Materials Horizons: From Nature to Nanomaterials

Materials are an indispensable part of human civilization since the inception of life on earth. With the passage of time, innumerable new materials have been explored as well as developed and the search for new innovative materials continues briskly. Keeping in mind the immense perspectives of various classes of materials, this series aims at providing a comprehensive collection of works across the breadth of materials research at cutting-edge interface of materials science with physics, chemistry, biology and engineering.

This series covers a galaxy of materials ranging from natural materials to nanomaterials. Some of the topics include but not limited to: biological materials, biomimetic materials, ceramics, composites, coatings, functional materials, glasses, inorganic materials, inorganic-organic hybrids, metals, membranes, magnetic materials, manufacturing of materials, nanomaterials, organic materials and pigments to name a few. The series provides most timely and comprehensive information on advanced synthesis, processing, characterization, manufacturing and applications in a broad range of interdisciplinary fields in science, engineering and technology.

This series accepts both authored and edited works, including textbooks, monographs, reference works, and professional books. The books in this series will provide a deep insight into the state-of-art of Materials Horizons and serve students, academic, government and industrial scientists involved in all aspects of materials research.

Review Process

The proposal for each volume is reviewed by the following:

1. Responsible (in-house) editor
2. One external subject expert
3. One of the editorial board members.

The chapters in each volume are individually reviewed single blind by expert reviewers and the volume editor.

Vinod K. Tiwari · Manoj K. Jaiswal ·
Sanchayita Rajkhowa · Sumit K. Singh

Click Chemistry

Springer

Vinod K. Tiwari
Department of Chemistry
Banaras Hindu University
Varanasi, Uttar Pradesh, India

Manoj K. Jaiswal
Department of Chemistry
Banaras Hindu University
Varanasi, Uttar Pradesh, India

Sanchayita Rajkhowa
Department of Chemistry
Haflong Government College
Haflong, Assam, India

Sumit K. Singh
Department of Chemistry
Banaras Hindu University
Varanasi, Uttar Pradesh, India

ISSN 2524-5384 ISSN 2524-5392 (electronic)
Materials Horizons: From Nature to Nanomaterials
ISBN 978-981-97-4595-1 ISBN 978-981-97-4596-8 (eBook)
https://doi.org/10.1007/978-981-97-4596-8

This Springer imprint is published by the registered company Springer Nature Singapore Pte Ltd.
The registered company address is: 152 Beach Road, #21-01/04 Gateway East, Singapore 189721, Singapore

Preface

The fields of '*Click Chemistry and Bioorthogonal Chemistry*' have recently been recognized with the 2022 Nobel Prize in Chemistry, awarded to Prof. Carolyn Bertozzi, Prof. Morten Meldal, and Prof. Karl B. Sharpless. Nowadays, this modular and regioselective protocol refers to a set of highly selective and efficient reactions that enable the rapid construction of complex molecules of high-rank applications. These reactions are characterized by their high yield, specificity, and modularity, making them ideal for wide applications ranging from drug discovery to bioconjugation and material design. The tool has been acknowledged as one of the most reliable synthetic tools discovered so far, and its scope is exponentially growing in almost all emerging fields including chemistry, biology, medicine, and materials science. Thus, we authors decided to bring this book to a wider readership.

This book covers twenty different chapters that illustrate the scope of modular click chemistry. Thus, the topics include the basics and advanced aspects of Click Chemistry starting from a brief introduction, the invention to its diverse applications. Chapter 1 discusses the general consideration of Click Chemistry and bioorthogonal chemistry. Starting from the invention of CuAAC click chemistry to its applications is covered in Chap. 2, while the scope of Ru-catalysis to give opposite regioisomer, i.e., 1,5-disubstituted 1,2,3-triazoles is presented in Chap. 3. Followed by metal-catalyzed 1,2,3-triazole forming click chemistry, the organocatalyzed route is presented in Chap. 4 followed by a catalyst-free method in Chap. 5. The strain-promoted azide-alkyne cycloaddition (SPAAC) reaction (known as bioorthogonal chemistry) and its notable applications in chemical biology is presented in Chap. 6. Thiol-ene Click Chemistry from invention to application is covered in Chap. 7 while SuFEx click tool with its ever-increasing scope in drug discovery and materials science is presented in Chap. 8.

The growing scope of *Click Chemistry* in various important fields such as peptide chemistry, dendrimer synthesis, polymer science, lipid modification, carbohydrate chemistry, nucleic acid chemistry, medicinal chemistry, and catalysis is well covered in Chaps. 9–16. An intramolecular version of Click Chemistry and its notable applications in macrocyclization as well as the synthesis of the biologically relevant fused ring systems are discussed in Chap. 17, while regioselective click chemistry under multicomponent reaction (MCR) and domino/tandem fashion are covered in Chap. 18. Followed by a discussion about 'Click Chemistry' starting from its discovery, generalization, and growing applications (Chaps. 2–18), a precise discussion about the post-functionalization of click products (Chap. 19) finally completes the whole theme of this book.

We authors sincerely thank our group members namely, Drs. Anand Agrahari, Mangal Singh Yadav, Anindra Sharma, Sunil Kumar, Nidhi Mishra, Abhishek Gupta, Faisal Ansari, Mayank Singh, and Vinay Pandey for their timely help during the preparation of this book. Our sincere thanks go to Prof. Richard Schmidt, Prof. Xi Chen, Dr. R. P. Tripathi, Prof. S. Bhattacharya, Prof. D. Basavaiah, Dr. Rakeshwar Bandichorr, Prof. Ganesh Pandey, Prof. Vinod K. Singh, and Prof. S. Chandrasekaran for their motivation and valuable suggestions. We also thank all investigators working on various aspects of click chemistry to make this topic popular.

In addition to the keen interest of journals such as *Chemical Reviews, Synthesis, Science of Synthesis, Current Organic Synthesis, Carbohydrate Research, and Chemistry Select* for their invitation to contribute to their respective special issues, we would like to convey special thanks to all who invited me to deliver lectures on '*Click Chemistry in Glycoscience*' during October 2022 to the end of 2023. The list includes the institutes like IIT-Delhi, IIT-Bombay, IIT-Gandhinagar, ILS-Bhuvneshwar, ICT-Mumbai, NIPER-Ahmedabad, NIT-Patna, BIT-Mesra, SVNIT-Surat, DTU-Delhi, SLIET-Longowal, and Amity Institute of Click Chemistry-Noida; universities, for example, University of Hyderabad, University of Lucknow, University of Mumbai, RTM Nagpur University, Gujrat University-Ahmedabad, Central University of Gujrat-Gandhinagar, BBA University-Lucknow, Bundelkhand University-Jhansi, MGCU-Motihari, Central University of Jammu, Cluster University-Jammu, Pt. Ravishankar Shukla University-Raipur, Dibrugarh University, Don Bosco University-Guwahati, IGNOU-Kohima, Cotton University-Guwahati, Central University of Panjab-Bhatinda, Akal University, Devi Ahilya University, SVV University-Indore, and Assam University-Silchar; and colleges like KKV College-Lucknow, Vidyawati College of Pharmacy-Jhansi, Sunbeam College for Women-Varanasi, Udai Pratap College-Varanasi, Harish Chandra College-Varanasi, Ambedkar College-Nagpur, Kirori Mal College-New Delhi, G. L. A. College-Daltonganj, and Haflong Government College-Assam. In addition, we would like to express our special thanks to pharmaceutical companies; Dr. Reddy's Laboratory-Hyderabad and Cadila Research Centre-Ahmedabad. We also thank The Indian Science Congress Association (ISCA), Indian Council of Chemists (ICC), Association of Carbohydrate Chemists and Technologists of India (ACCTI), Synergistic Training program Utilizing the Scientific and Technological Infrastructure (STUTI), Human Resource Development Centre (HRDC), and Asian Community of Glycoscience and Glycotechnology (ACGG) for their special consideration of our contribution on 'Click Chemistry in Glycoscience' in the form of invited lectures during their national meeting. At the end, I would like to express my deep sense of appreciation to the CSIR-New Delhi, SERB-New Delhi, SPARC, and IoE-BHU for funding our research on 'Click Chemistry in Glycoscience' and the authorities of BHU for all the support provided.

Last but not least, we would like to express our deep sense of appreciation to Springer Nature, for their keen interest and to the entire editorial team notably Ms. Priya Vyas, Mr. Suresh Dharmalingam, and Umamagesh A P for their consistent editorial support while maintaining the high standard of publication in bringing out this volume on *Click Chemistry*.

Varanasi, India — Vinod K. Tiwari
Varanasi, India — Manoj K. Jaiswal
Haflong, India — Sanchayita Rajkhowa
Varanasi, India — Sumit K. Singh
January 2024

Contents

About the Authors

Vinod K. Tiwari is Professor of Organic Chemistry at the Department of Chemistry, Banaras Hindu University (BHU, since 2005). He earned his M.Sc. degree in 1998 from BHU, Ph.D. from CSIR-Central Drug Research Institute, Lucknow (in 2004, Mentor: Dr. R. P. Tripathi), and had postdoctoral experience at University of Florida (Mentor: Prof. Alan R. Katritzky), University of California-Davis (Mentor: Prof. Xi Chen), and Guest Scientist at Universitat Konstanz-Germany (Mentor: Prof. Richard R. Schmidt). He has completed 11 projects, 16 Ph.Ds., 25 M.S. dissertations, and significantly contributed 179 peer-reviewed publications (Citations: 8321, h-index: 44, i10-index: 121), 8 patents, 25 book chapters, and 4 books. He is highly traveled Scientist and delivered > 265 invited lectures in India and abroad. His research is recognized by various academic/scientific societies including ISCA, CRSI, ICC, ICS, DST, UP-CST, IAPS, ACCT(I), BHU, Elsevier, ACS, RSC, Wiley, Thiem, Bentham, Holkar Science College, etc. At present, he is Hony. Secretary, Association of Carbohydrate Chemists and Technologists of India; Council Member of the Chemical Research Society of India (CRSI); and Editorial Board Member of Carbohydrate Research, Journal of Carbohydrate Chemistry, Trends in Carbohydrate Research, Journal of Chemistry, BVAAP (CSIR-NISCPR), Synthesis, Arkivoc, Current Organic Chemistry, Current Organic Synthesis, Molecules, Scientific Reports, etc. His current research is highly focused on '*Synthetic Carbohydrate Chemistry and Organic Synthesis*'.

Manoj K. Jaiswal is CSIR-Extended Senior Research Fellow at the Department of Chemistry, Institute of Science, Banaras Hindu University (BHU). He completed his doctoral research under the supervision of Prof. V. K. Tiwari at BHU. He earned his M.Sc. degree from Shri Baldev P.G. College, Varanasi, India, in 2014. He worked on the synthesis of drug candidates and active pharmaceutical ingredients (API) at Chembiotek Research International, Kolkata, and at CSIR-Central Drug Research Institute, Lucknow, before joining his doctoral research as CSIR-JRF in the laboratory of Prof. V. K. Tiwari at BHU. He qualified NET-CSIR-UGC with CSIR-JRF and GATE-2018. He significantly contributed to 18 peer-reviewed publications including *Chemical Reviews* and four book chapters of high repute.

His current research is mainly focused on '*Novel Glycosidation Methodology and Click Diversification in Glycoscience*'.

Sanchayita Rajkhowa is Assistant Professor of Chemistry at the Haflong Government College, Haflong (affiliated to Assam University), Assam. She earned her B.Sc. from Miranda House, Delhi (in 2006), M.Sc. degree from the University of Delhi (in 2008) and doctoral degree from North-Eastern Hill University (NEHU)-Shillong (in 2017). She qualified NET-CSIR-UGC with UGC-JRF in 2010 and worked at CSIR-NEIST, Jorhat. She taught Chemistry to UG/PG students at Department of Chemistry, Gauhati University, Guwahati (2017–2019), Jorhat Institute of Science and Technology (JIST)-Jorhat (2020–2021), and Royal Global University (RGU), Guwahati (2021–2022), before being appointed in the Haflong Government College-Haflong, Assam (2023). Dr. Rajkhowa significantly contributed 24 peer-reviewed publications in addition to several book chapters and five books of high repute. She was awarded 'CONIAPS-Young Scientist Award-2019' and Best Oral Presentation Award-2023 during an international conference at Don Bosco University-Guwahati and delivered over 21 invited lectures. At present, Dr Rajkhowa serves editor position of 'Journal of Chemistry' and her current research is focused on surface chemistry, ionic liquids, and carbohydrate chemistry.

Sumit K. Singh is Assistant Professor of Chemistry at Government P.G. College, Amarpatan, Satna, Madhya Pradesh, India. He was awarded his Ph.D. degree (in 2023) from Banaras Hindu University (BHU). He earned his B.Sc. (in 2015) and M.Sc. degrees in chemistry (in 2017) from BHU. He qualified GATE-2018 and NET-UGC-JRF and was selected for the prestigious Shyama Prasad Mukherjee (SPM) Fellowship. He significantly contributed to 21 peer-reviewed publications and 2 book chapters during his doctoral research under the guidance of Prof. V. K. Tiwari at the Department of Chemistry, BHU. His area of research interest is focused on the '*Click Chemistry-mediated Development of Novel Glycohybrids and their application in Catalysis*'.

Abbreviations

AAC	Azide-alkyne cycloaddition
ADME	Absorption, distribution, metabolism, and excretion
AFGPs	Anti-freeze glycoproteins
AIBN	Azobis-isobutyronitrile
AIE	Aggregation-induced emission
ANS	8-Anilino-1-naphthalenesulfonate
AR	Androgen receptor
ATRP	Atom transfer radical polymerization
AuNP	Gold nanoparticle
AZT	Azidothymidine
BARAC	Biaryl azacyclooctynone
BBB	Blood-brain barrier
Bn	Benzyl
BODIPY	Boron-dipyrromethene
BSA	Bovine serum albumin
BtRC	Benzotriazole ring cleavage
BTTES	2-[4-({bis[(1-*tert*-butyl-1*H*-1,2,3-triazol-4-yl) methyl] amino}methyl)-1*H*-1,2,3-triazol-1-yl]ethyl sulfate
Bz	Benzoyl
CAN	Ceric ammonium nitrate
CAR	Chimeric antigen receptor
CAs	Carbonic anhydrases
CCCTP	Cobalt-catalyzed chain transfer polymerization
CD	Circular dichroism
CHEF	Chelation-enhanced fluorescence
CMP	Cytidine-5′-Monophosphate
Con A	Concanavalin A
CRD	Carbohydrate recognition domain
CS	Chondroitin sulfate
CuAAC	Cu(I)-catalyzed azide-alkyne cycloaddition
DBU	Diazabicyclo[5,4,0]undec-7-ene

DCM	Dichloromethane
DDSs	Drug delivery systems
Dex	Dextran
DFT	Density functional theory
DIC	*N,N′*-Diisopropylcarbodiimide
DIPEA	*N,N*-Diisopropylethylamine
DMAP	4-Dimethylaminopyridine
DMPA	2,2-dimethoxy-2-phenylacetophenone
DNA	Deoxyribonucleic acid
DNJ	Deoxynojirimycin
DP	Degree of polymerization
DPV	Differential pulse voltammetry
DSC	Differential scanning calorimetry
DTBP	Di-tert-butyl peroxide
ECG	*Escherichia coli* galactosidase
ELPs	Elastin-like polypeptides
EMC	*N*-(ε-maleimidocaproyl)
Gal-3	Galectin-3
GalNAz	*N*-azidoacetylgalactosamine
GDP-fucose	Guanosine diphosphate *β*-L-fucose
GlcNAz	*N*-azidoacetylglucosamine
Glu	Glucose
GO	Graphene oxide
HA	Hemagglutinin
IAD	Intramolecular aglycon delivery
ICP-MS	Inductively coupled plasma-mass spectrometry
ICT	Internal charge-transfer
IRI	Ice recrystallization activity
ITC	Isothermal titration microcalorimetry
KLH	Keyhole limpet hemocyanin
Lac	Lactose
LCA	Lens culinaris agglutinin
MCR	Multicomponent reaction
MES	Maximum electroshock
MHC	Minimal hemolytic concentration index
MIC	Minimal inhibitory concentration
MNPs	Magnetic nanoparticles
MUNANA	2′-(4-Methylumbelliferyl)-α-D-N-acetylneuraminic acid
MW	Microwave
NaAsc	Sodium ascorbate
NanAs	*N*-acetylneuraminate lyases
NCL	Native chemical ligations
Neu5Ac	*N*-acetylneuraminic acid
NI	1,8-naphthalimide
NIS	*N*-iodosuccinimide

NS	*N*-hydroxy succinimide
PDI	Polydispersity index
PDT	Photodynamic therapy
PEG	Polyethylene glycol
PET	Photoinduced electron transfer
PFEx	Phosphorus Fluoride Exchange
PIFA	Phenyliodine(III) bis(trifluoroacetate)
PMAM	Polyamidoamine
PMDETA	Pentamethyldiethylenetriamine
PNA	Peanut agglutinin
PPI	Poly(propylene imine)
PR	HIV-1 protease
PTP	Protein tyrosine phosphatase
PTP1B	Protein tyrosine phosphatase 1B
PTZ	Pentylenetetrazole
QCM	Quartz crystal microbalance
RAFT	Reversible addition–fragmentation chain transfer
RCM	Ring-closing metathesis
RCY	Radiochemical yield
RNA	Ribonucleic acid
ROP	Ring-opening polymerization
RT	HIV-1 reverse transcriptase
RuAAC	Ru-catalyzed azide-alkyne cycloaddition
SAG	Sweet almond glucosidase
SAM	Self-assembled monolayers
SCSC	Single-crystal-to-single crystal
SEC	Size exclusion chromatography
SPR	Surface plasmon resonance
ST	Sialyl-transferase
SuFEx	Sulfur Fluoride Exchange
TAAC	Topochemical azide-alkyne cycloaddition
TBAA	Tetrabutyl ammonium azide
TBTA	Tris-(benzyltriazolylmethyl)amine
TCEP	Tris (2 carboxyethyl) phosphine
TcTS	*Trypanosoma cruzi trans*-sialidase
TEM	Transmission electron microscopy
TMSOTf	Trimethylsilyl trifluoromethane sulfonate
THPTA	Tris(3-hydroxypropyltriazolylmethyl)amine
TIPS	Triisopropylsilyl ether
$TMSCF_3$	(Trifluoromethyl)trimethylsilane
TPE	Tetraphenylethylene
TREN	Triaminoethylamine
TTMA	Tetraethylene glycol trimethyl ammonium
TTTA	Tris(4-trimethylsilylmethyl-1,2,3-triazolylmethyl)amine

Tyc	Tyrocidine
US	Ultrasound
WGA	Wheat germ agglutinin
β-CD	β-Cyclodextrin

Chapter 1
Click Chemistry and Bioorthogonal Chemistry: General Consideration from Discovery to Applications

1.1 Introduction

The amalgamation of two molecular components with diverse characteristics gives rise to a novel conjugate possessing intriguing collective traits derived from the parent entities, a phenomenon that has surged in popularity in recent decades (Fig. 1.1). *Just immediately after the invention of Cu(I)-catalyzed azide-alkyne cycloaddition reaction, commonly known as 'CuAAC—Click Chemistry'* [1, 2] by two distinct teams of researchers, Professor Karl Barry Sharpless at Scripps Research Institute at the USA and Professor Morten Meldal at the University of Copenhagen in Denmark in 2002, this regioselective and modular reaction for forming 1,2,3-triazoles from organic azides and terminal alkynes under Cu(I) catalysis has been extensively investigated across numerous burgeoning fields [3–17]. Regioselectively synthesized 1,4-disubstituted 1,2,3-triazole scaffolds are recognized as crucial heterocyclic cores in contemporary organic chemistry, medicinal chemistry, and materials science [3–6].

The CuAAC reaction boasts numerous notable attributes, including exceptional regioselectivity, high-to-excellent yields, ease of execution, short reaction times, modular characteristics, mild reaction conditions, and the use of inexpensive and easily available starting materials [7]. Additionally, the resulting regioselective 1,2,3-triazoles can function as amide bond isosteres, which are privileged functional groups in drug discovery and development [18]. The 'click' terminology signifies a method where molecules of two different functionalities can efficiently couple together, often through the biologically imperative linker, for example, 1,2,3-triazole. In contrast, the ruthenium-catalyzed azide-alkyne cycloaddition reaction, commonly referred to as 'RuAAC Click Chemistry', yields the exact opposite regioisomer, namely 1,5-disubstituted 1,2,3-triazoles as the sole product [8, 19].

Alternatively, a similar 1,2,3-triazole forming reaction via copper-free click chemistry was brilliantly introduced by applying the concept 'strain-promoted azide-alkyne cycloaddition (SPAAC)' that proceeds at a lower activation barrier and is well-known as Bertozzi's bioorthogonal chemistry tool [20]. The method is free

V. K. Tiwari et al., *Click Chemistry*, Materials Horizons: From Nature to Nanomaterials,
https://doi.org/10.1007/978-981-97-4596-8_1

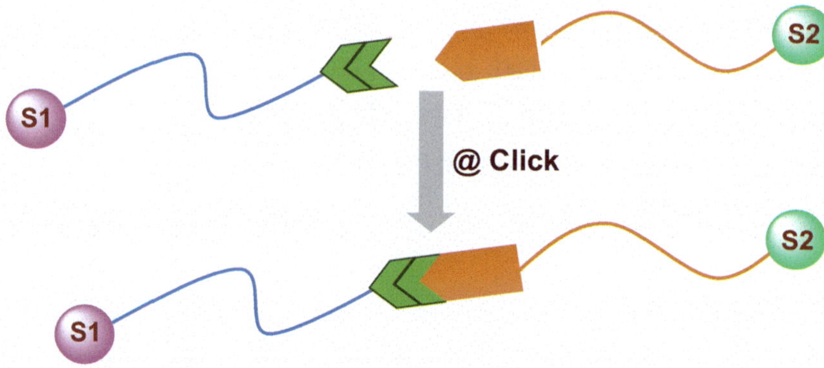

Fig. 1.1 Typical schematic representation of 'Click Chemistry' as a modular conjugation tool

of the use of cytotoxic transition metal catalysts and the absence of these exogenous metal catalysts makes the protocol even more suitable for in vivo applications in biological systems [21, 22]. The protocol has garnered the highest recognition with the awarding of the 'Nobel Prize in Chemistry for the year 2022' to Professors Carolyn R. Bertozzi, Morten Meldal, and Karl B. Sharpless for their remarkable contributions to the '*Development of Click Chemistry and Bioorthogonal Chemistry*'. Professor Carolyn R. Bertozzi has elevated the modular regioselective CuAAC tool to new dimensions by applying it within living organisms, a concept extensively investigated in chemical biology and drug development [21].

Over the past 22 years, significant efforts have been devoted to utilizing the regioselective CuAAC tool to create a diverse range of desired molecular architectures appended with simple-to-complex 1,2,3-triazole-appended molecular architectures [3–17]. In this context, we present this introductory chapter on '*Click Chemistry and Bioorthogonal Chemistry*' covering the background of the modular Click Chemistry with its variants such as CuAAC, RuAAC, SPAAC, SNAAC, thiol-ene, SuFEx, PFEx, inverse demand tetrazine-ene Click Chemistry starting from invention to applications. This chapter aimed to present a flavor of our book titled 'Click Chemistry' to a wide range of readership with a precise overview of almost all aspects of '*Click Chemistry and Bioorthogonal Chemistry*' from discovery to applications.

1.2 Sharpless's Click Chemistry Concept

Nobel laureate Prof. K. Barry Sharpless introduced the term 'Click Chemistry' for the first time in 1999, during the 217th American Chemical Society annual meeting [23a]. Subsequently, the term gained immense popularity within the scientific community. According to Professor Sharpless, for a chemical transformation to be considered a 'Click reaction', it must fulfill the following criteria:

> if the reaction is modular, easy to carry out, broad substrate scope, mild reaction condition, furnished products reasonably in high-to-excellent yields with ease of isolation preferably through avoid of chromatographic separation, generates only harmless by-products which can be free from the sole product without any difficulty, and also be regioselective and stereospecific (but not essentially to be enantioselective), moreover use of readily available starting materials or reagents and also the use of either solvent-free condition or environmentally benign solvent that can be removed easily [23b].

In this manner, Nobel laureate Professor K. B. Sharpless classified 'click reactions' into four primary categories (Scheme 1.1) as outlined below:

(a) Nucleophilic substitutions at saturated carbon atoms, encompassing ring-opening reactions of epoxides, aziridines, and related systems.
(b) Cycloaddition reactions, such as Diels–Alder $[4\pi_s + 2\pi_s]$ cycloaddition and Huisgen's $[3\pi_s + 2\pi_s]$ cycloaddition.

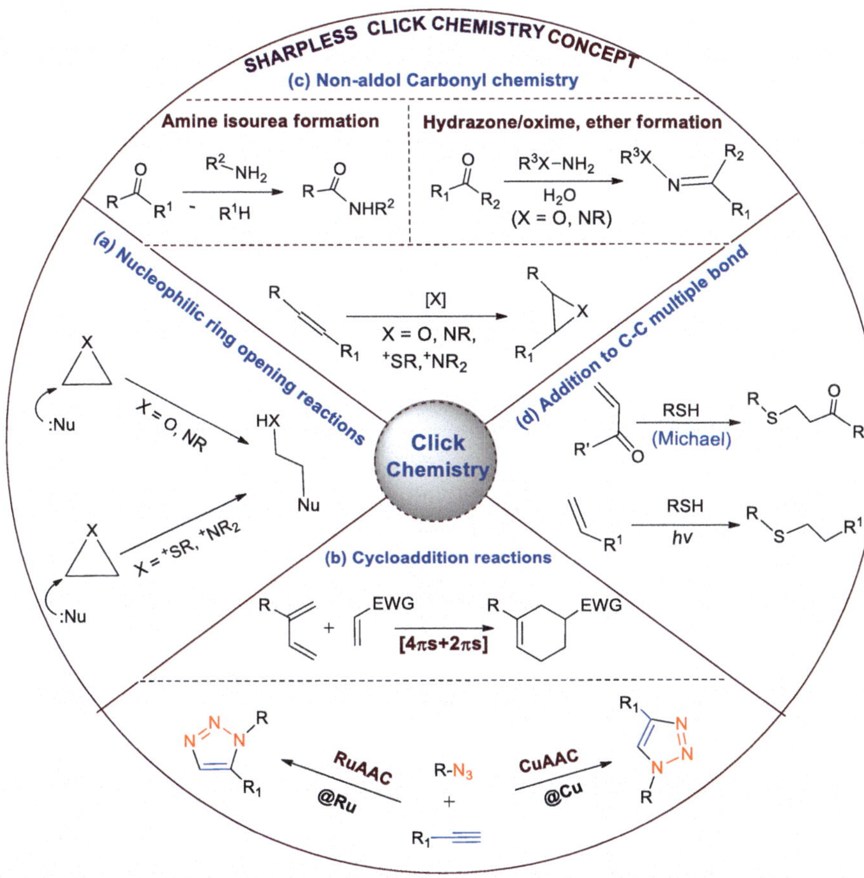

Scheme 1.1 Generalization of Sharpless's 'Click Chemistry' concept [9]

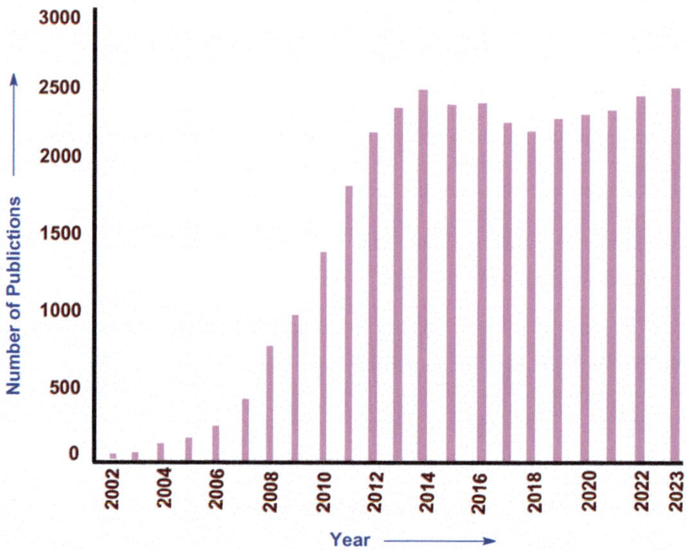

Fig. 1.2 Graphical presentation about progress on the contribution of 'Click Chemistry' during 2002–2023 [25]

(c) Non-aldol type functional group transformation of carbonyl and related systems; and
(d) Addition to carbon–carbon multiple bonds such as epoxidation, dihydroxylation, aziridination, aminohydroxylation, 1,4-conjugate additions, thiol-ene, and related click reactions.

A good number of important reviews dealing with the term 'Click Chemistry' have been reported in the literature [3–17, 24], which is indeed a wonderful reflection of the importance of this modular reaction in several emerging fields. The seminal progress indicates the enormous growth and flexibility of this regioselective tool of great synthetic importance. The progressive development since the invention of 'click chemistry' in 2002 and wide-ranging usage can be evidenced in surveys as the appearance of 'Click Chemistry' in CAS Sci-Finder search showed 32,699 hits that eventually demonstrating its diverse scientific and industrial applications (Fig. 1.2) [25].

1.3 Regioselectivity in 1,2,3-Triazole Forming Click Chemistry

Fortunately, all the stringent criteria outlined in Sharpless's Click Chemistry concept, as depicted in Scheme 1.1, align well with the Cu(I)-catalyzed azide-alkyne cycloaddition (CuAAC) reaction [1]. This CuAAC protocol is notably easy to execute, modular, efficient with a broad substrate scope, and yields high (often quantitative) product yields. Moreover, it is easily scalable, exhibits excellent regioselectivity and regiospecificity, and generates minimal by-products. These reactions are typically conducted in environmentally friendly solvents (such as aqueous media) or easily removable solvent systems, aided by the ready availability of both starting materials (alkynes/azides) and the catalyst [7]. Hence, the CuAAC reaction can be regarded as an ideal 'click' reaction and has been extensively investigated over the past two decades. Recently, it has garnered significant interest due to its remarkable applications across various scientific and technological disciplines. In contrast, Huisgen's catalyst-free thermal reaction of organic azides and terminal alkynes results in a mixture of regioisomeric 1,2,3-triazoles, which are challenging to separate into individual isomers in pure form [26, 27]. A similar cycloaddition reaction under Cu(I) catalysis (CuAAC) produces 1,4-disubstituted 1,2,3-triazoles as the exclusive product (Scheme 1.2) [1, 2]. The CuAAC method operates under mild reaction conditions, yielding regioselective 1,4-disubstituted 1,2,3-triazoles at room temperature and in reduced reaction times.

AAC reaction in the presence of Cu(II) salts and ascorbate as reducing agent is although the perfect method of choice for the regioselective 1,2,3-triazoles synthesis; however, it has still problematic in bioconjugation applications. Interestingly, similar AAC once performed in the presence of some suitable ligand such as *tris*[(1-benzyl-1*H*-1,2,3-triazol-4-yl)methyl]amine (TBTA), this has been effectively enhanced the CuAAC tool without damaging to the biological scaffolds [28]. As discussed above, CuAAC reaction exclusively afforded 1,4-disubstituted 1,2,3-triazoles with notable acceleration of kinetics than un-catalyzed condition. In contrast, utilizing a ruthenium catalyst resulted in the formation of 1,5-disubstituted 1,2,3-triazoles **11**, referred to as RuAAC that represents exactly an opposite regioisomer of CuAAC product **10** (Scheme 1.3) [18]. The ruthenium-catalyzed cycloaddition of azides to alkynes gives the complementary 1,5-disubstituted triazoles. While RuAAC can accommodate diverse functional groups, it has limited compatibility with various reaction media and tough steric requirements for the azide component, unlike CuAAC [8]. Several ruthenium complexes were well employed for the regioselective 1,5-disubstituted 1,2,3-triazole formation; however, the pentamethylcyclopentadienyl (Cp*) analogues, e.g., $Cp^{*}RuCl(PPh_3)_2$, was employed in most the cases to give the best results [8]. In contrast to the CuAAC reaction which was limited to terminal alkynes, the RuAAC variant is active with internal alkynes as well [18].

A very first example of 1,3-dipolar cycloaddition reaction of organic azide **8** and terminal alkyne **9** was reported by Michael [29]. The method was widely explored by R. Huisgen in 1960s ara and known as Huisgen's thermal 1,3-dipolar azide-alkyne cycloaddition (AAC) [26, 27]. Huisgen's [30] thermally induced 1,2,3-triazole synthesis was well utilized for the synthesis of diverse range of triazole-linked hybrid candidates of wide applications; however, the method has some serious limitations such as longer reaction time, required high-temperature condition and leads to the inseparable mixture of 1,4-(**10**) and 1,5-disubstituted 1,2,3-triazoles **11** (i.e., no regioselectivity) that complicate the purification process [31].

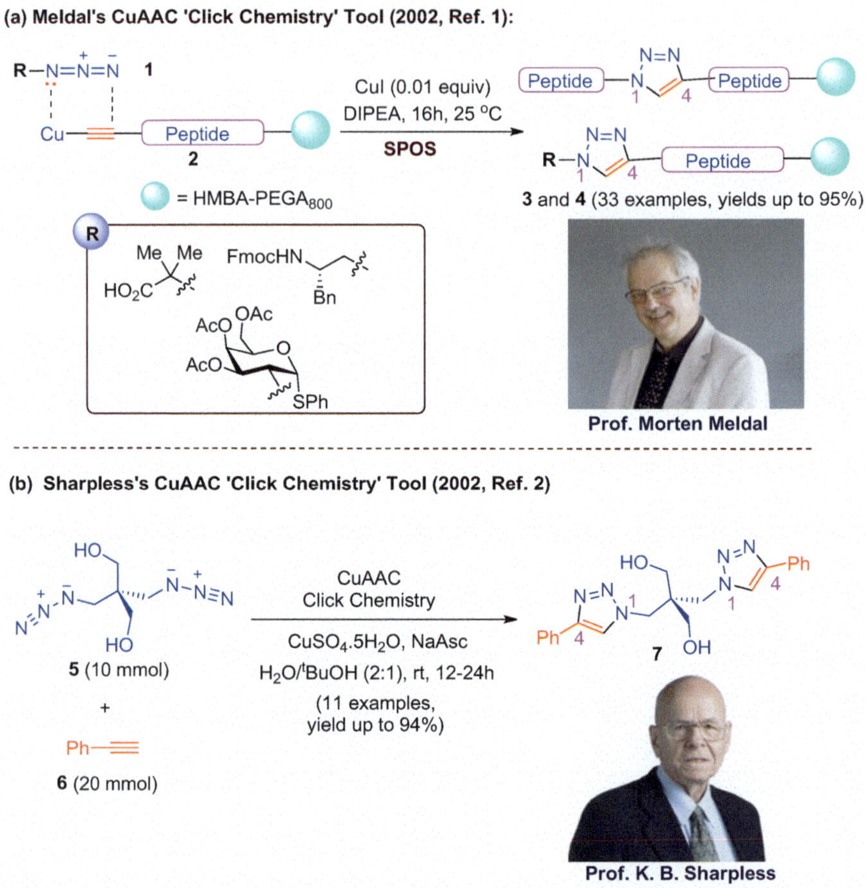

Scheme 1.2 CuAAC-mediated [3 + 2] azide-alkyne cycloaddition for the regioselective synthesis of 1,4-disubstituted 1,2,3-Triazoles CuI/DIPEA: Meldal's CuAAC Click tool; $CuSO_4 \cdot 5H_2O$/NaAsc: Sharpless's CuAAC Click tool (Nobel Prize in Chemistry for the year 2022)

To get the desired 1,2,3-triazole products from the molecules which are otherwise unstable under harsh condition through smooth route at room temperature with increased reaction rate (thus reduced reaction time) collectively expanded the widespread applications. Toward this end, Cella and Stefani [32] explored US-supported efficient click tool for CuAAC coupling of TMS-protected C-alkyne-glycoside **12** with organic azides **13** to give high yield of metabolic processes resistant trait of C-glycosyl-triazoles **14** (Scheme 1.4a).

Rufinamide **17**, a 1,2,3-triazole ring containing anti-convulsant drug, was developed in 87% yield in short span of time simply by reacting aryl bromide **15** with NaN_3, propiolamide **16** using recycled ionic liquid supported Cu(II) catalyst **18** in refluxing methanol for 3 h (Scheme 1.4b) [33]. In recent years, a number of copper complexes with PPh_3 have been explored as proficient CuAAC catalytic

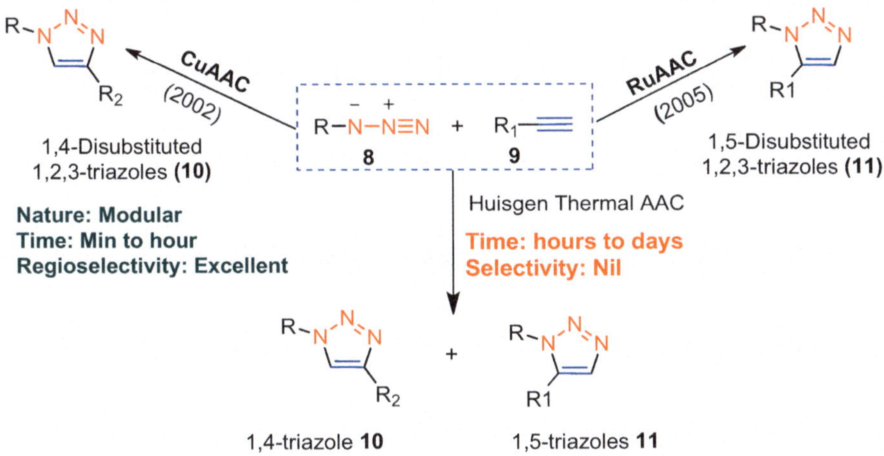

Scheme 1.3 Regioselectivity in 1,2,3-triazole forming CuAAC and RuAAC tool

systems for the regioselective 1,2,3-triazole forming click reactions. A notable example include $[(PPh_3)_2Cu(\mu\text{-tda})Cu(PPh_3)_2]0.6H_2O$ (tda = thiodiacetate anion). This highly stable dinuclear Cu(I) catalyst **21** displayed a significant catalytic activity for 1,4-disubstituted triazoles via CuAAC click chemistry [34]. For an example, click reaction of glycosylated azide **19** with terminal alkyne **20** in the presence of $[(PPh_3)_2Cu(\mu\text{-tda})Cu(PPh_3)_2]0.6H_2O$ (**21**) for 10 min time afforded the desired 1,4-disubstitute triazolyl glycoconjugate **22** in excellent yield (Scheme 1.4c) [34]. In addition to the notable advantages of catalyst **21** such as its high stability, easy preparation, and excellent recyclability, other features like easy work-up and short reaction time are the additional benefits. Further, this Click tool can also be performed under solvent-free, ligand-free, and base-free conditions and has wide substrate scope as the method was successfully applied in dendrimer synthesis [34–36].

Starting from corresponding amino compounds, a diazotransfer reaction has been widely used to give high yield of respective azide for click conjugation and both the steps can be performed under one-pot condition for an easy access of 1,2,3-triazoles [37]. Likewise, glycosyl-β-azido ester **19** required for the click conjugation with alkyne **25** under CuI/DIPEA catalysis to give biologically active glycohybrids **26** can also be obtained by the conjugate addition of ammonia on olefinic ester glycosyl olefinic ester **23** followed by metal-catalyzed diazotransfer reaction using imidazole-1-sulfonyl azide **24** in the presence of K_2CO_3 (as base) and $ZnCl_2$ under one-pot mild reaction condition (Scheme 1.4d) [38].

In addition to modular nature of click conjugation of alkynes with organic azides under catalytic condition (CuAAC or RuAAC) or catalyst-free condition, it is interesting to mention here that metal azides also undergo 1,3-cycloadditions with suitable dipolarophiles and the reactions in many such cases are more facile than the analogous organic azides. The first cycloaddition between an alkyne **28** and a metal-coordinated azide **27** group was first reported by Kreutzer et al. [39] in 1979. However,

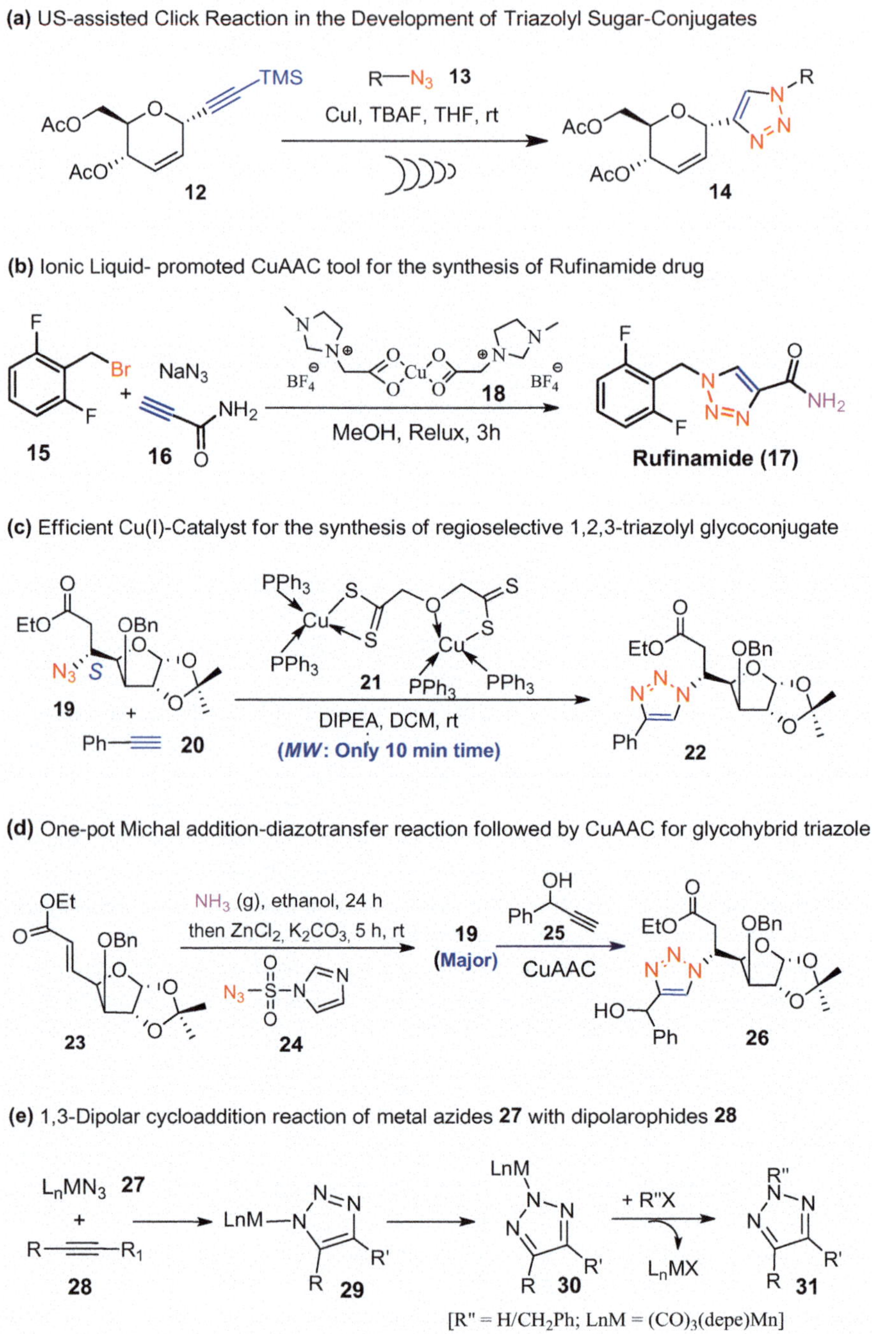

Scheme 1.4 Some notable 1,2,3-triazole forming coupling reactions under various reaction conditions

large numbers of reports available in the literature are much more recent. Usually, the triazolate ligand binds through N^2, and it is generally believed that the N^1-bonded isomer which is the kinetic product of the [3 + 2]-cycloaddition reaction subsequently isomerizes to the thermodynamically stable N^2-isomer **31** (Scheme 1.4e). The isomerization reaction (N^1 to N^2 bonding with metal) was studied using NMR spectroscopy by Chang and Lee [40] and Busetto et al. [41] independently. Notably, the cationic complexes such as $[(p\text{-cymene})Ru(N_3)(\text{diphosphine})]^+$do not undergo [3 + 2] cycloaddition with activated alkynes [42]. With terminal alkynes ligand exchange reaction is more preferred over cycloaddition because these alkynes do not behave as 1.3-dipolarophiles [43]. For several 1,2,3-triazolato complexes, the ligand from the metal ion is cleaved by electrophilic attack [44].

Both the CuAAC and RuAAC methods have notable advantages as similar cycloaddition reactions under thermal conditions afforded a mixture of both the regioisomeric triazoles. Copper-free click chemistry is based on an old reaction, reported by Wittig in 1961 which was the cycloaddition of cyclooctyne **31** with phenyl azide **32** proceeded like an explosion to give a 1-phenyl-4,5,6,7,8,9-hexahydro-1*H*-cycloocta[d][1,2,3]triazole, **33** a single sole product [45]. The metal-free cycloaddition of azide with cyclooctyne proceeds with unusually fast rate even at room temperature without the need for catalysis mainly because of the release of large amount of the ring strain (18 kcal/mol of ring strain) in the cyclooctyne molecule. The strained cyclooctynes react selectively with azides to form regioisomeric mixture of triazoles at ambient temperatures and pressures without the need for metal catalysis and no apparent cytotoxicity [46].

Bertozzi has taken up the Sharpless-Meldal CuAAC click tool to a higher level by integrating the concept of ‘Bioorthogonal Chemistry’, which operates within living organisms without disrupting normal biological functions [20] Her strain-promoted [3 + 2] azide-alkyne cycloaddition (SPAAC) for covalent modification of biomolecules in living systems (a concept recognized by her Nobel Prize) represents a remarkable advancement. This approach builds upon the Wittig concept, involving a metal-free [3 + 2] cycloaddition of azide **34** with a strained cyclooctyne **35** to yield functionalized 1,2,3-triazole **36** at room temperature (Scheme 1.5a). In Bertozzi’s SPAAC reaction, an azido-modified cell **38** reacts with an alkenyl fluorescent probe, allowing both molecules to coalesce and form a triazole residue **40** within a human cell (Scheme 1.5b) [20–22]. This coupling can occur within a biological system or even within the human body, offering significant advantages for tracking glycans in living systems. At first, Prof. Bertozzi fed cells with modified sugar having clickable azide on it and then the modified sugars were incorporated into glycans. Latter, an alkyne as fluorescent probe was then underwent SPAAC with modified sugar azide. At the end, the resultant florescent 1,2,3-triazoles were allowed to track the glycans on the cell surface (Scheme 1.5c).

Bertozzi’s ‘Bioorthogonal Chemistry’ tool has significant applications in chemical biology, notably in selectively labeling lipids and glycans within biological systems without apparent toxicity, as well as in labeling biomolecules in live mice [46]. This concept has facilitated the precise targeting of medicines to specific tissues, aiding

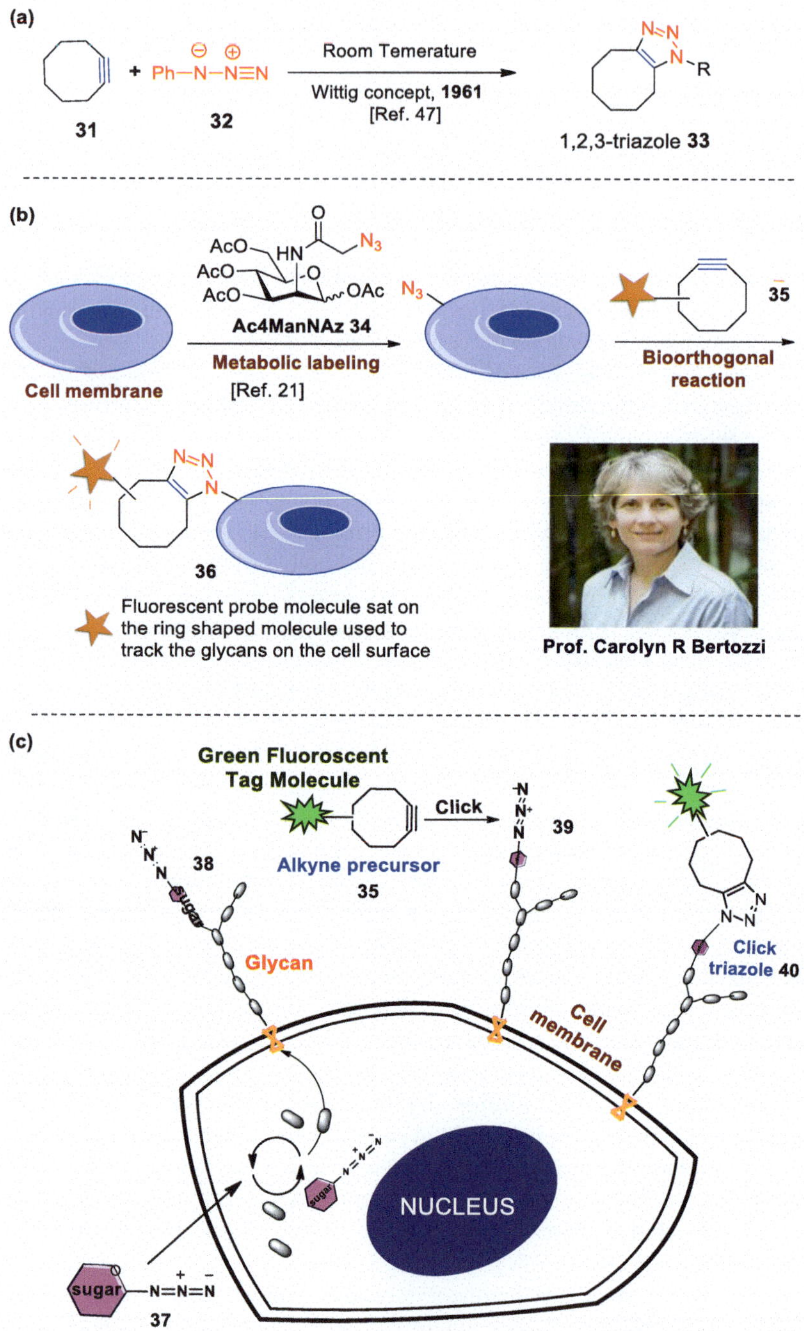

Scheme 1.5 **a** Strain-promoted azide-alkyne cycloaddition (**Wittig approach**); **b** strain-promoted azide-alkyne cycloaddition in living systems (**Bertozzi's Bioorthogonal approach, Nobel Prize in Chemistry for the year 2022**)

chemists in medicine development, and enabling the real-time monitoring of modifications to other metabolites within living systems. Some difluorinated cyclooctyne (DIFO) strained alkyne reagents have been extensively utilized in copper-free click chemistry for the site-selective labeling of biomolecules both in vitro and in vivo [20]. Additionally, biarylazacyclooctynone (BARAC) has been employed for live cell fluorescence imaging of azide-labeled glycans [47]. In situ 'click' cross-linking of azide-terminated photodegradable star polymers has been effectively managed using bifunctional fluorinated cyclooctynes [48]. Further, biorthogonal chemistry was explored to functionalize new platinum (IV) [PtIV] facilitating the installation of targeting moieties, delivery systems, and fluorescent reporters from a single precursor capable of releasing biologically active cisplatin [49]. The copper-free click tool has been employed for various applications, including the functionalization of gold nanoparticles with monovalent maleimide [50] and for the synthesis of chitosan and hyaluronan-derived biocompatible and biodegradable polysaccharide hydrogels of potential applications in soft-tissue engineering [51]. Furthermore, copper-free click chemistry is instrumental in developing radiotracers such as [^{64}Cu] DOTA-ADIBON$_3$-Ala-PEG$_{28}$-A20FMDV2, which finds notable application in positron emission tomography imaging of integrin $\alpha_v\beta_6$ expressing tumors [52]. It is also utilized for radioiodination, such as iodine radioisotope labeling of cyclooctyne-containing molecules [53]. In protein site-specific labeling, the strain-promoted azide-alkyne cycloaddition (SPAAC) of dibenzocyclooctyne-fluor 545 (DBCO-fluor 545) with an azide-bearing unnatural amino acid is employed [54]. In addition to widespread applications of click chemistry and strain-promoted bioorthogonal click reaction [3, 55, 56], a number of triarylphosphines and related systems has also been widely used under bioorthogonal fashion to tag azide-functionalized biomolecules in living systems [57].

The above-mentioned examples indicated for the regioselective triazole forming reactions includes, Cu(I)-catalyzed azide-alkyne cycloaddition (CuAAC) [1, 2], Ru-catalyzed azide-alkyne cycloaddition (RuAAC) [19], photoinduced click [58], topochemical AAC click [59], and Bertozzi bioorthogonal strain-promoted (SPAAC) click is now well highly explored tool for wide applications [20, 21]. The established SPAAC protocol has great scope in biology and medicine. Apart from metal-catalyzed (CuAAC and RuAAC) and SPAAC reaction, another green variant of click chemistry tool is the organocatalyzed synthesis of functionalized 1,2,3-triazole [60–62]. The method involves the reaction of carbonyl compounds **41** with azides **42** using catalytic amount of secondary amine, for example, DBU, or proline or pyrrolidine to give high-to-excellent yield of functionalized 1,2,3-triazoles **43** (Scheme 1.6a) [60–62]. Furthermore, functionalized 1,2,3-triazoles can be obtained from nitrostyrenes or related molecules by reacting them with azides under suitable catalysts [63]. For example, [3 + 2] cycloaddition reaction of nitroolefins **45** with azides **44** using Schreiner thiourea **I** catalyst which enable the required activation of nitroolefins through double hydrogen-bonding afforded high-to-excellent yield of 1,5-disubstituted 1,2,3-triazolyl glycoconjugates **46** (Scheme 1.6b) [64]. In addition to the excellent regioselectivity, operationally simple, wide substrate scope, and high

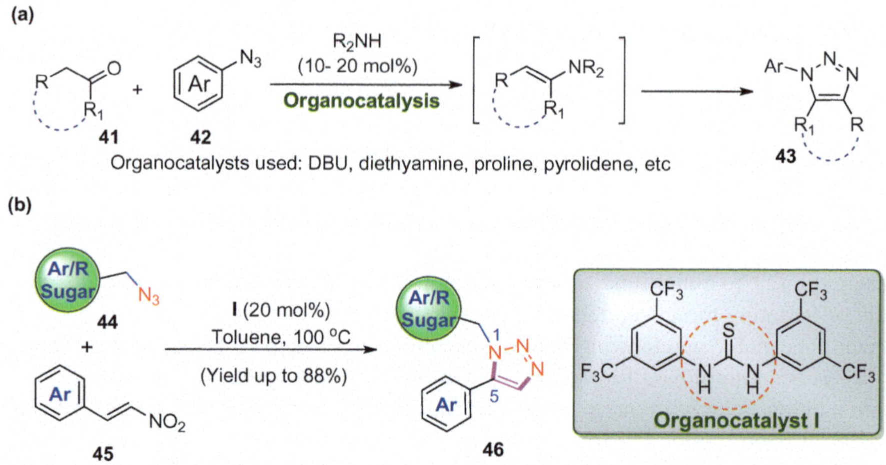

Scheme 1.6 Organocatalyzed variant of 1,2,3-triazole forming 'Click Chemistry'

reaction yield, straightforward metal-free, and acid-free condition are the notable features of this click method.

1.4 Method of Characterization of 1,4-Disubstituted-1,2,3-Triazole and 1,5-Disubstituted-1,2,3-Triazole

The [3 + 2] cycloaddition reaction of organic azide with terminal alkyne under Cu(I) catalysis (CuAAC) gave exclusively 1,4-disubstituted 1,2,3-triazoles as sole products [1, 2], while exactly opposite regiomer (i.e., 1,5-disubstituted 1,2,3-triazole) was obtained as sole product by utilizing the ruthenium-catalysis (RuAAC) [65]. Similar to the RuAAC, an enolate anions or acetylide anions on reaction with organic azides also gave 1,5-disubstituted 1,2,3-triazoles [66, 67]. The structural assignment of both the regioisomeric triazoles can be ascertained by using the spectral techniques like multidimensional NMR [68], NOE [69], and X-ray crystallographic analysis [70]. It is evident that both the regioisomer 1,2,3-triazole is hard to identify through its ^{1}H NMR spectra, and thus to resolve this issue, a fast and easy analytical technique is required to evaluate the regioselectivity in the click products. Creary et al. [68] used one-dimensional ^{13}C NMR spectroscopy using gated decoupling tool for rapid distinguishing between the isomeric 1,2,3-triazoles **10** and **11**. An expanded region of the ^{13}C NMR spectrum of isomeric triazoles obtained by CuAAC and organocatalytic route is depicted in Fig. 1.2. As per the trend, the C-5 carbon of 1,4-disubstituted triazole appeared at ~ 120 ppm, while the corresponding C-4 carbon signal of 1,5-disubstituted triazole was appeared at ~ δ 133 ppm. For an example,

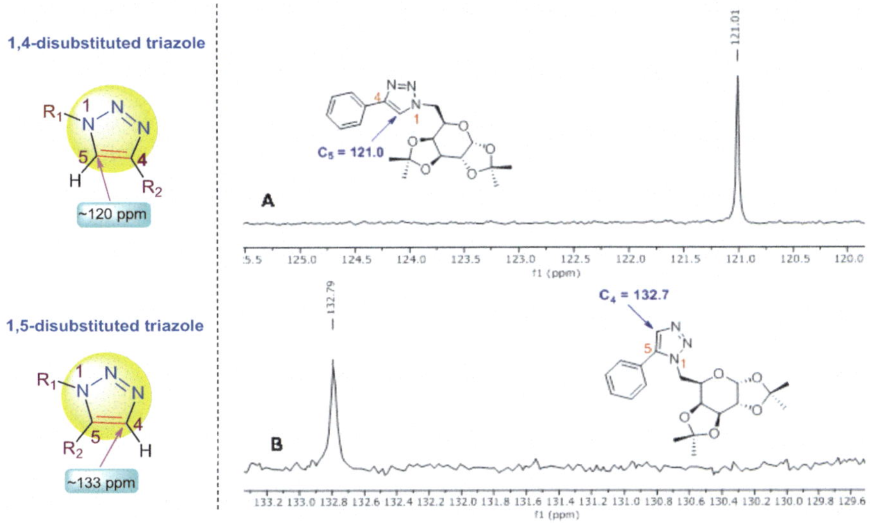

Fig. 1.3 Distinguishing between both the regioisomer 1,4-disubstituted 1,2,3-triazoles and 1,5-disubstituted 1,2,3-triazoles using one-dimensional ^{13}C NMR tool

the signal at δ 132.7 ppm corresponding to C_4 carbon in spectrum 1B (1,5-isomer) and δ = 121.0 ppm corresponding to C_5 carbon of spectrum 1A (1,4-regio isomer) prove the isomeric structure of 1,2,3-triazoles. Also, C_4 carbon signal of 1,5-isomer (spectrum 1B) has relatively large C-H coupling constant than C_5 carbon of 1,5-isomer (spectrum 1A). Further, the up-field shift of C5 signal of 1,4-disubstituted 1,2,3-triazole with respect to C-4 of 1,5-disubstituted 1,2,3-triazole was supported by computational studies (Fig. 1.3) [68, 71].

1.5 1,2,3-Triazoles as Isostere of Amides

Proteins, the most important biopolymer, consist of an amide bond (commonly known as peptide bond) that was obtained via acid–amine coupling of the COOH group of α-amino acid with the NH_2 group of another one. The amide group is utmost importance in medicinal chemistry as it is found in a majority of NCEs and clinically approved drug candidates [72]. Interestingly, a non-classical amide (CO–NH) isostere is considered as the click-mediated triazole heterocycle [18]. Some interesting features of click-mediated triazoles include its structural planarity, notable dipole–dipole interactions, chemical stability, and also hydrogen-bonding abilities which altogether could be comparable to the respective amide bond [73, 74]. Depend on the substitution pattern of both CuAAC- and RuAAC-mediated 1,2,3-triazole, this residue is known to mimic the conformations of respective *Z*- and *E*-amide. Thus, the structure of CuAAC-mediated 1,4-disubstituted triazole **10** is isosteric to a Z-amide **47**, while RuAAC-mediated resulting 1,5-disubstituted triazole **11** is known to mimic an E-amide **48** (Fig. 1.4) [72].

The distance between substituents, denoted as $R^1 \rightarrow R^2$, in triazole **10** is ~ 5.0 Å which is near to the value of 3.9 Å of Z-amide **47**. The lone pair available on *N*-3 of 1,4-disubstituted 1,2,3-triazole **10** serves as a HBA, akin to the C=O oxygen in the amide **47**. Additionally, the HBD capabilities of the *C*(5)-*H* bond of 1,4-disubstituted

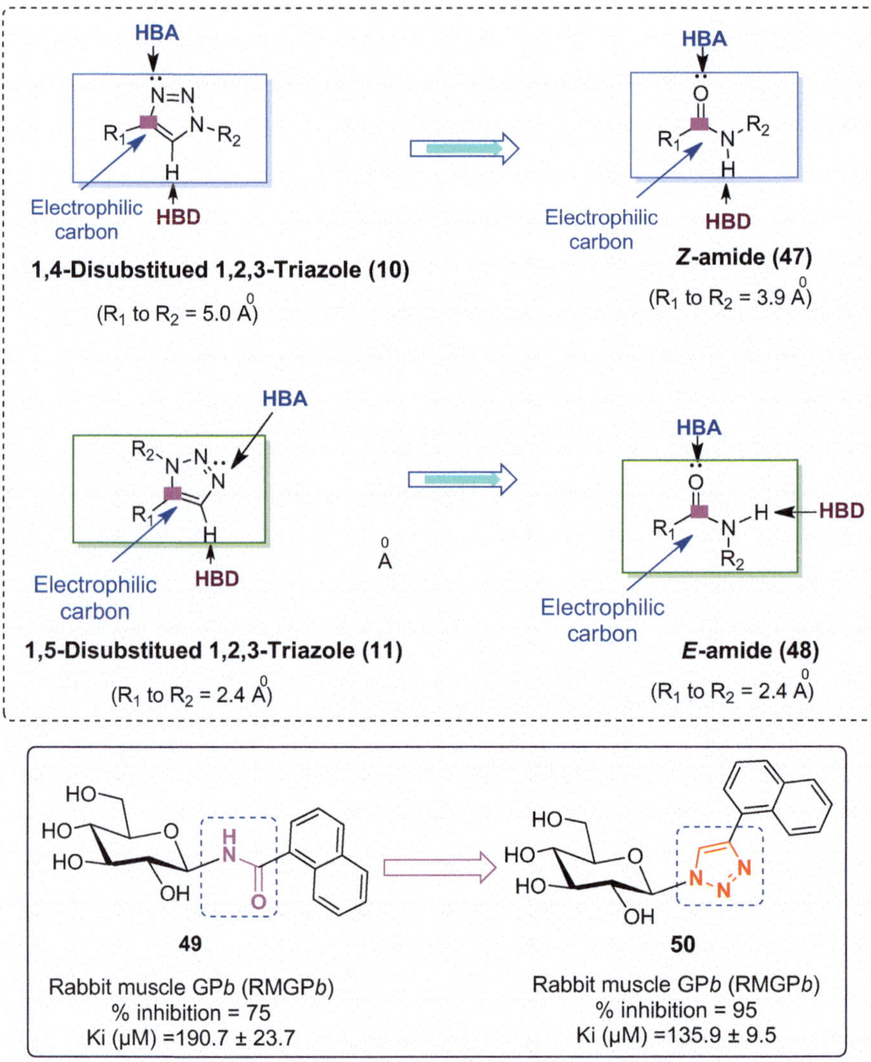

Fig. 1.4 Bioisosteric relationship of click-derived regioisomeric 1,2,3-Triazoles **10** and **11** with an Amides (HBD stands for hydrogen bond donor and HBA stands for hydrogen bond acceptor)

1,2,3-triazole **10** resemble those of the *N*–H of the amide bond. Moreover, the higher μ of the CuAAC triazole **10**, approximately 5.0 Debye, compared to an amide **47** (~3.8 Debye), imparts advantages to the triazole, allowing it to participate as both HBA and HBD (Fig. 1.4) [72].

A similar structural feature like the position of substituents and HBD or HBA sites present in 1,5-disubstituted 1,2,3-triazole is accounted for isostere with that of *cis*-amide. The 1,2,3-triazole exhibits resistance to cleavage by proteases, oxidation/reduction states, and hydrolysis, making it an attractive candidate as an amide surrogate. Consequently, the 1,2,3-triazole ring can effectively mimic the labile amide bond in vivo, offering the potential for developing protease-resistant peptidomimetics [18].

Because of the above-mentioned notable resemblances of CuAAC-mediated triazole heterocycle with that of peptide linkages, they displayed a wide range of pharmacological activities [4]. For instance, in the management of hyperglycemia in NIDDM type 2 diabetes, the CuAAC-inspired glycohybrid triazole **50** demonstrated little improved inhibitory potency against RMGP*b* enzyme compared to the amide **49** [75]. As evidenced from *in-silico* studies, this was revealed that both the glycohybrid triazole **50** and a sugar amide **49** efficiently bind to the catalytic site of RMGP*b* enzyme via resultant hydrogen bond interactions between the peripheral OH groups of glucopyranose ring and the residues finding the active site. This concept was further supported by X-ray studies confirming the relative binding of the triazole and an amide within the catalytic active site of the RMGP*b* enzyme. This result further envisaged the potential utilities of CuAAC-mediated triazole heterocycles as a bioisostere of an bioactive amide [75]. The bioisosteric relationship between amides and 1,2,3-triazoles is extensively explored for identifying novel lead candidates against various frontline diseases such as cancer, infectious disease, human immunodeficiency virus (HIV), and neurological disorders.

1.6 Other Important Variants of Click Chemistry (SuFEx, PFEx, Thiol-Ene, Tetrazine-Ene Click)

Just like the above-described metal-catalyzed azide-alkyne cycloaddition (CuAAC or RuAAC click chemistry) or organocatalyzed AAC or thermally induced catalyst-free AAC or strain-promoted AAC (SPACC, bioorthogonal chemistry) lead to the regioselective 1,2,3-triazole formation, a Sulfur(VI) Fluoride Exchange reaction, commonly known as '**SuFEx Click**' is identified as the next-generation variant of Click reactions (Scheme 1.7) [76]. This 'SuFEx Click' is also modular, rapid, robust, and reliable tool widely explored to access a diverse functional molecules of prime interest in drug development and material science [77].

Further after the successful innovation of SuFEx click, another newest toolbox of 'Click Chemistry' was recently introduced as the **Phosphorus Fluoride Exchange (PFEx) Click** by Prof. John Moses and Prof. K B Sharpless [78]. The PFEx variant

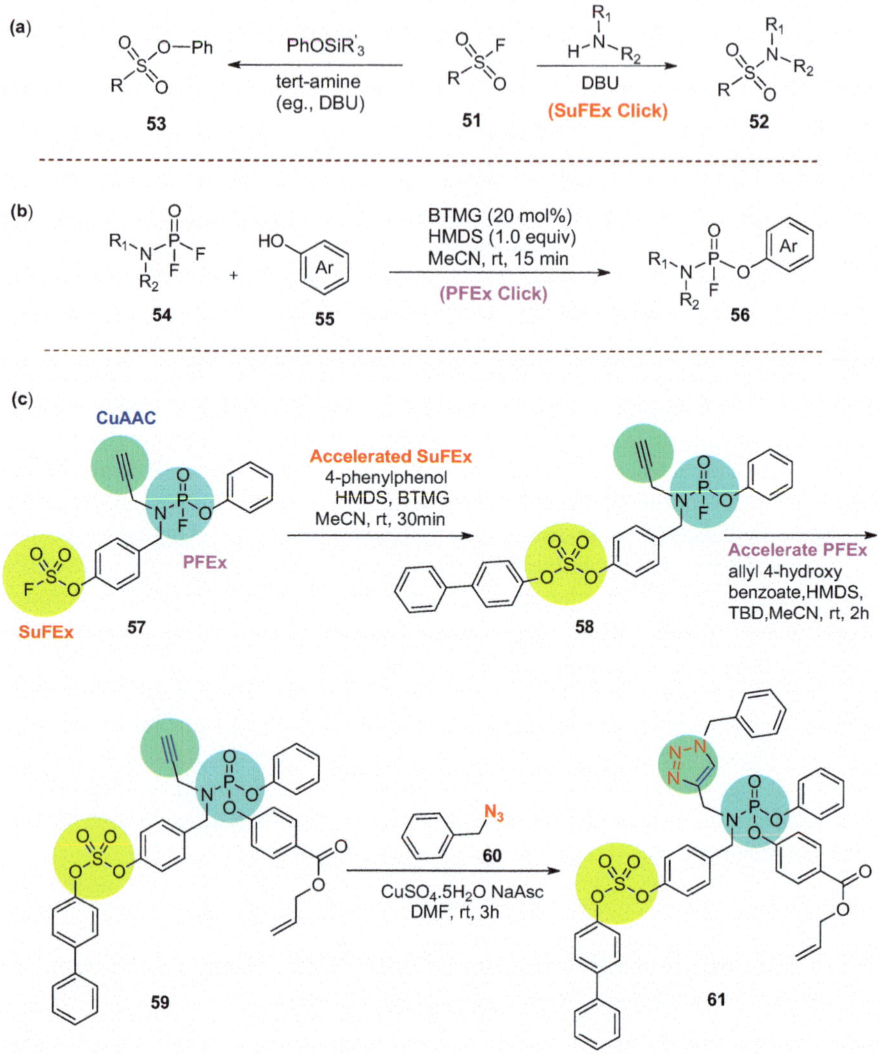

Scheme 1.7 Sulfur fluoride exchange (SuFEx) and phosphorus fluoride exchange PFEx tool as new variant of 'Click Chemistry'

of click chemistry was inspired by nature's proclivity for the well-known phosphorus connections which was prevalent in biomolecules serving as the connections in nucleic acids (DNA and RNA) and ATP as the nature cell currency. This PFEx tool because of the mild reaction condition and improved reactivity profile resulted by the shorter length and higher stability of P-F bonds than P-Cl bonds is considered to be even more advantageous and near perfect. PFEx reacts P-F containing molecules

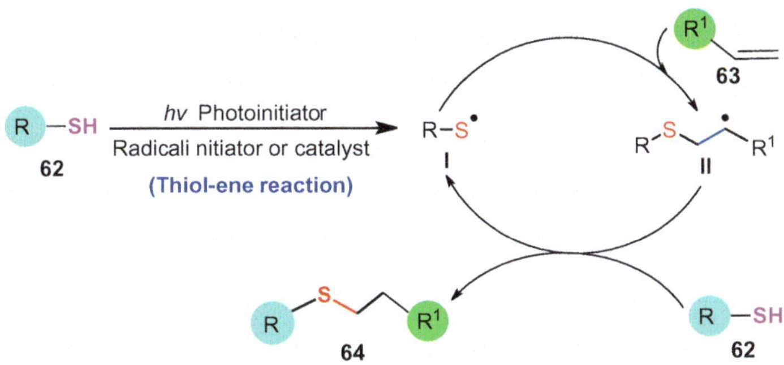

Scheme 1.8 Thiol-ene reaction as another variant of Click Chemistry

54 with nucleophile **55,** e.g., amines or alcohols to develop P-N or P-O-linked products **56**. The reaction is orthogonally to other variant of click chemistry like SuFFEx and CuAAC as shown in Scheme 1.8. The detailed information about the development and emerging scope of SuFEx click chemistry in various fields is discussed in Chap. 8.

Another variant of 'Click Chemistry' is the '**thiol-ene reaction**'. The method is an anti-Markovnikov addition of a thiol **62** to an alkene **63** to give high yield of respective thioether **64**. With the help of radical initiator, thiol **62** first resulted to thiol radical **I** that underwent anti-Markovnikov addition to alkene **63** and thiol-ene adduct radical **II** thus obtained on further reaction with another molecule of thiol gave the desired product **64** and again a thiol radical **I** that complete this process to be cyclic one (Scheme 1.8) [79]. The thiol-ene reaction is modular, stereoselective, high yielding and due to thermodynamic driving force it proceeds with high rate. This well-established method has enormous applications in organic synthesis for the development of polymers and dendrimers of great impact in biomedical and material sciences [80]. The detailed information about the invention to application of thiol-ene reaction and thiol-ene Click Chemistry in various emerging fields is provided in Chap. 7 of this book.

Another important variant of click chemistry is the '**Tetrazine-Alkene Click**' which is based on inverse-electron-demand Diels—Alder reactivity. This bioorthogonal click chemistry involves the reaction of tetrazines **65** with strained alkenes **66,** such as trans-cyclooctenes, and gives a stable product **68** (Scheme 1.9a). The reaction is efficient, high selectivity, and so fast that it can be completed within seconds at room temperature. Likewise, tetrazine reacts faster with a strained trans-cyclooctene (to release the associated ring strain) in an inverse-electron-demand Diels–Alder pattern and the resulting cycloaddition adduct underwent a retro [4 + 2] cycloaddition to give the desired product with great ease [81]. *Further, due to the* high ring strain in three- or four-membered cycloalkenes, they also considered an ideal alkene substrate and the tetrazine by virtue of having the additional nitrogens act as a good diene. Electron-donating substituents on the dienophile (strained ene) and electron-withdrawing substituents on the diene (tetrazine) accelerate the alkene-tetrazine inverse-demand Diels–Alder [82]. Such *alkene-tetrazine* click *ligation has great scope in bioorthogonal chemistry particularly* for the successful *imaging cellular DNA* [83].

Interestingly, there is an increased demand for the bond cleavage reactions that further improve the scope of click bioorthogonal chemistry ligation and facilitate novel tools for the in vivo probe activation and therapeutic delivery. The tetrazine/*trans*-cyclooctene click-to-release strategy has exemplary kinetics and adaptability

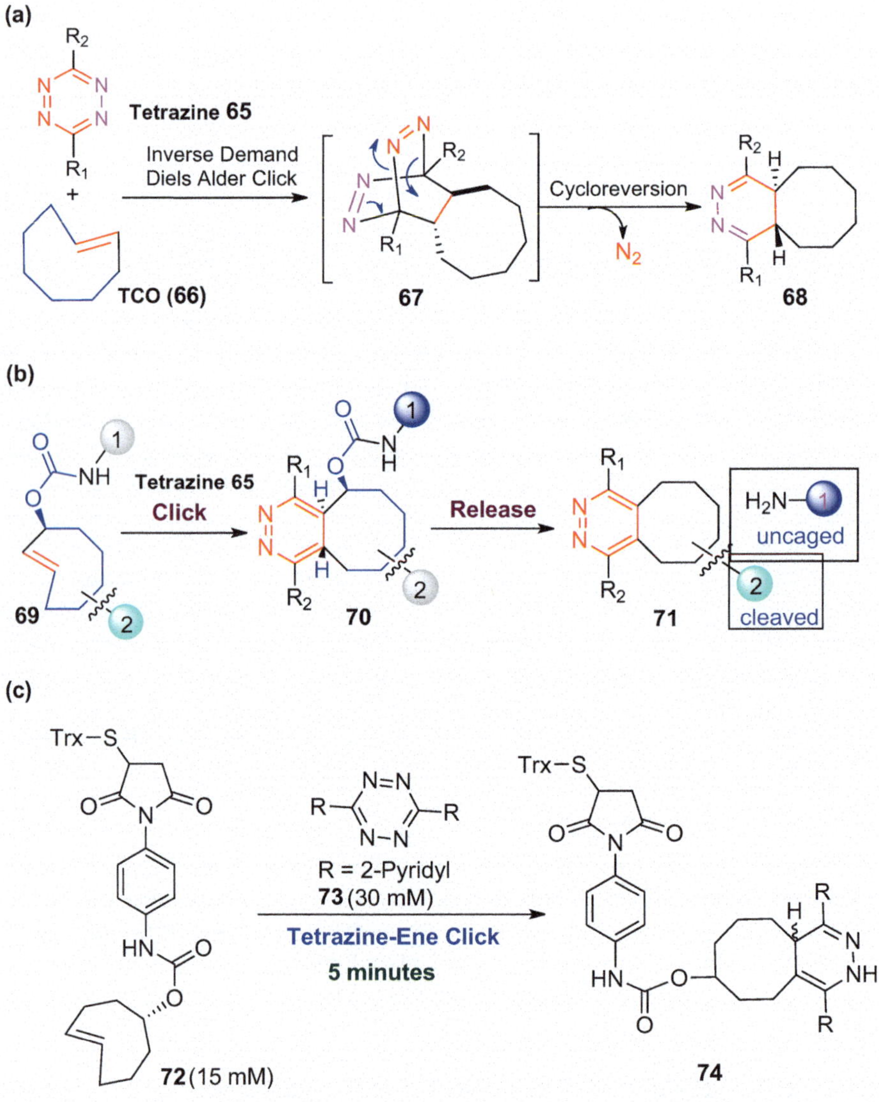

Scheme 1.9 Strain-promoted Tetrazine-Alkene Click Chemistry as bioorthogonal tool

with pH sensitivity and works well with acid-functionalized tetrazines with significantly enhancing the addition and eventually leading to dead-end isomer. A recent report demonstrated an expeditious way to overcome the issue that resulted in such dead end. The method works well with exceptional efficiency and complete with absolute release across the overall range of physiologic pH, potentiating drug delivery tool and moreover expand the dynamic range of bioorthogonal on/off control (Scheme 1.9b) [84].

This catalyst-free cycloaddition reaction of s-tetrazine **73** and trans-cyclooctene **72** proceeds with unusually fast reaction rates with tolerance of broad range of functional group present in organic solvents, aqueous media, cell media, or cell lysate and resulting compound **74** was isolated in high yield. The rate of the ligation of 3,6-di(2-pyridyl)-s-tetrazine and trans-cycloocotene is fast (k2 2000 M^{-1} s^{-1}), and this reactivity enables protein modification at low concentration (Scheme 1.9c) [85]. Similar to the Bertozzi's SPAAC tool, the Tetrazine-Alkene click method can also explored well in biological systems without interacting or interfering with other biological processes.

Another variant of strain-promoted click reaction enabled *rapid tunable reactions for bioorthogonal labeling* is the 'strain-promoted alkyne-nitrone cycloaddition (SPANC)' that involves the reaction of diaryl-strained-cyclooctynes, for example, dibenzylcyclooctyne (DIBO) with 1,3-nitrones to give high yield of *N*-alkylated isoxazolines [86]. SPANC-mediated isoxazoline product is although not as much stable in comparison with the SPAAC-mediated 1,2,3-triazole product. This can undergo rearrangements at biological conditions, despite the SPANC tool because of notably fast reaction kinetics [87], is quite useful for live cell labeling and also allows for multiplex labeling [88]. In short, the strain-promoted bioorthogonal tool allows the selective labeling and targeting of biomolecules in living organism and thus has great scope in chemical biology, particularly for in vivo imaging, and can also be used in drug delivery and thus very applicable in drug discovery and development.

1.7 Some Notable Features of 1,2,3-Triazole Forming Click

A copper-free azide-alkyne click reaction of azide-functionalized polyvinyl chloride **75** (PVC) with internal alkynes **76** was performed to develop 1,2,3-Triazole analogues of phthalate plasticizers **77** such as covalently linked 1,2,3-triazole to PVC such as di(2-ethylhexyl)-1*H*-triazole-4,5-dicarboxylate (PVC-DEHT), di(*n*-butyl)-1*H*-1,2,3-triazole-4,5-dicarboxylate (PVC-DBT) and di(methyl)-1*H*-triazole-4,5-dicarboxylate (PVC-DMT) (Scheme 1.10a) [89].

m-nitrophenyl α-galactosidase **79** and propargylated sugar **80** can be easily linked with an azido-functionalized dextran **78** through CuAAC-mediated 1,2,3-triazole linker and the resulting multivalent 1,2,3-triazole-appended glycan **81** displayed efficient as blood group dependence cholera toxin IC_{50} via inhibiting both AB_5 and GM1 toxins (Scheme 1.10b) [90].

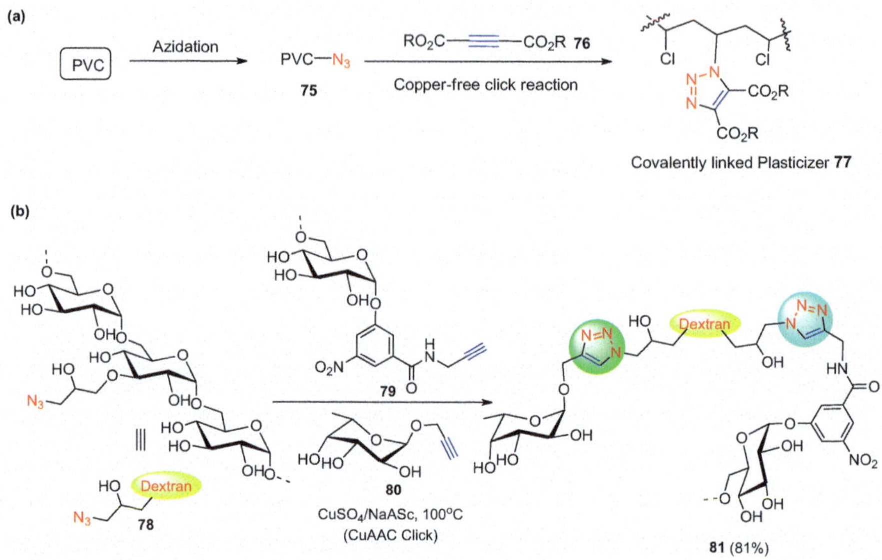

Scheme 1.10 **a** Copper-free azide-alkyne click chemistry for the synthesis of 1,2,3-triazole analogues of phthalate plasticizers. **b** Synthesis of multivalent glycopolymers as cholera toxin

Interestingly, the modular CuAAC tool was further explored to couple oligomannose-containing azides **85a** with sugar immunogen via 1,2,3-triazole linker and the resulting virus glycoconjugate **86** was used to target HIV-1. In step 1, the surface amino groups were acylated using excess of alkynyl *N*-hydroxysuccinimide ester **82** and then ligand **7** promoted CuAAC conjugation with azide **85a** was considered to give desired virus glycoconjugate **86** (Scheme 1.11) [91]. Interestingly, the Qβ glycoconjugate with oligomannose cluster was bound the antibody 2G12 with increased affinity and the tool has growing scope to develop carbohydrate-based vaccine for possible treatment of HIV.

The established virus glycoconjugate synthesis was further explored to highlight a very weak tumor-linked sugar antigen TACA on bacteriophage Qβ virus-conjugates **85b,c** (Scheme 1.12) [92]. Glycan microarray was used to renovate the selectivity of resulting antibodies with Tn antigens, and target IgG antibodies were detected to tether with the parental Tn antigens on human leukemia cells. This well-known virus-like tool has an efficient antigen delivery vehicle and can be used to discover an effective glycan-based anti-cancer vaccine.

Furthermore, the sequencing of a lactose-based probe can be achieved via CuAAC took required for the successful detection of the carbohydrate-binding protein interactions of galectin-1. For example, ethylene glycol linker was linked to the azide functionality at terminus pertaining to chemoselective CuAAC click ligation to meet the water-solubilizing properties and the resulting labeled proteins **91** can be easily in-gel visualization by 'clicking-on' a rhodamine to give **93** (Scheme 1.12) [93].

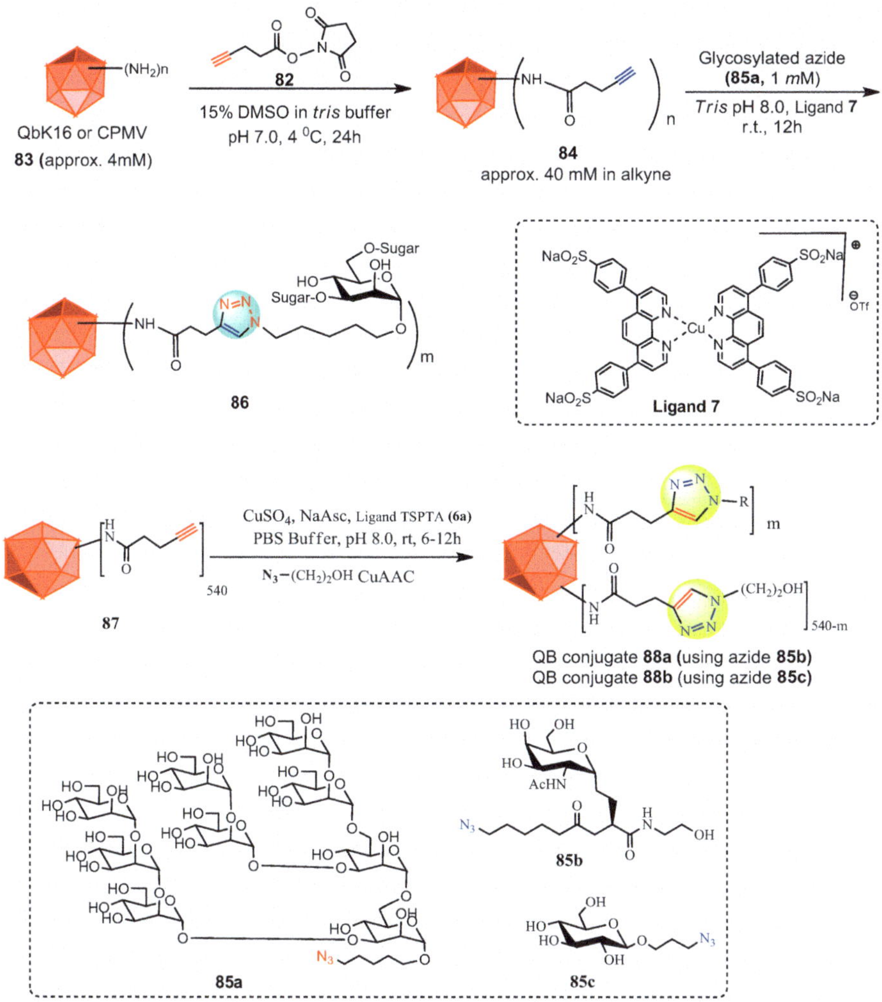

Scheme 1.11 Synthesis of virus glycoconjugate (**86** and **88**) by means of 'CuAAC Click Chemistry'

To attain the desired regio- and stereo (anomeric) control in glycosylation methodologies, intramolecular glycosylation methodology can be considered as powerful tool [94]. Schmidt and Tiwari in 2012 successfully explored click-mediated 1,2,3-triazole as rigid spacer in IAD intramolecular glycosylation to achieve the high selectivity in glycosylation product. Thus, glycosyl donor, e.g., phenyl 3,4,6-tri-*O*-benzyl-2-*O*-propargyl-1-thio-D-glucopyranoside **94** was first clicked with glycosyl azide **95** as part of *O*-(2-azidomethylbenzyl)-protected glycosyl acceptor under standard CuAAC condition (CuI/DIPEA in DCM) to give respective disaccharide triazole **96** in 96% yield. The *p*- PMB group was then removed by treating it with DDQ in

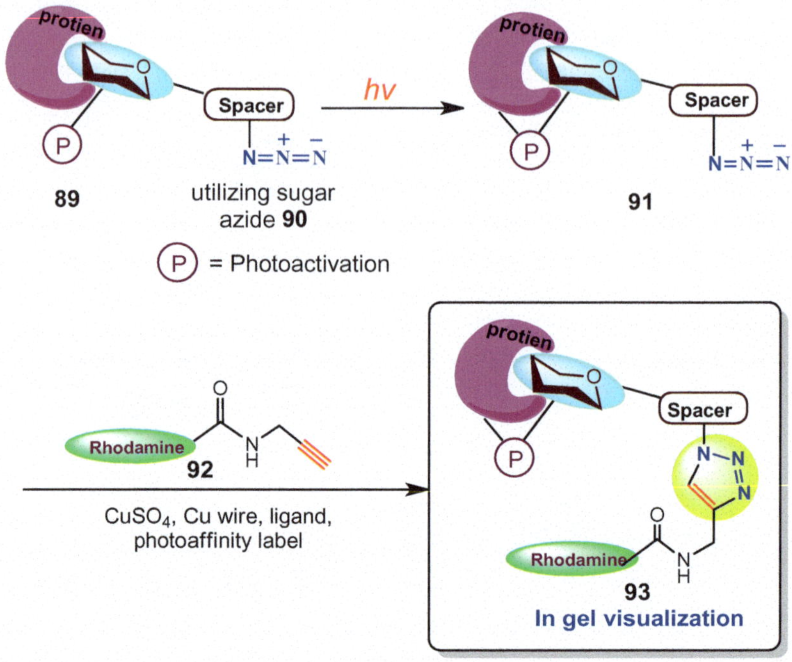

Scheme 1.12 CuAAC Click-mediated lactose-based probe for the recognition based on selective tagging of proteins ligation for in-gel visualization

CH_2Cl_2/H_2O at rt for 2 h and the resulting donor–acceptor system linked through CuAAC triazole **97** was subjected to intramolecular glycosylation through IAD using NIS/TfOTMS in DCM for 2 h to give the disaccharide-containing triazole-linked macrocyclic system **98** in 58% yield and >98% β-selectivity as a part of fifteen-membered ring formation (Scheme 1.13) [95]. Later on, Schmidt et al. successfully cleaved the 1,2,3-triazole rigid spacer from the resulting triazole-appended macrocycle disaccharide by applying Birch conditions (Na/Liq NH_3 at −75 °C) respective free disaccharide was obtained in high yield [96].

Carbohydrate-containing macrocyclic systems because of their unique properties like drug-likeness, ion complexation capacity, as well ligands for catalysis, and many more fascinating features have received considerable interests to the scientific community [97]. Toward this end, the versatility of the CuAAC click conjugation was well utilized for the development of complex cyclic molecular architectures. For example, octasaccharide macrocycle **100** was developed by utilizing Chen's one-pot multi-enzyme protocol followed by CuAAC click [98]. Thus, α-2,6-linked sialoside product **99** having azide at *C*-5 position of sialic acid and alkyne at the terminal end of galactose was obtained by OPME route which was then underwent facile macrocyclization under intramolecular click condition [CuI/DIPEA, MeCN/H_2O (1:1)] and afforded the 76-membered water-soluble macrocyclic product **100** in 78%

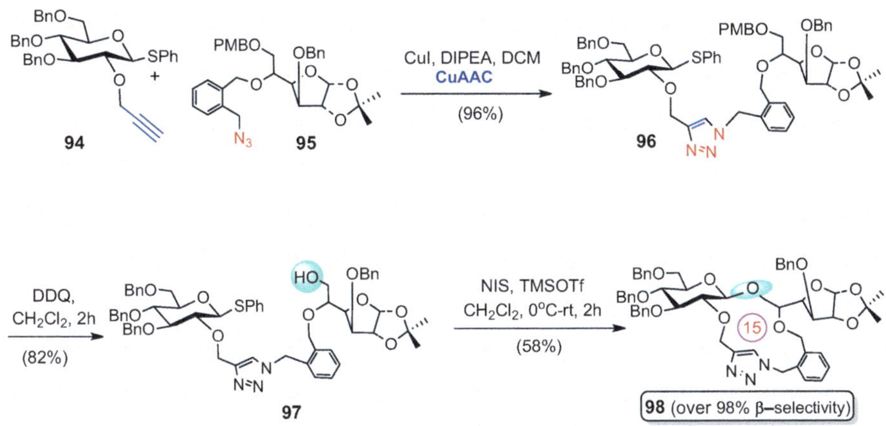

Scheme 1.13 CuAAC-mediated 1,2,3-triazole as the rigid spacer in intramolecular reaction

yield (Scheme 1.14) [98]. This hydrophilic molecule has potential use in the host–guest chemistry and the size-dependent interaction study with other hydrophobic molecules. Also, carbohydrates can modulate and facilitate several important cellular interactions in biological systems and can used for the study of the carbohydrate-protein interactions [99]. A number of carbohydrate-containing molecules including glycohybrids, glycoconjugates, neoglycoconjugates displayed promising biological activities and played very important role in biological systems [3, 99].

CuAAC conjugation in nucleic acid polymers is profitable when the cycloaddition was performed in the aqueous buffer in the presence of some organic solvents, e.g., DMSO, DMF, or NMP simply to avoid 'bury' the reactive groups. Notably, the CuAAC in 20% CH_3CN in buffer was as effective as performed in the presence of any ligand in the CuAAC labeling of nucleic acid strands. Once alkyne and azide modified strands were brought in close proximity by a complementary oligonucleotide, this resulted to act both the modified sequences as a template [100]. In the other hand, macrocycles can be used in the active template CuAAC reaction (known as AT-CuAAC) for the development of interlocked molecular machine systems. In AT-CuAAC, a metal ion bound within the macrocycle was used to organize fragments for the CuAAC click ligation as template AAC to capture the interlocked structure. For example, the templating macrocycle AT-CuAAC tool was successfully applied for the template synthesis of mechanically interlocked dinucleotide **103** by click conjugation of alkyne-modified cytosine **101** with an azide-modified thymine **102** under standard click catalysis (in the presence of a templating macrocycle **103** (Scheme 1.15) [101]. Similar AT-CuAAC strategy is widely utilized to develop a number of rotaxanes and catenanes [102].

This modular CuAAC tool was brilliantly used to couple azide **104** and alkyne **105** via 1,2,3-triazole linker under standard CuAAC conditions to give high yield of non-ionic xylo-nucleoside dimer that is composed of TL-t-T xL, TL-t-ABzxL, and TL-t-CBzxL **106** (Scheme 1.16a) [103]. Interestingly, CuAAC Click conjugation of

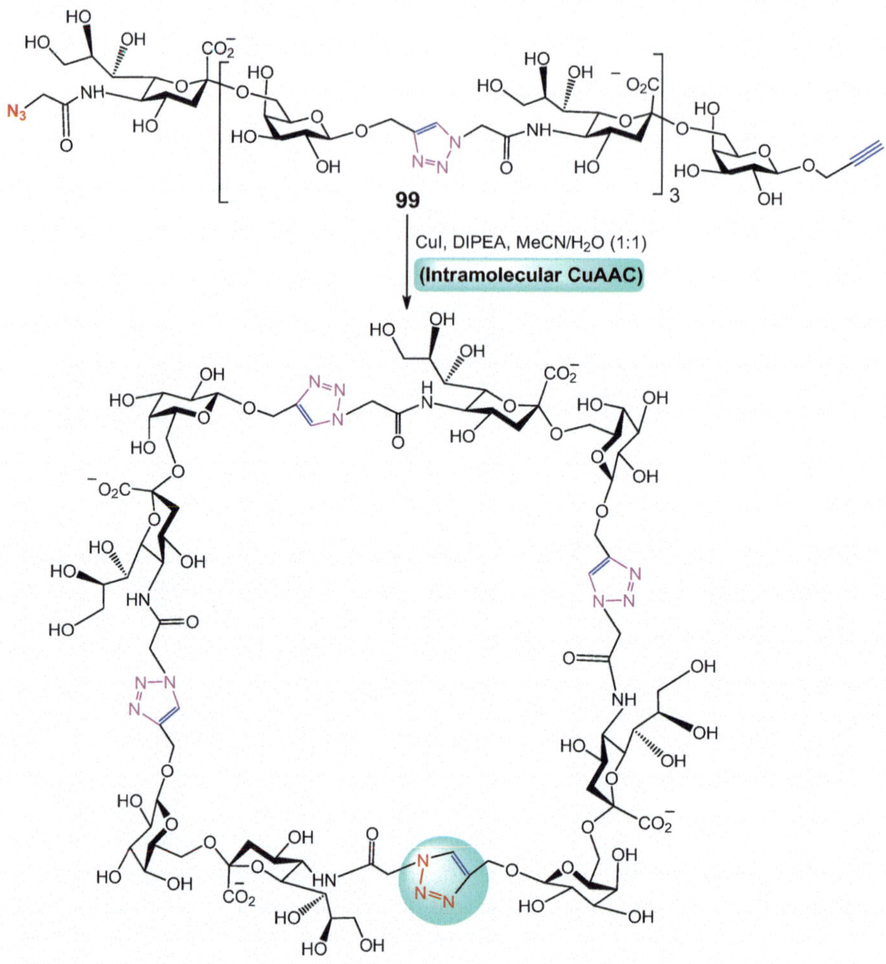

Scheme 1.14 Efficient synthesis of sialic acid-containing macrocycle using intramolecular CuAAC tool

two complementary oligonucleotides gave massive dsDNA catenanes which were composed of six helical turns, with T–T mismatches on either side of ten base pairs. Two strands that had been labeled with a 5′-alkyne and a 3′-azide were employed in a step-wise manner to provide the dsDNA catenane macrostructure. The non-templated CuAAC coupling of the first oligonucleotide was first cyclized via intramolecular fashion which was then served as a template for the second oligonucleotide's click ligation followed by cyclization (Scheme 1.16b) [100].

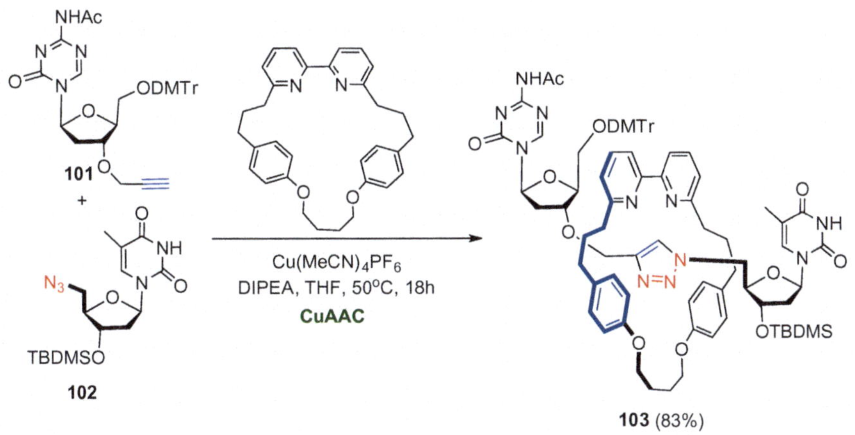

Scheme 1.15 Templating macrocycle for the synthesis of an interlocked dinucleotide **103** via AT-CuAAC tool

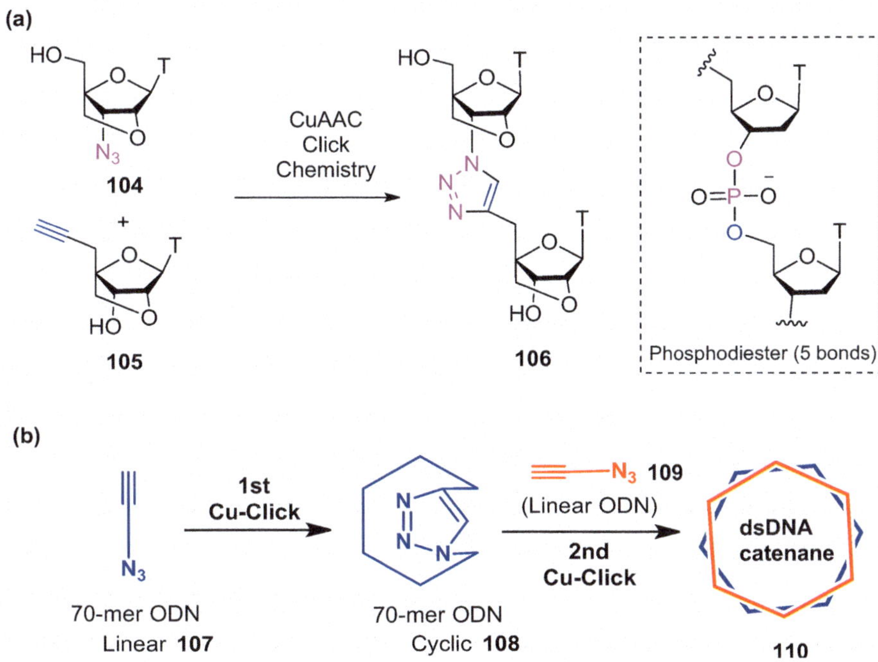

Scheme 1.16 Click Chemistry in synthesis of unnatural triazole-nucleosides and dsDNA catenane

The copper(I)-catalyzed 1,3-dipolar cycloaddition reaction, often referred to as 'CuAAC' or 'Click Chemistry', stands out as an exceptionally successful, adaptable, dependable, and versatile approach for the efficient and regioselective synthesis of 1,4-disubstituted 1,2,3-triazoles with diverse functionalizations. Carbohydrates, essential components of living cells, exhibit remarkable qualities, including structural diversity, biocompatibility, bioavailability, hydrophilicity, excellent ADME properties with the least toxicity [96, 97]. Consequently, there is a growing interest in exploring carbohydrates as versatile frameworks for the facile creation of a wide range of glycohybrids and glycoconjugates, enabling comprehensive chemical, biochemical, and pharmacological investigations [3]. This chapter focuses on the successful integration of CuAAC or 'Click Chemistry' into emerging fields related to nucleic acids, including their modifications through suitable linkage with 1,2,3-triazole, and explores their numerous applications in these emerging domains.

Click-derived 1,2,3-triazole serves as a most suitable spacer connecting to the two units to deliver water-soluble molecular architectures having prolific applications in catalysis, chemical biology, drug career, etc. Furthermore, this modular 'CuAAC Click Chemistry' tool has been widely explored to develop well-designed neoglyoconjugates (glycopolymer, glycopeptides, and glycodendrimers). For a representative example, Tiwari group used modular CuAAC to develop peripheral β-D-glucopyranose-coated porphyrin-cored dendrimer using the high-yielding convergent route (Scheme 1.17) [104]. Phenolic ester **111** was first propargylated to give compound **112**, which was then subjected to LAH reduction followed by chlorination and afforded high yield of respective alkyne **113**. Resulting alkyne **113** was coupled with glycosyl azide **114** under CuAAC condition followed by routine azidation to give azide-functionalized dendritic scaffold **115.** This azide dendron was click conjugated with alkyne 116 followed by azidation, and the resulting dendron 117 was finally clicked with the tetra-alkyne-functionalized porphyrin core **655** to give high yield of desired glycodendrimer **660** (Scheme 1.17) [104].

The similar convergent synthetic route was extended for the synthesis of porphyrin-cored hybrid dendrimers [105]. In addition to the absorption–emission behavior and photophysical properties, such porphyrin-cored dendrimer systems can be used in photodynamic therapy (PDT) and certainly have immense potential to affect biological events label to consider as possible therapeutic candidates for fatal diseases like cancer. Because of its highly facile, modular nature, and excellent regioselectivity in triazole synthesis, this CuAAC Click chemistry is widely explored in diverse range of dendrimer synthesis. In addition to the cationic or anionic dendrimers, click-mediated dendrimers are less toxic and have similar number of possible sites available to bind with an active species and thus its multivalence makes it therapeutically more potent [4].

Similar to the effect of metal complexation, binding of analytes, linker, and fluoroprobe, and reporter, the click-mediated 1,2,3-triazoles may take part in the detection of several chemical characteristics through sensing properties. Toward this end, a conformationally constrained (coumarin-pyrrolidinyl-triazolyl-bipyridyl) chemosensor was reported for the selective detection of Al^{3+} ions [106]. The synthetic route for triazole framework involves click conjugation of Boc-protected pyrrolidinyl

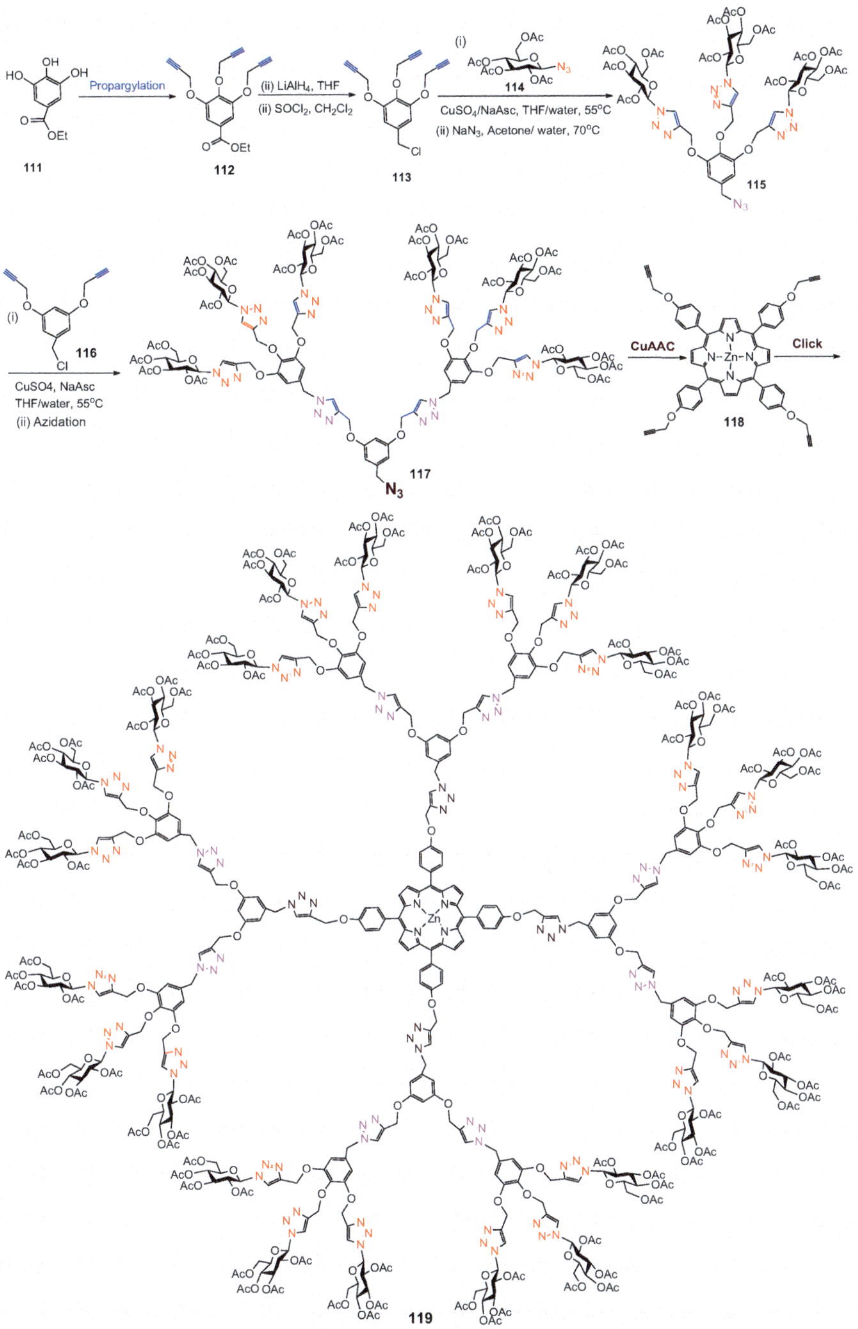

Scheme 1.17 Click-mediated porphyrin-cored glycodendrimer using the convergent route

azide **120** with bipyridyl-alkyne **121** followed by Boc deprotection produces triazole **122** (Scheme 1.18) [106]. Subsequent coupling with coumarin-3-carboxylic acid **123** furnished the desired 1,2,3-triazole framework **124** which serves as an ionophore by providing one of the cations coordinating sites together with bipyridyl moiety. The coumarin-bipyridyl complex **124** shows enhanced fluorescence with high sensitivity in the presence of Al^{3+} ions over the presence of most competing metal ions which was based on an internal charge-transfer (ICT) mechanism. Likewise, a sugar-derived *bis*-triazolyl chemosensor **125** was reported as promising candidate for the selective detection of Cu(II) ions [107]. A number of simple hybrid compounds were shown to be promising candidates for the efficient detection of chiral molecules, especially glucose in supramolecular systems, and could be used in the treatment of diabetes mellitus [108]. Interestingly, a fluorescent boronic acid hybrid **126** obtained through click chemistry displayed as a promising modular sensor for D-glucose (Scheme 1.18) [109].

Although a number of 1,2,3-triazole-appended molecules were displayed promising for the selective sensing of cations and anions, the prime objective should be the selective sensing of chiral analytes. An interesting representative example includes a simple fluorescent boronic acid 1,2,3-triazole hybrid **126** and was identified as a promising modular sensor for the successful detection of D-glucose in living system [109].

In recent years, click-mediated molecular hybridization through the combination of two active pharmacophoric units of either similar or different bioactive units into a single pharmacophoric framework with improved affinity and efficacy is highly

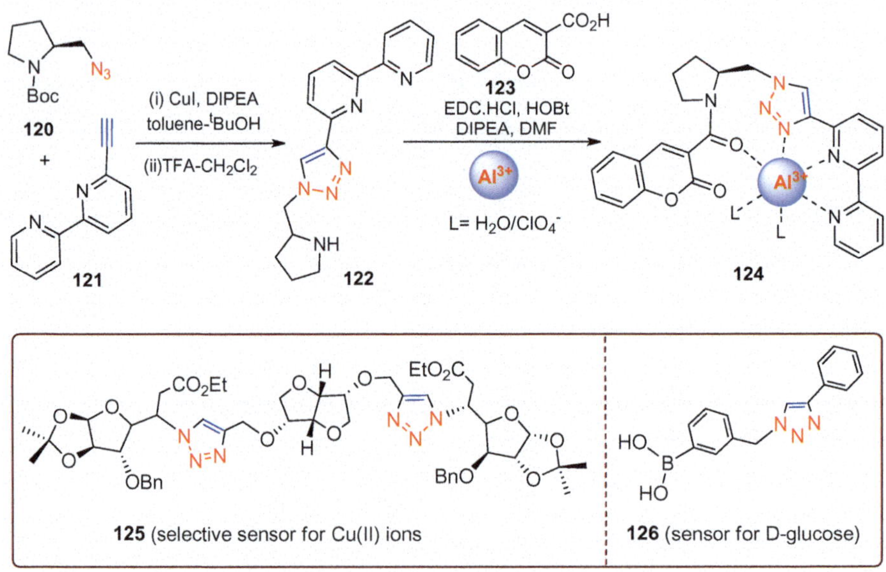

Scheme 1.18 Conformationally constrained (coumarin-triazolyl-bipyridyl) click **124** and α-amidoboronic acid **126** for the chiral sensing in supramolecular assembly

considered approach in medicinal chemistry [110]. This molecular hybridization tool having 1,2,3-triazole as a biocompatible linker is nowadays widely explored for the successful development of a novel, safe, and fast-acting drug candidates [111]. For example, a number of 1,2,3-triazole-based hybrid molecules (**127–132**) as potent drug candidates (Fig. 1.5a) are found in clinical trials for the treatment of several front-line diseases. Followed by Tiwari et al. explored CuAAC modular method to develop a water-soluble curcumin glycoconjugate **133** in order to potentiate conventional anti-microbials by inducing apoptosis-like phenomena among drug-resistant bacteria (Fig. 1.5b) [112]. This water-soluble curcumin glycoconjugate even sensitizes resistant gram-negative clinical isolates to the gram-positive specific antibiotic vancomycin, thereby expanding the anti-bacterial spectrum of curcumin.

A report demonstrated the total synthesis of an anti-seizure medication 'Rufinamide' using continuous-flow chemistry [113]. The approach is convergent and exhibits a [3 + 2] click cycloaddition reaction catalyzed by a copper tubing reactor. Each of the three chemical stages has several advantages and substantial benefits by being carried out in flow chemistry. Utilizing flow chemistry, a Rufinamide drug **17** was developed in an overall 92% yield with an average residence time of about 11 min (Fig. 1.5c).

Domino or tandem reactions have a notable feature as the processes avoid the isolation of intermediates involved in the synthetic steps and thus offer a highly efficient way to generate a diverse variety of 1,2,3-triazole-containing molecules of notable applications in medicinal chemistry. Thus, a sequential Lewis acid-catalyzed intermolecular Michael addition and an intramolecular azide-internal alkyne click cycloaddition under one-pot condition were applied for rapid synthesis of indole-based polyheterocycles using [114]. The method involves $Yb(OTf)_3$-mediated Michael addition of the 2-alkynyl indoles **135** with (*E*)-1-azido-2-(2-nitrovinyl)benzenes **134** in refluxing toluene to give respective azido-alkynes **136** which underwent thermally induced intramolecular click cycloaddition and afforded high yield of biologically relevant indole-based tetracyclic products **137** (Scheme 1.19a) [114].

Further, a thermally induced click reaction was used to develop carbohydrate-containing morpholine-fused triazoles **140** in two steps [115]. Glycosyl alkynes **138** on treatment with epichlorohydrin and NaN_3 in the presence of $CuSO_4.5H_2O$/NaAsc in H_2O at rt afforded azido alcohol **139** which on propargylation followed by thermally induced intramolecular click cyclization gave high yield of biologically relevant morpholine-fused [5,1-*c*] triazolyl glycohybrids **140** (Scheme 1.19b). Likewise, azido-alkyne **142,** obtained from isatin epoxides **141** using azidation (via regiospecific ring opening) followed by *O*-propargylation on thermally induced click cycloaddition in refluxing toluene afforded god-to-excellent yield of spiro-oxindole-tetracyclicmorpholine-fused-1,2,3-triazoles **143** (Scheme 1.19c). Spiro-fused glycohybrids **143** displayed potent anti-proliferative activity against the A549 cell line (IC_{50} = 1.87–4.36 μM) to the reference standards doxorubicin and 5-flourouracil [116]. The above-mentioned example on intramolecular azide-alkyne cycloaddition 'Click Chemistry' is nowadays widely utilized as modern tool for the cyclization

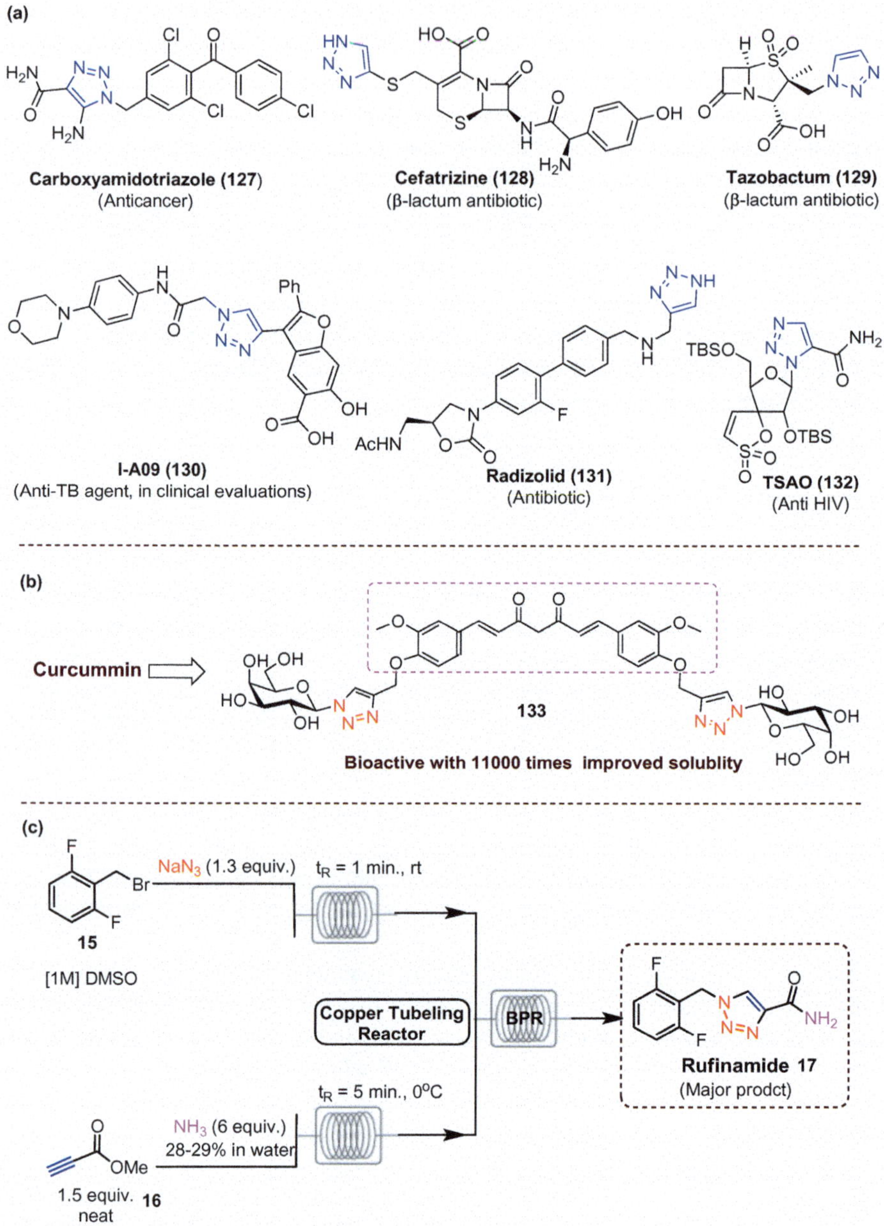

Fig. 1.5 Click chemistry-mediated 1,2,3-triazoles in medicinal chemistry

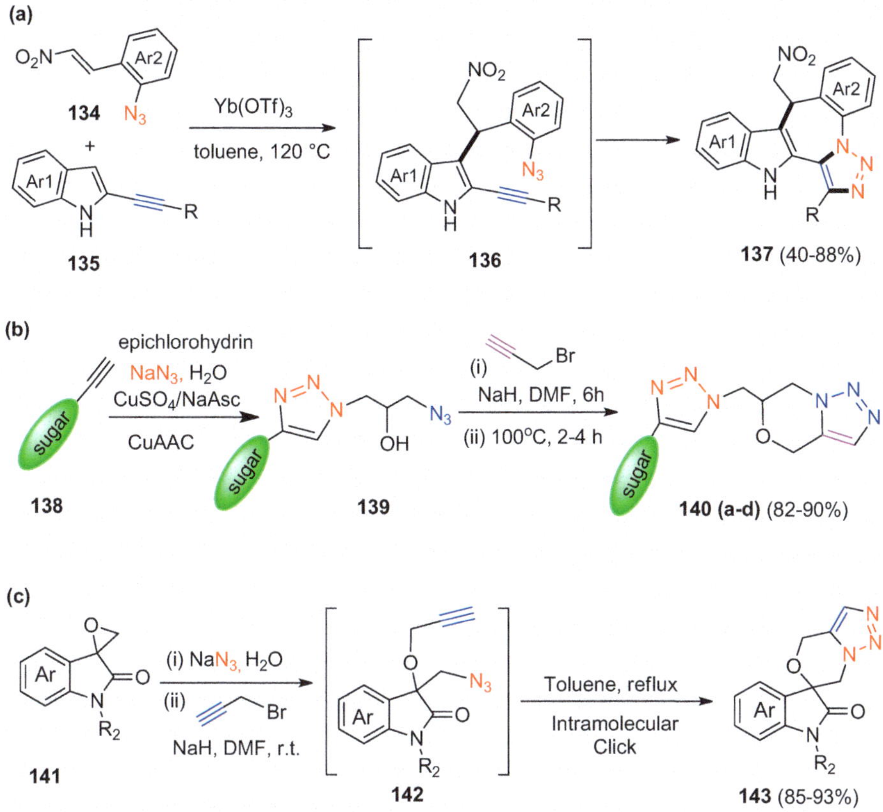

Scheme 1.19 One-pot expeditious synthesis of indole-based tetracyclic architecture **137** and spiro-oxindole-tetracyclicmorpholine-fused-1,2,3-triazole **143**

leading to the synthesis of number of fused-heterocycles, peptidomimetics, and macrocyclic architectures of wide applications [117].

The click-mediated 1,2,3-triazole-appended glycohybrids have demonstrated their effectiveness in catalyzing various chemical transformations as metal ligands as well as organocatalysts, despite the fact that there are very few instances of these compounds in the field of catalysis [118]. The presence of multiple hydroxylfunctional groups provides access to modify them to multiple binding sites, which further improves the catalytic efficiency even with very low catalytic loading, the chiral pool of the carbohydrate moiety contributes to the ability of glycosyl triazoles to be a good source to induce asymmetry in a reaction. The modular click protocol was utilized to develop a triazole-based pyrrolidine glycoconjugates **147** that was successfully explored as an organocatalyst for enantioselective Michael addition reaction of nitro olefins **145** with ketones **144** to afford selective addition product **146** in good yields and stereoselectivity (Scheme 1.20) [119].

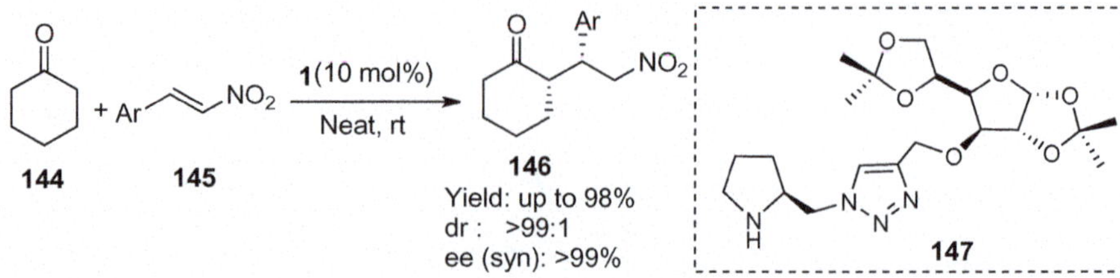

Scheme 1.20 Click-mediated organocatalyst **147** catalyzed stereoselective 1,4-conjugate addition

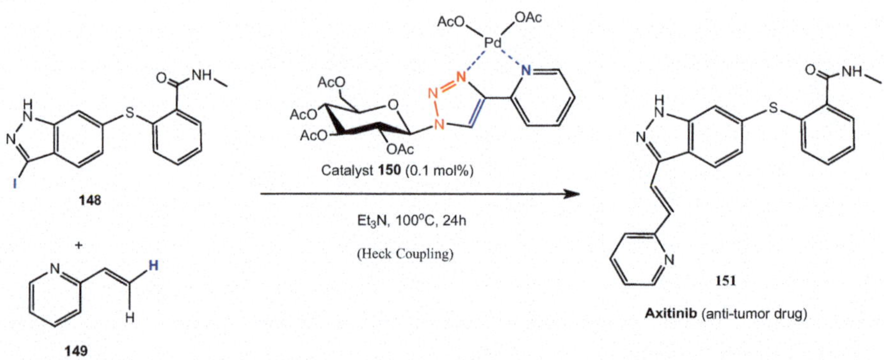

Scheme 1.21 CuAAC-mediated glycobrid 1,2,3-triazole Pd-complex **150** as promising catalyst in heck coupling leading to Axitinib, an anti-tumor drug

Further, a glycobrid 1,2,3-triazole in the form of a Pd-complex **150** was used as ligand for the synthesis of a well-known anti-tumor drug Axitinib **151** (Scheme 1.21). The coupling proceeds via Heck coupling of iodoarene **148** with olefin **149** in the presence of 0.01 mol% of catalyst under solvent-free condition as environmental benign condition [120].

Tiwari and co-workers explored the CuAAC and developed a number of triazolyl sugar analogues for various purposes [121]. The resulting 1,2,3-triazole ligands were explored as efficient metal ligands in various coupling reactions to achieve libraries of biologically important heterocyclic scaffolds as well as alkynes under mild reaction conditions. For example, the group developed click-mediated triazole-appended glycohybrid ligand **154** and reported the CuI/ligand **154** catalytic system as effective one for the intramolecular Ullmann type coupling reaction of *o*-halo aromatic amides or thioamides or benzimidines **152** that gave high yields of respective benzoxazoles or benzothiazoles or benzimidazoles **153** (Scheme 1.22a) [122]. Next the click-derived triazolyl glycohybrids ligand **155**/CuI catalytic system was explored in temperature-dependent Sonogashira and Glaser-Hay coupling reactions of terminal alkynes (Scheme 1.22b) [123]. Moreover, the group recently developed an array of mono- and bis-triazole-appended glycohybrids for the Sonogashira and Glaser-Hay coupling reactions of alkynyl sugars to attain various alkyne containing glycohybrids including glycosyl rods [124]. In addition to coupling reaction, the domino reaction began with metal-catalyzed coupling followed by cyclization were also demonstrated by the group for synthesis of various biological demanding quinazolinone and isoquinolinone analogues. The pyridyl glycosyl triazole ligand **156** was synthesized with the assistance of modular click reaction and explored in the CuI-catalyzed domino cross-coupling reaction of various *N*-substituted *o*-halobenzamides **157** with formamide and cyanamide to assemble 3-substituted quinazolinones **159** and 2-amino-3-substituted quinazolinones **160**, respectively (Scheme 1.22c) [125]. The group recently developed glycosyl triazole-based pyridinamide ligand **161** using click approach. The CuI/ligand **161** catalytic system was found to be efficient for the synthesis of dihydrophenanthridine-diones **165** and substituted isoquinolinones **166**, through the domino reaction of *N*-substituted *o*-halobenzamides **162** with dimedone **163** and active methylene compound **164**, respectively (Scheme 1.22d) [126].

Further, multicomponent reaction under domino fashion is identified as one of the prominent tools for the development of biologically potent scaffolds [127]. These emerging features may lead to the cyclization with great ease by variety of established methodologies including regioselective 1,2,3-triazole forming methodologies to generate diverse range of well-defined macrocyclic architectures as well fused ring systems. For example, 1,2,3-triazole containing appended macrocycle, for example, glucosyl spiro-pyrrolizidine grafted macrocycle **170** was obtained in 72% yield by reacting *N*-alkyl azidoisatin **169** with secondary amino acid, e.g., proline **168** and glucose-derived *O*-propargyl enone **167** in the presence of CuI/DIPEA via tandem one-pot MCR fashion (Scheme 1.23) [128]. This tandem double 1,3-dipolar click cycloaddition reaction is high yielding, stereoselective as well as regioselective and formation of spiromacrocycle involves the concurrent formation of five chemical bonds under one-pot fashion. The method has great scope and certainly can be explored further for the drug development. The detailed concept with their applications and challenges ahead is presented in Chap. 18.

Followed by 'Click chemistry', another powerful and reliable tool explored for the successful development of the radiolabeling of radiopharmaceutically biorelevant motifs is recognized as 'the CuAAC-to-chelation' approach [129]. Mindt's 'Click-to-chelation' tool involves the use of regioselective modular CuAAC conjugation of azide **171** and alkyne **172** followed by chelation with metal carbonyls including technetium and rhenium tricarbonyl cores to afford tridentate triazolyl metal chelators **174** under one-pot condition with higher specific activity, higher decay-corrected radiochemical yield (Scheme 1.24) [129]. The resulting candidates supportive for building tridentate chelating system bearing stable radio-metal complexes in the core can also be predictable with other significant biomolecules, including phospholipids, bombesin, thymidine, etc., for the notable biomedical applications including novel radiotracers in the real dynamics of required clinical use.

The growing impact and challenges ahead toward the post-functionalization of click-mediated regioselective 1,2,3-triazoles have been presented in separate chapter

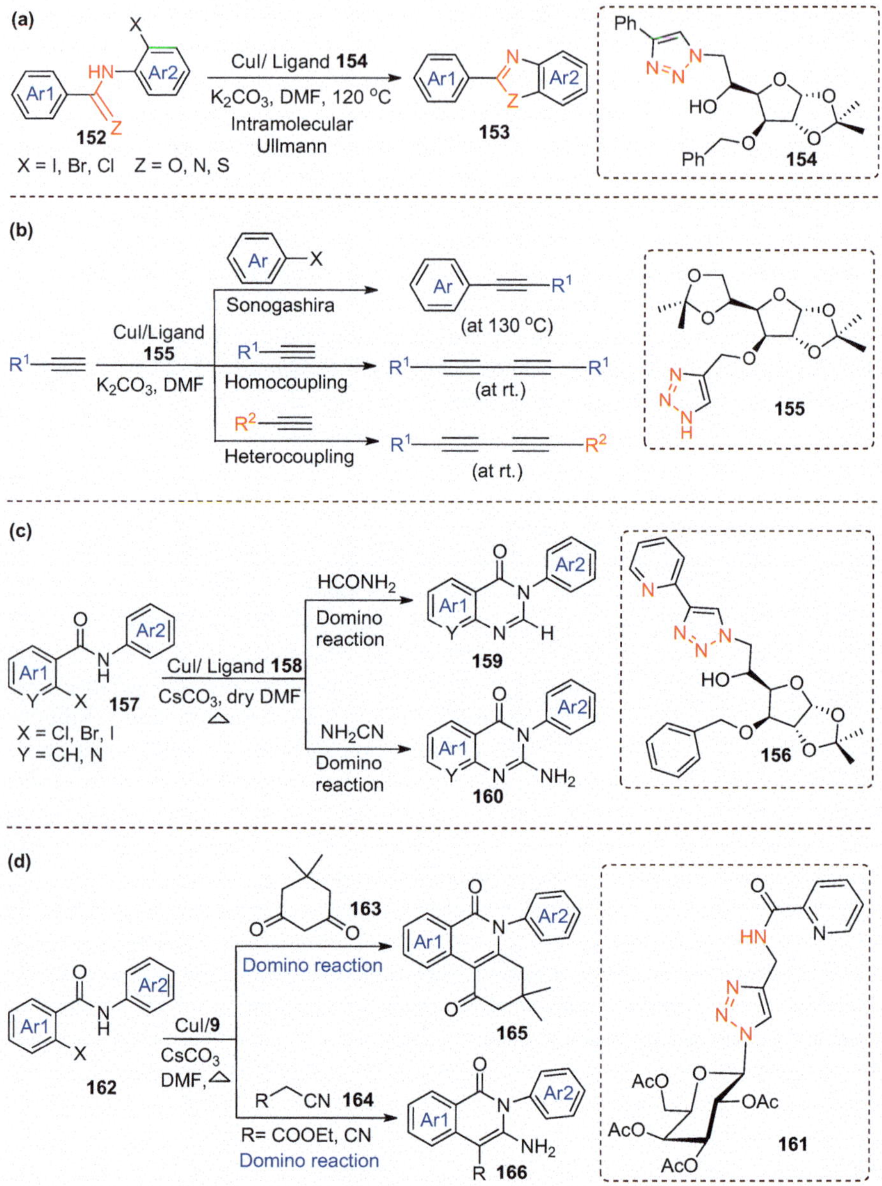

Scheme 1.22 Click-derived 1,2,3-triazole-appended glycohybrids/CuI-catalyzed reaction in organic synthesis

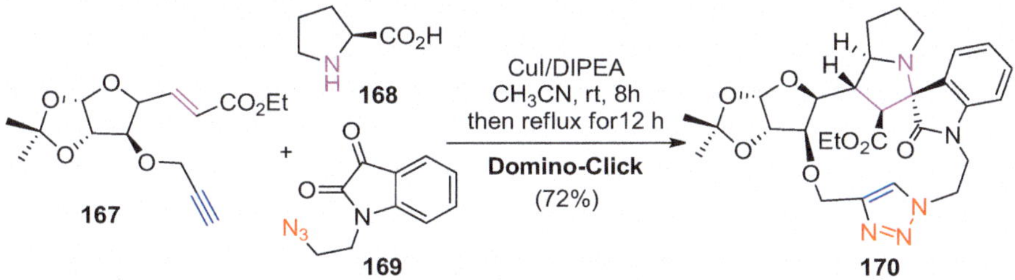

Scheme 1.23 Tandem one-pot double 1,3-dipolar cycloaddition reaction for the synthesis of spiro macrocycles **170**

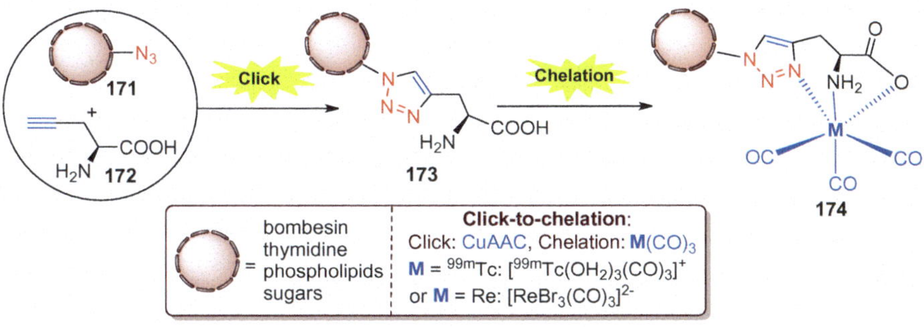

Scheme 1.24 Mindt's 'Click-to-Chelate' tool for the radiolabeling purpose

of this book (Chap. 19). The aforementioned applications of click reaction emphasized the importance and utility of various forms of 'Click Chemistry' is well-explored in modern organic synthesis. Among them, triazoles delivering click reactions have often been investigated in the field of organic synthesis for the development of an array of compounds including macromolecular architectures [3, 4]. In recent times, enthusiasm for the post-modification of click triazoles, generally referred to as interrupted click reaction, has steadily increased beyond the CuAAC. In this context, interrupted click cyclization is one of the most commonly used procedures for fashioning fused triazole architectures, that are 1,5-fused as well as 4,5-fused triazole cores in an intramolecular fashion using diverse metal catalysts [130–132]. In this endeavor, diverse range of compounds were synthesized, namely triazoloquinazolinones **176**, catalyzed by copper catalyst (Scheme 1.25a) [133], a programmable multicatalytic system of copper and palladium afforded fused polycyclic triazole **178** (Scheme 1.25b) [134], and 5-acyl triazoles **180** were constructed by use of a basic CuI/2,2′-bipyridine-based catalytic pathway (Scheme 1.25c) [135]. Moreover, a report on click-sulfenylation reaction leads to the formation of benzoxathiocino- or related benzothiazocino-fused triazole **182**, using post-functionalization approach was documented (Scheme 1.25d) [136]. These fused triazoles have particular applications in material chemistry, chemical biology, drug delivery, and development. Furthermore, functionalized 5-substituted 1,2,3-triazolyl analogues **182** and **184** of sulfur, selenium, and other heteroatoms were also constructed by employing a post-functionalization strategy (Scheme 1.25e, f) [137, 138]. In continuation, a significant tactic for ring opening of *N*-sulfonyl 1,2,3-triazole **187** was developed via denitrogenative route under rhodium catalysis for an easy access of α-sulfenylated imines (Scheme 1.25g) [139]. In recent year, 1,2,3-triazoles have been successfully explored as novel surrogates for the diazo compounds for an easy synthesis of diverse range of pharmaceutically important heterocyclic skeletons [140].

Moreover, regarding triazole ring cleavage, Tiwari et al. performed a ring-opening reaction of acyl or thioacyl bounded benz-fused 1,2,3-triazole **189** for the synthesis of biologically relevant molecules such as benzothiazoles **190** and related heterocycles

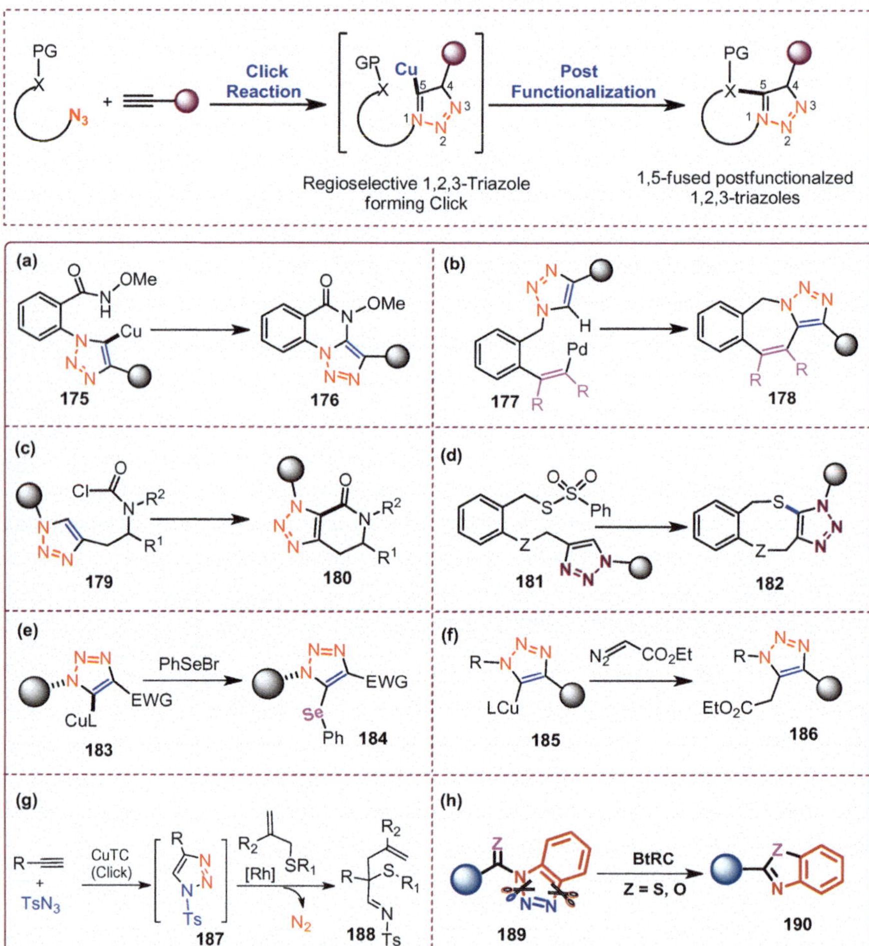

Scheme 1.25 Post-functionalization route for the synthesis of diverse fused and functionalized click triazole

under free-radical conditions (Scheme 1.25h) [141–143]. The growing impact and challenges ahead toward the post-functionalization of click-mediated regioselective 1,2,3-triazoles have been presented in separate chapter of this book (Chap. 19).

1.8 Conclusions and Future Perspective

Click chemistry is an immensely powerful tool for the robust and efficient covalent conjugation of two entirely different molecular entities. The method has found numerous applications across a wide variety of discipline of science and technology. The 1,2,3-triazoles framework possesses some unique salient characteristics such as high chemical stability, strong dipole moment, and hydrogen-bonding ability which motivated chemists to explore this modular process for an easy access of a diverse range of appealing molecules. The versatility and easy synthesis of both the required click azides and alkynes partners provide a basis for its wide acceptance in organic synthesis. Even azide can be masked as a primary amine and can be easily converted to azide through diazotransfer reaction; however, tandem triazole forming reaction under one-pot condition proceeds to give high-yield triazole conjugates. 1,2,3-triazole is an isostere of amide which serves as privileged pharmacophore in several commercial drugs, where 1,4-disubstituted 1,2,3-triazoles exhibited close similarity with Z amide while 1,5-disubstituted triazoles are comparable to E amide. The regioselective 1,2,3-triazole forming click chemistry through [3 + 2] azide-alkyne cycloadditions reaction has been extensively explored in almost all the emerging fields such as carbohydrate chemistry, nucleic acid chemistry, surface chemistry, nanotechnology, polymer chemistry, agriculture, peptidomimetics, drug discovery, material chemistry, sensing, and catalysis (Fig. 1.6). The compatibility of CuAAC reactions with biomolecules like carbohydrates, protein, and lipid further accelerates the development of bioorthogonal-mediated triazoles linked to bioinspired conjugates.

Thiol-ene and related click an important and well-established variant of 'Click Chemistry' have enormous applications for the synthesis of polymeric and dendrimer molecular architectures of great impact in biomedical and material sciences. Furthermore, the SuFFEx click chemistry and newest variant inspired by nature's proclivity is the PFEx tool for well-known phosphorus connections has also enormous impending impact in areas ranging from drug development to material sciences. Altogether, the SuFFEx and PFEx tool is orthogonally to significant variant of 1.23-triazole forming regioselective click chemistry (CuAAC or RuAAC Click).

The applicability of this robust methodology continues to expand exponentially in broad array of scientific domains including DNA modification, surface modification, glycan microarrays, labeling studies, carbohydrate-protein interactions, click-to-catalysis (as promising organocatalysts as well ligands in some important transformations) in several chemical reactions. In the similar way, selective and efficient covalent bond cleavage reactions can also have ample opportunities to tailor the composition and structure of complex macromolecular systems and such 'Clip

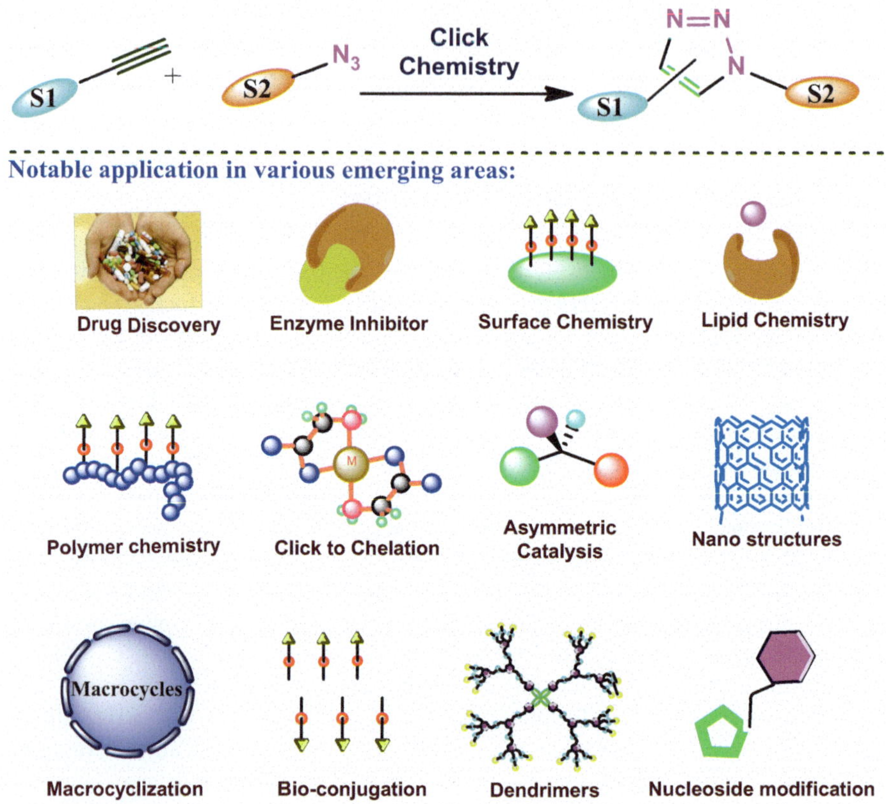

Fig. 1.6 Pictorial presentation of growing perspectives of regioselective 1,2,3-triazole forming click chemistry in various emerging areas

Chemistry' [143] tool is able to address next-generation challenges in chemistry and biology. The emerging aspects including basic as well as advance level information about the both synthetic and applied fields of click chemistry in various emerging areas discussed in the next nineteen chapters of this book titled 'Click Chemistry'.

References

1. Tornøe CW, Christensen C, Meldal M (2002) J Org Chem 67:3057–3062
2. Rostovtsev VV, Green LG, Fokin VV, Sharpless KB (2002) Angew Chem Int 41:2596–2599
3. Tiwari VK, Mishra BB, Mishra KB, Mishra N, Singh AS, Chen X (2016) Chem Rev 116:3086–3240
4. Agrahari AK, Bose P, Jaiswal MK, Rajkhowa SS, Singh AS, Hotha MN, Tiwari VK (2021) Chem Rev 121:7638–7956
5. Thirumurugan P, Matosiuk D, Jozwiak K (2013) Chem Rev 113:4905–4979
6. Chen MM, Kopittke PM, Jhao FJ, Wang P (2024) Trends Plant Sci 29:167–178

7. Meldal M, Tornoe CW (2008) Chem Rev 108:2952–3015
8. Johansson JR, Beke-Somfai T, Stålsmeden AS, Kann N (2016) Chem Rev 116:14726–14768
9. Fantoni NZ, Ei-Sagheer AH, Brown T (2021) Chem Rev 121:7122–7154
10. Fitzgerald PR, Paegel BM (2021) Chem Rev 121:7155–7177
11. Amblard F, Cho JH, Schinazi RF (2009) Chem Rev 109:4207–4220
12. Jha RK, Wooley KL, Nystrom AM, Burke DJ, Kade MJ, Hawker CJ (2009) Chem Rev 109:5620–5686
13. Levandowski BJ, Raines RT (2021) Chem Rev 121:6777–6801
14. Deb T, Tu J, Franzini RM (2021) Chem Rev 121:6850–6914
15. Worch JC, Stubbs CJ, Price MJ, Dove AP (2021) Chem Rev 121:6744–6774
16. Suazo KF, Park KY, Distefano MD (2021) Chem Rev 121:7178–7248
17. Albada B, Keijzer JF, Zuilhof H, Delft FV (2021) Chem Rev 121:7032–7058
18. Bonandi E, Christodoulou MS, Fumagalli G, Perdicchia D, Rastelli G, Passarella D (2017) Drug Discovery Today 22:1572–1581
19. Zhang L, Chen XG, Xue P, Sun HHY, Williams ID, Sharpless KB, Fokin VV, Jia GC (2005) J Am Chem Soc 127:15998–15999
20. Agard NJ, Prescher JA, Bertozzi CR (2004) J Am Chem Soc 126:15046–15047
21. Codelli JA, Baskin JM, Agard NJ, Bertozzi CR (2008) J Am Chem Soc 130:11486–11493
22. Chang PV, Chen X, Smyrniotis C, Xenakis A, Hu TS, Bertozzi CR, Wu P (2009) Angew Chem Int 48:4030–4033
23. (a) Sharpless KB (1999) Proceedings of the 217th American chemical society conference. Anaheim, CA, United States; (b) Kolb HC, Finn MG, Sharpless KB (2001) Angew Chem Int Ed 40:2004–2021
24. Witczak ZJ, Bielski R (2013) Click chemistry in glycoscience: new development and strategies. Wiley-VCH, New York
25. SciFinder: Chemical Abstract Service. http://scifinder.cas.org/. Accessed 3 Jan 2024)
26. Huisgen R (1963) Angew Chem Int Ed 2:565–598
27. Huisgen R (1963) Chem Int Ed 2:633–696
28. Chan TR, Hilgraf R, Sharpless KB, Fokin VV (2004) Org Lett 6:2853–2855
29. Michael A (1893) J Prakt Chem 48:94–95
30. Huisgen R (1989) Pure Appl Chem 61:613–628
31. Huisgen R (1984) 1,3-dipolar cycloadditions-introduction, survey, mechanism. In: Padwa A (ed) 1,3-dipolar cycloaddition chemistry. Wiley, New York, pp 1–176
32. Cella R, Stefani HA (2009) Tetrahedron 65:2619–2641
33. Saikia AA, Rao RN, Das S, Jena S, Rej S, Maiti B, Chanda K (2020) Tetrahedron Lett 61:152273
34. Sareen N, Singh AS, Tiwari VK, Kant R, Bhattacharya S (2017) Dalton Trans 46:12705–12710
35. Mishra KB, Tiwari N, Bose P, Singh R, Rawat AK, Mishra RC, Singh RK, Tiwari VK (2019) Chem Select 4:2644–2648
36. Agrahari AK, Kumar S, Sharma A, Tiwari VK (2021) J Chem 574:4209514
37. Meng G, Guo T, Ma T, Zhang J, Shen Y, Sharpless KB, Dong J (2019) Nature 574:86–91
38. Mishra A, Tiwari VK (2015) J Org Chem 80:4869–4881
39. Kreutzer P, Weis C, Boehme H, Kemmerich T, Beck W, Spencer C, Mason Z (1972) Naturforsch 27:74
40. Chang CW, Lee G-H (2003) Organometallics 22:3107
41. Busetto L, Marchetti F, Zacchini S, Zanotti V (2005) Inorg Chim Acta 358:1204
42. Singh KS, Kreisel KA, Yap GPA, Kollipara MR (2006) J Organomet Chem 691:3509
43. Agapiou K, Jones RA, Yang X (2010) Main Group Chem 9:41
44. Li GQ, Orchin M (1997) J Organomet Chem 535:43
45. Wittig G, Krebs A (1961) Chem Ber 94:3260–3275
46. Chang PV, Prescher JA, Sletten EM, Baskin JM, Miller IA, Agard NJ, Lo A, Bertozzi CR (2010) Proc Natl Acad Sci USA 107:1821–1826
47. Jewett JC, Sletten EM, Bertozzi CR (2010) J Am Chem Soc 132:3688–3690
48. Johnson JA, Baskin JM, Bertozzi CR, Koberstein JT, Turro NJ (2008) Chem Commun 26:3064

49. Pathak RK, McNitt CD, Popik VV, Dhar S (2014) Chem Eur J 20:6861–6865
50. Nieves DJ, Azmi NS, Xu R, Lévy R, Yates EA, Fernig DG (2014) Chem Commun 50:13157–13160
51. Fan M, Ma Y, Mao J, Zhang Z, Tan H (2015) Acta Biomater 12:2060–2068
52. Satpati D, Bauer N, Hausner SH, Sutcliffe JLJ (2014) Radioanal Nucl Chem 302:765–771
53. Jeon J, Kang JA, Shim HE, Nam YR, Yoon S, Kim HR, Lee DE, Park SH (2015) Bioor. Med Chem 23:3303–3308
54. Zhang G, Zheng S, Liu H, Chen PR (2015) Chem Soc Rev 44:3405–3417
55. Bird RE, Lemmel SA, Yu X, Zhou QA (2021) Bioconjugate Chem 32:2457–2479
56. Fairbanks BD, Macdougall LJ, Mavila S, Sinha J, Kirkpatrick BE, Anseth KS, Bowman CN (2021) Chem Rev 121:6915–6990
57. Heiss TK, Dom RS, Prescher JA (2021) Chem Rev 121:6802–6849
58. Kumar GS, Lin Q (2021) Chem Rev 121:6991–7031
59. Hema K, Sureshan KM (2019) Acc Chem Res 52:3149–3163
60. Ramachary DB (2014) Angew Chem Int Ed 53:10420–10424
61. Wang J (2011) Chem Eur J 17:3584–3587
62. Wang L, Peng S, Danence LJT, Gao Y, Wang J (2012) Chem Eur J 18:6088–6093
63. Kayet A, Pathak T (2013) J Org Chem 78:9865–9875
64. Jaiswal MK, Gupta A, Yadav MS, Pandey VK, Tiwari VK (2023) Chem Eur J 29:e202301749
65. Chandrasekeran S (2016) Click reactions in organic synthesis. Wiley, New York
66. Shashank AB, Karthik S, Madhavachary R, Ramachary DB (2014) Chem Eur J 20:16877–16881
67. Krasinski A, Fokin VV, Sharpless KB (2004) Org Lett 6:1237–1240
68. Creary X, Anderson A, Brophy C, Crowell F, Funk Z (2012) J Org Chem 77:8756–8761
69. Holzer W (1991) Tetrahedron 47:9783–9792
70. Rasmussen LK, Boren BC, Fokin VV (2007) Org Lett 9:5337–5339
71. Jaiswal MK, Gupta A, Ansari FJ, Pandey VK, Tiwari VK (2024) Curr Org Synth 21:513–558
72. Pedersen DS, Abell A (2011) Eur J Org Chem 13:2399–2411
73. Bauman A, Kluba CA, Vomstein S, Walter MA, Mindt TL (2013) Angew Chem Int Ed 52:8957–8960
74. Bachl J, Mayr J, Sayago FJ, Cativiela C, Díaz Díaz D (2015) Chem Commun 51:5294–5529
75. Chrysina ED, Bokor É, Alexacou KM, Charavgi MD, Oikonomakos GN, Zographos SE, Leonidas DD, Oikonomakos NG, Somsák L (2009) Tetrahedron Asymmetry 20:733–740
76. Dong J, Krasnova L, Finn MG, Sharpless KB (2014) Angew Chem Int Ed 53:9430–9448
77. Barrow AS, Smedley CJ, Zheng Q, Li S, Dong J, Moses JE (2019) Chem Soc Rev 48:4731–4758
78. Sun S, Homer JA, Smedley CJ, Chang QQ, Shapless KB, Moses JE (2023) Chem 9:2128–2143
79. Hoyle CE, Bowman CN (2010) Angew Chem Int Edn 49:1540–1573
80. Lowe AB (2010) Polym Chem 1:17–36
81. Liu F, Paton RS, Kim S, Liang Y, Houk KN (2013) J Am Chem Soc 135:15642–15649
82. Menzel JP, Feist F, Tuten B, Weil T, Blinco JP, Christopher BK (2019) Angew Chem Int Ed 58:7470–7474
83. Rieder U, Luedtke NW (2014) Angew Chem Int Ed Engl 53:9168–9172
84. Carlson JCT, Mikula H, Weissleder R (2018) Am Chem Soc 140:3603–3612
85. Blackman ML, Royzen M, Fox JM (2008) Am Chem Soc 130:13518
86. MacKenzie DA, Sherratt AR, Chigrinova M, Cheung LL, Pezacki JP (2014) Curr Opin Chem Biol 21:81–88
87. MacKenzie DA, Pezacki JP (2014) Can J Chem 92:337–340
88. Lang K, Chin J (2014) ACS Chem Biol 9:16–20
89. Earla A, Braslau R (2014) Macromol Rapid Commun 35:666–671
90. Haksar D, Van Ufford LQ, Pieters RJ (2020) Org Biomol Chem 18:52–55
91. Astronomo RD, Kaltgrad E, Udit AK, Wang SK, Doores KJ, Huang CY, Pantophlet R, Paulson JC, Wong CH, Finn MG et al (2010) Chem Biol 17:357–360

92. Yin Z, Aragones MC, Chowdhury S, Bentley P, Kaczanowska K, Mohamed LB, Gildersleeve JC, Finn MG, Huang X (2013) ACS Chem Biol 8:1253–1262
93. Ballell L, Alink KJ, Slijper M, Versluis C, Liskamp RMJ, Pieters RJ (2005) Chem Biol Chem 6:291–295
94. Jaiswal MK, Sharma A, Tiwari VK, Schmidt RR (2023) Recent advances in stereoselective intramolecular O-glycosylation. In: Tiwari VK (ed) Synthetic strategies in carbohydrate chemistry. Elsevier Inc., Amsterdam, pp 53–91
95. Tiwari VK, Kumar A, Schmidt RR (2012) Eur J Org Chem 274:2945–2956
96. Kumar A, Geng Y, Schmidt RR (2012) Eur J Org Chem 35:6846–6851
97. Chen A, Wang G (2023) Synthesis of carbohydrate containing macrocycles. In: Tiwari VK (ed) Synthetic strategies in carbohydrate chemistry. Elsevier Inc., Amsterdam, pp 407–516
98. Muthana S, Yu H, Cao H, Cheng J, Chen X (2009) J Org Chem 74:2928–2936
99. (a) Tiwari VK (2020) Carbohydrates in drug discovery and development. Elsevier Inc., Amsterdam; (b) Tiwari VK (2023) Synthetic strategies in carbohydrate chemistry. Elsevier Inc., Amsterdam
100. Kumar R, El-Sagheer A, Tumpane J, Lincoln P, Wilhelmsson LM, Brown T (2007) J Am Chem Soc 129:6859–6864
101. Acevedo-Jake A, Ball AT, Galli M, Kukwikila M, Denis M, Singleton DG, Tavassoli A, Goldup SM (2020) J Am Chem Soc 142:5985–5990
102. Denis M, Goldup SM (2017) Nat Rev Chem 1:61
103. Srivastava S, Singh SK, Sharma VK, Mangla P, Olsen CE, Prasad AK (2015) Nucleos Nucleot Nucl 34:388–399
104. Kushwaha D, Tiwari VK (2013) J Org Chem 78:8184–8190
105. Agrahari AK, Kumar S, Pandey MD, Rajkhowa S, Jaiswal MK, Tiwari VK (2022) Chem Sel 7:e202202273
106. Maity D, Govindaraju T (2010) Inorg Chem 49:7229–7231
107. Mishra A, Tiwari VK (2017) Chem Sel 2:9466–9471
108. Sun X, James TD (2015) Chem Rev 115:8001–8037
109. Scrafton DK, Taylor JE, Mahon MF, Fossey JS, James TD (2008) J Org Chem 73:2871–2874
110. Sharma A, Agrahari AK, Rajkhowa S, Tiwari VK (2022) Eur J Med Chem 238:114454
111. Bozorova K, Zhaoa J, Aisaa HA (2019) Bioorg Med Chem 27:3511–3531
112. Yadav S, Singh AK, Agrahari AK, Sharma K, Singh AS, Gupta MK, Tiwari VK, Prakah P (2020) Sci Rep 10:14204
113. Zhang P, Russel MG, Jamison TF (2014) Org Process Res Dev 18:1567–1570
114. Arigela RK, Mandadapu AK, Sharma SK, Kumar B, Kundu B (2012) Org Lett 14:1804–1807
115. Mishra KB, Tiwari VK (2014) J Org Chem 79:5752–5762
116. Senwar KR, Sharma P, Reddy TS, Jeengar MK, Nayak VL, Naidu V, Kamal A, Shankaraiah N (2015) Eur J Med Chem 102:413–424
117. Jaiswal MK, Tiwari VK (2023) Chem Rec 102:e202300167
118. Singh SK, Mishra N, Kumar S, Jaiswal MK, Tiwari VK (2022) Chem Sel 7:e202201314
119. Wang L, Liu J, Miao T, Zhou W, Li P, Ren K, Zhang X (2010) Adv Synth Catal 352:2571–2578
120. Shen C, Shen H, Yang M, Xia C, Zhang PA (2015) Green Chem 17:225–230
121. Tiwari VK (2021) Chem Rec 21:3029–3048
122. Mishra N, Singh AS, Agrahari AK, Singh SK, Singh M, Tiwari VK (2019) ACS Comb Sci 21:389–399
123. Mishra N, Singh SK, Singh AS, Agrahari AK, Tiwari VK (2021) J Org Chem 86:17884–17895
124. Singh SK, Kumar S, Yadav MS, Gupta A, Tiwari VK (2023) J Org Chem 88:13440–13453
125. Singh SK, Kumar S, Yadav MS, Tiwari VK (2022) J Org Chem 87:15389–15402
126. Singh SK, Kumar S, Yadav MS, Bhattacharya S, Tiwari VK (2023) SYNTHESIS 56:975–988
127. Witczak ZJ, Bielski R (eds) (2016) Domino and intramolecular rearrangement reactions as advances synthetic methods in glycoscience. Wiley, Amsteradm
128. Prasanna R, Purushothaman S, Raghunathan R (2014) Org Biomol Chem 12:9375–9383
129. Mindt TL, Struthers H, Brans L, Anguelov T, Schweinsberg C, Maes V, Tourwe D, Schibli R (2006) J Am Chem Soc 9:15096–15097

130. Brahma K, Achari B, Chowdhury C (2013) Synthesis 45:545–555
131. Larin EM, Lautens M (2019) Angew Chem 131:13572–13576
132. Reddy MN, Swamy KCK (2012) Eur J Org Chem 384:2013–2022
133. Selvaraju M, Sun CM (2014) Adv Synth Catal 356:1329–1336
134. Qureshi Z, Kim JY, Bruun T, Lam H, Lautens M (2016) ACS Catal 6:4946–4952
135. Larin EM, Lautens M (2019) Angew Chem Int Ed 58:13438–13442
136. Reddy RJ, Waheed M, Kumari AH, Krishna GR (2022) Adv Synth Catal 364:319–325
137. Sun P, Wang W, Lai J, Yan H, Tung C-H, Xu Z (1890) Org Chem Front 2023:10
138. Chen F-J, Mamidipalli P, Sabbasani VR, Liu H, Xia Y, Lee D (2021) Org Chem Front 8:6095
139. Yadagiri D, Anbarsan P (2013) Chem Eur J 19:15115
140. Akter N, Rupa K, Anbarasan P (2022) Chem Rev 122:13108–13205
141. Kumar D, Mishra A, Mishra BB, Bhattacharya S, Tiwari VK (2013) J Org Chem 78:899–909
142. Kumar D, Mishra BB, Tiwari VK (2014) J Org Chem 79:251–265
143. Shieh P, Hill MR, Zhang W, Kristufek SL, Johnason JA (2021) Chem Rev 121:7059–7121

Chapter 2
CuAAC 'Click Chemistry'-Mediated Synthesis of 1,4-Disubstituted 1,2,3-Triazoles

2.1 Introduction

The Cu-catalyzed 1,3-dipolar cycloaddition reaction of organic azides and alkynes to furnish 1,4-disubstituted triazoles (known as CuAAC or click reaction) has garnered significant attention from Huisgen's 1,3-dipolar cycloaddition to its regioselective improvements. Michael [1], in 1893, reported the initial instance of a 1,3-dipolar cycloaddition reaction utilizing organic azides and terminal alkynes to assemble 1,2,3-triazole. The chemistry was extensively explored by Huisgen in the 1960s era and became popular as Huisgen's [2–6] [3 + 2]-azide-alkyne cycloaddition reaction. The click reaction characterized by excellent regioselectivity with the assistance of a Cu-catalyst was revealed by a group of scientists represented by the Meldal and Sharpless [7–10]. The Cu(I)-catalyzed reaction is highly successful due to its excessive versatility, quantitative nature, robustness, insensitivity, and generality. It offers a reliable and modular strategy for the rapid and regioselective assembly of 1,4-disubstituted 1,2,3-triazoles. These triazoles have a wide range of applications, including orthogonal ligation reactions, bimolecular ligation, in vivo tagging, polymerization reactions for synthesizing long linear polymers, catalysis, organic synthesis, biology, medicinal chemistry, material sciences, and industrial applications [11–16]. Moreover, triazole-appended glycohybrids have great attention to exploring as metal ligands and organocatalysts [17–24]. Several articles have been reported on the topic of click chemistry, highlighting its significance in several fields and demonstrating the extensive growth and flexibility of this modular synthetic approach [25–31]. Herein, keeping in mind the growing impact of 1,4-disubstituted 1,2,3-triazole heterocycles and their future perspectives, this chapter covered the various approaches for Cu-catalyzed synthesis of triazolyl analogue **3** from terminal alkyne core **1** and azide component **2** (Scheme 2.1).

V. K. Tiwari et al., *Click Chemistry*, Materials Horizons: From Nature to Nanomaterials,
https://doi.org/10.1007/978-981-97-4596-8_2

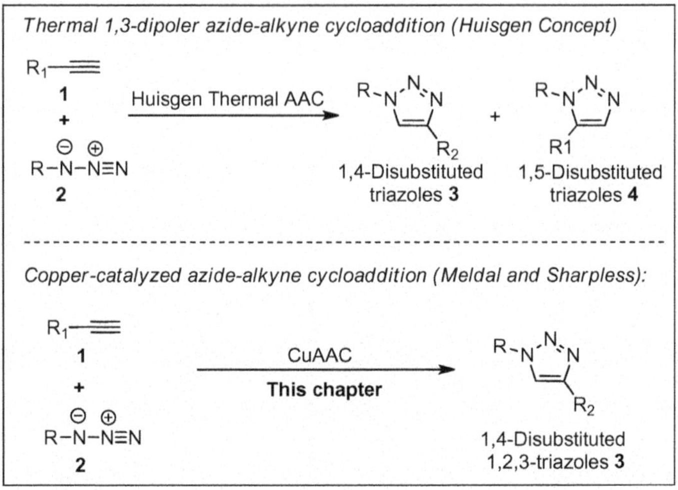

Scheme 2.1 Regioselective CuAAC click reaction for an easy synthesis of diverse 1,4-substituted 1,2,3-triazole analogues

2.2 CuAAC Catalysis and Common Additives

Prof. Morten Meldal [7] and Prof. Karl Barry Sharpless [8] established the regioselectivity in 1,2,3-triazole synthesis by utilizing copper catalysis in the new millennium, with flexible applications in several domains. The Cu-catalyzed click tool, with its advantages over uncatalyzed alkyne-azide coupling, has marked a new era in green chemistry. This method expanded the chemistry field to meet the needs of other parallel fields with diverse notable applications in chemistry, biology, medicine, and material science. Copper catalysis enhances reaction rate, yield, synthetic methodology ease, and cost-effectiveness, making it an environmentally friendly green approach. The commonly used methods for click reaction can be categorized into three types: (i) copper(I) source as such; (ii) obtaining in situ copper(I) from copper(II) system; and (iii) oxidizing the elemental copper to produce an oxidized Cu(I) source in the reaction medium.

The second approach was reported by K B Sharpless and is preferred in chemical synthesis, especially in living systems, due to its advantages with copper(II) salts like $CuSO_4$, $Cu(OAc)_2$, and related compounds [8]. The strategy offers the advantage of avoiding side redox reactions and maintaining an optimal copper(I) concentration throughout the procedure. An approach for the synthesis of 1,2,3-triazolyl analogue **6** falls in the 1st category was reported by M. Meldal in 2002 [7]. This is the first report of CuAAC click chemistry [7]. The group utilizes the CuI with DIPEA to develop compound **6** in from solid-supported alkyne substituent **5** and organic azide **2** under solid-phase conditions (Scheme 2.2) [7].

An alternative method was reported by K. B. Sharpless and Group in the same year for the regioselective synthesis of 1,2,3-triazole **9**, from terminal alkyne **7** and organic azide **8** with the assistance of $CuSO_4$ and a reducing agent. The approach

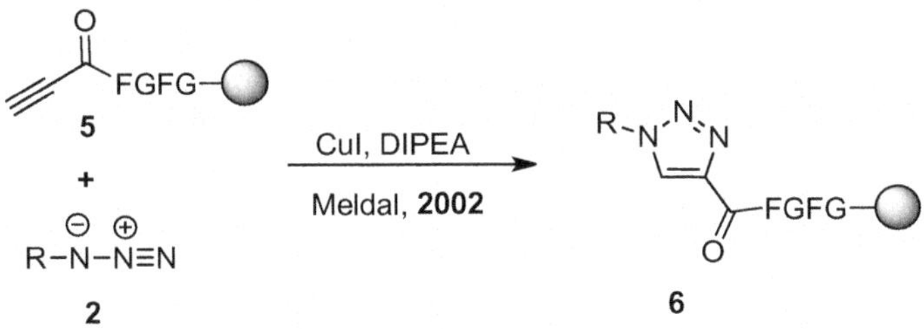

Scheme 2.2 Meldal's CuAAC catalytic (CuI/DIPEA in DCM) route for the regioselective 1,2,3-triazole forming click reaction

Scheme 2.3 Sharpless's CuAAC catalytic ($CuSO_4$/NaAsc in t-butanol/water) route for the regioselective 1,2,3-triazole forming click reaction

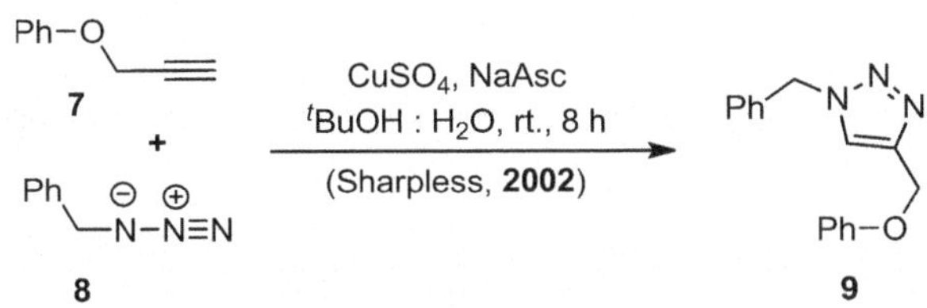

falls in category two, where Cu(II) ion undergoes a reduction in situ in the presence of sodium ascorbate (NaASc) in tertiary butanol and water to afford Cu(I) for active 1,3-dipolar cycloaddition reaction (Scheme 2.3) [8].

Water is the most accessible and appropriate solvent for the $CuSO_4$/NaAsc-catalyzed click technique. Nevertheless, this method is difficult due to the prolonged reaction time and the requirement for a larger catalytic quantity of $CuSO_4$ and NaAsc. Several other significant reducing agents, such as 3′,3″-phosphanetriyltripropanoic acid [32], and hydrazine [33], have been also employed and were found to be effective in the reaction system.

The use of reducing salts in biological systems can cause unwanted interactions with protein chains [32], potentially affecting biomolecules like DNA [34]. Light-mediated in situ synthesis of Copper(I) system addresses this issue [35]. Copper is utilized in various ways to assemble preferred triazolyl motif, including Cu-wire, Cu-nanoparticles [36], simple Cu-nanoparticles [37], and various amine salts of Cu^{2+} ions. An alternative method to get the triazole product with desired selectivity involves the use of copper(I) salts in their deoxygenated state, without the need for reducing agents. However, the presence of mixed aprotic solvents, such as THF, CH_3CN, CH_2Cl_2, toluene, is crucial for the reaction to occur. A report discusses the Cu-system, highlighting the modular click reaction using $Cu(acac)_2$ as a catalyst for the assembly of triazole-appended disaccharides, trisaccharides, or triazole-appended glycohybrids [38]. Common bases like NEt_3, 2,6-lutidine, DIPEA, and pyridine increase the reaction yield, while benzoic acid protonates copper triazolide intermediate and achieving the same goal [39, 40]. Various stabilizing ligands due to their significant catalytic activity such as phosphine-based complexes such as $Cu(PPh_3)_2Cl$ [41], $Cu(P(OMe)_3)_3Br$ [42], and $Cu(PPh_3)_3Br$ are frequently used in AAC [43, 44]. Copper particles immobilized on carbon nanotubes and graphene have been recognized as a reusable catalyst for the click reaction [45, 46].

Recent research has discovered that several amine-based hard ligands, connected to different heterocyclic donors including benzimidazole, benzothiazole, pyridyl, and click-constructed triazole, have proven to be successful in executing the process with in biological systems. Several traditional ligands, as well as triazole-based ligands from click chemistry, have been documented as transition metal ligands to enhance the catalytic efficiency of click reactions.

The ligand bound to the metal in the catalyst obstructs the oxidation of Cu(I) to Cu(II) by forming a complex with copper(I). Consequently, the inclusion of ligands improves the solubility of the catalyst and promotes catalytic conversion. Nitrogen-containing ligands play a crucial role in CuAAC processes, encompassing a wide range of molecules varying from small to high molecular weight polydentate ligands. It has been found that aliphatic amines have a higher reactivity compared to pyridine-based ligand, resulting in a faster reaction rate. This is attributed to their ability to facilitate the formation of Cu-acetylide, which is not as efficient with pyridine-based ligands (Fig. 2.1) [33].

Chan et al. [47] developed a unique intriguing approach to tackle catalyst toxicity issues using TBTA **10a**, as the leading Cu(I) stabilizing ligand for bioconjugation investigations. Although TBTA has low water solubility, it effectively prevents Cu(I) from undergoing oxidation and disproportionation, hence enhancing the catalytic activity in the reaction medium. Within this context, the ligand TTTA **10b** has gained recognition for its high catalytic efficiency in the CuAAC click reaction, particularly when subjected to identical reaction conditions [48]. The ligand THPTA **10c** has also gained fame for its role in the production of virus-like particles, specifically

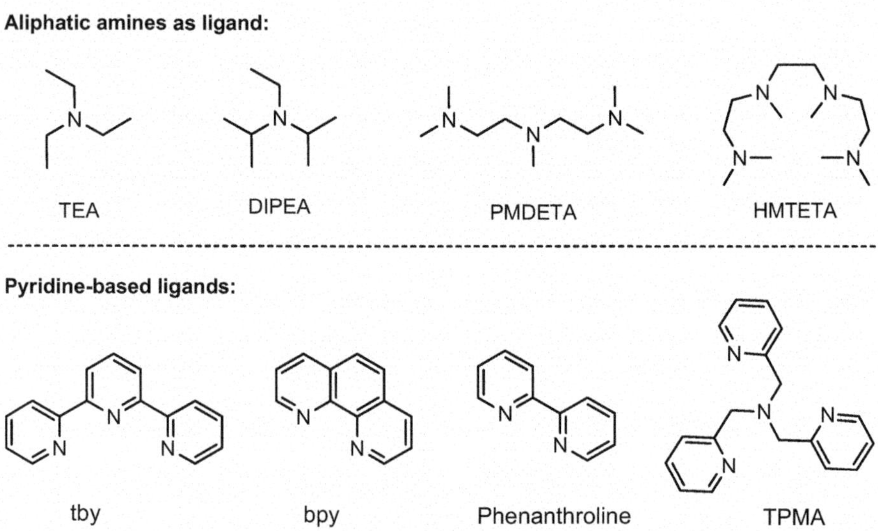

Fig. 2.1 Formidable amine and pyridine-based ligands for model CuAAC reaction

QB conjugates, which are employed in the creation of anti-cancer vaccines based on carbohydrates [49]. Two additional non-toxic and appealing ligands are BTTES **11a** and its propane-1-sulfonic acid derivative BTTPS **11b** (Fig. 2.2) [50]. The water-soluble $(BimC4A)_3$ **12** proved as quick and high-yielding in comparison to TBTA ligand in an aqueous solution as well [50]. Specifically, ligand **13**, which is based on sulfonated bathophenanthroline, has been widely utilized in bioconjugation tools because of its enhanced reactivity [51]. The researchers, Finn et al. developed a substantial biocompatible catalyst by combining ligand and copper in a ratio of 5:1. Additionally, they added aminoguanidine in order to reduce the amount of reactive oxygen species that were produced. When applied to a zebrafish model, the method primarily contributed to the acquisition of rapid in vivo imaging through the utilization of non-invasive imaging of fucosylated glycans in living systems [52]. The use of tris(benzimidazolylmethyl) amines (**14a,b**) as tris-triazolylamine ligands in such reactions resulted to proceed in a faster reaction rate [53, 54]. As a result, various chelating heterocycles, such as benzimidazole **15** [55], and indole **16** [56], were tested for the robust AAC click reaction and demonstrated good-to-excellent reaction yields of corresponding 1,2,3-triazoles with similar regioselectivity. Reiser et al. used fine-tuneable tris(triazolyl)methane as gram-scale promising ligands capable of producing Cu(I) complex **17a-c** (Fig. 2.2). The source of the copper catalyst and the different solvent media has a direct impact on the reaction outcome [57].

The triazolidine NHC ligands **18** were also discovered to be effective click catalysts for the assembly triazolyl analogue **3** from alkyne **1** and azide **2** (Scheme 2.4) [58]. On the other hand, it has been shown that complex **19**, which is an abnormal nitrogen heterocyclic carbene (Cu-ANHC) type copper(I) complex, is effective in catalyzing the CuAAC reaction for sterically hindered alkyne **20** and azide **8** as starting materials (Scheme 2.4) [59].

The ligand **22**, which is based on *S,N* coordination, demonstrated high efficacy in synthesizing glycodendrimers with a significant yield. This synthesis was achieved by reacting glycosyl azides with phenyl propargyl ethers in acetonitrile [60]. However, an array of copper complexes has been used and found to be excellently compatible with CuAAC click reaction to assemble a diverse range of triazolyl glycoconjugates. These complexes include Cu(I) thiobenzoate complexes $[CuSCOPh]_x$ **23**, [Cu(dppf)(SCOPh)] **24**, and $[Cu(bpy)(\mu\text{-}SCOPh)]_2$ **25** (Fig. 2.3) [61].

Metal–organic frameworks (MOFs), consisting of metal ions or clusters coordinated with suitable ligands, have become increasingly prominent in the field of catalysis. These frameworks possess a large surface area, a diverse range of functional groups, and metals referred to as coordination nanospace. The primary aspect of interest and significance to be explored for the rapid progress in heterogeneous catalysis is their ability to preserve their crystalline structure and regularity during each catalytic cycle. Fu et al. [62] developed a reduced Cu-MOF with strong catalytic activity, adequate stability, and excellent recyclability.

Ma et al. have developed two environmentally friendly metal–organic frameworks based on Cu(I), specifically $[Cu_4Cl_4L]\cdot CH_3OH\cdot 1\cdot 5H_2O$ **27** and $[Cu_4Br_4L]$ $\cdot DMF\cdot 0.5H_2O$ **28** utilizing functionalized thiacalix[4]arenes (L) **26** and halogen anions X − (X = Cl and Br) under solvothermal conditions. The developed MOFs

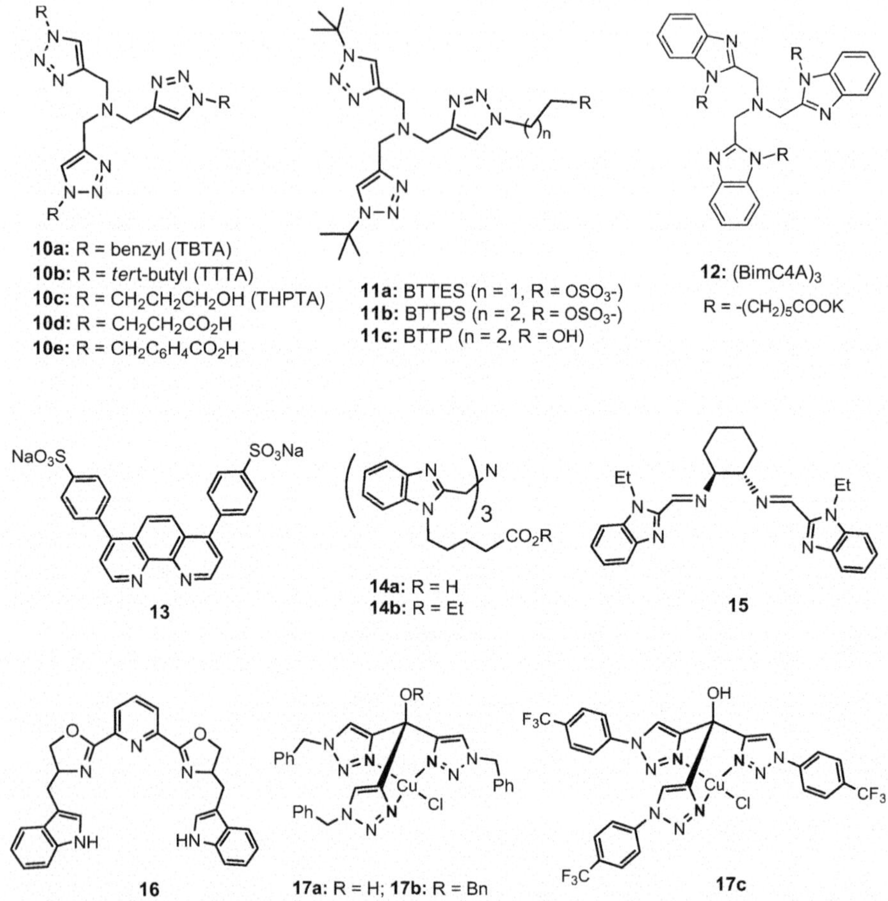

Fig. 2.2 Structure of *N*-based polydentate ligands along with catalysts for modular click reaction

effectively assist the cycloaddition of alkyne component **20**, sodium azide, and peroxide **29** to afford a high yield of desired triazolyl analogue **30** (Scheme 2.5) [63].

The click-bridged triazoles are made a group of highly stable analogue with one exception of the sulfonylated-triazoles. The influence of ring-chain isomerization in these compounds results in the assembly of equivalent *N*-acylated sulfonamides [64, 65]. In this study involving triazolyl analogues, the (CuTC) catalyst **31** was proposed to be utilized in click condition to attain the desired 1,2,3-triazole **33** from respective sulfonyl azide **32** (Scheme 2.6) [66].

In a recent scenario of research, copper complexed with PPh_3 has experienced extensive investigation and demonstrated competency in the CuAAC click reaction. The Bhattacharya and Tiwari research group developed an intriguing highly stable dinuclear Cu(I) catalyst $[(PPh_3)_2Cu(\mu\text{-tda})Cu(PPh_3)_2]0.6H_2O$ **34**. The developed

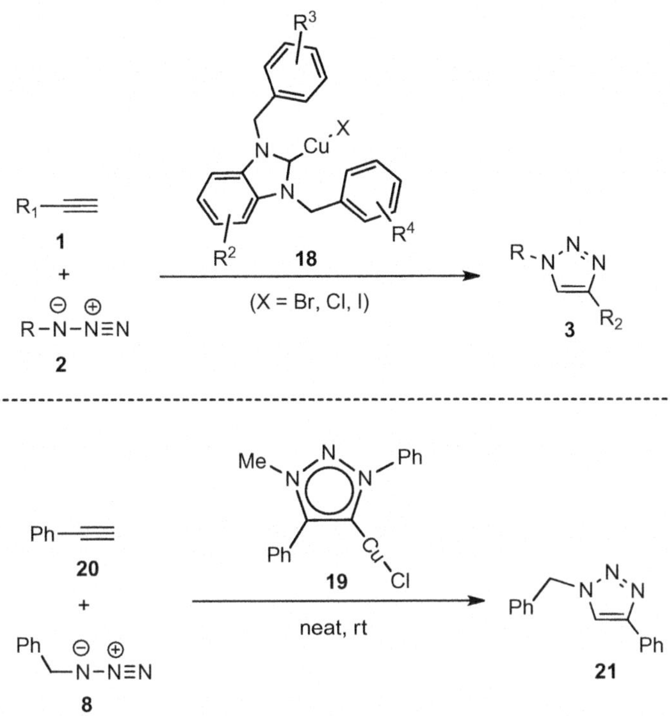

Scheme 2.4 Cu-based heterocyclic carbene-induced regioselective click chemistry

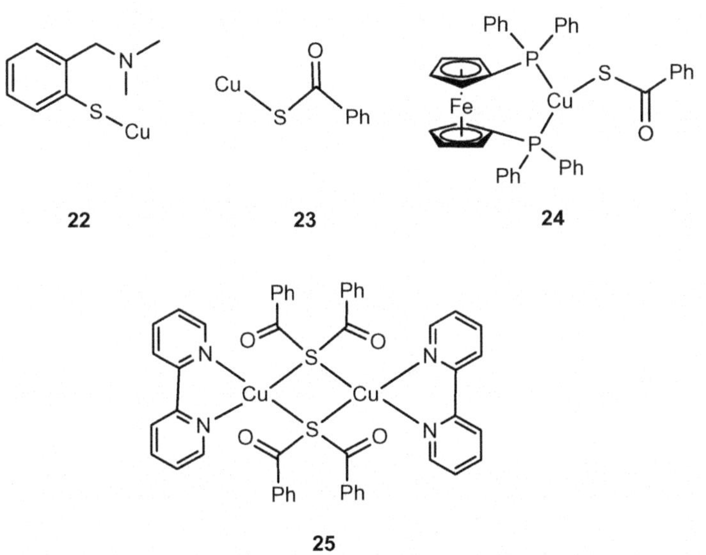

Fig. 2.3 Structure of the *S, N*-based ligand/Cu(I)-complexes for modular CuAAC click reaction

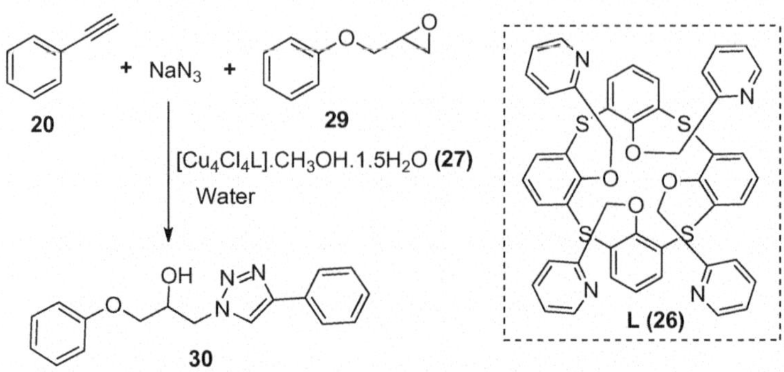

Scheme 2.5 CuAAC reaction in assistance of thiacalix[4]arenes (**L**) and anions-based MOF

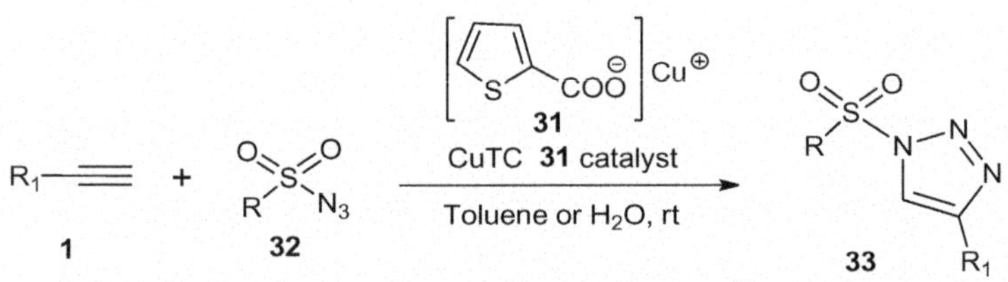

Scheme 2.6 Click reaction in the assistance of CuTC catalyst **31**

Cu-based complex **34** was found to be efficient in catalyzing modular click reactions of an array of sugar azide and terminal alkynes. The complex Cu(I)-based complex **34** showed notable catalytic properties for the efficient click conjugation of different glycosylated azides **35** and terminal alkynes **20** to afford excellent yields of desired 1,2,3-triazolyl glycoconjugates **36** in a regioselective manner (Scheme 2.7) [67]. Moreover, the same group developed hydrosulfide complexes of Cu(I) as a cubane-like tetramer with μ_3-SH-bridge [68] and Cu(I) furan-2-thiocarboxylate complexes which were displayed efficient CuAAC catalytic potency [69].

Singh et al. expanded upon the approach and assembly of a new heteroleptic Cu(I) β-oxodithioester complexes using PPh_3 ligands. The Cu(I) complexes served as a suitable catalyst in the CuAAC reaction, exhibiting a remarkable capability for catalytic transformation at room temperature in an open environment even without the presence of any base. The catalytic efficacy of resulting catalyst **37** was tested through the assembly of the library of 1,2,3-triazolyl glycoconjugates **39** utilizing sugar azide component **38** and an array of alkyne derivatives **1** (Scheme 2.8) [70].

The homoleptic Cu(I) complex **40** was developed and purified as a tetrafluoroborate salt (Fig. 2.4). The complex's structure was determined through the analysis of single-crystal X-ray diffraction data. It was shown that the complex is very stable, active, and selective toward the copper-catalyzed azide-alkyne cycloaddition. Furthermore, a range of alkynes and azides were examined to ascertain the degree of the substrate [71].

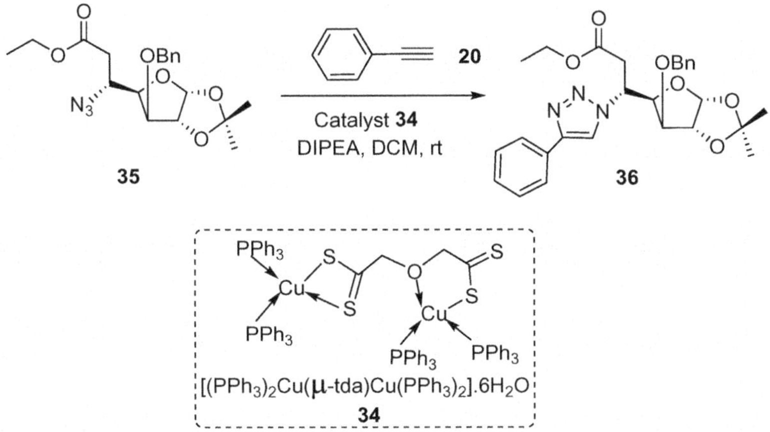

Scheme 2.7 Dinuclear Cu-complex **34** in CuAAC-mediated synthesis of glycohybrid triazoles

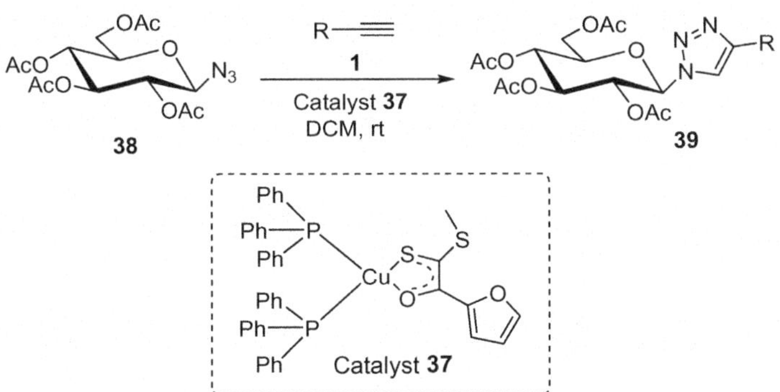

Scheme 2.8 Heteroleptic Cu(I) β-oxodithioester complex **37** in CuAAC-mediated synthesis of triazolyl glycoconjugates

Fig. 2.4 2,6-dimethylphenyl isocyanide-based Cu(I) catalyst for the CuAAC reaction

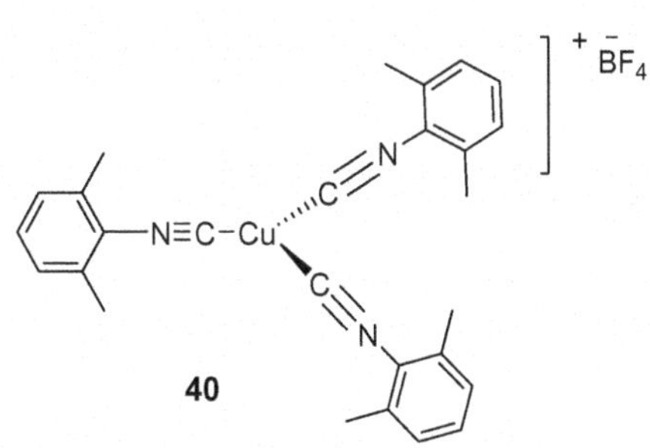

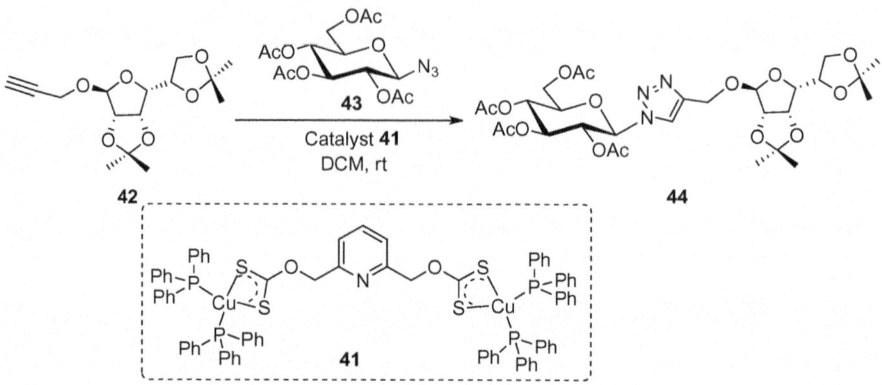

Scheme 2.9 Dinuclear dixanthate ligand as a promising click catalyst for the synthesis of triazolyl glycoconjugates

Singh et al. successfully assembled five heteroleptic copper(I) dithio complexes, which consisted of both dinuclear and mononuclear structures. The confirmation of these complexes was achieved by the use of single-crystal X-ray diffraction analysis. Furthermore, these complexes were tested for their effectiveness in the click reaction. The efficacy of dinuclear catalyst **41**, synthesized using the pyridyl linker-based dixanthate ligand, was demonstrated in the click reaction between alkynyl sugar scaffold **42** and glycosyl azide **43**, resulting in the formation of desired triazolyl glycoconjugates **44** (Scheme 2.9) [72].

Nembenna et al. reported a large thiourea stabilized Cu(I) catalyst **46** via the reaction of two equivalents of bulky thiourea **45** and $CuCl_2$ or CuCl as copper source in THF solvent. The bulky thiourea stabilized copper(I) halide complex **46** at 2 mol% catalytic loading efficiently catalyzes the 1,3-dipolar cycloaddition of azide **2** with alkyne **47**, to afford triazolyl analogue **48** (Scheme 2.10) [73]. The substituted thiourea and urea were also greatly utilized in the form of ligands as well as reductants in CuAAC [74, 75].

Khanam et al. [76] developed an intriguing thiocarbamide containing Cu(I) catalyst **49**, which was explored as the effective catalyst for the assembly of triazine core *N,N*-bis-sulfonylated dendrimers in high yields under liquid-assisted grinding CuAAC conditions (Fig. 2.5).

Bai et al. [77] developed four *bis*(2-(benzylthio)ethyl)amine (*SNS*) ligated copper complex; these complexes are well characterized by X-ray data analysis and found to be efficient for Cu-catalyzed azide-alkyne cycloaddition (Fig. 2.6).

A diverse range of heterogeneous Cu-catalysts are now being developed for varied purposes. These characteristics encompass recyclability in catalytic activities, a high reaction yield, minimal need for additive or ligands, and ease of handling. In particular, Cu-catalysts supported by amberlyte resin [78], a polymer [79], and zeolite [80], copper-based single-atom catalysts [81], were used in both usual and environmentally friendly chemical processes [82].

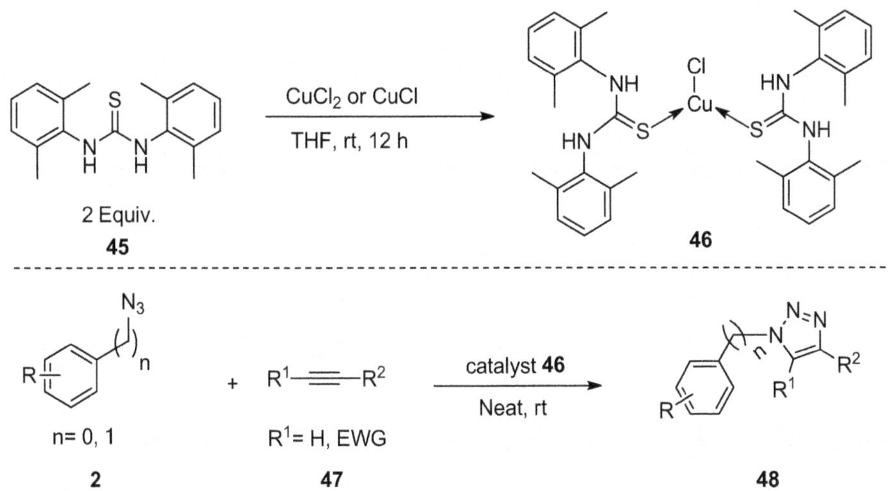

Scheme 2.10 Bulky thiourea stabilized Cu(I) catalyst-mediated synthesis of substituted 1,2,3-triazole analogue

Fig. 2.5 Thiocarbamide-based Cu(I) complex **49**

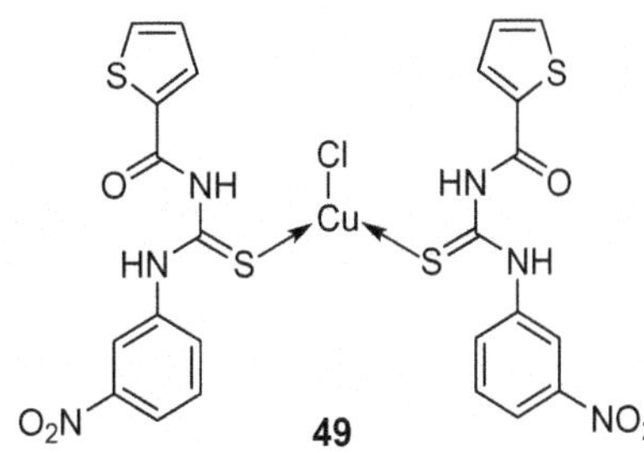

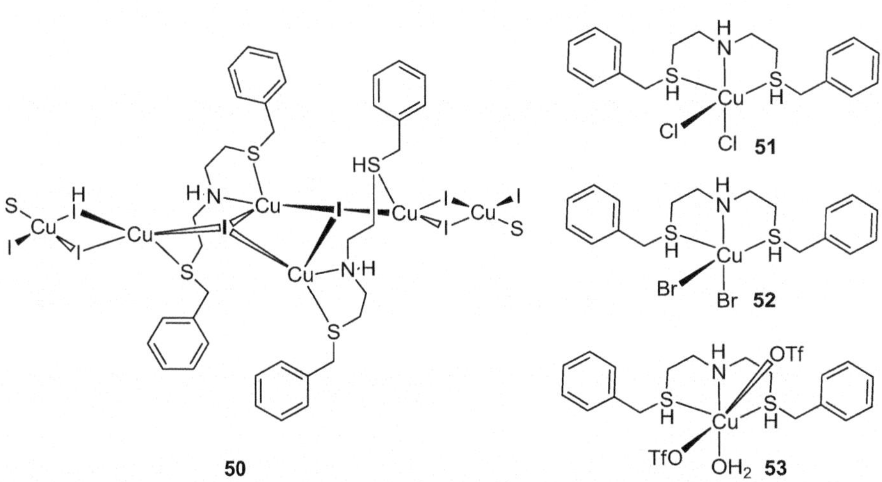

Fig. 2.6 Structure of *bis*(2-(benzylthio)ethyl)amine (*SNS*) ligated Cu-complexes

2.3 Plausible Mechanism of CuAAC 'Click Chemistry'

Numerous synthetic routes have been developed over the years to efficiently achieve the desired simple-to-complicated chemical configurations using either thermodynamically or kinetically favorable reactions, and among them, the regioselective 1,2,3-triazole forming CuAAC process is used more frequently than the other cycloaddition reactions. This modular strategy emphasizes the importance of the reaction system's mechanistic background, which points to a stepwise reaction that has been convincingly explained and demonstrated with valid experimental evidence [83–86]. The complex copper catalysis mechanism has contributed a significant rate of reaction, reaction yield, and regioselectivity in addition to offering an ambient platform for two sterically hindered or inaccessible molecular configurations, for which exploration of various additives like ligands in appropriate solvents has been well optimized and frequently used [87]. In a well-planned and strategic experimental expedition, it was shown that the ligands used in the experiment, in addition to the solvent systems, directly affect the formation of the copper complex and, thus, the probability that the reaction would occur as well as the enhanced reaction rate [88]. Although the assembly of copper complexes is a challenging process if the right conditions are met, the complex synthesis increases, and a cyclic reaction proceeds to give the desired product in a regioselective manner. An array of plausible mechanisms has been given for a proper explanation of click reaction in terms of the various steps involved. The most commonly accepted plausible approach is the one that can be used to satisfactorily explain click reactions and is well supported by kinetics and DFT calculation [84]. In various reports, the formation of one or two Cu species in a transition state has been significantly reported. The catalytic cycle of the click reaction began with the formation of the copper acetylide **I** intermediate from the Cu-alkyne complex, which on further complex coordination extension and reaction with azide **2**, results in the formation of complex **II**. Next, in complex **II** the nucleophilic nature of terminal *N*-3 favors attacking the *C*-4 position to get a stable spatial orientation [89] which leads to the rearrangement of complex **II** to complex **III**. Further, the reorder proceeds via *a* six-membered transition state [90], and the attack of lone pair of *N*-1 in metallacycle at the *C*-5 leads to the formation of a contracted as well as efficient smaller ring structured species **IV** [91]. In the final step of the catalytic cycle, intermediate reactive species **IV** undergoes a base or a solvent-mediated protonation to afford the desired 1,4-disubstituted-1,2,3-triazole product **3** as well as the facile dissociation of the copper complex takes place. The regenerated Cu-system is again involved in complexatation with terminal alkyne **1** to the formation of acetylide complex **I** and maintains the cyclic order of the catalytic route (Scheme 2.11) [92]. Ramesh and Chauhan [93] intriguingly cover the mechanistic perspectives of click chemistry in a book edited by S Chandrasekeran on title 'Click Chemistry in Organic Synthesis'.

Over the last decade, different research groups have focused their investigation on describing the most feasible and significant stepwise mechanism of click reaction using DFT simulations. The majority of investigators have determined that CuAAC

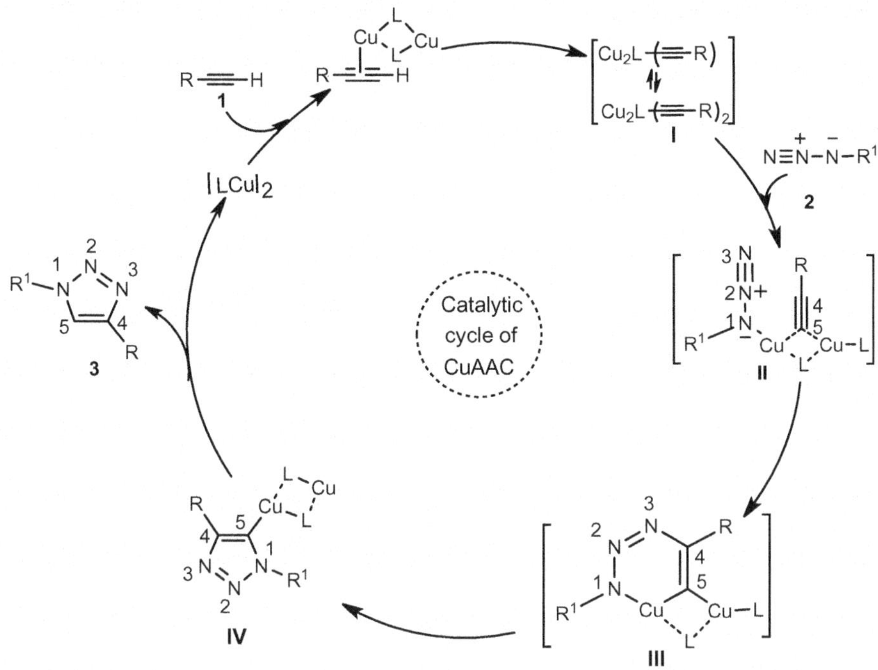

Scheme 2.11 Plausible mechanism of Cu(I)-catalyzed azide-alkyne cycloaddition (CuAAC)

chemistry occurs via the ligand replacement process during the catalytic cycle. In this cyclic mechanism, one of the ligands from Cu-acetylide **I** is replaced with azide functionality, resulting in the intermediate **II**, which has an exothermic energy value that varies depending on the pre-coordinated ligand molecule, for example, 0.7 kcal/mol for CH_3CN ligand and 2.0 kcal/mol for H_2O as ligand (Fig. 2.7). However, once the C-2 carbon of acetylide gains an electron from the N-3 atom of azide, the energy value increases to 12.6 kcal/mol for the H_2O ligand and 8.2 kcal/mol for the CH_3CN ligand. The energy value of the six-membered copper(III) metallacycle **III** grows further in the next step of rearrangement and reaches values of 14.9 kcal/mol for acetonitrile ligand and 18.7 kcal/mol for water ligand. On the other hand, the energy for an uncatalyzed reaction system is relatively high in comparison to the catalyzed one, i.e., 25.7 kcal/mol for CH_3CN ligand and 26.0 kcal/mol for H_2O ligand, demonstrating the benefit of the catalytic approach. The final step of ring contraction to obtain the triazolyl-Cu derivative **IV** from intermediate **III** had the lowest conversion energy of 3.2 kcal/mol for the H_2O ligand. In summary, the ligand exchange mechanism is the basic route of the click approach; however, it varies based on the ligand attached and the reaction system used [84].

In spite of the mechanistic studies and its DFT support about the catalytic cycle of Cu-catalyzed azide-alkyne cycloaddition [84], the straight proof of displayed mechanisms was lacking, so, to fulfill this rationale gap, Bertrand et al. performed

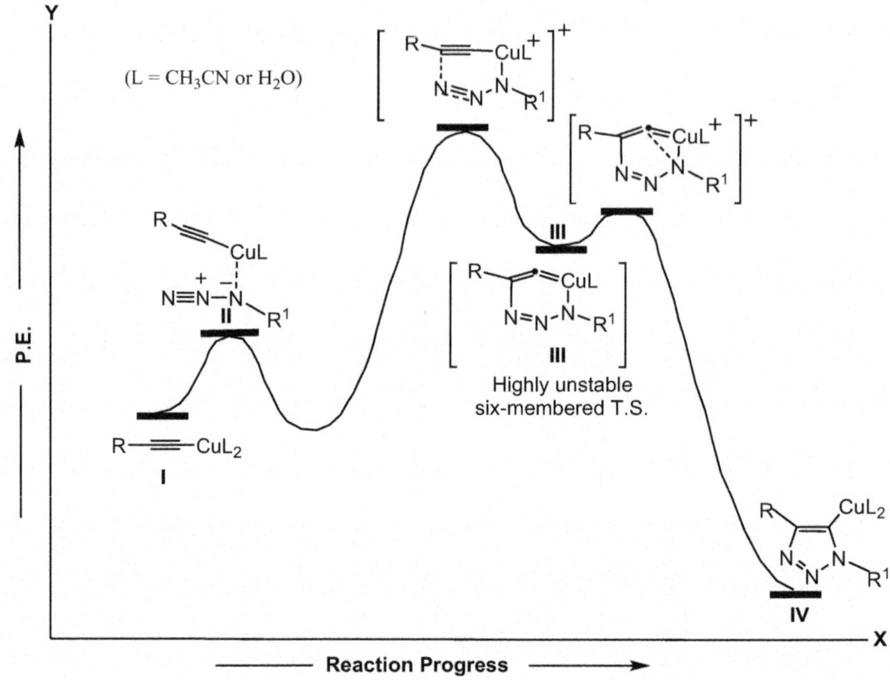

Fig. 2.7 DFT-based energy-profile diagram for the CuAAC click [84]. Reproduced from American Chemical Society, 2005

the CuAAC reaction of phenylacetylene **1** with benzyl azide **8** to assemble the desired 1,2,3-triazole analogue **13**. The authors reported the successful isolation of dinuclear complexes I(Cu_2) and II(Cu_2) which were never identified before (Scheme 2.12) [94]. Based on this kinetic examination, the authors suggested that the dinuclear complexes π,σ-*bis*(Cu-acetylide) I(Cu_2), and a 3,5-*bis*(metallated) triazole complex II(Cu_2) express catalytic activity with higher efficacy in comparison to the mononuclear complex, I(Cu) and II(Cu) [94].

2.4 NMR Spectroscopy-Based Characterization of 1,2,3-Triazoles

In addition to the notable features of CuAAC-mediated 1,4-disubstituted 1,2,3-triazoles in various fields, the other complementary regioisomers have also displayed a wide range of applications. The required 1,5-disubstituted 1,2,3-triazoles can typically be developed by using RuAAC [95]. Similar regioisomeric molecules can also be obtained exclusively in good yields under organocatalysis conditions [96, 97].

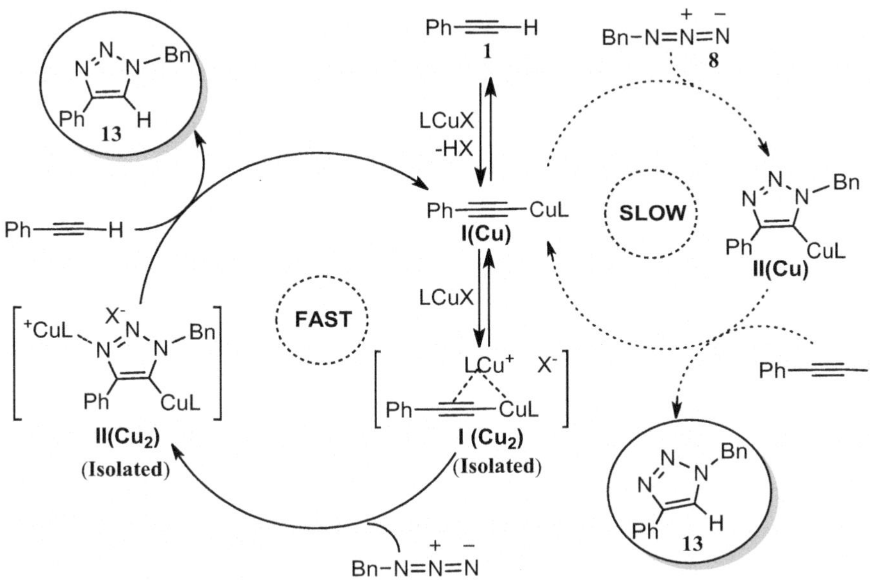

Scheme 2.12 Isolation of dinuclear complexes I(Cu_2) and II(Cu_2) during CuAAC reaction

There is a great demand to characterize both the individual 1,2,3-triazole regioisomer. One important tool to identify them is NMR. Creary et al. successfully demonstrated simple one-dimensional ^{13}C NMR spectroscopy with gated decoupling as a straightforward and effective methodology for quickly discriminating between the isomeric 1,2,3-triazoles **3** and **4**. Figure 2.8 depicts an extended of the signal of ^{13}C NMR spectrum of isomeric triazoles generated by a Cu-catalyzed synthesis and an organocatalytic reaction or through Ruthenium catalysis. The ^{13}C NMR signal at $\delta = 133.0$ ppm corresponding to C4 carbon of 1,5-isomer and ^{13}C NMR signal at $\delta = 120.0$ ppm corresponding to C5 carbon of 1,4-isomer prove the isomeric structure of triazole compounds. Further, decoupling experiments also identified that C_4 carbon signal of 1,5-disubstituted 1,2,3-triazole has a relatively larger C–H coupling constant than C_5 carbon of 1,4-substituted 1,2,3-triazole [98].

Fig. 2.8 ^{13}C NMR spectroscopy-based effective characterization of 1,4- and 1,5-disubstituted 1,2,3-triazole

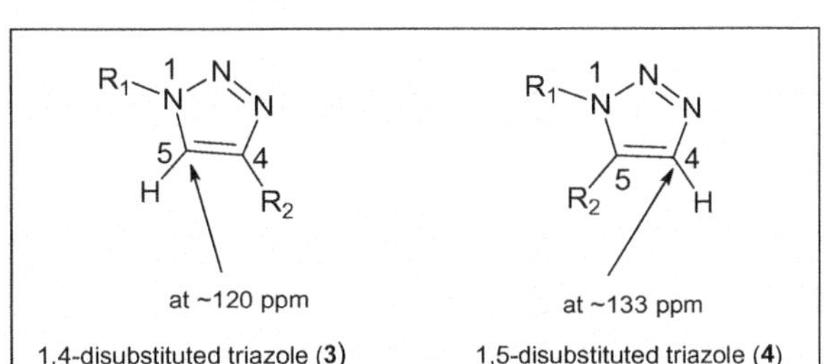

2.5 Various Representative Synthetic Approaches for CuAAC-Mediated 1,4-Disubstituted 1,2,3-Triazoles

In recent decades, the coupling of two molecular elements with different properties to afford a new conjugate with fascinating cumulative attributes of the mother entities has risen greatly. An array of novel conjugates, notably glycoconjugates, has demonstrated conventional biological activity, owing to the fact that the two components work in tandem to have a major influence [99, 100]. In this regard, CuAAC 'Click Chemistry' is the most adaptable, practical, and dependable approach widely explored in drug discovery for the benefit of human kind in the development of new medications. Fall et al. reported the assembly of biocompatible triazole ring in the side chain of CD ring of vitamin D analogue **57**. For this purpose, the authors utilize the $CuSO_4{\cdot}3H_2O$-mediated click reaction of azide **54** with ethylpropiolate **55** to give high yield of desired triazole **56** (Scheme 2.13) [101].

Considering the environmental benign green aspect, Praveena et al. reported a reaction of 2-chloro-3-(chloromethyl)-6-methoxyquinoline **58**, alkyne **59,** and NaN_3 in the presence of KI as a phase transfer catalyst and $CuSO_4{\cdot}5H_2O$/NaAsc as a source of copper(I) in water at room temperature. The reaction was carried out under one-pot CuAAC condition involving azidation in water under mild reaction conditions. The inclusion of 6 mol% TBAB as a phase transfer catalyst increases the reaction rate that completed in 8 min and generated the desired cytotoxic triazolyl product **60** in 93% yield (Scheme 2.14) [102].

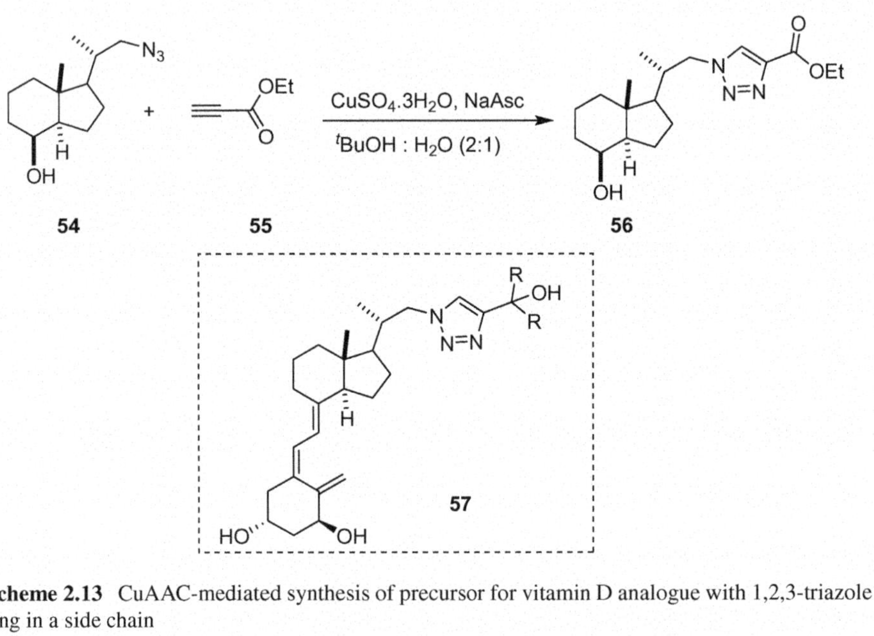

Scheme 2.13 CuAAC-mediated synthesis of precursor for vitamin D analogue with 1,2,3-triazole ring in a side chain

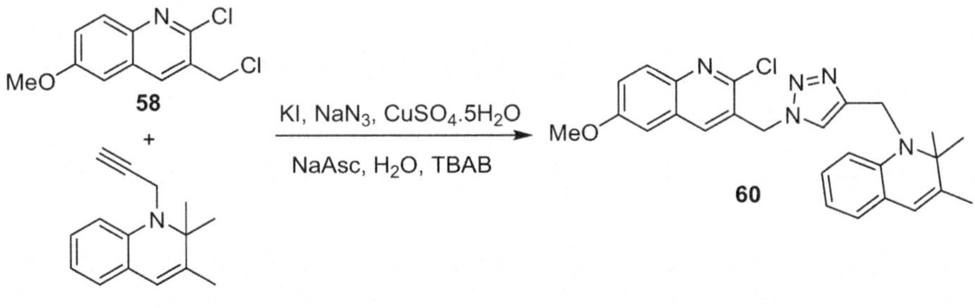

Scheme 2.14 $CuSO_4$-assisted one-pot regioselective synthesis of 1,4-disubstituted 1,2,3-triazole

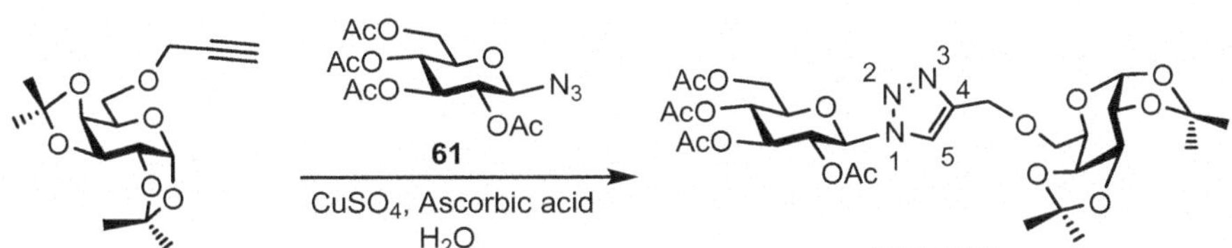

Scheme 2.15 $CuSO_4$-assisted synthesis of triazolyl glycohybrid

Norris et al. used $CuSO_4$/ascorbic acid in the water system for the coupling of glycosyl azide **61** with propargylated sugar **62** to give desired 1,2,3-triazole-linked disaccharide motif **63** in 84% yield (Scheme 2.15) [103].

After the invention of CuAAC using either CuI/DIPEA (Meldal Click catalyst) or $CuSO_4$/NaASc (Sharpless Click catalyst) in 2002, several copper nanoparticles (CuNPs) have also been explored with substantial efficiency in terms of ease, reaction yield, reaction time, and the desired regioselectivity to the similar AAC click conjugation. The first examples of CuNPs used in CuAAC to furnish regioselective 1,4-disubstituted 1,2,3-triazoles were independently reported by two different groups of researchers in 2005, Pachón et al. [104] and Orgueira et al. [105]. Rhothenberg et al. used copper nanoclusters to efficiently assembling of 1,4-disubstituted triazoles **3** from terminal alkyne **1** and organic azide **2**. However, Orgueira et al. utilized 10 mol % copper nanosize activated powder and 1.0 equiv of an amine hydrochloride salt for in situ generated the Cu(I) source during the AAC conjugation (Scheme 2.16) [104, 105].

Later on, a number of efforts were made to consider the notable features of supported nanoparticles over the unsupported NPs in AAC. The efficient CuNPs for this purpose include alumina [106], aluminum oxyhydroxide [107], montmorillonite [108], graphene oxide [109], metal–organic framework (MOF) [110], cellulose [111], SBA-15 [112], and zeolite [113]. Interestingly, Gupta et al. reported an intriguing scale-up synthesis of glycocohybrid **66** via AAC of glycosyl azide **64** with alkyne **65** with the assistance of a magnetic nanocomposite of GO-Fe_3O_4@CuO under sonochemical conditions (Scheme 2.17). The ease of the separation of GO-Fe_3O_4@CuO

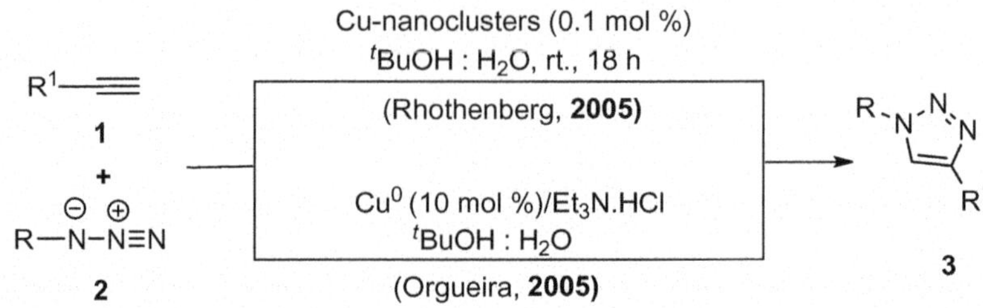

Scheme 2.16 Cu-nanocluster-mediated synthesis of 1,4-disubstituted 1,2,3-triazole ring

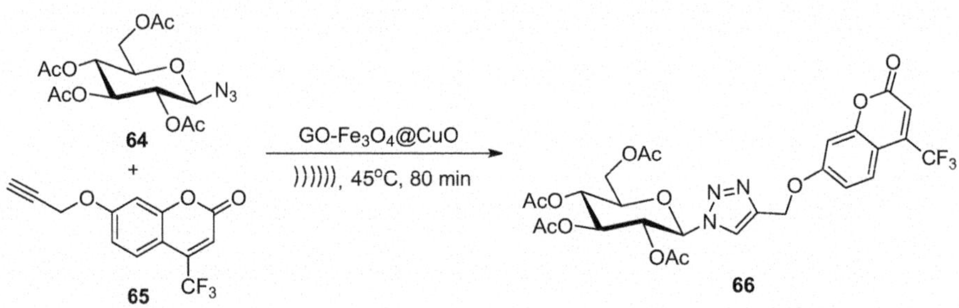

Scheme 2.17 Synthesis of potent anti-cancer glycohybrid triazole **66** in assistance of GO-Fe_3O_4@CuO under sonication condition

catalyst from the reaction mixture using an external magnet and also its reusability eight times without a substantial loss in yield are the key advantages of this modular CuAAC reaction. Further, the glycohybrid **66** had potential cytotoxic efficacy (IC_{50} = 8.3 μg/mL) in respect to cis-platin, a standard anti-cancer medication [114]. The role of click chemistry in drug discovery and development is presented in a distinct chapter of this book.

Several years of dedicated imprinting investigation for catalytic systems resulted in adaptability to create regioselective AAC under specific reaction conditions. However, due to the notable advantages of copper-inspired catalysts, they have been widely used in catalyst design in substrate-based needs [115]. PANFPABuBuX@CuX, a polymeric fiber-supported polyquaterniums@Cu(I) Click catalyst with exceptional stability, is used in efficient AAC in water with considerable yields and excellent functional group tolerance [116]. Lim et al. [117] designed solid-supported heterogeneous catalytic systems such as Cu(II)@pNIPAM-VP as aminopropyl-functionalized reverse phase silica-gel with an end-capped hydrophobic alkyl group and also Cu(I)@pNIPAM-VP as a thermo-responsive poly(*N*-isopropylacrylamide-co-4-vinylpyridine) that can switch between hydrophilic and hydrophobic properties as temperature dependence. Hayderi et al. designed Fe_3O_4@LDH@cysteine-Cu(I) nanoparticles **67** as a unique and recyclable catalytic system, in which copper (I)-cysteine complex immobilized on the surface of layered double hydroxide (LDH)-coated magnetite nanoparticles (Fig. 2.9). The

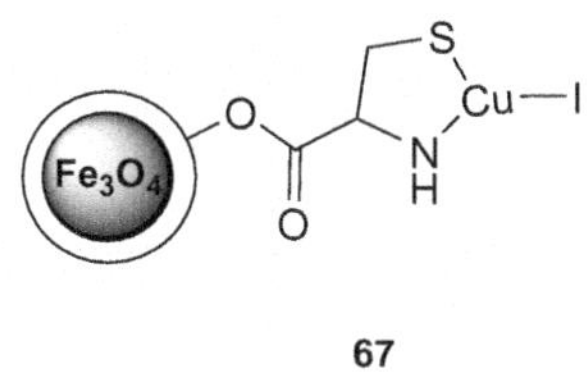

Fig. 2.9 Structure of Fe_3O_4@LDH@Cysteine–Cu(I) nanocatalyst in AAC conjugation

nanoparticle **67** exhibits high catalytic activity in similar AAC of organic halides and alkynes in the presence of choline azide in the reaction media [118]. Furthermore, this catalyst was easily recovered and reused five times without a decrease in the catalytic activity.

In addition to the above-mentioned CuNPs, the click-based NP substrate is also used to give a similar 1,2,3-triazole analogue of notable application in catalysis. For example, an intriguing self-catalyst oxazoline-based alkyne **68** was devised to proceed with regioselective CuAAC with silica NP-based azide **69** to high yield of give desired nanocomposite **70** (Scheme 2.18). Furthermore, the resulting nanocomposite **70** could catalyze the stereoselective benzoylation of geminal alcohols, indicating its synthetic utilities [119]. Kwak et al. [120] explored the Cu-catalyst immobilized on multiwalled carbon nanotubes modified with poly(3-carboxythiophene) to assist the regioselective CuAAC conjugation.

In the category of solid supports, nanostructured magnetic materials such as Fe_3O_4 in combination with functionally modified biomolecules work effectively to hold metal and metal ions. The copper anchored over these functionalized magnetic nanoparticles demonstrated exceptional CuAAC reaction efficiency. Sabaqian et al. [121] developed a copper-supported functionalized cellulose-based magnetic material (Fe_3O_4-Cellulose-NH_2-CuI) catalyst **71,** which was found to be effective in click reaction of azide and alkyne with some notable features such as low-cost, biodegradable support and efficient (Fig. 2.10).

Wang et al. [122] developed an intriguing approach for eco-friendly and highly efficient synthesis of triazole **73** from cycloaddition of terminal alkyne **72** and azide **8** with the assistance of phosphine-based $Cu(PPh_3)_2NO_3$ complex under solvent-free conditions (Scheme 2.19). The same group also utilizes the $[Cu(phen)(PPh_3)_2]NO_3$

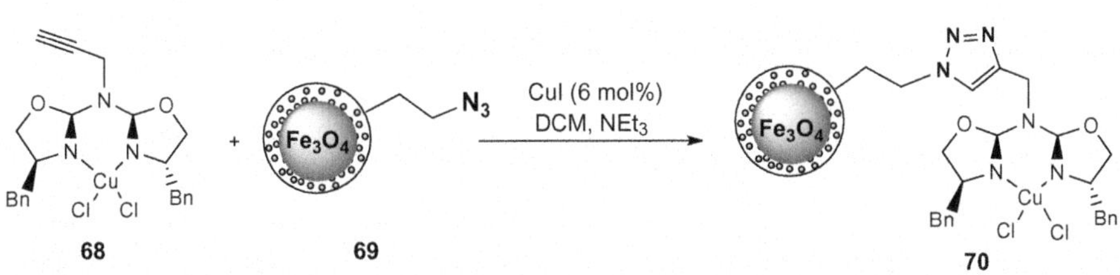

Scheme 2.18 Synthesis of nanocomposite **70** by exploiting the CuAAC click tool

Fig. 2.10 Structure of Fe_3O_4-Cellulose-NH_2-CuI catalyst **71**

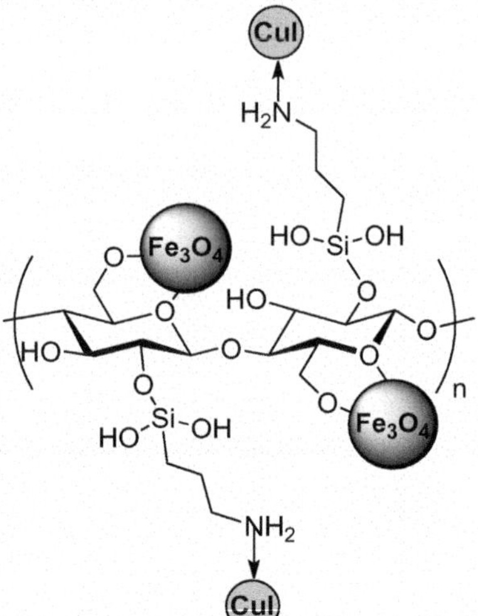

Fe₃O₄-Cellulose-NH₂-CuI (Catalyst **71**)

complex as the effective catalyst for the same reaction at 1 mol% of catalytic loading under solvent-free conditions with 2–25 min of reaction time [123].

González et al. reported the [$CuBr(PPh_3)_3$] catalyzed alkyne **74** azide **8** cycloaddition to attain the 1,4-substituted 1,2,3-triazoles **75**. The catalyst was found to be effective at 0.5 mol % catalytic loading at room temeperature without the addition of any additive (Scheme 2.20) [124].

Gonda and Novák [125] reported the $C_3H_7COOCu(PPh_3)_2$, a bis-triphenylphosphano-based complexes of copper(I)-carboxylates as efficient catalysts

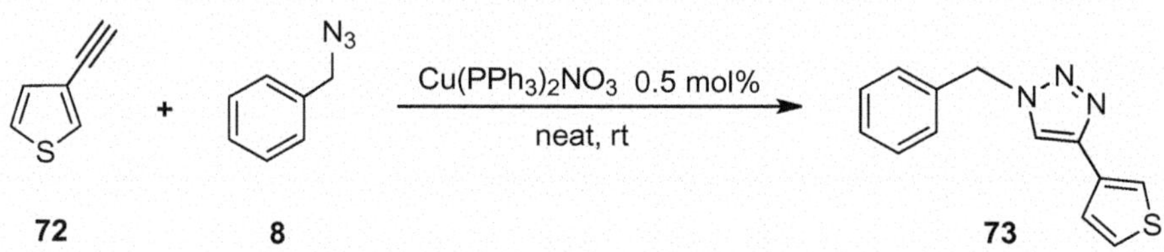

Scheme 2.19 Cu-triphenylphosphine-based complex-assisted synthesis of 1,4-disubstituted 1,2,3-triazole under solvent-free condition

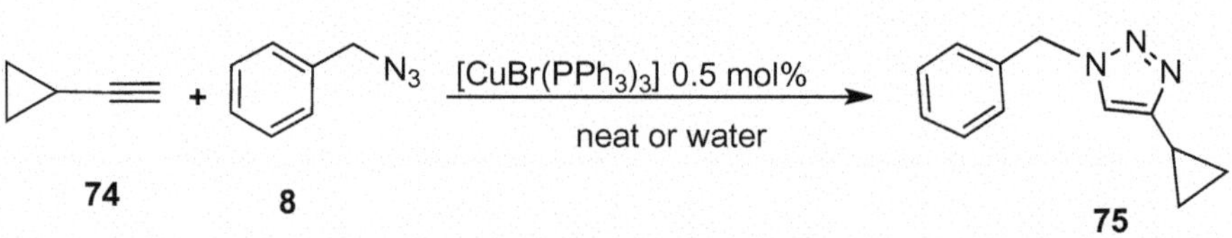

Scheme 2.20 [$CuBr(PPh_3)_3$]-mediated synthesis of 1,4-disubstituted 1,2,3-triazole

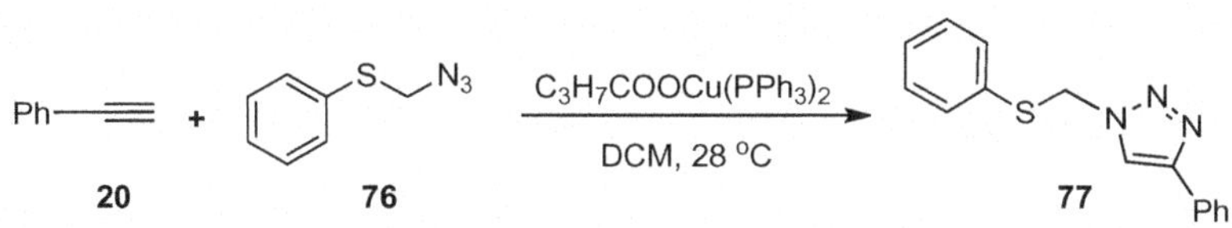

Scheme 2.21 The $C_3H_7COOCu(PPh_3)_2$-assisted synthesis of 1,4-disubstituted 1,2,3-triazole

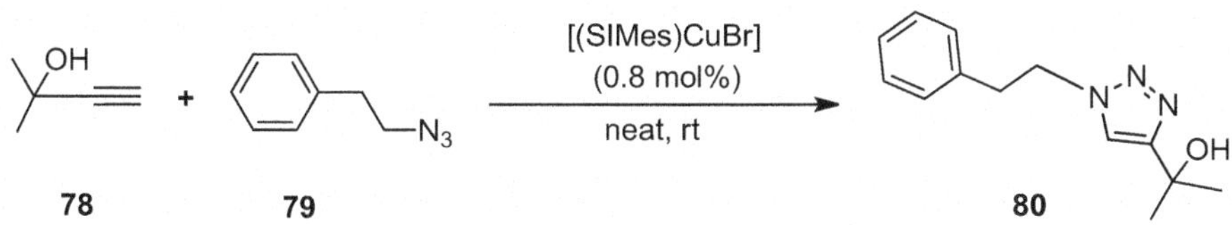

Scheme 2.22 [(SIMes)CuBr]-mediated synthesis of 1,4-disubstituted 1,2,3-triazole

for terminal alkyne **20** and azide **76** cycloaddition to attain corresponding triazole **77** under mild reaction condition (Scheme 2.21).

The pre-formed Cu(I) complexes with *N*-heterocyclic ligands (NHCs) in their structures are the most common catalysts in CuAAC processes. The Cu(I) center is more stable against oxidation in solution through the steric shielding from effective NHC ligand. Díez-González et al. [126] utilized the [(NHC)CuBr] complex (NHC = *N*-heterocyclic carbene) for the efficient click protocol. An array of alkyne **78** and azide **79** led to the formation of respective 1,2,3-triazole **80** with the assistance of [(SIMes)CuBr], (SIMes = *N,N′*-bis(2,4,6-trimethylphenyl)-4,5-dihydro-imidazol-2-ylidene) at room temperature (Scheme 2.22) [126].

Li et al. developed a series of ammonium salt-tagged SIPr-Cu(I) complexes **81**; these water-soluble derivatives successfully assist a three-component one-pot reaction for triazole synthesis in an aqueous medium. Furthermore, the system could be reused up to three times without substantial yield loss [127]. Gaulier et al. [128] used a similar approach to assemble water-soluble [(SIMes)CuI] derivatives **82** by substituting hydrophilic triazolyl-choline groups on the SIMes core (Fig. 2.11).

Zhu et al. performed a $Cu(OAc)_2$-mediated click cycloaddition reaction of propargyl alcohol with 2-picolylazide **83** in alcoholic media without the inclusion of any additional reducing agent. The reaction was carried out with 5 mol% $Cu(OAc)_2$ in ^{t}BuOH and gave the desired product **84** in 95% of yield in a very short period (within 1–2 min). The reduction of Cu(II) to Cu(I) by oxidation of alcohol to carbonyl compounds was confirmed by spectroscopic examination. The ability of 2-picolylazide molecule to form a chelate with copper(I) ions was attributed to its strong reactivity during the CuAAC cycloaddition (Scheme 2.23) [129].

Sahil et al. [130] developed two *bis*(imino)pyridine-copper(II) complexes **85** and **86** (Fig. 2.12) which were found to be effective in catalyzing the click reaction of azide and alkyne to assemble the corresponding 1,2,3-triazole.

Emami et al. developed an efficient Cu(II)-hydrazide coordination complex **87** based on silica-gel (Fig. 2.13) as a heterogeneous CuAAC catalyst for the successful

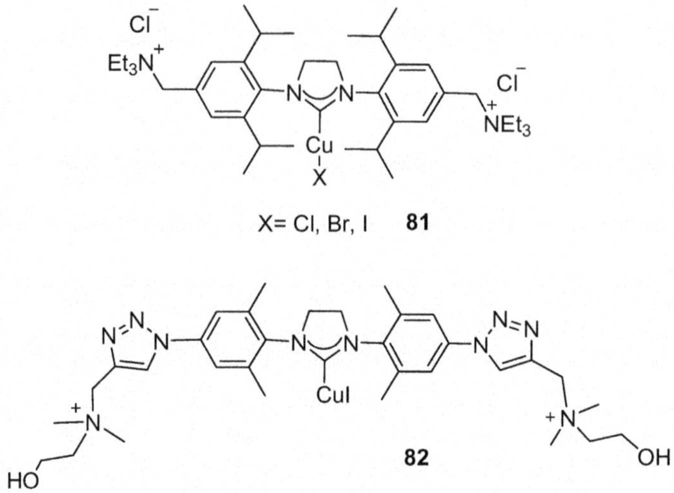

Fig. 2.11 First examples of water-soluble NHC-Cu(I)-complexes for CuAAC

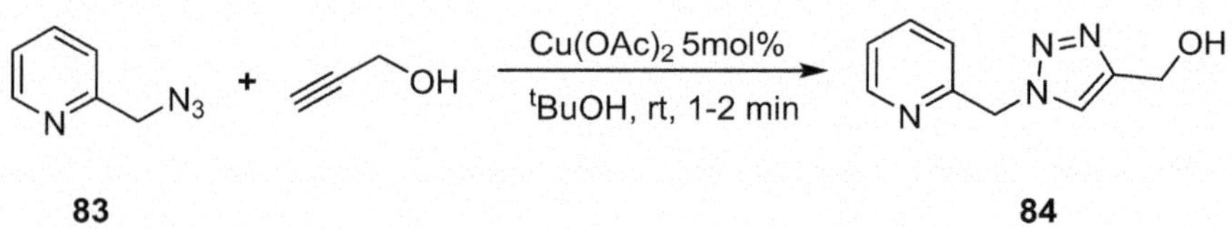

Scheme 2.23 $Cu(OAc)_2$-assisted synthesis of 1,4-disubstituted 1,2,3-triazole

Fig. 2.12 Structure of *bis*(imino)pyridine-copper(II) complexes

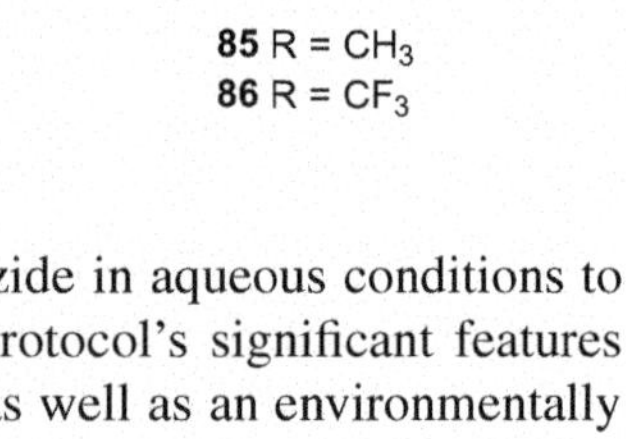

reaction of epoxides, terminal alkynes, and sodium azide in aqueous conditions to give high yields of β-hydroxy triazoles. The click protocol's significant features include sufficient catalyst stability and recyclability, as well as an environmentally friendly nature [131].

Konwar et al. [132] reported the readily available array of Cu-based catalytic systems to assembly a wide range of 1,4-disubstituted-*1H*-1,2,3-triazoles under mild conditions including Benedict's solution/Vitamin C catalyst system, Fehling

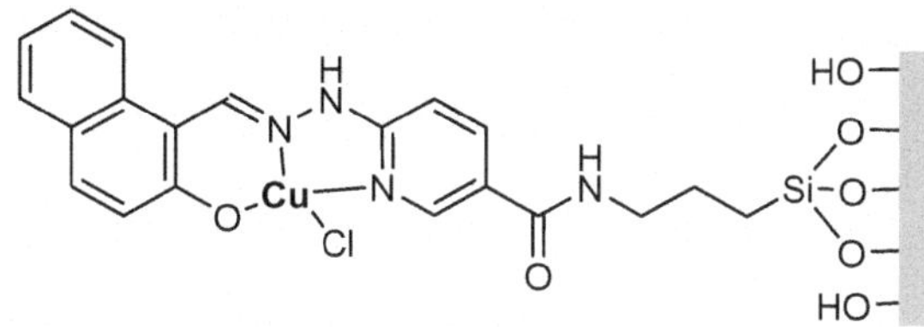

87

Fig. 2.13 Silica-gel supported Cu(II)-hydrazide coordination copper(II) catalyst **87** for click reaction

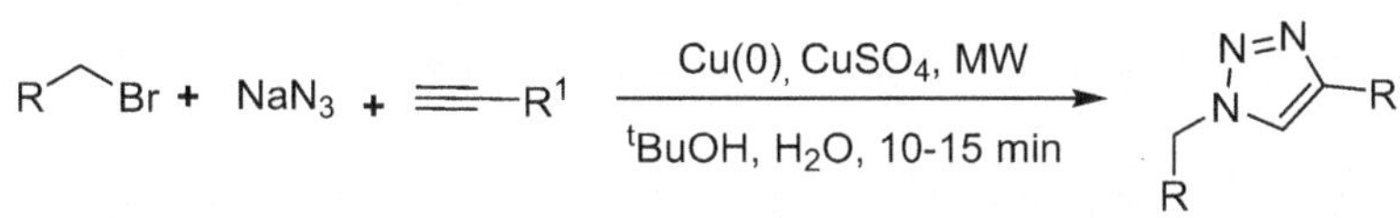

88 1 89

Scheme 2.24 Cu-assisted synthesis of 1,4-disubstituted 1,2,3-triazole in the presence of microwave

solution/DIPEA/hydrazine [133], $(DHQD)_2PHAL$ ligand-accelerated CuI catalytic system [134].

Interestingly, Appukkuttan et al. investigated microwave (MW)-assisted click reaction to develop an array of triazoles **89** with alkyl halides **88**, alkynes **1**, and NaN_3 in a single-pot. The MW-irradiation greatly promotes the reaction and in situ generation azide makes the protocol mild and safe (Scheme 2.24) [135].

Saikia et al. developed an intriguing ionic liquid-supported Cu(II) pre-catalyst **90** for effective 1,2,3-triazole synthesis under microwave conditions. The click reaction between azides **91** and alkynes **92** was carried out using this capable catalyst **90**, producing the desired 1,2,3-triazoles **93** in good-to-excellent yields. The authors further explore the efficient ionic liquid-supported Cu(II) catalyst in multistep synthesis of an antiepileptic drug candidate, Rufinamide **94** (Scheme 2.25) [136].

Metallic copper catalysis was greatly used to generate changes in cyclodextrins, which are a diverse group of ligands. As illustrated in Scheme 2.26, cyclodextrin-triazole derivatives **96** were attained through the reaction of azide **95** and alkyne **1** at elevated temperatures employing a trace quantity of metallic copper in DMF with the assistance of ultrasonic irradiation. Metal reactions are sonochemistry's favorite domain because the US promotes mechanical depassivation and improves both mass transfer and electron transfer from the metal to the organic acceptor [137].

Stefani et al. explored the click methodology for the synthesis of 1,4-disubstituted 1,2,3-triazole under ultrasound (US) irradiation. Thus, C-glycosyl-triazoles **98** obtained in good-to-excellent yield by reacting TMS-protected C-alkyne-glycoside **97** with organic azides **2** using CuI/TBAF in THF under US irradiation (Scheme 2.27). The protocol has advantages such as improved reaction rate and shorter reaction time in addition to approximately 1.6 times increased yield in contrast to the conventional approach and metabolic process resistance of C-glycoside analogues [138].

Nallapati et al. reported the synthesis of olanzapine-based 1,2,3-triazole using ultrasonic-assisted CuAAC. An alkyne **99** generated from olanzapine was subjected

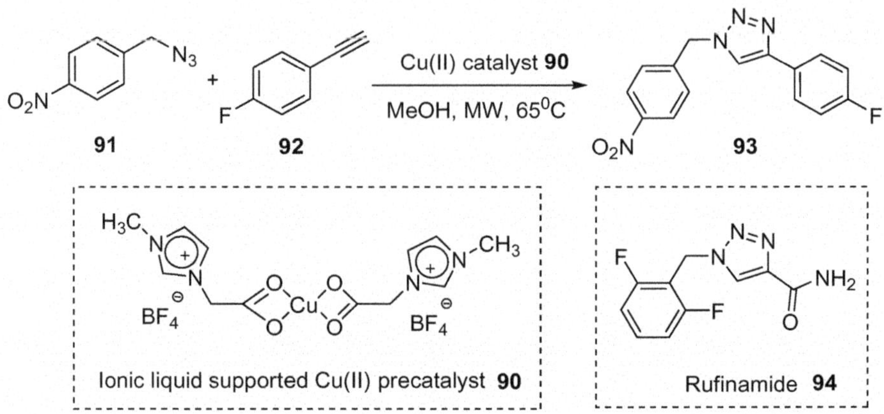

Scheme 2.25 Ionic liquid-supported Cu(II) catalyzes the synthesis of 1,2,3-triazoles with the assistance of microwave

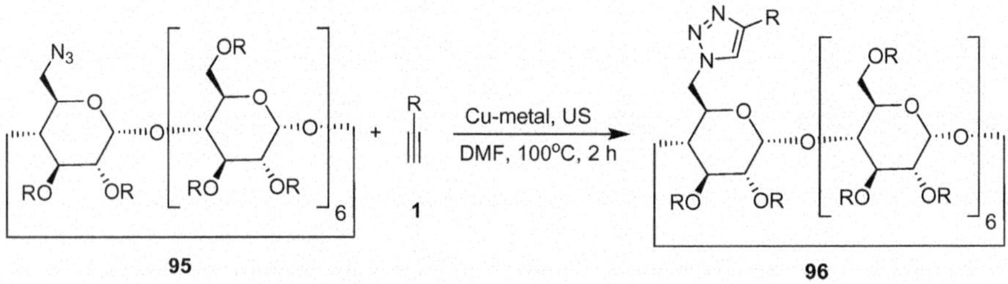

Scheme 2.26 *US*-assisted CuAAC of β-cyclodextrin azide and alkyne in the presence of catalytic loading of Cu-metal

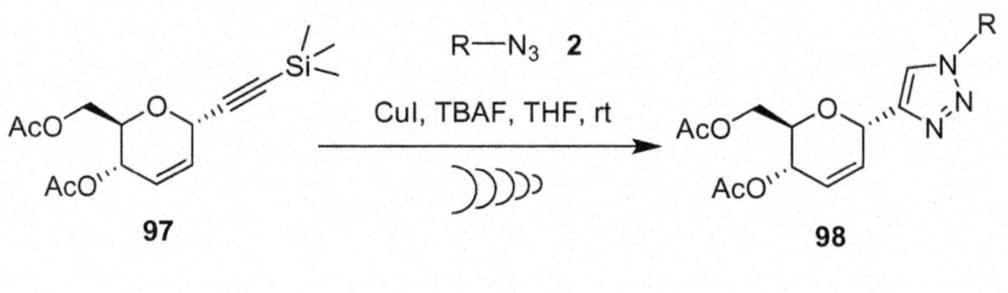

Scheme 2.27 *US*-assisted CuAAC to effectively develop triazolyl sugar-conjugates

to click with benzyl azide **8** in MeCN at 25 °C using CuI/DIPEA yielding triazole **100** with 89% yield in just 20 min. The inclusion of DIPEA as an external ligand resulted in an increase in rate conversion (Scheme 2.28) [139].

In contrast to the use of ultrasound (US) irradiation to facilitate the Cu-azide-alkyne cycloaddition of azide-alkyne to furnish respective triazole [138], some reports on retro-click (Click-to-Unclick) under similar conditions by mechanically

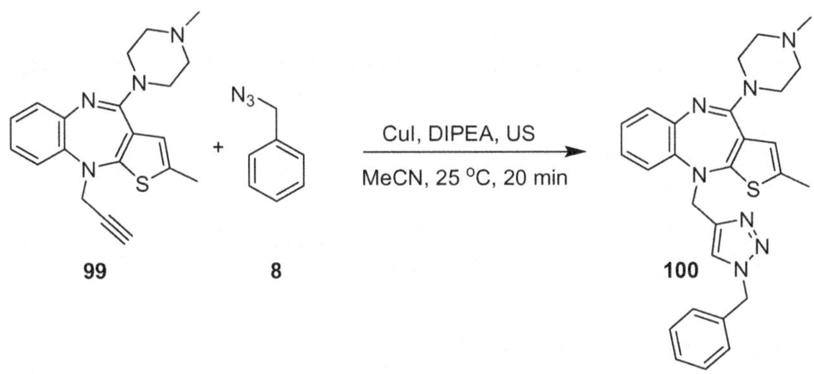

Scheme 2.28 *US*-assisted CuAAC of azide and alkyne in the presence of catalytic loading of CuI/ DIPEA

throwing a reaction into reverse, i.e., unclicking the click under mechanically facilitated click cycloreversions have been reported [140, 141]. Unfortunately, the analysis stated in 'Bielawski's 2011 Science publication' became contentious [142].

Nowadays, a flexible strategy is being sought in order to achieve a more logical and systematic structural component that includes a more favorable and effective photochemical procedure. Similarly, organic dyes were used as a Cu(II) reducing agent by photoredox reactions to get Cu(I) synthesis for attaining click reaction of azide-alkyne functionality. Furthermore, no organic solvent is utilized, and the amount of catalyst precursor used is substantially lower than in conventional methods. Konig et al. highlighted the first photoinduced CuAAC conjugation reaction via photoirradiation of riboflavin tetraacetate **101** in the presence of Et_3N as a base. Furthermore, the highly reducing riboflavin **102** reduces Cu(II) to Cu(I) to catalyze the click conjugation reaction between azide **2** and terminal alkyne **1**, resulting in high yields of regioselective 1,4-disubstituted 1,2,3-triazole-linked molecular scaffold **3** (Scheme 2.29) [143].

The photoreduction of Cu(OAc)2 to generate active Cu(I) species is indeed a crucial aspect of] CuAAC reactions. This process involves the use of light to reduce copper(II) acetate (Cu(OAc)2) to catalytically active Cu(I) species, which are essential for facilitating the CuAAC reaction. In traditional CuAAC reactions, a reducing agent such as sodium ascorbate or tris(2-carboxyethyl)phosphine (TCEP) is often employed to generate Cu(I) species from Cu(II) precursors. However, the use of photoreduction photoinduced Cu(I)-induced click conjugation of polyfunctional azides and alkyne scaffolds eliminates the need for additional reducing agents, simplifying the reaction setup in considerably easier ways [144]. The disodium salt of eosin Y (EY) enabled an efficient photochemical reduction of Cu(II) to Cu(I), which was greatly utilized as a suitable catalyst in the CuAAC reaction [145]. Vincent's group reported the synthesis of Cu(II)-tren ketoprophenate complexes **103,104** with benzophenone-like photosensitizing group as a counter anion and were found as an efficient photolatent CuAAC catalysts (Fig. 2.14). Upon UV irradiation, the catalyst

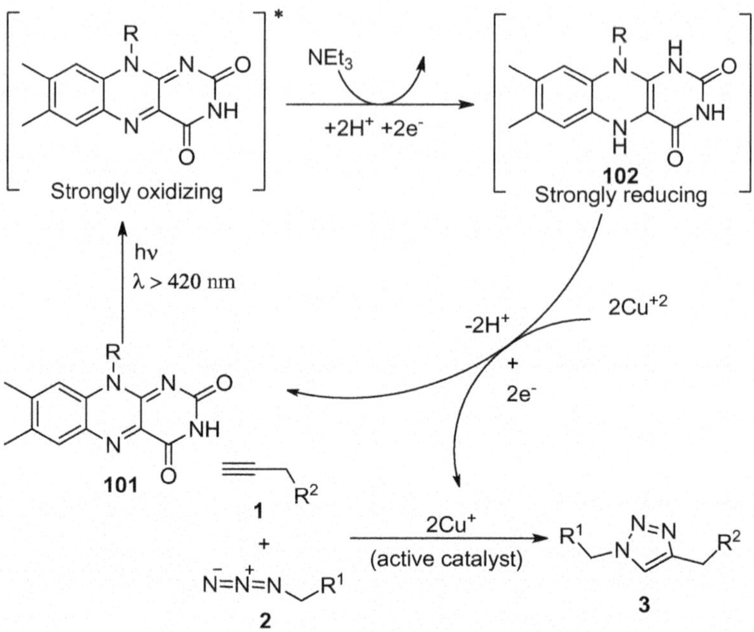

Scheme 2.29 Photoinduced CuAAC click reaction in the assistance of riboflavin substrate

was activated by a photoinduced electron transfer (PET) from photoreduced keto-prophenate to the Cu(II) center and the generated Cu(I) system efficiently assisted the click reaction [146, 147]. The group further developed a photolatent Cu(II) catalyst **105,** made up of an ionic pair produced by an octahedral coordinated Cu(II) dication with commercially available DMEDA ligand and 3-benzoylbenzoate as a counteranion and the resulting system **105** can be triggered by sunlight to proceeds CuAAC (Fig. 2.13) [148].

By harnessing photoreduction to generate Cu(I) species, efforts on photoinduced CuAAC reactions offer several advantages. These include improved temporal control over the reaction, as the reaction can be initiated and controlled by light irradiation. Additionally, photoreduction avoids the use of potentially toxic or interfering reducing agents, simplifying the reaction procedure and potentially enhancing the biocompatibility of the reaction for applications. Overall, the use of photoreduction to generate Cu(I) species for CuAAC reactions represents an important advancement in the field of bioconjugation and materials science, enabling the efficient and controlled formation of triazole linkages under mild conditions in biological systems.

Room temperature ionic liquids (RTILs) have been widely used in a variety of areas due to their superior physical and chemical properties, including organic transformation, biocatalytic synthesis, electrochemistry, and so on. RTILs due to their simplified reaction conditions and sustainable synthesis methods are a popular solvent in the pharmaceutical industry [149, 150]. As a result, these 'designer

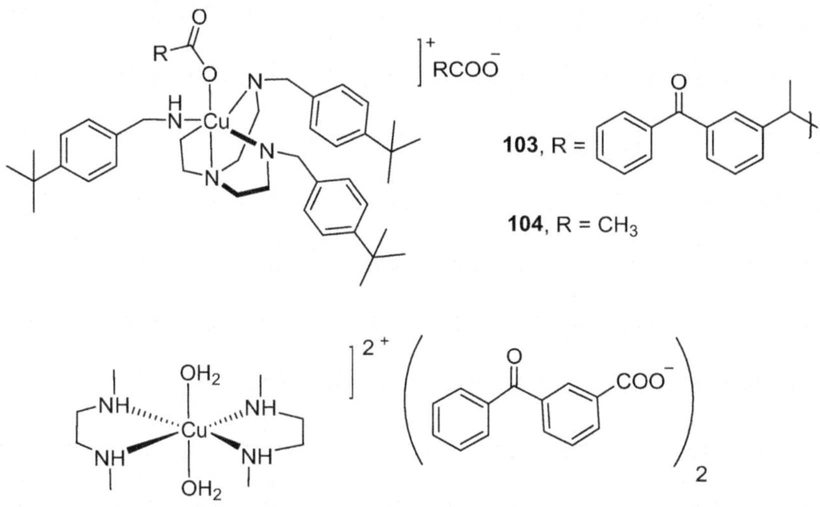

105

Fig. 2.14 First photolatent CuAAC catalyst

solvents' provide a green architectural platform for easy access to medicinally relevant scaffolds like 1,2,3-triazoles. An array of methods for affording 1,2,3-triazoles in ionic liquids are reported. Ali et al. devised an efficient protocol for the regioselective synthesis of 1,4-substituted 1,2,3-triazoles **107** via CuAAC conjugation of alkyne **1** and azide **106** with the assistance of [BMIM][OH] as green media in the absence of other organic solvents and external reducing agent, base, or ligand (Scheme 2.30) [151].

Furthermore, based on the above findings, the group also investigated the efficacy of this green catalytic medium in a one-pot, three-component cycloaddition reaction involving the in situ synthesis of organic azides from bromides **88** and NaN_3, which then react with alkynes **1** to afford the desired 1,4-substituted 1,2,3-triazole **3** in excellent yield and regioselectivity. This alternate multicomponent reaction approach was chosen to reduce the usage of organic azides in laboratory techniques due to their hazardous nature (Scheme 2.31).

Tavassoli research group reports the synthesis of immobilized Cu(II) containing ionic liquids on silica nanoparticles [SNIL-Cu(II)] as the catalyst. The synthesis was

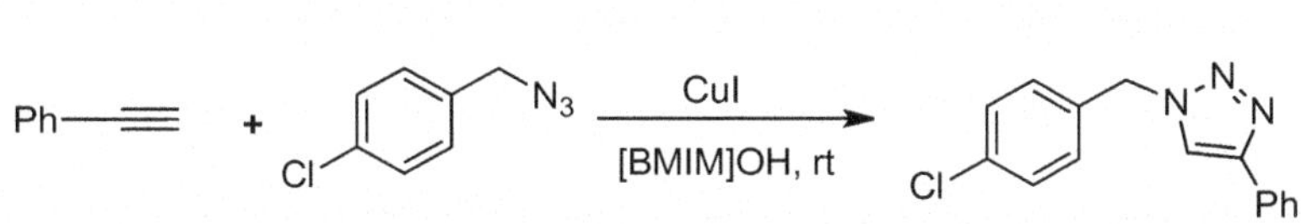

1 106 107

Scheme 2.30 CuI-catalyzed synthesis of 1,4-disubstituted 1,2,3-triazole in [BMIM]OH as green reaction media

Scheme 2.31 CuI-catalyzed one-pot, multicomponent synthesis of 1,4-disubstituted 1,2,3-triazole in the presence of [BMIM]OH

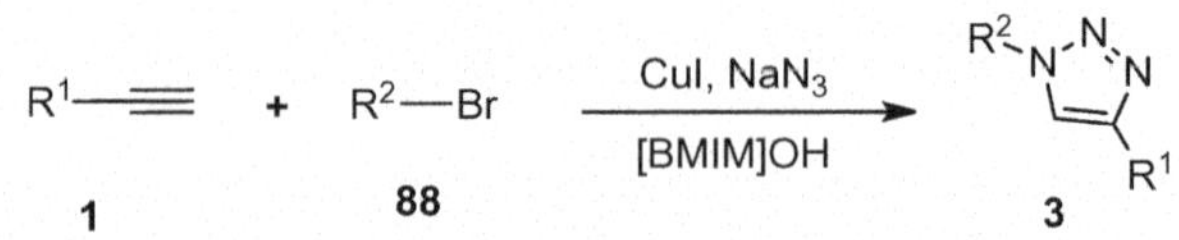

started through the surface-modified silica nanoparticles **108**, which was on treatment with 1,2-bis(4-pyridylthio)ethane **109** in DMF at 80 °C followed by methylation afford the silica nanoparticles immobilized ionic liquid **110**. This silica nanoparticle immobilized ionic liquid **110** was subsequently reacted with $Cu(OTf)_2$ in ethanol for 24 h at 78 °C to give the Cu(II) supported catalytic system **111**. This catalyst is a highly stable, active, reusable, and solid phase, with high activity in the 'one-pot' CuAAC of 1,4-disubstituted triazoles (Scheme 2.32) [152].

The use of various Cu-based hallo clay catalysts was reported to assist the efficient click reaction [153, 154]. Sadjadi et al. [155] reported the ionic liquid and cuprous sulfite containing halloysite nanoclay (Hal-IMI-SO_3Cu) **112** as an efficient catalyst for one-pot multicomponent reaction to attain the triazolyl analogues **114** of interest through the reaction of alkyne **1**, α-halo ketone **113** and sodium azide. In addition to the CuAAC conjugation reaction, the catalyst also assists the *N*- and *O*-arylations as well (Scheme 2.33).

Moreover, the halloysite nanotubes and Schiff base combination with deposited copper iodide nanoparticles was explored as a novel heterogeneous catalytic system for click reaction as well [156]. Furthermore, with environmental considerations in mind, research groups are using a variety of green and eco-friendly approaches to

Scheme 2.32 Synthesis of Cu(II) containing ionic liquids on silica nanoparticles [SNIL-Cu(II)]

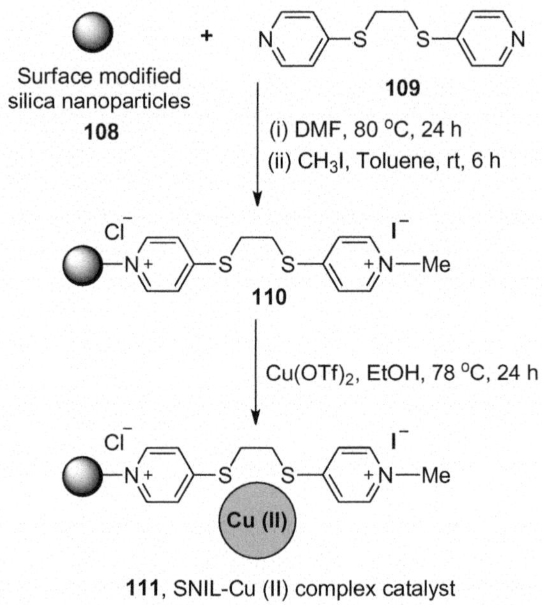

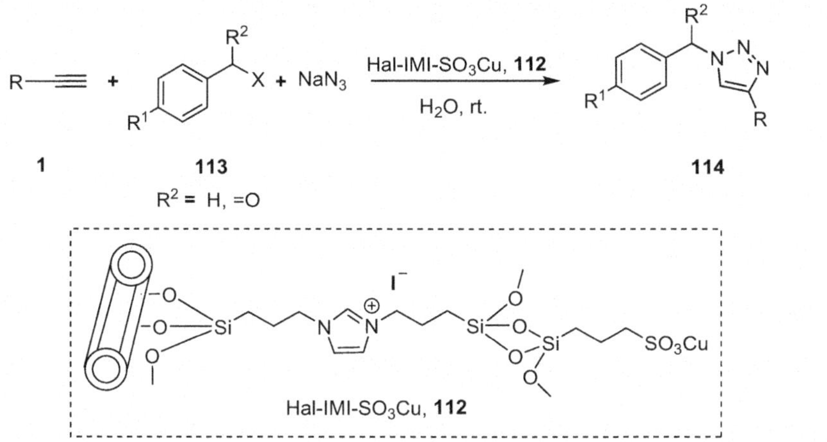

Scheme 2.33 Cu-based halloysite nanoclay-catalyzed one-pot, multicomponent synthesis of 1,4-disubstituted 1,2,3-triazole

synthesize triazolyl-based structural motifs. In the objective of assembly of a green catalytic medium, Wang et al. demonstrated a very simple and uncomplicated one-pot synthesis of 1,4-disubstituted-1,2,3-triazoles in [BMIM][BF4] with the assistance of amino acid ionic liquid (AAIL). In the synthetic approach at room temperature, tetrabutylammonium hydroxide [BMIM][OH] was combined with amino acids such as L-proline, trans-4-hydroxy-L-proline, and *N*, *N*-dimethylglycine to produce AAIL. Among all the developed AAIL, the AAIL **115** synthesized with tetrabutylammonium hydroxide and L-proline proved to be the most effective (Scheme 2.34) [157].

Flow chemistry, also known as continuous-flow chemistry or plug-flow chemistry, is a method of chemical synthesis where reactants are pumped through a reaction chamber where they mix and react under continuous-flow conditions. This method has grown in immense popularity over the last decade, with the discipline recently extending to encompass the use of copper flow reactors in bench-top organic synthesis. These reactors come in multiple configurations and offer several benefits over their batch reaction equivalents concerning yield and safety [158]. Smith et al.

Scheme 2.34 CuI/Amino acid ionic liquid (AAIL)-catalyzed synthesis of substituted 1,2,3-triazole

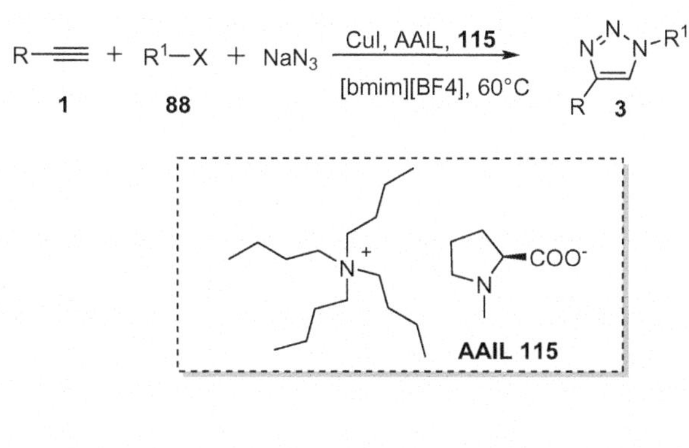

[159, 160] reported the first use of a modular flow reactor to carry out CuAAC reactions in 2007. Kappe research group investigated the click reaction by ICP-MS analysis in the continuous-flow strategy concerning the outcome of copper leaching from the catalyst over the reaction time and the composition of the reaction mixture, using a variety of copper metal sources as heterogeneous catalysts. The findings indicate that the reaction is based on a 'homogeneous' mechanism, with Cu_2O acting as an active species involving copper to assist the reaction [161]. Another intriguing example in this course is the flow method using a glass microchip (GMC) reactor outfitted with β-cyclodextrin to produce a good yield of the desired triazole via AAC of terminal alkynes with aromatic azides. This method works well and affords the desired product in a few minutes, providing a stable GMC surface and can be recycled [162]. Interestingly, another diverse approach to access more critical spatial arrangements is the solid-supported click strategy. The method has notable advantages such as robustness, insensitivity to different conditions, and the ability to use a wide range of polar or non-polar solvents [163]. With the need for synthetic routes and their application, the solid-support inspired click protocol is achieving diverse applications ranging from catalysis to pharmacologically relevant library genesis [164], but the possibility of homo-coupling via this strategy is still being investigated in the absence of a proper definition [165]. Jung et al. highlighted a glass microreactor with an inner layer functionalized with β-cyclodextrin **116** to carry out CuAAC reactions between aromatic azides **8** and alkynes **1**, by using 1,2,3-triazole linkers, the silane-modified reactor surface was connected to cyclodextrin rings. Separate inlets were used to introduce the alkyne/azide scaffold solution (in tert-BuOH) and the $CuSO_4$/NaAsc mixture (in water) into the microreactor. Aside from that, the 1,2,3-triazolyl-linked scaffolds **117** were developed in high yields from a variety of aromatic azides **8** and terminal (aromatic/aliphatic) alkynes **1** at optimal temperature using a low flow rate of 5 μL min^{-1} (Scheme 2.35) [162].

Despite the significant advantages of CuAAC click reaction over other synthetic protocols, the method has a few shortcomings. CuAAC clicks in a few cases because of unwanted oxidative coupling of Cu(I)-acetylides resulted in some trace amount of by-products such as bis-triazolyl analogues. This constraint could be mitigated by employing an inorganic base and air to get cross-coupling triazole as the sole product. Furthermore, the use of stoichiometric amounts of Cu(I) halides for AAC favors the 5-halogenated 1,2,3-triazoles; yet, a dimeric 5,5′-bis-triazole can be obtained as the sole product when an analogous approach is performed in excess CuCl/DMAP [166]. Precaution is required to use the amount of alkyne in AAC, as an alkyne once used in excess, triazole-alkyne oxidative coupling might occur and 2-alkylated tri-substituted triazoles can be obtained as the final product [167]. As per some investigation reports, the electrophiles present in the reaction medium could demetallize the Cu-triazole intermediate and may result in an undesirable product(s); for example, sulfonyl AAC coupling in an aqueous medium resulted in the undesired *N*-sulfonylatedamides, latter on similar AAC performed in presence of some suitable catalyst successfully deliver to high yield of desired 1,2,3-triazoles [50, 52, 168]. Some interrupted CuAAC reactions and triazole formation through different electrophiles are depicted

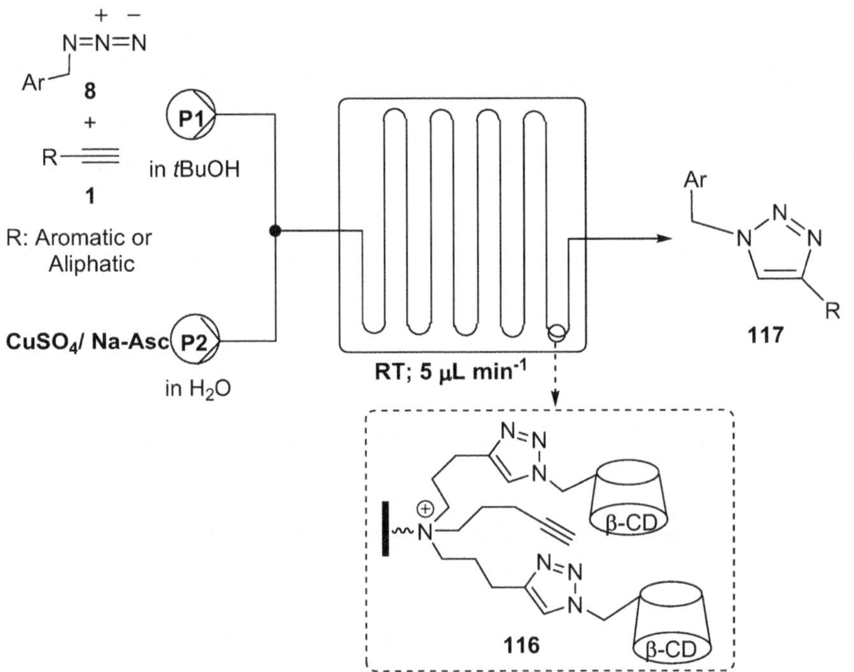

Scheme 2.35 CuAAC reaction through the β-cyclodextrin-functionalized microreactor

in Schemes 2.36, 2.37, and 2.38. The formation of the CuAAC product, the copper-triazole adduct **I** is essentially a reactive metal–organic molecule that can react with a wide range of electrophiles or even be employed in Burgess oxidative coupling to generate bis-triazoles 1**18** [167]. 5-Iodo triazole **121** was formed in the Cu-azide-alkyne cycloaddition coupling as a result of an electrophilic reaction with iodine in assistance of copper(I) species and the electrophilic triiodide ions [169, 170]. However, since then, an array of reactions known as interrupted CuAAC reactions have been characterized based on the use of the nucleophilic nature of copper-triazole adducts. The above reactions afford 1,4,5-substituted triazoles, **119–123** with the interruption step allowing for the introduction of a wide range of 5th-substituents (Scheme 2.36) [171–176]. The competitive copper-hydrogen exchange between the generated triazole and the residual alkyne has made it difficult to achieve high yields of triazole **3** in these reactions. Several ways have been devised to get around this, to outcompete the copper-hydrogen exchange an intramolecular reaction partner has been utilized which effectively gives rise to the elaborate chemical scaffolds [174, 177, 178], To eliminate any proton source, a pre-metallated alkyne or a strong base that allows complete deprotonation of the alkyne might be utilized [179], Selectively use of a second substituent on the alkyne, such as halogen or another leaving group,

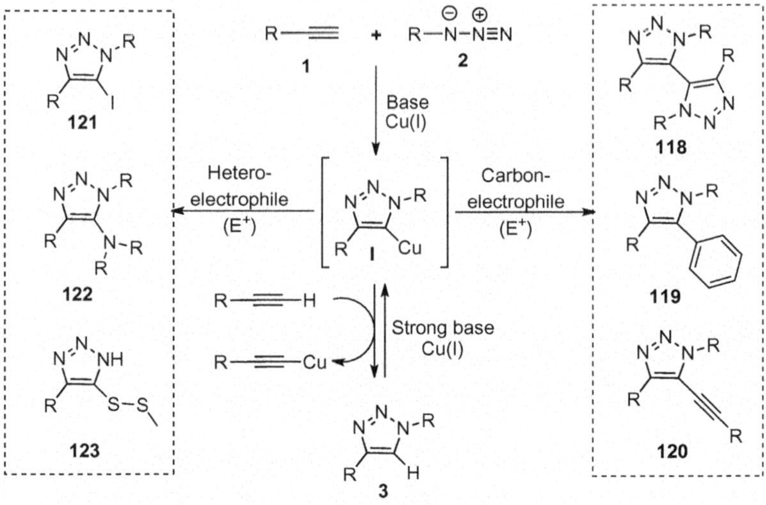

Scheme 2.36 CuAAC catalytic cycle showing cross-couplings of the Cu-triazole intermediate and applications in reactions with electrophiles

has proven to be extremely efficient [175]. Finally, a strong base can promote deprotonation or other C–H activation of the 5-position in the first produced 1,4-substituted 1,2,3-triazole, resulting in a triazole C5-nucleophile [180].

Electron-deficient triazoles **I** are easily transformed into other moieties through the elimination reaction, where N_2 elimination may lead to the formation of ketene imine **III**. These, in turn, react with a variety of nucleophiles to afford the compound **124** [181, 182]. In the presence of dirhodium tetracarboxylates, rhodium carbene intermediates **IV** can be further transformed into a variety of products **125,126** (Scheme 2.37).

An interruptive CuAAC reaction takes place when trifluoroacetic anhydride was added to the CuAAC reaction, the central triazole nitrogen was acylated, and a ring-opening–ring-closure cascade reaction ensued, affording 1,2,4-triazinones **127** instead of desired triazoles (Scheme 2.38) [183].

In addition to the above-mentioned Cu(I)-catalyzed regioselecive triazole forming AAC tool, some other homogeneous catalytic systems such as silver(I) salts have also been explored. The Ag-AAC method is comparable in terms of regioselectivity; however, the higher cost limits the use of CuAAC [184]. In addition, some other metal-induced click tools like Zn-AAC [185] and Pd-AAC [186] have also been employed for regioselective AAC; however, none of them can replace the Cu-AAC system in terms of regioselectivity, cost-effectiveness, ease of reaction, reactivity, and also wide availability. CuAAC coupling reaction in the presence of tris (triazolylmethyl) amin-based ligand, for example BTTES can make the protocol ever more rapid, biocompatible without apparent toxicity, and used successfully for noninvasive imaging of biomolecules in living organisms particularly fucosylated glycans during zebrafish early embryogenesis [187]. Since 2002, this modular CuAAC tool

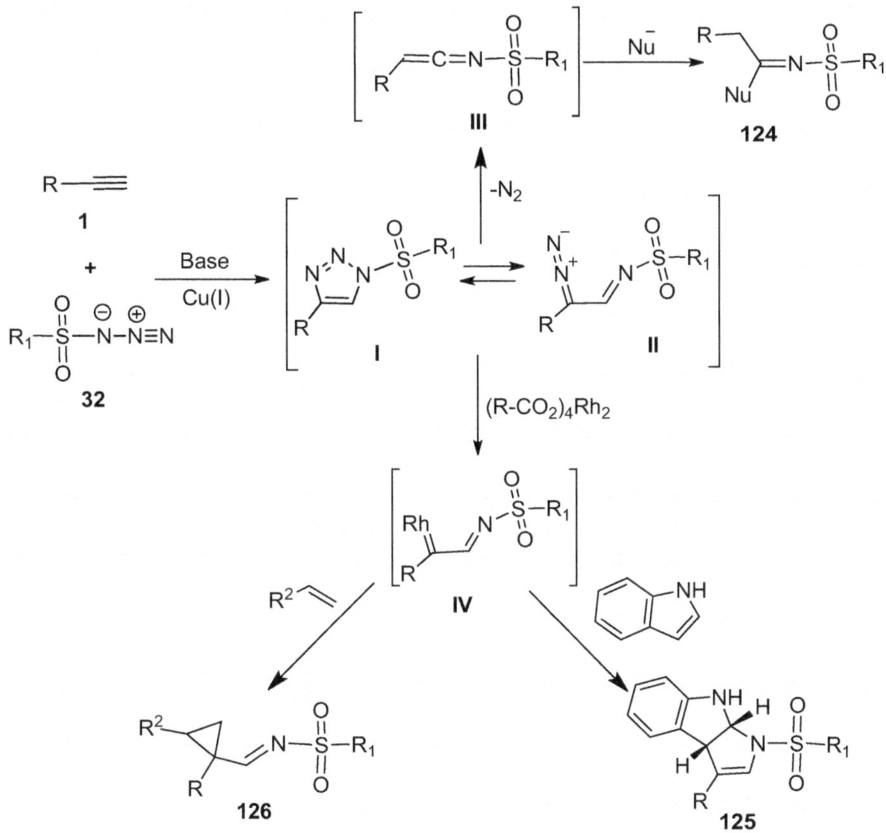

Scheme 2.37 Annulations and intermolecular cyclization reactions of electron-deficient, unstable sulfonyl triazoles via rhodium-interrupted disproportionation

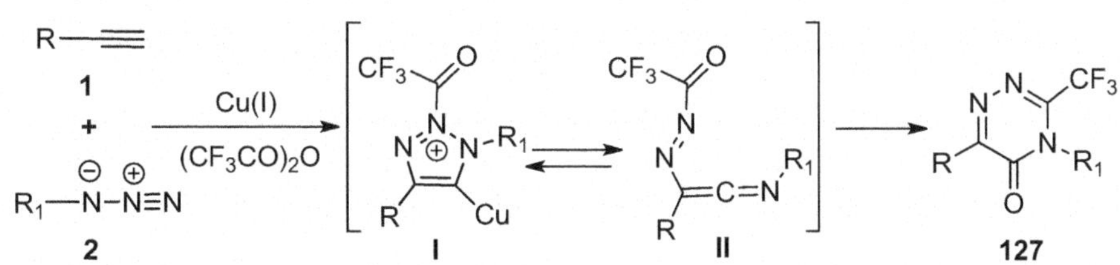

Scheme 2.38 Trifluoroacetic acid anhydride-assisted ring-opening–ring-closure cascade observed during CuAAC as a route to 1,2,4-triazinones

has received enormous applications in wide fields of growing interest ranging from medicinal chemistry to material science and chemical biology to catalysis. A detailed account of notable applications of CuAAC conjugation in various fields including catalysis, dendrimer synthesis, polymer science, peptide chemistry, lipid modification, nucleic acid chemistry, glycoscience/glycotechnology, and drug discovery and development is presented in a separate chapter of this book.

2.6 Conclusions and Future Outlook

Click chemistry is a powerful and versatile approach to chemical synthesis that has gained widespread popularity in various fields, including chemistry, biology, medicine, and materials science. Coined by Nobel laureate K. Barry Sharpless in 2001, the term 'click chemistry' refers to a set of highly selective and efficient reactions that enable the rapid and reliable construction of simple-to-complex molecular architectures. These reactions are characterized by their high yield, specificity, and modularity, making them ideal for applications ranging from drug discovery to bioconjugation and material design.

CuAAC 'Click chemistry' developed by Meldal and Sharpless, in 2002, encompasses a diverse array of highly efficient and selective chemical reactions that have revolutionized the field of chemical synthesis. These reactions are characterized by their modularity, reliability, and versatility, making them invaluable tools across a wide range of disciplines. This modular CuAAC reaction is one of the most adaptable organic reactions available to the scientific community at large, from biology to material sciences. The versatility and robustness of the reaction have led to application in almost all disciplines of chemistry. The outcome of the reaction is reliable and predictable, leading in all cases to the exclusive formation of 1,4-regioisomer of 1,2,3-triazole in high yield. The reaction is complex and has been demonstrated to involve clusters of Cu(I) and ligands. The extreme specificity for 1,4-triazole formation may be explained through simple six-membered transition states involving two Cu(1) atoms on a Cu-cluster. Notably, the CuAAC is highly selective, and the reaction conditions may be tailored to accommodate nearly all forms of protecting groups and reactive intermediates, an essential requirement for the growth of the reaction is the ongoing development of new ligands and conditions. The chapter included the various notable approaches for selective assembly of the substituted 1,2,3-triazole under various conditions and sources of Cu-catalyst. The development of novel Cu-AAC catalysts with potential efficacy and notable regioselectivity can enforce the wide utilities of azide-alkyne cycloaddition (AAC) in the emerging fields of human interest such as pharmaceutical, medical, biology, and green and sustainable chemistry.

References

1. Michael A (1893) J Prakt Chem 48:94–95
2. Huisgen R (1963) Angew Chem Int Ed Engl 2:565–598
3. Huisgen R (1963) Angew Chem Int Ed Engl 2:633–696

4. Huisgen R (1989) Pure Appl Chem 61:613–628
5. Huisgen R, Szeimies G, Moebius L (1967) Chem Ber 100:2494–2507
6. Huisgen R, Padwa A (ed) (1984) Wiley, New York, pp 1 – 176
7. Tornoe CW, Christensen C, Meldal M (2002) J Org Chem 67:3057–3064
8. Rostovtsev VV, Green LG, Fokin VV, Sharpless KB (2002) Angew Chem 114:2708–2711
9. Lee LV, Mitchell ML, Huang SJ, Fokin VV, Sharpless KB, Wong CH (2003) J Am Chem Soc 125:9588–9589
10. Whiting M, Muldoon J, Lin YC, Silverman SM, Lindstrom W, Olson AJ, Kolb HC, Finn MG, Sharpless KB, Elder JH, Fokin VV (2006) Angew Chem 118:1463–1467
11. Tiwari VK, Mishra BB, Mishra KB, Mishra N, Singh AS, Chen X (2016) Chem Rev 116:3086–3240
12. Agrahari AK, Bose P, Jaiswal MK, Rajkhowa S, Singh AS, Hotha S, Mishra N, Tiwari VK (2021) Chem Rev 121:7638–7956
13. Speers AE, Adam GC, Cravatt BF (2003) J Am Chem Soc 125:4686–4687
14. Beatty KE, Xie F, Wang Q, Tirrell DA (2005) J Am Chem Soc 127:14150–14151
15. Deiters A, Schultz PG (2005) Bioorg Med Chem Lett 15:1521
16. Kumar S, Yadav MS, Singh SK, Rajkhowa S, Tiwari VK (2023) SynOpen 7:145–153
17. Mishra N, Singh AS, Agrahari AK, Singh SK, Singh M, Tiwari VK (2019) ACS Comb Sci 21:389–399
18. Mishra N, Singh SK, Singh AS, Agrahari AK, Tiwari VK (2021) J Org Chem 86:17884–17895
19. Singh SK, Mishra N, Kumar S, Jaiswal MK, Tiwari VK (2022) Chem Select 7:e202201314
20. Singh SK, Kumar S, Yadav MS, Tiwari VK (2022) J Org Chem 87:15389–15402
21. Singh SK, Kumar S, Yadav MS, Gupta A, Tiwari VK (2023) J Org Chem 88:13440–13453
22. Singh SK, Kumar S, Yadav MS, Bhattacharya S, Tiwari VK (2024) SYNTHESIS 56:975–988
23. Mishra A, Mishra N, Tiwari VK (2016) Curr Org Synth 1:176–219
24. Wojaczynska E, Steppeler F, Iwan D, Scherrmann M-C, Marra A (2021) Molecules 26:7291
25. Meldal M, Tornøe CW (2008) Chem Rev 108:2952–3015
26. Saini PS, Singh G, Kaur G, Singh J, Singh H (2021) Mol Cataly 504:111432
27. Mosesa JE, Moorhousea AD (2007) Chem Soc Rev 36:1249–1262
28. Kolb HC, Sharpless KB (2003) Drug Disc Today 8:1128–1137
29. Sokolovaab NV, Nenajdenko VG (2013) RSC Adv 3:16212–16242
30. Kumar V, Lal KN, Tittal RK (2023) Catal Commun 176:106629
31. Mishra KB, Tiwari N, Bose P, Singh R, Rawat AK, Singh SK, Mishra RC, Singh RK, Tiwari VK (2019) ChemistrySelect 4:2644
32. Zhan WH, Barnhill HN, Sivakumar K, Tian H, Wang Q (2005) Tetrahedron Lett 46:1691–1695
33. Golas PL, Tsarevsky NV, Sumerlin BS, Matyjaszewski K (2006) Macromolecules 39:6451–6457
34. Liu PY, Jiang N, Zhang J, Wei X, Lin HH, Yu XQ (2006) Chem Biodiv 3:958–966
35. Natarajan P, Ferraudi G (1981) Inorg Chem 20:3708–3712
36. Lipshutz BH, Taft BR (2006) Angew Chem Int Ed 45:8235–8238
37. Molteni G, Bianchi CL, Marinoni G, Santo N, Ponti A (2006) New J Chem 30:1137–1139
38. Kushwaaha D, Pandey P, Kale RR, Tiwari VK (2012) Trends Carbohydr Res 4(3):45–53
39. Shao C, Wang X, Xu J, Zhao J, Zhang Q, Hu Y (2010) J Org Chem 75:7002–7005
40. Shao C, Wang X, Zhang Q, Luo S, Zhao J, Hu Y (2011) J Org Chem 76:6832–6836
41. Kamijo S, Jin T, Huo Z, Yamamoto Y (2004) J Org Chem 69:2386–2393
42. Perez-Balderas F, Ortega-Munoz M, Morales-Sanfrutos J, Hernandez-Mateo F, Calvo-Flores FG, Calvo-Asin JA, Isac-Garcia J, Santoyo-Gonzalez F (2003) Org Lett 5:1951–1954
43. Malkoch M, Schleicher K, Drockenmuller E, Hawker CJ, Russell TP, Wu P, Fokin VV (2005) Macromolecules 38:3663–3678
44. Wu P, Feldman AK, Nugent AK, Hawker CJ, Scheel A, Voit B, Pyun J, Frechet JMJ, Sharpless KB, Fokin VV (2004) Angew Chem Int Ed 43:3928–3932
45. Nia AS, Rana S, Dohler D, Jirsa F, Meister A, Guadagno L, Koslowski E, Bron M, Binder WH (2015) Chem Eur J 21:10763–10770
46. Alonso F, Moglie Y, Radivoy G (2015) Acc Chem Res 48:2516–2528

47. Chan TR, Hilgraf R, Sharpless KB, Fokin VV (2004) Org Lett 6:2853–2855
48. Gaetke LM, Chow CK (2003) Toxicology 189:147–163
49. Yin Z, Aragones MC, Chowdhury S, Bentley P, Kaczanowska K, Mohamed LB, Gildersleeve JC, Finn MG, Huang X (2013) ACS Chem Biol 8:1253–1262
50. Rodionov VO, Presolski SI, Gardinier S, Lim Y-H, Finn MG (2007) J Am Chem Soc 129:12696
51. Lewis WG, Magallon FG, Fokin VV, Finn MG (2004) J Am Chem Soc 126:9152–9153
52. Wang Q, Chan TR, Hilgraf R, Fokin VV, Sharpless KB, Finn MG (2003) J Am Chem Soc 125:3192–3193
53. Donnelly PS, Zanatta SD, Zammit SC, White JM, Williams SJ (2008) Chem Commun 324:2459–2461
54. Li F, Hor TSA (2009) Chem Eur J 15:10585–10592
55. Meng JC, Fokin VV, Finn MG (2005) Tetrahedron Lett 46:4543–4546
56. Ozkal E, Llanes P, Bravo F, Ferrali A, Pericas MA (2014) Adv Synth Catal 356:857–869
57. Liu M, Reiser O (2011) Org Lett 13:1102–1105
58. Touj N, Özdemir I, Yaşar S, Hamdi N (2017) Inorg Chim Acta 467:21–32
59. Nakamura T, Terashima T, Ogata K, Fukuzawa S (2011) Org Lett 13:620–623
60. Fabbrizzi P, Cicchi S, Brandi A, Sperotto E, Koten GV (2009) Eur J Org Chem 31:5423–5430
61. Joshi DK, Mishra KB, Tiwari VK, Bhattacharya S (2014) RSC Adv 4:39790–39797
62. Fu Q, Xie K, Tan S, Ren JM, Zhao Q, Webley PA, Qiao GG (2016) Chem Commun 52:12226–12229
63. Wang X-X, Yang J, Xu X, Ma JF (2019) Chem Eur J 25:16660–16667
64. Cho SH, Yoo EJ, Bae I, Chang S (2005) J Am Chem Soc 127:16046–16047
65. Cassidy MP, Raushel J, Fokin VV (2006) Angew Chem Int Ed 45:3154–3157
66. Raushel J, Fokin VV (2010) Org Lett 12:4952–4955
67. Sareen N, Singh AS, Tiwari VK, Kant R, Bhattacharya S (2017) Dalton Trans 46:12705–12710
68. Nayak M, Joshi DK, Kumar K, Singh AS, Tiwari VK, Bhattacharya S (2021) Inorg Chem 60:8075–8084
69. Kumar K, Agrahari AK, Pratap R, Tiwari VK, Bhattacharya S (2022) New J Chem 46:18471–18481
70. Kumari K, Singh AS, Manar KK, Yadav CL, Tiwari VK, Drew MGB, Singh N (2019) New J Chem 43:1166–1176
71. Ferraro V, Sole R, Bortoluzzi M, Beghetto V, Castro J (2021) Appl Organomet Chem 35:e6401
72. Anamika AK, Manar KK, Yadav CL, Tiwari VK, Drew MGB, Singh N (2019) New J Chem 43:8939–8949
73. Barman MK, Sinha AK, Nembenna S (2016) Green Chem 18:2534–2541
74. Wang S, Jia K, Cheng J, Chen Y, Yuan Y (2017) Tetrahedron Lett 38:3717–3721
75. Ali AA, Chetia M, Sarma D (2016) Tetrahedron Lett 15:1711–1714
76. Khanam S, Pandey SK, Rai SK, Verma D, Jasinski JP, Pratap S, Tewari AK (2017) Chem Select 2:6370–6374
77. Bai S-Q, Koh LL, Andy Hor T (2009) S. Inorg Chem 48:1207–1213
78. Girard C, Onen E, Aufort M, Beauviere S, Samson E, Herscovici J (2006) Org Lett 8:1689–1693
79. Sirion U, Bae YJ, Lee BS, Chi DY (2008) Synlett 15:2326–2330
80. Chassaing S, Kumarraja M, Sido ASS, Pale P, Sommer J (2007) Org Lett 9:883–886
81. Vilé G, Liberto GD, Tosoni S, Sivo A, Ruta V, Nachtegaal M, Clark AH, Agnoli S, Zou Y, Savateev A, Antonietti M, Pacchioni G (2022) ACS Catal 12:2947–2958
82. Thorwirth R, Stolle A, Ondruschka B, Wild A, Schubert US (2011) Chem Commun 47:4370–4372
83. Rodionov VO, Fokin VV, Finn MG (2005) Angew Chem Int Ed 44:2210–2215
84. Himo F, Lovell T, Hilgraf R, Rostovtsev VV, Noodleman L, Sharpless KB, Fokin VV (2005) J Am Chem Soc 127:210–216
85. Straub BF (2007) Chem Commun 12:3868–3870
86. Zhu L, Brassard CJ, Zhang X, Guha PM, Clark RJ (2016) Chem Rec 16:1501–1517

87. Hein JE, Fokin VV (2010) Chem Soc Rev 39:1302–1315
88. Presolski SI, Hong V, Cho S-H, Finn M (2010) G. J Am Chem Soc 132:14570–14576
89. Ahlquist M, Fokin VV (2007) Organometallics 26:4389–4439
90. Zhou Y, Lecourt T, Micouin L (2010) Angew Chem Int Ed 49:2607–2610
91. Nolte C, Mayer P, Straub BF (2007) Angew Chem Int Ed 46:2101–2103
92. Berg R, Straub BF (2013) Beil J Org Chem 9:2715–2750
93. Ramesh R, Chauhan P (2016) Click reactions in organic synthesis. In: Chandrasekeran S (ed), Wiley VCH, pp 1–24
94. Jin L, Tolentino DR, Melaimi M, Bertrand G (2015) Sci Adv 1:e1500304
95. Johansson JR, Beke-Somfai T, Stalsmeden AS, Kann N (2016) Chem Rev 116:14726–14768
96. Shashank AB, Karthik S, Madhavachary R, Ramachary DB (2014) Chem Eur J 20:16877–16881
97. Krasinski A, Fokin VV, Sharpless KB (2004) Org Lett 6:1237–1240
98. Creary X, Anderson A, Brophy C, Crowell F, Funk Z (2012) J Org Chem 77:8756–8761
99. Mishra N, Agrahari AK, Bose P, Singh SK, Singh AS, Tiwari VK (2020) Sci Rep 10:3586
100. Kushwaha D, Singh SK, Tiwari VK (2020) Carbo-click in drug discovery and development. In: Carbohydrates in drug discovery and development. Elsevier Inc., Amsterdam, pp 403–450
101. Suarez PL, Gándara Z, Gómez G, Fall Y (2004) Tetrahedron Lett 45:4619–4621
102. Praveena KSS, Murthy NYS, Ramarao EVVS, Akkenapally S, Kumar CG, Kapavarapu R, Pal S (2015) Bioorg Med Chem Lett 25:1057
103. Akula RA, Temelkoff DP, Artis ND, Norris P (2004) Heterocycles 63:2719–2725
104. Pachón LD, Van Maarseveen JH, Rothenberg G (2005) Adv Synth Catal 347:811–815
105. Orgueira HA, Fokas D, Isome Y, Chan PC-M, Baldino CM (2005) Tetrahedron Lett 46:2911–2914
106. Kantam ML, Jaya VS, Sreedhar B, Rao MM, Choudary BM (2006) J Mol Catal A Chem 256:273–277
107. Park IS, Kwon MS, Kim Y, Lee JS, Park J (2008) Org Lett 10:497–500
108. Borah BJ, Dutta D, Saikia PP, Barua NC, Dutta DK (2011) Green Chem 13:3453–3460
109. Sanka RVSP, Balaji K, Leterrier Y, Pandey S, Srivastava M, Srivastava A, Binder WH, Rana S, Michaud V (2019) Chem Commun 55:6249–6252
110. Jayaramulu K, Suresh VM, Maji TK (2015) Dalton Trans 44:83–86
111. Yu W, Jiang L, Shen C, Xu W, Zhang P (2016) Catal Commun 79:11–16
112. Fu F, Martinez A, Wang C, Ciganda R, Yate L, Escobar A, Moya S, Fouquet E, Ruiz J, Astruc D (2017) Chem Commun 53:5384–5387
113. Siuki MMK, Bakavoli M, Eshghi H (2019) Appl Organometal Chem 33(e4774):1–9
114. Jain Y, Kumari M, Singh RP, Kumar D, Gupta R (2020) Catal Lett 150:1142–1154
115. Gawande MB, Goswami A, Felpin FX, Asefa T, Huang X, Silva R, Zou X, Zboril R, Varma RS (2016) Chem Rev 116:3722–3811
116. Hu Q, Shi X-L, Chen Y, Wang F, Weng Y, Duan P (2019) J Ind Eng Chem 69:387–396
117. Lim M, Lee H, Kang M, Yoo W, Rhee H (2018) RSC Adv 8:6152–6159
118. Pazoki F, Salamatmanesh A, Bagheri S, Heydari A (2020) Catal Lett 150:1186–1195
119. Schatz A, Hager M, Reiser O (2009) Adv Func Mater 19:2109–2115
120. Kwak SJ, Shin US, Kim SH (2023) J Chem Sci 135:10
121. Sabaqian S, Nemati F, Heravi MM, Nahzomi HT (2017) Appl Organomet Chem 31:3660
122. Wang D, Li N, Zhao M, Shi W, Ma C, Chen B (2010) Green Chem 12:2120–2123
123. Wang D, Zhao M, Liu X, Chen Y, Li N, Chen B (2012) Org Biomol Chem 10:229–231
124. Lal S, Díez-González S (2011) J Org Chem 76:2367–2373
125. Gonda Z, Novák Z (2010) Dalton Trans 39:726–729
126. Díez-González S, Correa A, Cavallo L, Nolan SP (2006) Chem Eur J 12:7558–7564
127. Wang W, Wu J, Xia C, Li F (2011) Green Chem 13:3440
128. Gaulier C, Hospital A, Legeret B, Delmas AF, Aucagne V, Cisnetti F, Gautier A (2012) Chem Commun 48:4005
129. Brotherton WS, Michaels HA, Simmons JT, Clark RJ, Dalal NS, Zhu L (2009) Org Lett 11:4954

130. Sahil D, Yadav S, Dash C (2023) ChemistrySelect 8:e202301452
131. Emami M, Bikas R, Noshiranzadeh N, Kozakiewicz A, Lis T (2020) ACS Omega 5:13344–13357
132. Konwar M, Hazarika R, Ali AA, Chetia M, Khupse ND, Saikia PJ, Sarma D (2018) Appl Organometal Chem 32:e4425
133. Konwar M, Ali AA, Chetia M, Saikia PJ, Sarma D (2016) Tetrahedron Lett 57:4473–4476
134. Ali AA, Chetia M, Saikiab PJ, Sarma D (2014) RSC Adv 4:64388–64392
135. Appukkuttan P, Dehaen W, Fokin VV, Eycken EV (2004) Org Lett 6:4223–4225
136. Saikia AA, Rao RN, Das S, Jena S, Rej S, Maiti B, Chanda K (2020) Tetrahedron Lett 61:152273
137. Cintas P, Barge A, Tagliapietra T, Boffa L, Cravotto G (2010) Nat Protoc 5:607
138. Stefani HA, Silva NCS, Manarin F, Lüdtke DS, Zukerman-Schpector J, Madureira LS, Tiekink ERT (2012) Tetrahedron Lett 53:1742–1747
139. Nallapati SB, Sreenivas BY, Bankala R, Parsa KVL, Sripelly S, Mukkanti K, Pal M (2015) RSC Adv 5:94623
140. Leibfarth FA, Hawker CJ (2011) Science 333:1582–1583
141. Brantley JN, Wiggins KM, Bielawaski CW (2011) Science 333:1606–1609
142. Reaction DAU (2014) Chem Eng News 92(26):7
143. Ritter SC, König B (2006) Chem Commun 41:4694–4696
144. Sandmann B, Happ B, Vitz J, Hager MD, Burtscher P, Moszner N, Schubert US (2013) Polym Chem 4:3938–3942
145. Castro-Godoy WD, Heredia AA, Schmidt LC, Argüello JE (2017) RSC Adv 7:33967–33973
146. Harmand L, Cadet S, Kauffmann B, Scarpantonio L, Batat P, Jonusauskas G, McClenaghan ND, Lastécouères D, Vincent J-M (2012) Angew Chem Int Ed 51:7137
147. Harmand L, Lambert R, Scarpantonio L, McClenaghan ND, Lastécouères D, Vincent J-M (2013) Chem Eur J 19:16231
148. Beniazza R, Lambert R, Harmand L, Molton F, Duboc C, Denisov S, Jonusauskas G, McClenaghan ND, Lastécouères D, Vincent J-M (2014) Chem Eur J 20:13181
149. (a) Pârvulescu VI, Hardacre C (2007) Chem Rev 107:2615–2665; (b) Tiwari VK, Tripathi G, Rajkhowa S, Singh AK, Kumar A (2023) Green chemistry. Springer
150. Bose P, Agrahari AK, Singh SK, Singh AS, Yadav MS, Rajkhowa S, Tiwari VK (2021) Recent developments on ionic liquids-mediated synthetic protocols of biologically relevant five and six-membered heterocyclic skeletons. In: Brahmachari G (ed) Green synthetic approaches for biologically relevant heterocycles. Elsevier, Amsterdam, pp 301–364
151. Ali AA, Konwar M, Chetia M, Sarma D (2016) Tetrahedron Lett 57:5661–5665
152. Tavassoli M, Isfahani AL, Moghadam M, Tangestaninejad S, Mirkhani V, Baltork IM (2015) Appl Catal A Gen 503:186–195
153. Dar BA, Bhowmik A, Sharma A, Sharma PR, Lazar A, Singh AP, Sharma M, Singh B (2013) Appl Clay Sci 80–81:351–357
154. Bahri-Laleh N, Sadjadi S, Heravi MM, Malmir M (2018) Appl Organometal Chem 32:e4283
155. Sadjadi S, Heravi MM, Malmir M, Kahangi FG (2018) Appl Clay Sci 162:192–203
156. Daraie M, Bagheri D, Malmir M, Heravi MM (2021) Sci Rep 11:23658
157. Yan J, Wang L (2010) Synth 3:447–452
158. Baoa J, Tranmer GK (2015) Chem Commun 51:3037–3044
159. Smith CD, Baxendale IR, Lanners S, Hayward JJ, Smith SC, Ley SV (2007) Org Biomol Chem 5:1559
160. Smith CD, Baxendale IR, Tranmer GK, Baumann M, Smith SC, Lewthwaite RA, Ley SV (2007) Org Biomol Chem 5:1562
161. Fuchs M, Goessler W, Pilger C, Kappe CO (2010) Adv Synth Catal 352:323–328
162. Tahir MN, Qamar R, Adnan A, Cho E, Jung S (2013) Tetrahedron Lett 54:3268–3273
163. Amblard F, Cho JH, Schinazi RF (2009) Chem Rev 109:4207–4220
164. Alza E, Cambeiro XA, Pericas MA (2007) Org Lett 9:3717–3720
165. Ghosh KK, Ha HH, Kang NY, Chandran Y, Chang YT (2011) Chem Commun 47:7488–7490
166. Goyard D, Praly JP, Vidal S (2012) Carbohydr Res 362:79–83

167. Angell YL, Burgess K (2007) Angew Chem Int Ed 46:3649–3651
168. Yoo EJ, Bae I, Cho SH, Han H, Chang S (2006) Org Lett 8:1347–1350
169. Bock VD, Hiemstra H, van Maarseveen JH (2006) Eur J Org Chem 2006:51–68
170. Brotherton WS, Clark RJ, Zhu L (2012) J Org Chem 77:6443–6455
171. Ackermann L, Potukuchi HK (2010) Org Biomol Chem 8:4503–4513
172. Cheung KPS, Tsui GC (2017) Org Lett 19:2881–2884
173. Cui F-H, Chen J, Mo Z-Y, Su S-X, Chen Y-Y, Ma X-L, Tang H-T, Wang H-S, Pan Y-M, Xu Y-L (2018) Org Lett 20:925–929
174. Larin EM, Lautens M (2019) Angew Chem Int Ed Engl 58:13438–13442
175. Wei F, Zhou T, Ma Y, Tung C-H, Xu Z (2017) Org Lett 19:2098–2101
176. Zhou W, Zhang M, Li H, Chen W (2017) Org Lett 19:10–13
177. Cai Q, Yan J, Ding K (2012) Org Lett 14:3332–3335
178. Selvaraju M, Sun CM (2014) Adv Synth Catal 356:1329–1336
179. Yu XX, Xu J, Zhou Y, Song Q (2018) Org Chem Front 5:2463–2467
180. Wei F, Li H, Song C, Ma Y, Zhou L, Tung C-H, Xu Z (2015) Org Lett 17:2860–2863
181. Spangler JE, Davies HM (2013) J Am Chem Soc 135:6802–6805
182. Guo S, Dong P, Chen Y, Feng X, Liu X (2018) Angew Chem Int Ed Engl 57:16852–16856
183. Wu FF, Zhou W, Chen K, Chen W, Liu M, Wu H (2018) Adv Synth Catal 360:2435–2439
184. McNulty J, Keskar K (2012) Eur J Org Chem 28:5462–5470
185. Smith CD, Greaney MF (2013) Org Lett 15:4826–4829
186. Li J, Wang D, Zhang Y, Li J, Chen B (2009) Org Lett 11:3024–3027
187. Soriano D, Wang W, Jiang H, Besanceney C, Yan AC, Levy M, Liu Y, Marlow FL, Wu P (2010) J Am Chem Soc 132:16893–16899

Chapter 3
RuAAC 'Click Chemistry'-Mediated Synthesis of 1,5-Disubstituted 1,2,3-Triazoles

3.1 Introduction

The click reactions are robust, modular, and atom-efficient synthetic approach for the facile synthesis of 1,2,3-triazole analogues. The reaction was discovered by a group of scientists represented by the Meldal and Sharpless [1–4]. The 1,3-dipolar cycloaddition or click concept has drawn a lot of attention since the idea was introduced and has been essential applications in several fields including organic synthesis, biology, medicinal chemistry, and material sciences to industrial applications as corrosion inhibitors, agrochemicals, dyes, photostabilizers, optical brighteners [5–13]. The two regioisomeric triazoles, 1,4-disubstituted and 1,5-disubstituted 1,2,3-triazole, are the significant heterocyclic compounds of nitrogen-based scaffold, effectively important in various applications in the domain of organic syntheses. Moreover, the application the click-derived triazoles in various fields it's have efficient tendency a ligation scaffolds for the transition metal in transition metal-catalyzed coupling reactions as well as itself as an organocatalysts to enhance the rate of particular reaction [14–21]. In order to successfully attain the desired 1,5-disubstituted 1,2,3-triazole isomer, which is precisely the opposite regioisomer of the relevant Cu-variant of azide-alkyne cycloaddition, Fokin and co-workers introduced the use of ruthenium catalyst in 2005 [22]. In spite of various reports of its application growing exponentially, the ruthenium-catalyzed azide-alkyne cycloaddition (RuAAC) reaction is less frequently studied in contrast to the copper-catalyzed azide-alkyne cycloaddition (CuAAC). The 1,5-disubstituted 1,2,3-triazoles show a type of outstanding substance core present in many biologically significant compounds and functionalized materials within the diverse range of 1,2,3-triazoles. In moderate and catalytically ideal conditions, a number of functionalized triazoles were fashioned from the corresponding compounds in high-to-excellent yields for various applications [23–31]. In addition to RuAAC, the organocatalyzed path is also an effective tool for the synthesis of 1,5-disubstituted 1,2,3-triazoles, which we have comprise in next chapter of the book. Herein, keeping in the mind the various applications

V. K. Tiwari et al., *Click Chemistry*, Materials Horizons: From Nature to Nanomaterials,
https://doi.org/10.1007/978-981-97-4596-8_3

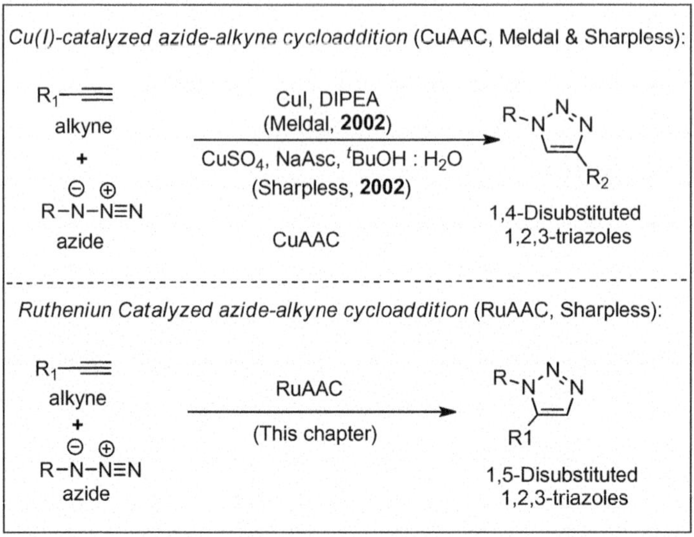

Scheme 3.1 The 1,4-regioselective to 1,5-regioselective click reaction for the synthesis of diverse triazole analogues

nitrogen-based 1,5-disubstituted 1,2,3-triazoles and their future perspectives; in this chapter, we have covered the Ru-catalyzed various devised approaches for synthesis of 1,5-disubstituted 1,2,3-triazoles (Scheme 3.1).

3.2 Ru-Catalyzed Synthesis of 1,5-Substituted 1,2,3-Triazoles

The efficient synthesis of 1,5-substituted 1,2,3-triazole compounds **3** from organic azides **1** and terminal alkynes **2** is covered in detail in this chapter along with an array of techniques that have been reported in the literature. Through the Ru-catalyzed protocol, we have delineated the function of catalysts and split the contents into several divisions based on the reaction mechanism and application of 1,5-disubstituted 1,2,3-triazole **3** in various fields (Scheme 3.2). The purpose of this chapter is to provide a thorough and organized account of the extremely selective assembly of 1,5-substituted 1,2,3-triazoles in assistance Ru-complex/catalyst.

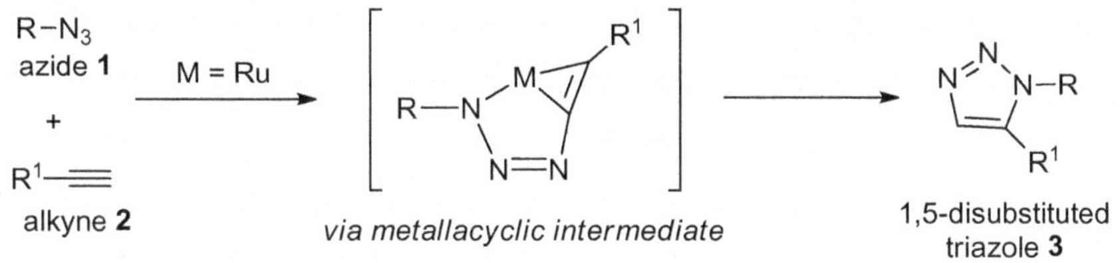

Some Notable Synthetic Applications

*Synthesis of 1,5-Disubstituted 1,2,3-Triazoles
*Peptide Side-Chain Mimetics
*Nucleoside/Nucleotide Analogues
*Synthesis of bioactive molecule

*Amide-Bond Mimetics
*Macrocycles
*Catalysis
*Glycomimetics

Scheme 3.2 The metallacyclic intermediate for regioselective assembly of 1,5-disubstituted 1,2,3-triazoles

3.2.1 RuAAC-Mediated Click Chemistry: Catalysis and Mechanism

In contrast to the cycloaddition of azide with internal alkyne under CuAAC conditions, the RuAAC-mediated regioselective triazole forming reaction is well documented in literature with both terminal and internal alkyne [22]. Folkin and co-workers devised an intriguing practical catalytic tool for the regioselective synthesis of 1,5-disubstituted triazoles **3** and 1,4,5-trisubstituted 1,2,3-triazoles **5** from cycloaddition of organic azides **1** separately with terminal alkyne **2** as well as internal alkynes **4** using catalytic amount of suitable ruthenium complexes. In the established protocol for easy access of various 1,5-substituted 1,2,3-triazolyl analogues, although a variety of ruthenium complexes were explored for efficient coupling, however few of them such as Cp*RuCl(PPh_3)$_2$, Cp*RuCl(COD), and Cp*RuCl(NBD) were found to be most active and suitable catalyst in such regioselective AAC systems. The Cp*RuCl(PPh_3)$_2$ or Cp*RuCl(COD) complexes were indeed to be the best catalyst to promote robust and regioselective cyclization of primary and secondary azides with a broad range of terminal alkynes with tolerance of several functional groups present therein and desired 1,5-disubstituted 1,2,3-triazoles were obtained in high yields with notable regioselectivity. The tertiary azides were significantly less reactive for the reaction. According to the reactivity order with both terminal and internal alkynes, this AAC reaction does not supposed to occur through a ruthenium acetylenide intermediate in contrast to CuAAC reaction that proceeds through in situ formation of copper acetylenide intermediate. In addition to the reactivity of terminal alkynes, both the ruthenium complexes can also promote the expeditious AAC reactions of organic azides with internal alkynes and provide high yield of fully functionalized 1,2,3-triazoles (Scheme 3.3) [32].

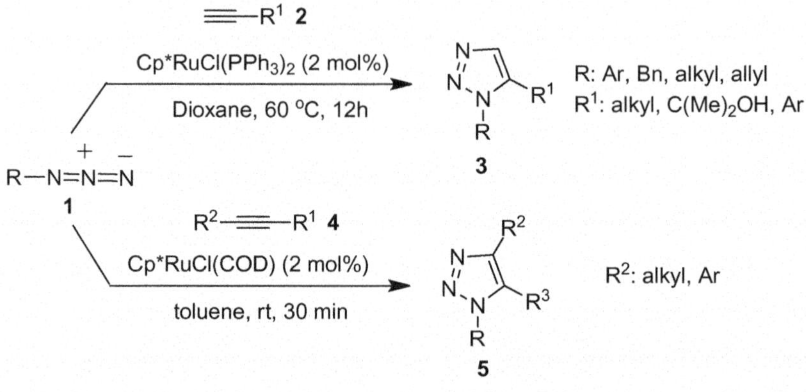

Scheme 3.3 RuAAC-induced diverse 1,2,3-triazoles obtained through AAC of terminal and/or internal alkynes and azides

The catalytic cycle of the RuAAC reaction proceeds with an activated intermediate **II** through substitution of two ligands by both the reactive click partners **1** and **2**. In the next step, the ruthenacycle intermediate **III** was obtained through the oxidative coupling of azide **1** and alkyne **2** and this step presumably control the regioselectivity of the AAC reaction. The new C-N bond formation takes place through the terminal N-atom of the azide **1** and the more electronegative as well less hindered C-atom of the alkyne **2** to give intermediate **IV**. At the end, the ruthenium complex **IV** was proceed to the reductive elimination to form the desired 1,5-triazole **3** with the regeneration of active Ru-catalyst **I** that completes the catalytic cycles (Fig. 3.1). Unlike the CuAAC reaction where copper acetylenide intermediate was formed in the catalytic cycle, the reactivity with both terminal and internal alkynes in RuAAC suggested that this modular reaction certainly not proceed via anticipated ruthenium acetylenide intermediate.

In comparison with the CuAAC, the alkyne in the RuAAC reaction is triggered through the enhanced nucleophilicity resulted by its π-interaction with the ruthenium core. Consequently, the metal-acetylide adduct could not formed that eventually extends the use of Ru-catalyzed AAC conjugation beyond cycloaddition with terminal alkyne. The substituents present on the propargyl residue usually influence the regioselectivity for internal changes, and this can be evaluated by its electronic characteristics, steric demand, and also the ability to form hydrogen bonds. The lowest-energy RuAAC path with plausible associated intermediate **I–V** is shown in Fig. 3.2 [22, 32].

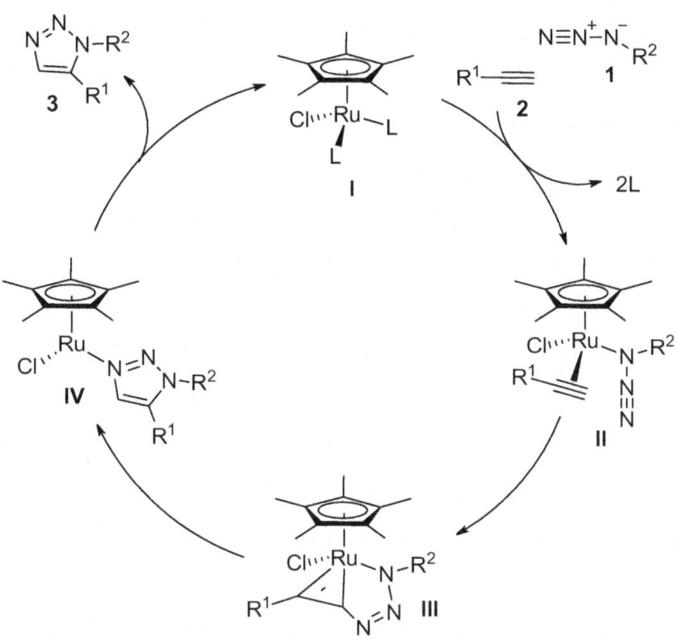

Fig. 3.1 Possible mechanistic approach for the RuAAC-mediated 1,5-disubstituted-1,2,3-triazole synthesis

3.2.2 RuAAC-Mediated Synthesis of 1,5-Disubstituted 1,2,3-Triazoles

Cintrat et al. reported the Cp*RuCl(PPh_3)$_2$-catalyzed cycloaddition of *N*-benzyl *N*-tosyl ynamide **6** with organic azides **1** to afford good yields of desired 1,5-disubstituted 1,2,3-triazoles **7**. The protocol has various notable features such as mild reaction conditions, excellent regioselectivity, and good-to-excellent yield (Scheme 3.4) [33].

Furthermore, Hawker et al. reported the convenient synthesis of 1,5-substituted 1,2,3-triazoles **9** via the cycloaddition reaction of various azides **1** with but-3-yn-1-yl methanesulfonate **8** in assistance of Cp*RuCl(PPh_3)$_2$ complex catalyst. Next, the developed 1,5-substituted 1,2,3-triazoles **9** were subjected for methane sulfonic acid elimination in the presence of sodium iodide and 1,8-diazabicyclo [5.4.0] undec-7-ene (DBU), which afforded the vinyl group-appended desired 1-substituted-5-vinyl 1,2,3-triazoles **10** in good yield. The vinyl-containing compound **10** acts as suitable monomer and demonstrated to polymerize under controlled conditions, yielding homopolymers and block copolymers with varied properties depending on the substituent and isomer orientation (Scheme 3.5) [34].

Wuest et al. used the Cp*RuCl(PPh_3) catalyst to develop a number of (aryl-1,2,3-triazole-1-yl)-methanesulfonylphenyl derivatives **12** through the cycloaddition of

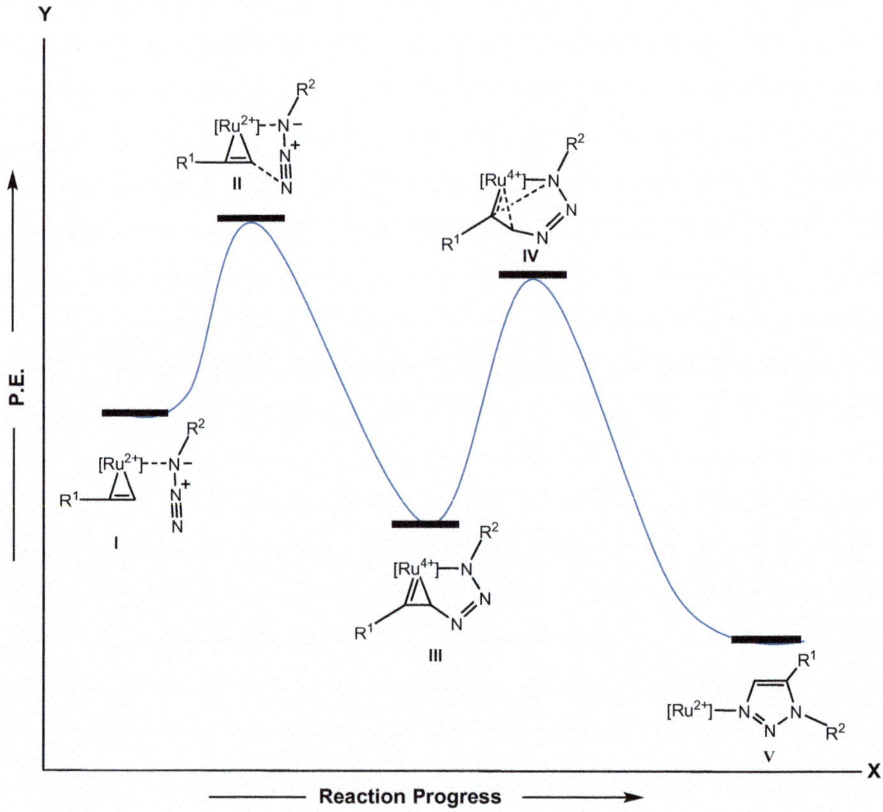

Fig. 3.2 DFT-based energy-profile diagram for the RuAAC reaction [22, 32]

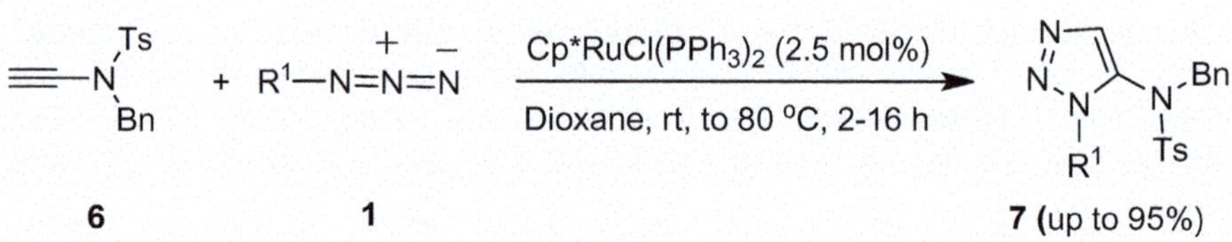

Scheme 3.4 RuAAC-catalyzed synthesis of 1,5-disubstituted 1,2,3-triazoles from ynamides and azides

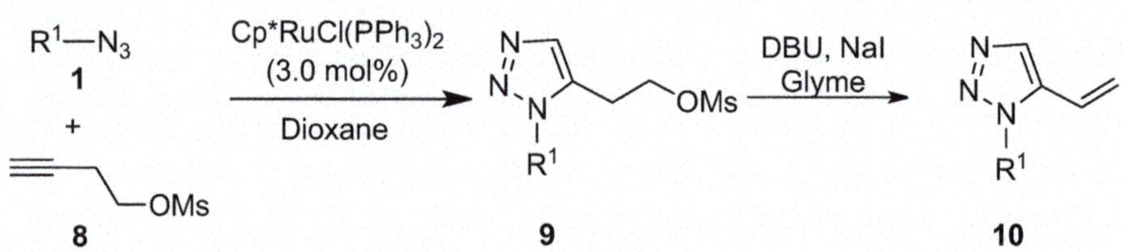

Scheme 3.5 RuAAC-mediated synthesis of 1-substituted-5-vinyl 1,2,3-triazoles as monomer for homopolymer and block copolymers

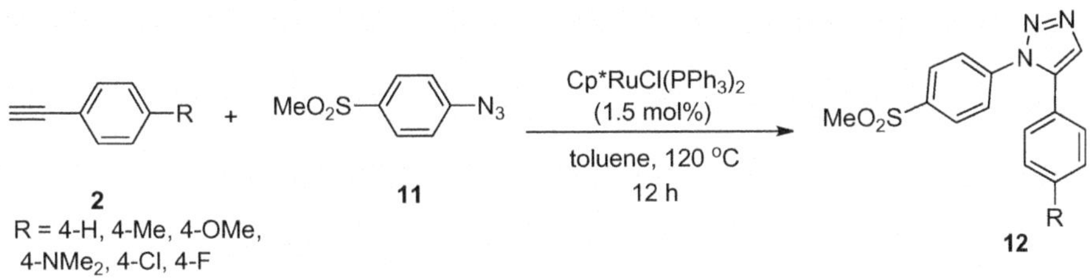

Scheme 3.6 RuAAC-catalyzed synthesis of (aryl-1,2,3-triazole-1-yl)-methanesulfonylphenyl derivatives

para-substituted phenyl acetylenes **2** and 1-azido-4-methane-sulfonylbenzene **11**. These 1,2,3-triazoles have beneficial characteristics for medicinal chemistry, and the author successfully screened them for in-vitro COX inhibitors assay and explored their suitable usage as in vitro COX inhibitors (Scheme 3.6) [35].

Astruc et al. reported a recyclable ruthenium(II) complex supported on magnetic nanoparticles, where a pentamethylcyclopentadienyl ruthenium complex was immobilized on iron oxide nanoparticles robust SiO_2/γ-Fe_2O_3 to develop a magnetically separable ruthenium catalyst $Cp^*(PPh_3)_2Ru/SiO_2/\gamma$-$Fe_2O_3$ **13**. The complex catalyst **13** was found to be reusable at least five times and is highly active and selective for the cycloaddition of organic azides **1** and alkynes **2** to afford desired 1,5-disubstituted 1,2,3-trizoles **14** in good to excellent yield (Scheme 3.7) [25].

There have been reports of a number of ruthenium(II) complexes acting as catalysts in azide-alkyne 1,5-selective cycloadditions, including $[Cp^*RuCl]_4$ and $[Cp^*RuCl(PPh_3)_2]$. The [Cp*RuCl(COD)] catalyst was further investigated because of its synthetic availability and stability. The catalyst [Cp*RuCl(COD)] can be used without difficulty at room temperature because the 1,5-cyclooctadiene (COD) ligand is more labile than other ligands. However, the $[Cp^*RuCl(PPh_3)_2]$ is extensively used for 1,5-selective triazole formation as well [32]. The presence of Cp ligand favors the ruthenium-catalyzed reaction and selectivity. Jia and Weinreb separately explore the use of $[Cp^*RuCl(PPh_3)_2]$ catalyst for the synthesis of 1,5-disubstituted 1,2,3-triazole compounds **15** and **16** utilizing organic azide **1** terminal alkynes **2** (Scheme 3.8) [22, 36].

In addition to Cp-containing Ru-complexes, non-Cp Ru-complexes were also reported as convenient catalytic systems for synthesis of 1,5-substituted triazolyl analogues. Numerous ruthenium azide complexes with Tp ligands were explored in synthetic chemistry. Lo et al. reported the intriguing complex $Tp(PPh_3)(EtNH_2)RuN_3$ (Tp = $HB(pz)_3$, pz = pyrazolyl) **17** as an efficient catalyst for the 1,5-selective cycloaddition reaction of benzyl azide **1** with terminal alkynes **2** to afford desired 1,5selective triazolyl compound **4** (Scheme 3.9). An array of functional groups can be tolerated in the 1,5-disubstituted 1,2,3-triazole synthesis catalyzed by this complex, and the reactions can take place in either organic or aqueous conditions [37].

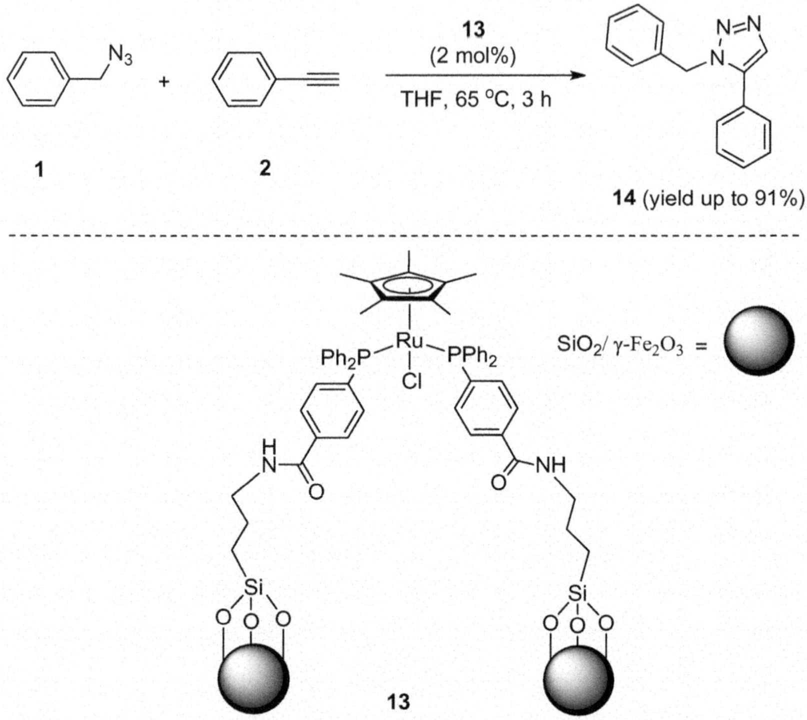

Scheme 3.7 Regioselective Ru-catalyzed synthesis of 1,5-substituted triazole from alkyne and azide

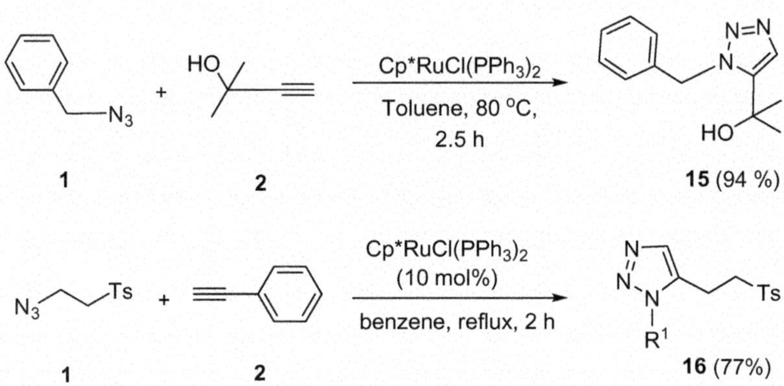

Scheme 3.8 [Cp*RuCl(PPh3)2]-catalyzed synthesis of 1,5-substituted triazole analogues from alkyne and azide

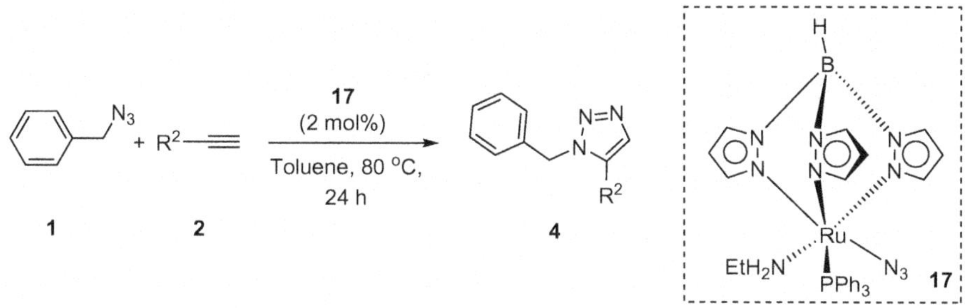

Scheme 3.9 Synthesis of 1,5-substituted 1,2,3-triazole in assistance of non-Cp-based Ru-complex

Recently, Iannazzo et al. explored the modular RuAAC protocol as a potent method for accessing 1,5-disubstituted and 1,4,5-trisubstituted triazole diazabicyclooctane (DBOs) with complete control over regioselectivity or selectivity toward position. The synthetic protocol involves the Cp*RuCl(COD)-assisted cycloaddition of diazabicyclooctane based-azide **18** with terminal alkyne **2** and internal alkyne **3** to afford 1,5-disubstituted **19** and 1,4,5-trisubstituted triazole diazabicyclooctane **20**, respectively. These developed substituted triazole analogues were subjected for biological activity against a hyperpermeable strain of Escherichia coli producing KPC-2 and found to be gave promising activities (Scheme 3.10) [38].

Although the microwave heating has been employed on numerous synthetic approaches to get faster synthesis of triazoles in the CuAAC reaction, it is significantly more relevant for the RuAAC reaction due to the high conversion in a short reaction time [39, 40]. Johansson et al. reported the ruthenium-catalyzed one-pot synthesis of 1,5-disubstituted triazole **23** using alkyl halide **21**, sodium azide, and an alkyne **22**. The reaction was assisted by microwave heating, which results shorter reaction time and good yield of the desired triazole analogue. In the synthetic route,

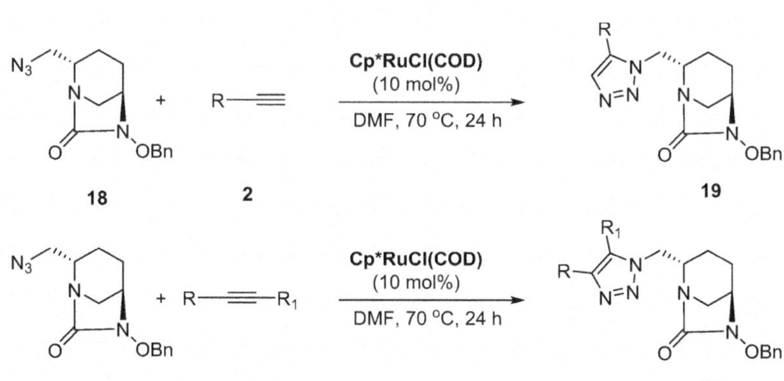

Scheme 3.10 RuAAC-mediated synthesis of 1,5-disubstituted and 1,4,5-trisubstituted triazole diazabicyclooctane (DBOs)

alkyl halide **21** is first treated with sodium azide in DMA with microwave heating to synthesize organic azide in situ through substitution reaction. Additionally, the cycloaddition product was achieved by microwave irradiating organic azides with the catalyst Cp*RuCl(PPh_3)$_2$ and an alkyne **22**. The secondary alkyl halides substrate, like cyclohexyl bromide, failed to afford any product in this experiment at the elevated temperature as well (Scheme 3.11) [41].

In a study by *Tamas* et al., the [Cp*RuCl_2]$_x$-catalyzed 1,5-selective cycloaddition of methyl 2-azidoacetate **24** and *N*-Boc-propargylamine **25** was harvested to synthesize 1,5-disubstituted 1,2,3-triazole amino acids **26**. The author reports a new set of peptidomimetics derived from δ-peptides, consisting of *n* units of a 1,5-disubstituted 1,2,3-triazole amino acid (5Tzl). In addition to existing triazole a chiral 1,5-disubstituted 1,2,3-triazole amino acid **27** is effectively synthesized as well by the [Cp*RuCl(COD)]-catalyzed 1,5-selective cycloaddition of chiral *N*-Boc-propargylamine **25b** with azide **24** (Scheme 3.12) [42–43].

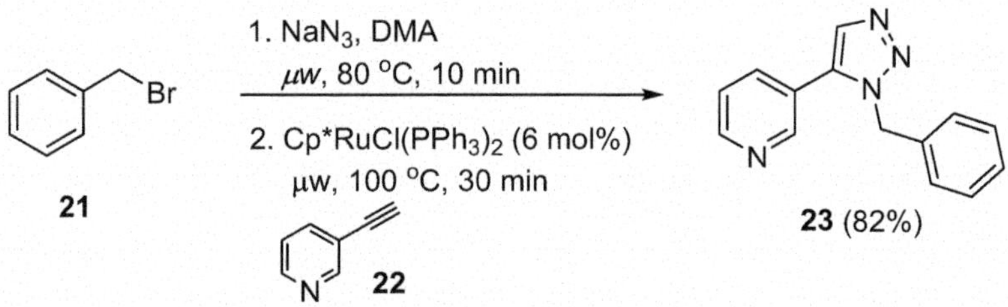

Scheme 3.11 Microwave-assisted sequential one-pot RuAAC reaction to develop 1,5-disubstituted triazole

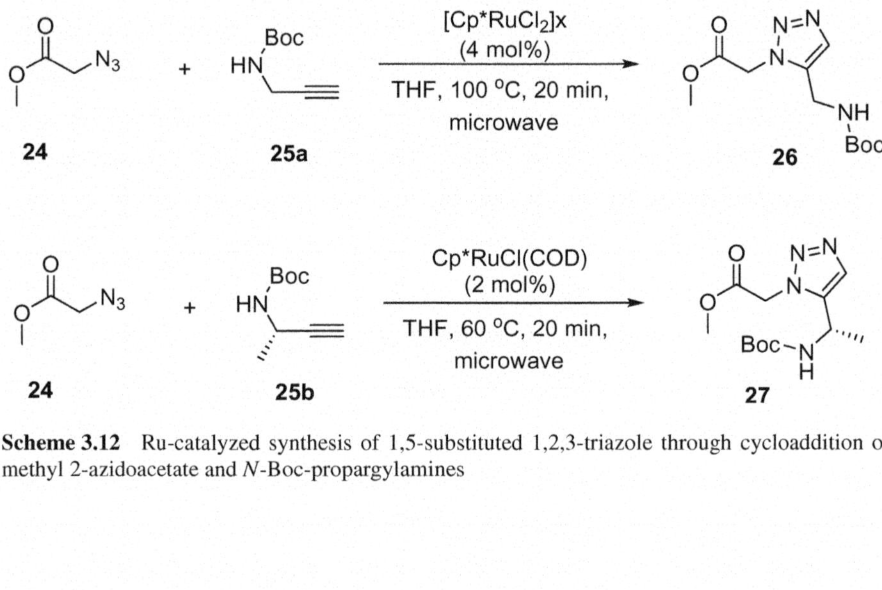

Scheme 3.12 Ru-catalyzed synthesis of 1,5-substituted 1,2,3-triazole through cycloaddition of methyl 2-azidoacetate and *N*-Boc-propargylamines

3.2.3 *1,5-Disubstituted 1,2,3-Triazoles as Mimetic of cis-Amide Peptidomimetics*

A large number of ruthenium catalysts for effective RuAAC coupling are now commercially available and their impact in various emerging fields of chemistry, biology, medicine, material science, etc., has been exponentially expanding. Thus, taking into account of the notable feature of RuAAC-mediated 1,5-substituted-1,2,3-triazoles, this section presents a concise survey of some representative strategies as its mimetic of *cis*-amide bonds. A large number of peptidomimetics displayed promising affinity for enzyme inhibition and ligand receptors of notable applications in drug discovery research [44, 45]. They were developed for particular purpose to replicate the chemical structure required to influence the biological action of peptides [44]. The RuAAC-mediated triazolo-amino acid (TzI) may underwent subsequent integration into a short peptoid to induce turn in an artificial peptide. Appella and co-workers developed 1,5-disubstituted 1,2,3-triazole amino acid (TzI) **31** in five synthetic steps. A ring-constrained Huygens AAC tool was implemented, and the resulting triazole was then inserted into a synthetic peptide sequence using established solid-supported peptide synthesis tool to give respective peptoid **32**. In the first step SN of propargyl amine with compound **28** gave the bromoacetylamide **29** in 67% yield which was subjected to azidation using NaN_3 in DMF followed by Huisgen's AAC afforded the respective bicyclic-triazole analogue **30**. The resulting compound **30** underwent ring-opening reaction by treating it in strong acidic medium (6 M HCl) followed by FMoc protection under basic condition gave the desired monomer, e.g., 1,5-disubstituted 1,2,3-triazole amino acid (TzI) **31**. The NMR analysis confirms that in phosphate buffer, the peptoid **32** adopts a hairpin-like turn conformational shape (Scheme 3.13) [46].

Buntkowsky and co-workers extended the similar tool and developed a triazole metallopeptide [47]. The resulting candidate is based on the nickel superoxide dismutase (NiSOD) peptide model and this can eventually copy the prolyl peptide bond found in biomimetics. Thus, AAC of azido acid **33** and amino alkyne **34** in the presence of catalytic amount of Cp*RuCl(COD) gave the desired 1,5-disubstituted triazole **35** which was then inserted into a synthetic peptide sequence using the similar polymer-supported peptide synthetic tool. The triazole amino acid analogue **36** thus obtained was subjected to react with $NiCl_2$ to give high yield of desired triazole metallopeptide architecture **37** (Scheme 3.14) [47].

3.2.4 *RuAAC-Mediated 1,5-Disubstituted 1,2,3-Triazoles in Peptide Side-Chain Mimetics*

One of the most frequent and well-researched post-translational modifications (PTMs) is protein phosphorylation. In the translational modifications, Kee et al. have developed promising, robust, and non-isomerizable phosphohistidine (pHis)

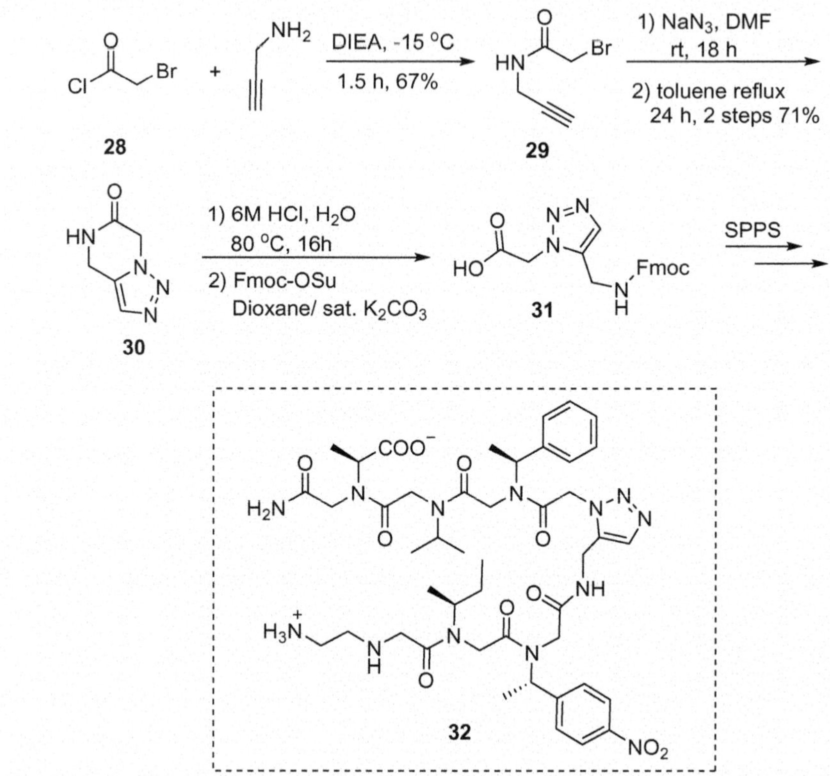

Scheme 3.13 Click-mediated triazolyl amino acid and it's in peptoid development

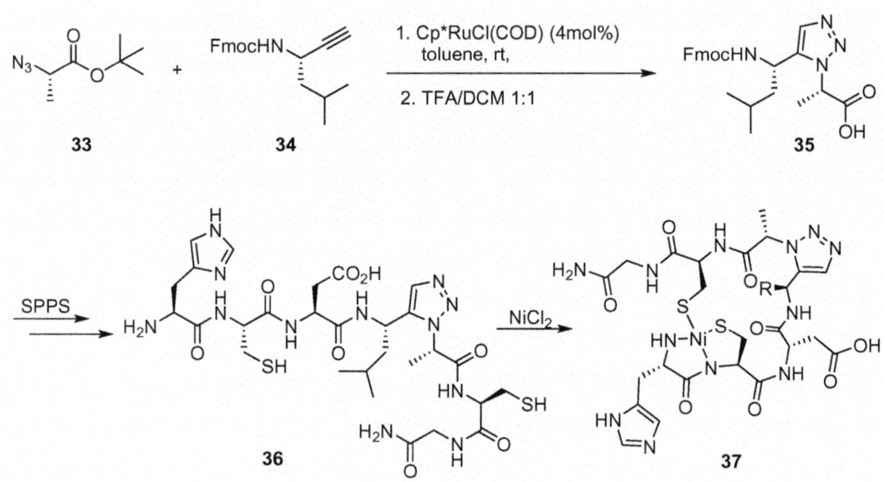

Scheme 3.14 Development of triazole metallopeptide **37** via RuAAC tool

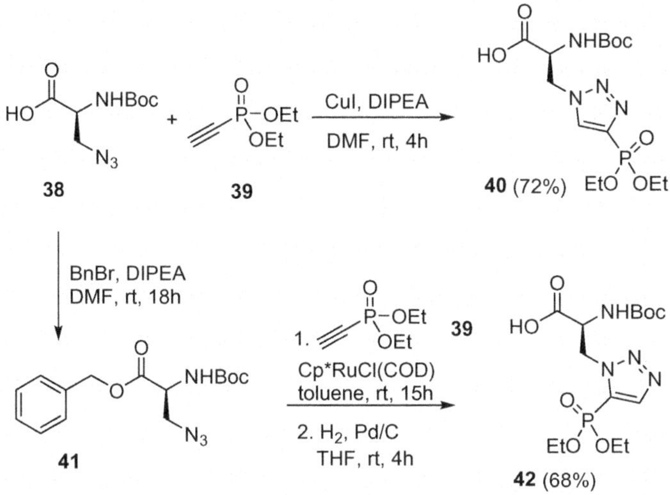

Scheme 3.15 Click-mediated synthesis of stable 1,4- and 1,5-disubstituted triazole-based pHis mimetics **40** and **42**

mimetics and phosphoryltriazolyalanines (pTza). The developed phosphohistidine mimetics and phosphoryltriazolyalanines were found to be efficient for the synthesis of pHis antibodies and partially synthetic pHis-containing proteins. Synthetic approach began with the reaction of protected azidoalanine **38** and **41** with diethyl ethylphosphonate **39** under both CuAAC and RuAAC conditions, yielding the required phosphoryltriazolylalanine (pTza) derivatives **40** and **42** in 72% and 68% of the reactions, respectively (Scheme 3.15) [48].

Ballet et al. have used intra- and intermolecular Huisgen cycloaddition processes to explore amino-triazolodiazepine (Ata) scaffolds as restricted histidine analogues. The ability of amino-triazolodiazepines to act as histidine mimics was demonstrated by replacing the His-Pro dipeptide section in angiotensin IV with the dipeptide mimic Ata-Gly and then evaluating the results in two inhibitory enzyme experiments. The Ata dipeptide mimics were assembled using two distinct ways. In the route **A** or intermolecular approach, RuAAC reaction of Fmoc-protected *N*-propargyl amine **43** with azido-Ala **44** affords the 1,5-substituted 1,2,3-triazole based-compound **45** in satisfactory yield, which on base-assisted removal of fluorenylmethyloxy-carbonyl and hydroxamate groups leads to the formation of compound **46**. The developed compound **46** via carbodiimide-induced cyclization afforded the anticipated Boc-amino-triazoloazepinone **47** (Boc-Ata) in satisfactory yield. In the route **B** intramolecular approach, *N*-propargyl amino acid **43,** undergo coupling reaction with azido-Ala **48** in assistance of coupling reagent, followed by thermal intramolecular cycloaddition afforded the Ata **47**. This method was deemed not enough because of the high level of epimerization and low yield of the desired triazole based compound (Scheme 3.16) [49]. A similar report introduced by Wales et al., they were utilized

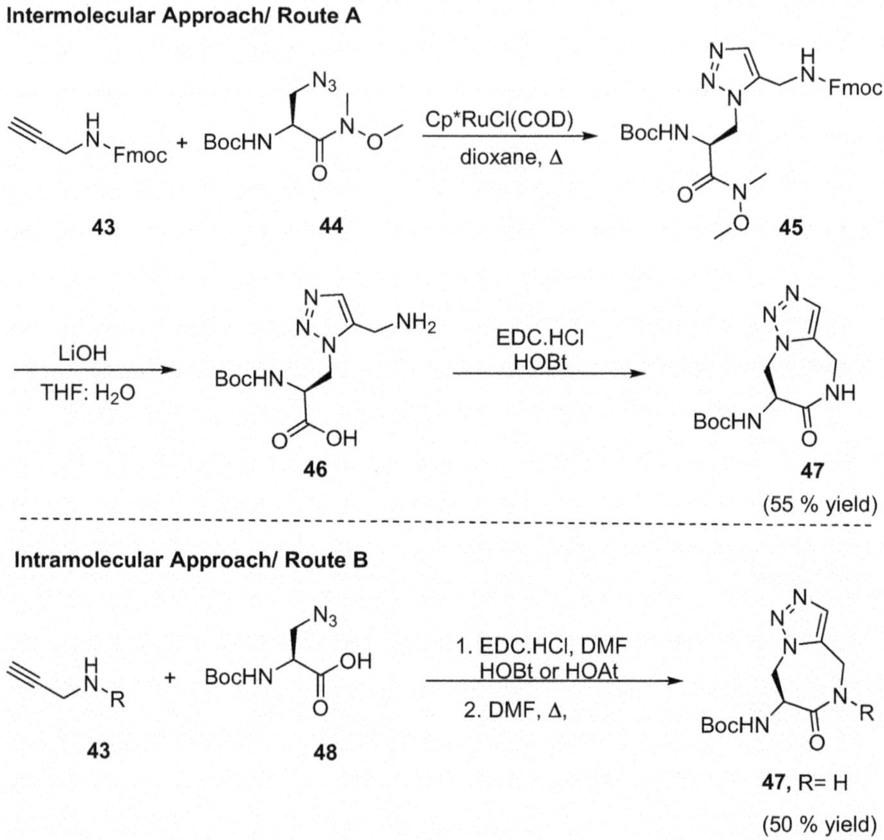

Scheme 3.16 RuAAC and thermal reaction for synthesis of dipeptide mimetic amino-triazolodiazepine (Ata) scaffold

both CuAAC and RuAAC for assembly of mono, di, and tripeptidomimetcs with a common 1,2,3-triazole ester isostere at the *C*-terminus [50].

3.2.5 RuAAC-Mediated Regioselective 1,5-Disubstituted 1,2,3-Triazole in Macrocycles

A macrocyclic triazole forming metal-catalyzed AAC approach with varied ring sizes is developed by Marcaurelle and colleagues from variety of azido alkyne substrates under intramolecular Ru-catalysis. For this purpose, some common scaffolds were first introduced, which were then subjected to various optimized reaction conditions to give high yield of variety of triazole containing n or n + 1 ring size macrocyclic architectures. After screening of several catalysts this was found that $[RuClCp^*]_4$ has a favorably better monomer/dimer ratio than the frequently used Cp*RuCl(PPh_3)$_2$. An excellent monomer-to-dimer ratio is achieved when the reaction is carried out at a higher temperature (80 °C) and lower concentration (2 mM). The optimization of RuAAC conditions on azido alkyne substrate **49**, which led to the macrocyclic triazole **50** in a 58% yield were explored for the numerous alternative structural frameworks (Scheme 3.17) [51].

The same group has extended this devised protocol to synthesize a sizable DOS library in their research for novel macrocyclic histone deacetylase inhibitors (HDACs), using a three-phase [52], diversity-oriented synthesis (DOS) process known as build/couple/pair (B/C/P). Macrocyclic triazole **52** was successfully synthesized in 78% yield by macrocyclizing azido-alkyne substrates **51** (2R, 5S, and 6R) generated from syn-aldol. The RuAAC cyclization methods yielded reliable macrocyclic triazoles that could regularly be carried out on gram scale (Scheme 3.18) [53].

Liskamp et al. explore the intramolecular RuAAC double macrocyclization to effectively developed vancomycin bicyclic mimics with a 1,5-triazole bridge. The

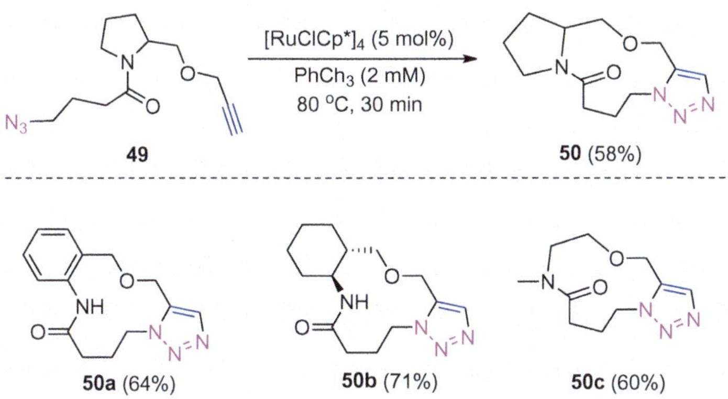

Scheme 3.17 RuAAC prompted substrate control intramolecular macrocyclization

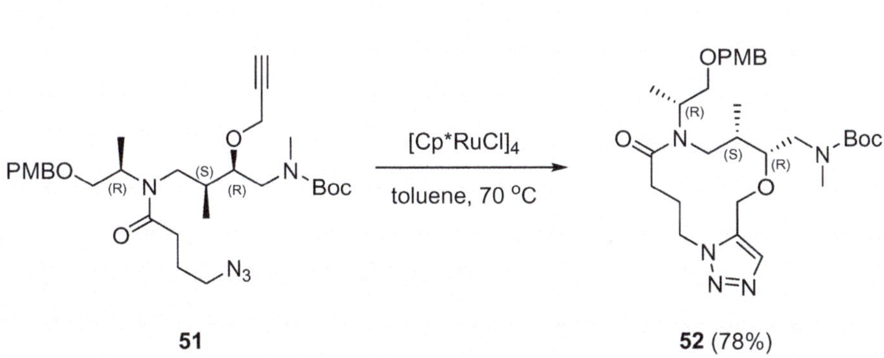

Scheme 3.18 RuAAC-mediated macrocyclization to achieve 12-membered triazole macrocycle **52**

Scheme 3.19 Intramolecular RuAAC double macrocyclization in the synthesis of vancomycin

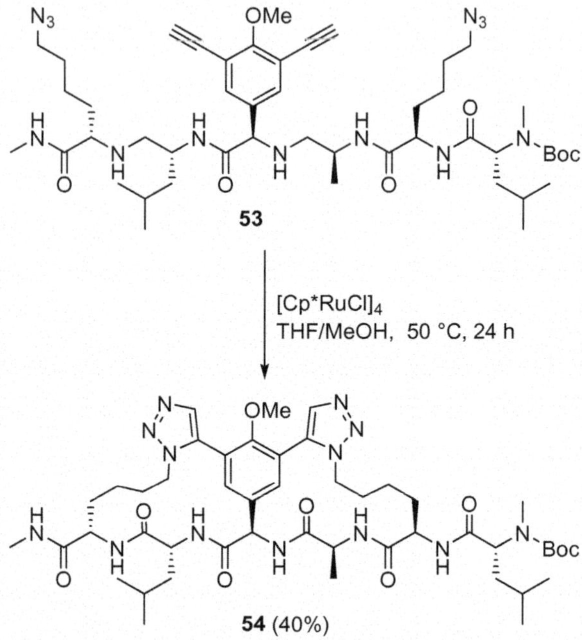

$[Cp^*RuCl]_4$-assisted cycloaddition reaction of linear bisazido alkyne **53** in THF/MeOH for 24 h at 50 °C affords the bicyclic bistriazole macrocycle **54** in 40% yield (Scheme 3.19) [54].

Zhang and colleagues reported the intriguing approach for effective synthesis of 1,4- and 1,5-disubstituted triazole-containing cyclic tripeptides that bear striking similarities to vancomycin mimics by profoundly utilizing both the modular CuAAC and RuAAC click cyclization approach. The stiffness of the related linear azido alkyne **55** (n = 1) leads to the assembly of both dimer and trimer as significant products, macrocyclization that leads to the formation of **56** (n = 1) by RuAAC was determined to be less promising. On the other hand, BOP-mediated macrocyclization (lactamization) with the longest side chain precursor **57** (n = 4) resulted in the assembly of monocycle **58** (n = 4) in a good yield. Additionally, the extent of macrocyclization was increased in the case of the shortest ring size systems (Scheme 3.20) [55].

Fang research group has developed a simple and highly productive efficient process for synthesizing triazoles fused to an oxazolidinone ring from a fused aziridine ring. The researchers used RuAAC macrocyclization to assembly the 11- to 12-membered 1,5-triazole fused macrocycles **60** and **61** using corresponding azido-alkyne oxazolidinone scaffold **59** (Scheme 3.21). The author suggested that the formation of 1,4-triazole dimer adduct under heating conditions leads the slightly decrease in the yield of macrocycle **61** [56].

Spring et al. explore the three-phase DOS build/couple/pair methodology for the synthesis of biologically relevant common macrocyclic peptidomimetic framework

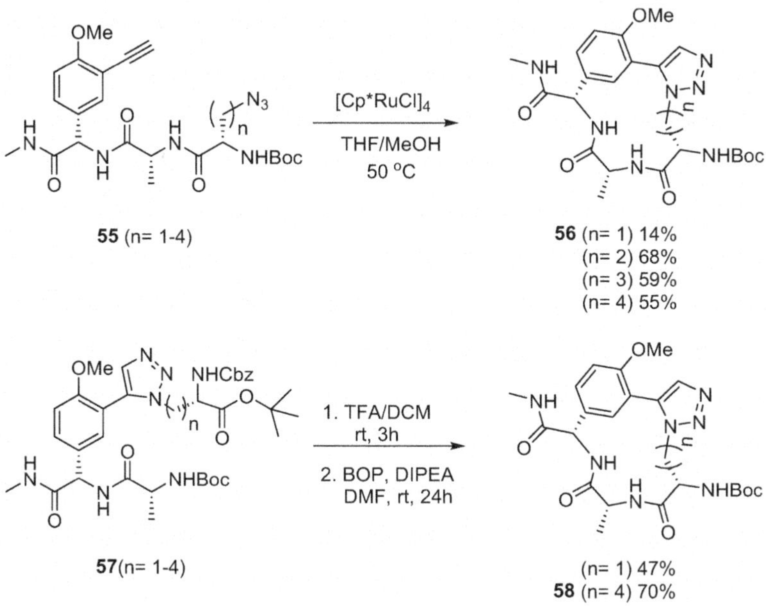

Scheme 3.20 RuAAC-mediated macrocyclization and BOP-mediated macrolactamization of azido-alkynes

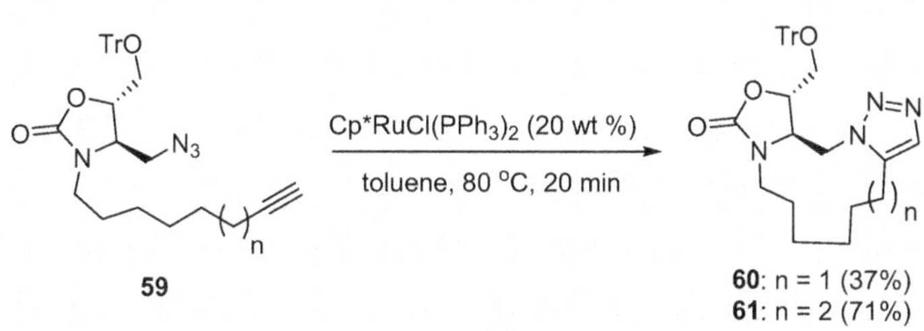

Scheme 3.21 RuAAC-inspired macrocyclization for the synthesis of triazole hybrid

65, **66**, and **67** using corresponding alkyne/azide-containing compound **62**, **63,** and **64** in the assistance of RuAAC reaction. A single stereoisomer was found in the ^{1}H NMR of each cyclization product under these idealized conditions, indicating that the 1,3-dipolar cycloaddition processes were entirely regioselective and did not result in epimerization of the chiral centers (Scheme 3.22). Additionally, a biological study conducted as a preliminary showed significant anti-bacterial activity against *Staphylococcus aureus* [57].

RuAAC was used by Krause et al. to macrocyclize and synthesize the C3-symmetric tris-1,5-triazole macrocycle. They carried out comprehensive NMR and ITC (isothermal titration calorimetry) tests to investigate the triazole macrocycle **71** anion-binding capabilities and revealed that it binds Cl, Br, I, and sulfate in both protic

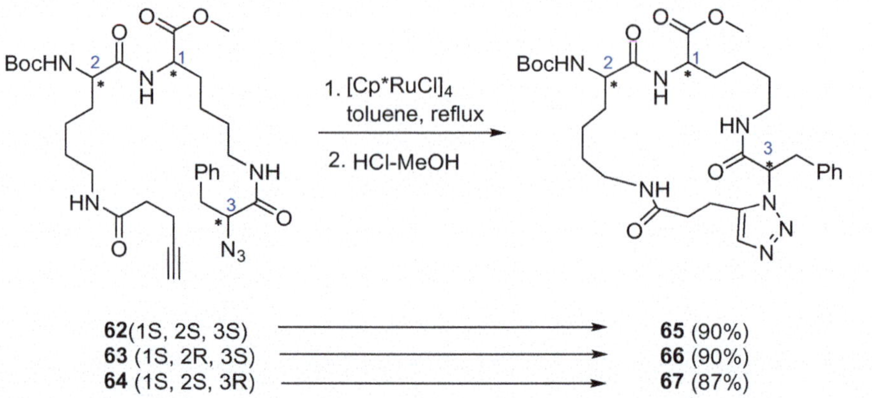

Scheme 3.22 DOS-mediated synthesis of triazole macrocycle in assistance of RuAAC macrocyclization

and aprotic solvents in a 1:1 ratio with effective affinity. In the synthetic methodology, the author used Ru-catalyzed cycloaddition of an azide analogue **68** with an alkyne **69** to afford the 1,5-substituted triazole **70** in good yield. The developed triazolyl compound **70** on two more sequential RuACC afforded the final C_3-symmetrical tris-1,5-triazole macrocycle **71** (Scheme 3.23) [58, 59].

3.2.6 RuAAC-Mediated 1,5-Disubstituted 1,2,3-Triazoles in Nucleoside/Nucleotide Analogues and Glycomimetics

Pradere et al. devised an intriguing approach to synthesize a variety of triazole-appended ribavirin analogues, which involves the substitution of the 1,3-disubstituted 1,2,4-triazole ring in ribavirin with 1,4- or 1,5-disubstituted 1,2,3-triazoles in assistance of Ru-catalyst. The triazole-appended-1,5-isomer of the ribavirin analogue **73** was achieved in good yield when RuAAC conditions were applied to a benzoyl-protected azido ribose **72** and different alkynes **2** using [Cp*RuCl(PPh_3)$_2$] as the catalyst in THF under microwave heating (Scheme 3.24). In this case, microwave irradiation greatly speeds up the reaction process compared to traditional thermal heating, yielding the necessary derivatives in good to excellent yields. Almost no 1,4-disubstituted isomer was generated, only the desired 1,5-regioisomer was obtained. The author subjected the developed triazoles to demonstrate its activities against the hepatitis C virus; screening showed that the synthesized triazoles had little action against the hepatitis C virus (HCV) [60].

The same group also describes the synthesis and antiviral activities of triazole compounds with a C5 substitution and a appended 2′-deoxyuridine structure greatly utilizing the RuAAC reaction under microwave irradiation. The reaction of

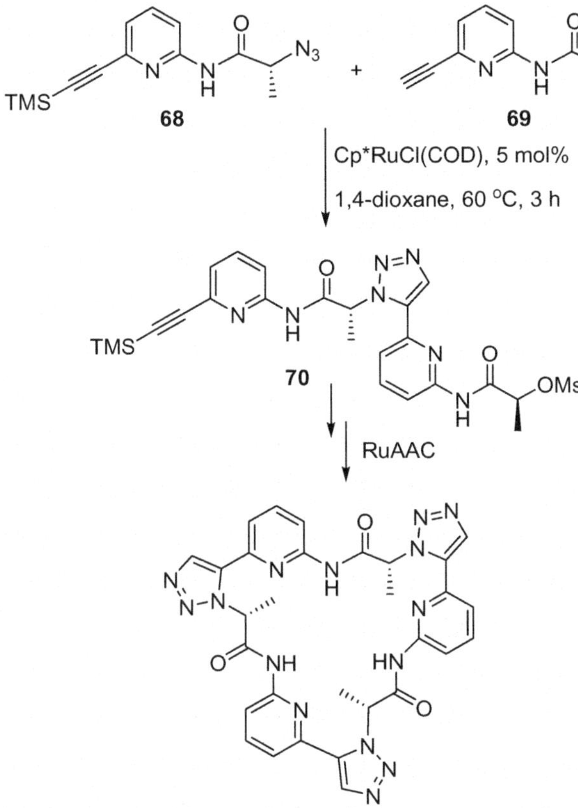

Scheme 3.23 Sequential RuAAC-mediated synthesis of anion binding tristriazole macrocycle

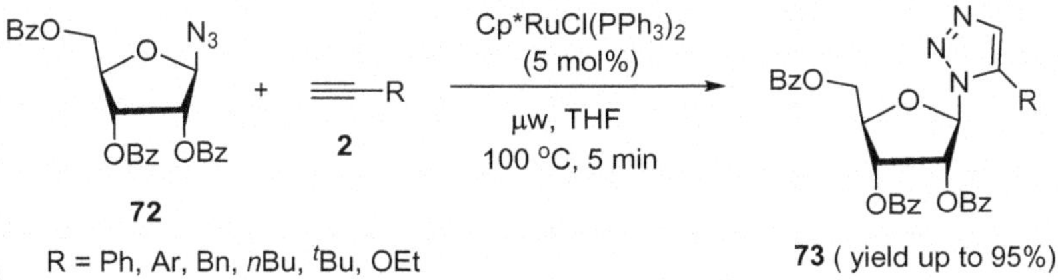

Scheme 3.24 RuAAC-assisted synthesis of triazole-modified analogues of potent ribavirin

alkenyl uridine **74** and organic azides in presence of 5 mol% catalytic loading of $Cp^*RuCl(PPh_3)_2$ afford the triazole-linked 2′-deoxyuridine compounds, which on deacetylation gave the final desired triazole-appended 2′-deoxyuridine compounds **75–78** in high yields (Scheme 3.25) [61].

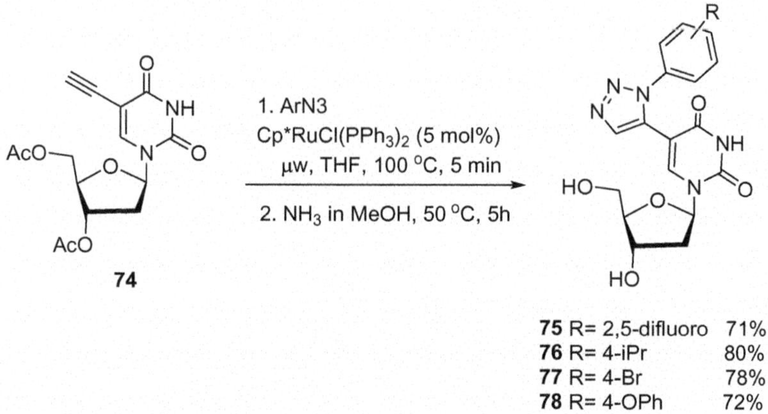

Scheme 3.25 RuAAC-mediated C_5 triazole substituted compounds of a 2′-deoxyuridine

A series of protection/deprotection strategies have been used by various research groups to examine the presumed therapeutic applications of 1,5-substituted-1,2,3-triazole isomers linked to carbohydrate moieties. The structures of a few novel developed triazole molecules **79a–d** summarized in Fig. 3.3 [62–65]

Recently, Brown et al. reported the synthesis of a novel 1,5-disubstituted triazole (TL2) dinucleotide phosphoramidite followed by its integration into oligonucleotides, and a comparison of its DNA polymerase replication competency with that of existing TL analogues under the assistance of RuAAC. In the synthetic route the RuAAC of alkyne **80** with azide **81,** afforded the dimer **82,** which afforded the deprotected compound **83** in the presence of TBAF as fluoride ion source. The final 1,5-TL phosphoramidite **84** was obtained by functional group transformation in compound **83** (Scheme 3.26) [66].

The RuAAC-mediated triazole motif linkers were extensively used by Crich and colleagues to build a substance that resembled a trisaccharide. The reaction was carried out in assistance of $[Cp^{*}RuCl]_4$ as the catalyst and microwave irradiation at 110 °C to activate the cycloaddition of both sugar-derived azido **85** and

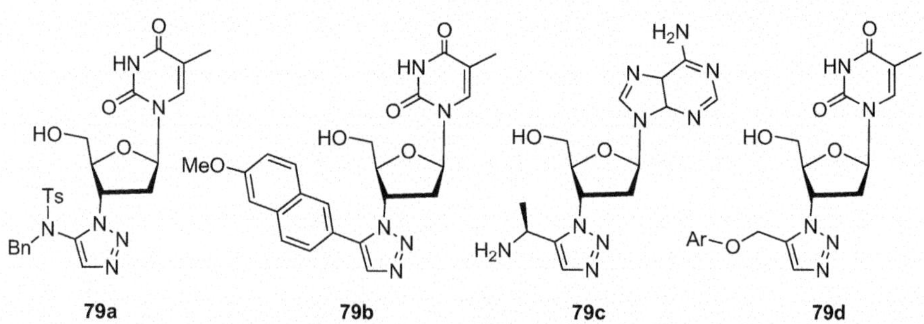

Fig. 3.3 RuAAC-mediated analogues of the nucleoside ribose or deoxyribose

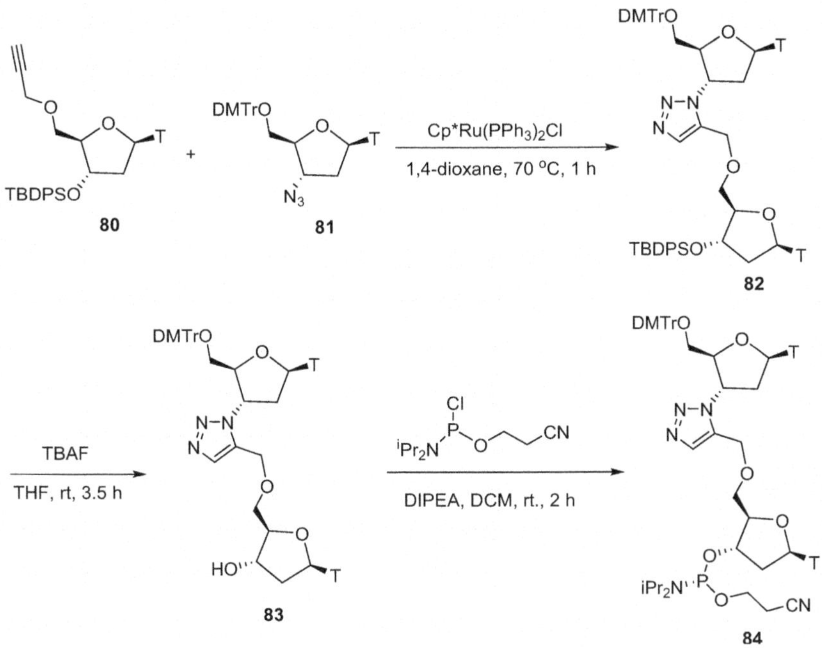

Scheme 3.26 RuAAC-mediated synthesis of 1,5-TL phosphoramidite **84**

alkenyl **86** residue yielded a 1,5-disubstituted triazole glycoconjugate **87** in 75% yield (Scheme 3.27). However, CuAAC was effectively used to easily synthesize the 1,4-triazole isomer as well [67].

Wiebe et al. used a similar technique to synthesize metabolically stable glycomimetic compounds by cycloadditioning acetylated galactose azide with fucosyl acetylene to achieve the designed 1,5-disubstituted 1,2,3-triazoles. The Cp*RuCl(COD) was used as a catalyst in DMA under microwave irradiation at 100 °C to react the sugar-containing azide residues **88** and alkyne analogue **89**.

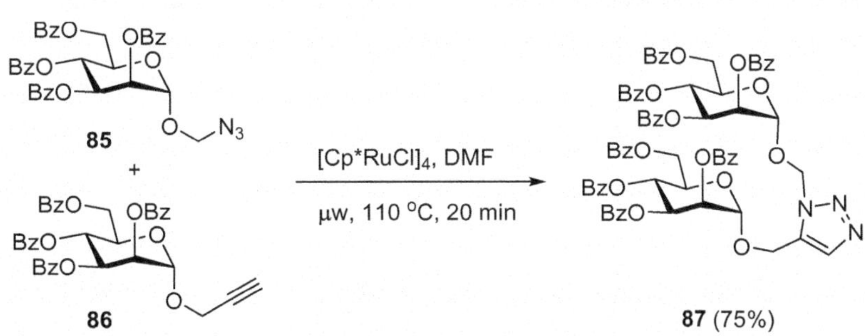

Scheme 3.27 RuAAC-mediated triazole-appended carbohydrate motif

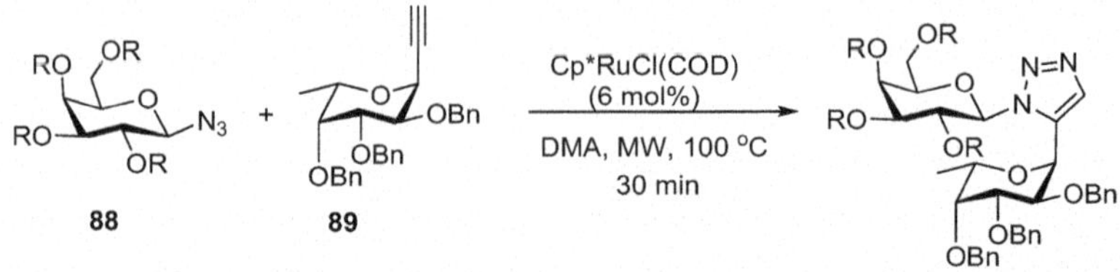

Scheme 3.28 RuAAC-inspired synthesis of triazole-containing sugar analogues using fucosyl acetylene with galactosyl azide

The desired 1,5-diglycosylated triazole **90** is achieved in yields of 60% for pivaloylated sugar azide and 76% for acetylated sugar azide. The 1,4-isomer of developed compound was also prepared by Cu-catalyzed modular click reaction (Scheme 3.28) [68].

In the effective synthesis of multivalent glycoclusters, the RuAAC reaction can be used as a crucial step [69]. Additionally, Poulsen and co-worker reported the intriguing synthesis of novel cancer therapies via the synthesis of glycoconjugate carbonic anhydrase (CA) inhibitors mediated by ruthenium-catalyzed azide-alkyne cycloaddition. The 5 mol% catalytic loading of [CpRuCl(COD)] catalyzes cycloaddition reaction, where sugar azides **91** and *p*-ethynyl-benzenesulfonamide **92** undergo the RuAAC to afford the designed 1,5-disubstituted triazole **93** in a highly regioselective manner (Scheme 3.29). In this study, the author successfully synthesized 10 molecules of 1,5-disubstituted 1,2,3-triazole benzene-sulfonamides, and their corresponding deacetylated triazoles were achieved by alkaline hydrolysis of these compounds. It has been demonstrated that a number of 1,5-disubstituted triazole acts as inhibitors to block CA IX with inhibition constants around 10 nM, among them compound **93** was found to be highly efficient [70].

Besra et al., reported the synthesis of the 6″-Triazole-substituted-α-galactosyl ceramides through the RuAAC click chemistry. The RuAAC-mediated cycloaddition of 6″-azido-6″-deoxy–galactosyl ceramide **94** with a variety of alkynes **3** resulted in the formation of a number of triazole-containing α-galactosyl ceramide (α-GalCer) analogues **95, 96** in good yields under 5 mol % catalytic loading of $Cp^*RuCl(PPh_3)_2$

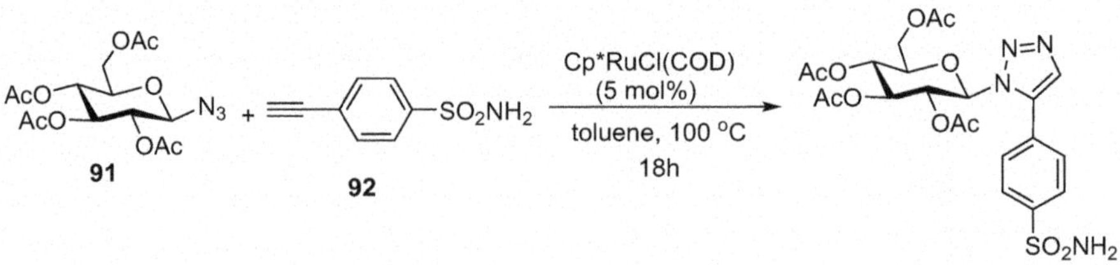

Scheme 3.29 RuAAC-mediated synthesis of glycoconjugate carbonic anhydrase (CA) inhibitors

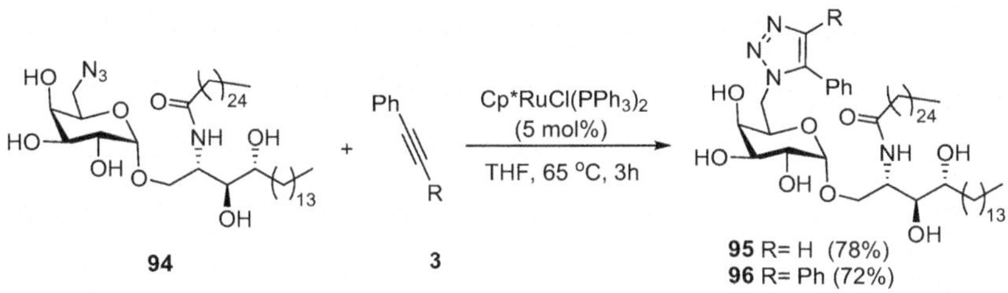

Scheme 3.30 RuAAC click-mediated efficient synthesis of triazole-containing α-galactosyl ceramide

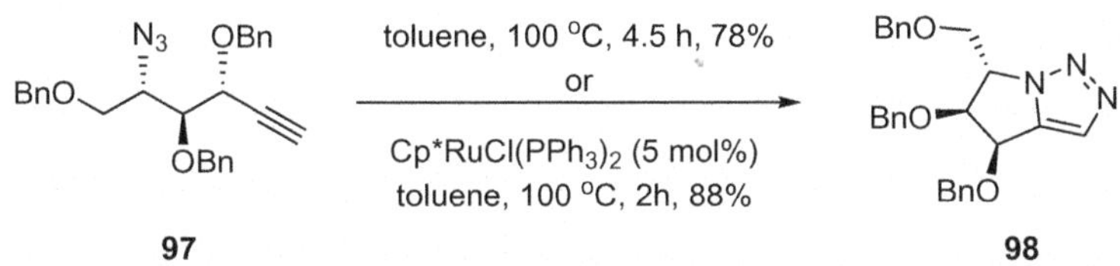

Scheme 3.31 Intramolecular RuAA-mediated synthesis of fused bicyclic pyrrolidine iminosugar hybrid

as Ru-based catalyst. The author subjected the developed triazole analogues for biological activities and found that in vivo testing of the α-GalCer analogues **95** and **96** revealed a favorable affinity to promote IL-2 production by marine iNKT hybridoma cells. Moreover, in the analysis the 1,5-regioisomer was found to be more effective than the comparable 1,4-regioisomer (Scheme 3.30) [71].

Furthermore, the fused bicyclic triazoles as a novel family of potent and effective α-glucosidase inhibitors were synthesized by Shaw et al., using the intramolecular 'click' reaction. Shaw et al. executed this feat superbly for the assembly of triazolyl iminosugars and greatly explored the thermal and RuAAC coupling of azide and alkyne approaches. In this study, the azido-alkyne precursor **97** undergoes thermal cycloaddition to afford bicyclic triazole **98** in 78% yields. However, RuAAC catalysis with 5 mol% catalytic loading of $Cp^*RuCl(PPh_3)_2$ leads to the increase of yield up to 88.3% (Scheme 3.31) [72].

3.2.7 *RuAAC-Mediated 1,5-Disubstituted 1,2,3-Triazoles in Catalysis and in Synthesis of Bioactive Molecule*

Luo, Cheng, and co-workers reported the asymmetric 1,4-addition of ketones and aldehydes **99** to nitroolefins **100** through organocatalysis in the presence of 1,4-disubstituted and 1,5-disubstituted 1,2,3-triazole based compounds as organocatalyst. The author employed modular CuAAC approach to develop an array of

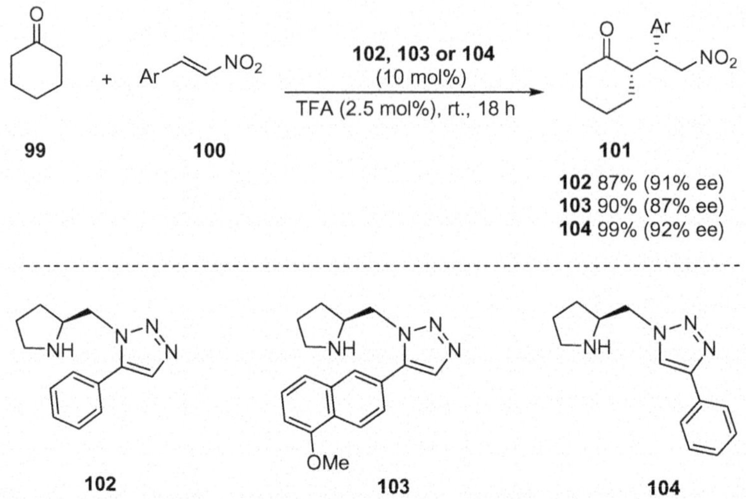

Scheme 3.32 Pyrrolidine-based triazolyl compound **102–104** as organocatalyst in asymmetric addition

1,4-disubstituted triazole for organocatalysis. In addition, two 1,4-isomer, two 1,5-disubstituted triazole were also synthesized to be used as organocatalyst and compare the catalysis results with 1,4-substituted one. As the effective organocatalyst, the 1,5-disubstituted 1,2,3-triazole **102** and **103** acts well in the asymmetric conversion to attain triazole analogues **101** shown in Scheme 3.32. The 1,4-disubstituted triazole **104** was found to be effective for the same with marginal differences in comparison to 1,5-disubstituted triazole **102** [73].

Takemoto et al. described the synthesis of trifunctional thioureas with a 1,2,3-triazole tether, where one of the functional groups is positioned somewhat far from the thiourea analogue. The ruthenium and copper catalyzed-Huisgen cycloadditions, respectively, afford two different types of regioisomeric catalysts with a 1,5- and 1,4-disubstituted triazole. Despite the vast range of functionalities that the RuAAC reaction can support, the thiourea sulfur atom has the potential to ligate to the ruthenium catalyst, which would deactivate it. In order to prepare catalysts of the type **105** (Fig. 3.4), a microwave-mediated RuAAC reaction with the catalyst $[Cp^*RuCl]_4$ was used, followed by coupling reactions to introduce the thiourea unit in the final step. This synthetic strategy involved introducing the thiourea unit in the last step. In the enantioselective conjugate addition of cyclohexanone to trans-β-nitrostyrene, screening of three of the catalysts produced up to 92% ee. A 2-furyl substituted nitroolefin was able to provide up to 98% ee utilizing these organocatalysts, expanding the range of aryl-substituted nitroolefins in a later investigation [74, 75].

In addition, a class of carbohydrate compounds known as 2,3,6-trideoxysugar triazole hybrids has been studied for potential medicinal uses, particularly as new broad-spectrum anti-bacterial medicines [76]. Csampai and co-worker explore the modular

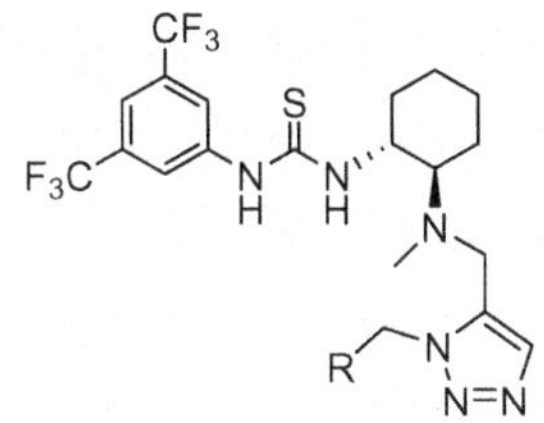

105a-e

105a, R = C_6H_5
105b = 2-OHC_6H_4
105c= 2-$HO_2CC_6H_4$
105d = (*R*)-2-pyrrolidynyl
105e = (*S*)-2-pyrrolidynyl

Fig. 3.4 Structure of developed trifunctional 1,5-disubstituted triazole–based organocatalyst for asymmetric conjugate addition

CuAAC and RuAAC protocols for the efficient synthesis of ferrocene-cinchona hybrid **108** as promising in vivo cytostatic efficacy toward the human tumor cell lines HT-29 colorectal adenocarcinoma and HepC2 hepatoma. The desired ferrocene-cinchona hybrid analogue **108** was achieved in 48% yield by Ru-assisted cyclization of the ferrocene-containing azido component **106** with the alkynyl cinchona alkaloids **107** using 5 mol% CpRuCl(COD) as catalyst (Scheme 3.33) [77].

Scheme 3.33 RuAAC-assisted synthesis of cytostatic ferrocene-cinchona triazole hybrid **108**

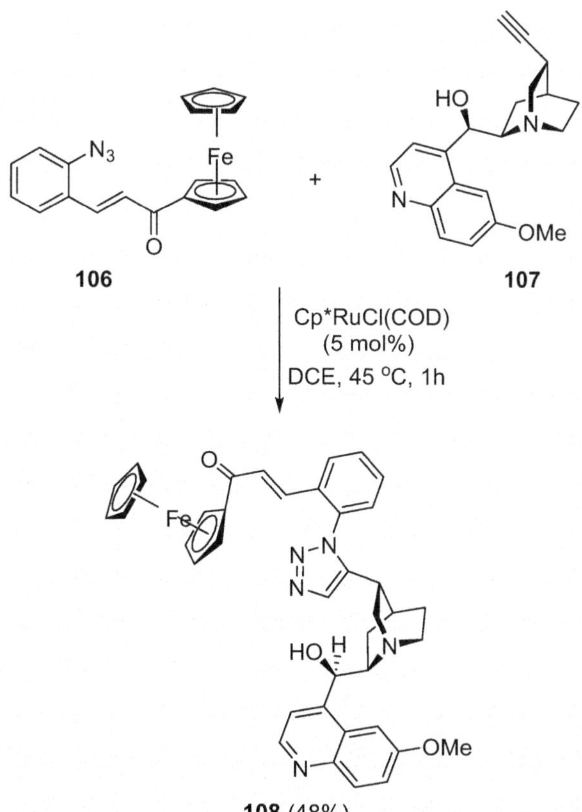

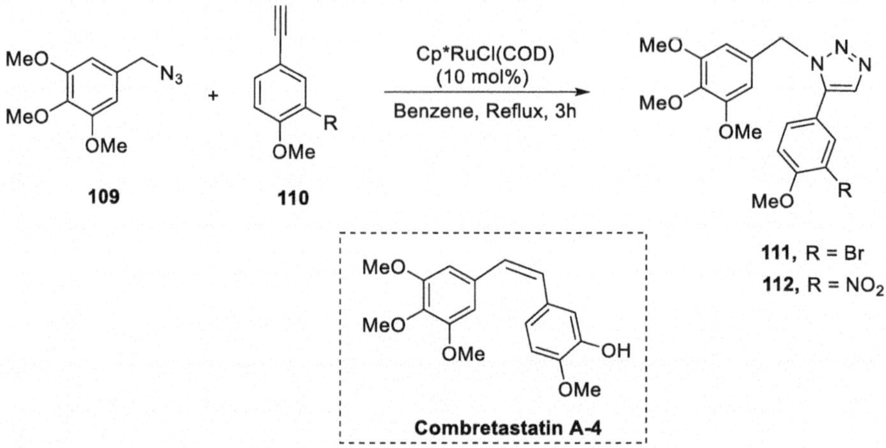

Scheme 3.34 RuAAC-assisted 1,5-substituted 1,2,3-triazole-containing combretastatin A-4 analogues

A cytotoxic natural substance called combretastatin A-4 was isolated from the South African tree Combretum caffrum bark. Hansen and co-workers previously created analogues of combretastatin in which the cis double bond, required for activity, had been substituted with a triazole ring in order to lock the two aromatic rings into a cis connection. This earlier study featured both 1,4- and 1,5-disubstituted 1,2,3-triazoles, where 1,5-regioisomers were obtained by using magnesium acetylides [78]. In a more recent work, combretastatin A-4 analogues **111** and **112** were developed in excellent yields using azide analogue **109** and alkyne analogue **110** in assistance of Cp*RuCl(COD) in refluxing benzene Scheme (3.34) [79].

The cycloaddition reaction of some selected metal acetylides with organic azides resulted in respective metallo-triazoles which subsequently underwent protonation to give the 1,5-disubstituted triazole adduct. Akimova et al. find scope in acetylide adducts with various metals to mediate the synthesis of desired 4-metalated 1,5-disubstituted 1,2,3-triazoles [80–82]. Sharpless et al. further modify the protocol with improved reaction yields in comparison to use of bromomagnesium acetylides. Scope of the cycloaddition reaction was further broadening once the intermediate 4-metalated triazole was trapped with some other electrophiles than proton and afforded good yield of the desired 1,4,5-trisubstituted 1,2,3-triazoles in regioselective manner [83]. However, the formation of several types of trace byproducts in the coupling reaction of metal acetylides with azides limit the scope in synthesis. For example, the reaction when performed for the longer time in the absence of strictly anaerobic conditions may gave 4-hydroxy 1,2,3-triazole adduct (**113**), 4-triazol (**114**) or 4-alkyne 1,2,3-triazole adducts (**115**) (Fig. 3.5) as a result of oxidative couplings.

A byproduct **116** can be obtained by coupling of the azide to the C-4 position of 1,2,3-triazole, when the organic azide is used in excess. Bromomagnesium acetylides when used as substrates, the 1,5-disubstituted-4-bromotriazoles **117** were isolated in trace amount [83]. For 1,5-disubstituted triazoles formation, the RuAAC tool is found to be more convenient than metal acetylides. Some Ru-catalysts such

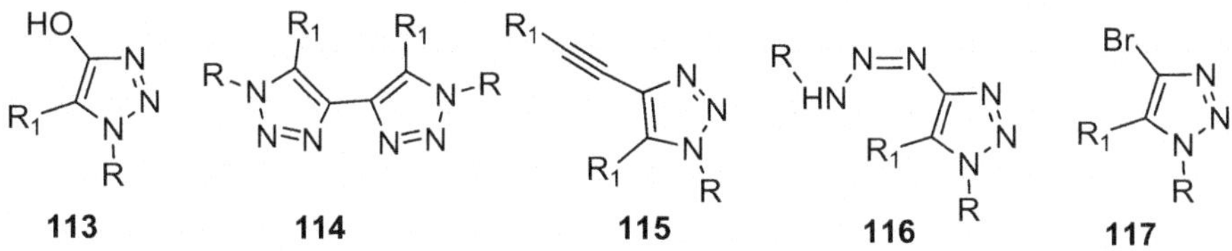

113 **114** **115** **116** **117**

Fig. 3.5 Structure of some byproducts obtained in metal acetylides and azides

as $Ru(OAc)_2(PPh_3)_2$, $RuCl_2(PPh_3)_3$, and $RuHCl(CO)(PPh_3)_3$) although favor the formation of 1,4-disubstituted 1,2,3-triazole formation like CuAAC mechanism [84], however use of appropriate hindered pentamethylcyclopentadienyl (Cp*) containing Ru-catalysis such as $[Cp^*RuCl_2]_2$, Cp* RuCl(COD), Cp*RuCl(NBD) (where COD = cyclooctadiene and NBD = norbornadiene), we can get 1,5-regioisomer of 1,2,3-triazole as the sole product. These sterically hindered pentamethylcyclopentadienyl (Cp*) derivatives of ruthenium are the most common catalyst explored in Ru-catalyzed azide-alkyne cycloaddition reaction (RuAAC) [31].

3.3 Conclusions and Future Outlook

In contrast to the copper-catalyzed modular triazole's 1,4-regioisomer triazole forming 'CuAAC click' approaches, the modular ruthenium variant RuAAC gives 1,5-disubstituted 1,2,3-triazoles as complementary opposite regioisomer. A number of Ru(II) complexes catalyze the 1,5-selectivity, but in order to attain the enhanced regioselectivity, a sterically hindered pentamethylcyclopentadienyl (Cp*) ligand and a chloride (Cl) atom-linked ruthenium system are typically needed. Since internal alkynes under RuAAC condition were reactive over respective CuAAC version and thus as per the mechanistic consideration, the RuAAC follows a different pathway than the CuAAC click. RuAAC has explored for the successful construction of well-defined macrocyclic architectures with 1,5-disubstitution pattern in triazoles. This is because, in general, the regioselectivity in smaller ring formation, RuAAC has more facile over to 1,4-disubstition pattern achieved through CuAAC route. The catalyst used in CuAAC is cheaper, and also, AAC condition is mild in comparison with RuAAC. Although both the tools have displayed tolerance of wide range of functional groups, RuAAC sometime gets complicated with substrates having acidic functional groups. The RuAAC-mediated 1,5-triazoles have promise in several fields of research including chemistry, biology, medicine and have increased potential as enzyme inhibitory activities.

A RuAAC-mediated derivative, however, has not yet appeared in a medicine that is commercially available. It is also interesting to access the resulting triazoles as mimics of the peptide bond or disulfide or as mimics of the α-helix, and various studies in this field have been reported. There is plenty of room for additional research of resulting 1,5-disubstituted triazoles in many growing areas of chemistry ranging from the development of assembly of novel materials or electronic devices to their application as suitable ligands in coordination chemistry as well catalysis. The development of newer Ru-catalysts with potential efficacy and selectivity can enforce the synthetic applicability of RuAAC in the growing field of pharmaceutical sciences, material sciences, green, and environment-sustainable synthetic approach which is the ultimate goal of organic synthesis.

References

1. Tornoe CW, Christensen C, Meldal M (2002) J Org Chem 67:3057–3064
2. Rostovtsev VV, Green LG, Fokin VV, Sharpless KB (2002) Angew Chem 114:2708–2711
3. Lee LV, Mitchell ML, Huang SJ, Fokin VV, Sharpless KB, Wong CH (2003) J Am Chem Soc 125:9588–9589
4. Whiting M, Muldoon J, Lin YC, Silverman SM, Lindstrom W, Olson AJ, Kolb HC, Finn MG, Sharpless KB, Elder JH, Fokin VV (2006) Angew Chem 118:1463–1467
5. Tiwari VK, Mishra BB, Mishra KB, Mishra N, Singh AS, Chen X (2016) Chem Rev 116:3086–3240
6. Agrahari AK, Bose P, Jaiswal MK, Rajkhowa S, Singh AS, Hotha S, Mishra N, Tiwari VK (2021) Chem Rev 121:7638–7956
7. Kumar SS, Kavitha HP (2013) Mini-Rev Org Chem 10:40–65
8. Matin MM, Matin P, Rahman MR, Ben Hadda T, Almalki FA, Mahmud S, Ghoneim MM, Alruwaily M, Alshehri S (2022) Front Mol Biosci 9:864286
9. Abd-Elaal AA, Aiad I, Shaban SM, Tawfik SM, Sayed A (2014) J Surfactants Deterg 17:483–491
10. Duan T, Fan K, Fu Y, Zhong C, Chen X, Peng T, Qin J (2012) Dyes Pigm 94:28–33
11. Abu-Orabi ST, Atfah MA. Jibril I, Mari'i FM, Ali AA-S (1989) J Heterocycl Chem 26, 1461–1468
12. Jiang Y, Li X, Li X, Sun Y, Zhao Y, Jia S, Guo N, Xu G, Zhang W (2017) Chin J Chem 35:1239–1245
13. Shi S, Wang Z, Deng Y, Tian F, Wu Q, Zheng P (2022) CCS Chem. 4:598–604
14. Mishra N, Singh AS, Agrahari AK, Singh SK, Singh M, Tiwari VK (2019) ACS Comb Sci 21:389–399
15. Mishra N, Singh SK, Singh AS, Agrahari AK, Tiwari VK (2021) J Org Chem 86:17884–17895
16. Singh SK, Mishra N, Kumar S, Jaiswal MK, Tiwari VK (2022) Chem Select 7:e202201314
17. Singh SK, Kumar S, Yadav MS, Tiwari VK (2022) J Org Chem 87(22):15389–15402
18. Singh SK, Kumar S, Yadav MS, Gupta A, Tiwari VK (2023) J Org Chem 88:13440–13453
19. Singh SK, Kumar S, Yadav MS, Bhattacharya S, Tiwari VK (2024) Synthesis 56:975–988
20. Mishra A, Mishra N, Tiwari VK (2016) Curr Org Synth 1:176–219
21. Wojaczynska E, Steppeler F, Iwan D, Scherrmann M-C, Marra A (2021) Molecules 26:7291
22. Zhang L, Chen XG, Xue P, Sun HHY, Williams ID, Sharpless KB, Fokin VV, Jia GC (2005) J Am Chem Soc 127:15998–15999
23. Coats SJ, Link JS, Gauthier D, Hlasta DJ (2005) Org Lett 7:1469–1472
24. Hong L, Lin W, Zhang F, Liu R, Zhou X (2013) Chem Commun 49:5589–5591
25. Wang D, Salmon L, Ruiz J, Astruc D (2013) Chem Commun 49:6956–6958
26. Koguchi S, Izawa K (2014) ACS Comb Sci 16:381–385
27. De Nino A, Maiuolo L, Costanzo P, Algieri V, Jiritano A, Olivito F, Tallarida MA (2021) Catalysts 11:1120
28. Zhao Y, Chai Z, Zeng Q, Zhang WX (2023) Molecules 28:1400
29. Johansson JR, Beke-Somfai T, Stålsmeden AS, Kann N (2016) Chem Rev 116:14726–14768
30. Taguchi R, Nakahata M, Kamon Y, Hashidzume A (2023) Polymers 15:2199

31. Jaiswal MK, Gupta A, Ansari AF, Pandey VK, Tiwari VK (2024) Curr Org Synth 21:513–558
32. Boren BC, Narayan S, Rasmussen LK, Zhang L, Zhao HT, Lin ZY, Jia GC, Fokin VV (2008) J Am Chem Soc 130:8923–8930
33. Oppilliart S, Mousseau G, Zhang L, Jia G, Thuery P, Rousseau B, Cintrat JC (2007) Tetrahedron 63:8094–8098
34. Nulwala H, Takizawa K, Odukale A, Khan A, Thibault RJ, Taft BR, Lipshutz BH, Hawker CJ (2009) Macromolecules 42:6068–6074
35. Wuest F, Tang X, Kniess T, Pietzsch J, Suresh M (2009) Bioorg Med Chem 17:1146–1151
36. Yap AH, Weinreb SM (2006) Tetrahedron Lett 47:3035–3038
37. Wang T-H, Wu F-L, Chiang G-R, He S-T, Lo Y-H (2014) J Organomet Chem 774:57–60
38. Bouchet F, Barnier J-P, Sayah I, Bagdad Y, Miteva MA, Arthur M, Ethève-Quelquejeu M, Iannazzo L (2023) ChemMedChem 18:e202300077
39. Kappe CO, Vander EE (2010) Chem Soc Rev 39:1280–1290
40. Barge A, Tagliapietra S, Binello A, Cravotto G (2011) Curr Org Chem 15:189–203
41. Johansson JR, Lincoln P, Norden B, Kann N (2011) J Org Chem 76:2355–2359
42. Johansson JR, Hermansson E, Nordén B, Kann N, Beke-Somfai T (2014) Eur J Org Chem 2014:2703–2713
43. Stålsmeden AS, Paterson AJ, Szigyártó IC, Thunberg L, Johansson JR, Beke-Somfai T, Kann N (2020) Org Biomol Chem 18:1957–1967
44. Trabocchi A, Guarna A (2014) Peptidomimetics in organic and medicinal chemistry. John Wiley & Sons Ltd.
45. Scott DE, Bayly AR, Abell C, Skidmore J (2016) Nat Rev Drug Discov 15:533–550
46. Pokorski JK, Jenkins LMM, Feng HQ, Durell SR, Bai YW, Appella DH (2007) Org Lett 9:2381–2383
47. Tietze D, Tischler M, Voigt S, Imhof D, Ohlenschlager O, Gorlach M, Buntkowsky G (2010) Chem Eur J 16:7572–7578
48. Kee JM, Villani B, Carpenter LR, Muir TW (2010) J Am Chem Soc 132:14327–14329
49. Buysse K, Farard J, Nikolaou A, Vanderheyden P, Vauquelin G, Sejer Pedersen D, Tourwe D, Ballet S (2011) Org Lett 13:6468–6471
50. Wales SM, Hammer KA, King AM, Tague AJ, Lyras D, Riley TV, Keller PA, Pyne SG (2015) Org Biomol Chem 13:5743–5756
51. Kelly AR, Wei JQ, Kesavan S, Marie JC, Windmon N, Young DW, Marcaurelle LA (2009) Org Lett 11:2257–2260
52. Nielsen TE, Schreiber SL (2008) Angew Chem Int Ed 47:48–56
53. Marcaurelle LA, Comer E, Dandapani S, Duvall JR, Gerard B, Kesavan S, Lee MD, Liu HB, Lowe JT, Marie JC et al (2010) J Am Chem Soc 132:16962–16976
54. Zhang JQ, Kemmink J, Rijkers DTS, Liskamp RMJ (2013) Chem Commun 49:4498–4500
55. Zhang JQ, Kemmink J, Rijkers DTS, Liskamp RMJ (2011) Org Lett 13:3438–3441
56. Fang F, Vogel M, Hines JV, Bergmeier SC (2012) Org Biomol Chem 10:3080–3091
57. Isidro-Llobet A, Murillo T, Bello P, Cilibrizzi A, Hodgkinson JT, Galloway W, Bender A, Welch M, Spring DR (2011) Proc Natl Acad Sci U S A 108:6793–6798
58. Krause MR, Goddard R, Kubik S (2011) J Org Chem 76:7084–7095
59. Krause MR, Goddard R, Kubik S (2010) Chem Commun 46:5307–5309
60. Pradere U, Roy V, McBrayer TR, Schinazi RF, Agrofoglio LA (2008) Tetrahedron 64:9044–9051
61. Montagu A, Roy V, Balzarini J, Snoeck R, Andrei G, Agrofoglio LA (2011) Eur J Med Chem 46:778–786
62. Chemama M, Fonvielle M, Arthur M, Valery JM, Etheve-uelquejeu M (2009) Chem.-Eur. J. 15:1929–1938
63. Sirivolu VR, Vernekar SKV, Ilina T, Myshakina NS, Parniak MA, Wang ZQ (2013) J Med Chem 56:8765–8780
64. Vernekar SKV, Qiu L, Zacharias J, Geraghty RJ, Wang ZQ (2014) MedChemComm. 5:603–608
65. Vernekar SKV, Qiu L, Zhang J, Kankanala J, Li HM, Geraghty RJ, Wang ZQ (2015) J Med Chem 58:4016–4028

66. Epple S, Modi A, Baker YR, Węgrzyn E, Traoré D, Wanat P, Tyburn AES, Shivalingam A, Taemaitree L, Sagheer AHE, Brown T (2021) J Am Chem Soc 143:16293–16301
67. Crich D, Yang F (2009) Angew Chem Int Ed 48:8896–8899
68. Wiebe C, Schlemmer C, Weck S, Opatz T (2011) Chem Commun 47:9212–9214
69. Cagnoni AJ, Varela O, Uhrig ML, Kovensky J (2013) Eur J Org Chem 972–983
70. Salmon AJ, Williams ML, Maresca A, Supuran CT, Poulsen SA (2011) Bioorg Med Chem Lett 21:6058–6061
71. Jervis PJ, Graham LM, Foster EL, Cox LR, Porcelli SA, Besra GS (2012) Med Chem Lett 22:4348–4352
72. Arora I, Kashyap VK, Singh AK, Dasgupta A, Kumar B, Shaw AK (2014) Org Biomol Chem 12:6855–6868
73. Luo SZ, Xu H, Mi XL, Li JY, Zheng XX, Cheng JP (2006) J Org Chem 71:9244–9247
74. Takasu K, Azuma T, Takemoto Y (2010) Tetrahedron Lett 51:2737–2740
75. Takasu K, Azuma T, Enkhtaivan I, Takemoto Y (2010) Molecules 15:8327–8348
76. Sharma S, Saquib M, Verma S, Mishra NN, Shukla PK, Srivastava R, Prabhakar YS, Shaw AK (2014) Eur J Med Chem 83:474–489
77. Kocsis L, Szabo I, Bosze S, Jernei T, Hudecz F, Csampai A (2016) Bioorg Med Chem Lett 26:946–949
78. Odlo K, Hentzen J, dit Chabert JF, Ducki S, Gani O, Sylte I, Skrede M, Florenes VA, Hansen TV (2008) Bioorg Med Chem 16:4829−4838
79. Odlo K, Fournier-Dit-Chabert J, Ducki S, Gani O, Sylte I, Hansen TV (2010) Bioorg Med Chem 18:6874–6885
80. Akimova G, Chistokletov V, Petrov A (1967) Zh Obshch Khim 3:2241−2247
81. Akimova G, Chistokletov V, Petrov A1 (1967) Zh Obshch Khim 3:968−974
82. Akimova G, Chistokletov V, Petrov A1 (1968) Zh Obshch Khim 4:389−394
83. Krasiński A, Fokin VV, Sharpless KB (2004) Org Lett 6:1237–1240
84. Liu PN, Li J, Su FH, Ju KD, Zhang L, Shi C, Sung HHY, Williams ID, Fokin VV, Lin ZY et al (2012) Organometallics 31:4904–4915

Chapter 4
Organocatalyzed 1,2,3-Triazoles Forming Click Chemistry

4.1 Introduction

Followed by Chaps. 2 and 3 dealing with copper (I)- and ruthenium-catalyzed regioselective 1,2,3-triazole forming click approaches through [3 + 2] azide-alkyne cycloaddition, we now move to the metal-free variant click approaches with an aid of organcatalysis [1, 2]. In this chapter, we have oriented to comprise an array of organocatalyzed approaches for synthesis of 1,2,3-triazolyl scaffolds. The 1,2,3-triazoles have various notable importance in pharmaceutical, materials, and synthetic chemistry which needs the new, efficient, mild, and non-toxic method for easy access of such 1,2,3-triazole-based functionalized compounds [1–5]. In numerous industrial uses, including agrochemicals, corrosion inhibitors, polymers, optical brighteners, photostabilizers, pigments, and metal chelators, they function as important synthetic intermediates [6, 7]. The 1,2,3-triazole heterocycles, amide-triazole bioequivalence was used to develop numerous advantageous pharmaceutical scaffolds with anti-HIV, bactericidal, and anticancer properties [8–10]. The carbohydrate-based analogues including triazolyl glycoconjugates or triazole-appended glycohybrids have received considerable attention to explore as metal catalyst to organocatalysts [11–18].

Organocatalyzed 1,2,3-triazole forming reactions are an important class of chemical reactions that involve the reaction of azides with alkynes using organic catalysts, gained significant attention in recent years due to their wide range of applications and the increasing demand for more sustainable and environmentally friendly chemical processes [19–21]. The development of new and efficient catalysts for these reactions has led to the discovery of novel and useful triazole compounds for a variety of applications. Unlike traditional metal catalysts, organic catalysts are generally more environmentally friendly, less expensive, and can be more selective in their reactions. In these reactions, triazole compounds are formed through the reaction of an azide compound with an alkyne or a diazo compound, the reaction is catalyzed by an organic

V. K. Tiwari et al., *Click Chemistry*, Materials Horizons: From Nature to Nanomaterials,
https://doi.org/10.1007/978-981-97-4596-8_4

molecule, which can be a variety of different compounds such as a amines, imidazole, proline, DBU, or various bases, and the reaction can proceed through a variety of mechanisms, depending on the specific catalyst and reaction conditions used [22–28]. In this chapter, we have comprised the efficient synthesis of 1,2,3-triazoles in assistance of organocatalysts through various modes and intermediates.

4.2 The Organocatalyzed Synthesis of Functionalized 1,2,3-Triazoles

4.2.1 Synthesis of 1,2,3-Triazole Through Iminium-Based Reactions

The organocatalytic route for the synthesis of substituted 1,2,3-triazoles generally involves two types of mechanisms such as enolate and enamine-based. Despite the above two mechanisms, the iminium ion-based mechanism is also reported in the literature. However, the iminium ion-based mechanism is less frequently reported for organocatalytically triazole synthesis. The synthesis of 1,4,5-trisubstituted triazoles from α,β-unsaturated ketones with the assistance of piperidine as an organocatalyst was described by Wang et al. In the report,10 mol% of piperidine was utilized to assist the conversion of many substituted unsaturated ketones to corresponding triazole with various azide partners such as aliphatic, aromatic, and heteroaromatic azides. The authors propose an iminium ion intermediate to describe the mechanistic path of the reaction. In the first step, piperidine catalyst reacts with the unsaturated ketone **1** to furnish the iminium intermediate **I,** which exists in equilibrium with dienamine structure **II** as well. The iminium species **II** and azide **2** undergo standard electron-demand cycloaddition to afford the triazoline intermediate **III**. The fully substituted 1,2,3-triazole **3** was achieved by hydrolysis of the iminium center of intermediate **III** to compound **IV** followed by air oxidation of the triazoline **IV** (Scheme 4.1) [29].

Later, the same group reported the synthesis of 1,4-disubstituted 1,2,3-triazole through DBU-catalyzed iminium ion-based 1,3-dipolar cycloaddition reaction of α, β-unsaturated ester **4** with azide **2**. The devised protocol was characterized by high regioselectivity and good yield of the desired triazole **5** (Scheme 4.2) [30].

4.2.2 Synthesis of 1,2,3-Triazole Through Enamine/Dienamine-Based Reactions

Since the early 1950s, extensive research has been done on the chemistry of preformed enamines, particularly their usage as enolate substitutes. Enamine catalysis, which is the catalysis of carbonyl transformations via enamine intermediates by using primary

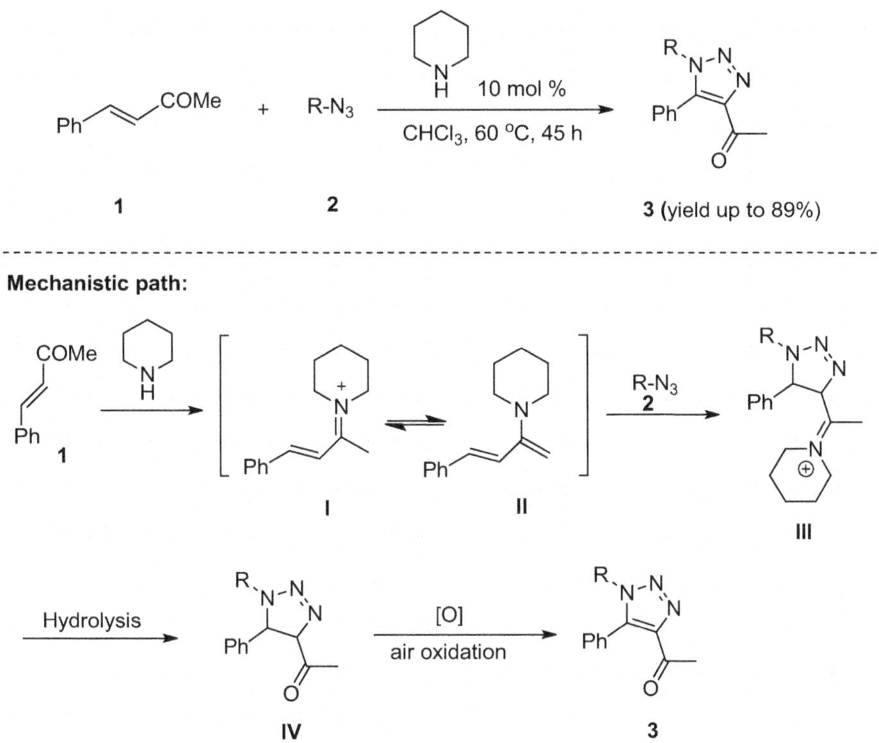

Scheme 4.1 Piperidine-catalyzed synthesis of substituted 1,2,3-triazoles via iminium-based intermediate

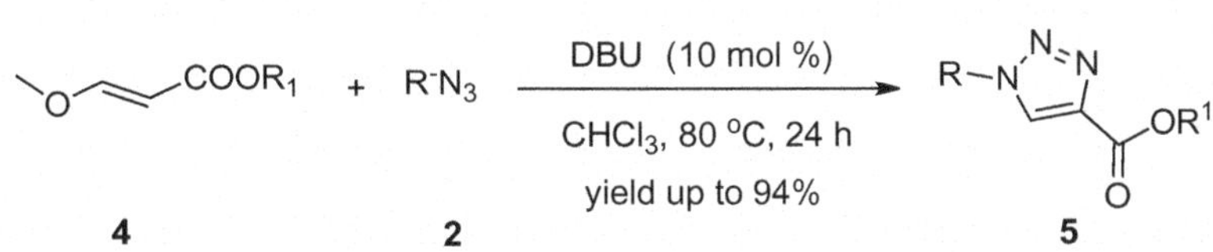

Scheme 4.2 DBU-catalyzed synthesis of substituted 1,4-disubstituted 1,2,3-triazole

and secondary amines as catalysts, has only recently come to be appropriately recognized as a potent method for asymmetric synthesis [31]. The reactions of isolated enamines analogues with organic azides afford substituted 1,2,3-triazoles have been reported from many years ago [32]. In this section, we have included the various cycloaddition reactions of catalytically derived enamines intermediate with azides to afford 1,2,3-triazoles. The enamine-based mechanism for the assembly of substituted 1,2,3-triazoles is one of the common pathways. The widespread substrates are carbonyl compounds and the organocatalyst is a secondary amine like proline, pyrrolidine, or diethylamine. The condensation of the carbonyl compounds **6** with the organocatalyst in the first stage of this mechanism results in the appropriate enamine intermediate **I**. Since this enamine is an olefinic partner with many electrons, it engages in an enamine-azide [3 + 2] cycloaddition reaction with various azides to produce a triazoline intermediate **II**. This intermediate **II** then goes through a conceivable 1,3-hydride shift to produce a triazoline intermediate **III**. The final step regenerates the organocatalyst to continue the catalytic cycle, and the intermediate **IV**, which is the zwitter ionic form of intermediate **III**, provides the desired 1,4,5-trisubstituted 1,2,3-triazole product **7** (Scheme 4.3).

In very initially toward the metal-free synthesis of substituted triazoles, Pocar et al. reported an intriguing approach where 1,3-diene-2-amines **10** were synthesized by conventional heating of morpholine **8** with ketal derivatives of methyl *n*-alkyl-ketones **9**. The developed dienes **10** were demonstrated as a click partner with aryl azides **2** to attain a substituted triazole skeleton, which on acidic hydrolysis gave the substituted 1,2,3-triazole **11** (Scheme 4.4) [33].

Vigevani et al. reported an interesting synthesis of triazole analogues **13** through the reaction of developed fused ring dienamine **12** with 4-nitrophenylazide **2**. The cycloaddition reaction of developed fused ring dienamine **12** with 4-nitrophenylazide

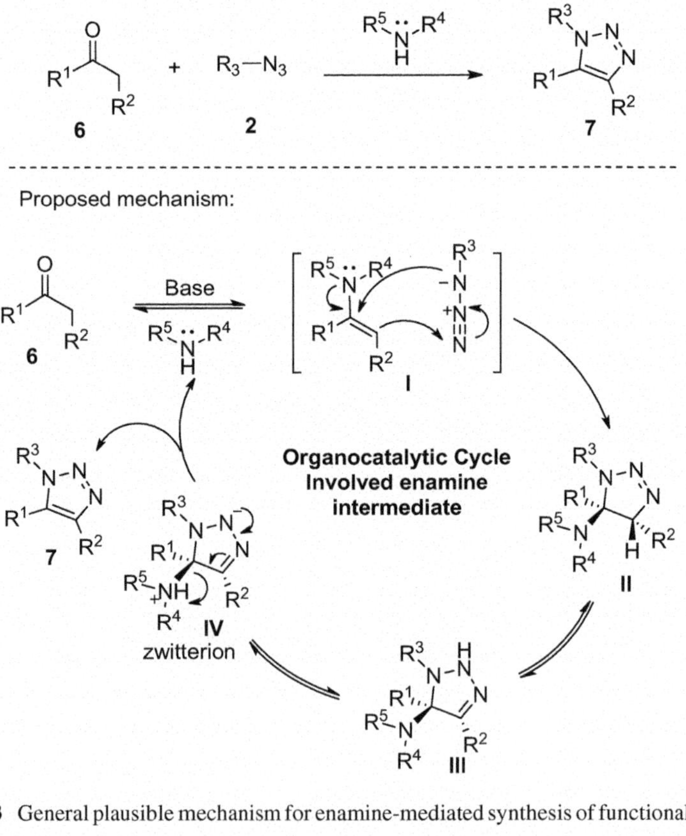

Scheme 4.3 General plausible mechanism for enamine-mediated synthesis of functionalized 1,2,3-triazoles

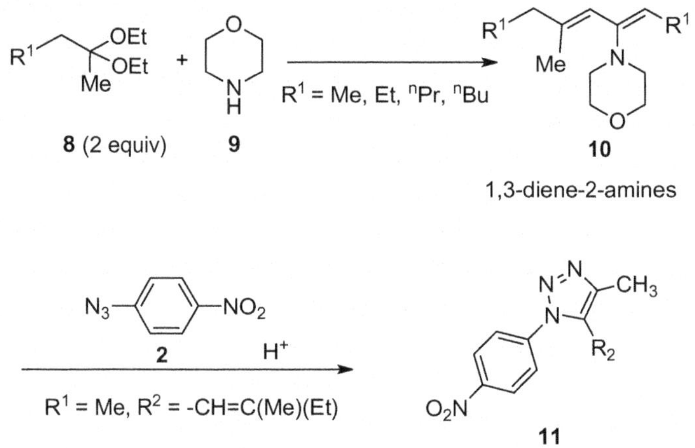

Scheme 4.4 Synthesis of 1,4,5-trisubstituted 1,2,3-triazole through the reaction of dienamines and aryl azide

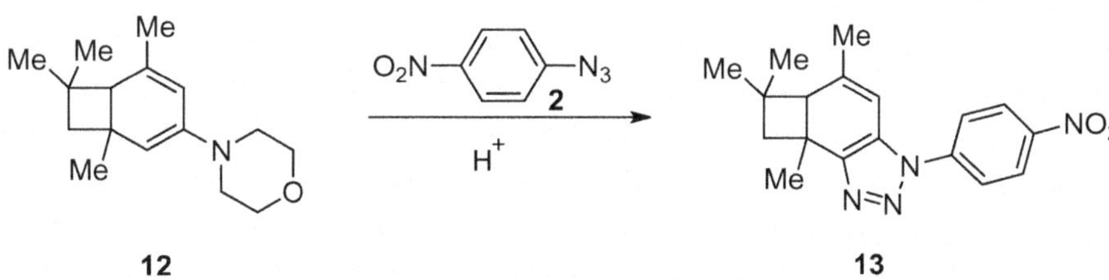

Scheme 4.5 Synthesis of triazole derivative using fused cyclic dienamine and aryl azide

affords the respective triazolinic adduct which on treatment with 25% acetic acid affords the desired triazole analogue **13** (Scheme 4.5) [34].

The cycloaddition of organic azides with a few conjugated enamines of the 2-amino-1,3-diene, 1-amino-1,3-diene, and 2-aminobut-1-en-3-yne types was studied by Gerhard Maas et al. The 2,3-dimethyl-2-morpholinobutene **14** was used as a conjugated enamine to synthesized 5-alkenyl-4,5-dihydro-5-morpholino-1*H*-1,2,3-triazoles **15** through the regioselective [3 + 2]-cycloaddition with a number of electrophilic azides **2** (Scheme 4.6) [35].

A proline-catalyzed interesting synthesis of fused triazoles from Hagemann's ester was revealed by Ramachary et al., in the assistance of proline as an amino acid catalyst tosyl azide **17** reacted with cyclic enone to afford high yields of *N*H-1,2,3-triazoles. In the catalytic path of the reaction organocatalyst on reaction with cyclic enone **16** generated the dienamine **I**, onto which the tosyl azide **17** undergoes cycloaddition followed by elimination of the proline, resulting in the formation of the desired triazole-containing compound **18** (Scheme 4.7) [36].

In the same report for the effective approach on amino acid-catalyzed synthesis of triazole analogue, the group also utilized the benzylamine in place of proline, and

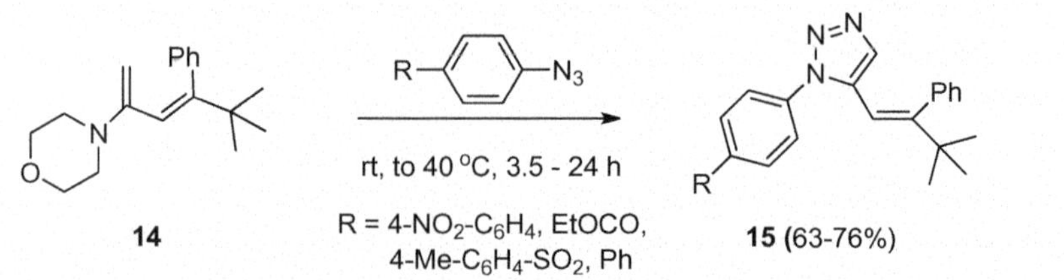

Scheme 4.6 Synthesis of 1,5-disubstituted 1,2,3-triazole through conjugated enamine

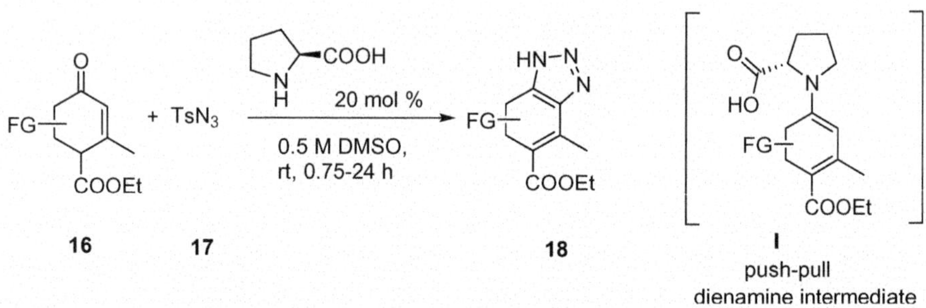

Scheme 4.7 L-proline-catalyzed synthesis of 1,2,3-triazole derivative through push–pull dienamine intermediate

α-diazo ketone compound **19** was formed rather than triazole **18** with the assistance of benzylamine as the catalyst. The authors suggested that the reaction precedes through the dienamine intermediate **II** (Scheme 4.8) [36].

Next, the Ramachary research group reported an intriguing multicomponent approach to synthesize trisubstituted 1,2,3-triazoles **21** by the reaction of cyclic enones **16** and aryl azides **2** using pyrrolidine as an amine-based catalyst. The reaction is characterized by the formation of an enamine intermediate and carried out in DMSO with 10 mol% of the pyrrolidine at room temperature and found to be suitable to afford triazoles **20** for both substituted and unsubstituted Hagemann's

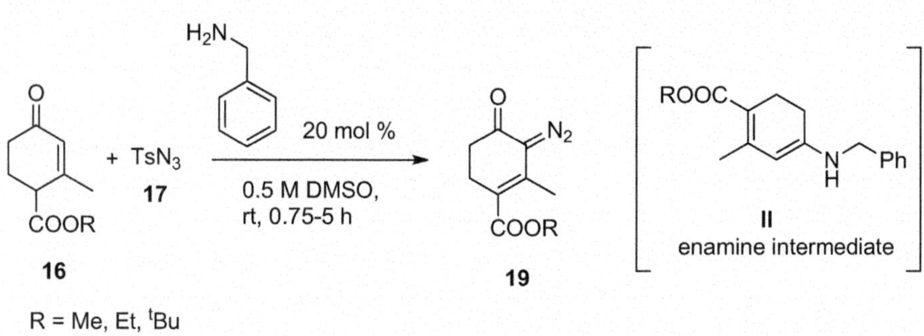

Scheme 4.8 Synthesis of α-diazoketone compound **19** in assistance of benzylamine as the catalyst

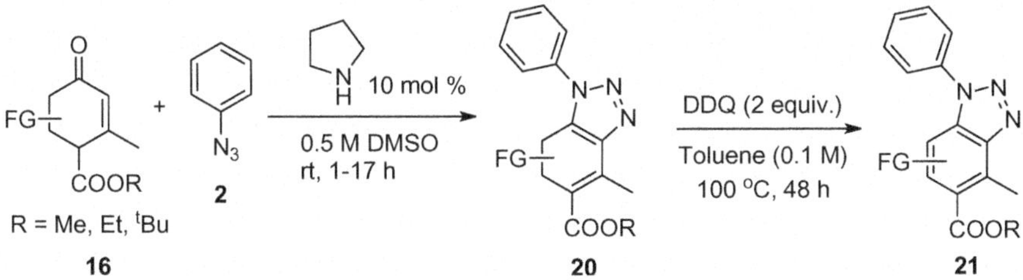

Scheme 4.9 Metal-free synthesis of benzotriazole derivative **21** with the assistance of pyrrolidine catalyst

esters **16**. The developed triazoles **20** on oxidation with 2,3-dichloro-5,6-dicyano-1,4-benzoquinone (DDQ) lead the formation of benzotriazole derivative **21** (Scheme 4.9) [37].

Wang et al. reported the regiospecific enamide–azide cycloaddition reaction for the synthesis of 1,4,5-trisubstituted triazoles from active methylene compounds such as aryl ketones or ketoesters and organic azides. An organic azide **2** when treated with active methylene compound **22** afforded the 1,4,5-trisubstituted 1,2,3-triazoles **23** in excellent yields. The reaction was catalyzed by 5 mol% catalytic loading of diethylamine in DMSO at 70 °C (Scheme 4.10). Several active methylene compounds **22** and substituted aryl azides **2** were found to be compatible with the devised methodology, which could also withstand a wide range of substituents. It is important to note that benzyl azide and aliphatic azide were also reported to provide the corresponding triazoles in a good yield. In the mechanistic part, active methylene compound **22** on condensation reaction with secondary amine as organocatalyst afforded the enamine intermediate **I,** which on regiospecific 1,3-dipolar cycloaddition reaction with organic azide **2** gave the triazoline **II**. The resulting triazoline **II** on 1,3-hydride shift affords the intermediate **III**, which on elimination step via zwitter ionic intermediate **IV**, affords the final 1,4,5-trisubstituted-1,2,3-triazole **23** in good-to-excellent yield (Fig. 4.1) [38].

Semenov and colleagues explore the synthetic approach of Wang et al. to develop a family of conformationally restricted 1,2,3-triazole analogues of combretastatin

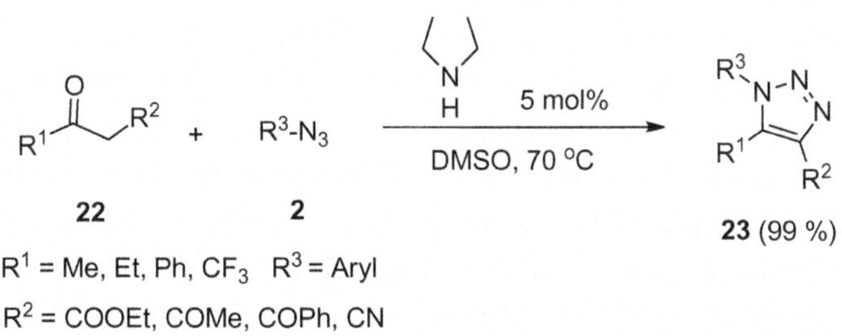

Scheme 4.10 Synthesis of 1,4,5-trisubstituted-1,2,3-triazoles **2** in assistance of diethylamine as the catalyst

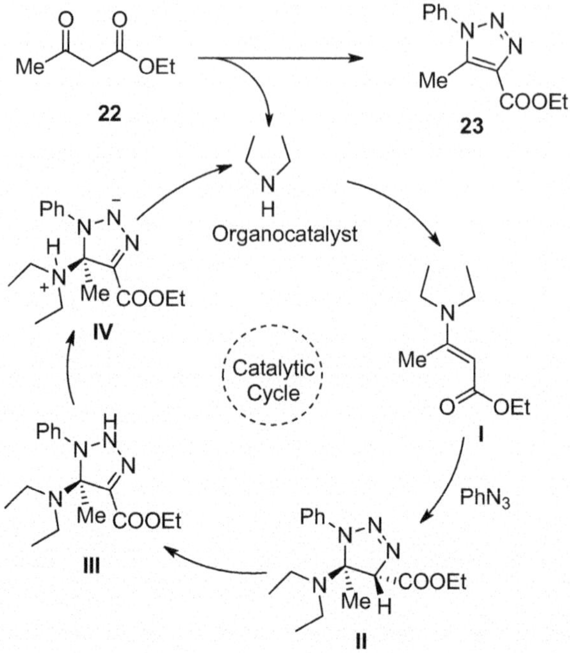

Fig. 4.1 Plausible reaction mechanism for the secondary amine-catalyzed synthesis of 1,4,5-trisubstituted-1,2,3-triazoles **23**

A4. They synthesized the corresponding 1,5-diaryl substituted triazoles **24** and **25** in good yields by treating various active methylene compound **22a,b** with aryl azides **2** in the presence of 5 mol% diethylamine as catalyst (Scheme 4.11). To investigate the relationship between structure and activity, several transformations were applied to these triazolyl products and found that the decarboxylated substances **26** (IC_{50}-20 nM) and **27**(IC_{50}-10 nM) derived from triazolyl analogue **25** were effective as potent antimitotic microtubule destabilizing agents even at nanomolar concentrations [39].

An organocatalytic method for the synthesis of trisubstituted-1,2,3-triazoles beginning with allyl ketones and azide was reported by Wang et al. (Scheme 4.12). The authors utilized a 2:1 combination of aryl azides **2** and allyl ketones **28** in DMSO with 10 mol% diethylamine as a catalyst for this regioselective devised approach. Aliphatic, aromatic, and heteroaromatic allyl ketones **28** all performed well in the reaction. Both aliphatic and aromatic azides **2** produced excellent yields of substituted triazoles **29** in this process. In the mechanistic part, the dienamine **I** was formed by condensation reaction of allyl ketone with diethylamine, which combines with aryl azide **2** to afford the intermediate **II**. The resulted intermediate **II** on 1,3-H-shift furnished the intermediate **III** which underwent intramolecular cycloaddition and gave the triazoline intermediate **IV**. In the final step, the desired triazole derivatives **29** were formed by the oxidation and hydrolysis of intermediate **IV** (Scheme 4.12) [40].

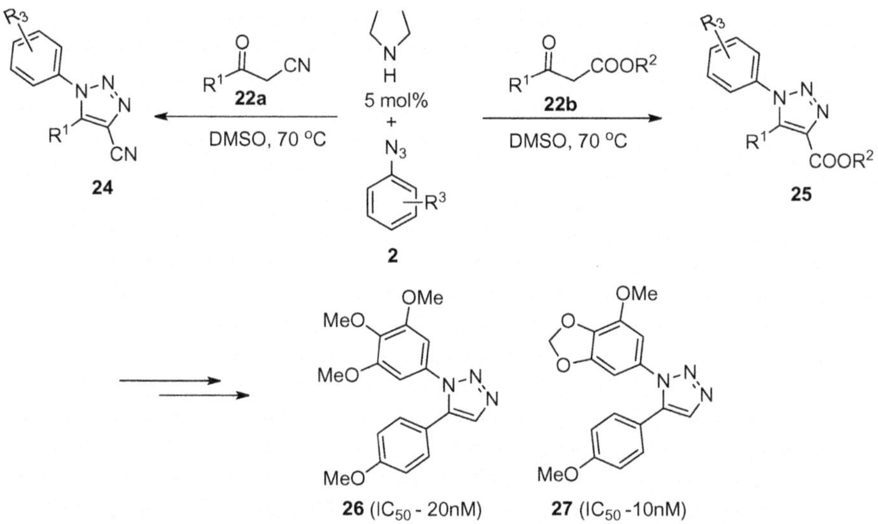

Scheme 4.11 Synthesis of 1,5-diaryl-1,2,3-triazole analogues of combretastatin A4 in assistance of diethylamine as the catalyst

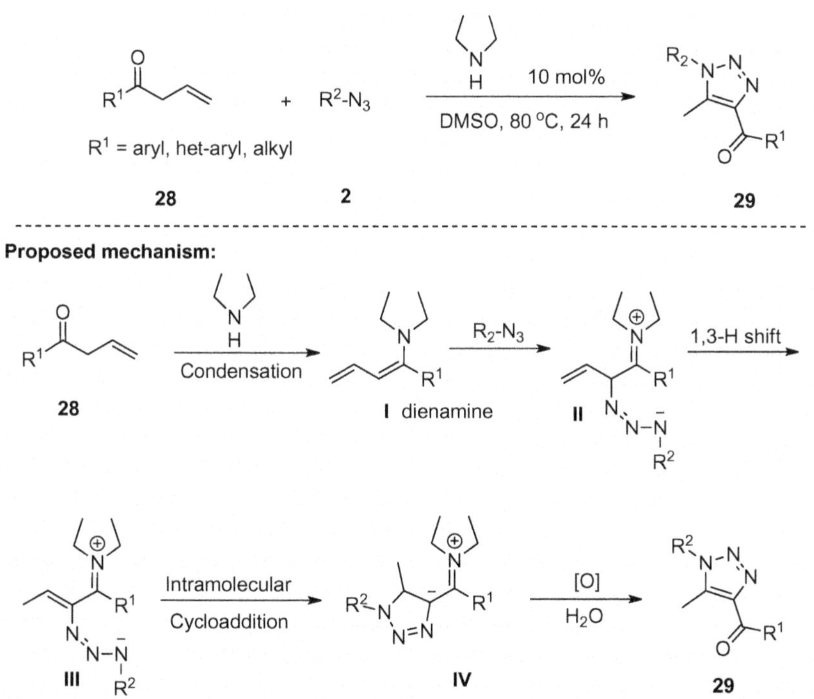

Scheme 4.12 Diethylamine-assisted synthesis of trisubstituted-1,2,3-triazoles using allyl ketones

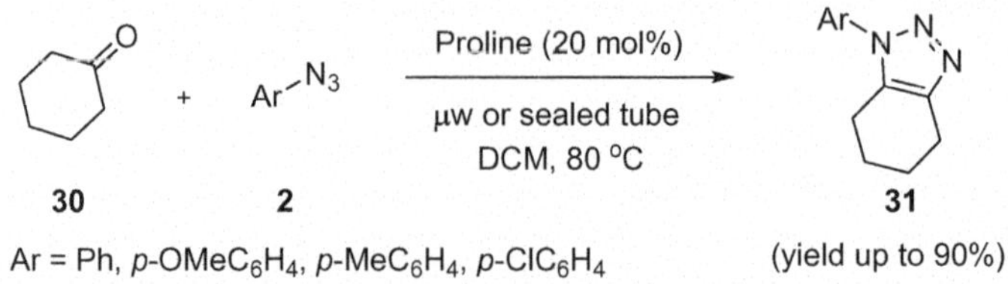

Scheme 4.13 Proline-catalyzed synthesis of 1,2,3-triazoles under microwave irradiation using unactivated ketones and aryl azide

An efficient organocatalytic procedure, complementary to Huisgen's metal-catalyzed triazole synthesis, was developed for synthesis of substituted 1,2,3-triazoles using unactivated ketones in assistance of different organocatalysts. The groups of Pons-Bressy and Wang reported a successful triazole synthesis starting from unactivated ketones [41, 42]. According to Pons-Bressy et al., the ideal conditions needed for the efficient conversion are 20 mol% proline as organocatalyst and a 2:1 ratio of unactivated ketone **30** and aryl azide **2** in DCM as the solvent (Scheme 4.13). Microwave irradiation at 80 °C was used to shorten the lengthy reaction time to 1 h, which turned out to be as high yielding of substituted 1,2,3-triazole **31**. Only aryl azides with electron-donating substituents were found to be effective in the devised protocol. Both cyclic and acyclic ketones were used in the reaction, which was highly regio- and chemoselective (Scheme 4.13) [41]. The Wang research group reported the enamine-catalyzed strategy for the synthesis of substituted triazoles **32** using unactivated ketones **30** and azides **2** at ideal catalytic loading as 20 mol% of pyrrolidine in DMSO as organic solvent (Scheme 4.14). They utilized the devised approach to develop the substituted triazole-based CB1 cannabinoid receptor antagonist precursor **35** through the reaction of the ethyl 3-(4-chlorophenyl)-3-oxopropanoate **33** with the dichlorophenyl azide **34** in the presence of 20 mol% pyrrolidine as catalyst in DMSO [42].

The Wang group reported the first organocatalytic synthesis of substituted 1,2,3-triazoles in aqueous media. They developed a prolinamide catalyst molecule with the right hydrophobic groups that could bind the reactants and compel the reaction to take place in hydrophobic pockets. These micelles might protect the enamine intermediate from water, leading to a reaction akin to those occurring in organic solvents. Experimental parts reveal that the prolinamide catalyst **37** with two *n*-octyl chains had the highest yield when synthesizing substituted triazoles in water (Scheme 4.15). When aryl azides **2** and both activated and inactive carbonyl compounds **36** were combined in water at 80 °C, good to outstanding yields of the corresponding trisubstituted triazoles **38** were produced. When compared to reactions conducted in organic solvents, the regioselective outcome for this environmentally friendly approach remained unaltered [43].

The same group provided a simple pathway from unsaturated aldehydes **39** to 4-alkenyl-1,2,3-triazoles **40** (Scheme 4.16). The author used diethylamine as a secondary amine catalyst and suggested that 10 mol% DBU is sufficient for this

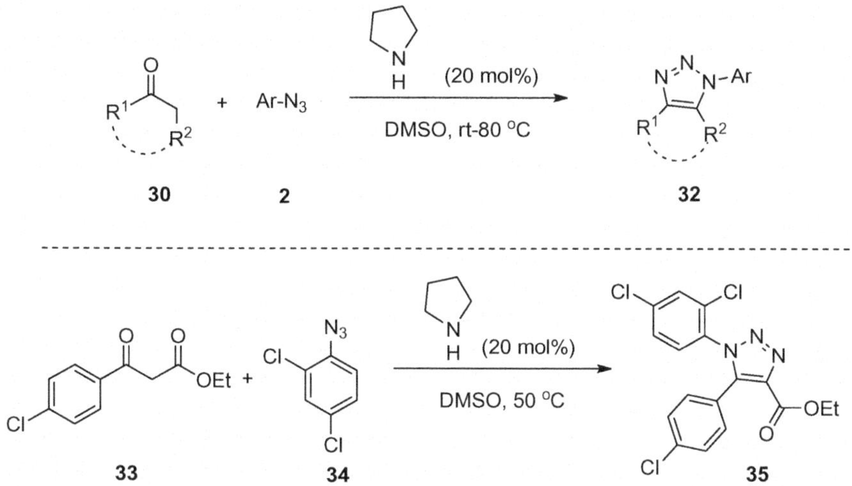

Scheme 4.14 Pyrrolidine-catalyzed synthesis of trisubstituted triazoles with unactivated ketones and azides

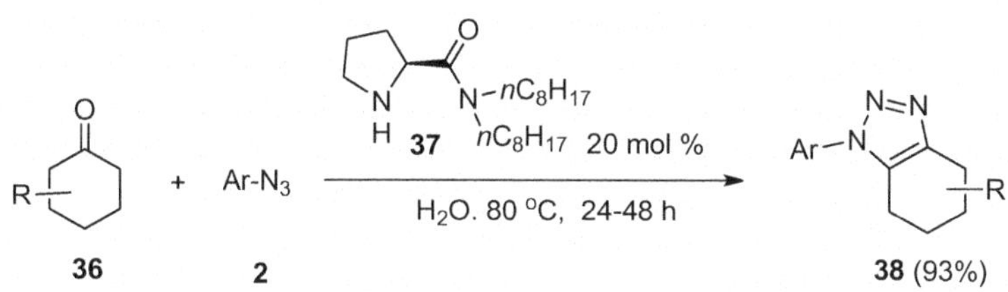

Scheme 4.15 Prolinamide-catalyzed synthesis of substituted triazole in aqueous media

reaction. The author reveals that in the absence of either a secondary amine catalyst or DBU, the yield was significantly reduced. The dienamine intermediate **III** was formed in situ as HOMO-raising dipolarophile and reacts with azide via catalytic inverse-electron-demand 1,3-dipolar cycloaddition process. It was discovered that the electron-donating or withdrawing substituents on the unsaturated aldehydes **39** and aryl azides **2** had no effect on the generality of the alkenyl triazole **40** synthesis [44].

The selenium-containing 1,2,3-triazoles have various importance in medicinal chemistry and material science as it acts as prevalent cores in biologically active compounds and functional materials. Considerable effort has been put into the design and development of suitable approaches for the synthesis of these scaffolds because of their special features [45]. The Alves-Paixao group developed an organocatalytic method to synthesize high yield of arylselanyl-1,2,3-triazole derivatives [46, 47]. The cycloaddition of active methylene compounds **41** with azidophenyl arylselenides **42** was performed in DMSO at room temperature in the presence of

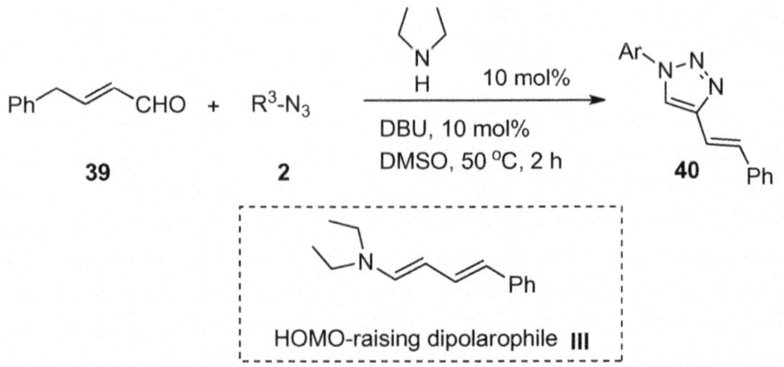

Scheme 4.16 Diethylamine, a secondary amine-catalyzed synthesis of 4-alkenyl-1,2,3-triazoles **32**

1–10 mol% of diethylamine to afford arylselanyl-1,2,3-triazole carboxylate derivatives **43**. Using this method, a variety of arylselanyl-1*H*-1,2,3-triazole-4-carboxylates were exclusively produced in good to outstanding yields involving *β*-enaminone–azide cycloaddition (Scheme 4.17) [46].

The same research group further expanded their work in 2013 by using *β*-oxo-amides **44** instead of β-ketoesters **41** which on cycloaddition reaction with azidophenyl arylselenides **45** under organocatalyzed click condition assemble the desired (arylselanyl)phenyl-1*H*-1,2,3-triazole-4-carboxamides **46**. The authors used DMSO and 1 mol% diethylamine as an organocatalyst in this reaction to develop the desired triazolyl compounds in good to exceptional yields (Scheme 4.18) [47].

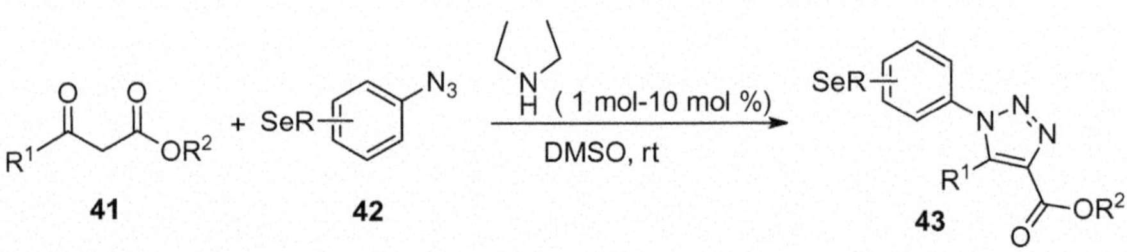

Scheme 4.17 Diethylamine, a secondary amine-catalyzed synthesis of arylselanyl-1,2,3-triazole derivatives

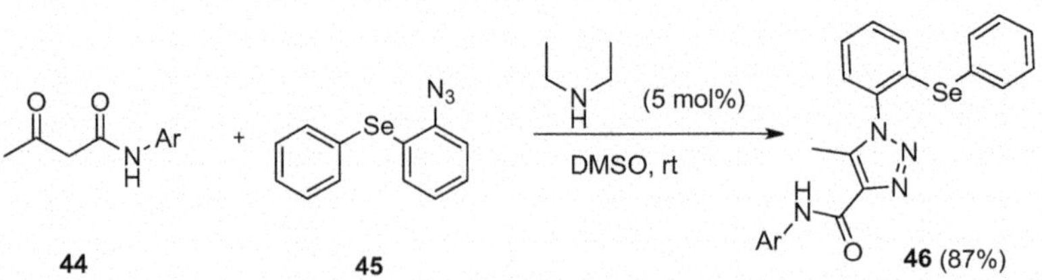

Scheme 4.18 Diethylamine-assisted synthesis of (arylselanyl)phenyl-1*H*-1,2,3-triazole-4-carboxamides

4.2.3 *Synthesis of 1,2,3-Triazole Through Enolate-Based Chemical Reactions*

Synthesis of 1,2,3-triazole through enolate-based mechanism is most common route for organocatalyzed synthesis of substituted 1,2,3-triazoles. The active methylene compounds, enolizable aldehydes, and carbonyl compounds bearing electron-withdrawing groups at the β-position were extensively used in the pathway, which involve synthesis of an enolate intermediate. The base-promoted hydrogen abstraction from suitable carbonyl compound **47** leads to the formation of enolate **I**, which on a concerted [3 + 2] cycloaddition reaction with various azides **2** afforded the 1,2,3-triazoline adduct **III**. Under ambient circumstances, the basic catalyst then encourages the quick removal of the water molecule from the adduct **III** to afford the desired substituted 1,2,3-triazole product **48** (Scheme 4.19). Ferreira et al. reported a fully detailed mechanistic study involving an organocatalyzed 1,3-dipolar cycloaddition via enolate intermediates and azide [48].

In 1902, Dimroth reported the elegant synthesis of 1,4-disubstituted 5-amino-1,2,3-triazoles **50** by the reaction of acetonitrile derivative **49** with organic azide **2** in assistance of sodium ethoxide. Reaction was carried out in boiling ethanol, and excellent yield was observed with aryl azide **2** [49a]. Reaction was sluggish for the benzyl and other alkyl azides. An improved result in yields was also obtained in assistance potassium t-butoxide as base (Scheme 4.20) [49b].

An intriguing synthetic approach for the synthesis of an array of 1-vinyl-1,2,3-triazoles **52** was developed by Hassner et al. The method involves the condensation of active methylene compounds, for example, acetonitrile derivative **49** with vinyl azides or their precursors, 3-haloalkyl azides **51**, in the presence of an alkoxide in methanol. The reaction efficiently work for the cycloaddition under optimized

Scheme 4.19 General plausible mechanism for enolate-mediated synthesis of 1,2,3-triazoles

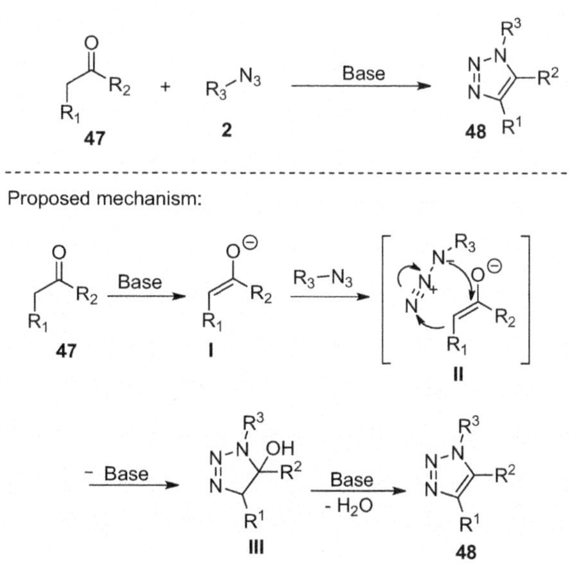

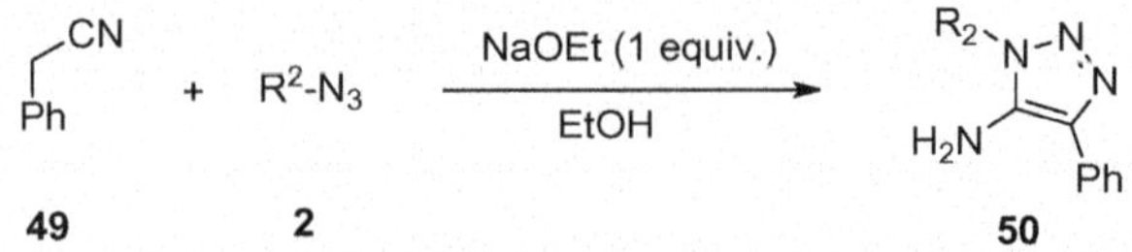

Scheme 4.20 Base promoted synthesis of 1,4-disubstituted 5-amino-1,2,3-triazoles

condition to afford 1-vinyl-1,2,3-triazoles scaffolds **52** in good-to-excellent yields (Scheme 4.21) [50].

In recent years, a number of organocatalytic methods have been discovered providing a simple access to functionalized triazoles with remarkable use. In this undertaking, Obushak et al. use the β-ketosulfones **53** and β-nitrile sulfones **55** as starting materials to demonstrate the synthesis of 1,2,3-triazoles **54** and **56,** respectively, by Dimroth cyclization. In this reaction, aryl azides **2** are combined with the sulfone reagents to furnished the 1,2,3-triazoles **54** and **56** in good yields (Scheme 4.22) [51].

Cui et al. reported an intriguing approach for the regioselective assembly of 1,2,3-triazoles **60**, employing iminoenolates and azides. In order to explore this proposal, a

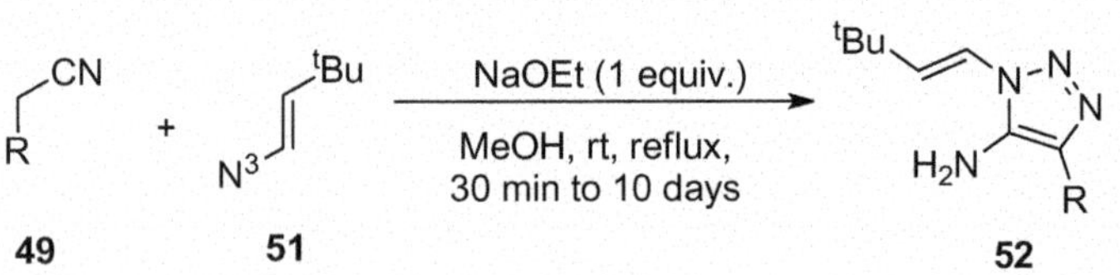

Scheme 4.21 Alkoxide-assisted synthesis of 1,4-disubstituted 5-amino-1,2,3-triazoles

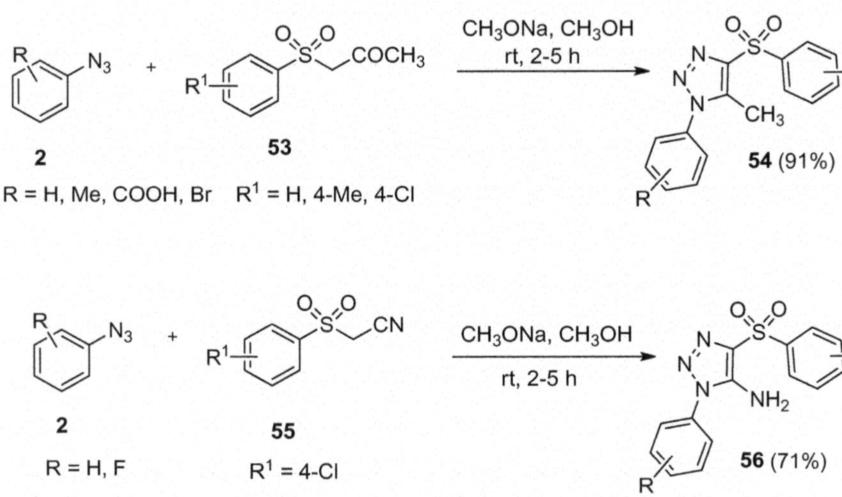

Scheme 4.22 Sodium methoxide-assisted synthesis of triazole-linked sulfones

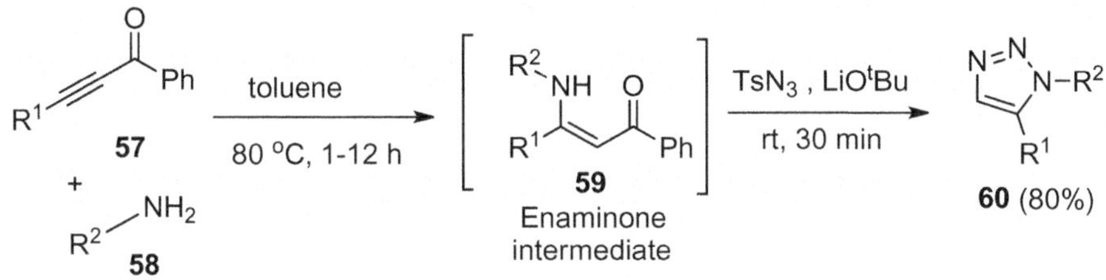

Scheme 4.23 Synthesis of 1,5-disubstituted 1,2,3-triazoles from primary amines and propynones-derived enaminone

novel approach that uses primary amine analogue **58**, substituted propynones **57**, and sulfonyl azides scaffold in a single step was devised (Scheme 4.23). This developed strategy yielded satisfactory results for amines and alkynes that were functionalized with diverse functional groups. In addition, it was feasible to efficiently synthesize the corresponding 1,5-disubstituted triazole **60** with high-to-excellent yields (71–99%) by employing several *N*-aryl enaminones **59** that were modified with electron-donating or electron-withdrawing groups. The initial stage in the likely process involves the occurrence of a Michael addition reaction between a primary amine and propynone, resulting in the formation of enaminone **59**. During the subsequent step, the compound **59** undergoes deprotonation in assistance of LiO^{t}Bu, leading to the creation of an iminoenolate intermediate. This intermediate then reacts with tosyl azide through a 1,3-dipolar cycloaddition, resulting in the development of the desired triazolyl scaffold **60** (Fig. 4.2) [52].

The enolate chemistry is an effective approach to assemble peptide chimaeras with a fixed cis-peptide mimetic configuration, since it has the potential to produce the desired 1,5-disubstituted triazoles. The Rademann research group reported a procedure for constructing peptide chimaeras with cis-locked peptide bonds containing 1,5-disubstituted triazoles under solid-phase synthesis. The process began with the C-acylation of solid-phase-appended phosphoranylidene **61** using fluorenylmethoxycarbonyl-protected leucine **62**, which afforded chain-elongated compound **63**. The compound **63** again subjected for chain elongation, with phenylalanine **64**, to afford enolate tautomers **A** and **B**. The synthesized phosphor ylide/phosphonium enolate tautomers **A** and **B** underwent cycloaddition reaction with different azides **2,** to afford 1,5-disubstituted triazole **65** (Scheme 4.24) [53].

Townsend research group explores the assembly of sugar-appended triazolyl analogue, 5-amino-l-(2,3,5-tri-*O*-acetyl-*β*-D-ribofuranosyl)-1,2,3-triazole-4-carboxamide **68** via cycloaddition reaction of cyanoacetamide **66** and sugar azide **67**. The developed compound by the group works a key intermediate in the assembly of biologically relevant 2-azacoformycin. In reaction the triazole **68** was synthesized with a 37% overall yield (Scheme 4.25) [54].

Bakulev group developed a traditional protocol for regioselective assembly of 4-acyl-5-trifluoromethyl triazoles **70** starting from 1,3-dicarbonyl compounds **69** at the same time. This process proceeds by cyclizing azide analogue **2** in assistance

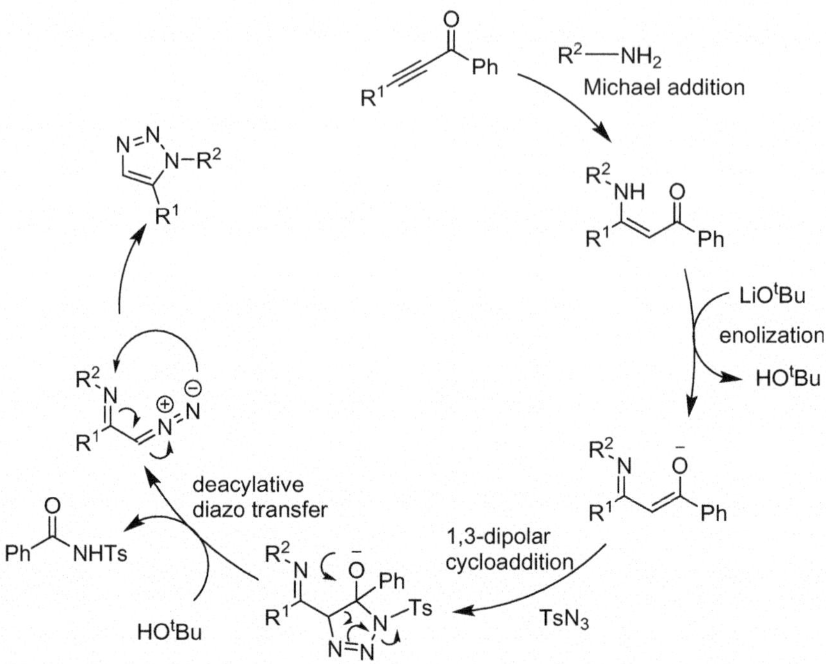

Fig. 4.2 Mechanism approach for assembly of triazolyl analogue **60**

of triethylamine with 1-trifluoromethyl-substituted 1,3-dicarbonyl compounds **69** to get triazolyl product **70** (Scheme 4.26) [55].

Ayoob Bazgir et al. reported the 1,1,3,3-tetramethylguanidine-catalyzed synthesis of 1,2,3-triazoles **73** via the cycloaddition reaction of azides **72** and CH acids **71** in ethanol at 30 °C. They used a straightforward, rapid, and organocatalyzed experimental method, under mild reaction conditions to produce good yields of 1,2,3-substituted triazole **73** (Scheme 4.27) [56].

The Ramachary group developed an efficient approach for synthesizing 1,4-disubstituted triazoles from enolizable aldehydes (Scheme 4.28). Earlier methods for synthesizing substituted triazoles from aldehydes were detailed; however, each of these protocols needs an excessive amount of base and is operated under challenging conditions [57–62]. The devised approach used 10 mol% catalytic loading of DBU to afford a range of disubstituted triazoles **75** from aldehydes **74** and azide component **2**. The reactions were completed in under 30 min at room temperature in DMSO. It was revealed that the reaction is applied to both electron-rich as well as electron-poor aromatic substituent-appended aldehydes. The catalytic cycle involves the reaction of catalyst's with the enolizable aldehyde, which furnished an enolate **I**. The approach involves a [3 + 2] cycloaddition and stepwise addition-cyclization to produce the triazoline intermediate **IV** from enolate **I** and azide **2**. A base-induced H_2O elimination as the final step affords triazolyl analogue **75** [63].

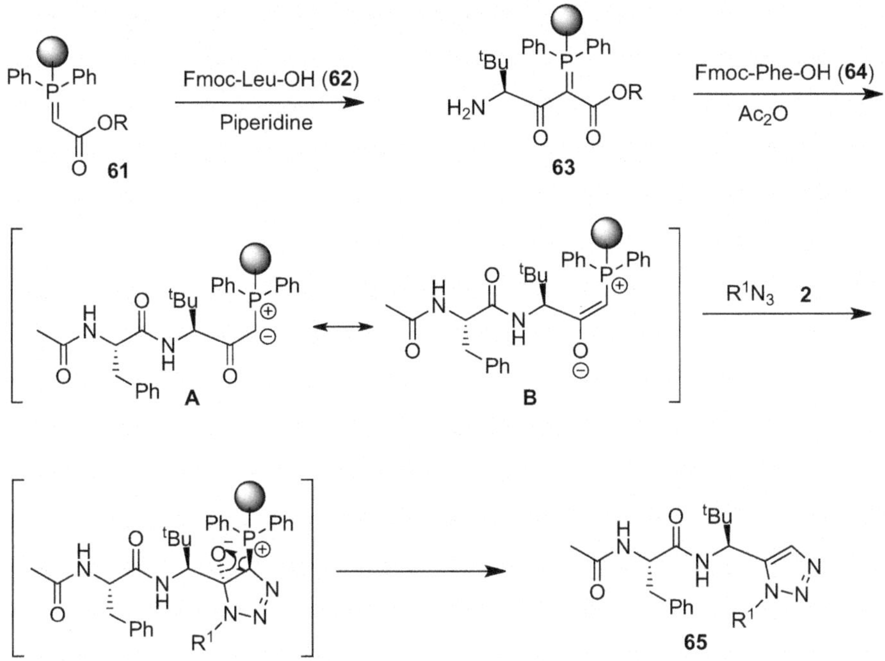

Scheme 4.24 Synthesis of 1,5-disubstituted triazoles-appended peptide chimaeras

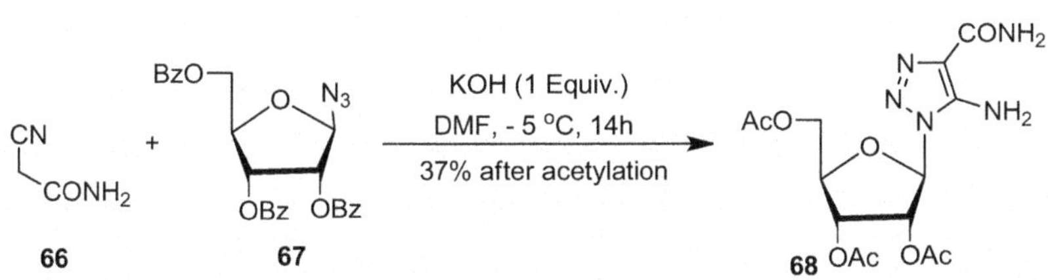

Scheme 4.25 Base-assisted cycloaddition of cyanoacetamide **66** and sugar azide **67**

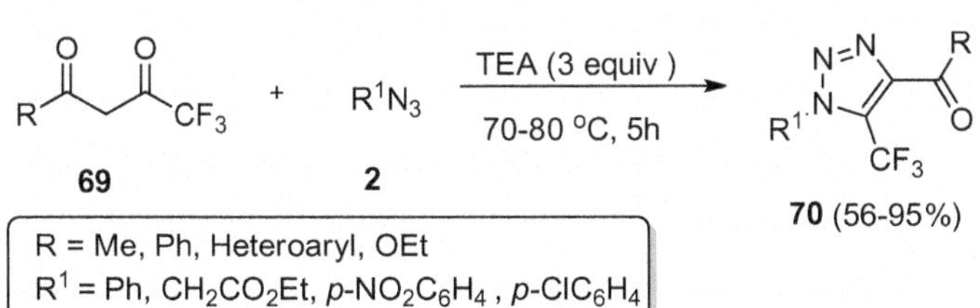

Scheme 4.26 TEA-assisted regioselective synthesis of 5-fluoroalkylated-1,2,3-triazoles

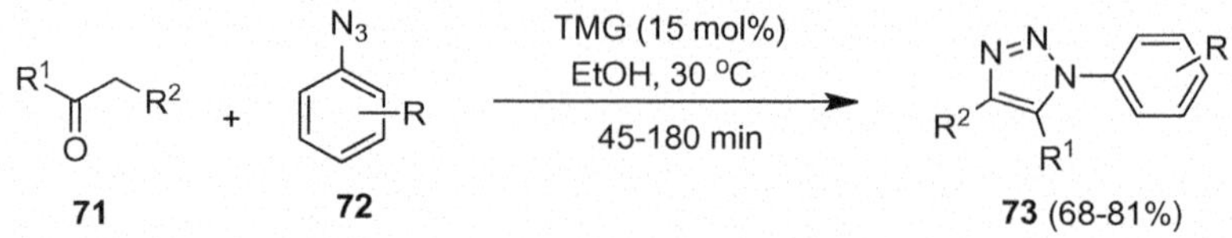

Scheme 4.27 TMG-assisted synthesis of triazole scaffolds **73**

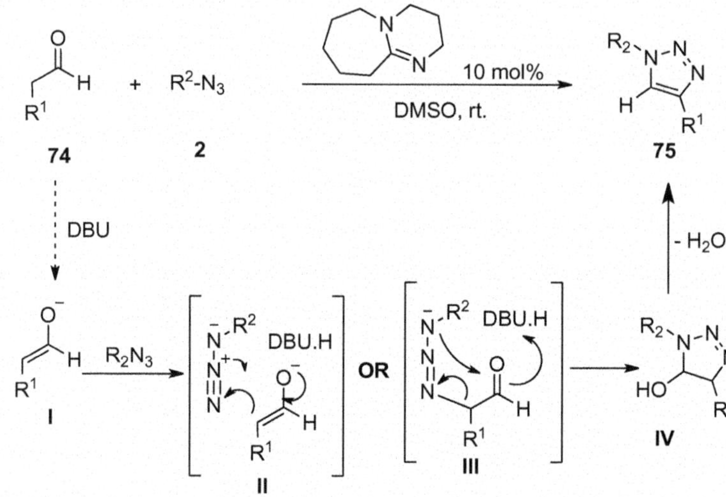

Scheme 4.28 Synthesis of substituted triazoles **75** in assistance of DBU

Furthermore, Ramachary and coworkers explored the methodology foe enolizable ketone **76** as well to assemble trisubstituted triazoles **77**. It was found that the reaction was universal with an array of aryl azides. Aliphatic azides were not examined in this reaction. The aryl group on the ketone's substituents was found to have no considerable impact on the reaction (Scheme 4.29) [64].

Wang and colleagues made a fascinating discovery regarding an organocatalytic intermolecular azide-zwitterion reaction under aerobic oxidative to assemble functionalized triazoles **79**. The reaction between an organic azide **2** and an activated alkene **78** was enabled by the presence of DBU, which acted as a catalyst. This novel

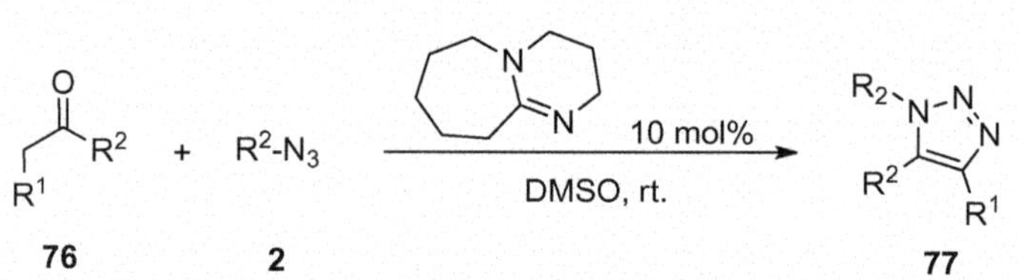

Scheme 4.29 DBU-catalyzed assembly of triazolyl **77** from enolizable ketone

approach is common for activated alkenes such as α,β-unsaturated esters, nitriles, ketones, amides, and aldehydes. Additionally, the reaction was successful with both aromatic and aliphatic azides. In the initial state of the plausible mechanism of the reaction, α,β-unsaturated ester was first reacted with DBU catalyst to furnished the required enolate (zwitterion) **I**. The zwitterion **I** on a Morita–Baylis–Hillman type reaction to the azide afforded the intermediate **II**. From intermediate **II**, three likely reaction paths were suggested. The first route utilized the S_N2 reaction of **II** to furnish intermediate **III**, that was oxidized into the triazolyl analogue **79**. The second pathway afforded the olefin intermediate **IV** by E2 elimination of DBU. The final triazole product **79** is formed via 6π electrocyclization followed by oxidative aromatization. The concluding approach was suggested to start with aerobic oxidation of intermediate **II** and proceeds to the intermediate olefin **V**, which was on 6π electrocyclization gave the intermediate **VI**, followed by subsequent elimination of DBU furnished the desired 1,4,5-functionalized 1,2,3-triazole **79** (Scheme 4.30) [65].

Alves et al. demonstrated the synthesis of fused [1,2,3]triazolo[1,5-a]quinolines using a one-pot sequential organocatalytic approach. The triazole intermediates **82** are created in the method by the reaction between active methylene compound **80** and o-carbonyl-substituted phenyl azide **81**.

The reactions were carried out using 20 mol% of DBU as a catalyst in DMSO and were able to tolerate an array of 1,3-dicarbonyl compounds (Scheme 4.31) [66].

In addition, the same group suggested a fascinating method for synthesis of fused triazole analogue **85**. The synthetic approach incorporates a two-stage process. In the first step, organocatalysis is used using 10 mol% of DBU. In the second step, an inorganic base, either 1.2 or 0.1 equiv. of potassium hydroxide, is employed. The reactions were conducted using O-carbonyl phenylazides **83** and β-keto amides **84** as reactants in DMSO as the solvent at a temperature of 70 °C (Scheme 4.32) [67].

The same group further extended the organocatalytic route to develop a number of derivatives of 1,2,3-triazoyl-zidovudine **88**. These compounds were demonstrated for primary antioxidant activity and synthesizing through cycloaddition reaction of zidovudine **86** with various functionalized keto compounds **87**, such as β-ketoesters, β-ketoamides, β-ketonitriles, and β-ketosulfones, in the presence of a 10 mol % of DBU in DMSO at 70 °C for 25 hours (Scheme 4.33) [68].

Ramachary et al. significantly contribute and explore the organocatalyzed route to assemble functionalized 1,2,3-triazole analogues. In addition to utilization of DBU in organocatalyzed reaction, the group also report a carbonate-catalyzed process to synthesize high yield of 5-amino-1,2,3-triazoles **91**. The title compounds **91** were obtained in good yields by reacting substituted acetonitriles **89** with azides **90** at room temperature for 0.5–2 h in the presence of 10 mol% Cs_2CO_3 (Scheme 4.34). With extremely strong regioselectivity, the 5-amino-1,2,3-triazoles **91** was formed. On the phenyl ring of the aryl acetonitriles, it was discovered that the reaction was tolerant of a variety of substituents. The products were produced in yields that ranged from good to outstanding using different substituted aryl azides [69]

Moreover, an enolate-based organocatalytic approach was developed to synthesize various substituted triazoles by the reaction of enones with azides. Reaction efficiently assisted by 20 mol% catalytic loading of DBU at room temperature to

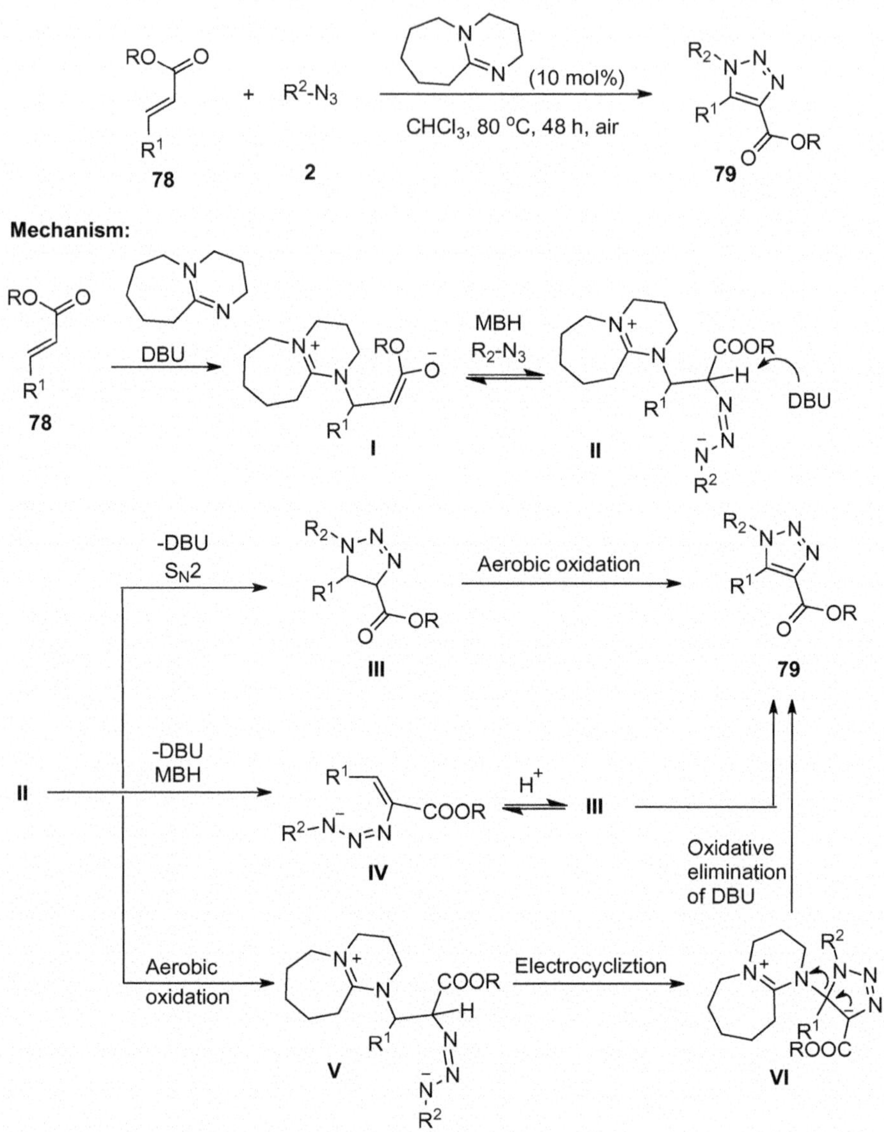

Scheme 4.30 DBU-catalyzed assembly of functionalized triazoles **79**

afford an array of vinyl-appended triazolyl scaffold **94** (Scheme 4.35). The developed compound **94** is preferred building blocks for a variety of applications from medicinal to synthetic chemistry [70].

Recently, Ramachary et al. developed a method using organic catalysts to combine azides and ketones in a [3 + 2] reaction. The study reports the cycloaddition of various 1-aryl-3-(arylthio)propan-2-ones or 1-alkyl-3-(arylthio)propan-2-ones **95**

Scheme 4.31 DBU-catalyzed assembly of functionalized triazoles **82**

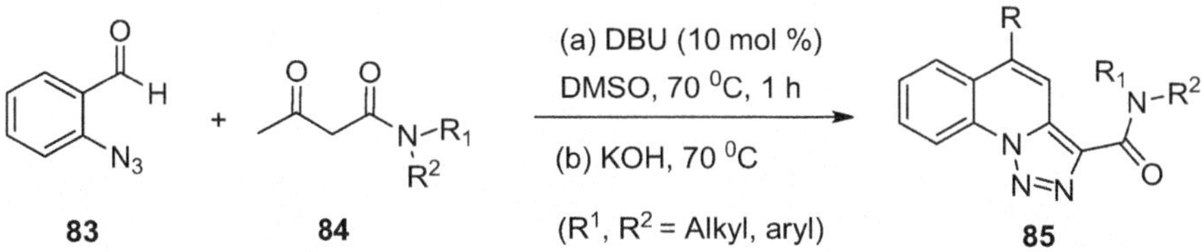

Scheme 4.32 DBU-catalyzed synthesis of [1,2,3]Triazolo[1,5-a]quinoline-3-carboxamides **85**

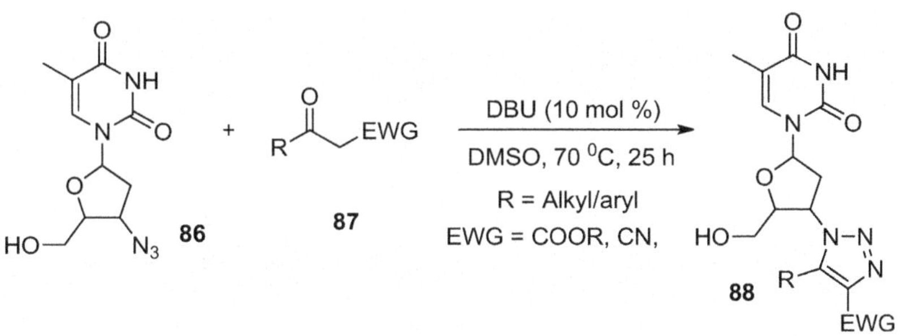

Scheme 4.33 DBU-catalyzed synthesis of 1,2,3-triazoyl-zidovudine **88** as primary antioxidant

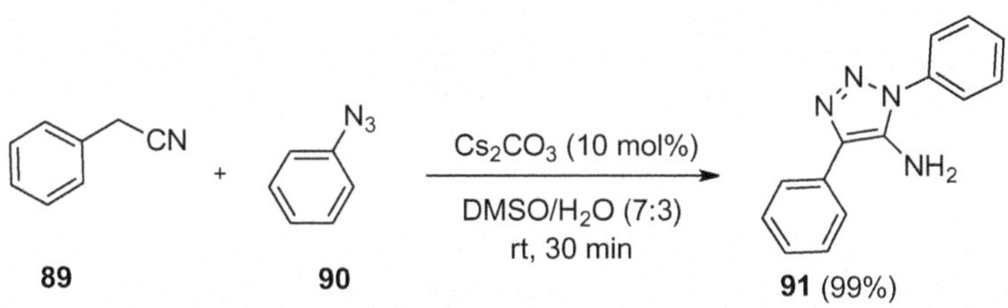

Scheme 4.34 Cs_2CO_3-catalyzed assembly of functionalized triazolyl analogue **91**

with different azide analogues **96**. This reaction takes place under normal conditions and results in the formation of the medically important compounds **97** and **98**, respectively (Scheme 4.36). The reaction has some notable features including regioselectivity and high yields/rates. In the beginning of the catalytic synthesis of substituted triazole, the thermodynamic enolate **I** resulting from the α-methylene

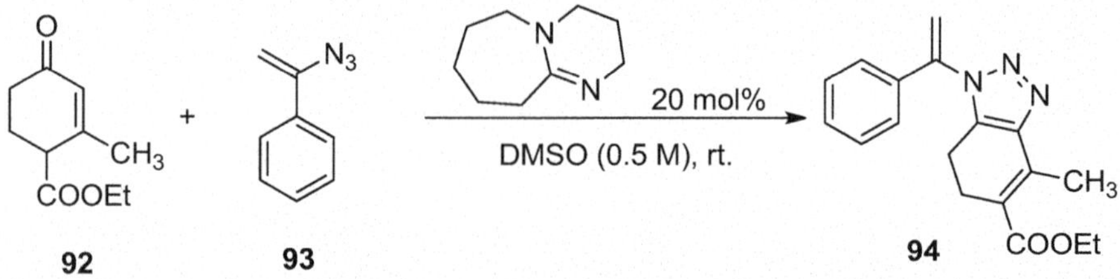

Scheme 4.35 DBU-catalyzed synthesis of functionalized C–N vinyl triazoles **40**

group as opposed to the ′α-methylene group is formed by the non-symmetric ketone **95** in the presence of DBU as the catalyst. The kinetic enolate **II** is produced as a result of a [1,3]-hydride shift if R1 is an aryl group, and it is thought that it is in dynamic equilibrium with the thermodynamic enolate. The kinetic enolate **II** combines with the azide **96** to make the intermediate **IV**, which then produces **97** following the removal of water when R1 is the aryl group. As and when the kinetic enolate **II** reacts, the equilibrium should also have been steadily shifting toward the right side, as shown by the 1,2,3-triazole **98** produced with a regioselectivity of > 20:1. In contrast, if R1 is an alkyl group, the thermodynamic enolate **III** instantly reacts with the azide **96** to form the intermediate **V**, which, when water is eliminated, yields the 1,2,3-triazole **98** (Fig. 4.3). The author suggests that the R^1 alkyl group is unable to cause the [1,3]-hydride shift during the enolization to produce the kinetic enolate **II** in situ. This may be because α-methylene is less acidic than the alkyl group, which may be the cause [71].

Koppula et al. explore to develop triazolyl pyrimidines **102** and **103** as effective medicinally important scaffolds in anticancer and EGFR inhibition. They were utilized [3 + 2] organocatalytic cycloaddition reaction of compound **99** with a number of azides **100** and **101** in assistance of catalytic concentrations of pyrrolidine (Scheme 4.37) [72].

Reddy et al. reported that the TBD catalyzes the cycloaddition of an array of activated carbonyl compounds **104** with vinyl azides **105** to assemble triazolyl analogue **106** (Scheme 4.38). The developed vinyl-substituted triazoles on hydrogenation reaction furnish the 1,4,5-trisubstituted *N*-alkyl-1,2,3-triazole as well [73].

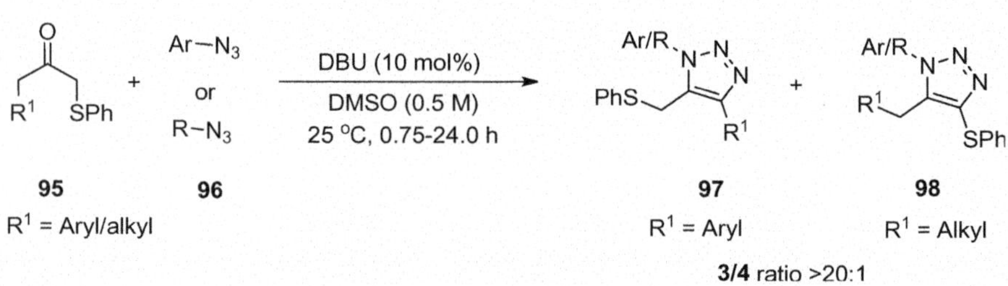

Scheme 4.36 DBU-catalyzed synthesis of functionalized triazoles **97** and **98**

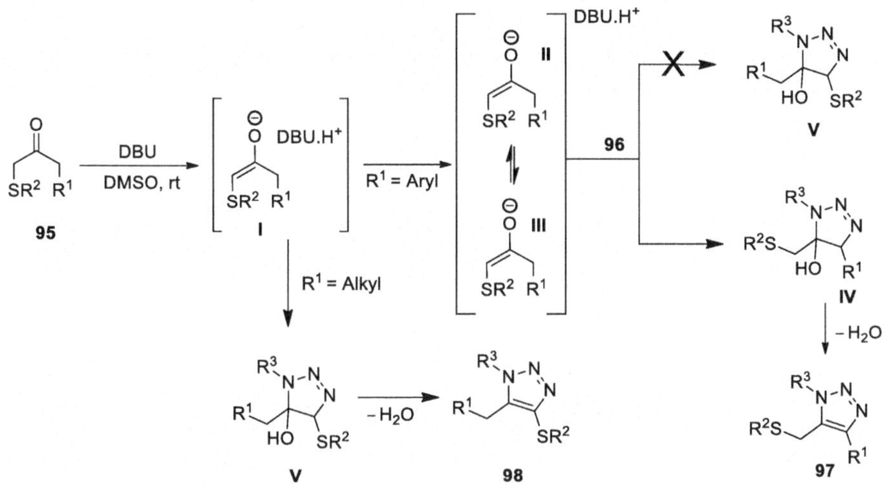

Fig. 4.3 Plausible mechanism for DBU-catalyzed synthesis of functionalized triazoles **97** and **98**

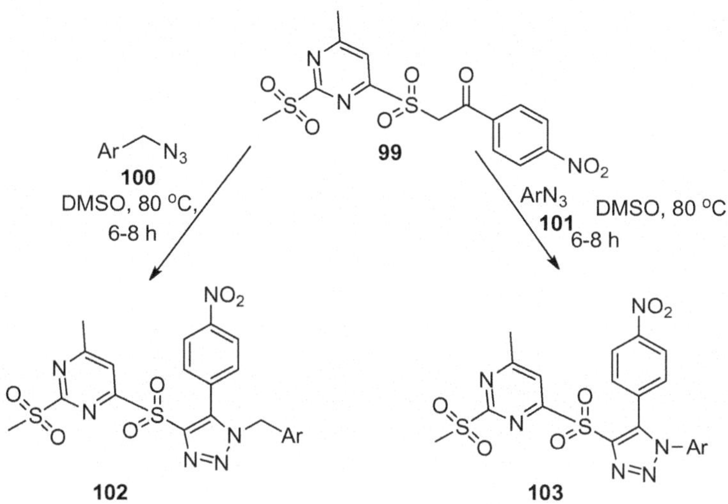

Scheme 4.37 DBU-assisted assembly of functionalized triazoles **102** and **103**

4.2.4 Synthesis of 1,2,3-Triazole Through Acetylide-Based Intermediate

The straightforward synthetic processes and excellent yields, metal-free [3 + 2] cycloaddition reactions, Diels–Alder reactions, and thiol-alkene radical addition reactions have become popular click reactions. Alkynes are acting as the most common dipolarophiles in metal-free 1,3-dipolar cycloadditions [74]. In the earlier

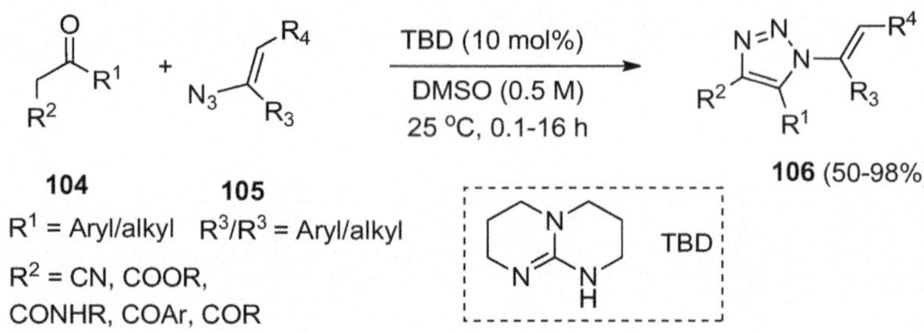

Scheme 4.38 TBD-catalyzed synthesis of*N*-vinyl-1,2,3-triazoles **106**

studies, only strained alkynes or high temperatures could be used to successfully complete the reactions involving alkynes [75]. Although ongoing research has led to alternate cyclization methods to assemble 1,2,3-triazoles, alkynes are still routinely used to synthesize these heterocyclic scaffolds in present scenario of synthetic chemist [76–78]. An effective method for synthesis of 1,5-disubstituted-4-(trimethyl-silyl)-1*H*-1,2,3-triazoles **109** was described by Schubert et al. In aqueous media, aromatic azides **107** and silylated alkynes **108** undergo a [3 + 2]-cycloaddition process. The trimethylsilyl (TMS) group in alkynes greatly favored the 1,5-isomer's regiopreference in the reaction (Scheme 4.39) [78].

Fokin et al. developed a methodology for exclusive formation of 1,5-disubstituted triazoles using alkynes, reaction accomplished formation of reactive acetylide intermediates in situ (Scheme 4.40) [79]. After screening a number of bases, it was revealed that NMe_4OH was the selected base due to the favorable yield. The range of the reaction revealed that aryl azides with sterically challenging ortho-substituents and aryl azides lacking electrons both provided lower yields. It is interesting that base-labile capabilities were found to be tolerated by this devised protocol. The reversible deprotonation of the terminal alkyne **43**, which results in the production of an aryl acetylide, starts the plausible catalytic cycle. The acetylide intermediate **I** afterward works as a nucleophile, attacking the azide component **2**, forming a

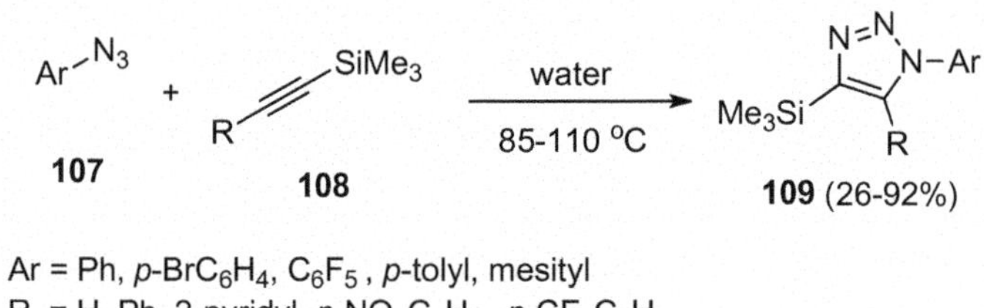

Scheme 4.39 Synthesis of 1,5-disubstituted-4-(trimethyl-silyl)-1*H*-1,2,3-triazoles **109** through [3 + 2] cycloaddition reaction using silylated alkynes

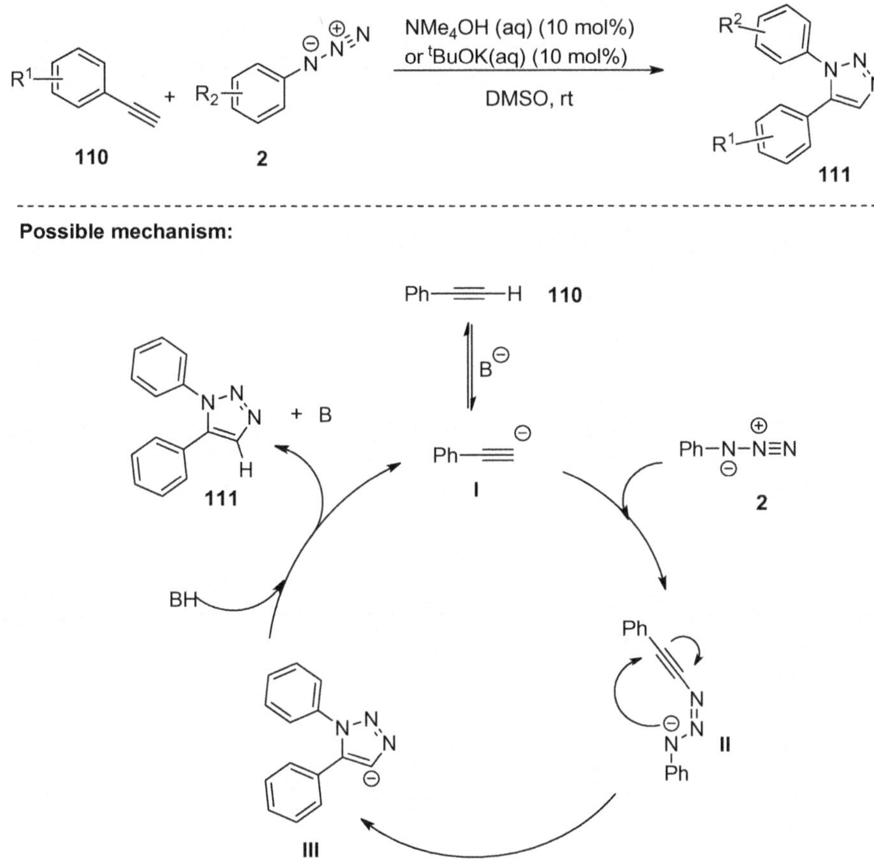

Scheme 4.40 Base-catalyzed synthesis of functionalized triazoles **111**

triazenide component **II**, which undergoes either a 6π-electrocyclization or a 5-endo-dig cyclization to form triazolyl anion **III**. The developed triazolyl anion **III** on protonation afforded the final product **44**.

Similarly to this approach, Lin et al. devised the assembly of 1,5-disubstituted 1,2,3-triazolyl analogue **113** through reaction of phenyl azide with vinyl halogenides **112** in presence of t-butoxide base in DMF (Scheme 4.41) [80].

4.2.5 Triazoles Via Miscellaneous Chemical Reaction

Tiwari et al. reported an interesting methodology to synthesize triazolyl glycoconjugates **116** using cycloaddition reaction of nitroolefins **114** with a range of azides **115** in assistance of Schreiner thiourea organocatalysts **I**. Using the devised

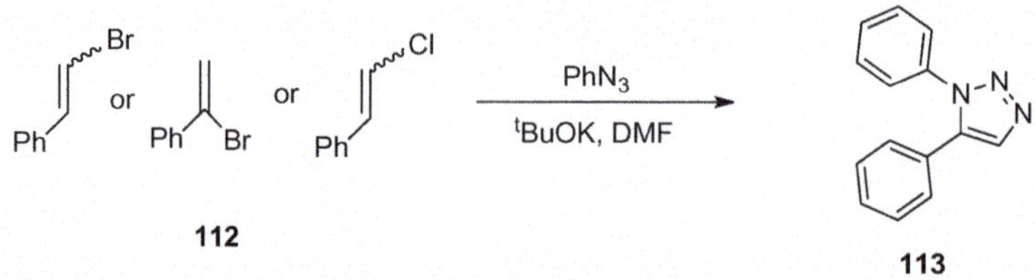

Scheme 4.41 Base-catalyzed synthesis of functionalized triazoles **113**

protocol, the group developed various triazolyl glycoconjugates including bis-triazolyl glycohybrids characterized **118** with the 1,4 and 1,5 substituted triazole as well (Scheme 4.42). The approach has some noticeable characteristics, such as excellent regioselectivity, operational simplicity, metal-free, straightforwardness, acid-free, broad substrate scope, and high yielding. The double hydrogen bonding provided by the Schreiner thiourea organocatalysts **I** allows for the necessary activation of nitroolefins [81].

Pathak et al. explore the organocatalyzed route for synthesis of various triazolyl glycoconjugates, and in the way, the group was developed an interesting metal-free approach for the regioselective synthesis of 1,5-disubstituted triazolylated pyranosides **121** and 1,5-disubstituted triazole-linked disaccharides. The exocyclic vinyl sulfones **120a** and **120b** derived from styrene epoxide and monotosylated glycerol were put through a reaction with a variety of C1, C2, C3, C4, and C6 azidopyranosides **119** in water at a high temperature, and the result was corresponding C1, C2, C3, C4, and C6 triazolylated monosaccharides **121a-f** in good-to-excellent yield (Scheme 4.43) [82].

The same group further enhanced and expanded their synthetic method to develop a metal-free, one-pot, three-component approach for the highly efficient synthesis of

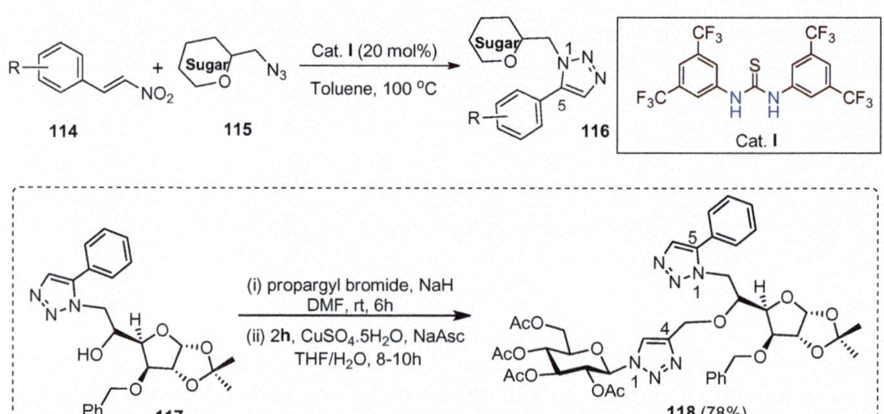

Scheme 4.42 Schreiner thiourea-catalyzed synthesis of 1,5-disubstituted 1,2,3-triazolyl glycoconjugates

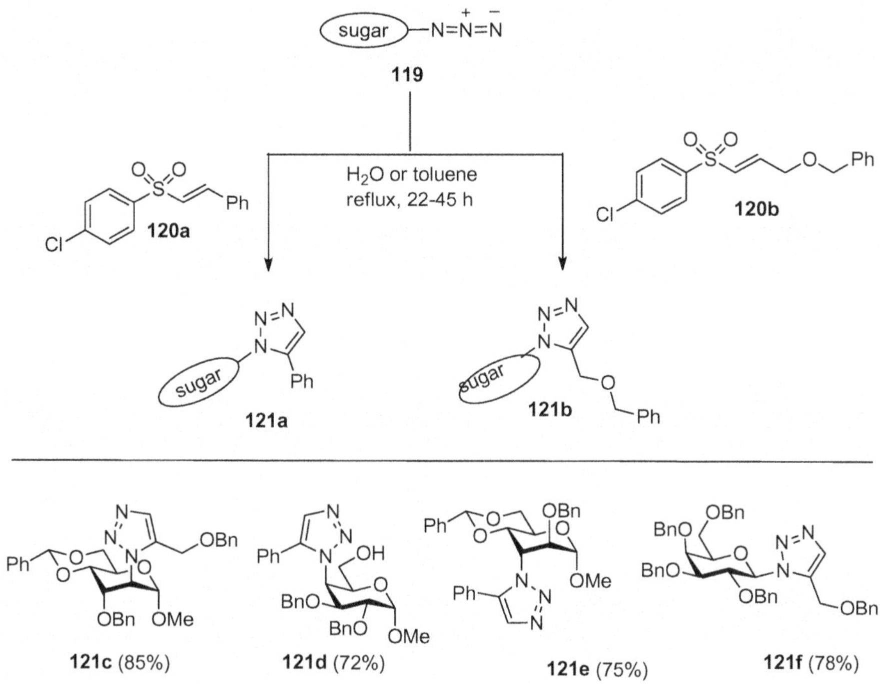

Scheme 4.43 Efficient regioselective synthesis of 1,5-disubstituted 1,2,3-triazolylated sugars

1,5-disubstituted triazoles. Three component reactions of vinyl sulfones, sodium azide, and alkyl bromides, tosylates, mesylates, or aryl amines are used in the process, which is carried out in DMF at 100 °C. In order to assemble the appropriate 1,5-disubstituted triazoles, the organic azides produced in situ from the interaction of sodium azide and alkyl halides were allowed to react with vinyl sulfones in a regioselective way [83]. Next, in subsequent investigation, the same group made noteworthy effort to extend a metal-free approach for the efficient assemblance of 1,5-disubstituted triazolyl glycoconjugates with great ease in aqueous medium. The author reports, the four different azido furanosides **122** undergo 1,3-dipolar cycloaddition reaction with vinyl sulfones **124** and vinyl sulfone containing carbohydrate molecules **123** in aqueous media under reflux condition to assemble corresponding 1,5-disubstituted 1,2,3-triazolylated carbohydrates **126** and **125** (Scheme 4.44). In addition to 1,5-disubstituted 1,2,3-triazolylated carbohydrates, the triazolyl-linked difuranosides were also prepared by the reaction of azidofuranosides **127** with vinyl sulfones derived from gluco- and allo furanosides under the similar optimized conditions [84].

Westermann and co-workers reported an azide-free synthesis of a library of 1,4,5-functionalized 1,2,3-triazoles. A variety of α,α-tosylhydrazones **128** could be utilized to react with a wide range of aromatic amines as well as aliphatic amines un presence of DIPEA at room temperature to afford high yield of functionalized triazole **129** (Scheme 4.45) [85].

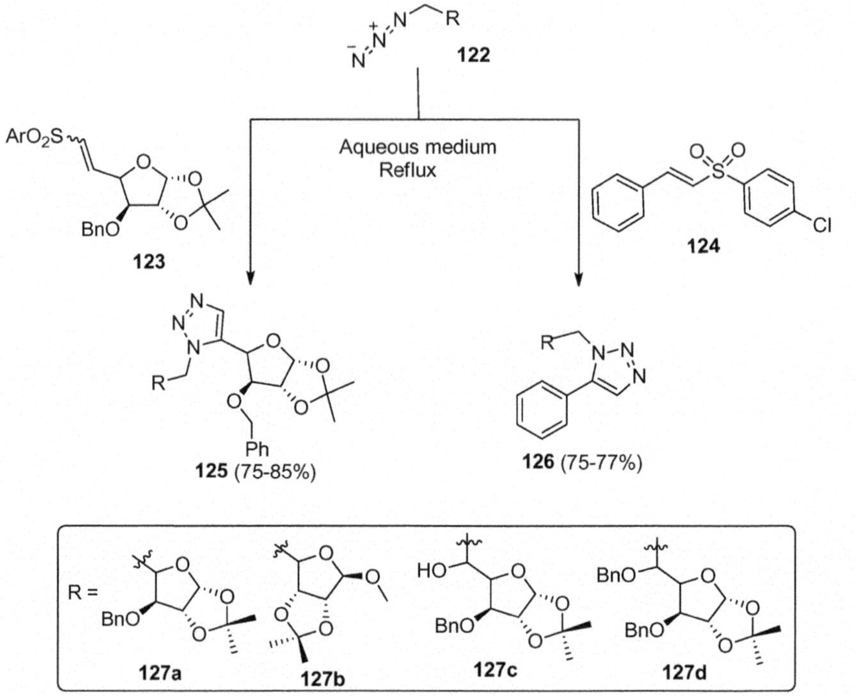

Scheme 4.44 Efficient synthesis of 1,5-disubstituted 1,2,3-triazolyl glycoconjugates **125** and **126** in aqueous medium

Scheme 4.45 Efficient synthesis of 1,5-disubstituted 1,2,3-triazole from α,α-tosylhydrazones

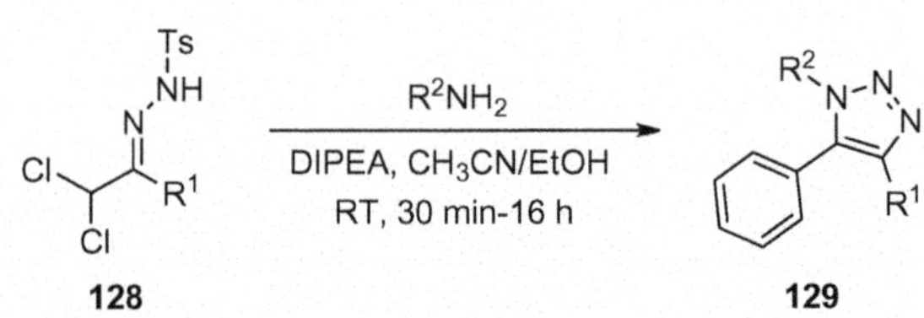

Guan et al. reported an azide-free technique for the synthesis of 1,4-disubstituted triazole **132**; they deploy *N*-tosylhydrazones along with I_2/TBPB to promote the reaction, which sets their strategy apart from the others (Scheme 4.46). The devised protocol showed broad functional group tolerance when applied to a variety of anilines **131** and *N*-tosylhydrazones **130**. The plausible mechanism for this transformation begins with the reaction of *N*-tosylhydrazone **130** with molecular iodine, which afford the intermediate **I,** through the loss of HI. HI is converted back to I_2 by TBPB in a concurrent catalytic cycle. The intermediate **I** then yields azoalkene **II**, which on reaction with toluidine gave **III**. Next, intermediate **III** via intramolecular cyclization gave **IV**. The ultimate 1,4-disubstituted 1,2,3-triazole **132** is produced by further oxidation of **IV** [86].

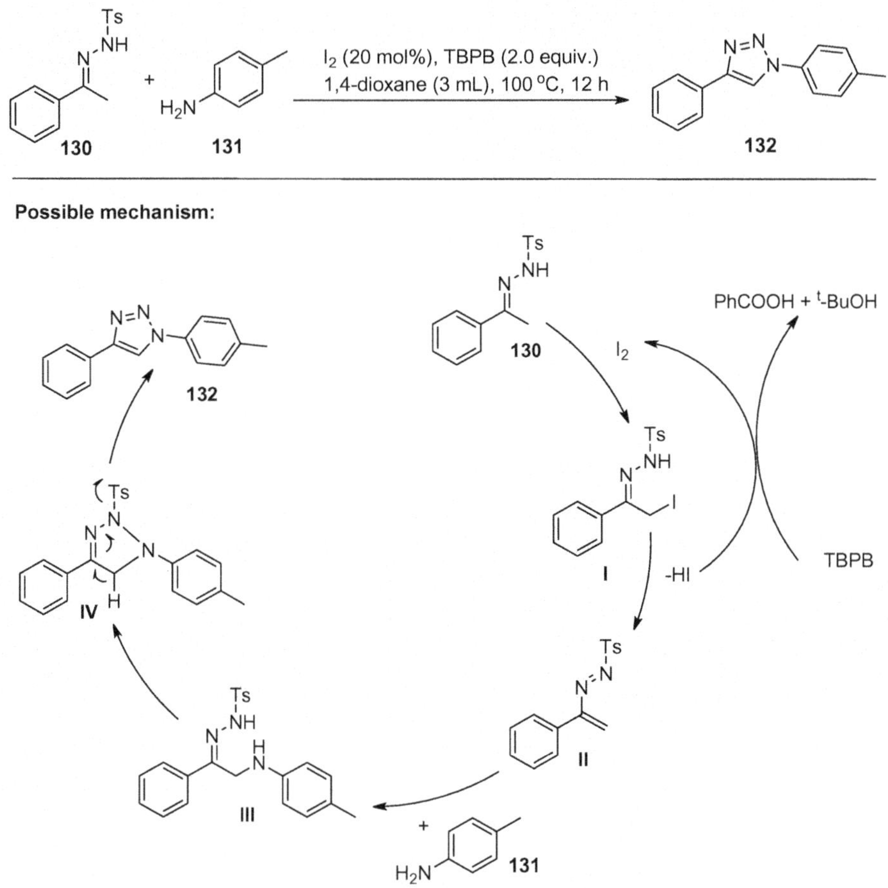

Scheme 4.46 Plausible mechanism and synthesis of 1,5-disubstituted 1,2,3-triazole from *N*-tosylhydrazones

4.3 Conclusions and Future Outlook

The copper(I)-catalyzed azide-alkyne cycloaddition reaction or click reaction has long been the preferred protocol for assembling the 1,4-disubstituted triazoles. Even though the CuAAC reaction is still extensively used in the synthesis of many 1,2,3-triazoles that are relevant to biology, new approaches have arisen and have proven to be effective replacements. The introduction of the strain-promoted azide-alkyne cycloaddition reaction, which commonly uses functionalized cyclooctynes as the preferred dipolarophile, has been a turning point in the development of so-called copper-free techniques. Other forms of reactions outside 1,3-dipolar cycloadditions have been researched in order to produce highly different triazoles, in addition to the use of numerous dipolarophiles (such as enamines, enolates, vinyl sulfones, peptidyl phosphoranes, and iminoenolates). These various dipholarophiles have the potential to develop into the dominant method for producing metal-free triazoles in the coming generation when used in so-called organocatalytic (3 + 2)-cycloaddition reactions. The most often studied dipolarophiles at the moment are enamines/enaminones and enolates because they can be easily made by condensation of amines and carbonyl compounds, which can be classified as activated or unactivated. The application of these dipolarophiles is confined to the synthesis of 1,4,5-trisubstituted 1,2,3-triazole, while showing considerable potential, and the majority of procedures still only allow the use of aryl azide. Although not based on the (3 + 2)-cycloaddition process, other promising techniques include the Sakai reaction, which involves the condensation of a primary amine and a tosylhydrazone that has been functionalized. The Sakai reaction, which typically offers high yields and high regioselectivities, is a fantastic alternative for the synthesis of 1,4-disubstituted triazole. Although intriguing new techniques for the synthesis of 1,2,3-triazoles have emerged in recent years, there is still much that can be done to optimize reaction conditions and substrate range. The latter will be especially important in the development of procedures that will allow reactions to occur under physiological settings, opening up new study fields.

References

1. Muller T, Braese S (2011) Angew Chem Int Ed 50:11844–11845
2. Tiwari VK, Mishra BB, Mishra KB, Mishra N, Singh AS, Chen X (2016) Chem Rev 116:3086–3240
3. Agrahari AK, Bose P, Jaiswal MK, Rajkhowa S, Singh AS, Hotha S, Mishra N (2021) Tiwari. Chem Rev 121:7638–7956
4. Hou J, Liu X, Shen J, Zhao G, Wang PG (2012) Expert Opin Drug Discov 7:489
5. Moses JE, Moorhouse AD (2007) Chem Soc Rev 36:1249
6. Lau YH, Rutledge PJ, Watkinson M, Todd MH (2011) Chem Soc Rev 40:2848–2866
7. Astruc D, Liang L, Rapakousiou A, Ruiz J (2012) Acc Chem Res 45:630–640
8. Soltis MJ, Yeh HJ, Cole KA, Whittaker N, Wersto RP, Kohn EC (1996) Drug Metab Dispos 24:799–806
9. Feng LS, Zheng MJ, Zhao F, Liu D (2021) Arch Pharm 354:e2000163
10. Rani A, Singh G, Singh A, Maqbool U, Kaur G, Singh J (2020) RSC Adv 10:5610–6563
11. Mishra A, Mishra N, Tiwari VK (2016) Curr Org Synth 1:176–219
12. Mishra N, Singh AS, Agrahari AK, Singh SK, Singh M, Tiwari VK (2019) ACS Comb Sci 21:389–399
13. Wojaczynska E, Steppeler F, Iwan D, Scherrmann M-C, Marra A (2021) Molecules 26:7291
14. Mishra N, Singh SK, Singh AS, Agrahari AK, Tiwari V (2021) K. J Org Chem 86:17884–17895
15. Singh SK, Mishra N, Kumar S, Jaiswal MK, Tiwari VK (2022) Chemistry Select 7:e202201314
16. Singh SK, Kumar S, Yadav MS, Tiwari VK (2022) J Org Chem 87(22):15389–15402
17. Singh SK, Kumar S, Yadav MS, Gupta A, Tiwari VK (2024) J Org Chem 88:13440–13453
18. Singh SK, Kumar S, Yadav MS, Bhattacharya S, Tiwari VK (2024) Synthesis 56:975–988
19. Opsomer T, Dehaen W (2021) Chem Commun 57:1568–1590
20. Chu C, Liu R (2011) Chem Soc Rev 40:2177–2188
21. Kharb R, Sharma PC, Yar MS (2011) J Enzyme Inhib Med Chem 26:1–21
22. John J, Thomas J, Dehaen W (2015) Chem Commun 51:10797–10806
23. Anebouselvy K, Ramachary DB (2016) Synthesis of substituted 1, 2, 3-Triazoles through organocatalysis. In: Click reactions in organic synthesis. Wiley VCH, pp 99–139

24. Ahmed M, Razaq H, Faisal M, Siyal AN, Haider A (2017) Synth Commun 47:1193–1200
25. Anebouselvy K, Ramachary DB (2018) In: Dienamine catalysis for organic synthesis. The Royal Society of Chemistry, pp 53–64
26. De Nino A, Maiuolo L, Costanzo P, Algieri V, Jiritano A, Olivito F, Tallarida MA (2021) Catalysts 11:1120
27. Shiri P, Amani AM, Mayer-Gall T (2021) Beil J Org Chem 17:1600–1628
28. Vala DP, Vala PM, Patel HM (2022) ACS Omega 7:36945–36987
29. Li W, Du Z, Zhang K, Wang J (2015) Green Chem 17:781–784
30. Li W, Zhou X, Luan Y, Wang J (2015) RSC Adv 5:88816
31. List B (2004) Acc Chem Res 37:548–557
32. Meilahn MK, Cox B, Munk ME (1975) J Org Chem 40:819–824
33. Bianchetti G, Croce PD, Pocar D (1965) Tetrahedron Lett 25:2039–2041
34. Bianchetti G, Pocar D, Croce PD, Gallo GG, Vigevani A (1966) Tetrahedron Lett 15:1637–1642
35. Brunner M, Maas G, Klarner FG (2005) Helv Chim Acta 88:1813–1825
36. Ramachary DB, Ramakumar K, Narayana VV (2008) Chem Eur J 14:9143–9147
37. Ramachary DB, Shashank AB (2013) Chem Eur J 19:13175–13181
38. Danence LJT, Gao Y, Li M, Huang Y, Wang J (2011) Chem Eur J 17:3584–3587
39. Demchuk DV, Samet AV, Chernysheva NB, Ushkarov VI, Stashina GA, Konyushkin LD, Raihstat MM, Firgang SI, Philchenkov AA, Zavelevich MP, Kuiava LM, Chekhun VF, Blokhin DY, Kiselyov AS, Semenova MN, Semenov VV (2014) Bioorg Med Chem 22:738–755
40. Li W, Du Z, Huang J, Jia Q, Zhang K, Wang J (2014) Green Chem 16:3003–3006
41. Belkheira M, Abed DE, Pons J-M, Bressy C (2011) Chem Eur J 17:12917–12921
42. Wang L, Peng SY, Danence LJT, Gao Y, Wang J (2012) Chem Eur J 18:6088–6093
43. Yeung DKJ, Gao T, Huang J, Sun S, Guo H, Wang J (2013) Green Chem 15:2384–2388
44. Li W, Jia Q, Du Z, Wang J (2013) Chem Commun 49:10187–10189
45. Alves D, Goldani B, Lenardão EJ, Perin G, Schumacher RF, Paixão MW (2018) Chem Rec 18:527
46. Seus N, Goncalves LC, Deobald AM, Savegnago L, Alves D, Paixao MW (2012) Tetrahedron 68:10456–10463
47. Seus N, Goldani B, Lenardao EJ, Savegnago L, Paixao MW, Alves D (2014) Eur J Org Chem, pp 1059–1065
48. Neto AC, Soares KC, Santos M da-S, Aímola TJ, Ferreira AG, Jardim GAM, Tormena CF, Paixão MW, Ferreira MAB (2022) Org Biomol Chem 20:6019–6026
49. (a) Dimroth O (1902) Ueber eine Synthese von Derivaten des 1.2.3-Triazols. Ber Dtsch Chem Ges 35:1029–1038; (b) Lieber E, Rao CNR, Rajkumar TV (1959) J Org Chem 24:134–135
50. Labbe G, Hassner A (1970) J Heterocycl Chem 7:361–366
51. Pokhodylo NT, Matiychuk VS, Obushak MD (2009) Synthesis 14:2321–2323
52. Cheng G, Zeng X, Shen J, Wang X, Cui X (2013) Angew Chem Int Ed 52:13265–13268
53. Ahsanullah P, Kuhne SR, Rademann J (2009) J Angew Chem Int Ed 48:5042–5045
54. Acevedo OL, Krawczyk SH, Townsend LB (1986) J Org Chem 51:1050–1058
55. Rozin YA, Leban J, Dehaen W, Nenajdenko VG, Muzalevskiy VM, Eltsov OS, Vasiliy BA (2012) Tetrahedron 68:614–618
56. Ahmadi F, Tisseh ZN, Dabiri M, Bazgir ACR (2013) Chim 16:1086–1090
57. Fusco R, Bianchetti G, Pocar GD (1961) Gazz Chim Ital 91:849–865
58. Bianchetti G, Croce PD, Pocar D, Gallo GG (1965) Rend Ist Lombardo Sci Lettere A 99:296–315
59. Almirante N, Forti LJ (1984) Heterocyclic Chem 21:1121–1123
60. Gilchrist TL, Gymer GE (1974) Adv Heterocycl Chem 16:33–85
61. Wamhoff H (1984) Compr. Heterocycl. Chem. 5:669–732
62. Di Nunno L, Scilimati A (1986) Tetrahedron 42:3913–3920
63. Ramachary DB, Shashank AB, Karthik S (2014) Angew Chem Int Ed 53:10420–10424
64. Shashank AB, Karthik S, Madhavachary R, Ramachary DB (2014) Chem Eur J 20:16877–16881
65. Li W, Wang J (2014) Angew Chem Int Ed 53:14186–14190

66. da Costa GP, Bach MF, de Moraes MC, Barcellos T, Lenardão EJ, Silva MS, Alves D (2020) Adv Synth Catal 362:5044
67. Da Costa GP, Blödorn GB, Barcellos T, Silva MS, Luque R, Alves D (2023) Eur J Org Chem 26:e202300604
68. Gomes CB, Balaguez RA, Larroza A, Smaniotto TA, Domingues M, Casaril AM, Silva MS, Rodrigues OED, Savegnago L, Alves D (2020) ChemistrySelect 5:12255
69. Krishna PM, Ramachary DB, Peesapati S (2015) RSC Adv 5:62062–62066
70. Reddy GS, Kumar AS, Ramachary DB (2020) Org Biomol Chem 18:4470–4478
71. Reddy GS, Reddy LM, Kumar AS, Ramachary DB (2020) J Org Chem 85:15488–15501
72. Velidandla JMR, Koppula SK (2023) Polycycl Aromat Compd. https://doi.org/10.1080/10406638.2023.2220864
73. Ramachary DB, Gujral J, Peraka S, Reddy GS (2017) Eur J Org Chem, 459–464
74. Becer CR, Hoogenboom R, Schubert US (2009) Angew Chem Int Ed 48:4900–4908
75. Ess DH, Jones GO, Houk KN (2008) Org Lett 10:1633–1636
76. Zhang H, Tanimoto H, Morimoto T, Nishiyama Y, Kakiuchi K (2013) Org Lett 15:5222–5225
77. Guggenheim KG, Toru H, Kurth MJ (2012) Org Lett 14:3732–3735
78. Kloss F, Kohn U, Jahn BO, Hager MD, Gorls H, Schubert US (2011) Chem Asian J 6:2816–2824
79. Kwok SW, Fotsing JR, Fraser RJ, Rodionov VO, Fokin VV (2010) Org Lett 12:4217–4219
80. Wu L, Chen Y, Luo J, Sun Q, Peng M, Lin Q (2014) Tetrahedron Lett 55:3847–3850
81. Jaiswal MK, Gupta A, Yadav MS, Pandey VK, Tiwari VK (2023) Chem Eur J 29:e202301749
82. Kayet A, Pathak TJ (2013) Org Chem 78:9865–9875
83. Dey S, Pathak T (2014) RSC Adv 4:9275
84. Kayet A, Dey S, Pathak T (2015) Tetrahedron Lett 56:5521–5524
85. Van Berkel SS, Brauch S, Gabriel L, Henze M, Stark S, Vasilev D, Wessjohann LA, Abbas M, Westermann B (2012) Angew Chem Int Ed 51:5343–5346
86. Cai Z-J, Lu X-M, Zi Y, Yang C, Shen L-J, Li J, Wang S-Y, Ji S-J (2014) Org Lett 16:5108–5111

Chapter 5
Catalyst-Free Thermally Induced 1,2,3-Triazole Forming Approaches

5.1 Introduction

In 2001, Professor Karl Barry Sharpless introduced the concept of 'Click Chemistry' which has received tremendous importance in recent years [1]. The conventional Huisgen's thermally induced method [2] afforded a mixture of 1,4- and 1,5-regioisomers which are difficult to isolate in their pure form. However, Sharpless [3] and Meldal [4] independently reported Cu(I)-catalyzed cycloaddition of azides **2** and terminal alkynes **1** (commonly known as CuAAC) for regioselective synthesis of 1,4-disubstituted-1,2,3-triazoles **3** (Scheme 5.1). This CuAAC reaction has several notable features including high regioselectivity, excellent yields, easy to perform, short reaction time, modular in nature, etc. However, a method for corresponding 1,5-disubstituted 1,2,3-triazole **4** isomer was accessed by Fokin and co-workers in 2005 by exchanging copper with ruthenium (RuAAC) [5]. The CuAAC reaction is more widely explored in the domain of organic synthesis, biology, medicinal chemistry, and material sciences as compared to RuAAC [6–15].

Further, a broad array of affordable and easily produced carbonyl derivatives, such as aldehydes, ketones, and nitriles, have shown notable advantages over the established CuAAC tool. The reactions of azides separately with alkenes or ketones are some pair of the recent alternate techniques for the construction of such 1,2,3-triazole and its derivatives [16, 17]. Bressy et al. reported an organoclick reaction in which microwave irradiation of unactivated ketones **5** with aryl azides **6** in the presence of proline afforded functionalized 1,2,3-triazole **7** in good yields (Scheme 5.2) [18]. Ionic liquids serve as green replacements for coupling reactions also along with being easily recyclable and reusable up to several times [19–21]. Recently, a report documented for diverse synthesis of 1,5-disubstituted triazole **10** via eliminative azide-olefin cycloaddition (EAOC) reaction using reusable homogeneous ionic liquid/iron(III) chloride [22]. In this chapter, we briefly demonstrate the basic aspect of metal-free click reaction in the development of cyclic or fused-1,2,3-triazoles

V. K. Tiwari et al., *Click Chemistry*, Materials Horizons: From Nature to Nanomaterials,
https://doi.org/10.1007/978-981-97-4596-8_5

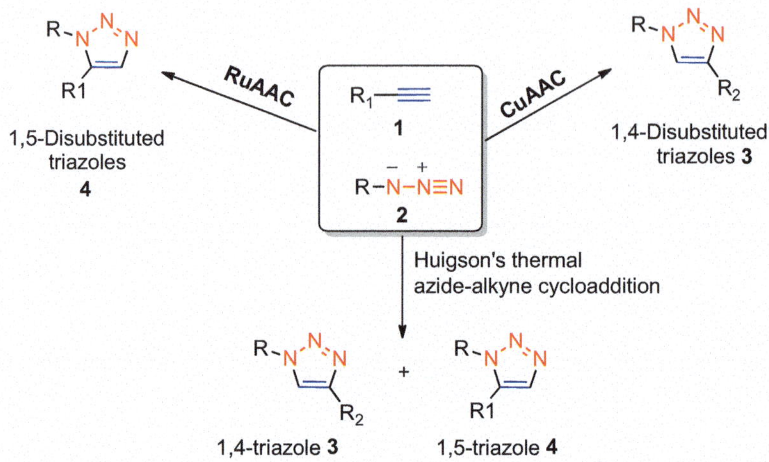

Scheme 5.1 Regioselective 1,2,3-triazole forming reaction under different catalytic conditions

with a wide range of applications in the domain of chemistry, chemical biology, and material sciences.

5.2 Greener 1,2,3-Triazoles Catalyst-Free Approaches

The presence of persistent hazardous metal traces in the product limited the scope of copper in CuAAC reaction for 1,4-triazoles and ruthenium, palladium, or nickel in RuAAC reaction for 1,5-triazoles synthesis. The widespread application of 1,4- and 1,5-triazoles [23–26] triggered to development of a greener approach for easy access to 1,2,3-triazoles in order to address this difficulty. Therefore, Pathak and co-workers describe the synthesis of 1,5-triazoles **16** simply by refluxing the vinyl sulfones **15** with sugar azide **12** in water (Scheme 5.3) [27]. Meanwhile, an interesting report on triazole formation in living systems was demonstrated by Bertozzi using strain-promoted click reaction [28, 29]. The modification of biomolecules through SPAAC reaction is a fantastic development of Wittig methodology that exemplifies the metal-free [3 + 2] cycloaddition of organic azide **2** with cyclooctyne **17** to produce corresponding triazole **18** at r.t [30]. Triazole residue **20** around a human cell was produced by the SPAAC reaction of azido-modified cell **19** with a fluorescent probe.

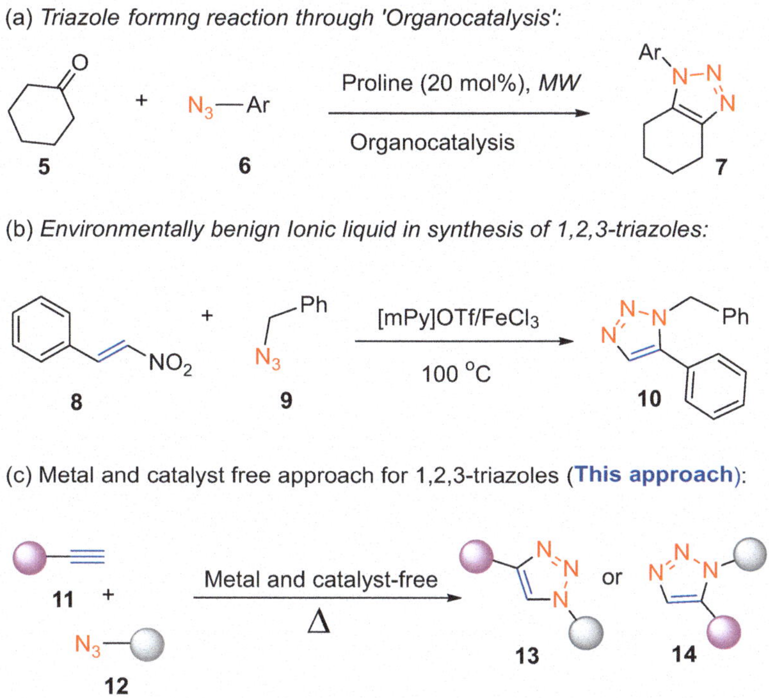

Scheme 5.2 Organocatalytic and metal-free approaches for functionalized 1,2,3-triazoles

5.3 Synthesis of 1,2,3-Triazoles Via Topochemical Azide-Alkyne Cycloaddition Reactions

Apart from the diverse application of modular CuAAC click reaction in solid-phase synthesis (CuAAC-SP) [31], topochemical azide-alkyne cycloaddition (TAAC) reaction has come to light as a structured and captivating method for the synthesis of crystalline biopolymer mimetic products with triazole attached. Topochemical reactions are solid-state phenomena that are solely controlled by the orientation of molecules in the crystal lattice [32]. These reactions result in very high yields of products that are both regio- and stereospecific owing to the molecular packing in the crystal lattice. Topochemical reactions are more appealing than solution-phase reactions due to a variety of fascinating features besides their remarkable specificity. Since solution-phase processes required the use of reactants, reagents, catalysts, and solvents, however, topochemical reactions did not require these materials as they are solvent and catalyst-free and even not require any chromatographic purification due to their high-yielding nature and specificity. Furthermore, the developed products structured in a crystalline order with fascinating characteristics are another interesting feature of topochemical processes (Fig. 5.1) [33]. Therefore, topochemical reactions are regulated both at the supramolecular level (packing) and at the

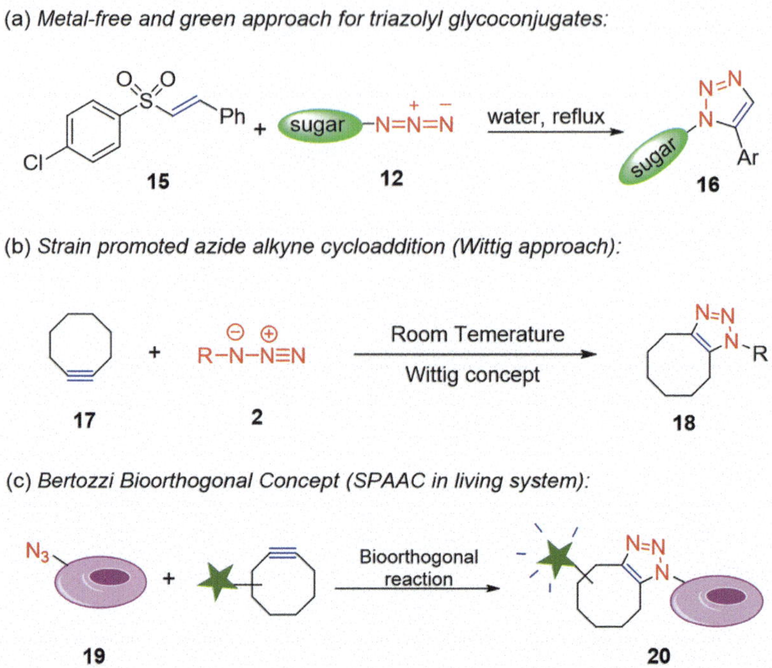

Scheme 5.3 Greener approach for regioselective formation of 1,2,3-triazoles

molecular level (regio-/stereospecificity). Triazole-linked biopolymer mimics were produced via the TAAC reaction that occurred either spontaneously in the crystals of monomers at ambient temperature or under heating. Furthermore, the X-ray diffraction (XRD) technique has been effectively used to establish the molecular alignment in the crystal structure and non-covalent interaction (NCI), such as the N–H…π interaction and N–H…O, using TAAC reaction [34].

In order to produce stereochemical control products in a single-crystal-to-single-crystal (SCSC) way, supramolecular chemistry and crystal engineering techniques are utilized to generate the desired crystal lattice for interpretation and facilitation of TAAC reactions [35]. In topochemical reactions, developing a monomer with an appropriate molecular packing for polymerization is the primary challenge. The primary feature for designing reactive monomer molecules is the introduction of hydrogen bonding as NCI for aligning molecules in an orientation favorable for TTAC reaction. Biomolecules that possess chirality, easy access, fascinating self-assembling patterns, and molecular diversity such as nucleosides, peptides, and carbohydrates have the ability to generate hydrogen bonds as monomers. Moreover, a number of chiral biopolymer mimics which have great use in chiral separation [36], asymmetric catalysis [37], and enantioselective crystallization [38] could be easily accessed using TAAC polymerization of biomolecules.

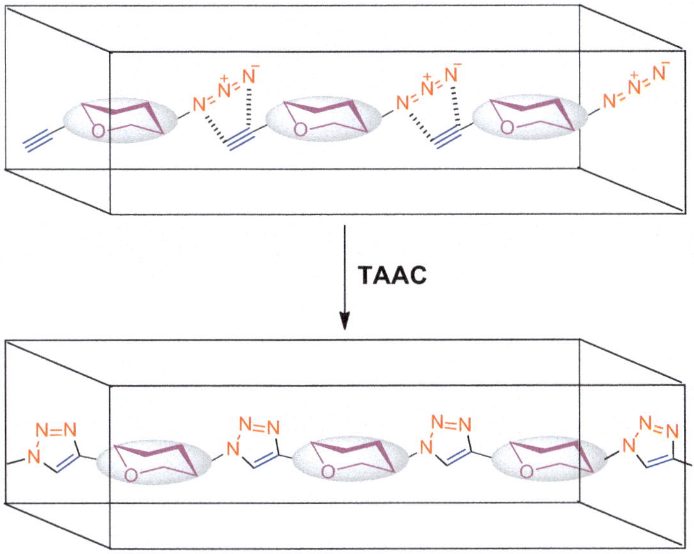

Fig. 5.1 Diagrammatic representation of triazole-appended sugar-derived crystal lattice of biopolymer through TAAC polymerizations

Toward this end, Sureshan and co-workers describe a topochemical click reaction for the synthesis of linear oligomers **22** and **23** through spontaneous regiospecific topochemical azide-alkyne cycloaddition of a monosaccharide analogue **21** (Scheme 5.4) [39]. The resulted 1,5-triazole-appended linear oligosaccharide analogues (**22**, **23**) were difficult to synthesize by conventional solution-state chemistry due to the preferential formation of cyclic oligomers. The reactant motifs (alkyne and azide) are entrapped in a reaction nanovessel formed by the crystal lattice; their arrangement which resembles a transition state facilitates the reaction, their relative orientation within the nanovessel determines the regiospecificity, and the linearity is determined by the crystal packing. A completely green click reaction is one that does not require the use of solvents, catalysts, or other means of activation, even if the click chemistry concept and green chemistry concepts overlap. Topochemical click reaction is consequently even more interesting because it is straightforward and proceeds spontaneously.

TAAC polymerization is a unique metal-free approach for the construction of triazole-appended products using either ordered crystal/gel as reaction medium [40]. Since crystallization is often difficult, therefore gelation is another process of self-assembly and as a result, the microstructures that form within the gels have an ordered arrangement. Topochemical reactions can also be facilitated by such ordered arrangement in gel microstructures [41].

In this order, Sureshan et al. exploited a crystal-to-crystal topochemical polymerization of unprotected monosaccharide **24** that produced a lattice-controlled highly homogeneous, enzyme-stable, 1,4-triazole-linked crystalline glycopolymer

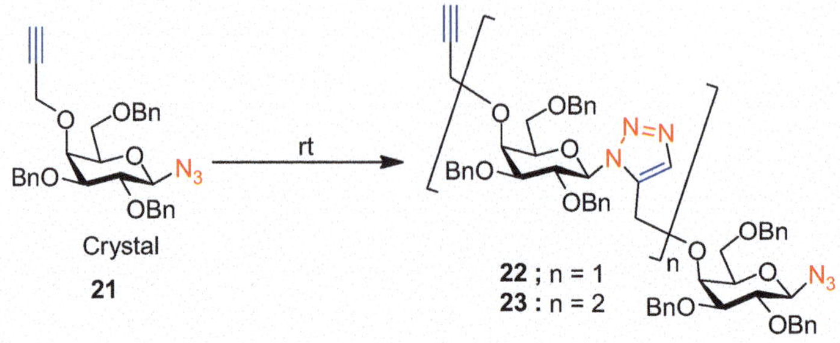

Scheme 5.4 Regiospecific topochemical azide-alkyne cycloaddition of **21** for the synthesis of linear oligomers **22** and **23**

25 regiospecifically in its crystal (Scheme 5.5) [42]. However, conventional solution-phase synthesis leads to o a heterogeneous mixture of polymers as a result of uncontrolled, non-selective, irregular, and incomplete polymerization. Unprotected, functionalized D-galactose monosaccharide **24** was exposed to temperature and time-dependent topochemical polymerization that resulted in a selective 1,4-triazole analogue of polysaccharide **25** with two hydroxyl groups at C_2 and C_3 positions oriented in such a way to the close proximity of adjacent molecule (Scheme 5.5). The crystal structure of pyranoside **24** exhibits zig-zag hydrogen bonding between its free hydroxyl group and molecules arranged in head-to-tail, resulting in the formation of 1,4-disubstituted triazole-attached polysaccharides **25**. This polymer could be chemically modified in order to produce a variety of functional polymers due to the availability of three hydroxy groups per monomer and the presence of an azide and an alkyne motif at either terminus of each polymer chain.

Further, the same group demonstrated the synthetic utility of miscible polymer blends in the preparation of triazole-linked polymers [43]. The proximal click partners of tetrol **26** exhibit self-shorted head-to-tail alignment, resulting in two distinct types of polymer chains by topochemical polymerization, each containing only one type of conformers. A head-to-tail conformer's orientation of complementary reactive azido-alkyne groups supports the synthesis of trans-triazole linkages **28**, while the other conformer's orientation favors the formation of cis-triazole linkages **27** (Scheme 5.6). Following thermal activation, cycloaddition was applied to both

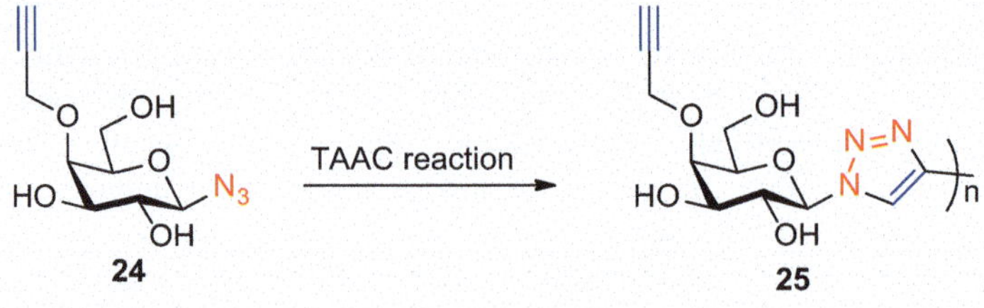

Scheme 5.5 TAAC-mediated synthesis of 1,4-triazole-appended linear polysaccharide **25** from unprotected monosaccharide **24**

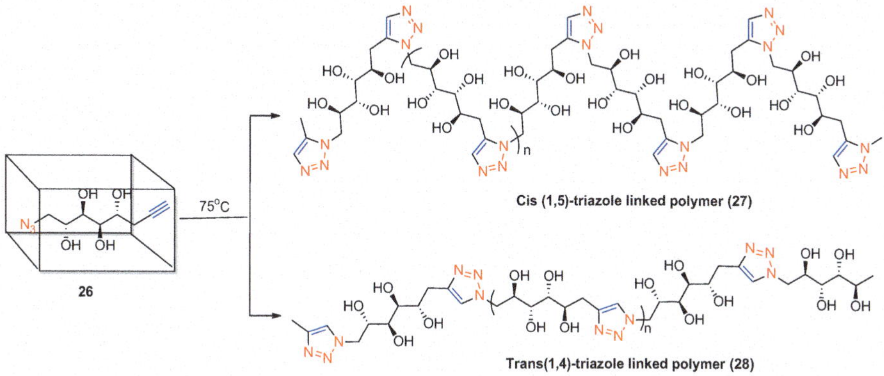

Scheme 5.6 TAAC-mediated synthesis of miscible polymer blends (**27, 28**) from azide and alkyne decorated tetrol **26**

types of self-sorted conformers, resulting in an equivalent quantity of two structurally distinct polymers that combined to produce an ideal polymer blend. During the polymerization process, each conformer forms an *H*-bond with four additional conformers of the same sort, resulting in a perfect polymer blend where every polymer chain is surrounded by new chains. Therefore, crystal molecular ordering in the pre-polymerized form is well-suited for producing a polymer blend. Thermal activation of both types of self-shorted conformers generates a solid-state miscible polymer blend that can be achieved architecturally differently using TAAC polymerization.

Moreover, considerable research has been attempted for a better understanding of the chemistry of other difficult parallel reactions, such as the TAAC polymerization of carbohydrates to develop triazolyl biopolymers. Toward this, Sureshan et al. greatly utilize TAAC to investigate the important effect by compelling azide and alkyne functionalized ribose derivative or multicomponent-derived non-identical reacting partners and their competing parallel reactions through its single-crystal X-ray diffraction analysis [44, 45].

Since TAAC is a stereoselective reaction and it was believed that the selectivity of the developed polymer is majorly influenced by the spatial orientation of reacting partners involved in polymerization. This property eventually can suggests the conformational restriction may lead to adopt the selectivity while conformational freedom leads to loss of the selectivity. Investigation revealed that the choice of conformational freedom of the reacting partner of **29** produced three competing reactions that proceeded in parallel, resulting in a dynamic equilibrium. A variety of triazolyl adducts were produced by the TAAC reaction upon thermal activation, where an equilibrium conformer produced a heterogeneous mixture of 1,5- and 1,4-triazolyl-linked products **33** and **34** respectively. However, the intermolecular TAAC reaction that produced the 1,5-triazolyl adduct **30** is expected to occur in the persistent arrangement or native conformation of the crystal (Scheme 5.7). Additionally, in a comparative study of **29** under CuAAC click protocol in the solution phase, only 1,4-triazolyl adduct **31** and **32** are produced by intermolecular cycloaddition, despite the fact that **30** was the only product formed by metal-free click process in the same condition.

The degree of potential for the TAAC reaction in the molecule functionalized with close proximal azide and alkyne depends on its spatial alignment of reacting partners and thus receives notable attention in the construction of 1,2,3-triazole link polymers in the solid state [46–48]. Sureshan et al. reported regiospecific dimerization of azido-alkyne **35** obtained from carbocyclic sugar inositol under mild thermal conditions. Compound **35** formed a 1,5-triazole-linked cyclic dimer **36** through topochemical 1,3-dipolar cycloaddition of azide and alkyne which pre-organized at C_2 and C_5 positions of the crystal structure **35** (Scheme 5.8) [49].

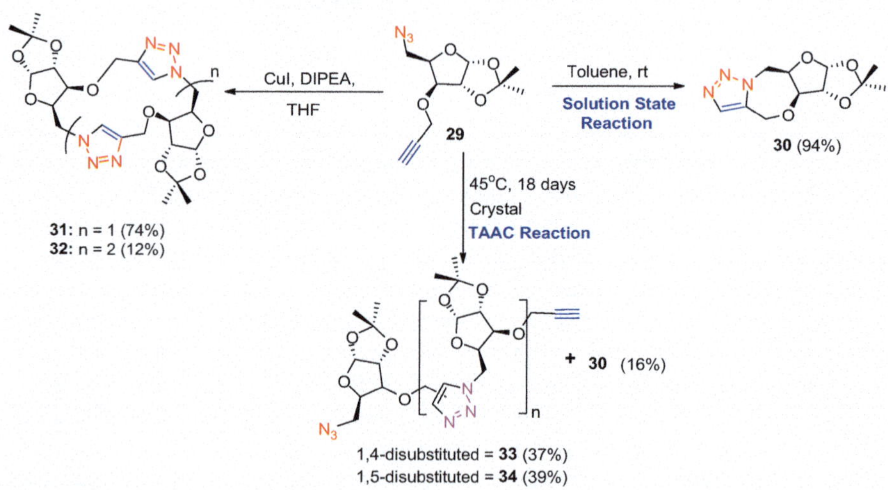

Scheme 5.7 Comparative study of **29** under TAAC polymerization and modular CuAAC click cycloaddition

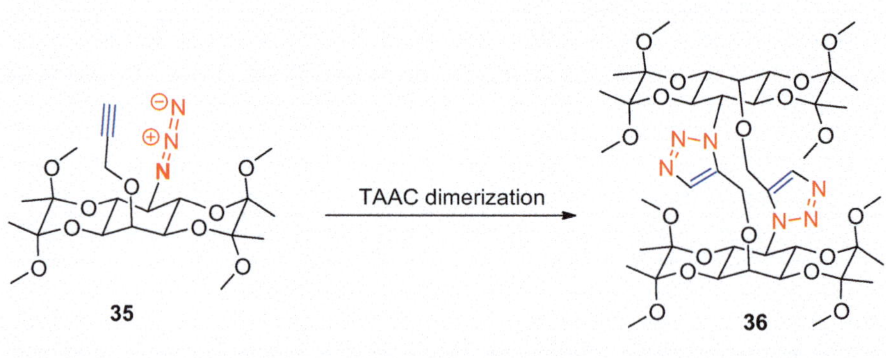

Scheme 5.8 Inositol derives azido-alkyne **35** afforded dimer adduct **36** under TAAC polymerization

Elevated temperature in TAAC polymerization is usually avoided as the progress of the reaction is determined by the rate of thermal activation of reacting molecule. The SCSC technique is employed to facilitate dimerization of the crystal by trapping surrounding water even though regular conditions are preferred for the TAAC process. Remarkably, the heating rate-dependent behavior of the crystal demonstrated significant benefits for the design of automated turn-on switches that activate the corrective electric circuit in the event of an abrupt rising temperature.

Thus, the carbohydrate-based topochemical azide-alkyne click reaction has emerged as a new promising approach for the design and regioselective/stereoselective synthesis of polymeric adducts. TAAC methodology enables the formation of a miscible polymer blend, which is effectively applied in the construction of adjustable switches. In light of greener, metal-free, and metal salt removal from the reaction mixture, TAAC polymerization has demonstrated a remarkable benefit over polymer synthesis by conventional Cu(I)-catalyzed click reaction. In order to produce 1,4 or 1,5-triazole adducts, the experimental challenge in the TAAC reaction is to develop a monomeric crystal in such a way that both reacting partners aligned in proximal orientation to one other. Peptides and nucleotides are two other macromolecules besides carbohydrates in which TAAC chemistry plays a significant role. The synthesis of monosaccharides was primarily inhibited due to the required protection and deprotection procedures in sugar chemistry before applying click conjugation. Conversely, the chemoselectivity and regioselectivity of the synthesized glycol-polymers were protected by click methods. Employing this reaction technique, practical and usable glycopolymers have been developed with a variety of topologies and carbohydrate residue. Additionally, it has proven to be an effective strategy for the development of glycol monomers and a promising class of vinyl-based triazolyl monomers.

5.4 Synthesis of Cyclic and Fused 1,2,3-Triazolyl Compounds

In order to increase the selectivity of the triazole forming process, Tiwari et al. reported thermally induced click reactions in carbohydrate molecules for the development of morpholine-fused triazole glycoconjugates **39a-d** in two steps [50]. In the presence of $CuSO_4.5H_2O$/NaAsc in H_2O at room temperature, sugar-derived terminal alkynes **37** react with epichlorohydrin and NaN_3 and furnished triazolyl azido alcohol **38** in good yields. Further alkynylation of **38** produces a suitable azido-alkyne motif which underwent intramolecular cyclization and resulted in a cyclic product **39** simply by heating the reaction mixture in DMF at 110 °C (Scheme 5.9). Furthermore, in the second investigation, authors addressed the development of physiologically relevant morpholine-fused [5, 1-c] triazole glycoconjugates from corresponding glycosyl azido alcohols without the usage of metals [51]. Notably, from similar sugar-based azido alcohol **39** scaffolds, CuAAC facilitates the quick synthesis of carbohydrate-derived oxazolodine-2-thiones [52], noscapine glycoconjugates [53, 54], and *bis*-triazolyl glycoconjugates [55].

Similarly, in a separate investigation regioselective synthesis of physiologically important spirocyclic 1,2,3-triazolo oxazine nucleoside **42** from azido alcohol **40** was also reported using click chemistry. This is an essential substance used extensively in chemical synthesis. Alkynylation of azido alcohol **40** in the presence of BEMP at 0 °C produces azido-alkyne **41** which subsequently underwent intramolecular click cyclization at elevated temperature to give spirocyclic glycohybrid adduct **42** in good yield (Scheme 5.10) [56].

Furthermore, Chattopadhyay and co-workers employed orthogonally protected sugars **43a,b** and developed biologically relevant 1,5-disubstituted triazole-fused glycohybrids **45a,b**. Thus, compounds **43a** and **43b** separately underwent reaction with 2-azidobenzyl bromide under basic condition to give respective azidos **44a** and **44b**, which were then proceed for 5,6-*O*-isopropylidene deprotection, $NaIO_4$

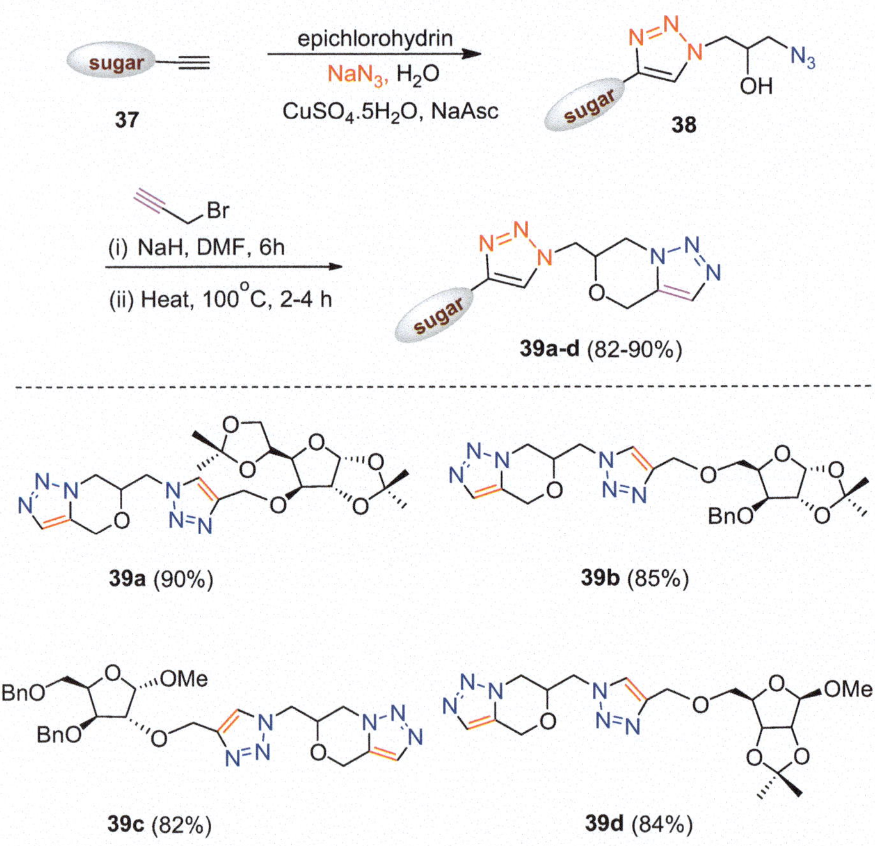

Scheme 5.9 Click-inspired synthesis of morpholine-fused triazoles **39a-d** using sugar-derived alkyne **37** and epichlorohydrin

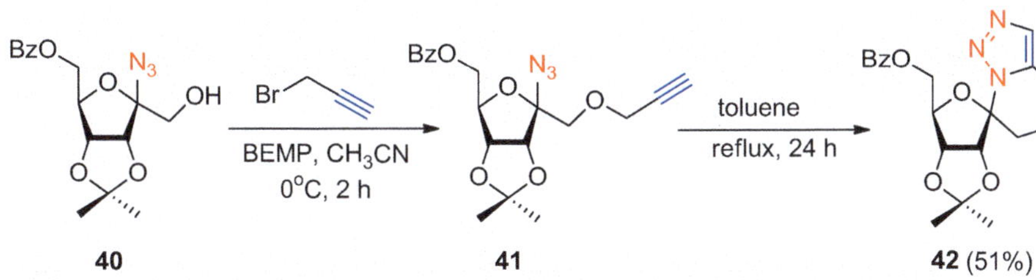

Scheme 5.10 Application of azido alcohol **40** for an access of spirocyclic glycohybrid **42**

oxidation, and the resulting aldehydes were then reacted with Bestmann reagent to give good yield of the required sugar alkynes **45a** and **45b**. At the end, alkynes **45a** and **45b** were separately heated at 120 °C in anhydrous DMF and gave good yield of triazole-fused polycyclic glycohybrids **46a** and **46b** via thermally induced intramolecular click chemistry (Scheme 5.11) [57].

Pericas and co-workers synthesized tricyclic 1,2,3-triazole **50** molecules using intramolecular 1,3-dipolar cycloaddition reaction. The epoxide **47** reacts first with NaN_3 in the presence of $LiClO_4$ in acetonitrile and reflux condition to give racemic trans-azido alcohol **48** with 90% yield. The azido alcohol **48** undergoes propargylation in the presence of propargyl bromide, NaH base, and tetrabutyl ammonium iodide catalyst in THF affording the desired alkynyl-azido compound **49** in 99%

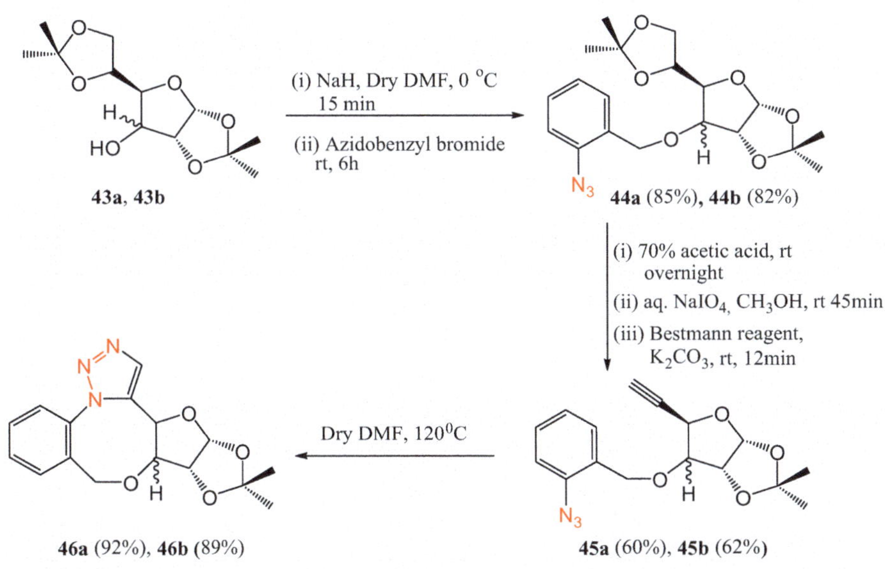

Scheme 5.11 Metal-free thermal azide-alkyne cycloaddition reaction for an easy access of 1,5-disubstituted triazole-fused polycycles **46a** and **46b** as sole product

Scheme 5.12 Click-inspired synthesis of structurally diverse, tricyclic 1,2,3-triazole **50**

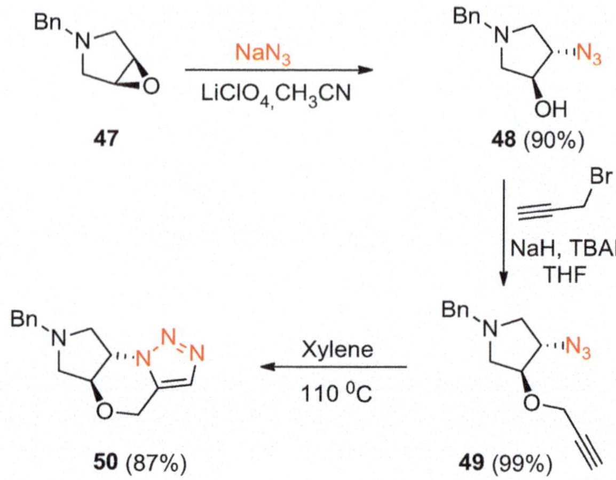

47 **48** (90%) **49** (99%) **50** (87%)

yield. Finally, the alkynyl-azido compound undergoes intramolecular click cycloaddition reaction in xylene/toluene under reflux to produce desired tricyclic compound **50** with 87% yield (Scheme 5.12) [58].

The macrocyclization of a range of azido-alkyne-based functionalized glycosides has been found to be most successfully achieved using the click chemistry approach. Hotha and co-workers reported a thermal-induced metal-free approach for construction of diverse 1,2,3-triazole-fused tetracyclic glycohybrid from functionalized sugar-based azido-alkyne using Huisgen's intramolecular click chemistry [59]. The resultant carbohydrate-containing macrocycles (**51a–o**) were easily produced as the only product in high-to-excellent yields and could serve as interesting molecular architecture with a variety of applications (Scheme 5.13).

Andrea and co-workers performed a very useful reaction for the stereocontrolled synthesis of fused 1,5-disubstituted triazoles **54a-d** from α-amino acid using intramolecular cycloaddition in microwave conditions. Initially, α-azido acid **52** was prepared by the reaction of amino acid and NaN_3, triflic anhydride, K_2CO_3, and $CuSO_4$. This azido acid **52** reacted with propargyl amine and DMTMM: 4-(4,6-dimethoxy[1,3,5] triazin-2-yl)-4-methylmorpholinium chloride gave the compound **53** which is the best model for intramolecular cyclization. Thus compound **53** finally heated at a temperature 160 °C cyclization in a mixture of acetonitrile and water solvent under microwave conditions afforded a desired compound 1,5-disubstituted triazole **54** (Scheme 5.14) [60].

Tiwari et al. nicely constructed 1,5-disubstituted 1,2,3-triazolyl glycoconjugates **57** using glycosyl azide **56** and nitrostyrenes **55** in the presence of TBAB in DMF with good yield (Scheme 5.15) [61]. The plausible mechanism involves the formation of triazoline intermediate **I** through a regioselective [3 + 2] cycloaddition of azides **56** with nitrostyrenes **55**. TBAB accelerates the rate of cycloaddition by enhancing the electrophilicity of the nitrostyrenes **54** at *β*-position through a weak ionic bond with

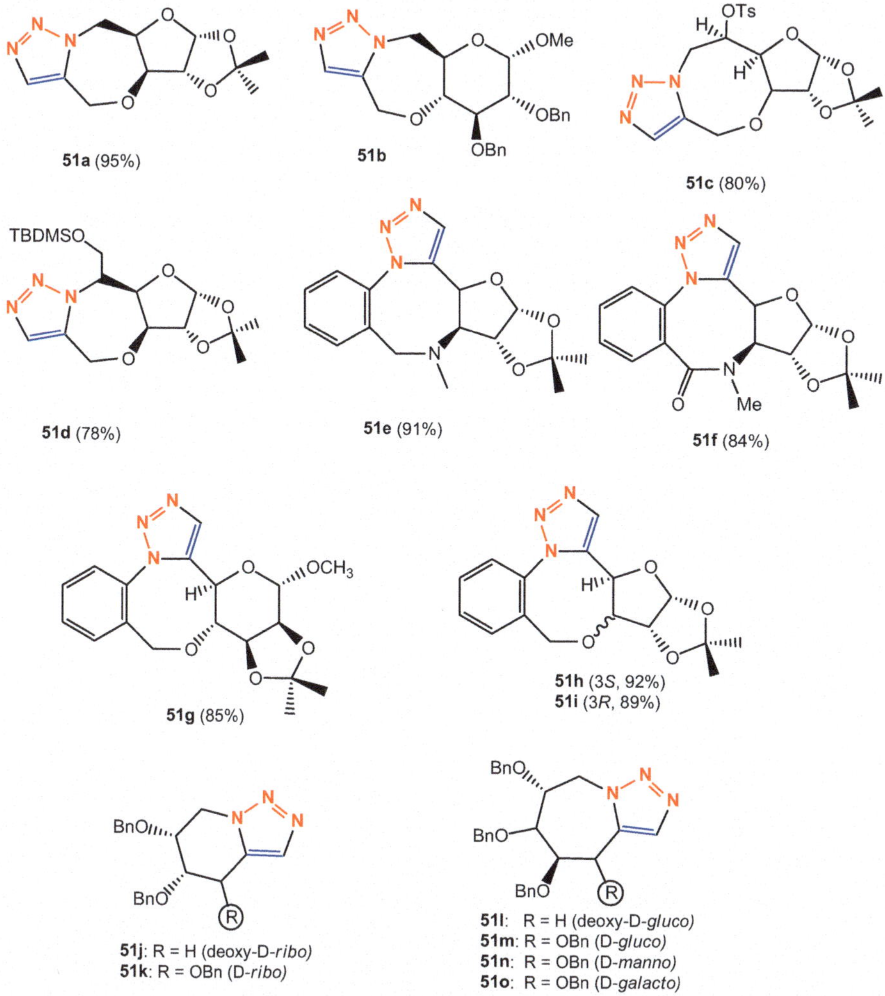

Scheme 5.13 Structure of some functionalized fused macrocycles **51a-o** from respective azido-alkynes through intramolecular 'click chemistry' under Huisgen's conditions

oxygen atoms of nitro group and ammonium ions. Subsequently, the triazoline intermediate **I** undergoes an elimination of HNO_2 to furnish the desired 1,5-disubstituted 1,2,3-triazolyl glycoconjugates **57** (Fig. 5.2).

Maiuolo and co-workers have developed a very facile and convenient method to synthesize 1,5-disubstituted 1,2,3-triazole **60** using eliminative azide-alkyne cycloaddition reaction using iron catalyst. Initially trans-nitrostyrene **58** treated with benzyl azide **59** to undergo cycloaddition reaction in presence of 1-methyl pyridinium trifluoromethanesulfonate ([mpy]OTf) at 100 °C temperature catalyzed by 20 mol % $FeCl_3$ (used as ionic liquid) in a 1:1 ratio of reagents and synthesized the

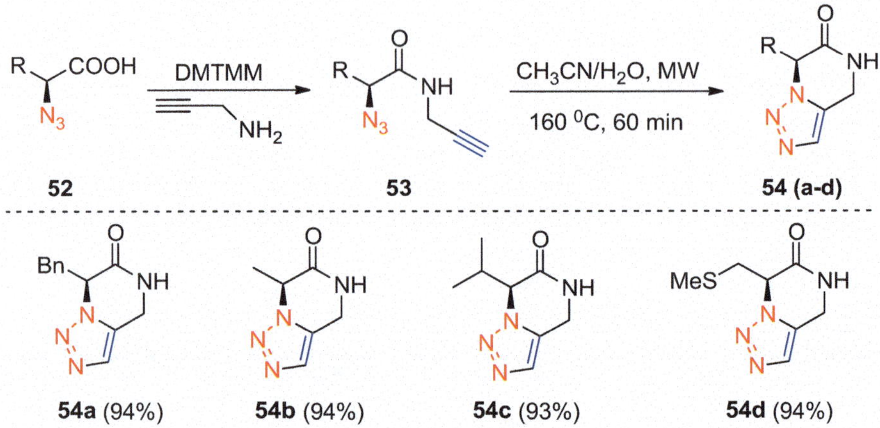

Scheme 5.14 *MW*-mediated intramolecular click cyclization of azido alkyne **53** derived from α-amino acids

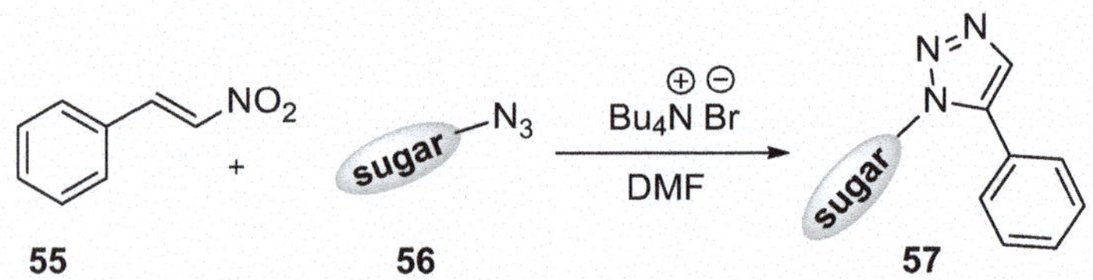

Scheme 5.15 TBAB-mediated synthesis of 1,5-disubstituted triazolyl glycoconjugates **57**

desired product 1,5-disubstituted triazole **60** in excellent yield (Scheme 5.16) [22]. There are a few examples of this reaction reported which are given below.

This instance of direct condensation between pyridine and methyl trifluoromethanesulfonate ([mPy]OTf) produced the catalyst $FeCl_3$, which was served as an ionic liquid, without the need for the traditional anionic exchange process that is employed to generate the majority of ionic liquids [62].

Pathak and co-worker reported construction of 1,5-disubstituted triazole **63** by utilizing the thermal cycloaddition of sugar azide **61** and vinyl sulfone **62**. A [3 + 2] cycloaddition of sugar azide **61** with vinyl sulfones **62** in toluene under reflux for 18 h produces the desired product 1,5-disubstituted triazole **63** in 79% yield (Scheme 5.17) [63].

Similarly, an extended protocol for the synthesis of 1,5-disubstituted triazole using vinyl sulfone and sugar modified vinyl sulfones with sugar azide under aqueous conditions was demonstrated by the same group. Refluxing sugar azide with vinyl sulfone in water undergoes [3 + 2] cycloaddition and produces 1,5-disubstituted triazole compounds as the sole product [64].

Further, Kumar et al. describe the synthesis of 1,5-disubstituted 1,2,3-triazole **66** using unsaturated acid **64** and aryl azide **65** in the presence of cupper triflate

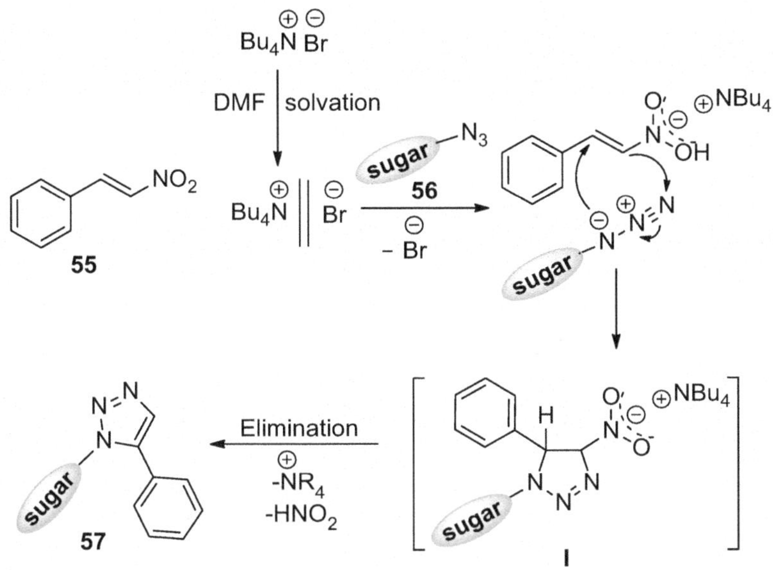

Fig. 5.2 Plausible mechanism for TBAB-catalyzed [3 + 2] cycloaddition reaction of azides with nitrostyrene

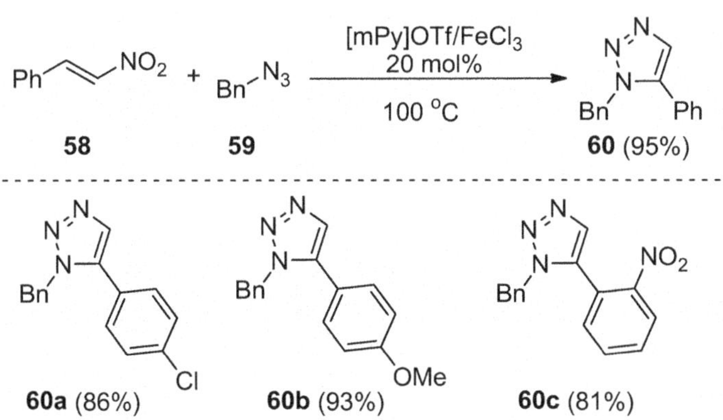

Scheme 5.16 Ionic liquid/iron(III) chloride-mediated synthesis of 1,5-disubstituted 1,2,3-triazoles **60a–c**

under open-air conditions. The cinnamic acid **64** reacted with aryl azide **65** in a mixture of solvent ascorbic acid and DMF (1:4 w/v ratio) in the presence of copper triflate catalyst at an elevated temperature 115 °C and afforded the synthesis of 1,5-disubstituted 1,2,3-triazole **66** in 90% yields (Scheme 5.18) [65].

In another investigation, aluminum oxide-catalyzed synthesis of 1,5-disubstituted 1,2,3-triazoles is reported. In this report, alkyl or aryl azide reacted with nitrostyrene

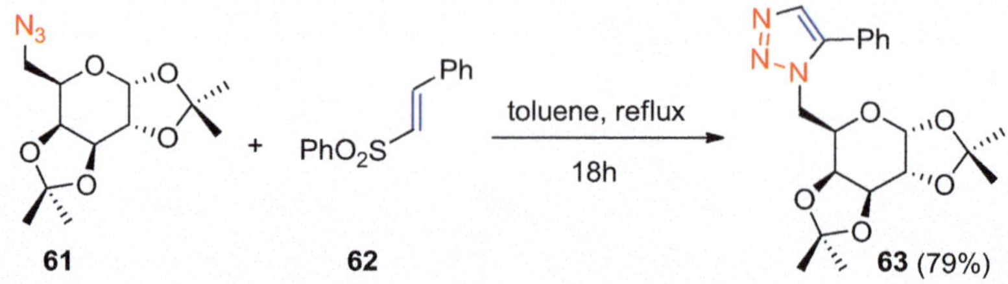

Scheme 5.17 Regioselective synthesis of 1,5-disubstituted triazole from reactions of vinyl sulfone with sugar azides

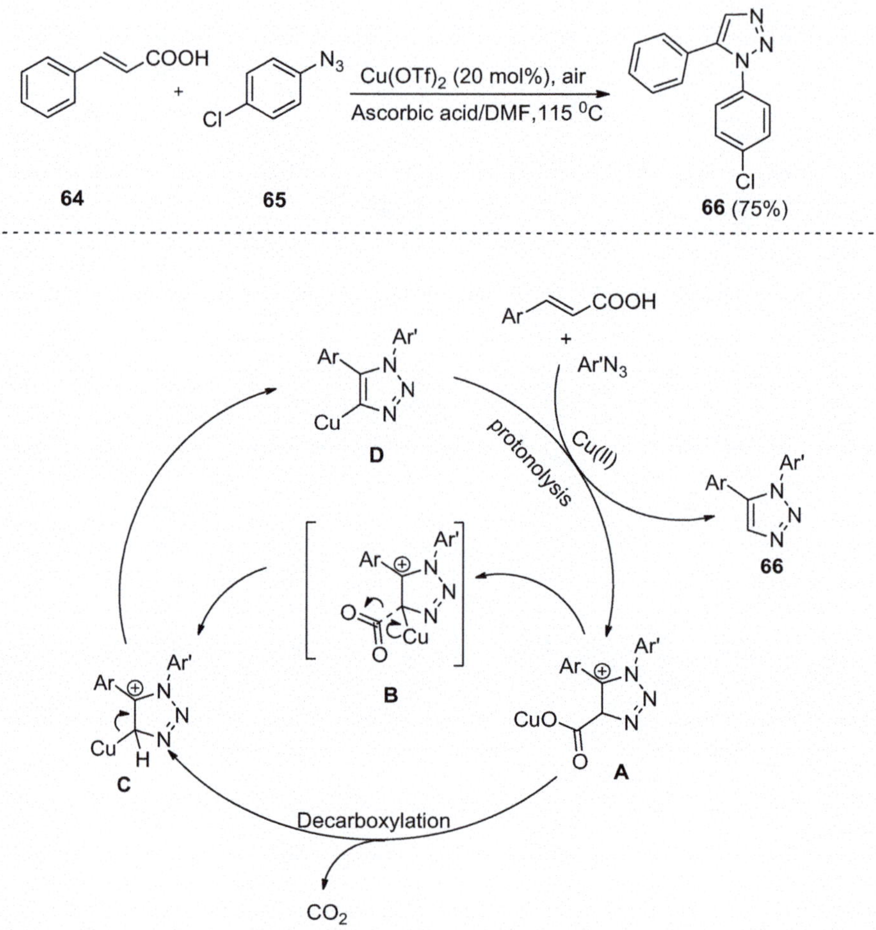

Scheme 5.18 Synthesis and proposed mechanism pathway of 1,5-disubstituted triazole **66** from unsaturated acid **64** with copper catalyst

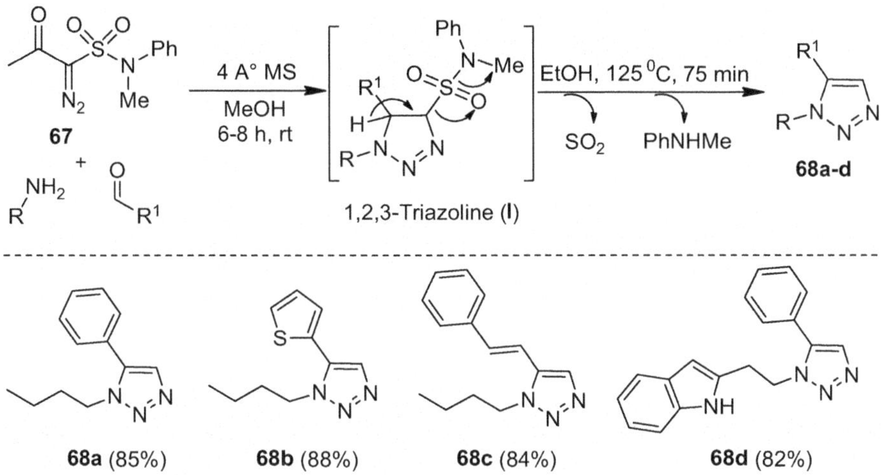

Scheme 5.19 Metal-free multicomponent method for synthesis of 1,5-disubstituted 1,2,3-triazole **68a–d**

in presence of Al_2O_3 in toluene solvent at temperature 100 °C and gave a satisfactory yield of desired product [66].

Furthermore, Krasavin et al. demonstrated a metal-free regioselective approach for the construction of 1,5-disubstituted 1,2,3-triazoles which involves the treatment of α-acetyl-α-diazomethane sulfonamides **67** with aliphatic primary amine and aromatic aldehyde in methanol [67]. The methodology involves the generation of 1,2,3-triazolines intermediate **I** through in situ deacetylation followed by [3 + 2] cycloaddition. Further, the subsequent elimination of SO_2 and PhNHMe in ethanol at 125 °C lead the formation of high yield of the desired 1,5-disubstituted 1,2,3-triazole **68** (Scheme 5.19).

The requirement for extremely high-dilution conditions for required macrocyclization over competing oligomerization has some limitations comparatively traditional routes [68]. Consequently, click-mediated continuous-flow chemistry presents itself as an intrinsic 'greener' alternative to classical syntheses and is simple to scale up through rate acceleration [69]. To make it simple to obtain biologically meaningful 12- to 22-membered triazole-appended macrocycles, Bogdan and co-workers effectively utilized the copper coil continuous-flow approach for intramolecular CuAAC-assisted macrocyclization (Scheme 5.20) [70]. Thus, in a copper coil flow reactor with continuous-flow conditions for 5 min at 150 °C, CuAAC-mediated macrocyclization of chiral amino alcohol azido-alkynes **69** in the presence of tris[1-tert-butyl-1H-1,2,3-triazolyl)methyl)amine (TITA, 0.1 equiv.) and DIPEA (2.0 equiv.) produced the desired macrocycle **70** in 73% yield. Surprisingly, the macrocyclization process accelerated over competing dimerization when it was carried out with an appropriate ligand and added base. Computational analysis confirmed the association to the strain energy in the macrocyclization product [71].

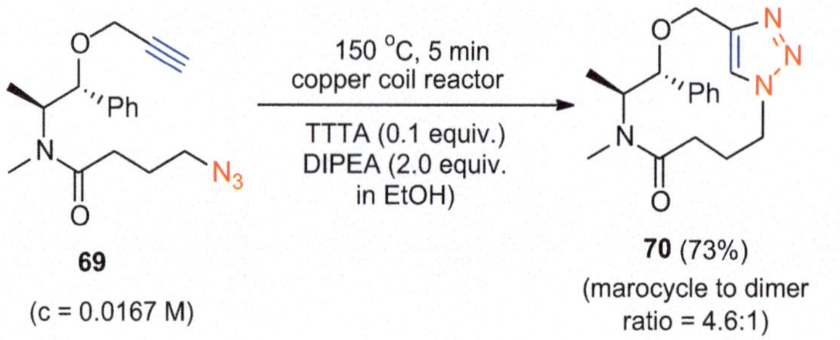

Scheme 5.20 Intramolecular click-induced macrocyclization using continuous-flow chemistry

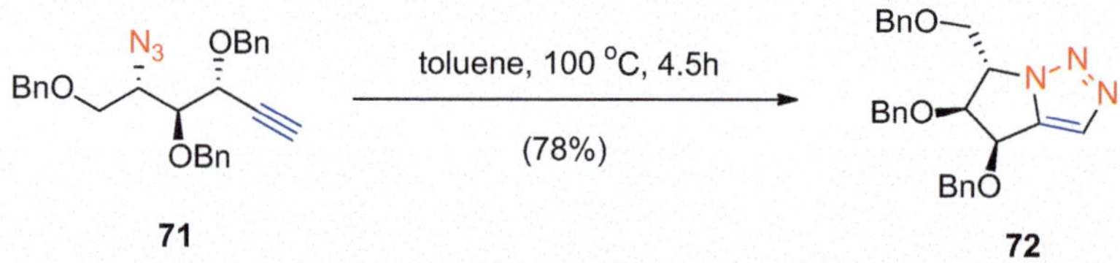

Scheme 5.21 Thermal-induced click cyclization for fused bicyclic pyrrolidine iminosugar hybrid **206**

Shaw and co-workers effectively employed intramolecular click chemistry in order to synthesize fused bicyclic triazoles, which demonstrated remarkable potency and selectivity as a novel family of α-glucosidase inhibitors [72]. The synthesis of triazolyl iminosugars was accomplished using both the thermal and RuAAC methods of coupling of azide and alkyne. The present study describes the formation of bicyclic triazole **72** in 78% yields upon thermal cycloaddition of azido-alkyne precursor **71** in 78% yields. The yield of the reaction is somewhat enhanced to 88%, when RuAAC catalysis is used (Scheme 5.21).

Apart from thermal-induced TAAC reaction and synthesis of carbohydrate-derived scaffolds of 1,2,3-trizoles, some reports were also documented for construction of biologically important triazole-fused heterocyclic scaffolds. Toward this, Basavaiah et al. presented a thermal-induced, metal-free, easy, and high-yielding process for the synthesis of triazole-fused benzoxazonines **77** from the Baylis–Hillman adduct **75** (Scheme 5.22) [73]. Therefore, the key Baylis–Hillman adduct **75** generated through DABCO-mediated reaction of *O*-propargylated salicylaldehyde **73** with alkyl acrylates **74** followed by treatment with acetyl chloride under basic condition. Further, treatment of **75** with sodium azide at room temperature furnished the required azido-alkyne motif **76**, which subsequently underwent click cyclization in toluene under reflux to produce biologically important triazole-fused nine-membered benzoxazonines skeleton **77** in good yield.

Using a one-pot, four-step synthetic process incorporating click cycloaddition, Banerji and co-workers reported the development of small triazolo-benzoxazepine

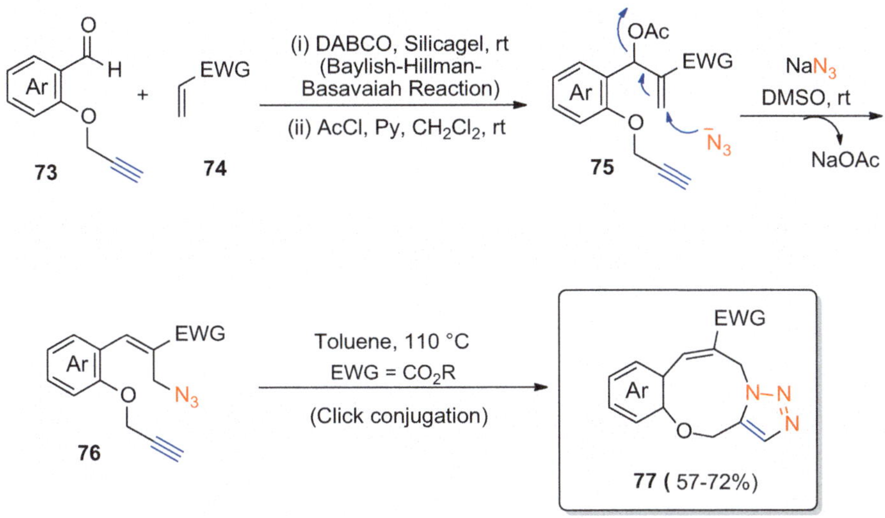

Scheme 5.22 Thermal-induced synthesis of tricyclic [1,2,3] -triazolo-[1,4]-benzoxazonine frameworks **77** from Baylis–Hillman–Basavaiah adduct **75**

scaffolds **83** that further evaluated against various cancer cell lines [74]. The corresponding propargyl ether **79** was initially produced in 95% yields by treating 2-amino benzyl alcohol **78** with propargyl alcohol in an acidic environment at 0 °C (Scheme 5.23). Moreover, a mixture of $Pd(PPh_3)_2Cl_2$, AgCl, and TEA in DMF at room temperature was used to functionalize the propargyl ether **79** with various aromatic iodides **80** via Sonogashira coupling. Substituted propargyl ethers **81** were next subjected to diazotization at 0 °C in the presence of $NaNO_2$ and HCl. Subsequently nucleophilic substitution with sodium azide afforded the corresponding azide **82**. Finally, silver chloride-catalyzed [3 + 2] click cyclization of azides **82** furnished the desire triazolo-benzoxazepine scaffolds **83** in moderate yields. A MTT assay was used to assess the biological efficacy of developed triazolo-benzoxazepine scaffolds **83** against three distinct cancer cell lines, including neuroblastoma Neura, liver cancer cell line Hep G2, and kidney cancer cell line Hek 293. The results of the screening showed that certain triazolo-benzoxazepine scaffolds had micromolar anti-cancer activity. Microscopic analyses utilizing phase contrast, fluorescence, and confocal images verified that one of the compounds **83** produced apoptosis within the synthesized triazolo-benzoxazepine scaffolds.

Following up on cascade click cycloaddition, a study revealed an easily synthesis of indole-based polyheterocycles in one pot by employing an intramolecular azide-internal alkyne click cycloaddition reaction and a sequential lewis acid-catalyzed intermolecular Michael addition [75]. The synthesis process for C-3 functionalize the indole includes the Michael addition reaction of the aryl 2-alkynyl indoles **85** with substituted (E)-1-azido-2-(2-nitrovinyl)benzenes **84** in toluene with the Lewis acid catalyst $Yb(OTf)_3$ generated desired azido-alkyne intermediates **86** (Scheme 5.24).

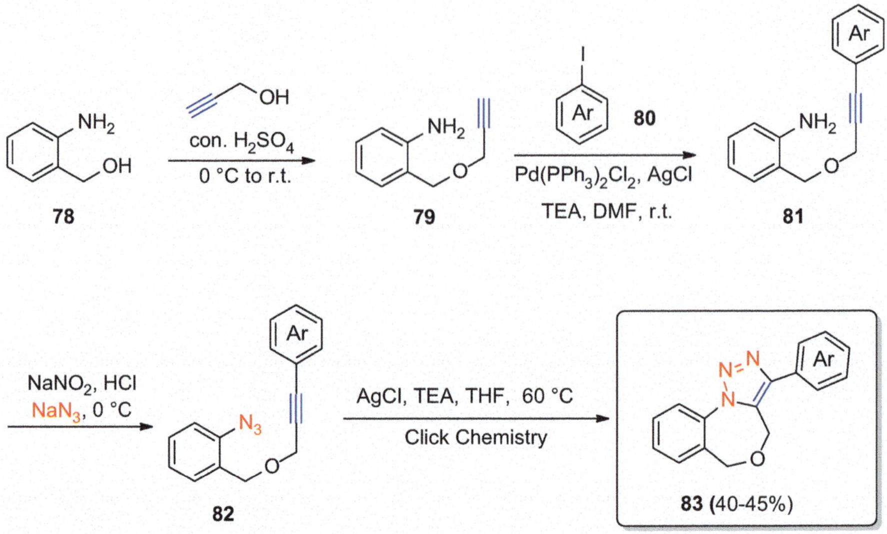

Scheme 5.23 Synthesis of triazolo-benzoxazepine **83** as anti-cancer agents

Finally, the required indole-based tetracyclic product **87** was formed by thermally induced click cycloaddition of corresponding azido-alkyne intermediates **86** at 120 °C in toluene. Additional research revealed that the Michael addition step occurs at a catalytic quantity of $Yb(OTf)_3$; however, the intramolecular click cycloaddition process completes thermally without a catalyst.

Comparably a one-pot, two-step cascade reaction methodology which is easy to operate was developed in order to construct physiologically potent fused 1,2,3-triazolo-heterocyclic scaffolds **91**, **94**, and **95** using 2-alkynyl aryl(heteroaryl) aldehydes **89**, **92**, and **93** and phenacyl azides **88** [76]. This metal-free synthetic technique uses piperidinium acetate to facilitate the Knoevenagel condensation of 2-alkynyl aryl(heteroaryl) aldehydes with phenacyl azides, resulting the adduct **90** which on click cyclization produces heterocyclic scaffolds **91**, **94**, and **95** (Scheme 5.25). The model reaction mediated by piperidinium acetate is shown in Scheme 5.25. For

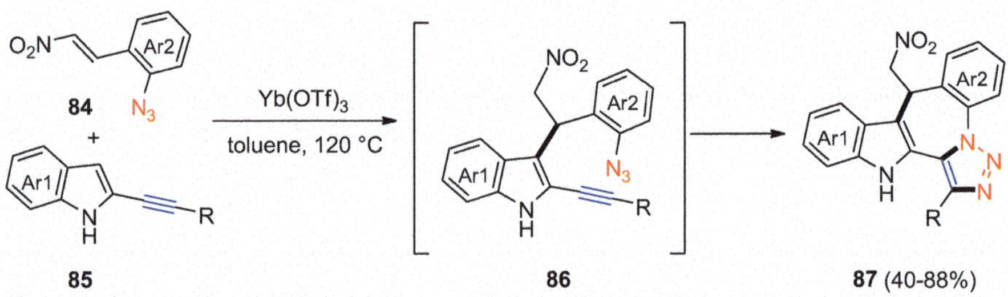

Scheme 5.24 Synthesis of indole-based tetracyclic architecture **87** via one-pot cascade reaction

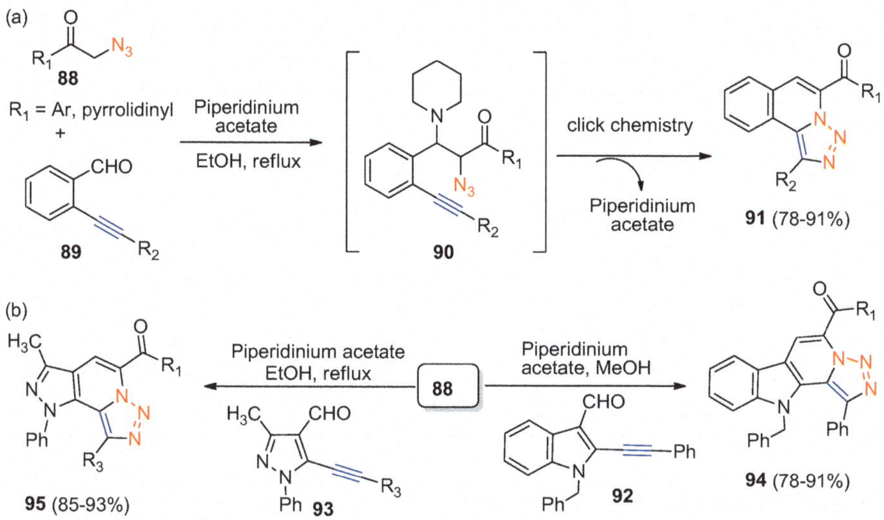

Scheme 5.25 Piperidinium acetate-mediated one-pot synthesis of 1,2,3-triazole-fused heterocycles **91**, **94**, and **95**

instance, the substrate **89** activated with piperidinium acetate followed by nucleophilic attack yielded the azido-alkyne intermediates **90**. Through a 1,3-dipolar azide/alkyne cycloaddition reaction in which piperidine is removed during the reaction, the Knoevenagel adduct **90** results in the formation of fused 1,2,3-triazole-heterocyclic scaffolds **91**. This protocol's key characteristics include high atom economy, metal-free environments, and scale-up synthesis.

Ning and co-workers demonstrated silver-catalyzed cascade hydroazidation and azide-alkyne click cycloaddition reaction of diverse diynes with $TMSN_3$ in presence of water for successful construction of pharmaceutically relevant 1,5-fused 1,2,3-triazole frameworks [77]. Under ideal reaction conditions, Ag_2CO_3-catalyzed hydroazidation of several terminal 1,6-diynes (Z = O, NR, and CR_6R_7) **96–98** with $TMSN_3$ provides the corresponding azido-alkyne intermediates **99–101** in good-to-high yields (Scheme 5.26). Further fused heterocyclic 1,2,3-triazoles units of morpholine **102**, piperazine **103**, and piperidine **104** were nicely achieved through click-inspired cycloaddition of azido-alkyne intermediates **99–101** in DMSO at 100 °C, respectively. Notable benefits of this synthetic process include easy access to starting materials, mild, high reaction efficacy, strong functional tolerance, and superior product yields.

Nekkanti and co-workers have used intramolecular azide-alkyne click cycloaddition to synthesize a wide range of poly-heterocyclic annulated 1,2,3-triazolo-fused-tetrahydro-β-carboline derivatives with good yields [78]. Synthesis of starting *O*-azido-benzaldehydes **105** from corresponding *O*-nitrobenzaldehydes via nucleophilic displacement with NaN_3 in HMPA at 20 °C is the first step in order to construct hexa-heterocycles. Furthermore, tryptamine **106** and *O*-azidobenzaldehydes **105**

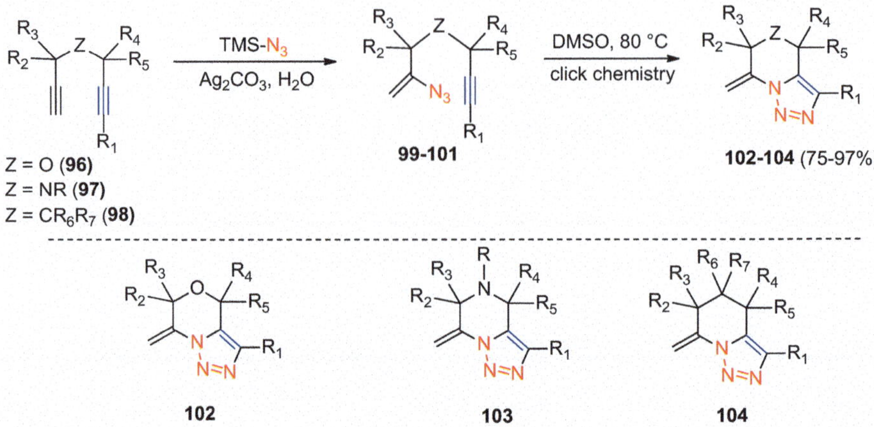

Scheme 5.26 Synthesis of 1,5-fused-triazole heterocycles **102–104** through tandem hydroazidation and click cyclization

were cyclized using the Pictet–Spengler method in DCM at 0 °C, and the desired alkynylated tetrahydro-*β*-carbolines intermediate **108** was produced with overall good yield through *N*-propargylation using propargyl bromide and K_2CO_3 in DMF (Scheme 5.27). As a result, the thermally induced click cycloaddition process of **108** in toluene provides the final 1,2,3-triazolo-fused-tetrahydro-*β*-carbolines **109** in good yields. Routine *N*-alkylation of *β*-carbolines **109** under basic conditions in anhydrous DMF resulted in an equivalent hexa-heterocyclic fused scaffold **110** in excellent yield.

Additionally, this well-established protocol was further utilized for the grafting of biologically important pentaheterocyclic annulated systems **114** in excellent as depicted in Scheme 5.28 [78]. Moreover, a tosylated analogue **112** which produces from L-tryptophan **111** over a series of reaction was underwent azidation followed by *N*-propargylation (by reacting with propargyl bromide using K_2CO_3 in DMF) afforded the high yield of the required azido-alkyne scaffolds **113**. Finally, thermal-induced click cycloaddition of **113** furnished the excellent yields of title pentaheterocyclic analogues **114** with great ease. Fascinatingly, the hexa-heterocyclic compound **115** exhibited significant in vitro cytotoxicity against human breast MCF-7 cell line with IC_{50} value of 6.45 ± 0.37 μM, whereas pentaheterocyclic compound **116** was identified as a potent cytotoxic agent against B16F10 cell line with an IC_{50} value of 4.01 ± 0.39 μM.

Similarly, tricyclic heterocycles 4*H*-benzo[*g*][1,2,3]triazolo[5,1-c][1,4]oxazocines **120** can also be synthesized using a one-pot cascade procedure that combines azide-alkyne click cycloaddition and Knoevenagel condensation and investigation of biological assay against several cancer cell lines was successfully reported by Sastry et al. [79] The reactive azido-alkyne intermediate **119** was generated using Knoevenagel condensation of alkylated aldehyde **117** and phenacyl azides **118** in methanol in presence of piperidinium acetate at rt (Scheme 5.29). At this stage, neither

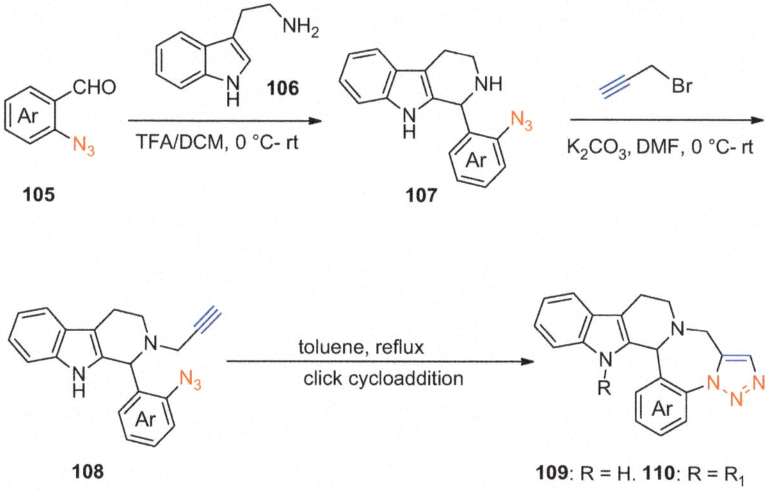

Scheme 5.27 Click-inspired synthesis of 1,2,3-triazolo-fused-tetrahydro-β-carboline **110** from *O*-azidobenzaldehydes **105**

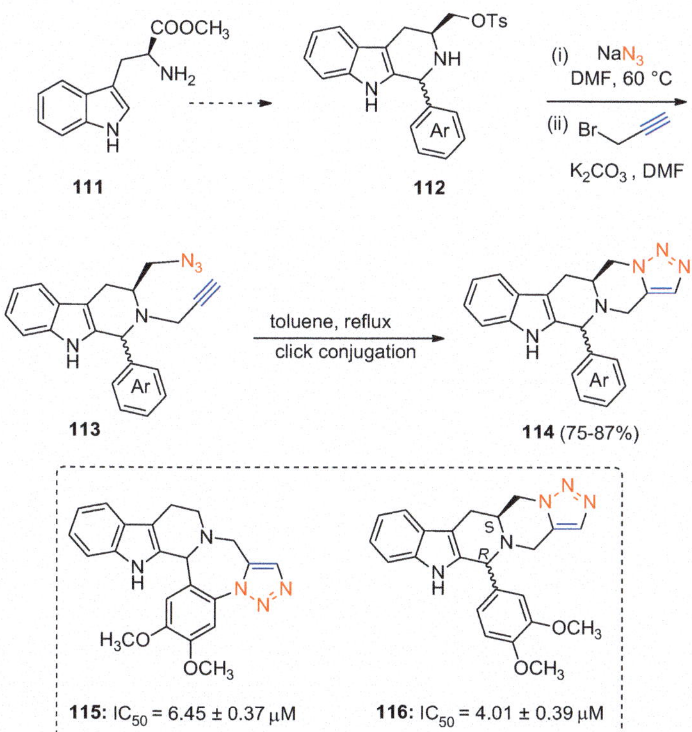

Scheme 5.28 Synthesis of triazolo-fused-tetrahydro-β-carboline **114** via thermal-induced click

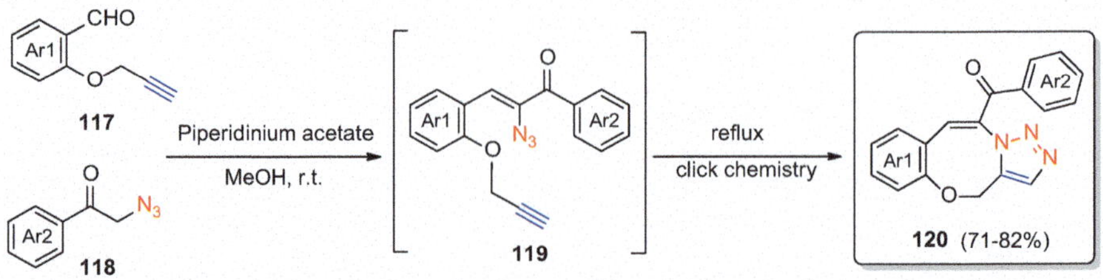

Scheme 5.29 One-pot cascade synthesis for tricyclic heterocycle framework **120**

intramolecular nor intermolecular cyclization was seen by the authors. Eventually, the same reaction mixture of **119** was subjected to reflux producing the title tricyclic heterocycles 4*H*-benzo[g][1,2,3]triazolo[5,1-c][1,4] oxazocines **120** via click conjugation in good yields. All new compounds were screened using a cytotoxicity assay to determine their anti-cancer potential against four distinct cancer cell lines such as MDA-MB-231 (a human breast adenocarcinoma epithelial cell line), DU145 (a human prostate cancer cell line), MCF-7 (a human breast cancer cell line), and HeLa (a human cervix cancer cell line). Based on the results of the assay, the tested compounds have an IC_{50} value of less than 13 μM, suggesting that they may be cytotoxic to the study cell lines. Furthermore, studies on fluorescence titration and viscosity show that these tricyclic compounds have a strong affinity for binding DNA.

Further in continuation of tandem reaction, a report is well documented for the development of biologically relevant tricycle triazolo[1,5-a][1,4] benzodiazepinones **125** through copper-catalyzed tandem Ullmann C-N coupling and click cyclization [80]. *N*-Propargylated anilines **123** and *O*-azidobenzylbromides **122** served as important starting materials in this one-pot protocol. The initial compound *O*-azidobenzylbromides **122** was obtained from *O*-toluidines in a two-step procedure in good yields. The formation of equivalent azide **121** from *O*-toluidines involves diazotization using $NaNO_2$/HCl, and a nucleophilic displacement with sodium azide. Furthermore, *O*-azidobenzylbromide **122** is produced in good yields by treating **121** with benzoyl peroxide and NBS (Scheme 5.30). Conversely, *N*-alkynylation of aniline with propargyl bromide in room temperature DMF with K_2CO_3 afforded *N*-alkynylated aniline **123**. Finally, azido-alkyne intermediate **124** obtained from copper-catalyzed Ullmann C-N coupling between *O*-azidobenzylbromides **122** and *N*-propargylated anilines **123** was subsequently underwent thermal-induced click cyclization in order to produce the triazolo[1,5-a][1,4] benzodiazepinone derivative **125** in 86–97% yields.

Further thermal-induced click cycloaddition was nicely utilized by Senwar and coworkers to construct spiro-oxindole-derived morpholine-fused-1,2,3-triazoles **130** from isatin spiro-epoxides **127** and evaluate their anti-proliferative activity on several human tumor cell lines [81]. The synthetic methodology for tetracyclic morpholine-fused-1,2,3-triazoles **130** requires generation of desired azido-alkyne intermediate **129**, which obtained from regiospecific ring opening of isatin epoxides **127** with sodium azide followed by *O*-propargylation (Scheme 5.31). Further, thermal-induced

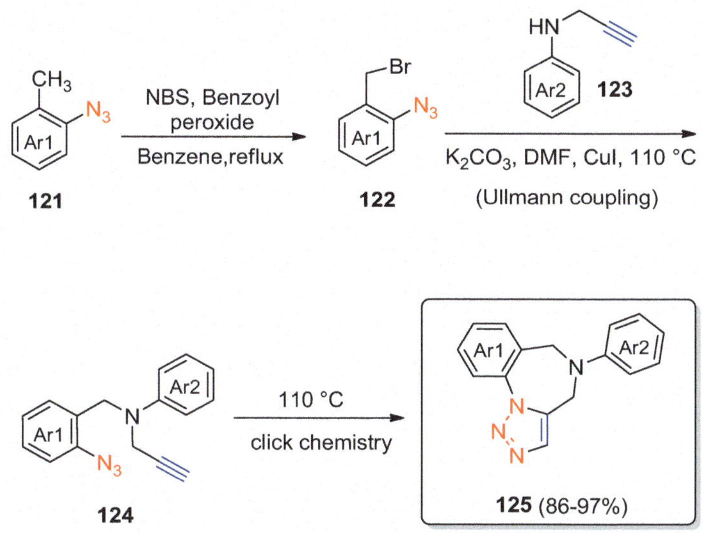

Scheme 5.30 Thermal-induced click cyclization for synthesis of triazolo[1,5-a][1,4]benzodiazepinones **125**

click cyclization of in situ generated azido-alkyne derivative **129** in toluene at reflux afforded the tetracyclic morpholine-fused-1,2,3-triazoles **130** in excellent yields. The treatment of *N*-alkylated isatins **126** with trimethylsulfoxonium ylide delivers the corresponding isatin epoxide **127** as depicted in Scheme 5.31. Moreover, the anti-proliferative effect of tetracyclic morpholine-fused-1,2,3-triazoles **130** was examined against several human tumor cell lines including breast (MCF-7), lung (A549), cervical (HeLa), and prostate (DU–145). Using IC_{50} values between 1.87 and 4.36 μM, which are comparable to the reference compounds doxorubicin and 5-flourouracil, the in vitro test reveals that specific spiro-derivatives effectively suppressed the proliferation of the A549 cell line. The examined triazoles demonstrate apoptotic traits in A549 cells, as indicated by flow chemistry analysis, and could potentially serve as important anti-cancer agents for lung adenocarcinoma A549 cancer cells.

A two-step, three-component protocol for the diastereoselective synthesis of tricyclic 4,6,7,8,8a,9-hexahydropyrrolo[1,2-a][1,2,3]triazolo[1,5-d]pyrazinesines **135** has been disclosed in order to construct fused triazoles via click cyclization and MCR coupling [82]. The essential propargylamines **134** are produced under solvent-free conditions by the $AuBr_3$-catalyzed coupling of cyclic amine **131**, aldehydes **132**, and terminal alkynes **133**. Comparing aliphatic aldehydes to aromatic aldehydes, the coupling reactivity of butyraldehyde, valeraldehyde, hexanal, and cyclohexane carboxaldehyde was found to be higher. Further, thermal-induced [3 + 2] dipolar cycloaddition of resulted propargylamines **134** in DMF at 100 °C afforded the tricyclic fused pyrazines **135** in good yields with excellent diastereoselectivity (>99% dr) (Scheme 5.32).

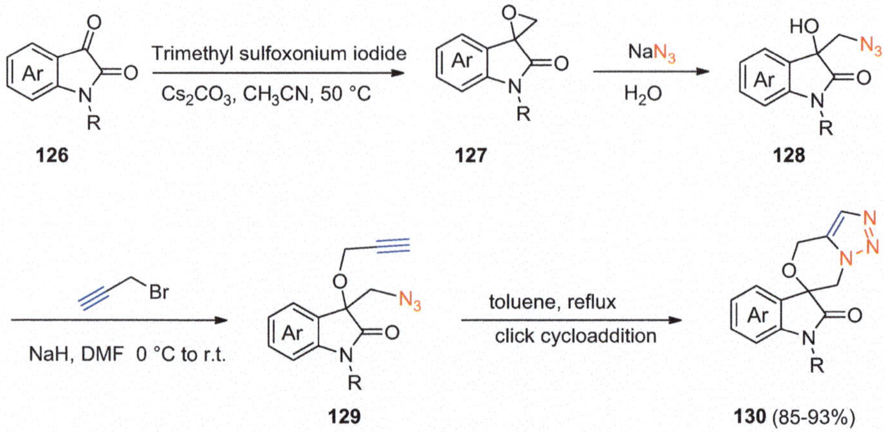

Scheme 5.31 Thermal-induced click cyclization for synthesis of spiro-oxindole-derived tetracyclicmorpholine-fused-1,2,3-triazoles **130**

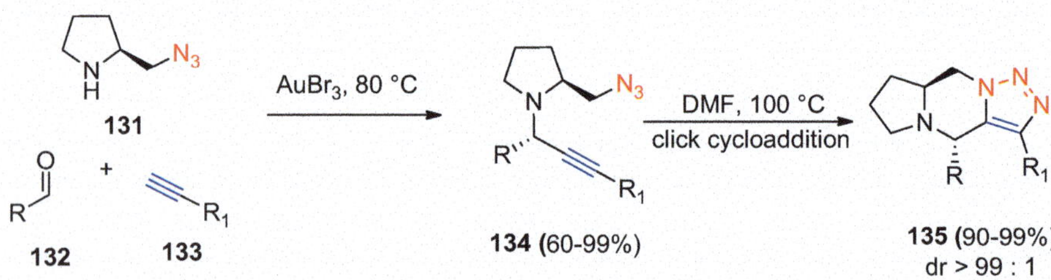

Scheme 5.32 Two-step three-component reaction for synthesis of tricyclic fused pyrazines **135**

5.5 Conclusions and Future Outlook

In contrast to the copper-catalyzed click reaction of azide with alkyne (CuAAC) which produces the 1,4-triazole isomer, catalyst-free click reaction offers direct access to structurally diverse cyclic and fused 1,2,3-triazole framework. Catalyst-free click reactions are useful tool for preparation of linear polymer via topochemical polymerization (TAAC), biologically important morpholine-fused [5, 1-c] triazolyl glycoconjugates, spirocyclic adduct, polycycles, and heterocyclic molecular architectures with great ease. Some catalyst and solvent-free solid-state reactions are driven by alignment of monomers in crystal lattice for polymer synthesis. Apart from that, azido-alkyne functionalized carbohydrate scaffolds effectively underwent intramolecular click cycloaddition in order to produce pharmacologically relevant variety of 1,2,3-triazoles under thermal conditions. Ionic liquids (ILs) recently received important attention for providing green and environmentally benign reaction media with great advantages in terms of yields and selectivity as compared to conventional conditions. This chapter briefly demonstrate the basic aspect of metal-free click reaction in development of novel functionalized 1,2,3-triazoles with a wide range application in domain of chemistry, chemical biology, and material sciences.

References

1. Kolb HC, Finn MG, Sharpless KB (2001) Angew Chem Int Ed 40:2004–2021
2. Huisgen R (1963) Angew Chem Int Ed 2:565–598
3. Rostovtsev VV, Green LG, Fokin VV, Sharpless KB (2002) Angew Chem Int Ed 41:2596–2599
4. Tornøe CW, Christensen C, Meldal M (2002) J Org Chem 67:3057–3062
5. Zhang L, Chen X, Xue P, Sun HHY, Williams ID, Sharpless KB, Fokin VV, Jia G (2005) J Am Chem Soc 127:15998–15999
6. Meldal M, Tornoe CW (2008) Chem Rev 108:2952–3015
7. Tiwari VK, Mishra BB, Mishra KB, Mishra N, Singh AS, Chen X (2016) Chem Rev 116:3086–3240
8. Agrahari AK, Bose P, Jaiswal MK, Rajkhowa S, Singh AS, Hotha S, Mishra N, Tiwari VK (2021) Chem Rev 121:7638–7956
9. Iha RK, Wooley KL, Nystrom AM, Burke DJ, Kade MJ, Hawker CJ (2009) Chem Rev 109:5620–5686
10. Thirumurugan P, Matosiuk D, Jozwiak K (2013) Chem Rev 113:4905–4979
11. Shieh P, Hill MR, Zang W, Kristufek SL, Johnson JA (2021) Chem Rev 121:7059–7121
12. Albada B, Keijzer JF, Zuilhof H, Delft FV (2021) Chem Rev 121:7032–7058
13. Fairbanks BD, Macdougall LJ, Mavila S, Sinha J, Kirkptrick BE, Anseth KS, Bowman CN (2021) Chem Rev 121:6915–6990
14. Witczak ZJ, Bielski R (2013) Click chemistry in glycoscience: new development and strategies. Wiley-VCH John Wiley & Sons, New York
15. Chandrasekaran S (2016) Click reactions in organic synthesis. Wiley-VCH, John Wiley & Sons, New York
16. Ramachary DB, Ramakumar K, Narayana VV (2008) Chem Eur J 14:9143–9147
17. Danence LJT, Gao Y, Li M, Huang Y, Wang J (2011) Chem Eur J 17:3584–3587
18. Belkheira M, Abed DE, Pons JM, Bressy C (2011) Chem Eur J 17:12917–12921
19. Yin D, Li C, Li B, Tao L, Yin D (2005) Adv Synth Catal 347:137–142
20. Rodriquez M, Sega A, Taddei M (2003) Org Lett 5:4029–4031
21. Aggarwal A, Lancaster NL, Sethi AR, Welton T (2002) Green Chem 4:517–520
22. Nino AD, Merino P, Algieri V, Nardi M, Gioia MLD, Russo B, Tallarida MA, Maiuolo L (2018) Catalysts 8:364
23. Barrow AS, Smedley CJ, Zheng Q, Li S, Dong J, Moses JE (2019) Chem Soc Rev 48:4731–4758
24. Mishra A, Agrahari AK, Tiwari VK (2021) Application of CuAAC in carbohydrates. In: Click chemistry volume (ed. Rutjes), Science of synthesis, pp 135–179
25. Jaiswal MK, Tiwari VK (2023) Chem Rec e202300167
26. Kumar GS, Lin Q (2021) Chem Rev 121:6991–7031
27. Kayet A, Pathak T (2013) J Org Chem 78:9865–9875
28. Agard NJ, Prescher JA, Bertozzi CR (2004) J Am Chem Soc 126:15046–15047
29. Chang PV, Chen X, Smyrniotis C, Xenakis A, Hu T, Bertozzi CR, Wu P (2009) Angew Chem Int Ed 48:4030–4033
30. (a) Wittig G, Krebs A (1961) Chem Ber 94:3260–3275; (b) Michael A (1893) J Prakt Chem 48:94
31. Castro V, Rodríguez H, Albericio F (2016) ACS Comb Sci 18:1–14
32. Hema K, Ravi A, Raju C, Pathan JR, Rai R, Sureshan KM (2021) Chem Soc Rev 50:4062–4099
33. Hema K, Sureshan KM (2019) Acc Chem Res 52:3149–3163

34. Biradha K, Santra R (2013) Chem Soc Rev 42:950–967
35. Li J, Tian X-Y, Zong L-P, Zhang Q, Zhang X-J, Marks R, Cosnier S, Shan D (2019) ACS Appl Mater Interf 11:32366−32372
36. Shen J, Okamoto Y (2016) Chem Rev 116:1094–1138
37. Megens RP, Roelfes G (2011) Chem Eur J 17:8514–8523
38. Medina DD, Goldshtein J, Margel S, Mastai Y (2007) Adv Funct Mater 17:944–950
39. Pathigoolla A, Gonnade RG, Sureshan KM (2012) Angew Chem Int Ed 51:4362–4366
40. Rondeau-Gagné S, Néabo JR, Desroches M, Larouche J, Brisson J, Morin JF (2013) J Am Chem Soc 135:110–113
41. Gronwald O, Shinkai S (2001) Chem Eur J 7:4329–4334
42. Pathigoolla A, Sureshan KM (2013) Angew Chem Int Ed 52:8671−8675
43. Hema K, Sureshan KM (2019) Angew Chem Int Ed 58:2754−2759
44. Hema K, Sureshan KM (2018) Cryst Eng Commun 20:1478–1482
45. Krishnan BP, Ramakrishnan S, Sureshan KM (2013) Chem Commun 49:1494–1496
46. Li W, Li X, Zhu W, Li C, Xu D, Ju Y, Li G (2011) Chem Commun 47:7728–7730
47. Krishnan BP, Rai R, Asokan A, Sureshan KM (2016) J Am Chem Soc 138:14824–14827
48. Pathigoolla A, Sureshan KM (2014) Angew Chem Int Ed 53:9522–9525
49. Ravi A, Sureshan KM (2018) Angew Chem Int Ed 57:9362–9366
50. Mishra KB, Tiwari VK (2014) J Org Chem 79:5752–5762
51. Mishra KB, Shashi S, Tiwari VK (2015) RSC Adv 5:86840–86848
52. Mishra KB, Agrihari AK, Tiwari VK (2017) Carbohydrate Res 450:1–9
53. Mishra KB, Mishra RC, Tiwari VK (2015) RSC Adv 5:51779–51789
54. Mishra KB, Tiwari N, Bose P, Singh R, Rawat AK, Mishra RC, Singh RK, Tiwari VK (2019) Chemistry Select 4:2644–2648
55. Mishra KB, Shashi S, Tiwari VK (2016) Trends Carbohydrate Res. 9(2):53–58
56. Dellisola A, McLachlan MMW, Neuman BW, Al-Mullah HMN, Binks AWD, Elvidge W, Shankland K, Cobb AJA (2014) Chem Eur J 20:11685–11689
57. Adhikary ND, Chattopadhyay P (2012) J Org Chem 77:5399–5405
58. Oliva AI, Christmann U, Font D, Cuevas F, Ballester P, Buschmann EH, Torrens A, Yenes S, Perica's MA (2008) Org Lett 10:1617–1619
59. Hotha S, Anegundi RI, Natu AA (2006) Tetrahedron Lett 46:4585–4588
60. Balducci E, Bellucci L, Petricci E, Taddei M, Tafi A (2009) J Org Chem 74:1314–1321
61. Mishra KB, Tiwari VK (2016) ChemistrySelect 1:3693–3698
62. De Nino A, Maiuolo L, Merino P, Nardi M, Procopio A, Roca-Lõpez D, Russo B, Algieri V (2015) ChemCatChem 7:830–835
63. Sahu D, Dey S, Pathak T, Ganguly B (2014) Org Lett 16:2100–2103
64. Kayet A, Dey S, Pathak T (2015) Tetrahedron Lett 56:5521–5524
65. Kumar N, Ansari MY, Kant R, Kumar A (2018) Chem Commun 54:2627–2630
66. Nanjundaswamy HM, Abrahamse H (2015) Synth Commun 45:967–974
67. Bubyrev A, Malkova K, Kantin G, Dar'in D, Krasavin M (2021) J Org Chem 86:17516–17522
68. Driggers EM, Hale SP, Lee J, Terrett NK (2008) Nat Rev Drug Disc 7:608–624
69. Otvos SB, Fulop F (2015) Catal Sci Tech 5:4926–4941
70. Bogdan AR, Jerome SV, Houk KN, James K (2012) J Am Chem Soc 134:2127–2138
71. Bogdan AR, James K (2011) Org Lett 13:4060–4063
72. Arora I, Kashyap VK, Singh AK, Dasgupta A, Kumar B, Shaw AK (2014) Org Biomol Chem 12:6855–6868
73. Basavaiah D, Reddy BS, Lingam H (2013) Tetrahedron 69:10060–10067
74. Banerji B, Pramanik SK, Sanphui P, Nikhar S, Biswas SC (2013) Chem Biol Drug Des 82:401–409
75. Arigela RK, Mandadapu AK, Sharma SK, Kumar B, Kundu B (2012) Org Lett 14:1804–1807
76. Maurya RA, Adiyala PR, Chandrasekhar D, Reddy CN, Kapure JS, Kamal A (2014) ACS Comb Sci 16:466–477
77. Ning Y, Wu N, Yu H, Liao P, Li X, Bi X (2015) Org Lett 17:2198–2201

78. Nekkanti S, Pooladanda V, Veldandi M, Tokala R, Godugu C, Shankaraiah N (2017) ChemistrySelect 2:7210–7221
79. Sastry KNV, Routhu S, Datta SG, Nagesh N, Babu BN, Nanubolu JB, Kumar CG, Maurya RA, Kamal A (2016) Org Biomol Chem 14:9294–9305
80. Majumdar KC, Ganai S (2013) Tetrahedron Lett 54:6192–6195
81. Senwar KR, Sharma P, Reddy TS, Jeengar MK, Nayak VL, Naidu V, Kamal A, Shankaraiah N (2015) Eur J Med Chem 102:413–424
82. Dhondge AP, Afraj SN, Nuzlia C, Chen C, Lee GH (2013) Eur J Org Chem, 4119–4130

Chapter 6
Bioorthogonal Click Chemistry: Invention to Applications in Living Systems

6.1 Introduction

The bioorthogonal concept was established by Prof. Carolyn R Bertozzi which after then attained remarkable importance in various areas [1, 2]. Within the framework, bioorthogonal chemistry refers to a set of reactions that occur within the (i) biological pH and temperature, without being affected by water, electrophiles or nucleophiles, oxidants, or reductants present in the biological system, (ii) the selective reaction should be sufficiently rapid and effective so that ligation can occur before the clearance of reactant from the studied organism, and (iii) product formed should be stable and should be easy to incorporate the reactant into biomolecules by metabolic or protein engineering [3, 4]. These reactions were invented to modify the biological systems in many ways that were previously impossible. Bioorthogonal reactions involve the incorporation of small functionalities like azides, that have excellent accessibility as metabolic precursors [5, 6]. Early 2000, bioorthogonal reactions were mainly classified as chemoselective ligation reactions involving, for example, carbonyl ligations [7–9], the Staudinger ligation [10], 1,3-dipolar cycloadditions [11], Click Chemistry has transformed chemical and biological sciences over the past two decades. This powerful and accessible technology enables the construction of complex molecules by scientists outside the elite synthetic circles. The history of click reaction started, in 1893, Michael first reported the 1,3-dipolar cycloaddition reaction between organic azides and alkynes, which yields the formation of 1,2,3-triazoles [12]. This reaction was further extensively explored by Huisgen during the 1960 ara [13]. The versatility and easy generation of azides and alkynes functionality, which remains stable during the reaction course, led to wide acceptance of this reaction in organic synthesis. Huisgen-thermal azide-alkyne cycloaddition (AAC) reaction was considered to generate many triazole-linked hybrid molecules for important application [14–19].

V. K. Tiwari et al., *Click Chemistry*, Materials Horizons: From Nature to Nanomaterials,
https://doi.org/10.1007/978-981-97-4596-8_6

However, this reaction has many limitations, such as long reaction time, high temperature, and non-regioselective triazole (both 1,4- and 1,5-disubstituted) formation. Building upon these foundational reactions, Sharpless and Meldal independently developed copper-catalyzed azide-alkyne cycloaddition (CuAAC) in early 2002 [11, 20]. By having the exceptionally fast kinetics, high yields, benign reaction conditions, copper-catalyzed azide-alkyne cycloaddition (CuAAC), rapidly accepted by the researchers in chemical biology for bioconjugation [21–23]. On the other hand, traditional chemoselective ligation synthesis of biomolecules was challenging under physiological conditions because of its metal toxicity [7]. Therefore, extending the click concept Carolyn R. Bertozzi pioneered copper-free click reaction compatible with living systems, enabling cell-surface engineering and glycan imaging [24]. Furthermore, Bertozzi group broadened the scope of biorthogonal chemistry by employing modified strain-promoted azide-alkyne cycloaddition reaction. The SPAAC (strain-promoted azide-alkyne cycloaddition reaction) proceeds via the release of ring strain within a cyclic octyne, eliminating the need for the catalyst and making it biocompatible with the biological system. However, both CuI-catalyzed [21, 22] and Cu-free click chemistry (SPAAC) are mainstays of chemical biology applications for beyond [25, 26], while other several powerful click reactions have emerged in recent years, including Staudinger ligation, inverse-electron-demand Diels–Alder cycloaddition Tetrazine-Alkene ligation, photoinduced ligation, etc. (Fig. 6.1) [27–29].

The first attempt to image the glycans over the cell surface was done by Bertozzi group in 1997. An unnatural derivative of mannosamine containing ketone was metabolically converted to corresponding sialic acid displaying the ketone group. Further, the ketone was ligated with hydrazide to image the surface [8]. Bioorthogonal chemistry enables organic synthesis reactions to be executed in the biological system without interfering with native biochemical processes, allowing for the site-specific modification and study of biomolecules in living systems [30]. The integration of bioorthogonal chemistry and biomedical nanotechnology has created new opportunities for accurate diagnosis and targeted treatment of cancerous tumors [31]. Biorthogonal click chemistry also facilitates robust spatiotemporal control over the integration of biochemical factors into the synthetic hydrogels to alter the cell behavior [32]. The presented workflow for high throughput reaction profile computation to screen for the bioorthogonal reaction, as well as for the development of novel machine learning models [33]. Bioorthogonal chemistry depicted the great potential across reproductive medicine by empowering tracking of biomolecules in gametes to comprehend development and function and targeted drug delivery [34]. The recent awardees of the 2022 Nobel Prize in Chemistry, Carolyn R. Bertozzi, Morten Meldal, and K. Barry Sharpless, provide the foundational impact of click chemistry and bioorthogonal chemistry across diverse disciplines. The pioneering work facilitated the simplicity, efficiency, and reliability of click reactions for synthesizing molecules and materials (Sharpless and Meldal) along with designing biocompatible reactions to visualize and probe biomolecules in living systems (Bertozzi's bioorthogonal concept) [35–37]. The joining of versatile biomolecules such as protein, antibodies, nucleic acids, lipids, and carbohydrates with molecules like fluorescent tags, and

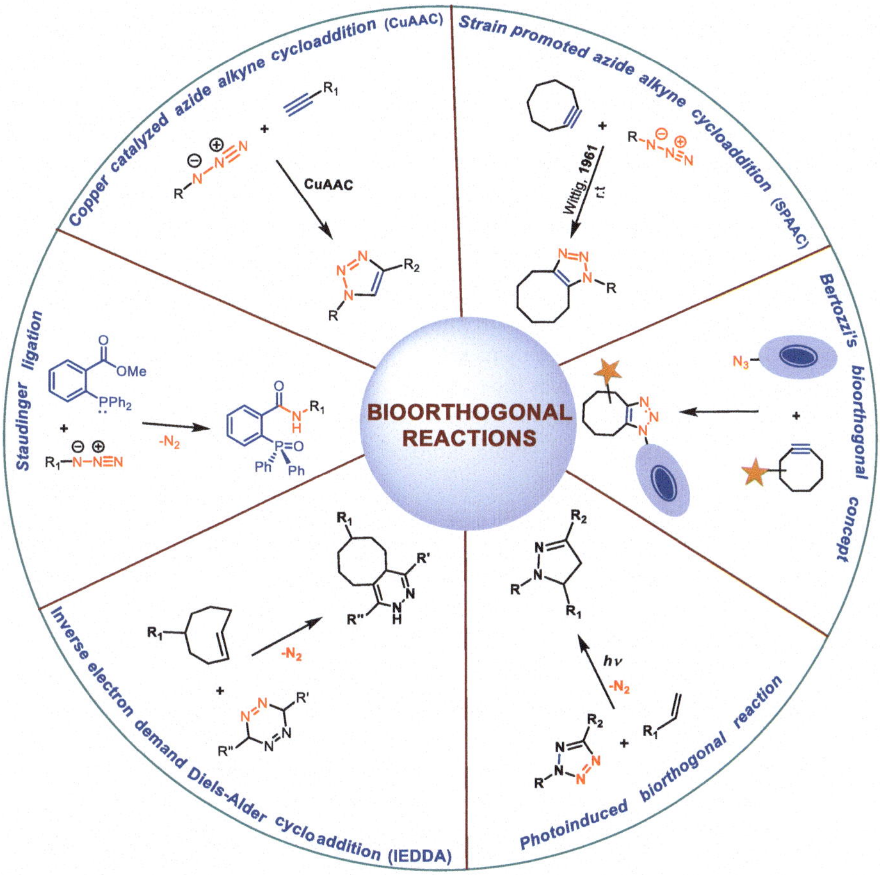

Fig. 6.1 Various types of ligation techniques for the biorthogonal reactions

drugs. This elevates the capacity of proteins to treat diseases both as therapeutics and as drug carriers by virtue of their complex functional properties. The bioconjugation tool has several notable pharmaceutical applications including drug delivery, biocompatibility, therapeutics, diagnostics, and pharmacologically [38].

This groundbreaking innovation laid the foundation for click chemistry to biological systems through the emerging field of bioorthogonal chemistry. Following the Nobel Laureates accomplishments, this chapter begins with a concise introduction to the development of bioorthogonal reactions and their rising significance. The discussion then outlines methods for incorporating bioorthogonal handles into biological systems including Staudinger ligation, CuAAC, copper-free click chemistry, and tetrazine ligation by highlighting successful applications in areas such as imaging, labeling, and targeted drug delivery [39].

6.2 Staudinger Ligation

Staudinger ligation was the earliest biorthogonal ligation reaction developed so far. In the classical Staudinger reaction, azides **1** selectively react with phosphines **2** to produce an iminophosphorane intermediate **3** (aza-ylide) (Scheme 6.1) [40]. In the presence of water, this resulting ylide immediately hydrolyzes to form amine **4** and corresponding phosphine oxide. The good thing about these ligating partners in classical Staudinger is unreactive to inside and on surface of cells. The Staudinger reaction implemented the azide, a small biocompatible group easily introduced into biomolecules. It reacts selectively with phosphines [41–43].

Since today, there have been many modifications done by several scientists, among which Bertozzi's modification was the most interesting, incorporating an electrophile trap like methyl easter **5** into the phosphine, where iminophosphorane **6** reacts with the ester group to generate the stable amide **7** linkage connecting two reactants (Scheme 6.2a) [10].

Highly specific chemical reactions that take place within cells can serve as useful tools for understanding biological processes or engineering novel cell interactions [44]. The authors demonstrated a chemical reaction that allows the selective formation of covalent bonds between complex biomolecules inside cells. The reaction was modeled on the Staudinger reaction, which joins azides and specially designed triphenylphosphines together via an amide bond. Additionally, the authors introduced azide groups into glycoproteins on the cell surface by incubation of azido sugars with **5**, which were then incorporated into glycoproteins through metabolism. Further, it reacted the azide-containing cell-surface glycoproteins **6** with a biotinylated triphenylphosphine **7**, forming stable modifications on the cell surface (Scheme 6.2b). Because this highly selective reaction avoids interfering with native biomolecules, the authors proposed that it could also be used inside cells to probe intracellular interactions in new ways [10].

The first well-tolerated bioorthogonal reaction by living beings was demonstrated by the Bertozzi group using modified Staudinger ligation in 2004. Azidoacetyl mannosamine **12** was administered for 7 days. On the 8th day, mice were euthanized and their splenocytes (rich in sialic acid) were harvested. Then, the presence

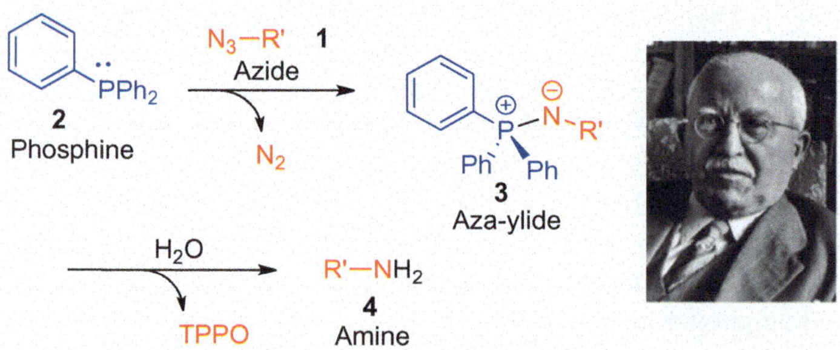

Scheme 6.1 Classical Staudinger reaction with reaction mechanism

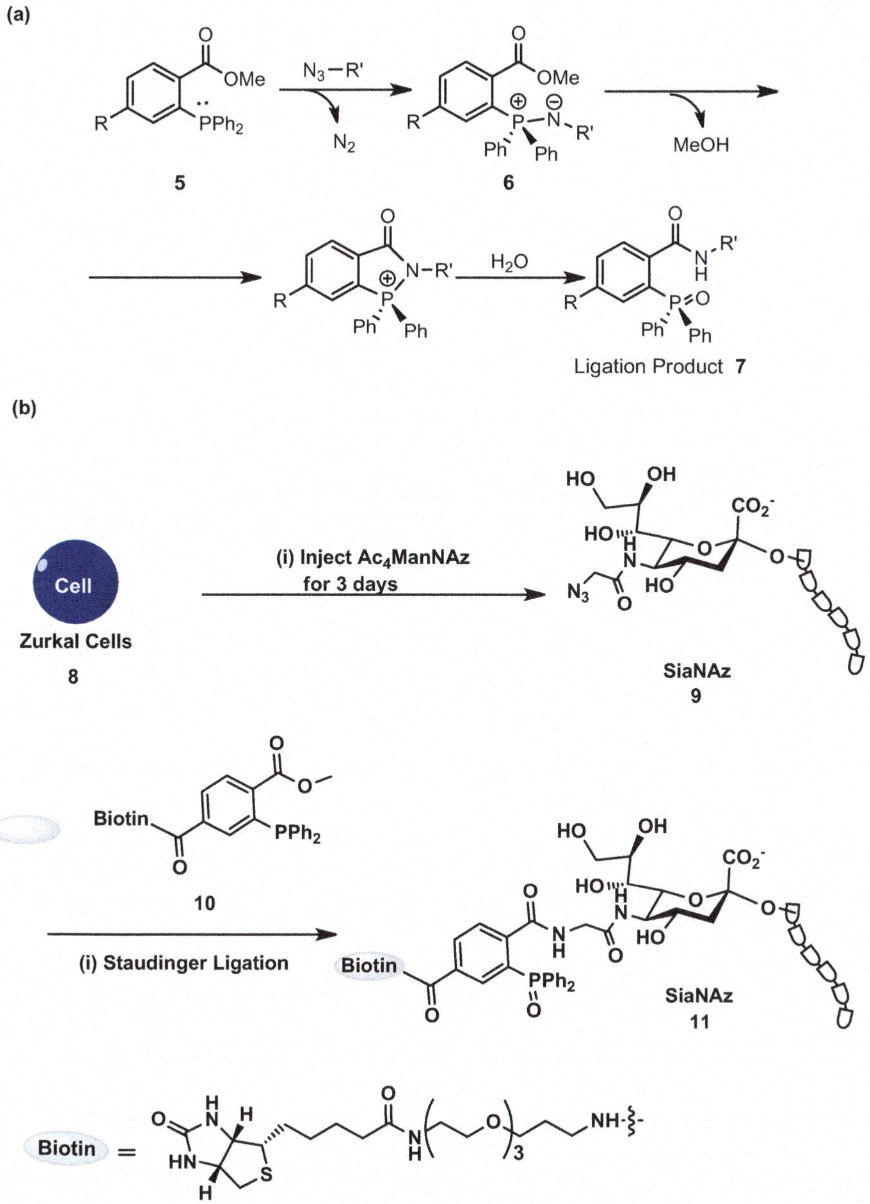

Scheme 6.2 **a** Mechanism of modified Staudinger ligation. **b** Metabolic glycan labeling using the Staudinger tool [10]

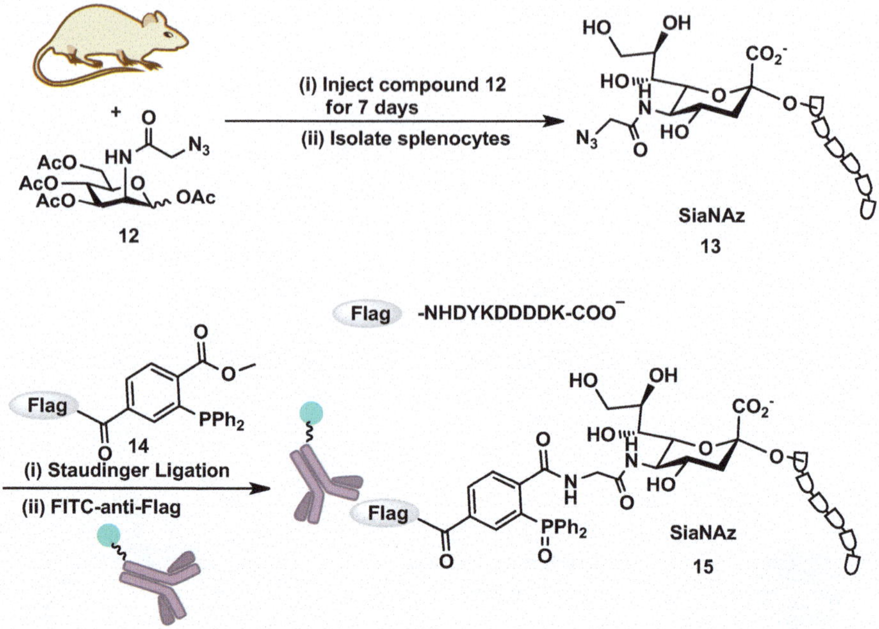

Scheme 6.3 First tolerated bioorthogonal reaction in live mice by employing Staudinger ligation [45]

of cell-surface azides **13** was reacted with phosphine probe **14** comprising a Flag peptide. Further, treatment with a fluorescein isothiocyanate-labeled anti-Flag antibody (FITC-anti-Flag) followed by cytometry analysis (Scheme 6.3). The obtained data confirmed that Ac_4ManNAz metabolized by the cells in the body of the mice which altered to its corresponding azido sialic acid and appeared on the surface of the cells [45].

Cell-surface glycan is involved in several important physiological processes such as cell–cell interaction, lymphocyte trafficking, cancer metastasis, and many more. Briendle et al. have employed the bioorthogonal metabolic labeling tool to identify cell-surface glycans and showcase the first time, fluorescence and radionuclide imaging of the sialylated glycans in a murine tumor model. Peracetylated azido-labeled mannosamine (Ac_4ManNAz) **16** was first injected intraperitoneally for the biosynthesis of azido sialic acid analogue **17** followed by its coupling with biotinylated phosphine **18** under standard Staudinger ligation condition. Moreover, the detection of biotin was done through succeeding intravenous injection of radiolabeled or fluorescent avidin analogue. After a day's administration of neutravidin when clicked with a fluorophore, there was substantial growth in tumor-to-tissue contrast that depended on the azido-labeled *N*-acetyl mannosamine. This enhanced contrast was utilized through optical imaging of SPECT (single electron computed tomography). By imaging altered glycoprotein could be used to track disease progression in patients (Scheme 6.4) [46].

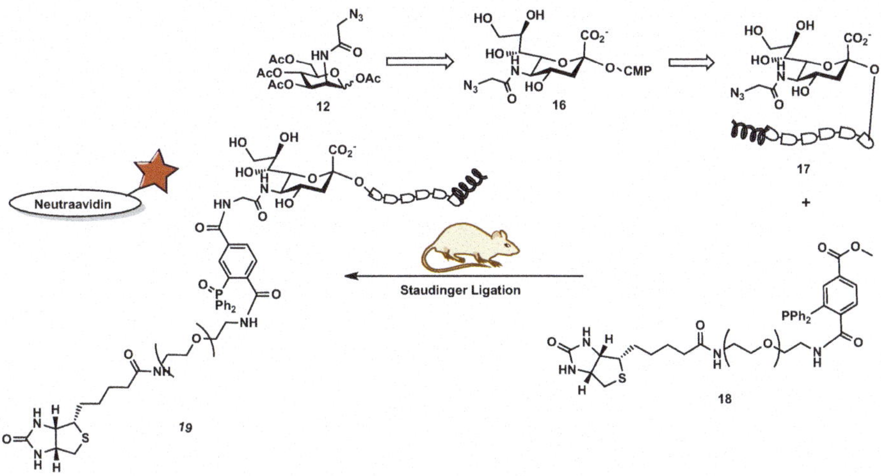

Scheme 6.4 Metabolic labeling of live mice by intraperitoneal injection of AC$_4$ManNAz [46]

Next, Ramstrom and Yan reported the fast Staudinger reaction between the perfluoroaryl azides (PFFAs) and aryl phosphines, with an enhanced rate constant of up to 18 Ml^{-} in CH_3CN/D_2O (1:1). The iminophosphorane ylide was found to be very stable toward hydrolysis, that's throw out the need of easier trap in Staudinger ligation. PFFA-derived acetylated mannosamine (**22a**), galactosamine (**22b**), and glucosamine (**22c**) were incubated with A549 cells **20** for three days, washed and clicked with phosphine-BSA-FITC (**23** or **24**). After then, labeling was analyzed by flow cytometry, which in turn afforded clear images with low background (Scheme 6.5). The Staudinger ligation has been superseded in many bioorthogonal chemistry applications by faster ligations [47].

However, the major drawback of the Staudinger ligation is the slow kinetics of this reaction, which leads to slow labeling. In contrast, electron-rich phosphine undergoes rapid oxidation under physiological condition. Albeit, the high utility of Staudinger ligation, there were many advancements in the biorthogonal chemistry was done by the research to enhance the rate of reaction. In the same context, it turns out that 1,3-dipolar azide-alkyne cycloadditions click chemistry Cu(I) is incredibly fast and can be utilized for the bioorthogonal reaction.

6.3 Copper-Catalyzed Azide-Alkyne Cycloaddition Reaction (CuAAC) Ligation

Prof. K. B. Sharpless, a Nobel laureate, first introduced the term "click chemistry" in 1999 at 217th American Chemical Society annual meeting. According to him, the reaction which meet certain criteria: easy to perform, be modular, has a

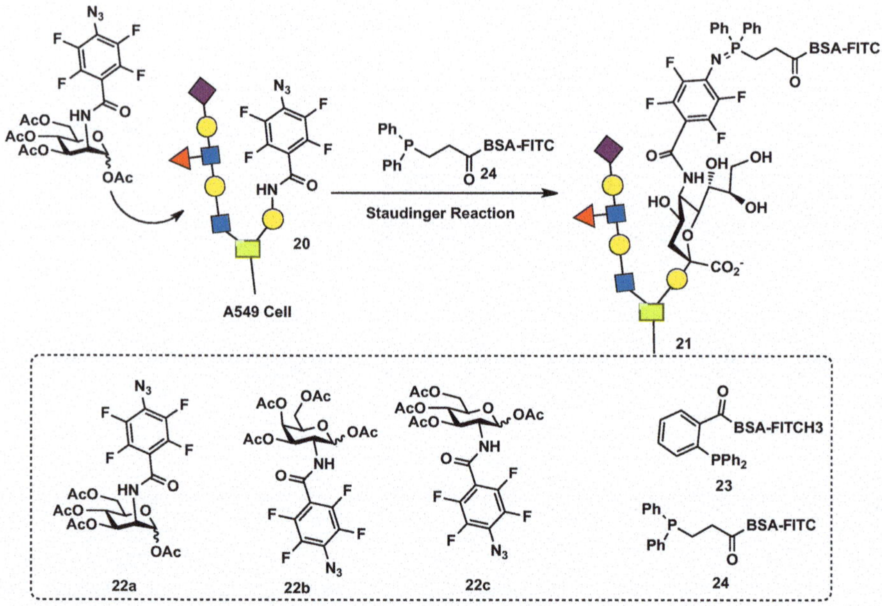

Scheme 6.5 Cell-surface labeling experiment by implementing modified Staudinger reaction [47]

broad substrate scope, and mild reaction condition, gives good to excellent yields, easy isolation (mostly without chromatography), producing non-toxic byproducts that can be easily removed, regioselective and regiospecific in nature, use readily available starting material and reagents, and use environmentally benign solvent. In 2002, Sharpless [11] and Meldal [20] independently discovered that a Cu(I) catalyst could afford 1,4-disubstituted triazolyl products in good to excellent yield. Owing to the remarkable acceleration of kinetics compared to the uncatalyzed reaction, they have revived interest in the AAC reaction [23, 48, 49]. The joining of versatile biomolecules with fluorescent tag candidates can be made easy by utilizing modular CuAAC click chemistry. This elevates the capacity of proteins to treat the diseases both as therapeutics and as drug carriers by virtue of their complex functional properties [50]. Bioconjugation has several pharmaceutical applications including drug delivery, biocompatibility, therapeutics, diagnostics, and pharmacologically [51–56]. However, the use of CuAAC with living things was challenging because the copper catalyst reactions are poisonous to living systems. The Cu(II) precursors used in CuAAC reactions cause oxidative damage to cells, while Cu(I) is readily oxidized to Cu(II), needing extra reducers like ascorbate which byproducts can also hurt to cells [57–60]. To make CuAAC technique, suitable for living things, a much effort has been made to stabilize Cu catalyst using various triazolyl ligands [61–63]. In this context, the tris(triazolyl)amine ligands have shown success by enhancing the reaction rate and by reducing the cytotoxicity of the copper catalyst. The fast rate of CuAAC has made it a commonly used reaction in biorthogonal chemistry.

The biological importance of the fucosylated glycoproteins, particularly associated with metabolic processes, has the significant interest. These glycoproteins are present over the glycoproteins in human T-lymphoma Jurkat cells. The authors employed the copper-catalyzed click chemistry to conjugate the azide-modified lysates **26a-c** with the biotin-based alkyne **27** in the presence of $CuSO_4$, TCEP (triazolyl ligands), and the western blot was analyzed with an a-biotin antibody horseradish peroxidase (HRP) conjugate (Scheme 6.6) [64]. It concluded that fucosylated glycoproteins can be labeled in humans cells using a specific compound **25c**. This selectivity suggests that salvage pathway enzymes have specific substrate tolerances.

Here, the authors described a chemical approach focused on the labeling and spotting fucosylated glycoproteins from cultured Bacteroidales. To obtain, Bacteroides were dosed with an alkyne-carrying fucose (**29**) for metabolically incorporation of an alkyne into the bacterial surface fucosylated glycoproteins. The alkyne-carrying glycoprotein **30** can then be clicked with biotinylated azide **31** by bioorthogonal

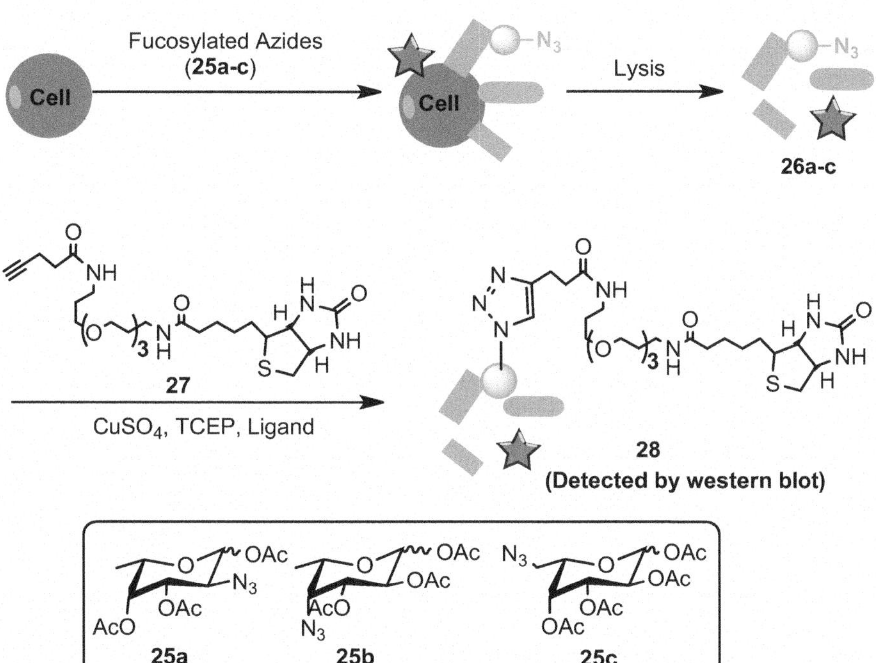

Scheme 6.6 Metabolic labeling of cellular glycoproteins with azido fucose entities for identification of glycoproteins [64]

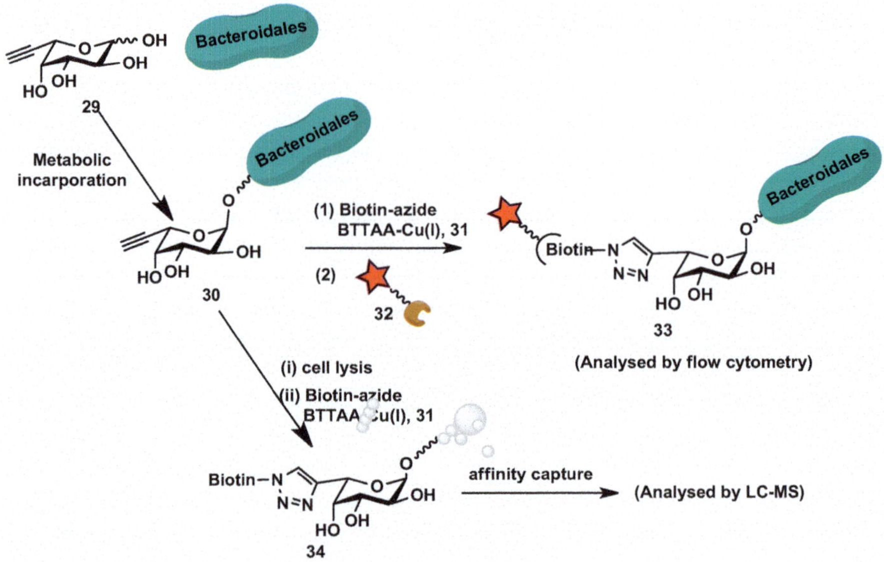

Scheme 6.7 Detection and labeling of fucosylated glycoprotein in Bacteroidales species using click tool [65]

(CuAAC) click chemistry followed by fluorescent tagging **32** to afford the fluorescent tagged glycoprotein **33** for the cytometry analysis. On the other hand, the alkyne-linked glycoprotein then specifically detected by LC–MS using azido probes by forming respective products **34** (Scheme 6.7). The precision and perceptiveness of this chemical strategy will allow for augmentation of fucosylated glycoproteins for the proteomic study [65].

To find the structure and function of fucosylated glycan, the authors developed an interesting fluorescent labeling tool, which demonstrates a comparison between tumor and normal cells. In this context, Wong and co-workers developed fluorogenic probes that utilize bioorthogonal chemical reactions to facilitate easy biomolecular imaging and detection [66]. The authors have incorporated a non-fluorescent precursor 4-alkynyl/azide-N-ethyl-1,8-naphthalimide **35** which triggered upon clicking with 6-modified (azido/alkyne) fucose analogues (**36a,b**) to generate a fucose-based triazole fluorescent probe **37** (Scheme 6.8). In addition, fluorescence microscopy studies then revealed intracellular localization of the glycoconjugate product (**38**). To identify the glycan-related biomarkers and therapeutics was the key goal of this strategy [67].

Bertozzi and Wu report established that the peracetylated *N*-(4-pentynoyl)mannosamine ($Ac_4ManNAl$) can be metabolically labeled sialic acids in cultured cells and mice with enhanced efficacy in comparison with $Ac_4ManNAz$. The authors synthesized $Ac_4ManNAl$ and injected in mice for seven days. Then, the mice were harvested and homogenized. To probe the existence of glycoprotein-linked alkynes **39** in the soluble fraction of homogenates, biotin-azide (**40**) was first

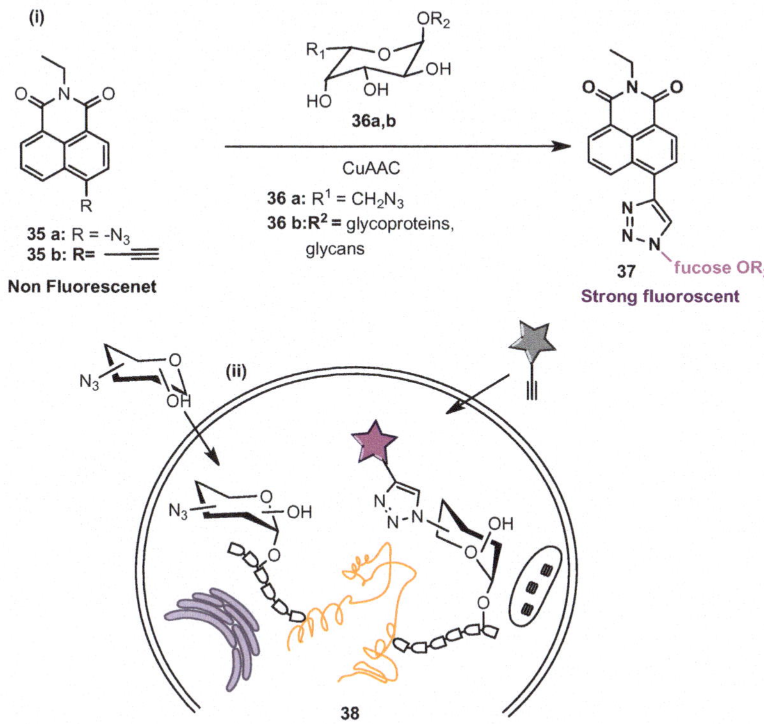

Scheme 6.8 Specific fluorescent labeling of fucosylated glycan in the cells using bioorthogonal click tool [67]

clicked using CuAAC technique followed by Western blot analysis (Scheme 6.9) [68].

Fluorogenic probes activated by bioorthogonal chemical reactions can enable biomolecule imaging in situations where unbound probes cannot be washed away. It had been anticipated by the authors that triazole formation would lower the aryl E-HOMO value relative to that of azide-functionalized substrate, which resulted in fluorescence turn-on. By taking this advantage, a highly selective fluorogenic azido-fluorescein analogue, 4-azidonaphthyl fluorescein (**42**), was developed to label protein-containing alkynes presented over the glycoproteins. A 3-day metabolic labeling of Chinese hamster ovary cells by using peracetylated *N*-(4-pentynoyl) mannosamine ($Ac_4ManNAl$) resulted to corresponding alkynyl sialic acid **41** which was then incorporated on the cell surface. The cells were fixed and then stitched with 4-azido-naphthylfluorescein **42** using $CuSO_4$, NaAsc and TBTA for 1 h, to investigate the imaging of cells without washing (Scheme 6.10). However, designing red-shifted fluorophores is interesting for future direction [69].

However, target identification in pharmacological research is heavily relies on activity-based probes, photoability labeling, or affinity chromatography [70]. In this

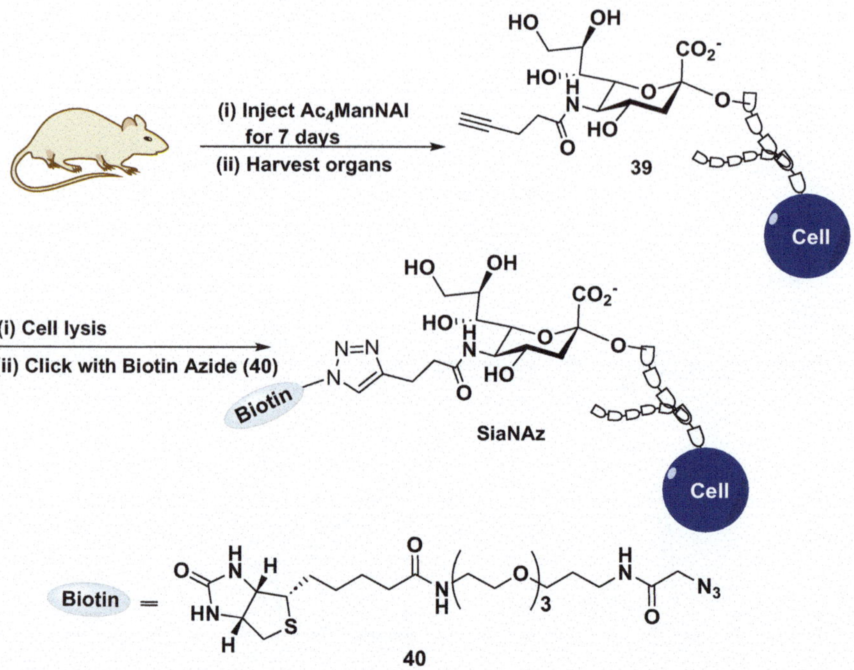

Scheme 6.9 Metabolic labeling of cellular glycans with $Ac_4ManNAl$ and detection using CuI-catalyzed CuAAC chemistry using TBTA ligand [68]

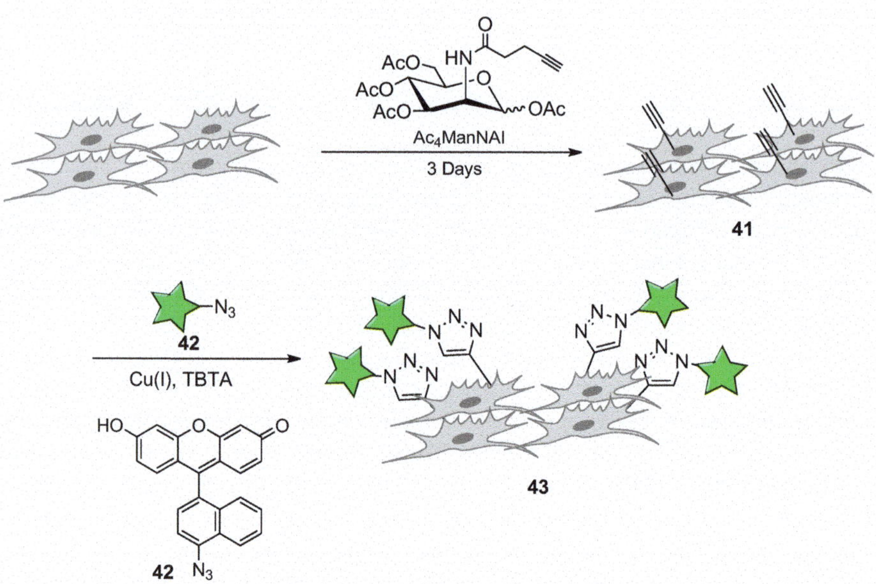

Scheme 6.10 Cell labeling by clicking with 4-azidonaphthylfluoresceinfluorogenic **42** [69]

context, Perrin et al. explored the click-to-kill approach while synthesizing a cytotoxic amanitin analogue for bioorthogonal conjugation. Additional studies revealed the propargyl-amide analogue had comparable cytotoxicity to natural amanitin [71].

Ballell and co-workers reported a multivalent photoaffinity probe to identify galectin-3 in a protein mixture. For this, lactose-derived azide was stitched with 3,5-dipropargyloxybenzoic acid under microwave heating for 20 min, giving the triazole product in good yield. The obtained acid motif was then coupled with propargyl amine, followed by deacetylation to afford the galectin-3 labeling agent (divalent probe **44**). It was further incubated with cell lysates rich in galectin-3 followed by irradiation at 366 nm. Finally, to visualize, bioorthogonally clicked with dye-azide **45** via a second click step using $CuSO_4$/TCEP under mild conditions (Scheme 6.11). This probe is selectively visualized galectin-3 mixtures and cell lysates which opens the possibility of future development. [72]

A similar approach was developed for the detection of galectin-1 carbohydrate-binding protein interaction. Ethylene glycol linkers were strategically attached to enhance the water-solubilizing properties. In addition, azide functionality **48** has

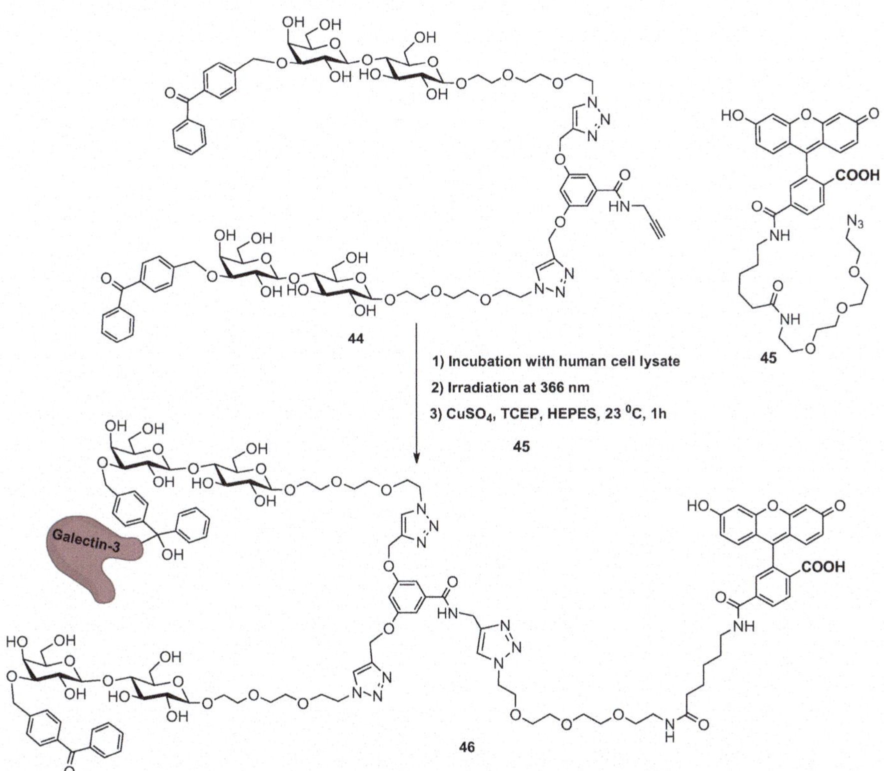

Scheme 6.11 Construction of a divalent photoaffinity probe for the selective labeling of galectin-3 [72]

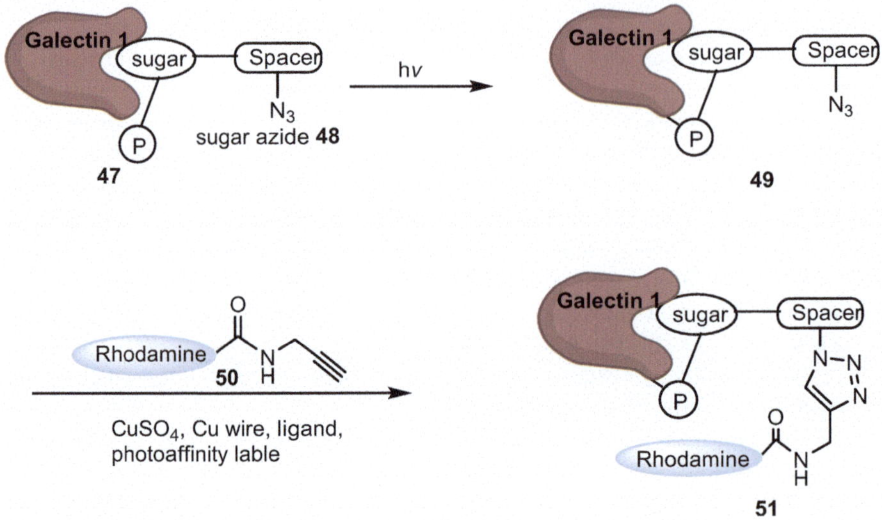

Scheme 6.12 Versatile approach for detection of labeled protein using click chemistry [73]

been incorporated at the terminus position for chemoselective ligation. The labeled protein **49** was then easily detected in gel by 'clicking on' a rhodamine tag **50** (Scheme 6.12). This straightforward as well versatile methodology enables simple detection and offers potential for further exploration [73].

Click chemistry has become a remarkable tool for non-invasive glycan imaging in zebrafish models, which enlightens the robust potential for rapid in vivo biomolecule imaging [74]. Ac_4ManNAz was incubated, then the incorporated as sia-azido glycan **52** over cell surface was clicked with alkyne probes **53a-b** rhodamine and biotin **53c** using $CuSO_4$, sodium ascorbate and the ligand tris(hydroxypropyltriazolyl)methylamine (THPTA) as a ligand to furnish scaffold **54**. This study highlights the reliability of cell-surface glycan labeling in mammalian cells without losing the cell viability (Scheme 6.13) [75].

With a similar approach, Soriano and colleagues resourcefully used the ligand BTTES and click condition to conjugate alkyne or azide glycan harboring cell surface with the respective biotin alkyne **55** or azide **56**, to obtain the biotin labeled cell surface **57** and **58** (Scheme 6.14). Then, it was stained with fluor 488-streptavidin and analyzed by flow cytometry. The authors reported that labeled glycan could be imaged in the outer layer of Zebrafish embryos in the early larval stage. This show-cause the versatility of their protocol for detecting cell-surface glycoconjugates in living cells [76].

Likewise, a robust and reliable method was developed for the labeling of cell-surface glycans using alkynated proteins **61** in the cell lysates. This method utilizes click ligation with ligands like BTTPS and BTTP coupled with picolyl azides. Compared to non-chelating azides, this approach showed a fivefold sensitivity in western blot assays. More interestingly, when an electron-donating picolyl azide

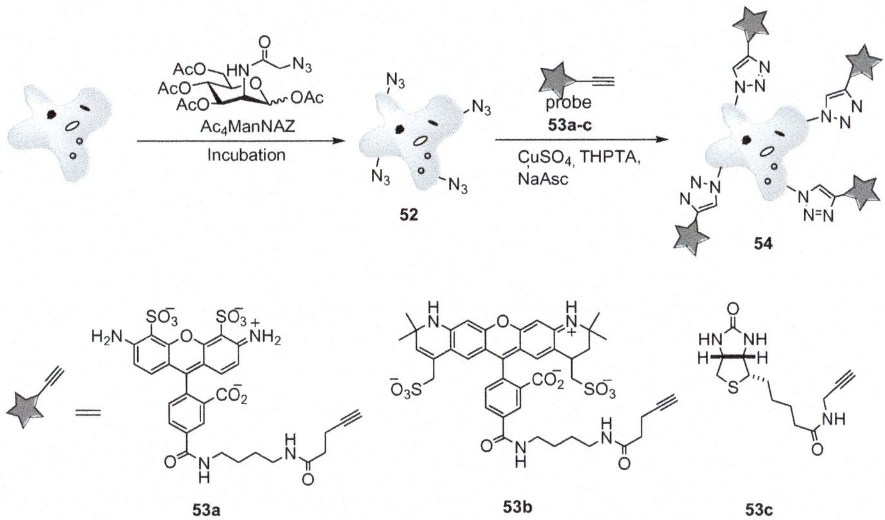

Scheme 6.13 Cell-surface labeling using the bioorthogonal copper click using triazolyl THPTA ligand [75]

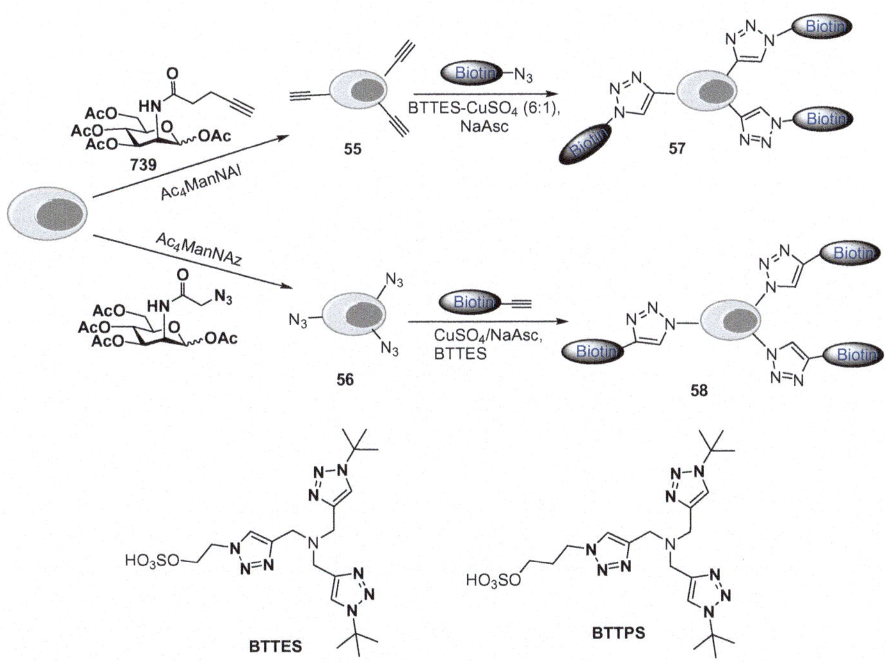

Scheme 6.14 Live cell identification of glycoconjugate using bioorthogonal click coupling [76]

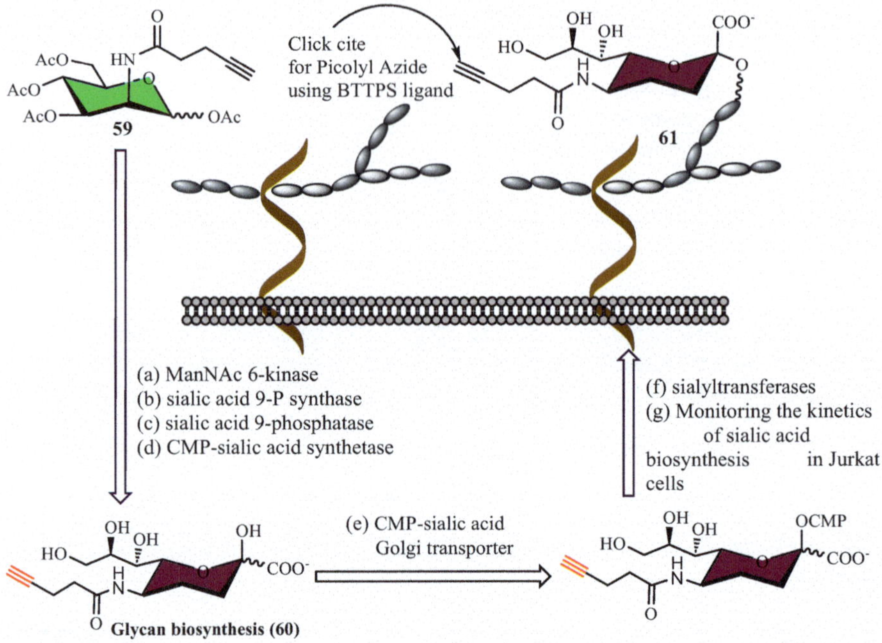

Scheme 6.15 Biosynthesis of modified sialic acid followed by labeling on cell surface using bioorthogonal chemistry [77]

was used for the click conjugation in living system, it imparted 20–38-fold higher efficiency with no detectable toxicity. In combination with BTTPS, the picolyl azide enabled the recognition click triggered cell-surface glycans *using* flow cytometry (Scheme 6.15), which allows the metabolic incorporation of glycans into cell surface with high specificity [77].

The one of the prominent techniques to prevent the tumor related diseases is anti-tumor vaccine. The ganglioside GM2 tumor associated carbohydrate antigen (TACA) **66** was chemically attached to alkyne functionalized virus like particle **62** derived from bacteriophage Qβ, [78] which ultimately incorporate 237 copies to each bacteriophage capsid **63**. In addition, the remaining free alkynes were quenched by using 3-azidopropan-1-ol **64** to result in the required Qb **65.** It is interesting to mention here, GM2 immobilized on Qβ through a thiourea linker elicited high titters of IgG antibodies, while clicked scaffold failed to induce enough anti-recognized GM2 positive tumor cells. The authors claimed the usefulness of bacteriophage being used as a suitable platform to enhance the antibody response (Scheme 6.16) [78].

Similarly, Finn and co-workers reported the click protocol to complete the chemoselective attachment of several significant molecules to cowpea mosaic virus (CPMV) in search of standard conjugation method [79]. CuAAC reactions are widely used to develop bioorthogonal probes for in vivo imaging and assessing biomolecular

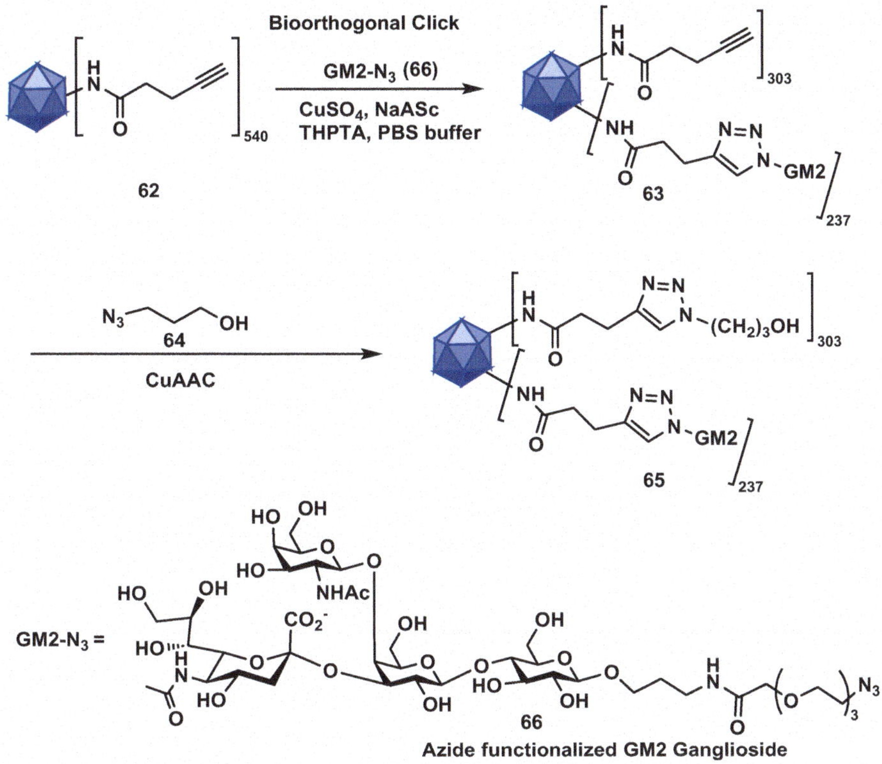

Scheme 6.16 GM2-Qβ **65** conjugates by using bioorthogonal click chemistry [78]

function. This strategy enables real-time detection and spatial/sequential tracking of targets in living systems. Numerous CuAAC-based fluorescent probes have been designed, especially for cell surfaces with glycans labeling [80] and studying microbial behavior. More applications in cell-surface engineering are expected in the future for further exploration of click chemistry [81]. Advancing click chemistry methodology requires developing biocompatible techniques and improving technological capabilities [82, 83]. However, the toxicity of the copper catalyst prevents its use for reactions involving live cells or organisms [48].

Cyclooctyne reagents can undergo cycloaddition with azides at rates similar to copper-catalyzed reactions. Hence, these cyclooctyne-azide cycloadditions are called ‘copper-free click chemistry’ [84]. Various cyclooctyne reagents have been utilized for bioorthogonal labeling of glycans on cultured cells and developing zebrafish [85], as well as for probing protein and phospholipid dynamics in live cells. This copper-free approach enables click chemistry applications in the presence of living systems [86, 87].

6.4 Strain-Promoted Azide-Alkyne Cycloaddition (SPAAC) Reaction: Copper-Free Click Chemistry

In 1961, Prof. Georg Wittig reported a reaction between the strained cyclooctyne **67** and phenyl azide **68**, which proceeded robustly to form the corresponding triazole **69** without the catalyst, which is also termed a copper-free cycloaddition reaction (Scheme 6.17) [88].

Copper-free click chemistry, also termed as strain-promoted azide-alkyne cycloaddition reaction (SPPAC), was first incorporated in biological system by Bertozzi [89–93]. The target was to eliminate the need for copper catalyst and improve the rate of slow kinetics of copper-free azide-alkyne cycloaddition reactions [94–96]. In SPPAC, the terminal alkyne is replaced with cyclic alkynes (mainly cyclooctynes). Cyclic octynes are strained due to sp-hybridized alkynyls carbons, which are normally oriented at the angles of 180°, which enhances the reaction rate by relieving the strain at the alkyne scaffold. There were many efforts have focused on improving cycloaddition kinetics by introducing electron withdrawing to cyclooctyne ring, which activates it toward nucleophile attack by the azide (Fig. 6.2). Further, Prof. Bertozzi, Baskin, and other co-workers developed the several modified cyclooctyne-based probes for copper-free click chemistry [97–99]. They tuned the reaction rate by modifying the cyclooctynes [100].

Recently, SPAAC protocol represents a potent bioorthogonal reaction for the cellular imaging [26]. In addition to the fast detection of azides in biological system, other important use of this copper-free non-toxic tool includes the quick molecular imaging and also the affinity tagging. The DIFO and its structural derivatives like

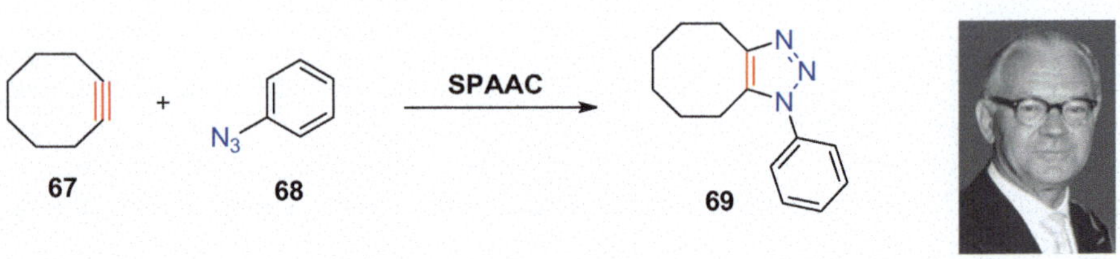

Scheme 6.17 Strain-promoted cycloaddition 'copper-free click chemistry' Wittig reaction [88]

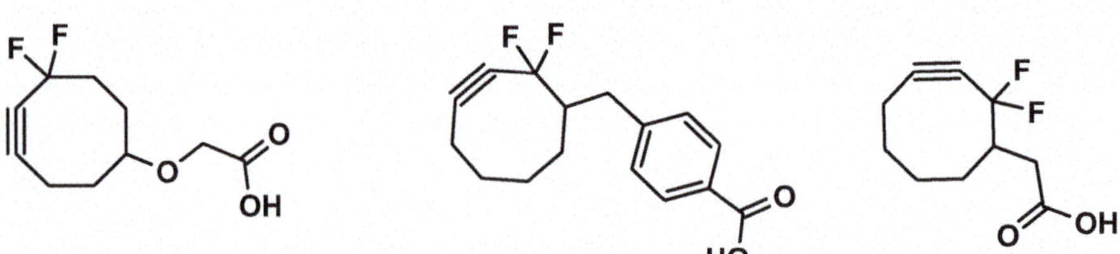

Fig. 6.2 Commonly used difluorinated cyclooctynes reagent **70–72** for copper-free click chemistry [100]

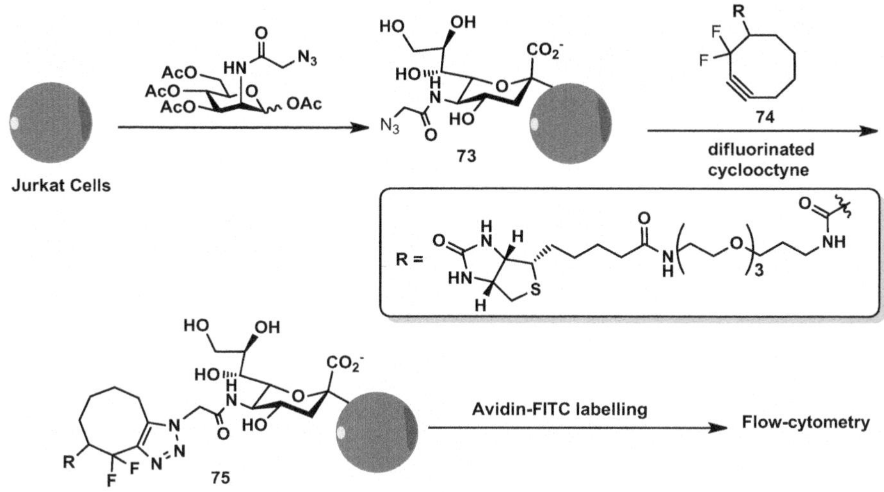

Scheme 6.18 Labeling of cell-surface glycan with azido sugar using copper-free click chemistry

biotinylated version **3** have recognized for their potential utilities for the labeling of cell-surface glycans with sugar azides under optimized biorthogonal conditions. For the required metabolic labeling of cell-surface glycans with azido sialic acid **73**, the Jurkat cells were first reacted with 25 μM per Ac_4Man*N*Az **12** for 3 days followed by washing and labeling with biotinylated reagent **74** either at different concentrations (0–10 μM) for I h or for different times (0–60 min) at 10 μM concentration. Altogether, the resulting cells were stained with FITC-avidin and analyzed by flow cytometry method. Both the second-generation DIFO reagents in bioorthogonal coupling with cell-surface azides showed concentration as well as time-dependent profiles. Importantly, this new DIFO reagent is found to be non-toxic to cells and also proceeds effectively for the selective tagging of azide-labeled biomolecules in living systems (Scheme 6.18).

It has been evident that such reagents have widespread applications in SPAAC chemistry for in vivo imaging and ex vivo labeling of biomolecules, and in developing new biocompatible materials [100–102].

Laughlin and Baskin demonstrated the invasive imaging of the sugar in the live cells of the zebrafish because these zebrafish are the very popular model organism to study the biology of vertebrate system having translucent advantage. Zebrafish embryos were treated with an azido N-azidoacetylgalactosamine **76** to metabolically label their cell surface with unnatural sugars and subsequently treated with fluorescent probe **77–79** by means of SPPAC, enabling the imaging of the glycan in vivo at subcellular resolution during development (Scheme 6.19). By employing a multicolor approach to detect glycans, the authors were able to conduct spatiotemporal mapping of glycan expression and transport. This study showcased patterns that would remain invisible using conventional molecular imaging technique [85].

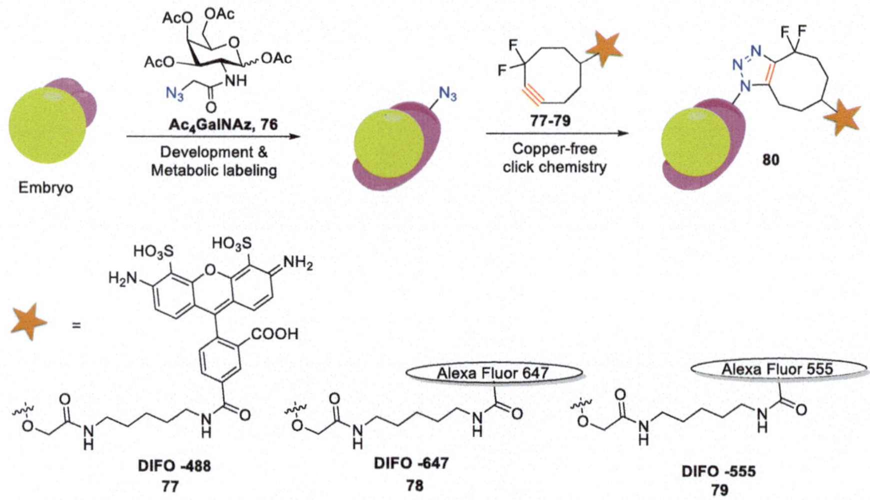

Scheme 6.19 Non-invasive imaging of glycans during zebrafish development by employing copper-free click chemistry [85]

Likewise, the same group reported an alternative bioorthogonal metabolic glycan labeling in live mice. Mice were incubated with per-acetylated N-azidoacetylmannosamine (Ac_4ManNAz) **12** to metabolically label cell surface with sialo-azides **80**. Further, injected with cyclooctynes (**81–85**) to couple with surface azides **80**. Analysis showed the successful labeling of glycoconjugate **86** isolated mouse spleen cells and tissue including intestine, heart, and liver without apparent toxicity. In addition, the cyclooctynes variants (**81–85**) displayed the differing labeling efficiency. Among the tested several cyclooctynes compounds for efficiency in labeling, DIFO-FLAG (**84**) displayed the highest intrinsic reactivity with azides and was most effective in tagging serum and tissue glycoproteins. DIMAC-FLAG (**83**) and ALO-FLAG (**85**) showed detected labeling compared to the OCT-FLAG (**81**) and MOFO-FLAG (**82**). In contrast to this, despite being a low intrinsic rate of PHOS-FLAG's has better labeling to splenocytes than DIFO-FLAG, which might be because of the low bioavailability of DIFO (Scheme 6.20). As the author stated, finding the optimized bioorthogonal reagent for in vivo application requires both intrinsic as well as pharmacokinetic possessions [104].

Li and co-workers reported an alternative strategy for the selective destruction of disease-relevant cells by the immune system. The strategy involves the installation of an antibody recruiting scaffold on the cell surface. The authors selectively delivered an azide-containing sialic acid precursor Ac_3ManNAz (**87**) to folate receptor overexpressing cells via folate conjugate and cleavable linker. Ac_3ManNAz metabolically labeled cell-surface glycans with azides **88**. These then served as anchors to incorporate the hapten L-rhamnose (Rha) using SPPAC with an azadibenzocyclooctyne-Rha reagent (DBCO-Rha). Testing in KB, HEK-293, and MCF7 cells depicted selective Rha recruitment on folate receptor-positive cells (Scheme 6.21). The surface Rha

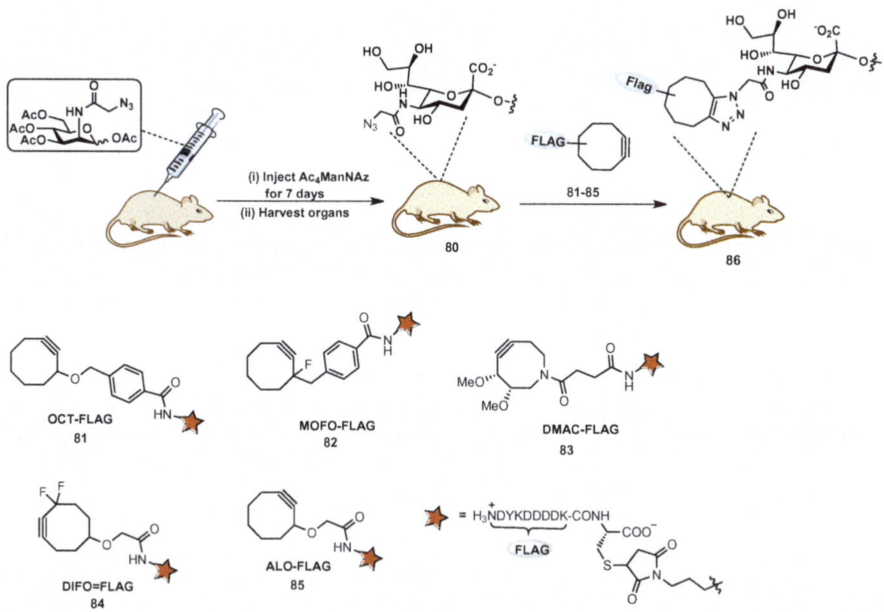

Scheme 6.20 Metabolic labeling of glycans in the live mice using the different cyclooctynes [104]

recruited anti-rhamnose antibodies, triggering complement-dependent cytotoxicity (CDC) and antibody-dependent cellular phagocytosis (ADCP) to vanish targeted cells [103].

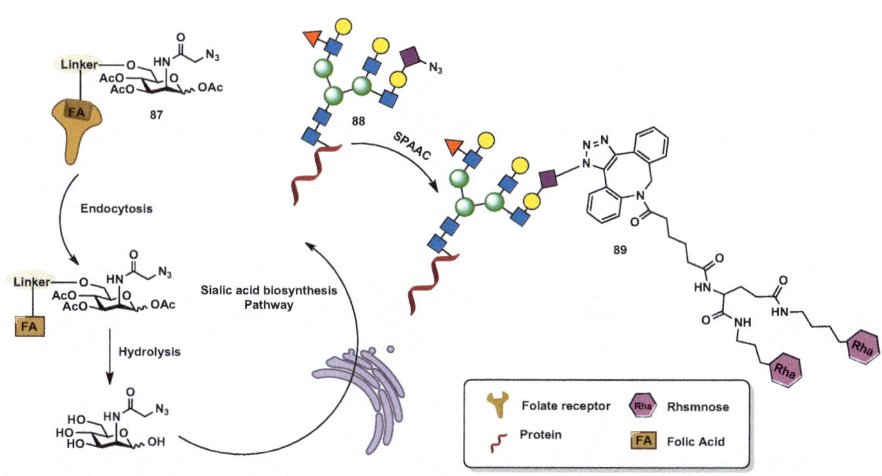

Scheme 6.21 Depiction of selective attachment process on Rha to the FR overexpressed cell surface by utilizing the bioorthogonal SPAAC click chemistry [103]

The authors introduced a novel in vivo active targeting tool for nanoparticles via strain-promoted copper-free click in living body. Targeted glycans, i.e., unnatural azido sialic acid **91** were generated artificially on the surface of tumor cell by intratumor injection of the starting material for example, tetraacetylated *N*-azidoacetyl-D-mannosamine **12**. This azide was then proceeded for the intravenously injection of DBCO conjugated PEGylated liposomes (known as DBCO-lipo) into the tumor mice required to the chemical conjugation with unnatural sialic acid present on cell surface by using copper-free strain-promoted bioorthogonal click tool. Interestingly, the resulting nanoparticles **92** via binding on the cell surface were well incorporated into cells that have notable viability required for targeted intracellular drug delivery. The optimized method could be highly fruitful for the targeted delivery of nanoparticles for imaging and drug delivery (Scheme 6.22) [105].

However, the issue of copper toxicity in click chemistry was solved by developing new ring-strained cyclooctyne derivatives that undergo cycloaddition with azides without a copper catalyst [84, 106]. On the other hand, many of these cyclooctynes have poor water solubility or require complex synthesis, limiting availability. Nonetheless, the copper-free click reaction between azides and cyclooctynes has been used for imaging cells and zebrafish embryos, despite the limitations of current cyclooctyne reagents.

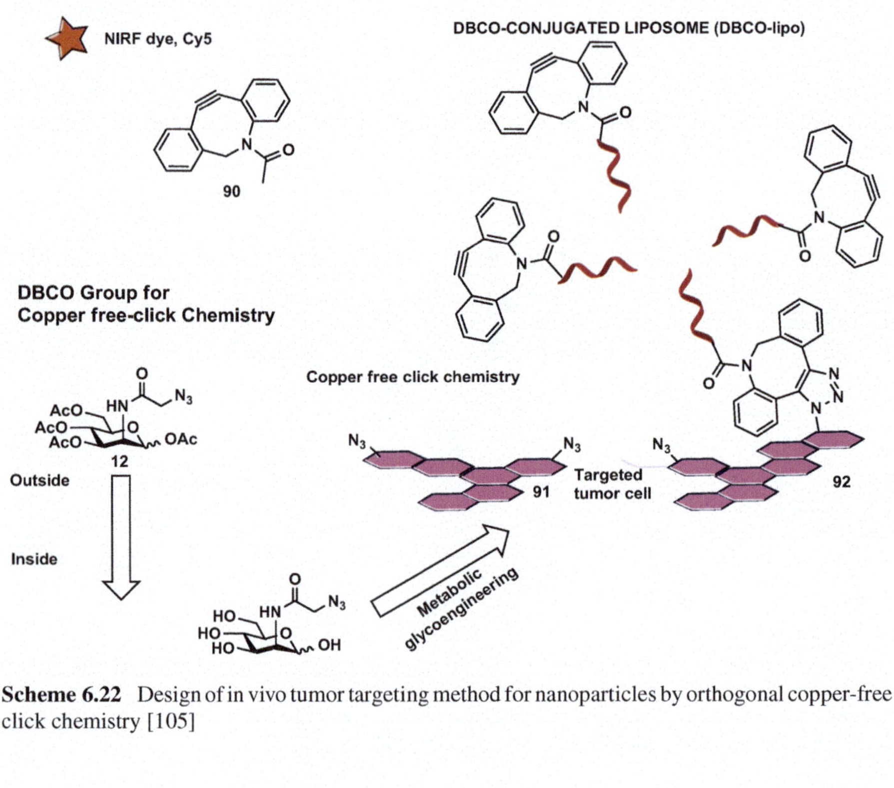

Scheme 6.22 Design of in vivo tumor targeting method for nanoparticles by orthogonal copper-free click chemistry [105]

6.5 Tetrazine Ligation 'Inverse-Electron-Demand Diels–Alder Cycloaddition Reaction'

Tetrazine ligation is an expeditious reaction and proceeds efficiently at the physiological pH in aqueous media. It is a reaction that starts through an inverse-electron-demand Diels–Alder cycloaddition between tetrazine and a strained alkene or alkyne [107]. At the beginning, it forms a bicyclic intermediate, then releases nitrogen and aromatizes to yield the pyridazine. Kinetics can be regulated by the structural modifications of the tetrazine and the dienophile [108, 109]. Electron deficient, fused ring and conformationally restricted tetrazine increase reactivity while for dienophiles, increasing ring strain and electron-withdrawing group enhance the reactivity. Along with that, the stable pyridazine product avails a handle for attaching the probes, labels, drugs, etc. Overall, tetrazine ligation is a versatile robust tool for the bioorthogonal reaction that enables chemical biology and biomedical applications at a broad aspect [110–112].

Tetrazine ligation was developed by Prof. Joseph Fox as an alternative to copper-free cyclooctyne-azide click reaction [107], which was further proceeded within the bodies of cancer patients by Hilderbrand and Devaraj [107, 113]. The Fox group utilized the inverse-electron-demand Diels–Alder reaction between strained trans-cyclooctene **94** and a tetrazine probe **93** (Scheme 6.23), producing highly reactive intermediates that rapidly decompose to nitrogen and stable dihydropyridazines **95** ligation product. Reaction rates are orders of magnitude faster than other's copper-free click approaches [109, 112].

Further, to overcome the problem of photoisomerization in trans-cyclooctene [114–118], Hilderbrand et al. used more stable norbornene with monoaryl tetrazines, though reaction rates are slower.

Wittmann et al. synthesized mannosamine derivatives with terminal alkene chemical reporter groups of varying chain lengths. They found the butenyloxycarbonyl derivative gave optimal cell-surface glycan labeling efficiency when evaluating metabolic oligosaccharide incorporation and subsequent tetrazine ligation detection [27]. Utilizing the concept of tetrazine ligation, Hilderbrand et al. utilized bioorthogonal tetrazine-norbornene cycloadditions for live cell labeling. Tetrazines rapidly and selectively react with strained norbornene dienophiles in aqueous media to generate dihydropyrazine products and nitrogen gas. Her2/neu receptors on live

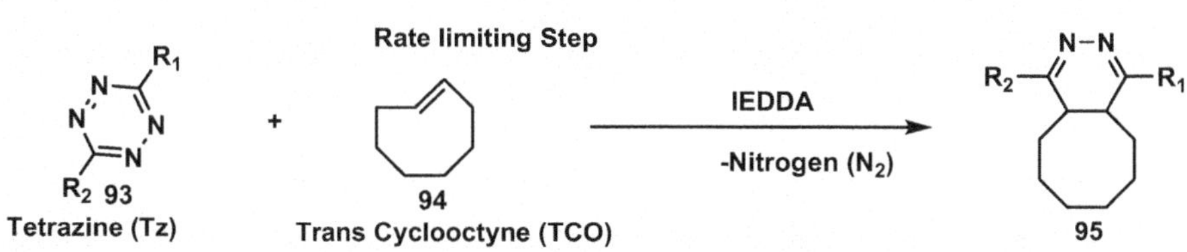

Scheme 6.23 Tetrazine ligation between *trans*-cyclooctene **94**, with biaryltetrazine **95**

breast cancer cells were first targeted with norbornene-modified monoclonal antibodies. These pretargeted antibodies were then rapidly and selectively labeled by tetrazine-conjugated near-infrared fluorophores. In this context, the FDA-approved monoclonal antibody trastuzumab (Herceptin) which binds to Her2/neu growth factor receptors [119] was labeled with norbornene and tetramethylrhodamine using standard coupling chemistry. Then, these antibodies **96** were incubated with human breast cancer SKBR3 cells, which overexpress Her2/neu, and then treated with 50 μM **3** in 10% FBS/HBSS solution for 30 min at 37 °C. After washing, the cells were imaged using both rhodamine **100** and NIR channels to detect antibody and tetrazine probe fluorescence **98**, respectively (Fig. 6.3). The high yield, speed, and selectivity of this chemistry indicate suitability for in vitro labeling and suggest the potential for in vivo pretargeted imaging applications [113].

Apart from these most common ligation techniques including Staudinger ligation, CuAAC, SPAAC, IEDDA, there are many other ligation methods have also been used like tetrazole ligation [120–122], oxime ligation [123–125], and isocyanide click reaction [126–128]. Apart from that, bioorthogonal chemistry reactions like Staudinger ligation, CuAAC, SPAAC, and IEDDA have become indispensable tools in radiochemistry and radiopharmaceutical sciences. They enable rapid, reliable radiolabeling, and bioconjugation for diverse applications from automated radiosyntheses to in vivo pretargeted imaging and therapy. Bioorthogonal chemistry has facilitated significant progress toward clinical translation of novel radiopharmaceuticals. Key priorities going forward are to continue advancing click-based radiochemistry

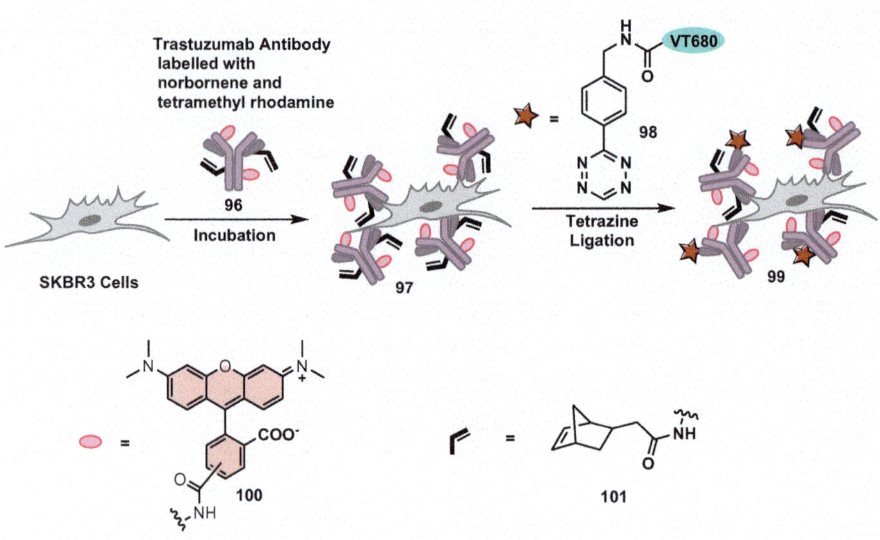

Fig. 6.3 Pre-targeting of breast cancer SKBR3 cells with norbornene and tetramethyrhodamine co-adorned trastuzumab followed by tagging with tetrazine-VT680 using biorthogonal chemistry [113]

into human studies and clinical use. Bioorthogonal chemistry has been studied as a method to control the release, localization, and formation of drugs in vivo.

6.6 Conclusions and Future Perspective

Bioorthogonal chemistry allows selective, high-yield labeling of glycans for imaging and other applications. Current efforts are focused on expanding the impact of bioorthogonal chemistry for studying glycans and other biomolecules in living systems. This is being accomplished by optimizing click reagents and reaction conditions. The continuing development of click ligations is transforming our capacity to probe and control intricate biological mechanisms. From a chemistry point of view, the development of bioorthogonal reactions comes with particularly stringent prerequisites. This reaction must establish a durable covalent connection between two functional groups that are biologically inert and ideally show no toxicity concern. This reaction should exhibit rapid kinetics even after low concentration of reactant, which is attainable at the physiological pH and temperature. Developing reactive partners with improved biological stability and faster reaction kinetics to enable labeling at low concentrations in in vivo is important. This could be paired with creating bioorthogonal reactions that proceed without catalysts, to reduce toxicity. Standardizing reagents with tunable pharmacokinetics and cell permeability would facilitate the use of these improved chemistries. Improving methods for incorporating small, non-perturbing bioorthogonal tags into biomolecules would further expand applicability. Generating labels in situ could minimize changes to analyze properties during labeling with these improved tags. Using native chemical linkages formed under mild bioorthogonal conditions would also help to maintain natural behavior. Finally, expanding photochemical bioorthogonal chemistries would better leverage spatial and temporal control with light for these advanced labeling approaches. Understanding the capabilities and constraints of several bioorthogonal reactions facilitates their combined use to study complex biological interactions. Despite rapid expansion, bioorthogonal chemistry still needs new reagents with fewer side reactions and robust efficiency. As this chapter discusses, biorthogonal tools have become invaluable for the precise detection, imaging, and therapeutic intervention of complex diseases such as cancer.

References

1. Kayser H, Kunstmann J, Schmidt RR, Kollmeier T, Veh RW, Cummings RD (1992) J Biol Chem 267:16934–16938
2. Prescher JA, Bertozzi CR (2006) Cell 126:851–854
3. Sletten EM, Bertozzi CR (2009) Angew Chem Int Ed 48:6974–6998
4. Lin FL, Hoyt HM, van Halbeek H, Bergman RG, Bertozzi CR (2005) J Am Chem Soc 127:686–2695

5. Ohtsubo K, Marth JD (2006) Cell 126:855–867
6. Prescher JA, Bertozzi CR (2005) Nat Chem Biol 1:13–21
7. Lemieux GA, Bertozzi CR (1998) Trends Biotechnol 16:506–513
8. Mahal LK, Yarema KJ, Bertozzi CR (1997) Science 276:1125–1128
9. Severinov K, Muir TW (1998) J Biol Chem 273:16205–16209
10. Saxon E, Bertozzi CR (2000) Science 287:2007–2010
11. Rostovtsev VV, Green LG, Fokin VV, Sharpless KB (2002) Angew Chem Int Ed 41:2596–2599
12. Michael A (1893) J Prakt Chem 48:94–95
13. Huisgen R (1963) Angew Chem Int Ed Engl 2:565−598
14. Huisgen R (1963) Angew Chem Int Ed Engl 2:633−696
15. Huisgen R (1989) Pure Appl Chem 61:613–628
16. Huisgen R, Szeimies G, Moebius L (1967) Chem Ber 100:2494–2507
17. Huisgen R, Padwa A (ed) (1984) Wiley, New York 1−176
18. Banahene N, Kavunja HW, Swarts BM (2022) Chem Rev 122:3336–3413
19. Bauer D, Cornejo MA, Hoang TT, Lewis JS, Zeglis BM (2022) Bioconjugate Chem 33:762–778
20. Tornoe CW, Christensen C, Meldal M (2002) J Org Chem 67:3057–3064
21. Agrahari AK, Bose P, Jaiswal MK, Rajkhowa S, Singh AS, Hotha S, Mishra N, Tiwari VK (2021) Chem Rev 12:7638–7956
22. Tiwari VK, Mishra BB, Mishra KB, Mishra N, Singh AS, Chen X (2016) Chem Rev 116:3086–3240
23. Kolb HC, Finn MG, Sharpless KB (2001) Anew Chem Int Ed 40:2004–2021
24. Saxon E, Armstrong JI, Bertozzi CR (2000) Org Lett 2:2141–2143
25. Agard NJ, Prescher JA, Bertozzi CR (2004) J Am Chem Soc 126:15046–15047
26. Baskin JM, Prescher JA, Laughlin ST, Agard NJ, Chang PV, Miller IA, Lo A, Codelli JA, Bertozzi CR (2007) Proc Natl Acad Sci USA 104:16793–16797
27. Späte A-K, Schart VF, Schçllkopf S, Niederwieser A, Wittmann V (2014) Chem Eur J 20:16502–16508
28. Schart VF et al (2019) ChemBioChem 20:166–171
29. Kozma E, Bojtár M, Kele P (2023) Angew Chem Int Ed 62:e202303198
30. Chen X, Varki A (2023) J Clin Invest 133:e169408
31. Zhang H, Wang J, Han R, Sun B, Luo C (2023) Chem Soc Rev 52:697–710
32. Azagarsamy MA, Anseth KS (2013) ACS Macro Lett 2:5–9
33. Stuyver T, Jorner K, Coley CW (2023) Sci Data 10:1–14
34. Sengupta P, Dutta S, Chhikara BS (2023) Chem Biol Lett 10:545
35. Bertozzi C (2023) ACS Cent Sc 9:558–559
36. Dube DH, Bertozzi CR (2003) Curr Opin Chem Biol 7:616–625
37. Bertozzi C, Philippidis A, Davies K (2023) GEN Biotechnol. 2:459–463
38. Wratil PR, Horstkorte R, Reutter W (2016) Angew Chem Int Ed 55:9482–9512
39. Agatemor C, Buettner MJ, Ariss R, Muthiah K, Saeui CT, Yarema KJ (2019) Trends Biotechnol 37:605–620
40. Staudinger H, Meyer J (1919) Helv Chim Acta 2:635–646
41. Gololobov YG, Kasukhin LF (1992) Tetrahedron 48:1353–1406
42. Wiley RA, Sternson LA, Sasame HA, Gillette JR (1972) Biochem Pharmacol 21:3235–3247
43. Kohn M, Breinbauer R (2004) Angew Chem Int Ed 43:3106–3116
44. Laughlin ST, Bertozzi CR (2007) Nat Protoc 2:2930–2944
45. Prescher JA, Dube DH, Bertozzi CR (2004) Nature 430:873–877
46. Neves AA, Stöckmann H, Harmston RR, Pryor HJ, Alam IS, Ireland-Zecchini H, Lewis DY, Lyons SK, Leeper FJ, Brindle KM (2011) FASEB J 25:2528–2537
47. Sundhoro M, Jeon S, Park J, Ramström O, Yan M (2017) Angew Chem Int Ed 56:12117–12121
48. Meldal M, Tornoe CW (2008) Chem Rev 108:2952–3015
49. Kolb HC, Sharpless KB (2003) Drug Discovery Today 8:1128–1137
50. Lutz JF, Zarafshani Z (2008) Adv. Drug Delivery Rev 60:958–970

51. Tanaka T, Kamiya N, Nagamune T (2005) FEBS Lett 579:2092–2096
52. Singh SK, Kumar S, Yadav MS, Tiwari VK (2022) J Org Chem 87:15389–15402
53. Van Berkel SS, Dirks AJ, Meeuwissen SA, Pingen DLL, Boerman OC, Laverman P, van Delft FL, Cornelissen JJLM, Rutjes FPJT (2008) ChemBioChem 9:1805–1815
54. Joshi NS, Whitaker LR, Francis MB (2004) J Am Chem Soc 126:15942–15943
55. Antos JM, Francis MB (2006) Curr Opin Chem Biol 10:253–262
56. Gauthier MA, Klok HA (2008) Chem Commun 2591−2611
57. Tron GC, Pirali T, Billington RA, Canonico PL, Sorba G, Genazzani AA (2008) Med Res Rev 28:278–308
58. Sharpless KB, Manetsch R (2006) Expert Opin Drug Discovery 1:525–538
59. Thirumurugan P, Matosiuk D, Jozwiak K (2013) Chem Rev 113:4905–4979
60. Hein JE, Fokin VV (2010) Chem Soc Rev 39:1302–1315
61. Chan TR, Hilgraf R, Sharpless KB, Fokin VV (2004) Org Lett 6:2853–2855
62. Gaetke LM, Chow CK (2003) Toxicology 189:147–163
63. Yin Z, Comellas-Aragones M, Chowdhury S, Bentley P, Kaczanowska K, BenMohamed L, Gildersleeve JC, Finn MG, Huang X (2013) ACS Chem Biol 8:1253–1262
64. Rabuka D, Hubbard SC, Laughlin ST, Argade SP, Bertozzi CR (2006) J Am Chem Soc 128:12078–12079
65. Besanceney-Webler C, Jiang H, Wang W, Baughn AD, Wu P (2011) Bioorg Med Chem Lett 21:4989–4992
66. Hsu TL, Hanson SR, Kishikawa K, Wang SK, Sawa M, Wong CH (2007) Proc Natl Acad Sci U S A 104:2614–2619
67. Sawa M, Hsu TL, Itoh T, Sugiyama M, Hanson SR, Vogt PK, Wong CH (2006) Proc Natl Acad Sci U S A 103:12371–12376
68. Chang PV, Chen X, Smyrniotis C, Xenakis A, Hu TS, Bertozzi CR, Wu P (2009) Angew Chem Int Ed 48:4030−4033
69. Shieh P, Hangauer MJ, Bertozzi CR (2012) J Am Chem Soc 134:17428–17431
70. Sumranjit J, Chung SJ (2013) Molecules 18:10425–10451
71. Zhao L, May JP, Dietrich DJ, Loonchanta A, Matinkhoo K, Pryyma A, Perrin DM, Blanc A (2015) ChemBioChem 16:1420–1425
72. Ballell L, van Scherpenzeel M, Buchalova K, Liskamp RMJ, Pieters RJ (2006) Org Biomol Chem 4:4387–4394
73. Ballell L, Alink KJ, Slijper M, Versluis C, Liskamp RMJ, Pieters RJ (2005) ChemBioChem 6:291–295
74. Wang Q, Chan TR, Hilgraf R, Fokin VV, Sharpless KB, Finn MG (2003) J Am Chem Soc 125:3192–3193
75. Hong V, Steinmetz NF, Manchester M, Finn MG (2010) Bioconjugate Chem 21:1912–1916
76. Soriano del Amo D, Wang W, Jiang H, Besanceney C, Yan AC, Levy M, Liu Y, Marlow FL, Wu P (2010) J Am Chem Soc 132:16893–16899
77. Jiang H, Zheng T, Lopez-Aguilar A, Feng L, Kopp F, Marlow FL, Wu P (2014) Bioconjugate Chem 25:698–706
78. Yin Z, Dulaney S, McKay CS, Baniel C, Kaczanowska K, Ramadan S, Finn MG, Huang X (2016) ChemBioChem 17:174–180
79. Gupta SS, Kuzelka J, Singh P, Lewis WG, Manchester M, Finn MG (2005) Bioconjugate Chem 16:1572–1579
80. Gutmann M, Memmel E, Braun AC, Seibel J, Meinel L (2016) Lühmann, T. ChemBioChem 17:866–875
81. Best MD (2009) Biochemistry 48:6571–6584
82. Gupta SS, Raja KS, Kaltgrad E, Strable E, Finn MG (2005) Chem Commun 34:4315–4317
83. Parker MFL, Flavell RR, Luu JM, Rosenberg OS, Ohliger MA, Wilson DM (2020) ACS Infect Dis 6:1587–1598
84. Ning X, Guo J, Wolfert MA, Boons GJ (2008) Angew Chem Int Ed Engl 47:2253–2255
85. Laughlin ST, Baskin JM, Amacher SL, Bertozzi CR (2008) Science 320:664–667
86. Fernandez-Suarez M et al (2007) Nat Biotechnol 25:1483–1487

87. Neef AB, Schultz C (2009) Angew Chem Int Ed Engl 48:1498–1500
88. Wittig G, Krebs AI (1961) Chem Ber 94:3260–3275
89. Jewett JC, Bertozzi CR (2010) Chem Soc Rev 39:1272–1279
90. Link AJ, Vink MK, Agard NJ, Prescher JA, Bertozzi CR, Tirrell DA (2006) Proc Natl Acad Sci U S A 103:10180–10185
91. Zou Y, Yin J (2008) Bioorg Med Chem Lett 18:5664–5667
92. Hur GH, Meier JL, Baskin J, Codelli JA, Bertozzi CR, Marahiel MA, Burkart MD (2009) Chem Biol 16:372–381
93. Nessen MA, Kramer G, Back J, Baskin JM, Smeenk LEJ, de Koning LJ, van Maarseveen JH, de Jong L, Bertozzi CR, Hiemstra H (2009) J Proteome Res 8:3702–3711
94. Davis DL, Price EK, Aderibigbe SO, Larkin MXH, Barlow ED, Chen R, Ford LC, Gray ZT, Gren SH, Jin Y, Keddington KS, Kent AD, Kim D, Lewis A, Marrouche RS, O'Dair MK, Powell DR, Scadden MIHC, Session CB, Tao J, Trieu J, Whiteford KN, Yuan KZ, Yun G, Zhu J, Heemstra JM (2016) J Org Chem 81:6816–6819
95. Anderton GI, Bangerter AS, Davis TC, Feng Z, Furtak AJ, Larsen JO, Scroggin TL, Heemstra JM (2015) Bioconjugate Chem 26:1687–1691
96. Poloukhtine AA, Mbua NE, Wolfert MA, Boons G-J, Popik VV (2009) J Am Chem Soc 131:15769–15776
97. Hart GW, Housley MP, Slawson C (2007) Nature 446:1017–1022
98. Agard NJ, Baskin JM, Prescher JA, Lo A, Bertozzi CR (2006) ACS Chem Biol 1:644–648
99. Sletten EM, Bertozzi CR (2008) Org Lett 10:3097–3099
100. Bach RD (2009) J Am Chem Soc 131:5233–5243
101. Johnson JJ, Baskin JM, Bertozzi CR, Koberstein JT, Turro NJ (2008) Chem Commun 3064–3066
102. Bird RE, Lemmel SA, Yu X, Zhou QA (2021) Bioconjugate Chem 32:2457–2479
103. Li S, Yu B, Wang J, Zheng Y, Zhang H, Walker MJ, Yuan Z, Zhu H, Zhang J, Wang PG, Wang B (2018) ACS Chem Biol 13:1686–1694
104. Changa V, Preschera JA, Slettena EM, Baskina JM, Millera IA, Agarda NJ, Loa A, Bertozzi CR (2010) PNAS 107(5):1821–1826
105. Koo H et al (2012) Angew Chem Int Ed 51:11836–11840
106. Codelli JA, Baskin JM, Agard NJ, Bertozzi CR (2008) J Am Chem Soc 130:11486–11493
107. Blackman ML, Royzen M, Fox JM (2008) J Am Chem Soc 130:13518–13519
108. Rossin R, van den Bosch SM, ten Hoeve W, Carvelli M, Verstehen RM, Lub J, Robillard MS (2013) Bioconjugate Chem 24:1210–1217
109. Selvaraj R, Fox JM (2013) Curr Opin Chem Biol 17:753–760
110. Karver MR, Weissleder R, Hilderbrand SA (2011) Bioconjugate Chem 22:2263–2270
111. Kamber DN, Liang Y, Blizzard RJ, Liu F, Mehl RA, Houk KN, Prescher JA (2015) J Am Chem Soc 137:8388–8391
112. Darko A, Wallace S, Dmitrenko O, Machovina MM, Mehl RA, Chin JW, Fox JM (2014) Chem Sci 5:3770–3776
113. Devaraj NK, Weissleder R, Hilderbrand SA (2008) Bioconjug Chem 19:2297–2299
114. Peplow M (2020) Click chemistry sees first use in humans, 21 Oct. https://cen.acs.org/pharmaceuticals/Click-chemistry-seesfirst-use/98/web/2020/10
115. Pigga JE, Rosenberger JE, Jemas A, Boyd SJ, Dmitrenko O, Xie Y, Fox JM (2021) Angew Chem Int Ed 60:14,975−14,980
116. Taylor MT, Blackman ML, Dmitrenko O, Fox JM (2011) J Am Chem Soc 133:9646–9649
117. Kamber DN, Nazarova LA, Liang Y, Lopez SA, Patterson DM, Shih H-W, Houk KN, Prescher JA (2013) J Am Chem Soc 135:13680–13683
118. Pipkorn R, Waldeck W, Didinger B, Koch M, Mueller G, Wiessler M, Braun K (2009) J Pept Sci 15:235–241
119. Lewis GD, Figari I, Fendly B, Wong WL, Carter P, Gorman C, Shepard HM (1993) Cancer Immunol Immun 37:255–263
120. Song W, Wang Y, Qu J, Madden MM, Lin Q (2008) Angew Chem Int Ed 47:2832−2835
121. Song W, Wang Y, Qu J, Lin Q (2008) J Am Chem Soc 130:9654–9655

122. Lim RKV, Lin Q (2011) Acc Chem Res 44:828–839
123. Dirksen A, Dirksen S, Hackeng TM, Dawson PE (2006) J Am Chem Soc 128:15602–15603
124. Rashidian M, Song JM, Pricer RE, Distefano MD (2012) J Am Chem Soc 134:8455–8467
125. Agten SM, Suylen D, Ippel H, Kokozidou M, Tans G, van de Vijver P, Koenen RR, Hackeng TM (2013) ChemBioChem 14:2431–2434
126. Kaur T, Wadhwa P, Bagchi S, Sharma A (2016) Chem Commun 52:6958–6976
127. Giustiniano M, Basso A, Mercalli V, Massarotti A, Novellino E, Tron GC, Zhu J (2017) Chem Soc Rev 46:1295–1357
128. Tu J, Svatunek D, Parvez S, Liu AC, Levandowski BJ, Eckvahl HJ, Peterson RT, Houk KN, Franzini RM (2019) Angew Chem Int Ed 58:9043−9048

Chapter 7
Thiol-Ene 'Click Chemistry': Discovery to Applications

7.1 Introduction

The concept of 'click chemistry' was established by Prof. K. B. Sharpless [1] in 2001 and was soon validated by Shapless [2] and Meldal [3] in 2002. The click chemistry has been widely acknowledged for its remarkable utility in producing novel molecules and materials of wider utilities in chemistry, biology, medicine, agriculture, and material science [4–14]. The wide compatibility of click chemistry is mainly because of its simplicity as it operates under mild reaction conditions, regioselectivity, and moreover, it offers rapid reaction rates and tolerates various functional groups as an additional notable characteristics of such modular conjugations [2–17]. As the Cu(I)-mediated azide-alkyne cycloaddition click reactions have gained great popularity, the thiol-based 'click' reactions have become a particular center of attraction among researchers due to its appealing features like metal-free nature and adherence to the essential characteristics of a click reaction [18, 19]. It is worth mentioning that while some thiol-based reactions may not truly stick to the click philosophy, the majority of them can be executed within reasonable times delivering high yields without the need for chromatographic purification [18].

The discovery of thiol-ene reaction is known long back in 1905 by Posner [20]; however, this method gained prominence in the late 1990s and early 2000s for its feasibility and wide range of synthetic utilities [19, 21, 22]. The term thiol-based 'click' reaction represents a range of reactions (Fig. 7.1), such as thiol-ene (also known as alkene hydrothiolation) [19, 23, 24], thiol-yne [25], thiol–isocyanate [26], thiol–epoxy [27], and thiol–bromo reactions [28]. In recent years, these reactions have typically applied in the development of polymers and well-defined dendrimeric architectures. These thiol-based reactions encompass various types including radical addition, nucleophilic addition, and nucleophilic substitution reactions [29]. The primary focus here lies in employing thiol-based 'click' reactions for constructing diverse topological structures, elucidating the reaction mechanisms, and exploring their utility in materials synthesis, specifically in the

V. K. Tiwari et al., *Click Chemistry*, Materials Horizons: From Nature to Nanomaterials,
https://doi.org/10.1007/978-981-97-4596-8_7

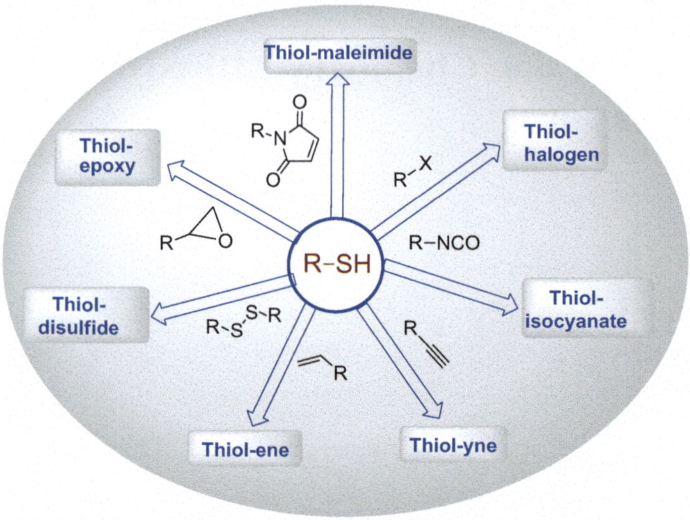

Fig. 7.1 Thiol-click-mediated reactions for construction of diverse molecular architectures

context of photoclick hydrogels [30]. This chapter highlights the notable features including mechanistic aspects and also the structural considerations and synthetic applications, also modification of materials through both the radical-mediated or base/nucleophile-initiated thiol-ene and thiol-yne click reactions.

7.2 Catalysis and Mechanism of Thiol-Ene and Thiol-Yne Click

In thiol-ene reaction, thiol **1** was reacted with an alkene **2** to form a thioether **3** (Scheme 7.1). The addition proceeds with an anti-Markovnikov addition of a thiol compound to an alkene. The thiol-ene reactions are modular, stereoselectivity high yielding and proceed with high rate due to thermodynamic driving force and therefore is widely accepted as another variant of click chemistry. The method is now well established one and displayed immense applications in organic synthesis particularly for the development of polymers and dendrimers and great scope in material and biomedical sciences [19, 31].

Thiol-ene additions are supposed to proceed via two different ways such as a free-radical addition (Scheme 7.2) or catalytic 1,4-conjugate addition (e.g., Michael addition, Scheme 7.3). A free-radical addition reaction is initiated either under photoinduced conditions (light) or thermal conditions using suitable radical initiators. The resulting thiyl radical species underwent propagation with an 'Ene' functionality (via an anti-Markovnikov addition) and give a carbon-centered radical. At the end,

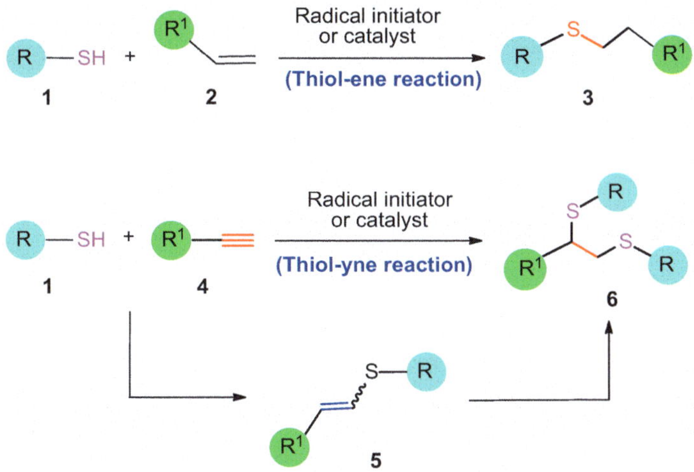

Scheme 7.1 Classical photoinduced or catalytic 'thiol-ene and thiol-yne' click reaction

a chain-transfer step removes a hydrogen radical from a thiol substrate and can afterward participate in the desired step of multiple propagation [31].

Interestingly, the propagation and chain-transfer (step-growth) and homopolymerization (chain growth) processes in the thiol-ene radical route can be explored effectively to develop a series of desired homogeneous polymer networks. Further, this method is quite useful as it proceeds rapidly with almost quantitative yield under ambient atmospheric conditions and thus can combine such advantages of radical-based photopolymerizations with the modular nature of click chemistry [31]. Depending on the thiol and ene functional groups, the resulting carbon-centered free-radical can undergo chain-growth polymerization for the successful development of a diverse range of uniform polymer networks [32].

Scheme 7.2 Photoinduced thiol-ene click reaction [33]

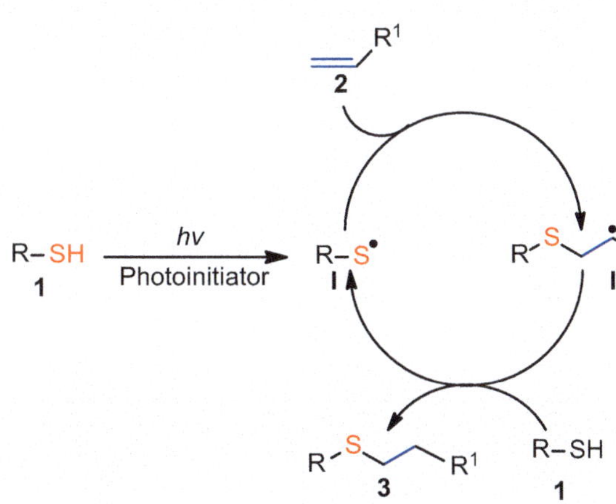

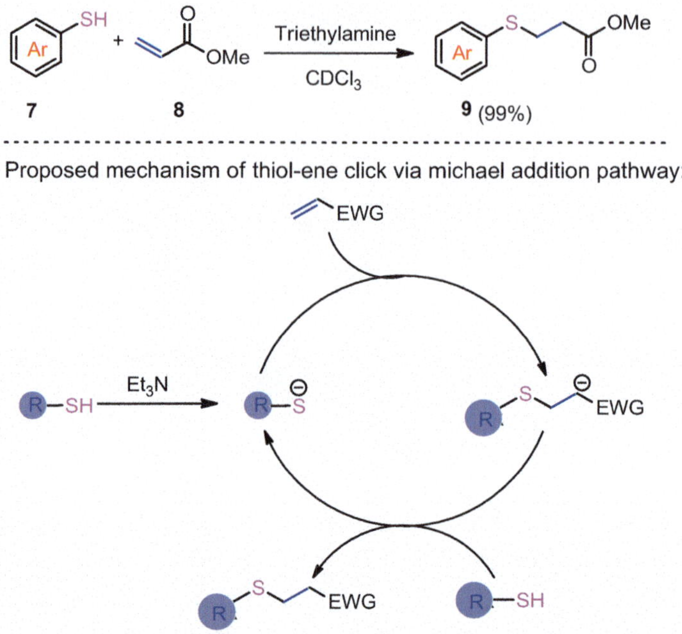

Scheme 7.3 Synthesis of adduct **9** with thiol and methyl acrylate via thiol-Michael reaction

The most possible mechanism is believed to proceed through the following steps: Thiol-ene radical reaction has occurred, and an addition product is obtained. In the first step, thiol radical **I** is formed with the help of radical initiator ruthenium catalyst and thiol. This formed radical **I** reacted with alkene to undergo anti-Markovnikov attack and yield thiol-ene adduct radical **II**. This further treated with another molecule of thiol and gave the desired product **3** and again a thiol radical **I**. This process is cyclic and completely underwent through radical initiator Scheme 7.2).

The way to get the similar desired addition product is believed to proceed through thiol-Michael click reaction, where Michael donors reacted with Michael acceptor in the presence of some catalyst (or in the absence of catalyst) to give 1,4-conjugate addition product without the involvement of free-radical conditions. For example, Northrop and co-workers experimented without hazardous metal via the thiol-Michael click reaction. In this route, thiophenols **7** (Michael donors) reacted with methyl acrylate **8** (Michael acceptors) in the presence of triethyl amine as a catalyst for an hour. Then, reaction mixture underwent a thiol-Michael addition reaction and gave the final product **9** in almost quantitative yield (Scheme 7.3) [34]. The method is highly applicable for an easy access of various materials including polymers, dendrimers, and other macromolecules.

The 'thiol-ene click' tool via radical route can be used for the high-yielding preparation of wide variety of molecules such as thioether, thiodendrimer and sulfur-linked peptides. However, in some cases, the radical resulting in thiol-ene click reaction may lead to the formation of unwanted by-products as radical works as chain propagation as well. Dondoni group established a similar method by reaction of thiol with an alkene in the presence of 2,2-dimethoxyphenylacetophenone (DMPA) as photoinitiator under ultra-light conditions and phosphine oxide for visible-light initiation, where the addition product was the desired thioether [35].

The '**thiol-yne click reaction**' is known to proceed via sequential propagation of a thiyl radical **I** with an alkyne followed by chain transfer of the radical to another thiol. Thus, the addition of a thiyl radical **I** to an alkyne **4** resulted in respective carbon-centered radical **II** (Scheme 7.4) that subsequently abstracted a hydrogen from another thiol **1** and led to the production of a vinyl sulfide **5** with the regeneration of a thiyl radical **I**. The vinyl sulfide **5** can undergo further addition of a second thiyl radical **I.** Each alkyne has to combine with two thiol residues to give a desired thioether **6**. Consequently, the alkyne has been established now as a difunctional nature in thiol-one step-growth polymerization reactions.

Bowman and co-workers performed an experiment using thiol-ene click reaction without any hazardous metal and synthesized the addition product in very beautiful way with the help radical-mediated reagent. In this reaction, terminal alkene **10** was treated in solvent MeOH/EGDA (1:1) with thiol **1** in the presence of DMPA photoinitiator and DBU. This reaction mixture underwent thiol-ene click reaction without any metal and gave the desired product **11** with good yield (Scheme 7.5) [36].

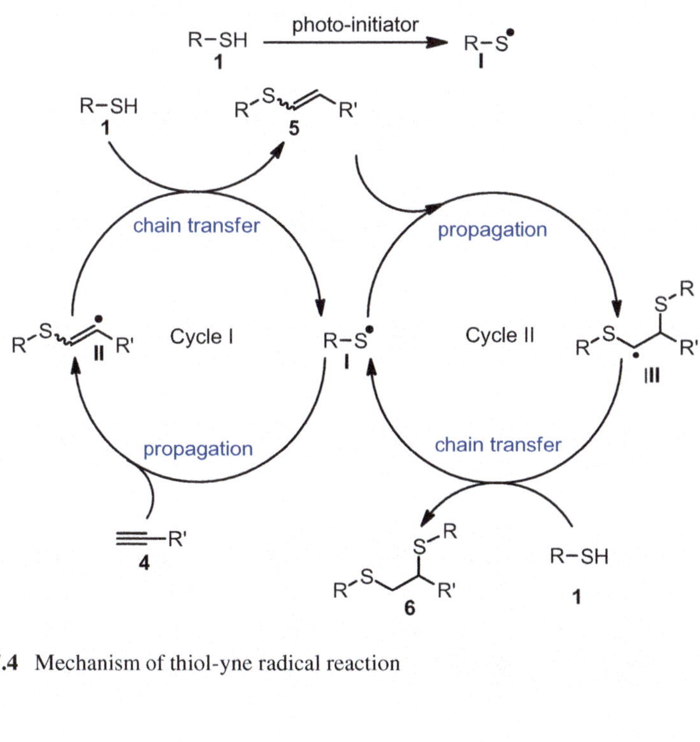

Scheme 7.4 Mechanism of thiol-yne radical reaction

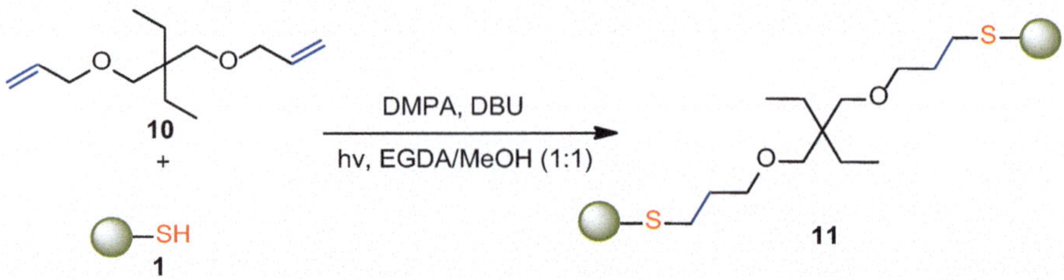

Scheme 7.5 Mechanism of thiol-ene click reaction in the presence of photoinitiator and light

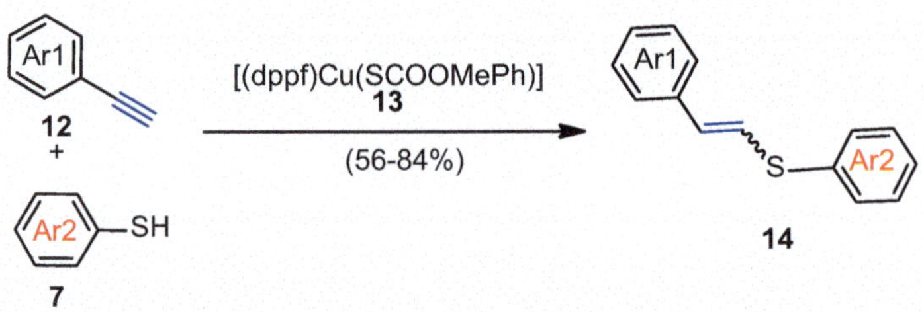

Scheme 7.6 Synthesis of vinyl sulfide **12** through thiol-yne reaction

Bhattacharya and co-workers synthesized a variety of Cu catalyst in lab, and by using this copper catalyst, vinyl sulfide is synthesized through thiol-yne reaction. In this reaction, aromatic terminal alkyne **12** reacted with aromatic thiol **7** in the presence of copper catalyst [(dppf)Cu(SCOOMePh)] **13** and heated the complete solution at a temperature 70 °C for several hours and underwent copper-catalyzed thiol-yne reaction to form vinyl sulfide **14** in good yield 54–84% (Scheme 7.6) [37].

Yoon and group established the C-S bond forming modular tool using transition metal catalyst as photoinitiator via thiol-ene radical reaction. In this method, benzyl mercaptan **15** reacted with styrene **16** in acetonitrile and a catalytic amount of ruthenium catalyst is added. Thus, the whole set up is exposed to blue LED for 2 h at room temperature that leading to formation of adduct **17** via thiol-ene radical reaction in excellent yield (Scheme 7.7) [38].

Dilman and co-worker performed an experiment based on thiol-ene click reaction which underwent though blue light irradiation and photoactive Lewis basic molecules such as acridine orange (catalyst X) or naphthalene-fused *N*-acylbenzimidazole (catalyst Y) is used as catalysts. There is a thought that the mechanism underwent though the steps which is initiated by a proton-coupled electron transfer within the complex between the thiol and the Lewis basic catalyst. *tert*-butyl thiol is added in cyclohexane under argon condition with allyl acetate **18** in the presence of small amount of catalyst **X** and blue light (LED) is irradiated for one hour at room temperature. Thus, this reaction underwent via thiol-ene radical reaction under proton-coupled electron transfer reaction and resulted to the formation of desired adduct **19** with 98% yield (Scheme 7.8) [39].

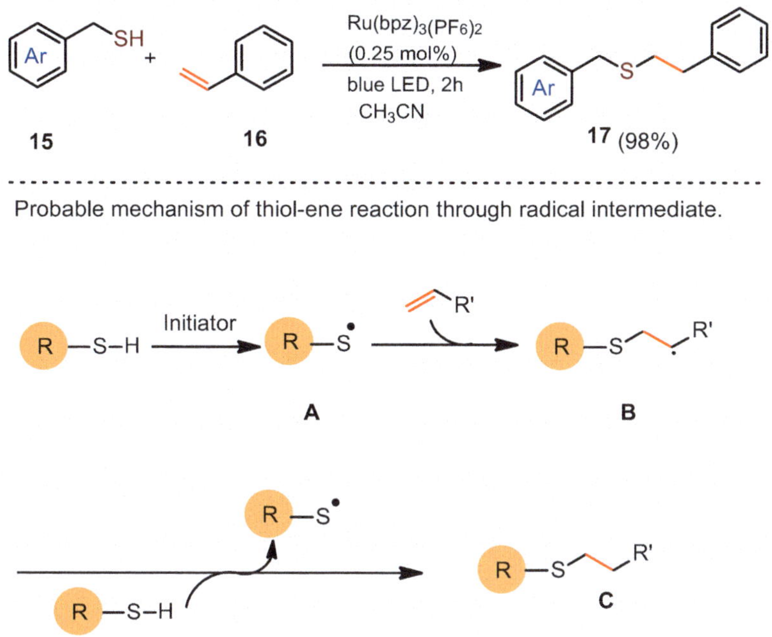

Scheme 7.7 Synthesis of adduct **17** via photocatalytic radical thiol-ene click reaction

Garrell et al. worked on thiol having trialkoxysilane functional group and alkene and formed C-S bond through thiol-ene radical reaction. Acrolein diethyl acetal **20** dissolved in chloroform and stirred with triethoxysilane **21** in the presence of DPAP and exposed to the ultra-violet light for several hours at room temperature then underwent through thiol-ene radical reaction. The addition product **22** is synthesized in very good yield (Scheme 7.9) [40].

Feng and co-workers developed the synthesis of Novel multifunctional alkoxysilanes by facile thiol-yne 'click' chemistry without any heavy and hazardous metal. In this work, alkyne **4** reacted with 3-mercaptopropyltrimethoxysilane **23** and DMPA in the reaction mixture is kept under a 100 W UV-light ($\lambda_{max} = 365$ nm) and irradiated for 20 min with gentle stirring to undergo thiol-yne click reaction and synthesized the desired adduct **24** with high conversion (97%) (Scheme 7.10) [41].

Bowman and his group used thiol-vinyl click reaction that is very powerful and popular reaction by which a large number of macromolecules such as polymers, dendrimers can be easily synthesized. First of all thiol **1** is treated with vinyl sulfone **25** in the presence of catalyst 1-mthyl imidazole and stirred for 55 min under neat conditions undergo via thiol-ene click reaction and synthesized the desired adduct **26** with excellent yield 96% (Scheme 7.11) [42].

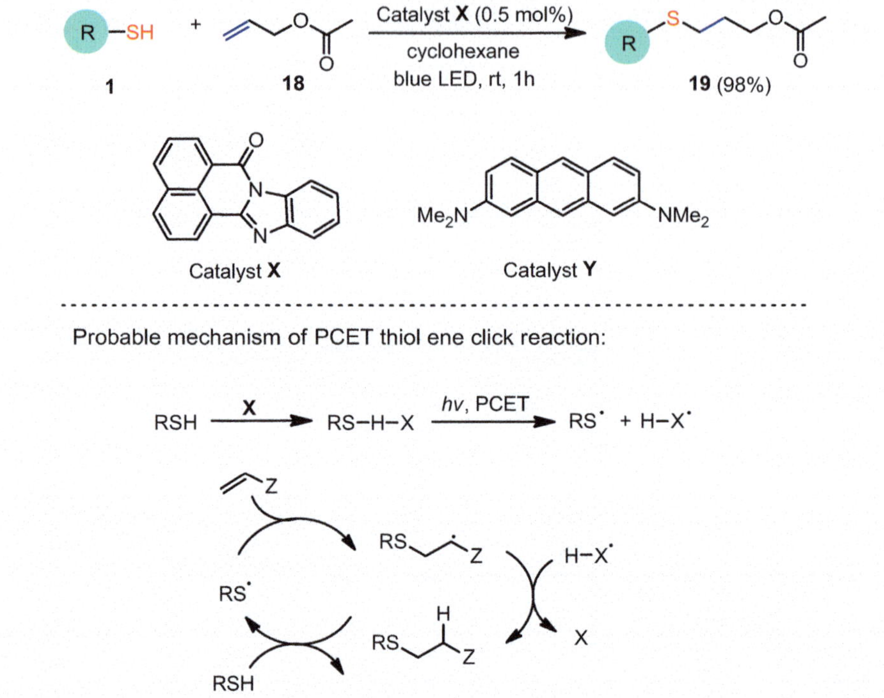

Scheme 7.8 Synthesis of adduct **19** via thiol-ene radical reaction

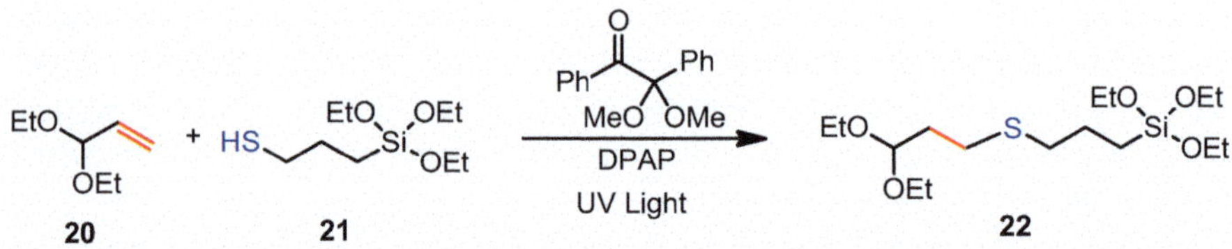

Scheme 7.9 Synthesis of diethyl acetal triethoxysilane **22** via thiol-ene radical reaction

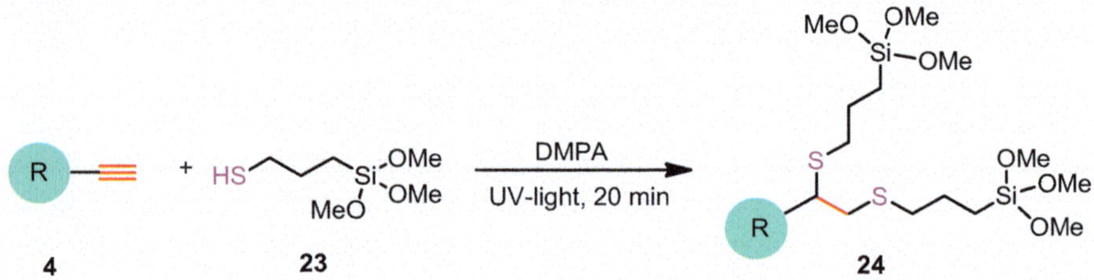

Scheme 7.10 Synthesis of multifunctional alkoxysilanes via thiol-yne click chemistry

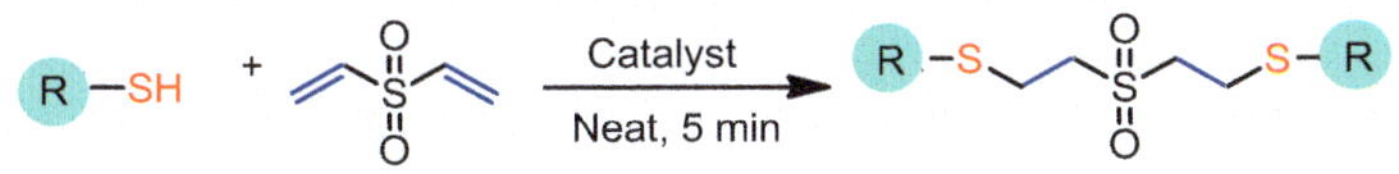

Scheme 7.11 Nitrogen center catalyzed addition reaction of thiol with vinyl sulfone

7.3 Applications of Thiol-Ene and Thiol-Yne in Organic Synthesis

Like standard CuAAC or RuAAC click chemistry [2–5], thiol-ene click and thiol-yne click chemistry as another variant of click is widely explored in variety of synthesis of simple to complex organic molecular architectures including carbohydrate-based molecules, peptides, polymer, nanocomposites, dendrimers, etc. [18, 19]. For example, Northrop et al. devised a simple way by experimentally the reactivity and selectivity of *N*-allyl maleimide **27** and *N*-propargyl maleimide **34** via thiol-ene/ -yne click reaction. In this reaction, *N*-allyl maleimide **27** reacted with methyl-3-mercaptopropionate **28** and *β*-mercapto ethanol **29,** respectively, in the presence of triethyl amine in $CDCl_3$ and underwent thiol-Michael reaction to give compound **30** and **32,** respectively. The compounds **30** and **32** were further treated with *β*-mercaptoethanol **29** and methyl-3-mercaptopropionate **28**, respectively, and DMPA (α,α-dimethoxy-α-phenylacetophenone) as catalytic radical initiator in CDC13 and underwent radical-mediated thiol-ene click reaction to give the final products **31** and **33**, respectively, with excellent yield (Scheme 7.12a) [43].

Likewise *N*-propargyl maleimide **34** underwent thiol-Michael reaction using triethyl ammine in $CDCl_3$ separately with methyl-3-mercaptopropionate **28** and *β*-mercaptoethanol **29** and afforded compound **35** and **37**, respectively. The resulting compounds **35** and **37** were further treated separately with β-mercaptoethanol **29** and methyl-3-mercaptopropionate **28**, in the presence of DMPA in $CDCl_3$ and underwent radical-mediated thiol-yne click reaction to give the final products **36** and **38** respectively in excellent yield (Scheme 7.12a,b) [43].

Cuadrado and co-workers explained the types of controlled addition either *β*–addition or α–addition by the development of ferrocene-linked molecules via thiol-ene and thiol-yne click reaction. First of all, vinyl ferrocene **39** was treated with thiol **1** and DMPA (2,2-dimethoxy-2-phenylacetophenone) and exposed the UV-light for several hour at room temperature and underwent different kind of addition product via photocatalyzed thiol-ene click reaction. As a result of β–addition, the desired product **40** is formed and α–addition leads to the formation of the desired product **41** with very good yield. When Ethynylferrocene **42** treated with thiol **1** in the presence of DMPA as photoinitiator and UV-light is irradiated for few hours at room temperature then undergoes different kind of addition via photocatalyzed thiol-yne click reaction. As a result of *β*–addition, the two stereoselective and regioselective stereoisomers *E*-isomer **43** or *Z*-isomer **44** were obtained depending upon the photoinitiator and

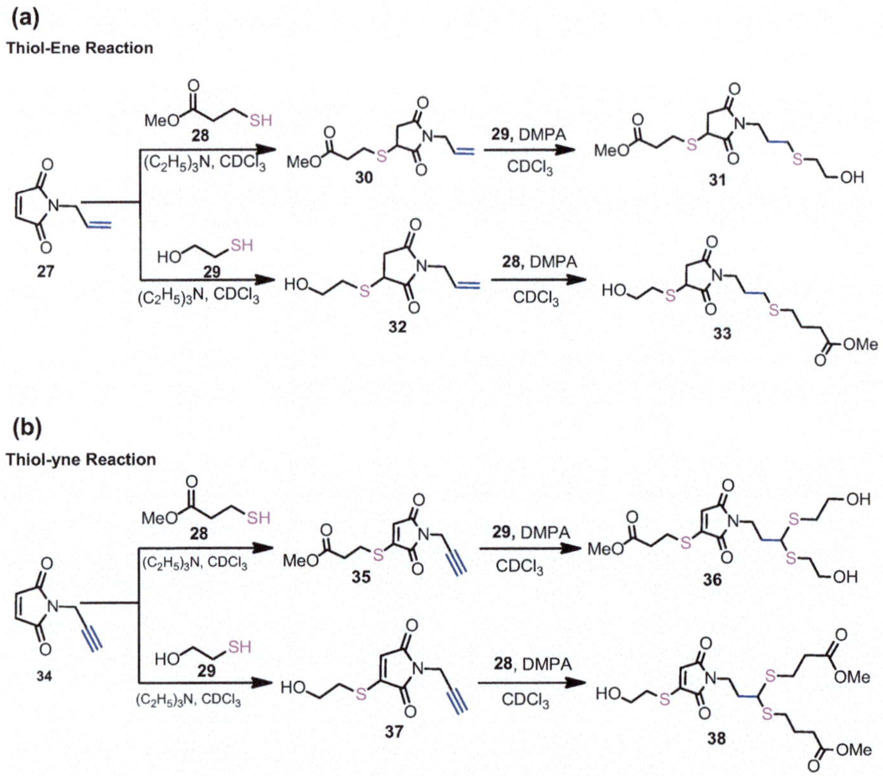

Scheme 7.12 **a** Selective base-initiated synthesis of thiol-ene products **30** and **32** followed by subsequent radical-mediated thiol-ene synthesis of products **31** and **33**. **b** Selective base-initiated synthesis of thiol-ene products **35** and **37** followed by subsequent radical-mediated thiol-yne synthesis of products **36** and **38**

solvent used in the controlled reaction. On the other hand, the α–addition of thiol **1** to alkyne **42** afforded to the formation of desired product **45** with very good yield (Scheme 7.13) [44].

Dondoni and co-workers devised a new methodology without any metal for synthesis of 1,6-linked disaccharide using moieties of sugar alkene, sugar thiol, and photoinitiator 2,2-dimethoxy-2-phenylacetophenone (DPAP) by photoinduced coupling thiol-ene click reaction. This sugar thiol **46** first treated with sugar alkene **47** in the presence of photoinitiator DPAP in DCM and irradiated UVA lamp (λ_{max} 365 nm, 4 × 15 W) several hours and afforded the adduct **48** thiol-ene click reaction with good yield and excellent diastereoselectivity (Scheme 7.14) [45].

Zhang and co-workers synthesized the eugenol-based polyols using thiol-ene click reaction that promote the green and environmental protection of polyurethane. In this method, groups took eugenol **49** and 2-mercaptoethanol **29** and stirred in the presence of 2-hydroxy-2-methylpropiophenone as photoinitiator. Thus, all the substrates were

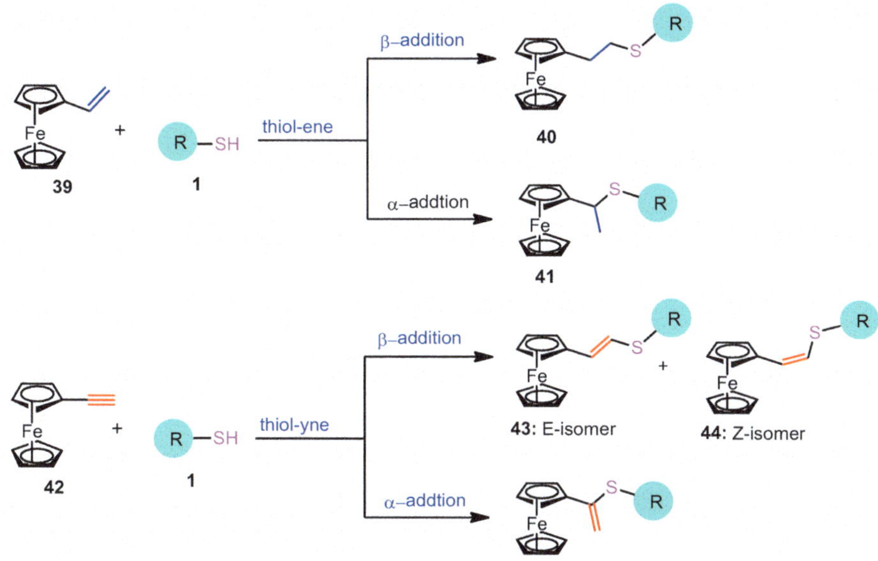

Scheme 7.13 Synthesis of thiol-ene/yne adduct of ferrocene link substituents

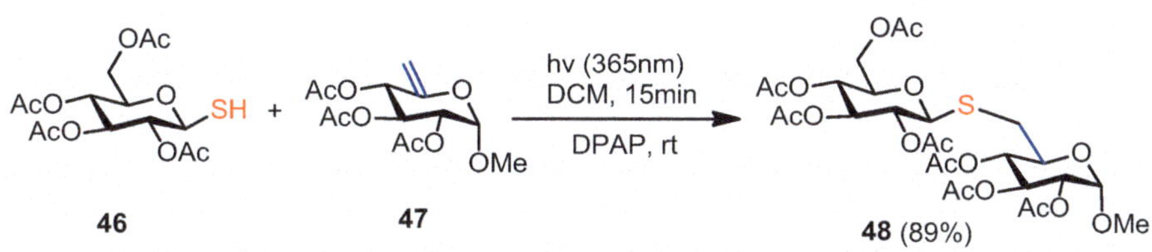

Scheme 7.14 Synthesis of 1,6-disaccharide of sugar thiol with sugar alkene

exposed to UV-light (100 W) for 3.5 h at room temperature and reaction underwent through thiol-ene click reaction under radical condition and gave the desired product **50** with good excellent yield (up to 96%). Similarly, eugenol **49** treated in separated sealed tubes with 6-hydroxy-1-hexanethiol **51** and 3-mercaptopropane-1,2-diol **53** in the presence of photoinitiator then both the solutions are irradiated by UV-light (100 W) for 3.5 h at room temperature and afforded the desired product **52** and **54** respectively with very good yield (Scheme 7.15) [46].

Hydrogen bonding plays an important role on stereo and regioselectivity of the product by controlling conformational position of attacking sites. Mal and groups select the different kind of thiol and amide having internal alkyne group and reported an article on synthesis of Z-selective anti-Markovnikov or Markovnikov product via thiol-yne click reaction depending upon the hydrogen binging between different functional groups. Thiophenol **11** is added the *N*,3-diphenylpropiolamide **55** under neat at room temperature and open air condition and led to the formation of Z-selective

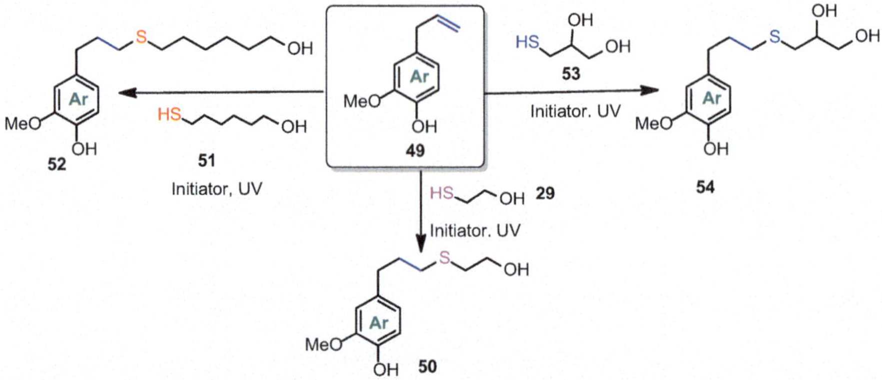

Scheme 7.15 Efficient synthesis of Eugenol-based polyols via thiol-ene click reaction

anti-Markovnikov product **56** through the transition state **TS-I** with excellent yield via thiol-yne click reaction by forming the *H*-bond between sulfur atom of thiophenol and proton of amide group present in *N*,3-diphenylpropiolamide **55** and when *N*,3-diphenylpropiolamide **55** treated with ortho-amino thiophenol **57** under neat at room temperature in open air condition and underwent through the thiol-yne click reaction and afforded the desired Z-selective Markovnikov adduct **58** with very good yield via transition state **TS-II** by forming the hydrogen bond between proton of amide group and nitrogen atom of amine group in ortho-amino thiophenol **57** (Scheme 7.16) [47].

Intramolecular thiol-ene reactions provide a variety of sulfur-containing heterocyclic. The radical-initiated thiol-ene reaction resulted to high yield of four-to-eight-membered rings and also well-defined macrocyclic architecture. The thiol-ene reaction via radical route favors anti-Markovnikov product, however, the regiochemistry of the intramolecular version depends on reaction condition as well substituent effects that serve the thermodynamically or kinetically favored cyclization product, respectively. The regioselectivity of intramolecular thiol-ene cyclization of a model reaction depicted in Scheme 7.17 depends on other factors and produces a mixture of 5-*exo* and 6-*endo* *C*-linked thiosugars [48]. The thiyl radical **60** underwent both 5-*exo* and 6-*endo* cyclization in order to produce pyranose **59** and furanose thiosugar **61,** respectively (Scheme 7.17). The ring closing mechanisms follow Baldwin's rules for such type of thiol-ene cyclization.

Similarly, a number of alicyclic and heterocyclic compounds were effectively accessed through intramolecular cyclization of thiyl and acyl thiyl via anti-Markovnikov thiol-ene reactions on 1,6-dienes, under photochemical conditions [49].

Carrilo and group performed an experiment of thiol-Michael addition (TMA) click reaction by using poly thiol and poly acrylate to synthesize polycyclic compounds, calixarene, etc. In this experiment, poly acrylate **62** reacted with poly thiols **63** in DCM in the presence of DBU as base for 85 min at a temperature of − 78 °C. This reaction underwent through TMA click reaction and gave the thiol-ene product

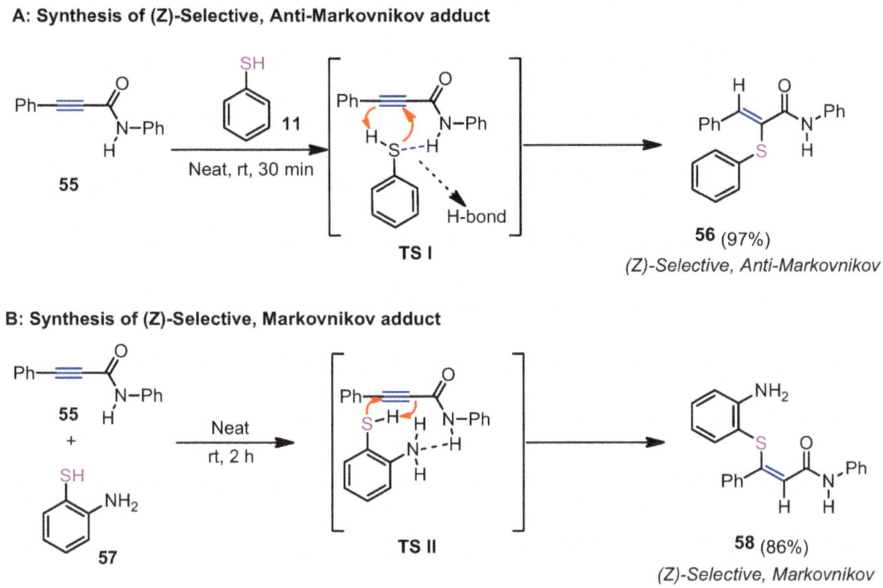

Scheme 7.16 Synthesis of (Z)-selective, anti-Markovnikov **56** and Markovnikov adduct **58** via controlled thiol-yne reaction

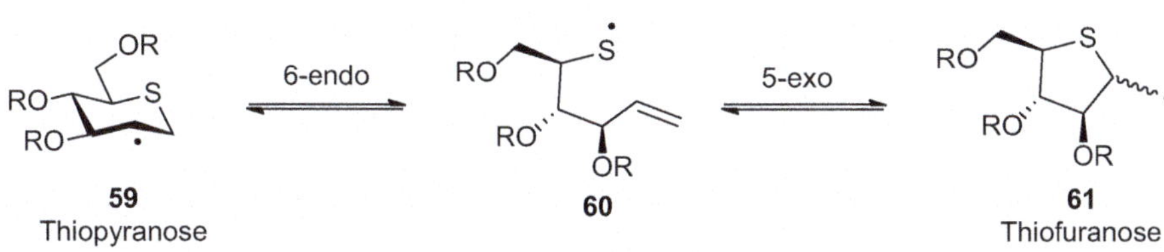

Scheme 7.17 Intramolecular ring closing thiol-ene reaction for synthesis of pyranose **59** and furanose thiosugar **61**

64 in one minute with slightly low yield of 50% at room temperature but after running the reaction for 85 min at − 78 °C compound **64** gave some more yield 55% (Scheme 7.18) [50].

Dove and groups also synthesized the cyclic polymer **70** from linear polymer by ethane-1,2-dithiol via thiol-ene click reaction. In first step, compound **65** treated with compound **66** and **67** and gave the linear polymer **68**. This formed product **68** further reacted with ethane-1,2-dithiol **69** to undergo thiol-Michael addition reaction and cyclized the linear polymer chain and afforded the desired product **70** with very good yield (Scheme 7.19) [51].

Li and group synthesized the macrocyclic and a rotaxane molecule via thiol-yne click reaction through simple dithiol coupling reaction. The macromolecule **71** treated with the compound **72** and formed the sulfuric macromolecule **73**. The

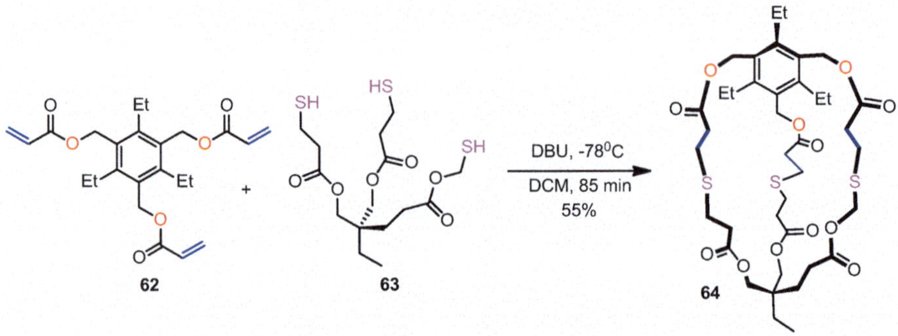

Scheme 7.18 Synthesis of calyx **64** through poly thiol and poly acrylate

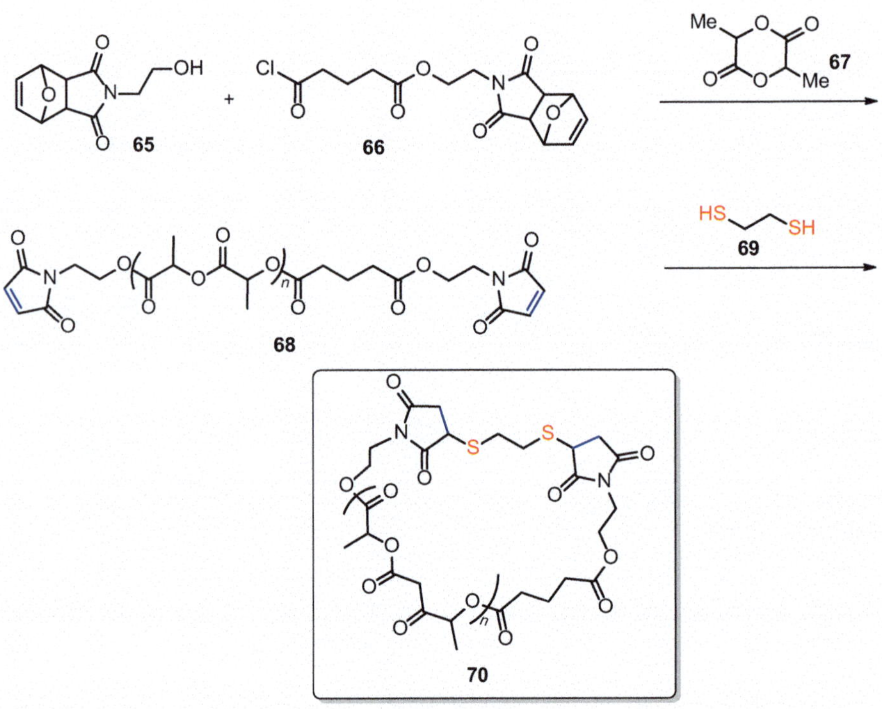

Scheme 7.19 Synthesis of cyclic polymer **70** via thiol-based click reaction

resulting compound **73** when reacted with alkylated ester compound **74** and NH_4PF_6 **75** (or secondary dialkylammonium ions) in DCM under irradiation of UV-rays for 20 min at room temperature, it undergo thiol-yne click reaction and afforded the desired macrocyclic product **76** with the yield 75% (Scheme 7.20) [52].

Lowe and co-workers introduced very beautiful way for development of sequential and chemoselective thiol-ene click in the presence of phosphine catalyst as an initiator

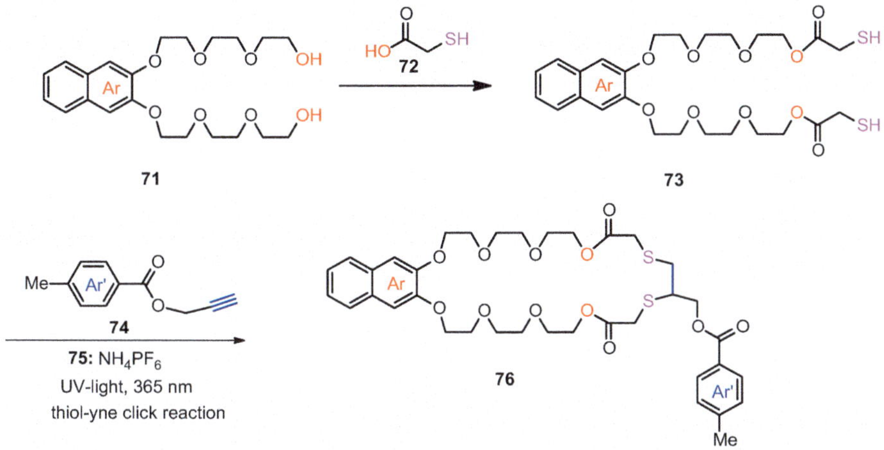

Scheme 7.20 Synthesis of sulfuric macrocyclic molecule via thiol-yne click reaction

and subsequently thiol-yne click via photocatalyzed photoinitiator DMPA. The group reported an experiment in which thiol **77** treated with the compound **78** (having both alkene and alkyne group) in the presence of dimethylphenylphosphine (Me_2PPh) as the initiator to undergo chemoselective thiol-ene click reaction under *nucleophile-initiated* conditions and synthesize the adduct **79** having terminal alkyne group with very good yield. This formed product **79** further underwent thiol-yne click reaction on reacting with 1-mercaptoglycerol **53** and DMPA as a photoinitiator by exposing the UV-light for 10 min at room temperature and afforded the addition product **80** with very good yield (Scheme 7.21) [53].

Many of the compounds have bridges of disulfide bonds such as peptide, polymer, and macromolecules and they can be modified and functionalized to vary their property. Bräse and his groups reported an article for rebridging through sulfur bond via thiol-yne click reaction. The compound **81** having disulfide bond in a ring underwent

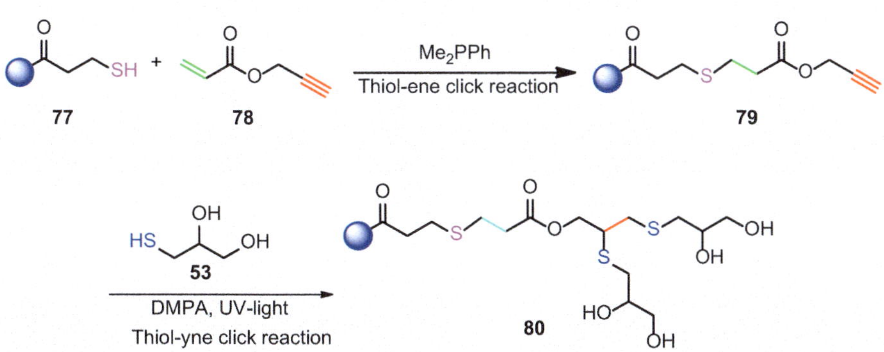

Scheme 7.21 Synthesis of sequential and selective thiol-ene/-yne click reaction

through reduction and broke the disulfide and gave the dithiol **82**. This formed dithiol **82** further reacted with alkyne **4** and DMPA as a photoinitiator and exposed the UV-light for several hours at room temperature to undergo photocatalyzed thiol-yne click reaction and synthesized the addition product **83** with very good yield (Scheme 7.22) [54].

Yoon et al. performed an experiment involving visible-light photoredox catalysis in thiol-ene reaction in the presence of transition metal polypyridyl phtotcatalysts. In this synthesis, *N*-Boc cysteine **85** treated with alkene, for example, styrene **84** in CH_3CN and a catalytic amount of ruthenium photoredox catalyst $Ru(bpz)_3(PF_6)_2$ **87** was added to the solution and blue light was irradiated for 90 min to form the desired click adduct **86** via thiol-ene click reaction with almost quantitative yield (Scheme 7.23) [55].

Li and co-workers developed a methodology using thiol-yne click reaction for cysteine modification triggered by sulfonium center and added the propargyl group to sulfonium center and reacted with thiol of amino acid or peptide chain. First of all,

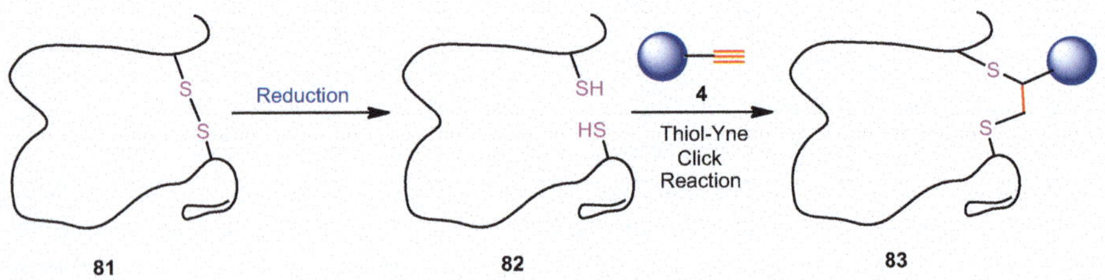

Scheme 7.22 Synthesis of sulfur inserted macrocyclic via thiol-yne click reaction

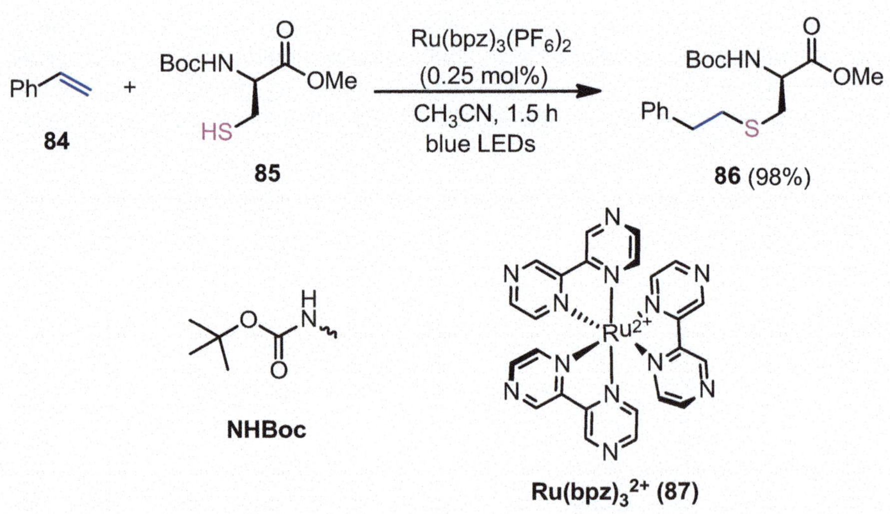

Scheme 7.23 Thiol-ene click reaction using photocataliytic radical

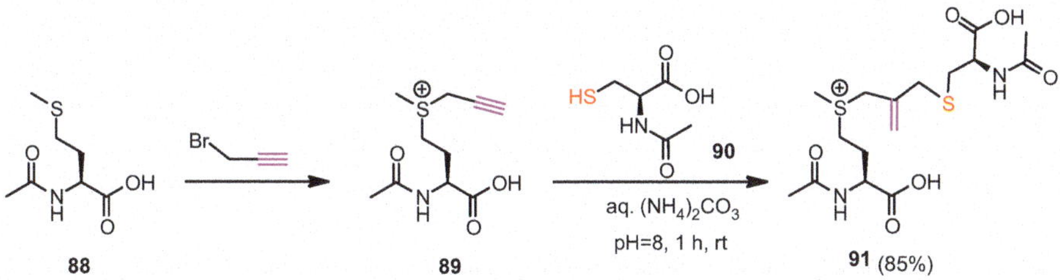

Scheme 7.24 Synthesis of sulfonium-derived product via thiol-yne click reaction

methionine **88** is propargylated on reacting with propargyl bromide and formed propargylated sulfonium center of methionine **89**. Then, the resulting propargylated sulfonium center **89** reacted with *N*-acyl cysteine **90** and aqueous ammonium carbonate at pH = 8 and stirred the solution for an hour at ambient temperature to undergo thiol-yne click reaction and synthesize the selective and controlled desired product **91** with excellent yield (Scheme 7.24) [56].

Scanlan and co-workers introduced the ligation reaction between an unprotected peptide thioester fragment and a second unprotected peptide possessing an *N*-terminal cysteine that results in formation of an amide bond is called native chemical ligation (NCL) and subsequently thiol-ene reaction. Peptide-1 **92** reacted with peptide-2 **93** and underwent NCL that formed the substitution product **94**. This synthesized product **94** was subsequently treated with alkene via thiol-ene reaction that led to formation of adduct **95** and further treated with PhSH to synthesize the thioester **96** with high yield (Scheme 7.25) [57].

Click chemistry, that is a library of efficient and reliable reactions, is very popular and powerful method for the synthesis and functionalization of various classes of bio and synthetic macromolecular systems. Wooley and group also used click chemistry for synthesizing the biodegradable and clickable polyphosphoester homopolymers via azide-alkyne cycloaddition and subsequently thiol-alkyne click reaction. 2-chloro-2-oxo-1,3,2-dioxaphospholane **97** is dissolved in THF and functionalized by the addition of triethylamine and alkyne **98** then stirred the complete solution for 12 h

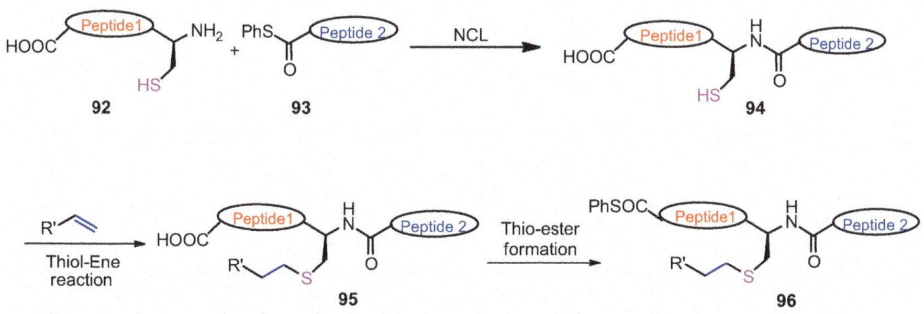

Scheme 7.25 Synthesis thioester via NCL and thiol-ene reaction

at 0 °C temperature that leads to the formation of substituted product **99**. This formed product **99** further reacted with phenol **100** and DBU (1,8-diazabicyclo[5.4.0]-undec-7-ene) and stirred for 5 min at room temperature and afforded a new type of reactive polyphosphoester bearing alkynyl groups **101** with very good yield. The formed product **101** subsequently reacted benzyl azide **102** CuBr and PMDETA after an hour triazole-linked polymer **103** with excellent yield via azide-alkyne click reaction and product **101** treated with thiol **104** and DMPA in methanol and UV-light exposed to the solution for 30 min at room temperature then synthesized the desired product **105** with very good yield via photocatalyzed thiol-yne click reaction (Scheme 7.26) [58].

In the domain of polymer synthesis, there are generally two main approaches for the synthesis of target polymers. One approach involves the use of thiol-ene or thiol-yne functionalized small molecules as monomer to produce the well-designed topological polymers. However, this approach has the drawback of limited control over the resulting polymer's molecular weight and molecular weight distribution. Nonetheless, it proves to be an expedient method for crafting functionalized networks or gels [59], particularly suitable for self-healing materials [60]. The alternate approach comprises post-polymerization modification of reactive precursor (co) polymers to construct complex (macro) molecules.

It is important to highlight that the significant benefits of this approach are a result of the diverse range of controlled polymerization (CLP) techniques that are

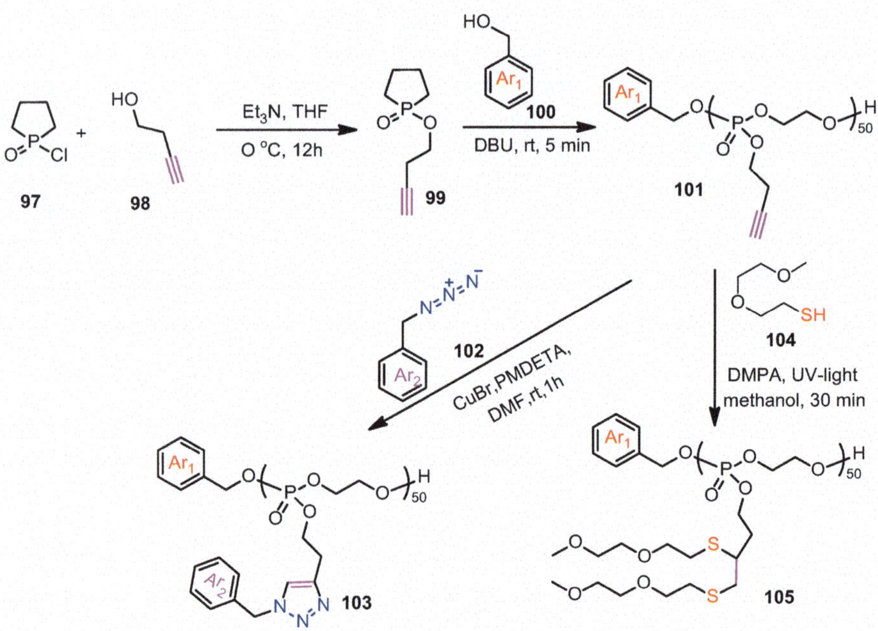

Scheme 7.26 Synthesis and functionalization of polymer chain via click reaction

available. One such technique, known as living radical polymerization (LRP) [61–63], is popular for its tolerance to various functional groups and its wide selection of polymerizable monomers, which allows for the synthesis of polymers with well-defined chain lengths, various structural designs, and specific end-group functionalities. Remarkably, the reversible addition fragmentation chain-transfer (RAFT) polymerization [64] technique can be employed to readily prepare thiol-terminated polymers from a wide range of monomers, followed by the removal of the chain-transfer agent that combined with CLP techniques and thiol-based 'click' chemistry, this approach offers an effective platform for constructing polymers in various structural designs.

Khan et al. used base-catalyzed thiol-epoxy click reaction for polymerization reaction to synthesize the linear polymer chains that have free hydroxyl group for further functionalization. 1,2–1,4-butanediol diglycidyl **106** ether and ethanedithiol **69** dissolved in THF and carried out with ice cold solution then added the lithium hydroxide (LiOH) as base. Thus, complete solution stirred for 6 h at 0 °C temperature. Therefore, polymerization reaction occurred via thiol-epoxy reaction and gave the desired linear polymer chain **107** with very good yield. The resulting product **107** has free hydroxyl group for functionalization (Scheme 7.27) [65].

Zhang et al. devised a new methodology for preparing of fully degradable brush using epoxide copolymerization and this methodology undergo through three steps alternating copolymerization of CO_2 and epoxides, thiol-ene click reaction, and ring-opening polymerization (ROP) of ε-caprolactone (ε-CL). First of all, 4-vinyl-cyclohexene-1, 2-epoxide **108** reacted with carbon dioxide in the presence of SalenCo **109** and underwent the copolymerization of CO_2 to give the compound **110** then treated with β-mercaptoethanol **29** in the presence of AIBN as radical initiator and synthesized the compound **111** via thiol-ene click reaction. Subsequently, the compound **111** reacted with ε-caprolactone in the presence of $Sn(Oct)_2$ and underwent ring-opening polymerization reaction and synthesized the final product **112** in good yield (Scheme 7.28) [66].

Scheme 7.27 Synthesis of linear polymer **107** via thiol-ene click reaction from epoxide **106**

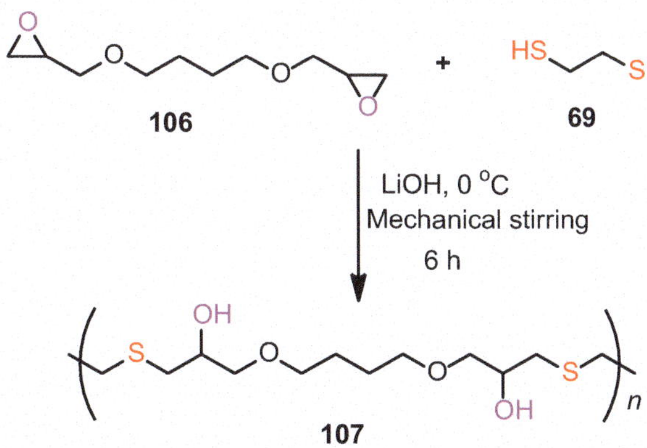

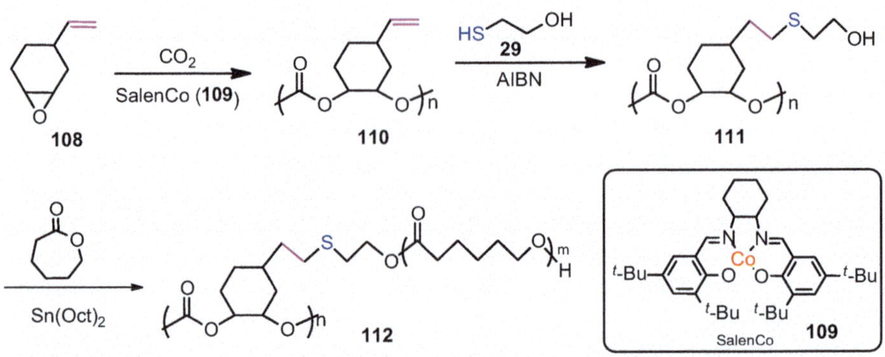

Scheme 7.28 Synthesis of degradable copolymer **112** with polycarbonate

Boyer et al. reported an article for functionalization and step-growth polymerization under aerobic condition via visible-light photocatalytic thiol-ene click reaction. In this report, $Ru(bpy)_3Cl_2$ acted as photoredox catalyst and *p*-toluidine as redox mediator. Alkene **113** treated with thiol in *N*-methyl-2-pyrrolidone and irradiated by visible light for several hours and underwent through photocatalytic thiol-ene click reaction and gave the desired adduct **114** with very good yield (Scheme 7.29) [67].

West and group gave a method for preparation of superhydrophobic cotton fabric using multiple steps via thiol-ene click reaction. In the first step, cellulose substrate (fabric) **115** was treated with aqueous solution of sodium carbonate for 10 min at 60 °C temperature and then fabric is neutralized with HCl because fabric was first treated with alkaline solution. After that fabric is stirred with cyanuric chloride **116** for an hour in chloroform at room temperature and thus resulted in cyanuric-linked cellulose compound **117 and** then this formed compound **117** turned to room temperature with cysteamine **118** and stirred for 24 h that results in the formation of compound **119** and washed with water and dried. Thus, the product **119** underwent thiol-ene radical reaction on reacting with AIBN in THF at room temperature for 10 min then reacted with alkene at temperature 80–120 °C from 30 to 120 min and afforded the desired product **120** in very good yield (Scheme 7.30) [68].

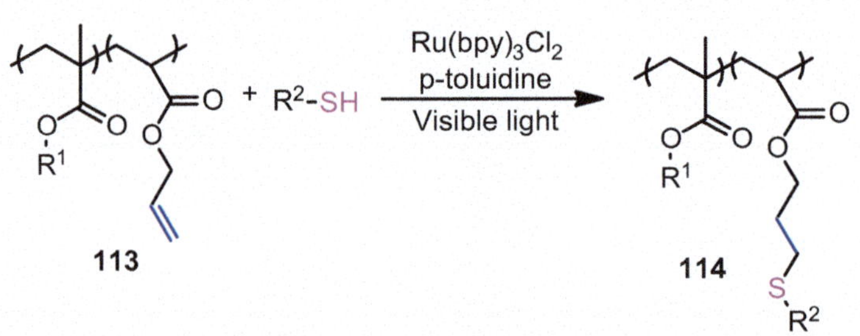

Scheme 7.29 Synthesis of thiol-ene adduct by photocatalytic process

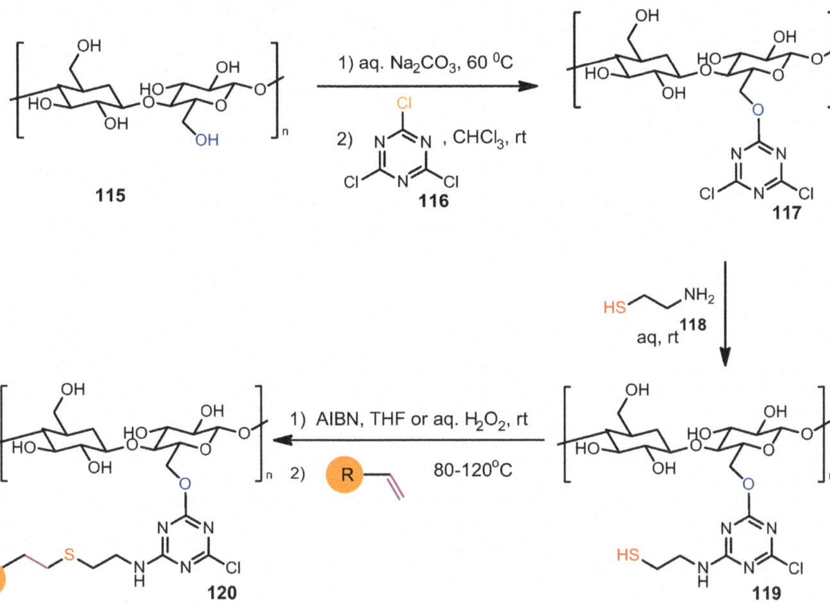

Scheme 7.30 Synthesis of super hydrophobic cotton fabric **120** from cellulose substrate **115**

Click chemistry is widely popular for synthesizing the dendrimers, macromolecules, and polymer synthesis. Durmaz and his groups reported an article for synthesis of controlled and stereoselective linear polymer via thiol-yne click reaction. The authors performed the experiment diverse condition with catalyst and without catalyzed. The dithiol **121** was treated with dialkyne **122** in DMF in the absence of catalyst and heated for several hours at 60 °C to undergo catalyst-free thiol-yne click reaction to give the linear polymer with good yield **123** without selectivity either E- or Z-isomer. Durmaz performed another experiment in the presence of catalyst in which dithiol **121 was** treated with dialkyne **112** and K_2PO_4 and NMP as a catalyst and heated the solution mixture for 24 h to 100 °C temperature and underwent stereoselective thiol-yne click reaction and afforded the Z-isomer **124** of linear polymer with very yield (Scheme 7.31) [69].

Dove and co-workers investigated the step-growth polymerization by using thiol-Michael addition reaction of dithiols and diynes and he reported the controlled and stereoselective polymer either E-isomer or Z-isomer with precise stereochemistry depending upon the presence of bases and solvent. Hexane 1,6-dithiol **125** was dissolved in chloroform added diynes **126** and DBU then heated to the solution for several hours and yielded very good the desired Z-isomer **127** via base-catalyzed thiol-yne click reaction (Scheme 7.32) [70].

Tang and his groups also succeeded in developing the polyhydrothiolation reaction and synthesized linear chain polymer via controlled thiol-yne reaction using diynes, dithiols, and transition metal catalysts. They found that polyhydrothiolation

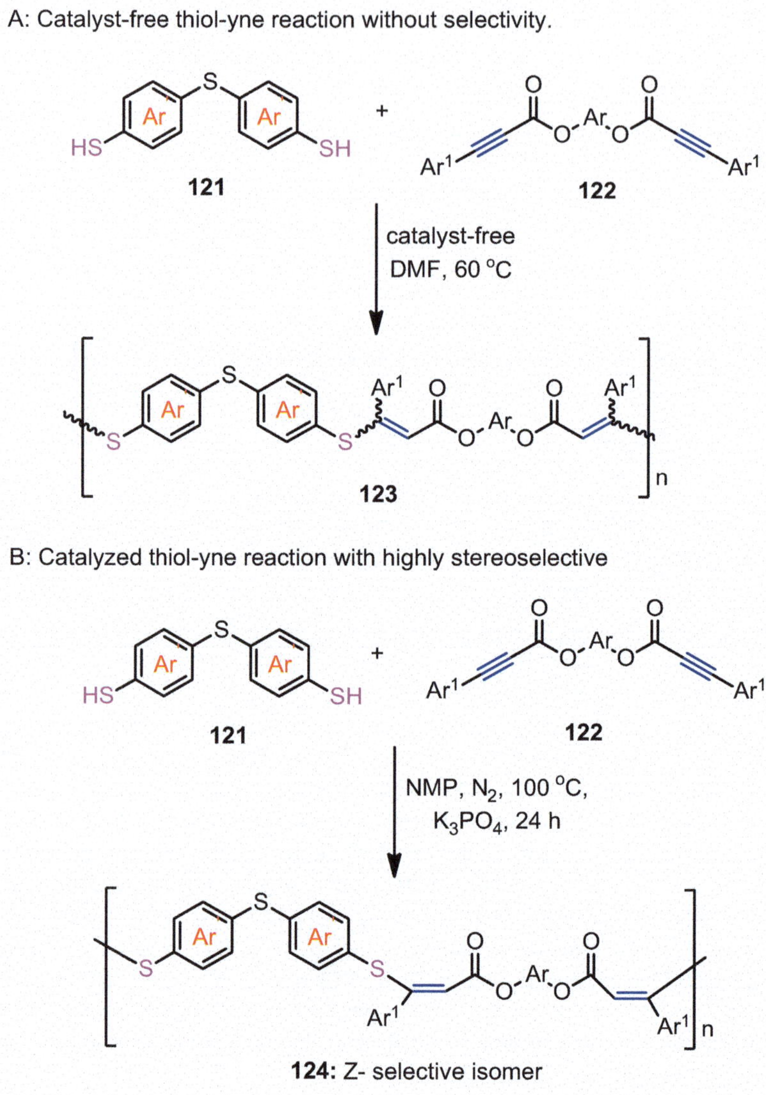

Scheme 7.31 Synthesis of thiol-yne adduct via catalyzed and catalyst-free pathway

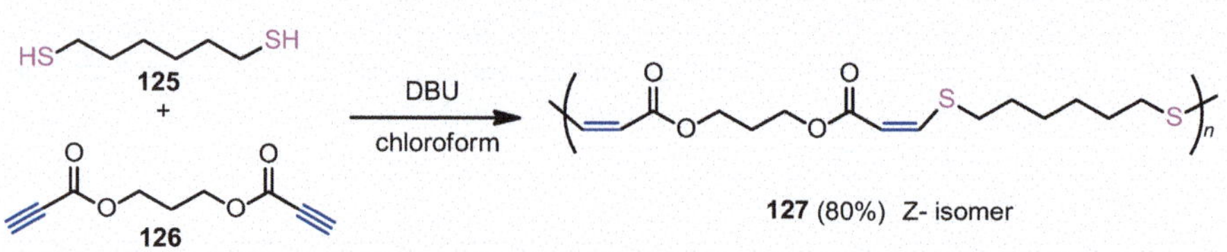

Scheme 7.32 Synthesis of controlled and stereoselective linear polymer via thiol-yne reaction

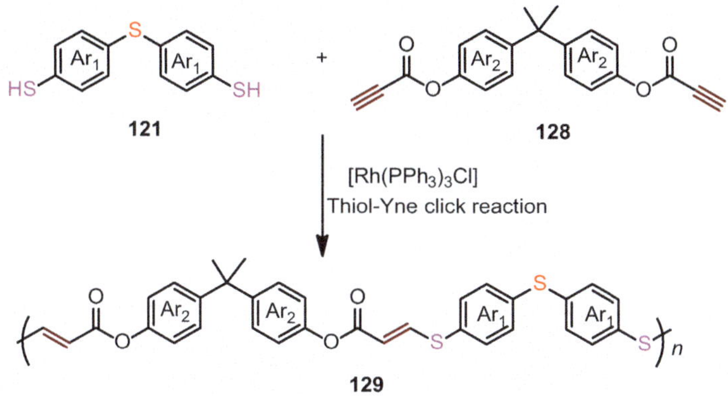

Scheme 7.33 Synthesis of regioselective linear polymer **129** via thiol-yne click reaction

reaction is stereoselective and regioselective and has no branched isomers, and this is well controlled by transition metal catalyst. The compound **121** was treated with compound **128** in the presence of ruthenium catalyst [Rh(PPh_3)$_3$Cl] and underwent controlled thiol-yne reaction and afforded the desired product **129** in very good yield (Scheme 7.33) [71].

In this methodology, Meier and co-workers reported novel step-growth polymerization reaction to synthesize the linear polymer of alkyne with dithiols through radical-mediated thiol-yne click reaction. Butane dithiol **130** was treated with 1-octyne **131** in the presence of radical initiator DMPA (2,2-dimethoxy-2-phenylacetophenone) and exposed the UV-light for 2 h at room temperature. Thus, the reaction proceeded very well and undergo thiol-yne click reaction and afforded the desired thiol-yne adduct **132** with yield 78%. When the similar reaction proceeded via thermally thiol-yne reaction in the presence of AIBN for 2 h, it produces 90% of the product **132** (Scheme 7.34) [72].

Zhang and co-workers have synthesized 1,2,4-triazole-thiol modified lignin-based adsorbent (LBA) by utilizing UV-light induced thiol-yne click reaction. Since LBA has good adsorption capacity and followed the second order reaction kinetics and works selectively for Cd adsorption. This LBA was made by two steps. In the first step, lignin (phenolic hydroxyl group) **133** was dissolved in an aqueous solution of NaOH at room temperature; then, propargyl bromide was added and the solution was

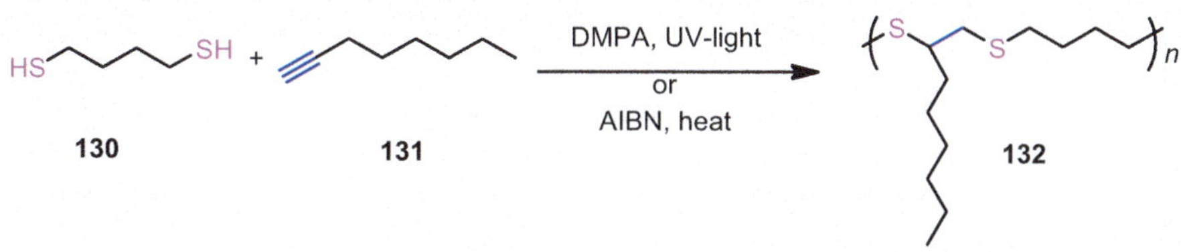

Scheme 7.34 Synthesis of linear polymer **132** of dithiol with alkyne via thiol-yne radical reaction

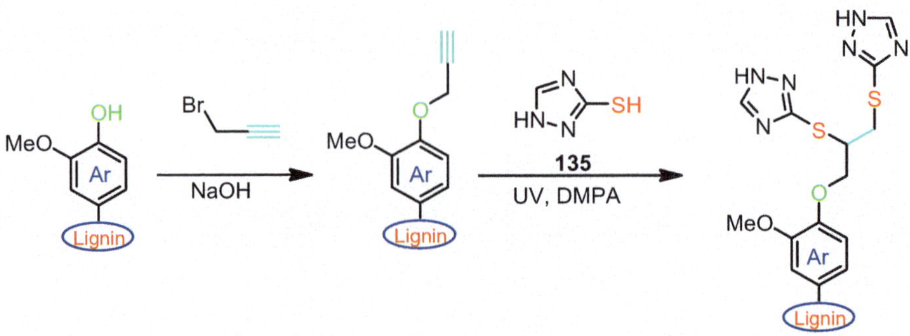

133 134 136

Scheme 7.35 Synthesis of lignin-based adduct **136** via thiol-yne click reaction

heated to 75 °C for 2 h led to the formation of alkylated lignin **134**. In the second step, 1,2,4-triazole-3-thiol **135** was added to the resulting product **134** in THF in the presence of DMPA and reaction was irradiated under a 365 nm UV lamp (150 W) for 30 min to give thiol-yne click product LBA **136** in 78% yield (Scheme 7.35) [73].

Stenzel and co-workers reported an article for the synthesis of copolymer via thiol-ene click reaction having carbohydrate side chain and subsequently developed the thermo-responsive glycomicelles as potential drug carrier. The compound **137** was treated with molecule **138** and gave the compound **139**. This resulting **139** product further reacted with AIBN and then added the 4-pentenic anhydride to the solution and raised the temperature to 80 °C for several hours that results in the formation of product **140**. The product **140** having alkene group reacted with Sugar-SH **141,** thus exposed UV-light at room temperature and underwent thiol-ene click reaction to synthesize the desired product **142** in excellent yield (Scheme 7.36) [74].

Wang and co-workers employed simple and universal route to synthesize the cyclic polymer (polyester) by copper catalyst click reaction of alkyne-azide cycloaddition of linear precursor and subsequently side chain functionalization of cyclized polymer through thiol-ene click reaction. The precursor **143** has both group alkyne and azide thus underwent copper-catalyzed click reaction with CuBr and PMDETA (pentamethyldiethylenetriamine) that results to the formation of 1,2,3-triazole adduct **144**. The resulting adduct **144** was treated with 2-mercaptoethylamine hydrochloride **145** and thiol **147** in the presence of photoinitiator DMPA and exposed the UV-light for an hour at room temperature and synthesized the addition product **146** and **148** respectively in very good yield via photocatalyzed thiol-ene click reaction (Scheme 7.37) [75].

The modular click nature of the thiol-Michael addition reaction is the key characteristic that makes it a highly valued tool in material sciences. This reaction is highly efficient and environmentally friendly and can be applied to a wide range of applications, from small molecule synthesis to surface modification of material coating to in situ polymer modification in biological systems [29]. Yogo et al. synthesized the cysteine-modified Fe_3O_4 nanoparticles (NPs) by functionalization

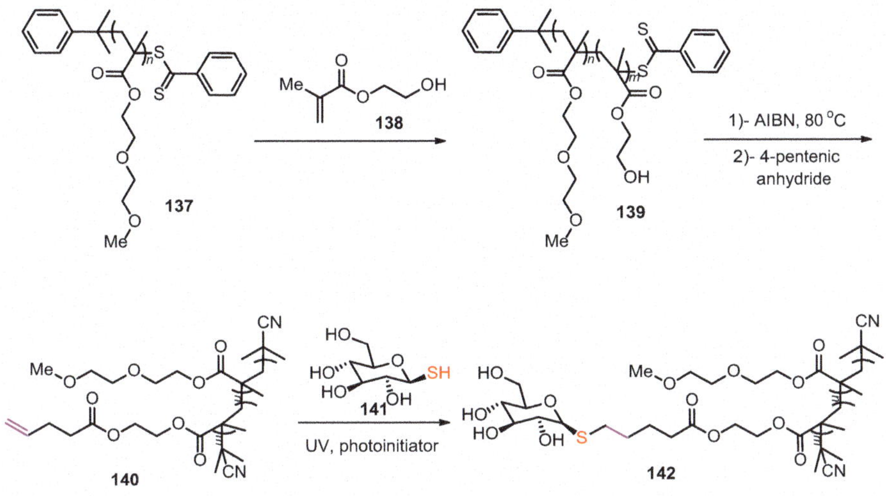

Scheme 7.36 Synthesis of sugar addition product via thiol-ene click reaction

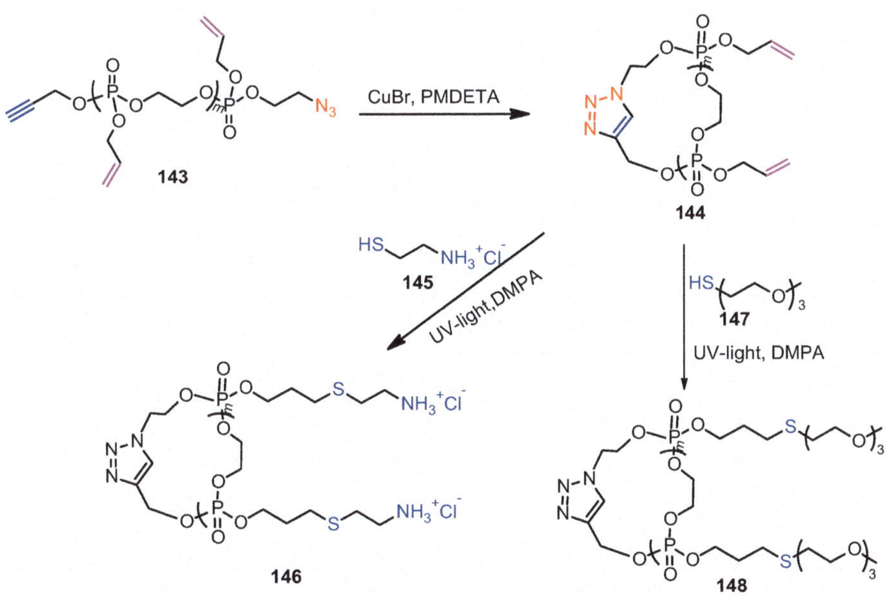

Scheme 7.37 Synthesis of cyclic polymers through thiol-ene click reaction

of allyl-functionalized Fe_3O_4 NPs (allyl-Fe_3O_4NPs) with cysteine using hydrolysis of iron(III) allylacetylacetonate **149** followed by thiol-ene click reaction. Iron(III) allylacetylacetonate **149** was dissolved in ethanol and added H_2O and N_2H_4 in the solution then refluxed at 80 °C for 24 h and after hydrolysis gave Allyl-Functionalized

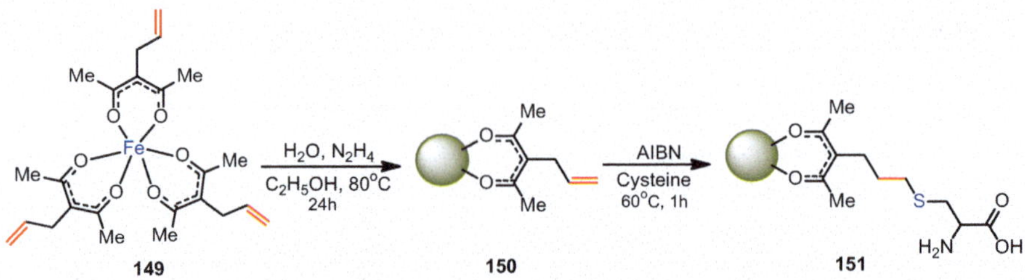

149 **150** **151**

Scheme 7.38 Synthesis of Cys-Fe_3O_4 nanoparticles by thiol-ene click reaction

Fe_3O_4 Nanoparticles (Allyl- Fe_3O_4 NPs) **150**. The resulting compound **150** further dissolved in ethanol with cysteine and radical initiator AIBN and heated to 60 °C for an hour and underwent thiol-ene click reaction and then finally synthesized Cysteine-functionalized Fe_3O_4 Nanoparticles (Cys-Fe_3O_4 NPs) **151** in very good yield (Scheme 7.38) [76].

Allard and his group made the functionalization of fullerene building block using thiol-ene click reation (thiol-maleimide reaction). Initially, the fullerene moiety having maleimide fragment **152** dissolved in dichloromethane and added *N*-Boc cysteine methyl ester [N-(*tert*-butoxycarbonyl)-L-cysteine methyl ester] **153** and triethylamine and stirred the solution at room temperature for 24 h and then afforded the fullerene additional product **154** via thiol-ene click reaction in 90% yield (Scheme 7.39) [77].

The orthogonal nature of thiol-ene coupling and anhydride-based esterification reactions were efficiently utilized in the synthesis of highly branched molecular architectures such as block copolymers, cross-linked materials, and dendrimers [78]. However, a number of applications of thiol-ene chemistry in synthesis of novel carbosilane-thioether dendrimers were greatly exploited in various report.[79, 80]. Cyclodextrin (CD) is a natural compound that made up of multisugar unit (cyclic oligosaccharides) in a ring. Cyclodextrin is used for drug delivery due to formation of inclusion complexes with guest molecules. Haddleton and groups synthesized the cyclodextrin-centered star polymers from per-6-thio-b-cyclodextrin [CD-$(SH)_7$] **155**

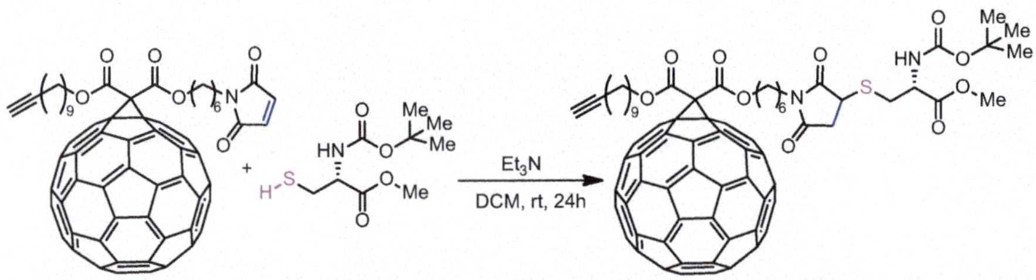

152 **153** **154** (90%)

Scheme 7.39 Formation of C–S bond between fullerene platform **152** and **154**

with vinyl terminated polymers **156** via thiol-ene click reaction. In this report, [CD-$(SH)_7$] **155** was treated with vinyl terminated polymer **156** and underwent thiol-ene click reaction and synthesized the desired adduct **157** in good yields (Scheme 7.40) [81].

Hawker and co-workers used thiol-ene click reaction and developed the dendrimer with the help of tris-alkene as a core and thiol by the formation of highly branch polymer via that has much different property from linear polymer and used in medicine and nanoengineering. 2,4,6-triallyloxy-1,3,5-triazine **158** was treated with 1-thioglycerol **53** and DMPA as photoinitiator, and then, UV-radiation exposed the solution for 30 min at room temperature to undergo thiol-ene click reaction. Thus, the desired thiol-ene adduct **159** is successfully synthesize in very good yield and this formed product underwent through several steps to produce the corresponding dendrimer **160** (Scheme 7.41) [82].

Thiol-yne click reaction was used in very beautiful way by Stenzel and co-workers to synthesize a functionalized dendrimer from tripropargyl ester of trimesic acid as a core of dendrimer. Tripropargyl ester **161** was treated with 1-thioglycerol **53** molecule in the presence of trace amount of photoinitiator DMPA and exposed by irradiation with UV lamp ($\lambda = 365$ nm) for ten minutes at room temperature. Thus, this reaction gave first generation of the dendrimer **162** with twelve hydroxyl groups via thiol-yne click reaction. Thus, resulting dendimer **162** subsequently reacted with acetylene anhydride **163** and triethylamine in DMF and further underwent thiol-yne reaction with 1-thioglycerol **53**. This afforded the synthesis of desired product poly-hydroxyl dendrimer **164** in good yield· (Scheme 7.42) [83].

Li and groups reported an article for the synthesis of dendrimers by using divergent approach and subsequently thiol-yne click reaction. Adipic acid **165** treated with two types of alkynes (aldo alkyne **166** and amido-alkyne **167**) in THF and raised the temperature to 40 °C and stirred the complete solution for 48 h then dendrimer core **168** is obtained in 92% yield. Further, the resulting core **168** further treated with 3-thio propionic acid **169** in the presence of DMPA in THF under UV-light is irradiated to the solution for 4 h at room temperature and underwent thiol-yne click reaction to synthesize the first generation of dendrimer **170** with 94% yield, therefore following the same process next generation of branched dendrimers or polymers synthesized (Scheme 7.43) [84].

Dondoni group exemplified the application of consecutive CuAAC click followed by 'thiol-ene click' coupling [35, 85] and develop high yield of desired calix[4]arene *S*-glycoside-based dendrimer **176** [23, 86]. The resulting dendrimer architecture have interesting structural feature where thiol residue placed at the upper rim and click-mediated triazole heterocycle incorporated at the lower rim on either calix[4]arene core. Thus, such calix[4]arene **172** having alkene and alkyne residue, respectively, placed on upper and lower rim was first CuAAC conjugated with galactose azide **173** using CuI/DIPEA in toluene at 25 °C for 26 h to give 56% of the respective triazole **174**, which was further allowed to 'thiol-ene click' with glucosyl thiol **175** using DMAP in CH_2Cl_2 for 60 min under photoinduced condition and gave high yield of the desired calix[4]arene *S*-glycoside **176a** (in 76%). At the end, acetate groups were

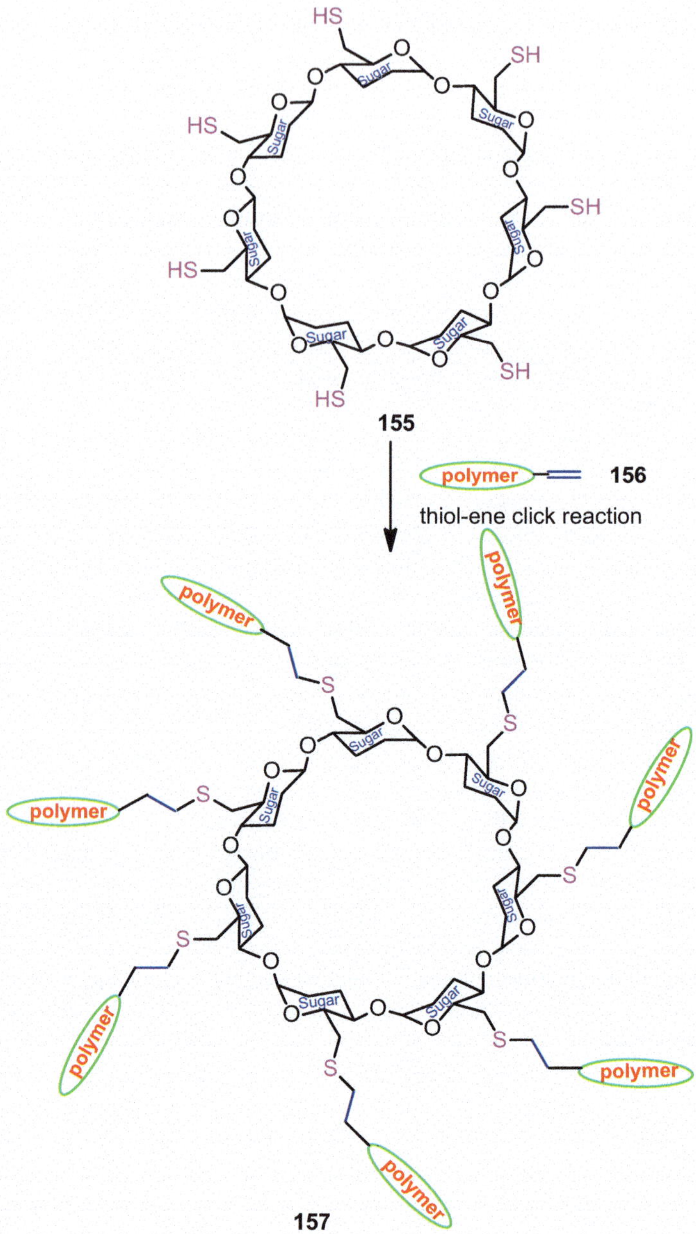

Scheme 7.40 Synthesis of cyclodextrin-centered polymer via thiol-ene click reaction

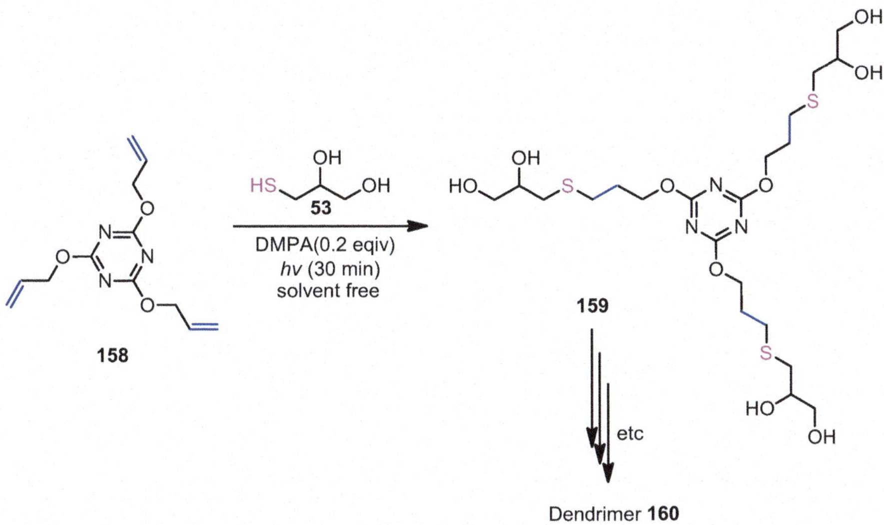

Scheme 7.41 Synthesis of dendrimer via thiol-ene click reaction

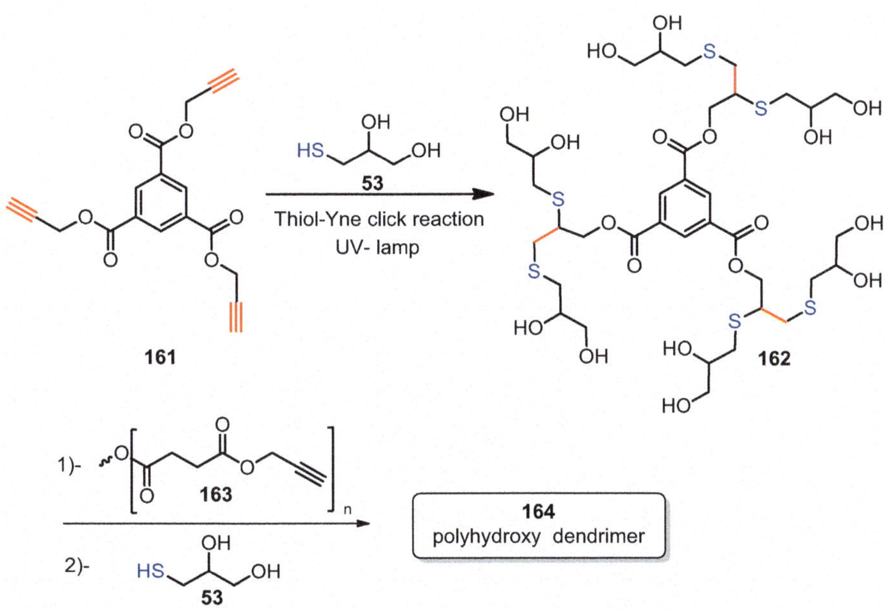

Scheme 7.42 Synthesis of dendrimer via thiol-yne click reaction

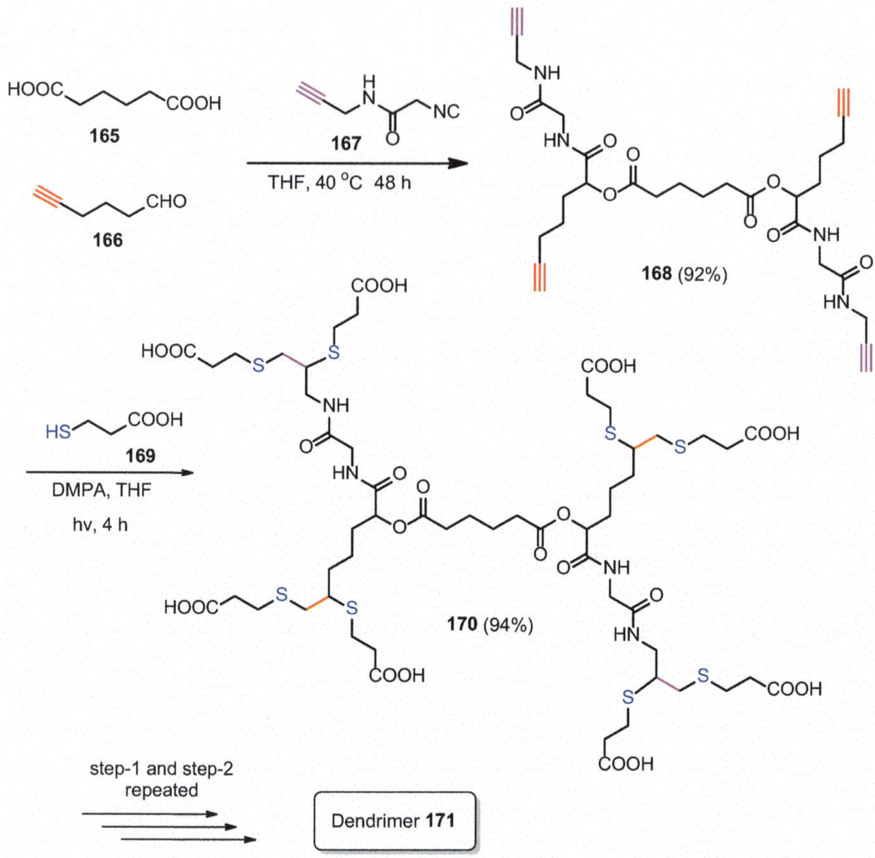

Scheme 7.43 Thiol-yne-mediated synthesis of dendrimeric architecture from dicarboxylic acid

successfully removed under Zamplen's conditions (NaOMe in MeOH) to give good yield of desired water-soluble glycodendrimer **176a** (Scheme 7.44).

In addition to the above-mentioned synthetic utilities, the thiol-ene click chemistry can be further explored for an anticipated functionalization of surface with desired spatial specificity and permitting the development of suitable photomasks [87]. The functionalization of surface through thiol-ene tool has great application in biotechnology and material science. Also, in continuation of the surface patterning, this modular tool can be utilized for the protein patterning on electron beam resist and the resulting nanostructures can permit direct protein functionalization of wide applications in material science [88].

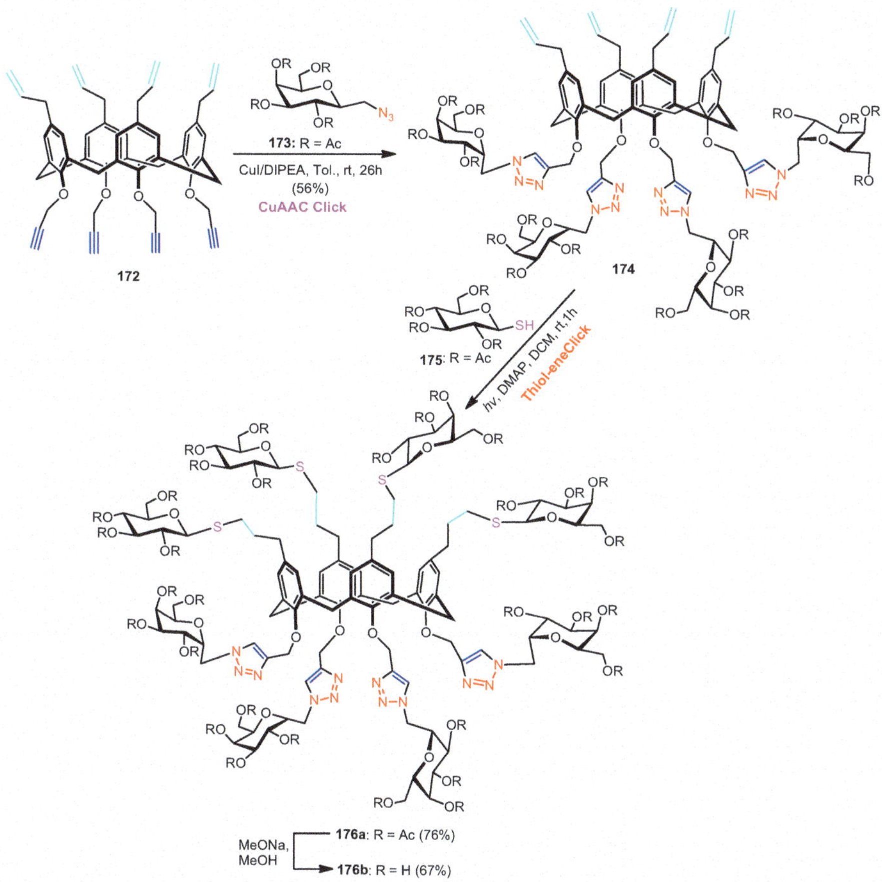

Scheme 7.44 Combined CuAAC-thiol-ene tool for easy access of calix[4]arene *S*-glycoside **176**

7.4 Conclusions and Future Perspective

This chapter summarizes the design and synthesis of diverse peptidomimetics, polymers and dendrimer molecular architectures via thiol-based 'click' chemistry. A number of polymer materials have been effectively synthesized using this simple, straight forward, and metal-free nature of reaction in combination with LRP and other click techniques. A large number of reactions, particularly those thiol-ene reactions triggered by UV-light, occur at room temperature and proceed swiftly. The photochemical, redox, thermal, and other pathways can all be effectively used to functionalize thiol-based click reactions due to the versatility of the thiol group. The study of polymer chemistry has found new applications due to advancements in visible-light catalysis. The thiol-based click chemistry in combination with visible light received more attention among researches across the globe toward the rapid development in stereoselective linear polymers. Furthermore, turning 'on' and 'off' the light is a simple process. It offers the possibility of programmed and patterned designing of new materials as it can be controlled both in terms of time and space. Since thiol group present in many biomolecules which may undergo thiol-based click chemistry to offer a broad range of applications for the development of functional molecules and materials of notable use in drug development and chemical biology. There are some drawbacks which limits the utility of thiol-based click reaction in some areas.

References

1. Kolb HC, Finn MG, Sharpless KB (2001) Angew Chem Int Ed 40:2004–2021
2. Tornøe CW, Christensen C, Meldal M (2002) J Org Chem 67:3057–3062
3. Rostovtsev VV, Green LG, Fokin VV, Sharpless KB (2002) Angew Chem Int Ed 41:2596–2599
4. Meldal M, Tornoe CW (2008) Chem Rev 108:2952–3015
5. Tiwari VK, Mishra BB, Mishra KB, Mishra N, Singh AS, Chen X (2016) Chem Rev 116:3086–3240
6. Agrahari AK, Bose P, Jaiswal MK, Rajkhowa SS, Singh AS, Hotha Mishra N, Tiwari VK (2021) Chem Rev 121:7638–7956
7. Kumar GS, Lin Q (2021) Chem Rev 121:6991–7031
8. Chen MM, Kopittke PM, Jhao FJ, Wang P (2024) Trends Plant Sci 29:167–178
9. Thirumurugan P, Matosiuk D, Jozwiak K (2013) Chem Rev 113:4905–4979
10. Kolb HC, Sharpless KB (2003) Drug Discov Today 8:1128–1137
11. Sharma A, Agrahari AK, Rajkhowa S, Tiwari VK (2022) Eur J Med Chem 238:114454
12. Jaiswal MK, Tiwari VK (2023) Chem Rec e202300167
13. Witczak ZJ, Bielski R (2013) Click chemistry in glycoscience: new development and strategies. Wiley, New York
14. Chandrasekeran S (2016) Click reactions in organic synthesis. Wiley, New York
15. Zhang L, Chen XG, Xue P, Sun HHY, Williams ID, Sharpless KB, Fokin VV, Jia GC (2005) J Am Chem Soc 127:15998–15999
16. Agard NJ, Prescher JA, Bertozzi CR (2004) J Am Chem Soc 126:15046–15047
17. Jaiswal MK, Gupta A, Yadav MS, Pandey VK, Tiwari VK (2023) Chem Eur J 29:e202301749
18. Worch JC, Stubbs CJ, Price MJ, Dove AP (2021) Chem Rev 121:6744–6774
19. Lowe AB (2010) Polym Chem 1:17–36
20. Posner T (1905) Ber Dtsch Chem Ges 38:646–657
21. Nilsson C, Simpson N, Malkoch M, Johansson M, Malmström E. J Polym Sci Part A: Polym Chem 46:1339–1348
22. Lowe AB (2014) Polymer 55:5517–5549
23. Dondoni A, Marra A (2012) Chem Soc Rev 41:573–586
24. Scanlan EM, Corce V, Malone A (2014) Molecules 19:19137–19151
25. Lowe AB, Hoyle CE, Bowman CN (2010) J Mater Chem 20:4745–4750
26. Hensarling RM, Rahane SB, LeBlanc AP, Sparks BJ, White EM, Locklin J, Patton DL (2011) Polym Chem 2:88–90
27. Jian Y, He Y, Sun Y, Yang H, Yang W, Nie J (2013) J Mater Chem C 1:4481–4489
28. Zhang Q, Anastasaki A, Li G-Z, Haddleton AJ, Wilson P, Haddleton DM (2014) Polym Chem 5:3876–3883
29. Nair DP, Podgórski M, Chatani S, Gong T, Xi W, Fenoli CR, Bowman CN (2014) Chem Mater 26:724–744
30. Shih H, Lin CC (2015) Biomacromol 16:1915–1923

31. Hoyle CE, Bowman CN (2010) Angew Chem Int Edn 49:1540–1573
32. Northrop BH, Coffey RN (2012) J Am Chem Soc 134:13804–13817
33. Sy Piecco KWE, Aboelenen AM, Pyle JR, Vicente JR, Gautam D, Chen J (2018) ACS Omega 3:14327–14332
34. Frayne SH, Murthy RR, Northrop BH (2017) J Org Chem 82:7946–7956
35. Dondoni A (2008) Angew Chem Int ed 47:8995–8997
36. Love DM, Kim K, Goodrich JT, Fairbanks FD, Worrell BT, Stoykovich MP, Musgrave CB, Bowman CN (2018) J Org Chem 83:2912–2919
37. Kumar K, Tiwari P, Kant R, Bhattacharya S (2022) New J Chem 46:17854–17864
38. Tyson EL, Ament MS, Yoon TP (2013) J Org Chem 78:2046–2050
39. Levin VV, Dilman AD (2019) J Org Chem 84:8337–8343
40. Tucker-Schwartz AK, Farrell RA, Garrell RL (2011) J Am Chem Soc 133:11026–11029
41. Zuo Y, Wang D, Zhang J, Feng S (2014) RSC Adv 4:62827–62834
42. Xi W, Wang C, Kloxin CJ, Bowman CN (2012) ACS Macro Lett 1:811–814
43. Stolz RM, Northrop BH (2013) J Org Chem 78(16):8105–8811
44. Enrıq́uez A, Gonzaĺ ez-Vadillo AM, Martıń ez-Montero I, Bruña S, Leemans L, Cuadrado I (2014) Organometallics 33:7307−7317
45. Fiore M, Marra A, Dondoni A (2009) J Org Chem 74:4422–4425
46. Hu Y, Tian Y, Cheng J, Zhang J (2020) ACS Sustain Chem Eng 8:4158–4166
47. Pramanik M, Choudhuri K, Chakraborty S, Ghosh A, Mal P (2020) Chem Commun 56:2991–2994
48. Scanlan E, Corcé V, Malone A (2014) Molecules 19:19137–19151
49. Lynch DM, Nolan MD, Williams C, Van Dalsen L, Calvert SH, Dénès F, Trujillo C, Scanlan EM (2023) J Org Chem 88:10020–10026
50. Perretti MD, Pérez-Márquez LA, Garcia-Rodriguez R, Carrillo R (2019) J Org Chem 84:840–850
51. Stanford MJ, Pflughaupt RL, Dove AP (2010) Macromolecules 43:6538–6541
52. Zhou W, Zheng H, Li Y, Liu H, Li Y (2010) Org Lett 12:4078–4081
53. Chan JW, Hoyle CE, Lowe AB (2009) J Am Chem Soc 131:5751–5753
54. Griebenow N, Dilmaç AM, Greven S, Bräse S (2016) Bioconjugate Chem 27:911–917
55. Tyson EL, Niemeyer ZL, Yoon TP (2014) J Org Chem 79(3):1427–1436
56. Hou Z, Wang D, Li Y, Zhao R, Wan C, Ma Y, Lian C, Yin F, Li Z (2020) J Org Chem 85:1698–1705
57. Markey L, Giordani S, Scanlan EM (2013) J Org Chem 78(9):4270–4277
58. Zhang S, Li A, Zou J, Lin LY, Wooley KL (2012) ACS Macro Lett 1:328–333
59. Shih H, Fraser AK, Lin CC (2013) ACS Appl Mater Interfaces 5:1673–1680
60. Hillewaere XKD, Teixeira RFA, Nguyen L-TT, Ramos JA, Rahier H, Du Prez FE (2014) Adv Funct Mater 24:5575–5583
61. Kamigaito M, Ando T, Sawamoto M (2001) Chem Rev 101:3689–3745
62. Chiefari J, Chong YK, Ercole F, Krstina J, Jeffery J, Le TPT, Mayadunne RTA, Meijs GF, Moad CL, Moad G, Rizzardo E, Thang SH (1998) Macromolecules 31:5559–5562
63. Matyjaszewski K, Xia J (2001) Chem Rev 101:2921–2990
64. Roth PJ, Boyer C, Lowe AB, Davis TP (2011) Macromol Rapid Commun 32:1123–1143
65. Brandle A, Khan A (2012) Polym Chem 3:3224–3227
66. Zhang JF, Ren WM, Sun XK, Meng Y, Du BY, Zhang XH (2011) Macromolecules 44:9882–9886
67. Xu J, Boyer C (2015) Macromolecules 48:520–529
68. Brown AS, Bozman ME, Hickman TJ, Hossain MI, Glover TG, West KN, West CW (2019) Ind Eng Chem Res 58:22534–22540
69. Daglar O, Luleburgaz S, Baysak E, Gunay US, Hizal G, Tunca U, Durmaz H (2020) Eur Polymer J 137:109926
70. Worch JC, Dove AP (2022) Acc Chem Res 55:2355–2369
71. Liu J, Lam JWY, Tang BZ (2009) Chem Rev 109:5799–5867
72. Türünç O, Meier MAR (2012) Polym chem 50:1689–1695

73. Jin C, Zhang X, Xin J, Liu G, Wu GM, Kong Z, Zhang J (2017) ACS Sustainable Chem Eng 5:4086–4093
74. Chen G, Amajjahe S, Stenzel MH (2009) Chem Commun 1198–1200
75. Yuan YY, Du JZ, Wang J (2012) Chem Commun 48:570–572
76. Hayashi K, Ono K, Suzuki H, Sawada M, Moriya M, Sakamoto W, Yogo T (2010) Chem Mater 22:3768–3772
77. Fensterbank H, Baczko K, Constant C, Idttalbe N, Bourdreux F, Vallée A, Goncalves AM, Meallet-Renault R, Clavier G, Wright K, Allard E (2016) J Org Chem 81(18):8222–8233
78. Montañez MI, Campos LM, Antoni P, Hed Y, Walter MV, Krull BT, Khan A, Hult A, Hawker CJ, Malkoch M (2010) Macromolecules 43:6004–6013
79. Rissing C, Son DY (2009) Organometallics 28:3167–3172
80. Son DY, Rissing C, Chen L, Andersson TE (2010) Polym Prepr 51:730–731
81. Zhang Q, Li GZ, Becer CR, Haddleton DM (2012) Chem Commun 48:8063–8065
82. Killops KL, Campos LM, Hawker CJ (2008) J Am Chem Soc 130(15):5062–5064
83. Chen G, Kumar J, Gregory A, Stenzel MH (2009) Chem. Commun 6291–6293
84. Deng XX, Du FS, Li ZC (2014) ACS Macro Lett 3:667–670
85. Witczak ZJ, Lorchak D, Nguyen N (2007) Carbohydr Res 342:1929–1933
86. Fiore M, Chambery A, Marra A, Dondoni A (2009) Org Biomol Chem 7:3910–3913
87. Jonkheijm P, Weinrich D, Köhn M, Engelkamp H, Christianen PCM, Kuhlmann J, Maan JC, Nüsse D, Schroeder H, Wacker R, Breinbauer R, Niemeyer CM, Waldmann H (2008) Angew Chem Int Ed 47:4421–4424
88. Shafagh R, Vastesson A, Guo W, van der WW, Haraldsson T (2018) ACS Nano 12 9940–9946

Chapter 8
SuFEx Click Chemistry: Discovery to Applications

8.1 Introduction

One of the most notable developments in enabling fast and reliable synthesis technologies during the last 22 years is identified as the regioselective 1,2,3-triazole forming reaction, commonly known as 'Click Chemistry' [1–4]. As the concept of 'Reactivity *vs* Selectivity', this is obviously acceptable thing that *'Reactivity is not everything; it's the only thing'* [5]. In addition to several benefits, 'Click Chemistry' has a notable feature and that is its excellent regioselectivity that just after the discovery by Meldal and Sharpless in 2002 opened a new window with wide opportunities in almost all emerging fields of science and technology. This modular method is widely explored in medicinal chemistry, carbohydrate chemistry, materials science, and bioconjugation, and is well recognized to felicitate the process of drug discovery and development research in academia and pharmaceutical industries [6–10].

After the 'Click Chemistry' concept coined by Prof. K. B. Sharpless in 2001 [9], the regioselective 1,2,3-triazole forming reaction using Cu(I)-catalyzed azide-alkyne cycloaddition (CuAAC) condition was discovered in 2002 by K B Sharpless [1] and Morten Meldal [11]. Latter on Ru-catalyzed azide-alkyne cycloaddition (RuAAC) [12], organo-catalyzed click [13–17], photo-catalyzed click [18], topchemical click [19], and strain-promoted click in the living system (known as, Bertozzi bioorthogonal chemistry) were developed and well established and explored in science and technology. The regioselective triazole forming click tool is modular, rapid, easy, and high-yielding and has become one of the most reliable protocols investigated so far having widespread applications in almost all fields such as carbohydrate chemistry [3], supramolecular chemistry [4], macrocyclization [20], dendrimers, biomolecular, material science [21, 22], surface chemistry [23], nanoparticle surface modification [24]. Agriculture, drug discovery and development [25], and technology. Soon after, the term 'click chemistry' emerged as a trendy that is not limited to regioselective triazole forming but also covered several other handy reactions such as thiol-ene

V. K. Tiwari et al., *Click Chemistry*, Materials Horizons: From Nature to Nanomaterials,
https://doi.org/10.1007/978-981-97-4596-8_8

coupling, thiol-yne coupling, oxime coupling, pyridyl sulfide reactions, ring-opening reactions, and the Sulfur (VI) Fluoride Exchange (SuFex) reactions [26–30].

Regioselective triazole formation under metal-catalyzed AAC reaction (CuAAC as well RuAAC) has notable features as thermally induced AAC resulted in a mixture of inseparable regioisomers (Scheme 8.1). Strain-promoted [3 + 2] azide-alkyne cycloaddition (SPAAC, Wittig approach) was brilliantly explored in 2004, where the SPAAC reaction of azido-modified cell **1** with alkenyl fluorescent probe **2** allows for the covalent linking of biomolecules together via a triazole residue surrounded by a human cell in the living system **3** (Bertozzi bioorthogonal chemistry) [31, 32]. The bioorthogonal coupling was performed in a biological system or even in the human body to track glycans in living systems. In addition to CuAAC/ RuAAC and SPAAC reaction, regioselective synthesis of functionalized triazoles using suitable organocatalysis for example, 20 mol% proline is well documented in the literature [33–36]. Recently in 2022, the Swedish Academy of Science recognized the field 'click chemistry and bioorthogonal chemistry' with Nobel Prize in Chemistry [37].

In addition to a modular triazole forming click chemistry, a Sulfur Fluoride Exchange reaction, commonly known as 'SuFEx Click' is also recognized as a modular and next-generation variant of click reactions. This tool is also triggered to a rapid, robust, and reliable one to access a family of diverse functional molecules of wide applications in various fields ranging from chemical biology to catalysis and medicinal chemistry to material science [29]. This chapter highlights the discovery, historic development, and emerging scope of SuFEx click chemistry in various fields.

8.1.1 Historic Development of SuFEx Connectors and SuFExable Molecular Plugins

A regioselective triazole forming Cu(I) catalyzed azide–alkyne cycloaddition CuAAC tool is just one of the identified versions of click reactions and other equally important methods included thio-ene click, thiol-yne-click as discussed in separate chapter of this book. Another important variant of click chemistry includes the Sulfur(VI) Fluoride Exchange (commonly known as, SuFEx Click Chemistry. This SuFEx click variant under metal-free conditions was introduced by the Sharpless Group at Scripps, in 2014 [38]. The chemistry of the higher organosulfur fluorides is not new as this was known back in the mid-1800s, while its incredible properties were again recognized after the seminal research of Steinkopf in the 1920s that finally and after a century led to the invention of SuFEx click chemistry [39].

Despite of Steinkopf's early investigations on the chemistry of aryl sulfonyl fluorides, the exploration of its higher valent sulfur–fluorides remains ignored. Although S–F bonds are more stable than the common S–Cl bonds and were known to tolerate harsh reaction conditions; despite they can be found reactive enough with several nucleophilic S–F exchange reactions with delicate control simply with an aid of desired activation (Scheme 8.2).

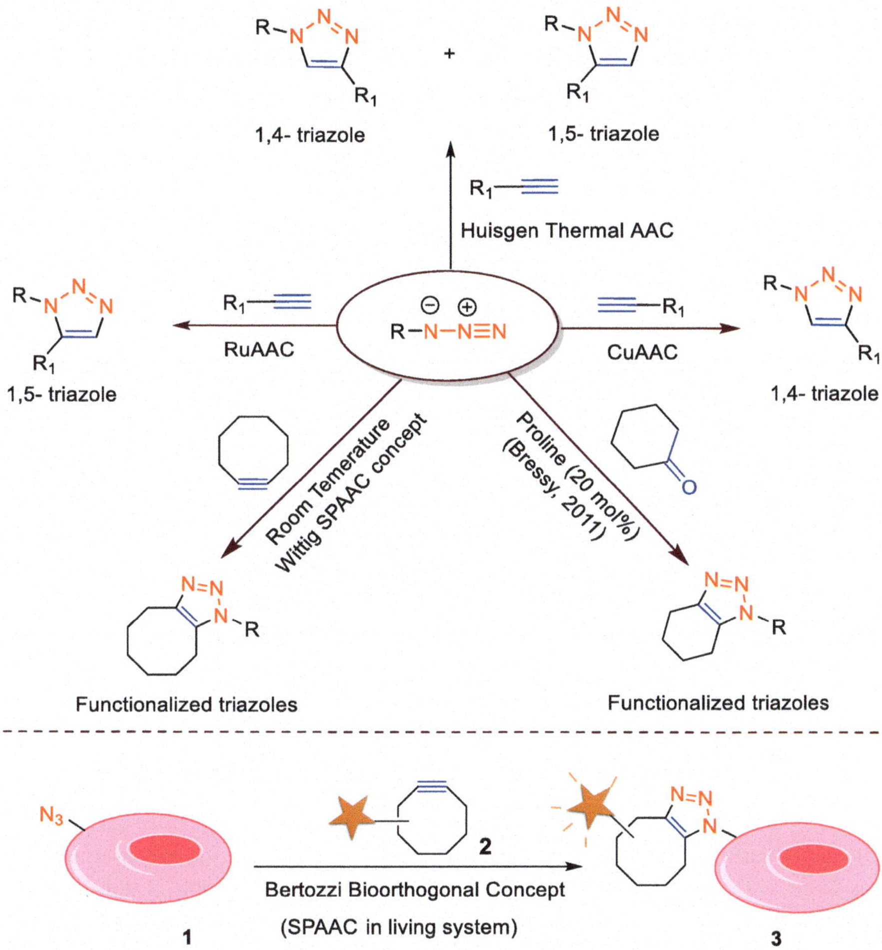

Scheme 8.1 Various 1,2,3-triazole forming click chemistry

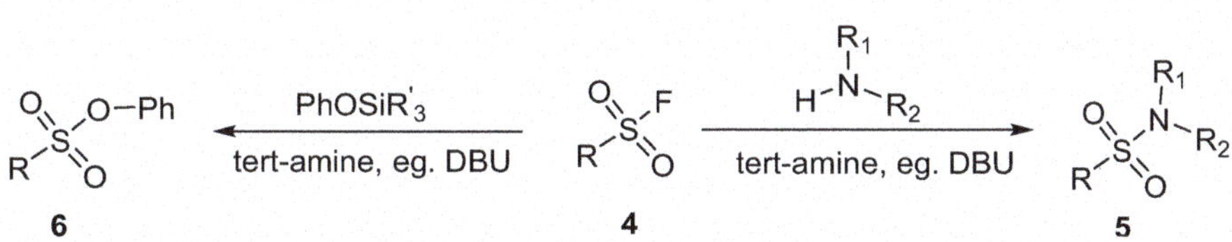

Scheme 8.2 Sulfur fluoride exchange (commonly known as, SuFEx) Click Chemistry (R stands for carbon and heteroatom-departure link such as sulfonyl fluoride, fluorosulfate, etc.)

This strange reactivity of higher valent sulfur–fluorides reproduced, where click-like reactivity is frequently observed and attributed to an individual combination of its kinetic stability and near-perfect reactivity of SuFEx as a reaction of near-perfect click variant.

8.1.2 Background and Historical Development

Due to space limitations and the structure of the book chapter, this is extremely difficult and challenging for us to cover all the relevant and impactful examples of SuFEx click chemistry and we authors deeply apologize to the original contributors whose work is not included herein. This chapter highlights the SuFEx chemistry a modular sulfur fluoride click tool and their versatile application in synthesis [40–43].

In contrast to CuAAC or RuAAC Click Chemistry, the SuFEx reactions are in general flexible and can proceed under metal-free conditions and are more diverse as the tool allows to form new bond via S–F exchange with various groups including aryl silyl ethers or amines or carbon nucleophiles. Such distinct characteristics make the SuFEx tool of wider scope to be explored in the biological system like Bertozzi Biorthogonal Chemistry and can also be well utilized in drug discovery research.

8.1.3 SuFEx Activation and Catalysis

The exact activation mechanism of SuFEx catalysis has not been fully established. It has been assumed that the exceptional activity is due to the conversion of fluoride ions from a strong covalent bond to a better-leaving group, which is stabilized by the interaction with the surrounding H^+ or R_3Si under stringent kinetic and spatial conditions. SuFEx is generally activated by tertiary amine-based catalysis such as Et_3N, DBU, etc., and is supposed to comprise bi-fluoride counterion species. Gembus et al. reported an efficient organocatalytic method for the interconversion of silyl ethers to sulfonate esters by using p-toluenesulfonyl fluoride and a catalytic amount of DBU in acetonitrile at room temperature [44]. The plausible mechanism involved the nucleophilic attack of DBU on sulfonyl fluoride **7** to form an activated aryl sulfonyl ammonium fluoride salt **8**, which may successively activate the silyl ethers **9** via the release of fluoride ion promoting aryl sulfonyl transfer to form the desired sulfonate product **10** (Scheme 8.3). It has been observed that the choice of catalysts mainly depends on the reactivity of the coupling partners and in exceptional cases, phosphazine-based catalysts such as BEMP (2-tertbutylimino-2-diethylamino-1,3-dimethylperhydro-1,3,2-diazaphosphorine) has also been used for this process. The order of the reactivity of these catalytic bases is directly proportional to their pKb values and follows the order: of $Et_3N < DBU < BEMP$.

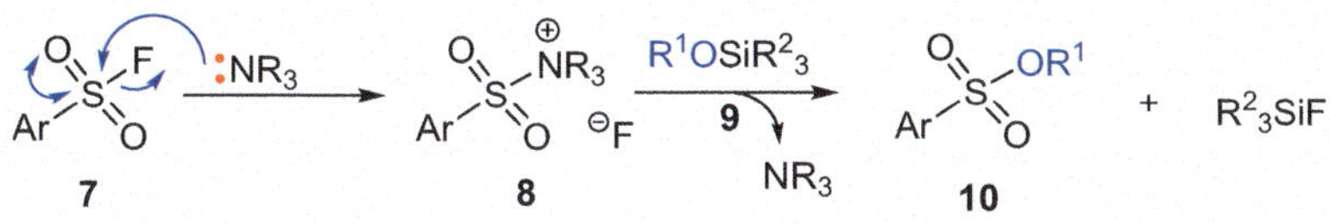

Scheme 8.3. Possible activation of sulfonyl fluorides with tertiary amine bases (Gembus's concept)

8.2 Synthesis of SuFEx Connectors and SuFExable Molecular Plugins

SuFEx click may able to join the universal native functional groups such as primary and secondary amines and phenols through distinct linked SuFExable hubs commonly known as molecular plugins, for example SO_2F_2 **11**, ethenesulfonyl fluoride (ESF) **12**, 1-bromoethene-1-sulfonyl fluoride (BESF) **13**, SOF_4 **14**, and etc. (Fig. 8.1). BESF is commonly referred to bromoethenyl sulfonylfluoride (Br-ESF) or more general, 1-bromoethene-1-sulfonylfluoride (1-Br-ESF).

8.2.1 Alkyl Sulfonyl Fluorides (RSO_2F) and Aryl Sulfonyl Fluorides (Ar-SO_2F)

In comparison with aryl sulfonyl chlorides, aryl sulfonyl fluorides were relatively less explored in organic synthesis. Sharpless and co-workers developed for the first-time aryl sulfonyl fluorides (Ar-SO_2F) by comprising a sulfur–carbon (S–C) bond installed into the native reaction modules and used them as suitable SuFExable connectors. The classical synthesis of aryl sulfonyl fluorides includes the direct HalEx reaction of a sulfonyl chloride with a suitable fluoride source in the presence of crown ether [45, 46]. Sharpless et al. employed the reaction of sulfonyl chlorides **15** in a biphasic

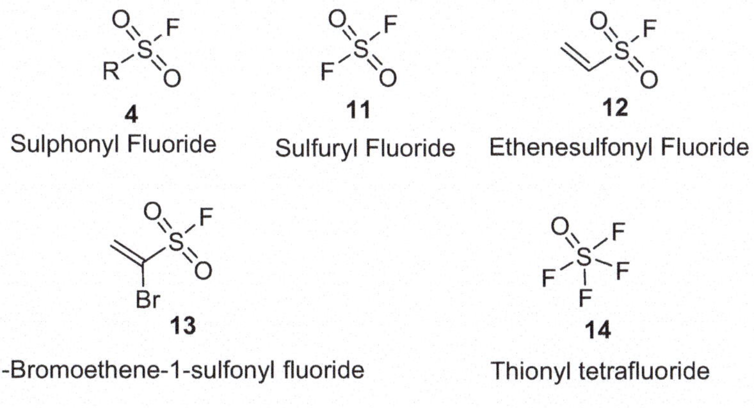

Fig. 8.1 Structure of some modular and robust SuFEx systems for click chemistry

Scheme 8.4 Synthesis of alkyl sulfonyl fluorides using potassium bifluoride and aryl sulfonyl fluorides under palladium catalysis

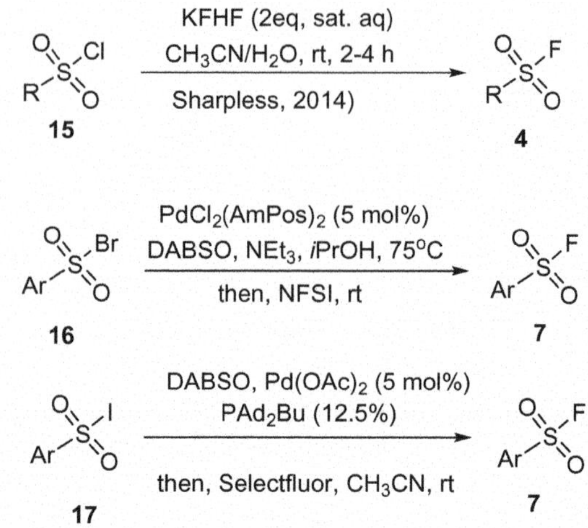

mixture of aqueous CH_3CN and saturated potassium hydrogen bifluoride (KFHF) for the practical synthesis of the desired alkyl sulfonyl fluorides **4** [38]. Furthermore, Barbasiewicz et al. performed the similar reaction in presence of the catalytic amount of phase transfer catalyst to enhance the reactivity of alkyl sulfonyl chlorides [47]. Interestingly, aryl bromide **16** or aryl iodide **17** under palladium catalysis with 1,4-diazabicyclo[2.2.2]octane bis(sulfur dioxide) (DABSO) were successfully converted to aryl sulfonyl fluorides. The reaction proceeds through a sulfonate salt intermediate which was then reacted with an electrophilic fluorine source to furnish high yields of respective aryl sulfonyl fluoride (Ar-SO_2F) (Scheme 8.4) [48, 49].

8.2.2 *Sulfuryl Fluoride (SO_2F_2)*

Sulfuryl fluoride (SO_2F_2) is a gas and was used as a pest control agent, which was later identified as a versatile SuFEx connector by Sharpless et al. in 2014. In the pioneer work, a wide range of aromatic fluorosulfates **7** were successfully obtained in good-to-excellent yields from the corresponding phenols **18** by treating them with SO_2F_2 in the presence of triethylamine in dichloromethane [38]. The method under the SuFEx condition is modular and equally useful for the easy and convenient synthesis of functional hetero-aromatic fluorosulfates. The simple and reliable tools were successfully demonstrated for their 'plugin' potential. Thus, the resulting aryl fluorosulfates **7** were efficiently reacted with aryl silyl ethers in the presence of catalytic amounts of DBU or BEMP or TAS-bifluoride at rt – 120 °C and afforded a high yield of respective sulfate-linked analogues **19**. Furthermore, Sulfuryl fluoride in reaction with a secondary amine gave a high yield of respective *N*-disubstituted sulfamoyl fluorides **20**. The resulting sulfamoyl fluorides **20** displayed notable stability than sulfonyl fluorides and thus required high-temperature conditions in a solvent having

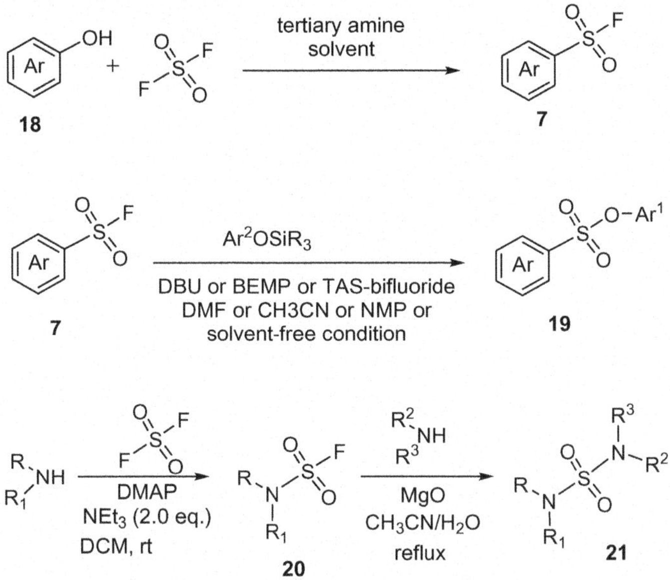

Scheme 8.5 Synthesis of diverse aryl fluorosulfates and their SuFEx reactivity with aryl silyl ethers and secondary amines

hydrogen-bonding ability for the further reaction with another secondary amine to give compound **21** (Scheme 8.5).

8.2.3 Sulfuryl Fluoride Surrogates

Fluorosulfuryl imidazolium salts offer themselves as an excellent surrogate of gaseous sulfuryl fluoride for delivering the F-SO_2^+ fragment to Nu-H acceptor groups. These salts are commercially available, benchtop stable solids, and are prepared by the reaction of 2-methylimidazole with sulfuryl fluoride, followed by the methylation with methyl triflate [50]. The enhanced reactivity, shorter reaction time, and better selectivity of these salts as compared to the SO_2F_2, provide the solution to the problems associated with SO_2F_2 such as difficulty in handling harmful SO_2F_2 gas, sluggish reaction with secondary amines to generate sulfamoyl fluoride with limited substrate scope and unsuccessful reaction with primary amines due to their instability under basic condition. Fluorosulfuryl imidazolium salt **22** reacts with primary amines under base-free conditions to give either mono- (NH-sulfamoyl fluoride) **23** or bis(fluorosulfonyl) imide **24** in excellent yield. AISF ([(acetylamino)-phenyl]imidodisulfuryl difluoride) **25** is another sulfuryl fluoride surrogate for delivering the F-SO2$^+$ fragment to phenol or secondary amine in the presence of DBU to afford corresponding fluorosulfate **26** or sulfamoyl fluoride **27** in excellent yield

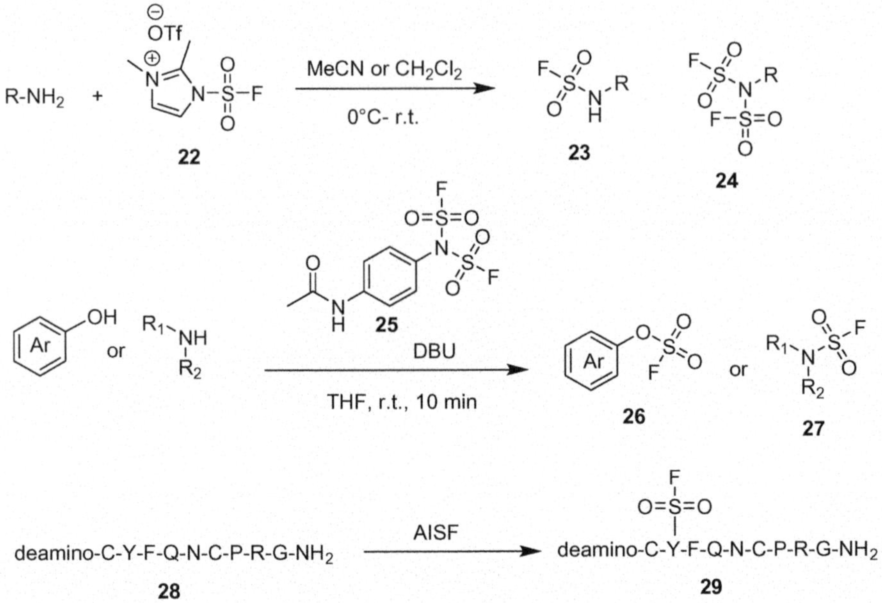

Scheme 8.6 Reactivity of various SO_2F_2 surrogates

(Scheme 8.6). The extended scope of AISF has been demonstrated by its ability to functionalize tyrosine residue of the peptide macrocycle **28** under aqueous condition [51].

8.2.4 Thionyl Tetrafluoride SOF_4

In 2017, the toolkit of available SuFEx connectors was further explored by utilizing trigonal bipyramidal thionyl tetrafluoride (SOF_4, **14**), a colorless gas and the first multidimensional click-linker [52]. In 1902, Moissan and Lebeau reported the synthesis of thionyl tetrafluoride, [53] which was further improved in 1960 by Smith and Engelhardt [54].

Soon after in 1961, Cramer and Coffman reported the synthesis of tetrahedral iminosulfur oxydifluorides **30** by reacting thionyl tetrafluoride with amines [55]. Likewise, Seppelt and Sundermeyer in the 1970s reported a primarily demonstration of silyl-mediated SuFEx chemistry using TMS-protected amines to afford a high yield of protected iminosulfur oxydifluorides **31** (Scheme 8.7) [56, 57]. For a long time gap of the above-mentioned early pioneering investigations, the chemistry of bipyramidal thionyl tetrafluoride SOF_4 remained unexplored. In 2017, Sharpless et al. reported the improved synthesis of iminosulfur oxydifluorides **30** by reacting SOF_4

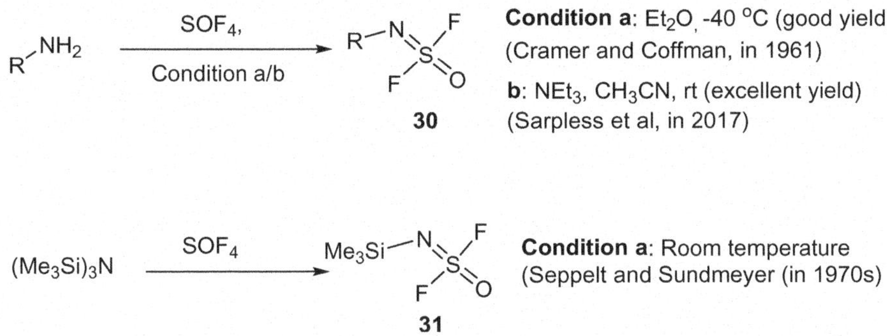

Scheme 8.7 Synthesis of iminosulfur oxydifluorides as manifestation of SuFEx chemistry

with a primary amine in the presence of triethylamine where the desired product was obtained in good-to-excellent yield with increased rates (Scheme 8.7) [52].

Iminosulfur oxydifluorides **30** possess two additional SuFExable handles, and by the selection of appropriate catalysis, each fluoride ion can be replaced by either phenols or amines or both to generate tetrahedral covalent departure vectors from the central sulfur (VI) hub (Scheme 8.8). It is important to note that the chemoselective preference of SOF_4 and SO_2F_2 are just opposite to each other toward the primary amines (favorably attacked by SOF_4) and phenols (favorably attacked by SO_2F_2). This chemoselective preference has been nicely demonstrated by Sharpless et al. through the reaction of 4-aminophenol with a 1:1 mixture of gaseous SO_2F_2 and SOF_4 in acetonitrile to yield a product **36** that contained both the fluorosulfate and iminosulfur oxidifluride SuFExable handles. The scope of iminosulfur oxidifluorides **37** has been further demonstrated by their reaction with limited amount of lithiated carbon-based nucleophiles (1.35 equivalent) to yield sulfonimidoyl fluorides **38**, which were further modified either by the reaction with lithiated carbon-based nucleophile, aryl sillyl ether, or secondary amine to produce the sulfoximines **39**, sulonimidate **40**, or sulfonimidamides **41**, respectively, via SuFEx click chemistry [58].

8.2.5 *Ethene Sulfonyl Fluoride (ESF)*

Ethene sulfonyl fluoride (ESF) and its derivatives are another class of functional SuFEx connectors, which deliver the additional mode of reactivity for the synthesis of stable linkages via the activated double bond. Zheng and Sharpless developed a process for the bulk synthesis of ESF from 2-chloroethenesulfonyl chloride in two steps (Scheme 8.9) [59]. ESF has two distinct pathways of reactivity site, i.e., SuFEx pathway at sulfonyl fluoride group and Michael addition pathway at the vinyl group, therefore ESF demonstrated orthogonal reactivity. Highly competent Michael addition reactions of ESF with amines, phenols, propargyl alcohols, and

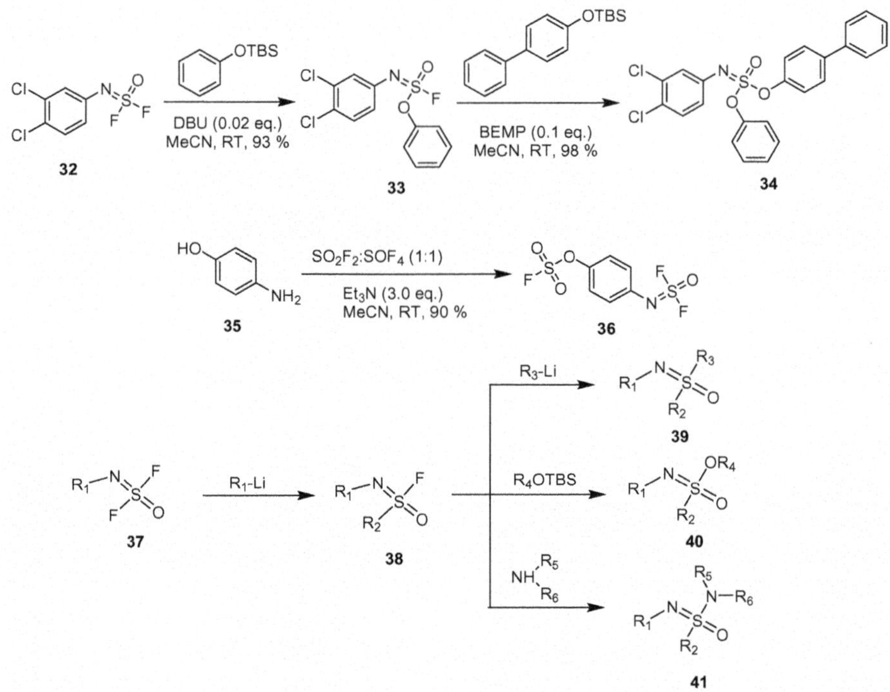

Scheme 8.8 Sequential and orthogonal reactivity of iminosulfur oxydifluorides

stabilized carbon nucleophiles have been reported, without touching the sulfonyl fluoride moiety (Scheme 8.9) [60, 38]. According to Sharpless, '*ESF is the most perfect Michael acceptor ever found*' and also acts as a dienophile and 1,3-dipolarphile.

8.2.6 1-Bromoethene-1-Sulfonyl Fluoride (BESF)

Incorporation of an additional reactive bromo-group in ESF may open a new window for the notable application of resulting 1-bromoethene-1-sulfonyl fluoride (BESF) in SuFEx click as a potent SuFEx hub. Vessiere et al. reported the first synthesis of 1-bromoethene-1-sulfonyl fluoride (BESF, **13**) from ESF which was the first bromination followed by the dehydrobromination of resulting 1,2-dibromoethane-1-sulfonyl fluoride (DBESF, **49**) under basic condition (Scheme 8.10) [61]. Further, BESF can readily be converted to ESF. Additional reactivity of the bromo group increases the efficacy of BESF as a potent SuFEx hub.

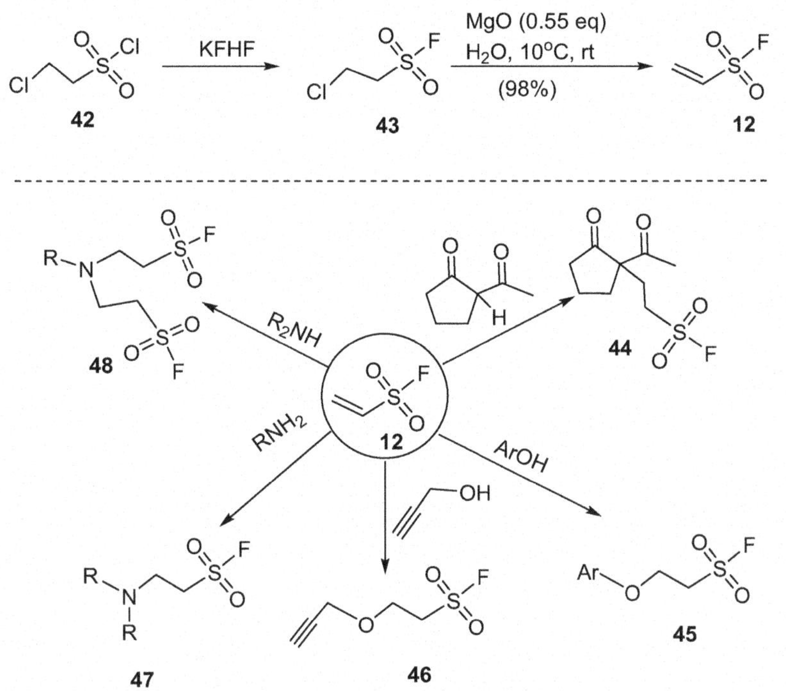

Scheme 8.9 Scale-up synthesis of ethene sulfonyl fluoride (ESF, **20**) and its reactivities with various nucleophiles

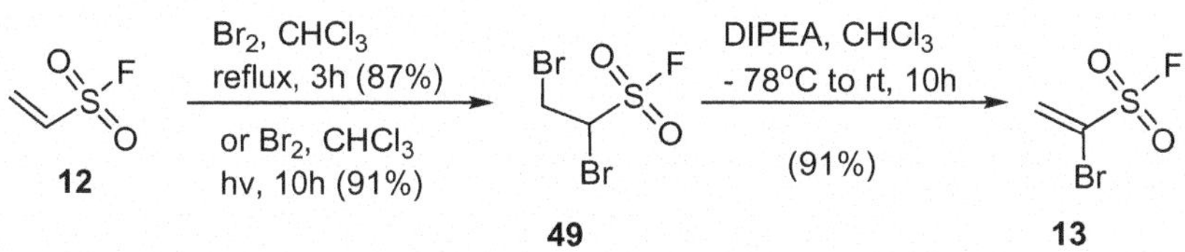

Scheme 8.10 Synthesis of BESF by bromination–dehydrobromination of ESF

8.3 Application of SuFEx Click Chemistry in Organic Synthesis

The application of SuFEx click chemistry in organic synthesis may be divided into two categories; (a) Installation of SuFEx building block to various substrates and (b) use of SuFEx building block as a functional reagent.

8.3.1 Installation of SuFEx Building Blocks

In 2016, Sharpless et al. reported a first efficient method for the synthesis of β-aryl ESF **51** by using Pd-catalyzed Heck–Matsuda cross-coupling reaction between ESF and arenediazonium tetrafluoroborates (Scheme 8.11) [62]. Due to the presence of activated C=C bond, and sulfonyl fluoride groups, these β-aryl ESF act as a bifunctional connector and can undergo further modification either by conjugate addition or by SuFEx pathway. The dual reactivity of β-aryl ESF was further demonstrated by their conjugate addition reaction with secondary amines, DBU catalyzed SuFEx transformation with aryl silyl ethers, and addition-substitution reaction with hydrazine to generate sulfonates **52**, aryl silyl ethers **53**, and cyclic sultams **54,** respectively (Scheme 8.11). Since β-aryl ESFs are approx. 4.5 times less electrophilic than ESF; therefore, Michael addition with C-nucleophile is very challenging. Recently, Michael addition reactions with C-nucleophiles such as diethylmalonate and dithiomalonate to β-aryl ESF have been reported by using tertiary amino-thiourea or water-accelerated phosphazene as a catalyst, respectively (Scheme 8.11) [63, 64].

Arvidsson et al. have reported the synthesis of β-aryl/heteroaryl ESF by using oxidative free boron Heck coupling reaction with ESF at room temperature, whereas Qin et al. reported similar Heck coupling by using DDQ or $AgNO_3$ as an oxidant (Scheme 8.12). Latter shapeless et al. developed a more general and efficient Heck-type coupling process to couple ESF and aryl/heteroaryl or vinyl iodides by using

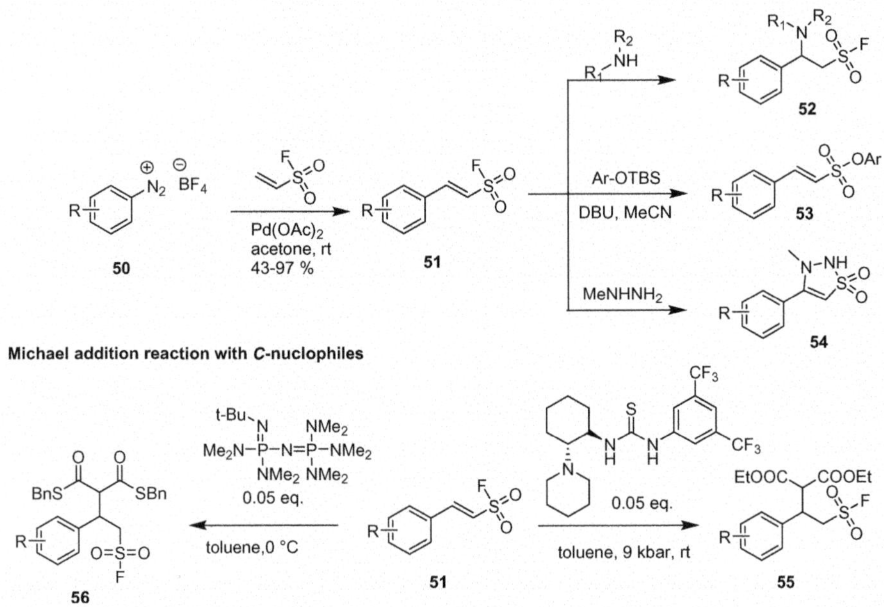

Scheme 8.11 Synthesis of β-aryl ESF by Matsuda–Heck coupling and their application as bifunctional connector

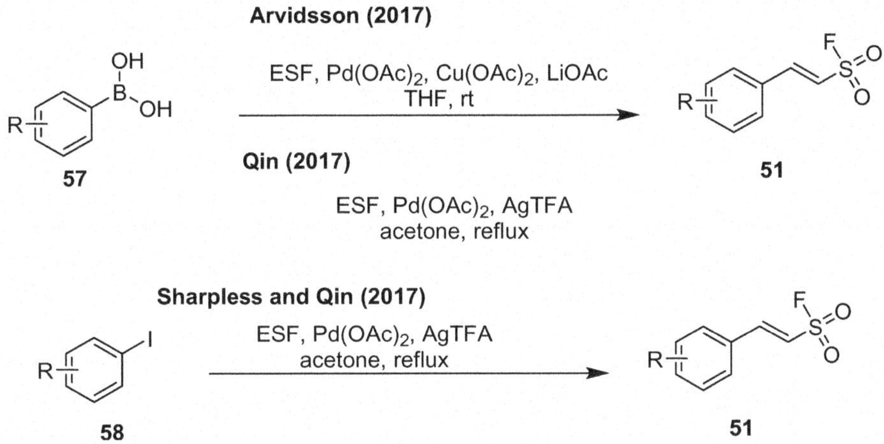

Scheme 8.12 Synthesis of β-aryl/heteroaryl ESF *via* Heck-mediated cross coupling

catalytic $Pd(OAc)_2$ with a stoichiometric amount of silver(I) trifluoroacetate to obtain β-aryl/heteroaryl or allyl ESF in good to excellent yield.

C–H bond activation is a highly significant, cost-effective, and environmentally friendly approach in organic chemistry, which directly facilitates the formation of C–C bonds by activating a C–H bond. Researchers have also explored the use of SuFEx reagents for activating C–H bonds. Quin et al. introduced Rh(III)-catalyzed C–H activation method for producing β-aryl ESF, which involves the oxidative coupling between *N*-methoxybenzamides and ESF, showcasing the amide group's capability as an ortho-directing entity (Scheme 8.13a) [65]. Selective ortho-functionalization of sp [2] C–H bonds in aryl or heteroaryl rings has been observed, resulting exclusively in *E*-stereoselective final products. Furthermore, it is important to mention that raising the amount of $AgSbF_6$ from 10 mol% to 1.0 equiv. results in lactonization as a consequence of the hydrolysis of the amine into acid. Later on, the same research group illustrated the effectiveness of ester, aldehyde, and ketone functionalities as directing groups in Ru-catalyzed ortho C–H bond functionalization with ESF. (Scheme 8.13b) [66, 67]. Various other functional groups, such as amides, phosphonates, oximes, sulfoxides, sulfides, and *N*-tert-butyl/benzyl nitrones, have been identified as ortho-directing groups for C–H functionalization with ESF [68, 69].

Yu et al. have also described the synthesis of β-aryl ESF through ligand-accelerated Pd(II)-catalyzed non-directive C–H alkenylation with ESF (Scheme 8.14) [70]. It was observed that the electron-deficient 2-pyridone ligand (**L1**) is essential for this conversion, especially when arenes are used as the limiting reagent.

Various cycloaddition reactions have been documented involving the double bond of different ethenesulfonyl fluorides (ESFs). Initially, Grygorenko et al. described the synthesis of pyrrolidine-3-sulfonyl fluorides **69** via a [3 + 2] cycloaddition reaction. This process involved the interaction between azomethine ylides **67**, which were obtained in situ from *N*-benzyl-1-methoxy-*N*-((trimethylsilyl) methyl)methanamine

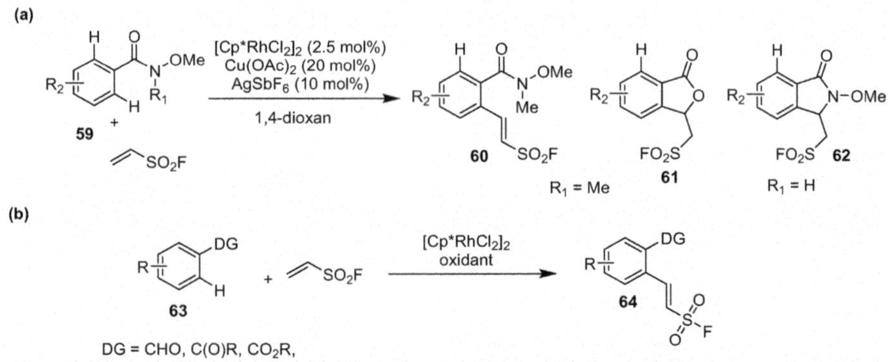

Scheme 8.13 Synthesis of β-aryl ESF by using Rh (III) catalyzed directive C–H bond activation

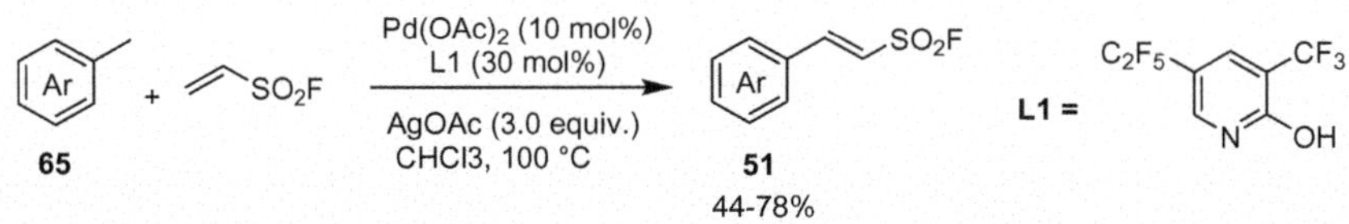

Scheme 8.14 Pd-catalyzed non-directive activation of C–H bonds for β-aryl ESF synthesis

66, and vinyl sulfonyl fluorides **68** (Scheme 8.15) [71]. Mykhailiuk et al. documented the [3 + 2] cycloaddition reaction between ESF and C_2F_5 or CF_3 substituted diazomethanes **70** to afford 3,5-substituted pyrazolines **71** in quantitative yield [72, 73]. Mayr and coworkers reported a similar [3 + 2] cycloaddition reaction between ESF and aryldiazomethanes, resulting in the synthesis of 3-sulfonyl fluoride substituted 5-arylpyrazolines with excellent yield (Scheme 8.15) [74].

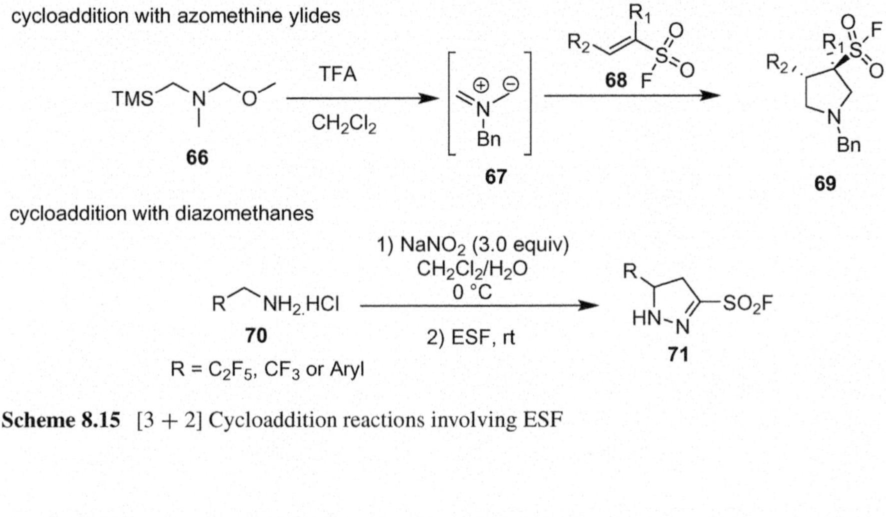

Scheme 8.15 [3 + 2] Cycloaddition reactions involving ESF

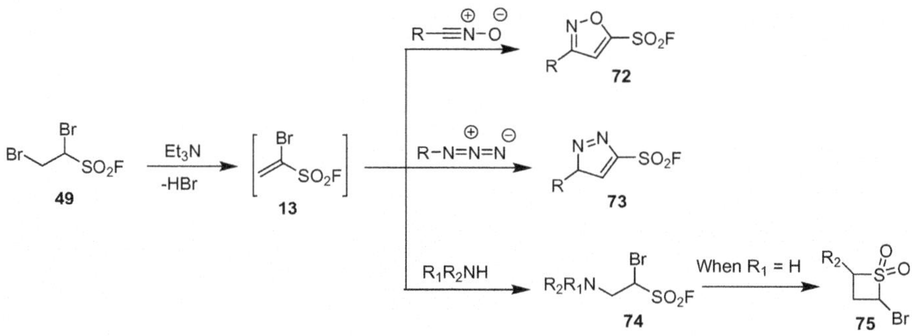

Scheme 8.16 In situ formation of BESF and its application

Moses and Fokin demonstrated a regiospecific [3 + 2] cycloaddition reaction using in situ generated BESF, derived from 1,2-dibromomethane-1-sulfonyl fluoride (DESF) **49** through dehydrobromination, with organic azides and nitrile oxides. This yielded SO_2F-substituted 5-membered heterocycles **72** and **73** in excellent yield (Scheme 8.16) [75, 76]. Under SuFEx conditions, it was observed that the S–F bonds in these products remained available for exchange reactions. Moreover, BESF was observed to undergo Michael addition reactions, resulting in 1,4-addition adduct with secondary amines **74** and 4-bromo-β-sultams **75** with primary amines.

Similar to [3 + 2] cycloaddition reactions, there have been limited reports on [2 + 2] and [4 + 2] cycloaddition reactions. Mykhailiuk et al. reported a photochemical [2 + 2] cycloaddition reaction between ESF and *N*-benzylmaleimide **76**, yielding functionalized 3-azabicyclo[3.2.0]heptane **77** in a 29% yield (Scheme 8.17) [77]. Initially, Daeniker and coworkers described a [4 + 2] cycloaddition reaction involving ESF with cyclopentadiene and cyclohexadiene. Later, in 1979, Hyatt and coworkers reported a Diels–Alder reaction between ESF and 2-methoxyfuran, resulting in adduct **81** with an 88% yield [78, 79].

A visible-light mediated decarboxylative fluorosulfonylethylation have been reported by Liao and coworkers for the synthesis of aliphatic sulfonyl fluorides **83** using activated carboxylic acid derivatives, *N*-hydroxy phthalimide esters, as alkyl radical sources and ESF as a radical acceptor (Scheme 8.18) [80]. Further diversification of alkylsulfonyl fluoride products were also explored through the SuFEx reactions and intramolecular cyclization.

8.4 Application of SuFEx Building Blocks as Chemical Tools

With the rapid advancement of SuFEx chemistry as a versatile tool for establishing novel connections, numerous methodologies have emerged. These methods focus not only on synthesizing substrates containing S–F bonds but also on utilizing them

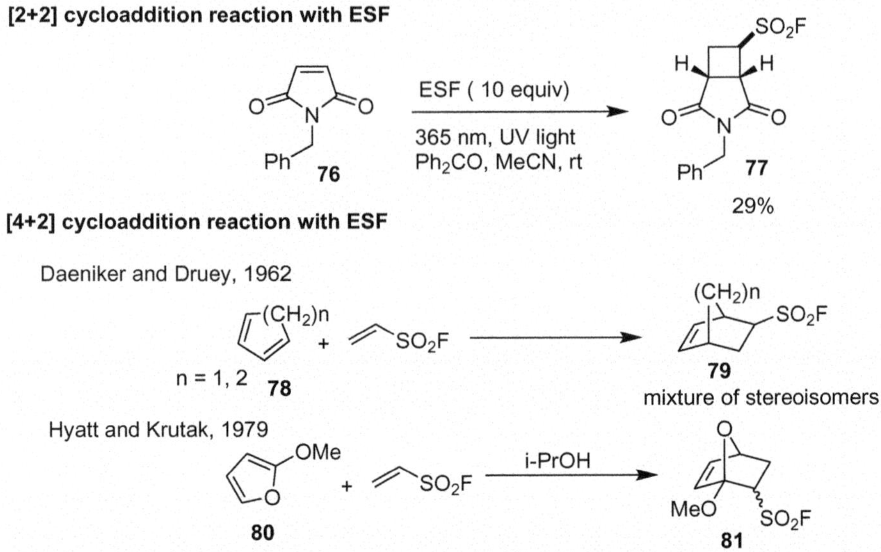

Scheme 8.17 [2 + 2] and [4 + 2] Cycloaddition reactions with ESF

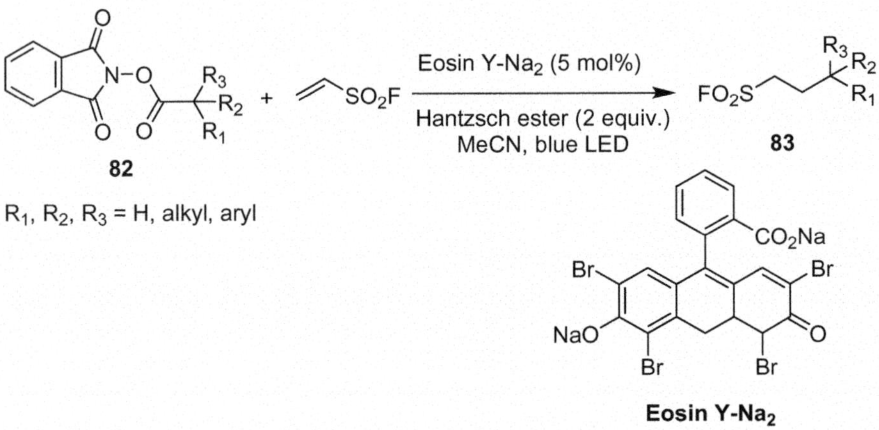

Scheme 8.18 Visible-light-mediated radical addition to ESF to activated carboxylic acid derivatives

as reagents. While this diverges from the essential 'molecular plugin' concept, the expansion of these advantageous techniques and conversions owes its success to the evolution of the SuFEx reaction family.

The incorporation of fluorine into various pharmaceuticals and agrochemicals is a common method to modulate their physicochemical properties, lipophilicity, acidity, and metabolic stability. Central glass described a method using SO_2F_2 as the fluorine source for deoxyfluorination of alkyl alcohols [81, 82].

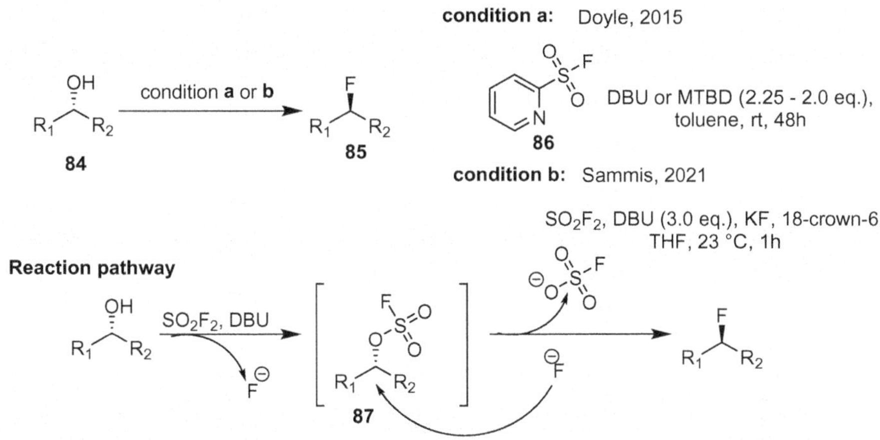

Scheme 8.19 SuFEx-mediated deoxyfluorination of alcohols

Doyle and coworkers have introduced a cost-effective deoxyfluorination reagent, 2-pyridinesulfonyl fluoride **86** (PyFluor), which fluorinates a wide array of alcohols **84** with minimal formation of elimination byproducts (Scheme 8.19) [83]. The reaction utilizes amidine or guanidine bases such as DBU or MTBD in excess, facilitating the rapid formation of intermediate sulfonate species **87**. Subsequently, these intermediates react slowly with free fluoride ions, yielding alkyl fluorides **85** in quantitative yield. Later, Sammis et al. have reported an efficient SO_2F_2-mediated deoxyfluorination reaction, effective for both primary and secondary alcohols with complete inversion of configuration [84].

The next application of SuFEx reagents is there use as an amide coupling reagent. Initially, Stienkopf explored the use of benzenesulfonyl fluorides as a reagent for the coupling between sodium salt of benzoic acid and aniline, but the reaction required harsh conditions with an inseparable mixture of products [85]. Latter, Yan and Coworkers have developed novel 5H-3-oxa-octafluoropentanesulfonyl fluoride **89** as a useful reagent for esterification, amidation and anhydridization with a limited substrate scope and mixed anhydride species **90** as a plausible reaction intermediate (Scheme 8.20) [86]. Further, Moses et al. have reported a novel method for the synthesis of sterically hindered and electron deficient amides in quantitative yield by using benzene-1,3-disulfonyl fluoride **92** as a cost-effective, powerful, and versatile coupling reagent [87]. The reaction proceed via acyl fluoride intermediates and the small size of the fluoride leaving group minimize the steric hindrance with the sterically hindered nucleophile. The industrial and pharmaceutical application of this SuFExAmide methodology was established by the synthesizing kinetoplastid proteasome inhibitor GNF6702 **98** in 77% yield, much better as compared to that previously reported 29.6% yield achieved by using HATU as a coupling reagent.

SuFEx reagents has also been demonstrated as an excellent diazotransfer reagent for the synthesis of azides from primary amines. Initially, Moses et al. have synthesized aryl sulfonyl azides **99** via base-activated sulfur fluoride–azide exchange, using

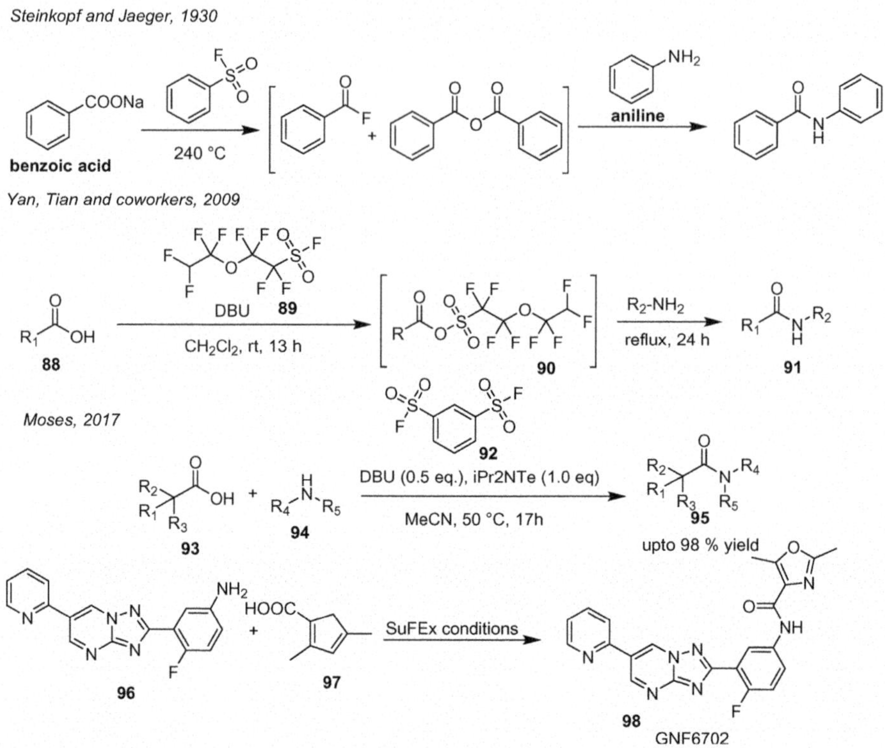

Scheme 8.20 Synthesis of amides through SuFEx-based chemistry

$TMSN_3$ as the azide source and DBU as a Lewis base (Scheme 8.21) [88]. Sharpless et al. have synthesized fluorosulfuryl azide **100** by reacting imidazolium fluorosulfuryl triflate salt **22** and sodium azide in biphasic system, which enables the synthesis of more than 1200 azides on 96-well plates in a safe and practical manner [89]. The extended scope of this reaction was further demonstrated by the quantitative azidation of amlodipine **101** or pazufloxacin mesylate at room temperature in just 5 min.

Sulfonyl fluorides can also be used as a difluorocarbene reagent. Dolbier et al. have used trimethylsilyl fluorosulfonyldifluoroacetate (TFDA) **103** as a difluorocarbene source for addition to relatively electron-deficient alkenes to produce *gem*-difluorocyclopropanes **104** in good yield (Scheme 8.22) [90]. Further Methyl 2,2-difluoro-2-(fluorosulfonyl)acetate (MDFA) **105** has been reported as an alternative to TFDA in terms of cost-effectiveness, safety, and ease of reaction albeit with considerably longer reaction time as compared to TFDA [91]. A non-corrosive and non-volatile hexamethyldisiloxane (HMDSO) could be used as the fluoride trap in place of trimethylsilyl chloride.

Aryl flurosulfates have also been reported as an electrophilic partner in the Buchwald-Hartwig C–N coupling reaction (Scheme 8.23). In 2016, Henly and

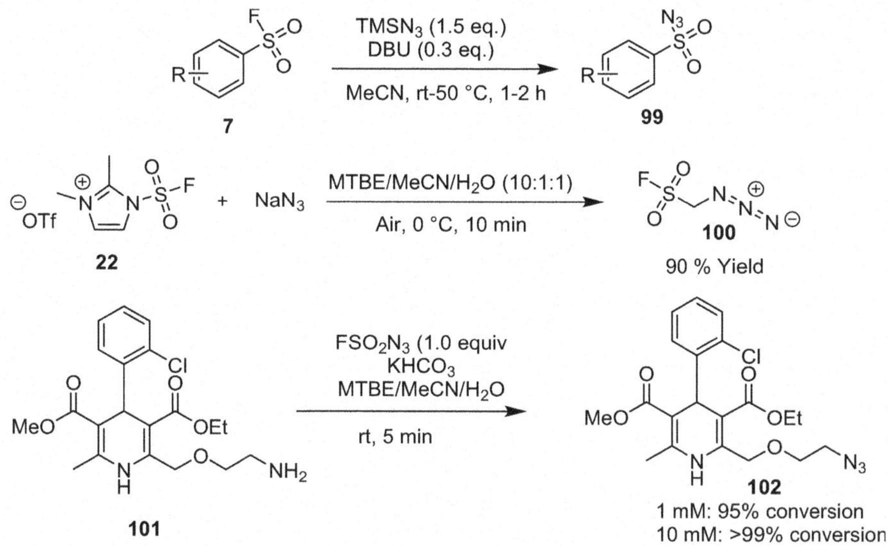

Scheme 8.21 Synthesis of sulfonyl azide through SuFEx-based chemistry and their application as a diazotizing agent

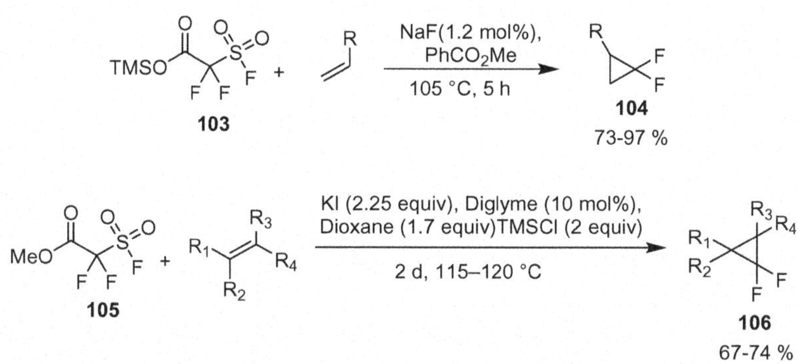

Scheme 8.22 Synthesis of gem-difluoropropanes by using either TDFA **103** or MFDA **105**

coworkers synthesized diarylamines **107** by utilizing Pd-catalyzed C-N cross-coupling reaction between (hetero)aryl fulorosulfate and anilines [92]. The reaction exhibited better reactivity with tolerance to both electron-rich and electron-deficient arylfluorosulfates. Later, Kim and coworkers reported a modified $Pd(OAc)_2$-catalyzed C-N cross coupling reaction for the amination of aryl fluorosulfates by using a diverse range of primary and secondary aniline derivatives under ligand-free condition.

Ding et al. have described the synthesis of aryl phosphonates **109** by employing of Ni and Pd-catalyzed cross coupling between the aryl sulfonyfluorides and phosphites (Scheme 8.24) [93]. The products were obtained in an excellent yield with

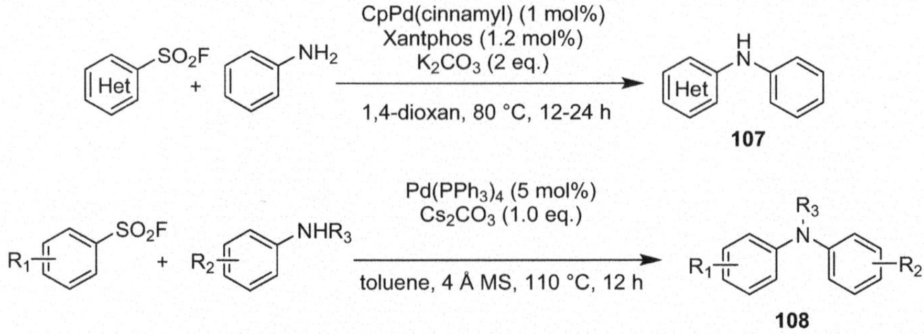

Scheme 8.23 Buchwald–Hartwig C–N coupling reaction of (hetero)aryl fluorosulfonates

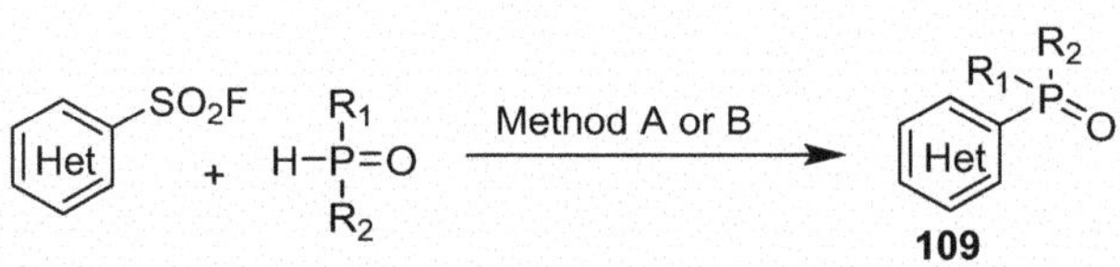

Method A: $Pd(OAc)_2$(5 mol%), DPEPhos (6 mol%), K_2CO_3 (2.0 eq.), THF, 80 °C. 8h

Method B: $NiCl_2$(dme)(5 mol%),XAntphos (6 mol%), Et_3N (2.0 eq.), Zn (1.5 eq.), DMF, 100 °C. 12h

Scheme 8.24 Ni- and Pd-catalyzed cross-coupling of (hetero)aryl fluorosulfonates and hydrogen phosphoryl compounds

broad functional group compatibility. All the three kinds of P(O)-H compounds (H-phosphonates, H-phosphinates, and secondary phosphine oxides) were applicable to these reactions. The same group also developed Ni based catalyst, i.e., $NiCl_2$(dme)/Xantphos/Et_3N/Zn combination in DMF for this interconversion to obtain the desired (hetero)aryl phosphonate derivatives in comparable yields.

Aryl, vinyl, and heteroaromatic halides are the most common electrophilic coupling partner in transition-metal-catalyzed cross-coupling reactions, but their high costs and environmental toxicity limit their use in industry. This led to the development of novel aryl electrophiles specially derived from phenols like sulfates. However, aryl triflate demonstrated a higher level of reactivity but the costly nature of reagents required to install the triflate group limits their use. Bristol Myers Squibb (BMS) prepared aryl fluorosulfates **110** by reacting phenols and fluorosulfonic anhydride in the presence of DIPEA and reported them as an effective electrophile for an alternative to triflates in various cross-coupling reactions such as Negishi, Stille, and alkoxy carbonylation reactions (Scheme 8.25) [94–96]. Though these coupling reactions are not SuFEx reactions because S-F exchange does not occur, these reactions are only possible due to the application of SuFEx in the synthesis of arylfluoro sulfates. Further Sharpless et al. have used SO_2F_2, AISF, and imidazonium for synthesizing aryl fluorosulfates to overcome the problem associated with sulfonic anhydride.

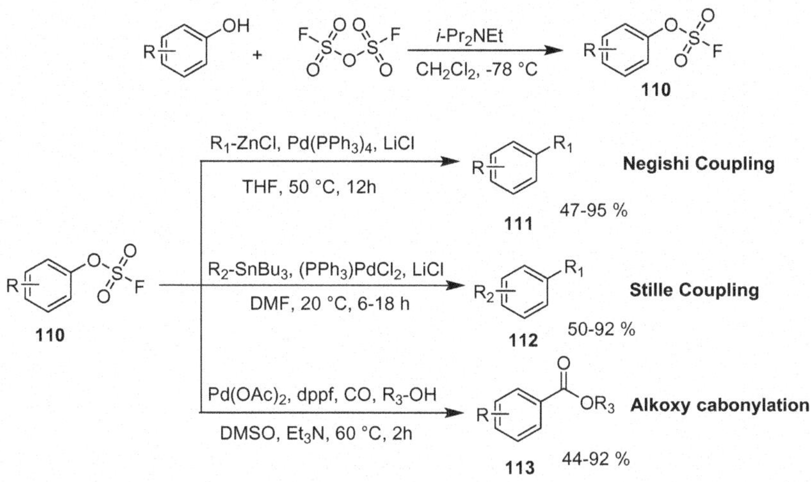

Scheme 8.25 Synthesis and application of (Hetero)aryl fluorosulfates in Negishi, Stille, and Pd-catalyzed alkoxy carbonylation reaction

Aryl fluorosulfates are recognized as highly efficient coupling partners in additive-free Suzuki–Miyaura cross-coupling reactions with boronic acids (Scheme 8.26) [97]. Sharpless et al. highlighted the significance of these reactions in synthesizing the COX-2 inhibitor Etoricoxib **119**, achieving an overall yield of 40.3% through an orthogonal approach [98]. Competitive experiments indicated that the relative reactivity of different leaving groups on substituted pyridines follows the order – Br > OTf > OSO_2F > Cl. Moreover, other competitive studies demonstrated that the relative reactivity of aryl fluorosulfonates is comparable to aryl triflates and aryl bromides, potentially surpassing the corresponding chloro-, tosyl-, and mesyl analogues.

Qin et al. have expanded the utility of these aryl fluorosulfates by developing a one-pot synthesis of carboxylic acids **120**, esters **121**, amides **122**, nitriles **123**, and deoxy-hydrogenated product **124** from phenols using Pd catalysis (Scheme 8.27) [99–101]. A recent report disclosed a Ni-catalyzed one-pot direct cross-coupling of aryl fluorosulfates with aryl bromides, yielding biaryls **125** in moderate to good yield with reasonable tolerance of diverse functional groups. Moreover, Sanford et al. demonstrated the effectiveness of SO_2F_2 as a deoxyfluorination reagent, enabling the conversion of phenols into fluoroarenes **126** through aryl fluorosulfonate intermediates [102].

Sammis's group reported sulfuryl fluoride as a novel reagent for 1,1-dihydrofluoroalkylation or trifluoroethylation of primary and secondary amines in the presence of DIPEA via in situ formation of trifluoroethyl fluorosulfate **128** as an intermediate (Scheme 8.28) [103]. It is noteworthy that under the standard reaction condition, other existing activating groups such as trifluoroethyl iodide, mesylates, and tosylates were completely unreactive, whereas the trifluroethyltriflates was more

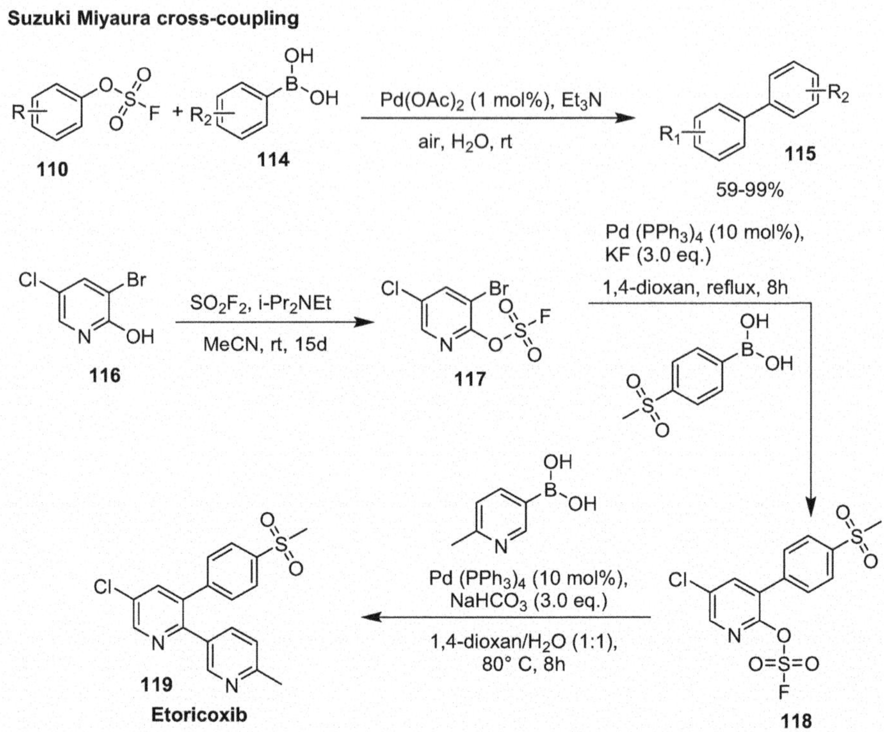

Scheme 8.26 Suzuki–Miyaura cross-coupling of aryl fluorosulfates without additives and its application to the chemoselective synthesis of Etoricoxib: A COX-2 inhibitor

reactive, but replacing the sulfuryl fluoride with triflic anhydride could not afford the anticipated product.

We can summarize that after the beginning of SuFEx click chemistry, the development of novel techniques to install the SuFExable functionality into modular building blocks has been rapidly expanded. Furthermore, sulfuryl fluorides and arylfluorosufates have been reinvigorated for their use as chemical reagents for their broad utilities as discussed above. The superiority of aryl fluorosulfates over triflate surrogates may ultimately find their most utility in transition metal catalyzed cross-coupling reactions.

8.5 Application of SuFEx Click Chemistry in Drug Discovery Endeavors

Late-stage functionalization (LSF) facilitates the quick diversification of drug or drug-like molecules during the latter stages of their synthesis, enhancing their physical and pharmaceutical attributes. Wu et al. described the application of SuFEx click

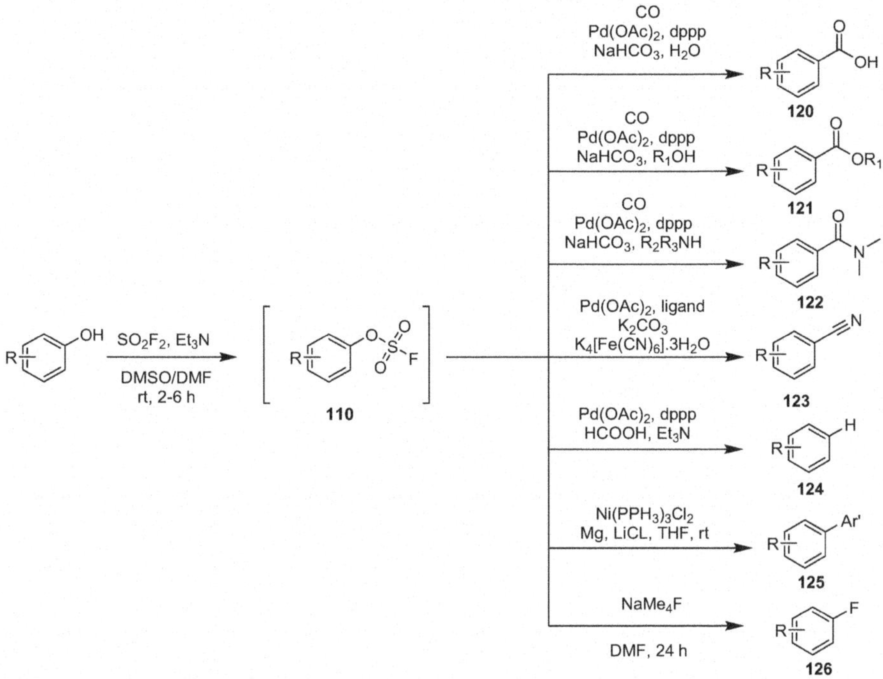

Scheme 8.27 Conversion of aryl fluorosulfates to benzoic acids, esters, amides, nitriles, deoxyhydrogenated products, and biaryl derivatives

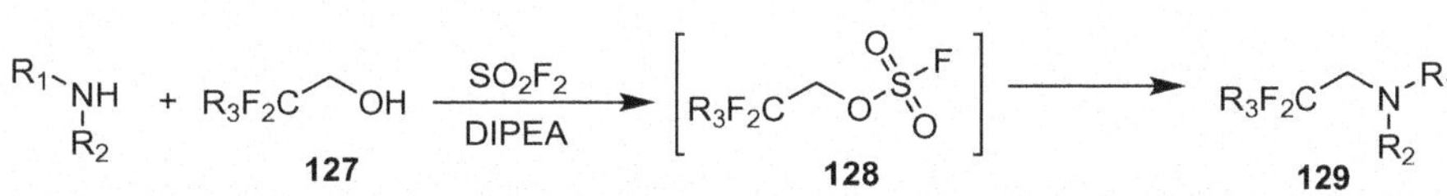

Scheme 8.28 1,1-dihydrofluoroalkylation or trifluoroethylation of amines via in situ formation of alkylfluorosulfates

chemistry to convert phenol-containing anti-cancer drugs or drug candidates into their corresponding arylfluorosulfate derivatives in situ within 96-well plates, employing SO_2F_2 (Scheme 8.29) [104]. Three of the in situ generated arylfluorosulfates demonstrated enhanced anti-cancer cell growth inhibition activities in comparison to their phenolic precursors. Notably, fluorosulfate analogue **131** of Fulvestrant **130** demonstrated significantly improved activity in down-regulating estrogen receptor expression in the ER^+ breast cancer cell line MCF7 compared to fulvestrant. Moreover, the fluorosulfate variant of Combretastatin A4 **133** showed a 70-fold enhancement in potency against the drug resistant colon cancer cell line HT-29 as compared to the combretastatin A4 **132**.

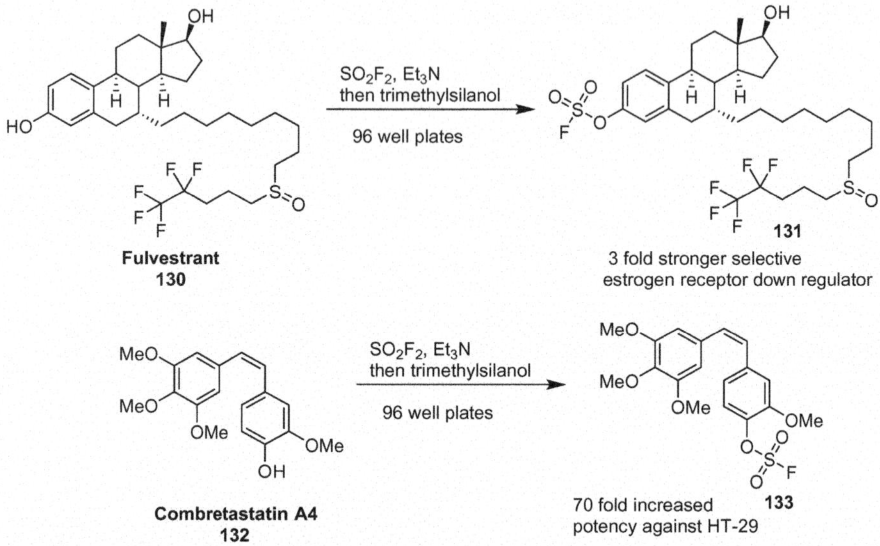

Scheme 8.29 Late stage fluorosulfonylation of fluvestrant and combretastatin

Conventional drug discovery typically involves screening compounds against isolated target proteins or using cell-based assays to identify potential drug candidates. In contrast, 'Inverse Drug Discovery' employs a different approach. It treats cells or cell lysates with compounds containing weak electrophilic functional groups. This method enables researchers to identify the proteins within the cell that interact with these compounds using affinity purification mass spectrometry. By doing so, this technique provides insights into potential drug targets within complex biological systems, opening up new avenues for drug discovery. Sharpless and coworkers investigated the latent electrophilic nature of aryl fluorosulfates in inverse drug discovery due to their efficient reactivity with intracellular lipid binding proteins (iLBPs). These iLBPs possess binding sites with specific arrangements of functional groups on their side chains, enabling them to catalyze the SuFEx reaction with a Tyrosine residue located at the binding site [105]. Three different arylfluorosulfates **134**, **135,** and **136** having intermediate complexity were synthesized and bound with the specific proteins (Fig. 8.2). An alkyne unit in each molecule facilitates successive identification via chromatography–mass spectrometry and CuAAC-enabled pull-down of labeled proteins. These aryl fluorosulfates selectively react with lysine and tyrosine side chains, targeting distinct groups of human proteins without significant overlap. Eleven protein targets were additionally validated through covalent reaction with a specific aryl fluorosulfate and recombinant protein, resulting in the formation of a singly labeled adduct. The researchers suggested that this approach has the potential to identify novel covalent probes and with further medicinal chemistry efforts to develop covalent or noncovalent drug candidates.

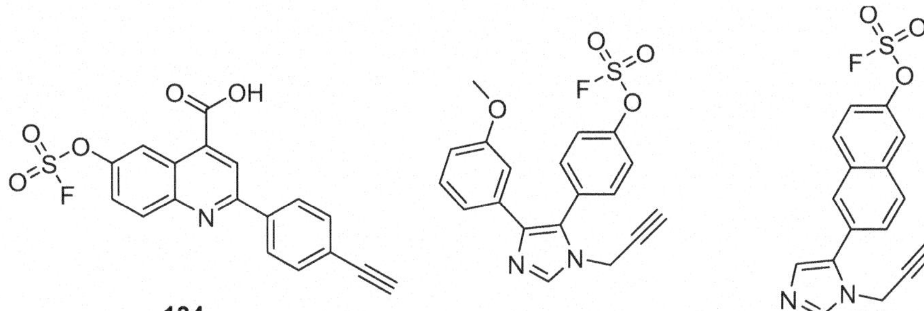

134 135 136

Fig. 8.2 Chemical probes used to establish inverse drug discovery approach

Human neutrophil elastase (hNE), a member of the serine protease superfamily is the key target for the development of anti-inflammatory agents. In the search of selective covalent hNE inhibitors, Sharpless et al. screened different categories of SuFExble probes such as aryl or alkyl or vinyl sulfonyl fluorides and aryl fluorosulfates to exploit the 'sleeping beauty' phenomenon of sulfur fluoride compounds. Evaluation of these SuFExable compounds toward various assays yields benzene 1,2-disulfonyl fluoride **137** as an initial hit, which were further derivatized to 2-triflyl benzenesulfonyl fluoride **138** (IC50 = 1.1 μM) and **139** (IC50 0.24 mM) with ~ 200 and 833-fold enhanced selectivity as compared to the homologous hNE inhibitor, cathepsin G (Scheme 8.30) [106, 107]. It is important to note that the optimized probe only covalently bonded to the active hNE protein, but not its denatured form.

In addition to the development of novel covalent inhibitors via the installation of flurosulfate probes, SuFEx click chemistry has also been explored for the synthesis of several drugs or drug-like molecules. Daclatasvir (Brand name: Daklinza) is a non-structural 5A (NS5A) inhibitor, that exhibits high anti-viral activity and is used in combination with NS5B inhibitor, sofosbuvir for the treatment of genotype 1 or 3 chronic HCV. Several symmetric biaryl sulfate analogues of daclatasvir, prepared by using phenols and sulfonyl diimidazole, have exhibited high anti-viral activities and good additive effects with sofosbuvir, without any cytotoxic effect [108]. Non-symmetrical biaryl sulfate analogues of daclatasvir, the next generation of HCV drugs used for the treatment of multi-genotype strain could not be synthesized by using the above synthetic strategy. SuFEx click chemistry offers the solution for the

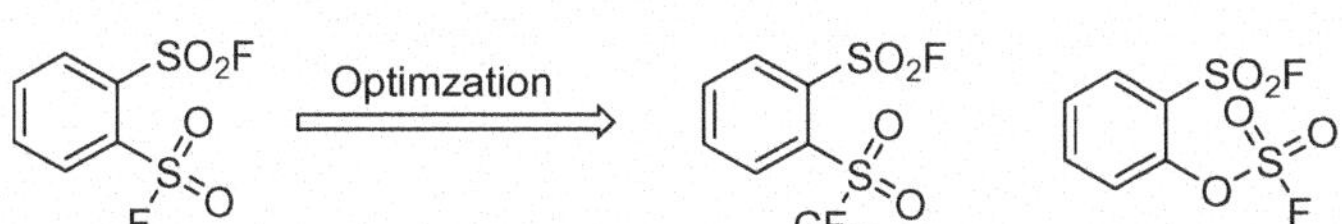

137 Initial Hit

138 **139**

Optimized compounds

Scheme 8.30 Optimization of benzene-1,2-disulfonylfluoride a covalent modifier of hNE

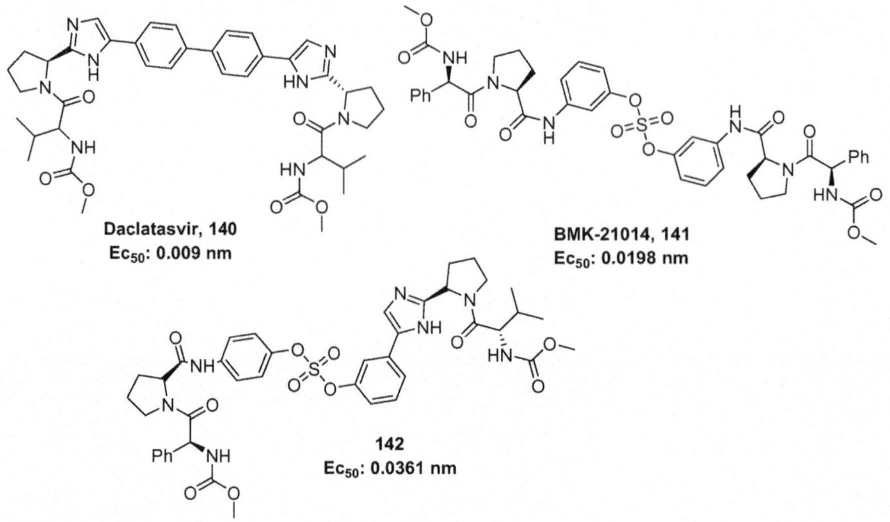

Fig. 8.3 SuFEx-mediated synthesis of homo- and cross-coupled biaryl sulfate analogues of daclatasvir

synthesis of both symmetrical as well as non-symmetrical biaryl sulfate-based NS5A inhibitors with suitability over a broad range of substituted phenols [109]. Utilizing this approach, researchers synthesized and tested a library of biaryl sulfate analogues, achieving impressive potency with two-digit picomolar EC50 values for inhibitors possessing both symmetric **141** and nonsymmetric **142** structures (Fig. 8.3). Sharpless et al. reported biocompatible SuFEx modification of SOF_4-derived iminosulfur oxydifluorides under aqueous buffer environment, which functions at low concentration, mild pH, and temperature conditions [110]. The SuFEx biomolecular conjugation of iminosulfur oxydifluorides to amine-tagged single-stranded DNA and BSA proteins effectively illustrated this concept, highlighting the promising applications of SuFEx click chemistry in biology. Therefore, the advancement of biocompatible SuFEx click reactions represents a valuable expansion of the click chemistry toolkit.

8.6 Application of SuFEx Click Chemistry in Material Science

The core concept of industrial processes, characterized by highly efficient reactions with minimal purification, resonates with the principles of click chemistry. Similarly, in the polymer industry, simple modular units are combined to yield new materials with desirable properties. The modular nature and orthogonal reactivity of SuFExable groups allow them to serve as the foundational backbone of polymers or be

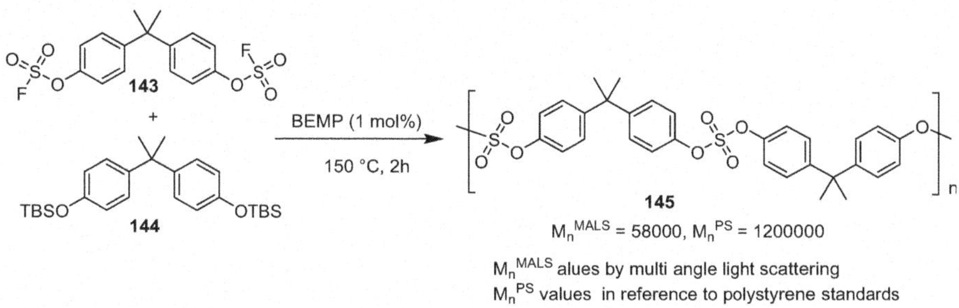

Scheme 8.31 SuFEx-mediated synthesis of polysulfates

easily incorporated as side chains onto monomers prior to assembly. This approach facilitates post-polymerization modification of polymers.

Sharpless and Fokin pioneered the practical application of SuFEx click chemistry in the field of materials science by combining bis(aryl fluorosulfates) **143** with bis(aryl silyl ethers) **144**, utilizing DBU, BEMP, or CsF as catalysts [111]. High-molecular weight polysulfates **145** with low polydispersity were obtained in quantitative yield through the crystallization of crude material in methanol. Bis(aryl silyl ethers) containing the tert-butyl dimethylsilyl (TBS) group emerged as optimal monomers, with low catalyst loadings yielding polymers of substantial molecular weights (Scheme 8.31). Further investigations revealed that among BPA-copolymers with similar structures, the polymer molecular weight correlates with the electronic properties of the silyl monomer. Introducing either electron-donating or withdrawing groups led to lower molecular weight polymers. These polymers exhibited superior resistance to chemical degradation compared to their polycarbonate counterparts, along with outstanding mechanical, optical, and oxygen-barrier properties.

Wu et al. synthesized bisalkyl/aryl sulfonyl fluorides **146** from the direct reaction between ESF and the required amine and used this for the copolymerization with the bisphenolsilyl ether **144** [112]. Due to the deprotonation of alkyl sulfonylfluorides, subsequent dehydrofluorination and termination of polymerization, DBU proved to be ineffective for this polymerization. A bifluoride salt $[Ph_3P{=}NPPh_3]^+[HF_2]^-$ was found to be suitable for this polymerization and a quantitative conversion was observed within 10 min with PDI ranging between 1.2 and 1.7. Exothermic polycondensation reaction between bisalkylsulfonyl fluorides **147** and bisphenyl silyl ethers **144** were also scaled up with 0.5 mol reaction making 245 g of polysulfate **148** (M_n^{PS} = 15 kDa, PDI = 1.2) (Scheme 8.32).

Polymer brushes offer a distinctive architecture for modifying surface functionalities because of their typical physicochemical properties. However, developing novel brushes imposes limitations on the monomers that can be directly grown from the surface. Various forms of click chemistry have been employed to efficiently modify polymer brushes, but they encounter challenges in incorporating unprotected reactive moieties into the original monomer before polymerization. SuFEx chemistry provides a solution to this inherent issue by utilizing robust sulfonyl fluorides and fluorosulfates moieties, enabling the formation of polymer brushes with pendant

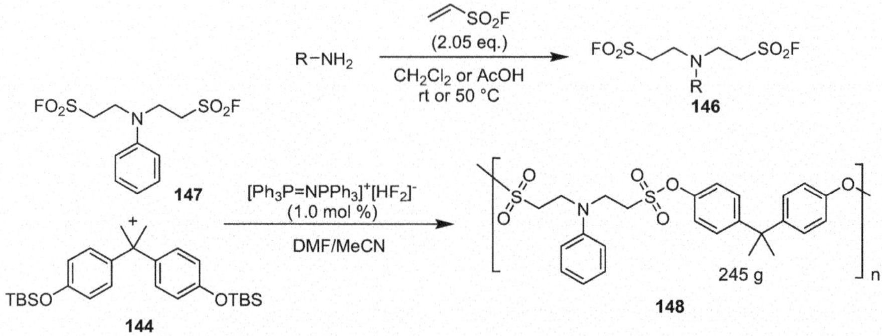

Scheme 8.32 Synthesis of bisalkyl sulfonyl fluorides and bifluoride salt promoted large-scale synthesis of polysulfonate

SuFExable groups. Locklin and coworkers demonstrated the post polymeric modification of the synthesized polymeric brushes via SuFEx modification with TBS protected alcohols (Scheme 8.33) [113]. Further detailed studies have been performed by reacting three different polymeric brushes having aryl sulfonyl fluorides **149**, aryl fluorosulfates **150**, and alkyl sulfonyl fluorides **151**, with three silyl ether derivatives (aryl, alkyl, benzyl) using different catalysts. The reactivity order follows the trend: Aryl sulfonyl fluorides > alkyl sulfonyl fluorides > aryl fluorosulfates, and aryl silyl ethers exhibited lower reaction rate as compared to the corresponding alkyl species.

Another important application of SuFEx click chemistry is surface functionalization, which is achieved by two complementary approaches, pseudo-Graft to' and

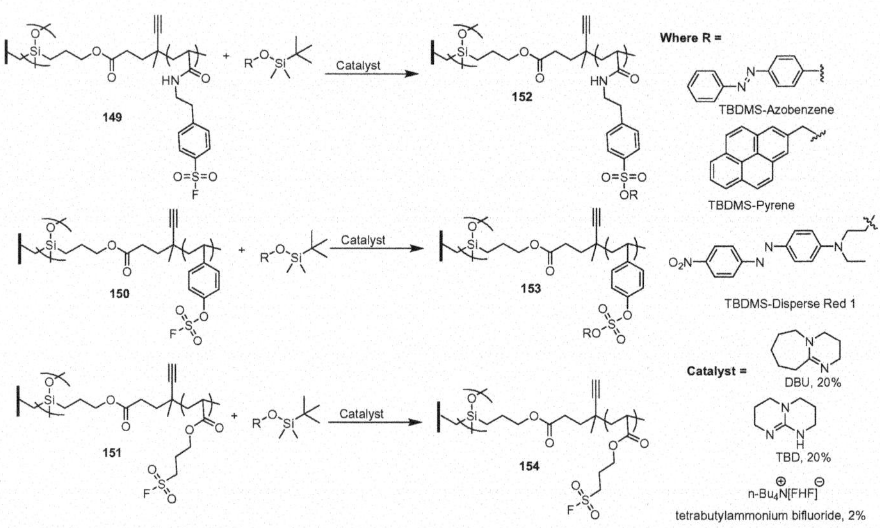

Scheme 8.33 SuFEx post-polymerization modification on aryl sulfonyl fluoride, aryl fluorosulfonate, and alkyl sulfonyl fluoride polymer brushes

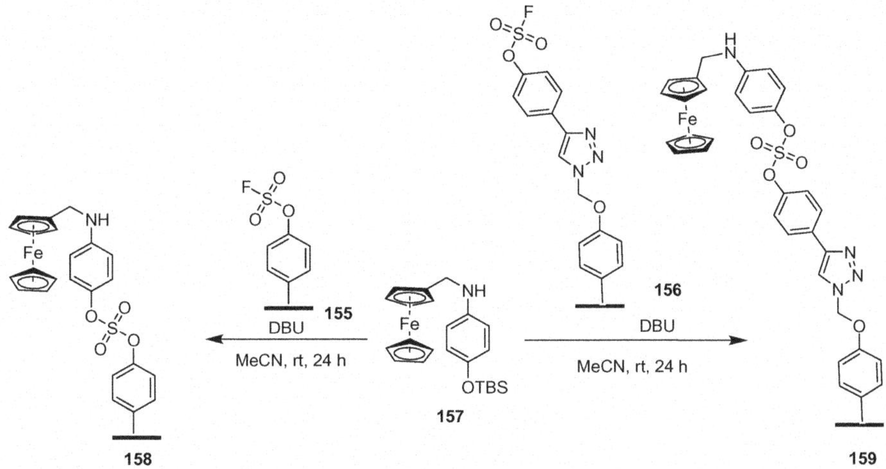

Scheme 8.34 Surface modification of carbon fibers by SuFEx click chemistry

pseudo-Graft from'. In the former approach, functionalized surface was fluorosulfonylated with SO_2F_2, whereas the latter approach involved the direct installation of aryl fluorosulfate on the surface by employing electrochemical deposition. Moses et al. reported the modification of the carbon fiber surface via SuFEx reactions [114]. Two different surfaces **155** and **156** were treated with ferrocenyl silyl ether **157** in the presence of DBU to yield surface **158** with sulfate functionality and **159** with both triazole as well as sulfate functionality. It has also been established by CV experiments that sulfate linker exhibited superior electrochemical stability as compared to the 1,2,3-triazole linkers (Scheme 8.34).

8.7 Conclusion and Future Outlook

SuFEx, introduced by Sharpless is a new generation of click chemistry, which is grounded based on the inertness of the sulfonyl fluoride functional group under many reaction conditions, but easily undergoes S-F exchange reaction with a range of nucleophile to give stable S-C, S–O and S–N covalent linkages under certain conditions. This chapter highlights the discovery and synthesis of SuFEx hub followed by their synthetic application and their utilities in drug discovery and material sciences. The number of applications of SuFEx are growing continuously and become a 'go-to' technology for the synthesis of diverse range of molecules. However, a complete mechanistic understanding of SuFEx reactions is necessary to exploit the true potential of SuFEx click chemistry.

References

1. Rostovtsev VV, Green LG, Fokin VV, Sharpless KB (2002) Angew Chem Int Ed 41:2596–2599
2. Meldal M, Tornøe CW (2008) Chem Rev 108:2952–3015
3. Tiwari VK, Mishra BB, Mishra KB, Mishra N, Singh AS, Chen X (2016) Chem Rev 116:3086–3240
4. Agrahari AK, Bose P, Jaiswal MK, Rajkhowa S, Singh AS, Hotha S, Mishra N, Tiwari VK (2021) Chem Rev 121:7638–7956
5. Sharpless KB (2018) Adaptation of the famous quote by UCLA Bruins football coach Henry Russell ("Red") Sanders: 'Winning Isn't Everything; It's the Only Thing. In: 15th national meeting on fluorine chemistry; Shanghai Institute of Organic Chemistry, November 2018
6. Suazo KF, Park KY, Distefano MD (2021) Chem Rev 121:7178–7248
7. Fantoni NZ, El-Sagheer AH, Brown T (2021) Chem Rev 121:7122–7154
8. Fitzgerald PR, Paegel BM (2021) Chem Rev 121:7155–7177
9. Kolb HC, Finn MG, Sharpless KB (2001) Angew Chem Int Ed 40:2004–2021
10. Moses JE, Moorhouse AD (2007) Chem Soc Rev 36:1249–1262
11. Tornøe CW, Christensen C, Meldal M (2002) J Org Chem 67:3057–3064
12. Zhang L, Chen XG, Xue P, Sun HHY, Williams ID, Sharpless KB, Fokin VV, Jia GC (2005) J Am Chem Soc 127:15998–15999
13. Ramachary DB, Ramakumar K, Narayana VV (2008) Chemistry 14:9143–9147
14. Ramachary DB, Shashank AB, Karthik S (2014) Angew Chem Int Ed Engl 53:10420–10424
15. Belkheira M, El Abed DE, Pons JM, Bressy C (2011) Chemistry 17:12917–12921
16. (a) Wang L, Peng S, Danence LJT, Gao Y, Wang J (2012) Chemistry 18:6088–6093; (b) Danence LJT, Gao Y, Li M, Huang Y, Wang J (2011) Chemistry 17:3584–3587
17. Sangwan R, Javed A, Dubey PKM (2017) Chem Select 2:4733–4743
18. Kumar GS, Lin Q (2021) Chem Rev 121:6991–7031
19. Hema K, Sureshan KM (2019) Acc Chem Res 52:3149–3163
20. Pasini D (2013) Molecules 18:9512–9530
21. Iha RK, Wooley KL, Nyström AM, Burke DJ, Kade MJ, Hawker CJ (2009) Chem Rev 109:5620–5686
22. Mishra A, Agrahari AK, Tiwari VK (2021) Application of CuAAC in carbohydrates in click chemistry volume. In: Rutjes (ed) Science of synthesis, pp 135–179
23. Liang L, Astruc D (2011) Coords. Chem Rev 255:2933–2945
24. Li N, Binder WH (2011) J Mater Chem 21:16717–16734
25. Thirumurugan P, Matosiuk D, Jozwiak K (2013) Chem Rev 113:4905–4979
26. Hoyle CE, Bowman CN (2010) Angew Chem Int Ed Engl 49:1540–1573
27. Massi A, Nanni D (2012) Org Biomol Chem 10:3791–3807
28. Worch JC, Stubbs CJ, Price MJ, Dove AP (2021) Chem Rev 121:6744–6776
29. Barrow AS, Smedley CJ, Zheng Q, Li S, Dong J, Moses JE (2019) Chem Soc Rev 48:4731–4758
30. Rodionov VO, Fokin VV, Finn MG (2005) Angew Chem Int Ed 117:2250–2255
31. Agard NJ, Prescher JA, Bertozzi CR (2004) J Am Chem Soc 126:15046–15047
32. Agard NJ, Baskin JM, Prescher JA, Lo A, Bertozzi CR (2006) A.C.S. Chem Biol 1:644–648
33. Ramachary DB, Reddy GB (2006) Org Biomol Chem 4:4463–4468
34. Jaiswal MK, Gupta A, Yadav MS, Pandey VK, Tiwari VK (2023) Chem Eur J 29:e202301749
35. Jalani HB, Karagoz AC, Tsogoeva SB (2017) Synthesis 49:39–41
36. Thomas J, John J, Parekh N, Dehaen W (2014) Angew Chem Int Ed 53:10155–10159
37. Bertozzi CACS (2023) Cent Sci 9:558–559
38. Dong J, Krasnova L, Finn MG, Sharpless KB (2014) Angew Chem Int Ed 53:9430–9448
39. Steinkopf W (1927) J Prakt Chem 117:1–82
40. Revathi L, Ravindar L, Leng J, Rakesh KP, Qin H-L (2018) J Org Chem 7:662–682
41. Martín-Gago P, Olsen CA (2019) Angew Chem Int Ed Engl 58:957–966

42. Fattah TA, Saeed A, Albericio F (2018) J Fluor Chem 213:87–112
43. Chinthakindi PK, Arvidsson PI (2018) Eur J Org Chem 3648–3666
44. Gembus V, Marsais F, Levacher V (2008) Synlett 1463–1466
45. Bianchi TA, Cate LA (1977) J Org Chem 42:2031–2032
46. Davies W, Dick JH (1931) J Chem Soc 2104–2109
47. Talko A, Barbasiewicz M (2018) ACS Sustain Chem Eng 6:6693–6701
48. Davies AT, Curto JM, Bagley SW, Willis MC (2017) Chem Sci 8:1233–1237
49. Tribby AL, Rodríguez I, Shariffudin S, Ball ND (2017) J Org Chem 82:2294–2299
50. Guo T, Meng G, Zhan X, Yang Q, Ma T, Xu L, Sharpless KB, Dong J (2018) Angew Chem Int Ed 57:2605–2610
51. Zhou H, Mukherjee P, Liu R, Evrard E, Wang D, Humphrey JM, Butler TW, Hoth LR, Sperry JB, Sakata SK, Helal CJ, Am Ende C (2018) W. Org Lett 20:812–815
52. Li S, Wu P, Moses JE, Sharpless KB (2017) 5 Int Angew Chem Int Ed 56(11):2903–2908
53. Moissan H, Lebeau P (1902) Ann Chim Phys 145–178
54. Smith WC, Engelhardt VA (1960) J Am Chem Soc 82:3838–3840
55. Cramer R, Coffman DD (1961) J Org Chem 26:4010–4014
56. Seppelt K, Sundermeyer W (1970) Angew Chem Int Ed Engl 9:905
57. Sundermeyer W, Roland A, Seppelt K (1971) Angew Chem Int Ed Engl 10:419–420
58. Gao B, Li S, Wu P, Moses JE, Sharpless KB (2018) Angew Chem Int Ed 57:1939–1943
59. Zheng Q, Dong J, Sharpless KB (2016) J Org Chem 81:11360–11362
60. Wipf P, Aslan DC, Southwick EC, Lazo JS (2001) Bioorg Med Chem Lett 11:313–317
61. Champseix A, Chanet J, Etienne A, Le Berre A, Lemasson JC, Napierala C, Vessiere R (1985) Bull Soc Chim Fr 463–472
62. Qin H-L, Zheng Q, Bare GAL, Wu P, Sharpless KB (2016) Angew Chem Int Ed 55:14155–14158
63. Kopyt M, Tryniszewski M, Barbasiewicz M, Kwiatkowski P (2023) Org Lett 25:6818–6822
64. Park JH, González-Montiel GA, Cheong PH, Bae HY (2023) Org Lett 25:1056–1060
65. Wang S-M, Li C, Leng J, Bukhari SNA, Qin H-L (2018) Org Chem Front 5:1411–1415
66. Wang S-M, Moku B, Leng J, Qin H-L (2018) Eur J Org Chem 4407–4410
67. Li C, Wang SM, Qin HL (2018) Org Lett 20:4699–4703
68. Ncube G, Huestis MP (2019) Organometallics 38:76–80
69. Wang TT, Zhao LM (2022) Chem Commun 58(79):11099–11102
70. Chen XY, Wu Y, Zhou J, Wang P, Yu JQ (2019) Org Lett 21:1426–1429
71. Mykhalchuk VL, Yarmolchuk VS, Doroschuk RO, Tolmachev AA, Grygorenko OO (2018) Eur J Org Chem 2870–2876
72. Mykhailiuk PK (2014) Chem Eur J 20:4942–4947
73. Slobodyanyuk Y, Artamonov OS, Shishkin OV, Mykhailiuk PK (2014) Eur J Org Chem 2487–2495
74. Jangra H, Chen Q, Fuks E, Zenz I, Mayer P, Ofial AR, Zipse H, Mayr H (2018) J Am Chem Soc 140:16758–16772
75. Smedley CJ, Giel MC, Molino A, Barrow AS, Wilson DJD, Moses JE (2018) Chem Commun 54(47):6020–6023
76. Thomas J, Fokin VV (2018) Org Lett 20:3749–3752
77. Skalenko YA, Druzhenko TV, Denisenko AV, Samoilenko MV, Dacenko OP, Trofymchuk SA, Grygorenko OO, Tolmachev AA, Mykhailiuk PK (2018) J Org Chem 83:6275–6289
78. Krutak JJ, Burpitt RD, Moore WH, Hyatt JA (1979) J Org Chem 44:3847–3858
79. Daeniker HU, Druey J (1962) Helv Chim Acta 45:1972–1981
80. Xu R, Xu T, Yang M, Cao T, Liao S (2019) Nat Commun 10:3752
81. Central Glass Company (2012) U.S. Pat. 8304576B2
82. Central Glass Company (2014) U.S. Pat. 8835669B2
83. Nielsen MK, Ugaz CR, Li W, Doyle AG (2015) J Am Chem Soc 137(30):9571–9574
84. Lee C, Lai J, Epifanov M, Wang CX, Sammis GM (2021) J Fluor Chem 251:109888
85. Steinkopf W, Jaeger P (1930) J Prakt Chem 128:63–88
86. Yan Z, Tian W, Zeng F, Dai Y (2009) Tetrahedron Lett 50:2727–2729

87. Smedley CJ, Barrow AS, Spiteri C, Giel MC, Sharma P, Moses JE (2017) Chem Eur J 23:9990–9995
88. Barrow AS, Moses JE (2016) Synlett 1840–1843
89. Meng G, Guo T, Ma T, Zhang J, Shen Y, Sharpless KB, Dong J (2019) Nature 574:86–89
90. Tian F, Kruger V, Bautista O, Duan JX, Li AR, Dolbier WR, Chen QY (2000) Org Lett 2(4):563–564
91. Eusterwiemann S, Martinez H, Dolbier WR (2012) J Org Chem 77:5461–5464
92. Hanley PS, Clark TP, Krasovskiy AL, Ober MS, O'Brien JP, Staton TS (2016) A.C.S. Catal 6:3515–3519
93. Zhang JW, Guan C, Zhao Y, Ding C (2021) Eur J Org Chem 810–813
94. Roth P, Sapino C (1991) Tetrahedron Lett 32:4073–4076
95. Roth P, Thomas JA (1992) Tetrahedron Lett 33:1959–1962
96. Roth GP, Fuller CE (1991) J Org Chem 56:3493–3496
97. Liang Q, Xing P, Huang Z, Dong J, Sharpless KB, Li X, Jiang B (2015) Org Lett 17:1942–1945
98. Zhang JT, Li S, Wu P, Moses JE, Sharpless KB (2016) Chem Eur J 22:5692–5697
99. Fang WY, Leng J, Qin HL (2017) Chem Asian J 12:2323–2331
100. Fang W-Y, Huang Y-M, Leng J, Qin H-L (2018) J Org Chem 7:751–756
101. Zhao C, Fang W-Y, Rakesh KP, Qin H-L (2018) Org Chem Front 5:1835–1839
102. Schimler SD, Cismesia MA, Hanley PS, Froese RDJ, Jansma MJ, Bland DC, Sanford MS (2017) J Am Chem Soc 139:1452–1455
103. Epifanov M, Foth PJ, Gu F, Barrillon C, Kanani SS, Higman CS, Hein JE, Sammis GM (2018) J Am Chem Soc 140:16464–16468
104. Liu Z, Li J, Li S, Li G, Sharpless KB, Wu P (2018) J Am Chem Soc 140:2919–2925
105. Mortenson DE, Brighty GJ, Plate L, Bare G, Chen W, Li S, Wang H, Cravatt BF, Forli S, Powers ET, Sharpless KB, Wilson IA, Kelly JW (2018) J Am Chem Soc 140:200–210
106. Zheng Q, Woehl JL, Kitamura S, Santos-Martins D, Smedley CJ, Li G, Forli S, Moses JE, Wolan DW, Sharpless KB (2019) Proc Natl Acad Sci USA 116:18808–18814
107. Smedley CJ, Zheng Q, Gao B, Li S, Molino A, Duivenvoorden HM, Parker BS, Wilson DJD, Sharpless KB, Moses JE (2019) Angew Chem Int Ed 58:4552–4556
108. You Y, Kim HS, Bae IH, Lee SG, Jee MH, Keum G, Jang SK, Kim BM (2017) Eur J Med Chem 125:87–100
109. You Y, Kim HS, Park JW, Keum G, Jang SK, Kim BM (2018) RSC Adv 8:31803–31821
110. Liu F, Wang H, Li S, Bare GAL, Chen X, Wang C, Moses JE, Wu P, Sharpless KB (2019) Angew Chem Int Ed 58:8029–8033
111. Dong J, Sharpless KB, Kwisnek L, Oakdale JS, Fokin VV (2014) Angew Chem Int Ed 53:9466–9470
112. Wang H, Zhou F, Ren G, Zheng Q, Chen H, Gao B, Klivansky L, Liu Y, Wu B, Xu Q, Lu J, Sharpless KB, Wu P (2017) Angew Chem Int Ed 56:11203–11208
113. Brooks K, Yatvin J, Kavoliov M, Crane GH, Horn J, Averick S, Locklin J (2018) Macromolecules 51:297–305
114. Randall JD, Eyckens DJ, Stojcevski F, Francis PS, Doeven EH, Barlow AJ, Barrow AS, Arnold CL, Moses JE, Henderson LC (2018) Chem Phys Chem 19:3176–3181

Chapter 9
CuAAC 'Click Chemistry' in Synthesis of Peptides and Protein Conjugates

9.1 Introduction

1,2,3-Triazoles have been identified as the most important heterocyclic compounds of nitrogen-based substances, particularly due to their numerous applications in the fields of organic synthesis, chemical biology, medicinal chemistry, and material sciences [1–4]. Di-substituted 1,2,3-triazole derivatives are generally produced by a 1,3-dipolar cycloaddition reaction of organic azides and terminal alkynes. Nevertheless, the regioselective synthesis of 1,4-disubstituted 1,2,3-triazoles via CuAAC click has been recognized as the most important chemical transformation with an extensive range of applications throughout the past 22 years [5–9]. As significant amide bond isosteres with distinctive chemical characteristics, 1,2,3-triazoles are frequently utilized as active pharmaceutical ingredients (API) in pharmaceutical products [10, 11]. Apart from that, in 2005, Fokin et al. introduced the use of ruthenium catalyst for the production of 1,5-disubstituted 1,2,3-triazole isomer which exactly an opposite regioisomer of corresponding CuAAC click conjugation (Scheme 9.1) [12, 13]. The CuAAC conjugation is more extensively studied than corresponding ruthenium-catalyzed azide-alkyne cycloaddition (RuAAC), however, reports of its applicability are growing exponentially. In addition to its noteworthy biological significance [4]. 1,2,3-triazole rings can act as important ligands or intermediaries in several important chemical transformation events such as the Glacer, Sonogashira, and various C-hetero atom coupling reactions [14–17]. Moreover, the properties of molecules can be greatly enhanced by post-functionalization of the 1,2,3-triazole framework, in which a series of C–C and C–N bonds are used to selectively add a substituent on three different locations. In the design and production of bioactive materials for medicinal purposes, peptides the diverse sequence of amino acids that make up the functional modules of proteins have grown in significance [18–20].

The main focus of this chapter is to introduce the impact of CuAAC reaction in the development of proteins and peptides. Since there have been many reported uses for peptides and proteins over the years, we have only included a brief technique

V. K. Tiwari et al., *Click Chemistry*, Materials Horizons: From Nature to Nanomaterials,
https://doi.org/10.1007/978-981-97-4596-8_9

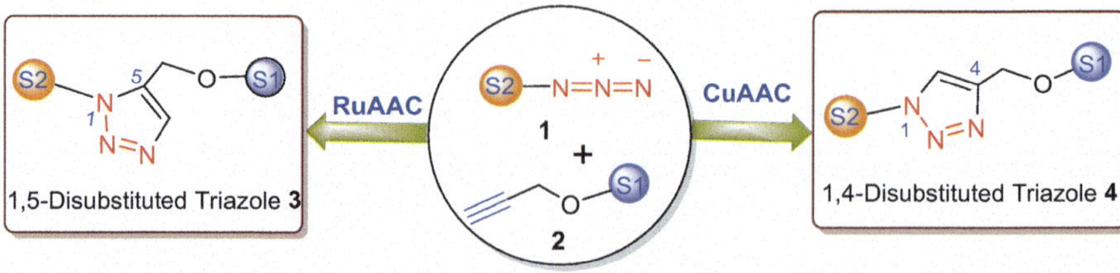

Scheme 9.1 Regioselectivity in 1,2,3-triazole synthesis under different catalytic conditions

for creating peptidomimetics here rather than a thorough underprivileged of all the examples that have been published. A variety of methods have been developed for conjugating peptides with either an alkyne or an azide functional group on one of the side chains or at one of the terminals with molecules in the widest sense that have the complementary functional group. One such method is the click reaction.

9.2 CuAAC Click-Inspired Peptide and Glycopeptide Conjugation Strategies

Triazoles synthesized through CuAAC protocol act as amide isosteres, i.e., they are good mimics of peptide or amide bonds because of their polar nature and hydrogen bond-forming capabilities [21, 22]. In research describes the creation of well-oriented nanotube assemblies using peptide chains that comprise macrocyclic building blocks. Chattopadhyay et al. synthesized two distinct regioisomeric triazole attached macrocycles, **5** and **6**, using cis-β-furanoids and β-alanine as suitable frames (Fig. 9.1) [23]. Because the functional groups on macrocycles **5** and **6** have different orientations, molecules are compelled to pack and self-assemble differently, resulting in the generation of nanotubes with distinct polarity.

The CuAAC protocol receives great attention to conjugating peptides [24] with sugar moiety which renders peptides resistant to in vivo hydrolysis [25]. A wide spectrum of peptide–triazole buildings were constructed from synergistic interaction between sugar-based alkynes/azides and the corresponding peptidic azide/alkynes using the established click chemistry technique [26]. Toward this end, a number of pseudo peptide conjugates were established by Scolastico and groups by using click chemistry tools [27]. Additionally, utilizing click-based cycloaddition, glycoconjugates originating from β-lactams were produced that bore a resemblance to Ser–Glu dipeptide O-glycoside hybrids **7**, lectin antagonists that have β-turn as a non-repetitive structural part. Compound **8** was predicted by the authors to bind easily to lectin or the carbohydrate recognition domain (CRD) that adorns human E-selectin (Scheme 9.2) [28].

In order to peptide synthesis, a report documented toward synthesis of triazole-appended glycoaminoacids with heterocyclic moiety using modular CuAAC click

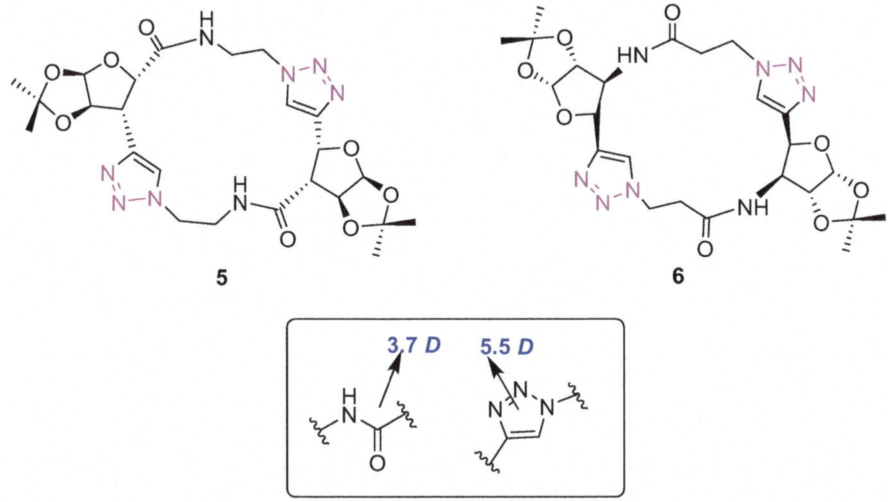

Fig. 9.1 Chemical structures of two different regioisomeric cyclic peptidomimetic macrocycles **5** and **6**

Scheme 9.2 CuAAC click chemistry for synthesis of β-lactam based glycohybrids **8**

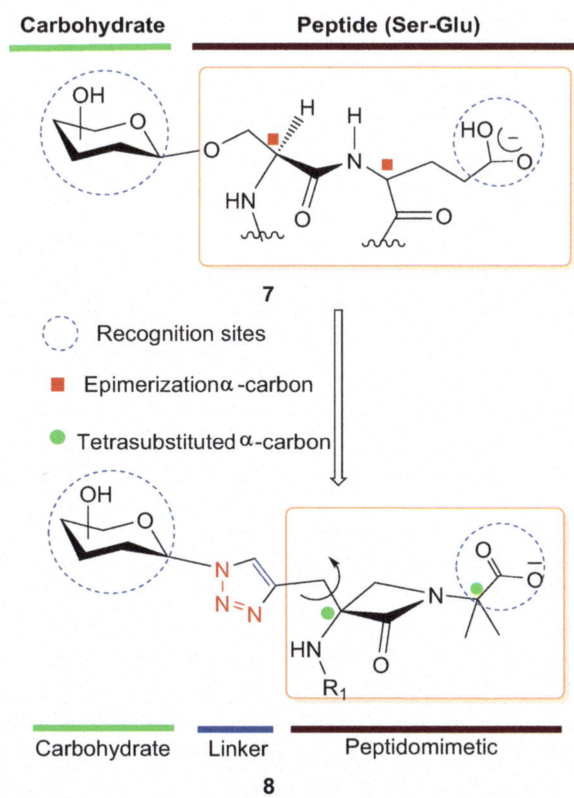

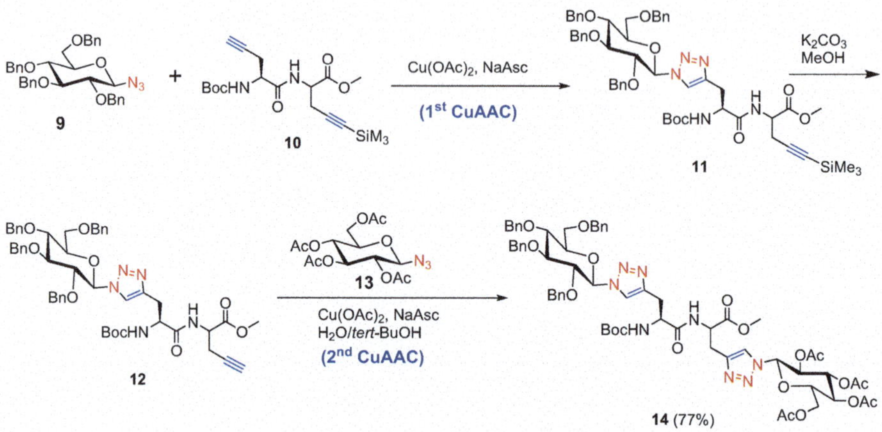

Scheme 9.3 Synthesis of triazole-linked diglycosylated peptide **14** via step-wise click reaction

reaction between the corresponding azido and acetylene moieties [28]. The step-wise introduction of carbohydrate moiety through click conjugation was achieved from protected diacetylenic dipeptide **10**. Initially first click of dipeptide **10** with tetra-*O*-benzyl azido glucoside **9** resulted in the conjugation of a single carbohydrate moiety **11** leaving the trimethylsilyl function fully unaffected. Subsequently, desilylation with potassium carbonate in methanol produces free terminal alkyne **12** which further underwent a second click with azide **13** in the presence of Cu(I) catalyst and afforded the respective triazole-linked diglycosylated peptide **14** in an overall 77% yield (Scheme 9.3).

A number of investigations have been carried out on glycoconjugate mimetics, or neoglycoconjugates understand the synthetic bonding between the aglycone and carbohydrate moieties [29]. Toward the above strategies, a new protocol was reported for the convergent synthesis of neoglycoconjugate **17** using a combination of click chemistry and chemical ligation [30]. Lee and co-workers elaborated this unique strategy of combination of two protocols and this was achieved by changing the sequence of synthesis. From three unprotected building blocks (peptides **15** and **16** and a sugar azide), an effective one-pot synthesis of a neoglycopeptide **17** containing two sugars has been achieved through the successful combination of native chemical ligation and click chemistry. Alternatively, click cycloaddition before native ligation (**18** and **19**) offers the reverse strategy for neoglycopeptide **17** (Scheme 9.4). The two ligation chemistries can be used in any order and are completely compatible. This has given chemists accessibility to a large variety of neoglycopeptides based on triazoles that contain even more than 20 amino acids.

The viability of click reaction in conjunction with native chemical ligation (NCL) strategy has been demonstrated by Macmillan and Blanc's use of chemoselective methods to replace alkyl bromides with side chain cysteine thiols. Additionally, the study suggested that the synthesis of neoglycopeptides did not require a specific

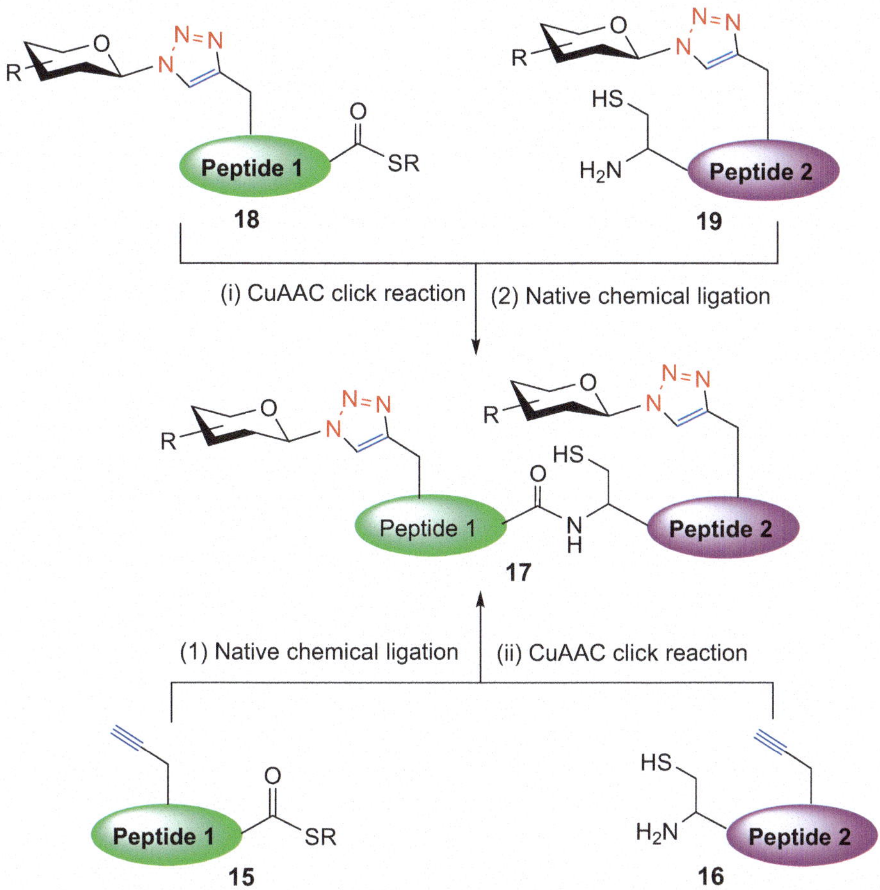

Scheme 9.4 A one-pot approach to neoglycopeptides **15–19** using orthogonal native chemical ligation and click chemistry

order for the production of thioether and triazole rings [31]. Two important pharmacodynamics parameters such as water solubility and biological activity can be addressed with the conjugation of non-ribosomal peptides and polyketides with carbohydrates via macrolactonization. Walsh and co-workers reported a chemoenzymatic approach for the synthesis of carbohydrate-modified cyclic peptide antibiotics [32]. The author developed thirteen head-to-tail cyclized tyrocidine derivatives **21** by utilizing a thioesterase domain from the decapeptide tyrocidine synthetase **20**, including one to three propargylglycines at positions 3–8 (Scheme 9.5). Then, CuAAC click conjugation of these cyclic peptides with 21 azido sugars afforded the corresponding glycopeptide variants of the antibiotic **22**. The anti-bacterial and hemolytic assay investigation of some glycopeptides suggested better biological efficacy of the developed glycopeptides in comparison with natural tyrocidine.

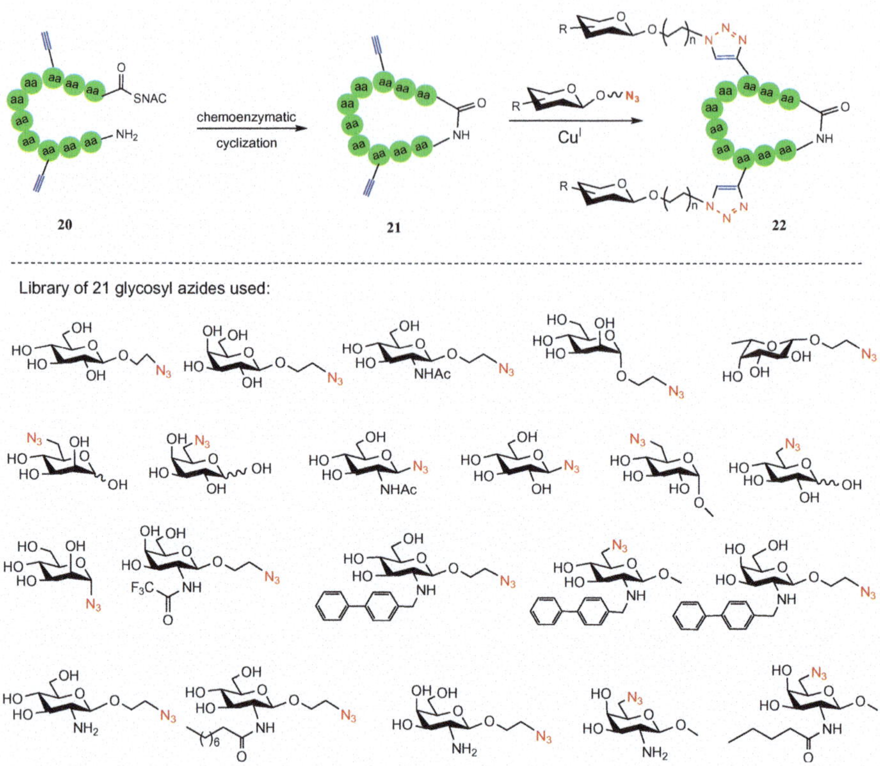

Scheme 9.5 The development of a library of carbohydrate-modified cyclic peptide antibiotics using CuAAC click conjugation

The click conjugation tool has been a more effective technique in enhancing the chemotherapeutic index of various well-established drugs. Among these, tyrocidine (Tyc) systemic applicability is limited by its potential to insert into eukaryotic membranes and lyse human red blood cells. Therefore, the introduction of various carbohydrate motifs into tyrocidine to develop the modified Tyc-derivatives like **23** and **24** (Fig. 9.2) which significantly enhance the therapeutics index glycopeptide. Furthermore, minimal hemolytic concentration (MHC) and minimal inhibitory concentration (MIC) assay screening of these conjugates against bacteria suggested that, at a concentration of 3.0 μM the glycopeptide **23** and **24** have sixfold better therapeutic index than native tyrocidine with enhanced biocompatibility [32].

There are several naturally occurring antibiotics have macrocycles formed from carbohydrates or glycosylated cyclic peptides [33]. Thus, by utilizing modular CuAAC click conjugation and ring-closing metathesis (RCM) methods, a range of glycosylated cyclic peptide mimics and peptidomimetic were effectively generated [23]. Using glycosyltransferases, Thorson and colleagues used a chemoenzymatic method to modify the naturally occurring antibiotic vancomycin, resulting in the production of azide-functionalized vancomycin analogue **25**, which was the intended

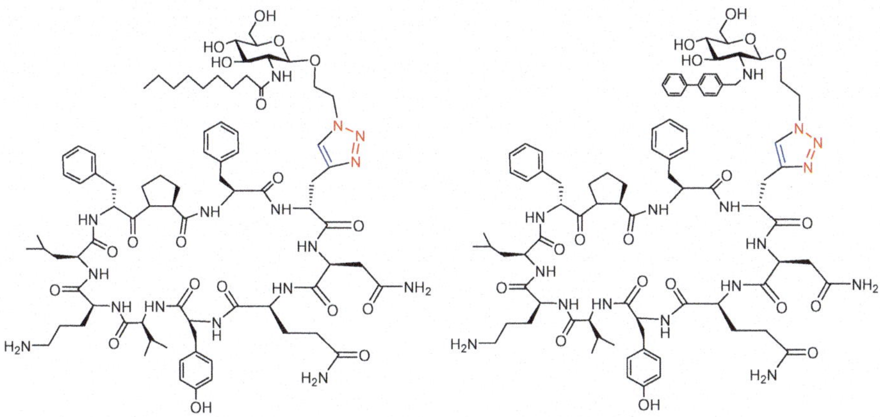

Tyc4PG-14 (**23**) Tyc4PG-15 (**24**)

Fig. 9.2 Glycopeptide structures (**23,24**) with enhanced chemotherapeutic potential

outcome. Furthermore, a library of triazole-linked compounds **26a–l** was produced by the regioselective click conjugation of **25** with a variety of alkynes (Scheme 9.6) [34]. In the absence of a CuI catalyst, non-terminal alkynes react to form a 1:1 mixture of 1,4- and 1,5-triazole. The carboxylic acid derivative produced from vancomycin antibiotics using the click technique was twice as effective as vancomycin itself against antibiotic-resistant bacteria.

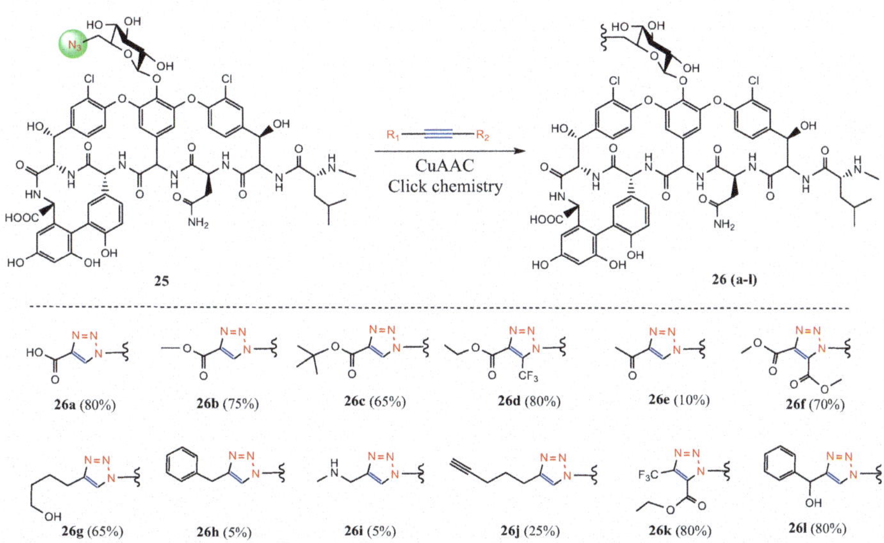

Scheme 9.6 Click chemistry inspired chemoenzymatic synthesis of vancomycin derivatives **26(a–l)**

Pramlintide is a synthetic neuroendocrine hormone analogue that is used in combination with insulin or insulin analogues to treat individuals with type I and type II diabetes. Brimble and colleagues described the production of mono- and di-glycosylated pramlintide analogues, as well as their effects on the $AMY_{1(a)}$ receptor. The introduction of *N*-glycosylated amino acids into the pramlintide sequence yields native *N*-linked glycomimetic, whereas modular CuAAC click cycloaddition yields 1,2,3-triazole attached glycomimetics. Furthermore, adding one or more GlcNAc moieties to positions 21 and 35 of the natural pramlintide maintained $AMY_{1(a)}$ receptor activation. Remarkably, $AMY_{1(a)}$ activity did not distinguish native *N*-linked glycomimetics (**30**–**32**) from 1,2,3-triazole-linked glycomimetics (**27**–**29**), indicating that click versions may serve as biological stand-ins for native *N*-glycosides (Fig. 9.3). Moreover, the physicochemical and pharmacokinetic properties of pramlintide **1** may be further enhanced by click chemistry-mediated *N*-glycosylation.

Despite the many benefits of CuAAC protocol for organic and biological systems, its toxicity due to free copper ions remains a threat to living things. The development of a single turnover system with a histidine moiety was demonstrated by Finn and colleagues through the use of a peptide-based accelerating ligand that was capable of cellular expression. The ligand-free reaction was significantly accelerated, suggesting that the protocol is biocompatible [35].

An interesting synthesis of arginine–glycine–aspartate (RGD) derivatives (glycosylated cyclic peptide **35**) was carried out by Rutjes and co-workers. The group exploited both solid-phase and solution-phase synthetic procedures for selective targeting of integrin expressed in tumors [36]. The conjugation of C-terminus amino acid Fmoc-glycine **33** and trityl resin was performed in the presence of BOP/DIPEA

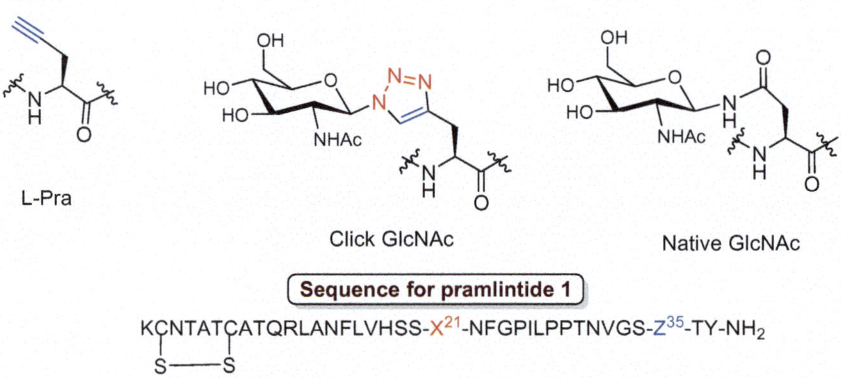

27. X^{21} = Click GlcNAc; Z^{35} = Asn
28. X^{21} = Asn; Z^{35} = Click GlcNAc
29. X^{21} = Click GlcNAc; Z^{35} = Click GlcNAc
30. X^{21} = Native GlcNAc; Z^{35} = Asn
31. X^{21} = L-Pra; Z^{35} = Native GlcNAc
32. X^{21} =Native GlcNAc; Z^{35} = Native GlcNAc

Fig. 9.3 Click chemistry inspired synthesis and amylin receptor activity of triazole-linked GlcNAc (**27**–**29**) and native *N*-linked GlcNAc (**30**–**32**) glycomimetics of pramlintide **1**

in DCM to prevent racemization. Further, Fmoc-L-^{N}TGA(Ac)4-OH was applied for the elongation of pentapeptide **34** and standard Fmoc chemistry involving HOBt/DIPCDI-mediated coupling. Furthermore, after completion of synthesis a sequential reaction involving cleavage, cyclization, and deprotection of cyclic arginine–glycine–aspartate peptide leading to the formation of cyclic glycosyl peptide derivative c(RGDy-^{N}TGA) **35** with good yield (Scheme 9.7). Although, the developed cyclic RGD derivatives exhibited favored uptake in the hypodermic tumors, in which the triazole-linked derivative **35** showed maximum tumor-to-blood ratio over the other peptides. Furthermore, this research demonstrated the utility of triazole motifs as a potent alternative to amide bonds in the synthesis of glycosylated peptides, possessing wide pharmaceutical applications.

Payne and co-workers describe the synthesis of dendrimer having four copies of glycopeptide belonging to MUC1 eicosa peptide tandem repeat sequence using solid-phase peptide chemistry in combination with click chemistry [37]. The desire tetrapeptide GSTA motif **37** possessing azido functionality at the threonine and

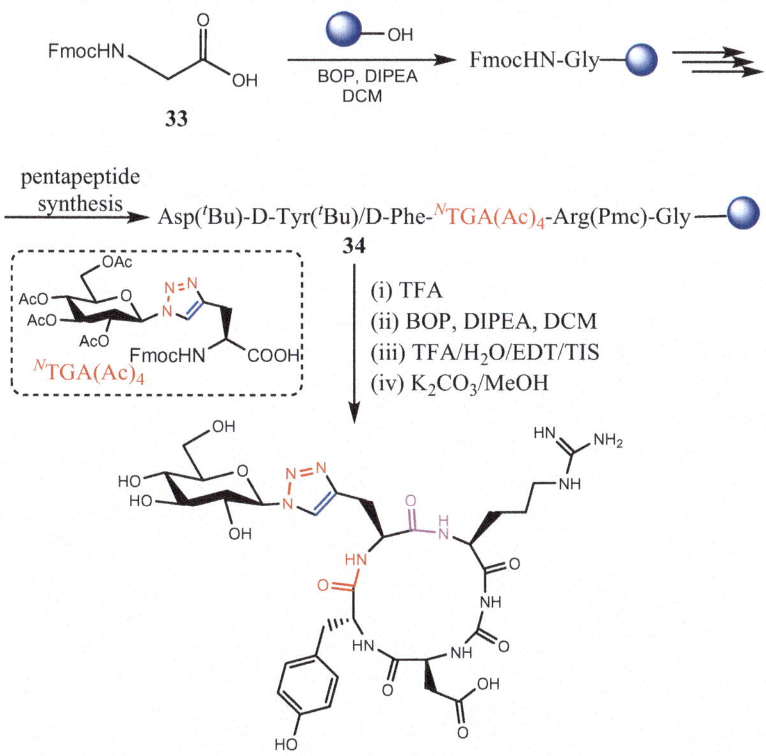

Scheme 9.7 Click chemistry-inspired synthesis of triazole-linked cyclic c(RGDy-^{N}TGA) peptide **35**

serine residues containing TN-antigen was thus constructed by adapting the solid-phase peptide synthesis approach. Therefore, CuAAC click-inspired coupling of azido functionalized glycopeptide **37** with appropriate tetra-alkyne-derived polyamidoamine **36** derived from PAMAM in the presence of $CuSO_4$/NaAsc in *t*-BuOH/H_2O (1:1) afforded the title multivalent glycopeptide dendrimer **38** having four glycosylated GSTA fragment conjugated via 1,2,3-triazole linkage (Scheme 9.8). The synthetic protocol of glycopeptide fragments has now emerged as a rapid and efficient way for the construction of a library of peptide dendrimers.

To facilitate the synthesis of biologically important modified unsymmetrical bis-triazolyl peptidomimetics, the amide bond and its isosteric triazole moiety can be put together in a one-pot chemoselective process. This is achieved by utilizing bifunctional alkynylated amines in a metal-free, multi-component CuAAC reaction [38]. The methodology has not been well investigated in the field of carbohydrate-based peptidomimetics synthesis, despite the method's significant prospects for synthetic applications.

Langle and co-workers used a modular copper(I)-catalyzed azide-alkyne cycloaddition (CuAAC) technique to produce a series of cysteine-containing glycopeptides with good yield [39]. This was accomplished by functionalizing peptides by propargylating the cysteine residue. The resulting product was then stitched together with appropriate free or protected azido-carbohydrates such as galacto, manno, and glucose sugars to produce a range of glycopeptides. Therefore, by utilizing the above approach authors developed *bis*-glycopeptide **41** by employing modular click conjugation of alkyne-functionalized peptide **40** with *bis*-6-azido trehalose **39** in 65% yield (Scheme 9.9). Further, in vitro platelet aggregation analysis of a set of 'glyco-RGD' peptides revealed that the IC_{50} value was lowered by pseudo glycosylation of the adhesion sequence and thus could improve the in vivo pharmacokinetic properties.

A common motif presents in anti-freeze glycopeptides (AFGPs) isolated from Antarctic fish linked to the hydroxyl group of ser/ thr (ser/thr-*O*-linked glycosides) is commonly termed as Tn-antigen. Toward this end, Brimble et al. developed click-inspired analogues of anti-freeze glycopeptides and 'Tn-antigen mimic' under microwave irradiation [40]. CuAAC click conjugation of azido alanine containing peptides **42** and **43** with alkyne functionalized pyranoside **44** afforded the glycopeptides **45** and **46** under *MW* irradiation under the catalysis of $CuSO_4$/NaAsc and in the presence of $H_2O/^{t-}$BuOH (1:1) ambient for 10 min (Scheme 9.10). Highly pure corresponding AFGPs scaffolds **47** and **48** were accomplished in good yield by deacetylating the resulting glycopeptide using 1 M CH_3ONa in methyl alcohol (pH = 11.3). This microwave irradiation protocol facilitated the efficient and prompt synthesis of anti-freeze glycopeptides, creating the way for the production of a broader spectrum of neoglycopeptides with significant biological applications.

Furthermore, same group performed the coupling reaction of carbohydrate-cored triazole-tethered glycoclusters to MUC1 peptides under click chemistry conditions in order to synthesize multivalent neoglycoconjugates of MUC1 [41]. The synthetic methodology for triazole-linked neoglycotetrasaccharide clusters **50, 51,** and **52** having azido functionality involves the click conjugation of **49** with appropriate glycosyl azide in the presence of NaN_3 in aqueous MeCN under reflux for

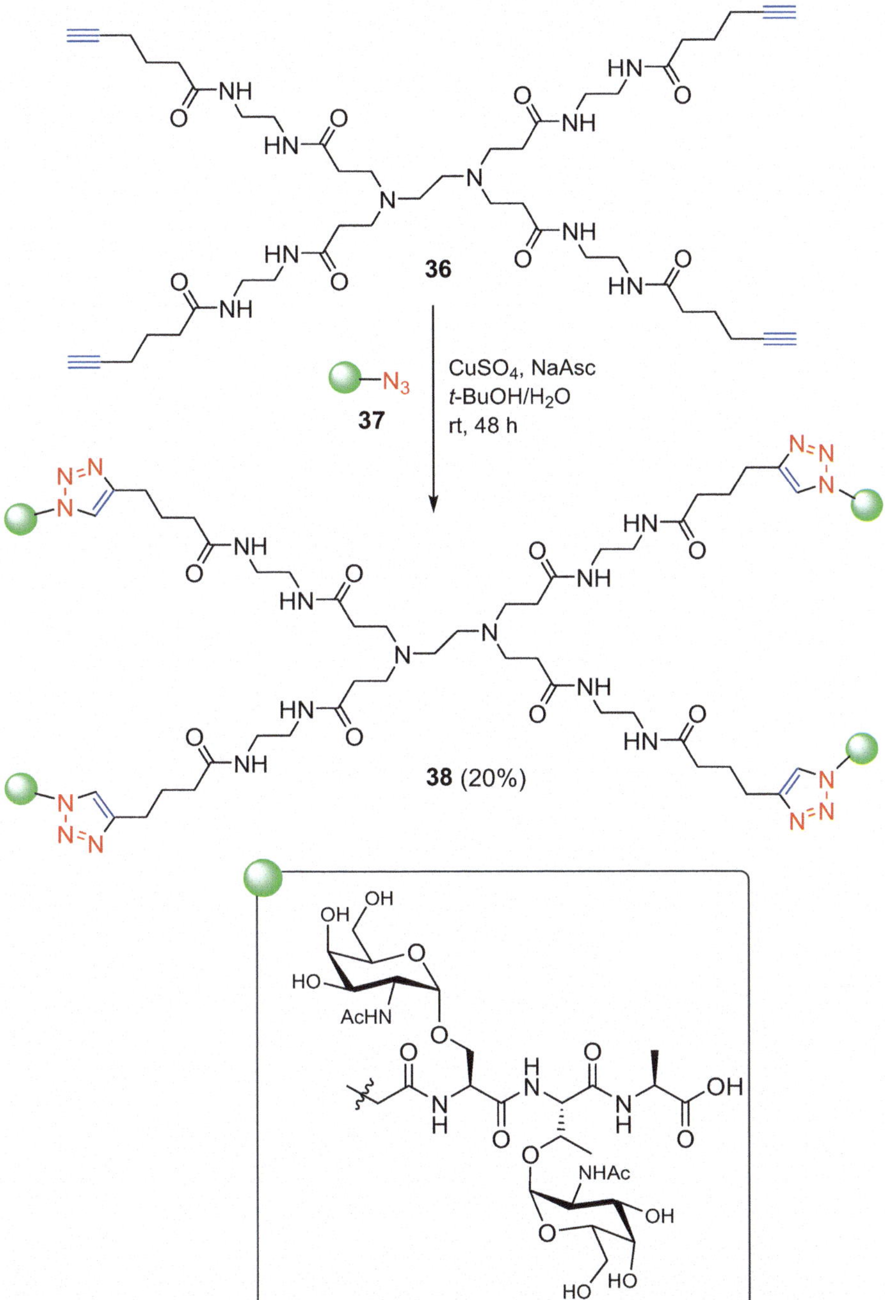

Scheme 9.8 Synthesis of MUC1 glycopeptide dendrimer **38** under CuAAC click strategy

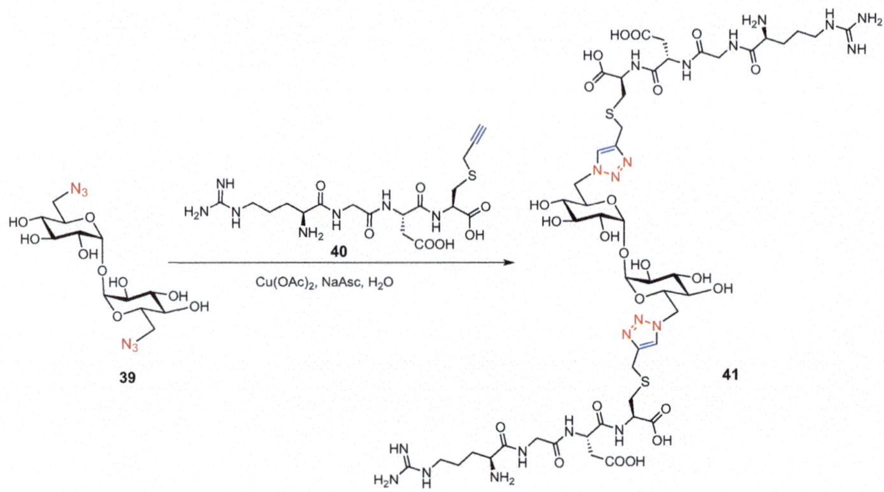

Scheme 9.9 CuAAC click chemistry inspired synthesis of *bis*-glycopeptide **41**

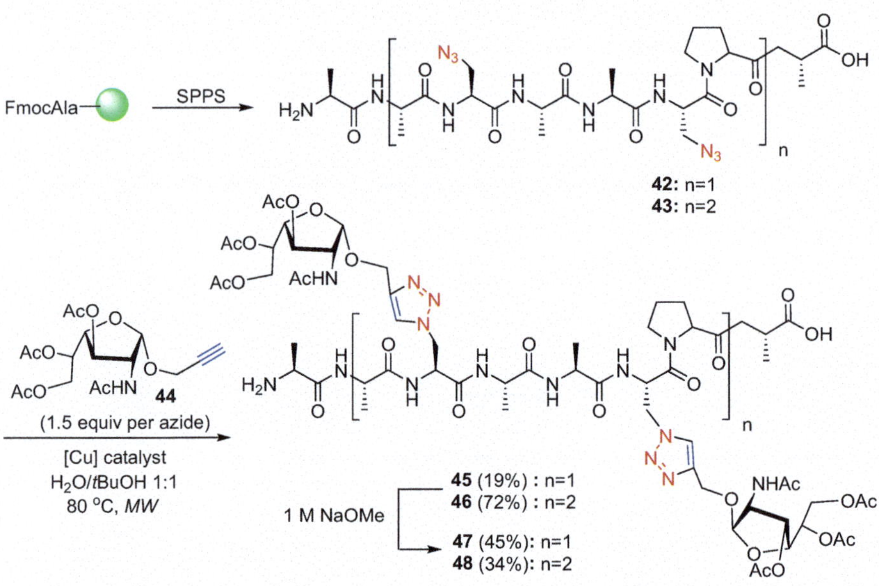

Scheme 9.10 Click chemistry-inspired synthesis of Tn-antigen mimetics **47** and **48**

5 h (Scheme 9.11). Further, all three sugars within neoglycoclusters underwent deacetylation producing respective triazole clusters having GlcNAc at its center (**53, 54,** and **55**). Additionally, GlcNAc-centered neoglycotetrasaccharide clusters were appropriately coupled with propargylated MUC1 peptide **56** under the catalysis of $CuSO_4 \cdot 5H_2O$/TCEP which was premixed under heating conditions before addition

to the peptide. This reaction was performed at pH 7.2 under microwave conditions at 50 °C and yielded the proposed multivalent neoglycoconjugates **57a–c**, consisting of biologically important glycopeptides, in excellent yields (Scheme 9.11).

Chandrasekaran and co-workers have nicely utilized click chemistry concept for the synthesis of sugar conjugated peptides and unusual amino acids [42]. In this order, the authors have constructed lysine-derived trialkynyl derivative Poc-Lys(Poc)-OPrp **59** and tera-alkyne derivatives **61**, which subsequently underwent click conjugation with C_6-azido glucose derivative **58** in the presence of in situ generated Cu(I) to

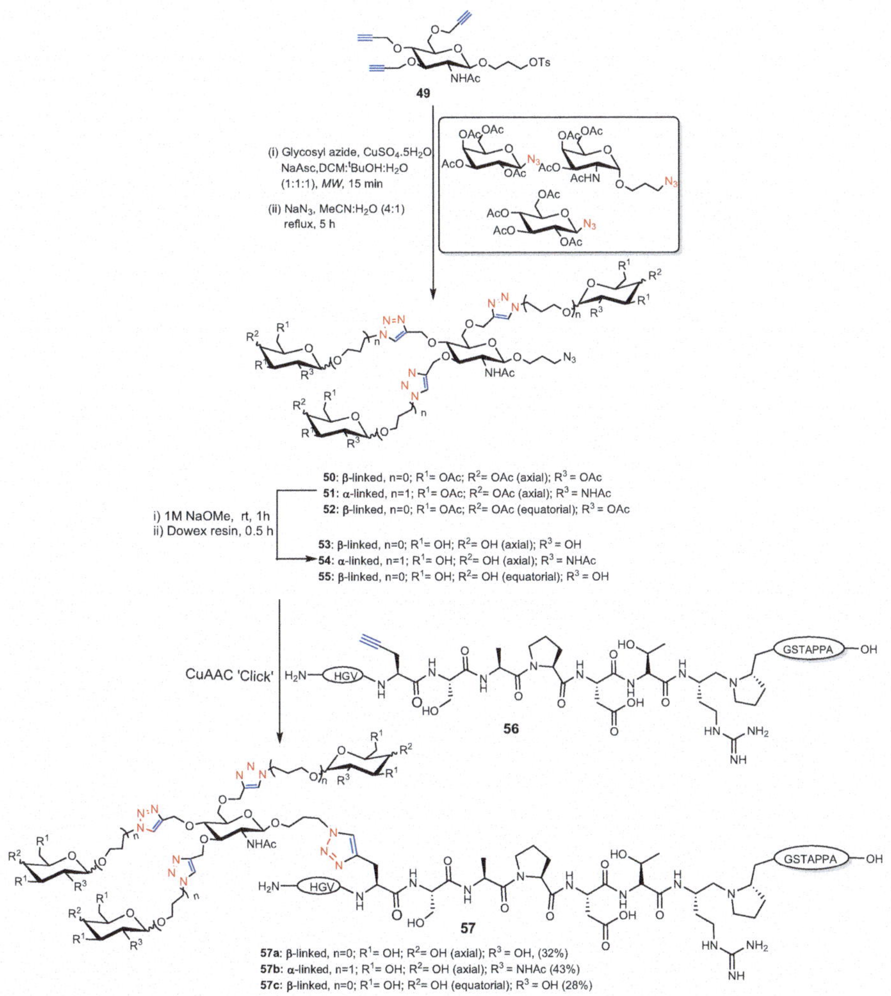

Scheme 9.11 Click chemistry-inspired synthesis of multivalent neoglycoconjugates **57a–c** of MUC1by the conjugation of carbohydrate-centered glycoclusters **50–55** and MUC1 peptides **56**

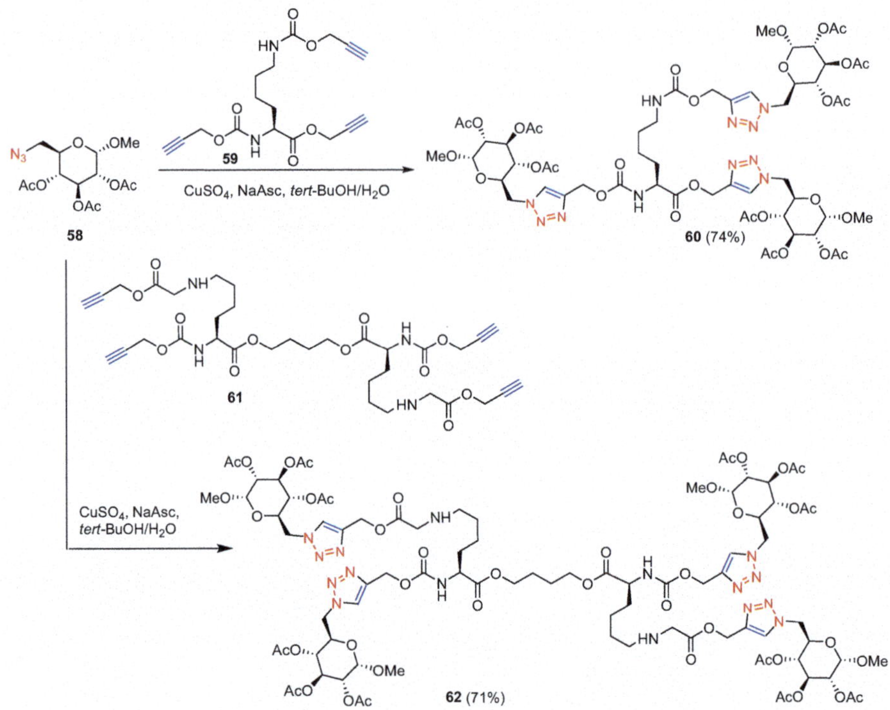

Scheme 9.12 Synthesis of tri-and tetra-glycoconjugates of Lysine using click chemistry

produce the corresponding tri- and tetra-glucose conjugated peptides **60** and **62** with good yields (Scheme 9.12).

In this continuation, same group developed lysine-based tripeptide analogues as inhibitors of Sir2 activity with various sugar entities [43]. Toward this end, click chemistry inspired coupling of tripeptide (Boc-Phe-Lys(Poc)-Val-Ome) with diverse glycosyl azide such as galactose, glucose, and mannose to furnish the corresponding glycopeptides (**63–65**) in good yield (Fig. 9.4). Moreover, these synthesized glycopeptides were tested against the *P. falciparum* Sir2.

Similarly, in another investigation, click chemistry-based coupling of inert (Dex) and bioactive (Lam and HA) polysaccharide with stimuli-responsive ELPS was carried out for the synthesis of notable polysaccharide-block-ELP copolymers [44]. For click conjugation, polysaccharide underwent successful conversion into azido scaffolds **66–68** facilitated by a bifunctional-*N*-methoxyamine linker. These scaffolds were then coupled with alkyne-functionalized ELP **69** under CuAAC click condition, resulting in the production of corresponding polysaccharide-b-ELP copolymers **70**, **71**, and **72** (Scheme 9.13). This work offers new insights into the advancement of thermo-sensitive bioactive self-assemblies, giving rise to innovative approaches for constructing 'elegant' and biologically effective nanocarriers

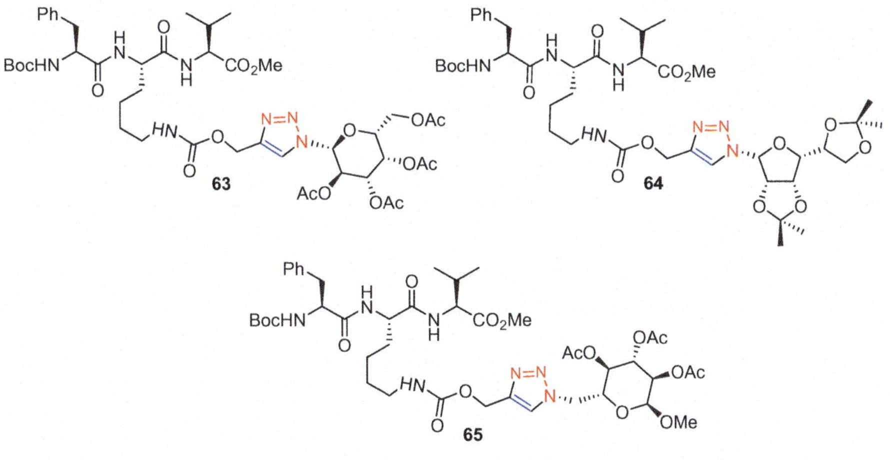

Fig. 9.4 Chemical structures of triazole-appended lysine-based tripeptides **63–65** conjugated with different sugars

through the fusion of bioactive polysaccharides with stimuli-responsive ELPs having applications in drug delivery, biomaterial, and receptor recognition.

Fascinatingly, a large number of deep-sea Teleost fish possess C-linked anti-freeze glycoproteins (AFGPs), which protect them from cryoinjury. Ben et al. reported a solid-phase synthesis of C-linked triazole-containing AFGP analogues using convergent click tool. The methodology involves coupling of protected galactosyl azides with resin bound peptides **73, 74** in presence of cupric acetate and NaASc under microwave irradiation in DMF. Subsequent cleavage of the Wang resin followed by deacetylation furnished the desired AFGP analogues **75–78** (Scheme 9.14) [45]. Additionally, using the splat cooling experiment, triazole-containing AFGP analogues **75–78** were examined for its ability to inhibit ice recrystallization activity (IRI). The assay results suggested that analogue **75** did not possess IRI activity, whereas other analogue exhibited moderate IRI activity. Conversely, the triazole is widely recognized for its ability to mimic the structure of an amide linkage; however, the amide bond was more important for the AFGP analogue effective IRI activity than the triazole. In addition, a hydrophobic spacer closer to sugar residue as well as the position of triazole linkage is very important for such activity.

The cyclopeptide-based glycocluster has received great attention due to its capacity for assessing the impact of diversity in sugar–lectin interactions. Renaudet and group explored the CuAAC reaction and demonstrated the synthesis of glyco-cyclopeptides comprising of α-D-mannopyranosyl, β-D-galactopyranosyl, and α-L-fucopyranosyl groups. In this order, the coupling of tetra azido-glycopeptides **79** with diverse sugar alkynes **80a–c** using Cu-micro powder as catalyst in $^{t-}$BuOH/ NH_4OAc for 3 h afforded the successful synthesis of the tetravalent glycocyclopeptides **81a–c** (Scheme 9.15) [46]. Furthermore, the potential of tetravalent glycopeptide **81a–c** containing triazole pharmacophores to inhibit *Ulex europaeus* (UEA-1)

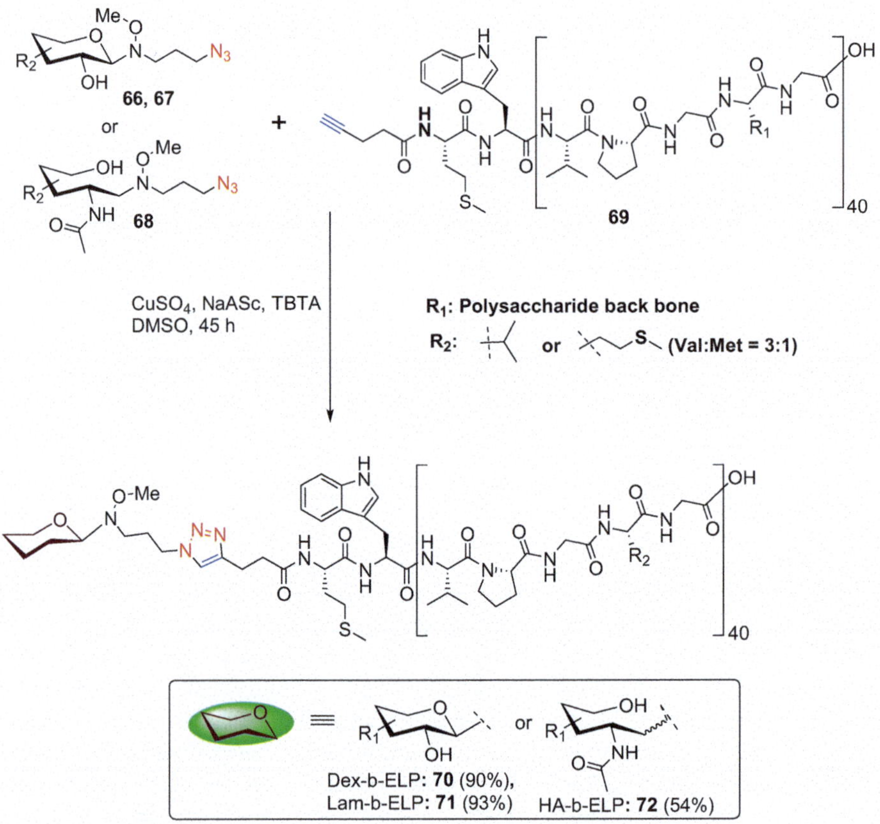

Scheme 9.13 Click-inspired synthesis of polysaccharide-block-ELP copolymers

and *Canavalia ensiformis* (ConA) was evaluated. Remarkably, in comparison with their oxime counterparts, the chemical exhibiting α-fucosylated glycoclusters showed less inhibition.

Michael addition of ammonia to glycosyl olefinic ester followed by diazotransfer reaction using imidazolyl sulphonyl azide can be done in one-pot fashion and the resulting glycosylated-β-azido ester was successfully coupled with terminal azide under CuAAC condition to give high yield of biologically relevant glycosylated-β-1,2,3-triazolo conjugates, [47] a mimic to biologically potent amide [8]. Similar in situ access of clickable azides was successfully explored through diazo transfer reactions [48]. Mancheno co-workers explored direct C(sp3)-Hazidation followed by CuAAC conjugation for the straightforward access of click-inspired triazole conjugates [49]. Thus, the direct C(sp3)-H azidation of unprotected tetrasaccharides **82** as reducing sugar furnish in situ corresponding azido glycoside which further underwent click conjugation with peptide **83** having the clickable propargyl glycine at side chain in one-pot fashion and produces the corresponding complex glycopeptide **84** with an improved way (Scheme 9.16).

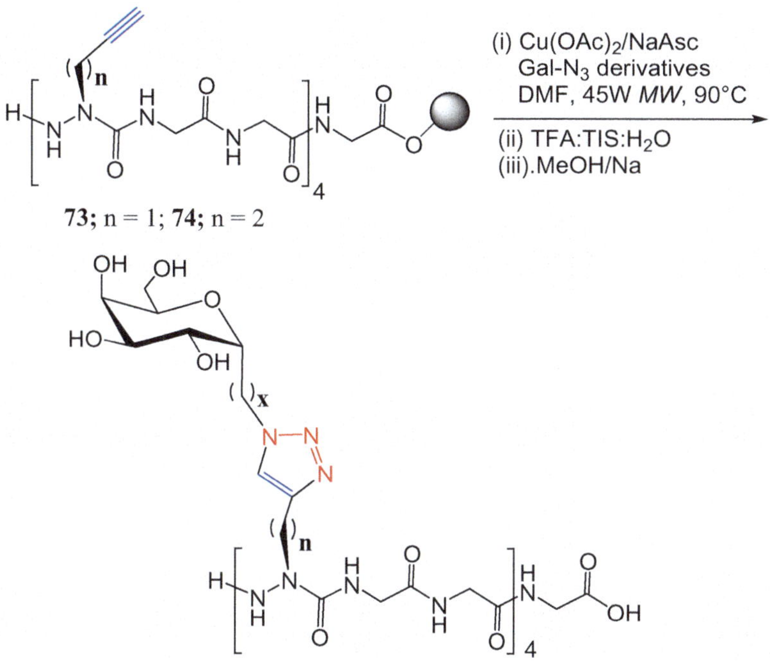

73; n = 1; **74**; n = 2

75 : x = 0; n = 1; **76** : x = 1; n = 1
77 : x = 2; n = 1; **78** : x = 1; n = 2

Scheme 9.14 Synthesis of C-linked triazole-comprising AFGP analogues using click methodology

In 2004, Rutjes and co-workers describe a highly productive method for synthesizing two varieties of triazole-appended glycopeptides using protected building blocks [50]. The development of stable glycopeptide mimics both α- and β-triazole linkages **87** involves coupling of mono/disaccharide azides **85** with mono/dipeptide alkynes **86** using $Cu(OAc)_2$/NaAsc as catalyst and H_2O/*tert*-butanol mixture at optimum temperature. Conversely, to provide another class of glycopeptides **90**, mono/disaccharide alkynes **88** were stitched to mono/dipeptide azide **89** using the identical reaction protocol (Scheme 9.17). Apart from that, a similar investigation was further exploited by the same group to develop a triazole-linked c(RGDy-^{N}TGA) scaffold comprised of glycosylated cyclic arginine–glycine–aspartate derivatives by adopting the same concept.

In this endeavor, a report was documented by Ghorai et al. in which they utilized a modular intramolecular CuAAC dimerization reaction of functionalized azido-alkynes (**95, 96**) and achieved an excellent synthesis of triazole-based peptidomimetic macrocycles (**97**, **98**) which belongs to the class of pseudo cyclo-β-peptides (Scheme 9.18) [51]. An azide derived from D-glucose **91** was taken as substrate which was first converted into the corresponding acyl azide **92** on reaction with DPPA/Et_3N followed by the treatment of the acyl azide with *N*-hydroxy

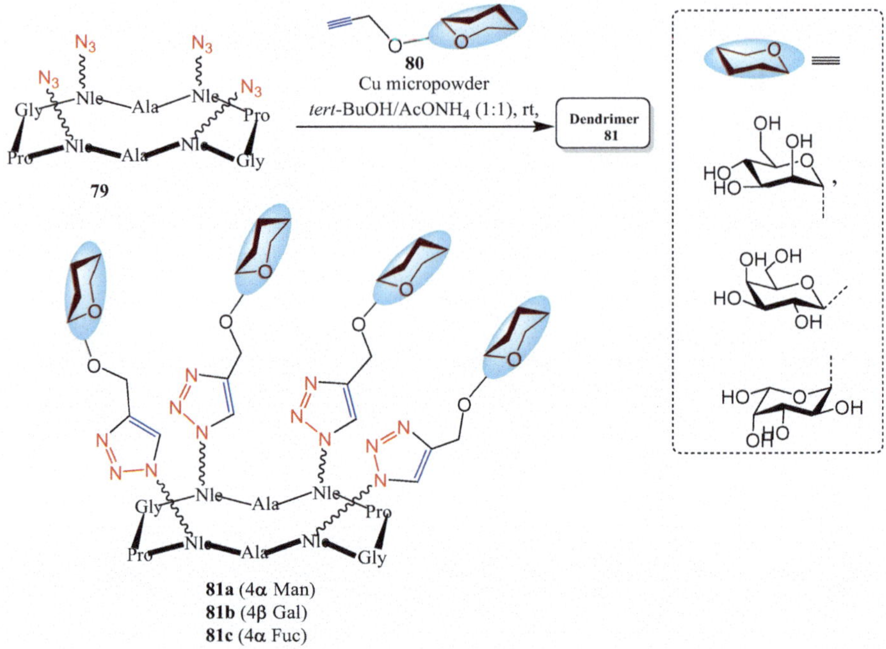

Scheme 9.15 Click chemistry-inspired synthesis of tetravalent glycocyclopeptides

succinimide which generates carbamate **93** via Curtius rearrangement. A standard coupling reaction of carbamate **93** with *N*-methylated amino alkyne **94** in the presence of DIPEA as base furnished the desired azido-alkyne **95** and **96** in good yields. Linear azido-alkynes **95** and **96** were subjected to intermolecular followed by intramolecular click macrocyclization reaction in the presence of CuI/DIPEA and *Tris*-(benzyltriazolylmethyl) amine (TBTA) as a catalyst and as a result, the triazolo-urea-based peptidomimetic macrocycles **97** and **98** were obtained as desired products. Also, it was observed that the triazole-based macrocycles **97** and **98** can self-assemble in a controlled chiral fashion and finally transform into a nanorod molecular structure (Scheme 9.18).

Interestingly, A peptidomimetic macrocycle comprising of triazole/amide **105** is synthesized by same group using a triazole/amide oligomer derived from a cis-furanoid sugar triazole amino acids. Despite its uniform backbone chirality, this macrocycle exhibits a conformation similar to the cyclic peptides based on D- and L-α-amino acids [52]. The authors emphasize the significance of triazole-di-*β*-peptide isostere **101** in their study, derived from the coupling of cis-furanoid azido derivative **99** with *N*-Cbz-protected cis-furanoid homopropargyl amine **100**. This coupling reaction achieved a high yield utilizing CuSO4/NaAsc along with TBTA in a mixture of t-BuOH and H_2O. Compound **101** possesses both ester and Cbz-protected amine functionality in two separate furanose fragments. Compound **101** separately underwent basic hydrolysis and Cbz deprotection to get free acid **102** and free amino

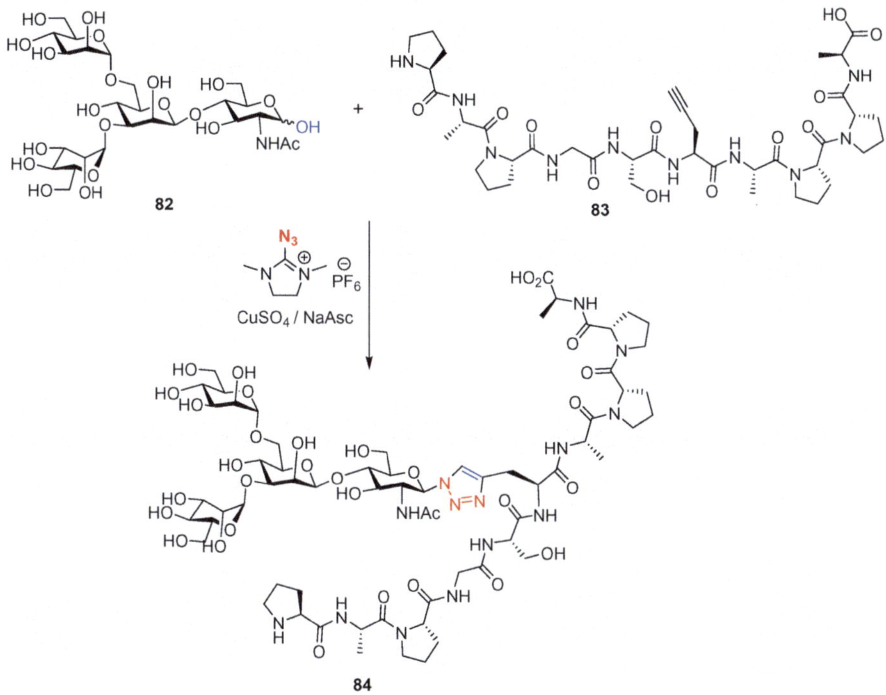

Scheme 9.16 Synthesis of glycopeptide **84** from fully unprotected tetrasaccharides **82** and peptide **83** through one-pot C(sp3)-H azidation and CuAAC click coupling techniques

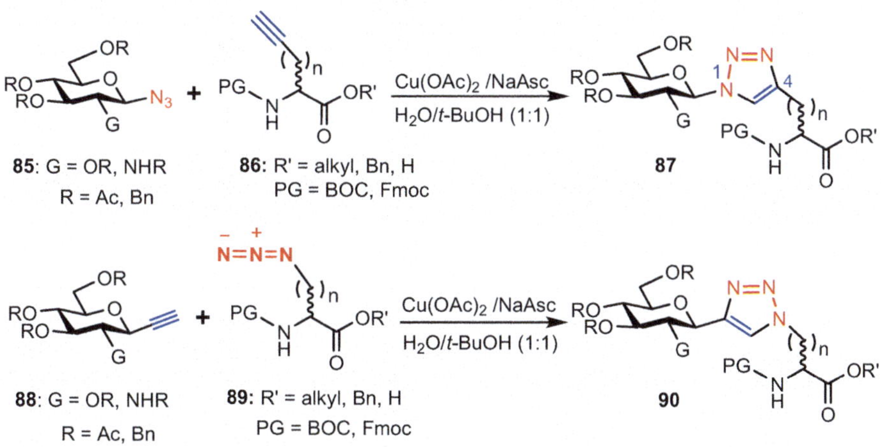

Scheme 9.17 Click-inspired synthesis of triazole-appended glycopeptides (**87**, **90**) by coupling of either sugar azide with peptide alkyne or vice versa

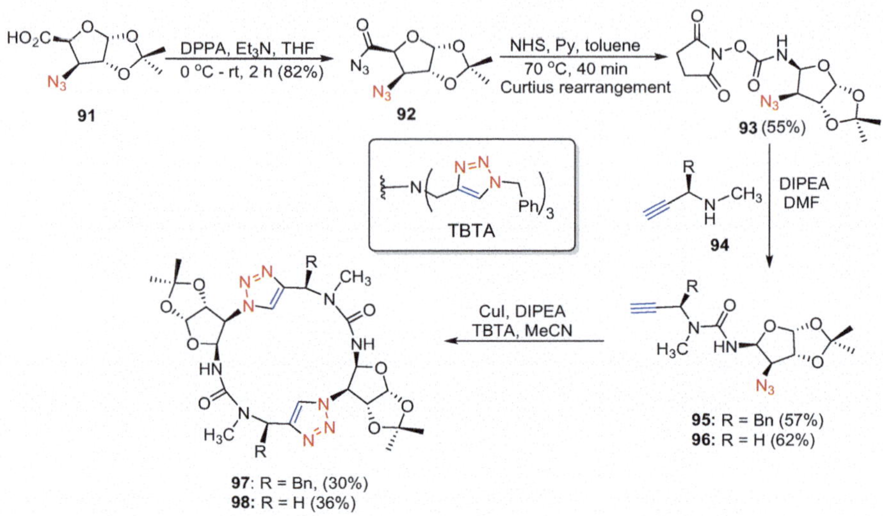

Scheme 9.18 Synthesis of triazole/urea-based peptidomimetic macrocyclic compounds **97** and **98** using click protocol

ester intermediate **103,** respectively (Scheme 9.19). Further acid, amine coupling of both the acid **102** and amine **103** intermediates using EDC/HOBt/DIPEA catalytic system to furnish tetramer **104** as a sole product. Initially, hydrolysis of the ester group in linear tetramer **104** was performed, succeeded by the removal of the Cbz protecting group. Subsequently, macrocyclization was achieved through intramolecular acid–amine coupling, facilitated by activating the acid group with pentafluorophenol. Finally, the desired amide-triazole macrocycle 105 was obtained through consequent hydrogenation (Scheme 9.19).

The resulting peptidomimetic macrocycle **105** underwent an extraordinary mode of self-assembly that included parallel stacking of an *H*-bond between amide-NH and carbonyl oxygen (C=O) and an anti-parallel hydrogen bond between amide-NH and N2/N3 of the triazole obtained through CuAAC. In this process, a tubular nanostructure representing two rotamers of the pseudo cyclic β-peptides was formed. Although the macrocycles including pseudo cyclic β-peptides are highly relevant, the ion-binding characteristics were still not studied.

A number of independent research groups, such as Wong, [53] Abell, [54] Horne, [55] Bock, [56] Dhavale, [57] Springer, [58] and James [59] were developed similar peptidomimetic-based macrocyclic systems and successfully explored as model illustrations in the area of artificial ion channels and other growing applications. Apart from that, hydrogen-bonded tubular assemblies based macrocyclic peptides have been nicely exploited [60].

Furthermore, intramolecular click macrocyclization was explored by Nilsson et al. for the efficient development of biologically relevant hybrid novel glycopeptide-containing macrocycles by attaching suitable amino acids with carbohydrates [61].

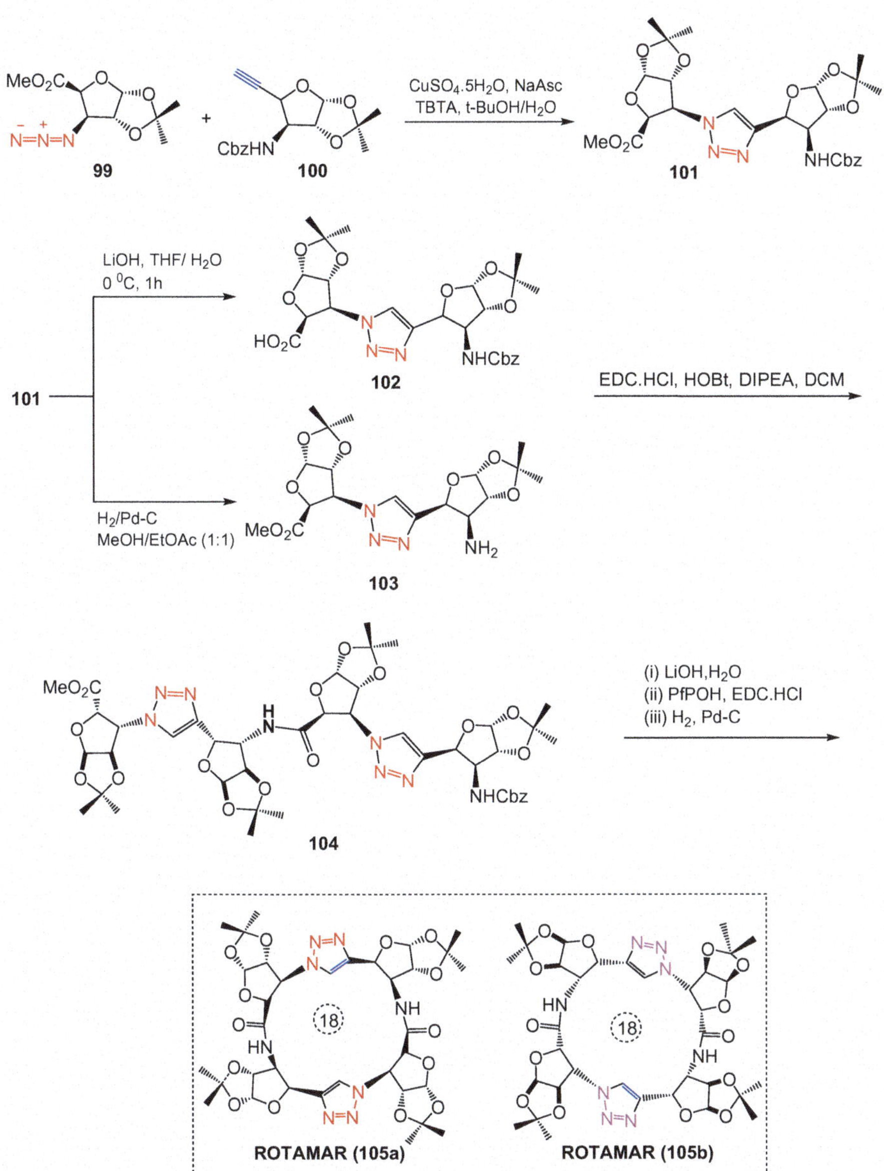

Scheme 9.19 Click chemistry-inspired synthesis of peptidomimetic macrocycles represents two rotamers of the pseudo cyclic β-peptides

Initially, dipeptides **106** and **107**were coupled independently with glycosyl azide **108** in the presence of DIC/HOBt in such a fashion to give rise to the glucosamine scaffolds **109** and **110** comprising of clickable azido group and terminal alkyne functionality. In the next step, intramolecular macrocyclization conditions (CuI/DIPEA at 45 °C in MeCN) were employed to cyclodimerize the compounds **109** and **110** to furnish macrocyclic products **111** and **112** having C_2 symmetry in excellent yields (Scheme 9.20). The developed macrocycles find application as artificial water receptors due to the possibility of variable conformations arising from enhanced tunable flexibility because of the 1,2,3-triazole component in the macrocyclic structure. The same group also applied the above-described methodology to efficiently synthesize a fluorescent macrocycle. This involved an acid–amine conjugation succeeded by two consecutive intramolecular CuAAC reactions.

Similarly, sugar-based macrocyclic systems were synthesized using CuAAC macrocyclization and used for stereochemical characteristics in biological systems, including complexation to metals [62, 63]. Likewise, Ziegler and group showcased a macrocyclization technique based on click chemistry, employing propargylated and homopropargylated 6-azido-6-deoxy-glycosides as effective substrates. This approach successfully produced macrocycles containing carbohydrates [64]. Likewise, Nillson and group contributed significantly toward construction of a fluorescent macrocycle derived from sugars, achieved by employing acid–amine conjugation succeeded by two sequential CuAAC reactions [65]. Thus, glycohybrid amide **115**

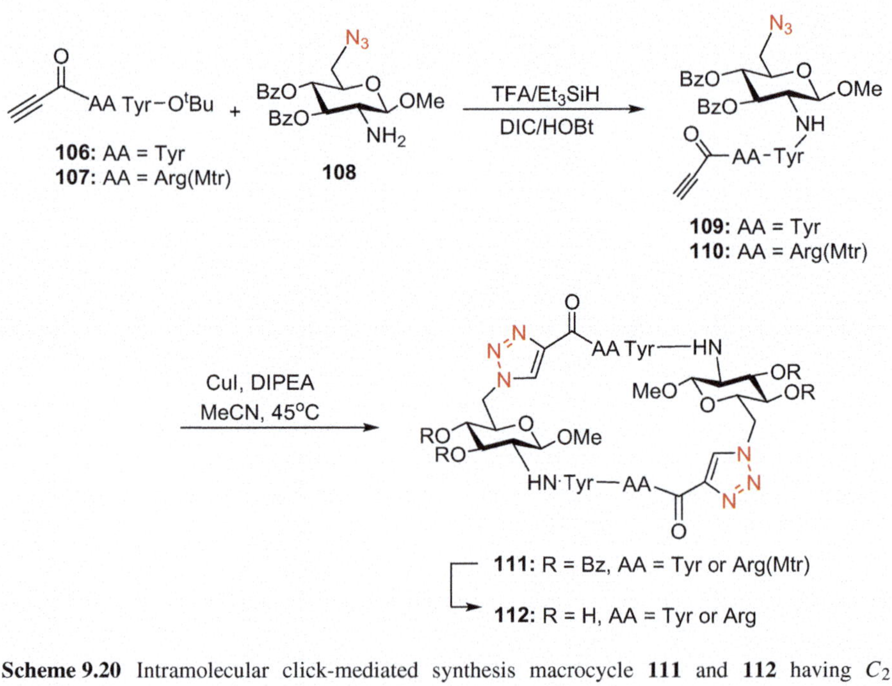

Scheme 9.20 Intramolecular click-mediated synthesis macrocycle **111** and **112** having C_2 symmetry

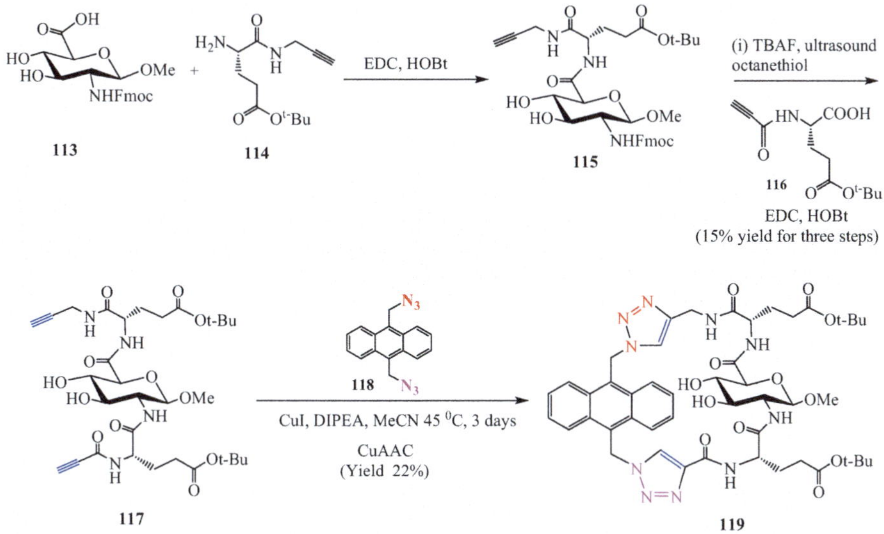

Scheme 9.21 CuAAC click-inspired synthesis of sugar-based fluorescent macrocycle **119**

was obtained from the acid/amine coupling of F-moc-protected sugar amino acid **113** with propargylated glutamic acid-derived amine **114** in the presence of EDC/HOBt (Scheme 9.21). Further, the resulting glycohybrid amide **115** underwent Fmoc removal, then its acid–amine coupling with *N*-propargylated glutamic acid derivative **116** using DIC/HOBt, resulting in the clickable glycohybrid **117**. Finally, a macrocyclization reaction inspired by CuAAC was performed by treating *bis*-propargyl analogue **117** with anthracene diazide **118** in the presence of CuI/DIPEA in CH_3CN solvent at 45 °C for 3 days in extremely diluted condition (0.22 mM) to furnish the 22% yield of the sugar-comprising fluorescent macrocycle **119** (Scheme 9.21).

Modular CuAAC click tool was effectively used to construct well-defined and precise glycoclusters from several other established important macrocyclic systems, such as porphyrins, [66, 67] cyclopeptides, [68] and cyclic peptoids [69]. (Scheme 9.22). CuAAC click chemistry tool was effectively employed for synthesis of new class of cyclopeptide-based oligomannose hetero clusters **121** from corresponding alkyne **120**. Moreover, the click conjugation alkyne cyclic peptide **120** with oligomannose (tetrasaccharide) clusters in presence of $CuSO_4$/NaAsc in $^{t-}$BuOH–H_2O system produces the desire template-assembled oligomannose cluster **121** (Scheme 9.22). The developed glycocluster displayed significant binding affinity to human antibody 2G12 using surface plasmon resonance (SPR) technology [70]. Therefore, the multivalency mechanism demonstrated that the recognition of carbohydrates is not limited to lectins only. It may also stimulate the action of carbohydrate-directed antibodies against HIV-1.

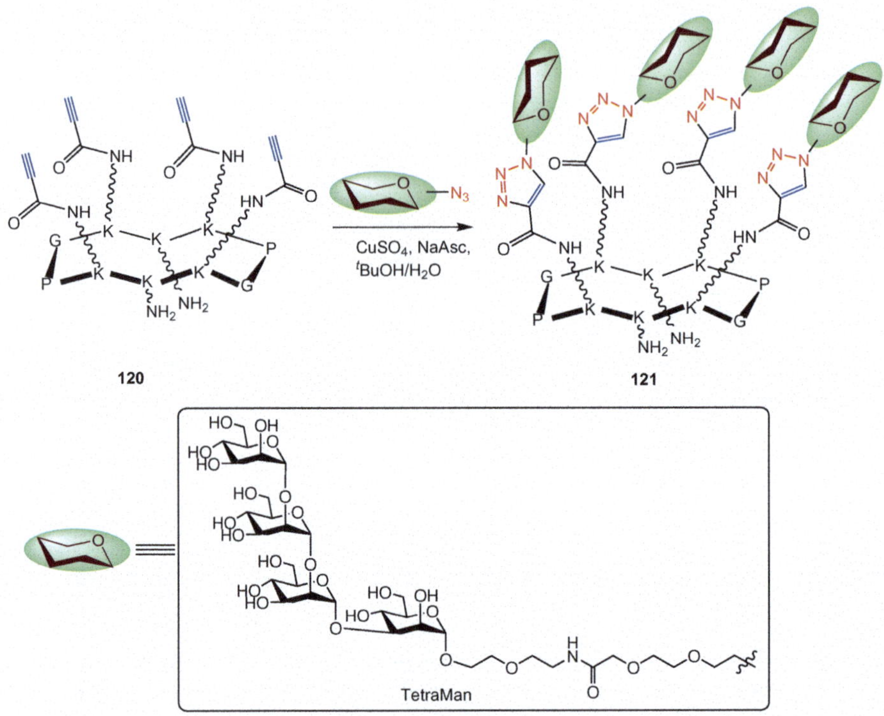

Scheme 9.22 Click-inspired synthesis of template-assembled oligosaccharide clusters **121**

9.3 Conclusions and Future Outlook

The CuAAC click chemistry is a tremendously adaptable reaction because of its mild reaction conditions, compatibility with functional groups, and ability to react with reaction media like water. Additionally, the procedure has been effectively used to conjugate fluorescently tagged proteins and peptides, polymers, saccharides, and radiolabels arranged covalently and side-specifically. The reaction can take place in organic solvents, water, or even physiological conditions, depending on the solubility of the substrates. Apart from these advantages, the CuAAC reaction has become an important conjugation reaction in the chemical biology domain. The traces of copper catalyst are sometimes difficult to remove and thus causes adverse effect on in vivo activity due to toxicity of metal ion. In addition to conjugating peptides, the CuAAC click reaction has been effectively used to change the peptide backbone by utilizing the functional similarities between amide and triazole groups [71]. The amide functional group is substantially replaced by the 1,2,3-triazole ring, which results in peptides that are more stable metabolically. However, due to slight difference in conformation, the characteristics of amino acids will also change. We have discussed the click reaction's versatility in this chapter and its past achievements in chemical modification of peptides and proteins. Due to vast variety of applications,

CuAAC 'click' reaction is expected to remain the most effective in future years, convenient and reliable synthetic techniques for creating materials functionalized with peptides that imitate the cooperative control of biological processes in living organisms.

References

1. Sharma A, Agrahari AK, Rajkhowa S, Tiwari VK (2022) Eur J Med Chem 238:114454
2. Lauria A, Delisi R, Mingoia F, Terenzi A, Martorana A, Barone G, Almerico AM (2014) Eur J Org Chem, pp 3289–3306
3. Lau YH, Rutledge PJ, Watkinson M, Todd MH (2011) Chem Soc Rev 40:2848–2866
4. Thirumurugan P, Matosiuk D, Jozwiak K (2013) Chem Rev 113:4905–4979
5. Kolb HC, Finn MG, Sharpless KB (2001) Angew Chem Int Ed 113:2056–2075
6. Tiwari VK, Mishra BB, Mishra KB, Mishra N, Singh AS, Chen X (2016) Chem Rev 116:3086–3240
7. Agrahari AK, Bose P, Jaiswal MK, Rajkhowa S, Singh AS, Hotha S, Mishra N, Tiwari VK (2021) Chem Rev 121:7638–7956
8. Tiwari VK (2021) Chem Rec 21:3029–3048
9. Jaiswal MK, Tiwari VK (2023) Chem Rec, e202300167
10. Kushwaha D, Dwivedi P, Kuanar SK, Tiwari VK (2013) Curr Org Synth 10:90–135
11. Mishra A, Agrahari AK, Tiwari VK (2021) Application of CuAAC in carbohydrates. In: Click chemistry volume, science of synthesis, pp 135–179
12. Zhang L, Chen XG, Xue P, Sun HHY, Williams ID, Sharpless KB, Fokin VV, Jia GC (2005) J Am Chem Soc 127:15998–15999
13. Jaiswal MK, Gupta A, Ansari FJ, Pandey VK, Tiwari VK (2024) Curr Org Synth 21:513–558
14. Mishra N, Singh SK, Singh AS, Agrahari AK, Tiwari VK (2021) J Org Chem 86:17884–17895
15. Singh SK, Kumar S, Yadav MS, Tiwari VK (2022) J Org Chem 88:15389–15402
16. Singh M, Singh AS, Mishra N, Agrahari AK, Tiwari VK (2019) ACS Omega 4:2418–2424
17. Mishra N, Singh AS, Agrahari AK, Singh SK, Singh M, Tiwari VK (2019) ACS Comb Sci 21:389–399
18. Lowik DWPM, Leunissen EHP, van den Heuvel M, Hansen MB, van Hest JCM (2010) Chem Soc Rev 39:3394–3412
19. Altunbas A, Pochan DJ (2012) Top Curr Chem 310:135–167
20. Tanga W, Becker ML (2014) Chem Soc Rev 43:7013–7039
21. Pasini D, Ricci M (2007) Curr Org Synth 4:59–80
22. Chandrasekhar S, Rao CL, Nagesh C, Reddya CR, Sridhar B (2007) Tetrahedron Lett 48:5869–5872
23. Ghorai A, Achari B, Chattopadhyay P (2016) Tetrahedron 72:3379–3387
24. Hackenberger CPR, Dawson PE, Chen Y-X, Hojo H (2020) J Org Chem 85:1328–1330
25. Davies BJ (2002) Chem Rev 102:579–601
26. Huo J, Hu H, Zhang M, Hu X, Chen M, Chen D, Liu J, Xiao G, Wang Y, Wen Z (2017) RSC Adv 7:2281–2287
27. Arosio D, Btoli M, Manzoni L, Scolastico C (2006) Tetrahedron Lett 47:3697–3700
28. Kuijpers BHM, Groothuys S, Hawner C, Dam J, Quaedflieg PJLM, Schoemaker HE, van Delft FL, Rutjes FPJT (2008) Org Proc Res Devlop 12:503–511
29. Groothuys S, Kuijpers BHM, Quaedflieg PJLM, Roelen HCPF, Wiertz RW, Blaauw RH, Delft FLV, Rutjes FPJT (2006) Synthesis 18:3146–3152
30. Lee DJ, Mandal K, Harris PWR, Brimble MA, Kent SBH (2009) Org Lett 11:5270–5273
31. Macmillan D, Blanc J (2006) Org Biomol Chem 4:2847–2850
32. Lin H, Walsh CT (2004) J Am Chem Soc 126:13998–14003

33. Wessjohann LA, Ruijter E, Garcia-Rivera D, Brandt W (2005) Mol Divers 9:171–186
34. Fu X, Albermann C, Jiang J, Liao J, Zhang C, Thorson JS (2003) Nat Biotechnol 21:1467–1469
35. Aioub AG, Dahora L, Gamble K, Finn MG (2017) Bioconjugate Chem 28:1693–1701
36. Kuijpers BHM, Groothuys S, Soede AC, Laverman P, Boerman OC, van Delft FL, Rutjes FPJT (2007) Bioconjugate Chem 18:1847–1854
37. Chun CKY, Payne RJ (2009) Aust J Chem 62:1339–1343
38. Niu TF, Gu L, Wang L, Yi WB, Cai C (2012) Eur J Org Chem, pp 6767–6776
39. Lamandé-Langle S, Collet C, Hensienne R, Vala C, Chrétien F, Chapleur Y, Mohamadi A, Lacolley P, Regnault V (2014) Bioorg Med Chem 22:6672–6683
40. Miller N, Williams GM, Brimble MA (2009) Org Lett 11:2409–2412
41. Lee DJ, Yang SH, Williams GM, Brimble MA (2012) J Org Chem 77:7564–7571
42. Chandrasekaran S, Ramapanicker R (2016) Chem Rec 17:63–70
43. Chakrabarty SP, Ramapanicker R, Mishra R, Chandrasekaran S, Balaram H (2009) Bioorg Med Chem 17:8060–8072
44. Xiao Y, Chinoy ZS, Pecastaings G, Bathany K, Garanger E (2020) Biomacromol 21:114–125
45. Capicciotti CJ, Trant JF, Lectere M, Ben RN (2011) Bioconjugate Chem 22:605–616
46. Bossu I, Berthet N, Dumy P, Renaudet O (2011) J Carbohydr Chem 30:458–468
47. Mishra A, Tiwari VK (2015) J Org Chem 80:4869–4881
48. Meng G, Guo T, Ma T, Zhang J, Shen Y, Sharpless KB, Dong J (2019) Nature 574:86–91
49. Brandhofer T, Ozdemir A, Gini A, Mancheno OG (2019) Chem Eur J 25:4077–4086
50. Kuijpers BHM, Groothuys S, Keereweer AR, Quaedflieg PJLM, Blaauw RH, van Delft FL, Rutjes FPJT (2004) Org Lett 6:3123–3126
51. Ghorai A, Reddy SK, Achari B, Chattopadhyay P (2014) Org Lett 16:3196–3199
52. Ghorai A, Gayen A, Kulsi G, Padmanaban E, Laskar A, Achari B, Mukhopadhyay C, Chattopadhyay P (2011) Org Lett 13:5512–5515
53. Brik A, Alexandratos J, Lin YC, Elder JH, Olson AJ, Wlodawer A, Goodsell DS, Wong CH (2005) Chem Bio Chem 6:1167–1169
54. Pehere AD, Sumby CJ, Abell AD (2013) Org Biomol Chem 11:425–429
55. Horne WS, Olsen CA, Beierle JM, Montero A, Ghadiri MR (2009) Angew Chem Int Ed 48:4718–4724
56. Bock VD, Speijer D, Hiemstra H, Maarseveen JHV (2007) Org Biomol Chem 5:971–975
57. Burade SS, Saha T, Bhuma N, Kumbhar N, Kotmale A, Rajamohanan PR, Gonnade RG, Talukdar P, Dhavale DD (2017) Org Lett 19:5948–5951
58. Springer J, de Cuba KR, Calvet-Vitale S, Geenevasen JAJ, Hermkens PH, Hiemstra H, van Maarseveen JH (2008) Eur J Org Chem 2008:2592–2600
59. Chouhan G, James K (2013) Org Lett 15:1206–1209
60. Ulijn RV, Smith AM (2008) Chem Soc Rev 37:664–675
61. Billing JF, Nilsson UJJ (2005) Org Chem 70:4847–4850
62. Ajay A, Sharma S, Gupt MP, Bajpai V, Kumar HB, Kaushik MP, Konwar R, Ampapathi RS, Tripathi RP (2012) Org Lett 14:4306–4309
63. Potopnyk MA, Jarosz S (2013) Sweet sucrose macrocycles *via* a click chemistry route. In: Witczak ZJ, Bielski R (eds) Click chemistry in glycoscience, new developments and strategies. John Wiley and Sons, pp 235–252
64. Pietrzik N, Schmollinger D, Ziegler T (2008) Beil J Org Chem 4:30. https://doi.org/10.3762/bjoc.4.30
65. Argen JKM, Grundberg HE, Billing F, Nilsson UJ (2006) Synthesis 18:3141–3145
66. Locos OB, Heindl CC, Corral A, Senge MO, Scanlan EM (2010) Eur J Org Chem 6:1026–1028
67. Garcia G, Naud-Martin D, Carrez D, Croisy A, Maillard P (2011) Tetrahedron 67:4924–4932
68. Chen YX, Zhao L, Huang ZP, Zhao YF, Li YM (2009) Bioorg Med Chem Lett 19:3775–3778
69. Cecioni S, Faure S, Darbost U, Bonnamour I, Parrot-Lopez H, Roy O, Taillefumier C, Wimmerová M, Praly JP, Imberty A et al (2011) Eur J Org Chem 17:2146–2159
70. Wang J, Li H, Zou G, Wang XL (2007) Org Biomol Chem 5:1529–1540
71. Yule LR, Bower RL, Kaur H, Kowalczyk R, Haya DL, Brimble MA (2016) Org Biomol Chem 14:5238–5245

Chapter 10
Click Chemistry in Dendrimer Synthesis

10.1 Introduction

The term 'dendrimer' was first coined by D. A. Tomalia, which made up of two Greek words: 'dendros' (which translates to 'tree' and is associated with their shape) and 'meros' (which translates to 'part' and is related to their chemical composition) [1]. A group of macromolecules known as dendrimers are highly tailored over the core and have a very low polydispersity index. In addition to their interesting architectures, dendrimers have greatly utilized in broad array of protein mimics, nanomedicine, sensors, light-harvesting devices, and catalytic systems [2–4]. In recent years, it has become noticeable that carbohydrate-protein interactions, which are critical molecular recognition processes in cell biology, are mediated by a lipid-like protein [5, 6]. Dendrimers classification is on the basis of their linkages to form the enormous architecture of different generations [7]. Typically construction of dendrimers involve three different building blocks, i.e., (i) core unit, (ii) interior multiple branching layers, and (iii) peripheral groups (sugar or hetero group). The synthesis of dendrimers has followed usually two routes, namely (i) divergent approach and (ii) convergent approach. A typical presentation of both the synthetic routes is outlined in Fig. 10.1.

'Click Chemistry' concept was introduced by Prof. K B Sharpless in 2001 [8]. Soon after in 2002, Prof. K B Sharpless and Prof. Morten Meldal individually established the concept of Cu(I)-catalyzed azide-alkyne cycloaddition (CuAAC), a most popular reaction known so far. This regioselective cycloaddition technique enables the construction of highly specific, reliable, and biocompatible 1,2,3-triazoles at room temperature (Scheme 10.1) [9, 10]. Cu(I)-catalyzed CuAAC which is quick, regioselective, simple to purify, and produces high yields of triazoles discovered one of the most popular and reliable tools studied [11–13]. Meldal studies revealed that click reaction is an efficient tool to couple azides and alkynes in a 1,4-disubstituted manner and their application in broad array of science [14]. The resulted 1,2,3-triazole motif readily produced in high to excellent yields. This orthogonal azide-alkyne coupling

V. K. Tiwari et al., *Click Chemistry*, Materials Horizons: From Nature to Nanomaterials,
https://doi.org/10.1007/978-981-97-4596-8_10

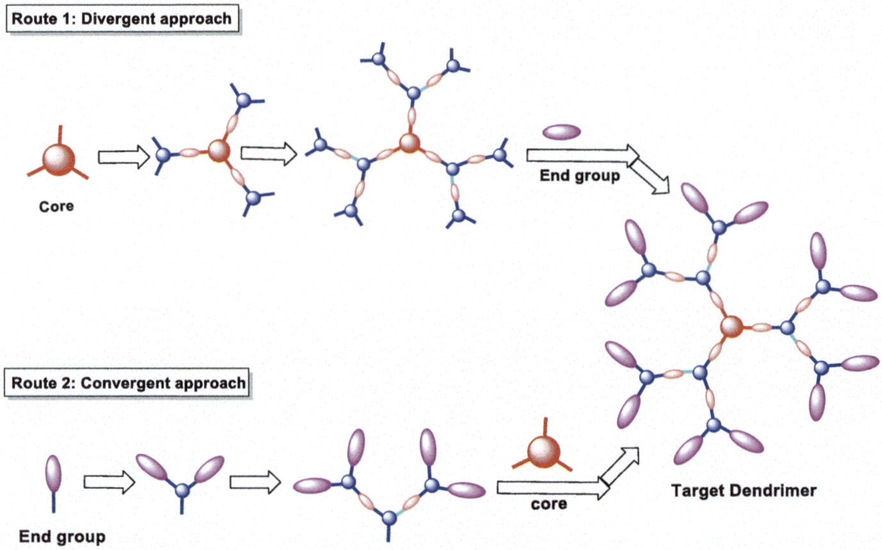

Fig. 10.1 Two convenient approaches for the synthesis of dendrimers

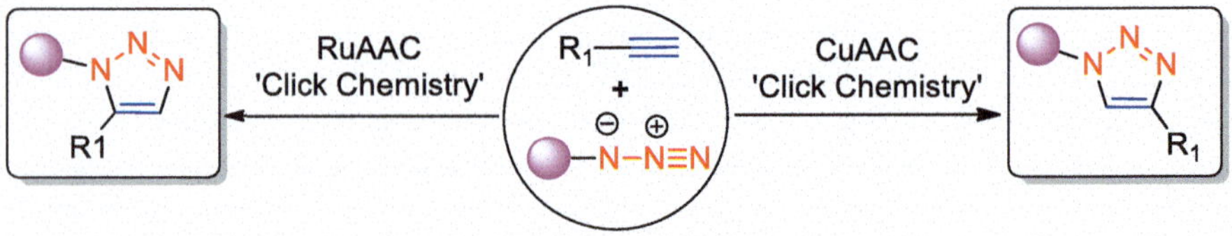

Scheme 10.1 Regioselective synthetic approach for the construction of 1,2,3-triazoles

is well-discovered to construct diverse clusters, polymer, peptides, dendrimers, etc. [15–19].

10.2 Classification of Dendrimers on the Basis of Their Linkages

Dendrimers are classified on the basis of their linkages, that is ether, amide, ester, etc., namely PPI, BOLTRON, PAMAM, polyether, polyester, and polysiloxane dendrimers of different generations as depicted in Fig. 10.2 [7].

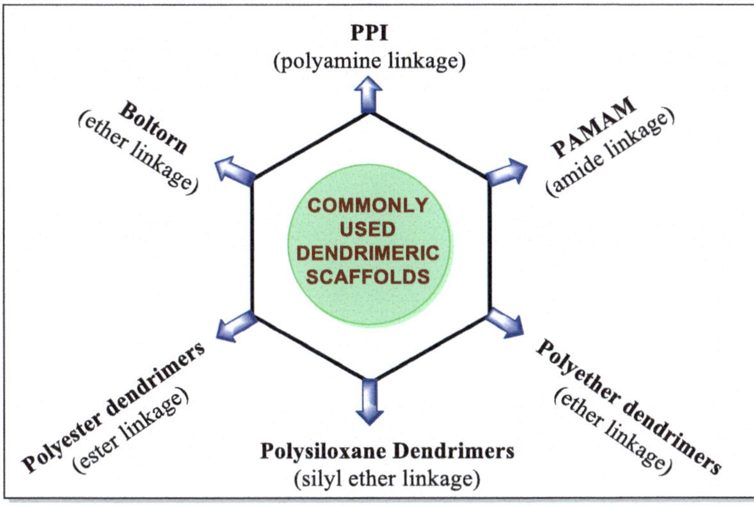

Fig. 10.2 Commonly used dendrimers on the basis of linkages

10.2.1 Classification on the Basis of Occurrence of Sugar Residue

Basic structure of glycodendrimers consists of three components: (i) end group (ii) branching layer, and (iii) core unit. The term 'glycodendrimers' is used to describe dendrimers that are attached to carbohydrates known as glycodendrimers evidence from the literature. It can be categorized as carbohydrate-coated, carbohydrate-based, and carbohydrate-cantered glycodendrimers depending on the presence of sugar units (Fig. 10.3).

10.2.2 Modern Method to Construct Dendritic Architectures

Modern methods for synthesis of dendrimer can be classified typically in two categories:

(i) double-stage convergent method and
(ii) onion peel strategy.

10.2.2.1 Double-Stage Convergent Method

This is most common and frequently used method for synthesis dendrimer molecules. Both approaches, i.e., convergent and the divergent are involved in this tactic of synthesis of dendrimer. The convergent approach involves in construction of dendron

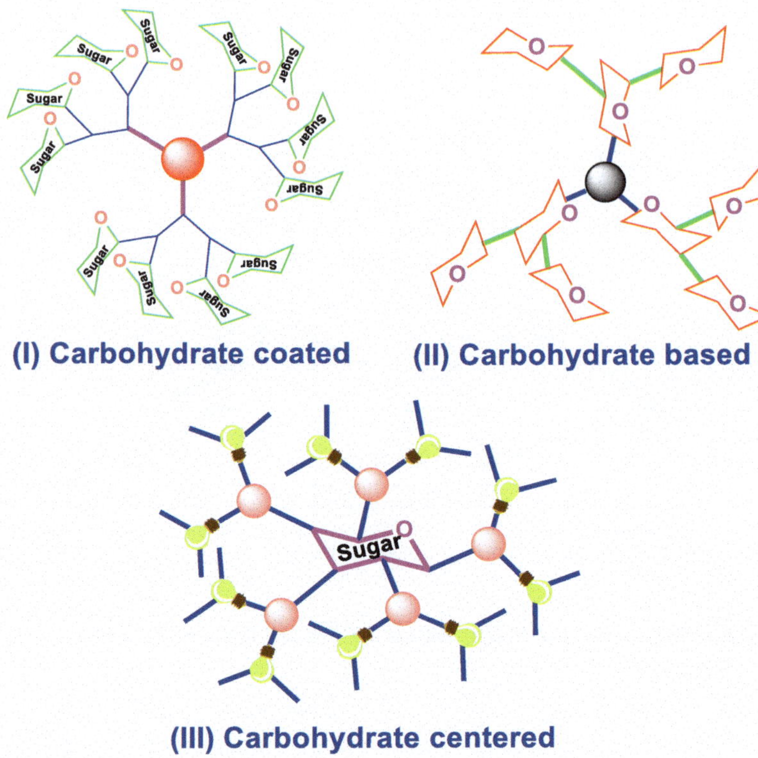

Fig. 10.3 Classification of glycodendrimers on the basis of sugar presence

molecule while, the hypercore units are produced through divergent approach separately. Finally, click-inspired coupling of both dendron moiety with hypercore afforded a low-generation multifunctional dendrimers (Fig. 10.4) [20, 21].

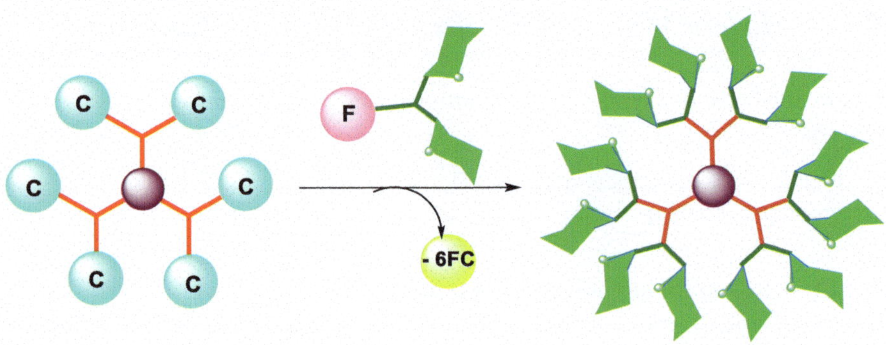

Fig. 10.4 Double-stage convergent technique (F = functional group, C = coupling groups)

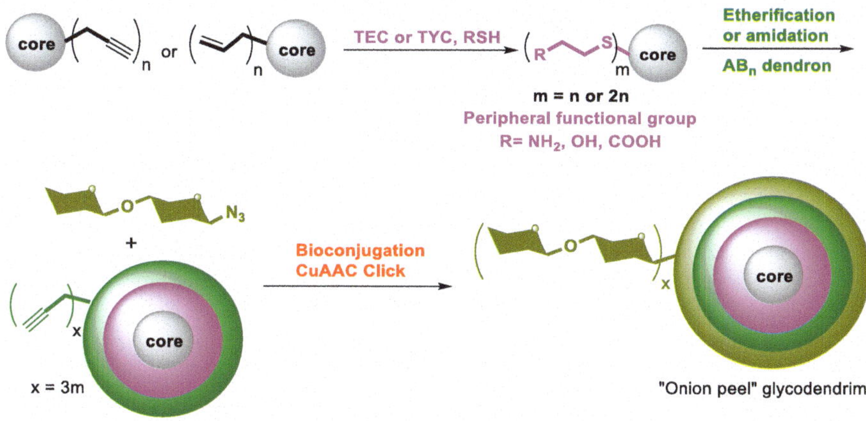

Fig. 10.5 Synthesis of sugar-coated 'onion peel' dendritic structure via divergent method

10.2.2.2 Onion Peel Strategy

A novel approach for creating functional glycodendrimers that involves assimilating various families of building blocks with orthogonal functionality at each generation. At each stage of dendritic growth, the advantage of using such a technique may result in the targeted hydrophilicity/hydrophobicity and flexibility/rigidity. The onion peel method offers an alternative to the established methodologies for the quick and efficient synthesis of dendritic scaffolds. The representative process for creating glycosylated structures is described in (Fig. 10.5) [22].

10.3 CuAAC Click-Mediated Construction of Diverse Dendrimer and Glycodendrimer Architectures

A group of macromolecules known as dendrimers are highly tailored over the core and have a very low polydispersity index. In addition to their architectures, dendrimers have greatly applied in broad array of protein mimics, nanomedicine, sensors, light-harvesting devices, and catalytic arrangements [2–4]. In recent years, it has become obvious that carbohydrate-protein interactions, which are critical molecular recognition procedures in cell biology, are facilitated by a lipid-like protein. Besides, Du et al. made significant effort to develop a C-3 symmetric (1–6)-*N*-acetyl-*β*-D-glucosamine octadecasaccharide **1** via convergent synthesis approach. Thus, the authors utilized Cu(I)-catalyzed azide-alkyne cycloaddition (CuAAC) reaction to amalgamate hexasaccharide-based clickable azide with alkyne functionalized 1,3,5-tri-*o*-propargylated benzene core in order to produce the desired C-3 symmetric (1–6)-*N*-acetyl-*β*-D-glucosamine octadecasaccharide in good yield (Fig. 10.6) [23].

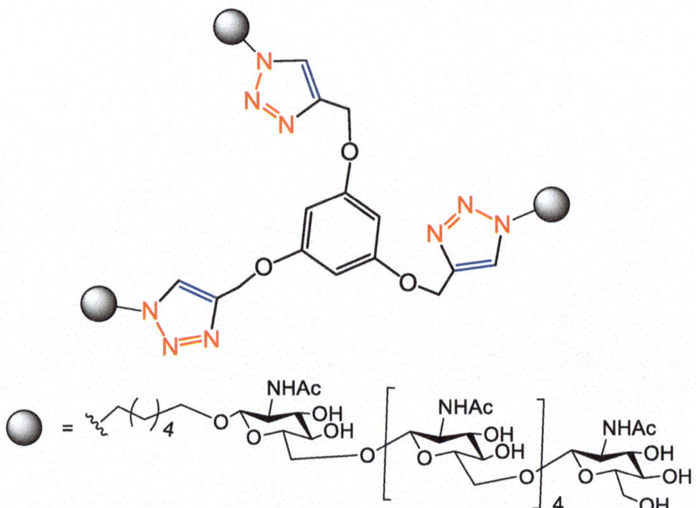

1

Fig. 10.6 Structure of *C*-3 symmetric octadecasaccharide glycocluster **1**

Gouin et al. in 2007 aimed to synthesize of carbohydrate-cored multimannosides **4** and **5** using a CuAAC click reaction. The alkyne precursor **2** and azide precursor **3** were developed in number of steps followed by protection and deprotection strategies. The azide scaffolds **3** were clicked with mannosyl alkynes **2** regioselectively under copper catalysis to furnish desired multivalent mannoside-based glycocluster anomers **4** and **5**, owing important cluster effect (Scheme 10.2) [24].

Further, design of triazolyl connections with the cyclodextrin core was made by 'CuAAC Click Chemistry,' as well as exocyclic triazolyl link to cyclodextrin, it is evident from the literature, the biomedical industry has seen numerous applications for the click-tailored cyclodextrins [25]. Cyclodextrin cored dendrimers have a possible use in drug delivery systems due to their ability to encapsulate substances, particularly nucleotides. Sugar-tethered oligorotaxanes with 1,2,3-triazole fraction at the periphery of the cyclodextrin core in addition to flexible threading ratios were additionally established via click conjugation of α-CD/decane pseudo rotaxanes under typical catalytic system [26]. Besides, using the same principle of click chemistry, *β*-CD-cored glycodendrimers were formed. For example a series of *β*-CD glycodendrimers **8a–8e** and **9a–9e** were obtained by click with *β*-CD resulting azide **6a** and **6b** with numerous appropriate dendron **7** such as alkyne having Boc-protection (Scheme 10.3) [27]. Under typical reaction conditions, both the protective functionality (for instance, *N*-Boc- and -OAc) from glycodendrimers **8a–8e** was successfully removed.

In year 2011, Compain et al. by applying the similar synthetic protocol, AB_3 type dendron **10** was conjugated with heptakis (2,3-di-*O*-methyl-6-*O*-propargyl) cyclomaltoheptaose **11** by using the typical click reaction condition under microwave irradiation to synthesized 21-valent iminosugar **12a** in excellent yield. Further, the complex

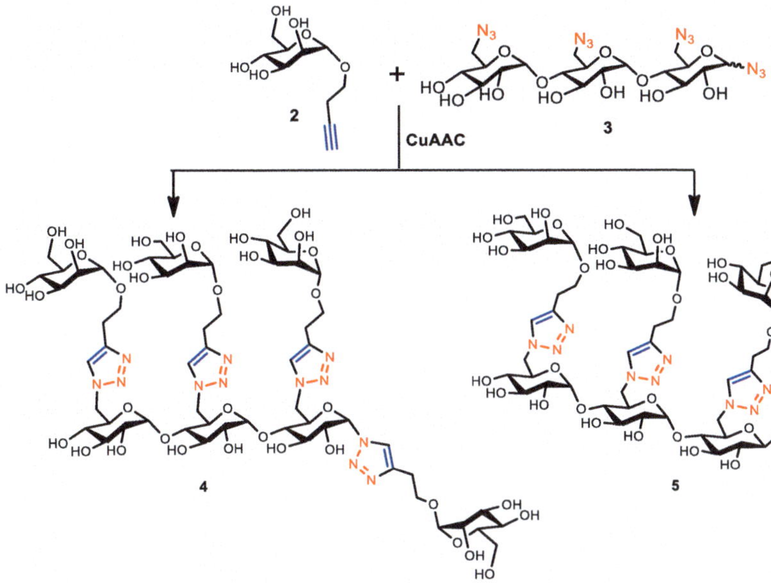

Scheme 10.2 Structure of click-inspired mannoside-based glycoclusters **4** and **5**

12a that was deacetylated utilizes amberlite IRA 400 ($^{-}$OH) anion exchange resin in CH_3OH/H_2O with ratio 1:1 to afford the targeted 21 peripheral amino sugar scaffold **12b** (Scheme 10.4) [28]. The biological examination of polyvalent iminosugar **12b** gave interesting results.

Zhang et al. report the synthesis of triazole-appended glycocluster via click approach. In order to produce the per-substituted *β*-CD-(Man)$_7$ glycocluster **14**, the Cu-catalyzed cycloaddition of CD-based azide **6b** and alkyne-functionalized mannose **13** leads the formation of desired glycodendrimer in assistance of CuBr/ bpycatalyst in DMSO as the solvent (Scheme 10.5) [29].

Fullerene is an extremely fascinating chemical from both a synthetic and biological point of view due to its electrical and structural characteristics. Compain et al. reported a methodology, in which employing a fullerene core decorated with a *N*-alkylated residue at *N*-1-deoxynojirimycin at the periphery to achieve multivalency on a variety of glycosidases [30]. Authors imagined that these scaffolds would be crucial for assessing the effect of multivalency on the regulation of inhibitory selectivity for various types of glycosidases in addition to inhibiting glycosidases. The desilylation of complex **15** in situ with fluoride ion source, TBAF provide the respective hexa adduct having twelve alkynes at periphery which on further CuAAC reaction with benzyl-protected iminosugar **16** and iminosugar **17**, in the presence of $CuSO_4.5H_2O$/NaAsc resulted to assemble glycodendrimers **18** and **19** in 78 and 83% yield, respectively (Scheme 10.6).

Benzotriazole and their analogues are an important tool in organic synthesis as precursor along with numerous biological properties [31]. Regarding this, Tiwari

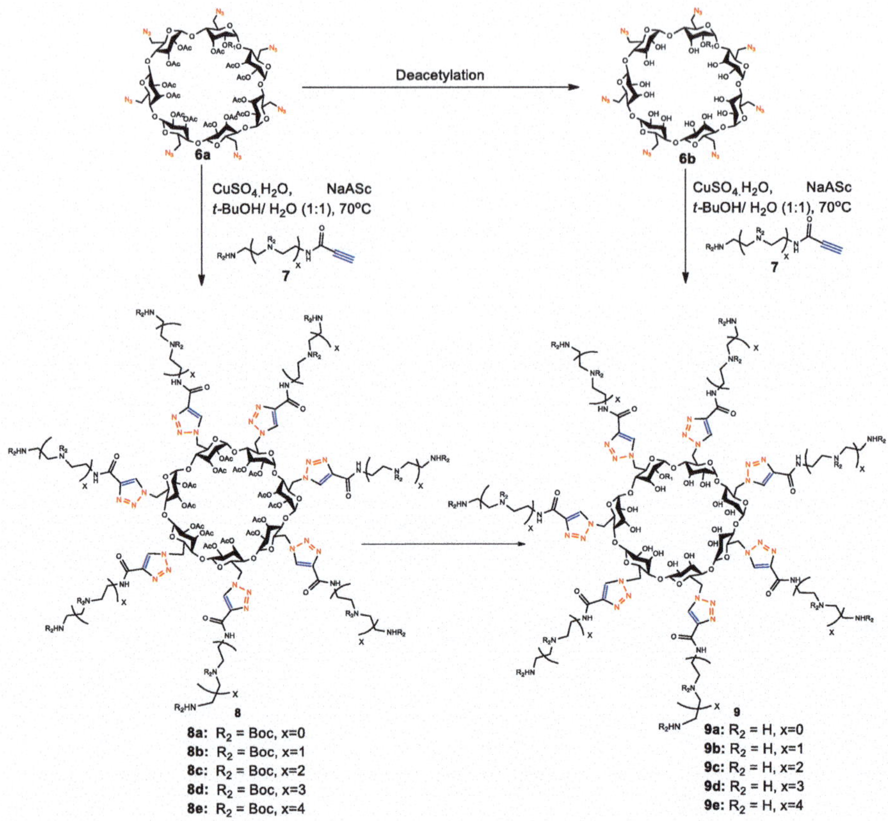

Scheme 10.3 Synthesis of β-CD-cored glycodendrimers **8** and **9** using click chemistry

et al. developed p-tert-butylcalix[4]arene-tethered benzotriazolyl dendrimers by using CuAAC click reaction. In these dendrimers include *N*-1, *N*-2-type azide-functionalized two-fold compounds **21** and **22**, was click with calix[4]arene core **20** to produce 'G(1)' generation dendrimers **23** and **24,** respectively. On the other hand, when the similar reaction was performed separately with *N*-1, *N*-2 benzotriazole-appended azide **25** and **26,** it afforded high yield of desired 'G(0)' generation dendrimer **27** and **28,** respectively (Scheme 10.7) [32]. The therapeutic potential of these dendrimers has also been evaluated, both in vitro and in vivo. These synthesized calix[4]arene-tethered benzotriazolyl dendrimers were characterized by typical spectroscopic analysis including NMR (1H and 13C), MS, and SEC data. Synthesized benzotriazolyl dendrimers compound **28** has been found to be the most effective. It has the ability to prevent bacteria and biofilm formation in drug-resistant and slime-producing organisms without causing cytotoxicity to eukaryotic systems.

Tiwari et al., demonstrated and explored the Cu-catalyzed click strategy for the synthesis of calix[4]arene-cored glycodendrimers separately, which contain galactose, mannose, and lactose units at the periphery. In the synthesis part the author

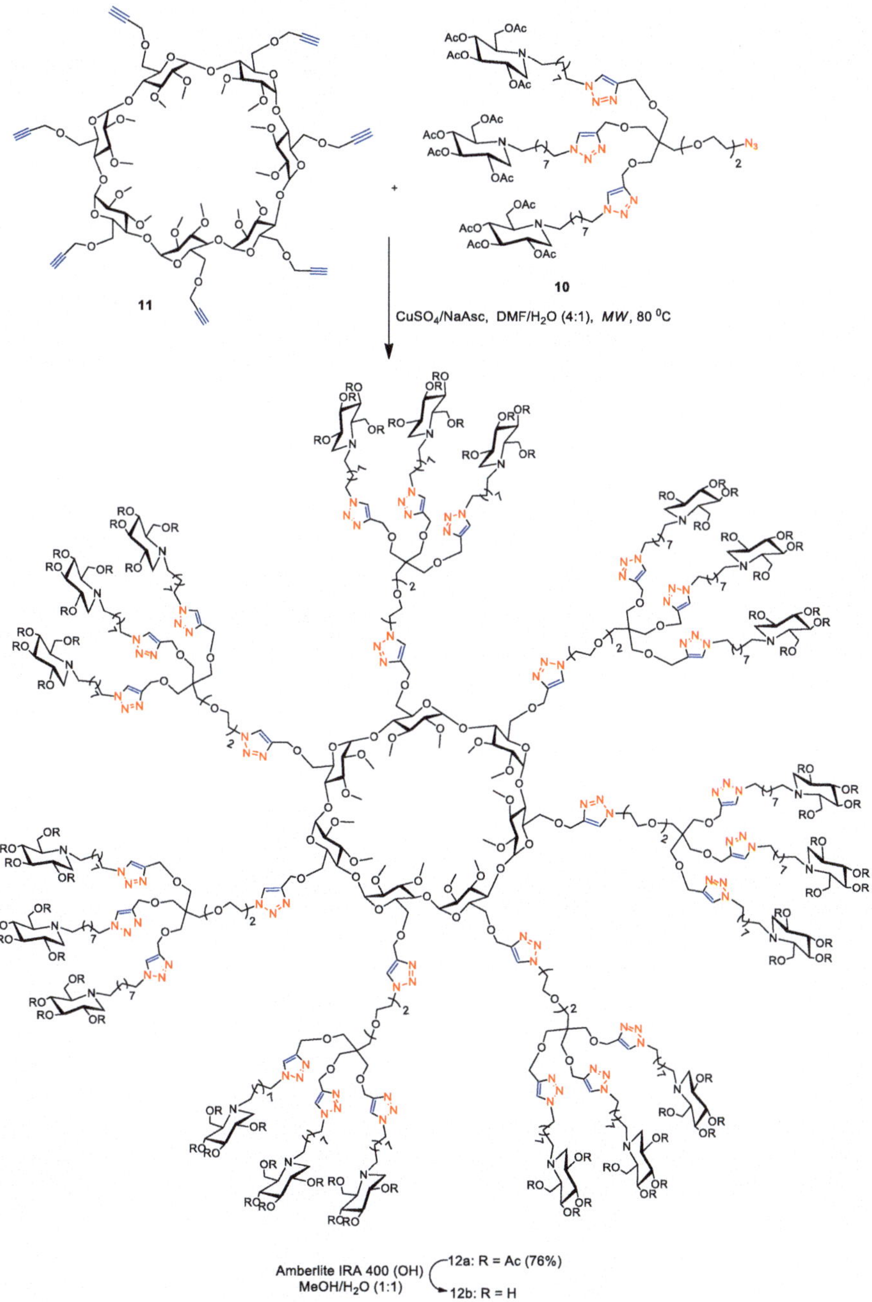

Scheme 10.4 Synthesis of 21-valent iminosugar-tethered dendrimer using CuAAC chemistry

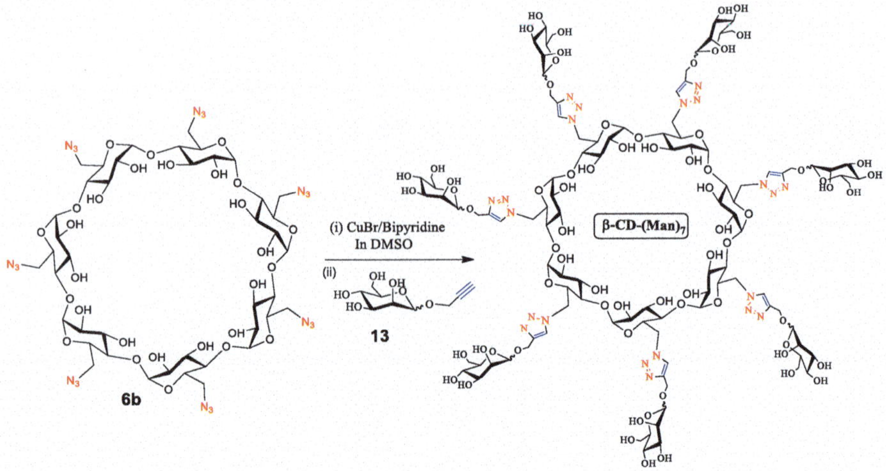

Scheme 10.5 Synthesis of a mannose glycocluster based on cyclodextrin **14** using the click protocol

utilizes different types of sugar azide **29** which was click with alkyne-functionalized compound **30** to produce the chlorine-fuctionalized dendron **31**. These dendrons that were on azidation with NaN_3 in dry DMF solvent for 8 h at 80 °C afford the azide-functionalized dendron **32**, which was on click reaction with calix[4]arene **20** afforded the desired glycodendrimers **33** (Scheme 10.8) [33, 34].

Recently the same group successfully developed G_2 generation galactosylated glycodendrimer in assistance with Cu-catalyzed click modular click approach, in which azide-containing dendron **34** on 1,3-dipolar cycloaddition reaction with alkyne-appended calix[4]arene **20** core afforded the desired G_2 generation galactosylated glycodendrimer **35** (Scheme 10.9) [34].

The fact that the anomeric carbon–sulfur bond is not sensitive to chemical or enzymatic conditions is one of the most intriguing structural characteristics of calix[4]arene S-sialoside compounds. The synthesis of these compounds commonly makes use of the well-known sialoside *N*-acetyl-neuraminic acid (Neu5Ac) [35]. Dondoni and colleagues have developed calix[4]arene-based sialyl clusters to investigate host–guest interactions in the context of viral infections. The Neu_5Ac-adorned clusters often prevent sugar–protein recognition, which is a key defence mechanism against the polyoma virus or BK virus (BKV), [36, 37] that causes severe nephropathy and, eventually, organ failure [38, 39]. Keeping in the mind the various importance of calix[4]arene, Roy and group develop an interesting approach for the synthesis of triazolylcalix[4]arene-*S*-sialosides **38** adorned at the top rim and calix[4]arene-S-sialosides **40** decorated at the bottom rim. In current methodology first of all propargyl thiosialoside **37**, which underwent to CuAAC with azido-calix[4]arene **36** scaffold under the assistance of copper(I) to furnish triazolylcalix[4]arene-S-sialosides **38** [40]. A tetravalent calix[4]arene-based S-sialoside cluster **38a** with a flexible *N*-propyltriazole appendage was synthesized in assistance of devised approach through

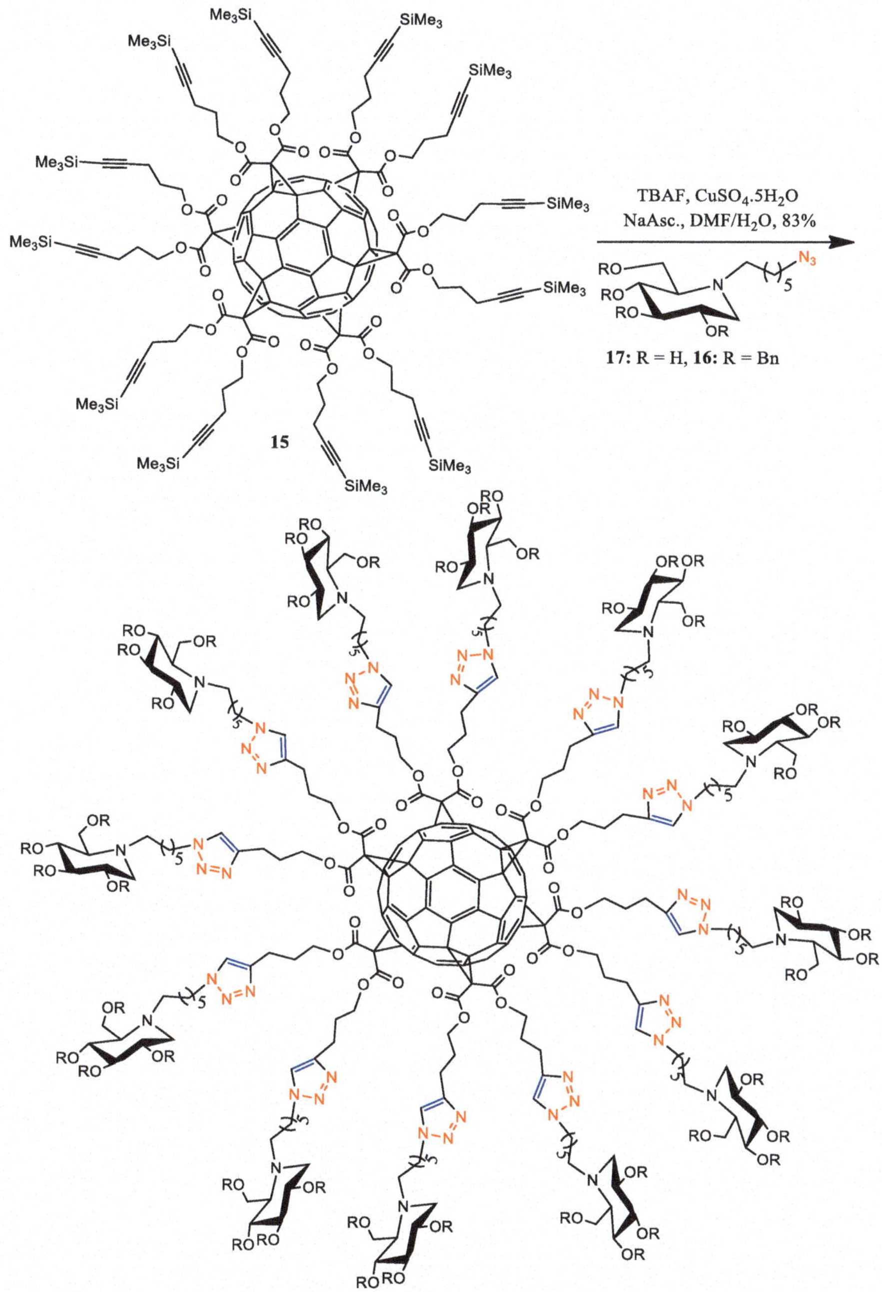

19 (83%) : R= H, **18** (78%) : R = Bn

Scheme 10.6 Synthesis route for fullerene iminosugar-tethered dendrimers using CuAAC protocol

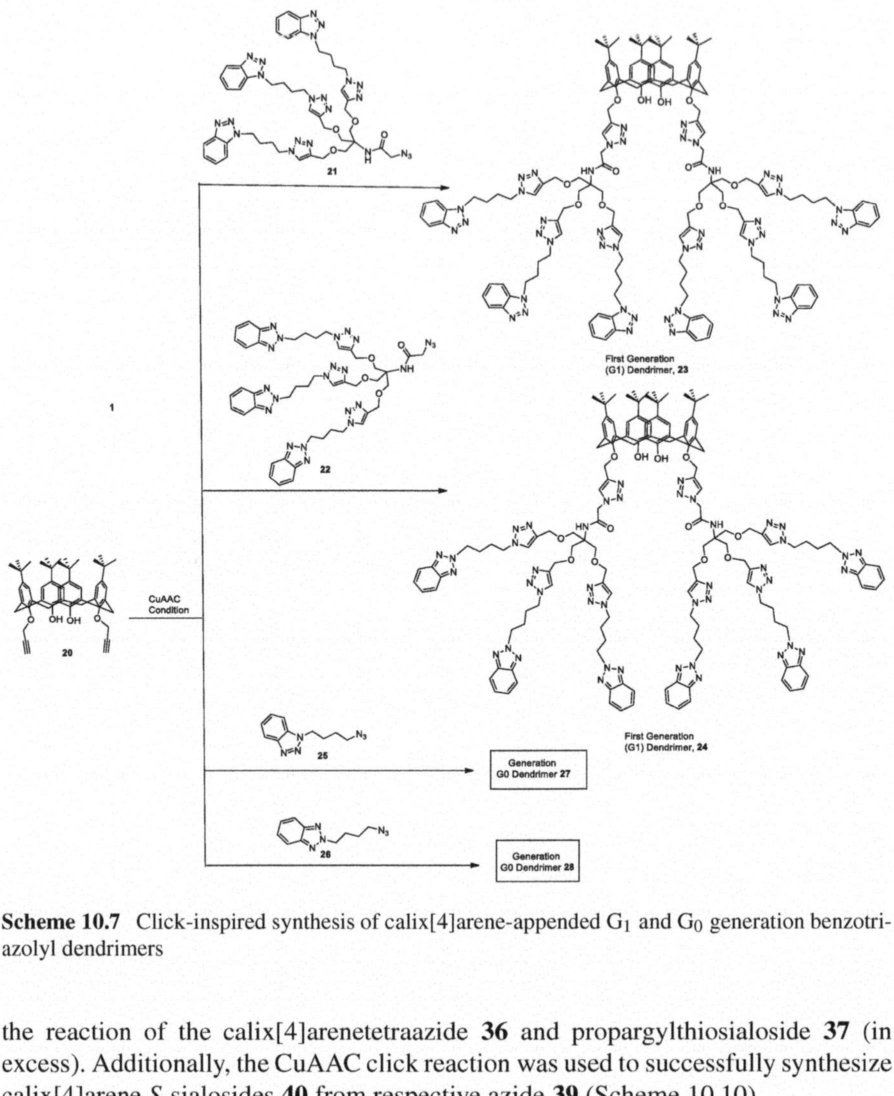

Scheme 10.7 Click-inspired synthesis of calix[4]arene-appended G_1 and G_0 generation benzotriazolyl dendrimers

the reaction of the calix[4]arenetetraazide **36** and propargylthiosialoside **37** (in excess). Additionally, the CuAAC click reaction was used to successfully synthesize calix[4]arene *S*-sialosides **40** from respective azide **39** (Scheme 10.10).

In order to achieve the desired 1,2,3-triazole linked sialocluster compound **42**, Dondoni et al. used modular CuAAC click chemistry [41]. The synthesis was started using compound **37** and were treated with octa-azide calyx[4]arene **41** in assistance with CuI as a catalyst along with DIPEA to develop the intriguing acetylated octavalent sialocluster, which has a total of eight sialyl scaffolds at the upper and lower rim of the calix[4]arene. Further the obtained compound was subjected for deacetylated to produce the pure octavalent sialo cluster **42** in 61% yield, which, indicate the unprecedented usefulness of the regioselective click approach (Scheme 10.11) [41].

Dondoni and Marra successfully studied the highly facile Cu-catalyzed azide-alkyne cycloaddition reaction or click chemistry and thiol-ene coupling to develop an

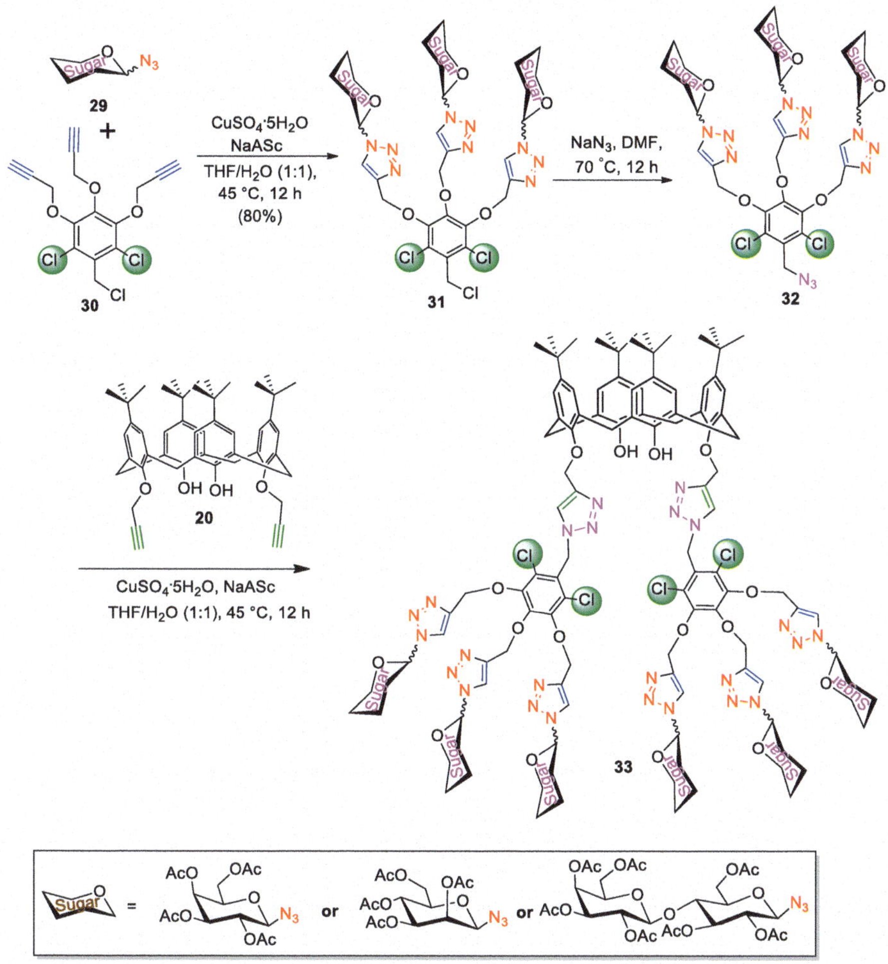

Scheme 10.8 Click-inspired synthesis of calix[4]arene-appended G_1 generation glycodendrimers of galactose, mannose, and lactose

intriguing dendrimer structure which is a calix[4]arene *S*-glycoside-based dendrimer with a triazole heterocycle at the lower and an upper rim on a calix[4]arene scaffold [42–44]. Alkene and alkyne appended calix[4]arene **43** on the upper and lower rim were first combined with galactose-derived azide **44** in toluene at room temperature for 26 h to afford the respective triazole **45** in 56% yield. The developed triazole **45** was subsequently subjected to a thiol-ene reaction with glucosyl thiol **46** in the presence of DMAP in dichloromethane under photo-induced conditions for 60 min to produce calix[4]arene S-glycoside **47a** in a 76% yield. Compound **47a** was successfully treated with sodium methoxide in methanol to obtain deacetylated product **47b** (Scheme 10.12) [45].

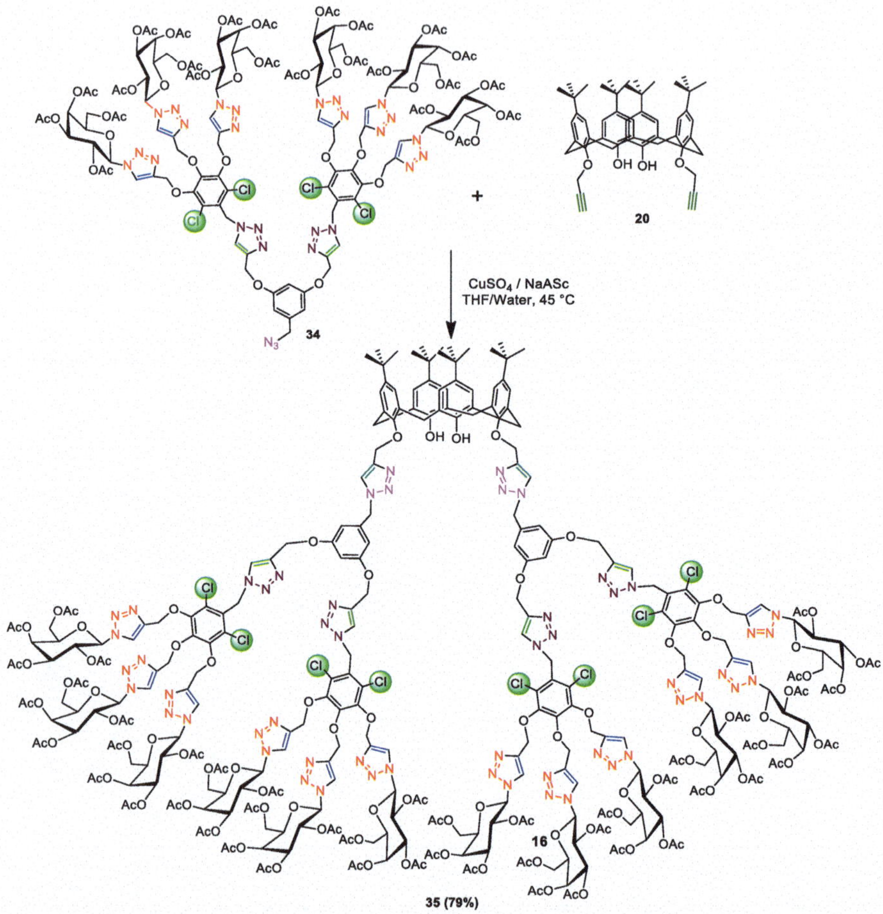

Scheme 10.9 Click-inspired synthesis of calix[4]arene-appended, G_2 generation galactosylated glycodendrimer

Tiwari et al. have successfully developed an interesting 18 galactose-coated dendrimer using the high yielding, modular CuAAC approach. In the reaction path, the synthesized azide dendron **48**, was on click reaction in the presence of CuI/ DIPEA in anhydrous dichloromethane with hexapropargylated benzene core to **49** gave the desired glycodendrimers **50**, which was characterized as18 galactosylated at the peripheral (Scheme 10.13) [46].

To explore the glycodendrimer chemistry and its application, Tiwari and co-workers introduced the synthesis of G1 generation glycodentromer by utilizing the 'double stage convergent technique'. In order to achieve this, the initial step involved the synthesis of a dentromer core **51**, and a dendritic wedge **52** via a successive high-yielding reaction steps. Additionally, the dendritic scaffold **52** was linked with non-valent alkyne core **51** in assistance of copper(I)-catalyzed azide/alkyne cycloaddition

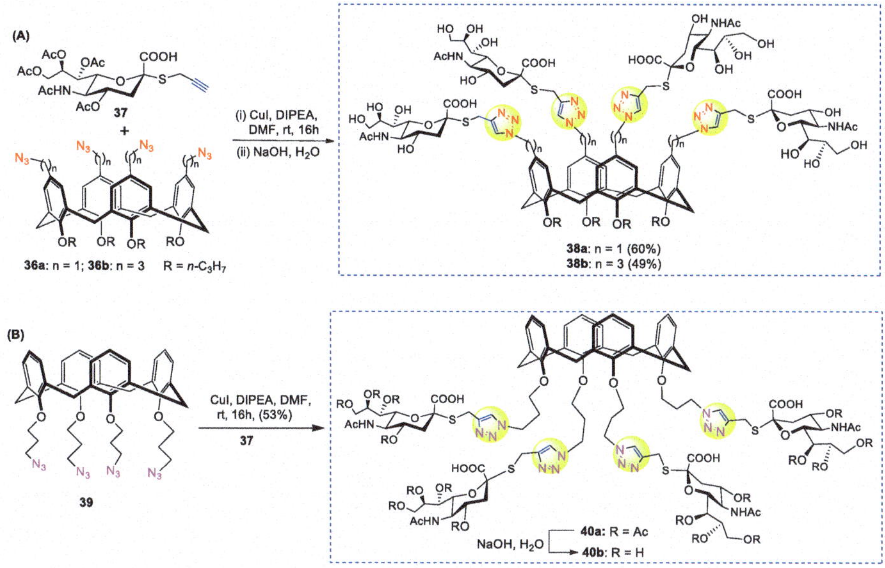

Scheme 10.10 Synthesis of calix[4]arene *S*-sialosides **40** decorated at the lower rim and triazolyl-calix[4]arene *S*-sialosides **38** decorated at upper rim

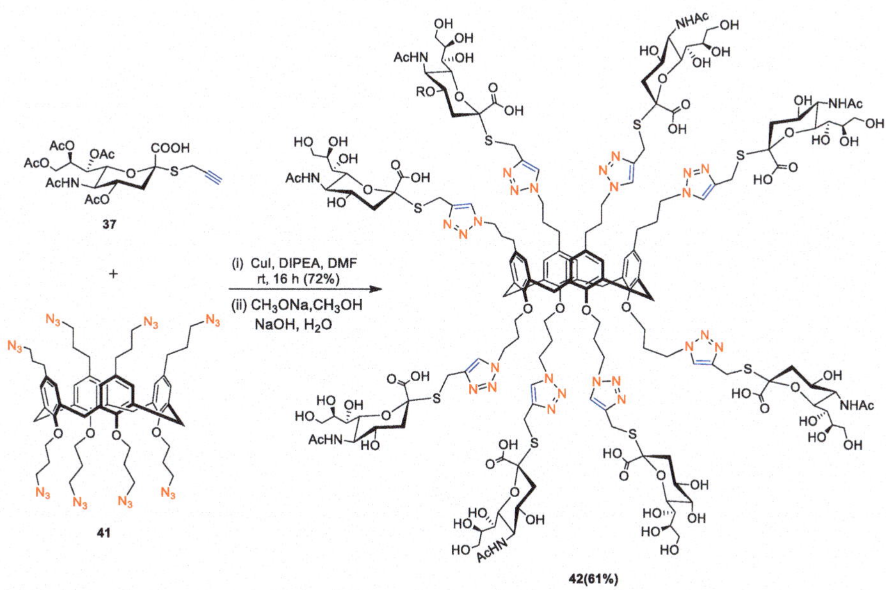

Scheme 10.11 Click-inspired synthesis of calix[4]arenes *S*-sialosides furnish at lower as well as upper rim

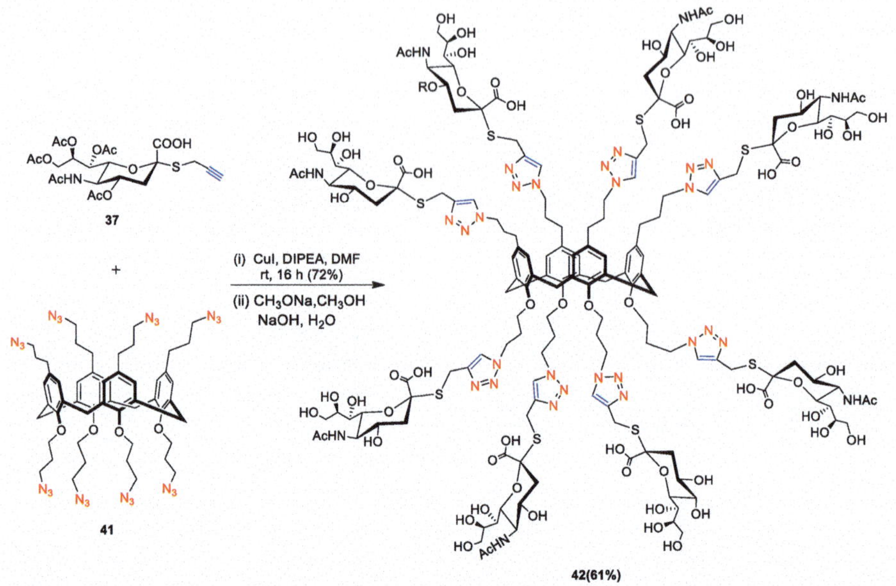

Scheme 10.12 Synthesis of calix[4]arene S-glycoside architecture utilizing combined CuAAC chemistry, thiol-ene coupling

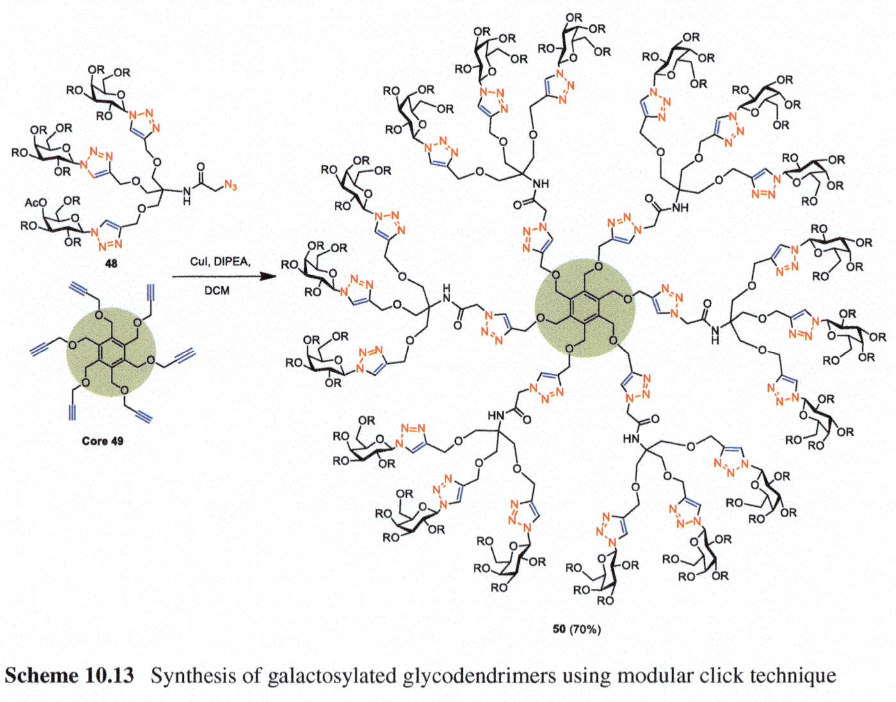

Scheme 10.13 Synthesis of galactosylated glycodendrimers using modular click technique

to afford the desired 27-peripheral-D-galactopyranoside tethered glycodentromer **53**. The synthesized glycodendrimers were identified by using NMR, MALDI-TOF MS, GPC data, etc. (Scheme 10.14) [47].

Roy et al. used the high yielding 'CuAAC Click Reaction' as the key technique to generate a series of $G_{(0)}$ generation glycodendrimers that had rigid, 'star-shaped' hexaphenyl benzene cores with mannosylated and lactosylated scaffolds at the outer

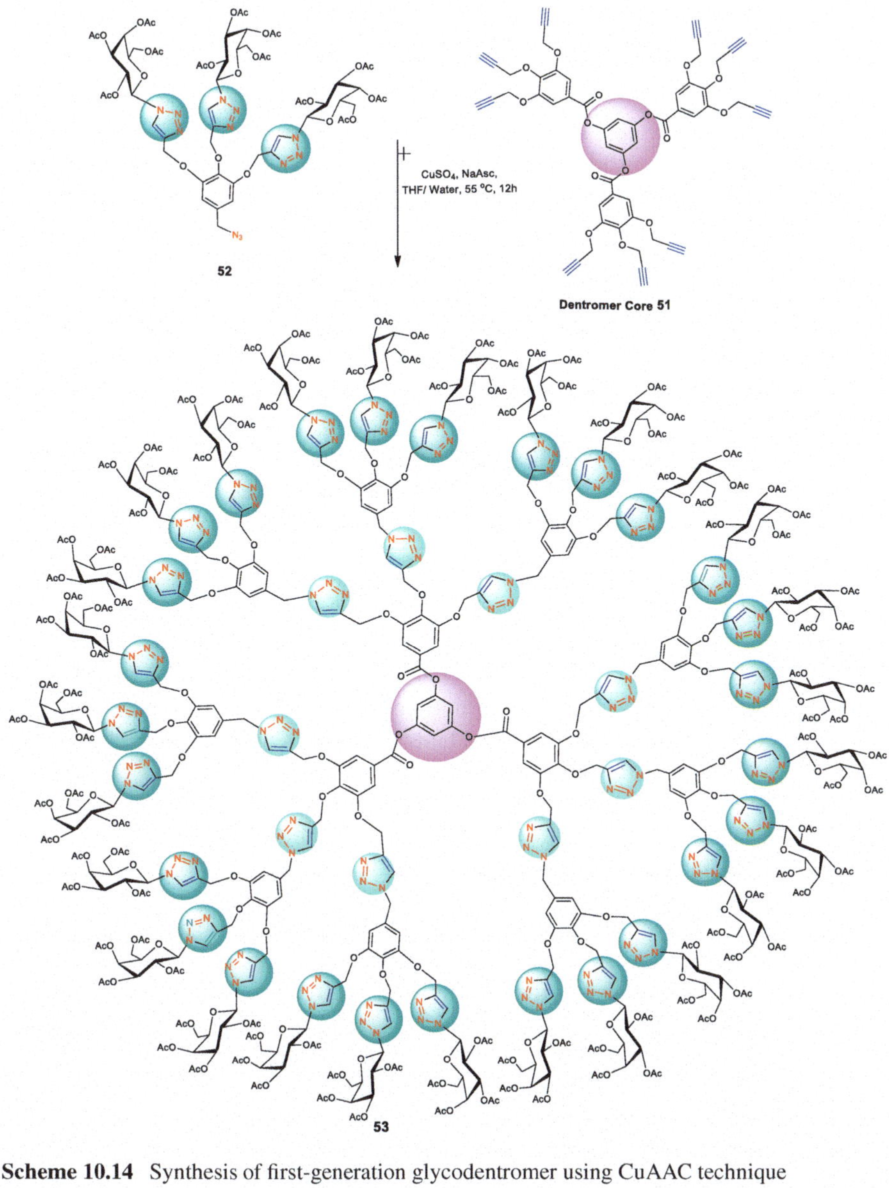

Scheme 10.14 Synthesis of first-generation glycodentromer using CuAAC technique

sphere. On the other hand of the authors developed peripheral mannopyranoside dendrimer **54** in 48% yield. Similarly, researchers have investigated the use of a similar click reaction to synthesize a new type of glycodendrimer. In this study, the sulfurated benzene core was combined with tris-derived alkynes to develop stable polypropargylated scaffolds. This compound were conjugated with sugar azide, specifically α-D-mannopyranoside azide, under CuAAC conditions, resulting in the development of dodeca- and octadeca-mannopyranoside, well-distinct benzene-cored glycodendrimers **54** (Fig. 10.7) [48, 49]. Although no biological research or catalytic activity were shown, the synthesized glycodendrimers are unquestionably intriguing.

Next, another approach for the synthesis of glycodendrimers was devised in which the multivalency concept demonstrates its scope beyond carbohydrate-lectin interaction and processes like glycomimetic-enzyme inhibition. In this affection, to achieve the multivalent iminosugar clusters under mild reaction conditions, authors profusely exploited the modular click chemistry [50]. A halide-based functionalized iminosugar **55** was synthesized and used as click partner under microwave-assisted CuAAC reaction condition to furnish glycodendrimer **58** [28]. The compound **55** was subjected for azidation in presence of sodium azide to afford AB_3 type synthon **56** containing azide functionality. Furthermore, authors used the second subsequent click by grafting of AB_3 synthon on the alkyne-functionalized-tetra propargyl core unit **57** in assistance of $CuSO_4$/NaAsc in DMF/H_2O (4:1) at 80 °C, which lead to the development of desired dodecavalent iminosugar **58a** in 56% yield. The developed dodecavalent iminosugar **58a** on *O*-deacetylation in acidic catalyst, amberlite IRA-400 (OH^-) anion-exchange resin gave the desired iminosugar cluster **58b** in good yield (Scheme 10.15) [51, 52].

Similarly, Sharpless et al. described synthesis of extremely beautiful di-block unsymmetrical glycodendrimer **59** having the both mannose-binding parts and coumarin fluorescent binding parts which demonstrated for the dual function, appreciation as well as recognition. By utilizing the most advanced click technique (CuAAC), an alkyne-functionalized dendritic moiety carrying a coumarin-based fluorophore was coupled with deprotected mannose-derived azide in the presence of THF/water to afford desired asymmetrical high-yield glycodendrimer **59** (Fig. 10.8) [53]. On the other hand, it is found that the developed asymmetrical glycodendrimer works with the appropriate receptors to prevent haemagglutination as a result of quantitative alteration at one end. In contrast to this, it is easier to distinguish coumarin units, which may be helpful for recognition. Conclusively, due to its dual properties the asymmetrical well-organized dendrimer **59** was found to be 240 times more effective than the monomeric mannose unit at inhibiting haemagglutination.

In continuation to synthesis of hybrid dendrimers, Ardabevskaia et al. made significant effort to develop hybrid dendrimers **62** composed of flexible carbosilane core **60** and rigid aromatic shell **61** by utilizing azide-alkyne click conjugation [54]. The CuAAC coupling of azide-functionalized carbosilane core with monoethynyl rigid hexaphenylbenzene (HPB) dendrons in the presence of CuI, TEA in dioxane at 105 °C afforded the desire hybrid dendrimer in good yields (Scheme 10.16). The thermogravimetric (TGA) and phase analysis of dendrimer suggested the similar

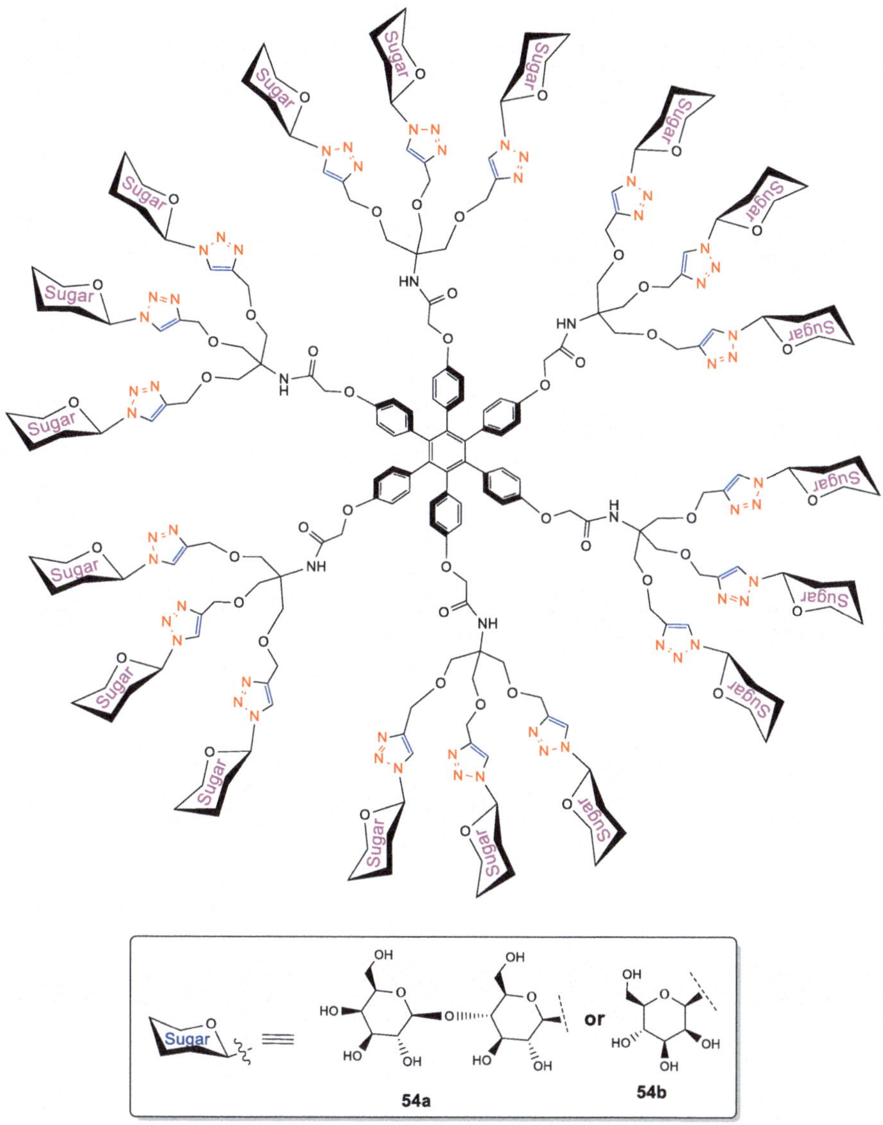

Fig. 10.7 Utilizing a modular CuAAC conjugation strategy, for construction hexaphenyl benzene-centered glycodendrimers **54**

performance of G_2 generation of dendrimer regardless rigid and flexible blocks of carbosilane core. The DSC and SWAXS investigation confirmed the propensity of crystallization and packing of crystal structure after annealing procedure.

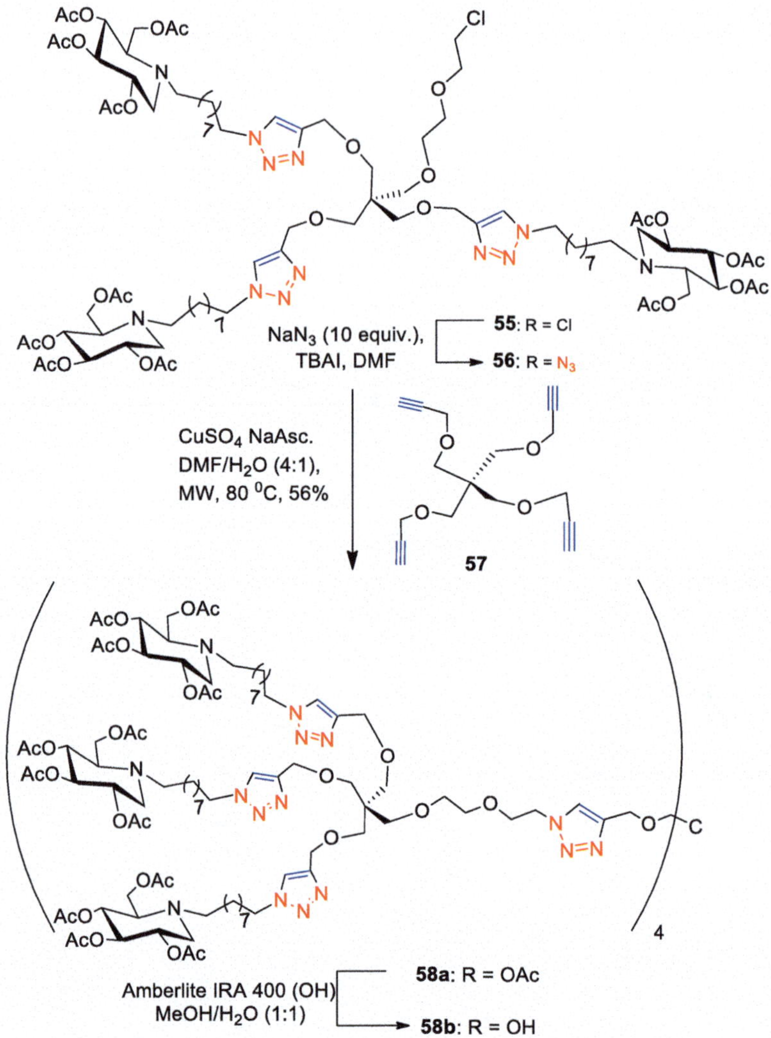

Scheme 10.15 Synthesis of dodecavalent iminosugar clusters synthesized by using *MW*-assisted CuAAC methods

Astruc et al. brilliantly used modular CuAAC tool to develop interesting glycodendrimer 243(G_3) **63** having xylopyranoside residue at the periphery of dendrimer (Fig. 10.9) [55].

Tiwari et al. in order to generate a range of novel chlorine-containing cyclen-cored glycodendrimers **66** of high-sugar tethers at low generation that has auspicious therapeutic potential. Synthesis of a unique hypercore unit **64** contained chlorine and twelve alkynyl functionalize group. The unit **64** was derived from the cyclen scaffold and confirmed by the study of its single crystal X-ray data. Furthermore, for the

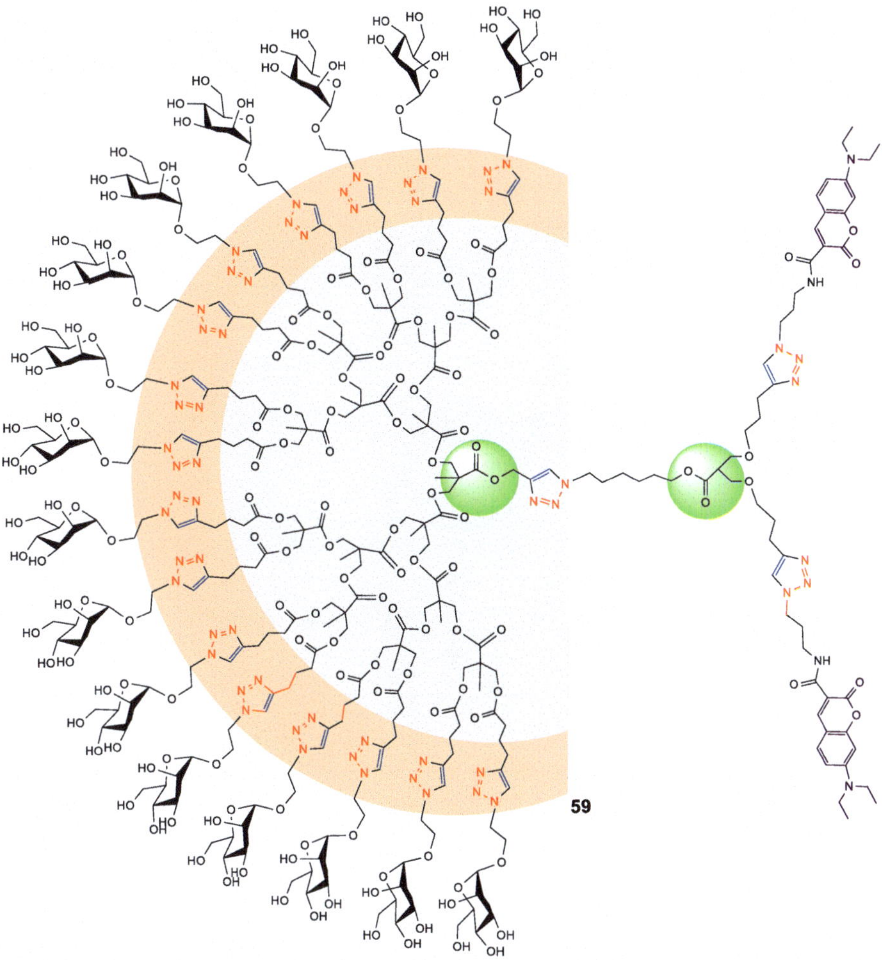

59

Fig. 10.8 Synthesis of irregular glycodendrimer **59** containing coumarin as fluorophores and triazolyl-mannose at the outer sphere

synthesis of 36-galactose-coated G1 generation glycodendrimer **66** and a diversity of unique 12-sugar-coated G0 generation glycodendrimers **65** were produced by utilizing the modular CuAAC approach in good-to-high yield (Scheme 10.17). The structures of synthesized glycodendrimer architectures have been well-explained by wide-ranging spectral analysis including NMR (^{1}H & ^{13}CNMR), HRMS, MALDI-TOF MS, UV–Vis, IR, and size exclusion chromatogram (SEC) data [56].

Tiwari et al. explore expeditious and modular CuAAC 'click' protocol for the assembly of porphyrin-cored hybrid glycodendrimers of G_1 and G_2 generation. In the synthetic approach the group synthesized the hybrid glycodendrimer **70** (G_1 generation) and **71** (G_2 generation), in assistance of $CuSO_4$/NaAsc using alkyne-appended core **67** and fusing with azide-based dendron **68** and **69**, respectively.

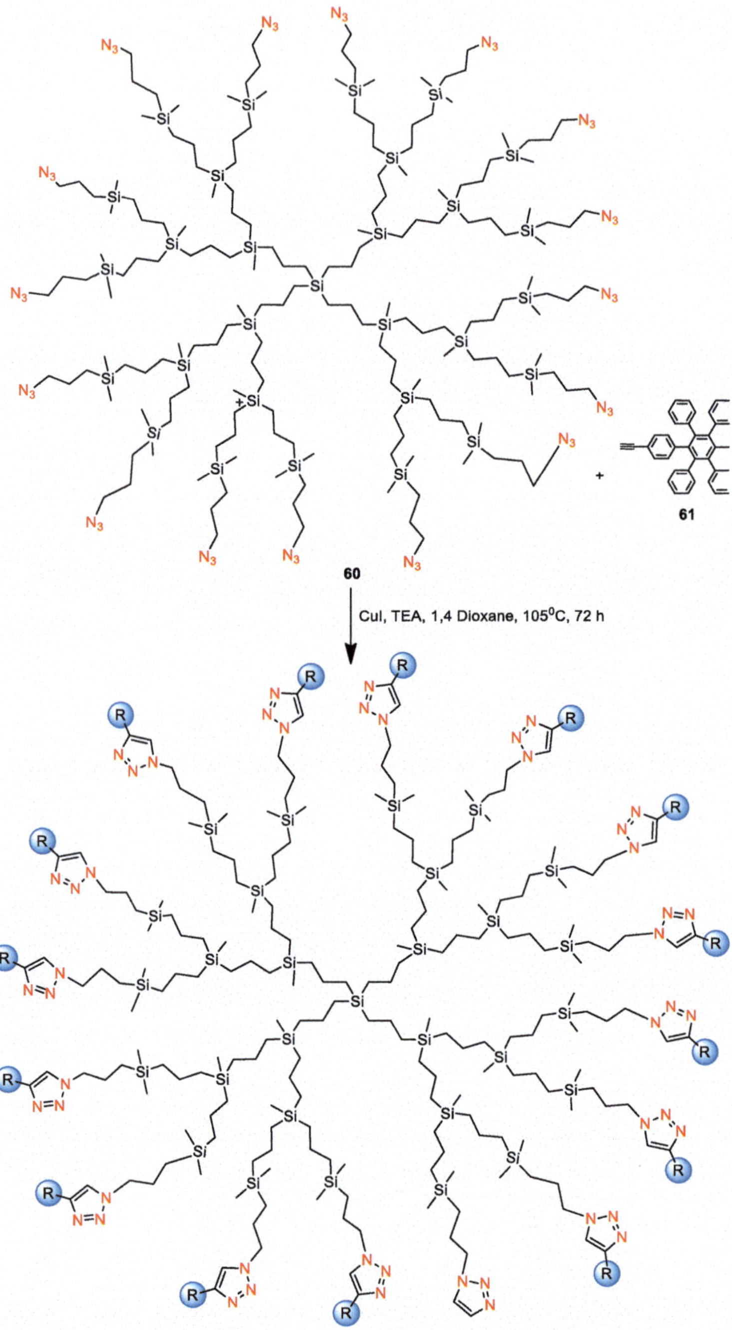

Scheme 10.16 Click-inspired synthesis of carbosilane-cored hybrid dendrimers **62**

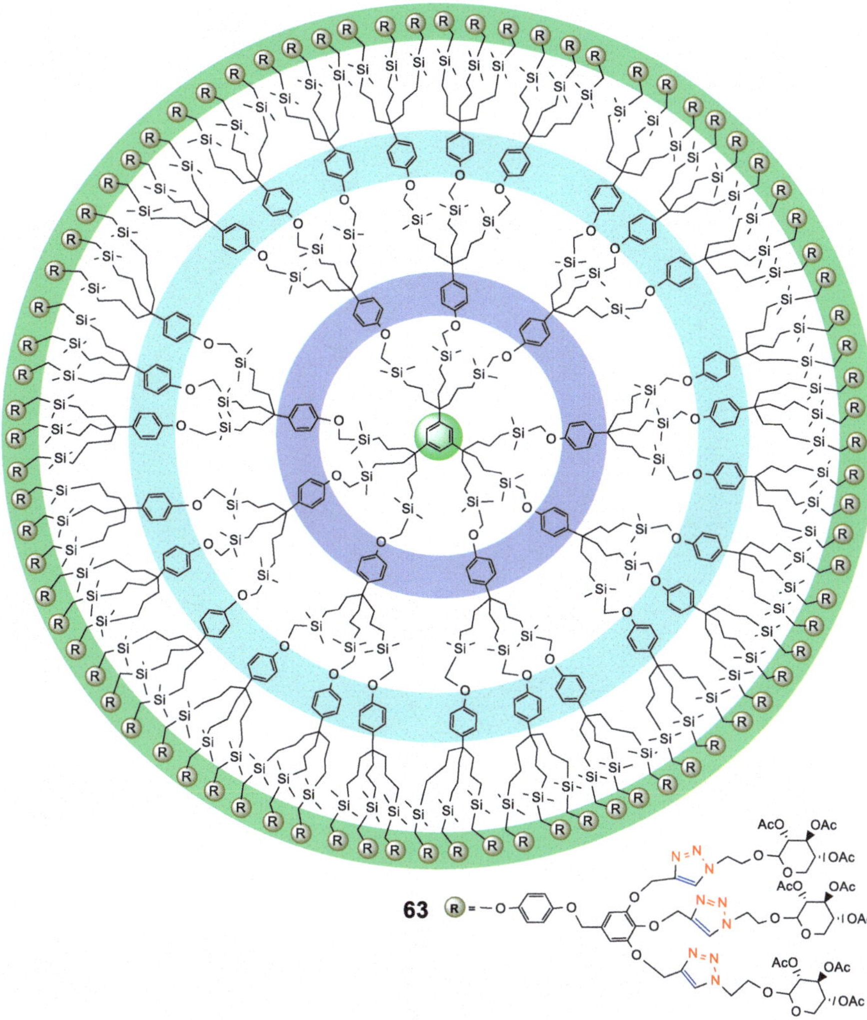

Fig. 10.9 Synthesis of G_3-243-xylopyranoside glycodendrimer by CuAAC 'click chemistry'

The developed glycodendrimers of G_1 and G_2 generation are well-characterized by various spectroscopic analyses including NMR (^1H, ^{13}C, and ^{19}F), MS, MALDI-TOF, IR, and SEC and evaluated for their sensing ability for arrangement of metal ions. Interestingly, the resulting glycodendrimers displayed notable sensing for Cu(II) ion with solvatochromism properties (Scheme 10.18) [57, 58].

Hasegawa et al. reported an efficient intriguing approach for synthesis of glycodendrimer, the approach characterized by some notable feature such as easy, quantitative, and one-pot introduction of eight β-lactoside-modules onto a porphyrin-core in assistance of Cu + -catalyzed chemoselective coupling (click chemistry) between

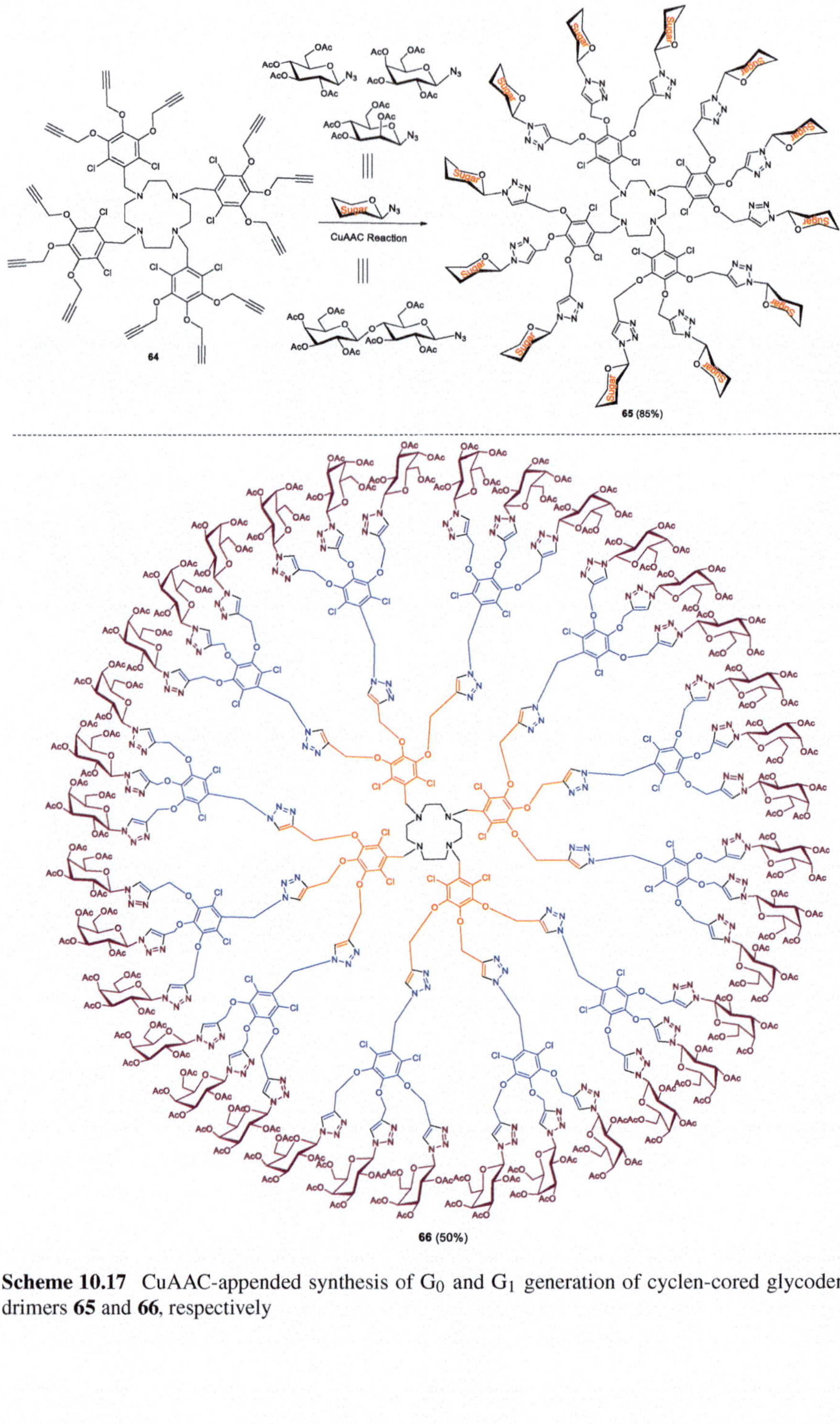

Scheme 10.17 CuAAC-appended synthesis of G_0 and G_1 generation of cyclen-cored glycodendrimers **65** and **66**, respectively

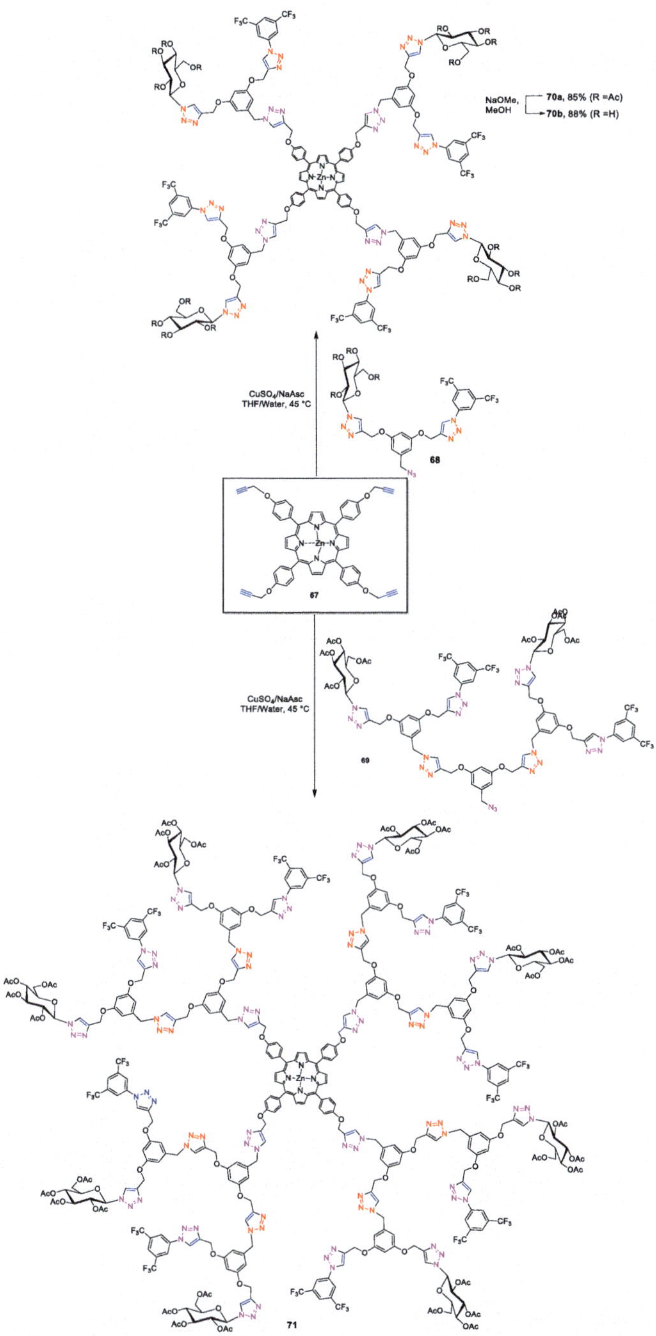

Scheme 10.18 Synthesis of G_1 and G_2 generation of porphyrin-cored hybrid glycodendrimers

a porphyrin carrying eight alkyne-terminals **73** and *β*-lactosylazides **72** to obtain porphyrin-based glycocluster **74** in good yield (Scheme 10.19) [59].

Moving forward, Mukhopadhyay et al., successfully explored the click approach to develop metal-centered glycodendrimer. They were utilizing the comparable tetra-alkyne-functionalized porphyrin core and applying it to develop porphyrin-cored

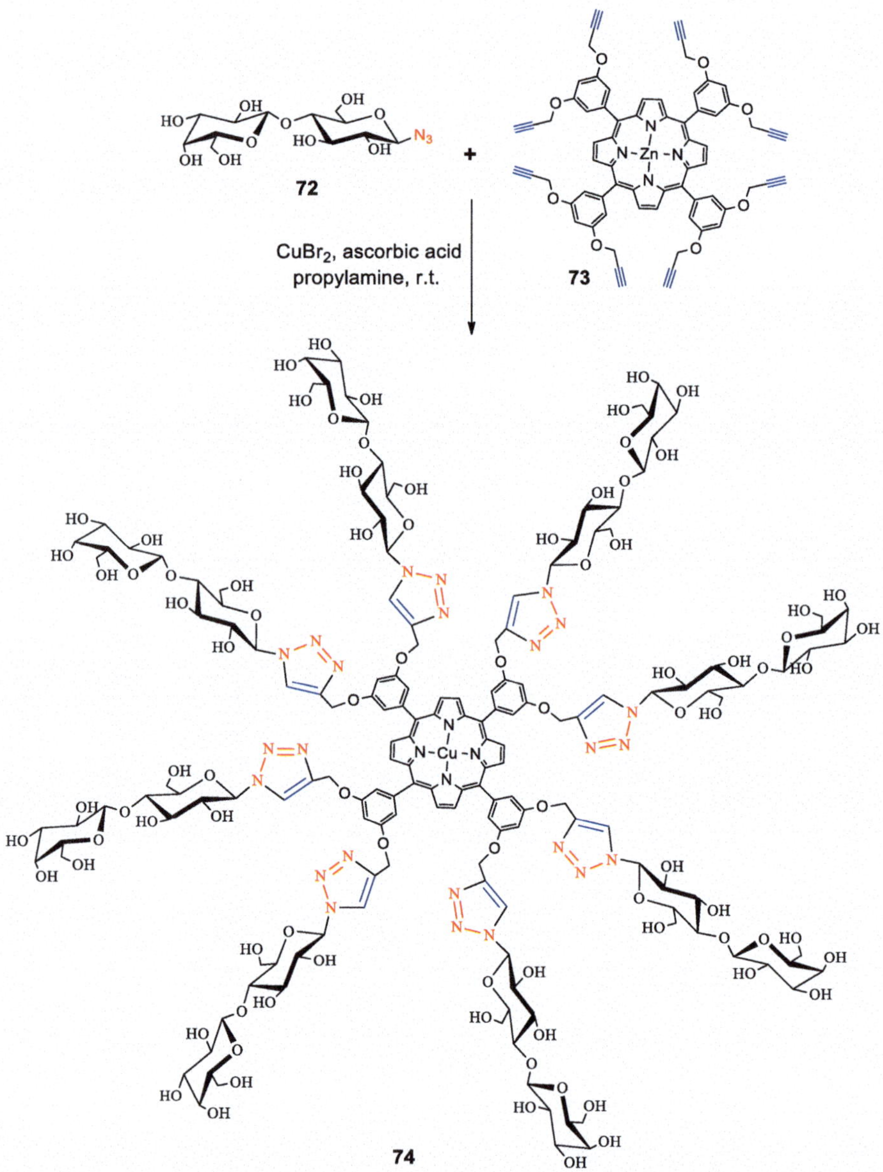

Scheme 10.19 Synthesis of octa-*β*-lactosylated porphyrinatocopper using modular click reaction

glycodendrimers using various dendritic wedges. In the assistance of the simple click reaction concept, the alkyne-appended core unit **75**, second-generation dendritic wedges **76**, were stitched together to develop 36 peripheral α-D-mannose-coated dendrimers **77a**, **b** (Scheme 10.20). While using the fluorescence characteristic of porphyrin-based glycodendrimers, the authors have also investigated lectin sensing of the developed dendrimer compound [60].

It has been demonstrated that the surface modification of the gallic acid-triethylene glycol dendrimer by particular ligands is a highly effective strategy for diagnostics and medication issue. As a result, it is important to examine how peripheral alteration of gallic acid-triethylene glycol (GATG) dendrimers affects their absorption and intracellular transport in living cells. To achieve this, the surfaces of $[G_3]$-N_3 dendrimer (**78a**) and PEG_{5000}-dendritic block copolymer 'PEG-$[G_3]$-N_3' (**78b**) were functionalized using a variety of physicochemical features that are biologically relevant. The twenty-seven azide groups that surround these dendrimers easily click functionalize, displaying their efficient conjugation in accordance with the choice for the interaction with cell surfaces. In order to produce **80a**, dendrimer $[G_3]$-N_3 (**78a**) was combined with the alkynated fluorescien (FITC) **79a** derivative using the standard click tool CuAAC ($CuSO_4$/sodium ascorbate in t-BuOH/H_2O) in outstanding yields. In this continuation azide functionality of dendrimer **80a** was reacted with lactosyl alkyne **79b** to give the desired glycodendrimer $[G_3]$-FITC-Lac **81a** with good yield (77%). In similar technique was used to develop, PEG-$[G_3]$-FITC-Lac **81b** (86%) yield through PEG-$[G_3]$-N_3 a series of reaction (Scheme 10.21) [61].

Toward the divergent synthesis of glycodendrimer comprising with a variety of building blocks at each layer of the dendritic growth, Roy et al. used an '***Onion Peel Strategy***'. The glycodendrimers produced by utilizing this technique differ significantly from the traditional dendritic family, which is typically made up of repeated nano synthons with limited end groups. In this respect, the authors have created the glycodendrimer **87** with an aliphatic backbone and 36 LacNAc (at the surface) by employing the aforementioned method. To do this, the initial step required the photo addition of mercaptopropionic acid to hexapropargylated cyclotriphosphazene **82** using thiol-yne click (TYC), which produced pure derivative **83** which develops targeted hypercore **85** after treatment with hydroxylated AB_3 dendron **84** via carbodiimide-mediated esterification. Finally, using the CuAAC technique, the lactosamine azide **86** and hypercore **85** were amalgamated to generate the required glycodendrimer **87** (Scheme 10.22) [62].

Another efficient approach was developed by Roy and co-workers, in the report group utilize azide functionality **88** as a click partner to react with hexapropargylated cyclotriphosphazene core **89** to give the 12 peripheral glycodendrimer **90**, which on subjected for deacetylation to afford the compound **91** (Scheme 10.23) [63].

Zhang et al. synthesized the multifunctional alkyne core **92**, were click with azido-conjugated functionalized mannose and glucose precursors **93a**, **b**, respectively, they are quite effective despite the presence of multisite interactions and steric hindrance among the carbohydrate terminals. Glycodendrimers **94a**, **b** were deacetylated by

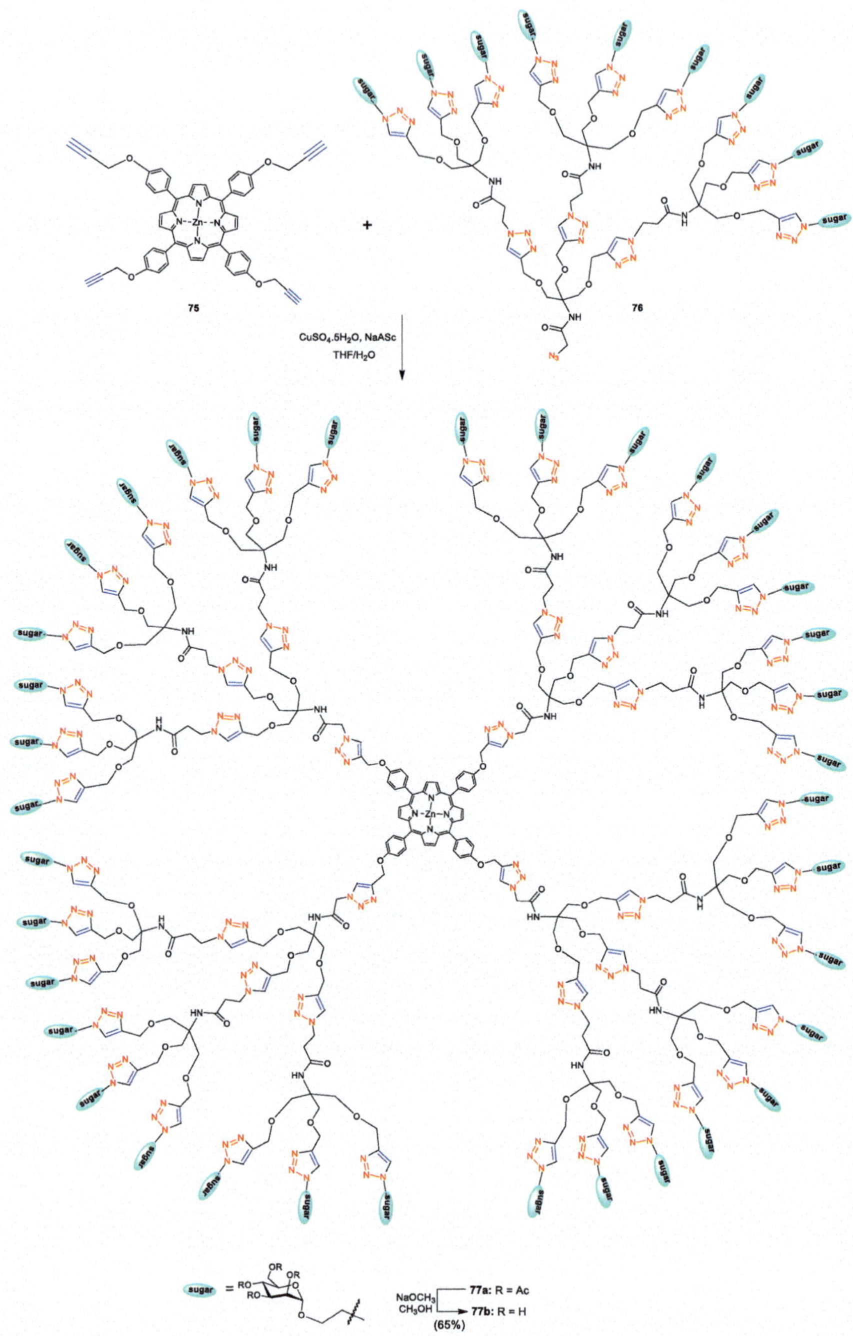

Scheme 10.20 Structure of porphyrin-cored 36-mannosylated glycodendrimers

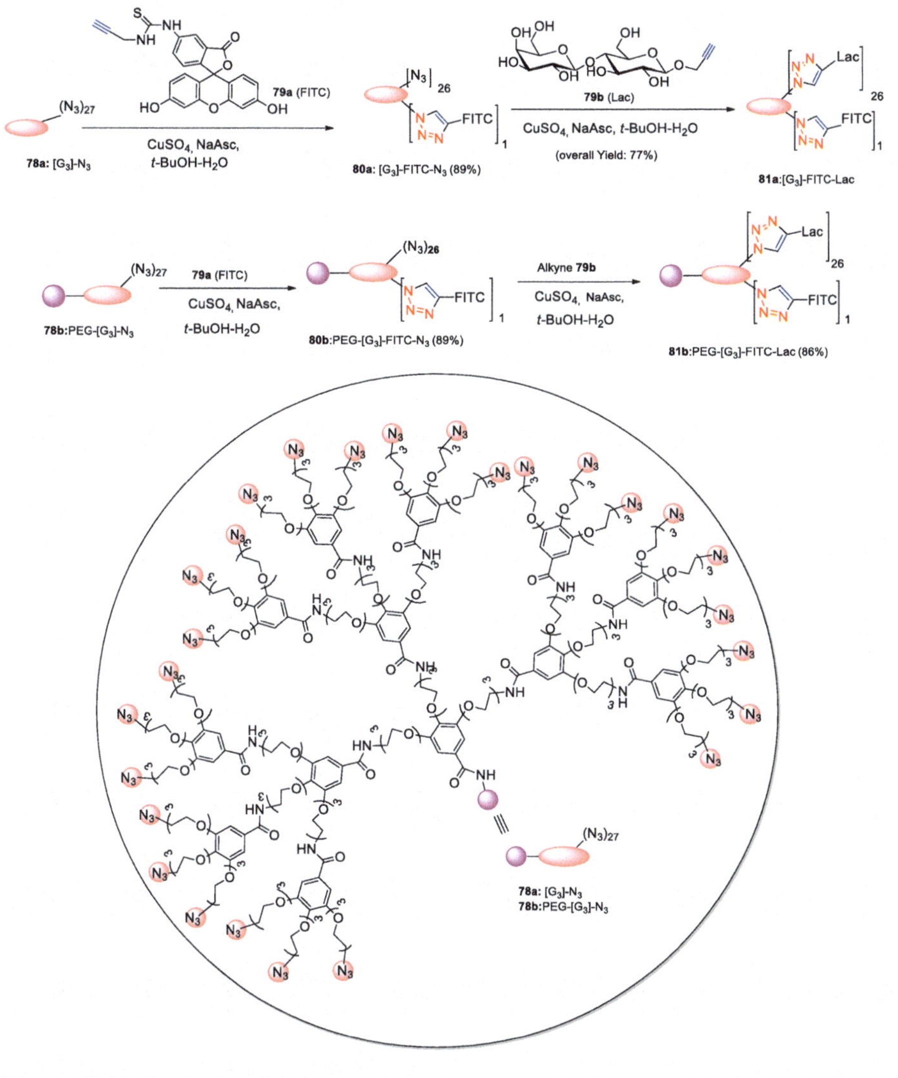

Scheme 10.21 Synthesis of biologically applicable [G3]-FITC-Lac and Peg-[G3]-FITC-Lac glycodendrimer

using lithium hydroxide in CH_3OH/H_2O to obtain the bola-amphiphilic glycodendrimers **95a** and **95b**, which was bearing mannose and glucose terminals, respectively (Scheme 10.24) [64].

Seeberger et al. designed and explored a glycodendrimer-based Ru-chelates as supplementary step in the 'Click-to-Chelation' approach. The process of synthesis involves an amide-based glycodendron **97** click with an alkyne functional core **96** in the presence of $CuSO_4$ and NaAsc in THF/H_2O (1:1), to succeed an Ru(II)-appended glycodendrimer **98** as a good ligand (Scheme 10.25) [65].

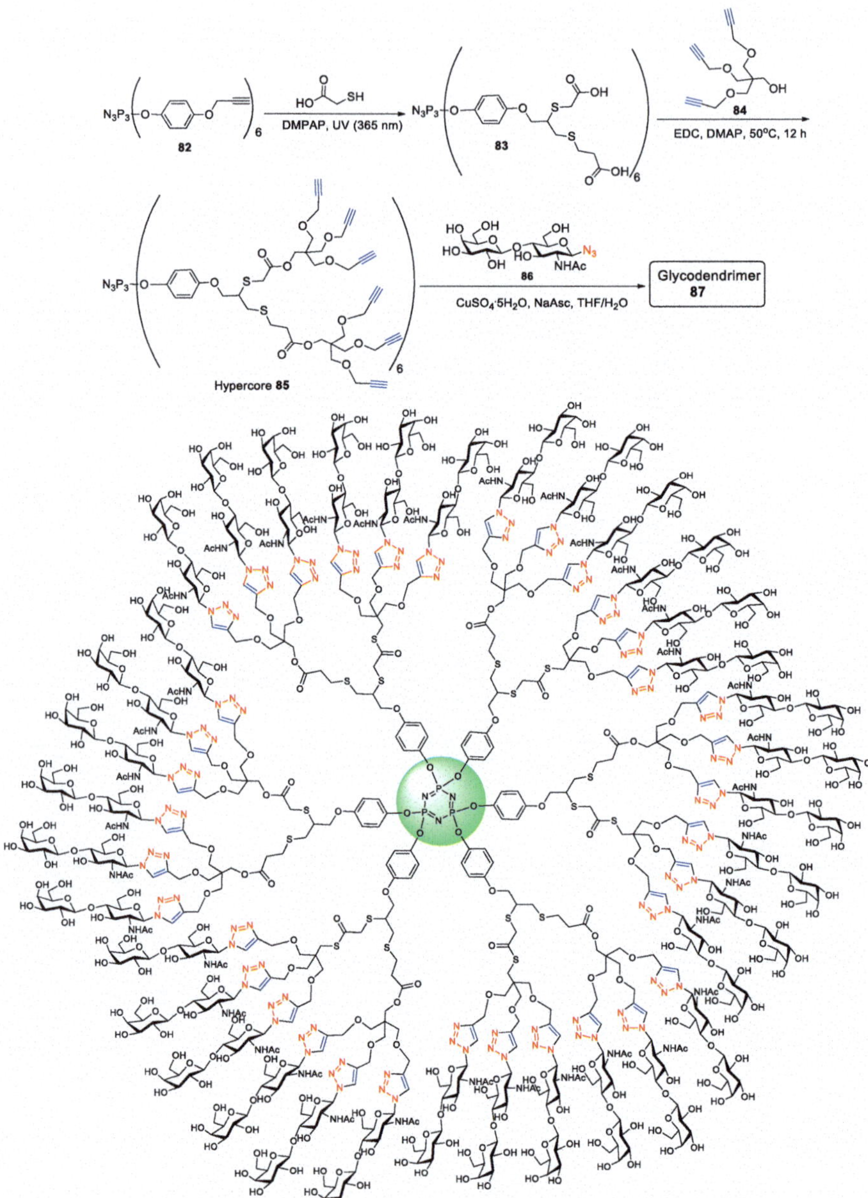

87

Scheme 10.22 Synthesis and development of 36LacNAc peripheral glycodendrimer **87**

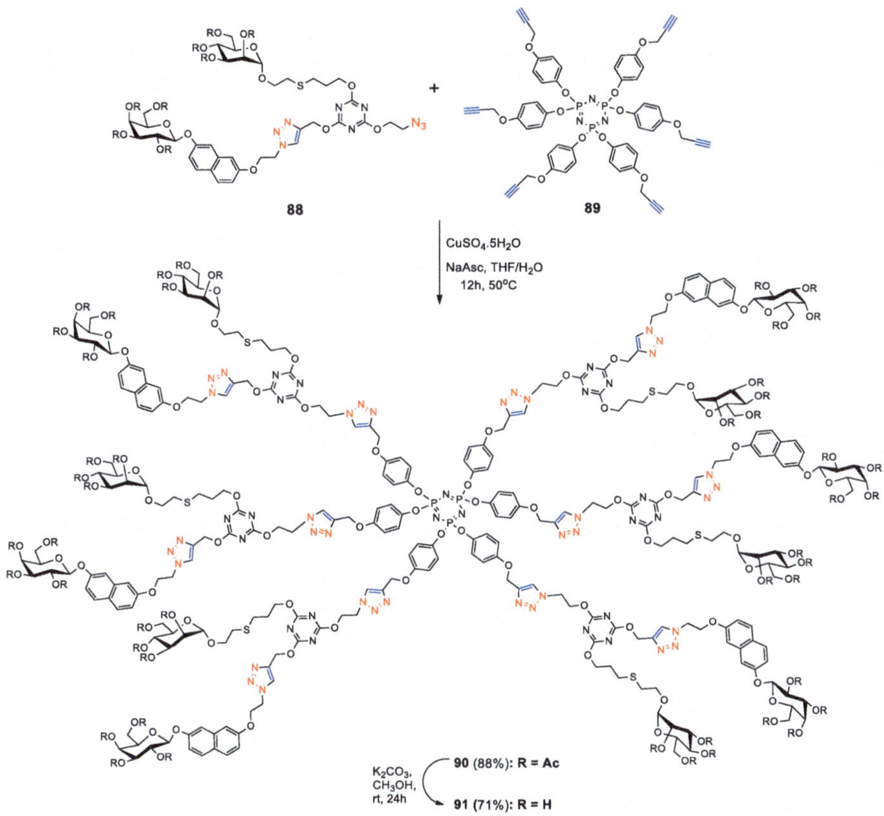

Scheme 10.23 Utilizing CuAAC technique to afford hetero-bi-functional glycodendrimers

Remarkably, Michael Gottschaldt and co-workers attempted to synthesis of pyridyl triazole-altered sucrose ligand in a seven-step synthesis utilizing *D*-glucose as the protection group for *D*-fructose and our synthesis beginning from commercially available sucrose which was complexes with $Ru(bpy)_2Cl_2$ precursor of a sucrose-based Ru complex with general formula $[Ru(bpy)_2(L)]Cl_2$ was successful in a good yield. Acid-catalyzed cleavage of *D*-glucose and pyridyl triazole(pyta)-altered *D*-fructose were achieved of the equivalent Ru complex. Elemental analysis confirmed the molecule's stability, and cytotoxicity reached confirmed that it was safe up to 100 mmol x L-1 against the L929, MDA-MB-231, or MCF-7 cell lines [66]. A progressive strategy has been demonstrated in macromolecular study and their development to obtain seven paramagnetic arms comprising β-cyclodextrin click cluster **101** as MRI contrast agent. In this syntheses acetylated per-azido-cyclodextrin core **6** and protected DTTA-hexyne dendron (**99**) were clicked in the presence of CuSO4/NaAsc as the standard CuAAC condition to yield **100a**, which was deprotected and hydrolyzed to produce **100b**. It was ultimately subjected to titration with $GdCl_3$ in order to provide a high yield of the required *β*-cyclodextrin click cluster **101**

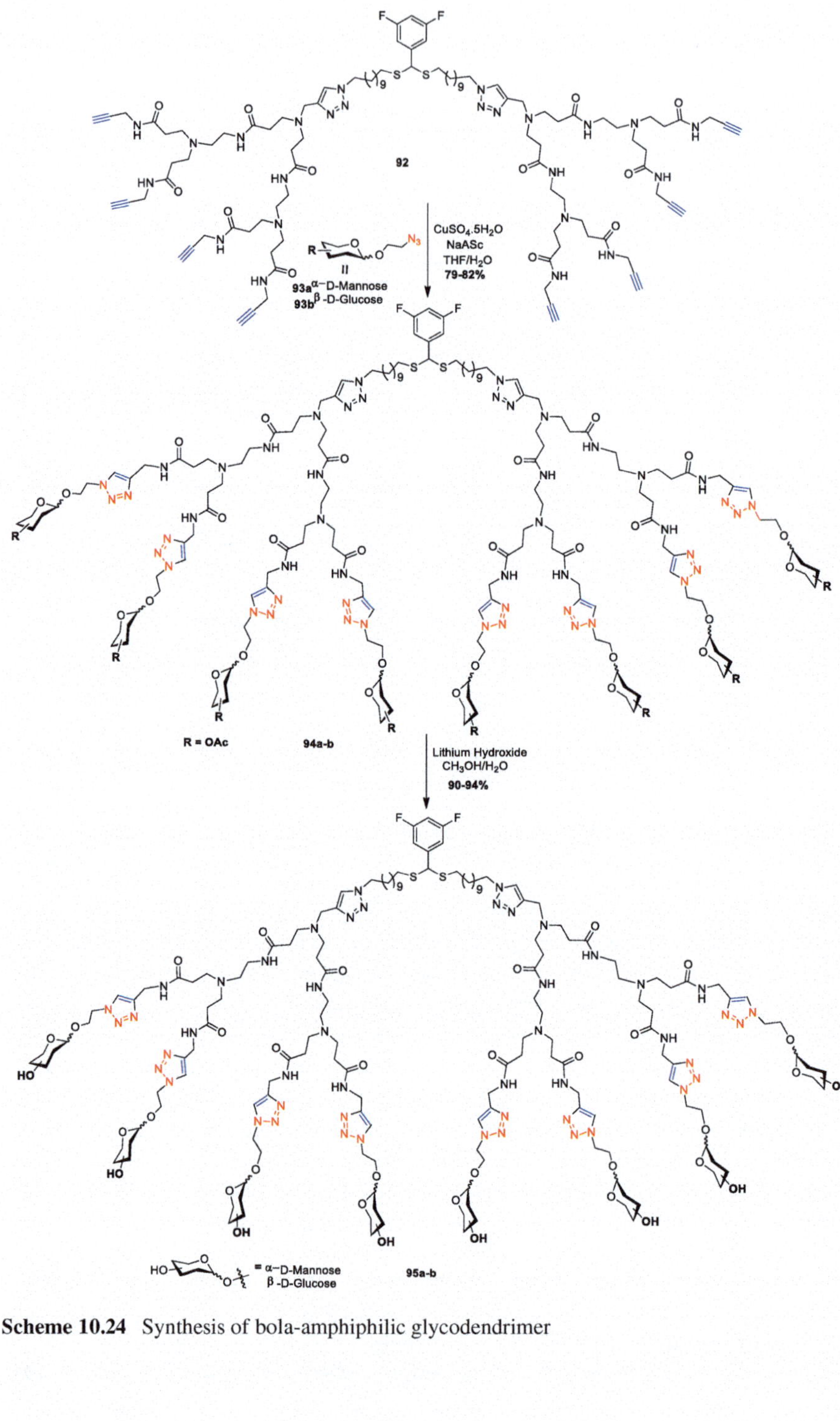

Scheme 10.24 Synthesis of bola-amphiphilic glycodendrimer

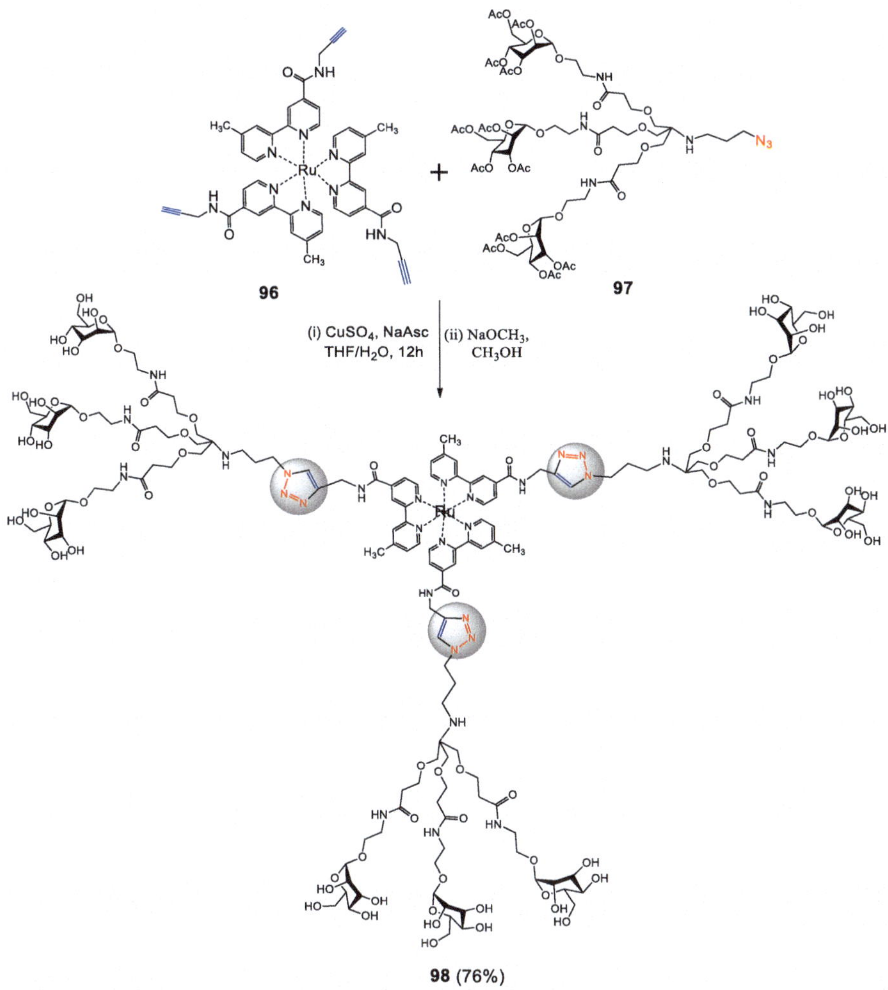

Scheme 10.25 Click reaction-facilitated sugar-based dendrimer-Ru chelate (**1593**)

(Scheme 10.26) [67]. The synthesized *β*-cyclodextrin CuAAC click cluster **100b** includes two water exchange positions at each arm, and assay investigations have shown that the cluster **101** exhibits a noticeably high-relaxivity profile (43.4 mM-1 s-1 at 9.4 T) at high magnetic field and also improves contrast on a human MRI scanner. Another benefit is the hydrophobicity connected to the key characteristics of *β*-cyclodextrin, which supports its candidature for the effective scaffold as receptor-specific targeted molecules.

Recently, it was reported that how gold nanoparticle decorated glycodendrimers were synthesized via a click reaction. The group synthesized the glycodendrimer Au-DSNPs **106** through click reaction pathway. Primarily, 1,3-bis(azidomethyl) phenol **102** on clicking with propargyloxy-acetylated glucose furnished acetylated

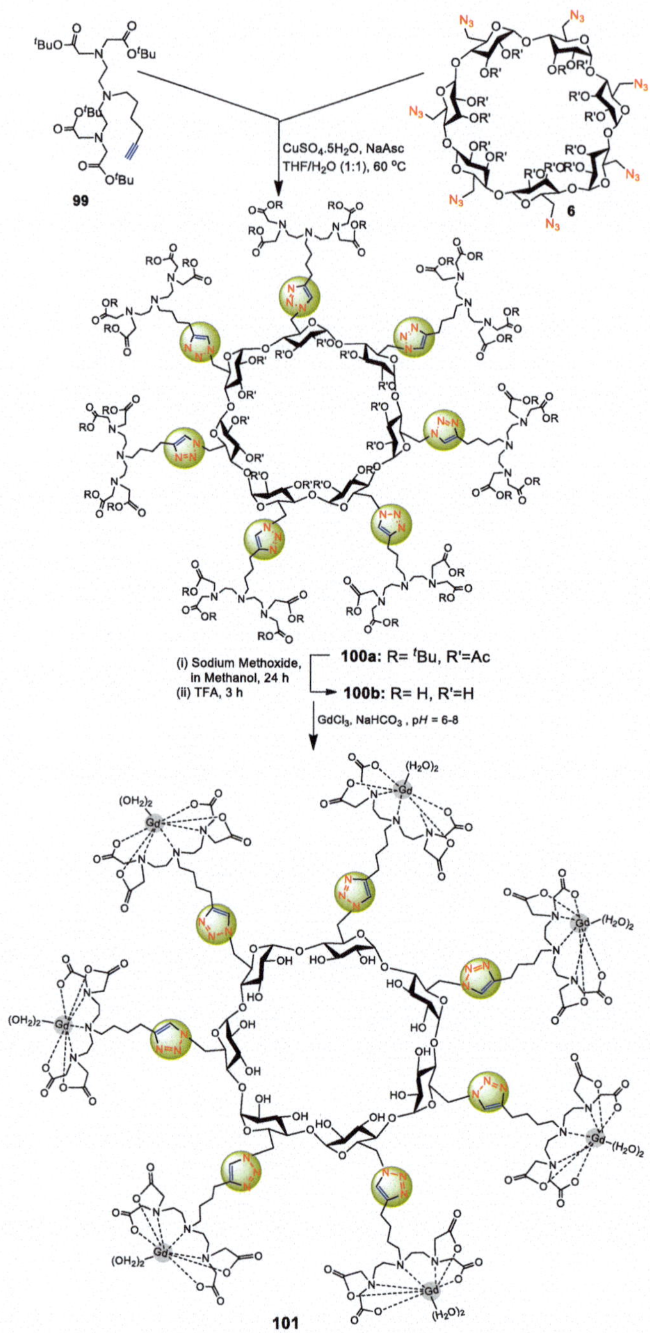

Scheme 10.26 Development of paramagnetic β-cyclodextrin click reaction-mediated Cluster **101**

compound **103**, followed by propargylation using propargyl bromide in DMF to afford respective dendron **104** in 72% yield. The resulted dendron **104** followed by a second CuAAC reaction with azide **105** to give desired nanoparticle-decorated glycodendrimers **106** (Scheme 10.27) [68]. UV–visible spectroscopy has been employed for assessing the catalytic activity of Au-DSNPs in the reduction of 4-nitrophenol to 4-aminophenol using $NaBH_4$ in an aqueous media.

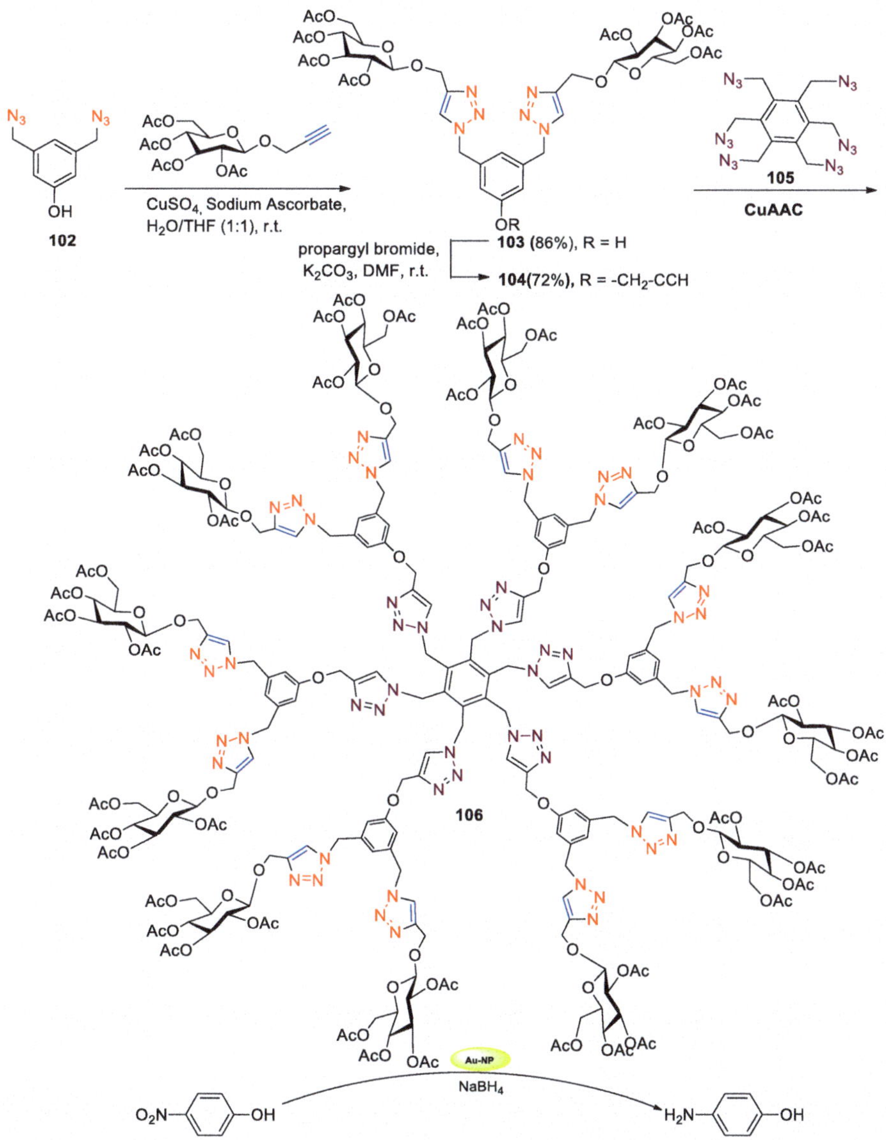

Scheme 10.27 Construction and catalytic properties of Au-DSNPs glycodendrimers **106**

10.4 Conclusions and Future Outlook

In summary, this chapter constitutes a series of dedicated strategic and synthetic applications of homo and hetero bifunctional dendrimers by using conventional 'CuAAC' click chemistry. Two important synthetic approaches namely, convergent and divergent were discussed for their construction and we discuss the section of literature on important topics and applications of CuAAC in dendrimers and glycodendrimers. For this, we express regret to other key contributors whose work has not yet been covered. In accumulation to a few of the selected examples which are utilizing CuAAC are several other applications correspondingly significant and they will be covered in successive reviews. Of particular interest are the contributions dealing with multivalent anti-adhesins against numerous pathogenic agents and microarrays. Several reviews have previously covered glycodendrimers in general and a few have been cited in this work. However, we concentrated our efforts to provide information utilizing CuAAC for synthesis and application of dendrimer and glycodendrimer. This chapter serves as a base for the importance and usefulness of click reaction in dendrimer synthesis to the researchers working in construction of giant molecules, macrocycles, and polymer chemistry.

References

1. Bosman AW, Janssen HM, Meijer EW (1999) Chem Rev 99:1665–1688
2. Jha RK, Wooley KL, Nystrom AM, Burke DJ, Kade MJ, Hawker CJ (2009) Chem Rev 109:5620–5686
3. Amblard F, Cho JH, Schinazi RF (2009) Chem Rev 109:4207–4220
4. Binder WH, Sachsenhofer R (2008) Rapid Commun 29:952–981
5. Lis H, Sharon N (1998) Chem Rev 98:637–674
6. Aebi M, Helenius A (2001) Science 291:2364–2369
7. Caminade A-M, Turri C-O, Laurent R, Ouali A, Beatrice D-N (2011) John Wiley & Sons
8. Kolb HC, Finn M, Sharpless KB (2001) Angew Chem Int Ed 40:2004–2021
9. Tornøe CW, Christensen C, Meldal M (2002) J Org Chem 67:3057–3062
10. Rostovtsev VV, Green LG, Fokin VV, Sharpless KB (2002) Angew Chem, Int Ed 41:2596–2599
11. Chandrasekeran S (2016) Click reactions in organic synthesis. Wiley-VCH, John Wiley & Sons, New York
12. Fantoni NZ, El-Sagheer AH, Brown T (2021) Chem Rev 121:7122–7154
13. Fairbanks BD, Macdougall LJ, Mavila S, Sinha J, Kirkptrick BE, Anseth KS, Bowman CN (2021) Chem Rev 121:6915–6990
14. Mendal M, Tornoe CW (2008) Chem Rev 108:2952–3015
15. Tiwari VK, Mishra BB, Mishra KB, Mishra N, Singh AS, Chen X (2016) Chem Rev 116:3086–3240
16. Agrahari AK, Bose P, Jaiswal MK, Rajkhowa S, Singh AS, Hotha S, Mishra N, Tiwari VK (2021) Chem Rev 121:7638–7956
17. Agrahari AK, Mishra A, Tiwari VK (2021) Science of synthesis, pp 135–179
18. Yadav MS, Jaiswal MK, Kumar S, Singh SK, Ansari FJ, Tiwari VK (2022) Carbohyd Res 521:108674
19. Zhang L, Chen X, Xue P, Sun HHY, Williams ID, Sharpless KB, Fokin VV, Jia G (2005) J Am Chem Soc 127:15998–15999

20. Wooley KL, Hawker CJ, Fréchet JMJ (1991) J Chem Soc Perkin Trans 1:1059–1076
21. Miller TM, Kwock EW, Neenan TX (1992) Macromolecules 25:3143–3148
22. Sharma R, Kottari N, Chabre YM, Abbassi L, Shiao TC, Roy R (2014) Chem Commun 50:13300–13303
23. Chen Q, Yang F, Du YG (2005) Carbohydr Res 340:2476–2482
24. Gouin SG, Vanquelef E, Fernandez JMG, Mellet CO, Dupradeau FY, Kovensky J (2007) J Org Chem 72:9032–9045
25. Jia Z, Singh RK, Wang D, Witczak ZJ, Bielski R (eds) (2013) Wiley, pp 271–292
26. Chwalek M, Auzely R, Fort S (2009) Org Biomol Chem 7:1680–1688
27. Srinivasachari S, Fichter KM, Reineke TM (2008) J Am Chem Soc 130:4618–4627
28. Decroocq C, Rodríguez-Lucena D, Russo V, Barragán TM, Mellet CO, Compain P (2011) Chem Eur J 17:12825–12831
29. Zhang Q, Collins LSJ, Chen G, Wallis R, Mitchell DA, Haddleton DM, Becer CR (2014) J Am Chem Soc 136:4325–4332
30. Compain P, Decroocq C, Iehl J, Holler M, Hazelard D, Barrag TM, Mellet CO, Nierengarten JF (2010) Angew Chem, Int Ed 49:5753–5756
31. (a) Yadav MS, Singh SK, Agrahari AK, Singh AS, Tiwari VK (2021) Synthesis 53:2494–2502. (b) Yadav MS, Singh AS, Agrahari AK, Mishra N, Tiwari VK (2019) ACS Omega 4:2418–2424
32. Agrahari AK, Singh AK, Singh AS, Singh M, Maji P, Yadav S, Rajkhowa S, Prakash P, Tiwari VK (2020) New J Chem 00
33. Kumar S, Yadav MS, Maddeshiya T, Asthana S, Jaiswal MK, Tiwari VK, Pandey MD (2023) Org Mat 5:112–117
34. Kumar S, Yadav MS, Singh SK, Rajkhowa S, Tiwari VK (2023) SynOpen 7:145–153
35. Bose P, Agrahari AK, Singh AS, Jaiswal MK, Tiwari VK (2020) Carbohydrates in drug discovery and development. Tiwari VK (ed). Elsevier Inc., pp 213–266
36. Hale AD, Bartkeviciute D, Dargeviciute A, Jin L, Knowles W, Staniulis J, Brown DWG, Sasnauskas K (2002) J Virol Methods 104:93–98
37. Gee GV, Dugan AS, Tsomaia N, Mierke DF, Atwood WJ (2006) Glycoconjugate J 23:19–26
38. Dugan AS, Eash S, Atwood W (2006) J Infect Dis 8:62–67
39. Randhawa P, Brennan DC (2006) Am J Transplant 6:2000–2005
40. Gan Z, Roy R (2002) Can J Chem 80:908–916
41. Marra A, Moni L, Pazzi D, Corallini A, Bridi D, Dondoni A (2008) Org Biomol Chem 6:1396–1409
42. Dondoni A, Marra A (2012) Chem Soc Rev 41:573–586
43. Witczak ZJ, Lorchak D, Nguyen NA (2007) Carbohydr Res 342:1929–1933
44. Dondoni A (2008) Angew Chem, Int Ed 47:8995–8997
45. Fiore M, Chambery A, Marra A, Dondoni A (2009) Org Biomol Chem 7:3910–3913
46. Agrahari AK, Singh AS, Singh AK, Mishra N, Singh M, Prakas P, Tiwari VK (2019) New J Chem 43:12475–12482
47. Agrahari AK, Singh AS, Mukharjee R, Tiwari VK (2020) RSC Adv 10:31553–31562
48. Chabre YM, Brisebois PP, Abbassi L, Kerr SC, Fahy JV, Marcotte I, Roy R (2011) J Org Chem 76:724–727
49. Chabre YM, Pepin CC, Placide V, Shiao TC, Roy R (2008) J Org Chem 73:5602–5605
50. Mollard A, Zharov I (2006) Tricarboranyl pentaerythritol-based building block. Inorg Chem 45:10172–10179
51. Papp I, Dernedde J, Enders S, Haag R (2008) Chem Commun 44:5851–5853
52. Kuijpers BHM, Dijkmans GCT, Groothuys S, Quaedflieg PJLMR, Blaauw H, Delft FLV, Rutjes FPJT (2005) Synlett 20:3059–3062
53. Wu P, Malkoch M, Hunt JN, Kaltgrad RVE, Finn MG, Fokin VV, Sharpless KB, Hawker CJ (2005) Multivalent, bifunctional dendrimers prepared by click chemistry. Chem Commun 46:5775–5777
54. Ardabevskaia SN, Chamkina ES, Krasnova IYu, Milenin SA, Sukhova EA, Boldyrev KL, Bakirov AV, Serenko OA, Shifrina ZB, Muzafarov AM (2022) Int J Mol Sci 23:1546

55. Camponovo J, Hadad C, Ruiz J, Cloutet E, Gatard S, Muzart J, Bouquillon S, Astruc DJ (2009) Org Chem 74:5071–5074
56. Agrahari AK, Jaiswal MK, Yadav MS, Tiwari VK (2021) Carbohyd Res 508:108403
57. Agrahari AK, Kumar S, Pandey MD, Rajkhowa S, Jaiswal MK, Tiwari VK (2022) Chem Select 7:e202202273
58. Kushwaha D, Tiwari VK (2013) J Org Chem 78:8184
59. Okada M, Kishibe Y, Ide K, Takahashi T, Hasegawa T (2009) Int J Carbo Chem, Article ID 305276, 9
60. Das R, Mukhopadhyay B (2016) Tetrahedron Lett 57:2129–2131
61. Rajakumar P, Anandhan R (2011) Eur J Med Chem 46:2687–2695
62. Sharma R, Naresh K, Chabre YM, Rej R, Saadeh NK, Roy R (2014) Polym Chem 5:4321–4331
63. Bagul RS, Hosseini M, Shiao TC, Saadeh NK, Roy R (2017) Polym Chem 8:5354–5366
64. Zhang W, Dhumal D, Zhu X, Ralahy B, Ellert-Miklaszewska A, Wu J, Laurini E, Yao Y, Kao C, Iovanna JL, Pricl S, Kaminska B, Xia Y, Peng L (2022) Chem Eur J, e202201400
65. Kikkeri R, Liu X, Adibekian A, Tsaiab YH, Seeberger PH (2010) Chem Commun 46:2197–2199
66. Prohl M, Moser PD, Czaplewska JA, Hoffmann P, Bus T, Traeger A, Goerls H, Schubert US, Gottschaldt M (2017) Carbohydr Res 446:19–27
67. Bryson JM, Chu WJ, Lee JH, Reineke TMA (2008) Bioconjugate Chem 19:1505–1509
68. Kannan A, Rajakumar P (2015) RSC Adv 5:46908–46915

Chapter 11
Click Chemistry in Polymer Science

11.1 Introduction

The most prominent advantage of glycopolymers in the scientific world is the development of a broad range of multifunctional glycomaterials. Apart from this, 1,2,3-triazole linker has played a fundamental role in stitching polymer chains with the several carbohydrate units and inclusion of such multivalent biocompatible scaffolds has provided the new dimension to polymerization technique for being a distinctive flair [1, 2] On the other hand, there are many ligations techniques, such as atom transfer radical polymerization (ATRP), [3, 4] cobalt-catalyzed chain transfer polymerization (CCTP), [5] ring-opening polymerization (ROP), [6] reversible addition fragmentation chain transfer (RAFT), [7, 8] ring-opening metathesis polymerization (ROMP), and topochemical azide-alkyne cycloaddition (TAAC) [9, 10] to conjoin the sugar moieties to polymers. The click-generated triazole is more entrusted in terms of pre-polymerization and post-polymerization tool [11]. As evident from the literature, post-functionalization is more convenient in terms of unpreferred interactions among insightful polymerization pathways [12]. Multivalent glycomaterials (glycoclusters, polymers, glycodendrimers, and nano particles) showcased the remarkable affinity toward sugar-binding proteins (lectins) owing to their glycoside cluster effect. This affinity has decisive potential in understanding of basic glycobiology and also shows wide applications from diagnostics and targeted delivery to anti-infectives. Apart from this, nitrogen-containing heterocycles such as triazoles have displayed diverse application in pharmaceutical industry, therefore synthesis of 1,2,3-triazoles are important [13, 14]. In addition, 1,4 and 1,5-disubstituted 1,2,3-triazoles constructed through copper-catalyzed azide-alkyne cycloaddition (CuAAC) and ruthenium-catalyzed azide-alkyne cycloaddition (RuAAC) (Scheme 11.1), [15] respectively have shown the profound utility outside the basic synthesis [16, 17]. The aim of this review is to show that glycomaterials has a remarkable impact on glycoscience and this will also introduce the reader to some significant new strategies to achieve it. Remarkable gain in selectivity can only be possible by employing the

V. K. Tiwari et al., *Click Chemistry*, Materials Horizons: From Nature to Nanomaterials,
https://doi.org/10.1007/978-981-97-4596-8_11

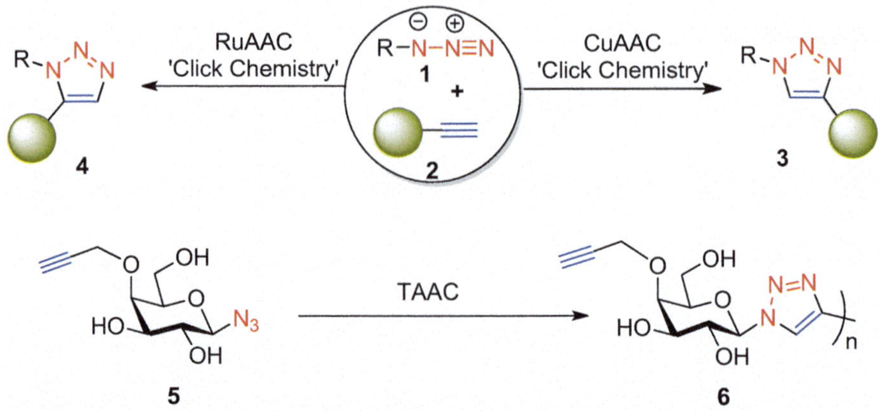

Scheme 11.1 Regioselective formation of 1,2,3-triazole under different catalytic conditions

importance of multivalency; glycans, steric effects, and tune of the linker chemistry. Glycomaterials could be used to understand that the media affects the performance in biological environments [18–20].

11.2 1,2,3-Triazole-Linked Glycopolymer

The use of glycopolymers in science has significantly increased over the last two eras, which has stimulated the development of a wide variety of multifunctional glycopolymers. Curdlan **7** was converted into 6-azido-6-deoxycurdlan (CUR-N_3) **8** using a high-yielding and demonstrated process. Initially, curdlan was altered into (CUR-Br) by utilizing PPh_3, DMF, and LiCl; further it was azidated by using sodium azide to create CUR-N_3. These azides were grafted onto the carbohydrate-derived alkynes (**9a–c**) using the modular CuAAC technique to produce the carbohydrate-appended curdlans (**10a–c**) (Scheme 11.2) [21]. Additionally, the newly synthesized carbohydrate-containing curdlanes can interact with polycytosine to form stable macromolecular compound consisting of two polysaccharide constituents and one polycytosine strand that are well-built and have a selective affinity for sugar-binding proteins (lectins). As a result, the carbohydrate-based curdlanes can be used as a new class of glycoclusters.

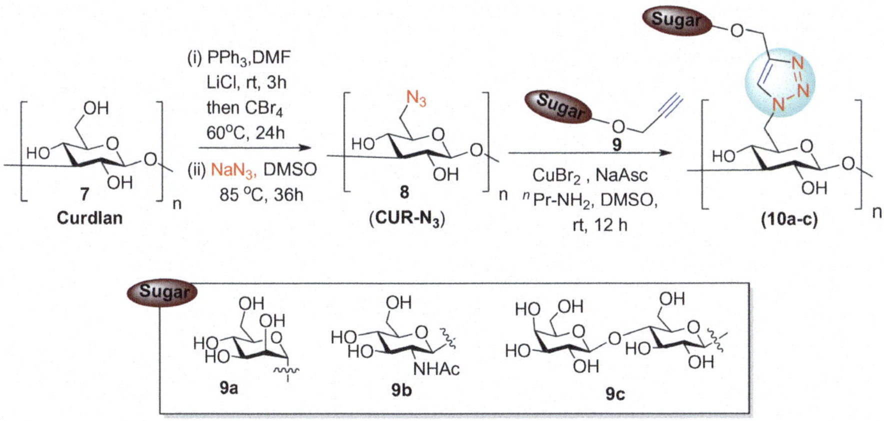

Scheme 11.2 Utilization of click technique for the synthesis of carbohydrate-appended curdlans **10a–c**

11.3 1,2,3-Triazole-Appended Glycopolymers

Haddelton and co-workers highlighted the prospect of the modular click method in the construction of maleimide-ended neoglycopolymers (PDI = 1.20–1.28) along with numerous D-mannose epitopes using ATRP polymerization path (Scheme 11.3). In route **I**, alkyne methacrylate **11a** was clicked with mannose azide **12** to form mannose-derived monomer **13**, which underwent polymerization to afford the desired neoglycopolymer **14** [22]. Post-polymerization involves the post-translational modification of pre-synthesized. Alternatively, the authors have used the TMS-protected propargyl methacrylate monomer **11b** to avoid the radical transfer to the alkyne bond during radical polymerization (Scheme 11.4). After the synthesis of alkyne-functionalized polymer **15**, it was coupled to the mannosyl azide **12** to furnish the mannose-appended glycopolymer **14** with no appreciable change in its polydispersity index.

The potential tool for the synthesis of a series of glycomaterials is further extended by one pot method for alkyne-azide cycloaddition reaction between functionalized methacrylate **11a** and azido-mannose **12** coupled, followed by radical polymerization to afford glycopolymer **16** (Scheme 11.5) [23]. Although, these methods are quite interesting and may be implemented in the industry to achieve the significant useful glycomaterials.

Catalytic chain transfer polymerization (CCTP) is a prominent technique, this results in ω-end functional polymers with the desired properties, such as increased chain-end integrity and regulated molecular weight. To produce the probable glycopolymers with a wide range of applications, sequential exploitation of CCTP with thio-Michael addition and the click technique has been implemented. In this continuation, Haddleton and groups have achieved the 'double-clickable' alkyne-functionalized macromonomer **18** with the consecutive implementation of catalytic

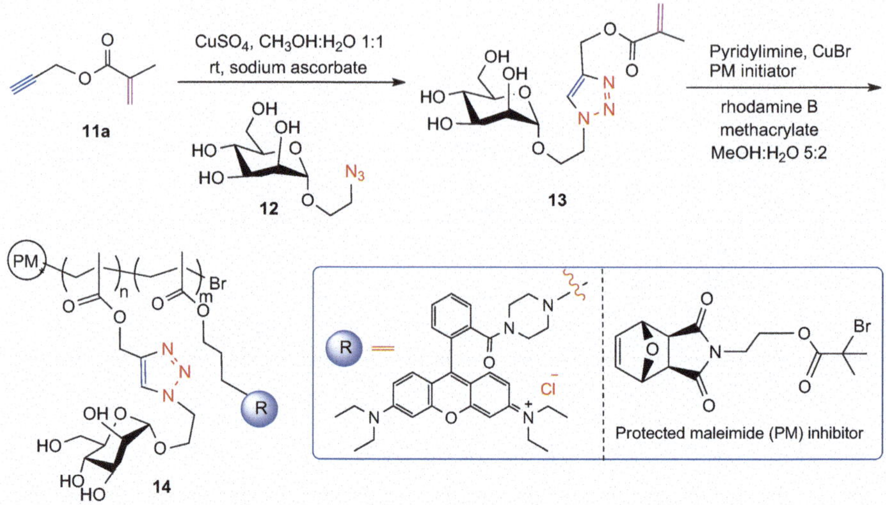

Scheme 11.3 Synthesis of triazole-appended neoglycopolymer using ARTP

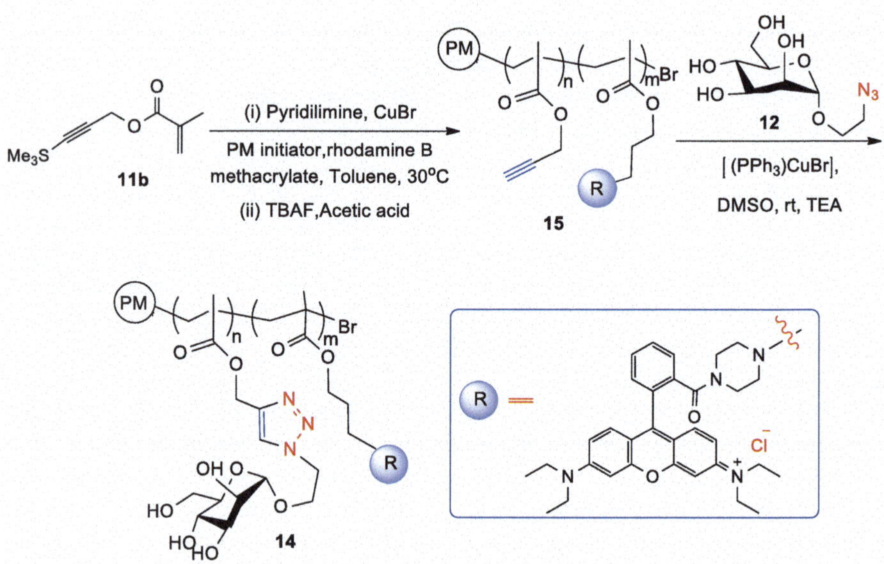

Scheme 11.4 Synthesis of 1,2,3-triazole-appended neoglycopolymer **14** using ARTP followed by CuAAC

chain transfer polymerization (CCTP) and thio-Michael addition to scaffold **11b**. Thereafter, glycosyl azides were allowed to stitch with polymer (**18a**, **b**) by using CuBr to furnish the required series of glycopolymers **19a–c** containing galactose, mannose, and cellobiose, respectively (Scheme 11.6) [6]. Moreover, glycopolymers

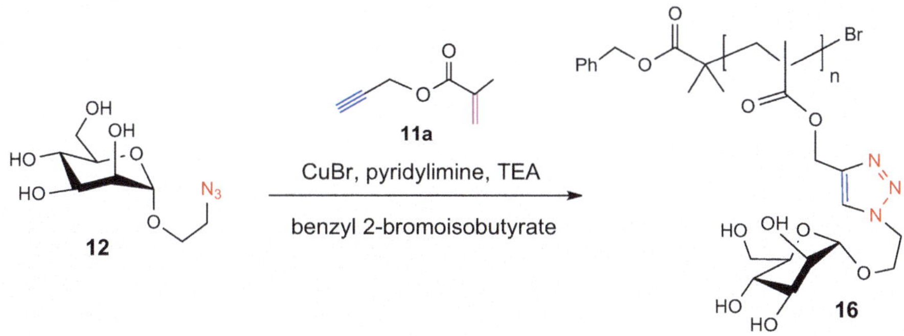

Scheme 11.5 Synthesis of 1,2,3-triazole-appended neoglycopolymer **16** using ARTP

having gal and man have depicted a significant affinity toward the RCA I and Con A, respectively.

Click tool is a flexible method that can be used to produce bioactive stable glycomonomers and glycopolymers. Another polymerization technique, i.e., reversible addition fragmentation chain transfer (RAFT) was implemented

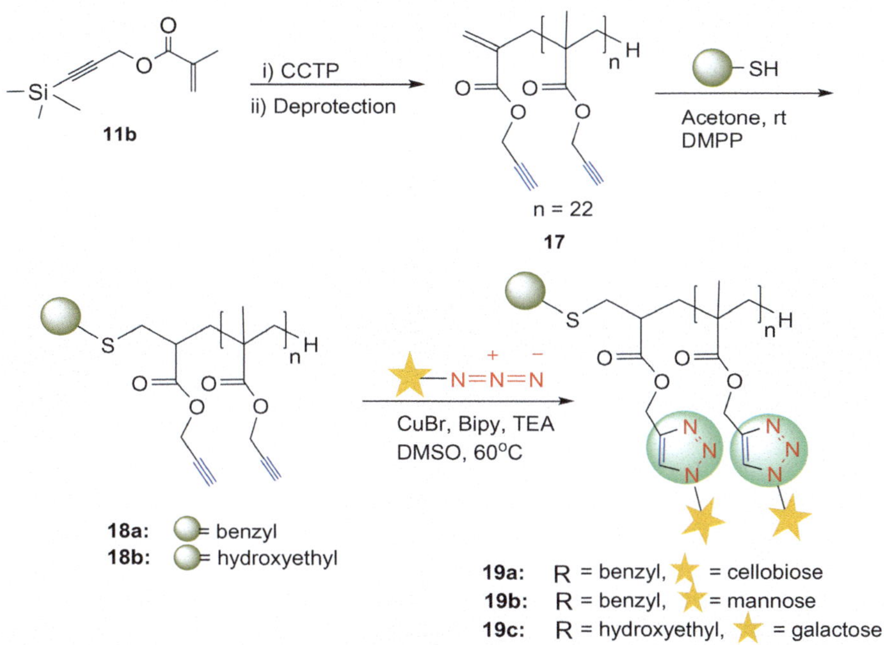

Scheme 11.6 Synthesis of triazole-linked glycopolymer **19a–c** by employing CCTP and click

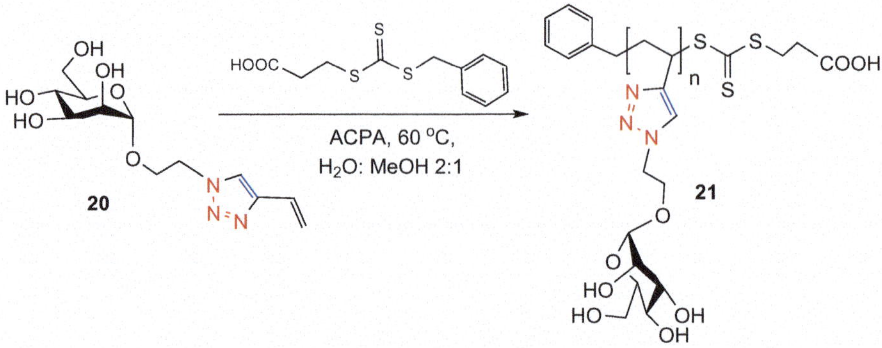

Scheme 11.7 Development of triazole-appended glycopolymer **21**

to generate the suitable functionality at α- and ω-position in the desired polymers. For this purpose, a click-inspired method mediated by reversible addition fragmentation chain transfer (RAFT) has been used to synthesize several types of glycopolymers. In this context, Stenzel et al. has synthesized a triazole appended mannose-based glycomonomer **20** by click conjugation. The resulting compound was then further undergoing polymerization under RAFT condition (i.e., 3-benzylsulfanylthiocarbonylsulfanyl propionic acid) to furnish the respective glycopolymer **21** (Scheme 11.7) [7] Biological investigation of glycopolymer **21** revealed the enhanced interaction with ConA with PDI of 1.16.

Moreover, glycopolymer containing fluorescent moieties have prompted significant biological applications, including biosensing, sugar-lectin interactions, and chemical biology. In this regard, Stenzel and co-workers have constructed block polymers of PDEGMA using RAFT technique [5]. Similarly, Han and co-workers have prepared the fluorescent-conjugated poly-fluorene **22** coated with lactopyranosyl glycopolymer **23** using triazole ligation and subsequent Suzuki coupling polymerization. Moreover, based on fluorescence spectroscopy the authors also claimed that the developed polymer was first used for the metal-mediated sugar-sugar interaction (Scheme 11.8) [24].

Haddleton et al. demonstrated the use of living radical polymerization (LRP) to create innovative, distinctive alkyne-functionalized polymers with a restricted molecular weight dispersion. In this study, the authors have underlined the synthesis of mannose and galactose-based glyocopolymers **25** by clicking mannose and galactose-based azides with alkyne-containing homopolymer **24** (Scheme 11.9) [25]. This 'co-clicking' tool successfully achieved the importance of guarded radical copolymerization along with the increased dexterous post-modification operation; leading to formation of multivalent glycopolymers.

In order to create polyphosphazene glycopolymers, Huang et al. extensively utilized the exciting idea of click chemistry. A standard click tool was used to bind alkyne-functionalized poly(dichlorophosphazene) to carbohydrate-based azides [26]. Reineke group has reported the synthesis of biocompatible triazole-appended

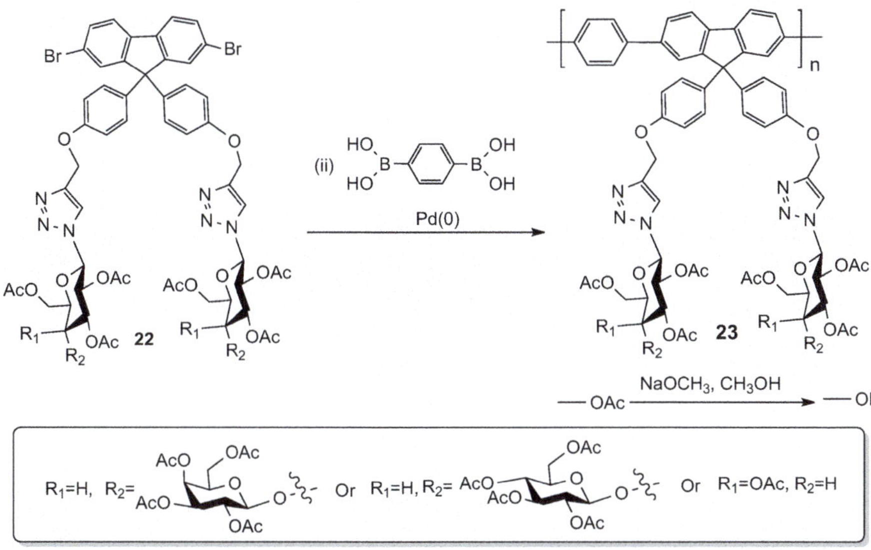

Scheme 11.8 Development of lactopyranosyl-coated fluorescent-conjugated polyfluorene glycopolymer

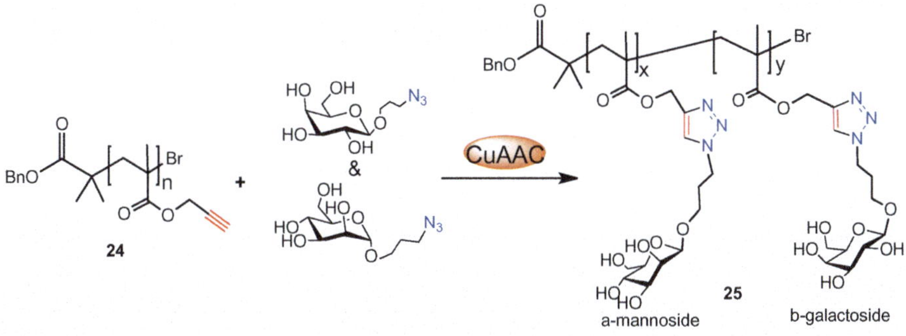

Scheme 11.9 Development of neoglycopolymer **25** by means of 'co-clicking' with mannose and galactose-based azides

glycopolymer. In order to achieve glycopolymer **28a–c**, the authors coupled the dialkyne derivatives **27a–c** and diazido disaccharide monomer **26** in equal molar ratio under standard click condition followed by deacetylation and then de-BOC using TFA in acetic acid (Scheme 11.10) [27]. Apart from this, the biological evaluation of the resulting glycopolymers **28a–c** revealed its biocompatibility as well as ability to deliver pDNA into HeLa cells.

Nguyen et al. have developed a bivalent polymerizable glycan monomer which has capability to polymerize in to the targeted neoglycopolymer **30**. To construct the polymerizable glycan monomer, the authors coupled the ROMP-capable rigid linker

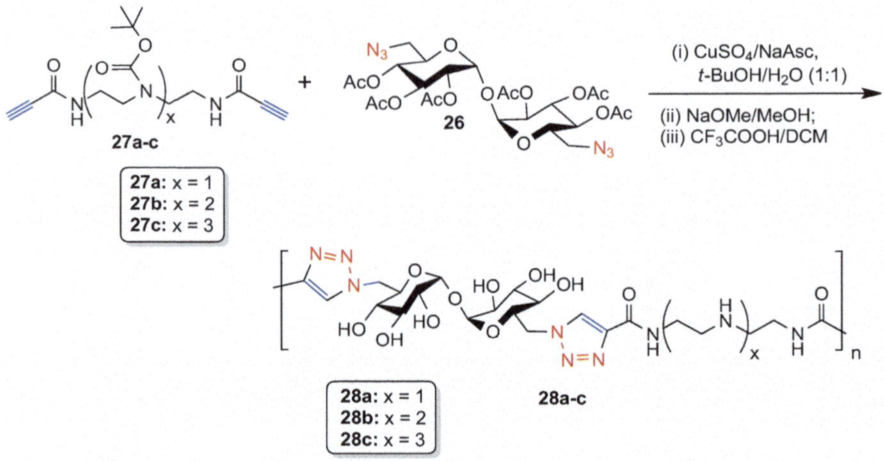

Scheme 11.10 1,2,3-triazole-appended polymerization toward development of glycopolymers

having alkyne functionality with man-α-(1–2)man disaccharide **29** unit of the D1 arm of $Man_9GlcNAc_2$ of HIV gp120 (Scheme 11.11) [28].

It's fascinating that Finn and co-workers created flexible polyvinyl chloride (PVC) tubing containing azide, which allows for various surface functionalization via azide-alkyne cycloaddition processes [29]. Hatano and Matsuoka have conveniently achieved the required glycopolymer using styrene-modified water-soluble sugar monomer **31**. The authors successfully polymerized the GlcNac monomer **31** to construct the corresponding white powdery triazole-appended glycopolymers **32** and also studied their biological interaction against wheat germ agglutinin (WGA) (Scheme 11.12) [30].

Multivalent-glycosylated materials (polymers, surfaces, and particles) frequently exhibit strong carbohydrate affinity to the binding proteins based on the glycoside effect. This affinity increase has applications ranging from diagnostics, biosensors, and targeted distribution to anti-infectives and basic glycobiology. This viewpoint emphasizes the issue of selectivity of binding, which is less frequently addressed due to the reductionist nature of glycomaterials and the promiscuity of many lectins binding protein. The utilization of macromolecular features to overcome this difficulty, including as architecture, heterogeneous ligand display, and the incorporation of non-natural glycans, is described in terms of selectivity of binding by the Richards and Gibson. By targeting secondary sites, heterogeneity improves affinity/selectivity, although this is not the only way of action. Steric shielding from a nonbinding partner is another method of action for heterogeneity gains (and hence potentially selectivity). The click-derived glycopolymer, α-mannosylated glycopolymer **33** on dilution with β-galactose side chains reduced DC-SIGN binding (Fig. 11.1a) [31]. Alternatively, as observed with mannosylated gold nanoparticles **34** for reducing FimH-driven adhesion, heterogeneity may play just a minor effect (Fig. 11.1b) [32, 33] through dilution with galactose unit **35**.

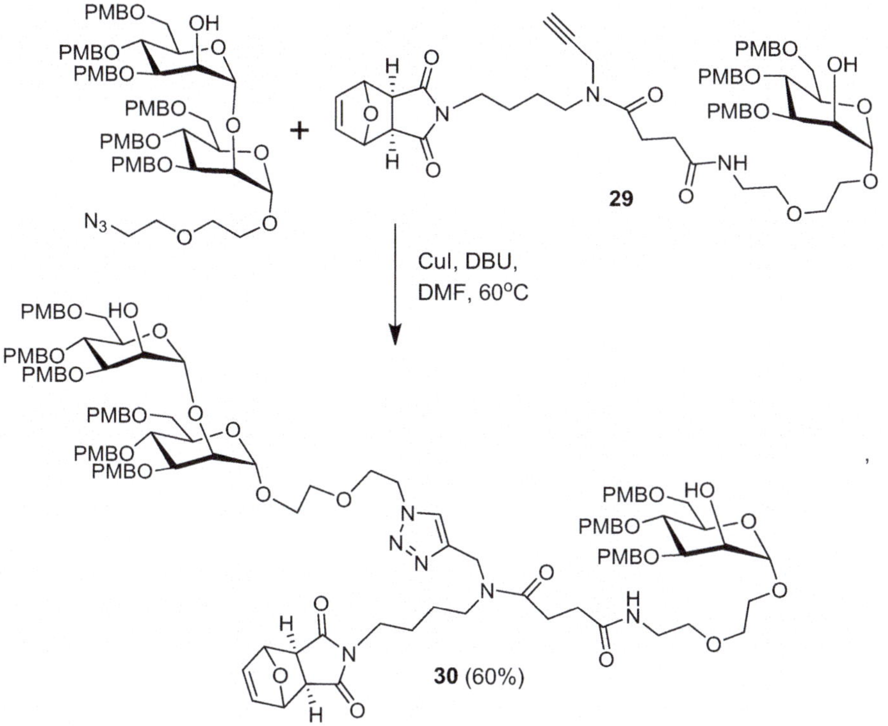

Scheme 11.11 Synthesis of 1,2,3-triazole-linked glycan monomer

Scheme 11.12 Synthesis of triazole-appended glycomaterial from styrene-manipulated sugar monomer

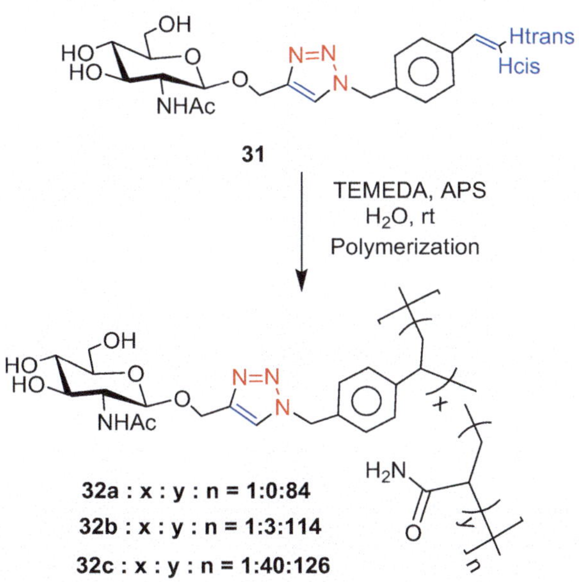

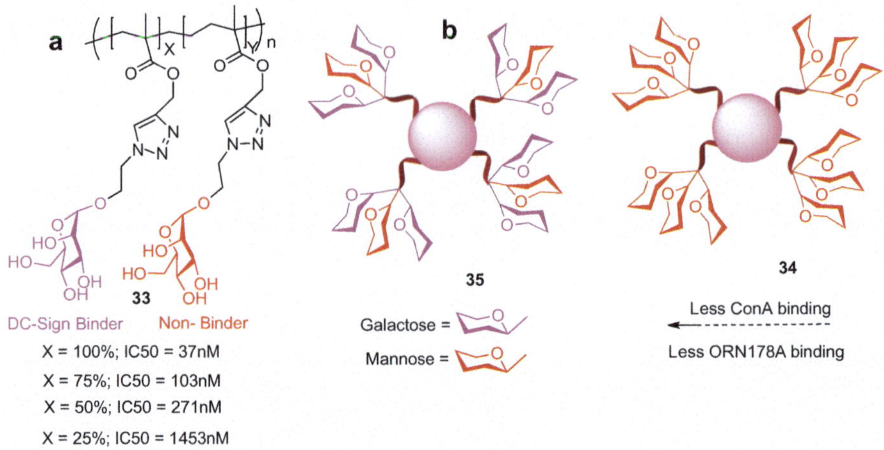

Fig. 11.1 Glycan heterogeneity and targeting binding affinity: **a** Mannosylated glycopolymer **33** displays reduced inhibitory activity *vs* DC-SIGN on dilution by galactose reveals the connectivity between heterogeneity and binding affinity; **b** mannosylated glycoparticles **34** with reduced affinity toward Con A and ORN178 (*E.* coli) as galactose is presented **35**

As a result of their biocompatibility and environmentally benign properties, glycopolymer-based green electronics are surprisingly quite popular. In order to create a resistive memory device with a high on/off current quotient that could be operated often, Karakawa and groups used modulated carbazolyl-adorned cellulose as the electret layer [34]. In extension Chen and colleagues described the synthesis, morphology, and OFET characteristics of malto-heptose-based diblock copolymers such as maltoheptose-block-poly(1-pyrenylmethyl methacrylate (MH-b-PPyMA, **39**) and related supramolecules with MH(4Py-Acceptor-4Py)x-b-PPyMA. The formation of polymers PPyMA-OSi^{i}Pr$_3$ (**36a–c**) from the t-Bu-P4-catalyzed GTP utilizing TIPS-MTS as an initiator was the first step in the synthesis of the required glycopolymer MH-b-PPyMA (**39**). Using TBAF, the polymer (**36**) triisopropylsilyl group was removed, resulting in PPyMA-OH (**37a–c**), in the next steps which, when combined with 6-azidohexanoic acid was produced by using EDC/DMAP in dichloromethane to produce the azide analogue PPyMA-N$_3$ (**38a–c**). Further, the azide-functionalized compound **38** was clicked with two equivalents of alkyne-functionalized maltoheptose (MH) in the occurrence of CuBr/PMDETA to provide the final required glyco-diblock copolymer, MH-b-PPyMA, **39a–c** (Scheme 11.13) [35]. The results of this work provided concrete proof that the ratio of acceptor to donor in block copolymer-based supramolecules can play a critical role in controlling the memory distinctiveness.

Kolender and co-workers have designed the synthesis of α-azide-ω-alkynyl ester monomers to produce the hydrolytically degradable poly(ester-triazoles) by applying the α-azido-ω-alkyne ester **43** as a monomer. The first step in making monomer was brominating *D*-galactone-1,4-lactone **40**, which after being treated with sodium azide and then alkaline hydrolysis, produced *O*-isopropylidene-protected galactonic acid

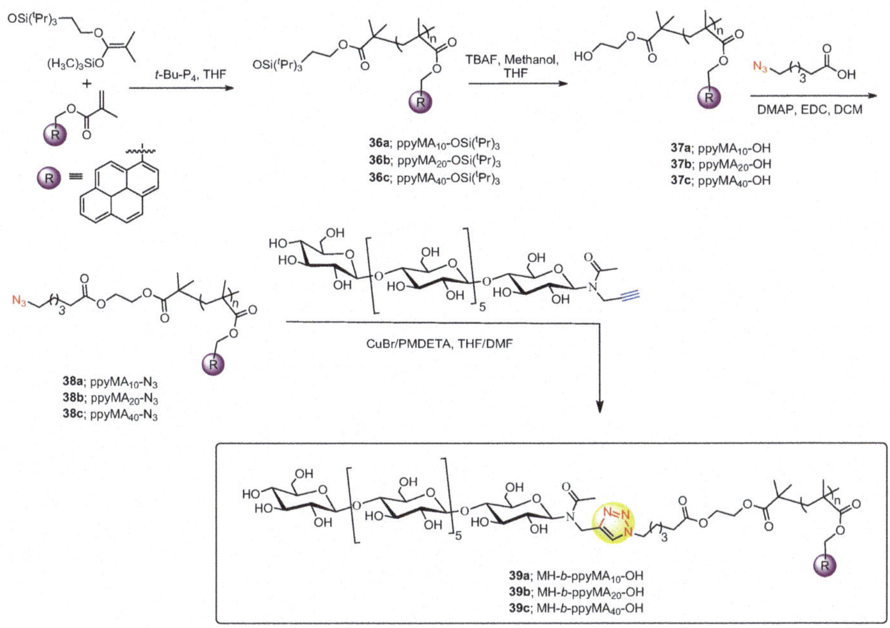

Scheme 11.13 Development of MH-b-diblock co-polymers with π-conjugated pyrene scaffolds

42. In the next steps in order to produce the required monomer **43** with a yield of 68% overall, the compound **42** esterification with propargyl alcohol was carried out. The obtained monomer **43** undergoes click polymerization in optimized condition to provide poly(ester-triazoles) with an inconstant amount of cyclic-oligomer. Therefore, click polymerization of monomer **43** in bulk generated mostly poly(ester-triazole) **44**, however cyclic compounds (**46–48**) have succeeded when the reaction was passed out with the diluted solution of monomer (Scheme 11.14) [36]. The poly(ester-triazole) **45** was eventually created without using copper as a catalyst. While random distributions of triazole rings were observed in the polymer backbone during metal-free polymerization, Cu-catalyzed click polymerization produced 1,4-regioselective triazolyl products. The obtained polymer easily hydrolyzed under high acidic or basic conditions, and the breakdown products were also examined.

The heterogeneous glycopolymers can be constructed with a variety of sugar motifs that can be used to improve the affinity and selectivity to a particular receptor, which also has a great significance in the glycopolymer area. Bearing in mind the importance of heterogeneity, in 2019, Liu and co-workers conducted a variety of studies to create a library of homo, hetero, as well as a block and random glycopolymers. First, from the molecule cis-5-norbornene-exo-2,3-dicarboxylic anhydride, pre-monomer **50** was created through a sequence of processes. Using the appropriate reagent, the compound was converted into the necessary intermediate **49**, which was then propargylated to produce the desired dialkyne-scaffold **50**. In this

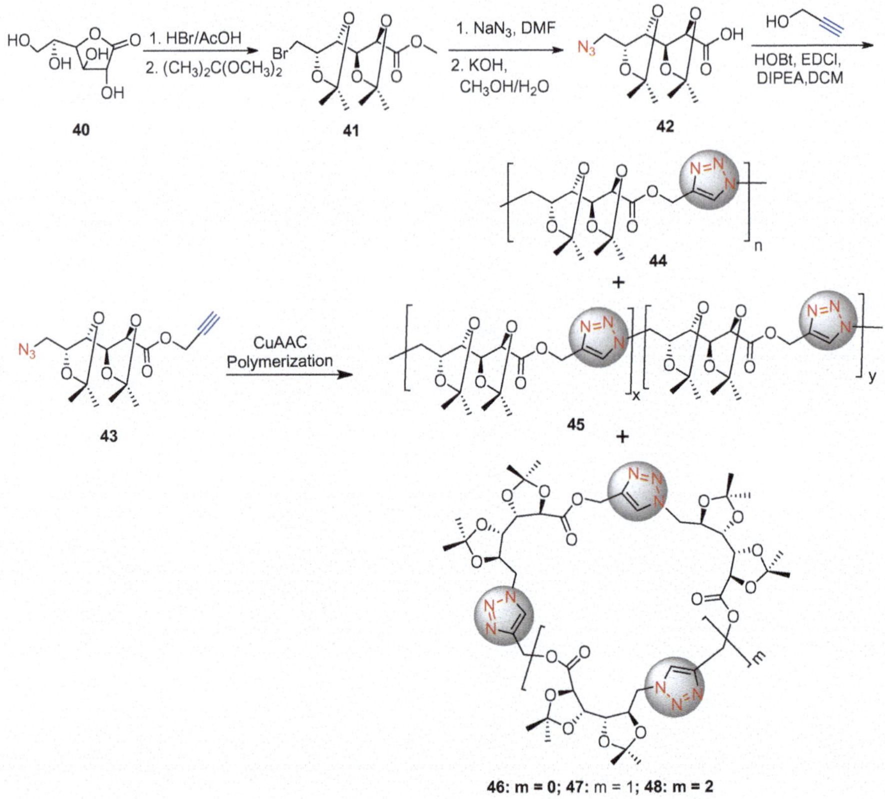

Scheme 11.14 Development of polyester-triazoles (**44**–**48**) using CuAAC click technique

regard, the authors developed a straightforward method to produce different acetyl-protected homodisaccharide monomers (**51a**–**54a**). Using the CuAAC technique, several carbohydrate units (*β*-D-glucose, *α*-D-glucose, *β*-D-galactose, and *α*-D-mannose) were incorporated into a single pre-monomer **50** to create monosaccharide monomer (**51b**, **c**) and heterodisaccharide monomer (**52b**–**54b**) (Scheme 11.15) [37]. Additionally, these monomers were polymerized under the standardized ring-opening metathesis polymerization (ROMP) conditions to produce the desired acetyl-protected glycopolymer, which upon ester hydrolysis in the presence of $NaOCH_3/CH_3OH$ condition to give corresponding homo, hetero, and block-random glycopolymer was obtained. In addition, it has been discovered that heteroglycopolymers, such as poly(Man-*β*-Glu) and poly(Man-*β*-Gal), containing both *α*-mannose and *β*-glucose or *β*-galactose have an affinity for concanavalin A (Con A) that is more than five times greater than that of poly(man-alkyne), a monoglycopolymer, and that is approximately a thousand times greater than that of methyl-*α*-mannose. It has been determined that heteroglycopolymers with non-binding sugar scaffolds

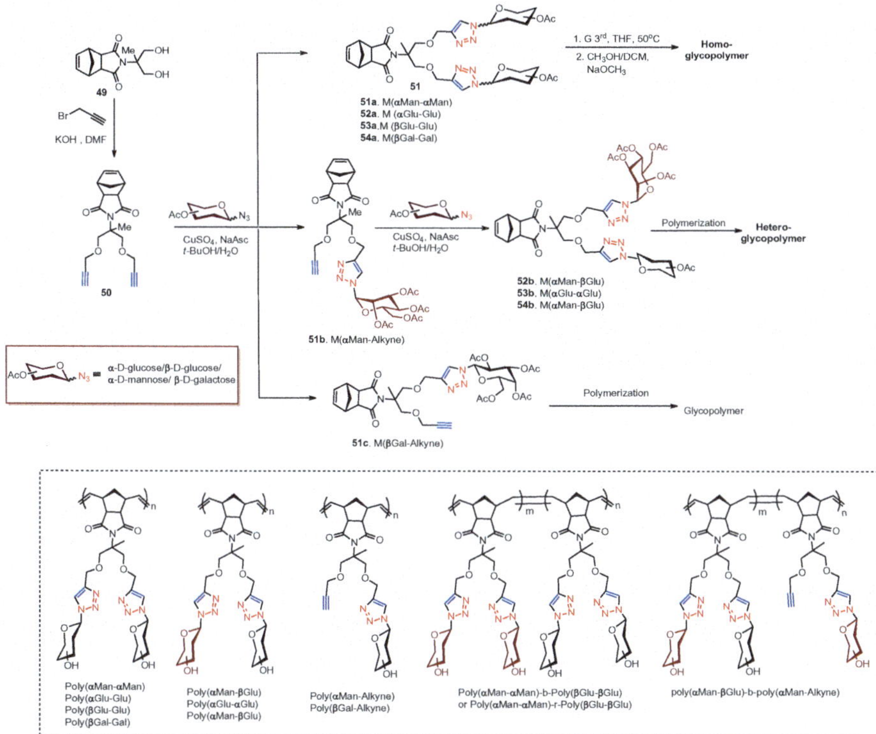

Scheme 11.15 Development of glycomonomers and their corresponding homo, heteroglycopolymers using CuAAC and ROMP approach

have demonstrated better selectivity for Con A than monoglycopolymers due to the synergistic impact.

Additionally, for the study of biological signal transmission, glycopolymers have developed into remarkably flexible glycan scaffolds. In this work, different interfacial glycopolymers built onto gold foil or nanoparticles are assessed for the alleged interaction between glycan and proteins. In order to improve the effectiveness and accessibility of this transducer interface, a combined protocol of copper(0)-mediated living radical polymerization (Cu(0)-LRP) and 'CuAAC' click reaction approach was used to generate a library of novel glycopolymer precursors with altered scaffold structures and pyranose ligands. The formation of poly (GMA) **57** moiety, which is made up of compounds **55** and **56** utilizing the (Cu(0)-LRP) method, is the first step in the synthesis of glycopolymers (Scheme 11.16) [38]. In the next steps compound **57** was on azidation in the presence of NaN_3, NH_4Cl in DMF at 50 °C to afford the compound **58**, i.e., poly(azide), it was clicked with different sugar-alkynes (glucose, galactose, and mannose) while utilizing CuBr and bipyridyl in DMF to produce the appropriate poly(glucose), poly(galactose), and poly(mannose), glycopolymers (**59a–59c**). Additionally, the glycopolymer brush (GlyB) surfaces harboring various

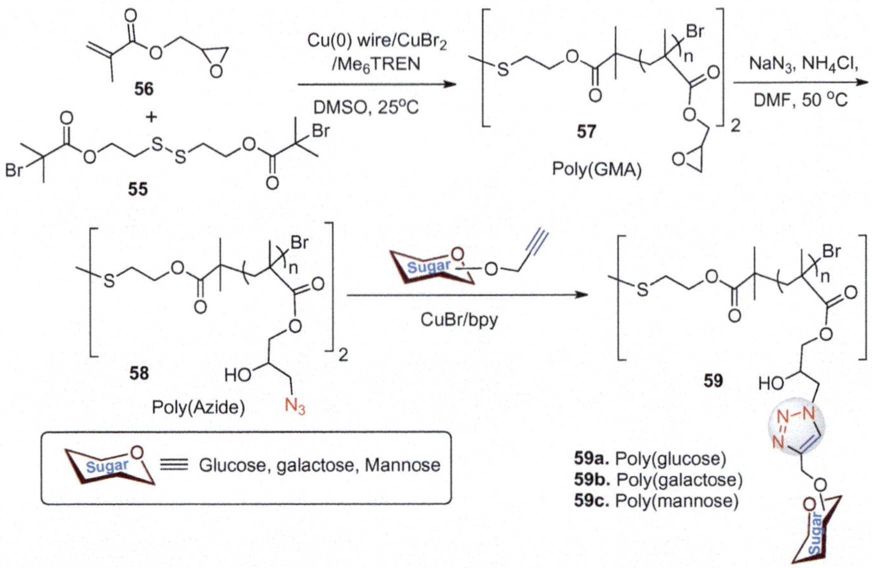

Scheme 11.16 Synthesis of poly(glucose), poly(galactose), and poly(mannose) glycopolymers utilizing CuAAC 'click chemistry'

saccharides (glucose, galactose, and mannose) were examined to determine whether or not they adhered to concanavalin A, peanut agglutinin (PNA), and bovine serum albumin. Among them, a proof-of-concept aggregation-mediated sensing system was created using the high affinity between poly(galactose) and PNA.

According to Frechet et al., a pH-sensitive system designed for immune-modulation using mannosylated dextran nanoparticles has been reported. Azido-AcDEX nanoparticles **60** were first suspended in HEPES buffer (Conc = 0.1 M, pH = 8.0), and after that, they were reacted with mannoside alkyne **61** in the presence of CuSO4/NaAsc and tris(benzimidazolylmethyl)amine $(BimC4A)_3$ **62** to produce a good yield of the desired man-AcDEX particle **63**. Interestingly, the click technique becomes more helpful in terms of an incremental increase in reaction rate when using the Cu-stabilizing ligand **62** (Scheme 11.17) [39]. Surprisingly, an immune modulatory system with dextran-based pH-sensitive material could be created using the well-known method [40]. Also, the author clearly explained how mannose-inspired monosaccharide-targeted cellular consumption by DCs results in better MHC I antigen than the untargeted entity.

The azide-alkyne click polymerization synthesis, which was Cu(I)-catalyzed, solely produced the 1,4 regioselective isomer with virtually quantitative yield, as is clear from the literature. In mark contrast, Tang et al. have developed a powerful Ru(II)-catalyzed azide-alkyne click polymerization tool that is tuneable through the use of appropriate ligands present on the Ru(II) catalysts. The authors used the catalyst $RuH_2(CO)(PPh_3)_3$ to make a novel 1,4-regioregular polytriazole **64** with strong

Scheme 11.17 Ligand $(BimC_4A)_3$-assisted click-inspired development of mannosylated AcDEX particle **63**

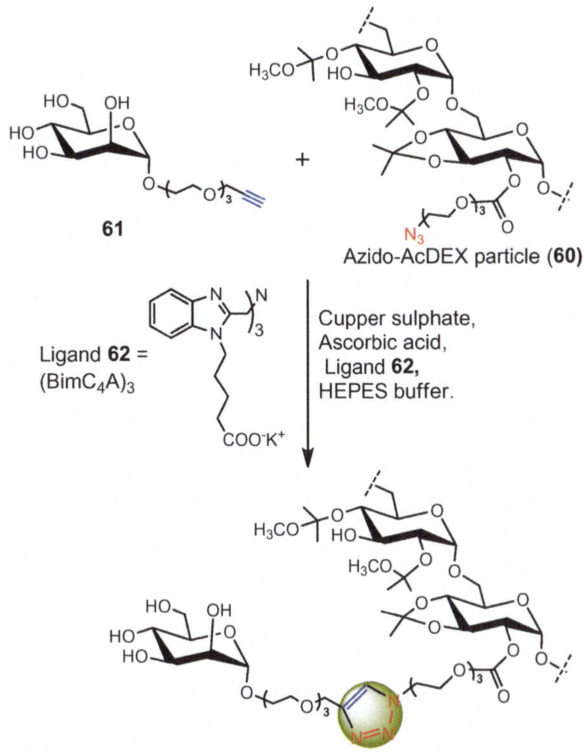

regioselectivity (up to 100%), while the catalyst $Cp^*Ru(PPh_3)_2Cl$ produced a 1,5 regioregularpolyazole **65** with complete regioselectivity (Scheme 11.18) [41].

Zhang et al. have used CuAAC click tool for the conjugation of azide-functionalized cyclodextrin (CDs) with mannosylated alkynes followed by living radical polymerization and developed cyclodextrin core mannose cluster **66** as star glycopolymers (Fig. 11.2) [42]. The resulting clycodextrin-mannose star glycopolymers **66** portrayed an excellent adhesive property toward human transmembrane lectin. Further the inhibition potency impedes the binding of HIV protein gp120 to DCSIGN at minimal concentrations. The authors have showcased the usage of sugar units with CD-core glycopolymer **66** and this would lead to a potential inhibitor of adhesion of gp120 with CRDs of DCSIGN. Astonishingly, the glycopolymer showed brilliant loading capacity of anti-cancer and anti-HIV drugs due to the presence of multiple saccharide units. Driven by the outcomes, the wider explorations of the application of lead molecules were conducted to counteract HIV and ineffective drug release.

Further, the aggregation-induced enhancement (AIE) fluorescence behavior of tetraphenyl ethylene (TPE) scaffold was broadly investigated for the conformational mobility of the fluorophore upon aggregation. Qin and co-workers prepared

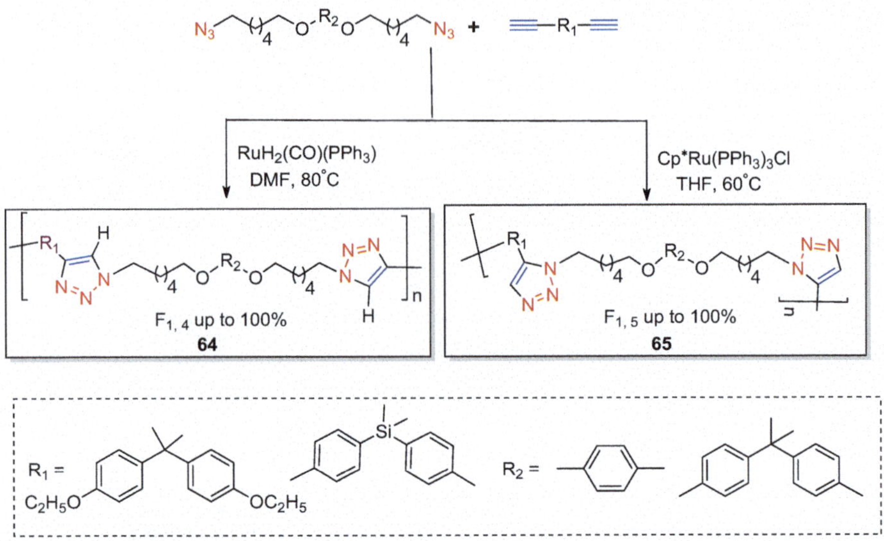

Scheme 11.18 Ligand control Ru(II)-catalyzed click polymerization reactions leading to regiodivergent 1,4-disubstituted 1,2,3-triazole-linked and 1,5-disubstituted 1,2,3-triazole-linked polymers

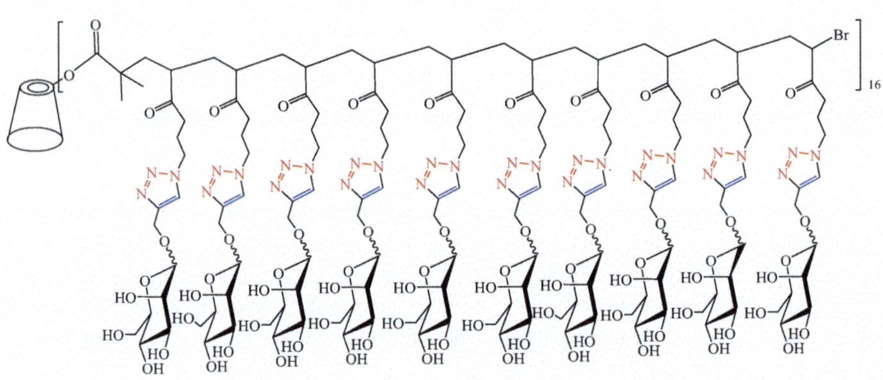

Fig. 11.2 CuAAC click chemistry-inspired structure of star-shaped glycopolymer **66**

tetraphenyl ethylene (TPE) polymer scaffold **69** with reaction of alkyne **68** with azide-fictionalized compound **67** via CuAAC (Scheme 11.19). These polymers 69 demonstrated the expected aggregation-induced enhancement (AIE), which resulted in a precipitation process that increased the fluorescence quantum yield by approximately 100 times [43]. Additionally, these nanoaggregates in solution showed effective quenching in response with aromatic compounds rich in nitro groups, like TNT, suggesting that they could be valuable sensors for nitro-aromatic explosives.

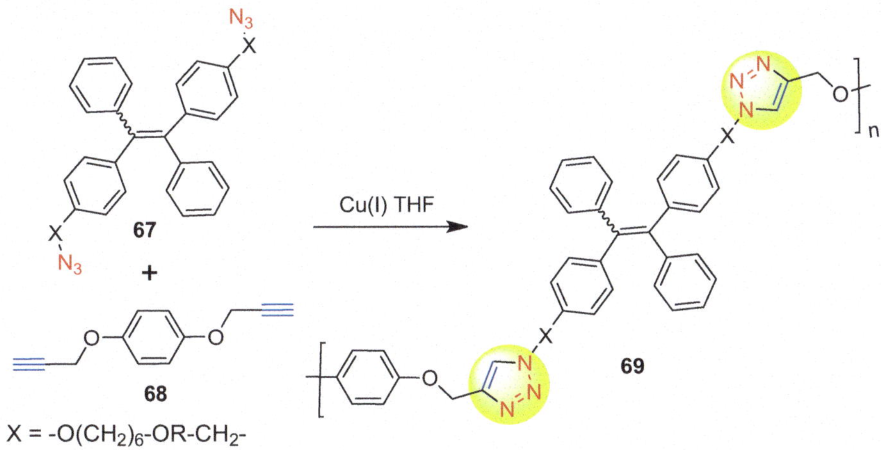

Scheme 11.19 Polymer incorporating tetraphenyl ethylene via CuAAC; polymer displays AIE

Qui and co-workers applied the click reaction for the synthesis of cyclic poly (N-isopropylacrylamide), c-PNIPAM **73** and study the solubility behavior as effect on the LCST. First of all, they prepare the precursor of c-PNIMMA **73** from azido-derivatized RAFT agent (AIP) **72** which was prepared from one pot aminolysis followed by thiol-Michael addition on trithiocarbonate end group **71** in the presence of propargyl acrelate and n-butylamine as well, the trithiocarbonate end group **71** was formed by azido derivative **70** (Scheme 11.20) [44]. At high-dilution condition, cyclization of c-PNIMMA **73** was going well with good yield.

In year 2008, Qin and group came with new approach and demonstrated that hyper-branched polymers (HBPs) **76** could be prepared by using 1,3,5-tris(4-ethynylphenyl) amine **74** and an appropriate diazide **75** via CuAAC route (Scheme 11.21) [45]. Inevitably, this method required careful monitoring of both stoichiometry and conversion to avoid the generation of cross-linked products. The group also showed that how the presence of residual azide groups in these systems allows the polymer to be photo-cross-linked by the photolysis of the azide, which produces reactive nitrenes that may help to cross-links through a C–H insertion process. For this reason, these HBPs were employed as photo-patternable coatings.

In order to polymerize styrene and vinyl acetate independently, Quemener et al. constructed two unique reversible addition fragmentation chain transfer (RAFT) agents. One of the RAFT agents added an azide unit **77** at the end of polyvinyl acetate chain and the other side alkyne unit **78** to the end of the polystyrene chain. The copper-catalyzed azide-alkyne cycloaddition reaction among these two polymers facilitated the development of a polvstyrene-block-polyvinyl acetate **79**, which is usually a difficult block copolymer to construct, due to the substantial variations in the reactivity's of the two monomers (Scheme 11.22) [46].

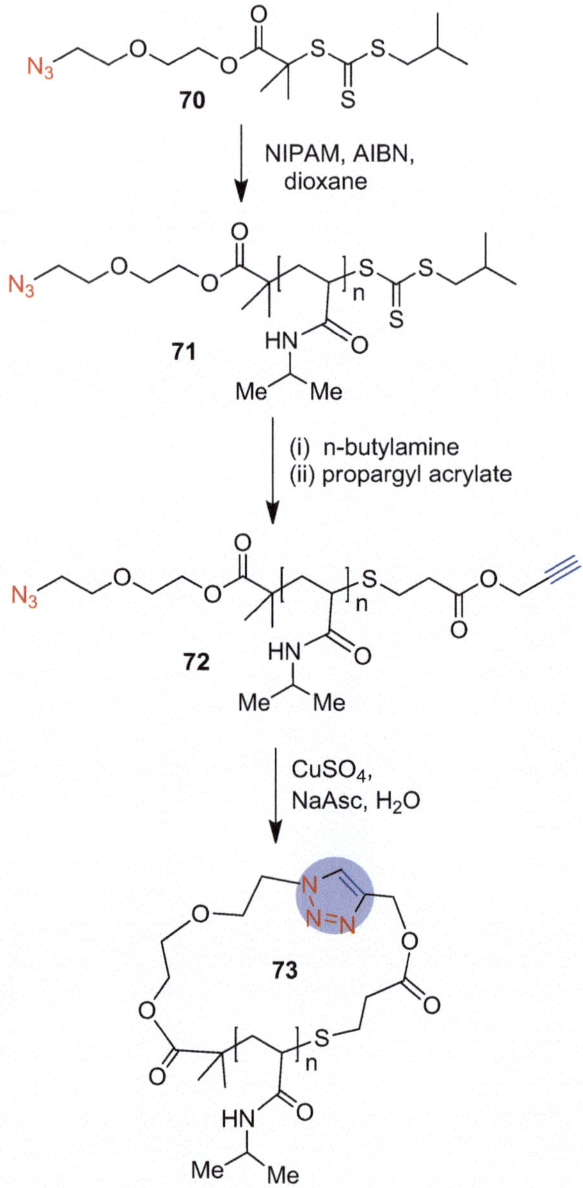

Scheme 11.20 Synthesis of heterotechelic PNIPAM via CuAAC reaction

Porous organic polymers are one of the most fascinating scaffolds in chemistry and many chemists have focused their attention on this field driven by the impressive variety of structures in metal–organic framework solids (MOFs) and covalent organic frameworks (COFs). These interesting frameworks have been investigated over the past few decades, particularly regarding their potential applications in gas

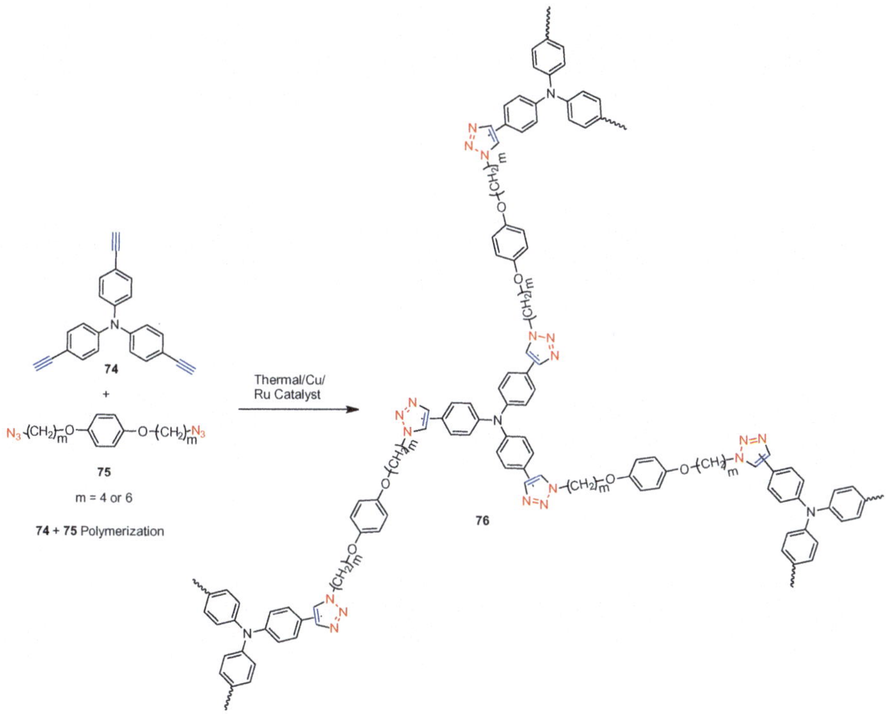

Scheme 11.21 Reaction with trialkyne and diazide via CuAAC to give hyperbranch polymer **76**

adsorption and catalysis. Numerous approaches have been explored to develop porous organic polymers and conjugated microporous polymers (CMPs). These materials are essentially highly interconnected polymers constructed from rigid multifunctional monomers. Notably, Holst and group were the pioneers in utilizing the CuAAC reaction for the synthesis of porous polymeric **82**. The group employed tetrakis(4-azidophenyl) methane **80** and tetrakis(4-ethynylphenyl) methane **81** as starting materials in the presence of $CuSO_4$ and ascorbic acid, to obtain three-dimensional cross-linked polymeric materials (Scheme 11.23) [47].

Star polymers are the most basic form of branched polymers, characterized by the connection of multiple linear chains to a single molecule with multiple binding sites. These star polymers can have various numbers of arms, ranging from three to four to even twelve [48]. The most efficient method for producing such polymers typically involves utilizing a living polymerization technique, like anionic polymerization, and concluding the process with a nucleophilic substitution reaction on a multifunctional core. With the introduction of (CRPs) and the copper-catalyzed azide-alkyne cycloaddition (CuAAC) reaction, Gao and Matyjaszewski pioneered a novel approach by combining ATRP and click chemistry to synthesize star-branched polymers **85**. As previously explained, ATRP can be easily employed to transform the terminal bromide, usually introduced at one chain end during ATRP,

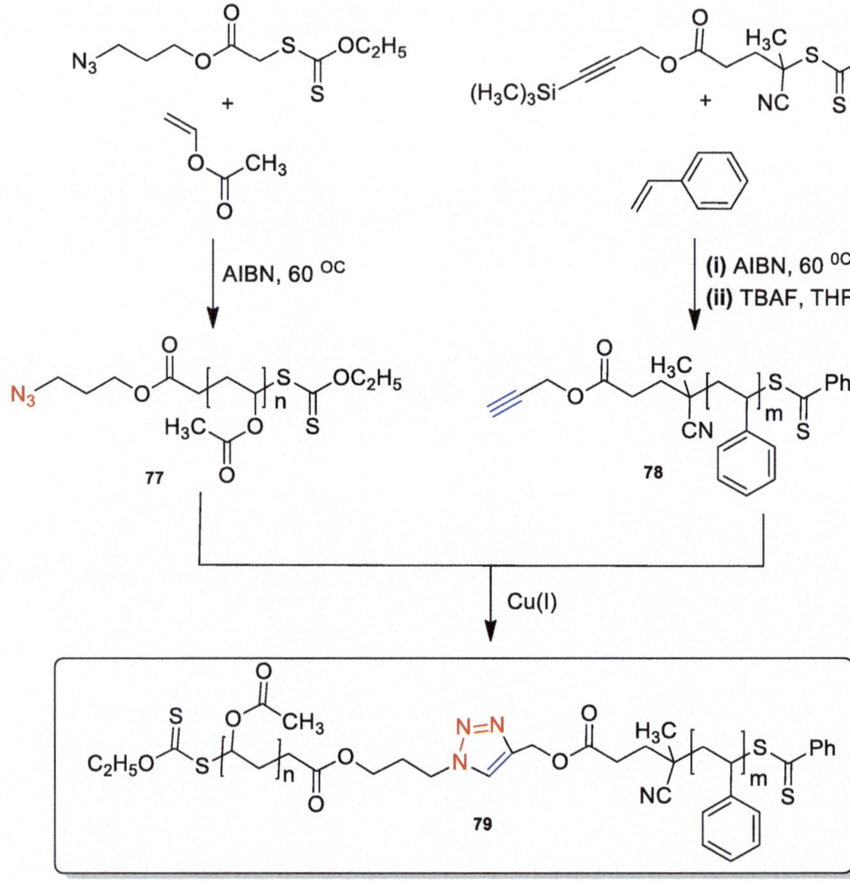

Scheme 11.22 Polystyrene-block-polyvinyl acetate copolymer synthesized by CuAAC reaction and RAFT polymerization

into an azide. Subsequently, this PS-azide **84** was reacted with diverse core molecules containing multiple propargyl groups **83**, one example being the pentynoyl ester of pentaerythritol, resulting in a tetra-arm polystyrene polymer **85** (Scheme 11.24) [49].

Scheel et al. used modular CuAAC tool to produce a new hyperbranched poly([1,2,3]-triazole)s from various AB_2 monomers. Thermal polymerization of the compound 3,5-bis(propargyloxy)benzyl azide **86** at room temperature produced 1,4- and 1,5-disubstituted poly([1,2,3]-triazole) **87a**, **b** and under the CuAAC reaction condition produced only 1,4-disubstituted poly([1,2,3]-triazole) **87a**. But products form in thermal condition and they are fully soluble (Scheme 11.25) [50]. The products are characterized by using NMR study.

An alternative chain polymerization method can be utilized instead of relying on vinyl polymerization to create the different blocks. In one of the early instances, Agut group nicely explored ring-opening polymerization (ROP) of γ-benzyl-L-glutamate

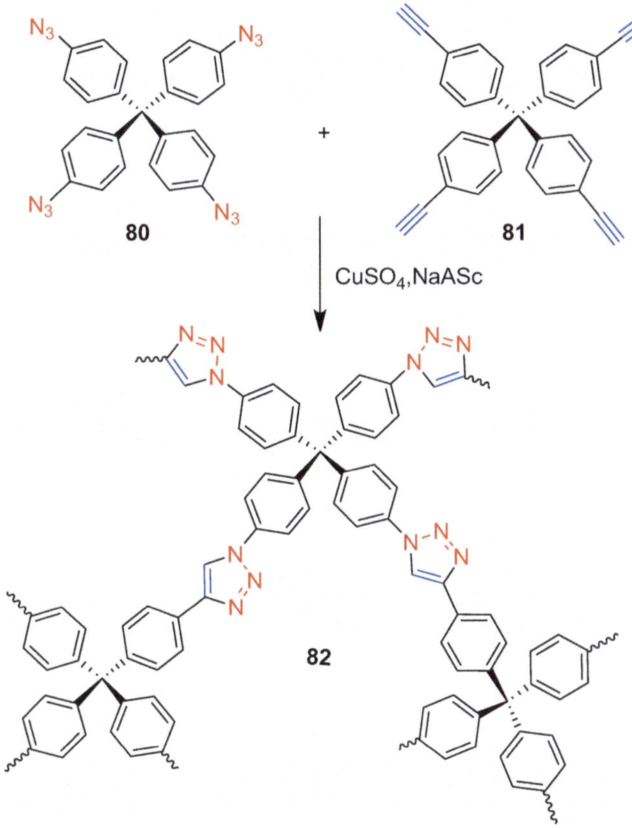

Scheme 11.23 CuAAC-inspired synthesis of porous organic polymer from tetrakis(4-azidophenyle) methane and tetrakis(4-ethynylphenyl) methane

N-carboxyanhydride (NCA) **89** in presence of amine initiator **88** and accomplished the synthesis of desired azide as rod-coil block copolymer **90** (Scheme 11.26). Another alkyne block copolymer **93** was obtained via Cu-catalyzed ATRP of 2-(dimethylamino)ethyl methacrylate **92** with an alkyne-bearing initiator **91** [51]. Both the azide and alkyne functionalities remained unaffected during the course of polymerization, i.e., ROP of NCA **89** with an amine **88** as well the ATRP of the vinyl monomer **92**. Finally, the CuAAC conjugation of poly (benzyl glutamate) azide **90** with alkyne terminated polymer PDMAEMA **93** to give the desired diblock copolymer **94** with excellent transformation rates (Scheme 11.26). A small excess of component **93** was used in reaction which could be readily eliminated through the use of an alumina column. Poly (benzyl glutamate) azide **90** is known to adopt a rigid helical conformation, while alkyne component **93** is recognized for taking on a random-coil conformation. Finally, the removal of benzyl via hydrolysis of the benzyl ester afforded a water-soluble polyglutamic acid diblock copolymer which imparts both temperature- and pH-sensitive characteristics.

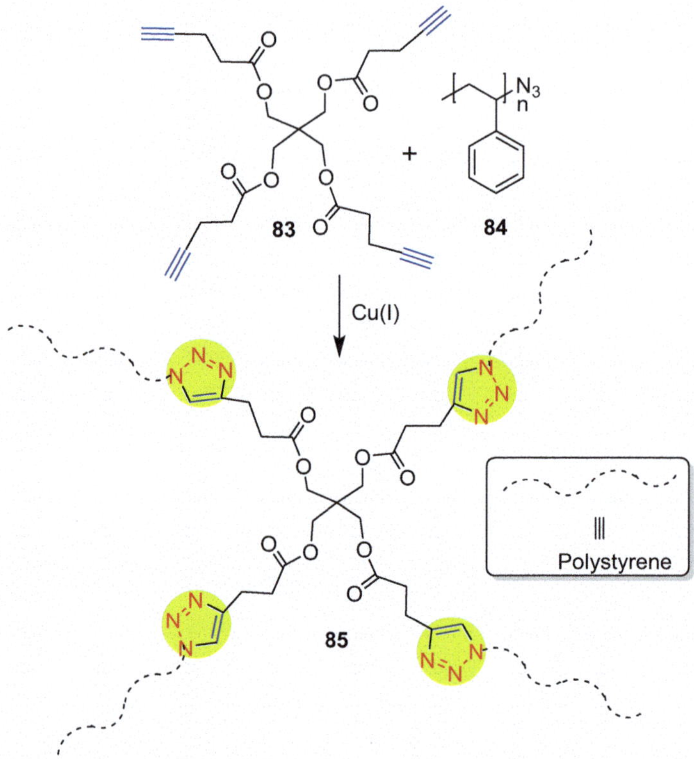

Scheme 11.24 CuAAC-mediated synthesis of four-arm PS-star polymer from PS-azide with a core bearing multiple propargyl groups

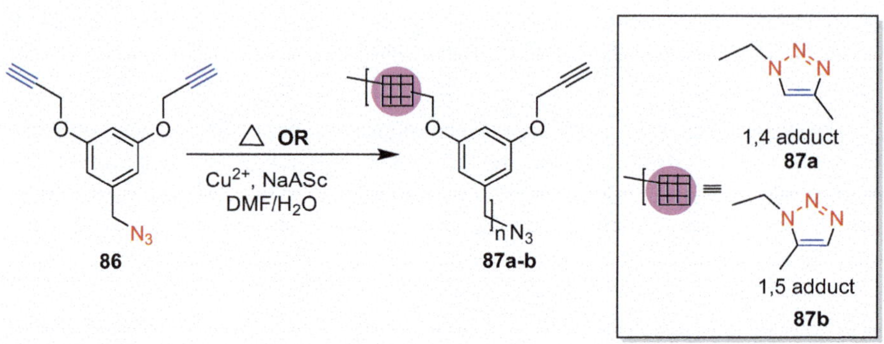

Scheme 11.25 Hyperbranched polymer is directly synthesized using the CuAAC reaction with AB_2 monomer

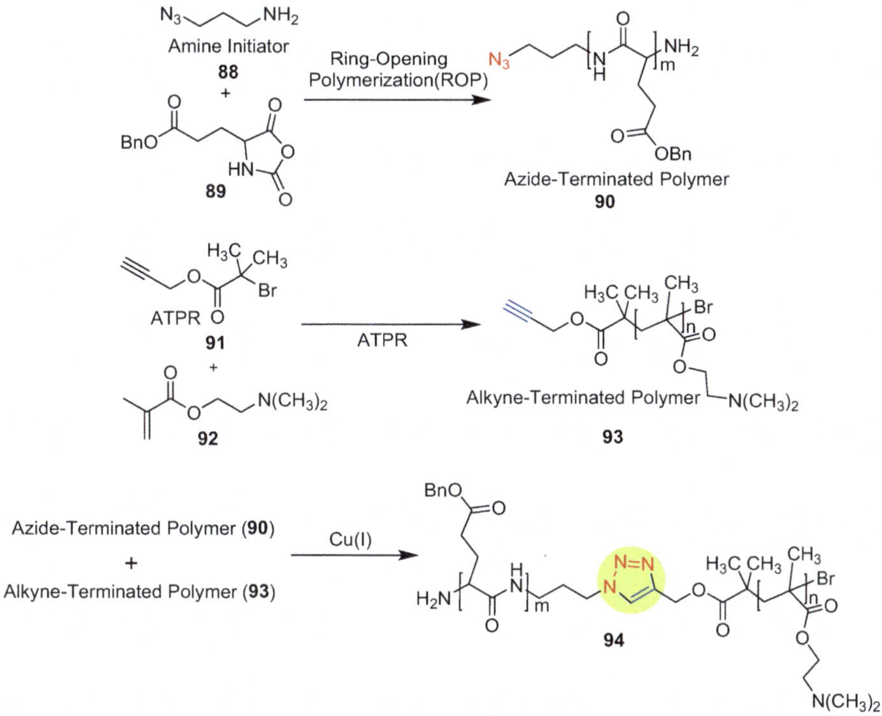

Scheme 11.26 Preparation of rod-coil block copolymer through combination of ROP, ATPR, and CuAAC reaction

Huisgen [3 + 2] dipolar cycloaddition procedure catalyzed by copper(I) was used to create π-conjugated block copolymers from terminal azide-functionalized polystyrenes (PS) **96** and alkyne-functionalized poly(3-hexylthiophene) (P3HT) **97**. Using an initiator with the azide function **95** that was particularly designed, the functionalized α-azido-PS homopolymer **96** was formed using atom transfer radical polymerization (Scheme 11.27). By utilizing CuAAC as a modular and regioselective tool, a new tri-block copolymer which contain (P3HT) central block [1,2,3]-triazoles **98** has been developed in good yield [52].

Nandivada et al. employed show that the polymer coatings generated under carefully regulated circumstances of monomer ethynyl paracyclophane **99** vaporization/pyrolysis experienced very little inadvertent cross-linking, alkyne-functionalized polymer **100** films that are deposited from the gas phase allow spatially well-defined immobilization of biomolecules, such as biotin, finished the 1,3-dipolar cycloaddition click reaction (CuAAC) with azide derivatives of biotin to give biotin derivative polymer containing triazole units **101** (Scheme 11.28) [53]. Applications for these biofunctional surfaces include biosensors and diagnostics.

Polymer scientists have always been fascinated with cyclic polymers because their lack of chain ends presents a number of fascinating conformational challenges and

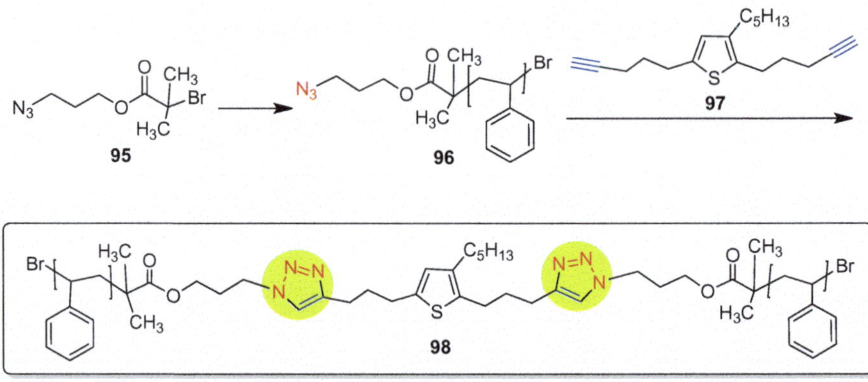

Scheme 11.27 Click-inspired synthesis of ABA triblock copolymer containing P3HT central block reaction between dialkyne derivative of P3HT and azide-terminal polystyrene

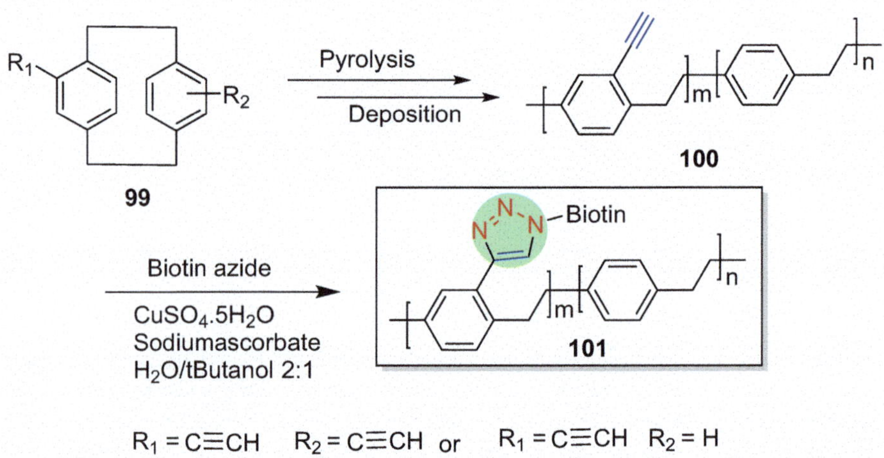

Scheme 11.28 Click-inspired immobilization of biomolecules between biotin-based azide-ligand and the polymer

features of diffusion and translocate [54]. The preceding sections have made it clear that the employment of CRP allows for the creation of heterotelechelic polymers with distinct functional groups at both chain ends through appropriate initiator structure selection. Subsequently its able to incorporate functional groups that have the potential to interact intramolecularly to produce cyclic structures. The effectiveness of the CuAAC reaction in conjunction with ATRP for the synthesis of cyclic polymers was initially shown by Laurent and Grayson [55]. As the ATRP initiator, propargyl 2-bromoisobutyrate **102** was used to create polystyrene **103** with a benzylic bromide at one end and a propargyl group at the other, which with sodium azide, was easily

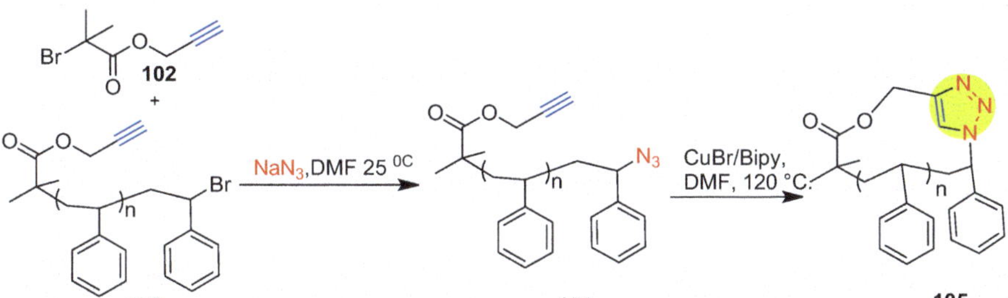

Scheme 11.29 Click-mediated synthesis of cyclic-PS from a heterotechelic propargyl-PS-azide

converted in quantitative yield to an azide **104** (Scheme 11.29). The CuAAC reaction would normally result in the creation of both higher molecular weight polymers **105** from such a heterotelechelic PS (propargyl-PS-azide) **104**, by straightforward step-growth method using an intrachain reaction and cyclic polymers through an interchain reaction.

This method was expanded upon by Eugene and Grayson to create a cyclic diblock copolymer of PS **105** and polymethylacrylate (PMA) **109** [56]. Using the Opsteen and van Hest-developed trimethylsilyl-protected ATRP initiator **106** (Scheme 11.30) [57]. The TMS-protected PMA-PS-Br **107** was created by polymerizing methyl acrylate and then styrene. It was subsequently converted to the necessary alkyne-PMA-PS-azide **108** in two step: first one is, the bromide was converted to the azide, and then the TMS group was deprotected using TBAF. Here, the authors used the CuAAC pathway to prepare cyclic diblock copolymer of PS **105** and polymethylacrylate (PMA) **109** in good yield. Some other groups also prepared cyclic diblock copolymer such as polyethylene oxide, polycaprolactone and so on via CuAAC pathway [58, 59].

11.4 Conclusions and Future Perspectives

The application of the CuAAC reaction in polymer science has proven highly beneficial, yielding a wealth of novel structures and strategies to enhance the synthesis of previously investigated systems. The rapid and chemoselective character of CuAAC, its compatibility with various reaction conditions, and its tolerance for a wide range of functional groups have facilitated its successful application across a diverse array of scenarios. The essential methods for preparing polymers, functional agents, or substrates incorporating alkyne or azide functionality are fundamental to the effectiveness of CuAAC protocol. As evidenced by prior discussions, polymer scientists have utilized CuAAC for diverse purposes, including synthesis of novel cyclic structures, forming cross-linked hydrogels, developing cross-linked functional coatings, engineering biocompatible substrates for tissue applications, creating porous polymeric systems for catalysis, and more. The potential for automated synthesis through

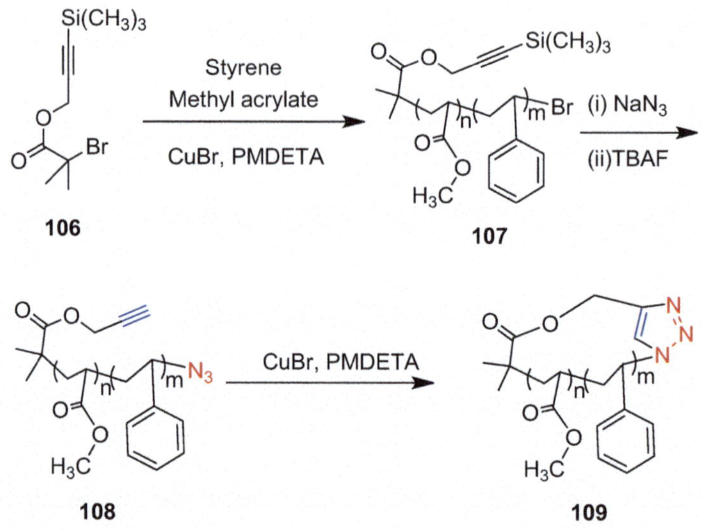

Scheme 11.30 Development of cyclic diblock copolymer of PMA and PS by an amalgamation of ATRP CuAAC reaction

the combination of CuAAC and flow synthesis represents a particularly innovative demonstration of the applicability of CuAAC. While CuAAC in itself has opened up and will continue to offer promising opportunities for further exploration, we believe that the principal objective lies in the pursuit of an almost perfect chemical reaction that could make significant contributions in the realms of polymers and material science.

References

1. Kempe K, Krieg A, Becer CR, Schubert US (2011) Chem Soc Rev 41:176–191
2. Zhang L, Chen X, Xue P, Sun HHY, Williams ID, Sharpless KB, Fokin VV, Jia G (2005) J Am Chem Soc 127:15998–15999
3. Geng J, Lindqvist J, Mantovani G, Haddleton DM (2008) Angew Chem, Int Ed 47:4180–4183
4. Semsarilar M, Ladmiral V, Perrier S (2010) Macromolecules 43:1438–1443
5. Kumar J, McDowall L, Chen G, Stenzel MH (2011) Polymer Chem 2:1879–1886
6. Nurmi L, Lindqvist J, Randev R, Syrett J, Haddleton DM (2009) Chem Commun 19:2727–2729
7. Hetzer M, Chen G, Barner-Kowollik C, Stenzel MH (2010) Macromol Biosci 10:119–126
8. Ning X, Rui W, Fu-Sheng D, Zi-Chen L (2009) J Polym Sci, Part A: Polym Chem 47:3583–3594
9. Castro V, Rodríguez H, Albericio F (2016) ACS Comb Sci 18:1–14
10. Hema K, Ravi A, Raju C, Pathan JR, Rai R, Sureshan KM (2021) Chem Soc Rev 50:4062–4099
11. Huo J, Hu H, Zhang M, Hu X, Chen M, Chen D, Liu J, Xiao G, Wang Y, Wen ZA (2017) RSC Adv 7:2281–2287
12. Huang D, Liu Y, Qin A, Tang BZ (2018) Recent advances in alkyne-based click polymerizations. Polym Chem 9:2853–2867
13. Agalave SG, Maujan SR, Pore VS (2011) Chem—Asian J 6:2696–2718

14. Keri RS, Patil SA, Budagumpi S, Nagaraja BM (2015) Chem Biol Drug Des 86:410–423
15. (a) Tornoe CW, Christensen C, Meldal M (2002) J Org Chem 67:3057–3064. (b) Rostovtsev VV, Green LG, Fokin VV, Sharpless KB (2002) Angew Chem, Int Ed 41:2596–2599
16. (a) Agrahari AK, Bose P, Jaiswal MK, Rajkhowa S, Singh AS, Hotha S, Mishra N, Tiwari VK (2021) Chem Rev 121:7638–7956. (b) Tiwari VK, Mishra BB, Mishra KB, Mishra N, Singh AS, Chen X (2016) Chem Rev 116:3086–3240
17. Johansson JR, Beke-Somfai T, Stalsmeden AS, Kann N (2016) Chem Rev 116:14726–14768
18. Walczyk D, Bombelli FB, Monopoli MP, Lynch I, Dawson KA (2010) J Am Chem Soc 132:5761–5768
19. Wan S, Kelly PM, Mahon E, Stöckmann H, Rudd PM, Caruso F, Dawson KA, Yan Y, Monopoli MP (2015) ACS Nano 9:2157–2166
20. Mandal J, Ramakrishnan S (2016) Azide-alkyne click reaction in polymer science. In: Chandrasekharan S (ed) Click reaction in organic synthesis. Wiley-VCH Verlag, Germany
21. Hasegawa T, Numata M, Okumara S, Kimura T, Sakurai K, Shinkai S (2007) Org Biomol Chem 5:2404–2412
22. Geng J, Mantovani G, Tao L, Nicolas J, Chen G, Wallis R, Mitchell DA, Johnson BRG, Evans SD, Haddleton DM (2007) J Am Chem Soc 129:15156–15163
23. Nagarsenkar A, Prajapti SK, Guggilapua SD, Birinenib S, Kotapallib S, Ummanni S, Babu BN (2016) Med Chem Commun 7:646–653
24. Chen Q, Cui Y, Zhang TL, Cao J, Han BH (2010) Biomacromol 11:13–19
25. Ladmiral V, Mantovani G, Clarkson GJ, Cauet S, Irwin JL, Haddleton DM (2006) J Am Chem Soc 128:4823–4830
26. Huang X, Huang XJ, Yu AG, Wang C, Dai ZW, Xu ZK (2011) Macromol Chem Phys 212:272–277
27. Srinivasachari S, Liu Y, Zhang G, Prevette L, Reineke TM (2006) J Am Chem Soc 128:8176–8184
28. Sletten ET, Svec RL, Nguyen HM (2015) Tetrahedron Lett 56:3473–3476
29. Beveridge JM, Chenot HM, Crich A, Jacob A, Finn MG (2018) Langmuir 34:10407–10412
30. Matsuoka K, Kurita A, Koyama T, Hatano K (2014) Carbohydr Polym 107:209–213
31. Becer CRR, Gibson MIMI, Geng J, Ilyas R, Wallis R, Mitchell DADA, Haddleton DMDM (2010) J Am Chem Soc 132:15130–15132
32. Toraskar S, Gade M, Sangabathuni S, Thulasiram HV, Kikkeri R (2017) Chem Med Chem 12:1116–1124
33. Richards S-J, Gibson MI (2021) JACS Au 1:2089–2099
34. Karakawa M, Chikamatsu M, Yoshida Y, Azumi R, Yase K, Nakamoto C (2007) Rapid Commun 28:1479–1484
35. Sun HS, Chen Y, Lee W-Y, Chiu Y-C, Isono T, Satoh T, Kakuchi T, Chen W-C (2016) Polym Chem 7:1249–1263
36. Rivas MV, Petroselli G, Erra-balsells R, Varela O, Kolender AA (2019) RSC Adv 9:9860–9869
37. Liu Z, Zhu Y, Ye W, Wu T, Miao D, Deng W, Liu M (2019) Polym Chem 10:4006–4016
38. Li J, Tian X-Y, Zong L-P, Zhang Q, Zhang X-J, Marks R, Cosnier S, Shan D (2019) ACS Appl Mater Interfaces 11:32366–32372
39. Cui L, Cohen JA, Broaders KE, Beaudette TT, Frechet JMJ (2011) Bioconjugate Chem 22:949–957
40. Beaudette TT, Cohen JA, Bachelder EM, Broaders KE, Cohen JL, Engleman EG, Frechet JMJ (2009) J Am Chem Soc 131:10360–10361
41. Huang D, Liu Y, Qin A, Tang BH (2019) Macromolecules 52:1985–1992
42. Zhang Q, Collins LSJ, Chen G, Wallis R, Mitchell DA, Haddleton DM, Becer CR (2014) J Am Chem Soc 136:4325–4332
43. Qin A, Lam JWY, Tang L, Jim CKW, Zhao BZ (2009) Macromolecules 42:1421–1424
44. Qiu XP, Tanaka F, Winnik FM (2007) Macromolecule 40:7069–7071
45. Qui A, Lam JWY, Jim CKW, Zhang L, Yan J, Haussler M, Liu J, Dong Y, Liang D, Chen E, Jia G, Tang BZ (2008) Macromolecules 41:3808–3822
46. Quemener D, Davis TP, Kowollik CB, Stenzel MH (2006) Chem Commun, 5051–5053

47. Holst JR, Stockel E, Adams DJ, Cooper AI (2010) Macromolecules 43:8531–8538
48. Hadjichristidis N (1999) J Polym Sci, Part A: Polym Chem 37:857–871
49. Goa H, Matyjaszewski K (2006) Macromolecules 39:4960–4965
50. Scheel AJ, Komber H, Voit BI (2004) Macromol Rapid Commun 25:1175–1180
51. Agut W, Taton D, Lecommandoux S (2007) Macromolecules 40:5653–5661
52. Urien M, Erothu H, Cloutet E, Hiorns RC, Vignau L, Cramail H (2008) Macromolecule 41:7033–7040
53. Nandivada H, Chen HY, Bondarenko L, Lahann J, Angew Chem Int Ed 45:3360–3363
54. Semleyen ER (2000) Kluwer Academic Publishers. Tezuka, Y.; World Scientific
55. Laurent BA, Gyayson SM (2006) J Am Chem Soc 128:4238–4239
56. Eugene DM, Grayson SM (2008) Macromolecules 41:5082–5084
57. Opsteen JA, van Hest JCM (2005) Chem Commun, 57–59
58. Hoskins JN, Grayson SM (2008) Macromolecules 42:6406–6413
59. Zhang B, Zhang H, Li Y, Hoskins JN, Grayson SM (2013) ACS Macro Lett 2:845–848

Chapter 12
Click Chemistry in Lipid Modification

12.1 Introduction

Lipids constitute an essential class of bio-organic compounds that play an important role in various cell functions including cell energy storage, membrane-traffic regulation and permeability, signaling, transport of material, absorption of vitamins, and making of hormones [1]. Amphipathic lipid molecules administer the protein functions also by the phenomenon of protein-membrane binding [2]. These vital aspects make lipid derivatives appropriate for mimicking cell membrane environments [3], imaging, gene transport, and drug delivery [4, 5]. The tagging of lipids with suitable labeling agents creates a pathway to achieve extraordinary results by mimicking the lipid functions along with the controlled functionalization of the membrane assembly. In this view, the regioselective Cu(I)-catalyzed azide-alkyne variant of click chemistry (CuAAC) offers a facile and apt technique to conjugate the lipids with other moieties for the desired functionalization of membrane surface [6–8]. A [3+2] cycloaddition reaction between terminal alkynes **1** and organic azides **2** in the presence of a Cu(I) catalyst, largely known as CuAAC or Click reaction is acknowledged to deliver high yields of 1,4-disubstituted 1,2,3-triazoles **3** with consummate regioselectivity under mild reaction conditions (Scheme 12.1) [9–11]. The opposite regiomer, i.e., 1,5-disubstiituted 1,2,3-triazoles can be selectively obtained once the cycloaddition reaction was performed in the presence of suitable ruthenium catalyst (RuAAC) [11]. The CuAAC-mediated modular triazole forming tool has enormous applications in various fields including glycoscience of diverse applications [12–14]. Easy access to clickable azide or alkyne derivatives of lipids and sugar moieties and the tolerance and compatibility of other functional groups with the click conditions make this strategy fascinating toward the development of glycolipid analogs with a 1,4-disubstituted 1,2,3-triazole link in-between. This chapter includes a brief discussion of the application and progress of CuAAC click reaction in lipid-modification, its advantage in this area and the associated challenges.

V. K. Tiwari et al., *Click Chemistry*, Materials Horizons: From Nature to Nanomaterials,
https://doi.org/10.1007/978-981-97-4596-8_12

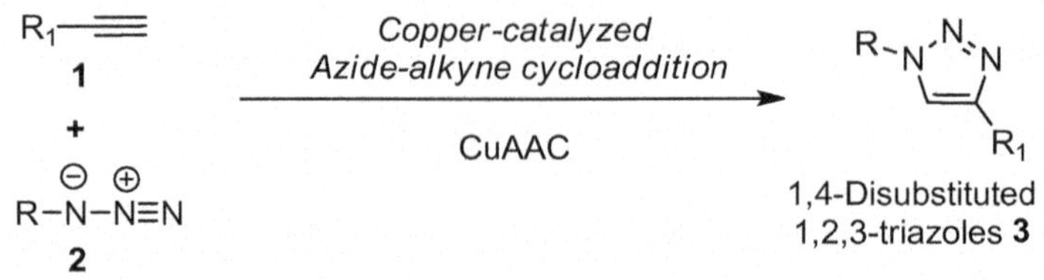

Scheme 12.1 The regioselective click reaction for the synthesis of diverse 1,4-substituted 1,2,3-triazole analogues

12.2 Click Chemistry for the Functionalization of Membrane Surface

The Cu(I) catalyzed click reaction was employed by Schuber et al. for the coupling of an azide derivative of α-thiomannose **4** with a glycerolipid derived terminal alkyne **5** to tag D-mannose-residue preformulated liposome-surface. The use of regular Cu-catalysts would result in the increased size of liposome to about 50% by membrane disruption. For this purpose, they used a copper chelation strategy through the involvement of a water-soluble ligand bathophenanthroline disulfonate to maintain the membrane integrity (Fig. 12.1a) [15]. This strategy not only helped to maintain membrane integrity but also improved the reaction yield with complete conversion within 24 h under mild reaction conditions. The membrane integrity was evaluated by the dynamic light scattering (DLS) method during the reaction, as well as, fluorescence dye-based assay. Noticeably, agglutination of the lectin Con A was found in the process of binding target proteins to the mannose-tagged liposomes. The Staudinger ligation tactic was further explored by Sun et al., while tagging a lactose-based azide **6** with a specific cholesterol liposome containing 5 mol% of lipid **7** (Fig. 12.1b) [16]. DLS and fluorescence dye-based assay throughout the ligation process again assured the maintenance of membrane integrity. Boxer and co-workers demonstrated an effective technique that uses a DNA-templated click reaction to covalently bind lipid vesicles to a supported lipid bilayer to account DNA-mediated-membrane binding interactions and fusion. This work used a sense/antisense DNA-templated Cu-catalyzed azide-alkyne cycloaddition, click reaction to immobilize intact liposomes onto supported lipid bilayers (SLBs). This orthogonal approach to liposome anchoring improved control over DNA-binding exchanges. To achieve this, liposomes bearing a PE analog with the complementary 24-mer sequence ending in an alkyne moiety, as well as a separate azide-tagged PE derivative and a DNA–lipid conjugate comprising a 24-mer attached to PE, were affixed onto SLB, shown as **8**. In order to maintain liposome–liposome binding contacts and fusion, the liposomes additionally included a second set of sense/antisense DNA–lipid conjugates. In order to create DNA sense/antisense liposome–SLB-binding interactions, labeled liposomes and SLBs were incubated with high concentrations of salt. Additionally, copper sulfate, sodium ascorbate, and the copper-ligand TTMA were added to the mixture to ensure covalent attachment via CuAAC. The triazole linkage kept liposomes bonded to the surface even when the salt concentration was reduced to allow DNA–DNA interactions. Then, in order to initiate interliposome binding through the DNA strands on the membrane surface, the salt concentration was raised once again. Different methods were employed to change the liposome–SLB connection point, including alternating the azide and alkyne concentrations between the liposome and SLB as well as the conjugation of DNA-lipid conjugates with alkynes and terminal azides shown as **9** (Fig. 12.1c). Notibly, the authors pointed out that in order to obtain significant conjugation yields, the CuAAC reaction had to be tampered with during liposome deposition. This method allowed for the direct observation of liposome diffusion and liposome–liposome binding in this system, where the liposomes were restricted to movement in a parallel plane by SLB attachment. Fluorescence microscopy was employed via fluorophore-tagged liposomes [17].

Additionally, Kros et al., reported characterizing click-chemistry-based membrane functionalization using a Förster resonance energy transfer (FRET) device. The authors explored phosphatidylethanolamine (PE) analogue **10** with an alkyne functional group which was then subjected to liposomes made of PC and lissamine-rhodamine-tagged PE (Fig. 12.2a). Moreover, in the reaction of lysine analogue **11** having both an azide and NBD (FRET donor) label in the presence of CuBr (0.5 equiv.) catalyst, FRET activation was detected through rhodamine (FRET acceptor) emission signal than control in the absence of copper. The reaction was completed after approximately 4 h as confirmed by Fluorescence analysis. FRET activation was attained by enhancement upon reaction with lysine derivative **11** that carried both a clickable azide and an NBD (FRET donor) label, employing CuBr (0.5 equiv.) and enhancement of the rhodamine (acceptor) emission signal in comparison to copper-deficient control groups [18]. In the absence of a copper catalyst, several strained cyclooctynes were demonstrated to be effective for click chemistry; for this reason, the aza-dibenzocyclooctyne (ADIBO) system because of its robustness, desirable kinetic properties, and simplicity in synthetic preparation were used. Liposome immobilization on streptavidin-coated microplates has been performed to evaluate the conjugation. A fluorescent NBD-PE conjugate and azido-TEG-glycerolipid **12** was added to liposomes followed by incubation of the ADIBO-TEGbiotin reagent **13** (Fig. 12.2b) [19–21]. In order to get around problems with liposome stability and lipid phase segregation, Cai et al., reported clickable polymerized liposomes for in vivo delivery applications. Here, alkyne-tagged PE analog **14** containing polydiacetylene moieties in the acyl chains were exposed to radiation to access polymerized liposomes (Fig. 12.2c).

CuAAC conjugations still required a copper-chelating ligand to increase yields and prevent the degradation of the unsaturated lipids that come from polymerization, even if polymerized liposomes were more stable. Because of its weak water solubility, TBTA was ineffective for the copper-chelating ligand, and bathophenanthroline disulfonate was avoided because of worries about the vulnerability of its copper complex to air oxidation. The rate of CuAAC was reported to increase twelve times when the scientists used a tris-TEG-triazole based ligand. The optimal conditions were ascorbic acid (6 mM), ligand, and copper sulfate (0.3 mM) for conjugation

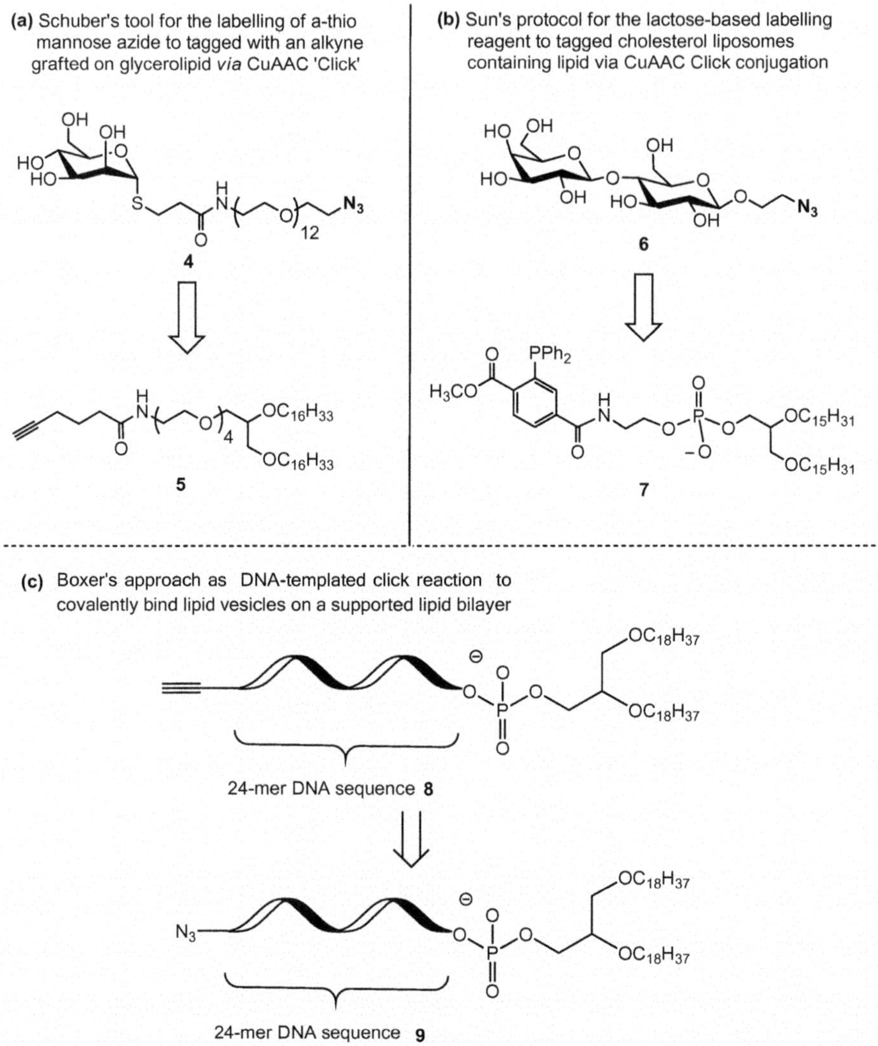

Fig. 12.1 **a** Schuber's approach for the labeling of α-thiomannose azide to tagg with an alkyne-grafted on glycerolipid lipids using CuAAC; **b** Sun's approach for the lactose-based labeling reagent to tagged cholesterol liposomes containing lipid; **c** Boxer's reported the DNA-templated click reaction to covalently bind lipid vesicles to a supported lipid bilayer

with fluorogenic azidocoumarin. The conjugation was completed in one hour under these conditions. The predicted yield of this reaction was 23%, however as only the lipids on the outer leaflet of the bilayer should react; this could be an underestimate due to fluorescence quenching. Under these conditions, Cu-catalyzed azide-alkyne cycloaddition was utilized to add a RGD peptide as an integrin-targeting ligand,

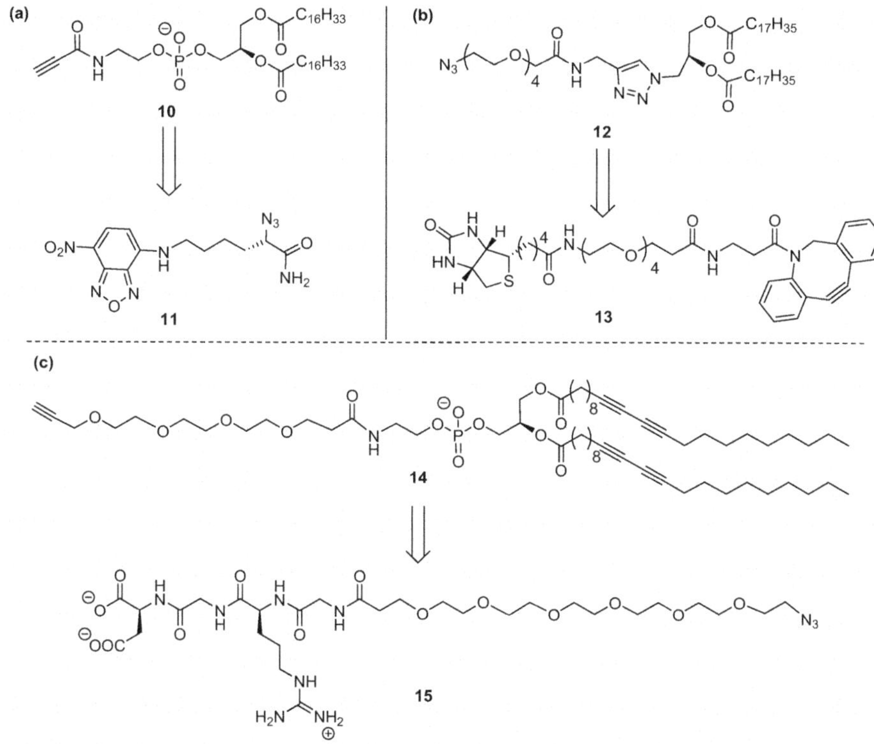

Fig. 12.2 **a** Click-chemistry-based membrane functionalization; **b** an assay using liposome immobilization; **c** CuAAC-mediated tagging

through reaction with azide component **15**, and a fluorescent label to the polymerized liposomes, and fluorescence imaging established successful delivery to cell membranes (Fig. 12.2c) [22].

12.3 Click Chemistry in the Synthesis of Functionalized Lipid Analogues

Solaiman et al. employed modular Cu-catalyzed azide-alkyne click reaction to develop multiple dimer-lipid analogues. They reported a reaction sequence in which azido triglycerol analogue **16** was first derived from respective 17-azido sterearic acid. This azido derivative was clicked with alkyne-functionalized glycolipid **17** in an aqueous mediumin the presence of $CuSO_4.5H_2O$ and sodium ascorbate at room temperature which afforded high triglycerol derivative **18** in high yield (Scheme 12.2) [23].

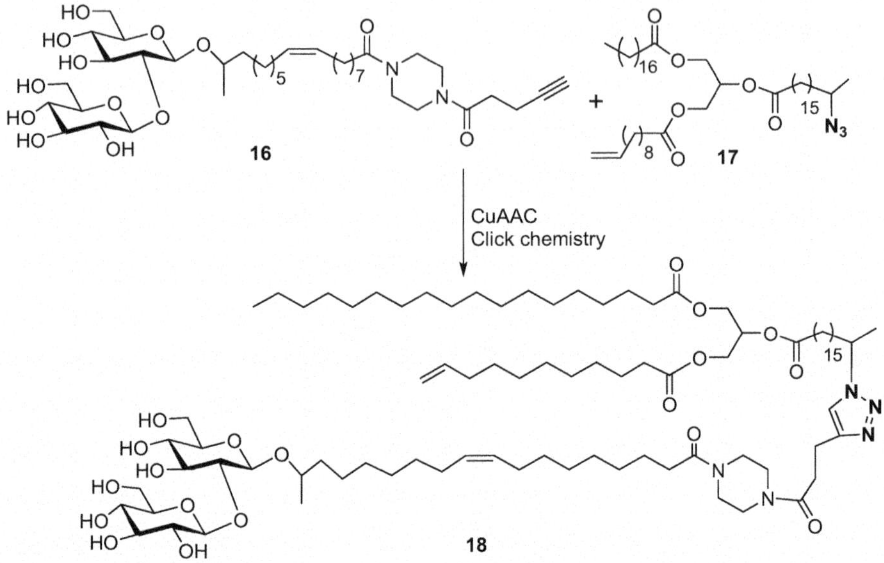

Scheme 12.2 CAAC-mediated tethering of tagged lipid component to result of respective dimer-lipid analogue **18**

Guo and co-workers did their bit to add a new chapter in click-inspired synthesis of highly functionalized complex glycolipids with potential applications. Their strategy followed the development of analkyne derivative of a completely protected analogue of monophosphoryl lipid **19** having a significant role in endotoxicity because of the activation of toll-like receptor. The CuAAC conjugation of alkyne-grafted glycolipid analogue **19** with a sialic acid-containing trisaccharide-based azide **20** in the presence of CuI/DIPEA in THF/MeOH at room temperature for 24 h afforded respective 1,2,3-triazole linked glycolipid product **21** in 70% yield. Despite of the apparent steric hindrance of glycolipid **19**, the regioselective click conjugation with azide **20** was quite satisfactory and expediency in term of reaction yield and ease of reaction proceeded. The resulting modified triazole analogue **21** could have the potency to explore possible immunostimulants to combat sepsis. The desired click conjugate product **21** (Scheme 12.3) [24].

Click-chemistry-based derivatization takes use of the bioorthogonal character of this reaction by labeling and visualizing lipids in situ within the diverse surroundings of biological samples (Scheme 12.4a). Lipids are active substances that travel through cellular membranes and change constantly through metabolic pathways. Each specific lipid's location is controlled by mechanisms like enzyme modification and diffusion, which influence both spatially and temporally. These processes affect key biological functions like fusion and signaling. Schultz and Neef intriguingly used synthetic alkyne-tagged PA analogs in a recent study to image the dynamics of this lipid in both fixed and live cells [25].

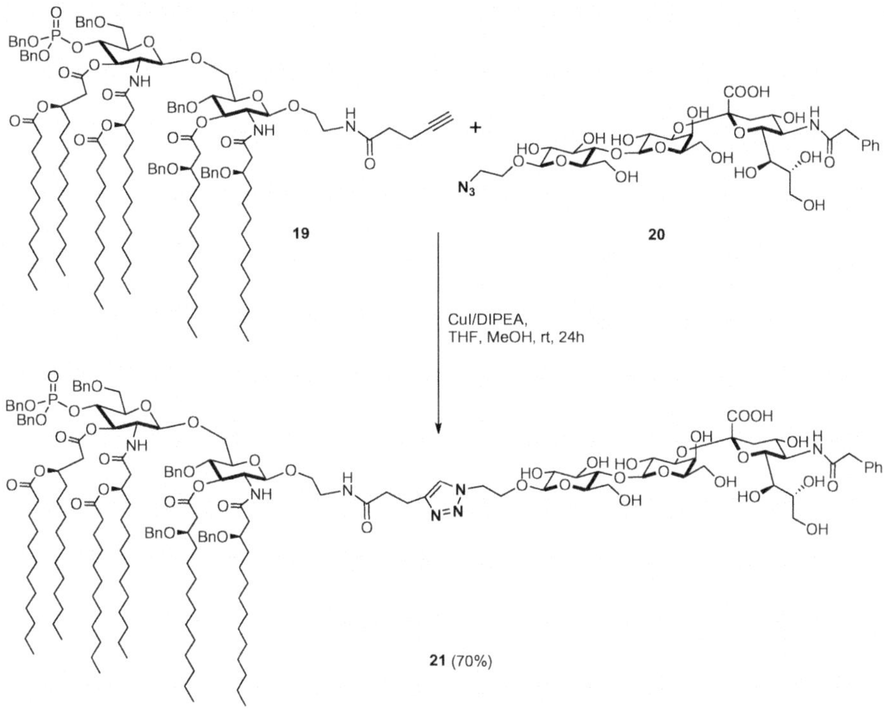

Scheme 12.3 Functionalization of alkyne-grafted monophosphoryl lipid in assistance of CuAAC/click reaction

Salic and colleagues took a different method to in vivo phospholipid tagging by means of an alkyne-tagged precursor which was through metabolic pathways transformed into the target lipid. Propargylated choline **22** was assembled in this way to be converted into alkyne-tagged PC lipid analogs **23** once injected into cells. Cell membranes were observed to be tagged after incubation with propargylated choline **22** followed by fluorescence tagging with CuAAC click conjugations. Such conjugation was influenced by the addition of detergent and inhibited by the application of phospholipase C which indicated that the labeled species were indeed PC lipid analogs. The resulting tagged PC derivatives were found in various cellular membranes, such as the plasma membrane, mitochondria, Golgibody, and ER (Scheme 12.4b) [26].

The engagement of relevant lipid analogues in non-covalent protein-lipid-binding interactions generally influences the protein function and also localization [27, 28]. A suitable probe-based technique could be used to efficiently detect proteins with binding capacity and adaptability to specific lipids [29]. The azide and alkyne functionalities in the bioorthogonal Cu-azide-alkyne cycloaddition process are compelling tags for the post-modification of labeled proteins. The Click chemistry has been actively investigated over the last twenty years to generate a number of

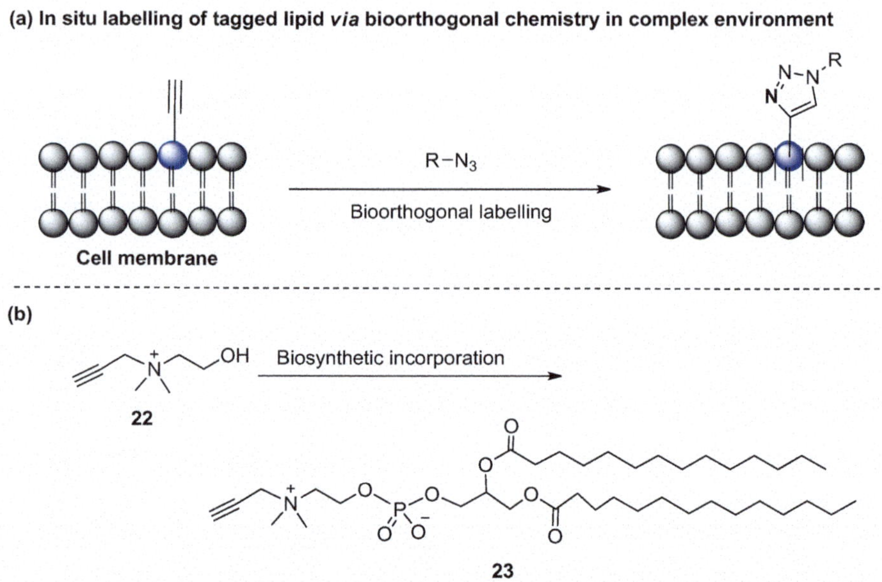

Scheme 12.4 **a** An *in-situ* labeling of tagged lipid component in complex environment using Bioorthogal concept; and **b** structure of developed analogue in in-situ labeling

intriguing lipids that have proven to be extremely effective as active probes in medicinal applications [30]. Phosphoinositidelipids are known for their critical regulatory role in cell physiology, and their derivatives are recognized as critical chemical tools for elucidating protein-signals and binding activities, as well as the activation of numerous proteins [31]. Best et al., concentrated their research on the production of appropriate probes in response to certain lipids, such as PA and PIPns. For efficient probe development, the authors were able to successfully develop headgroup conjugates following all seven amino conjugates of phosphatidylinositol polyphosphate isomers [32, 33]. Lipids PIPns were chosen first because they are known to bind and then activate protein targets, which is an unusual behavior in a variety of illnesses. The efficacy of the PH domain labeling of protein kinase B was assessed by the authors using the gel fluorescence scanning method. In which the probes in the series, phosphatidyl inositol-3,4,5-triphosphate **24** was determined to be the most promising and effective moiety. Subsequently, they conducted more research and asserted that the labeling phenomena involving probe **24** was the result of unambiguous activity of binding interactions. Once after explicit labeling, the probe was then analyzed for the target protein classification in cancer cell extracts. At the end, the enrichment of probe-labeled proteins was determined using proteomics based on MS. As a result, probe **24** was finally recognized to completely bind to the target proteins. The result could be considered for specific labeling and recognition of a total of 265 target proteins. Further, 44 predictable phosphatidylinositol-(3,4,5)-triphosphate proteins can also be well established (Fig. 12.3) [32, 33]. Kroon and colleagues used functionalized PC probes **25a–b** to describe those proteins that can efficiently bind to this lipid [34, 35]. In one study Cravatt and Tully recently explored the PC activity probes **26a–b** to examine phospholipases [36]. Other phospholipids have also been identified as potential candidates for activity probe development. Bong and Bandyopadhyay developed PS probes **27** required to identify proteins that bind to this lipid (Fig. 12.3) [37].

Likewise, biocompatible carbohydrate-based azide or alkyne required to connect with other lipid analogues via stable 1,2,3-triazole linker as lipid-rafts is explored to develop molecular architecture. To accomplish this, a protected oligosaccharide **28** was used to assemble corresponding alkynes having oligosaccharide-derived linker **29** as needed supplementary common spacers. The process involved selective removal of NH-Cbz group protection from **28**, under Pd/C-H_2 catalyzed hydrogenolysis followed by protection by carbamate protected alkyne in the presence of $NaHCO_3$ in methanol/water (1:1). The resulting spacer molecule **29** can be used as a probe for glyco-lipid-fluorophore conjugates as well as glyco-lipid-protein conjugates by tying together with other variable linker lengths such as tri-, penta-, and heptasaccharide (Scheme 12.5) [38].

Similarly, it is very simple to investigate the modular CuAAC conjugation approach to allow the homogenous coupling of orthogonally protected carbohydrates, such as the hydrophilic end and the counter component as the hydrophobic end. CuAAC surfactant conjugation primarily avoided hydrophobic contamination of surfactant when low MW alcohols were used, resulting in the CuAAC-intervene triazole fascinatingly connecting the two surfactant domains, and thus may recommend the formation of pure surfactants [39]. In order to obtain greater yields of the corresponding carbohydrate-appended lipids **31** and **33**, a mechanochemical ball milling methodology was lastly investigated for the click mediated cycloddition of propargylated carbohydrate **30** and **32** with azides produced from long chain hydrocarbons employing a copper catalyst under solvent-free conditions. It's interesting to note that the assembled carbohydrate-anchored lipids were able to self-assemble in solvents according to different patterns of carbohydrate shape and size coupled to the lipid residue via the liker 1,2,3-triazole. Furthermore, the temperature, solvent type, and lipid architecture all affect how well they self-assemble (Scheme 12.6) [40].

Labrada et al., reported a sequence of two-step high yielding synthesis of an array of neoglycolipids using a combination Ugi reaction followed by CuAAC. By using the Ugi-four component reaction of organic amino azide, para formaldehyde, corresponding isocynide, and fatty acids, azido containing ceramide mimics **34** as clickable partner were achieved. For instance, under Ugi-4CR conditions, 2-azidoethyyl amine, $(CH_2O)_n$, $C_{12}H_{25}NC$, and $C_{13}H_{27}COOH$ produced a high yield of ceramide-based azide **34a** (Scheme 12.7) [41].

Now after the successful synthesis of ceramide-based azides **35**, they were subjected for the click reaction with an array of glycosyl alkynes **36** derived from respective sugar analogue such as monosaccharides as well as from trisaccharides. The modular click reaction were takes place in assistance of $Cu(OAc)_2$ /NaAsc in

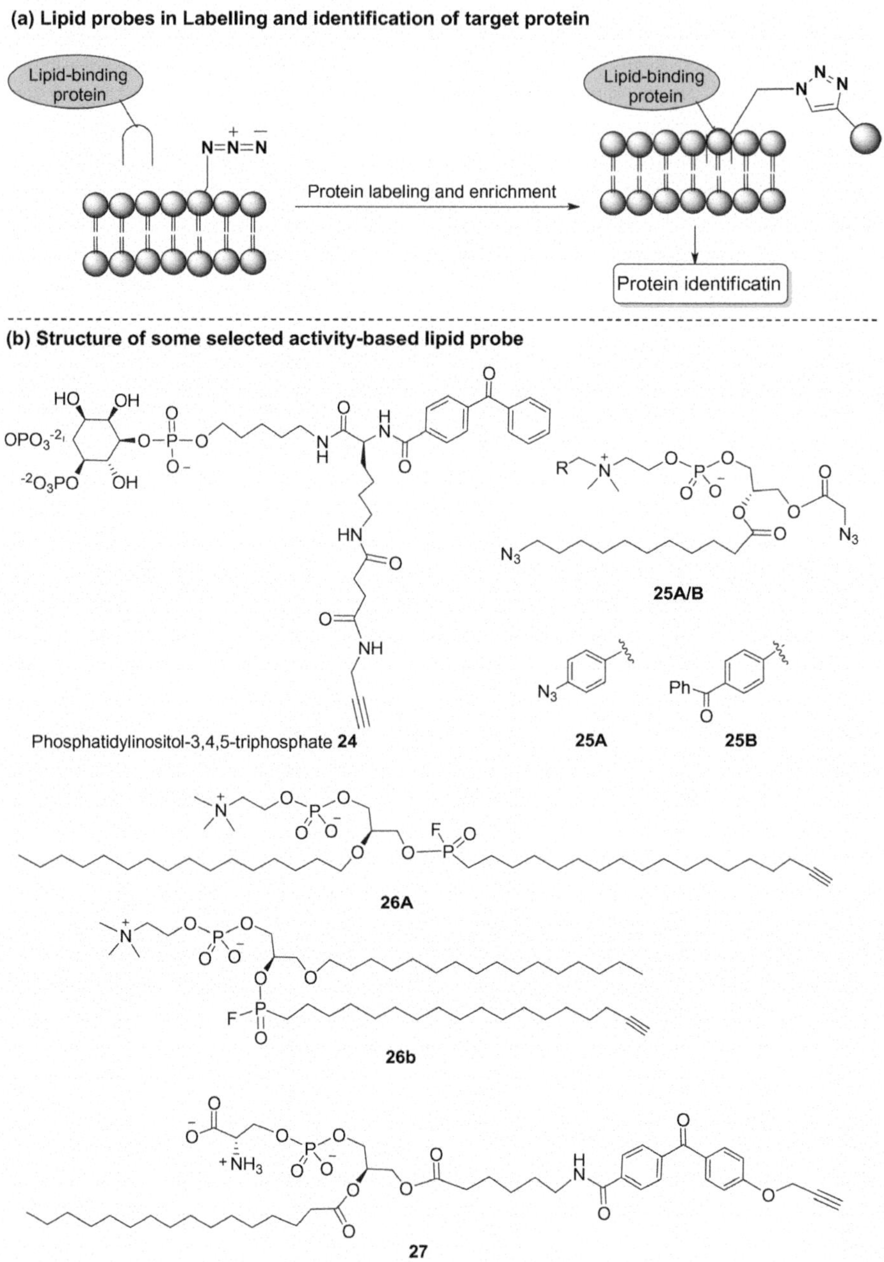

Fig. 12.3 The use of lipid activity probes for tagging and identifying target proteins with various examples of lipid activity probes

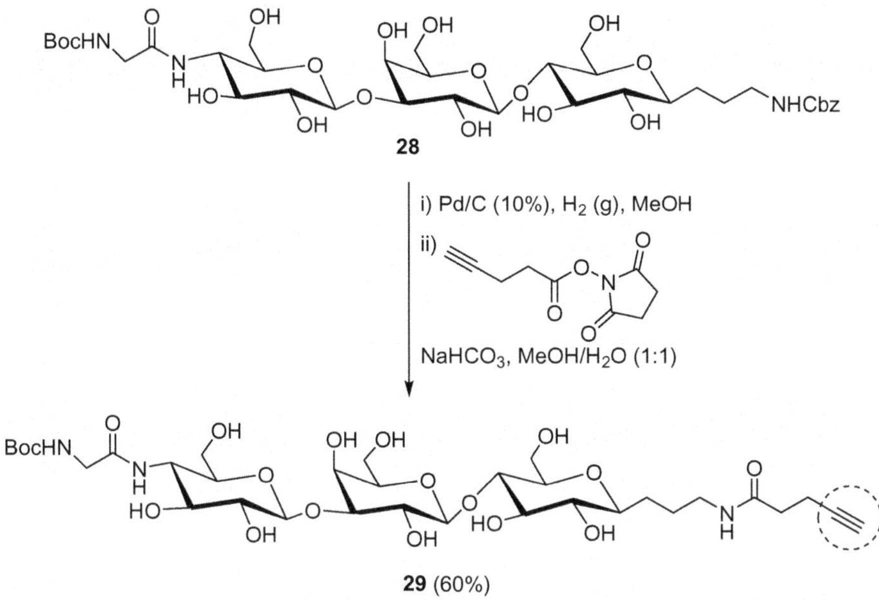

Scheme 12.5 Synthesis of glyco-lipid-fluorophore alkyne probe for CuAAC Click reaction

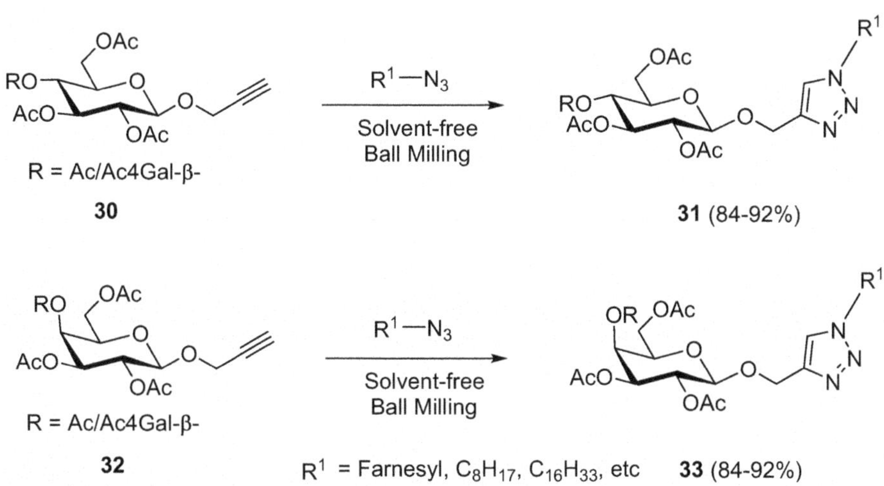

Scheme 12.6 Click-mediated synthesis of carbohydrate-appended triazole-linked lipids under ball milling solvent-free condition

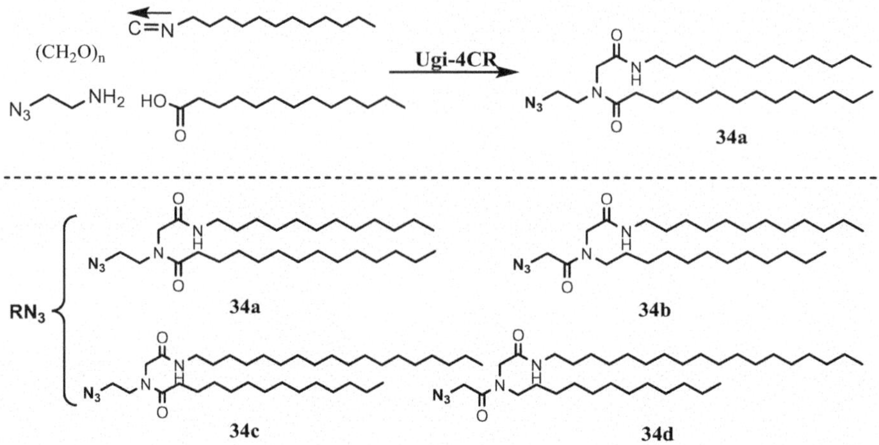

Scheme 12.7 The Ugi-four component reaction-mediated synthesis of azido-containing ceramide mimics **34** as a clickable analogue

THF/H_2O under mild condition to furnish the desired triazole-based neoglycolipids **37a–f** in high yield (Scheme 12.8) [41].

Similarly, a double-lipidic scaffold with a clickable alkyne residue **38** was assembled using the Ugi four-component approach. These lipid aklynes were subjected for click conjugation with glycosyl azides **39a,b** under modular click conditions to afford neoglycolipids **40a–d** (Scheme 12.9) [41]. It is interesting to note that both steps are practical, easy to follow, and have a broad range of substrates. For this reason, the protocol can be regarded as a highly effective, flexible, simple, and useful way to produce a variety of neoglycolipids with complex architectures that include lipid hybrids and carbohydrate with 1,2,3-triazole. The dynamic NMR-based charactaerization has validated the cis–trans isomerization of the tertiary amide bond in the desired neoglycolipids **40a–d**.

Triazoles are comparises a class of isoesters of amides, and since the neoglycolipids share a close structural relationship with a class of naturally occurring glycosphingolipids as well as several complex synthetic glycolipids with intriguing anti-cancer properties. In this way neoglycolipids constitute an important claas of biological significance analogues. Even though the developed protocol may provide a fresh perspective for very easy access to complex glycolipids, the clinical application of these molecules is still pending. This will hopefully motivate researchers to overcome the obstacles in the way of clinical practice, which will be important in the future for drug discovery and development. In order to successfully track the bacterial glycosyl hydrolase activity, Reichardt et al., investigated CuAAC Click coupling of azide-functionalized carbohydrates with a MALDI active surface. They also developed an enzymatic test technique for this purpose [42]. Zilio and coworker reported in their initial study that 2-azidoethyl lactose, which is derived from peracetylated lactose was an appropriate probe for the effective immobilization on alkyne-modified indium tin oxide (ITO) in assistance of $CuSO_4.5H_2O$/L-ascorbic acid as the standard CuAAC

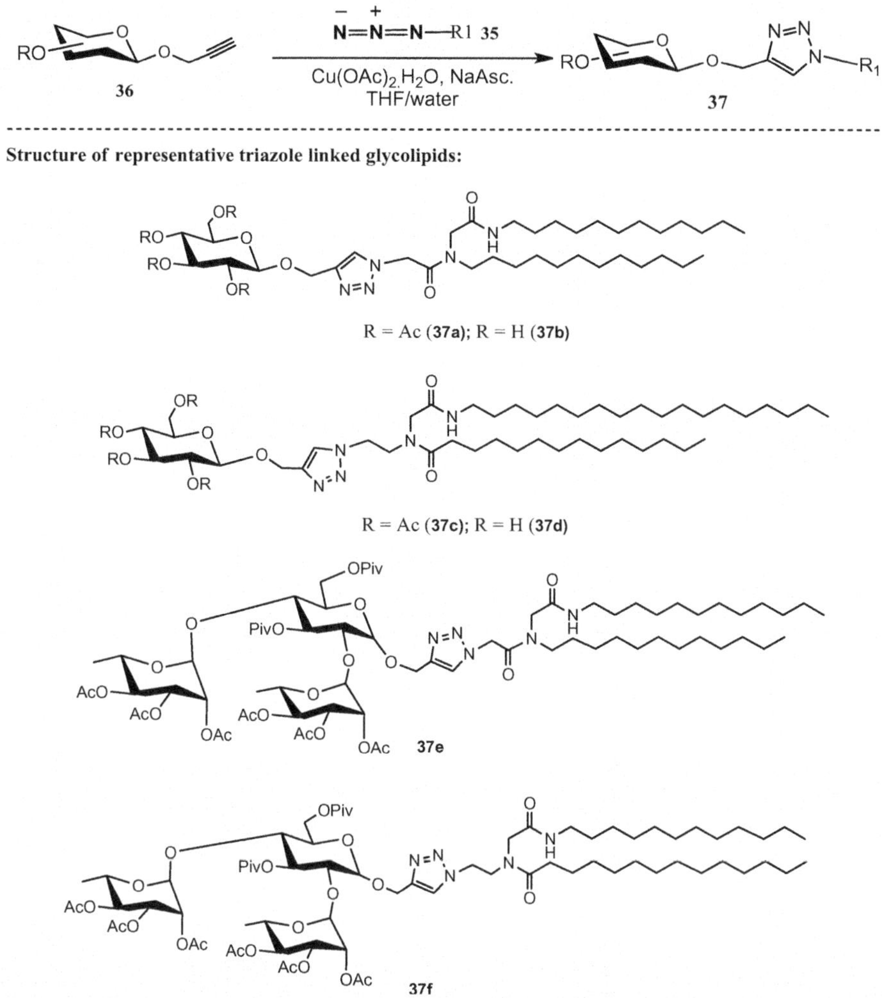

Scheme 12.8 CuAAC-inspired synthesis of triazole-appended glycolipids **37a–f** from respective lipid azides **35**

condition [43, 44]. Thus, in order to successfully establish an unnatural substrate for the study of bacterial β-galactosidase enzyme activities, the authors first attached a suitable carbohydrate-based probe, such as 2-azidoethyl lactose as linker, to the clickable alkyne partner comprises with bacterial cultures under standard coupling conditions. Afterward, they used mass spectrometry (MALDI-TOF) [44]. Indium tin oxide (ITO) slides to have the necessary alkyne-functionalized bidentate 1,2-sn-dipalmitoyl glycerol, they were simply functionalized using a carbamate linker under conventional coupling conditions. In the synthetic approach to achieve the desired alkoxycarbonylation product, such as alkyne-carbamate linker **42**, two consecutive

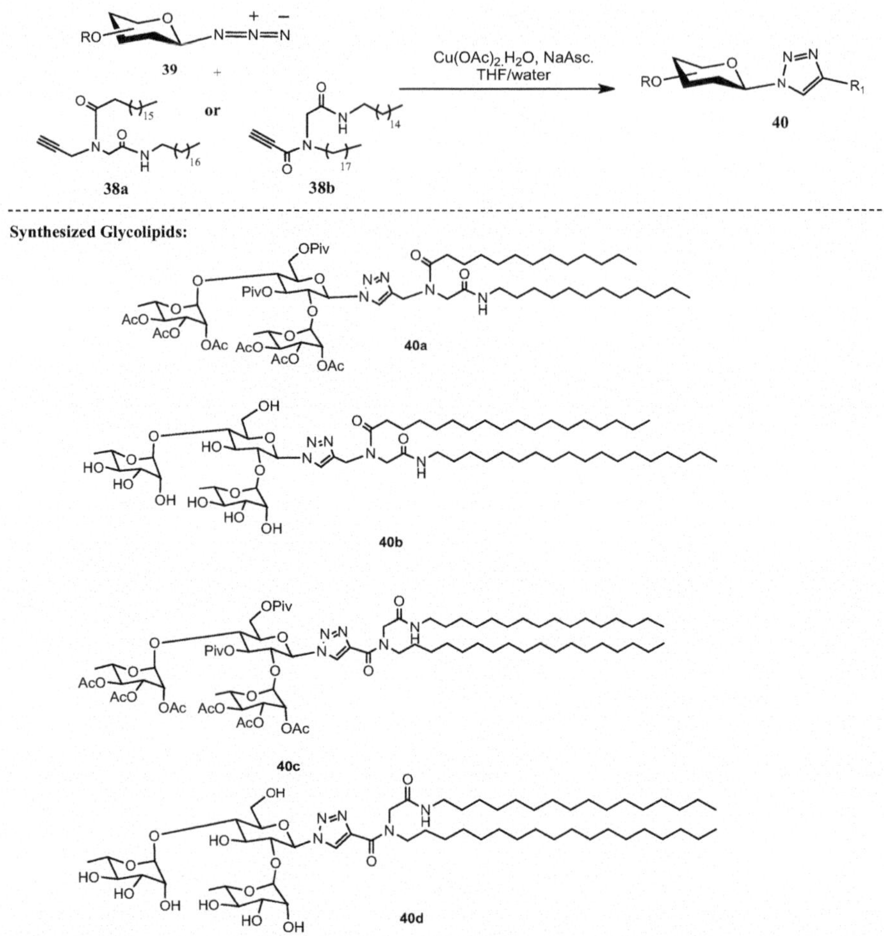

Scheme 12.9 CuAAC-mediated synthesis of 1,2,3-triazole-appended glycolipids **40a–d**

reaction were performed including reaction of 1,2-sn-dipalmitoyl glycerol **41** with *N*,*N*′-disuccinimidyl carbonate (DSC) in the presence of triethylamine followed by propargylation of respective product with propargyl amine under basic conditions (Scheme 12.10). However by employing surface-based MALDI-TOF examination of β-galactosidase activity, the non-covalent immobilization of the alkyne-carbamate linker via hydrophobic interactions on indium tin oxide slides may be suitable for studying the carbohydrate probe conjugate. It's interesting to note that the experiment showed that the ratio of digested to non-digested lactose probe may be used to measure for bacterial hydrolase activity. The protocol for evaluating the potential susceptibility of bacterial cultures to a specific antibiotic was further investigated by the authors, and as a result, it may be considered a promising tool in clinical laboratories for the accurate assessment of the activities of several health-related bacterial glycosidase in humans.

Additionally, the metabolic labeling under Cu-azide-alkyne cucloaddition reaction of a clickable alkynated myristic acid derivative was performed well and allowed for the regulation of lipid containing proteins in both the insect and host life phases of *T. brucei*. Moreover, to assess the particular aptitude to incorporate distinct fatty acids into proteins in biological systems, there is a great deal of room for further chemistry with other longer chain fatty acids, such as palmitate derivative [45]. The additional examples in term of chemistry were provided, in order to support and highlights the careful consideration of how microorganisms obtain phospholipids, cholesterol,

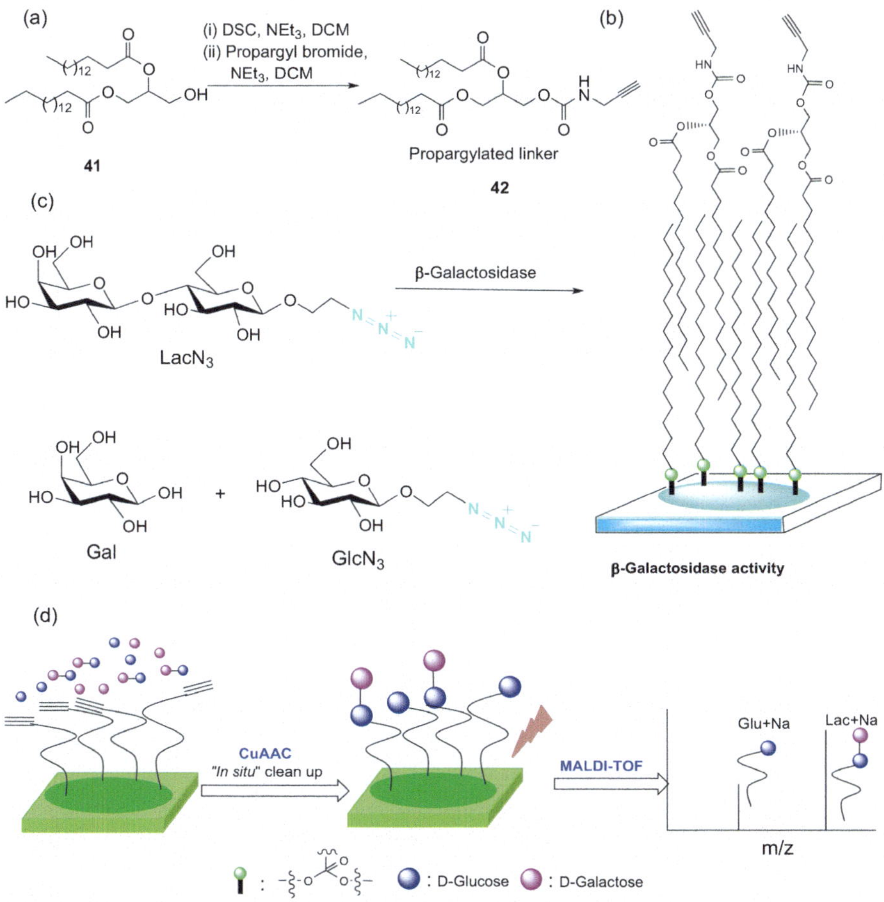

Scheme 12.10 **a** Synthesis of propargylkated linker from 1,2-sn-dipalmitoyl glycerol; **b** β-galactosidase assisted hydrolysis of 2-azidoethyl lactose; **c** Structure of ITO slide alkyne-modification; **d** CuAAC conjugation of $LacN_3$ on alkyne-modified ITO slides followed by MALDI-TOF analysis to evaluate β-galactosidase activity

and carbohydrates from the host,. Numerous significant processes, including phosphorus uptake, nitrogen fixation, and aluminum detoxification, depend heavily on the bacterial and fungal species that are found in soil. Researchers from UC Davis developed a synthetic pathway for a variety of glycolipids (derived from both plant and microbial sources) in order to establish a workable method for the extraction of metabolites from host tissue because the transient nature of chemical activity necessitates high detection to accumulate plant tissues for the purpose of tea cultivars in surroundings [46]. The Cu-catalyzed modular alkyne azide cycloaddition approach was used by the authors to start their investigation with the goal of in-vitro metabolic labeling, which is required to assess the effectiveness of chemical profiling and interactions in host tissue with a measurement goal in femtomolar detection limits. Consequently, *H.* pylori was first exposed to react with 17β-([3″-azidopropoxy)-5-androsten-3-β-ol **43** in a microaerobic reaction at room temperature to achieve **44**. Four days later, the bacterial cells were harvested and extracted, and the resulting extracts were then used to successfully metabolically label CGds through CuAAC coupling with individual fluorescent alkyne **45** in assistance of CuSO4/sodium ascorbate with tris(benzyltriazolylmethyl)amine for one hour to assemble various conjugated product **46** (Scheme 12.11) [47]. The investigation concluded that the microorganisms require some building blocks from the host, such as carbohydrates, phospholipids, and cholesterol in the glycosides form, which are then transported back to the host to allow symbiotic connections to form.

A variety of simple to complex glycolipids are easily prepared using the Cu-catalyzed alkyne azide cycloaddition reactions; it is obvious to conclude that their ease of development from corresponding azide/alkyne scaffolds certainly opens up a new area to use them as valuable tools for various purposes ranging from carbohydrate-based surfactant to biomedical applications. The impact of these click-inspired glycolipids has been demonstrated for significant application in biomedical science, specifically for the characterization of proteins in tumor cells. Additionally, while the established MALDI-TOF concept for studying bacterial β-galactosidase enzyme activities only identified one concept of enzyme activity, it is highly likely that the quantification of multiple glycosyl hydrolase activities with clinical relevance could be achieved in a single experiment through the use of multiple substrate probes in a multiplexed version. Despite the small number of researchers in this subject, the notable biomedical uses of click-inspired glycolipids may stimulate additional researchers worldwide to collaborate in this important and promising topic. It's interesting to note that some click-inspired glycolipids that have been assembled thus far may be investigated as appropriate linkers for spacer molecules and lipid rafts [48]. Many synthetic approaches have been employed in response to the need for different types of therapeutic molecules to treat a wide range of illnesses, and click chemistry has emerged as a flexible approach for assembling new and improved molecular arrangements. An intriguing approach exemplified the utility of modular click chemistry includes the synthesis of triazolyl α-GalCer moieties (**49a–f**) by Lee et al., where the click partner azide substituted lipid **48** derived from D-ribo-phytosphingosine **47** was subjected for the click reaction with terminal alkyne functionality having long chain of aliphatic with an incremental mannerism in the

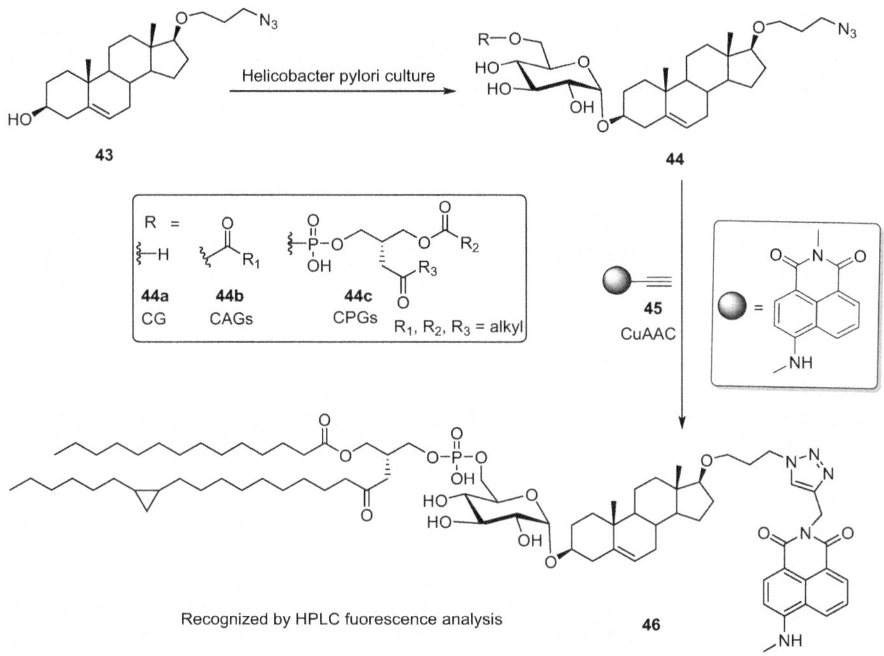

Scheme 12.11 CuAAC-mediated metabolic labeling of Helicobacter pylori CGds through coupling reaction of 17β-([3''-azidopropoxy)-5-androsten-3-β-ol with fluorescent alkyne [47]

chain length (Scheme 12.12) [49]. The α-GalCer, an agonistic antigen of the T cell receptor under natural killer cells (NKT), is the source of the revealed class, which was tested for cytokine effectiveness and shown to be a promising immune modulator. Additionally, the study demonstrated the impact of an elevated lipid chain on stimulation for a stronger Th2 cytokine response. Additional clinical research is still needed to fully understand the pharmacophore's potential for the lead optimization process in immunomodulator drug development research.

Nonetheless, Slamova et al. effectively investigated the MW-assisted click tool to synthesize triazolyl β-2-acetamidoGlcNAc-2-deoxyglucosides. Based on the bioassay analysis of the developed glycohybrids authors revealed that the compounds are hydrolyzed by *Talaromycesflavus*CCFβ-*N*-acetylhexosaminidase and are recognized as prospective inhibitors of NKT cells inhibitors [50]. Consequently, click chemistry has left a lasting legacy in the development of scaffolds for HIV therapy, as well as a path toward greater adaptability in the clicking of sugars with alkyne activity to generate triazole rings. In the field of biomedical imaging, spin sugar inspired radiolabeling has grown in acceptance over time. This method stands out among other similar approaches due to its link with the click approach, which offers advantages like simple synthesis protocols, a shorter reaction time, a high radiochemical yield, and excellent specificity. This method is incredibly effective, exhibiting excellent

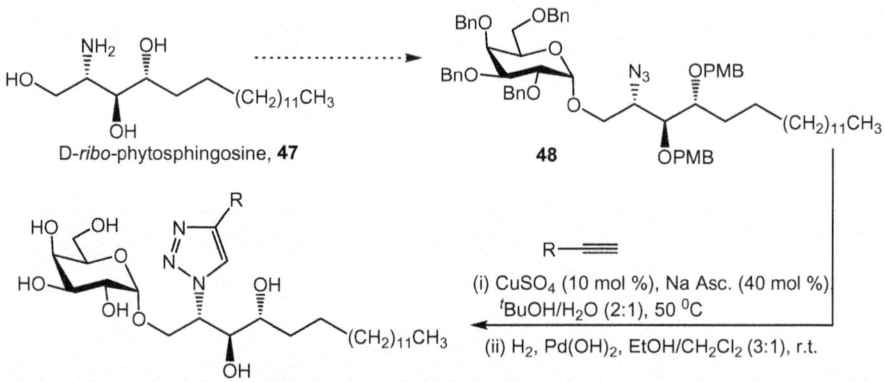

49a: R = -$(CH_2)_5CH_3$ (62%); **49b**: R = -$(CH_2)_6CH_3$ (56%)
49c: R = -$(CH_2)_{15}CH_3$ (50%); **49d**: R = -$(CH_2)_{22}CH_3$ (54%)
49e: R = -$(CH_2)_{23}CH_3$ (55%); **49f** : R = -$(CH_2)_{24}CH_3$ (51%)

Scheme 12.12 The triazole containing α-GalCer analogues synthesis under modular CuAAC Click Chemistry

selectivity and matching the necessary qualities of the equipment in ^{18}F carbohydrate probes to support medical science. However, the majority of the designing and development of radio tracers is still concentrated on including the ^{18}F labeled carbohydrate scaffold. In this manner, further research into the other effective non-toxic radiomolecules should be done in order to find a further development using carbohydrate motifs. Rao et al., intriguingly explored the modular Cu-azide-alkyne cycloaddition reaction to assemble a non-toxic cardanol derived glycoconjugate **53** for the specific utilization of its possible interactions with the lipid membrane [51]. In synthetic approach the 3-*O*-propargylated-1,2;5,6-di-*O*-isopropylidene glucofuranose **50** as alkyne unit was treated with hydrogenated cardanol-based azide **51** in assistance $CuSO_4$ (10 mol%) and sodium ascorbate (20 mol%) in DCM/H_2O (1:1) as solvent at room temperature for 10 h gives the triazole **52** in 92% yield, which was subjected for the acetonide deprotection with CF_3COOH in water (1:1) at room temperature for 10 h to achive the final desired cardanol derived glycoconjugate **53** (Scheme 12.13) [52]. The gelation properties of the resulting triazolyl cardanol glycoconjugate **53** were assessed to explore applicability. Remarkably, it creates four different types of gels in which hydrogel are formed from aqueous/protic polar solvents, whereas organogel is formed from non-polar solvents. Additionally, this glycohybrid successfully creates gel from both petroleum and vegetable oils. Even at a concentration of 0.03%, the glycoconjugate **53** was produced as a stable thermally reversible supergelator in a 1:1 mixture of H_2O and MeOH. While most comparable carbohydrate-based gels hydrolyze easily in acidic media, this gel is sufficiently stable in both basic and acidic environments. It was observed that this hydrogel's intrinsic fluorescence responded in the direction of a gel–sol transition.

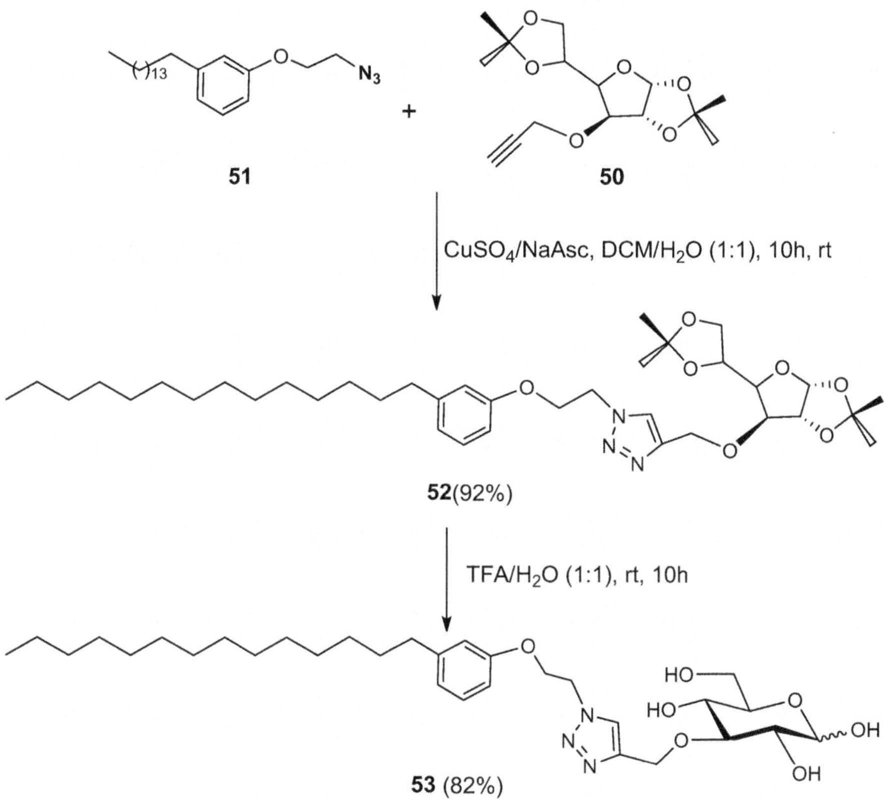

Scheme 12.13 Click-inspired synthesis of Cardanol-based Glycoconjugate **53**

For a clear understanding of the solubilization as well as the concentration-dependent membrane perturbation and the thermo-tropic phase transition progression of particular lipids, such as 1,2-*di*-palmitoyl-glycero-3-phosphocholine MLVs induced by cardanol derived glycoconjugate **53**, the authors used fluorescence molecular probe analysis and DSC. The outcomes seemed to indicate the high ability of the cardanol-based glycoconjugate to disturb the lipid bilayer membrane's structure [52]. Similarly, prior research for monosaccharide carbohydrate-based low molecular weight organo-gelators; Wang et al. intriguingly illuminated peracetylated disaccharide based organo-gelators **54, 55** by explicitly examining lactose and maltose residues (Fig. 12.4) [53]. The nature of glycosylation of monomer unit and substituent present on 4-position of triazole ring revealed the mechanism of action of disaccharide residues toward molecular self-assembling for their successful gelation. The mechanism of action of disaccharide residues toward molecular self-assembling for their effective gelation was described by the nature of glycosylation of monomer unit and substituent present on 4-position of triazole ring. The author used click chemistry to develop an array of maltose and lactose derivative analogues and tested their gelation capabilities in various organic solvents and aqueous solvent system. An experiment found that peracetylated maltose carrying β-triazole derivatives were superior to the comparable peracetylatedlactosyl-triazole derivatives for tested solvents. Such gelation phenomenon is caused by a configurational difference between two disaccharide molecules. Dynamic NMR spectroscopy demonstrates the self-assembly properties of disaccharide sugars and confirms that triazolyl cycle influences molecular assembling via non-covalent interactions such as hydrogen bonding and π-π-stacking. This concept was clarified by examining two individual compounds, **54a** and **55a**, whose dynamic NMR in two different solvents, $CDCl_3$/DMSO-d_6, demonstrated the gelation potential of maltosyl triazole derivatives. Optical microscopy and rheology analysis were used to investigate the gel formation of numerous maltose-derived triazole-appended glycoconjugates **55b** and **55c**, which were identified as more effective gelators and may be useful in drug immobilization and release (Fig. 12.4).

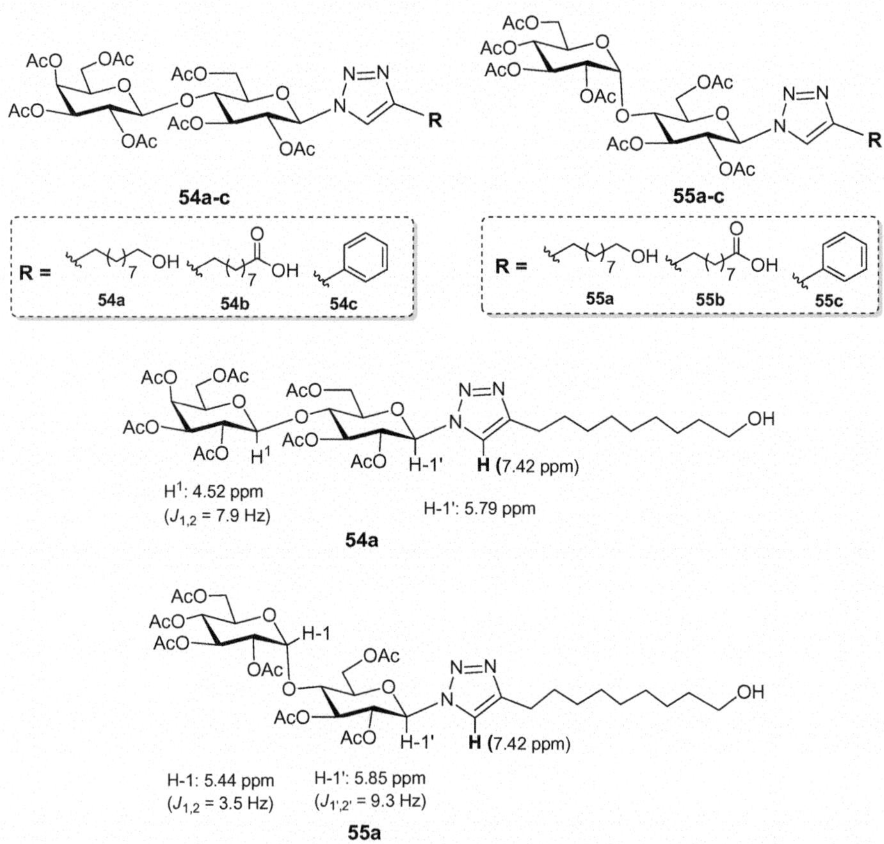

Fig. 12.4 Click-inspired developed Peracetylated Lactose **54a–c** and Maltose-based Gelators **55a–c**

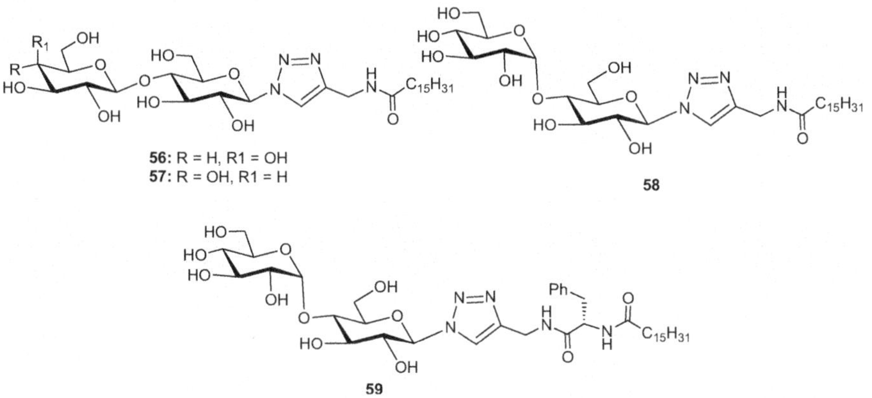

Fig. 12.5 Click-mediated Disaccharide Sugar-based Amphiphiles **56–59** as Supramolecular Hydrogelators

A series of low molecular weight sugar amphiphiles **56–59** were synthesized utilizing biocompatible disaccharides and hydrophobic linear chains via modular CuAAC 'Click Chemistry' to examine their liquid crystalline and gel formation capabilities (Fig. 12.5) [54, 55].

All of these triazole appeneded glycolipids **56–59** displayed stable hydrogel at room temperature. The electron microscopy has been used to analyze the fibrillar structure of supramolecular hydrogels that are formed by glycolipids self-assembling in a solvent. A torsion in the fibrillar structure was analyzed using circular dichroism (CD), which was most likely caused by the chiral supramolecular arrangement of amphiphiles. Thermal analysis of the hydrogel using differential scanning calorimetry (DSC) revealed the sol–gel transition temperature. Chiral supramolecular organization of self-assembled amphiphiles revealed homogeneous hydrogel formation at minimal concentrations of 1 or 0.5 weight percent twisted ribbons in lactose and cellobiose-derived glycohybrid triazoles **56** and **57**. These amphiphiles have been identified as excellent possibilities for tissue engineering matrix applications. These amphiphilic hydrogels' biocompatibility and degradability have been shown to be important for cell encapsulation and controlled drug release.

Lipids in the cell membrane have been found to be important unit in cell signaling processes. Their function as site-specific ligands in protein-membrane binding interactions, which lead to the cellular membrane anchoring of peripheral proteins, is one of their key roles. In general, these activities control the location and function of proteins and have been connected to the initiation of disease states as well as regular physiological processes. Therefore, it is crucial to clarify the specifics of interactions at the molecular level, including the placement of receptor binding domains, multivalency in binding, and lipid-binding specificities and affinities. Smith et al., developed azido-tagged lipid analogues as easily functionalizable lipid probe scaffolds specifically for this use. In the intriguing studies synthesis and design of the first structure of this kind, called diacylglycerol analogue **60**, which has an azide tag attached to the lipid headgroup at position sn-1. After the successful synthesis of azido compound DAG scaffold **60**, derivatize this core in order to demonstrate the capability of generating an array of probes **62a–d** in a single handy step, the 1,3-dipolar cycloaddition procedure was utilized to add fluorescent **62a–d** and photoaffinity **62e** groups onto **60** with alkyne reagents **61a–e**, which were obtained from propargylamine in a single step (Scheme 12.14). Azide-alkyne cycloaddition's moderate conditions are very helpful for assembled functionalized DAGs since they prevent the acyl migration byproducts. Many complex phospholipids and glycolipids with numerous Fermi resonance energy transfer (FRET) pairs are among the fluorophores that have been integrated. As a result, these DAG probes will function as reliable sensors for research into binding interactions and as models for the creation of similar probes for more intricate phospholipids and glycolipids [56].

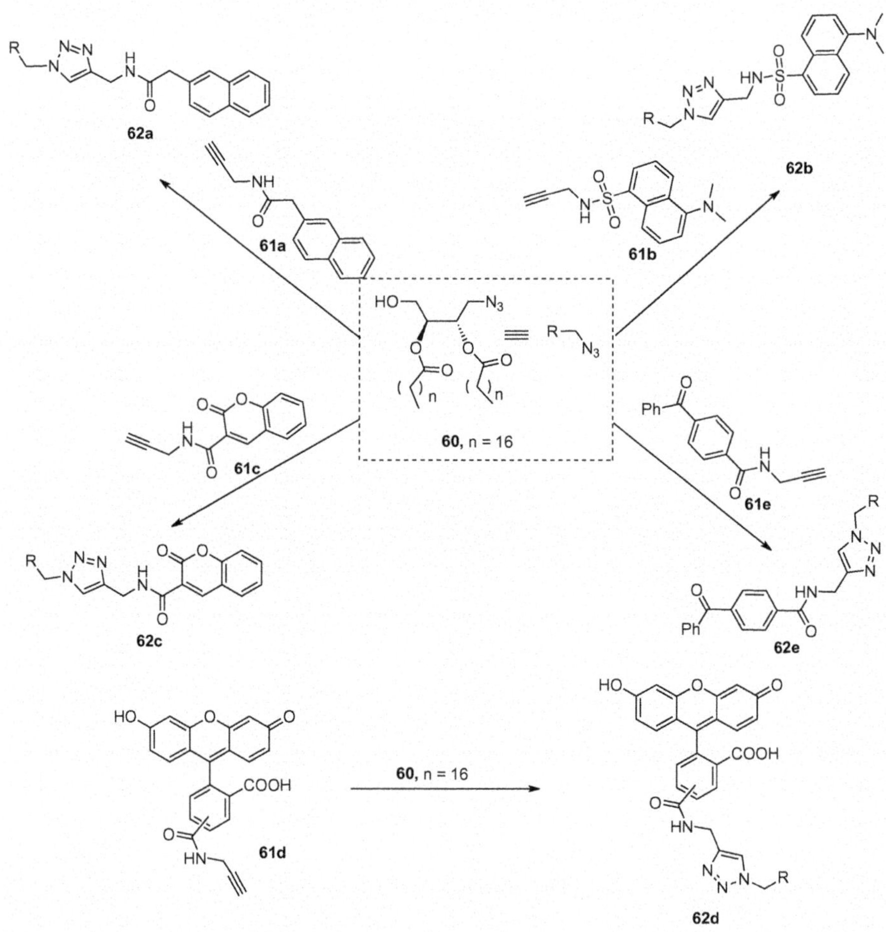

Scheme 12.14 Click-mediated derivatization of azido-DAG scaffold **60** to introduce various group

Moreover the same group reported a productive modular method that uses a late-stage introduction of reporter groups to synthesize a variety of PA probes. This technique was used to create fluorescent and photoaffinity-tagged PA probes, as well as a bifunctional probe with an azide serving as a secondary handle for purification and a photoaffinity moiety. In order to ascertain whether these PA analogs could replicate the original lipid's protein-binding characteristics, each one was added to vesicles for binding experiments employing the C2 domain of PKCa, a recognized PA receptor.

Each of the compounds in these investigations showed binding characteristics similar to those of synthetic PA, suggesting that they are viable probes for efficiently examining the roles played by PA in cellular processes. In the synthetic approach for the successful synthesis of analogue of phosphotriester-protected PA **63**, the scaffold was then explored to produce several derivatized PA probes. The authors utilized the modular click tool to connect probe **63** to a variety of alkynes **64a–e** to give high yield of respective phosphotriesters **65a–e** (Scheme 12.15) Bromotrimethylsilane was used for the successful removal of phosphotriester groups and desired PA probes **66a–e** were obtained in pure form and high yield. This method could be efficiently used to develop selective PA probes linked with a variety of fluorescent dyes including naphthyl (**66a**), dansyl (**66b**), coumarin (**66c**), rhodamine (**66d**), and also a photoaffinity tag benzophenone **66e** [57].

The author explores the concept of activity-based protein profiling (ABPP) for probe development. The ABPP concept utilizes the substrate as probes for a

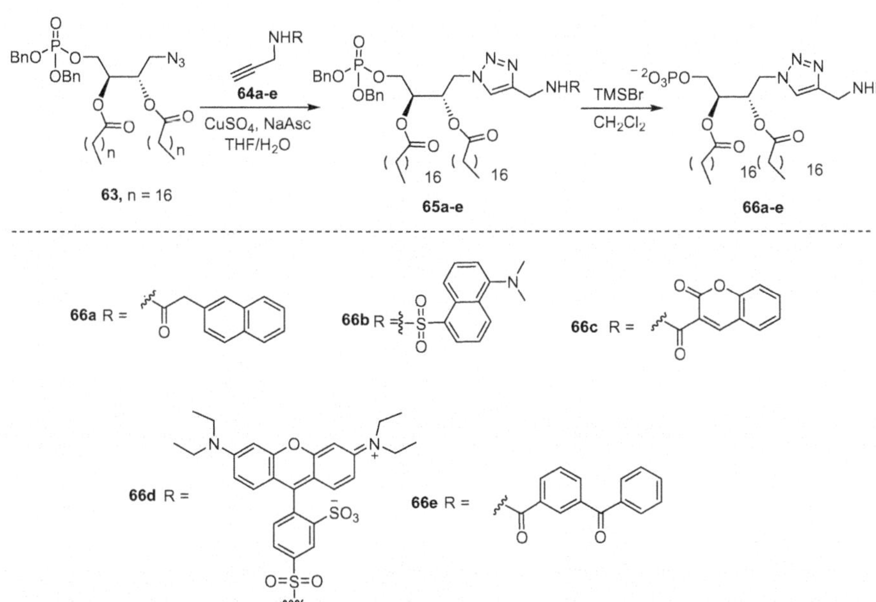

Scheme 12.15 Click-mediated derivatization of protected PA analogue **63** to develope PA probes **66**

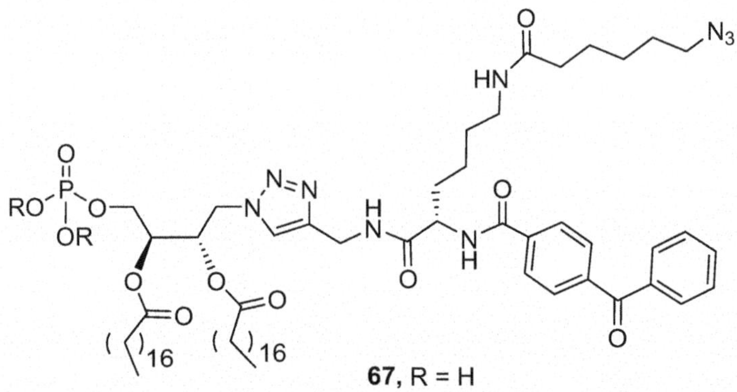

67, R = H

Fig. 12.6 Structure of a PA probe **67**, contains both a photo-cross-linking group benzophenone and azide as a secondary tag residue

mechanism-based effective tool required to the collective labeling of enzyme as molecular target as per the associated activity. A PA probe (**67**, Fig. 12.6) contains benzophenone as a photo-cross-linking group to cognate receptors and azide as a secondary tag for purification of effectively cross-linked proteins, is considered to fish out PA-binding receptors from a complex molecular architecture.

The robost tool is required for the charactrerization of membrane-based processes. In the way, an azide-tagged lipid anchor **68** can be easily resulted to respective 1,2,3-triazole architecture via click chemistry and can further used to functionalize the membrane surface. The first target was biotin-conjugate **70b**, achieved via the click conjugation of azide **68** with alkyne core **69b,** to immobilize the entire liposome. However, after immobilization, a tetraethylene glycol (TEG)-based tether can be established between the biotin and lipid components for the effective space of liposomes from the microplate surface. Fluorescein **70a** as a fluorophore-tagged anchor can also assembled through the click approach using azide-tagged modular lipid anchor **68** and alkyne core **69a to** detect functionalized liposomes during investigation (Scheme 12.16) [58].

Sheng et al., reported the modular "Click" reaction approach to assemble an array of sterol-based cationic lipids **75–78**. In the synthetic approach, alkynyl-bearing sterol hydrophobes **71** and **72** were synthesized. After that, the as-prepared alkynyl-bearing sterol hydrophobes **71** and **72** were reacted separately to azide-bearing buiding blocks **73** and **74** in assistance of $CuSO_4.5H_2O$/sodium ascorbate in THF/H_2O, at 80 °C for 2 h to afford Boc-protected sterol-based cationic lipids which on Boc-deprotection in acidic medium affords the final desired two cholesterol-based cationic lipids **75** and **76** and two lithocholate-based lipids **77** and **78** as the final products with the isolation yields of 62–79% (Scheme 12.17). These developed cationic lipids were shown to have effective pDNA binding and loading capabilities. It was discovered that the molecular structure components, such as sterol skeletons and headgroups, have a significant influence on the physico-chemical characteristics of their pDNA lipoplexes, including particle sizes, morphologies, and surface charges. Nevertheless,

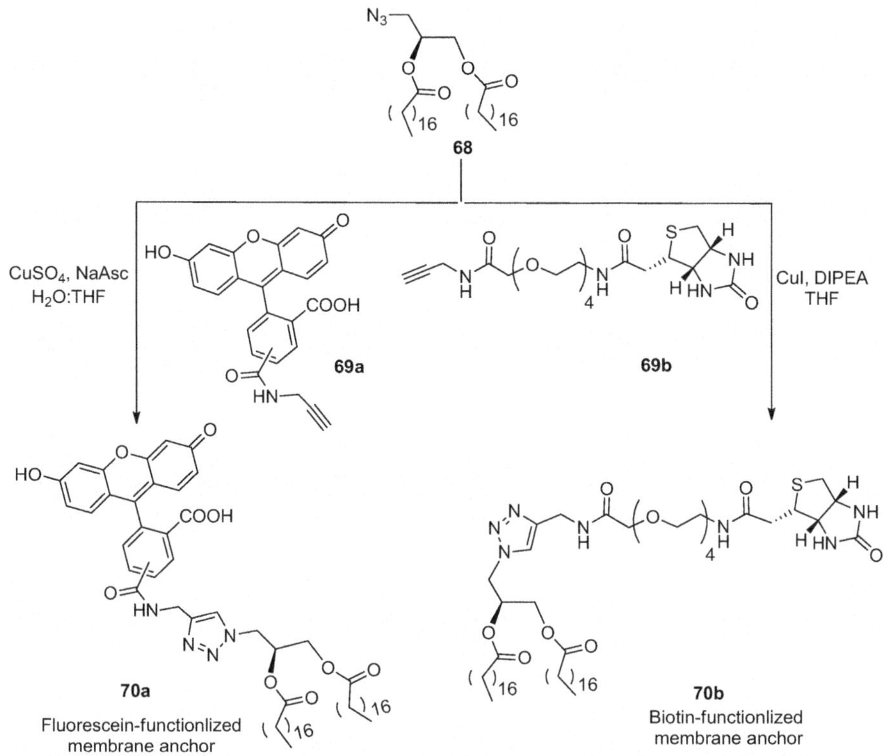

Scheme 12.16 CuAAC-mediated derivatization of azide scaffold **68** to develop functionalized membrane anchors **70**

it was shown that these sterol-based cationic lipids showed remarkably varied cytotoxicity and luciferase gene transfection effectiveness, both of which were highly dependent on their molecular structures. It's interesting to note that lipids **75** and **76** that contained cholesterol had a significantly higher transfection capacity (7–10^4 times) than lipids **77** and **78** that contained lithocholate. This suggests that cholesterol could be used as a more effective moiety than lithocholate for creating new lipid gene carriers [59].

Zhang et al., developed a library of triazole containing ionizable lipids **80**, using the biorthogonal CuAAC reaction. In the modular approach triazole was biocompatiblely introduced into ionizable lipid structures **79**, as shown by a cell survival experiment. Then, **80** were further used to get LNPs ready for the transport of luciferase mRNA (Scheme 12.18) [60].

The anti-cancer vaccines consisted of MUC1 glycopeptide and immune-stimulating lipopeptide ligands receive some practical issues in the peptide conjugation. To overcome the associoated common problem, a modular click tool successfully demonstrated the linking of lipopeptide lysine with MUC1 glycopeptide [61]. Therefore, click conjugation of multivalent functionalized lipopeptide alkyne **81** with

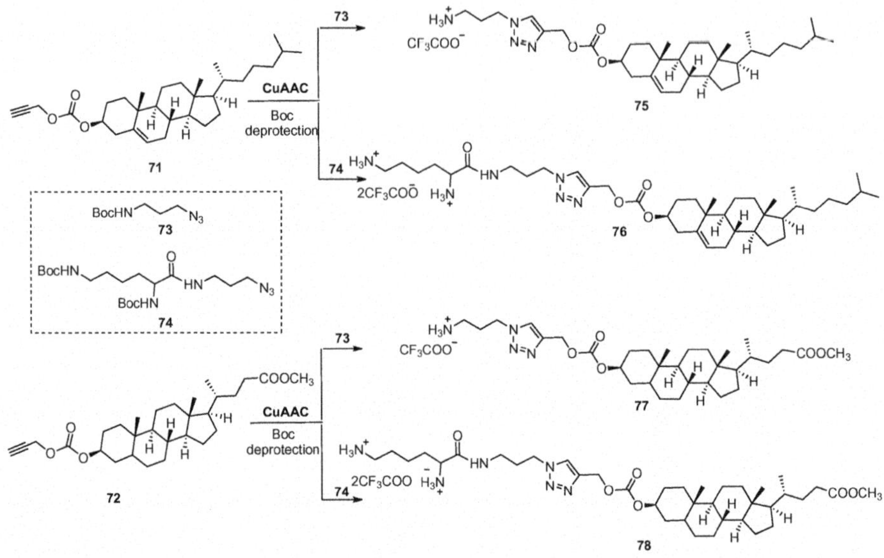

Scheme 12.17 CuAAC-mediated synthesis of steroids based cationic lipids

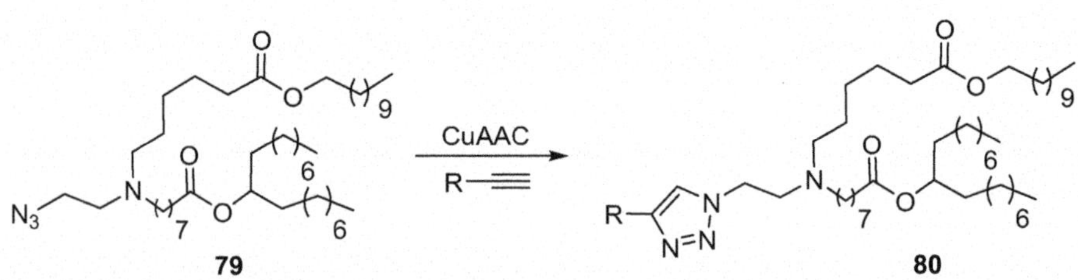

Scheme 12.18 CuAAC-mediated synthesis triazle appended lipid **80**

the glycopeptide azide **82** in presence of $Cu(OAc)_2$/ NaAsc furnished the desired oligovalent glycopeptide-lipopeptide antigens **83a–d** in moderate yield (30–40%) (Scheme 12.19). As both the lipopeptides and glycopeptides were water-soluble, thus the coupling was performed in deoxygenated water. Further, the azido-glycopeptide was used in eight times excess required for the convenient conjugation. Additionally, it has been noted that oligovalent lipopeptide-glycopeptide conjugates were significantly more effective than monovalent analogue in activating strong immune responses.

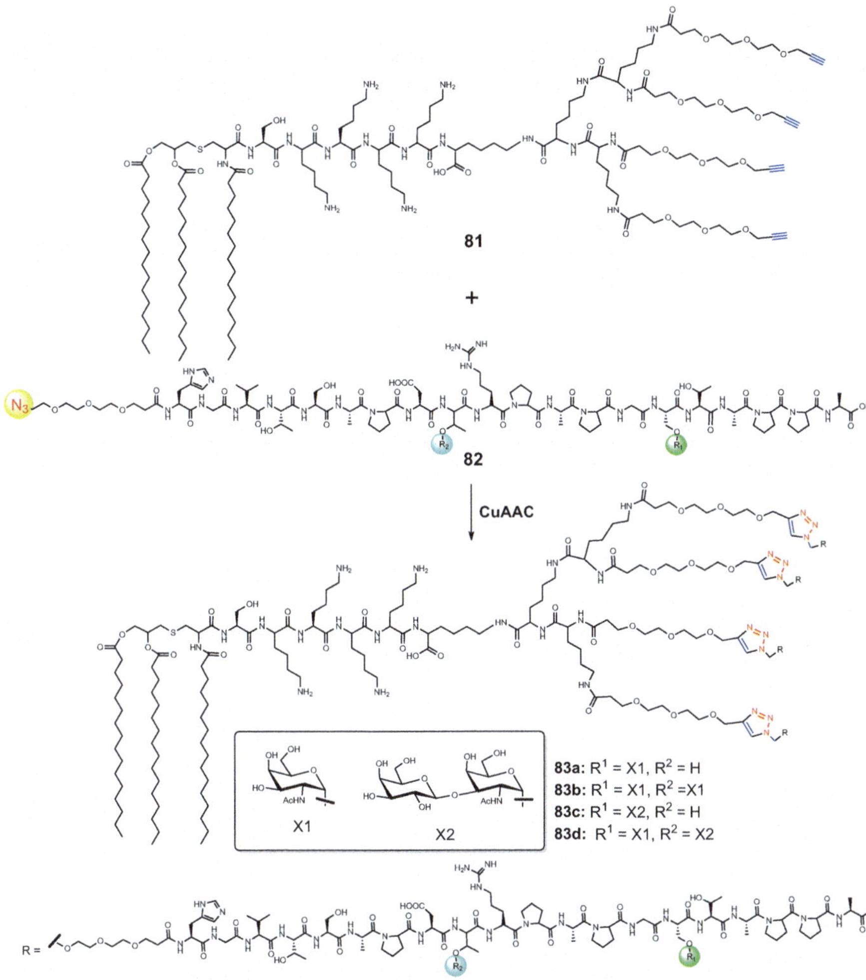

Scheme 12.19 CuAAC Click-inspired synthesis of tetra glycopeptides **83a–d** with a lipopeptide

12.4 Conclusions and Future Perspective

The modular CuAAC click conjugation of diverse biologically relevant scaffolds including sugars and other important biomolecules with lipids as imperative structural component known to communicate the resulting biocompatible molecules required to explicit cell behavior is covered in this chapter. Fundamental cellular components such as lipids, sugars, and others can be conjugated with other biologically relevant components including glycans, peptides, proteins, polymers, etc., can be easily conjugated by utilizing the regioselctive and modular CuAAC tool. The coupling method via triazole forming reaction can also be incorporated in absence of metal mediated coupling and for this purpose bioorthoganl click chemistry were also applied well. The resultant lipid containing 1,2,3-triazole based molecular architectures have some notable features notably its biocompatibility, bioactivities, and displayed a number of fascinating properties. Moreover, they also explored in modulation of cellular activities.

References

1. Subramaniam S, Fahy E, Gupta S, Sud M, Byrnes RW, Cotter D, Dinasarapu AR, Maurya MR (2011) Chem Rev 111:6452–6490
2. Fahy E, Subramaniam S, Murphy RC, Nishijima M, Raetz CR, Shimizu T, Spener F, van Meer G, Wakelam MJ, Dennis EA (2009) J Lipid Res 50:9–14
3. Chan YHM, Boxer SG (2007) Curr Opin Chem Biol 11:581–587
4. Brochu H, Polidori A, Pucci B, Vermette P (2004) Curr Drug Deliv 1:299–312
5. Mintzer MA, Simanek EE (2009) Chem Rev 109:259–302
6. Best, M. D. Witczak, Z. J.; Bielski, R., *Ed.; John Wiley and Sons,* **2013**, 79–105.
7. Best MD, Rowland MM, Bostic HE (2011) Acc Chem Res 44:686–698
8. Suazo KF, Park KY, Distefano MD (2021) Chem Rev 121:7178–7248
9. Rostovtsev VV, Green LG, Fokin VV, Sharpless KB (2002) Angew Chem 114:2708–2711
10. Tornoe CW, Christensen C, Meldal M (2002) J Org Chem 67:3057–3064
11. Zhang L, Chen XG, Xue P, Sun HHY, Williams ID, Sharpless KB, Fokin VV, Jia GC (2005) J Am Chem Soc 127:15998–15999
12. Tiwari VK, Mishra BB, Mishra KB, Mishra N, Singh AS, Chen X (2016) Chem Rev 116:3086–3240
13. Agrahari AK, Bose P, Jaiswal MK, Rajkhowa S, Singh AS, Hotha S, Mishra N, Tiwari VK (2021) Chem Rev 121:7638–7956
14. Mishra A, Agrahari AK, Tiwari VK (2021) Application of CuAAC in carbohydrates. In: Click chemistry volume, science of synthesis, pp 135–179
15. Hassane FS, Frisch B, Schuber F (2006) Bioconjug Chem 17:849–854
16. Zhang HL, Ma Y, Sun XL (2009) Chem Commun 21:3032–3034
17. Van Lengerich B, Rawle RJ, Boxer SG (2010) Langmuir 26:8666
18. Cavalli S, Tipton AR, Overhand M, Kros A (2006) Chem Commun 30:3193
19. Kuzmin A, Poloukhtine A, Wolfert MA, Popik VV (2010) Bioconjugate Chem 21:2076
20. Debets MF, van Berkel SS, Schoffelen S, Rutjes F, van Hest JCM, van Delft FL (2010) Chem Commun 46:97
21. Bostic HE, Smith MD, Poloukhtine AA, Popik VV, Best MD (2012) Chem Commun 48:1431
22. Kumar A, Erasquin UJ, Qin GT, Li K, Cai CZ (2010) Chem Commun 46:5746–5748
23. Zerkowski JA, Nunez A, Strahan GD, Solaiman DKYJ (2009) Am Oil Chem Soc 86:1115–1121
24. Tang SC, Wong QL, Guo ZW (2010) Chem Eur J 16:1319–1325
25. Neef AB, Schultz C (2009) Angew Chem Int Ed 48:1498
26. Jao CY, Roth M, Welti R, Salic A (2009) Proc Natl Acad Sci USA 106:15332
27. Cho W, Stahelin RV (2005) Ann Rev Biophys Biomol Struct 34:119–151
28. Lemmon MA (2008) Nat Rev Mol Cell Biol 9:99–111
29. Cravatt BF, Wright AT, Kozarich JW (2008) Annu Rev Biochem 77:383–414
30. Alam S, Alves DS, Whitehead SA, Bayer AM, McNitt CD, Popik VV, Barrera FN, Best MD (2015) Bioconjugate Chem 26:1021–1031
31. Di Paolo G, De Camilli P (2006) Nature 443:651–657
32. Gong D, Bostic HE, Smith MD, Best MD (2009) Eur J Org Chem 4170–4179
33. Rowland MM, Bostic HE, Gong D, Speers AE, Lucas N, Cho W, Cravatt BF, Best MD (2011) Biochemistry 50:11143–11161

34. Gubbens J, Ruijter E, de Fays LEV, Damen JMA, de Kruijff B, Slijper M, Rijkers DTS, Liskamp RMJ, de Kroon A (2009) Chem Biol 16:3
35. Gubbens J, de Kroon A (2010) Mol Biosyst 6:1751
36. Tully SE, Cravatt BF (2010) J Am Chem Soc 132:3264
37. Bandyopadhyay S, Bong D (2011) Eur J Org Chem 4:751
38. Fyrner T, Magnusson K, Nilsson KPR, Hammarstrom P, Aili D, Konradsson P (2012) Bioconjugate Chem 23:1333–1340
39. Sani FA, Heidelberg T, Hashim R, Farhanullah (2012) Colloids Surf B 97:196−200
40. Tyagi M, Kartha KPR (2015) Carbohydr Res 413:85–92
41. Perez-Labrada K, Brouard I, Mendez I, Perez CS, Gavín JA, Rivera DG Eur J Org Chem 3671−3683
42. Serna S, Ercibengoa M, Marimón JM, Reichardt N-C (2018) Anal Chem 90:12536–12543
43. Chernyak AY, Sharma GVM, Kononov LO, Krishna PR, Levinsky AB, Kochetkov NK.; Rama Rao AV Carbohydr Res 223:303−309
44. Zilio C, Bernardi A, Palmioli A, Salina M, Tagliabue G, Buscaglia M, Consonni R, Chiari M (2015) Sens Actuators B 215:412–420
45. Wright MH, Paape D, Price HP, Smith DF, Tate EW (2016) ACS Infect Dis 2:427–441
46. Schombs M, Park FE, Du W, Kulkarni SS, Gervay-Hague J (2010) J Org Chem 75:4891–4898
47. Jan H-M, Chen Y-C, Shih Y-Y, Huang Y-C, Tu Z, Ingle AB, Liu S-W, Wu M-S, Gervay-Hague J, Mong K-KT et al (2016) Chem Sci 7:6208–6216
48. Ancajas CF, Ricks TJ, Best MD (2020) Chem Phys Lipids 232:104971
49. Lee T, Cho M, Ko SY, Youn HJ, Baek DJ, Cho WJ, Kang CY, Kim S (2007) J Med Chem 50:585–589
50. Slamova K, Marhol P, Bezouska K, Lindkvist L, Hansen SG, Kren V, Jensen HH (2010) Bioorg Med Chem Lett 20:4263–4265
51. Rao SPH.; Kamalraj M, Swain J, Mishra AK (2014) RSC Adv 4:12175
52. Swain J, Kamalraj M, Surya Prakash Rao H, Mishra AK (2015) J Mol Struct 1081:124−127
53. Okafor IS, Wang G (2017) Carbohydr Res 451:81–94
54. Clemente MJ, Fitremann J, Mauzac M, Serrano JL, Oriol L (2011) Langmuir 27:15236–15247
55. Clemente MJ, Romero P, Serrano JL, Fitremann J, Oriol L (2012) Chem Mater 24:3847–3858
56. Smith MD, Gong D, Sudhahar CG, Reno JC, Stahelin RV, Best MD (2008) Bioconjugate Chem 19:1855–1863
57. Smith MD, Sudhahar CG, Gong D, Stahelin RV, Best MD (2009). Mol Biosyst 962–972.
58. Losey EA, Smith MD, Meng M, Best MD (2009) Bioconjugate Chem 20:376–383
59. Sheng R, Luo T, Li H, Sun J, Wang Z, Cao A (2013) Bioorg Med Chem 21:6366–6377
60. Wang Y, Si X, Feng Y, Feng D, Xu X, Zhang Y (2023) Molecules 28:4046
61. Cai H, Huang Z-H, Shi L, Zhao Y-F, Kunz H, Li Y-M (2011) Chem Eur J 17:6396–6406

Chapter 13
CuAAC 'Click' in Carbohydrate Chemistry

13.1 Introduction

During the last 22 years, Cu(I)-catalyzed azide-alkyne cycloaddition (CuAAC) reactions have been considered the most essential chemical transformation tool in organic synthesis which is well explored in glycosciences for the regioselective synthesis of a diverse range of 1,4-disubstituted 1,2,3-triazole-appended glycohybrids, glycoconjugates, and neoglycoconjugates (glycopolymers, glycopeptides, glycodendrimers, etc.) of a wide range of applications in chemical biology, catalysis, drug development, and other emerging areas [1, 2]. The 1,3-dipolar cycloaddition reaction of organic azides **2** with terminal alkynes **1** (usually known as 'CuAAC Click Chemistry') for regioselective synthesis of 1,4-disubstituted 1,2,3-triazole **3** was extensively explored by Huisgen [3]-Meldal [4]-Sharpless [5] (Scheme 13.1). This regioselective tool is recognized as a powerful strategy in organic synthesis for adjoining two entirely different building blocks of simple-to-complex molecular architectures [6–9].

Ruthenium-catalyzed azide-alkyne cycloaddition reaction was established in 2005 [10, 11] that regioselectively furnished 1,5-disubstituted triazole **4**. Meanwhile, the strain-promoted azide-alkyne cycloaddition (SPAAC) concept emerged for the chemical transformation of biomolecules in biological systems [12]. To develop a triazole pharmacophore surrounded by a human cell, Bertozzi et al. employed an alkynyl fluorescent probe in conjugation with an azido-modified cell in an SPAAC reaction [13]. This coupling is very useful for tracking glycans in living organisms. CuAAC click tool offers a versatile, successful, and reliable protocol for coupling two different molecular entities and thus largely explored in a broad array of science and technology, from medicinal chemistry to catalysis [14–17]. Carbohydrates are an essential component of living cells and possess several interesting characteristics, including their extensive structural and functional diversity, biocompatibility, hydrophilicity, excellent ADME features, and minimal toxicity [18]. In light of the aforementioned facts, there is a greater desire to inspect sugars as flexible scaffolds

V. K. Tiwari et al., *Click Chemistry*, Materials Horizons: From Nature to Nanomaterials,
https://doi.org/10.1007/978-981-97-4596-8_13

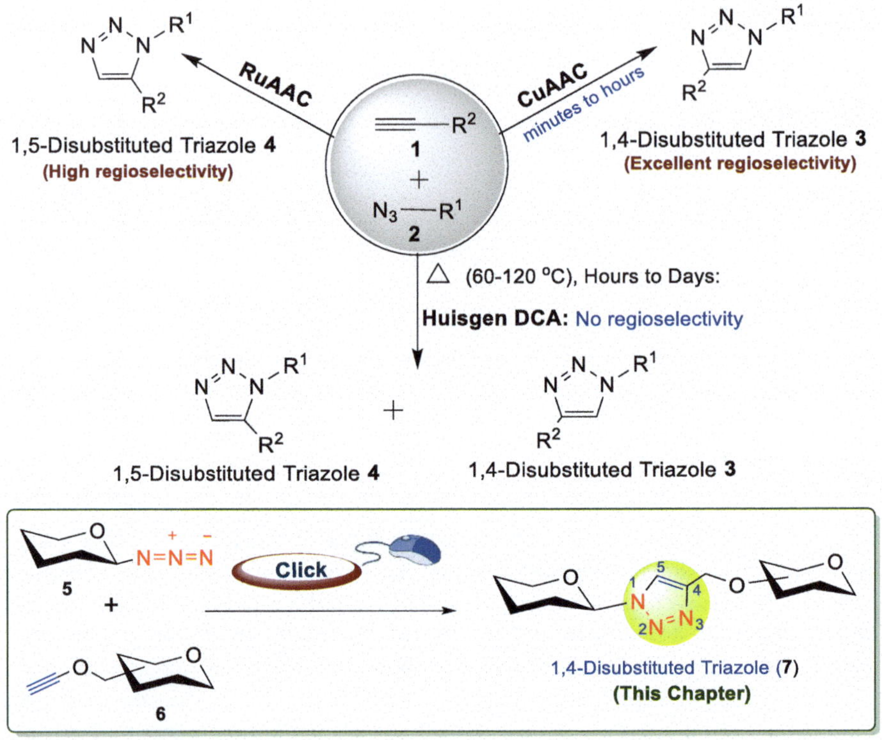

Scheme 13.1 Metal and thermal induced [3 + 2] azide-alkyne cycloaddition reaction in deciding regioselectivity

for the construction of an array of glycans and distinct glycoconjugates for their chemical, biochemical, and pharmacological investigations [19–22].

This chapter briefly highlights the notable applications of modular CuAAC 'Click Chemistry' for the development of regioselective 1,2,3-triazole glycoconjugates with their growing impact in various emerging fields of basic and applied science. We hope that the readers may find this chapter well suitable for their teaching and research on various emerging disciplines of science and technologies. We provide a brief overview with future perspectives of CuAAC click chemistry in glycoscience and their diverse applications. We express our regret to the readers in case their contributions are not included in our chapter due to limitations concerning page number.

13.2 Preparation of Carbohydrate-Based Azide and Alkynes for CuAAC Click Coupling

Although most organic reactions are known to be influenced by the stereoelectronic effect of the substituents, however, click-inspired azide-alkyne cycloaddition (CuAAC) of sugar azide with terminal alkyne is a common and preferred tool in glycoscience [23]. Matyjaszewski et al. demonstrated the steric and electronic effects on the rate of organic reaction and suggested that organic azides with an attachment of electron-withdrawing group underwent a faster rate of click cycloaddition [24]. The α-carbonyl-alkynes were found to be more reactive than alkylated alkynes in click cycloaddition, however, the click counter coupling partner terminal alkynes are slightly less reactive. The synthesis of organic azides can be achieved using several well-documented high-yield synthetic processes nevertheless, owing of the low molecular weight and toxic nature of azides, their practical adaptation has been limited in both academia and industry for quite a while.

In addition, carbohydrate-based azides show required persistence at room temperature and also found to be inert toward a broad variety of reaction conditions. It is usually preferred to introduce the required clickable azide and alkyne groups at the anomeric position of carbohydrates.

Paulsen et al. reported the preparation of glycosyl azides **9** from penta-*O*-acetyl-aldohexopyranose **8** using trimethylsilyl azide under acidic conditions, i.e., BF_3 or $SnCl_4$ in good yields (Scheme 13.2a) [25]. The clickable azide group can be simply generated in carbohydrate moiety at specific positions by producing suitable leaving groups namely halides, tosyl, triflate, mesyl, and other similar groups. Therefore, tosylation of compound **10** in the presence of pyridine produces a sulfonate derivative (stable and can be isolated at r.t.) which subsequently underwent azidation on treatment with NaN_3 in anhydrous DMF at 80 °C resulted in the 6-azido GlcNAc derivative **11** in overall 86% yield (Scheme 13.2b) [26]. Further, NaN_3 mediated oxirane ring opening of carbohydrate-based epoxides effectively produces corresponding diaxial azido alcohol. Therefore, a similar ring opening of glycosyl epoxide **12** with NaN_3 in DMF/H_2O at 120 °C furnishes respective 2-azido sugar **13** in 76% yield (Scheme 13.2c) [27]. This fascinating epoxide-ring opening protocol offers another valuable technique to introduce the clickable azido group at specific positions of orthogonally protected saccharides. Apart from the above protocols, some report describes the direct transformation of NH_2 group to azido functionality through a diazo-transfer reaction [28]. Toward this end, Vasella and co-workers made a significant effort to introduce the azido group in carbohydrate scaffolds through diazo-transfer of respective amino sugars without affecting the stereochemistry at that stereo center [29]. The authors treated the glucosamine **14** with triflyl azide in the presence of $CuSO_4$ and K_2CO_3 to achieve the corresponding azido sugar, which subsequent acetylation furnished a high yield of acetate-protected azido sugar **15** (Scheme 13.2d). A one-pot protocol including Michael addition followed by diazo-transfer was documented by Tiwari et al. for easy access to glycosyl-*β*-azido ester

[30]. Thus, olefinic ester **16** was treated with ethanolic ammonia to produce a diastereomeric mixture glycosyl β-amino ester, which on diazotization with imidazolyl-1-sulphonyl azide **109** in the presence of K_2CO_3 and $ZnCl_2$ furnished high yield of glycosyl azide **17** (Scheme 13.2e).

Kotik and co-workers have demonstrated the enzymatic mode of selectivity for the successful insertion of azide functionality into carbohydrate scaffold [31]. Rutin **19**, a β-rutinosylated flavonoid was readily converted into rutinosyl β-azide **20** using wild-type rutinosidases (as glycosyl donor) (Scheme 13.3). However, the opposite anomeric selectivity was achieved by treating rutin **19** with catalytic nucleophile mutant of sodium azide.

Following the installation of clickable azide at specific orthogonally protected glycoside positions, it is also essential to produce alkyne functionality at the desired position of carbohydrate moiety. Thus, Koester group reported Sonogashira–Hagihara coupling of 1-iodinated glycal with several aromatic and aliphatic alkynes for preparation of alkynyl *C*-glycoside in good yields [32]. Sonogashira–Hagihara reaction of 1-iodoglycal **21** with TMS-alkyne in the presence of Pd-catalyst and CuI under basic conditions at room temperature afforded **22** in 79% in 12 h (Scheme 13.4a). Further, the silyl group of protected alkyne can be easily removed by treating with tetra-butyl ammonium fluoride under mild conditions to furnish the desired terminal alkyne **23**. Likewise, glycosylation of glucal **24** with propargyl alcohol in the presence of $CeCl_3$ and L-proline as a chiral ligand in toluene furnished exclusively alkynylated glycoside **25** with α-selectivity in 98% yield (Scheme 13.4b) [33].

Furthermore, an interesting glycosyl alkyne **28** was obtained in good yields from the corresponding aldehyde using Bestmann ohira reagent (dimethyl (1-diazo-2-oxopropyl) phosphonate) under basic conditions in methanol at rt (Scheme 13.4c) [34]. The reactive aldehyde molecule **27** was obtained from corresponding diol **26** through NaIO4-mediated oxidation of glycosyl diol and subsequent periodinate oxidation in dry DCM at rt for 4 h. Similarly, a clickable *O-p*ropargyl group can be regioselectively installed in the both equatorial as well as the axial hydroxyl group of various monosaccharides **29** by selecting a suitable ligand to afford a high yield of desired glycosyl alkyne **30** and **31** with complete regioselectivity (Scheme 13.4d) [35]. Excellent regioselectivity, mild reaction conditions, wide substrate scope, and also the readily availability of catalysts and reagents are the noteworthy advantages of this straightforward method.

Interestingly, a variety of azide transfer processes were effectively investigated in situ to provide clickable azides for convenient access and potential application in click reaction [36]. Mancheno et al. reported a Cu-Catalyzed one-pot protocol for the facile synthesis of click-inspired triazole conjugates via C(sp3)-*H* activation with azidation and CuAAC conjugation [37]. Thus, in situ Csp(3)-*H* azidation of unprotected tetrasaccharide reducing sugar **32** followed by click conjugation with alkynyl peptide **33** furnish a high yield of desired complex glycopeptide **34** (Scheme 13.5).

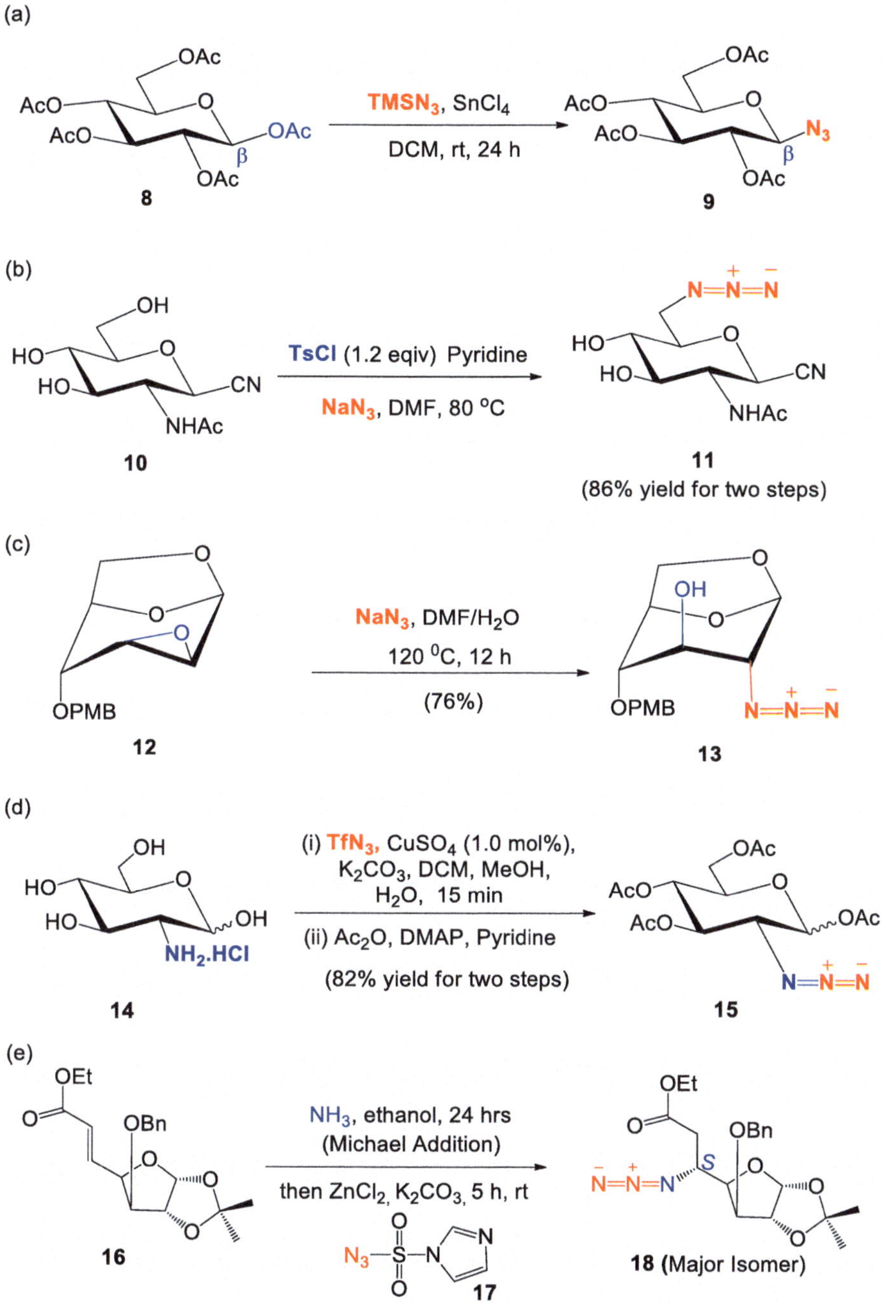

Scheme 13.2 Common approaches for the synthesis of different glycosyl azides

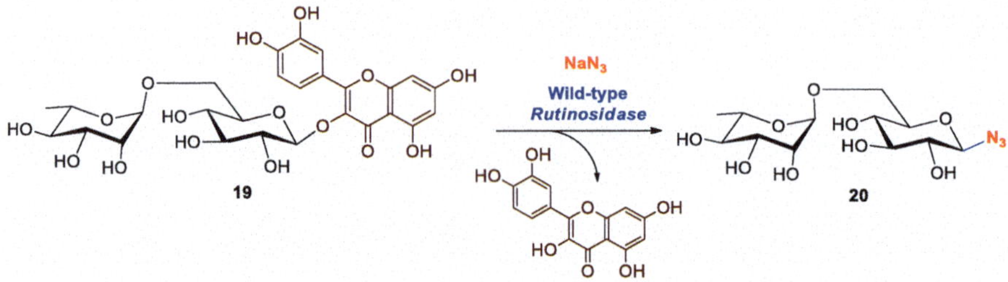

Scheme 13.3 Enzymatic mode of selectivity for the insertion of azide functionality

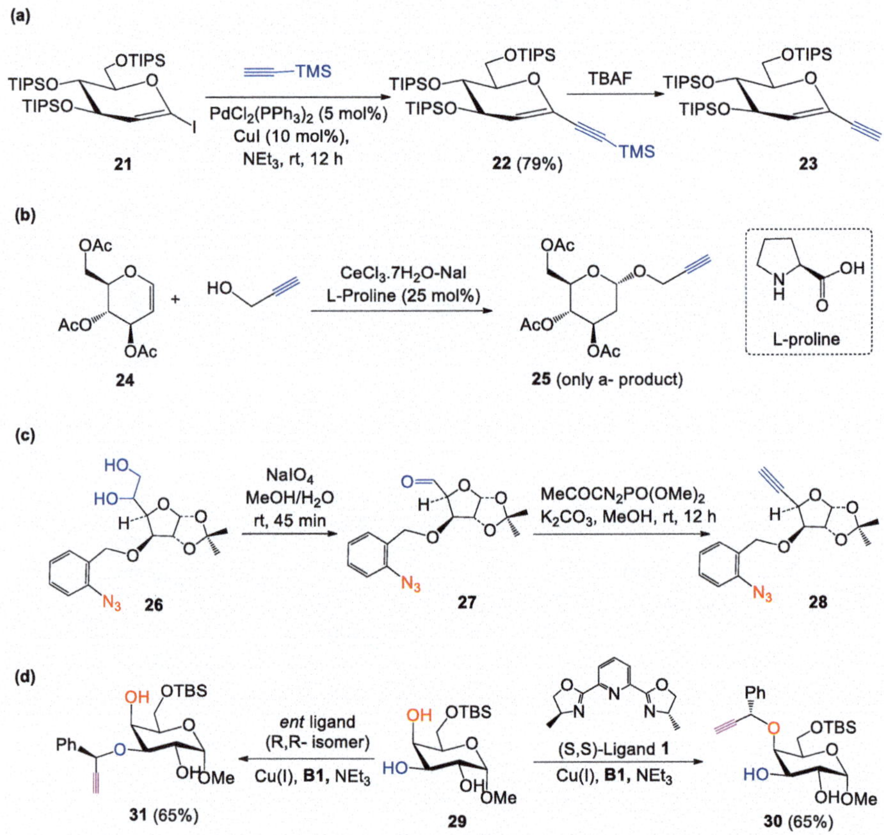

Scheme 13.4 Some common approaches for regioselective synthesis of different glycosyl alkynes

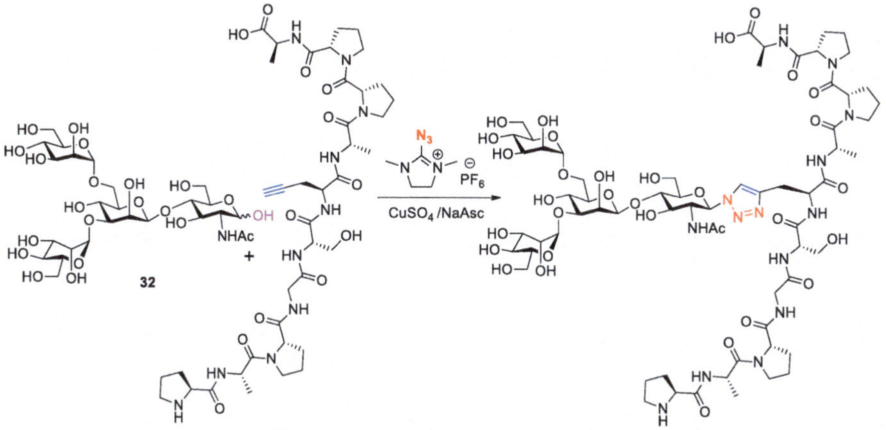

Scheme 13.5 One-pot azidation followed by CuAAC click conjugation of fully unprotected tetrasaccharide **32**

13.3 Cu-Catalyzed 'Click Chemistry' for Easy Access to Diverse 1,2,3-Triazole-Linked Glycohybrids and Glycoconjugates

In a recent study, a number of copper complexes displayed promising affinity toward [3 + 2] cycloaddition reaction. Toward this end, Bhattacharya and Tiwari reported the catalytic efficacy of a highly stable dinuclear Cu(I) catalyst $[(PPh_3)_2Cu(\mu\text{-}tda)Cu(PPh_3)_2]0.6H_2O$ (tda = thiodiacetate anion) for CuAAC click reaction [38]. The complex **35** displayed promising results for click reaction of various glycosylatyed azide **18** and diverse terminal alkynes **36**, which produces regioselective 1,2,3-triazolyl conjugates **37** in excellent yields in 10 min (Scheme 13.6). Wide substrate scope, solvent-free, base-free, recyclable, easy work-up, short reaction time, and high stability of the catalyst **35** are some notable advantages of the reaction. This conjugation can be regarded as environmentally friendly because the catalyst was recycled twice and performed effectively in base- and solvent-free conditions.

A conceivable method to exploit the potential of biomolecules is to couple the characteristics of two distinct molecular entities of different properties to produce unique glycoconjugates with simple-to-complex architectures [39]. The CuAAC click reaction of azide-alkyne has emerged as a promising synthetic tool for this purpose under mild reaction conditions [16, 40, 41]. A similar CuAAC click conjugation of 1-Azido-1-deoxy-2,3,4,6-tetra-*O*-acetyl-*β*-D-glucopyranose **9** with 6-propargyl-1,2–3,4-di-*O*-ispropylidene-α-D-galactopyranose **38** in water in the presence of $CuSO_4$/ascorbic acid afforded 1,4-triazole-linked disaccharide **39** (Scheme 13.7a) [42]. Campo et al. describe the synthesis of triazole appended sialyl-mimetic neoglycoconjugates using click conjugation between azido sugars (galactopyranose, glucopyranose) and alkyne-functionalized sialic acid precursors and screened for their TcTS

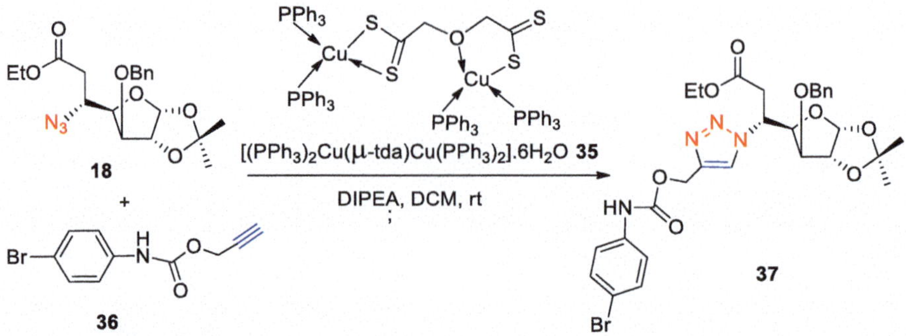

Scheme 13.6 A stable dinuclear Cu(I) catalyst **35** for synthesis of triazolyl glycoconjugate **37**

inhibitory activity [43]. The modular click conjugation of 1-N3-Gal **40** with the corresponding alkyne-based 2-propynyl-sialic acid **41**, afforded desired 1,2,3-triazole sialylmimetic **42** (Scheme 13.7b). Compound **42** was screened at 1.0–0.5 mM concentrations which exhibited high TcTs inhibitor. Also, hybrid compound **42** was found as most powerful inhibitor with IC_{50} in the 260 μM range.

Further, α-sialic acid azide has been used as a substrate for the construction of 1,2,3-triazolyl analogues of sialic acid using modular click chemistry protocol. In this order, Weiwer and co-workers demonstrated the synthesis of disaccharide **45** from click conjugation of α-azido sialic acid **43** and *C*-8 terminal alkyne-functionalized molecule **44** in the presence of 10% $CuSO_4$, and 50% NaAsc in 1:2:1 mixture of H_2O/tBuOH/DCM (Scheme 13.8). Additionally, a similar reaction of akynyl functionalized scaffold **46** with azido sialic acid donor **43** produces dendrimeric architectural molecule **47** under the same catalytic conditions. The developed triazolyl derivatives

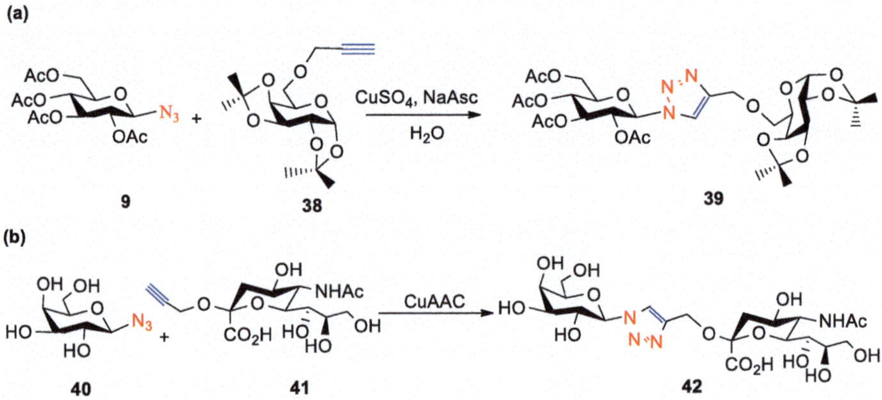

Scheme 13.7 CuAAC 'click chemistry' inspired synthesis of triazole-linked disaccharide analogue **39** and sialoconjugate **42**

were inspected for neuraminidase inhibitory activity using a 96-well plate fluorescence assay. The biological assay results suggested that multivalent derivatives such as **45** (IC_{50} = 17 μM) and dendrimer **47** (IC_{50} = 20 μM) were recognized as potential neuraminidase inhibitors [44].

Likewise, a similar click protocol was reported for development of ferrocene containing triazolyl glycconjugates **50** by treating ferrocenyl azide **49** with terminal alkyne-functionalized sugar compound **48** under modular click condition (Scheme 13.9). The resulting ferrocenyl glycoconjugate **50** has been designed to be used in electrochemical biosensing techniques in order to explore the interaction between proteins and carbohydrates [45].

Many natural product mimetics were effectively synthesized by utilizing the CuAAC click protocol. Toward this end, Tiwari and co-workers made significant efforts to develop triazole-linked ethisterone glycoconjugates **53** via CuAAC coupling of azido-sugars **52** with alkynyl ethisterone **51** [46]. In order to compete for androgen receptor (AR) binding, multivalent and divalent triazole bridging ethisterone conjugates were developed, which proven to be effective against prostate

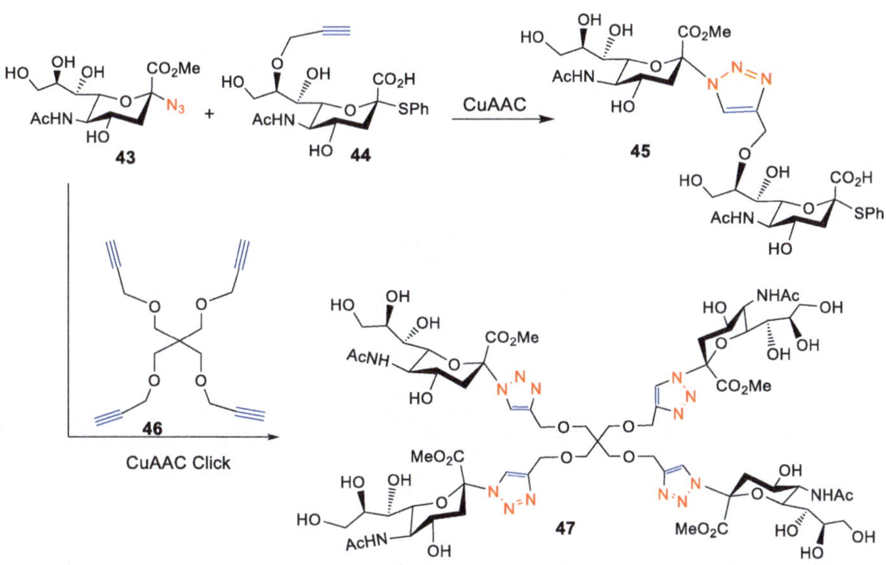

Scheme 13.8 Click chemistry inspired synthesis of sialic acid-based nuraminidase inhibitors

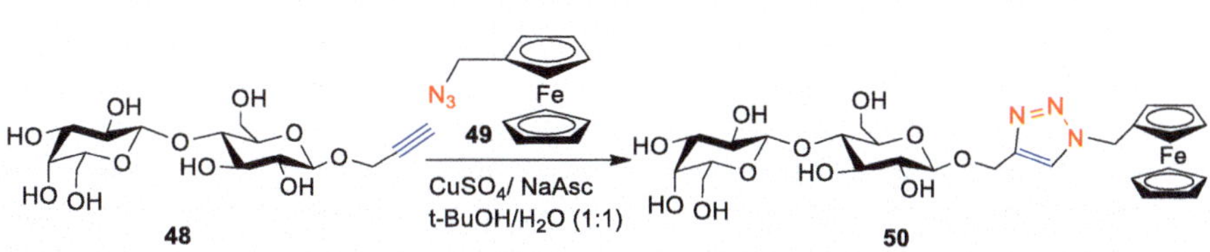

Scheme 13.9 Synthesis of triazole appended ferrocene-carbohydrate conjugates

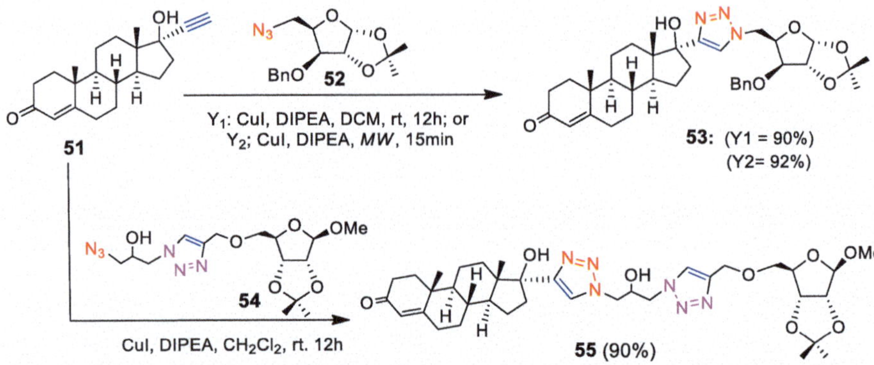

Scheme 13.10 CuAAC click chemistry inspired synthesis of ethisterone glycoconjugates

cancer. Therefore, several sugar azido alcohols **54** were synthesized which subsequently clicked with ethisterone **51** to produce 1,2,3-triazolyl ethisterone glycoconjugate **55** of potential therapeutic values in excellent yield (Scheme 13.10) [47].

Alkaloids are regarded as the most valuable and promising scaffolds in pharmacology, with a wide range of applications in contemporary drug discovery and development. It was discovered that relocated noscapine, which was initially used as an anti-tussive medication but was later converted to an anti-cancer agent, was 100 times more effective than the original noscapine at the 7-hydroxyl position [48]. Carbohydrates owing to their multivalency effect have been demonstrated to enhance the affinity toward the targets in various biological phenomena [21]. Therefore, Tiwari et al. greatly utilized click chemistry to tether alkaloids with sugar-hybrids for the development of potent therapeutics. The alkyne-functionalized second-generation noscapine scaffold **56** was produced by demethylation of (C-7) group under NaN_3/NaI followed by alkynylation. Further, triazole appended noscapine conjugates **57** has been achieved by coupling of compounds **56** with various glycosyl azides **9** (Scheme 13.11) [49]. Thus a series of second-generation noscapine glycohybrid **57** at the C-7 position have been developed. The method works well for producing new noscapine conjugates with enhanced chemotherapeutic efficacy.

In continuation to development of triazolyl glycoconjugates, Tiwari et al. have successfully introduced a triazole skeleton at the β-position (C-5) of glycosyl ester. A one-pot conjugate addition reaction of ammonia to glycosyl olefinic ester followed by a diazo-transfer reaction produces the desired glycosyl-β-azido ester **58** and **18** (as minor and major isomer). Afterward, click conjugation of these developed glycosyl-β-azido ester with a diverse range of terminal alkynes (**59** and **51**) produces 1,4-triazole-linked glycoconjugates **60** and **61** in high yields (Scheme 13.12) [30].

Click chemistry inspired one-pot MCR reactions have regarded invaluable tool for the synthesis of rare and multi-step compounds that serve as precursor to potential fused heterocycle sytems. Toward this ends, Tiwari and co-workers made significant efforts to develop a two-step synthetic protocol for morpholine-fused [5,1-c]-triazoles

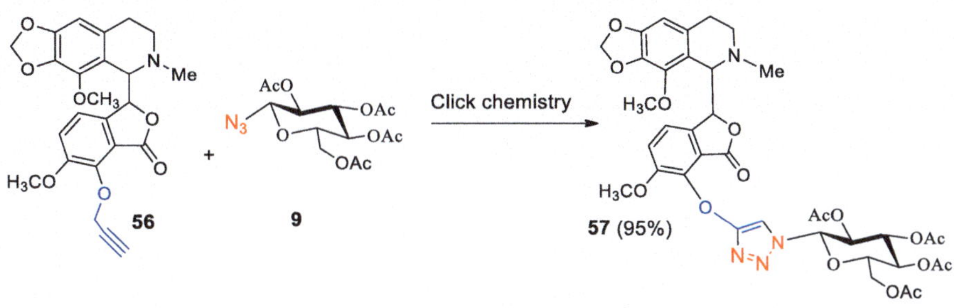

Scheme 13.11 CuAAC click-inspired synthesis of noscapine glycoconjugates **57**

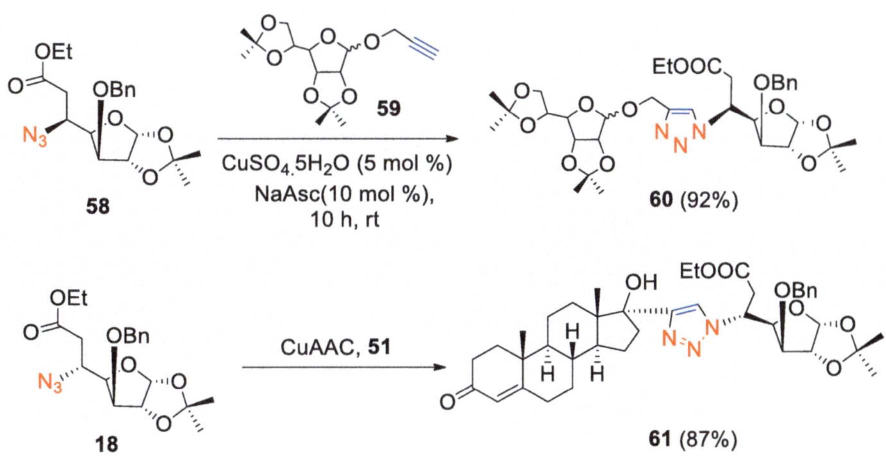

Scheme 13.12 Click ispired synthesis of triazolyl glycoconjugates

from glycosyl alkynes. This one-pot methodology involves the formation of desired triazolyl *azido* alcohol intermediate **63** from various sugar alkynes **62** via oxirane ring opening of epichlorohydrin in the presence of NaN_3 and subsequent azidation of chlorohydroxy triazoles. Further, propargylation of azido alcohol **63** followed by subsequent thermal induced and metal-free intramolecular cyclization produces the morpholine-fused[5,1-c]-triazoles **64** in good-to-excellent yield (Scheme 13.13) [50].

Likewise, Tiwari et al. have further explored modular CuAACclick protocol and develop a water-soluble curcumin glycoconjugate **68** containing 1,4-triazole linker from curcumin **65** in three synthetic steps. The synthetic methodology involves propargylation of curcumin followed by click conjugation with galactosyl azide **66** and finally deacetylation under basic conditions furnished the corresponding glycoconjugate **68** (Scheme 13.14) [51]. Interestingly, curcumin-attached glycohybrid **68** broadens the antibacterial spectrum significantly by sensitizing resistant gram-negative clinical isolates to the gram-positive specific antibiotic vancomycin.

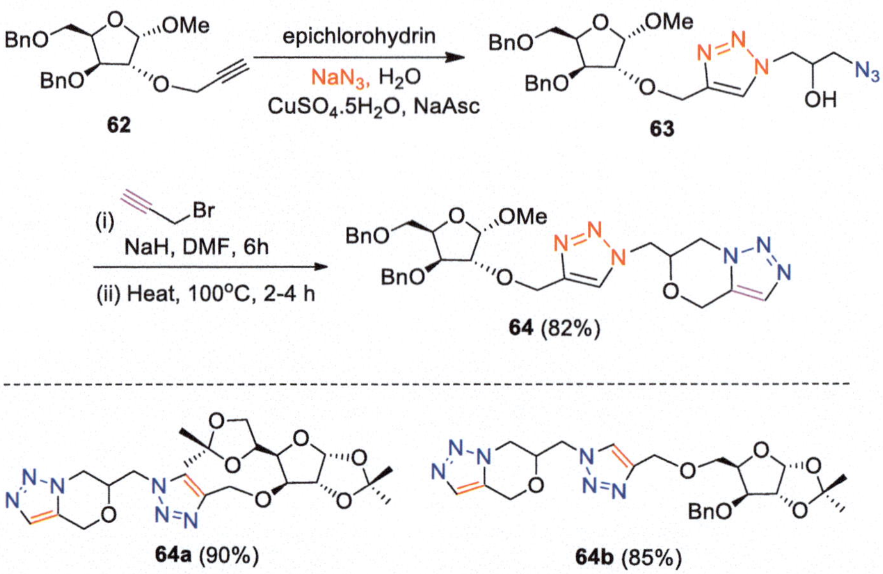

Scheme 13.13 Click-inspired synthesis ofmorpholine-fused triazolyl glycoconjugates

The higher water solubility of curcumin glycoconjugate (> 11,000 times more water-soluble) was thought to amplify the effects of conventional anti-microbials by causing drug-resistant bacteria to exhibit apoptosis-like behavior.

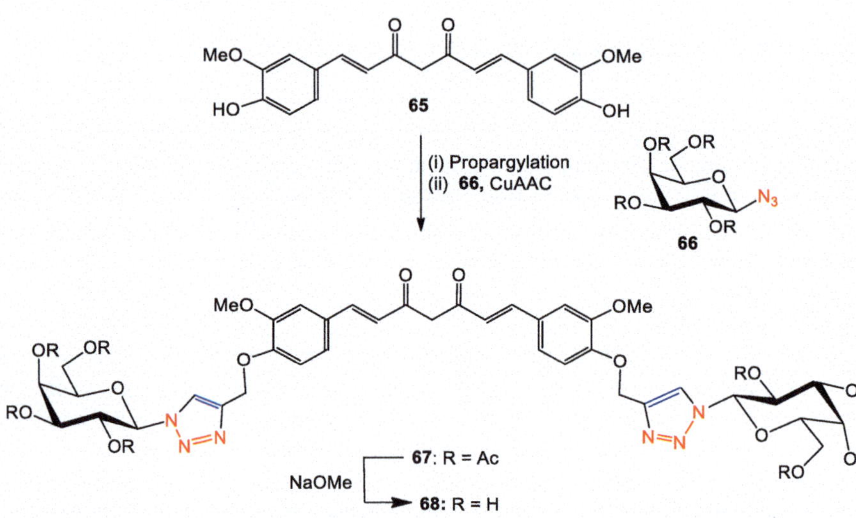

Scheme 13.14 Synthesis of water-soluble curcumin glycoconjugate **68** using click chemistry

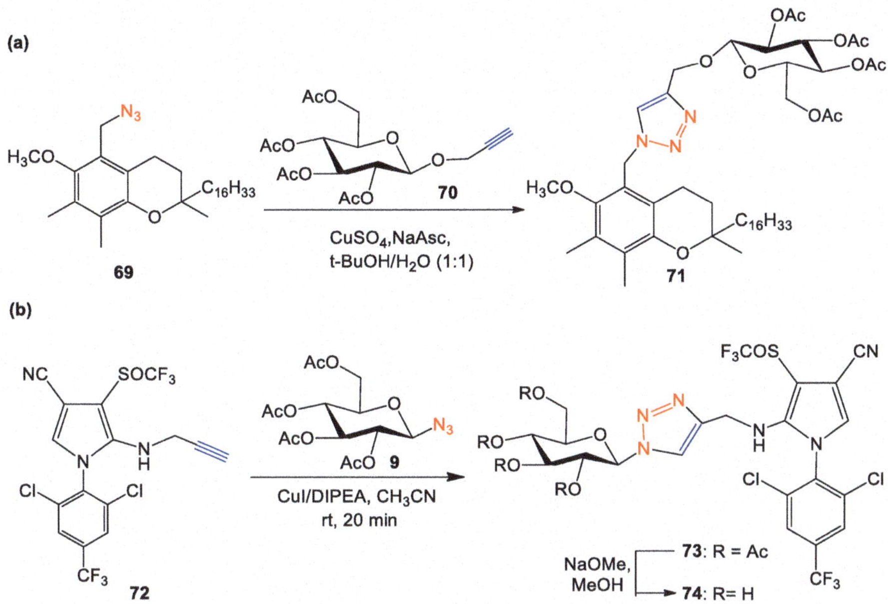

Scheme 13.15 **a** click-inspired synthesis of α-tocopherol derived triazolyl glycoconjugate **71** and **b** glucose-fipronil conjugate **74** as potential insecticidal activity

Similarly, the construction of α-tocopherol derived triazolyl glycoconjugate **71** involves click conjugation of azide **69** with alkyne **70** that displayed radical scavenging activities similar to α-tocopherol (Scheme 13.15a) [52]. Conversely, under typical conditions, the glycoconjugate produced by similarly click conjugation of α-tocopherol alkyne with sugar azide was unstable and sensitive to air.

Likewise, CuAAC click coupling of non-phloem-mobile insecticide containing alkynyl functionality **72** sugar azide **9** produces glucose-fipronil conjugate **73** that turns it into phloem-mobile insecticide **74** (GTF) under basic treatment (Scheme 13.15b). The phloem-mobility test on *Ricinus communis* was used to ascertain the mobility of GTF inside the sieve tubes. Conversely, GTF conjugate **74** showed less insecticidal activity against *Pzlutellaxylostella* larvae in their third instar [53].

In continuation of construction of 1,4-triazole compounds, Tiwari et al. employed click tool for development three distinct series of triazolyl *O*-benzyl quercetin glycoconjugates and tested for anti-leishmanial activity against the promastigotes and amastigotes of *Leishmania donovani*. alkynylation of benzylated quercetin at three distinct positions furnished alkyne fuctionalized *O*-benzylquercetin **75** that effectively underwent click conjugation with a variety of glycosyl azide **76** (Scheme 13.16a). The monosaccharide attached to 3,7,3',4'-*O*-tetra benzyl quercetin skeleton **77** displayed the strongest in vitro action against promastigotes and intra-macrophage amastigote forms of *L. donovani* among all the produced triazolyl glycoconjugates [54].

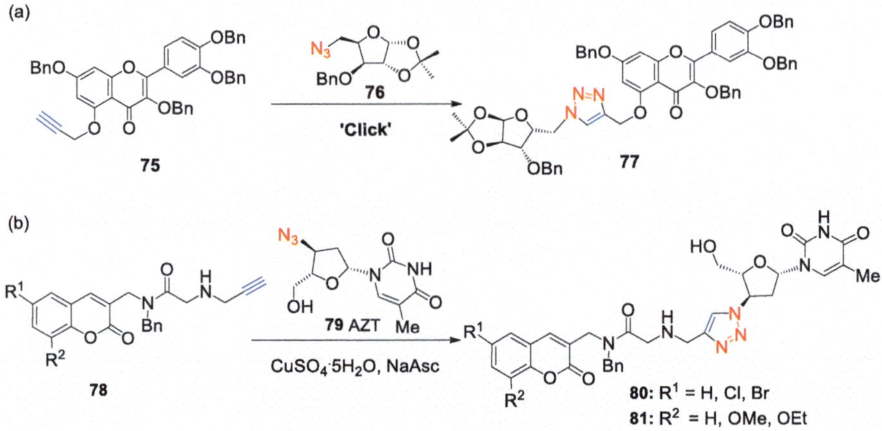

Scheme 13.16 **a** Synthesis of anti-leishmanial triazolyl *O*-benzylquercetin glycoconjugates; **b** click-inspired synthesis of dual HIV-1 PR/HIV-1 RT inhibitor

A coumarin-based alkyne **78** was click conjugated with the HIV-1 RT inhibitor azidothymidine (AZT) **79** to develop a variety of dual HIV-1 PR/HIV-1 RT inhibitors **80** and **81,** respectively, (Scheme 13.16b). The resulting triazoles inhibit HIV-1 PR and HIV-1 RT at 50 μM by 52–60% and 94–99%, respectively. These analogues provide a compelling strategy for the synthesis of such glycoconjugates with dual inhibitory properties [55].

Sometimes, small changes in the molecule cause a significant shift in the bioactivity profile. Selectivity and affinity of glycosidases were found to be significantly impacted by multivalency. The trivalent imino sugar-based glycoclusters **84** was develop via azide-alkyne click reaction of azido DNJ motif **82** and alkyne-functionalized scaffold **83** (Scheme 13.17). The trivalent glycocluster **84** exhibited lower inhibition activity for *β*-glucosidase and α-mannosidase while results overturned in the case of amyloglucosidase and α-mannosidase [56].

Triazolyl glycoconjugates that have been 1,4-disubstituted showed remarkable results in suppressing PTP enzymes and are being investigated as possible therapeutic agents. Microwave irradiated click conjugation of glycosyl alkyne **85** with amino acid azide **86** followed base hydrolysis produces the desire 1,4-triazolyl glycoconjugate **87** (Scheme 13.18a). The resulting triazolyl acid **87** was identified as a new inhibitor of PTP1B and CDC25B enzymes, exhibiting selectivity multiple times greater than that of other analogous PTPs [57].

A report documented by Thorson and co-workers demonstrated the identification of in vitro antiproliferative activity of perillyl alcohol glycosides in addition to their ability to inhibit S6 ribosomal protein phosphorylation. Therefore, 6-alkylated perillyl glucoside **88** underwent click conjugation with 3-azido propanol **89** to produce 6'-triazole perillyl glucoside **90** in good yield (Scheme 13.18b) [58].

Further, CuAAC tool greatly clicked glycosyl azide **92** with α-acetamido-α-propargyl-*β*-lactam **91** for the generation of hybrid glycopeptide *β*-lactam

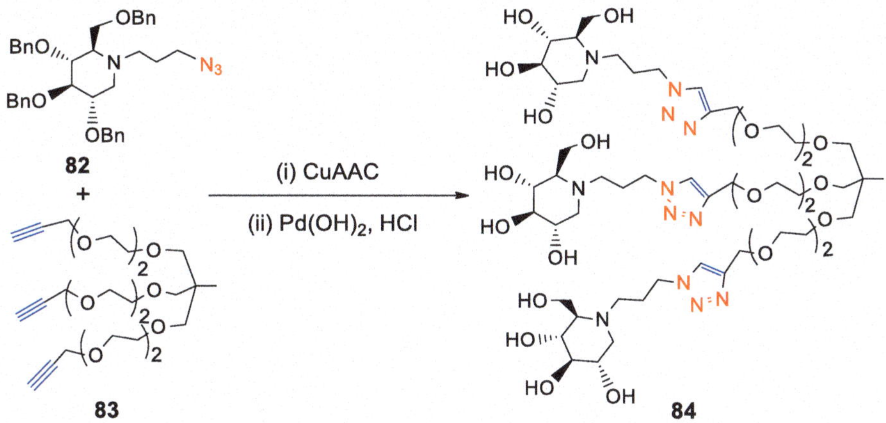

Scheme 13.17 Click-inspired synthesis of trivalent imino sugar **84**

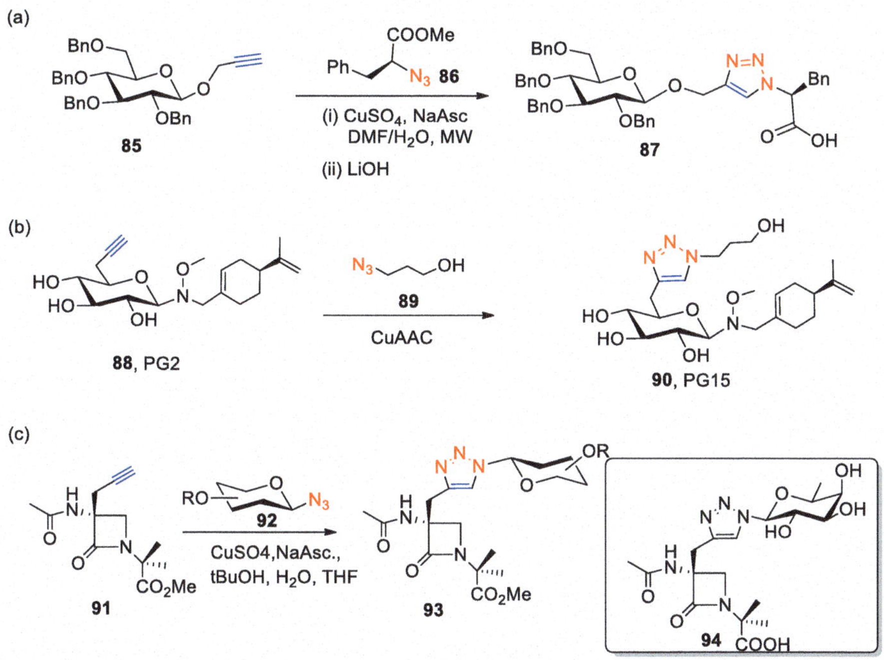

Scheme 13.18 Synthesis of triazolyl glycoconjugate **87**, **90,** and **93** as selective PTP1B inhibitor, antiproliferative activity, and lectin inhibition, respectively, using click chemistry

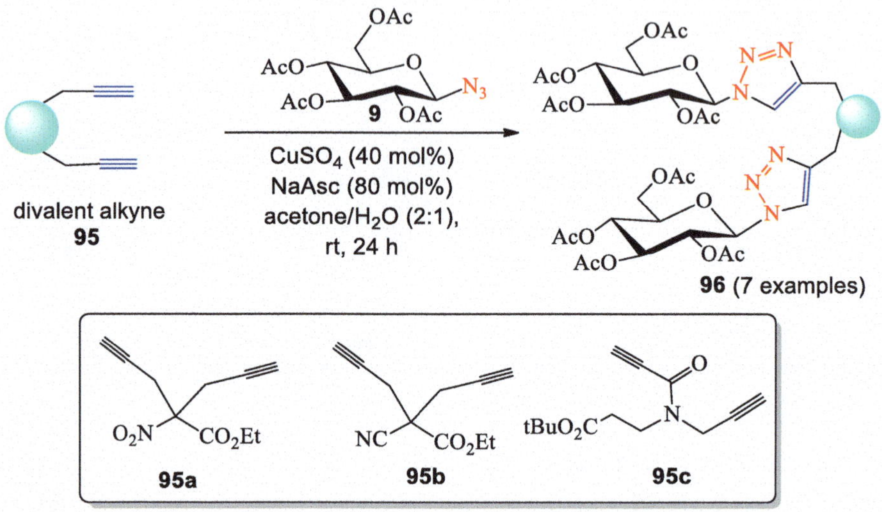

Scheme 13.19 Click-inspired synthesis of triazole-linked divalent glycoamino acid mimetics

mimetic **93** which bind lectins or carbohydrate recognition domains in selectins (Scheme 13.18c). After undergoing a conformational shift around a rotatable triazolylmethylene molecule, *Ulex Europaeus Lectin*-1 (UEL-I) was bound to the glycohybrid **94**, which was derived from *β*-1-azido-L-fucose [59]. UEL-I, a model fucose-binding lectin, was used as a receptor in the NMR/docking technique to determine the adjustable shape of glycomimetics.

Sahoo et al. successfully develop triazole appended divalent glycoconjugates **96** using click reaction of alkyne-functionalized divalent building blocks **95a-c** with per-*O*-acetylated glycosyl azides **9** (Scheme 13.19). The resulting glycoconjugates **96** selective transforms from nitro- and cyano- substituents to α- or β-amines, which are used in glycopeptide mimics synthesis having potential biological applications [60].

The click-mediated azobenzene glycoconjugate displayed exciting photochromic behavior where 1,2,3-triazole-linked glycoconjugates **97** and **98** were interchanged between two isomeric states (*E*- and *Z*-geometry) to alter the spatial position of the multivalent glycoazobenzenes (Scheme 13.20) [61]. Also, the multivalency effect was noticed in photoisomerisation and the concept may be used to develop photo-switchable multivalent glyco-arrays for wide biological applications.

Sen et al. reported one-pot solvent and ligand-free protocol for construction of 1,4-disubstituted triazolyl *N*-carboxamides **101** from glycosyl amines **99** in presence of alumina as solid support (Scheme 13.21a). Microvave assisted domino reaction involves Ullmann coupling followed by click reaction for the preparation of pharmalogically relevant 1,2,3-triazole adducts [62].

Membrane-bound isozymes CA IX and CA XII which are carbonic anhydrase (CA) enzymes, support a pH-regulating system that permits hypoxic tumor cell

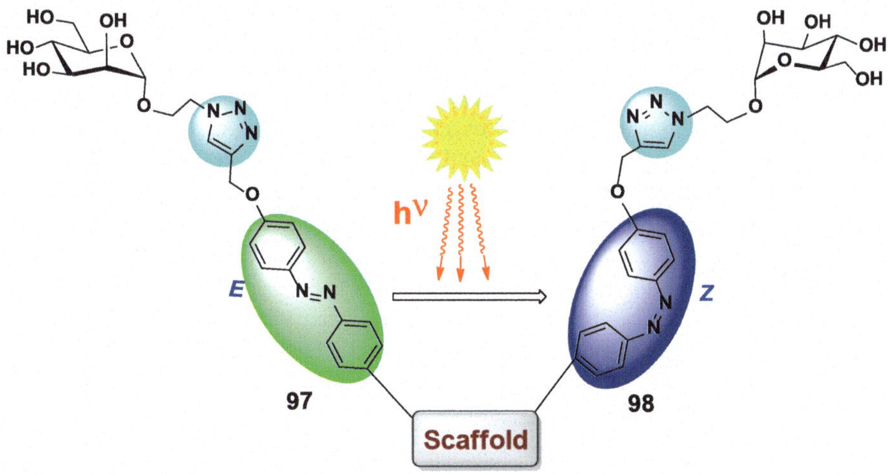

Scheme 13.20 Azole-based glycohybrids **97** and **98** with triazole attached in both isomeric E- and Z-forms had different photochromic characteristics

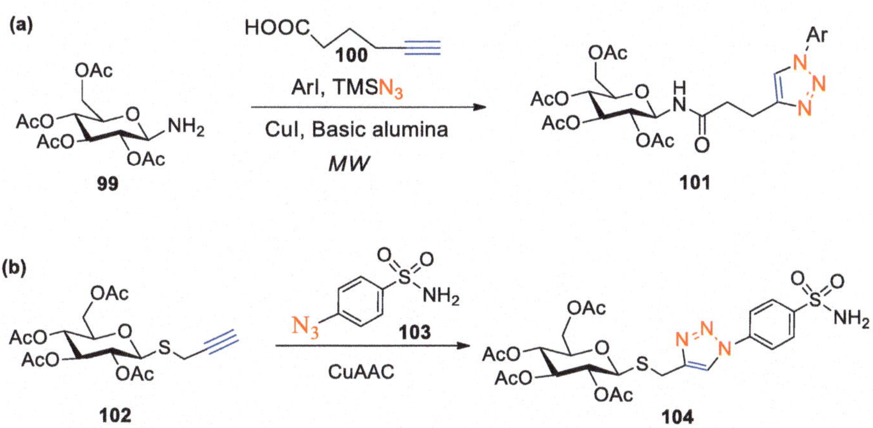

Scheme 13.21 Click chemistry inspired synthesis of triazolyl *N*-carboxamides (**a**) and carbonic anhydrase inhibitors (**b**)

survival and growth. It has been suggested that CA IX and XII could be used as targets in the development of novel hypoxic cancer treatments. Poulsen et al. report the synthesis of novel small molecules based on carbohydrates which serve as CA inhibitors. In this order, a library of small molecules based on carbohydrates was developed and tested for CA inhibitory activity in cell-based model (Scheme 13.21b). Among all synthesized compounds glycoconjugate **104** was found to be a promising CA IX and XII inhibitor within narrow range K_i values (5.3–11.2 nM) [63].

The administration of ManT inhibitors could be effective in treatment of tuberculosis as *M tuberculosis* cell envelope is biosynthesized by the mycobacterial α-(1,6)-mannosyltransferase enzyme. A new class of (1,6)-oligomannose analogues with triazole as a bridge can be generated by click chemistry [64]. Thus, CuAAC click coupling of sugar azide **105** with alkyne **106** followed by deactylation successfully produces glycoconjugate **107** (Scheme 13.22). TLC analysis of a broad range of radioactive lipids integrated into the membranes revealed 95% suppression of ManT activity at 1 mM concentration of the mycobacterial membranes [^{14}C]-Man from GDP [^{14}C]-Man. Similarly, click conjugation was used to generate triazole-linked mannose conjugates, which are composed of two or three mannose epitopes and showed significant inhibitory efficacy against ManT. Nevertheless, it was discovered that hexa- and octa-mannosides **108**, with IC_{50} values ranging from 0.14–0.22 mM, were effective ManT inhibitors [65].

Development of triazole appended sugar-based selectin inhibitors via azide-alkyne click reaction is a promising strategy to treat inflammatory diseases by preventing selectin-leukocyte interaction through selectin-antagonists. Azide-alkyne click conjugation of seven different alkynes **110** with sLex-based azide **109** successfully produces triazolyl glycoconjugate **111** (Scheme 13.23a) [66]. The in vitro screening of developed glycoconjuateagainst E-selectin show that only unsubstituted triazole **111** and the azido compound **109** exhibited slightly improved activity for E-selectin. Other substituted compounds of **111** showed similar activity to sLex.

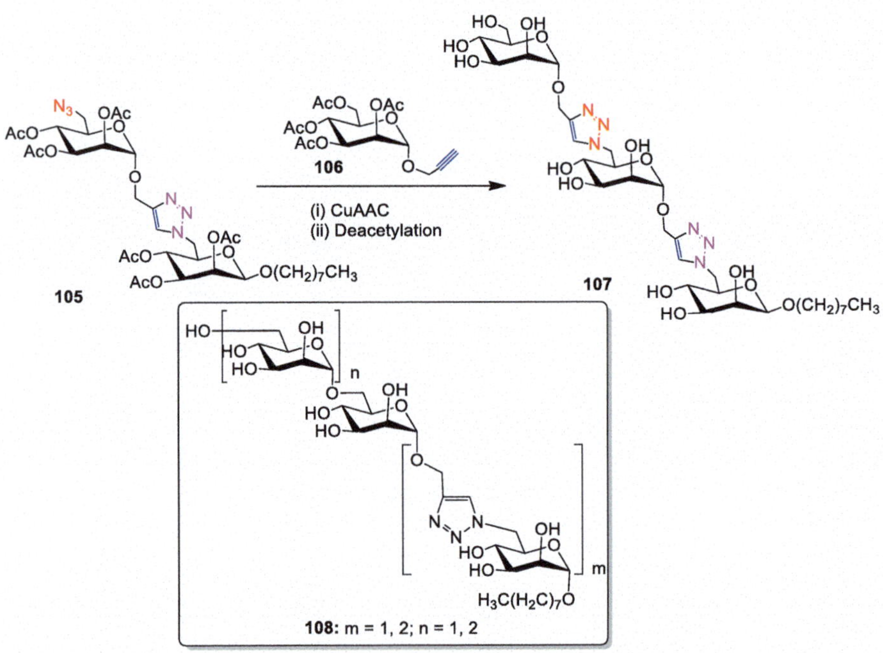

Scheme 13.22 Synthesis of triazole-linked 1,6-α-D-oligomannosides against ManT activity

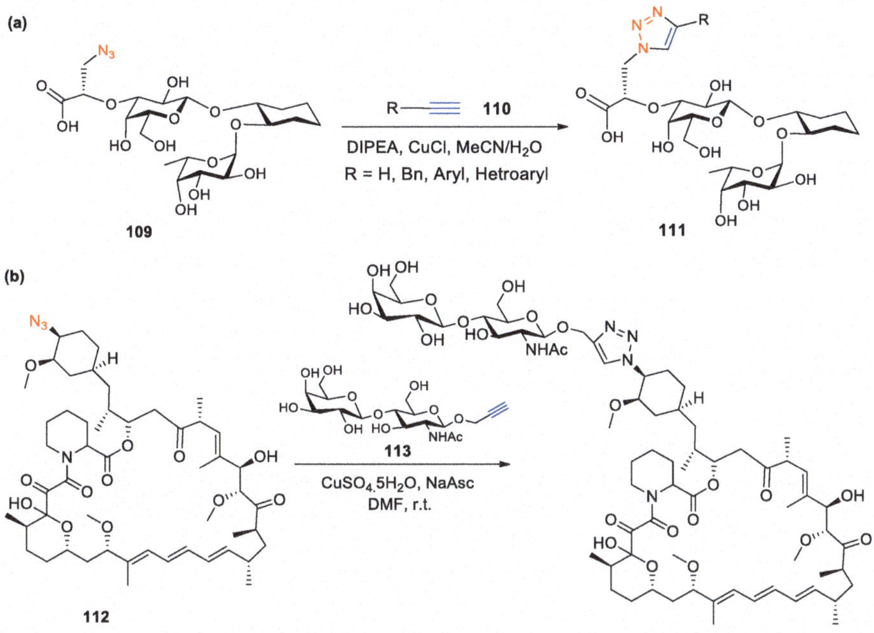

Scheme 13.23 **a** CuAAC click derived modification of E-selectin antagonists **111** based on sLex and **b** rapamycin triazolyl glycoconjugate **114**

Dondoni and co-workers successfully installed azido group on C-40 secondary hydroxyl group of rapamycin **112**, which on subsequent click conjugation with various unprotected propargyl-*O*-and S-glycosides and other glycosyl alkye (for example alkyne **113**) afforded high yield of triazole-bridged rapamycin glycoconjugates **114** (Scheme 13.23b) [67].

Roy et al. describe the synthesis of galactoside and lactoside bearing triazole and tested them by inhibition of hemagglutination assay at a concentration of 1 mM. The assay results suggested that monovalent galactose derivative **116** exhibited maximum inhibitory properties of 1250 μM whereas trivalent lactoside **115** displayed inhibitory activity of 20 μM for galectins (Fig. 13.1) [68]. Copper(I)-mediated [3 + 2] cycloaddition reaction of alkynyl derivatives of amino acids joined with 2-azidoethyl *β*-D-galactopyranosyl-(1 → 4)-*β*-D-glucopyranoside produces a panel lactoside clusters of diverse valency. The glycocluster **117** was recognized as a significant galectin-1 inhibitor with a K_d value as low as 3.2 μM. In comparison to natural disaccharide ligand lactose, compound **117** demonstrated a relative efficacy of 30 per lactose unit [69].

Starting from azide-alkyne scaffolds, the click-mediated macrocyclization can be emplemented via intramolecular CuAAC or intermolecular followed by intramolecular Click, where both the CuAAC may be proceed either simultaneously or may

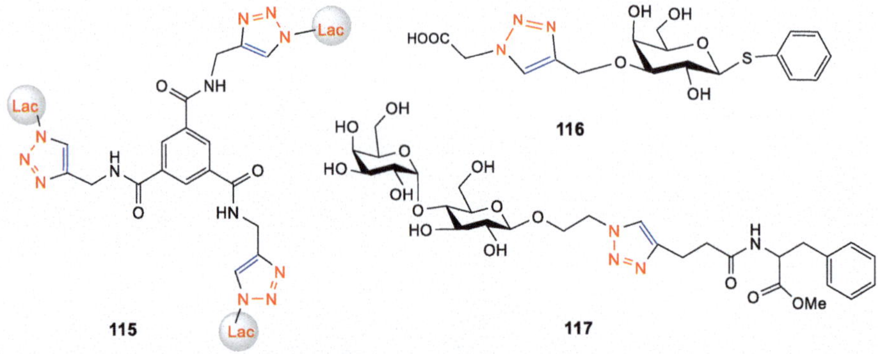

Fig. 13.1 CuAAC click-inspired Galectin (S-lectin) inhibitors (**115–117**)

occurs on the priority basis as the intramolecular cyclization. On the other hand, macrocyclization via CuAAC derived triazole may followed by other convenient cyclization route like intermolecular Ru-catalyzed ring closing metathesis (RCM) or through CuAAC followed by intramolecular glycosylation as part of IAD or intramolecular aldol reaction as convenient ring closing method. Thus, intramolecular aldol condensation was used to deliver a high yield of macrocyclicenone **122** [70]. For this purpose, the desired 1,2,3-triazole **120** was obiabed via click coupling of glucose propargyl analogue **119** and azide **118**. The resulting compound underwent for isopropylidene removal followed by $NaIO_4$ oxidation to hive respective aldehyde **121**, at the end proceed to intramolecular aldol condensation to give high yield of desired sugar-based macrocycle **122** (Scheme 13.24).

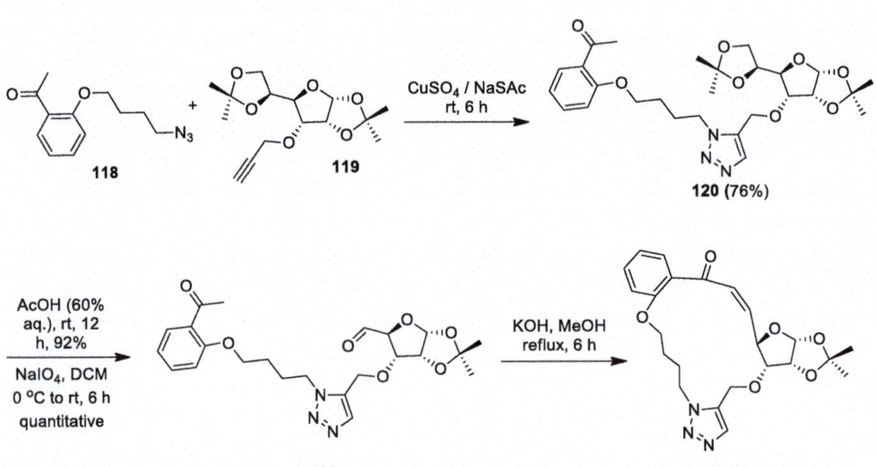

Scheme 13.24 Triazole containing sugar macrocycles **122** cyclized by intramolecular aldol condensation

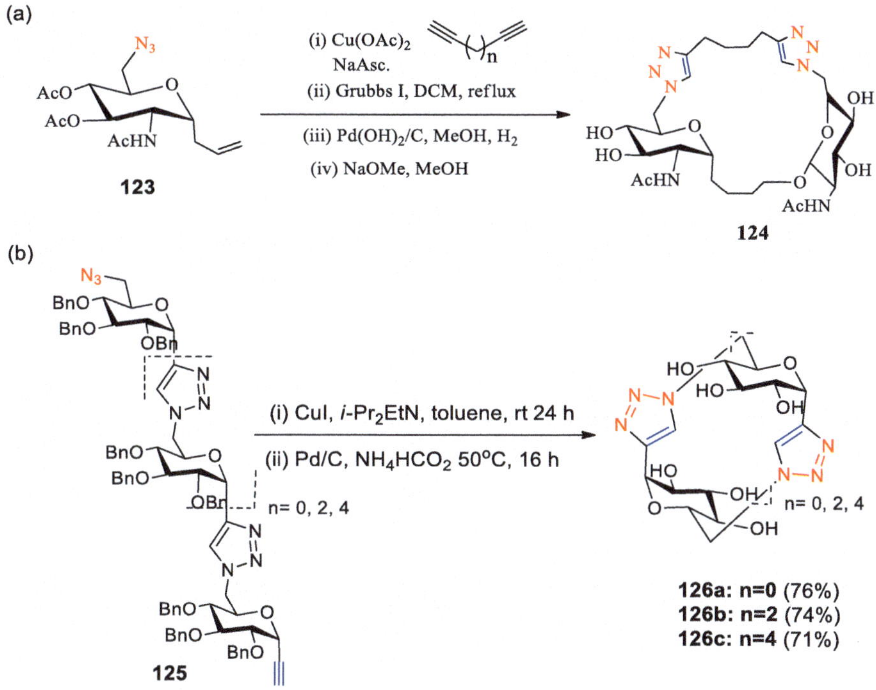

Scheme 13.25 **a** Synthesis of macrocycle **124** via a click dimerization; **b** synthesis of *Cn*-symmetric triazole-linked cycloglucopyranosides **126**

In order to synthesize macrocyclic scaffold **124** from the corresponding 6-azido moiety **123** and 1,7-octadiyne, Dorner and Westermann describe a cascade reaction that included click dimerization and ring-closing metathesis (Scheme 13.25a) [71]. This cascade transition, which uses several precursors, is a potent method for developing molecular complexity.

Conte et al. have identified a convergent method for development of Cn-symmetric triazole-linked cycloglucopyranosides through Cu(I) catalyzed intramolecular click cycloaddition reaction. Bifunctionalized linear oligomers **125** have been generated via Cu(I) catalyzed intramolecular cyclization to cyclodextrin-like Cn-symmetric macromolecules with two, four, or six *C*-glucopyranoside residues **126a-c** (Scheme 13.25b) [72].

Tiwari et al. utilized the intramolecular variant of modular CuAAC click macrocyclization tool and developed a *bis*-1,2,3-triazolyl vanillin glycoconjugate **129** in 78% yield by reacting *bis* deoxy-azido sugar **127** with *bis*-alkyne **128** in the presence of CuI/DIPEA at room temperature (Scheme 13.26) [73]. The reaction time can be reduced to 10 min once the CuAAC-mediated macrocyclization was performed under microwave irradiation at 100 °C.

The oligosaccharides **130** that contain sialic acid and have azide and alkyne functionalities at both ends, such as sialoside, was obtained through a one-pot three

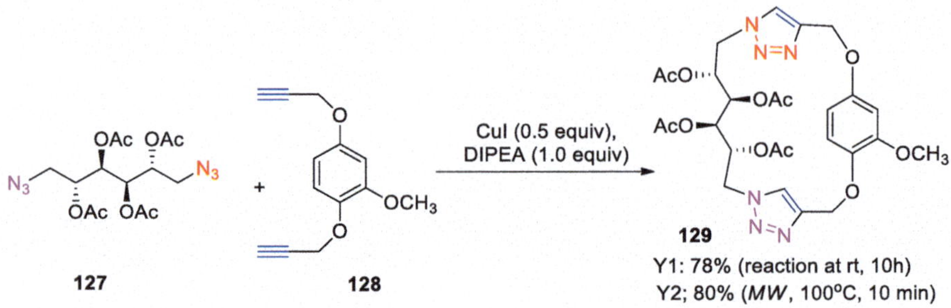

Scheme 13.26 CuAAC click macrocyclization for *bis*-1,2,3-triazolyl vanillin glycoconjugate **129**

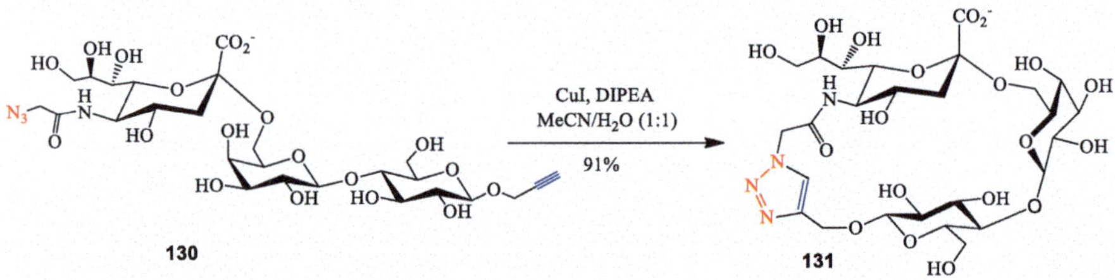

Scheme 13.27 Chemoenzymatic synthesis sialic acid-containing macrocycle **131**

enzyme chemoenzymatic strategy which on consecutive click cyclization in presence of CuI/DIPEA resulted a good yield of sialic acid-containing macrocycle **131** (Scheme 13.27) [74].

Intramolecular aglycon delivery (IAD) emerge as prominent approach for attaining high selectivity in glycosidation processes [75]. In this order, Schmidt et al. effectively develop an IAD technique that uses the triazolylmethyl moiety as a reasonably rigid spacer system in intramolecular glycosylations along with the o- and m-xylylene moiety. Therefore, the necessary donor–acceptor coupled triazole containing disachharide **134** was obtained by click conjugation of glycosyl alkyne **132β** with *O*-(2-azidomethylbenzyl)-protected acceptors **133** using CuI/DIPEA (Scheme 13.28) [76]. Moreover, a perfect α-selectiv 15-membered macrocycle **135α** was achieved in 55% yield using NIS/TMSOTf-mediated IAD of disachharide **134**.

Lewandowski and Jarosz clicked lutidine-derived dialkyne **137** linkers with sucrose-derived azide **136**, subsequently activate terminal free hydroxyl groups to produce a sucrose-derived C2-symmetric macrocyce **139** by macrocyclizing with ethylene diamine [77]. Template **140** was used to synthesize the final macrocycle with a 20% yield, however in the absence of **140** no product was detected (Scheme 13.29). Similarly, the same procedure was used to develop amino acid templated assemblies of macrocycles produced from sucrose [78].

It has been proposed that studies of the interactions between carbohydrates and proteins can benefit from the glycoclusters that are tightly attached to a calixarene that has undergone conformational rigidification [79]. For instance, CuAAC click

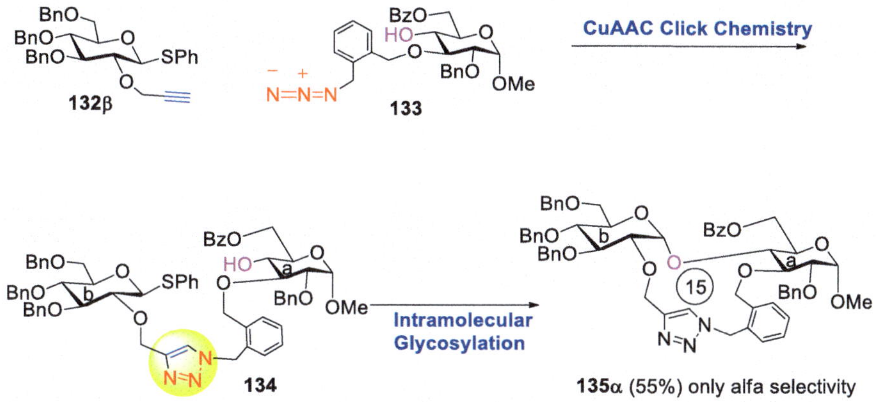

Scheme 13.28 Click generated triazole serve as rigid spacer in intramolecular glycosidation

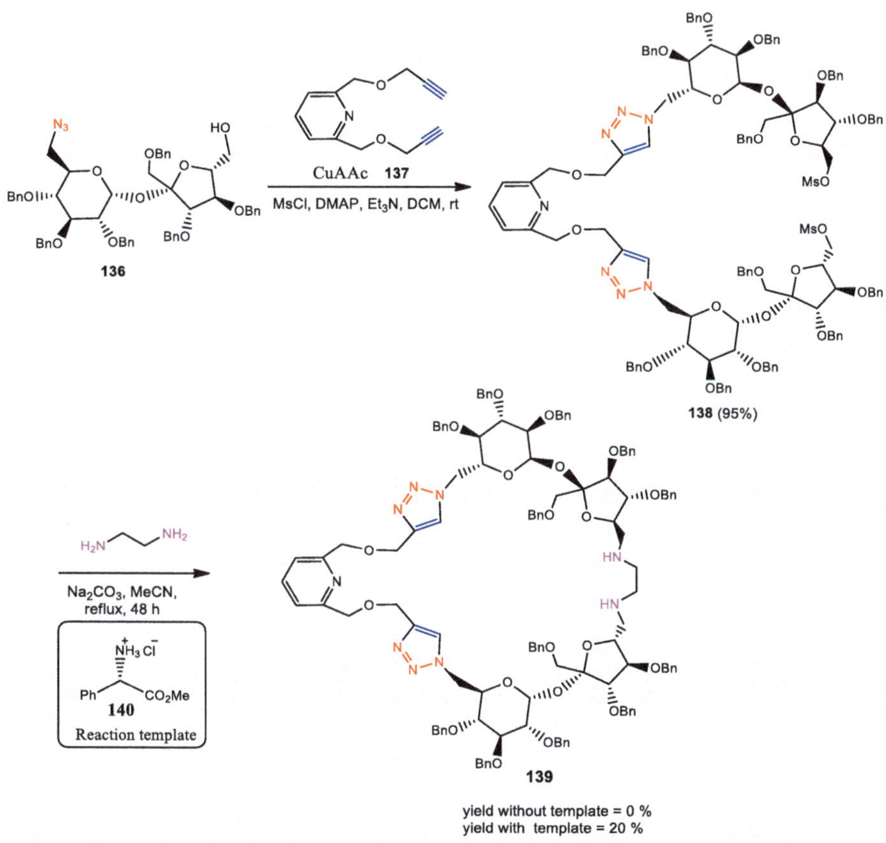

Scheme 13.29 Amino acid tempelated macrocyclization to access sucrose-derived macrocycle **139**

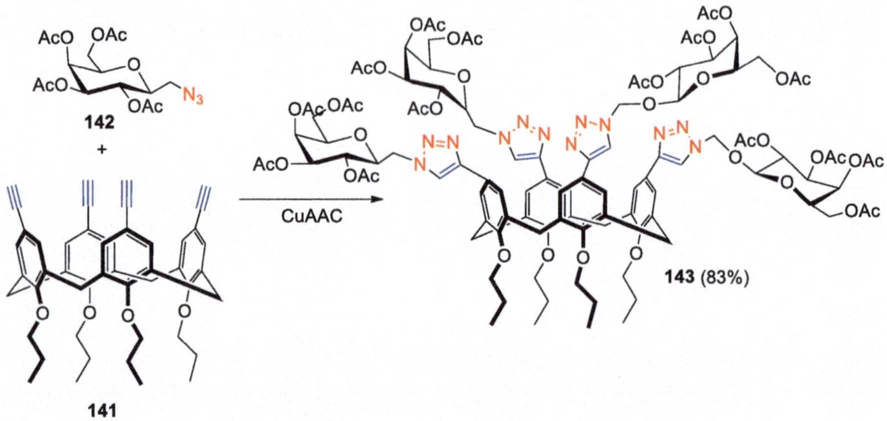

Scheme 13.30 CuAAC click-inspired synthesis of calix[4]arene glycocluster **143**

conjugation of calix[4]arene moiety **141** with sugar azide **142** furnished triazole-linked calix[4]arene *O*-glycosides **143** (Scheme 13.30) [80]. This [3 + 2] azie-alkyne reaction proceeds with high yield and exclusively furnished 1,4-disubstituted triazole. Similarly, a sequence of partially and fully *O*-galactosylated calix[4]arene clusters in cone, partial cone, and 1,3-alternate conformations were synthesized using a similar method [81].

Glyconanoparticles comprises a good biomimetic model of saccharides exhibition at the cell surfaces. The presence of sugar units at surfaces, significantly enhance the interaction those taking place in natural ambient conditions [82]. Dondoni et al. describe the synthesis of a glyconanoparticle **147** using click conjugation. For this purpose, a calixarene-coated TiO_2 nanoparticle **145** was develop from calix[4]arene **144** by treatment with TiO_2 in acetone at room temperature. Additionally, a CuAAC click tool in presence of CuSO4/NaAsc in methanol/water was used to clicked *O*-propargylated sialoside **146** with calixarene-coated TiO_2 nanoparticle **145** in order to develop the required glyconanoparticle **147** (Scheme 13.31) [83]. The azide-alkyne cycloaddition reaction was monitored using the infrared spectroscopy, while the loading of glycocluster per gram of TiO_2 was determined using thermogravimetric analysis (TGA).

Glycopeptides are a class of naturally occurring compounds that have several biological applications. A convenient method of obtaining physiologically relevant glycopeptides is made possible by 1,2,3-triazole, which functions as a crucial linker between peptide and carbohydrates [84]. Therefore, Rutjes and co-workers describe that a combination of $Cu(OAc)_2$ and Na-ascorbate mediated click coupling of mono and dipeptide azides or alkynes with mono or disaccharide azides and alkynes in water/^{t}BuOH afforded a range of α- and β-triazole-linked glycopeptides **150** (Scheme 13.32) [85]. The production of glycosylated cyclic arginine-glycine-aspartate derivatives **151**, which specifically target αvβ3 integrin, was further extended by the authors using a triazole linker [86].

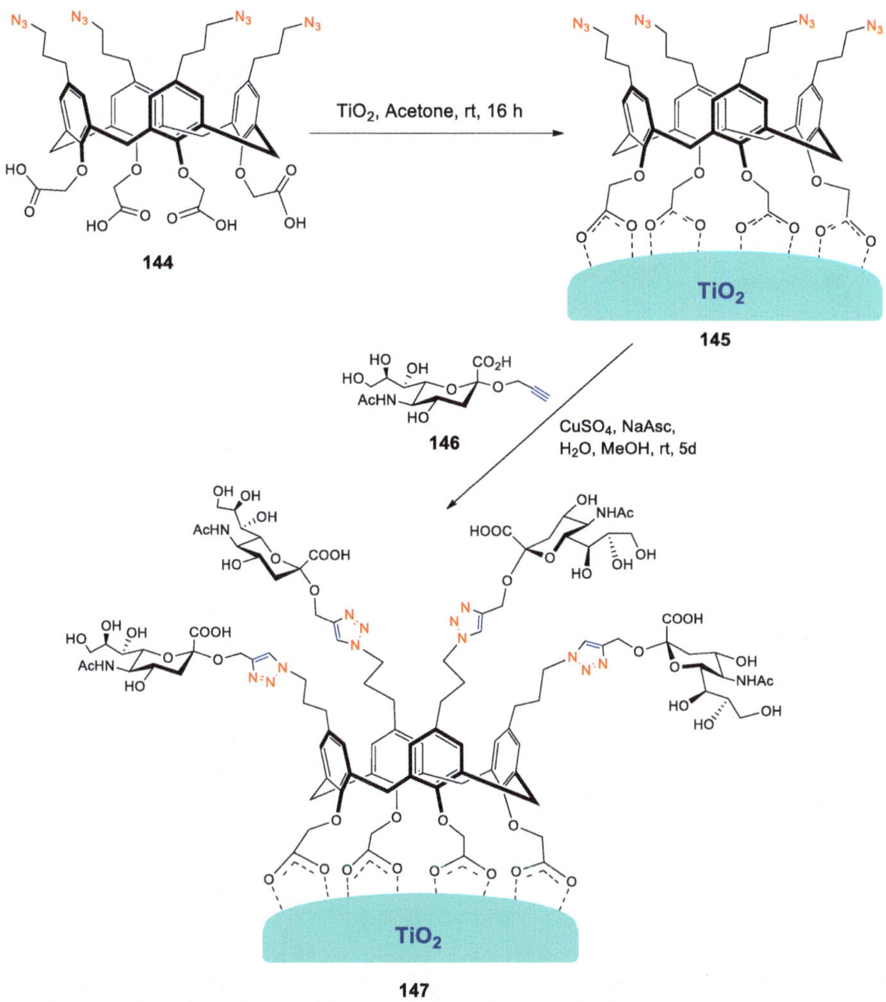

Scheme 13.31 Calix[4]arenes-based glycocluster **147** is immobilized on TiO_2 nanoparticle

Building a library of highly carbohydrates decorated scaffolds is thought to be very useful for a number of objectives, particularly in supramolecular chemistry where the weaker and reversible non-covalent interactions of triazoles have been extensively studied [87, 88]. Dondoni reported construction of triazole glycoclusters **155** and **156** having four triazoles glycosides by utilizing click protocol. The targeted tetravalent *O*-benzyl or *O*-acetyl triazole-*C*-glycosides **155** and **156a** were obtained by click conjugation of tetrakis(4-ethynylphenyl)-adamantane **152** with azidomethyl *C*-glycosides **153,154** separately in presence of CuI and DIPEA in anhydrous toluene at 80 °C for 18 h (Scheme 13.33) [89]. Finally, deacetylation of polyfunctionalized

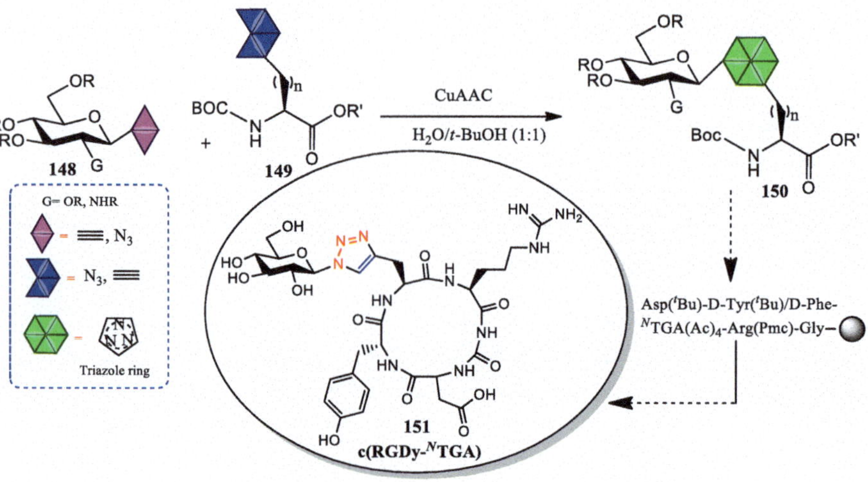

Scheme 13.32 CuAAC click-inspired synthesis of cyclic glycopeptide **151**

glycoconjugates (**155, 156a**) to furnish the desired supramolecular architecture **156b** under standard reaction conditions.

The significance of the CuAAC reaction is revealed by a class of molecular structures that reveal a legion of saccharides due to its enoromous practicality and expanding horizon. Such class of molecular architechtures namely glycoclusters, play important role in the study of carbohydrate-lectin interaction forbilogically relevant aspects. A number of hydrolysable *N*-glycosyl triazole derivatives such as

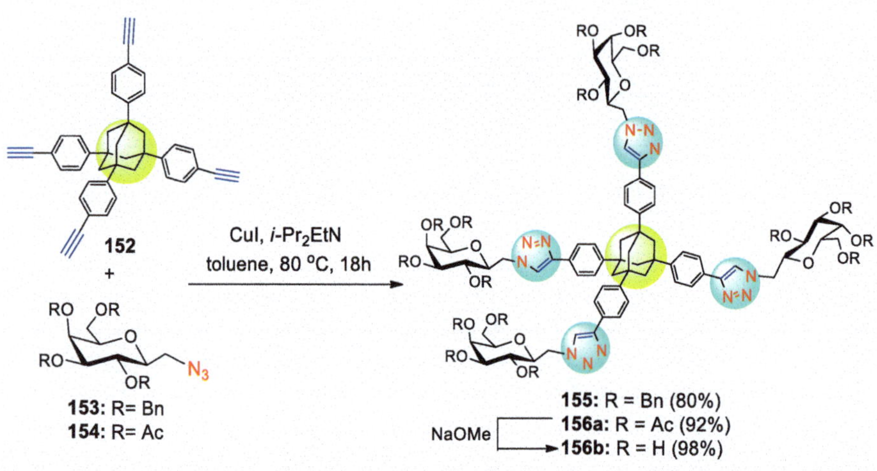

Scheme 13.33 CuAAC click-inspired synthesis of glycoclusters **155, 156** having adamantane core

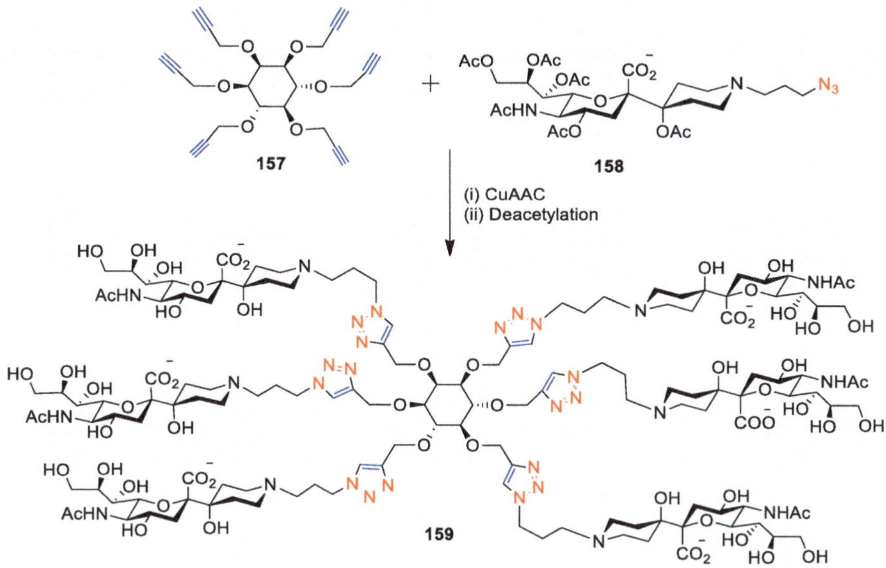

Scheme 13.34 Click chemistry inspired synthesis of multivalent *C*-sialoside monomer **159**

*C*3-symmetric (1–6)-*N*-acetyl-*β*-D-glucosamine octadecasaccharide [90], second-generation iminosugar clusters using trivalent dendrons [91], fullerene glycoconjugates [92] were effectively achieve through modular click conjugation of different glycosyl azides with polyalkyne-functionalized core.

Papin and co-workers demonstrated the construction of *C*-sialoside cluster **159** by exploiting click conjugation of diverse alkynyl carba-sugars **157** with sialic acid-basedazide **158** with an expectation to generate a glycocluster that interact with diverse biological targets (Scheme 13.34) [93].

Ligeour et al. develop three mannose-derived galactoclusters (**162–164**) using CuAAC click reaction and evaluate their affinity toward LecA [94]. Three distinct galactoclusters were produced as galactosyl azide **161** underwent click conjugation with the free hydroxyl groups of mannose sugar **160** that had been functionalized with alkynyl groups (Scheme 13.35). All synthesized clusters were evaluated for their inhibitory capability toward biofilms, and the results indicated that two glycoclusters showed a promising affinity toward LecA, with Kd values of 157 and 194 nM from ITC measurements at a concentration of 10 μM.

Tetrameric C-lectin 'DC-SIGN' regulates the chelation process with protein molecules, with interbinding site distances of 40 Å. The combined effect of statistical rebinding and chelation in succession for a specific multivalency effect could be possible in a very dense structure with many epitopes. Ordanini and co-workers demonstrated DC-SIGN antagonists by combining one selective monovalent glycomimetic ligand with trivalent dendrons separated by a rigid molecular rod of controlled length [95]. SPR analysis of glycodendrimer **168** (IC_{50} 5 μM, rp/n = 9) suggest that trivalent ligand exhibited promising affinity for DC-SIGN however,

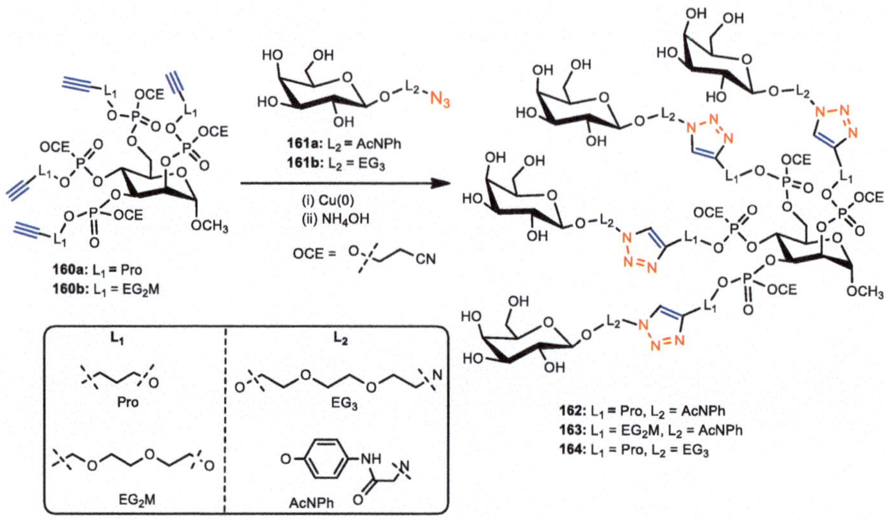

Scheme 13.35 Synthesis of biologically active mannose centered tetragalactose clusters

results for bivalent glycodendrimer (IC_{50} 8 μM and rp/n = 17) show moderate affinity at particular lengthen spacer. Modular CuAAC click protocol effectively used for construction of such rod-based glycodendrimer **168** by coupling of rod-based alkyne **166** with dendron **167** (Scheme 13.36).

Based on promising results of iminosugar cluster-based dendrimer architecture, Compain et al. explored the cyclopeptoid core for the synthesis of a higher generation of iminosugar cyclopeptoid conjugates using CuAAC tool [96]. Therefore using Cu(I) mediated azide/alkyne cycloaddition, a series of DNJ clusters **171a-f** were develop by attaching dendritic scaffold **170** on the surface of cyclic *N*-propargyl α-peptoids **169** (Scheme 13.37) [97]. Moreover, the 36-valent inhibitor **171a** was investigated by the use of native ESI–MS, AUC-SV, EM, and inhibition assay. The glycocojugate describe a noteworthy inhibition at value of 4747 rp/n. The combined action of cluster formation and chelation may be responsible for this figure. Comparing this synergistic effect to other designed molecules, the activity is substantially more efficient. Lesser activity is produced by geometrical restriction as chelation mostly happens in the porphyrin core dendrime.

The click method proved to be an effective means of producing glycodendrimers, a class of monodisperse, highly branching nanostructured polymers with a broad range of applications, from drug delivery systems to material science, biomedicine, and catalysis [98–100]. The modular click protocol was exploited for the synthesis of giant glycodendrimer embodying 243 peripheral xylopyranoside moieties [101] and star shaped G(0) dendrimers which displayed lactosyl glycosides at perypheryl of rigid hexaphenyl benzene core [102].

Wang et al. reported link spacer controlled supramolecular chirality based on the self assembly of the perylene bisimide-glycoconjugates. The mannose modified

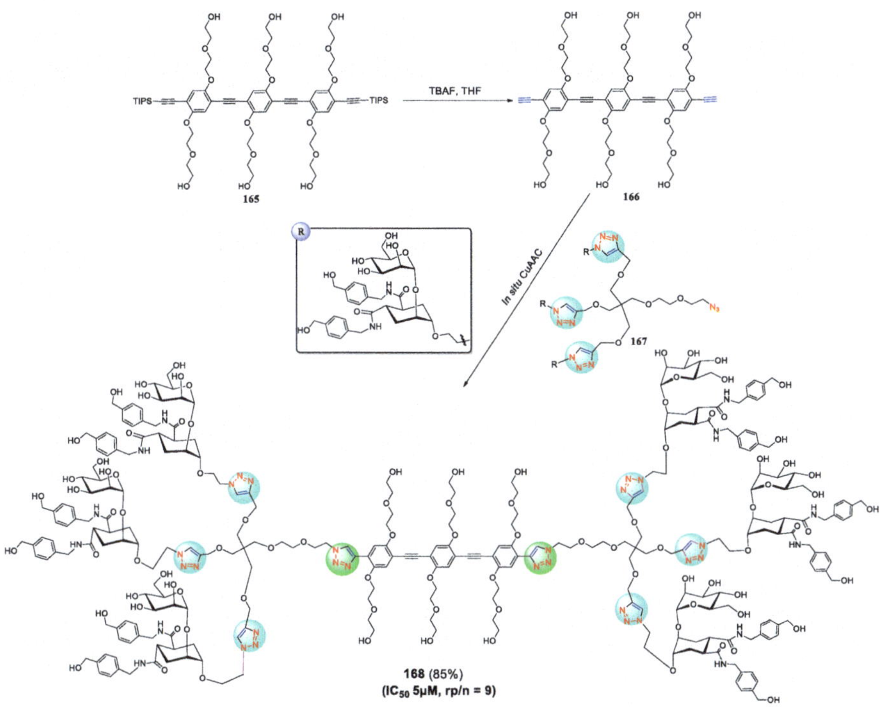

Scheme 13.36 Construction of rod-based glycodendrimer **168** as promising DC-SIGN antagonists

perylene bisimide conjugate obtained through click conjugation of **172** with sugar azide **173** in presence of $CuSO_4.5H_2O$ under 1:1 mixture of THF/water. Interestingly, the conjugate **174** with triazole linker exhibited right-handed chirality, in aqueous solution (Scheme 13.38) [103].

Glycopolymers are a common replacement for costly natural oligosaccharides since they can rapidly generate defined macrocarbohydrate characteristics due to their unique structural and chemical properties. In this regard, the CuAAC click reaction has emerged as an additional tool for this kind of polymerization, offering an opportunity to incorporate polymer chains that are functionalized with several glycosides [104, 105]. The CuAAC reaction in conjunction with the RAFT approach was used to develop several glycopolymers. For instance, Stenzel et al. developed the unique 4-vinyl-1,2,3-traizole monomer **176** using CuAAC and allowed it to react under RAFT polymerization process to produce polymer **177** (Scheme 13.39) [106]. Similarly, the synthesis of poly(diethylene glycol methyl ether methacrylate) block polymers was investigated, wherein the hydroxyl end groups were functionalized with terminal alkynes and subsequent reaction with galactosyl azide under click conditions resulted in the formation of a thermoresponsive polymer, which inhibited the toxin ricin [107].

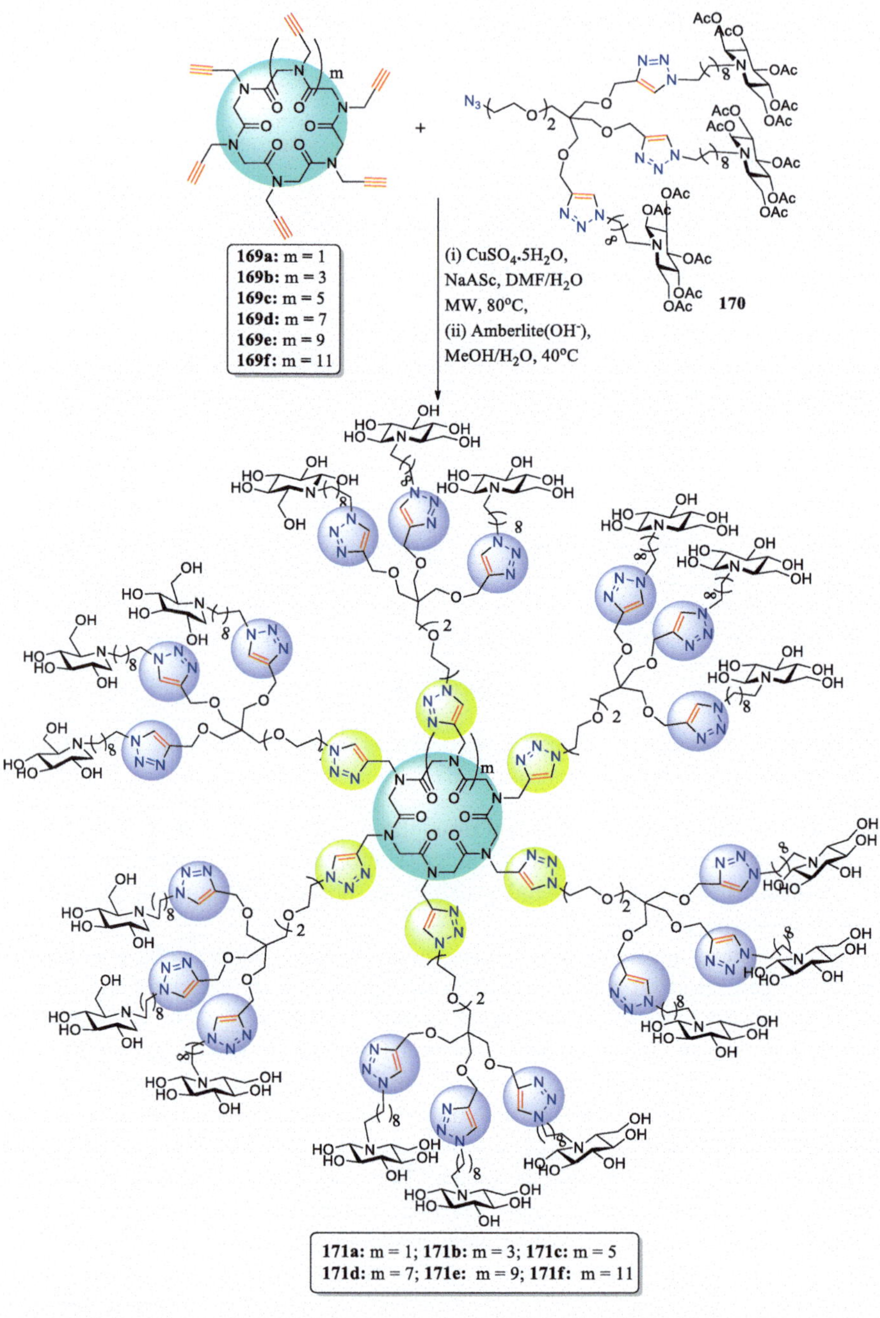

Scheme 13.37 CuAAC click-inspired synthesis of DNJ clusters **171a-f** (multivalent inhibitor) display a multivalency effect of 4747 rp/n for Jb-αMan as a result of aggregation

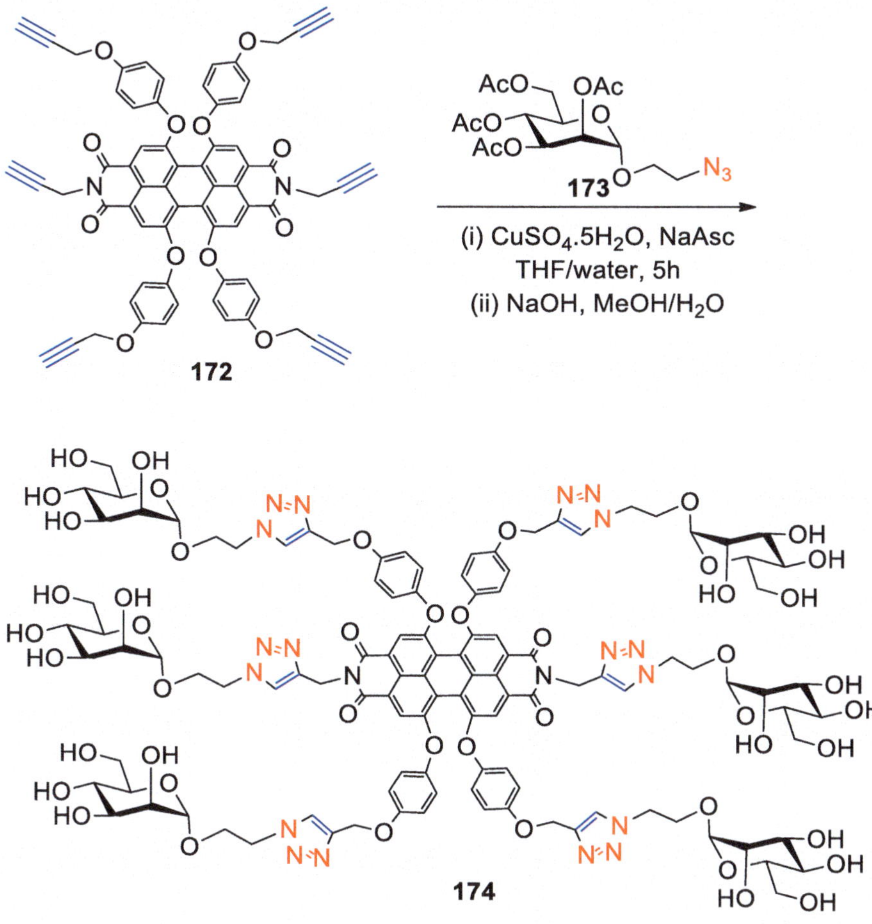

Scheme 13.38 Supramolecular chirality based on self assembly of triazole appended perylene bisimide-glycoconjugate

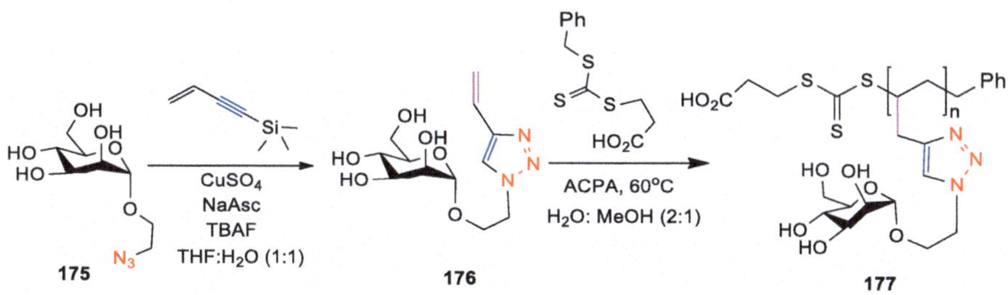

Scheme 13.39 Combined CuAAC and RAFT technique used for synthesis of polymer **177** from monomer 4-vinyl-traizole

The clickable oligomannose compounds **178a,b** (Man_8, and Man_9) were consider as potential building blocks for the synthesis of virus glycoconjugates, for example, a sugar immunogen to target HIV-1. The synthetic strategy involves CuAAC conjugation of **178a** with large excess of alkynyl group resulted the glycoconjugates **179** which on further acylation the surface amino groups using excess of alkynyl *N*-hydroxysuccinimide ester **180** and then clicking with azides **178b** under standard click condition furnished the desire virus glycoconjugate **182** (Scheme 13.40) [108]. Surprisingly, it was found that Qβ glycoconjugates containing oligomannose clusters exhibited a strong affinity for binding the antibody 2G12. The developed click protocol has been considered as effective strategy for the development of an interesting sugar-based vaccine component for HIV.

Carbohydrate microarrays, a technique of assembling sugar moieties in bulk over immobilized surface, has drawn a lot of interest as a possible tool for comprehending recognition events based on interactions between carbohydrates and proteins or lectins [109]. CuAAC proved to be an effective strategy for grafting multiple copies of alkyne/azide-functionalized saccharides on immobilized solid surfaces [110] and polymeric surfaces, despite the availability of advanced detecting tools [111]. Due to

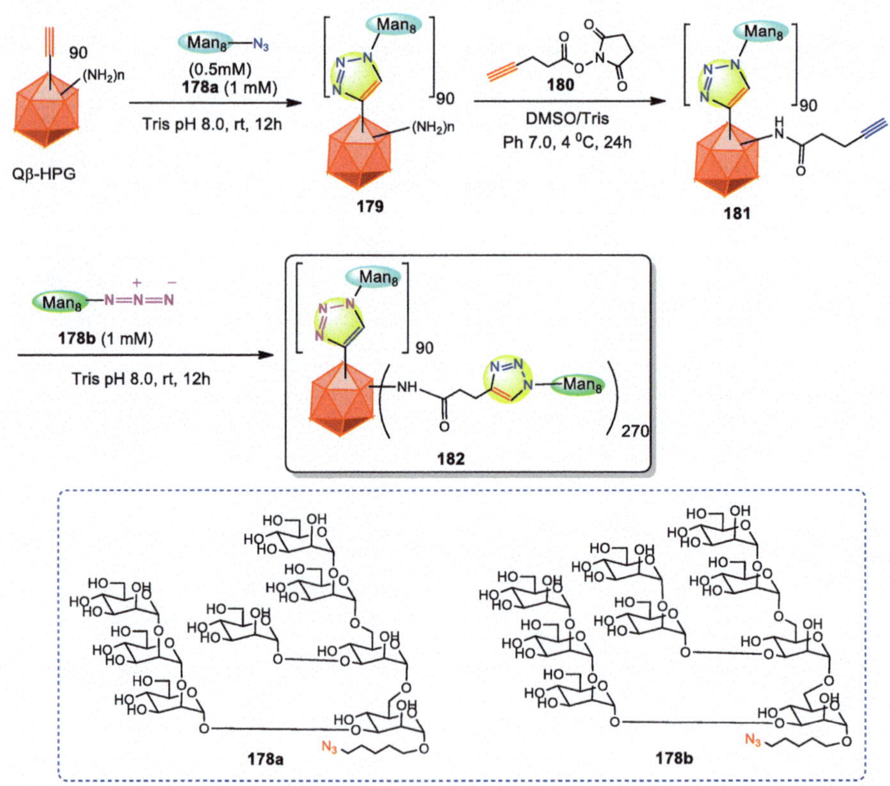

Scheme 13.40 CuAAC-mediated synthesis of Qβ conjugtae **182**

its involvement in host recognition/adhesion and biofilm development, the fucose-specific lectin PA-IIL could turn out to be a useful therapeutic target. As a result, a number of fucosylated glycoclusters were reported using click chemistry in conjunction with DNA solid-phase synthesis. Microwave induced click conjugation of solid-supported alkynyl **183** with sugar azide **184** produces solid-supported mannose platform **185**, which on linking with alkyne **186** followed by labeling and ammonia treatment afforded Cy3-oligonucleotide **187** (Scheme 13.41) [112]. Further, clicking of alkyne **187** with fucosylated azide **188** and treating with aqueous ammonia produce triazole-linked fucosylated glycocluster **189**. The influence of the linker arm on biological activity was addressed by the evaluation of binding assays against PA-IIL for these fucomimetics, and it was found that the mannose-based core performed better than the glucose- and galactose-based cores.

Tiwari et al. made significant effort to develop D-glucose derived chemosensors **193** and **194** for selective detection of Cu^{2+} ions by means of fluorescence quenching using fluorogenic dual CuAAC protocol [113]. *Bis*-propargylated subtrates **190** and **191**, obtained from D-glucose in a series of high-yielding standard reaction, underwent CuAAC coupling with a non-fluorescent 3-*azido*-coumarin **192** in order to

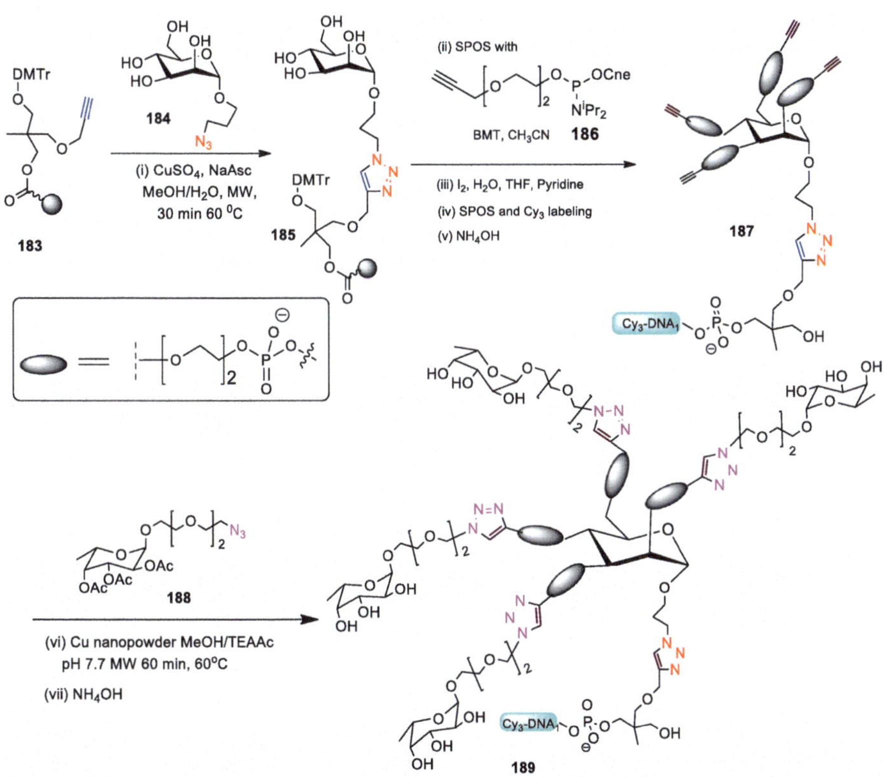

Scheme 13.41 Crown-like tetra-fucosylated glycocluster based on mannose core

produce *bis*-triazolyl probes **193** and **194** (Scheme 13.42). Structure of both probes **193** and **194** were assigned by variance in their relative positions of triazolocoumarin arms linked through the phenylene ring. The detection limit of glycosyl probes **193** and **194** were calculated to 6.99 lM and 7.30 lM, respectively.

Nguyen reported the synthesis of 1,2,3-triazole glycoconjugate **196** bearing BODIPY backbone with three glucosidesepitopes present at meso and boron atom position using click chemistry tool (Scheme 13.43) [114]. In vitro analysis suggested that triazole BODIPY glycoconjugate **196** is highly permeable against specific cell and non-toxic to human epithelial type 2(HEp2) cells either in presence or absence of light. Furthermore, as the amount of sugar residues has increased, the biological uptakes of these molecules have significantly increased as well.

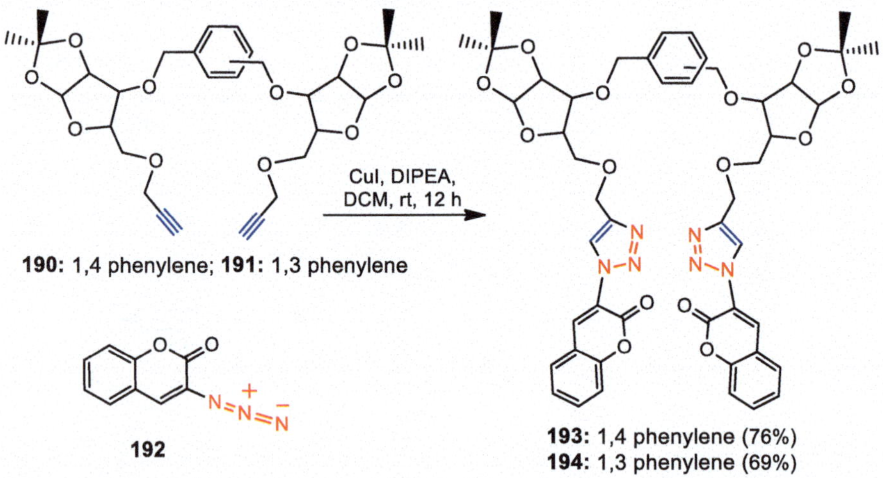

Scheme 13.42 Click-inspired synthesis of *bis*-triazole chemosensors **193** and **194** for selective detection of Cu^{2+} ions

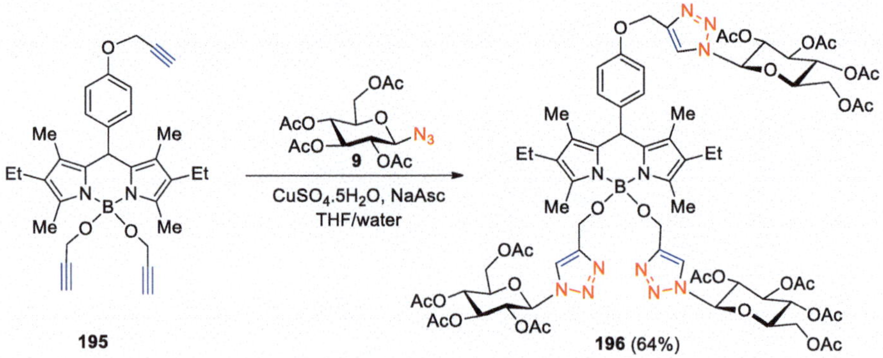

Scheme 13.43 Click ispired synthesis of glucosylated BODIPY glycoconjugates **196**

The diverse role of lipids in metabolism, drug delivery, imagine, gene transfer, and development of system that imitate membrane enviroments have compelled researchers to investigate the diverse role of lipid useful characteristics. Some valuable glycolipid analogsuch as multiple dimer lipids [115] and a complex glycolipid analog [116] were synthesized via click chemistry. In addition, carbohydrate-anchored triazole-linked lipids that fascinatingly self-assemled in different patterns (size and shape) was also documented [117]. Further a four-components Ugi reaction reported for construction of a novel glycolipid comprising of carbohydrate/triazole/lipid hybrid architectures [118]. Toward this end, a trisaccharide alkyne **197** was clicked with azido functionalized lipid **198** in presence of $Cu(Ac)_2$/NaAsc in THF/H_2O to produces triazole appended neoglycolipids **199** in good yields (Scheme 13.44).

There is a fascinating and well-established method of complexing metals, including radionuclei, without using triazole-*N*. Clicking an alkyne **200** with azide **201** generated a strong lanthanide chelate **202** which possess functionalized pyridine-tetraacetic ligands bound to glucoside and maltoside frameworks (Scheme 13.45) [119].

In a similar fashion, the mannose and galactose-containing triazolyl glycoconjugates were synthesized having distinct carbohydrate-lectin binding affinities with functionalized oligothiophene and Con A interaction [120]. Additionally, given click conditions, were implemented for the construction of glucoside framework appended with Tc-chelating agents and Rh- complexes [121]. In continuation, another route developed by the Mindt group for the construction of triazole chelators involving

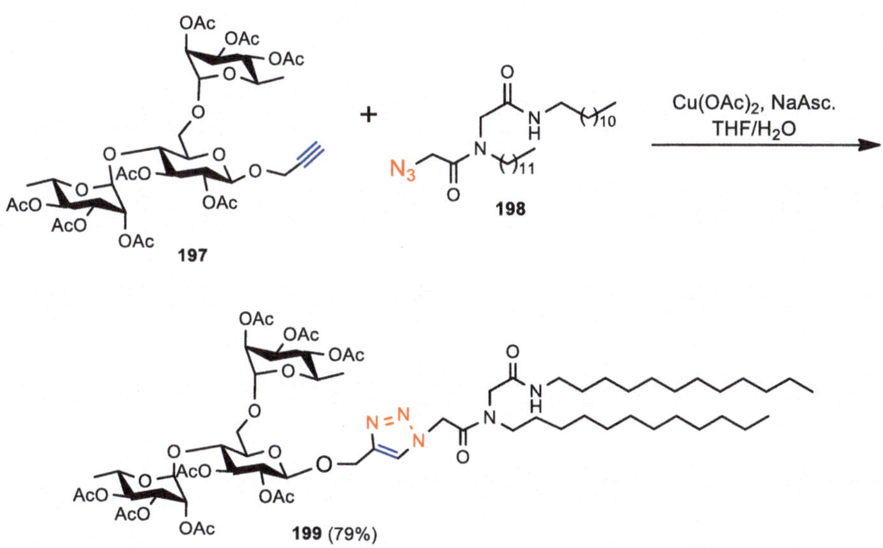

Scheme 13.44 Construction of triazole containing glycolipids using click reaction

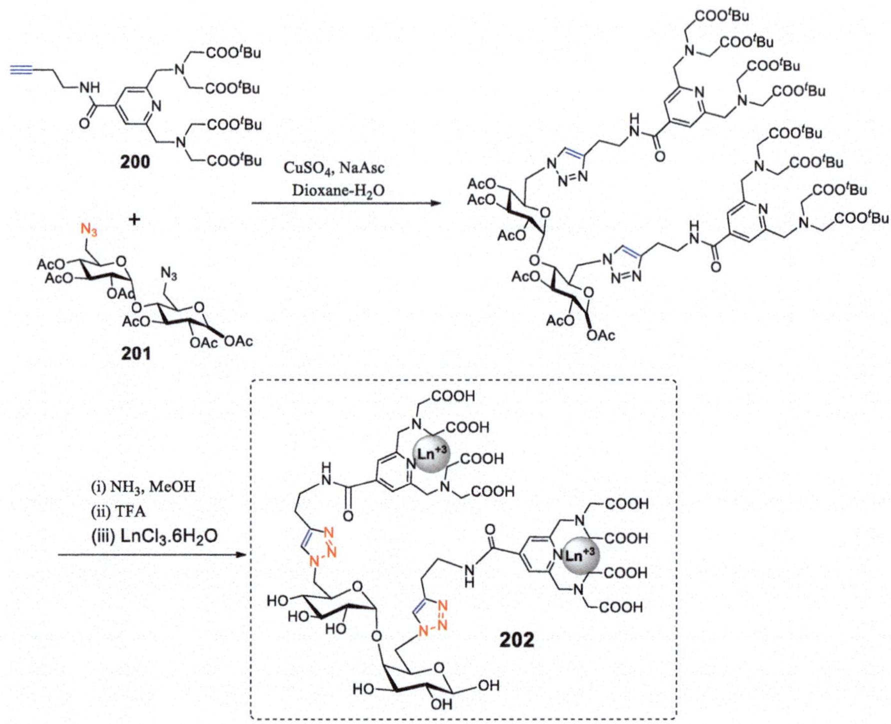

Scheme 13.45 Construction of pyridine-tetraacetic acid glycoconjugates **202** using click protocol

the reaction of L-propargyl glycine and azide-functionalized D-galactose in association with $[^{99m}Tc(CO)_3]^+$ [122]. Kim and the group implemented the carbo-click strategy for the development of ^{18}F-labeled triazole-bounded glycoconjugates, these developed glycoconjugates act as an intriguing positron emitter with a half-life of 109.8 min [123, 124]. The successful radiolabeling was achieved using this devised protocol and subsequently patented by another research group [125]. For the in vitro assessment of the PET radiotracer, a click-inspired labeling technique was investigated toconstruct 4-[(2-[^{18}F]fluoroethyl)-1-(β-D-glucopyranosyl)]-1*H*-1,2,3-triazole **205** in significant yield from glycosyl azide **203** in association with ^{18}F labeled alkyne functionality **204**. A similar ^{18}F-labeled scaffold **207** was achieved by treating peracetylated glycohybrid triazole **206** with n-Bu4N[18F]F under optimized reaction conditions (in acetonitrile at 100 °C for 20 min), accompanied by acetate removal at room temperature for 15 min under conventional NaOMe/MeOH conditions (Scheme 13.46).

Moreover, Fischer and co-workers used a similar strategy to produce [^{18}F]fluorodeoxyglucosyl folate **210** utilizing a CuAAC-mediated triazole ring formation reaction of folate alkyne **209** and2-deoxy-2-[^{18}F]fluoroglucopyranosylazide **208** with significant RCY (25% d.c.) and AS (90 GBq/mol) (Scheme 13.47) [126]. The embedded glucose moiety made a

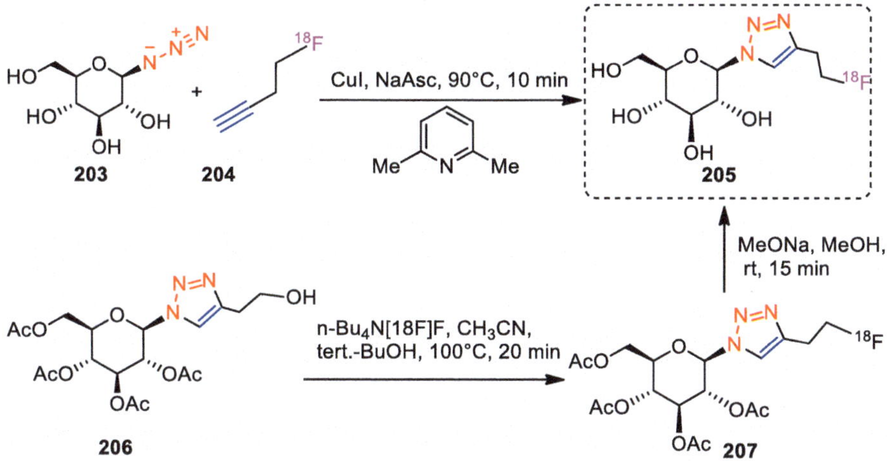

Scheme 13.46 Click reaction for the development of radiolabeled 4-[(2-[^{18}F]fluoroethyl)-1-(β-D-glucopyranosyl)]-1*H*-1,2,3-triazole **205**

noteworthy contribution to the improved hydrophilicity of the developed radiotracer, which displayed selective and high uptake for FR-positive tumors (10.03 ± 1.12% ID/g, 60 min p.i.) in the in vivo investigations using the KB tumor-bearing mice.

Another notable instance related to the development of radiotracers is the labeled triazolyl-glycohybrid **213**, which has been constructed from acetylated or free

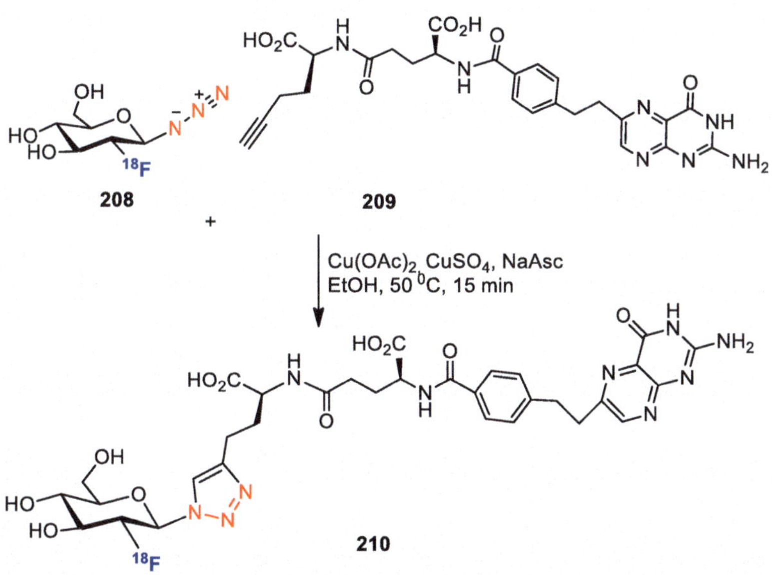

Scheme 13.47 Click-mediated ^{18}F-glycoconjugation for the synthesis of respective triazole **210**

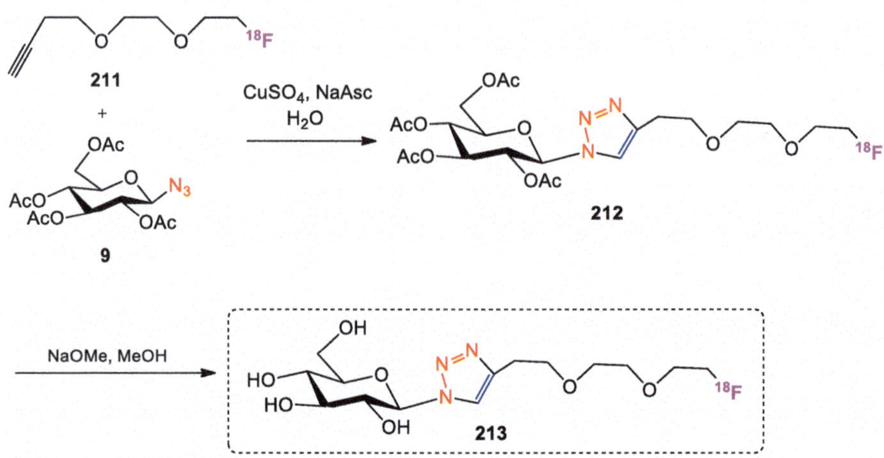

Scheme 13.48 Development of labeled triazolyl-glycohybrid **213** under click condition

azido-glucose and labeled acetylinic derivative **211**, using standard click condition ($CuSO_4$/NaAsc in ^{t}BuOH/H_2O system at rt). The subsequent removal of acetate group from resulting triazole appended glycohybrid **212** using NaOMe in methanol, furnished compound **213**, exhibiting acceptable RCY values (Scheme 13.48) [127].

Whenever the 18F-radiolabeled glucose analogue, or 2-deoxy-2-[18F]fluoro-D-glucose[18F]FDG, was established, it underwent numerous advancements as a radiotracer and nowadays a well-known chemo- and neuro-selective metabolic imaging tool for PET. An interesting article describes the development of a class of β-configured [^{18}F]FDG analogue **216** by coupling [18F] lebeled terminal alkyne **215** with aglycon azide functionality **214** in a facile two-step one-pot approach utilizing a CuAAC tool, resulting in a significant yield (Scheme 13.49) [128]. Similarly, a click conjugation of [18F] lebelledazide **218**, with alkyne **217**, produced an excellent yield of [^{18}F]FDG triazole **219** that was also *β*-configured. The *β*-configured precursors with related [^{18}F]FDGs showed RYC of 77–88%, while the Fmoc-protected amino acids, Fmoc-N-(propargyl)-glycine, and Fmoc-3-azido-L-alanine, showed a parallel range of RYC (75–83%) and produced the appropriate ^{18}F-fluoroglycosylated amino acid conjugates.

In recent years, chiral ionic liquids (ILs) that are produced from naturally occurring chiral precursors have found extensive application [129]. Regarding this, glucose-bounded 1,2,3-triazolium ionic liquid **221** was synthesized through a regioselective cycloaddition reaction of glucose alkyne with a glucose azide under copper catalysisand subsequently quaternized with methyl iodide (Scheme 13.50). The resulting ionic liquid **221** was employed as a reusable solvent and ligand in the copper(I)-catalyzed amination of aryl halides with aqueous ammonia [130]. This identical procedure, which involves clicking *β*-propargyl xyloside with phenyl azide followed by deacetylation and methylation was used to synthesize D-xylose-based ionic liquids [131].

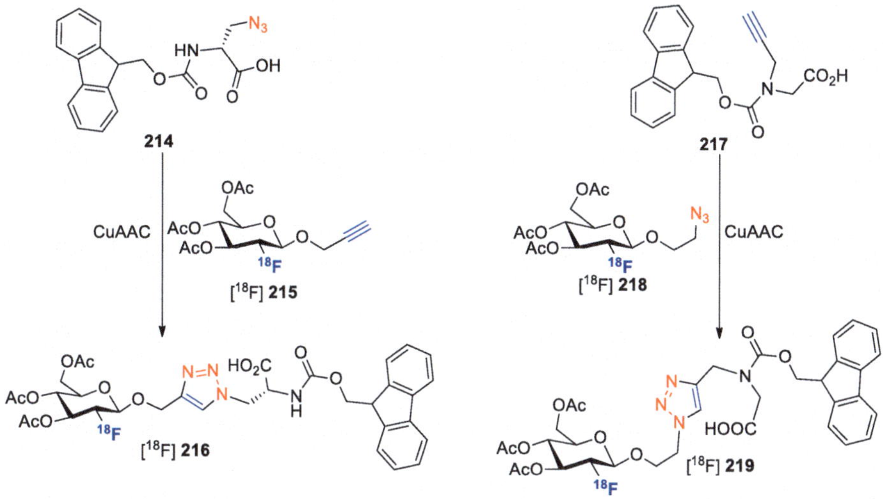

Scheme 13.49 CuAAC-mediated route for construction of [^{18}F]FDGs **216** and **219** as ^{18}F-fluoroglycosylation decorum convenient for PET therapy

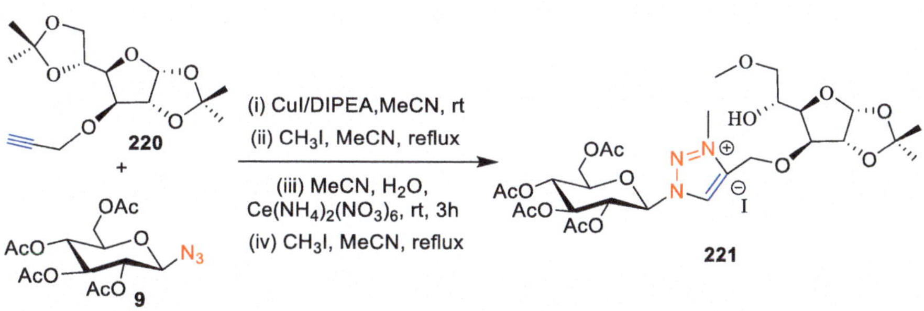

Scheme 13.50 Synthesis of 1,2,3-triazolium ionic liquid **221** linked to glucose

13.4 Conclusions and Future Perspective

CuAAC, also known as Click Chemistry, is a potentially useful method for efficiently and simply accessing complex molecular-level structures in good yields by treating organic azides with terminal alkynes in presence of Cu(I). Following Sharpless and Meldal's remarkable discovery in 2002, the 1,2,3-triazole synthesis by CuAAC carbo-click received a lot of attention and remained persistent in organic chemistry. This conjugation technique has a wide range of applications in drug discovery, surface chemistry, polymer chemistry, material chemistry, radio labeling, cell labeling, and plant science due to the relative inertness of the *N*-heterocyclic triazole ring and its compatibility. Also, molecular hybridization of a variety of pharmacophoric unit with 1,2,3-triazole unit may provide effective and safe and novel drug

Fig. 13.2 CARBO-CLICK together can-do miracles

candidates [132]. In particular, the regioselective 'CARBO-CLICK tool' [1, 2, 133, 134] provided a dependable path for design and developing a range of biocompatible multivalent neoglycoconjugates, including glycopolymers, glycopeptides, glycodendrimers, glycoclusters, glycomacrocyles, macromolecules, and glycan microarrays and known to do miracles in diverse range of applications (Fig. 13.2).

Click chemistry was successfully used to generate a number of carbohydrate-based ligands, ionic liquids, and 1,2,3-triazolyl organocatalysts. These products serve as asymmetric catalytic tools and offer new possibilities for the synthesis of asymmetric compounds derived from sugars. Moreover, the generation of solid-supported organocatalysts does not yet address by carbo-click, which requires more considerable investigation. A number of noteworthy characteristics of this CuAAC reaction such as outstanding regioselectivity, high-to-excellent yields, ease of execution, short reaction times, modular nature, mild conditions, and easy access to starting materials make it ideal for the pharmaceutical industry's large-scale synthesis of glycohybrids. Despite the fact that the click tool has many applications in glycoscience, especially in chemical biology and medicinal chemistry, more research is needed to fully understand the use 'carbo-click' in catalysis with focus on an asymmetric approach. This chapter covers a literature assessment on the development of functionalized 1,2,3-disubstituted triazoles through 'carbo-click' protocols during the last decades.

References

1. Tiwari VK, Mishra BB, Mishra KB, Mishra N, Singh AS, Chen X (2016) Chem Rev 116:3086–3240
2. Agrahari AK, Bose P, Jaiswal MK, Rajkhowa S, Singh AS, Hotha S, Mishra N, Tiwari VK (2021) Chem Rev 121:7638–7956
3. Huisgen R (1963) Angew Chem Int Ed 2:565–598
4. Tornoe CW, Christensen C, Meldal M (2002) J Org Chem 67:3057–3062
5. Rostovtsev VV, Green LG, Fokin VV, Sharpless KB (2002) Angew Chem Int Ed 41:2596
6. Kolb HC, Finn MG, Sharpless KB (2001) Angew Chem Int Ed 40:2004–2021
7. Meldal M, Tornoe CW (2008) Chem Rev 108:2952–3015
8. Thirumurugan P, Matosiuk D, Jozwiak K (2013) Chem Rev 113:4905–4979
9. Amblard F, Cho JH, Schinazi RF (2009) Chem Rev 109:4207–4220
10. Zhang L, Chen X, Xue P, Sun HHY, Williams ID, Sharpless KB, Fokin VV, Jia G (2005) J Am Chem Soc 127:15998–15999
11. Johansson JR, Beke-Somfai T, Stalsmeden AS, Kann N (2016) Chem Rev 116:14726–14768
12. Agard NJ, Prescher JA, Bertozzi CR (2004) J Am Chem Soc 126:15046–15047
13. Chang PV, Chen X, Smyrniotis C, Xenakis A, Hu TS, Bertozzi CR, Wu P (2009) Angew Chem Int Ed 48:4030–4033
14. Jaiswal MK, Tiwari VK (2023) Chem Rec 23:e202300167
15. Jaiswal MK, Gupta A, Ansari FA, Pandey VK, Tiwari VK (2024) Curr Org Synth 21:513–558
16. Witczak ZJ, Bielski R (2013) Click chemistry in glycoscience: new development and strategies. Wiley-VCH John Wiley & Sons, New York
17. Chandrasekeran S (2016) Click reactions in organic synthesis. Wiley-VCH John Wiley & Sons, New York
18. Tiwari VK (2020) Carbohydrates in drug discovery and development. Elsevier Inc.
19. Pérez S, Tvaroška I (2014) Adv Carbohydr Chem Biochem 71:9–136
20. Tripathi RP, Upadhyaya K, Singh K, Tiwari VK (2015) Trends Carbohydrate Res 7(4):1–21
21. Bose P, Jaiswal MK, Singh SK, Singh RK, Tiwari VK (2023) Carbohydrate Res 527:108804
22. Tiwari VK (2023) Synthetic strategies in carbohydrate chemistry. Elsevier Inc.
23. Liang L, Astruc D (2011) Coord Chem Rev 255:2933–2945
24. Golas PL, Matyjaszewski K (2010) Chem Soc Rev 39:1338–1358
25. Paulsen H, Gyorgydeák Z, Friedmann M (1974) Konformationsanalyse. III. Chem Ber 107:1568–1578
26. Sicherl F, Wittmann V (2005) Angew Chem, Int Ed 44:2096–2099
27. Arndt S, Hsieh-Wilson LC (2003) Org Lett 5:4179–4182
28. Cavender CJ, Shiner VJJ (1972) Org Chem 37:3567–3569
29. Vasella A, Witzig C, Chiara JL, Martin-Lomas M (1991) Helv Chim Acta 74:2073–2077
30. Mishra A, Tiwari VK (2015) J Org Chem 80:4869–4881
31. Kotik M, Brodsky K, Halada P, Javurkova H, Pelantova H, Konvalinkova D, Bojarova P, Kren N (2021) Catalysis Commun 149:106193
32. Koester DC, Werz DB (2012) Beil J Org Chem 8:675–682
33. Gupta MR, Thakur K, Khare NK (2018) Carbohydr Res 457:51–55
34. Adhikary ND, Chattopadhyay P (2012) J Org Chem 77:5399–5405
35. Li RZ, Tang H, Wan L, Zhang X, Fu Z, Liu J, Yang S, Jia D, Niu D (2017) Chem 3:834–845
36. Meng G, Guo T, Ma T, Zhang J, Shen Y, Sharpless KB, Dong J (2019) Nature 574:86–91
37. Brandhofer T, Ozdemir A, Gini A, Mancheno OG (2019) Chem Eur J 25:4077–4086
38. Sareen N, Singh AS, Tiwari VK, Kant R, Bhattacharya S (2017) Dalton Trans 46:12705–12710
39. Singh Y, Spinelli N, De Francq E, Dumy P (2006) Org Biomol Chem 4:1413–1419
40. Kushwaha D, Dwivedi P, Kuanar SK, Tiwari VK (2013) Curr Org Synth 10:90–135
41. Tripathi RP, Dwivedi P, Sharma A, Kushwaha D, Tiwari VK (2013) Triazolyl glycoconjugates in medicinal chemistry. In: Witczak ZJ, Bielski R (eds) Click-chemistry in glycoscience. John Wiley & Sons, Inc., Hobken, New Jersey, pp 293–324.

42. Akula RA, Temelkoff DP, Artis ND, Norris P (2004) Heterocycles 63:2719–2725
43. Campo VL, Sesti-Costa R, Carneiro ZA, Silva JS, Schenkman S, Carvalho I (2012) Bioorg Med Chem 20:145–156
44. Weiwer M, Chen CC, Kemp MM, Linhardt RJ (2009) Eur J Org Chem, 2611–2620
45. Casas-Solvas JM, Vargas-Berenguel A, Capitan-Vallvey LF, Santoyo-Gonzalez F (2004) Org Lett 6:3687–3690
46. Kumar D, Mishra KB, Mishra BB, Mondal S, Tiwari VK (2014) Steroids 80:71–79
47. Mishra KB, Mishra BB, Tiwari VK (2014) Carbohydrate Res 399:2–7
48. Mishra RC, Karna P, Gundala SR, Pannu VR, Stanton A, Gupta KK, Robinson MH, Lopus M, Wilson L, Henary M et al (2011) Biochem Pharmacol 82:110–121
49. Mishra KB, Mishra RC, Tiwari VK (2015) RSC Adv 5:51779–51789
50. Mishra KB, Tiwari VK (2014) J Org Chem 79:5752–5762
51. Yadav S, Singh AK, Agrahari AK, Sharma K, Singh AS, Gupta MK, Tiwari VK, Prakah P (2020) Scientific Rep 10:14204
52. Singh AK, Gopu K (2010) Tetrahedron Lett 51:1180–1184
53. Yang W, Wu HX, Xu HH, Hu AL, Lu ML (2011) Agric. Food Chem 59:12534–12542
54. Dwivedi P, Mishra KB, Mishra BB, Singh N, Singh RK, Tiwari VK (2015) Glycoconjugate J 32:127–140
55. Olomola TO, Klein R, Mautsa N, Sayed Y, Kaye PT (2013) Bioorg Med Chem 21:1964–1971
56. Diot J, Garcia-Moreno MI, Gouin SG, Mellet CO, Haupt K, Kovensky J (2009) Org Biomol Chem 7:357–363
57. Yang JW, He XP, Li C, Gao LX, Sheng L, Xie J, Shi XX, Tang Y, Li J, Chen GR (2011) Bioorg Med Chem Lett 21:1092–1096
58. Nandurkar N, Zhang J, Ye Q, Ponomereva L, She Q, Thorson JS (2014) J Med Chem 57:7478–7484
59. Palomo C, Aizpurua JM, Balentova E, Azcune I, Santos JI, Nez-Barbero JJ, Canada J, Miranda JI (2008) Org Lett 10:2227–2230
60. Sahoo L, Singhamahapatra A, Kumar K, Loganathan D (2013) Carbohydr Res 381:51–58
61. Chandrasekaran V, Lindhorst TK (2012) Chem Commun 8:7519–7521
62. Pal R, Sarkar S, Chatterjee N, Sen AK (2013) Tetrahedron Lett 54:5642–5646
63. Morris JC, Chiche J, Grellier C, Lopez M, Bornaghi LF, Maresca A, Supuran CT, Pouyssegur J, Poulsen SA (2011) J Med Chem 54:6905–6918
64. Lo Conte M, Marra A, Chambery A, Gurcha SS, Besra GS, Dondoni A (2010) J Org Chem 75:6326–6336
65. Polakova M, Bel M, Miku K, Lattov E, Perreault H (2011) Bioconjugate Chem 22:289–298
66. Titz A, Patton J, Smiesko M, Radic Z, Schwardt O, Magnani JL, Ernst B (2010) Bioorg Med Chem 18:19–27
67. Moni L, Marra A, Skotnicki JS, Koehn FE, Gharbia MA, Dondoni A (2013) Tetrahedron Lett 54:6999–7003
68. Giguere D, Patnam R, Bellefleur MA, St-Pierre C, Sato S, Roy R (2006) Chem Commun 22:2379–2381
69. Tejler J, Tullberg E, Frejd T, Leffler H, Nilsson UJ (2006) Carbohydr Res 341:1353–1362
70. Srinivas A, Rao EK (2021) Acta Chim Slov 68:404–413
71. Dorner S, Westermann B (2005) Chem Commun, 2852–2854
72. Conte ML, Grotto D, Chambery A, Dondoni A, Marra A (2011) Chem Commun 47:1240–1242
73. Dwivedi P, Mishra KB, Pritika MBB, Tiwari VK (2017) Glycoconj J 34:61–70
74. Muthana S, Yu H, Cao H, Cheng J, Chen XJ (2009) Org Chem 74:2928–2936
75. Jung KH, Muller M, Schmidt RR (2000) Chem Rev 100:4423–4442
76. Tiwari VK, Kumar A, Schmidt RR (2012) Eur J Org Chem, 2945–2956
77. Lewandowski B, Jarosz S (2011) Synth Commun 41:2161–2168
78. Lewandowski B, Jarosz S (2010) Org Lett 12:2532–2535
79. Dondoni A, Marra A (2010) Chem Rev 110:4949–4977
80. Dondoni A, Marra A (2006) J Org Chem 71:7546–7551

81. Vecchi A, Melai B, Marra A, Chiappe C, Dondoni J (2008) Org Chem 73:6437–6440
82. Dondoni A (2020) Synlett 31:1361–1371
83. Moni L, Rossetti S, Scoponi M, Marra A, Dondoni A (2010) Chem Commun 46:475–477
84. Dondoni A, Giovannini PP, Massi A (2004) Org Lett 6:2929–2932
85. Kuijpers BHM, Groothuys S, Keereweer AR, Quaedflieg PJLM, Blaauw RH, van Delft FL, Rutjes FPJT (2004) Org Lett 6:3123–3126
86. Kuijpers BHM, Groothuys S, Soede AC, Laverman P, Boerman OC, van Delft FL, Rutjes FPJT (2007) Bioconjugate Chem 18:1847–1854
87. Xu L, Li Y, Li Y, Asian J (2014) Org Chem 3:582–602
88. Ramkumar R, Anbarasan P (2020) Cu-catalyzed click reactions. In: Anilkumar G, Saranya S (eds) Copper catalysis in organic synthesis. Wiley-VCH, Chapter 9
89. Chen Q, Yang F, Du YG (2005) Carbohydr Res 340:2476–2482
90. Forman A, Pfoh R, Eddenden A, Howell PL, Nitz M (2019) Org Biomol Chem 17:9456–9466
91. Tarnus C, Bodlenner A, Compain PA (2014) Eur J Org Chem, 1866–1872
92. Isobe H, Cho K, Solin N, Werz DB, Seeberger PH, Nakamura E (2007) Org Lett 9:4611–4614
93. Papin C, Doisneau G, Beau JM (2008) Chem Eur J 15:53–57
94. Ligeour C, Vidal O, Dupin L, Casoni F, Gillon E, Meyer A, Vidal S, Vergoten G, Lacroix JM, Souteyrand E, Imberty A, Vasseur JJ, Chevolot Y, Morvan F (2015) Org Biomol Chem 13:8433–8444
95. Ordanini S, Varga N, Porkolab V, Thépaut M, Belvisi L, Bertaglia A, Palmioli A, Berzi A, Trabattoni D, Clerici M et al (2015) Chem Commun 51:3816–3819
96. Lepage L, Meli A, Bodlenner A, Tarnus C, De Riccardis F, Izzo I, Compain P (2014) Beil J Org Chem 10:1406–1412
97. Lepage ML, Schneider JP, Bodlenner A, Meli A, De Riccardis F, Schmitt M, Tarnus C, Nguyen-Huynh NT, Francois YN, Leize-Wagner E et al (2016) Chem Eur J 22:5151–5155
98. Ortega-Muñoz M, Morales-Sanfrutos J, Perez-Balderas F, Hernandez-Mateo F, Giron-Gonzalez MD, Sevillano-Tripero N, Salto-Gonzalez R, Santoyo-Gonzalez F (2007) Org Biomol Chem 5:2291–2301
99. Zhang S, Zhaob Y (2010) Macromolecules 43:4020–4022
100. Papp I, Dernedde J, Enders S, Riese SB, Shiao TC, Roy R, Haag R (2011) ChemBioChem 12:1075–1083
101. Camponovo J, Hadad C, Ruiz J, Cloutet E, Gatard S, Muzart J, Bouquillon S, Astruc D (2009) J Org Chem 74:5071–5074
102. Chabre YM, Brisebois PP, Abbassi L, Kerr SC, Fahy JV, Marcotte I, Roy R (2011) J Org Chem 76:724–727
103. Wang K, Han D, Cao G, Li X (2015) RSC Adv 5:47728–47731
104. Kempe K, Krieg A, Becer CR, Schubert US (2011) Chem Soc Rev 41:176–191
105. Geng J, Mantovani G, Tao L, Nicolas J, Chen G, Wallis R, Mitchell DA, Johnson BRG, Evans SD, Haddleton DM (2007) J Am Chem Soc 129:15156–15163
106. Hetzer M, Chen G, Barner-Kowollik C, Stenzel MH (2010) Macromol Biosci 10:119–126
107. Kumar J, McDowall L, Chen G, Stenzel MH (2011) Polymer Chem 2:1879–1886
108. Astronomo RD, Kaltgrad E, Udit AK, Wang SK, Doores KJ, Huang CY, Pantophlet R, Paulson JC, Wong CH, Finn MG et al (2010) Chem & Biol 17:357–360
109. Cecioni S, Imberty A (2009) Vidal S (2015) Chem. Rev 115:525–561; Song EH, Pohl NL (2009) Curr Opin Chem Biol 13:626–632
110. Fazio F, Bryan MC, Blixt O, Paulson JC, Wong CH (2002) J Am Chem Soc 124:14397–14402
111. Michel O, Ravoo BJ (2008) Langmuir 24:12116–12118
112. Gerland B, Goudot A, Pourceau G, Meyer A, Dugas V, Cecioni S, Vidal S, Souteyrand E, Vasseur JJ, Chevolot Y, Morvan F (2012) Bioconjugate Chem 23:1534–1547
113. Kushwaha D, Singh RS, Tiwari VK (2014) Tetrahedron Lett 55:4532–4536
114. Nguyen AL, Griffin KE, Zhou Z, Fronczek FR, Smith KM, Vicente MGH (2018) New J Chem 42:8241–8246
115. Zerkowski JA, Nunez A, Strahan GD, Solaiman DKYJ (2009) Am Oil Chem Soc 86:1115–1121

116. Tang SC, Wong QL, Guo ZW (2010) Chem Eur J 16:1319–1325
117. Tyagi M, Kartha KPR (2015) Carbohydrate Res 413:85–92
118. Labrada KP, Brouard I, Mendez I, Perez CS, Gavín JA, Rivera DG (2014) Eur J Org Chem, 3671–3683
119. Gouin SG, Roger M, Leygue N, Deniaud D, Julienne K, Benoist E, Picard C, Kovensky J, Galaup C (2012) Bioorg Med Chem Lett 22:2684–2688
120. Mandal S, Das R, Gupta P, Mukhopadhyay B (2012) Tetrahedron Lett 53:3915–3918
121. Benoist E, Coulais Y, Almant M, Kovensky J, Moreau V, Lesur D, Artigau M, Picard C, Galaup C, Gouin SG (2011) Carbohydr Res 346:26–34
122. Mindt TL, Struthers H, Brans L, Anguelov T, Schweinsberg C, Maes V, Tourwe D, Schibli R (2006) J Am Chem Soc 9:15096–15097
123. Van Swieten PF, Leeuwenburgh MA, Kessler BM, Overkleeft HS (2005) Org Biomol Chem 3:20–27
124. Kim DH, Choe YS, Lee I, Choi JY, Choi Y, Lee KH, Kim BT (2007) J Label Compd Radiopharm 50:S34–S40
125. Fischer CR, Muller C, Reber J, Müller A, Kramer SD, Ametamey SM, Schibli R (2012) Bioconjugate Chem 23:805–813
126. Maschauer S, Einsiedel J, Haubner R, Hocke C, Ocker M, Hübner H, Kuwert T, Gmeiner P, Prante O (2010) Angew Chem, Int Ed 49:976–979
127. Sirion U, Kim HJ, Lee JH, Seo JW, Lee BS, Lee SJ, Oh SJ, Chi DY (2007) Tetrahedron Lett 48:3953–3957
128. Elgland M, Nordeman P, Fyrner T, Antoni G, Peter K, Nilsson R, Konradsson P (2017) New J Chem 41:10231–10236
129. Rajkhowa S, Kale RR, Sarma J, Kumar A, Mohapatra PP, Tiwari VK (2021) Curr Org Chem 25:2542–2578
130. Jha AK, Jain N (2013) Tetrahedron Lett 54:4738–4741
131. Ferlin N, Gatard S, Nguyen A, Nhien V, Courty M, Bouquillon S (2013) Molecules 18:11512–11525
132. Sharma A, Agrahari AK, Rajkhowa S, Tiwari VK (2022) Eur J Med Chem 238:114454
133. Mishra A, Agrahari AK, Tiwari VK (2021) Application of CuAAC in carbohydrates. In: Click chemistry volume, science of synthesis, pp 135–179
134. Tiwari VK (2021) Chem Rec 21:3029–3048

Chapter 14
Click Chemistry in Nucleic Acids

14.1 Introduction

Connecting two molecules of different functional groups with diverse properties to create a new functional compound with combined attributes is increasingly becoming popular in recent years [1, 2]. Interestingly, a great number of novel hybrid molecules exhibit significant biological potential due to the collaborative action of the constituents contained therein [3]. Just after the discovery of 'Click Chemistry' in 2002 by Prof. Meldal and Prof. Sharpless [4, 5], this versatile and highly steady copper(I)-catalyzed azide-alkyne 1,3-dipolar cycloaddition (CuAAC) is identified as one of the most reliable approaches for advancing drug discovery process and ultimately benefiting the human kind [6–9]. The modular nature, high reaction yield, scalability, user-friendly, and biocompatibility of CuAAC click reactions attracted considerable interest in chemistry, biology, and medicine [10, 11]. The click-derived 1,2,3-triazoles are proving valuable scaffolds in pharmaceutical innovation as well as material science [12].

The Cu(I)-catalyzed click reaction yields an exclusive 1,4-triazole product [4, 5], while opposite regioselectivity was found once the azide-alkyne reaction was performed in the presence of ruthenium catalysis [13, 14]. The metal-free variant including organocatalysis or catalyst-free triazole forming click chemistry is further explored as one of the emerging tools of diverse application [15–17]. The regioselective 1,2,3-triazole forming click chemistry through [3 + 2] azide-alkyne cycloaddition reactions has been extensively explored in carbohydrate chemistry for diverse applications [18]. This modular conjugation chemistry is equally explored in nucleic acid chemistry [19–24]. that enabled investigators working across to generate a wide variety of biocompatible tools of diverse applications in chemistry, biochemistry, biology, biotechnology, and medicine. This chapter highlights the growing applications of modular click chemistry tools applied in nucleic acid chemistry including the modification of nucleoside architecture, with their notable uses in diagnostic tools,

V. K. Tiwari et al., *Click Chemistry*, Materials Horizons: From Nature to Nanomaterials,
https://doi.org/10.1007/978-981-97-4596-8_14

drug development, etc. The purpose of this chapter is to provide a handy overview of the scope of the modular click tool in nucleic acid chemistry.

14.2 CuAAC 'Click' for the Chemical Modification of Nucleosides

In addition to the notable applications of click chemistry in various conjugations with other molecular systems such as polymers, peptides, lipids, and carbohydrates, the regioselective 1,2,3-triazole forming click chemistry further has notable use in nucleoside, nucleotides, and nucleic acids. For example, this enables the diverse facile labeling of oligonucleotides with small-molecular probes [25] and found an attractive tool to join the oligonucleotide sequences (eg. single strands, double strands, or complementary strands) [26, 27] and circular nucleic acids through their cyclization [28–30]. In addition, the modular click route has been successfully used for the metabolic labeling of nucleic acids [31] and the resulting biocompatible nucleic acid scaffolds have found wide applications as PCR [32–35]. Furthermore, the method may be used to generate oligonucleotide-functionalized surfaces, DNA nanostructures, and nanosensors [36–40].

Prasad and colleagues have effectively utilized a modular CuAAC approach to achieve a significant yield of non-ionic xylo-nucleoside dimers **3**. These dimers are composed of TL-t-T xL, TL-t-ABzxL, and TL-t-CBzxL via triazole linker and were formed in good yield by coupling of azide **1** and alkyne **2** under standard CuAAC conditions as illustrated in Scheme 14.1 [41].

Brown *et. al.* developed a novel and highly efficient protocol based on a non-enzymatic DNA strand ligation technique in combination with polymerase chain reaction (PCR) amplification for the synthesis of DNA strands containing an unnatural T-triazole-T linkage. The versatile CuAAC Click chemistry has been employed for the DNA ligation between oligonucleotides with 3′-AZT **4** and 5′-propargylamido dT **5** and amplification was achieved efficiently by polymerase chain reaction (PCR) using various distinct polymerases to obtain the desired oligonucleotides containing an AZT-based T-triazole-T linkage **6**. Also, the DNA sequencing of PCR amplicons has been performed which displays the presence of a single thymidine at the ligation site (Scheme 14.2) [42].

The reason behind this inaccurate amplification was due to the inflexibility of the amide bond, which could potentially force the substituents to adopt the *trans*-form. This structural change might lead to the deviation of the thymine nucleobase, interfering with the proper incorporation of deoxyadenosine triphosphate (dATP) during the replication process (Fig. 14.1) [42].

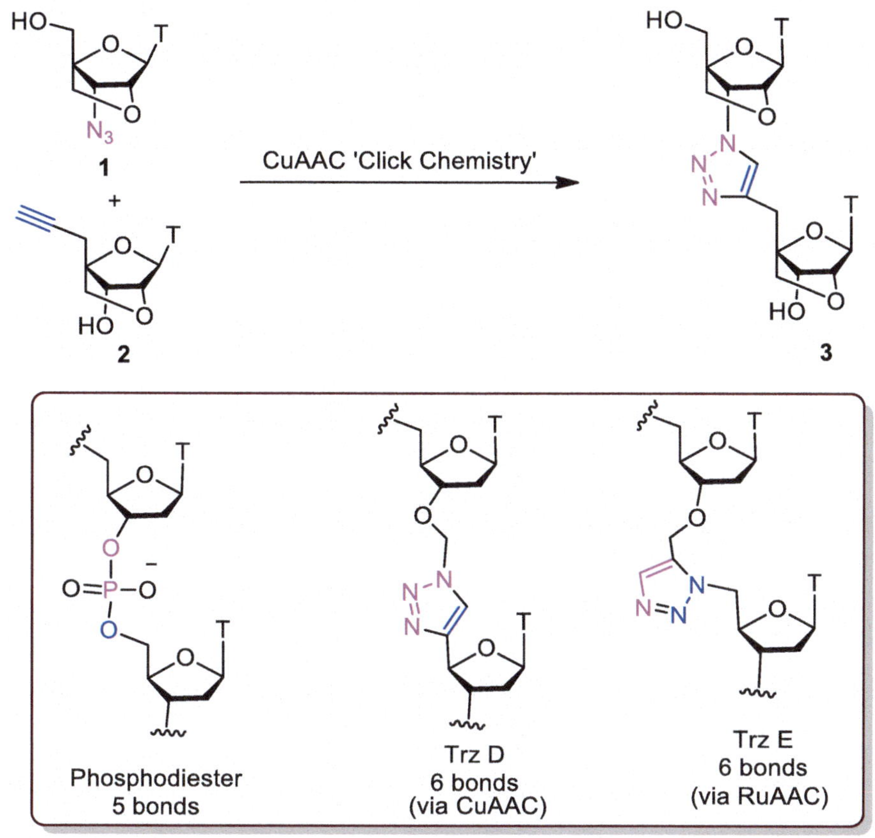

Scheme 14.1 CuAAC-triggered synthesis of triazole-linked-modified nucleoside derivative **3**

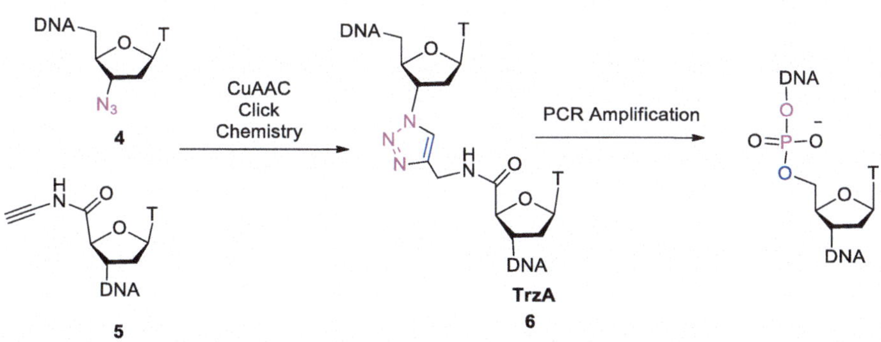

Scheme 14.2 CuAAC Click chemistry-based DNA ligation of alkyne and azide oligonucleotides combined with PCR amplification

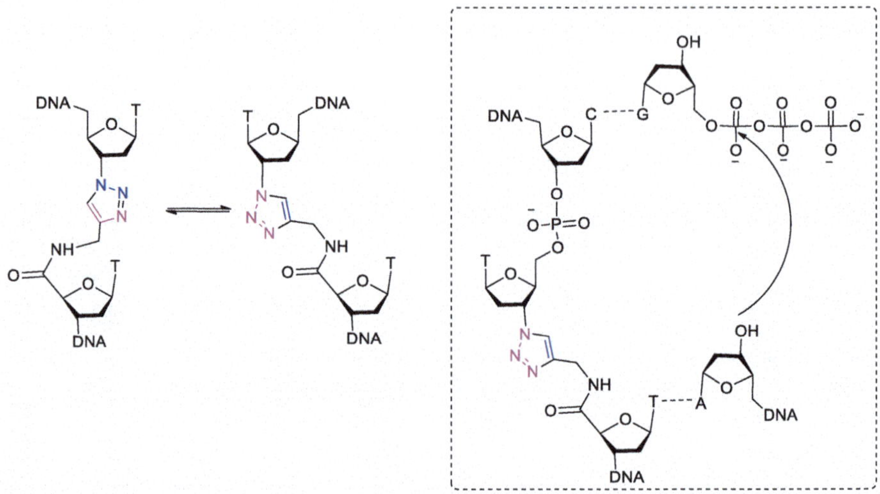

Fig. 14.1 Possible *cis–trans* isomerization at the amide bond leading to the skipping of one of the thymine bases during replication

14.3 Click-Mediated Biocompatible 1,2,3-Triazole as Alternative Phosphate Linkage in DNA

Isobe and co-workers reported a highly efficient and selective route for the design and synthesis of a new triazole-linked analogue of DNA; oligothymine ^{TL}DNA using the established copper-catalyzed Huisgen cycloaddition approach. This protocol for the first time employed for chain elongation, proceeded efficiently, and has proven useful in the field of oligomer synthesis. Also, the artificial oligonucleotide synthesized via this methodology formed a stable double strand with the complementary strand of natural DNA.

The solid-phase synthesis of ^{TL}DNA was carried out by loading the terminal monomer **123** on a poly(ethyleneglycol)-polystyrene-based resin. The first coupling reaction went smoothly to produce the 2-mer ^{TL}DNA **124** in quantitative yield. In the next step, a copper-mediated desilylation furnishes the unprotected acetylene derivative **125**. The repetition of the deprotection/elongation reactions under the optimized reaction conditions on solid support followed by HPLC purification affords the 10-mer ^{TL}DNA **126** (Scheme 14.3) [33].

Triazole-appended scaffolds are a significant class of biocompatible in drug delivery and development, are aforementioned. In this regard, Brown et al. reported the interesting synthesis of azido analogue **14** in which, the initial step involves the assembly of the 3′-alkyne **12** and **13** followed by its transformation into the 5′-azide **14**. This approach allows for the modification of oligonucleotides on solid support with either 5′-azide, 3′-alkyne, or both. A second-generation triazole linkage (Triazole B, Trz B) is devised and created through click ligation of 5′-azide and

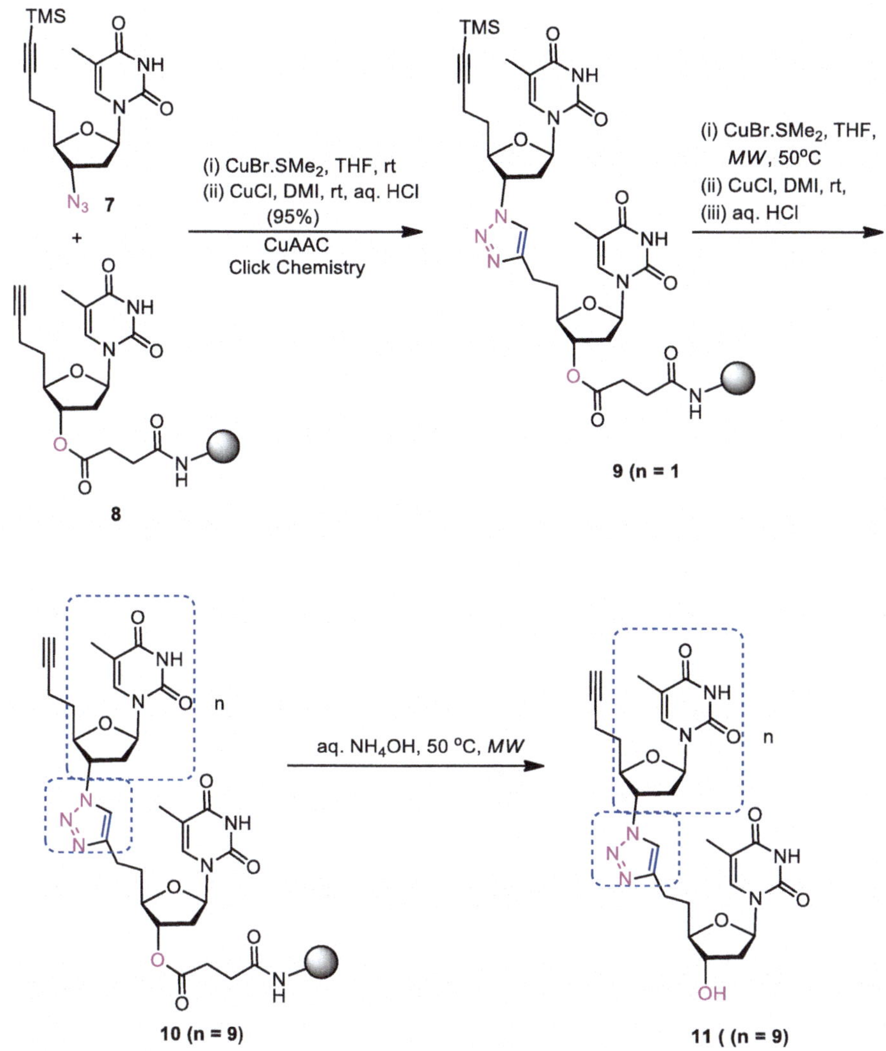

Scheme 14.3 Click chemistry-mediated solid-phase synthesis of 10-mer ^{TL}DNA

3′-propargyl-modified oligonucleotides to improve the resemblance to the phosphodiester backbone. This linker has a synthetic benefit as it was prepared from oligonucleotides which, in turn, were obtained solely by solid-phase phosphoramidite method. Notably, both linear 81- and 300-mer DNA strands, including one or two Trz B backbone modifications, were effectively amplified in vitro using thermostable DNA polymerases with or without proofreading activity (Scheme 14.4) [31].

Carell and group further explored the technique of solid-phase synthesis and utilized phosphoramidite derivatives of these two nucleosides for labeling numerous

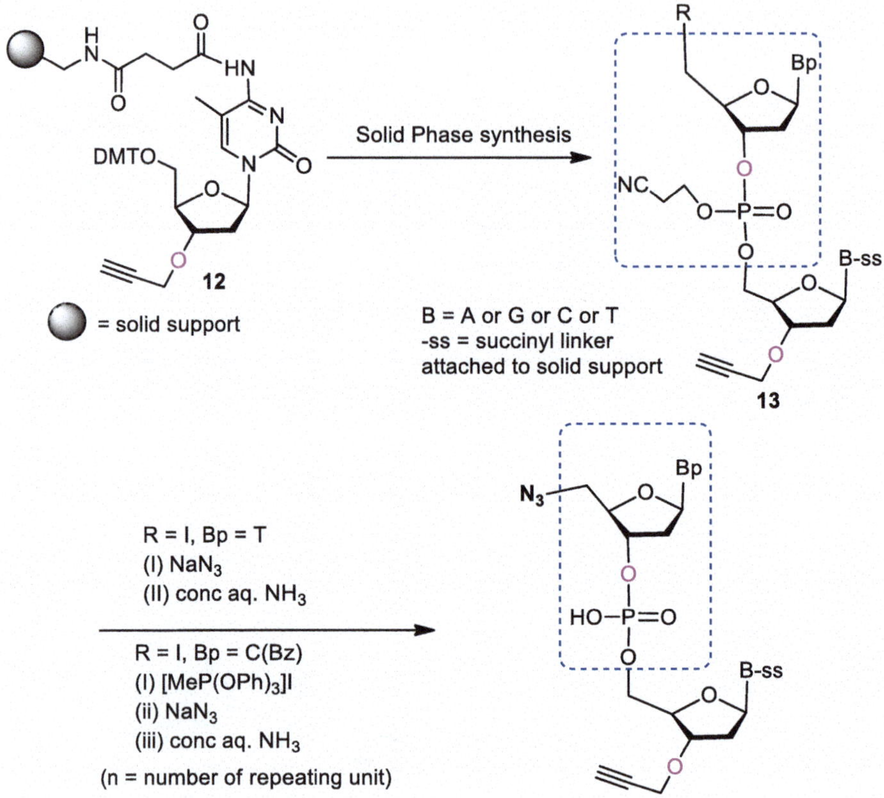

Scheme 14.4 Synthesis of azido analogue **14**

positions within 16-mer oligonucleotides (ONs) through solid-phase synthesis. The initial synthetic process involved preparing monomers through Sonogashira coupling of the alkyne group to the C5 position of a deoxyuridine, followed by the reaction of the 5′-dimethoxytrityl-nucleoside with 2-cyanoethyl *N,N,N′,N′*-tetraisopropylphosphorodiamidite. Also, the group used diverse azide-modified reporter groups to evaluate the effectiveness of the click labeling reaction (Fig. 14.2) [24].

Peng et al. presented the copper-mediated multicomponent synthesis of DNA-bounded 1,2,3-triazole **19**. The synthesis was achieved in one pot by reacting aryl borates with TMS-N_3, subsequently followed by regioselective click [3 + 2]-cycloaddition reaction in assistance with the binuclear macrocyclic nanocatalyst Cu(II)-β-cyclodextrin (Scheme 14.5). Additionally, a DNA-compatible macrocyclization reaction was performed in one pot under an optimized reaction condition. Based on qPCR analysis and Sanger sequencing data, the reported optimized reaction procedure causes no discernible damage to DNA [43].

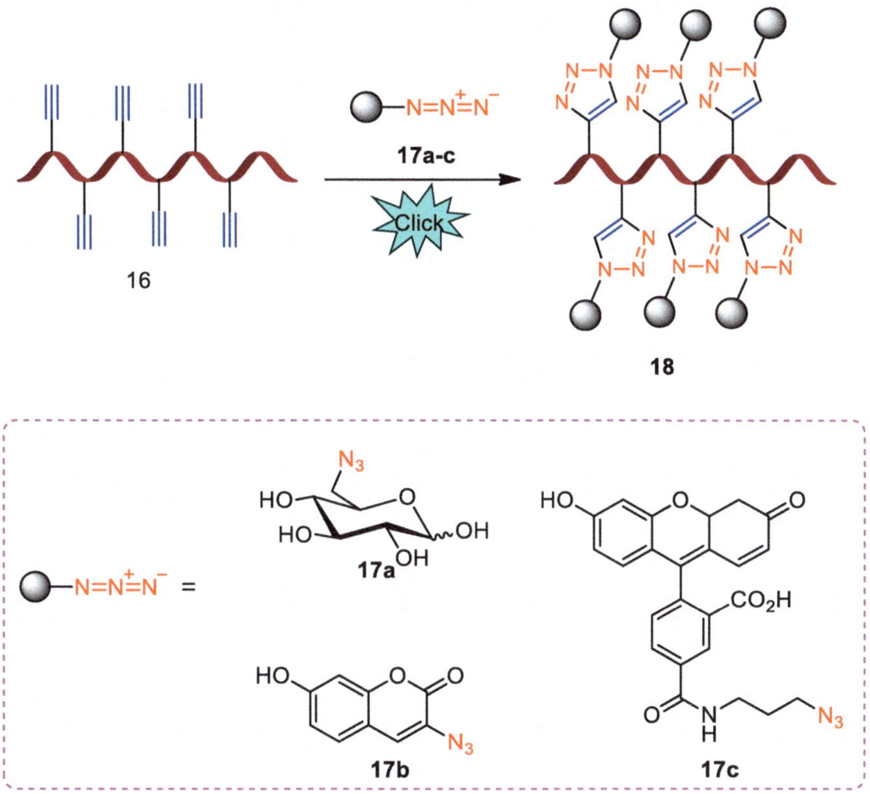

Fig. 14.2 Click-labeling approach for oligonucleotides

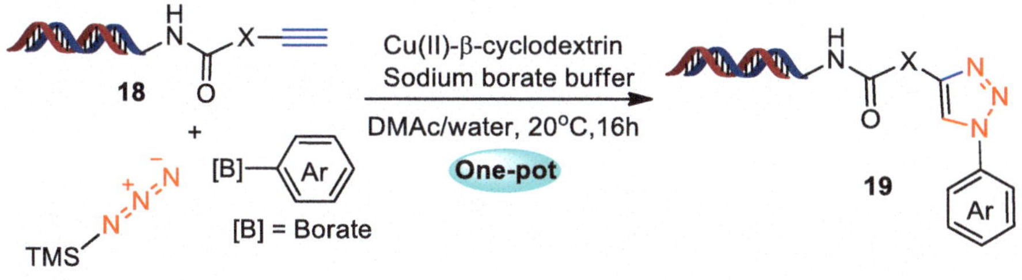

Scheme 14.5 Peng approach for the copper-mediated multicomponent synthesis of DNA-bounded 1,2,3-triazole **19**

The CuAAC protocol is although of limited utility in tagging oligonucleotides with multiple small molecules due to the challenge of controlling the introduction of distinct labels at specific alkyne-modified positions through conventional methods. To address this limitation and precisely guide the incorporation of fluorophores at predetermined sites, a DNA sequence was modified with an unprotected alkyne and

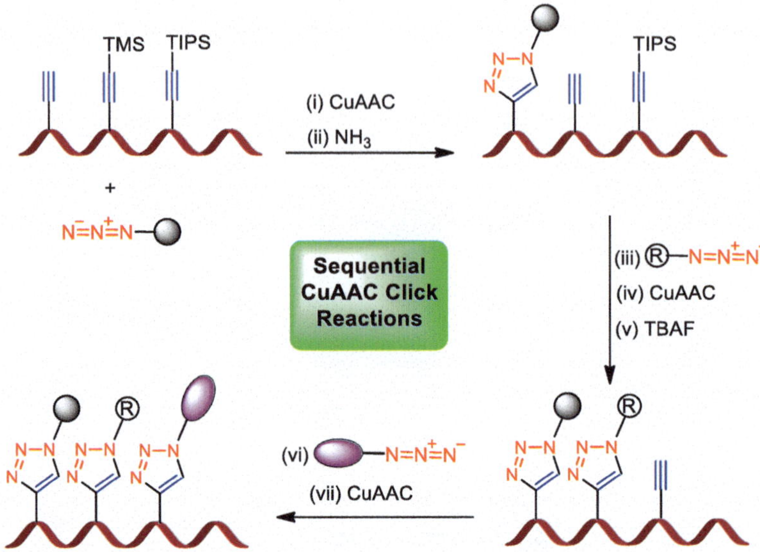

Scheme 14.6 Sequential CuAAC strategy for the labeling of oligonucleotides

two orthogonally protected octadiynyl groups (Scheme 14.6). Standard solid-phase synthesis facilitated the introduction of alkyne modifications into oligonucleotides, allowing three distinct fluorescent dyes to be efficiently integrated either directly on the resin or in solution post-oligonucleotide deprotection.

Initially, the free alkyne was coupled to a stable azide-modified fluorophore directly on the resin. Subsequent treatment with ammonia led to the removal of the trimethylsilyl (TMS) protecting group from the second alkyne and enabled the addition of a second fluorophore at specific positions in the modified strand. Throughout this protocol, the protection of the third alkyne group with triisopropylsilyl (TIPS) remained intact, making it available for deprotection using tetrabutylammonium fluoride, followed by labeling with a third fluorophore. A similar strategy has recently been employed to introduce multiple distinct fluorophores at predetermined locations along an oligonucleotide strand [44].

In similar fashion, further investigations pertaining to linking of modified nucleosides with other systems including coumarin and carbohydrates via CuAAC were carried out [45]. Triazole-appended coumarin scaffolds are supposed to enhance the biological and photophysical properties because of the inherent pharmacological activities and photophysical properties possessed by coumarin scaffolds. Considering this, Prasad and group developed a convenient and high-yielding synthesis of a library of N_1-(2′-*O*,4′-C-methylene-β-D-ribofuranonucleoside-3′-yl)-C4-(coumarin-7-oxymethyl)-1,2,3-triazoles **22** under CuAAC conditions. The coupling reaction of 3′-azido-3′-deoxy-2′-C-methyleneuridine derivative **20a** and **20b** as azide precursor with 7-propargyloxy coumarins **21a-d** as the alkyne coupling partner was performed in the presence of Na_2SO_4/NaAsc (Scheme 14.7) [46].

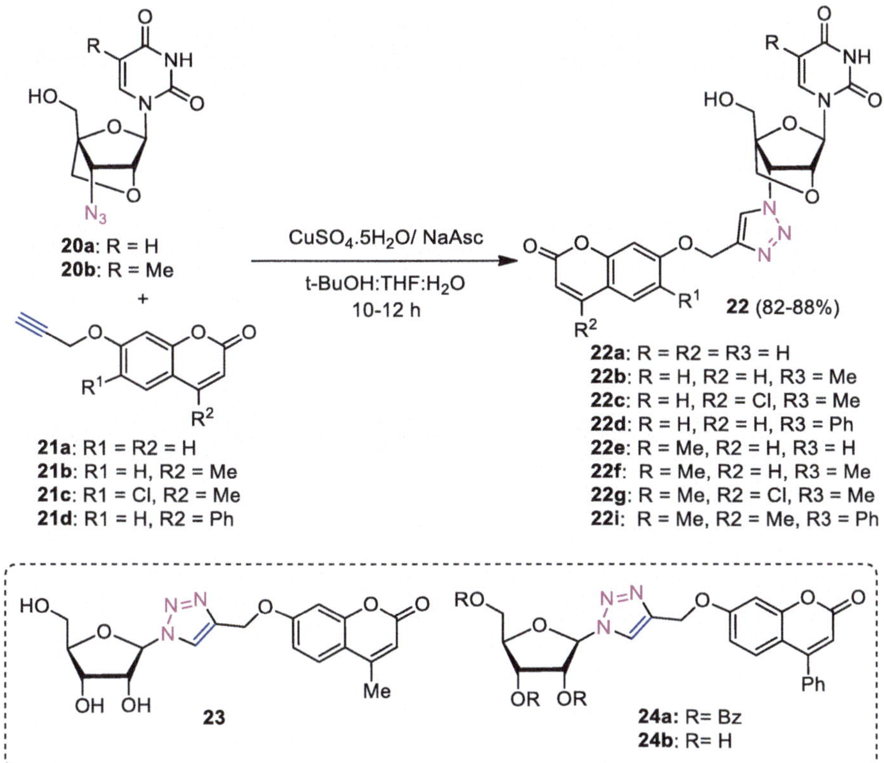

Scheme 14.7 CuAAC in linking of modified nucleosides using coumarin and carbohydrates

The resulting triazole products exhibited a Stokes shift of over 70 nm in their photophysical investigation.

The β-D-ribofuranosylcoumarinyl triazolyl glycohybrids hybrids were tested against *M. tuberculosis* and a multidrug-resistant clinical isolate strain, where compounds **23** and **24a,b** [47] displayed significant sensitivity against both the reference strain H37Rv and the multidrug-resistant clinical isolate, surpassing the activity of standard first-line anti-tubercular drugs by a significant margin. Furthermore, a thorough investigation indicated that these molecules have the potential to inhibit bacterial InhA and DNA gyrase B enzymes, thus disrupting the cell wall and thereby demonstrating anti-mycobacterial properties (Scheme 14.7).

The same group further extended their work of coumarin-triazolyl nucleoside and came up with a synthetic methodology for efficient synthesis of N1-(*β*-D-ribofuranosyl)-C4-(coumarin-4''-yl)-1,2,3-triazoles (**27a-h**). The group utilized 1-azido-1-deoxy-2,3,5-tri-*O*-benzoyl-β-D-ribofuranose **25** and 4-ethynylcoumarins (**26a-h**) as starting materials under standard click catalysis, i.e., $CuSO_4.5H_2O$ (0.4 equiv.)/NaAsc (0.9 equiv.) in *t*-butanol/THF/water (1:1:1 ratio) at room temperature (Scheme 14.8) [47]. Further, the obtained products were treated with sodium

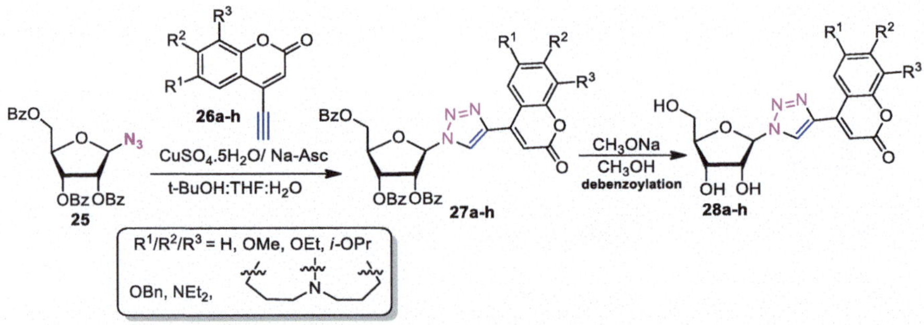

Scheme 14.8 'CuAAC Click Chemistry'-induced synthesis of coumarin-triazolyl glycosides **28a-h**

methoxide in methanol at room temperature to undergo debenzoylation to deliver a good yield of the desired coumarinlinked glycoside products (**28a-h**).

A facile synthetic approach for functionalizing the 2-position of uridine has been discovered, in which starting with 2'-azido-2'-deoxyuridine **29**, a Cu-catalyzed cycloaddition with alkynes **30** was utilized to attach lipid groups, to the uridine moiety via a 1,2,3-triazole unit as examples of recognition activities. These lipidated 2'-triazolyl-2'-deoxyuridines **31a-c** compounds are attractive prospects for the construction of functionalized oligonucleotides with biochemical and medicinal applications (Scheme 14.9) [47].

Astakhova and group pioneered the development of oligonucleotides inspired by proteins. They created these oligonucleotides **33** from a propargylated oligonucleotide **32**. This was achieved using click protocol in the presence of Cox 17. Cu(I), a human copper-binding chaperone (Scheme 14.10) [48]. The synthesis of triazole products **33**, was accomplished in a completely selective manner and with remarkable simplicity. Notable advancements in the reaction process include improved biocompatibility, the utilization of extremely gentle labeling conditions with aerated solutions, and the need for low copper concentrations to facilitate successful coupling. Additionally, the resulting oligonucleotides attached to triazole groups displayed a strong affinity for complementary nucleic acid targets, highlighting their utility as diagnostic tools and therapeutic probes.

Ermolat and group used CuAAC tool to establish a fascinating nucleoside mimic candidate of furo[2,3-b]pyrazine nucleoside derivative **36**. Ribofuranosyl azide **34** and pyranzide species **35** were employed as precursors in the presence of Cu(I)-catalyst to undergo Click reaction to produce the desired 1,2,3-triazolyl nucleoside product in excellent yield (Scheme 14.11) [49]. Several unnatural nucleoside analogues have been synthesized using a similar modular approach.

There has been a growing emphasis on innovation in the synthetic approach to antisense oligonucleotides. This involves chemically modifying nucleosides to regulate or inhibit the expression of specific genes, and it has become a prominent focus in the development of advanced therapeutics and diagnostic tools [50]. In the field of drug development, particular attention is given to functionalized-oligonucleotide

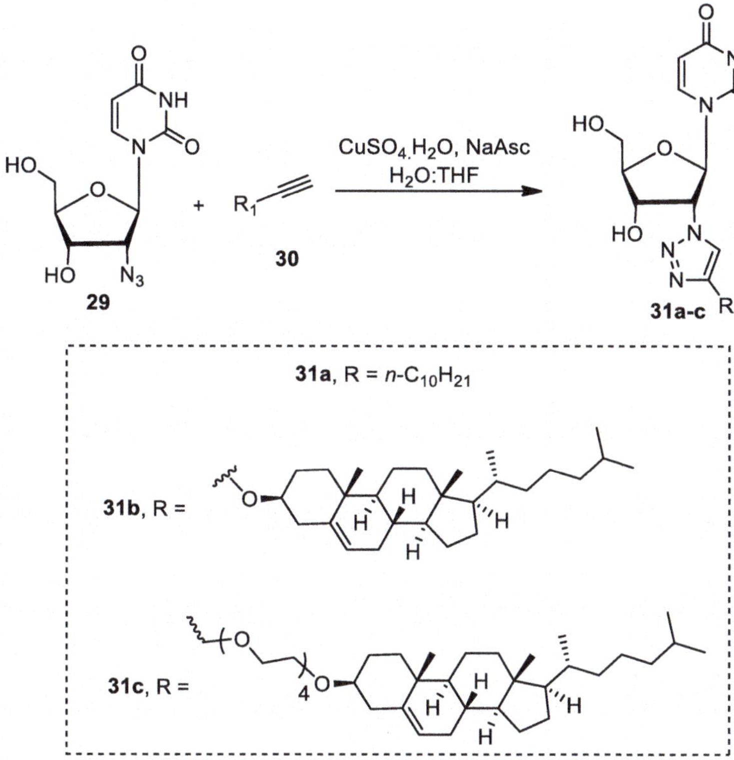

Scheme 14.9 Click-mediated synthesis of 2'-Triazolyl-2'-deoxyuridines **31a-c**

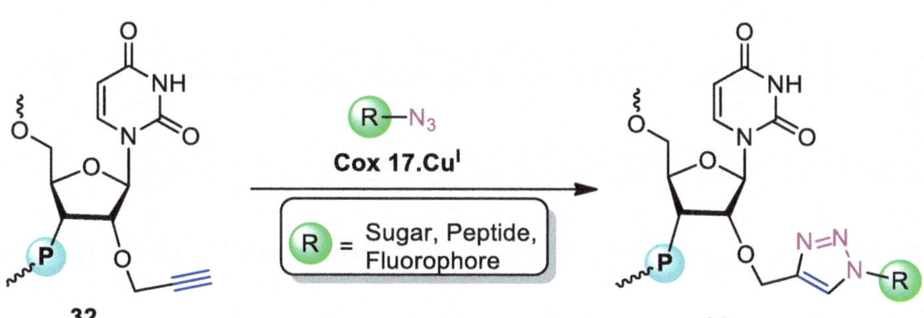

Scheme 14.10 Bioclick route for the functional oligonucleotides in the presence of human Cu-binding chaperone [48]

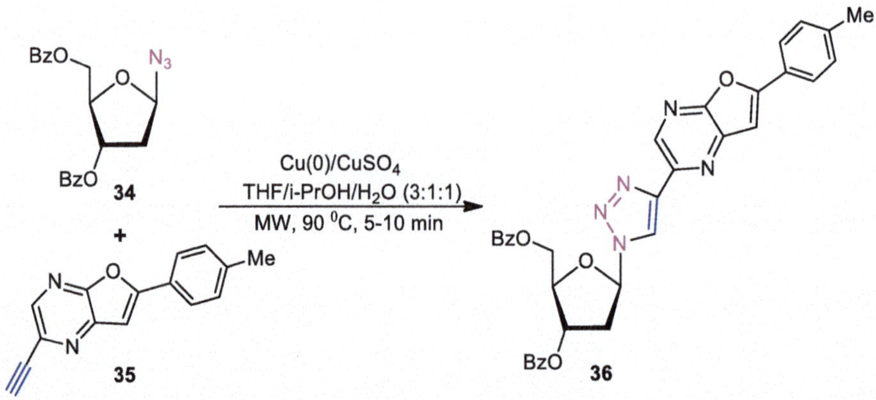

Scheme 14.11 'Click Chemistry' CuAAC-mediated synthesis of 1,2,3-triazolyl nucleoside analogue **36**

derivatives with chemically customized phosphate backbones, as these offer potential due to their ability to overcome the typical limitations of conventional phosphodiester linkages, potentially providing solutions for a variety of diseases [51]. Additionally, in the current context, there is a well-structured exploration of triazolyl scaffolds inspired by click chemistry for a broad spectrum of applications [52, 53].

Lucas group used regioselective click tool to develop a high yield of 1,4-disubstituted triazolyl-3′-5′ thymidine dimer as well as a pentamer [54]. Isobe et al. extended this modular tool to produce a 1,4-disubstituted triazole-linked 10-mer analogue of thymine DNA, which exhibited a significantly higher melting temperature compared to the standard DNA sequence [55]. Moreover, biocompatible triazole linkage in both DNA and RNA can be accurately interpreted by polymerases [31].

Nucleic acids conjugates play a substantial and diverse role, particularly in the context of drug development, diagnostic tools, and drug delivery systems. This biocompatible, efficient, and cost-effective tool is widely used in nucleic acids for the various purposes including labeling, ligation, modification as well cycllization of oligonucleotides and moreover DNA-encoded chemistry [19, 56]. Notably, Micura and colleagues successfully developed RNA conjugates tagged with a fluorescent marker [57]. This achievement centered around the utilization of an azido group in place of the traditional alkyne group for inverse click labeling, offering the necessary fluorescent tag. Employing a strategic click approach, they effectively linked 2′-*O*-(2-azidoethyl) RNA **37** with the fluorescent rhodamine alkyne derivative F545 **38** in the presence of $CuSO_4$/NaAsc in an acetonitrile/water solution, resulting in a high yield of dye-labeled RNA **39** (Scheme 14.12) [58]. This innovative method has significantly enhanced the importance and broadened the scope of Staudinger ligation in the field of chemical biology.

Within the context of siRNA modification methods, the same research team adopted a strategic approach, resulting in the creation of three RNA duplexes carrying a fluorescent label known as F545. The significance of this study has been expanded

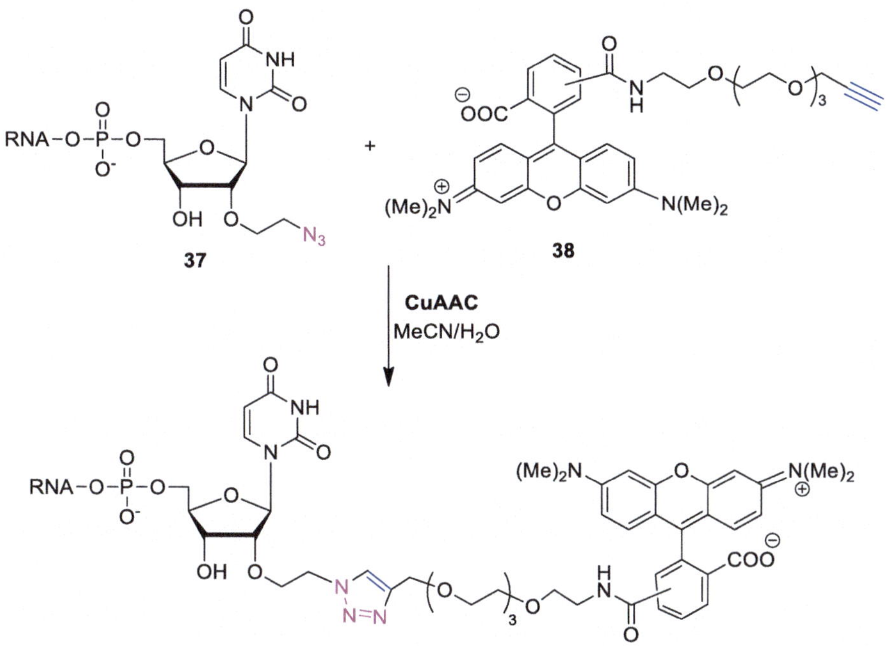

Scheme 14.12 CuAAC click-mediated synthesis of 1,2,3-triazole-tethered RNA conjugate **39**

through the growth of opportunities for bioconjugation using the efficient 'Click Chemistry' tool [59].

Similarly, Pujari and colleagues devised a modular method for linking DNA strands using the regioselective click technique. They connected two complementary oligonucleotide strands, one composed of 2′-O-propargylated 2-aminoadenosine **40** and the other of propargylated 5-methyluridine nucleoside **41**, with a diazido compound in the presence of copper wires in an acetonitrile/water solvent mixture. This process yielded triazole-appended cross-linked DNA strands **42**, achieving appreciable yield as depicted in (Scheme 14.13) [60]. The reaction was conducted in a stepwise fashion to prevent potential side reactions, such as the formation of homo-DNA duplexes during the cycloaddition. To obtain a selectively mono-functionalized azido derivative from alkyne **40**, an excess amount of bis-azide (in a 1:15 alkyne/azide ratio) was employed in an H_2O/CH_3CN solvent system. Next, the second click step was executed to produce the desired cross-linked strands **42**, involving the cycloaddition of the intermediate azide and propargylated 5-methyl uridine nucleoside **41**. These resultant cross-linked DNA conjugates hold promise for various applications, particularly in the field of anti-cancer therapy.

Cobb and group extensively explored the applicability of azido alcohols in modern organic synthesis and employed them as precursors for efficient synthesis of biologically important spirocyclic triazolooxazine nucleosides. The reaction was carried out

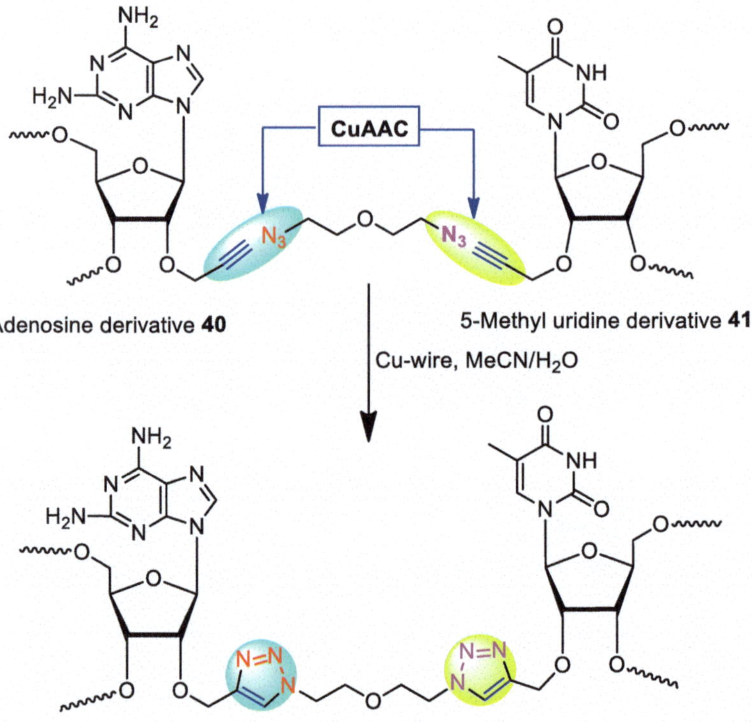

Scheme 14.13 Construction of triazole bridged cross-linked DNA strands **42** using CuAAC methodology

with glycosyl azido alcohol **43** which was first propargylated using BEMP in acetonitrile at 0 °C for 2 h and an azido-alkyne intermediate **44** thus obtained was then underwent intramolecular 1,3-dipolar azide-alkyne cycloaddition reaction to afford good yield of biologically relevant spirocyclic glycohybrid analogue **45** (Scheme 14.14) [61].

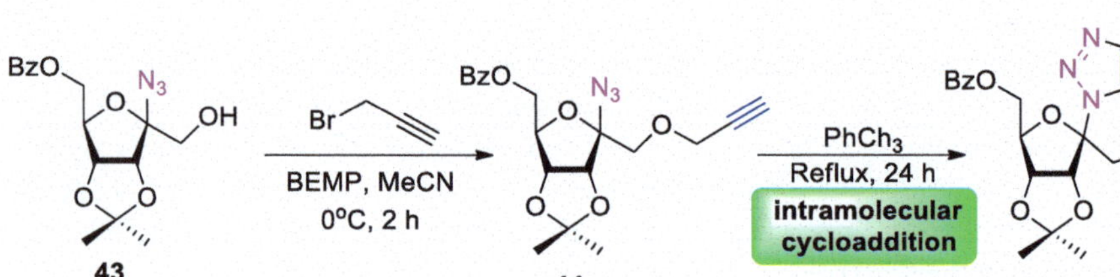

Scheme 14.14 Synthesis of spirocyclic glycohybridproduct **45** from azido alcohol precursor **43** in intramolecular fashion [61]

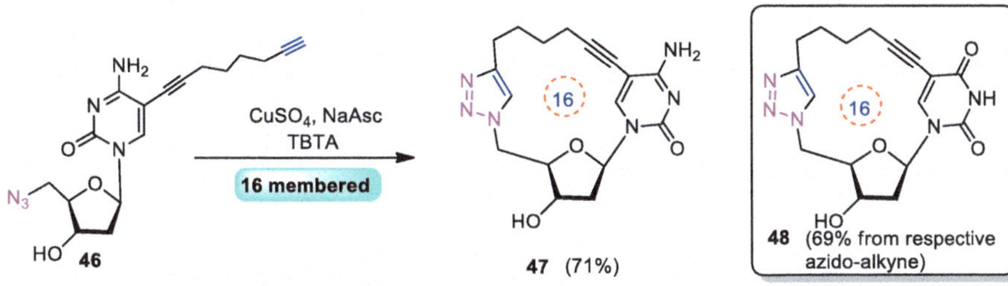

Scheme 14.15 Synthesis of 16-membered nucleoside macrocycles **47** and **48** through intramolecular CuAAC reaction

Likewise, Luzzio and co-workers exemplified carbocontiguous azidoalkynyl oxazole precursors which in presence of $CuSO_4$/NaAsc underwent intramolecular triazole forming click to give high-to-excellent yield of respective ring-closure product, for example, endo-fused-1,4-disubstituted-1,2,3-triazole-appended macrocyclic system [62]. Seela group used similar intramolecular variant of CuAAC-induced click to give good yield of biologically relevant triazole-linked nucleoside macrocycles.[63] The authors treated azido alkyne **46** $CuSO_4$/NaAsc and TBTA, where an azide present at the 5'-position of deoxyribose sugar and an octadiynyl group at the C-5 position of nucleobase underwent click-mediated macrocyclization to give desired macrocycles **47** in 71% yield (Scheme 14.15) [63]. Interesting to note that the resulting macrocycle displayed Watson–Crick recognition sites that were liberally accessible with great ease.

Click chemistry has been employed in the creation of compact cyclic DNA frameworks, designed for use in drug-DNA binding studies and potential therapeutic applications. Utilizing the CuAAC reaction, the 5′-alkyne and the 3′-azide group of hairpin oligonucleotides possessing a hexaethylene glycol loop in the hairpin were conjugated to construct an artificial mini-duplex structure (Fig. 14.3) [64].

Yamada and colleagues employed a *MW*-induced click method to devise oligoribonucleotide glycoconjugates, which have been applied in high-throughput screening. The synthetic procedure involved the click reaction of monomers **50** and **51** to generate a triazole-appended oligonucleotide intermediate. Subsequently, a solution-phase 'click coupling' was performed, followed by a similar conjugation onto an immobilized oligonucleotide motif having a propargyl group, resulting in the triazole-tethered oligonucleotide **52** (Scheme 14.16) [65]. These siRNA conjugates effectively suppressed the expression of the luciferase gene, responsible for bioluminescence in the transformed HeLa cell line. Consequently, this click-mediated approach for producing oligoribonucleotide glycoconjugates using orthogonally protected solid-supported alkyne-oligonucleotides has proven to be a highly efficient method for site-specific high-throughput oligonucleotide conjugate synthesis, with potential applications in the targeted molecular arrangements [65].

In similar fashion, a post-synthesis tool under microwave irradiation (*MW*) condition was utilized to connect a porphyrin molecule possessing an azido functional

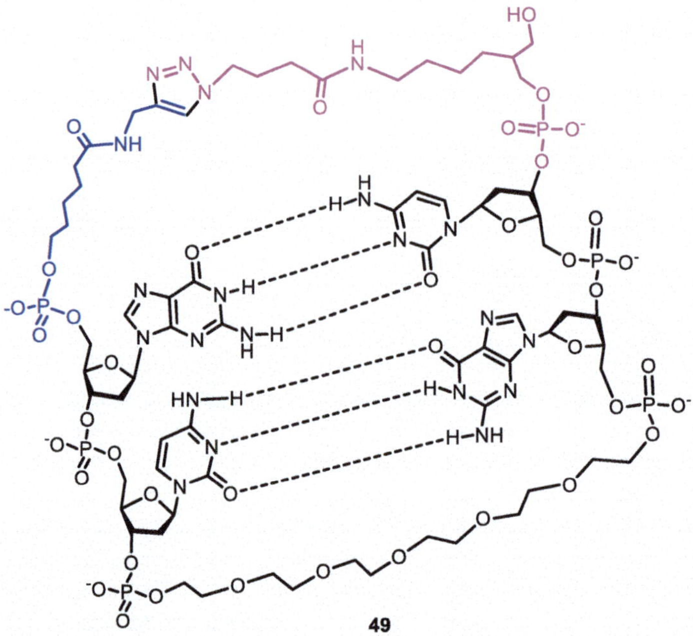

49

Fig. 14.3 Structure of cyclic GC/GC-dinucleotide mini-duplex **49**

group **53** to an oligonucleotide attached to 2′-*O*-propargylated uridine or adenosine **54**. The resulting triazole-linked derivative **55**, achieved in a high yield, can be described as an artificial DNA-chromophore [66]. The use of a click chemistry approach has been harnessed for assembling the components of 2′-deoxyuridine derivatives, ultimately yielding porphyrin-based oligonucleotides at room temperature. This approach further led to the creation of a targeted dendrimer scaffold [67]. Additionally, the porphyrin-embedded oligonucleotides were configured as four oligonucleotides, each featuring a porphyrin core **53**, using a CuI/DIPEA catalytic system at room temperature (Scheme 14.17) [68]. Therefore, the click chemistry strategy demonstrates substantial potential for the development of molecular architectures based on artificial DNA-chromophores with diverse applications.

An intriguing development by Neef and colleagues involved the creation of a glyco-complex labeled with a bioorthogonal chemical. They employed a modular approach based on the CuAAC click reaction. The resulting dF-EdU-based fluorophore hybrids could detect Herpes Simplex Virus-1 infected cells. This detection was likely due to the phosphorylation of dF-EdU by a viral enzyme. CuAAC's contribution to this field was, without a doubt, remarkable [69]. A similar click approach was used for the synthesis of an interlocked catenane ring system [70]. The intricate catenane architecture consists of twenty-four single-stranded DNA and possesses the advantage of maintaining structural stability even at high temperatures, reaching up to 95 °C [70]. This topologically stable structure exhibits nanoscale precision in pure water.

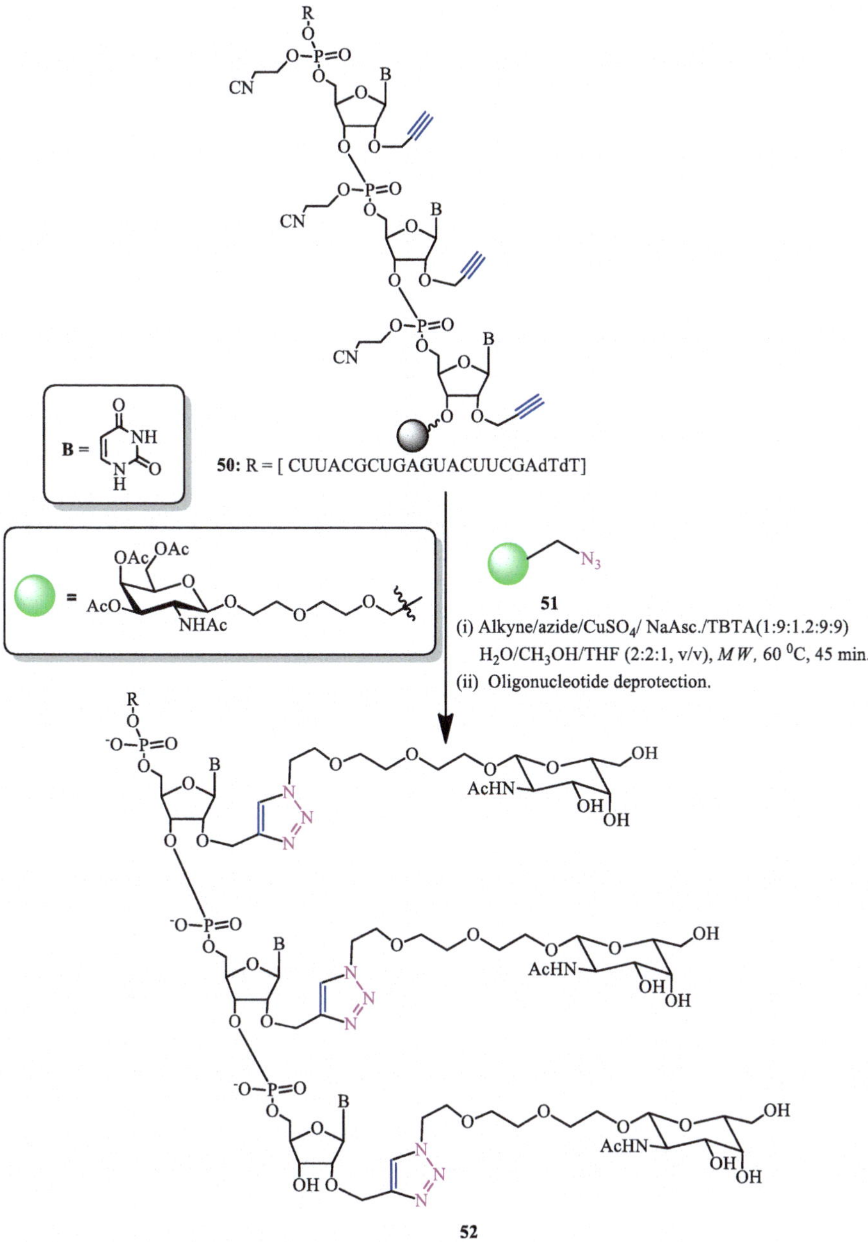

Scheme 14.16 CuAAC Click chemistry-mediated development of 1,2,3-triazole-linked oligonucleotide **52**

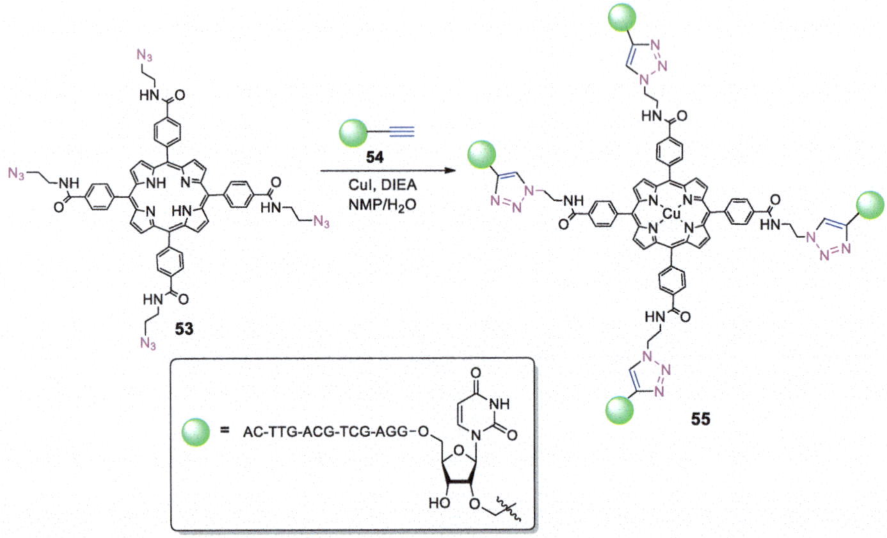

Scheme 14.17 Synthesis of triazole-tethered porphyrin–oligonucleotide hybrid **55** induced by CuAAC click [68]

Morvan and colleagues took into account the benefits of both combinatorial and click-combined-phosphoramide chemistry techniques and embarked on the creation of a set of oligonucleotides tagged with carbohydrates (**61a-c**). This was achieved by starting with the propargylated compound **56** on a solid support and employing a pivotal CuAAC coupling reaction between compounds **58a-c** and galactose-derived azides **59a, b** conducted under standard coupling conditions as depicted in Scheme 14.18 [71]. These multivalent glycomolecules, combined with saccharide-labeled oligonucleotides, hold promise for diverse applications in the realm of both drug development research and drug delivery systems.

Morvan and group undertook a noteworthy investigation using a phosphoramidite strategy, employing repetitive click chemistry, to create an oligonucleotide-based dendrimer **70** equipped with sixteen peripheral galactose molecules (Scheme 14.19) [72]. This method involved a series of synthetic steps, including CuAAC-mediated coupling of 1-*O*-propargyl-α-D-mannopyranoside to the polymer-bound azide **62** and yielding a solid-supported mannosylated scaffold **63**. Subsequent reaction of scaffold **63** with pentyn-4-yl of 6-bromohexyl phosphoramidites **64** produced a desired oligonucleotide intermediate **65** in excellent yield. The second click coupling was employed to deliver the respective glycocluster-labeled oligonucleotide **67** which was used for 3'-end labeling of the oligonucleotide **70** by CuAAC coupling with a galactose-derived sugar azide **69** under standard Click conditions [72]. The resulting DNA-based microarray having multivalent galactosylated clusters shows promise for various topological applications. The bioassays demonstrated that glycoclusters with a mannose core efficiently recognized LecA bio-macromolecules, making them a potential candidate for biolabeling purposes [73].

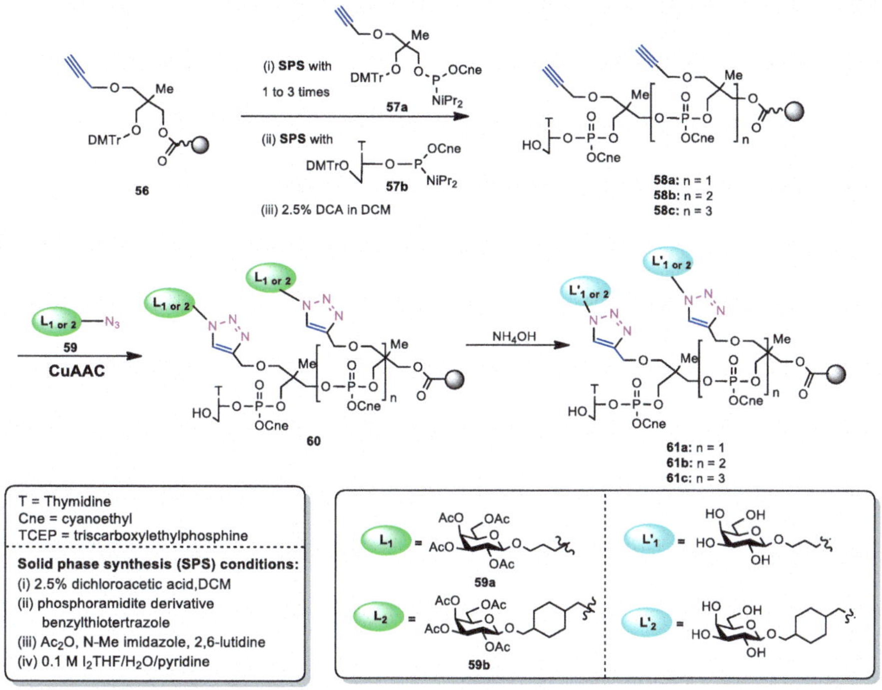

Scheme 14.18 Construction of multiple labeled oligonucleotides through solid-supported click conjugation technique

Moni and group successfully employed modular CuAAC tool to craft a range of intriguing 1,2,3-triazole-appended C-glycocluster nucleotide analogues such as the synthesized hybrid molecules **73** and **74** (Scheme 14.20) [74]. These triazole-appended C-glycoclusters **73** and **74** were straightforward obtained in high yield from the corresponding C-glycosides **71** and **72**. These glycosides featured at lower rim-functionalized azido-tether calix[4]arene systems and were subjected to CuAAC conjugation with corresponding alkyne partner.

The research team further performed a DNA-based microarray technique to assess the binding affinity of the resulting multivalent sugar-nucleotide framework **73** and **74** against PA-IL and RCA 120 [75]. Unfortunately, none of the developed hybrid oligonucleotides exhibit any significant binding affinity for PA-IL. However, a strong binding affinity for RCA 120 is demonstrated. Furthermore, in the context of immobilized hybrid oligonucleotides, the IC_{50} value determined the minimum lactose concentration to inhibit half of the RCA 120 was comparable with the monovalent and trivalent galactosides and compounds **73** and **74** exhibited a less pronounced cluster effect. Consequently, the study concludes that the specific spatial arrangement plays a remarkable role in demonstrating such multivalency.

Several effective strategies have been successfully employed over a period to create more potent oligonucleotides. Associated key challenges revolve around

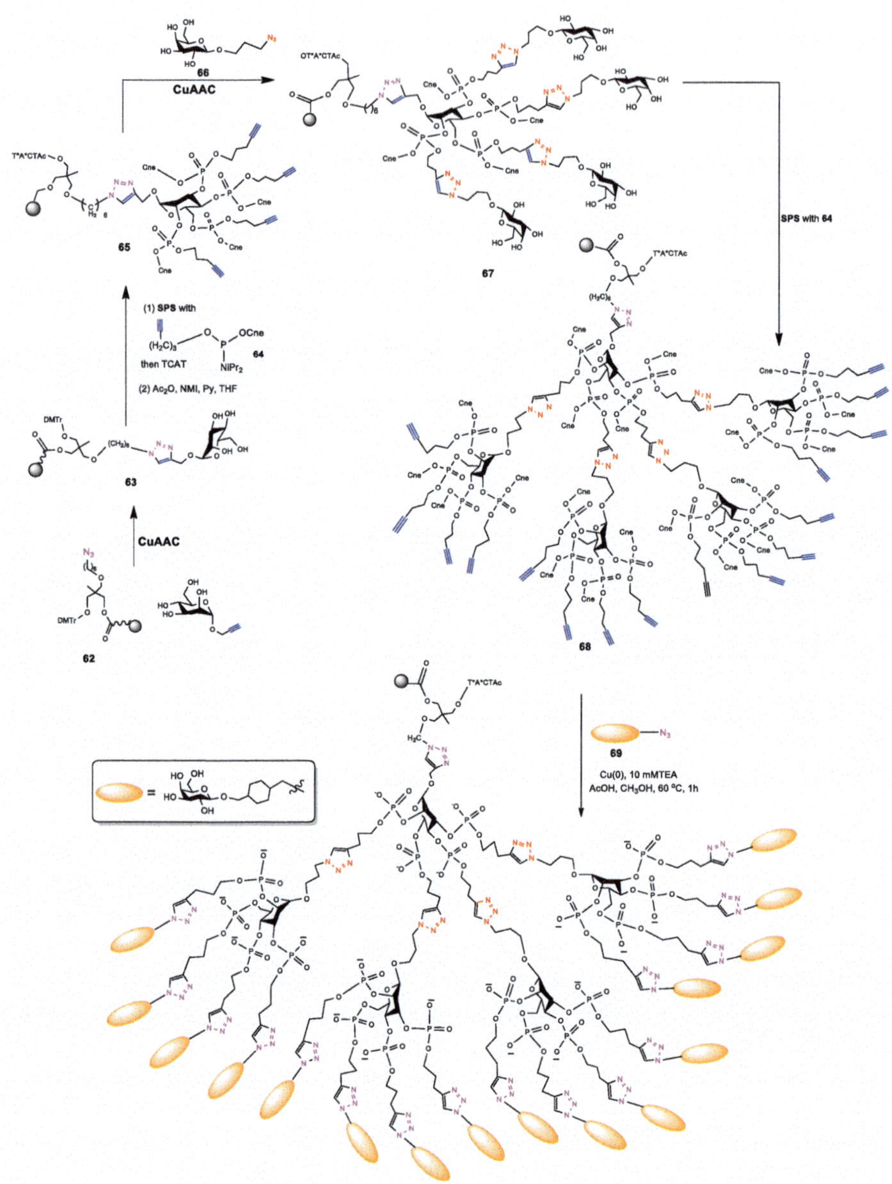

Scheme 14.19 CuAAC click-inspired construction of mannose centered triazole-linked oligonucleotide glycodendrimer macrostucture

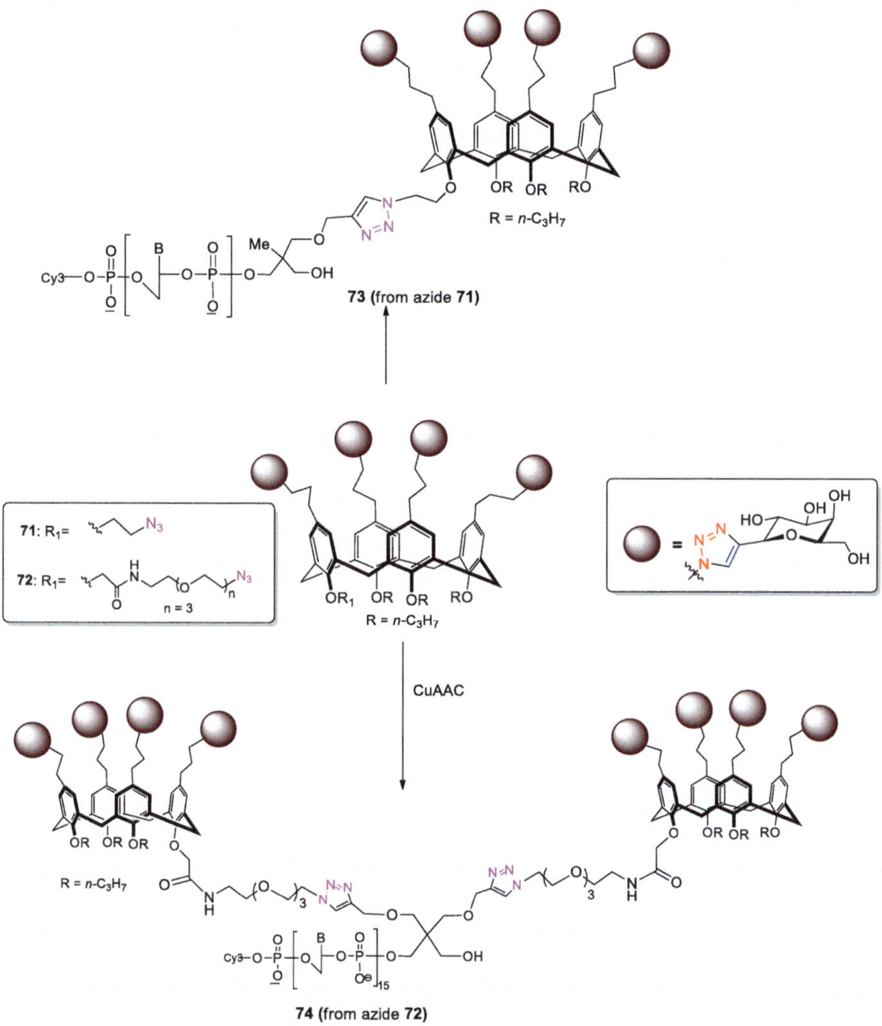

Scheme 14.20 Calix[4]arene-based multivalent glycocluster containing functionalized-oligonucleotide hybrids **73** and **74** via CuAAC tool [75]

stabilization of the polyanionic system and addressing vulnerability to nuclease-induced cleavage. To overcome the associated limitations and enhance their biomedical applications, innovative functionalities are being incorporated into nucleic acid structures.

Moreover, there is a growing interest to develop an expedient technique for glycoconjugation enabling a diverse range of homogeneous sugar-linked biomolecules such as glycopolymers, glycoproteins, and nucleic acids. This interest arises from the necessity to consider the precise role of individual glycan in biological systems [76]. Throughout this journey, click chemistry has repeatedly demonstrated its great utility and continues to offer available potential with the right efforts. Also, in nucleic acid therapy, hybrid molecules that interact with DNA to target specific genes are of utmost importance. Considering this, Fantoni and group further introduced a Click-mediated DNA-cleavage system that utilizes an artificial metallo-nuclease for the oxidative DNA cleavage [77].

In summary, CuAAC-mediated 1,2,3-triazole-linked hybrid glycoconjugates have served as a significant source of stimulation for creating novel bioconjugates, biorelevant supermolecular molecular structures, mutating or gene-silencing agents, multivalent architectures, and have unlocked numerous possibilities in various developing fields.

14.4 Click Chemistry-Inspired Glycan Microarrays in DNA Technology

The structural diversity of complex glycans poses a major challenge and by utilizing the glycan microarrays tool, investigation in glycan-binding proteins (GBPs) at a high-throughput level can be greatly supportive to understanding the role of glycoconjugates in disease progression [78, 79]. In general, the glycan microarrays can be employed either through non-covalent immobilization or covalent immobilization method and involve the immobilization of various glycans onto a high-density solid surface in an organized manner that allows for the rapid assessment of multiple binding interactions [80–84]. Among the various covalent immobilization methods required for the functionalization of the surface, for example use of some expeditios reactions like ring opening of epoxide-functionalized surfaces with amine or free glycans with amino-oxy or hydrazide surfaces [85–88], or thioether formation, amide bond formation, etc., regioselective CuAAC of azide with alkyne-functionalized surfaces like multivalent attachment of glycosyl azides onto alkyne-grafted solid surfaces receiving considerable importance for the successful immobilization of glycan microarrays. Wong group demonstrated a highly effective click approach for in situ immobilizing sugars on a solid polystyrene microtiter plate surface [89]. Further, Ravoo et al. used CuAAC to establish a microcontact printing technique to create glycan microarrays [90]. Norberg photoclick tool is used for effective covalent immobilization for a specific polymeric surface [91].

Toward click-inspired glycan microarrays in DNA technology, Gerland et al. investigated the properties of the fucose-binding protein PA-IIL. Pseudomonas aeruginosa (PA) is recognized for causing chronic respiratory infections. The fucose-binding protein PA-IIL plays a crucial role in adhering to host cells and facilitating biofilm formation, making it a potential therapeutic target. In pursuit of this objective, Gerland group employed a combined CuAAC and DNA-solid-phase approach and created a library of sixteen fucosylated glycoclusters [92]. The author first performed a MW-assisted CuAAC reaction involving a propargylated solid substrate **75** and a mannose-based azide **76** in a methanol/water to give a polymer-supported mannose scaffold **77**. Subsequently in the presence of BMT in acetonitrile, the designed alkynated phosphoramidite linker **78** gets incorporated into the solid-supported mannose scaffold **77**. This process enabled oligonucleotide elongation, followed by Cy3 labeling. At the end of the sequence, the product underwent treatment with ammonium hydroxide to yield the desired Cy3-oligonucleotide linker **79**. The resulting mannose-core linker **79** comprised of four clickable alkyne groups, which then underwent MW-induced CuAAC coupling with fucosylatedazide **80** in the presence of Cu-nanopowder in a methanol/TEAAc at 60 °C to afford high yield of desired fucosylated glycocluster **81** (Scheme 14.21). Due to solubility issues arising from the introduction of non-polar residues during two consecutive CuAAC steps, the resulting glycocluster **81** was underwent for deacetylation using aqueous ammonia to produce a water-soluble fucosylated glycocluster **82**.

The efficacy of the resulting fucomimetics in binding assays against PA-IIL was then assessed. They employed a glycan microarray to identify the antenna-like mannose-core fucomimetic **81**, which featured four fucosyl residues and exhibited

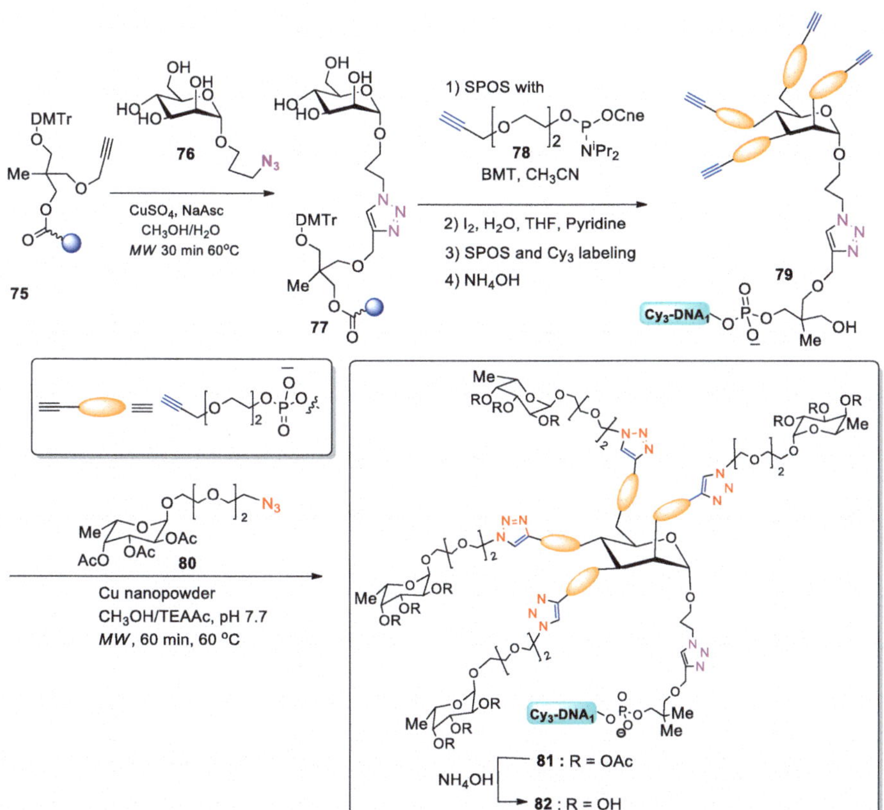

Scheme 14.21 Synthesis of mannose-core-based crown-type tetra-fucosylated glycocluster through click-induced CuAAC approach [92]

the most potent binding with PA-IIL. Furthermore, the importance of mannose-based cores with linker arms in a variety of carbohydrate-centered fucosylated glycoclusters was examined and found to be superior to glucose- and galactose-based cores [92].

Gerland and group conducted a noteworthy study involving glycan microarrays [70]. They employed an expanded glycan microarray approach, where they synthesized a range of structurally diverse multivalent galactosylated clusters, consisting of twenty-five members at the nanoscale, possessing potential topological qualities. These multivalent synthons were derived from the mannose core (**83a-c**) and were separately conjugated with azides **84a, b**underCuAACclick protocol conditions. Subsequently, the reaction with ammonium hydroxide resulted in an excellent yield of the corresponding triazole-appended glycoclusters **85a-d** (Scheme 14.22). The group also evaluated these glycoclusters for their binding interactions with LecA, and the octa glycocluster with a mannose core exhibited a notably enhanced binding affinity compared to other glycoclusters with glucose or galactose units [70]. This substantial 1361-fold increase in selective potency clearly establishes the superiority of the mannose core and holds the potential to open new opportunities in the field of chemical biology.

Similarly, Ligeour and group explored the next-generation DNA-based glycocluster, by incorporating pentofuranose and mannitol oligonucleotide as appropriate core in motif [93]. To construct the desired molecular framework, they developed a small DNA sequence on the polymer-supported alkyne **86** followed by its click

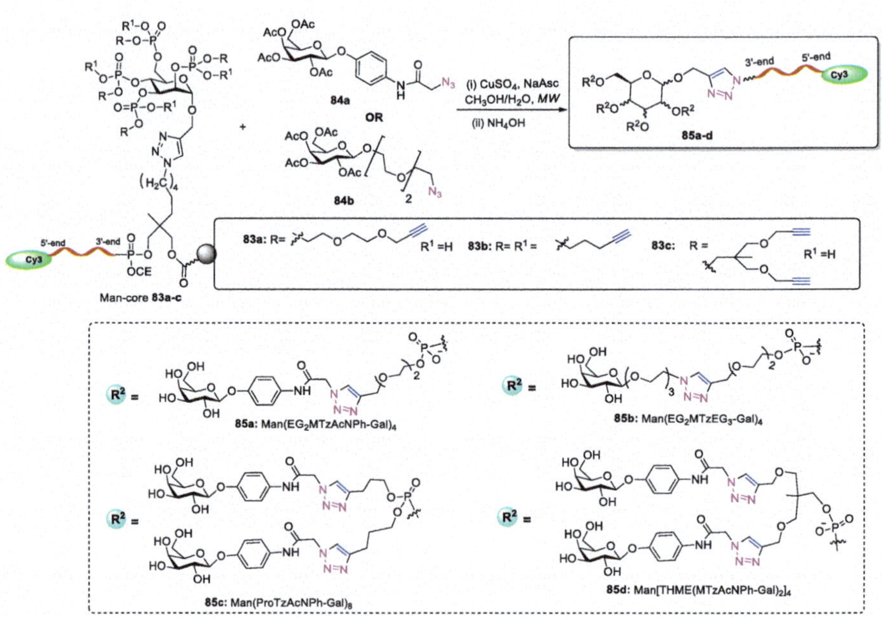

Scheme 14.22 CuAAC-inspired development of multivalent triazole-appended glycocluster

coupling with mannitol azide **87a,b** under microwave condition using $CuSO_4$/NaAsc to give triazole-tethered glycodendrimers in excellent yields. Further, acetate removal was performed on the oligonucleotide beads under treatment with ammonia, then subsequent elongation followed by desired labeling with the Cy3 fragment. At the end, this on reaction with ammonia finally produced high yield of desired decaglycoclusters **88** and **89** (Scheme 14.23) [93]. The resulting glycoclusters were displayed significant affinity-based potency against lectin A, lectin B, and BambL variation. However, in silico molecular investigation confirm that the topology as well the nature of linker was found to be a key factor in achieving the required high affinity in the context of multivalency phenomena. This established versatile methodology holds significant promise for generating novel and selectively patterned glycoclusters in the future.

14.5 Click Chemistry-Inspired Bioactive 1,2,3-Triazole-Appended Nucleoside Analogues

A substantial quantity of sialic acids and their derivatives are known to demonstrate potential inhibitory effects on sialyltransferase enzymes. Zou et al. carried out the synthesis of potential inhibitors **94** and **95** for bacterial sialyltransferases by coupling *N*-acetylated azido sialyl derivatives **90** with alkyne functionalities (**91a, b**) followed by removal of protection using TFA/H_2O (Scheme 14.24). According to the SAR study, the molecule's binding affinity and specificity are crucial for its pharmacological activity. Increasing hydrophilicity can have a positive impact on the molecule's binding affinity by introducing substituents that create charge interactions. In biological testing against α-(2,3)-sialyltransferase Cst-06, Cst-I of *Campylobacter jejuni*, sialyl hybrids **94** and **95** exhibited less than 50% inhibitory activity [94].

A collection of triazolyl sugar-nucleotide derivatives has been developed by Nishimura and group by using the established click protocol. These analogues serve as effective inhibitors of glycosyltransferases and display a high degree of specificity for the enzyme's active site [95]. Compound **96** and **97** have demonstrated significant potency and selectivity for two essential classes of human FucTs (Fig. 14.4). It is worth noting that although both compounds incorporate the same azido sugar nucleotide, they exhibit distinct molecular recognition patterns due to their different inhibition mechanisms of action in R1,3-FucT and R1,6-FucT.

Amid global anti-viral drug emergency, various research groups have embarked on a collaborative effort to build an extensive synthetic library of molecules aimed at identifying pharmacophores for lead compound generation. In this context, glycohybrid molecules have emerged as potential versatile candidates with less cytotoxicity. Notably, Chu and group adopted an effective approach to produce cyclopentenol (+) motifs, which are novel anti-viral candidates. This involved the integration of a ring-closing metathesis (RCM) reaction with the useful application of a second-generation Grubbs catalyst, followed by a click-mediated cycloaddition reaction with

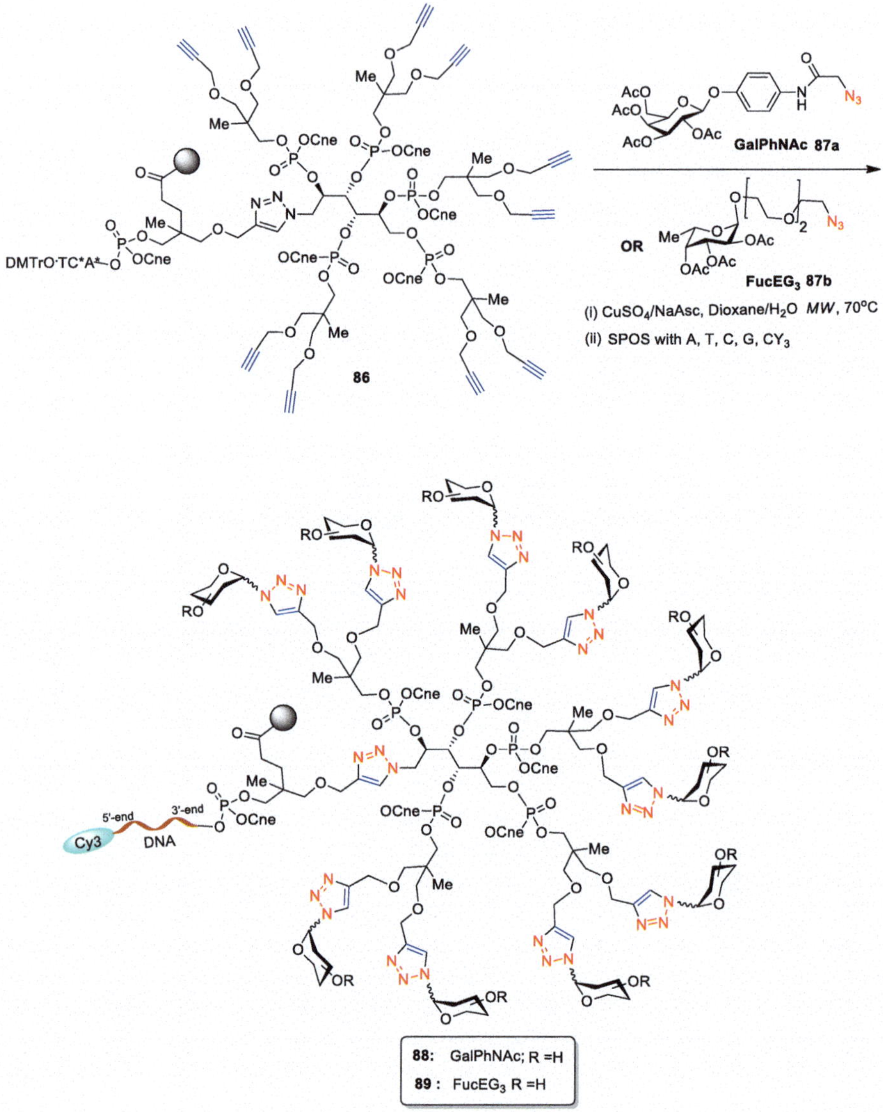

Scheme 14.23 CuAAC 'Click Chemistry'-based synthetis of lectin interactive decaglycoclusters [93]

five-membered ring heterocycles [96]. Within this development, a series of glyco-hybrid NPA derivatives **98–100** (Fig. 14.5) were introduced to a diverse screening panel. Among these compounds, **99** exhibited significant potential against vaccinia (EC50: 0.4 μM, SI > 750), as well as moderate activity against cowpox (EC50: 39 μM, SI > 7.7) and SARSCoV (EC50: 47 μM, SI > 2.1).

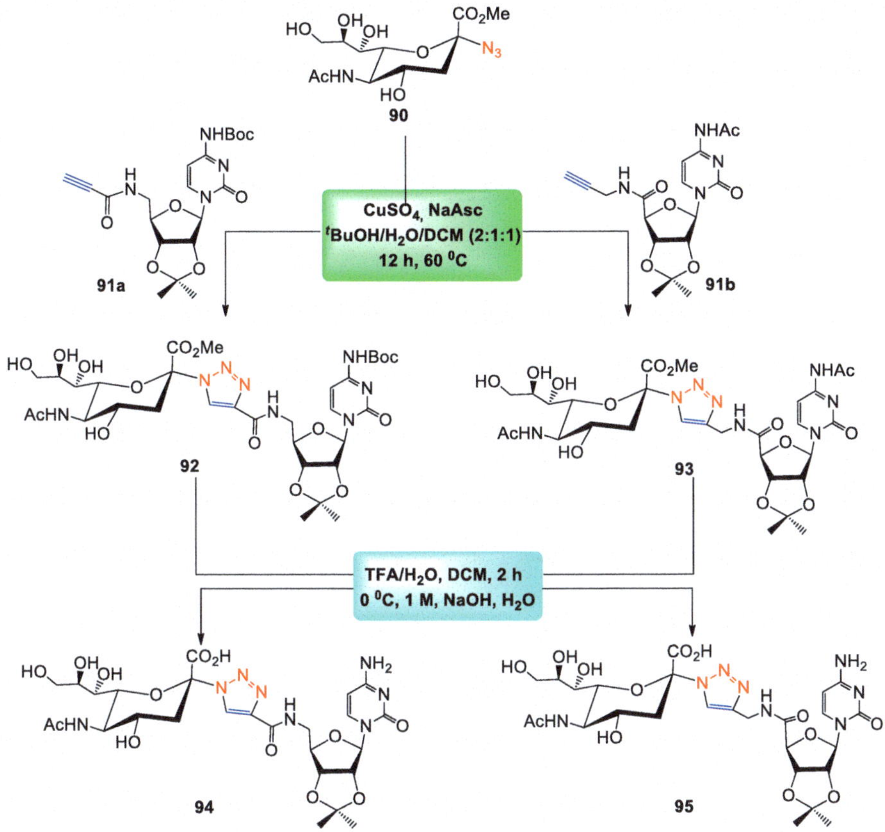

Scheme 14.24 Efficient synthesis of nucleoside-appended sialosides **94** and **95** as potentsialyltransferase inhibitors [94]

The efficient CuAAC click reaction serves as a versatile tool for Olomola and group, enabling the creation of a new category of hybrid compound **103**. This compound combines two prominent classes of pharmacologically active compounds: coumarins **101**, a recognized inhibitor of HIV-1 protease, and azidothymidine **102**, known as an inhibitor of HIV-1 reverse transcriptase (Scheme 14.25) [97]. This innovative series underwent screening for inhibition of both HIV-1 protease and HIV-1 reverse transcriptase. The results of the study revealed that this compound series exhibited bioactivity against both enzymes, with an efficiency ranging from 52 to 60% for the former and 94–99% for the latter at a concentration of 50 μM. Further development and optimization through clinical trials are necessary to transform this compound into a potential lead drug molecule [97].

In another work, Slamova and team conducted a thorough exploration of microwave-induced click chemistry technique for synthesizing triazolyl β-2-acetamidoGlcNAc-2-deoxyglucosides. In bioassay assessments of these newly developed glycohybrids, it was observed that these compounds were susceptible

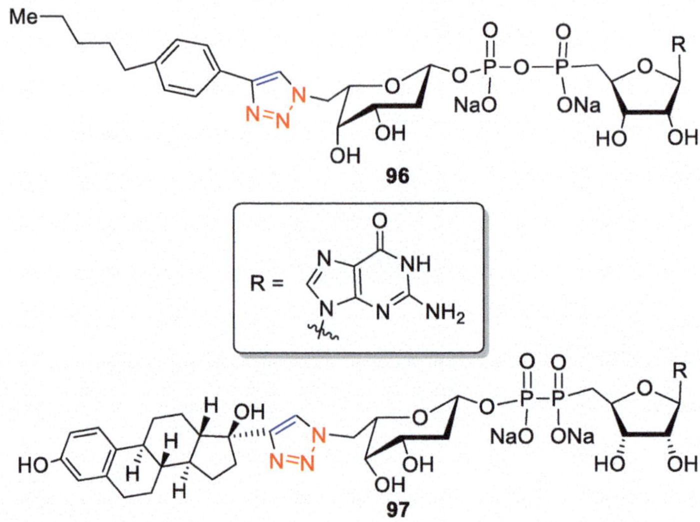

Fig. 14.4 Click-based synthesis of triazolyl glycohybrids possessing sialyltransferase inhibition activity

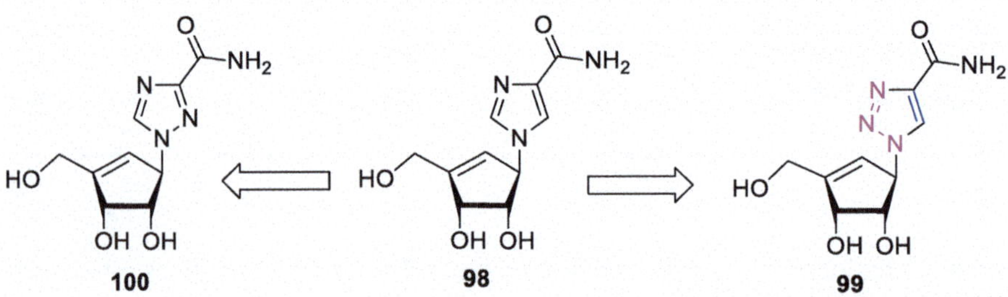

Fig. 14.5 Demonstration of the substituent variation effects on the activity of glycoconjugates showing triazole containing scaffolds as potential anti-viral agents

to hydrolysis by *Talaromyces flavus* CCF β-*N*-acetylhexosaminidase, marking them as promising inhibitors of NKT cells [98]. Consequently, click chemistry has made significant contributions to scaffold design for HIV treatment, introducing greater versatility in coupling sugars with alkyne functionality to form triazole rings. These intricate scaffolds inherently possess specific 3D spatial orientations that contribute to their potency.

A severe respiratory disorder known as ‘COVID-19’ has been identified, resulting from an infection with SARSCoV-2 [99]. This condition has developed into a global pandemic, causing significant illness and loss of life [100]. The virus’s relentless spread is because of the absence of a vaccine and effective treatments. Consequently, there is an urgent need to explore various potential therapeutic and preventive strategies to curb the disease’s continued expansion. Numerous commendable efforts

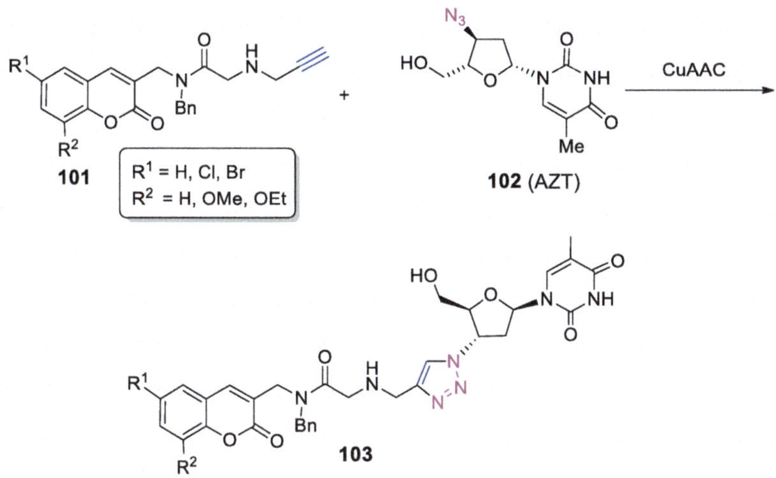

Scheme 14.25 CuAAC-mediated coumarin glycohybrid **103** as potent HIV-1 PR/HIV-1 RT inhibitor [97]

have been undertaken to develop a potent agent to combat this virus. One noteworthy endeavor involves the use of a selective CuAAC click approach to synthesize seven-carbon quantum dots (CQDs). These CQDs were designed and studied for their potential to treat human coronavirus HCoV-229E infections, representing a significant advancement (Scheme 14.26) [100].

This represents the first generation of anti-viral CQDs produced through hydrothermal carbonization of ethylenediamine and citric acid as carbon precursors, subsequently modified with boronic acid ligands **106**. They exhibited a concentration-dependent ability to inactivate the virus, with an EC50 value of 52 ± 8 μg/mL [101]. A second-generation variant **107**, derivatized through substitution with 4-aminophenylboronic acid, showed an even lower EC50 value of 5.2 ± 0.7 μg/mL. The underlying mechanism of action appears to involve inhibiting the entry of HCoV-229E by engaging with the virus's entry receptors. Additionally, there is evidence of unclear inhibition of viral replication. In a related context, the sugar comprised molecule remdesivir **108**, also known as GS-5734, has undergone trials in various clinical and model studies and has shown promise as an effective treatment for severe cases of COVID-19 [102].

Similarly, Ay and group carried out further CuAAC conjugation of azido furanoses and propargylated nucleobases to give respective triazolyl nucleoside glycohybrids as potential anti-cancer agents [103]. The nucleoside-based glycohybrids **109–112** (Fig. 14.6) demonstrated significant cytotoxicity against cell lines including MDA-MB-231, Hep3B, PC-3, SH-SY5Y, and HCT-116, where they were particularly effective in inhibiting the growth of Hep3B cells within the range of 66.1–92.8 μM and could be a promise as potential anti-cancer agents worth further exploration.

Furthermore, Halay et al. used similar efficient CuAAC catalytic system to connect pyrano-sugar azide with nucleobase, possess alkyne functionality, and

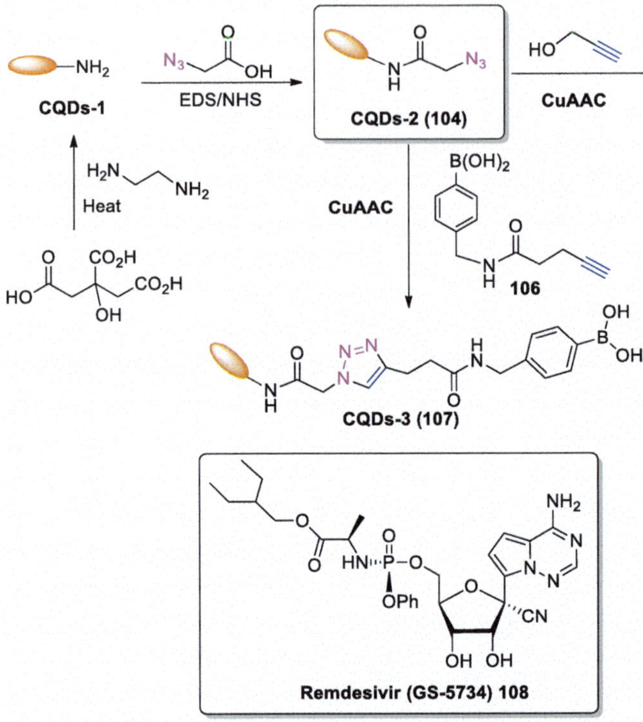

Scheme 14.26 Utilization of CuAAC click chemistry in the battle against the COVID-19

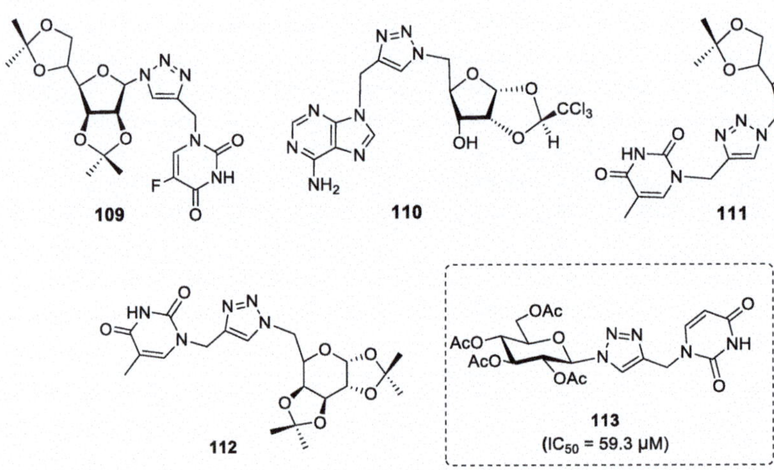

Fig. 14.6 Development of nucleoside-based glycohybrids **109–112** via CuAAC possessing anti-cancer property

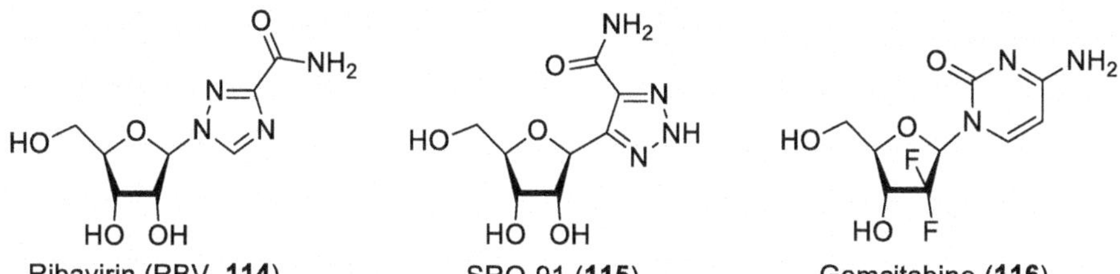

Fig. 14.7 Structure of recognized and pipeline compounds; ribavirin, SRO-91, and gemcitabine [105]

develop respective triazolyl nucleoside **113** [104]. This nucleoside exhibited the potent activity against the Hep3B cell line (IC50 = 59.3 μM) and the in vitro experiments unequivocally demonstrated the potential of this pharmacophore as a valuable candidate for anti-cancer drug development.

The effectiveness of anti-metabolite and nucleoside-based drugs in the market has motivated numerous research groups to direct their efforts toward developing similar compounds with reduced cytotoxicity and fewer side effects. A similar strategy was embraced by Lubin-Germain and their group, who sought to create triazolyl glycohybrids inspired by ribavirin (**114–116**) (Fig. 14.7). These glycohybrids were assessed for their anti-proliferative properties against a panel of nine cancer cell lines, including MDAMB231, MDAMB435, B16F10, Caco2, HuH7, DU145, SNB75, C6, and A431, using MTT assay [105]. Both the designed compound and the standard drug exhibited remarkable potency and responsiveness against all the tested cell lines. Additionally, the triazolyl derivative displayed superior IC_{50} values falling within the range of 0.5–9.5 mM when compared to the original motif. Furthermore, this molecule demonstrated an increased sensitivity and selectivity toward ovarian cancers.

Reddy et al. developed uridine glycohybrids to recognize the significance of nucleobase-containing triazoles. The analogues were subsequently assessed for their effectiveness in combating cancer and bacterial infections, demonstrating superior potency compared to the standard anti-cancer drug, cisplatin. Notably, compound **117** (IC50: 11.34 μM) (Fig. 14.8) displayed strong activity against the MCF-7 cancer cell line. Furthermore, compound **118** (IC_{50}: 7.93 μM) exhibited notably enhanced efficacy against the HeLa cell line. Also, the bacterial assay confirmed that compound **118** possessed remarkable anti-bacterial activity against *E. coli* and *B. subtilis*. Consequently, this pharmacophore emerged as a versatile compound with distinct biochemical attributes, demonstrating a broad range of cytotoxic capabilities, hence presenting a promising avenue for further exploration in molecular development [106].

In addition to the above-described bioactivity associated with CuAAC-mediated nucleosides, few of the related 1,2,3-triazole-appended glycohybrids displayed immuno-modulatory activity and cyto-interactive activity. Glycosyltransferases,

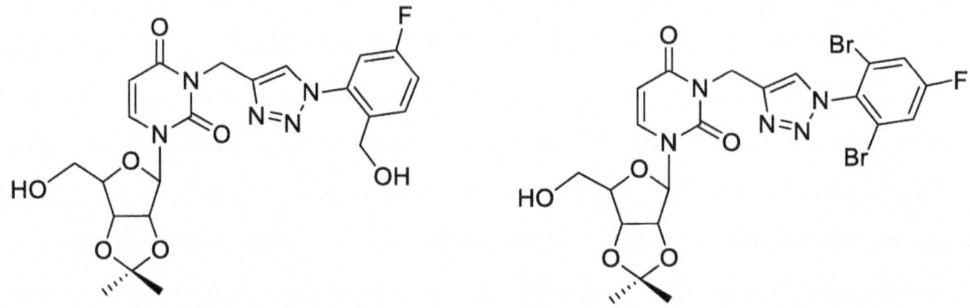

117 (IC_{50} = 11.34 μM) **118** (IC_{50} = 7.93 μM)

Fig. 14.8 Uridine glycoconjugates **117–118** derived via CuAACas potential anti-cancer activity

found in various cellular organelles, play vital role in construction of oligosaccharide chains by transferring a monosaccharide from an activated sugar donor to an acceptor through the formation of glycosidic bonds. Within this enzyme, galactosyltransferase subfamily consists of various enzymes, including β1-4-, β-1-3-, β-1-6-, α1-3-, and α1-4-galactosyltransferases, which are associated with several congenital glycosylation disorders. Each of these enzymes uses UDP-α-D-Gal as the sugar donor, but they generate different types of linkages, such as β1-4-, β1-3-, β1-6-, α1-3-, and α1-4-linkages, as indicated by their names. Each subfamily also has other extra members [107]. As a result, these enzymes represent important targets for pharmacophoric lead generation and drug development. In a specific study conducted by Gawolek and the group, a class of triazolyl glycoconjugates **119–122** was designed as substrate donor type inhibitors for galactosyltransferase. These inhibitors were created through the cycloaddition of clickable azide and alkyne-functionalized groups [107]. In the newly constructed analogues, the pyrophosphate moiety generally found in natural UDP substrates was replaced with an amide unit or a 1,2,3-triazolyl moiety and an amide unit. Additionally, the anomeric oxygen atom in the natural substrates was transformed to nitrogen atoms intentionally in the designed analogues to enhance their stability against enzymatic hydrolysis. Also, the sugar head group in these compounds has the opposite configuration compared to the natural uridine diphosphate substrate. This results in no activity against β4-Gal-T, underlining the importance of linkers connecting sugar and uridine for optimal binding at the enzyme's active site. Keeping this in view, a different set of molecules was designed, which included suitable linkers between the sugar unit and the nucleoside with appropriate length and flexibility. The glycohybrids **121** and **122** (Fig. 14.9) were synthesized through a coupling of 1-azido sugar with propargylated uridine derivatives using standard CuAAC protocol. On the other hand, the reaction of propargylated β-*O*-glycosides with 5'-azido uridine under similar conditions delivered another set of related glycohybrids **123** and **124**. Fascinatingly, it was observed that the inhibitory activity against β-1,4-galactosyltransferase I was significantly improved by the replacement of the pyrophosphate with an O-methylene triazole unit in compounds **123** and **124**.

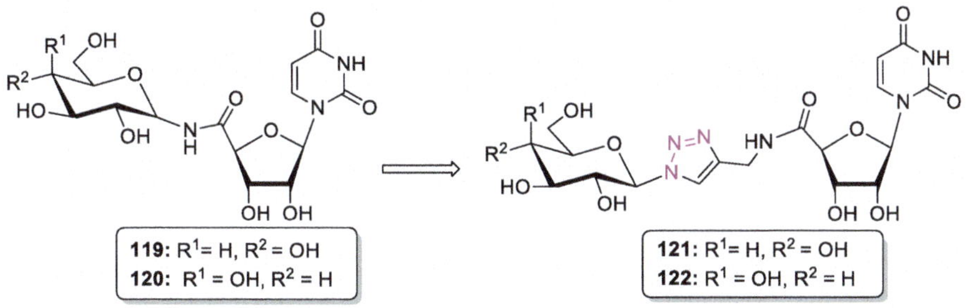

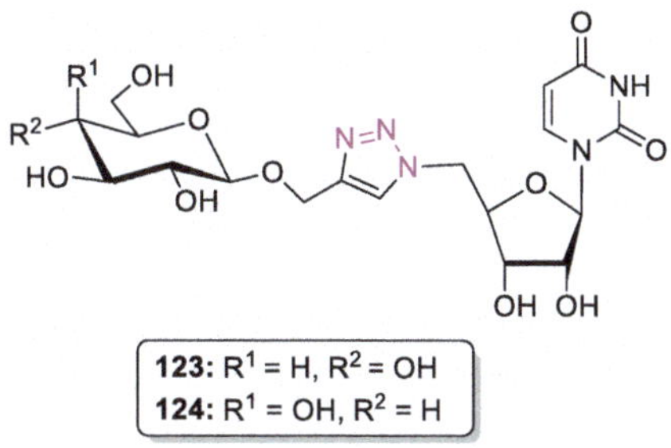

Antitumor and galactosyltransferase inhibitory activity of glycohybrids119-124:

Glycohybrids	% inhibition at 0.8mM	IC_{50} (mM)
119	1.3 ± 0.8	-
120	8 ± 0.2	-
121	1.5 ± 1.1	-
122	3 ± 0.5	-
123	48 ± 2.4	0.72
124	14 ± 2.6	-

Fig. 14.9 CuAAC chemistry-based glycohybrids **119–124** as anti-tumor agents [104]

The authors also assessed the inhibitory potential of resulting glycohybrids **119–124** against commercially accessible β-1,4-galactosyltransferase I (β4GalT), where glycohybrid **123** exhibited promising enzyme inhibition activity ($IC_{50} = 0.72$ mM). Also, the MTT assay confirmed the anti-tumor potency with glycohybrids [108]

14.6 Click Chemistry-Inspired Triazole-Linked Nucleosides as Radio Tracer Agents

In pursuit of enhanced complexity, a newly developed radiotracer **126**, inspired by thymidine, was synthesized with a notable RCY value of 75%. Starting from the uridine alkyne analogue **125,** a novel thymidine-inspired radiotracer **126** was obtained in appreciable RCY value of 75% by treating it with 2-[^{18}F]fluoroethylazide under standard CuAAC condition (CuSO4/NaAsc) at 80 °C in water for 10 min (Scheme 14.27) [109]. The resulting thymidine 1,2,3-triazole analogue on in vivo metabolism analysis exhibited notable stability in human plasma with fine selectivity for A431 tumors.

Likewise, a silylated thymidine analogue **127** was obtained in high yield by reacting thymidine-derived azide **128** with dialkyl silylated alkyne **127** in presence of $CuSO_4$/NaAsc in *t*-butanol/water system at 100 °C for 10 min. The resulting 1,2,3-triazole **129** was then underwent for the fluorination with K[^{18}F]F-K_{222} in DMSO/AcOH at 60 °C for 15 min to give the desired fluorinated radio tracer **130**.

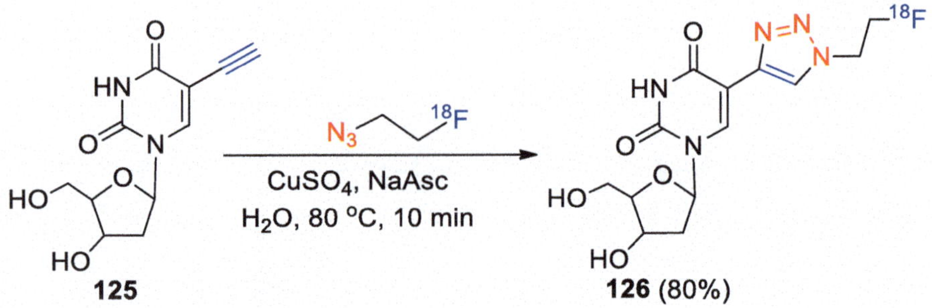

Scheme 14.27 CuAAC-mediated labeled thymidine 1,2,3-triazole **126**

Interestingly, a complementary analogue of 3'-silylated thymidine precursors can be developed via amalgamation of dialkyl silylated azide and thymidine-derived alkyne, and the alternative tool was found to be reliable for [^{18}F]- FLT radiotracer (Scheme 14.28) [110, 111].

Application of click chemistry in fluorescent modification has been done by the amalgamation of blue phenoxazinium dye **131** and red coumarin dye **13**) with alkyne-modified nucleotides photosynthetically. An array of DNA duplexes was modified by attaching dyes at the 2'-position of uridine (Fig. 14.10) [112].

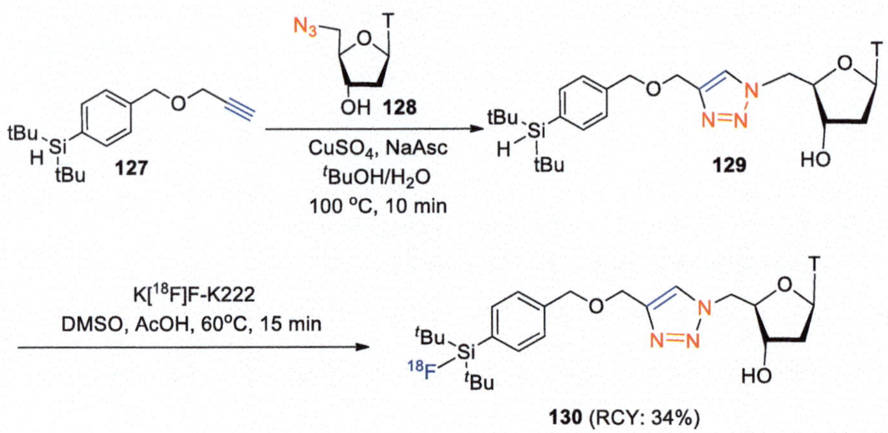

Scheme 14.28 Click-mediated selective ^{18}F-labeled silylated thymidine analogue **130**

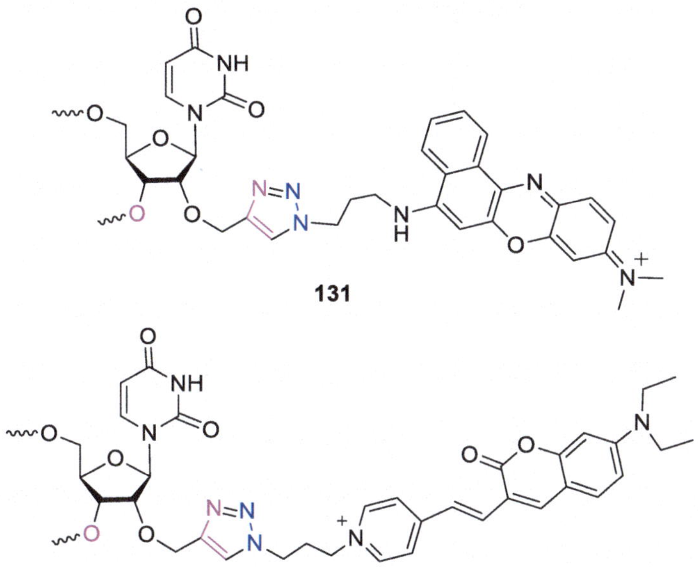

131

132

Fig. 14.10 Click chemistry in fluorescent modification

14.7 Cellular Metabolic Labeling of Nucleic Acids (DNA and RNA)

Arabinofuranosyl-5-ethynyluridine analogues **133–135** [113] and modified azido uridine analogues **136** and **137** were investigated for their role in cellular metabolic labeling of DNA [30, 114]. Among them **133** displayed least impact on the function of genome in order to inhibit cellular toxicity while in azido derivatives the compound **137** is more effective (Fig. 14.11a). The modified deoxynucleosides just discussed were limited only for the incorporation of DNA. In order to labeled RNA alkyne-modified nucleosides likewise ethynyl uridine **138** [115] and ethynyl adenosines **139,140** [116, 117] were tested with outcome of interesting results. The adverse effects of alkyne nucleosides mainly on RNA sequencing create a limit for their use. In order to resolve this problem, a reverse labeling using azido derivatives **141–143** [118–120] was tested and introduced effectively in RNA for SPAAC coupling (Fig. 14.11b).

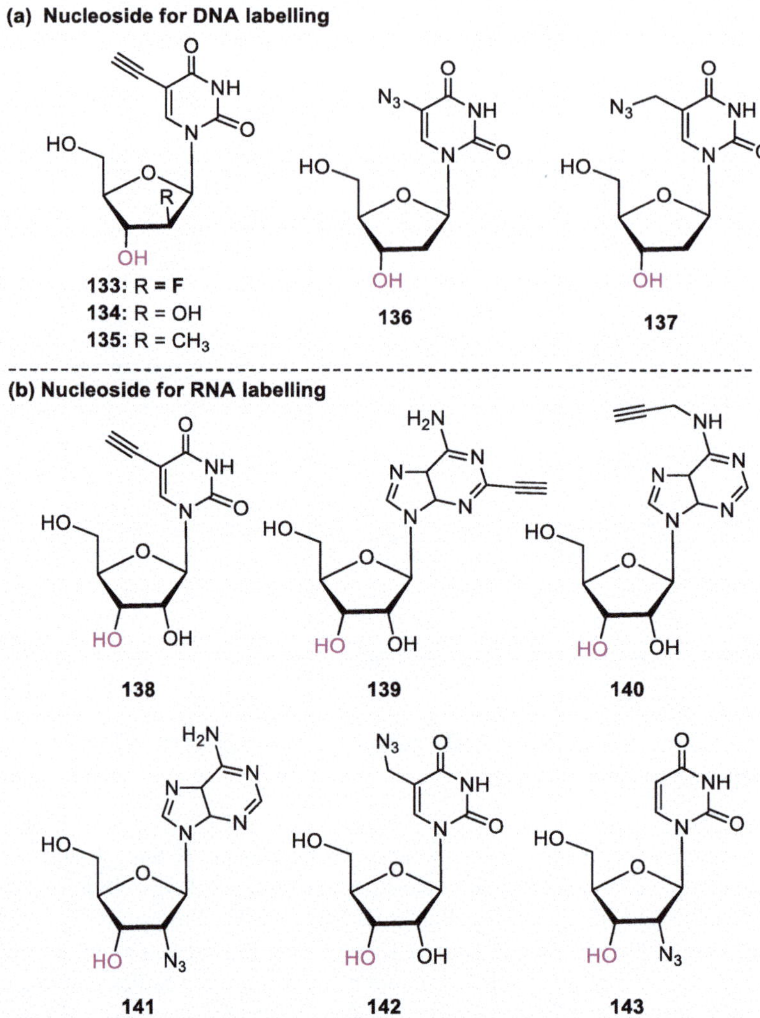

Fig. 14.11 Modified azide and alkyne-bounded nucleoside for DNA and RNA labeling

14.8 CuAAC in Assembling of Nucleic Acid

In order to effectively cross-link DNA strands, Seela and colleagues established a template-free approach that involved clicking two oligonucleotide strands over a bifunctional azide linker. It is possible to produce single-stranded DNA, duplexes, and multistranded structures using the 'bis-click' process, demonstrated by Seela and co-workers. A bis-click construct, for instance, hybridized with the complementary strands to produce a four-stranded DNA structure made up of two cross-linked duplexes (Fig. 14.12) [121].

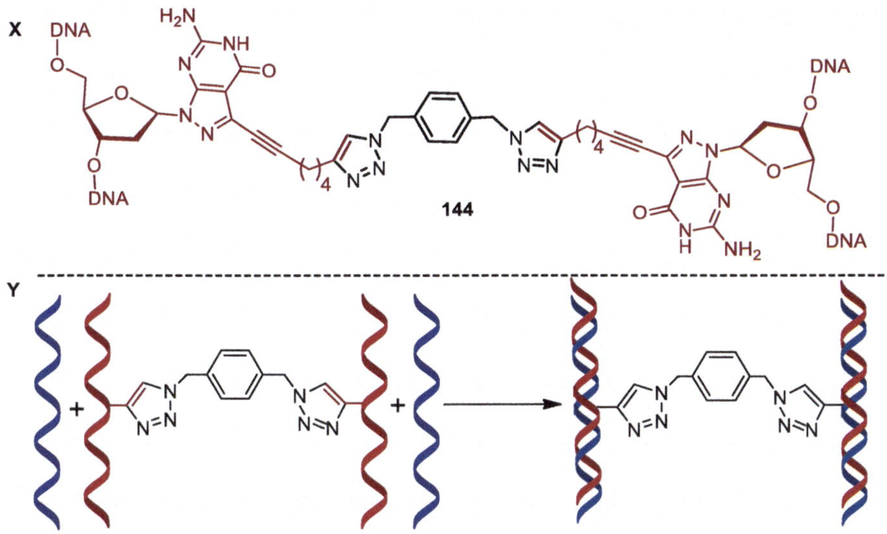

Fig. 14.12 Bis-click approach for the development of a four-stranded DNA molecule **Y** constituted of two cross-linked duplexes from cross-linker **X**

Furthermore, by conjugating two complementary oligonucleotides, the CuAAC process was employed to produce massive dsDNA catenanes **145**. The catenane was composed of six helical turns, with T-T mismatches on either side of ten base pairs. The base mismatches were employed as hinges to enable the duplex to take on a cyclic shape and be flexible. Two strands that had been labeled with a 5′-alkyne and a 3′-azide were used to construct the macrostructure in a stepwise manner. By employing the application of non-templated CuAAC chemistry, the first oligonucleotide was cyclized and served as a template for the second oligonucleotide's click ligation and cyclization (Scheme 14.29) [25].

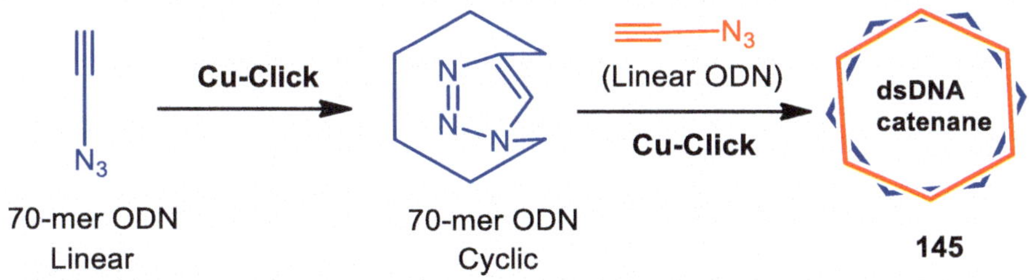

145

Scheme 14.29 Construction of dsDNA catenanes using CuAAC tool

14.9 Conclusions and Future Perspective

Owing to its versatility, modularity, and compatibility with various synthetic conditions and readily availability of substrates, the azide-alkyne cycloaddition click chemistry techniques have significantly influenced to a number of research domains with their notable applications in chemistry, biology, biochemistry, medicine, agricultures, and biotechnology. The effectiveness and flexibility of AAC chemistrycome into play especially, CuAAC and RuAAC as well as a metal-free variant the strain-promoted azide-alkyne cycloaddition (SPAAC) when applied to the domain of nucleic acids. Utilizing AAC chemistry with nucleic acids enables efficient and cost-effective labeling, ligation, and cyclization of oligonucleotides compared to traditional chemical and enzymatic approaches. Currently, click chemistry with oligonucleotides has evolved into a well-established field with a broad range of protocols and valuable modifications. This chapter discusses a broad range of applications, highlighting the substantial impact of azide-alkyne cycloadditions in the area of nucleic acid research. Furthermore, future progress in click chemistry with nucleic acids is expected to be carried out through comprehensive studies on modified CRISPR/Cas systems for gene editing and the development of novel artificial nucleic acid analogues. In addition, oligonucleotide therapeutics hold the promise of introducing entirely new classes of medicines for challenging diseases, with potential applications in diverse scientific fields including biotechnology, nanotechnology, and materials chemistry.

References

1. Singh Y, Spinelli N, De Francq E, Dumy P (2006) Org Biomol Chem 4:1413–1419
2. Virta P, Katajisto J, Niittymaki T, Lonnberg H (2003) Tetrahedron 59:5137–5174
3. Agrahari AK, Bose P, Jaiswal MK, Rajkhowa S, Singh AS, Hotha S, Mishra N, Tiwari VK (2021) Chem Rev 12:7638
4. Tornøe CW, Christensen C, Meldal M (2002) J Org Chem 67:3057
5. Rostovtsev VV, Green LG, Fokin VV, Sharpless KB (2002) Angew Chem Int Ed 41:2596–2599
6. Thirumurugan P, Matosiuk D, Jozwiak K (2013) Chem Rev 113:4905–4979
7. Chen MM, Kopittke PM, Jhao FJ, Wang P (2024) Trends Plant Sci 29:167–178
8. Witczak ZJ, Bielski R (2013) Click chemistry in glycoscience: new development and strategies. Wiley-VCH, New York
9. Ustinov AV, Stepanova IA, Dubnyakova VV, Zatsepin TS, Nozhevnikova EV, Korshun VA (2010) Russ J Bioorg Chem 36:401–445
10. Jha RK, Wooley KL, Nystrom AM, Burke DJ, Kade MJ, Hawker CJ (2009) Applications of orthogonal "click" chemistries in the synthesis of functional soft materials. Chem Rev 109:5620–5686
11. Yi G, Son J, Yoo J, Park C, Koo H (2018) Biomater Res 22:13
12. Zhang L, Chen X, Xue P, Sun HHY, Williams ID, Sharpless KB, Fokin VV, Jia G (2005) J Am Chem Soc 127:15998–15999
13. Johansson JR, Beke-Somfai T, Stålsmeden AS, Kann N (2016) Chem Rev 116:14726–14768
14. Danence LJT, Gao Y, Li M, Huang Y, Wang J (2011) Chem Eur J 17:3584–3587
15. Belkheira M, Abed DE, Pons JM, Bressy C (2011) Chem Eur J 17:12917–12921

16. Jaiswal MK, Gupta A, Yadav MS, Pandey VK, Tiwari VK (2023) Chem Eur J 29:e202301749
17. Tiwari VK, Mishra BB, Mishra KB, Mishra N, Singh AS, Chen X (2016) Chem Rev 116:3086
18. Amblard F, Cho JH, Schinazi RF (2009) Chem Rev 109:4207–4220
19. Fantoni NZ, El-Sagheer AH, Brown T (2021) Chem Rev 121:7122–7154
20. Amblard F, Sari O, Boucle S, Khalil A, Schinazi RF (2016) Modification of nucleosides, nucleotides, and nucleic acids using Huisgen's [3+2] azide-alkyne cycloaddtion: Opening Pandora's box. In: Chandrasekeran S (ed) Click reactions in organic synthesis, pp 309–336
21. Sharma VK, Mangla M, Singh SK, Prasad AK (2024) Curr Org Synth 21:436–455
22. Merkel M, Peewasan K, Arndt S, Ploschik D, Wagenknecht HA (2015) ChemBioChem 16:1541–1553
23. El-Sagheer AH, Brown T (2010) Chem Soc Rev 39:1388–1405
24. Gierlich J, Burley GA, Gramlich PME, Hammond DM, Carell T (2006) Org Lett 8:3639–3642
25. Kumar R, El-Sagheer A, Tumpane J, Lincoln P, Wilhelmsson LM, Brown T (2007) J Am Chem Soc 129:6859–6864
26. Kocalka P, El-Sagheer AH, Brown T (2008) ChemBioChem 9:1280–1285
27. El-Sagheer AH, Kumar R, Findlow S, Werner JM, Lane AN, Brown T (2008) ChemBioChem 9:50–52
28. El-Sagheer AH, Brown T (2010) Proc Natl Acad Sci USA 107:15329–15334
29. Paredes E, Das SR (2011) ChemBioChem 12:125–131
30. Salic A, Mitchison TJ (2008) Proc Natl Acad Sci U S A 105:2415–2420
31. El-Sagheer AH, Sanzone AP, Gao R, Tavassoli A, Brown T (2011) Proc Natl Acad Sci U S A 108:11338–11343
32. Fujino T, Yamazaki N, Isobe H (2009) Tetrahedron Lett 50:4101–4103
33. Isobe H, Fujino T, Yamazaki N, Guillot-Nieckowski M, Nakamura E (2008) Org Lett 10:3729–3732
34. Sanzone AP, El-Sagheer AH, Brown T, Tavassoli A (2012) Nucleic Acids Res 40:10567–10575
35. Rozkiewicz DI, Gierlich J, Burley GA, Gutsmiedl K, Carell T, Ravoo BJ, Reinhoudt DN (2007) ChemBioChem 8:1997–2002
36. Fischler M, Simon U, Nir H, Eichen Y, Burley GA, Gierlich J, Gramlich PME, Carell T (2007) Small 3:1049–1055
37. Fischler M, Sologubenko A, Mayer J, Clever G, Burley G, Gierlich J, Carell T, Simon U (2008) Chem Commun 44:169–171
38. Lundberg EP, El-Sagheer AH, Kocalka P, Wilhelmsson LM, Brown T, Norden B (2010) Chem Commun 46:3714–3716
39. Lundberg EP, Plesa C, Wilhelmsson LM, Lincoln P, Brown T, Norden B (2011) ACS Nano 5:7565–7575
40. Srivastava S, Singh SK, Sharma VK, Mangla P, Olsen CE, Prasad AK (2015) Nucleos Nucleot Nucl 34:388–399
41. Srivastava S, Maikhuri VK, Kumar R, Bohra K, Singla H, Maity J, Prasad AK (2018) Carbohydr Res 470:19–25
42. El-Sagheer AH, Brown T (2009) J Am Chem Soc 131:3958–3964
43. Qu Y, Wen H, Ge R, Xu Y, Gao H, Shi X, Wang J, Cui W, Su W, Yang H, Kuai L, Satz AL, Peng X (2020) Org Lett 22:4146–4150
44. Gramlich PME, Warncke S, Gierlich J, Carell T (2008) Angew Chem, Int Ed 47:3442–3444
45. Srivastava S, Bimal D, Bohra K, Singh B, Ponnan P, Jain R, Basil MV, Maity J, Thirumal M, Prasad AK (2018) Eur J Med Chem 150:268–281
46. Maikhuri VK, Bohr K, Srivastava S, Kavita, Prasad AK (2019) Synth Commun 49:3140–3147
47. Kaczmarek O, Scheidt HA, Bunge A, Föse D, Karsten S, Arbuzova A, Huster D, Liebscher J (2010) Eur J Org Chem, 1579–1586
48. Nabo LJ, Madsen CS, Jensen KJ, Kongsted J, Astakhova K (2015) Chem. Bio. Chem. 16:1163–1167
49. Ermolat DS, Metha VP, Van der Eycken EV (2007) QSAR Comb Sci 26:1266–1273
50. Snead NM, Rossi JJ (2012) Nucleic Acid Ther 22:139–146

51. Jain ML, Bruice PY, Szab'o IE, Bruice TC (2012) Chem Rev 112:1284–1309
52. Varizhuk A, Chizhov A, Florentiev V (2011) Bioorganic Chem 39:127–131
53. Varizhuk AM, Kaluzhny DN, Novikov RA, Chizhov AO, Smirnov IP, Chuvilin AN, Tatarinova ON, Fisunov GY, Pozmogova GE, Florentiev VL (2013) J Org Chem 78:5964–5969
54. Lucas R, Neto V, Bovazza AH, Zerrouki R, Granet R, Kravsz P, Champavier Y (2008) Tetrahedron Lett 49:1004–1007
55. Isobe H, Fujino T (2014) Chem Rec 14:41–51
56. Fitzgerald PR, Paegel BM (2021) Chem Rev 121:7155–7177
57. Santner T, Hartl M, Bister K, Micura R (2014) Bioconjugate Chem 25:188–195
58. Fauster K, Hartl M, Santner T, Aigner M, Kreutz C, Bister K, Ennifar E, Micura R (2012) ACS Chem Biol 7:581–589
59. Weisbrod SH, Marx A (2007) Chem Commun 18:1828–1830
60. Pujari SS, Leonard P, Seela F (2014) J Org Chem 79:4423–4437
61. Dellisola A, McLachlan MMW, Neuman BW, Al-Mullah HMN, Binks AWD, Elvidge W, Shankland K, Cobb AJ (2014) Chem Eur J 20:11685–11689
62. Patil PC, Luzzio FAT (2017) Tetrahedron 73:4206–4213
63. Liu J, Leonard P, Müller SL, Daniliuc C, Seela F (2018) Beil J Org Chem 14:2404–2410
64. El-Sagheer AH, Kumar R, Findlow S, Werner JM, Lane AN, Brown TA (2008) Chem Bio Chem 9:50–52
65. Yamada T, Peng CG, Matsuda S, Addepalli H, Jayaprakash KN, Alam R, Mills K, Maier MA, Charisse K, Sekine M, Rajeev KG et al (2011) J Org Chem 76:1198–1211
66. Stephenson AWI, Bomholt N, Partridge AC, Filichev VV (2010) Chem Bio Chem 11:1833–1839
67. Stephenson AWI, Partridge AC, Filichev VV (2011) Chem Eur J 17:6227–6238
68. Clave G, Chatelain G, Filoramo A, Gasparutto D, Saint-Pierre C, Le Cam E, Pietrement O, Guerineau V, Campidelli S (2014) Org Biomol Chem 12:2778–2783
69. Neef AB, Pernot L, Schreier VN, Scapozza L, Luedtke NW (2015) A bioorthogonal chemical reporter of viral infection. Angew Chem, Int Ed 54:7911–7914
70. Cassinelli V, Oberleitner B, Sobotta J, Nickels P, Grossi G, Kempter S, Frischmuth T, Liedl T, Manetto A (2015) Angew Chem, Int Ed 54:7795–7798
71. Pourceau G, Meyer A, Vasseur JJ, Morvan F (2008) J Org Chem 73:6014–6017
72. Pourceau G, Meyer A, Chevolot Y, Souteyrand E, Vasseur JJ, Morvan F (2010) Bioconjugate Chem 21:1520–1529
73. Gerland B, Goudot A, Ligeour C, Pourceau G, Meyer A, Vidal S, Gehin T, Souteyrand VE, Vasseur JJ, Chevolot Y et al (2014) Bioconjugate Chem 25:379–392
74. Moni L, Pourceau G, Zhang J, Meyer A, Vidal S, Souteyrand E, Dondoni A, Morvan F, Chevolot Y, Vasseur JJ et al (2009) Chem Bio Chem 10:1369–1378
75. Chevolot Y, Bouillon C, Vidal S, Morvan F, Meyer A, Cloarec JP, Jochum A, Praly JP, Vasseur JJ, Souteyrand E (2007) Angew Chem, Int Ed 46:2398–2402
76. Pergolizzi G, Dedola S, Field RA (2016) Carbohydr Chem 42:1–46
77. Fantoni NZ, McGorman B, Molphy Z, Singleton D, Walsh S, El-Sagheer AH, McKee V, Brown T, Kellett A (2020) ChemBioChem 21:3563–3574
78. Muthana SM (2020) Glycan microarray: towards drug discovery and development. In: Tiwari VK (ed) Carbohydrates in drug discovery and development, pp 213–266, Chapter 6
79. Cummings RD, Molinaro JH, Song X, Yu H, Chen X, Lasanajak Y, MTappert M, Air GM, Tiwari VK, Cao H, Chokhawala HA, Zheng H, Smith DF (2011) J Biol Chem 286:31610–31622
80. Song EH, Pohl NL (2009) Curr Opin Chem Biol 13:626–632
81. De Paz JL, Seeberger PH (2012) Methods Mol Biol 808:1–12
82. Rillahan CD, Paulson JC (2011) Annual Rev Biochem 80:797–823
83. Lonardi E, Balog CI, Deelder AM, Wuhrer M (2010) Natural glycan microarrays. Exp Rev Proteomics 7:761–774
84. Park S, Gildersleeve JC, Blixt O, Shin I (2013) Carbohydrate microarrays. Chem Soc Rev 42:4310–4326

85. Calarese DA, Lee HK, Huang CY, Best MD, Astronomo RD, Stanfield RL, Katinger H, Burton DR, Wong CH, Wilson IA (2005) Proc Natl Acad Sci USA 102:13372–13377
86. Manimala JC, Li Z, Jain A, VedBrat S, Gildersleeve JC (2005) Chem Biochem 6:2229–2241
87. Lee MR, Shin IF (2005) Org Lett 7:4269–4272
88. Zhang Y, Luo S, Tang Y, Yu L, Hou KY, Cheng JP, Zeng X, Wang PG (2006) Anal Chem 78:2001–2008
89. Fazio F, Bryan MC, Blixt O, Paulson JC, Wong CH (2002) J Am Chem Soc 124:14397–14402
90. Michel O, Ravoo BJ (2008) Langmuir 24:12116–12118
91. Norberg O, Deng L, Yan M, Ramstrom O (2009) Bioconjugate Chem 20:2364–2370
92. Gerland B, Goudot A, Pourceau G, Meyer A, Dugas V, Cecioni S, Vidal S, Souteyrand E, Vasseur JJ, Chevolot Y et al (2012) Bioconjugate Chem 23:1534–1547
93. Ligeour C, Dupin L, Angeli A, Vergoten G, Vidal S, Meyer A, Souteyrand E, Vasseur J-J, Chevolot Y, Morvan F (2015) Org Biomol Chem 13:11244–11254
94. Kumar R, Nasi R, Bhasin M, Khieu NH, Hsieh M, Gilbert M, Jarrell H, Zou W, Jennings HJ (2013) Carbohydr Res 378:45–55
95. Hosoguchi K, Maeda T, Furukawa JI, Shinohara Y, Hinou H, Sekiguchi M, Togame H, Takemoto H, Kondo H, Nishimura SI (2010) J Med Chem 53:5607–5619
96. Cho JH, Bernard DL, Sidwell RW, Kern ER, Chu CK (2006) J Med Chem 49:1140–1148
97. Olomola TO, Klein R, Mautsa N, Sayed Y, Kaye PT (2013) Bioorg Med Chem 21:1964–1971
98. Slamova K, Marhol P, Bezouska K, Lindkvist L, Hansen SG, Kren V, Jensen HH (2010) Bioorg Med Chem Lett 20:4263–4265
99. Gorbalenya AE, Baker SC, Baric RS, Groot RJ, Drosten C, Gulyaeva A, Haamans BL, Lauber C, Leontovich AM, Neuman BW et al (2020) Nat Microbiol 5:536–544
100. Łoczechin A, Seron´K, Barras A, Giovanelli E, Belouzard S, Chen Y-T, Metzler-Nolte N, Boukherroub R, Dubuisson J, Szunerits S (2019) ACS Appl Mater Interfaces 11:42964–42974
101. Kupferschmidt K, Cohen J (2020) Science 367:610–611
102. Wang Y, Zhang D, Du G, Du R, Zhao J, Jin Y, Fu S, Gao L, Cheng Z et al (2020) Lancet 395:1569–1578
103. Halay E, Ay E, Şalva E, Ay K, Karayıldırım T (2018) Chem Heterocycl Comp 54:158–166
104. Halay E, Ay E, Şalva E, Ay K, Karayıldırım T (2017) Nucleos Nucleot Nucl 36:598–619
105. Sabat N, Migianu-Griffoni E, Tudela T, Lecouvey M, Kellouche S, Carreiras F, Gallier F, Uziel J, Lubin-Germain N (2020) Eur J Med Chem 188:112009
106. Thatipamula RK, Narsimha S, Battula K, Chary R, Mamidala E, Reddy NVJ (2017) Saudi Chem Soc 21:795–802
107. Hanßke B, Thiel C, Lübke T, Hasilik M, Höning S, Peters V, Heidemann PH, Hoffmann GF, Berger EG, Körner C (2002) J Clin Invest 109:725–733
108. Gawolek GP, Plesniak M, Komor R, Wyrostek AB, Erfurt K, Szeja W (2017) Bioorg Chem 72:80–88
109. Ackermann U, O'Keefe G, Lee ST, Rigopoulos A, Cartwright G, Sachinidis JI, Scott AM, Tochon-Danguy HJ (2011) J Label Compd Radiopharm 54:260–266
110. James D, Escudier J-M, Amigues E, Schulz J, Vitry C, Bordenave T, Szlosek-Pinaud M, Fouquet E (2010) Tetrahedron Lett 51:1230–1232
111. Schulz J, Vimont D, Bordenave T, James D, Escuddier J-M, Allard M, Szlosek-Pinaud M, Fouquet E (2011) Chem Eur J 17:3096–3100
112. Berndl S, Herzig N, Kele P, Lachmann D, Li XH, Wolfbeis OS, Wagenknecht HA (2009) Bioconjugate Chem 20:558–564
113. Neef AB, Luedtke NW (2011) Proc Natl Acad Sci U S A 108:20404–20409
114. Neef AB, Luedtke NW (2014) ChemBioChem 15:789–793
115. Jao CY (2008) Salic. Proc Natl Acad Sci U S A 105:15779–15784
116. Curanovic D, Cohen M, Singh I, Slagle CE, Leslie CS, Jaffrey SR (2013) Nat Chem Biol 9:671–673
117. Grammel M, Hang H, Conrad NK (2012) Chem Bio Chem 13:1112–1115
118. Nainar S, Beasley S, Fazio M, Kubota M, Dai N, Correa IR, Spitale RC (2016) ChemBioChem 17:2149–2152

119. Nainar S, Cuthbert BJ, Lim NM, England WE, Ke K, Sophal K, Quechol R, Mobley DL, Goulding CW, Spitale RC (2020) Nat Methods 17:311–318
120. Wang DY, Zhang Y, Kleiner RE (2020) J Am Chem Soc 142:14417–14421
121. Xiong H, Seela F (2011) J Org Chem 76:5584–5597

Chapter 15
Growing Opportunities of Click Chemistry in Drug Development

15.1 Introduction

1,2,3-Triazoles, a fundamental class of N-heterocycle is considered one of the widely explored fundamental scaffolds in medicinal chemistry that play an important role in serving physiologically active drug candidates. Cu(I)-catalyzed 1,3-dipilar cycloaddition reaction of organic azides and terminal alkynes (CuAAC or Click Chemistry) [1, 2] has been identified as one of the most fascinating and highly efficient strategies for the rapid construction of diversely functionalized 1,4-disubstituted 1,2,3-triazole hybrids and their conjugates of well-known pharmacological activities [3–11]. Interestingly, the opposite regioselectivity in 1,2,3-triazoles was observed when the similar azide-alkyne cycloaddition was performed under ruthenium catalysis (RuAAC click) [12], and this chemistry has again diverse applications in medicinal chemistry [13]. Similar 1,2,3-triazole analogues can also be obtained by use of organocatalysis [14, 15] or catalyst condition [16]. Both the metal-mediated azide-alkyne 1,2,3-triazole forming click chemistry (CuAAC and RuAAC) were widely explored in an intramolecular fashion to give a high-to-excellent yield of triazole-fused as well macrocyclic molecular scaffolds of chemotherapeutic potential [17]. Interestingly, Bertozzi's strain-promoted azide-alkyne 1,2,3-triazole forming click reaction (SPAAC, bioorthogonal chemistry) works well in biological systems under catalyst-free conditions without interfering or interacting with the stuff found in the living system and thus have more impact in chemical biology and drug discovery and development [18–20].

The significance of the CuAAC tool lies in product 1,4-disubstituted 1,2,3-triazole, which itself is a potent pharmacophore and shows chemical and electronic resemblance with amide bond [21]. The compatibility of click reaction with almost all the natural and unnatural scaffolds has enabled the easy access of diverse hybrids and conjugates of chemotherapeutic potential and thus this way greatly speeded up the process of drug discovery. This chapter briefly present about the recent development

V. K. Tiwari et al., *Click Chemistry*, Materials Horizons: From Nature to Nanomaterials,
https://doi.org/10.1007/978-981-97-4596-8_15

of biologically potent 1,2,3-triazoles and their conjugates or hybrid analogues and their growing application in drug discovery and development.

15.2 Importance of 1,2,3-Triazole Scaffolding in Medicinal Chemistry

In recent years, molecular hybridization has been identified as one of the widely explored tools in medicinal chemistry for the successful development of novel, safe, and fast-acting drug candidates [6, 22, 23]. It involves the combination of two active pharmacophoric units of either similar bioactive units or different bioactive units into a single framework with improved affinity and efficacy. Two pharmacophoric units can be joined either by a linker or by framework integration strategies. It has been generally observed that the resulting hybrid molecules displayed better efficiency with minimal toxicity as compared to the parent drug candidates [8]. Considering this concept, several heterocyclic moieties including various azoles are being explored for the development of novel lead molecules for the possible treatment of frontline diseases.

Among various azoles, like imidazole, pyrazole, triazole, tetrazole, and pentazole, click-mediated 1,2,3-triazoles have been more widely explored as pharmaceutically important scaffolds than other related heterocycles [22]. This scaffold possesses some unique salient features as high chemical stability, a strong dipole moment. In general, 1,2,3-triazoles are stable to hydrolysis under acidic or basic conditions and redox conditions as well as were inert to metabolic degradation. The lone pair on N-3 of the 1,2,3-triazole simulates the O-atom of the carbonyl group and the electrophilic nature of C-4 of 1,2,3-triazole heterocycle resembles the carbonyl carbon of the amide group. Also, the C-H bond of the triazole possesses a similar hydrogen donor tendency to the N–H bond of the respective amide [24, 25]. Therefore, in addition to the π–π stacking interactions, both the 1,4-disubstituted 1,2,3-triazoles and 1,4-disubstituted 1,2,3-triazoles can form H-bonds and they considered as mimetic of the amide bond, similar to a *Z*- and *E*-amide, respectively [26].

Because of notable resemblances of 1,2,3-triazole with peptide linkage, they exhibit a wide spectrum of pharmacological activities and some of the representative commercially available potent drug molecules [21] that contain 1,2,3-triazole moiety in their core structure is listed in Fig. 15.1.

The representative potent candidates (**I–VI**) include carboxyamido-triazole (**I**, anticancer agent), Tazobactum (**II**, β-lactum antibiotic), I-A09 (**III**, an anti-TB agent in clinical evaluations), Rufinamide (**IV**), Radezolid (**V**, an antibiotic), and a non-nucleoside reverse transcriptase inhibitor as anti-HIV agents, tert-butyldimethylsilyl spiro amino oxathiole dioxide (TSAO, **VI**), etc. Several 1,2,3-triazole-linked conjugates displayed notable pharmacological activities. For example, the bioactivity of marketed ketolide telithromycin **1** was further improved via linking with 1,2,3-triazole spacer using click tool and antibiotic Solithromycin **2** thus obtained is

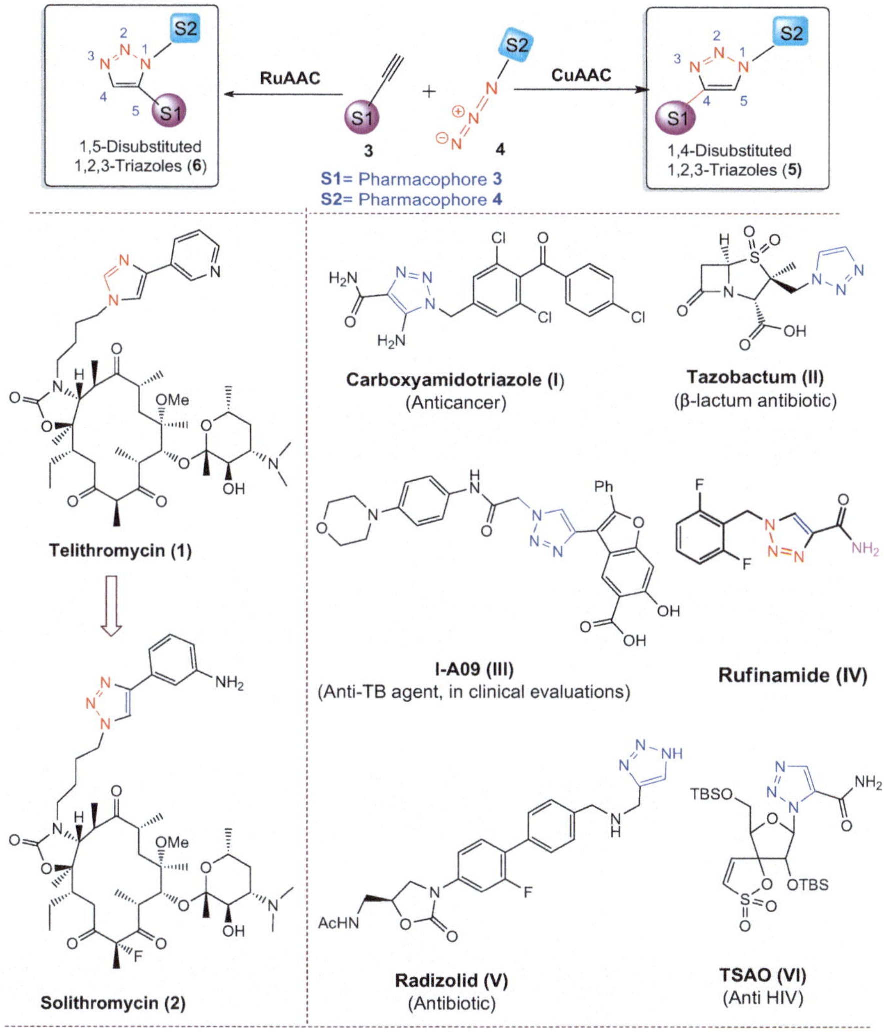

Fig. 15.1 Click-mediated 1,2,3-triazole functionalization required to improve their potency and led to even more potent drug candidates

now under phase 3 clinical trial. It has 1,2,3-triazole-phenylamine in the side-chain terminal on the contrary of imidazole-pyridinyl in telithromycin. The associated side effects with telithromycin **1** such as nausea, diarrhea, headache, anxiety, and muscle weakness were notably reduced in the antibiotic Solithromycin (Fig. 15.1) [27, 28]. The various biological activities such as anti-diabetic, anti-viral, anti-cancer, anti-tubercular, etc. associated with the click chemistry-mediated 1,2,3-triazoles and their analogues is presented in separate sections of this chapter.

15.3 1,2,3-Triazole Hybrids as Anti-diabetic Agents

The breakdown of glycosidic linkages in molecular arrangements in glycans frequently occurs in several cellular processes and thus through modulation of the pathways of cell cycles may result to the development of novel inhibitors of glycosidases and glycosyl transferases as promising chemotherapeutics to cure diabetes [29], viral infections [30], cancer [31], and other diseases. Toward this end, click-derived 1,2,3-triazole-appended hybrids or conjugates of sufficient stability in the physiological pH have been developed to influence the molecular systems required for specific glycosidase inhibitors. Once sugar is linked to amide functionality, the resulting hybrid scaffolds not only find resistance to proteases and enhanced thermal stability, but they modulate receptor binding and signaling phenomena. Since click-mediated triazoles are the bioisoesters of amides, and therefore, the 1,2,3-triazole-linked sugars have been considered privileged scaffolds in medicinal chemistry.

Glycohybrid 1-β-D-glucosyl-4-phenyl triazole **7** showed appreciable inhibition for *E.* galactosidase and 1-β-D-galactosyl-4-phenyl triazole **8** displayed improved inhibitory activity against bovine liver galactosidase [32]. The glycohybrids **9** and **10** displayed promising activities against enzymes glycogen phosphorylase, α-glucosidase, and glucose-6-phosphatase inhibitory activities [33] The glycohybrid **11** is identified as a potential α-glycosidase enzyme inhibitor that displayed 20-fold more potency than acarbose **12**, a standard naturally occurring anti-diabetic drug and can be further explored as anti-diabetic therapeutics (Fig. 15.2). The bioactivity was further supported by docking studies on a homology model of yeast maltase MAL12 [34].

Iminosugar (for example, DNJ **13**) and their derivatives, e.g., *N*-hydroxyethyl analogue, miglitol **14** (marketed as Glyset), miglitol, etc.) is an approved drug for the treatment of type II diabetes and a *N*-butyl derivative (NBDNJ) miglustat **15** (marketed as Zavesca in phase II trials for the cure of cystic fibrosis) is an effective drug for the treatment of Gaucher disease. This important sugar once linked with click-mediated 1,2,3-triazole pharmacophore, the resulting hybrids may serve as a promising scaffold in drug development for diabetes [35–37]. Considering the advantage of DNJ-based glycopolypeptide action against α-mannosidase, [36] a iminosugar-based glycohybrid **16** (Fig. 15.3) was developed which displayed improved potency against α-glycosidase inhibition ($IC_{50} = 1.15\ \mu M$) than DNJ ($IC_{50} = 1.67\ \mu M$) [37].

DNJ–admantyl hybrid **17** showed potent α-glucosidase inhibition and was found to disrupt the viral glycoprotein processing with increased glucosylation on bovine viral diarrhea virus BVDV E2 species [38]. Also, deoxynojirimycin–triazole hybrid **18** ($IC_{50} = 0.063 \pm 0.006$ mM) displayed competitively enhanced α-glucosidase inhibitory activity than standard drug, 1-deoxynojirimycin DNJ **13** ($IC_{50} = 0.155 \pm 0.015$ mM) [39]. Likewise, multivalent iminosugars **19** and **20**, a 1,2,3-triazole-linked analogue of glycosidase inhibitor 1-DNJ with oligoethylene spacer displayed promising glycosidase inhibitory activities (Fig. 15.3) [35].

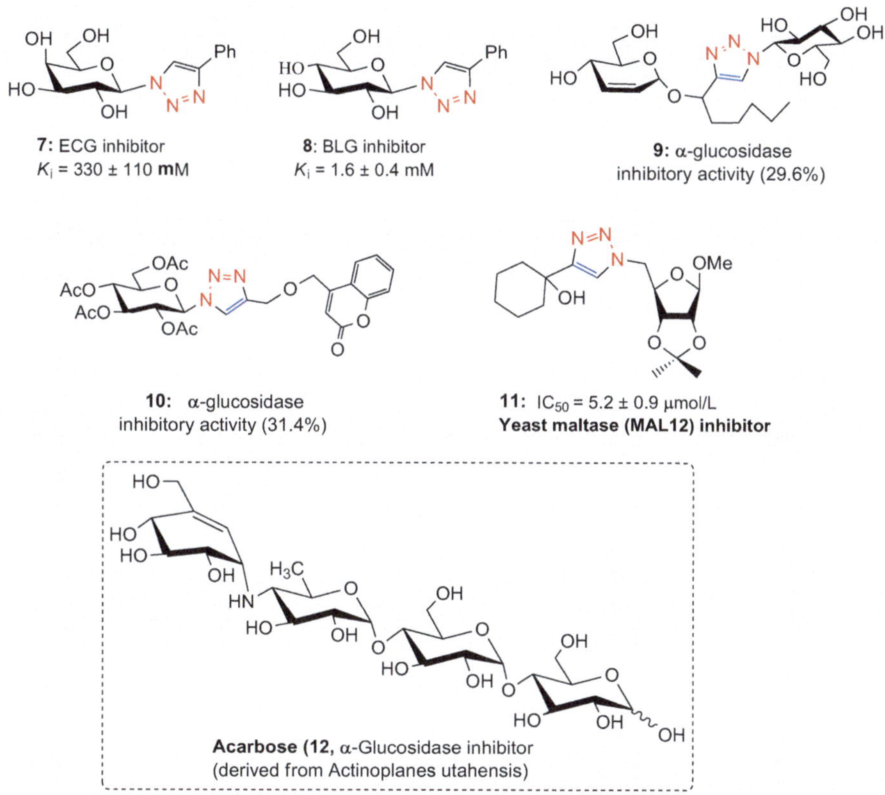

Fig. 15.2 α-Glucosidase inhibitory activity of some potent glycohybrid 1,2,3-triazoles (**7–11**)

Sugar-based triazolium functionalized motifs displayed potent inhibitor of glycogen phosphorylase [40]. 1,2,3-Triazole-oleanolic acid-appended **21** (IC_{50} = 1.14 μM) showed notable glycogen phosphorylase inhibitory activity [41]. The glycohybrid **22** (IC_{50} = 26 μM) and pentacyclic triterpene glycohybrid **23** showed moderate phosphorylase inhibition against RMGPa (Fig. 15.4) [42].

4-Phenyl-*N*-(*β*-D-glucopyranosyl)-1*H*-1,2,3-triazole-1-acetamide **24** (IC_{50} = 13.6 μM) displayed potent glycogen phosphorylase inhibitory activity with Ki = 17.9 μM [43]. Further exploration lead to the glycohybrid **25** having improved glycogen phosphorylase inhibitory potency, [44] while glycohybrid **26** have moderate potency than the standard drug (Fig. 15.5) [45]. Some simple glyco hybrids **27–29** displayed similar inhibitory potency of the rabbit muscle glycogen phosphorylase inhibitory although further clinical investigation required to reach the most potent inhibitor [46]. Also, a series of xylose- and 5-thioxylose-based 1,2,3-triazoles **30–32** displayed glycogen phosphorylase inhibitory activity [47]. Likewise, *β*-D-glucopyranosyl functionalized glycohybrids **33** and **34** were identified as promising

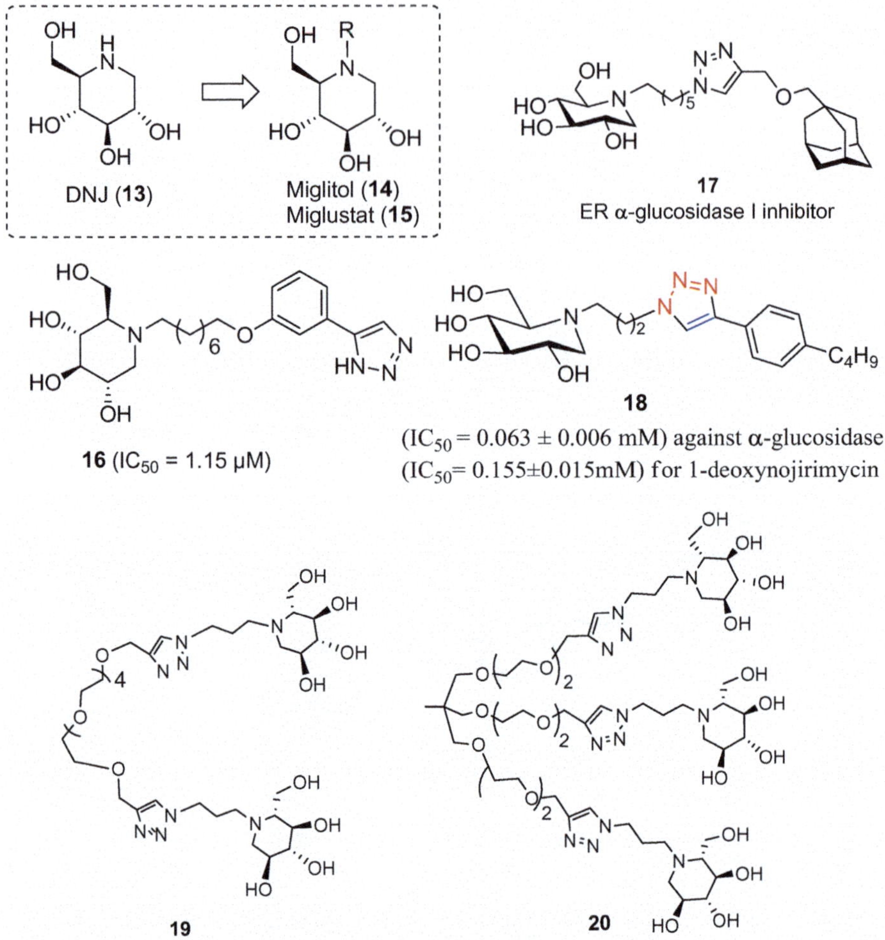

Enzyme	1-DNJ (12)	19	20
α- Galactosidase	-	105 ± 5	225 ± 5
β- Galactosidase	42	760 ±10	790 ± 10
β- Glucosidase	47	70 ± 2	400 ± 10
α- Mannosidase	270	90 ± 3	35 ± 2
β- Mannosidase	-	772 ±10	--
Amyloglucosidase	2.1	20 ± 2	11 ± 1

Fig. 15.3 DNJ-inspired 1,2,3-triazole-appended glycohybrids **16–20** as potent anti-diabetic agent

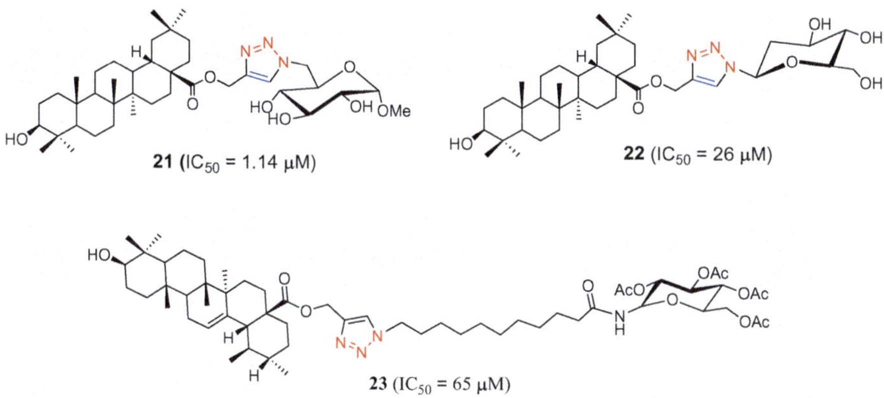

Fig. 15.4 Pentacyclic triterpene glycohybrids (**21–23**) as bivalent glycogen phosphorylase inhibitors

glycogen phosphorylase inhibitor in rabbit muscle at a moderate level (Fig. 15.5) [48].

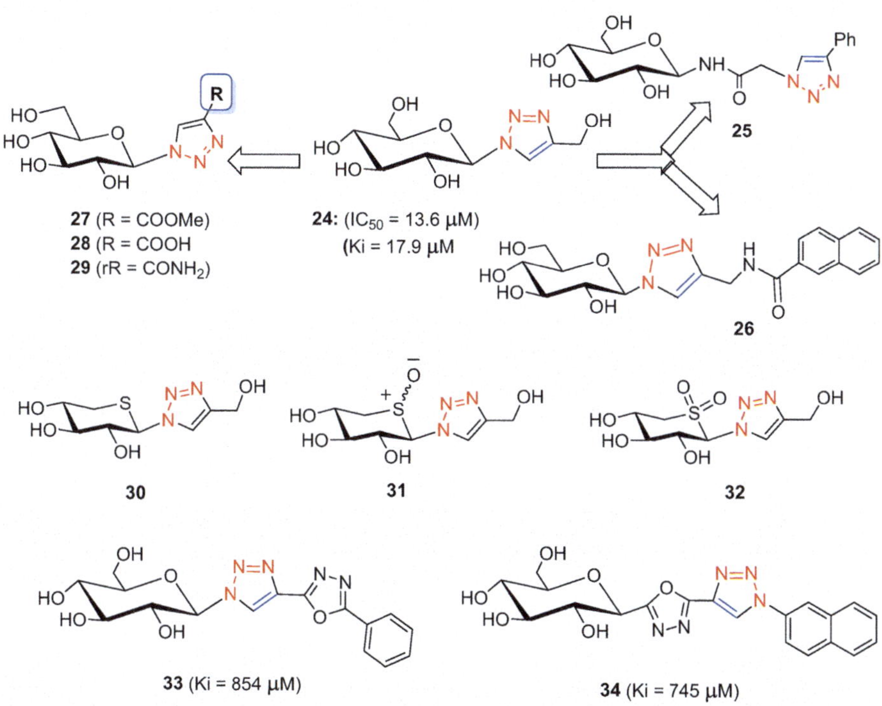

Fig. 15.5 Simple glycoconjugates **24–34** as potent glycogen phosphorylase inhibitors

For the drug development against type 2 diabetes mellitus, the protein tyrosine phosphatase 1B (PTP1B) is also well considered and several efforts are focused on structure–activity relationship (SAR) to achieve the lead optimization. 1,2,3-Triazole-appended glycohybrid **35** was found to be the most potent PTP inhibitory activity [49]. Also, benzyl 6-triazolo(hydroxy)benzoic glucoside **36** displayed potent PTP1B inhibitory activity [50]. These pharmacophore PTP1B inhibitors are more selective over LAR (leukocyte antigen-related tyrosine phosphatase), TCPTP (T-Cell PTP), SHP-1(PTP-1 having SH-2), and SHP-2. Phenylalanine-aryl *C*-glycoside **37** ($IC_{50} = 5.6\ \mu M$) and tyrosine-aryl *C*-glycoside **38** ($IC_{50} = 5.5\ \mu M$) displayed potent inhibitory activity against PTP1B [51]. The 1,2,3-triazole-linked glycohybrids **39** ($K_i = 14.4\ \mu M$) and **40** ($K_i = 8.7\ \mu M$) displayed promising inhibitory activity against PTP1B, where glycoconjugate **40** ($IC_{50} = 31.0 \pm 9.6\ \mu M$) was identified as a potent inhibitor against TCPTP enzyme (Fig. 15.6) [52].

Further, triazolyl α-ketocarboxylic acids derived glycoconjugate **41** exhibited promising PTP1B inhibitory activity [53]. The glycohybrid **42** ($IC_{50} = 5.1\ \mu M$) displayed notable potency as the selective inhibitor of PTP1B [54]. Further, the dimeric pharmacophoric module of *β*-CD-glycosyl 1,4-dimethoxy-arene 1,2,3-triazole glycohybrids **43** and **44** (($IC_{50} = 0.62$–$0.88\ \mu M$) identified as the potent inhibitor of PTP1B [55]. Interestingly, *O*-benzylated glycosyl analogue showed better and selective inhibitory potency against PTP1B (Fig. 15.6) [56].

The enzyme *O*-GlcNAcase is known for the selective hydrolysis of *O*-linked *β*-*N*-acetyl glucosamine and influence the function of specific proteins essential to control the diabetes type-2. The triazolyl glucopyranosides **45–47** (Fig. 15.7) on MTT assay and Western Blot assay has been identified as potent inhibitor of OG type-A inhibitor [57].

Because of the resemblance of 1,2,3-triazole residue with peptide linkages, several triazole-based hybridization scaffolds exhibit a wide range of pharmacological activities. For a representative example, the 1,2,3-triazole analogue **49** is identified as the better inhibitor of rabbit muscle unphosphorylated glycogen phosphorylase (RMGP*b*) than respective amide analogue **48** (Fig. 15.8) and can be explored as lead candidate to control the hyperglycemia in diabetes type-2 [58].

The above-mentioned examples on 1,2,3-triazole-appended conjugates or their hybrid analogues, this is obviously easy to draw a conclusion that these scaffolds have enormous potential to explore in drug development against diabetes. The bioisosteric relationship between amide and 1,2,3-triazole can be well utilized for generation of novel lead against this disease. Further, the variation on carbohydrate residue like carbasugar, thiosugar, or iminosugars, there should be a great scope to explore them via permutation and combination with other aglycon scaffold with 1,2,3-triazole as linker for the development of effective drug molecules.

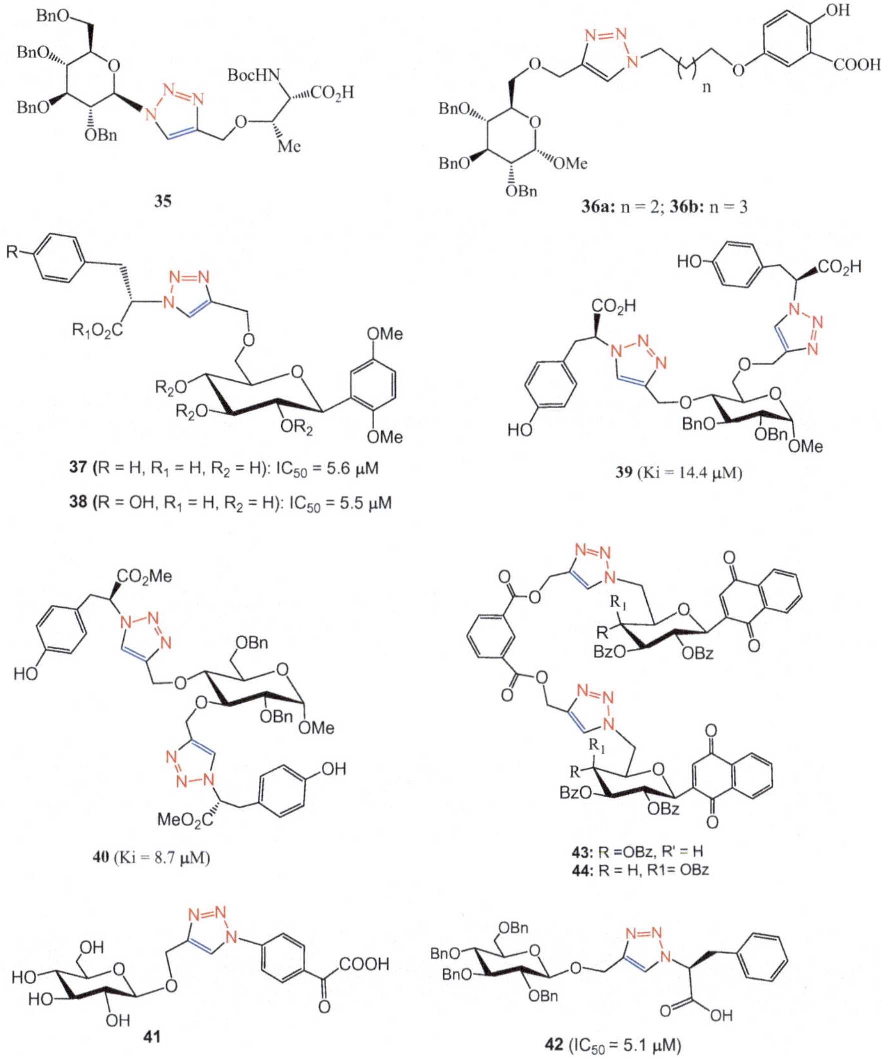

Fig. 15.6 Some simple glycohybrid 1,2,3-triazoles **35–44** as potent protein tyrosine phosphatase (PTP1B) inhibitors

15.4 1,2,3-Triazole Hybrids and Conjugates as Anti-cancer Agents

Click-mediated 1,2,3-triazoles have shown notable application in autophagy-dependent apoptosis in cancer therapy [23]. A 1,4,5-trisubstituted triazolo analogue **50** having the inhibition property against breast cancer cell growth. Based on biological evidence, compound **50** suppresses the growth of breast cancer cells in vitro and

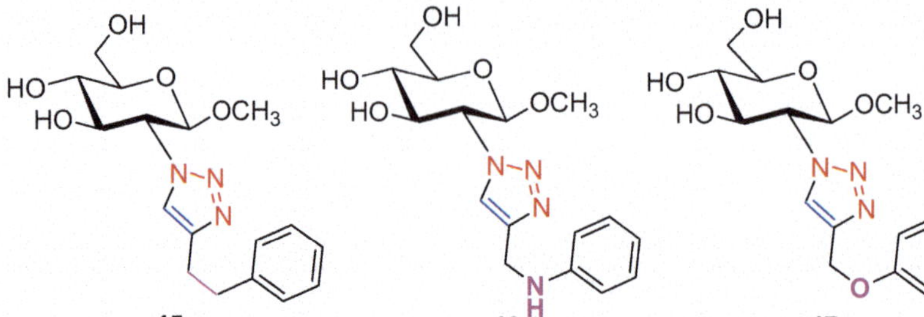

45 46 47

Fig. 15.7 Structure of 1,2,3-triazole-linked glucopyranosides as potent OGA inhibitor

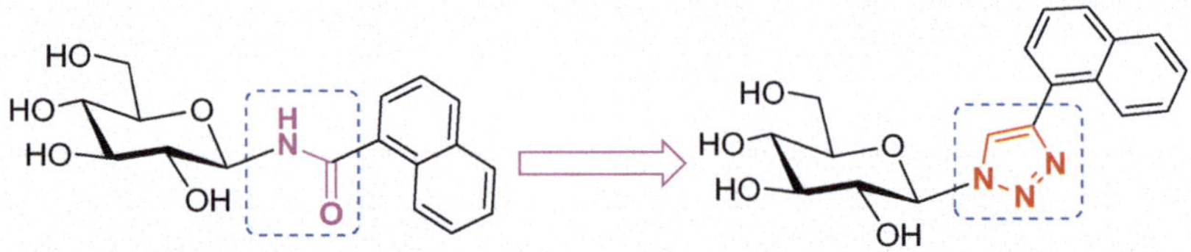

48

Rabbit muscle GP*b* (RMGP*b*)
% inhibition = 75
Ki (μM) =190.7 ± 23.7

49

Rabbit muscle GP*b* (RMGP*b*)
% inhibition = 95
Ki (μM) =135.9 ± 9.5

Fig. 15.8 Resembles of amide linkage to 1,2,3-triazole, biologically potent and stable linker

in vivo by triggering autophagy-mediated apoptosis and may lead to the development of a viable therapeutic approach for the treatment of breast cancer (Fig. 15.9) [59].

Moreover, click chemistry was employed for the development of **Histone Deacetylase** (HDAC) inhibitors **51–54**. Among them, compound **51** was found slightly potent in screening cancer cell line. On the other hand, an interesting result with compound **51** was observed and demonstrated a tenfold potency relative to tumor cells in in vitro safety testing, demonstrating considerably less toxicity to

Fig. 15.9
1,2,3-Triazole-bounded anti-cancer agent **50** for the treatment of breast cancer

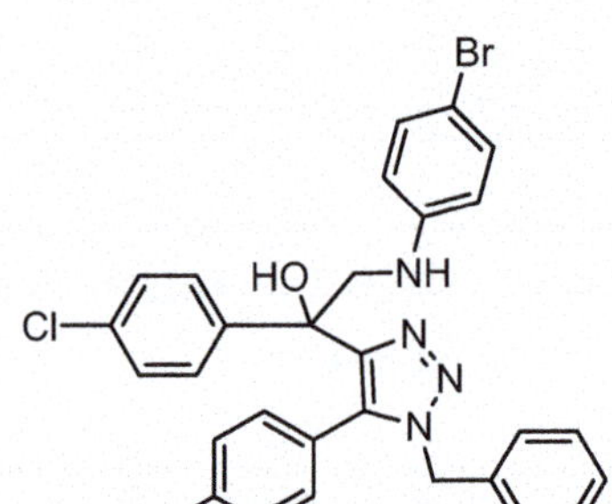

50

non-transformed cells than the commercially available medication, suberoylanilide hydroxamic acid (SAHA) (Fig. 15.10) [60a].

A 1,2,3-triazole-appended branched hydroxamate histone deacetylase (HDAC) inhibitor **54** displayed strong growth inhibition ($GI_{50} < 1\ \mu M$) when tested against three tumor cell lines—the human liver carcinoma cell line HepG2, the human adenocarcinomic alveolar basal epithelial cells A549, and the breast cancer cell line MDA-MB-231 [60b].

Employing the click reaction HDAC8 Inhibitors **55** and **56** were synthesized and found that compound **55** (IC_{50} value of 0.070 μM) demonstrated greater inhibitory action against the known HDAC8 inhibitor PCI-34058 ($IC_{50} = 0.31\ \mu M$). Additionally, it became apparent that these inhibitors had growth-inhibitory effects on T-cell lymphoma and neuroblastoma cells ($GI_{50} = 3–80\ \mu M$), which reveals that HDAC8 inhibitors may have anti-cancer properties. These findings led to the development and

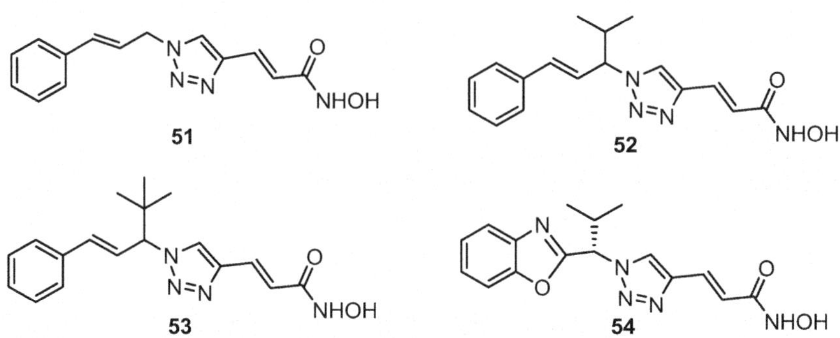

Compound	Tumor cell lines	IC_{50} (nM)
TSAHA	HDACs	67
51	HDACI	104±30
52	HDACs	22
53	HDACs	18
54	HDACs	7

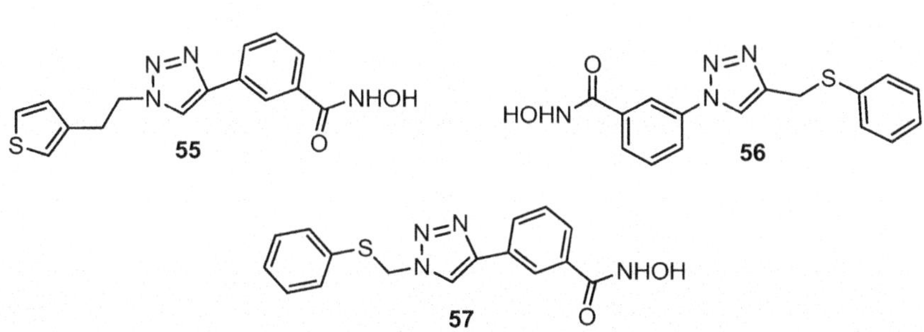

Fig. 15.10 1,2,3-Triazole-linked analogues **51–57** as potent HDAC inhibitors

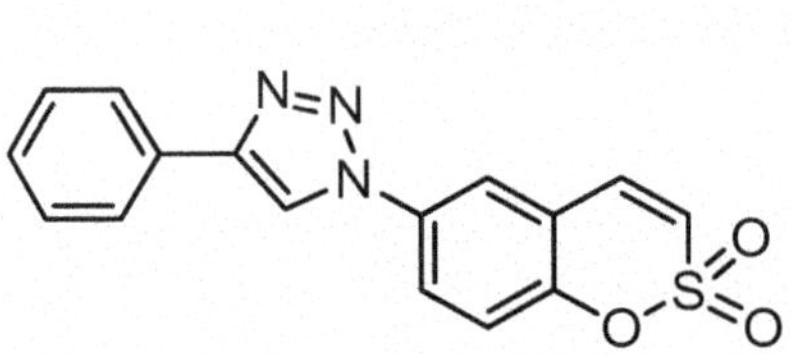

1,4-substituted triazole **58**

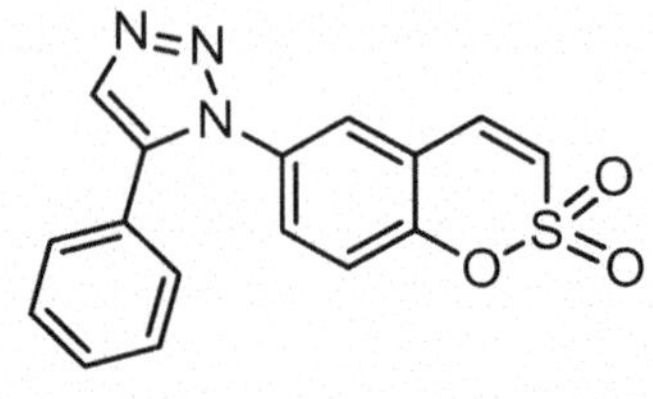

1,5-substituted triazole **59**

Fig. 15.11 Developed active agents **58** and **59** against human CAs IX and XII

synthesis of derivatives of **56** that are HDAC8-selective inhibitors of these, triazole **57** (IC_{50} = 0.053 μM), exhibited the highest potency (Fig. 15.10) [61].

In addition, a number of 1,2,3-triazole-linked conjugates or their hybrid analogues were identified as potent **Carbonic anhydrase inhibitor**. The coumarin-based conjugates were known for their carbonic anhydrase inhibitory activity and thus sulfocoumarin derivatives appended with triazole was developed by Supuran and group, where compound **58** showed potent activity against human CAs IX and Cas XII (IC_{50} = 29–32 nM range) [62]. Motivated from results, in preceding year the same group developed 5-substituted 1,2,3-triazole **59** under ruthenium catalysis. The compound **59** showed the activity against humanCAs IX (67.1 nM) and Cas XII (9.5 nM) with the value of Ki in nanomolar range (Fig. 15.11) [63].

A report has been documented in which use of human acetylcholinesterase (AchE) as an enzyme template for the development of huprine-based inhibitors were demonstrated [64]. For click tool (in situ), Huprine X (IC_{50} = 0.75 nM) and Huprine Y (IC_{50} = 0.78 nM) found to be appropriate catalytic site binder, for instance, respective triazolo derivative **62** was developed via the association of azido analogue of Huprine **60** and alkyne **61** (Fig. 15.12) [65].

The intriguing structural heterogeneity of glycohybrids serves an array of functions and has been dubbed 'glycocode' in glycobiology. The term glycocode refers to the processes of anomerization, isomerization, functional substitution, glycosidation, and other biological functions. Furthermore, the restriction of glycosidic linkage flexibility results in a number of conformational analogues with unique thermodynamically favorable configurations. This phenomenon is known as the third dimension in sugar code, and it explains the glycohybrid molecular recognition process. Furthermore, glycohybrids affect the physiochemical characteristics of the extracellular matrix through charge density or hydrophilicity. However, some low molecular weight glycoconjugates, such as eleutherobin, are potential chemotherapeutics. Eleutherobin is a marine diterpene glycoside with unusual microtubule stabilizing properties comparable to paclitaxel and significant cytotoxic activity that is now being tested in the drug development process. The identification of appropriate binding sites and their favorable chemical structures on target proteins to serve as templates for the synthesis of ligands, or the random combination of chemical groups to assemble a diverse molecular library, have become the predominant

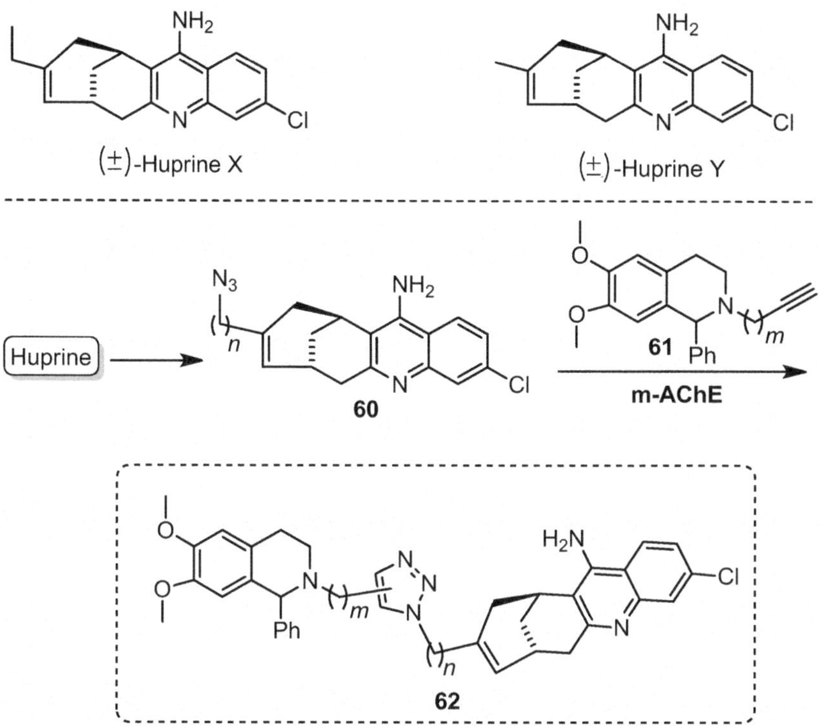

Fig. 15.12 Application of Huprine in click reaction as catalytic site binder

approaches to drug design since 2000 [66]. Hybridization of bioactive candidates with other molecular motifs is becoming a more well-known and effective approach for drug development to target an array of disease. In addition to the possibility of better medications to treat illnesses like respiratory infections, heart diseases, cancer, malaria, tuberculosis, and AIDS, it has the ability to address a variety of trial failure drawbacks [67–69]. The first mention of new carbohydrate-appended conjugates in literature dates back to 1994, Chamberlin et al. develop an acyclovir prodrug for use in chemotherapy [70]. Nevertheless, the modular click approach is unquestionably a more comprehensive, intriguing, and investigated synthetic method for the synthesis of glycohybrids. Through the use of 1,2,3-triazolyl spacers, the CuAAC has greatly aided in the extemporization of an enormous library of bioactive entities that bridge appropriate glycon and aglycon. The triazole unit contributes to the bioactivity of the molecule by increasing the protein interaction, which in turn satisfies stable and secure linkage between glycol hybrid molecular building components and remain an active part of drug discovery and development activities. Over the past 20 years of ongoing finding of gap, rationale or research, numerous research groups have made enormous efforts that have assembled adaptable click-inspired anti-cancer scaffolds. The carbonic anhydrase, a member of the rapidly expanding enzyme family, has a regulatory role in oncogenesis. The CA-IX subfamily inhibits the growth of cells by

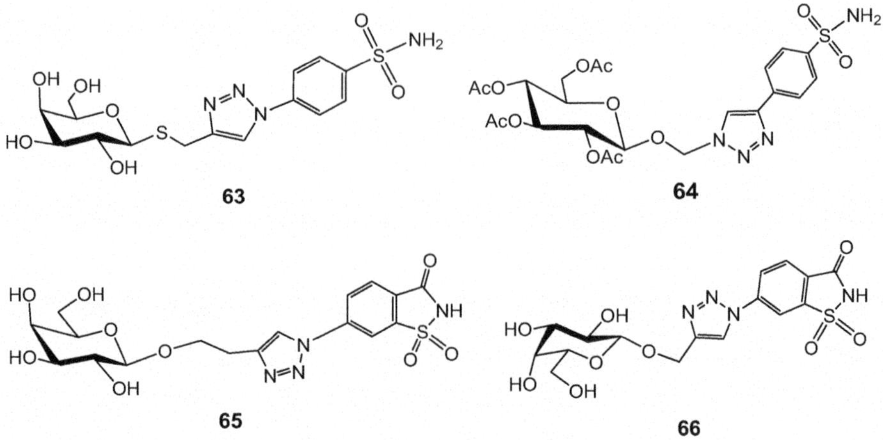

Fig. 15.13 Some sulfonamide-appended glycoconjugates as potential CA inhibitors

affecting the von Hippel–Lindau vHL0–hypoxia-inducible factor (HIF). As a result, the macromolecule received enough attention to with stand strong antagonist library advancement [71].

The first click assisted synthesis of CA inhibitor was reported in 2007. Followed by, an array of research groups enforced to develop the molecular library. The developed molecular analogues found to be efficient for a range of activities including the glycoconjugated benzene sulfonamides were human carbonic anhydrase isozymes inhibitors and the compounds **63, 64** was reported as the highly potent, specifically selective for the CA-IX enzyme (Fig. 15.13) [72]. Furthermore, some ruthenium-catalyzed benzenesulfonamide-based triazole-appended glycohybrid was developed as CA inhibitors [73]. Moreover, extend the same approach saccharin as a cyclic secondary sulfonamide invested triazolyl glycohybrids **65**, **66** are revealed up as CA inhibitors [74, 75].

1,2,3-Triazole-appended glycohybrid have intriguing activities in addition to sulfonamide-appended glycoconjugates, other cell lines, such as HeLa, MCF-7, and PM-3, were also examined and reported as potential glycohybrids. Some triazolyl glycohybrids of 18-β-glycyrrhetinic acid compound **67** for HeLa cell line [76], novobiocin triazolyl glycohybrids with coumarin unit **68** for MCF-7 [77], cyclopamine glycohybrids compound **69** for lung cancer [78], 8-hydroxyquinoline glycohybrids **70** for ovarian cell line [79], 6′-triazole perillyl glucoside **71** against A549 and PC3 cell lines [80], L-glucurono-γ-lactone hybrid **72** as potent tubulin depolymerization inhibitor have been reported in the literature (Fig. 15.14) [81].

However, there have been reports of various cell line active glycohybrids being corrected. Flavonoid-enhanced glycohybrids, for example, shown strong anti-cancer activity against three cancer cell lines, Hela, HCC1954, and SK-OV-3 (triazolyl

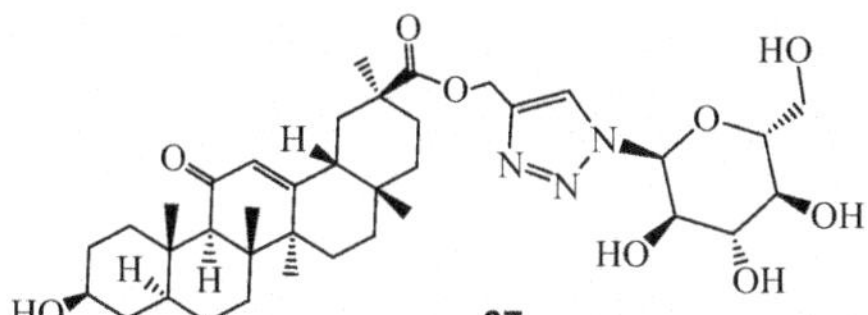

67

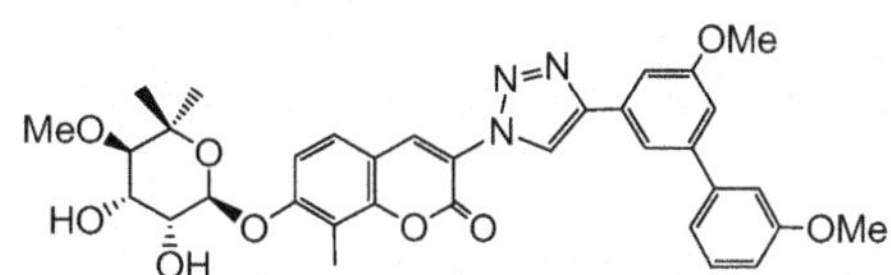

68

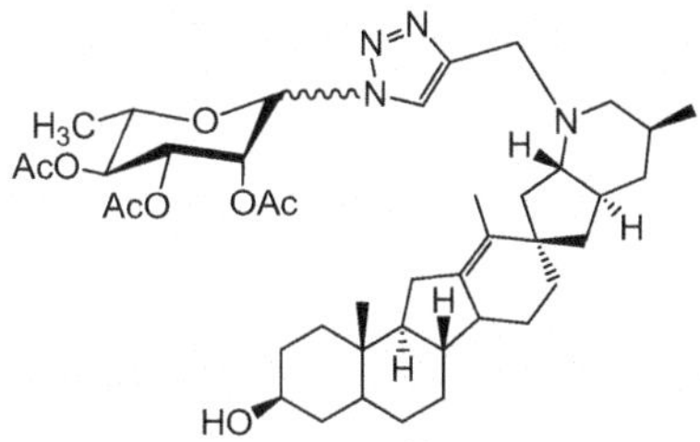

69

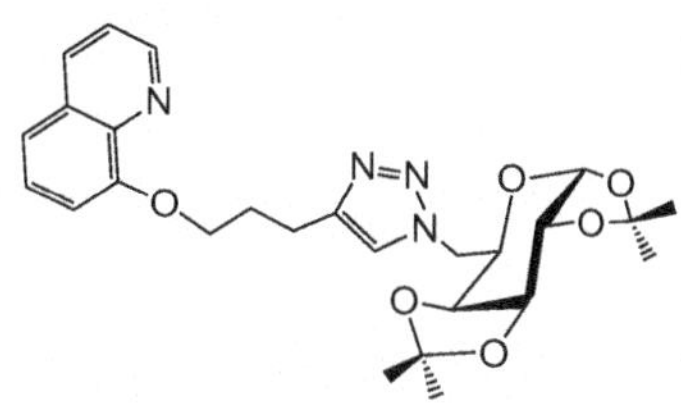

70

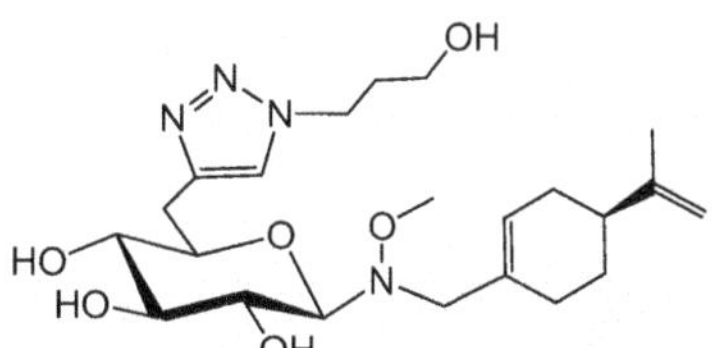

71

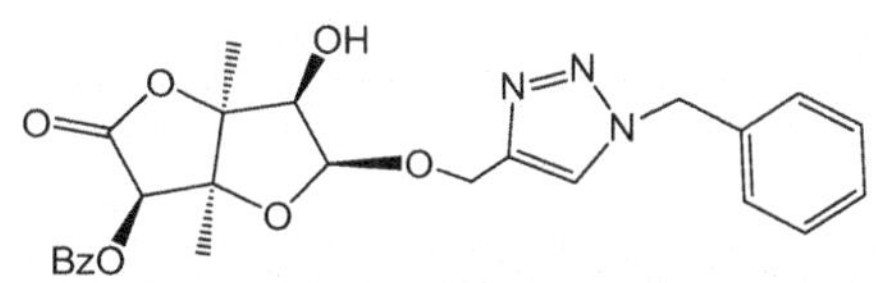

72

Fig. 15.14 1,2,3-triazole-appended glycohybrids **67–72** having versatile cell line activity

analogues **73, 74**) [82]. The nucleoside glycohybrids were tested against MDA-MB-231, Hep3B, PC-3, SH-SY5Y, and HCT-116 cell lines; however, they were particularly effective against Hep3B (triazolyl analogue **75**) [83]. Rapamycin–sialoside hybrid **76** demonstrated consistent biochemical impact (Fig. 15.15) [84].

Furthermore, indole-based glycohybrid **77** and oxindole-based glycohybrid **78** were effective against the DU145 cell line. Furthermore, cell cycle research confirms that the mechanism of action occurs during the sub-G1 phase of DU145 cells in a dose-dependent manner [85]. Interestingly, triazolyl analogues **77** and **78** can cause cell growth arrest by inducing apoptosis via the mitochondrial-mediated intrinsic mechanism and the pro-proliferative pathway (Akt/PKB). However, the 8-hydroxyquinoline based triazolyl glycohybrid **79** was more active than quinaldines and ten times more active than the parent motif against HeLa, HCT116, MCF7, U251, Hs683, PANC-1, and AsPC-1 cancer cell lines [86]. Through caspase activation and autophagy, glycohybrid **80** demonstrated promising anti-leukemic activity in multiple hematopoietic cell lines (Fig. 15.16) [87].

Remarkably, ferrocene chalcogeno glycoconjugates **81** and **82** furnished with selenium showed favorable activity against A549, HeLa, MDA-MB-231, and MCF-7 [88]. After that, hederacolchiside a glycohybrid compound **83** was shown to be

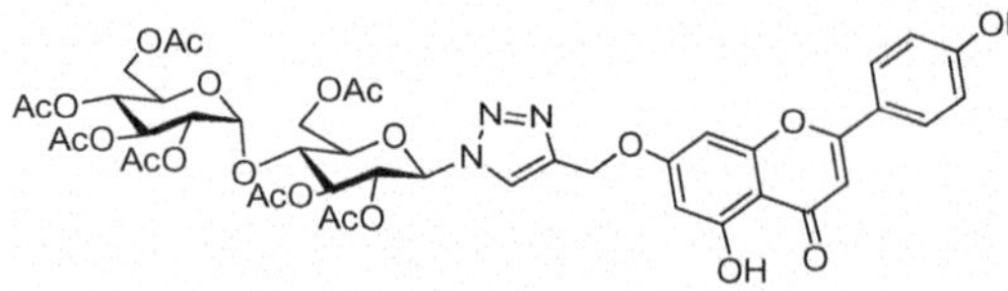

73 (IC_{50} = 4.67 μM)

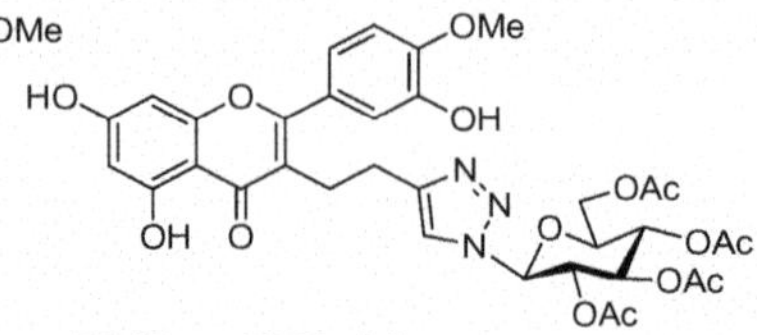

74 (IC_{50} = 30.56 μM)

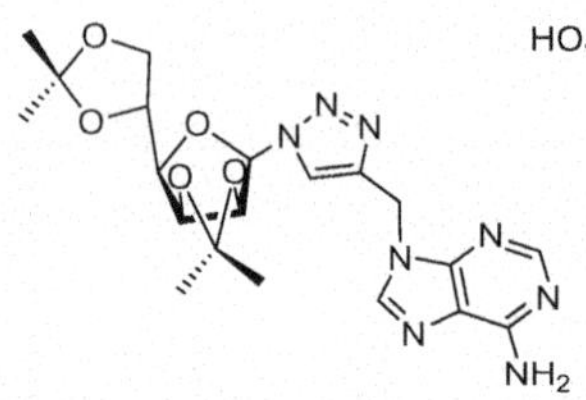

75

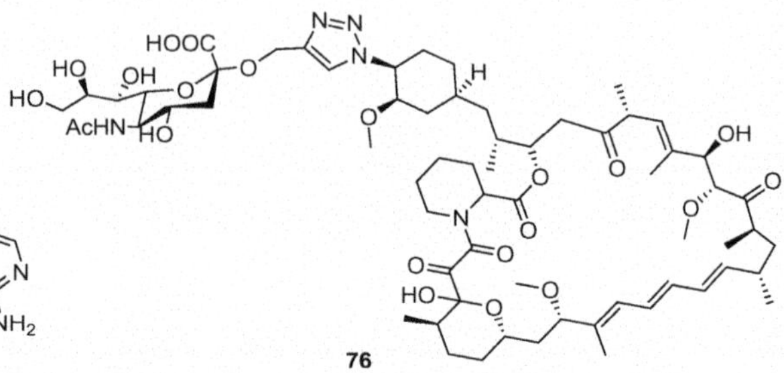

76

Fig. 15.15 Structure of some versatile anti-cancerous triazole-appended glycohybrids

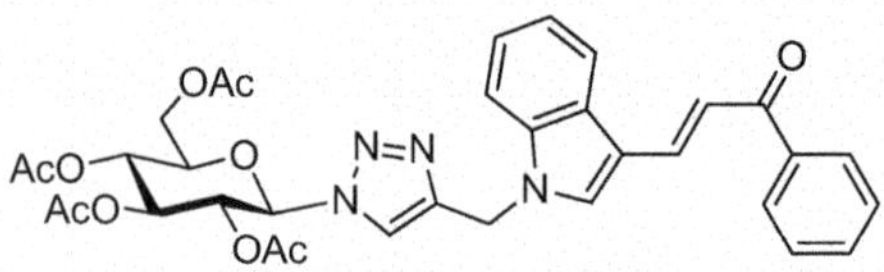

77

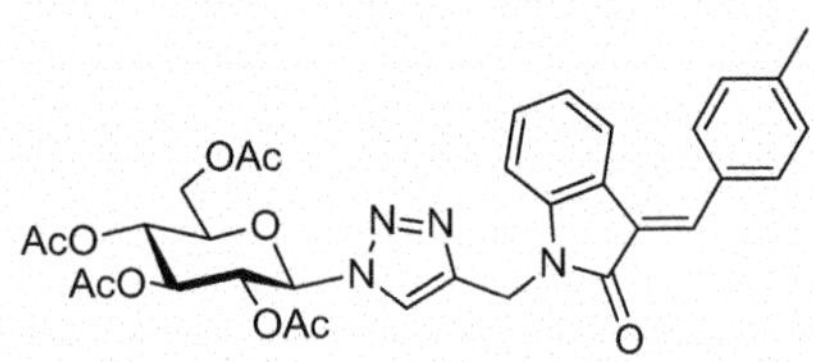

78

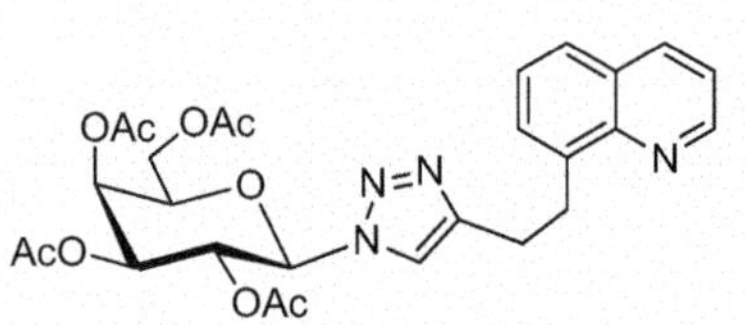

79

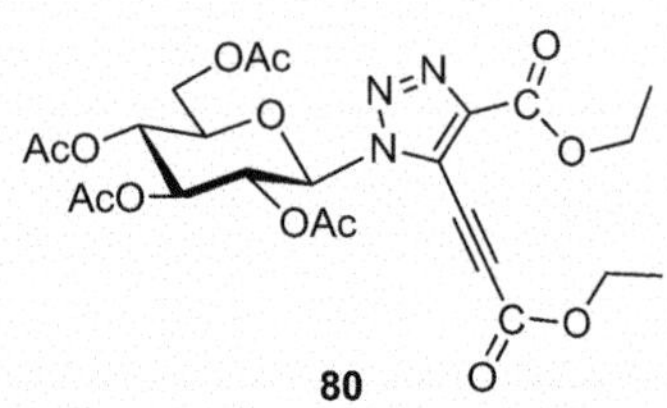

80

Fig. 15.16 Structure of various cell line active glycohybrids **77–80** as promising anti-cancer agent

effective against PC3, HT29, HepG2 A549, HL60, and U937, among other examined cell lines [89]. Additionally, cell cycle analysis and the apoptosis assay verify that the mechanism of action is to suppress cell proliferation by triggering apoptosis during the G1 and S stages of the cell cycle. Over the past ten years, noscapepine has been recognized as an anti-cancer pharmacological agent. It is a natural substance with dual uses. The second-generation noscapine triazolyl glycochybrids, with an emphasis on the C-7 position of the benzofenone ring, have been synthesized and subjected for biological activities by Tiwari et al. The noscapine-galactose hybrid

84 showed the most encouraging efficacy against the Hela cell line. The effectiveness of the glycoconjugate in the molecule's powerful nature was determined by additional docking investigation [90]. Moreover, to the analysis of anti-cancer activities of click-mediated triazolyl analogues, it is interesting to note that compound **85**, dimeric podophyllotoxin glycohybrid, exhibited in vitro anti-cancer activity against several cancer cell lines, including HL-60, SMMC-7721, A-549, MCF-7, and SW-480 (Fig. 15.17) [91].

An array of triazolyl glycoconjugates were assembled by adaptable substitution from well-known sulfonamides to saponin targeting from CA to microtubules to create a canvas for cancer-relevant pharmacophore discovery, molecular library genesis, and drug sculpting. Novel click-inspired glycohybrids are predicted to establish themselves as more reliable and biocompatible anti-cancer drugs in future years.

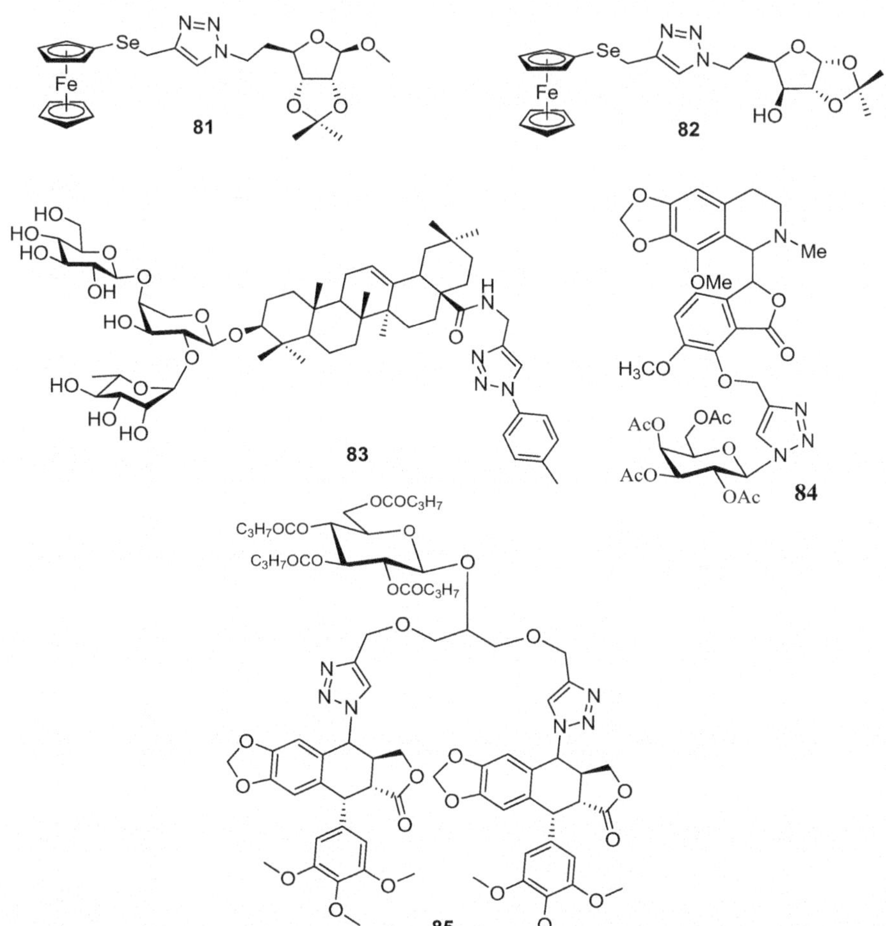

Fig. 15.17 Structure of click-mediated triazolyl glycohybrids **81–85** as anti-cancer analogues

Fig. 15.18 Structure of the thiosugar naphthalene diimide conjugate **86**

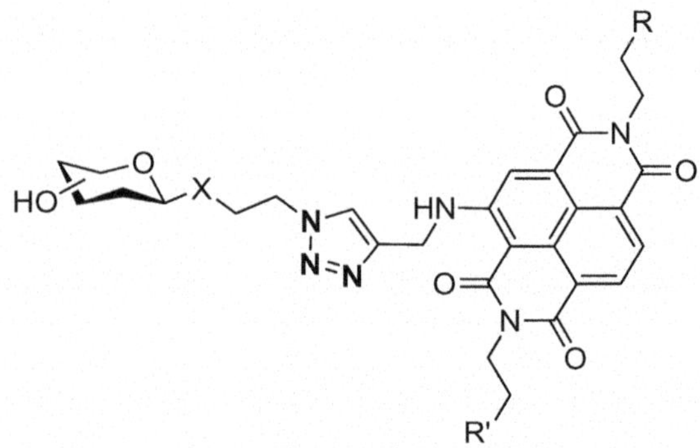

86

Another study found that at sub-μM concentrations, selectively developed thiosugar naphthalene diimide conjugates **86** have strong growth inhibition effect against colon cancer cells. As a result, the glycohybrid has the potential for high selectivity against human fibroblast controls (Fig. 15.18) [92].

Liu et al. identified a family of DNA-interacting anti-cancer drugs in the form of bowl-shaped corannulene sugar conjugates **87a–c**. These corannulene-bowl analogs block DNA transcription and show specific biological and pharmacological effects (Fig. 15.19) [93]. The 7-ethyl-10-hydroxycamptothecin glycohybrid **88** displayed potent anti-tumor activity against HCT116 human colon cancer cells. Among the series, glucose and galactose conjugates had the best anti-tumor effectiveness. Furthermore, the type of sugar connected to glycoconjugated influenced its anti-tumor efficacy [94].

The cytotoxic effects of 4β-(1,2,3-triazol-1-yl) podophyllotoxin's carbamate analogues were assessed against the HeLa, HCT-8, HL-60, and A-549 human cancer

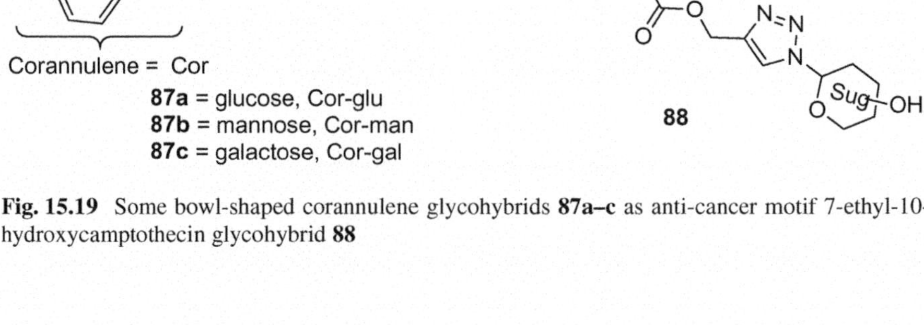

Fig. 15.19 Some bowl-shaped corannulene glycohybrids **87a–c** as anti-cancer motif 7-ethyl-10-hydroxycamptothecin glycohybrid **88**

cell lines. A triazolylmethyl carbamate **90** analogous to podophyllotoxin's carbamate originated from podophyllotoxin, **89** (an anti-microtubule agent) [95], exhibited the most effective cytotoxicities, inducing cell cycle arrest in the G2/M phase, followed by apoptosis, and inhibiting microtubule formation in A-549 cells. This molecule may additionally inhibit the DNA topoisomerase II (Fig. 20a) [96] Interestingly, a podophyllotoxin analogue 4β-1,2,3-triazolyl-podophyllotoxin glycohybrid **92** displayed potent anti-cancer activity than the standard drug etoposide **91** against five major human cancer cell lines (Fig. 20b) [97]. Moreover, this 1,2,3-triazole-appended sugar-linked molecular arrangement in podophyllotoxin scaffold contributed more improved chemical stability under the physiological conditions.'

Employing click chemistry, an array of 1,2,3-triazole bounded derivatives of oleanolic acid have been developed. Results based on anti-proliferative evaluation demonstrated that some of the compounds displayed good anti-cancer activity against five cancer cell lines (HeLa, HepG2, HCT116, A375-S2, and HTT1080). Compound **93** had the strongest inhibitory action against HTT1080 (human fibrosarcoma) cells among these derivatives, and it also markedly caused apoptosis (Fig. 15.21) [98].

Comprising of 1,2,3-triazole-appended limpet hemocyanin (KLH) polypeptide carriers resulted to an apparently activated with oligosaccharide antigen selective anti-cancer vaccines **94** (Fig. 15.22) [99]. The glycosyl-mimetic reside over the cancer cells and potential to explore in order to get the required antigenicity for in vivo antibody trigger.

Rapamycin, a well-established mTOR inhibitor after further modification resulted to 2-acetamido-glucoside **95** and rapamycin sialoside **96** (Fig. 15.23), a promising biochemical influence in particular for anti-cancer and anti-viral efficacy [84].

As a result, development of more pharmacological target-based molecular scaffold for the possible cure of cancer is needed to reduce adverse effects and adopt more focused therapy and Click Chemistry may provide a remarkable tool to achieve such targets in future.

15.5 1,2,3-Triazole Hybrids and Conjugates as Anti-viral Agents

Sialic acids present at the terminal glycan chains of the mammalian cell surface is a well-known targets of many receptors that play a crucial role in numerous biological events and therefore further structural modification in sialoside molecular arrangement can serve as potent drug candidates with improved efficacy, particularly to cure viral disease [100–102]. 2,3-Dehydroneuraminic acid (DANA, **98**), a synthetic analogue Neu5Ac **97** is a moderate inhibitor of neuraminidase [103]. Later more potent and selective neuraminidase inhibitors zanamivir analogue **101** [104], and oseltamivir analogue **102**, [105] were developed for the treatment of viral infection caused by influenza A and B virus (Fig. 15.24).

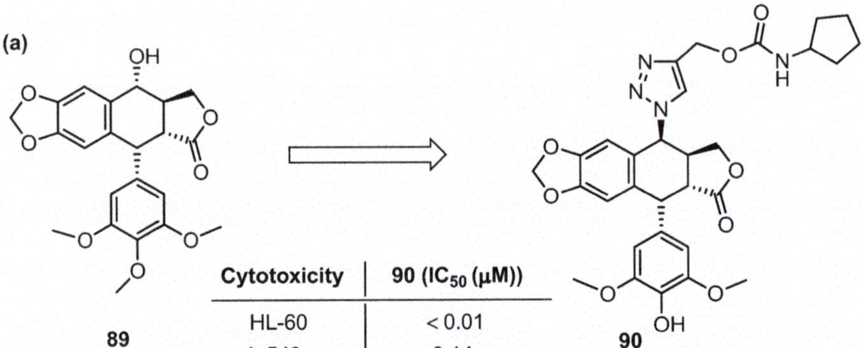

Cytotoxicity	90 (IC_{50} (μM))
HL-60	< 0.01
A-549	0.14
HeLa	0.51
HCT-8	0.05

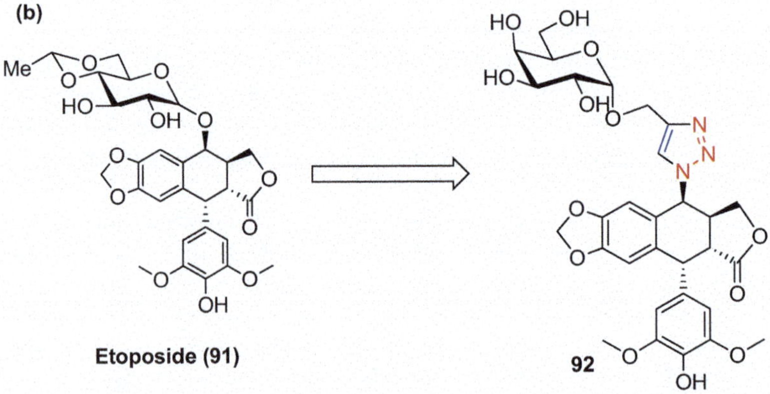

IC50 (μM)	Glycohydbrid 91	Etoposide	Cis-platin
HL-60	2.85	0.31	1.17
SMMC-7721	3.99	8.12	6.43
A-549	4.07	11.92	9.24
MCF-7	7.28	32.82	15.86
SW480	5.52	17.11	13.42

Fig. 15.20 Click-derived triazolylmethyl carbamate **90** and etoposide-based glycohybrid **92** as anti-cancer agent

DANA-based neuraminidase inhibitors have received tremendous interest in recent years, for example, C9 and *N*-5Ac positions of DANA analogue **103** displayed more potency against mammalian NEU3 enzyme than the standard reference [106]. The glycohybrid **104** (Fig. 15.24) [107] displayed potent and selective inhibition of VCNA and hNeu2 in the successive screening of two-phase. A glycosyl *C*-4 triazolyl analogue **101** of zanamivir showed moderate inhibitory activity

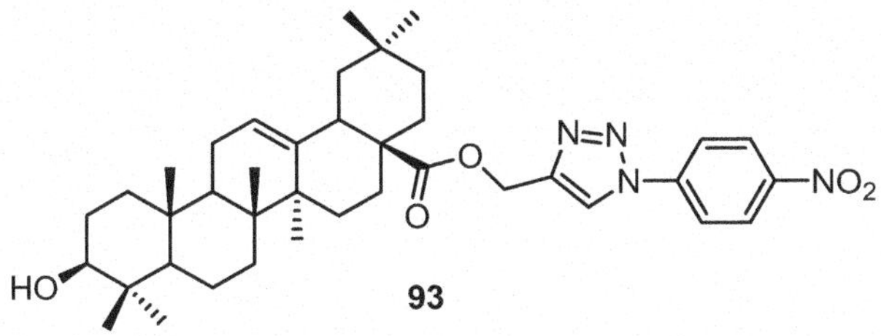

93

Cell lines	93 (IC_5 in μM)	5-fluorouracil (IC_{50} in μM)
HeLa	10.85	26.18
HepG2	24.15	67.64
HCT116	12.28	35.16
A375-S2	4.97	90.74
HT1080	3.51	25.46

Fig. 15.21 Study of oleanolic acid-appended 1,2,3-triazole with different human cancer cell line

concerning the standard drug [108]. Acetamido-3,5-dideoxy-2-(4-octyl-1*H*-1,2,3-triazol-1-yl)-D-glycero-α-D-galacto-non-2 ulopyranosidic acid **102** displayed potent neuraminidase inhibitory activity than parent molecule Neu5Ac2en ($IC_{50} = 28$ μM) (Fig. 15.24) [109]. The Concept of cluster effect was exemplified with Oseltamivir, where a divalent oseltamivir glycohybrid exhibited potent neuraminidase inhibitory activity [110]. A oligovalent *C*-sialoside cluster displayed notable interaction with pharmacologically important receptor candidates [111].

Poly-alkyne functionalized core once clicked with sugar azide led to tetravalent glycocluster **105** conjugated model carrier protein BSA as a promising vaccine candidate for O1-specific polysaccharides (O-SP) of *Vibrio cholera* (Fig. 15.25) [112]. The designed conjugates are structurally single-ended conjugation models and are utilized to ascertain well-defined building blocks for the vaccines along with the non-cross-linked molecular arrangement.

Several sialic acid-containing glycans displayed promising *Trypanosomacruzi trans*-sialidase (TcTS) inhibitory activities, a well-known choice of macromolecular target to get potent therapeutics [113]. The glycohybrid **106** exhibits the significant in vitro as TcTS inhibitory activity with respect to DANA [114]. The glycohybrid **107** (Fig. 15.26) displayed promising TcTS inhibition than DANA. Also, this glycohybrid was identified as potent trypanocidal anti-parasite activity and can be explore to have a potent candidate against Chagas diseases [115].

The glycohybrid **108** ($IC_{50} = 5.47$ mM against influenza A/WSN/33 virus) displayed significant potency against anti-influenza in vitro activity (Fig. 15.27) [116]. Compound **109** displayed promising competitive inhibitory action with $K_i = 185.6$ μM against human *O*-GlcNAcase enzyme [117].

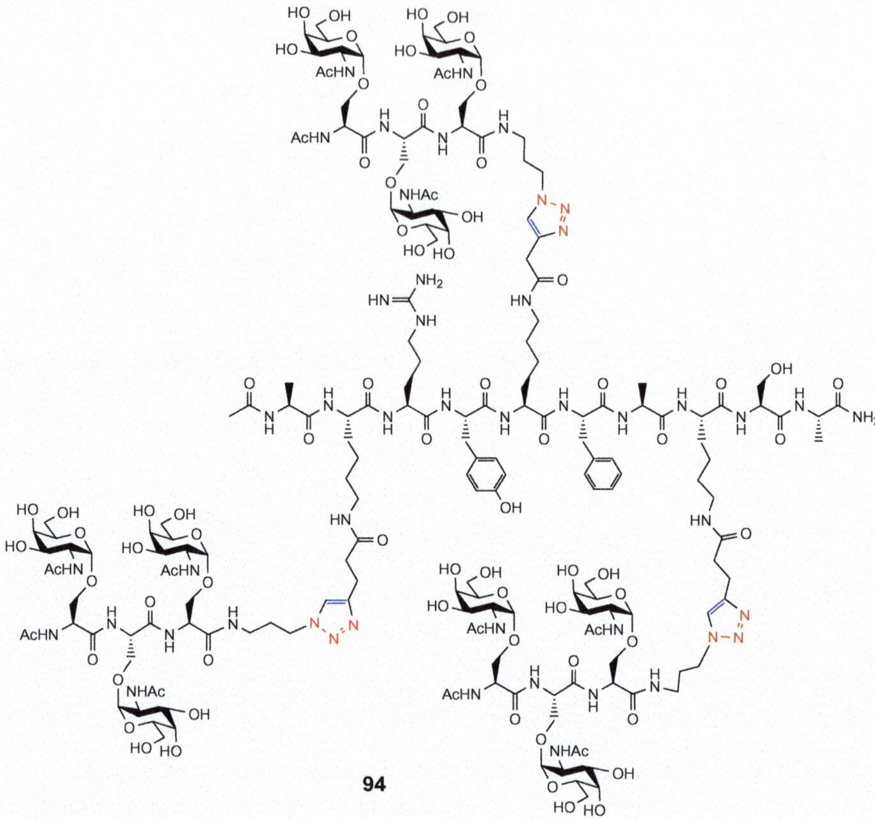

94

Fig. 15.22 Structure of Danishefsky's KLH mimetic anti-cancer vaccine required for anti-genicity for in vivo antibody

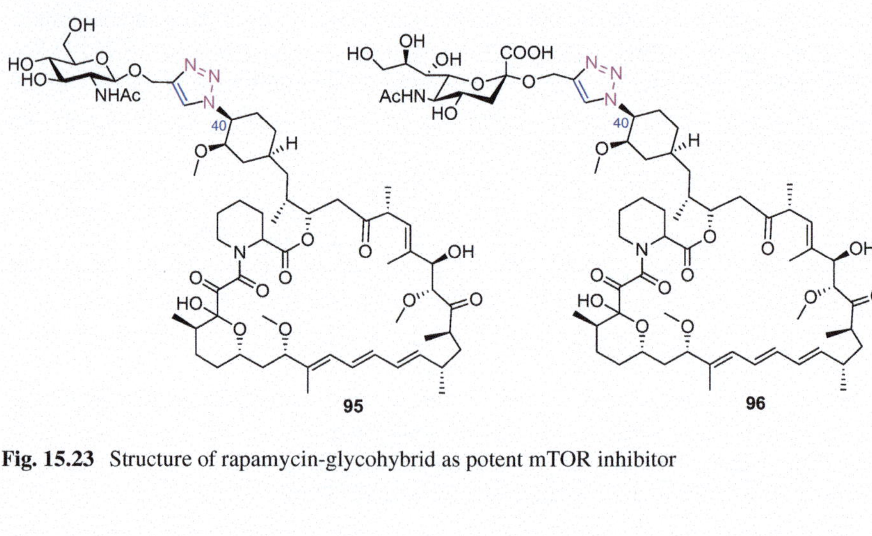

95 96

Fig. 15.23 Structure of rapamycin-glycohybrid as potent mTOR inhibitor

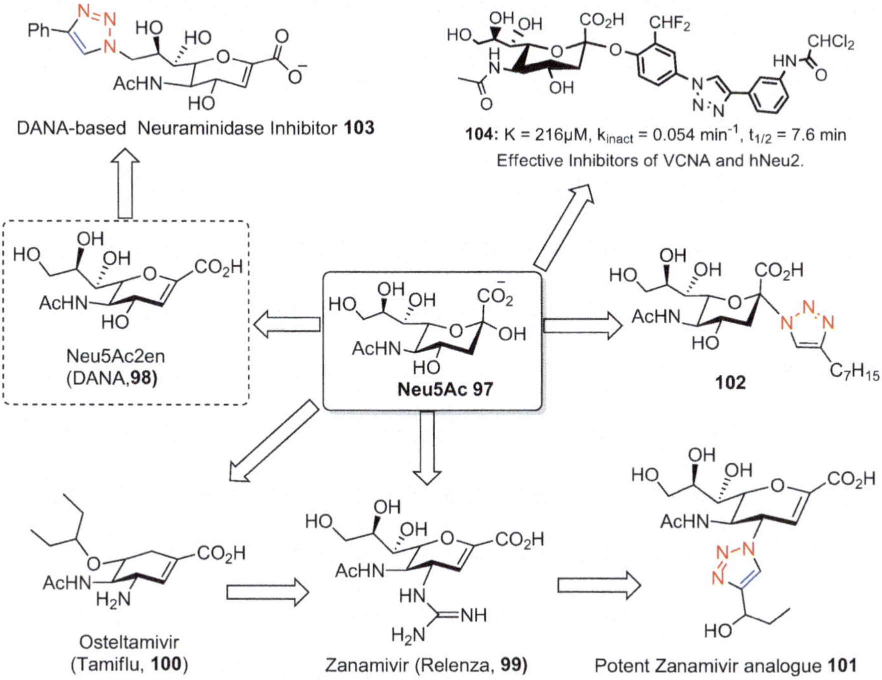

Fig. 15.24 Structure of DANA (**98**), zanamivir (Relenza, **99**), and oseltamivir (Tamiflu, **100**) as well-known neuraminidase inhibitors and their 1,2,3-triazole analogues (**101–103**)

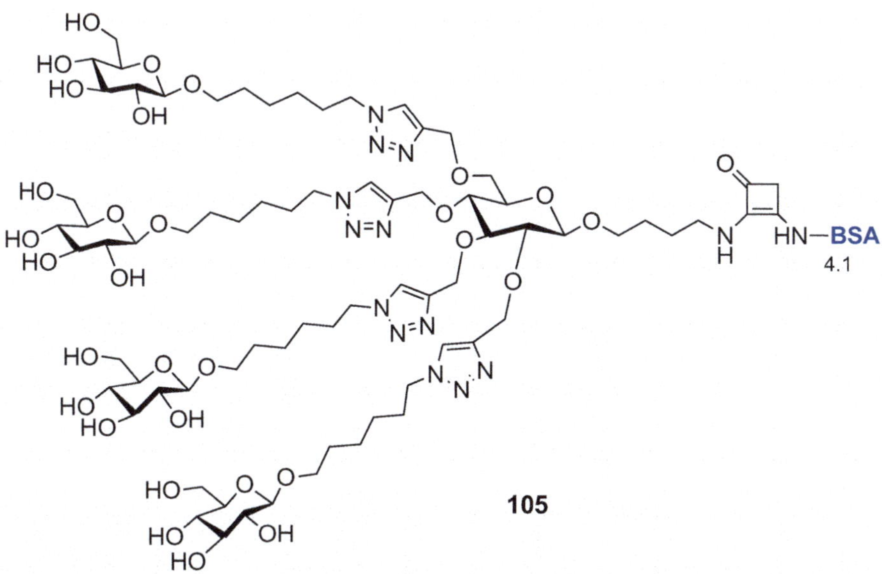

Fig. 15.25 Versatile core as subcarriers for multivalent immunogens

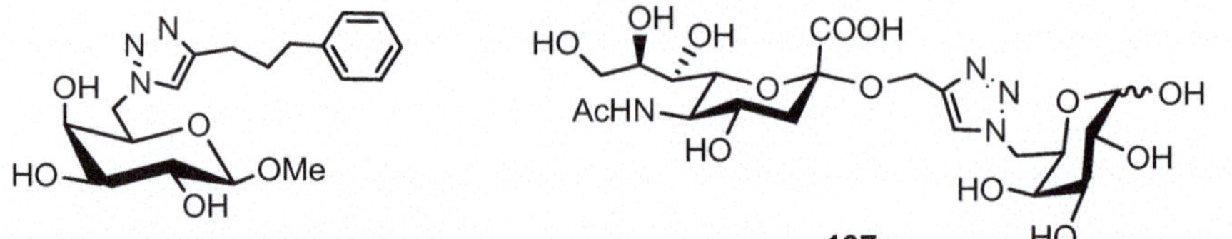

106 **107**

Fig. 15.26 Click-inspired potent glycohybrids (**106, 107**) with TcTS inhibitory activity

Fig. 15.27 Structure of potent against influenza glycohybrids **108, 109**

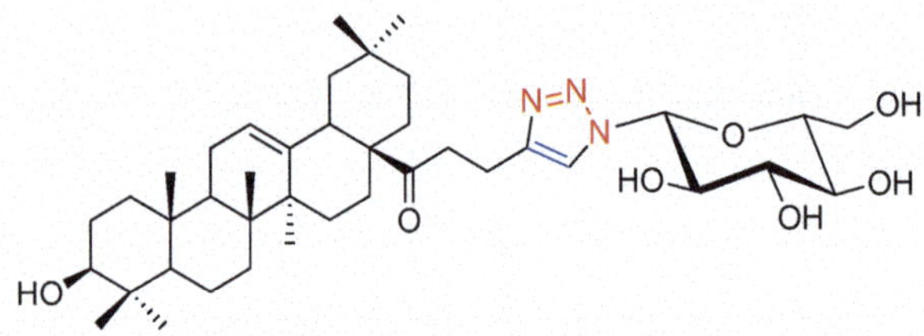

108 (IC_{50} = 5.47mM, Potent Anti-influenza Agent)

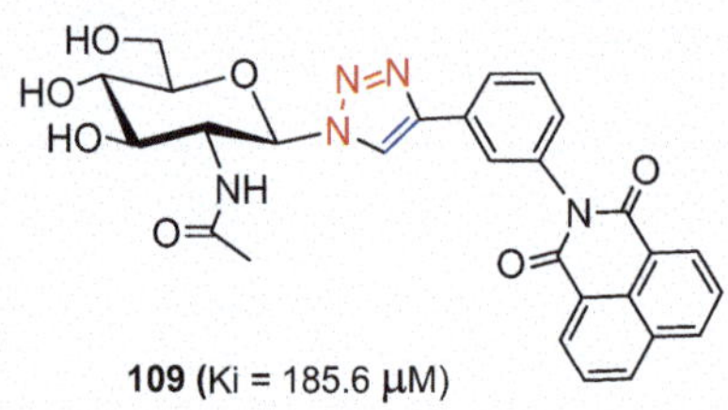

109 (Ki = 185.6 μM)

Several 1,2,3-glycohybrids displayed potent HIV-1 Inhibitory activity**.** TSAO derivative 2′,5′-bis-*O*-(tert-butyldimethylsilyl)]-3′-spiro-5″-(4″-amino-1″,2″-oxathiole-2″,2″-dioxide) was found to be more potent and selective inhibitor against HIV-1 [118]. Further modification by replacement of cyclohex-4-ene-1,3-dione with 1,2,3-triazolyl carbamoyl pharmacophore resulted to more potent TSAO analogue [119]. The glycohybrid **110** (Fig. 15.28) exhibited an improved selectivity index and potent against HIV-I than TSAO-T with no cytotoxicity at 250 μM [119]. Glycohybrid **113** displayed HIV-1 protease inhibitor and reverse transcriptase inhibitory activity with efficiency of 52–60% and 94–99% at 50 μM, respectively [120].

Neomycin dimer benzimidazole conjugate target numerous binding sites on HIV TAR [121], and thus, the neomycin glycohybrid dimer **114** (Fig. 15.29) with having shorter linker lengths developed which showed higher affinity for the HIV TAR RNA with (IC_{50} = 36–99 nM) than Tat-TAR and neomycin IC_{50} of 56 nM and 417 ± 115 nM, respectively [122].

A sialic acid-containing glycan **115** (Fig. 15.30) obtained via glycosylation of enfuvirtide, an *anti-HIV* peptide displayed 15-fold extended pharmacokinetic efficacy in rat with potent anti-viral activity [123].

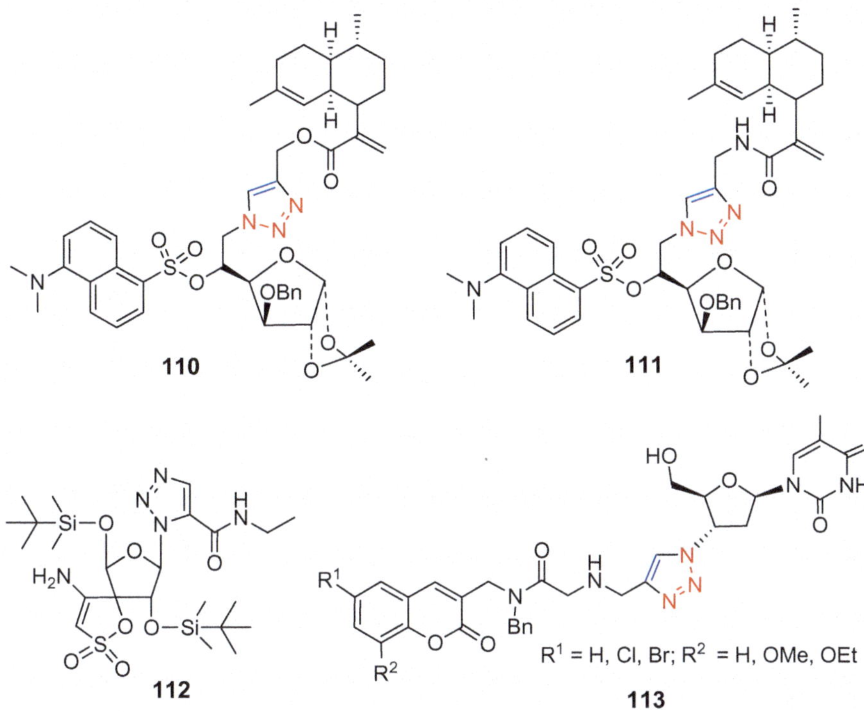

Fig. 15.28 1,2,3-Triazole-linked TSAO derivatives (**110–113**) as promising HIV-1 inhibitor

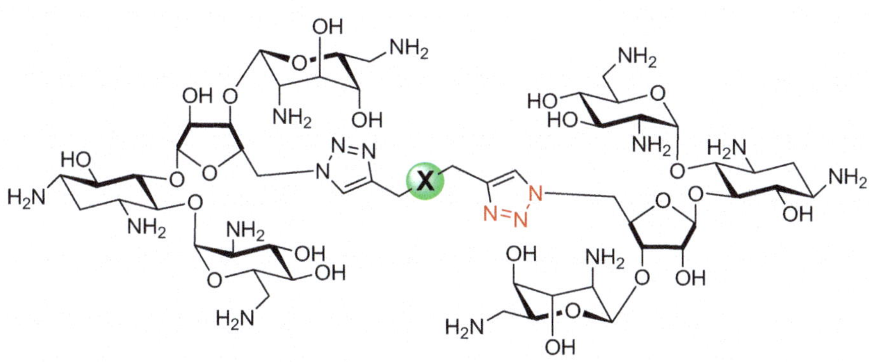

Fig. 15.29 Glycohybrid dimers **114** for targeting HIV TAR RNA

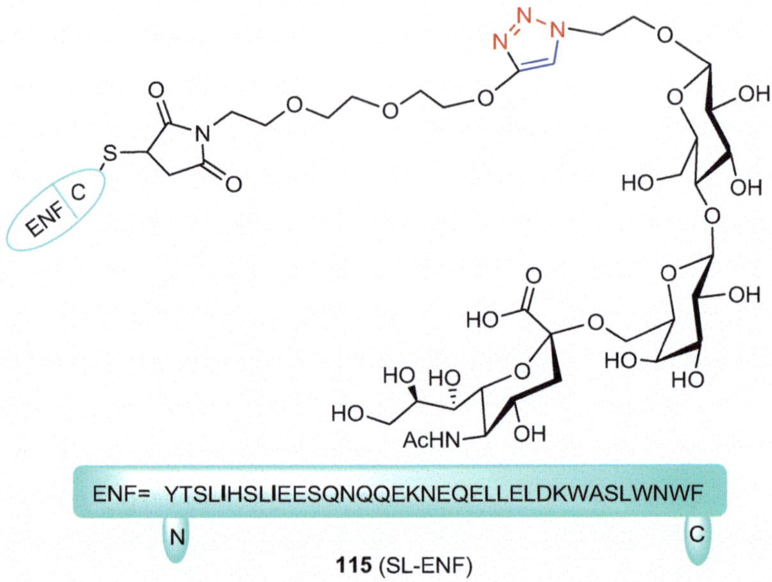

Fig. 15.30 CuAAC click-inspired sialylated peptidyl glycan **115** as anti-HIV

'COVID 19', caused by SARS-CoV-2 infection is a severe form of respiratory disease [124, 125]. This evolved into a pandemic and causing notable morbidity and mortality [126]. Many efforts were made to develop a potent candidate to fight against this virus where seven carbon quantum dots (CQDs) was developed. This first-generation anti-viral CQDs successfully used to cure human coronavirus HCoV-229E infections (Fig. 15.31) [126]. Further, post-modified with boronic acid ligand was active with an EC_{50} value of 52 ± 8 μg/mL [125]. Followed by this, a second-generation analogue **116** with 4-aminophenyl boronic acid was found EC_{50} lower to $5.2 \pm 0.7\ \mu g\ mL^{-1}$. The molecule work via the inhibition of HCoV-229E. Remdesivir **117** also known as GS-5734 is under trial against COVID-19 [127].

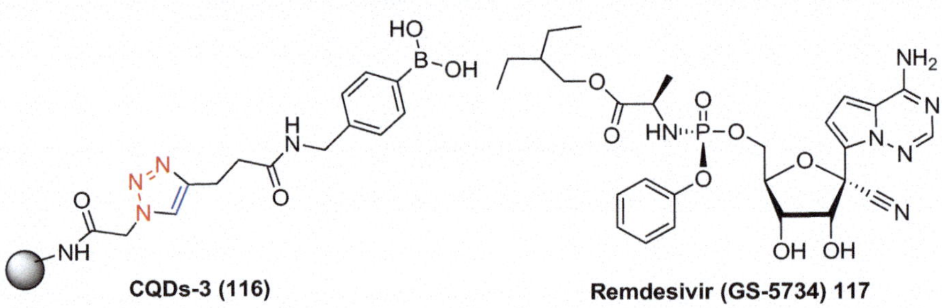

Fig. 15.31 Click-mediated CQDS conjugate to fight against COVID-19

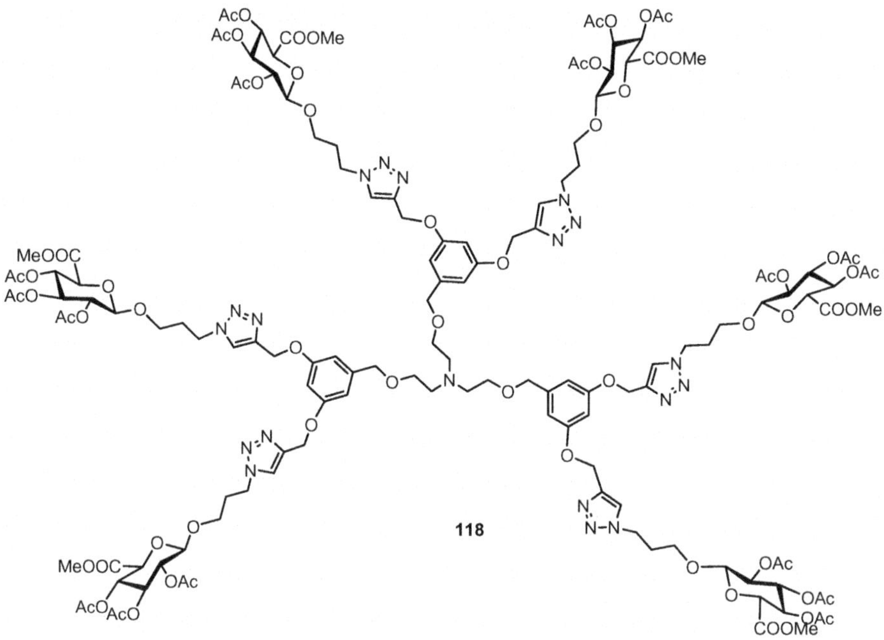

118

Fig. 15.32 1,2,3-Triazole-linked multivalent glycodendrimer **118** promising agent against dengue virus

As per the required components to increase the specificity and affinity, 1,2,3-triazole linked multivalent glycodendrimer can be used as valuable tools for examination of studying carbohydrate – protein interactions. Toward this end, click-mediated glycodendrimer **118** (Fig. 15.32) with flexible scaffold displayed notable binding interaction with the dengue virus envelope protein ($KD = 0.487$ μM). The result was supported by molecular modeling and its surface plasmon resonance studies where the binding constant value of dendrimer **118** was higher than the respective rigid tetravalent glucuronic acid dendron. This scaffold may offer for the possible development of new active candidates to fight against the dengue virus [128].

15.6 1,2,3-Triazoles as Anti-tubercular Agents

Despite of the advances made in the treatment and management, tuberculosis still remains as the major public health problem [129]. The reasons behind the widespread of disease are as follows: emergence of multidrug resistant (MDR) tuberculosis strains, latency and long duration treatment, financial burden in developing countries, and scarcity in new drug candidates with novel mechanism of action. One of the initial steps for development of therapeutic agents is the identification of lead molecule followed by lead optimization, which after careful SAR studies may define

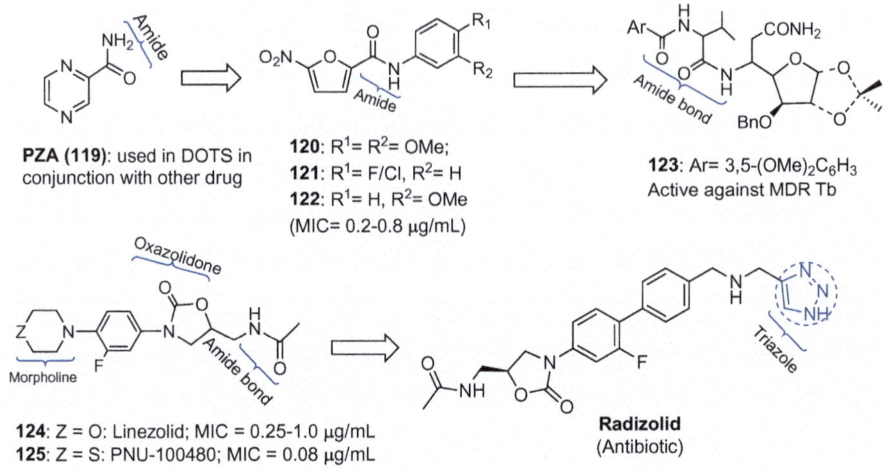

Fig. 15.33 Amides as anti-tubercular agents

the key elements for maximal enzyme inhibitory activities. Previous researches reveal that numerous simple amides (**119–125**) exhibit excellent anti-tubercular activities (Fig. 15.33).

Since 1,2,3-triazole are isostere with amides and therefore a number of triazole conjugates or their hybrid analogues were developed as possible effective candidate to fight against tuberculosis [130]. Such novel analogues presumably block lipid biosynthesis required to inhibit the growth of mycobacteria or similar mode of action of the first-line anti-tubercular drug isoniazid (INH). A benzofuran salicylic acid-based 1,2,3-triazole I-A09 **126** (Fig. 15.34) is now under clinical trial for the possible treatment of *M.* tuberculosis (TB) [131]. This way suitably functionalized hybrid 1,2,3-triazoles have been considered as a promising scaffold to be explored medicinal chemistry to find a novel and safe lead against tuberculosis.

The 1,4-disubstituted triazoles that can mimic the amide bond, are known to display a wide range of biological activities including excellent anti-tubercular properties [132]. 1,2,3-Triazole ring embodied in such molecules not just serve as a passive connecting unit rather interplay with suitable biological targets and provide stability

Fig. 15.34 Structure of I-A09, a potent anti-TB agent in clinical evaluation

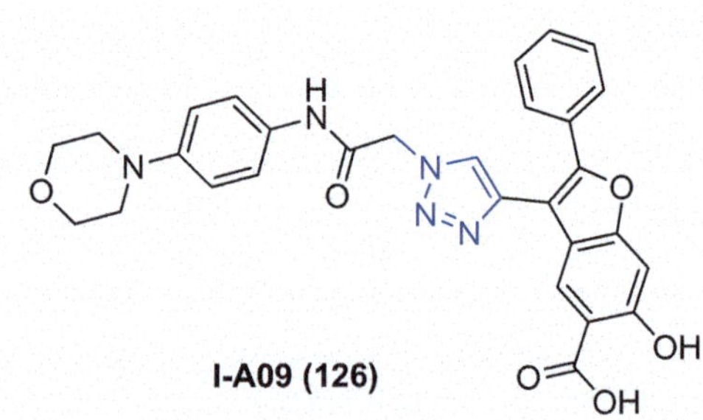

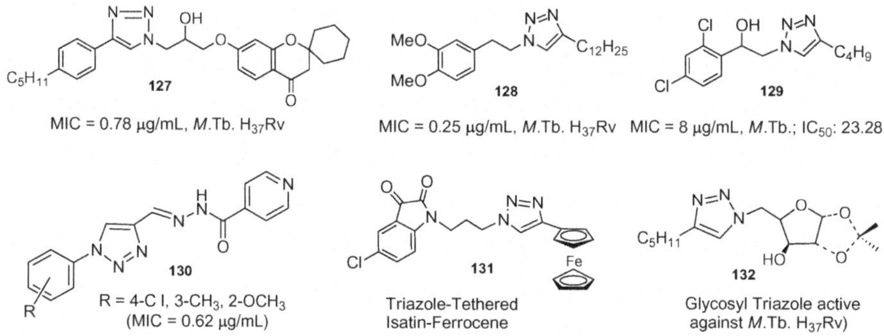

Fig. 15.35 Some 1,4-disusbstituted triazoles active against *Mycobacterium tuberculosis*

toward metabolic degradations. This when coupled to carbohydrates having structural diversity, optimum pharmacokinetics, low toxicity, and predominant role in different cellular processes, may offer an easy access to library of compounds with potential against frontline diseases including tuberculosis [133].

The CuAAC approach introduces 1,2,3-triazole skeleton to diverse building blocks of notable anti-tubercular activity profile, for example, a chromone–triazole conjugate **127** with MIC of 0.78 μg/mL against *M.* Tuberculosis $H_{37}Rv$ shows greater potency than the ethambutol [134]. A triazole bearing long alkyl chain **128** exhibits promising anti-tubercular activity with MIC of 0.25 μg/mL against *M.* tuberculosis $H_{37}R_V$ strain [135]. Likewise, hydroxy-triazole **129** exhibits two-fold anti-tubercular activity compared to econazole, thus appears promising for optimization as *M.* Tb. specific agents [136]. The triazolyl isonicotinoyl hydrazides **130** display activity at MIC values 2.5-0.62 μg/mL against susceptible $H_{37}Rv$ strain while exhibiting low cytotoxicity against liver and kidney cells, thereby providing a high therapeutic index [137]. Similarly, isatin–ferrocene–triazole conjugate **131** [138], and triazolyl glycoconjugate **132** [139] showed promising anti-tubercular activity (Fig. 15.35).

Modification of delamanid via 1,2,3-triazole as hybridization approach resulted to 1,2,3-triazolyl-6-nitro-2,3-dihydroimidazooxazole **133** of potent anti-tubercular activity against *MTB* H37Rv (MIC = 0.23 μM). This hybrid **133** exhibited synergistic to additive effect G with first-line anti-tuberculosis drug (Fig. 15.36) [140]. Triazole–imidazo[2,1-b][1,3,4]thiadiazole hybrid **135** displayed anti-tubercular activity (MIC: 3.125 μg/mL) against *MTB*-H37Rv (ATCC27294) strain employing Agar dilution method which is comparable to the ethambutol (MIC = 3.125 μg/mL) [141]. Further, click-mediated structural modification imidazo[1,2-a]-pyridine in Zolpidem lead to potent hybrid analogues **134** with MIC 1.56 μg/mL as compared to the standard Zolpidem (MIC: 3.12 μg/mL) and Pyrazinamide (MIC: 3.12 μg/mL) [142]. Hybrid **136** (MIC: 24.72 μM, $IC_{50} = 1.01 \pm 0.32$ μM) displayed potent anti-Tb activity against *M. pantothenate synthetase* and MTB (Fig. 15.36) [143].

Coumarin pharmacophore on molecular hybridization with 1,2,3-triazole moiety may deliver novel potential anti-TB agents with better potency, for example, hybrid

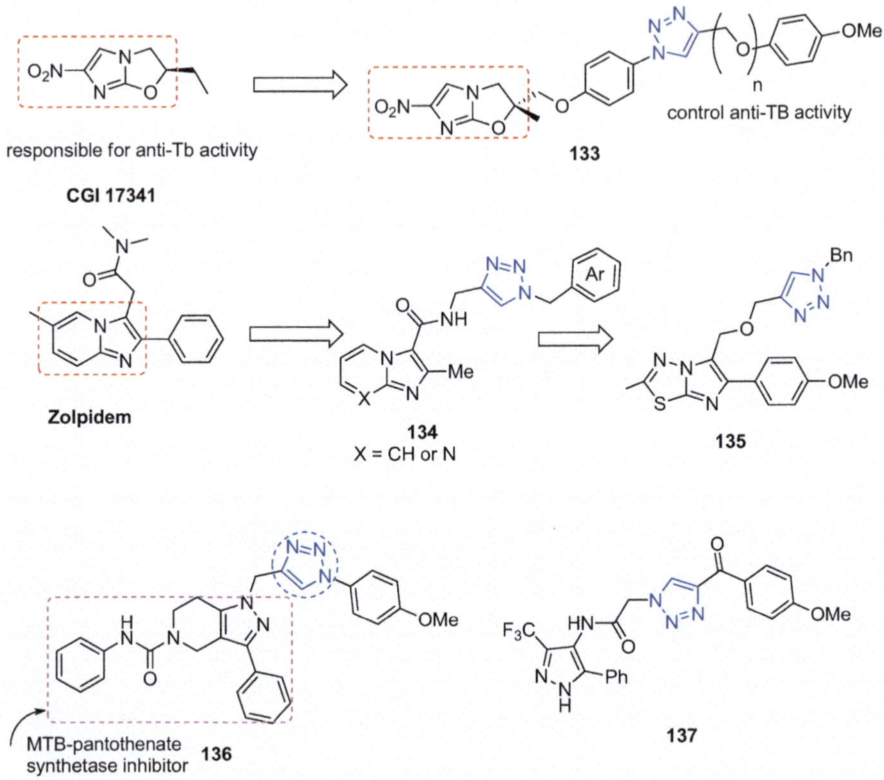

Fig. 15.36 Chemical structures of 1,2,3-triazole–azole hybrids with anti-tubercular activity

138 (MIC 6.25 μg/mL) is potent and equivalent to the streptomycin (MIC: 6.25 μg/mL) [144] and *bis*-chromenyl triazole hybrid **139** (MIC: 0.2–12.5 μg/mL) displayed potent activity against *M*TB strain H37Rv (Fig. 15.37) [145]. 1,2,4-Triazolone hybrid **140** displayed MIC 1.6–12.5 μg/mL displayed potency against *M. tuberculosis* H37Rv (MTB; ATCC-27,294) strain, where the docking results validate its binding of the active motifs against InhA with active site InhHa-148G [146]. Compound **141** (MIC: 50 μg/mL) was commensurate with RIF (MIC: > 32 μg/mL) and potent than **INH** (MIC: > 128 μg/mL) against MDR-TB strain [147]. Compounds **142** (MIC = 4.4 μM) displayed promising anti-TB activity as compared to Ethambutol (MIC = 9.7 μM) (Fig. 15.37) [148]. Compound **143,144** (MIC 1.56 μg/mL and lipophilicity clogP < 5) displayed potent anti-TB activity and the result was further supported by molecular docking with pantothenate synthetase (PS) enzyme of MTB (Fig. 15.37) [149].

1,2,3-Triazole–chromene conjugate **145** (MIC 12.5 μg/mL) displayed high binding affinity with *M. tuberculosis* DprE1 [150]. 1,2,3-Triazole-fused spiro-chromenes **146** (MIC 1.56 μg/mL) displayed potent against *M. tuberculosis* H37Rv

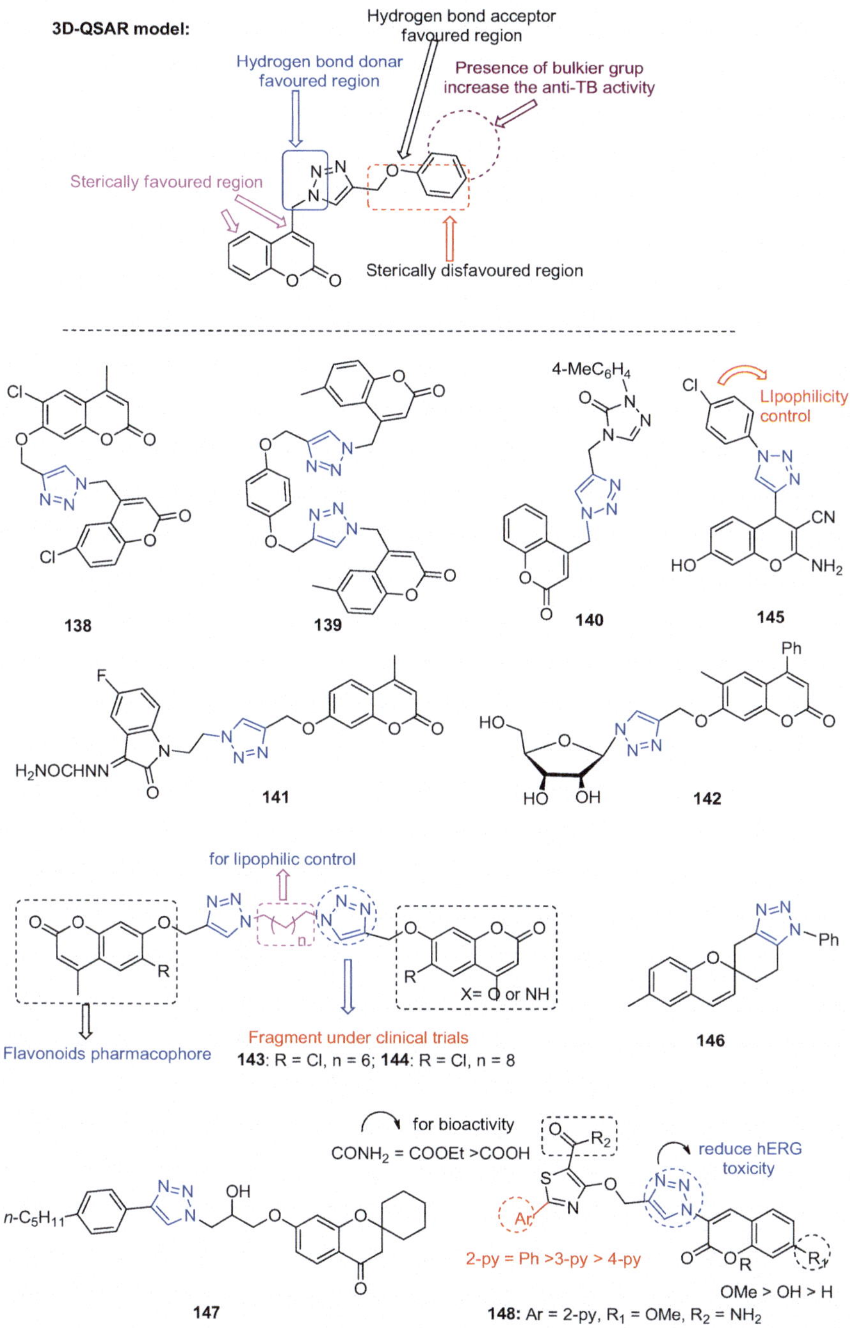

Fig. 15.37 Chemical structures of anti-tubercular-1,2,3-triazole–coumarin hybrids

and was equivalent to ethambutol (MIC 1.56 μg/mL) and thus can be further explored to develop effective anti-tubercular agents [151].

A spirochromone-1,2,3-triazole hybrid **147** (MIC: 0.78 μg/mL) against *Mtb*-H37Rv comparable to rifampicin (MIC: 0.2 μg/mL) and ethambutol (MIC: 1.56 μg/mL) [134]. 1,2,3-Triazole linked with the coumarin **148** (MIC 6.5 μM) displayed notable Gyr-B inhibitory activity ($IC_{50} = 0.44$ μM against *M. semgmatis* GyrB and *M. Tb* H37Rv strain) and equally correlated with supercooling inhibitory potency (Fig. 15.37) [152].

Isoniazid (INH), a standard first line anti-tuberculosis drug and 1,2,3-triazole are bacterial cell wall inhibiter and thus their hybridization may provide a lead with improved anti-TB efficacy [153]. The concept was widely explored and a number of INH–triazole hybrids were developed in recent years. Isoniazid-1,2,3-triazole conjugate **149** (MIC = 1.56 μg/mL) was active against *MTB*-H37RV strain with minimal cytotoxicity against RAW 264.7 cell lines and has the excellent binding affinity toward *DNA gyrase and* INHA (Fig. 15.38) [154]. Lapachone-based isoniazid-triazole **150** (MIC value 12.5 μg/mL) [155] Isoniazid-1,2,3-triazole conjugate **151** ($MIC_{99} = 0.195–0.39$ μM) is almost two-fold potent against *M. tuberculosis* H37Rv than the isoniazid ($MIC_{99} = 0.39$ μM) (Fig. 15.38) [156]. INH-1,2,3-triazole hybrid **152** displayed improved potency (MIC 0.62–2.5 μg/mL with minimal cytotoxicity (MDL50 > 1000 μg/mL) against the liver cells (hepatoma HepG2) and kidney cells (BGM) and also higher selectivity index [157] A potent isoniazid–triazole hybrid **153** (MIC = 0.78 μg/mL) have no toxicity against RAW 264.7 cell line and InhA as possible molecular target supported by molecular docking studies [158]. Further optimization of such potent hybrids can be useful to develop potent anti-Tb lead candidates. More sincere efforts are needed to explore 1,2,3-triazole-based hybrid or their conjugates to have NCE as a lead to fight against tuberculosis.

Bedaquiline (known as TMC 207), a diaryl quinoline-based US-FDA-approved anti-tubercular drug inhibit the mycobacterial ATP synthase. The molecular hybridization of this scaffold with 1,2,3-triazole is resulted to more potent hybrid candidates. For example, 1,2,3-triazole hybrid **154** (MIC = 0.625 μg/mL) was as potent as the RIF (MIC = 0.5 μg/mL) and INH (MIC = 0.7 μg/mL) against the MTB-H37Rv [159]. Amide group decrease lipophilicity and favor the adsorption and permeability of the resulting hybrid analogue (Fig. 15.39).

Quinoline–triazole hybrid **155** displayed 58.20 and 80.79% reduction in RLU, respectively, at concentration 5 and 25 μg/mL against *M. Tb.* H37Rv using RIF as standard (87% reduction in RLU of 2 μg/mL conc) [160]. Quinoline-1,2,3-triazole **156** had minimal cytotoxicity against normal VERO cell lines and displayed excellent potency (MIC: 1.6 μg/mL) against *M. tuberculosis* H37Rv using pyrazinamide (MIC: 3.12 μg/mL), streptomycin (MIC: 6.25 μg/mL) and ciprofloxacin (MIC: 3.12 μg/mL) as standard control [161]. Hybrid **157** (MIC: 1.56 μg/mL) was more potent and found minimal cytotoxicity against HEK-293 T cell lines (Fig. 15.39) [162].

Benzotriazoloquinolinyl spiro-oxindolopyrrolizidine hybrid **158** (MIC = 1.13 μM) against *M. tuberculosis* H37Rv strain as compared to ethambutol (7.64 μM) [163]. Hybrid **159** displayed 52.34% inhibition of *M. tuberculosis* H37Ra dormant strain as compared to the rifampicin with 99.5% inhibition [164]. Hybrid **160** (MIC

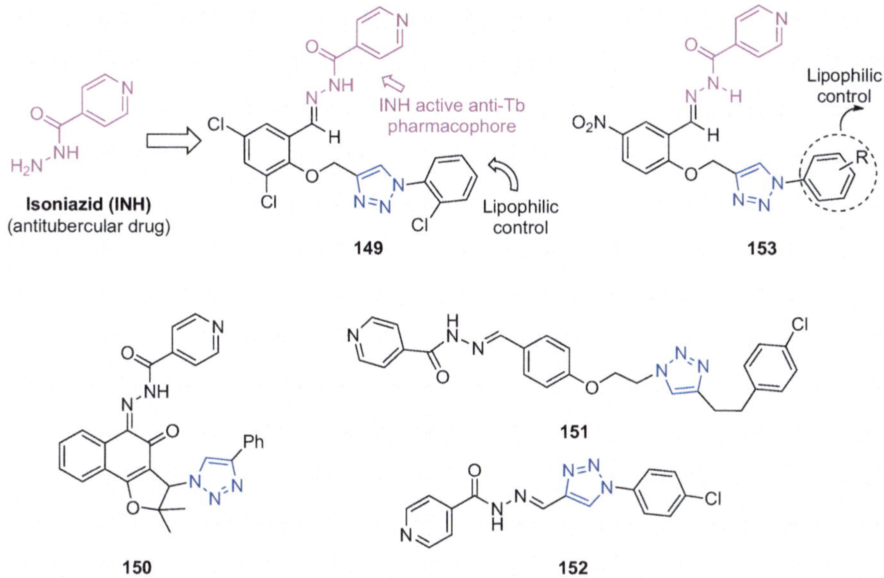

Fig. 15.38 INH-1,2,3-triazole hybrid analogues as potent anti-tubercular activity

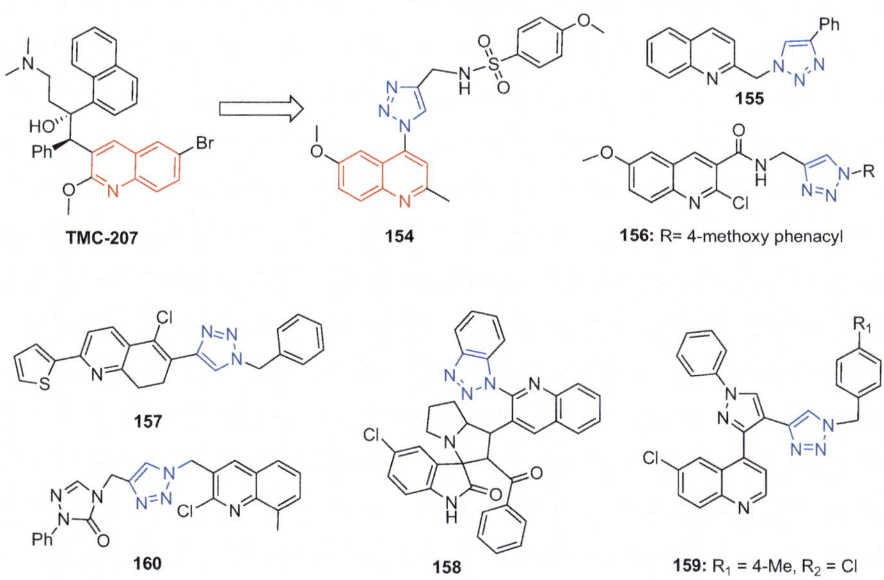

Fig. 15.39 Structure of some 1,2,3-triazole–quinoline hybrids as potent anti-tubercular agents

= 1.6 μg/Ml) displayed similar potency in comparison to isoniazid (MIC: 1.60 μg/mL), pyrazinamide (MIC: 6.25 μg/mL), and ciprofloxacin (MIC: 3.12 μg/mL) with therapeutic index (SI = 233.58–633.00) (Fig. 15.39) [165].

A number of **fluoroquinolones** such as ciprofloxacin (CPFX), ofloxacin, gatifloxacin, moxifloxacin, and levofloxacin are known for their notable anti-tubercular activity by inhibiting topoisomerase IV enzyme and DNA gyrase in bacterial species. Permeation of these drug candidates through bacterial cell wall is affected by increasing the lipophilicity at the C-7 position and this was well considered through its molecular hybridization with 1,2,3-triazole to further improve its potency against tuberculosis. For example, 1*H*-1,2,3-triazole-tethered 8-OMe ciprofloxacin isatin hybrid **161** displayed potent activities against *M. tuberculosis* H37Rv and MDR TB with MIC varied in-between 0.20–3.12 μg/mL and 0.25–8.0 μg/mL, respectively (Fig. 15.40) [166]. Ciprofloxacin (CPFX), 8-OMe-CPFX and RIF were used as standard with MIC value 3.12, 1.56, and 0.39, respectively, against MTB-H37Rv and 4.0, 2.0 and 64, respectively, against MDR-TB strain. SAR analysis confirms that the halogen at C-5 (i.e., R_2) position of satin increases the bioactivity. A further modification resulted to isatin-(thio) semicarbazide/oxime hybrid **162** (MIC: 0.39 μg/Ml) which was comparable activity with RIF (MIC: 0.39 μg/mL) against MTB H37Rv strain and more potent than CPFX (MIC: 3.12 μg/mL) and 8-OMe CPFX (MIC: 1.56 μg/mL) [167]. Moreover, the hybrid **162** displayed adequate toxicity (CC50: 16–64 μg/mL) in VERO cell lines than standard 8-OMe CPFX with CC50: 16 μg/mL. CPFX-isatin-1*H*-1,2,3-triazole hybrid **163** displayed (MIC: 0.39–50 μg/mL against MTB H37Rv strain) and conclusively changing of 8–OMe in CPFX with hydrogen has not improved the bioactivity profile [168]. A amide tethered ciprofloxacin-1,2,3-triazole-isatin hybrid **164** (MIC: 0.5 μg/mL) displayed similar activity to INH (0.12 μg/mL), RIF (0.25 μg/mL), and CPFX (1.0 μg/mL) against MTB-H37Rv strain. The hybrid **164** was more potent against MDR-TB (MIC: 0.12 μg/mL) than INH (MIC: > 64 μg/mL), RIF (MIC: > 64 μg/mL) and CPFX (MIC: 2.0 μg/mL), and thereby offer a novel mechanism of action on MDR-TB (Fig. 15.40) [169]. 1*H*-1,2,3-triazole-tethered gatifloxacin isatin conjugate **165** had more lipophilicity than gatifloxacin and displayed potent activity against H37Rv and MDR-TB (MIC: 0.10–8.0 μg/mL) [170]. GTFX-triazole analogue **166** (MIC: 0.025 μg/mL) displayed improved potency than INH (MIC: 0.05 μg/mL), GTFX (MIC: 0.78 μg/mL) and RIF (MIC: 0.39 μg/mL) against *MTB* H37Rv strain [171]. A gatifloxacin-1,2,3-triazole–isatin hybrids **167** (MIC: 1.0 μg/mL, CC50: 64 μg/mL) displayed improved potency and minimal toxicity than reference GTFX [172].

Considering the activity of Moxifloxacin (MXFX, a drug in clinical trial phase III), a number of related moxifloxacin–triazole–isatin analogues were reported for their potent anti-tubercular activities. For example, hybrid **168** displayed further improved activity profile against *M. tuberculosis* H37Ra and MDR-TB strains [173]. Likewise, moxifloxacin-acetyl-1,2,3-1*H*-triazolemethylene-isatin derivatives **169a–d** (MIC: 0.12–0.25 μg/mL) were developed and displayed potent activity against *M. tuberculosis* H37Rv and were cytotoxic against mammalian VERO cell (Fig. 15.40) [174]. Thus, through the suitable conjugation of quinolone and isatin via biocompatible 1,2,3-triazole linker may resulted to their significant improvement of

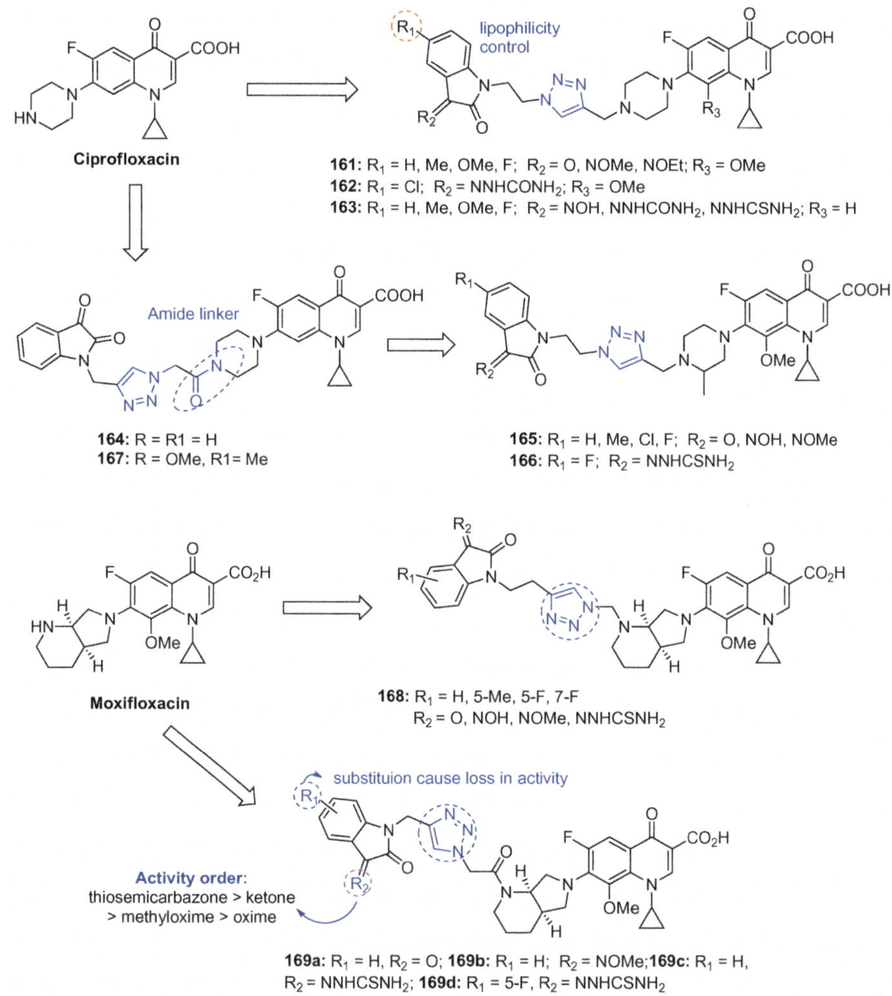

Fig. 15.40 Chemical structures of 1,2,3-triazole-quinolone hybrids

anti-tubercular activities and after further optimization may use for the successful development of novel mechanism based drug candidates to fight against tuberculosis.

Further, metals once after incorporation to the drug like molecules for example, ferrocene may result to some notable alteration like physical and pharmacological properties, tenability, redox states, and its control release. This pharmacophore (known as metalo-pharmaceuticals) for example, ferrocene 1,2,3-triazole hybrids like β-lactam-ferrocene-1,2,3-triazole hybrid **170** (MIC: 10 μg/Ml) [175], and 1H-1,2,3-triazole-tethered isatin-ferrocene **171** (Fig. 15.41) has significant impact in drug discovery against tuberculosis [138]. And hybrids **172a,b** (MIC: 27 μg/mL)

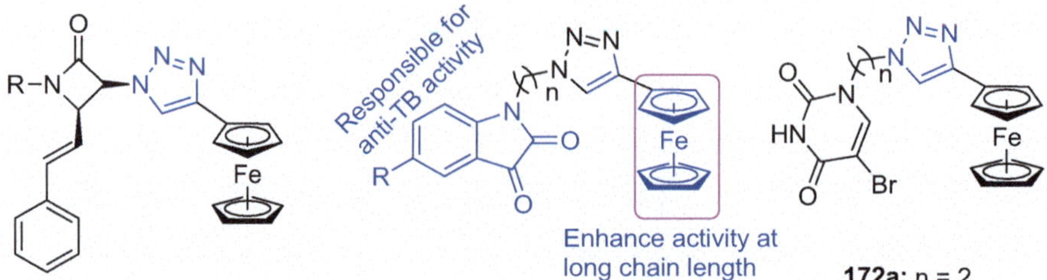

Fig. 15.41 Chemical structures of 1,2,3-triazole-ferrocene hybrids **170–172**

after insertion of a ferrocene in uracil–ferrocenyl conjugate resulted to the improvement in bioactivity against *M. tuberculosis and also were* non-cytotoxic even at the concentration of 100 μg/mL [176].

15.7 1,2,3-Triazoles as Antibacterial Agents

Boronic acid transition-state inhibitors are well-recognized as β-lactamase inhibitors. In this endeavor, the click tool has been utilized by Prati and group for the development of an array of 1-amido-2-triazolyl ethaneboronic acid analogues, where the 1,2,3-triazole hybrid **173** displayed prominent anti-bacterial activity (Fig. 15.42) [177].

Curcumin due to the associated limitations mainly poor solubility and instability is not proper to use for the medication. Toward this end, a click-mediated curcumin glycoconjugate **174** (Fig. 15.43) was developed which has notable water-solubility (almost 11,000 times more water-soluble than its native counterpart) [178]. This curcumin-linked glycohybrid **175** (with MIC 32 and 64 μg/ml, respectively, against MRSA and MDR *Klebsiella* isolates) was sensitized resistant gram-negative clinical isolates to the gram-positive specific antibiotic vancomycin and displayed potent anti-bacterial activity [178, 179].

Fig. 15.42 Click tool for the development of potent β-lactamase inhibitor **173**

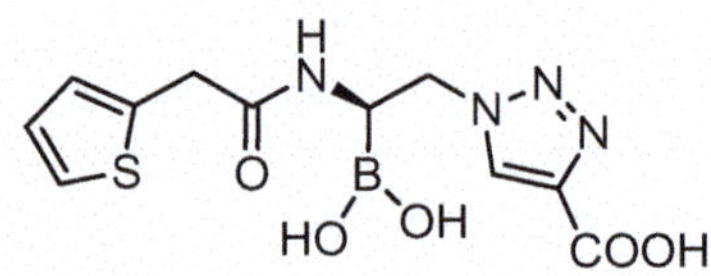

173

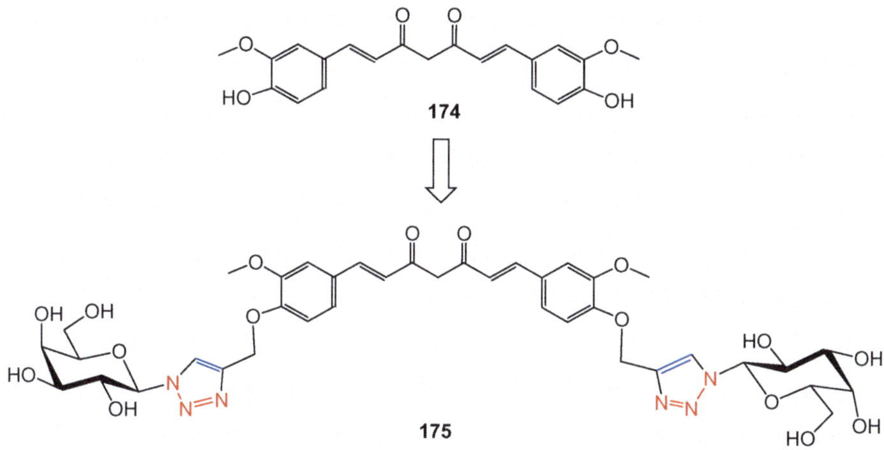

Fig. 15.43 Click-mediated water-soluble curcumin glycoconjugate **175** of potent anti-bacterial activity

15.8 1,2,3-Triazoles as Anti-leishmanial Activity

Visceral Leishmaniasis (VL, also known as kala azar caused by *Leishmania donovani*) threatens over 350 million people worldwide and kills 60,000 annually [180]. Despite many efforts, still no effective drugs is available to complete cure of leishmaniasis as existing one is still treated with old antimony-based drugs that are associated with significant side effects and therefore, there is an urgent need to develop well-tolerated inexpensive drugs to fight against this deadly disease [181]. To identify potent anti-leismanial leads, our group conjugated quercetin, a potent flavonoid, and resulting triazole-linked *O*-benzylquercetin glycoconjugates (**176** and **177**) displayed promising anti-leishmanial activity against the promastigotes and intramacrophase amastigote forms of *L. donovani* with the IC_{50} values of 7.76 and 6.08 μg/mL, respectively (Fig. 15.44) [182]. Further, noscapine, a naturally occurring biologically active opium alkaloid was glycoconjugated, where 1,2,3-triazole-linked noscapine glycoconjugates **178** ($IC_{50} = 8.36$ μM) and **179** ($IC_{50} = 15.9$ μM) displayed significant inhibition of parasite growth against *L. donovani* promastigotes comparable to standard drug miltefosine ($IC_{50} = 10.4$ μM) and none of them exhibited cellular cytotoxicity against mouse macrophages [90].

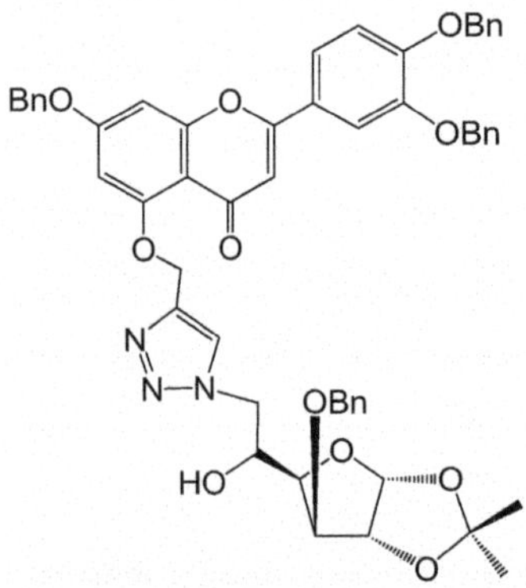

176: IC_{50} = 7.76 mg/mL for promastigote
IC_{50} = 6.08 mg/mL for amastigote

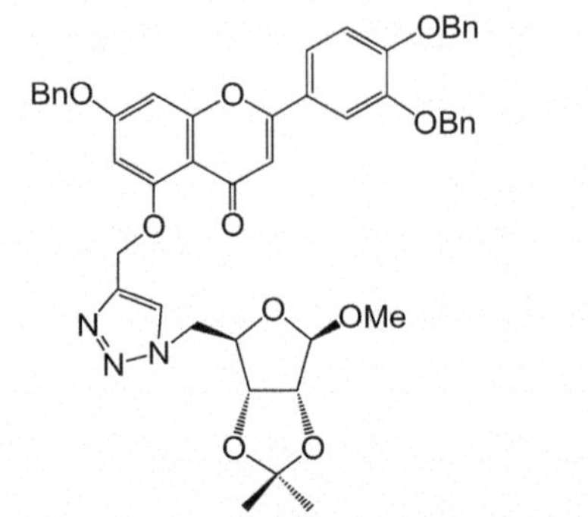

177: IC_{50} = 9.92 mg/mL for promastigote
IC_{50} = 7.65 mg/mL for amastigote

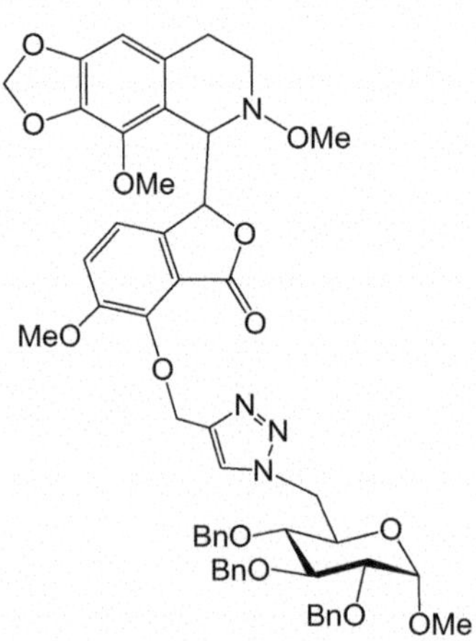

178: IC_{50} = 8.36 mM

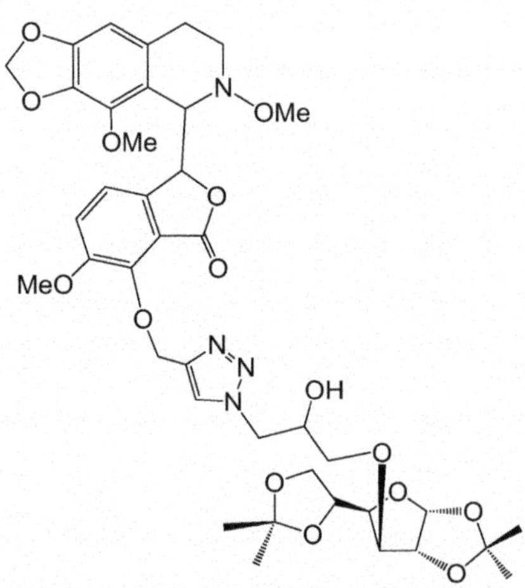

179: IC_{50} = 15.9 mM

Fig. 15.44 Click-mediated potent anti-leishmanial triazole-linked *O*-benzylquercetin (**176, 177**) and noscapine glycoconjugates (**178, 179**)

15.9 1,2,3-Triazoles as Potential Lead Candidates for Miscellaneous Activities

The efficacy of 1,2,3-triazoles as HIF prolyl hydroxylase 2 (HIF-PHD2) inhibitory activity in fluorescence polarization assay for the treatment of renal anemia is well explored. For example, the activity of the HIF-PHD inhibitor **180** ($IC_{50} = 62.23$ nM) was about tenfold higher than that of roxadustat ($IC_{50} = 591.4$ nM) (Fig. 15.45). Additionally, it has the ability to restore the hemoglobin levels of mice with cisplatin-induced anemia (120 g/L) to normal levels (160 g/L) without exhibiting apparent toxicity [183].

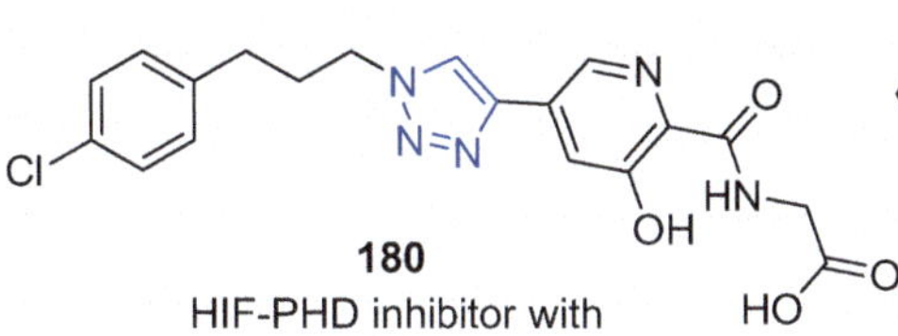

180
HIF-PHD inhibitor with
IC_{50} value of 62.23 nM

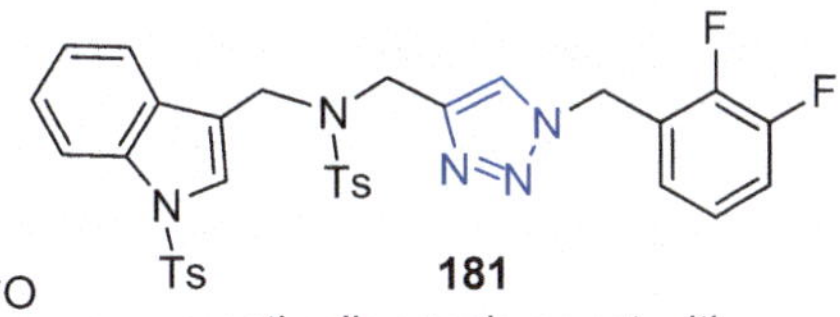

181
anti-adipogenic agent with
an IC_{50} value 1.67 μM

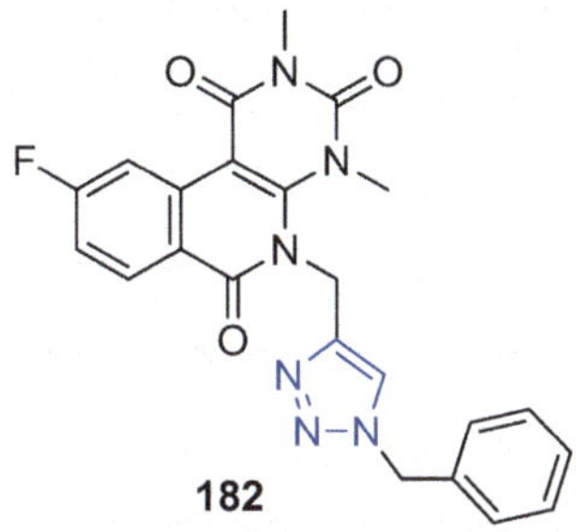

182
antioxidant potency with an IC_{50}
value 6.02 μM than the positive
control drug Trolox

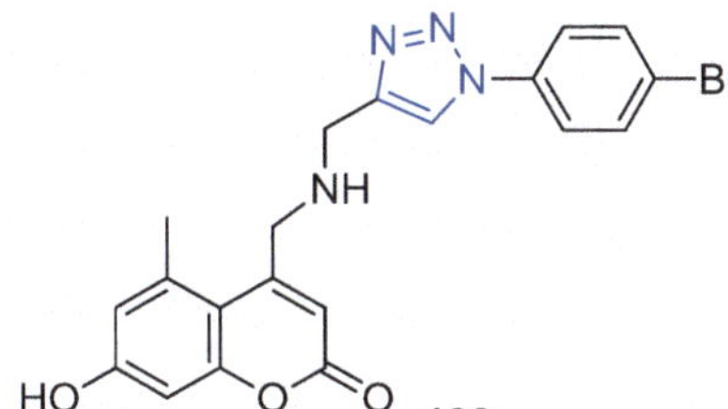

183
TNF-α (62%) and
IC_{50} value in LPS-stimulated U937
cells was 8.01 μM

Fig. 15.45 Some 1,2,3-triazole-appended molecules **180–183** serve as effective lead candidates

Further, click-mediated synthesis of indole–triazole hybrid **181** was effectively reported for potent anti-adipogenic and anti-dyslipidemic activity with an IC_{50} of 1.67 μM, which was more promising than that of its parent compound [184]. The mechanistic investigations suggested that triazole hybrid **181** exhibited anti-adipogenic properties by activating the Wnt/β-catenin pathway, and adipogenic genes such as *PPARγ* and *C/EBPα*.

A novel one-pot click methodology reported for the synthesis of 1,2,3-triazole-appended pyrimido[4,5-*c*]isoquinolines toward an evaluation of their antioxidant activity [185] In vitro anti-oxidant investigation using the DPPH method demonstrated that compound **182** (IC_{50} = 6.02 μM) displayed 1.94 fold greater antioxidant potency in comparison with Trolox. Furthermore, pro-inflammatory cytokine assay of a triazole–coumarin hybrid **183** displayed IC_{50} value 8.01 μM in lipopolysaccharide-stimulated U937 cell line and significantly suppressed by 62% of tumor necrosis factor (TNF)-α at 10 μM [186].

Abelson (Abl) tyrosine kinase, in its cellular form, is closely related to the Src family of tyrosine kinases. Chronic myelogenous leukemia is directly linked to Abl tyrosine kinase. It has been suggested that click chemistry may be used to quickly synthesize Abl tyrosine kinase inhibitors by combining the head region of imatinib with a 1,2,3-triazole ring and an appropriate spacer linked to the N1-position. In enzymatic screening, compounds **185** and **186** exhibited superior inhibition against **protein tyrosine kinase** (Src kinase), with an efficacy remarkably analogous to that of the standard drug imatinib **184** (Fig. 15.46) [187].

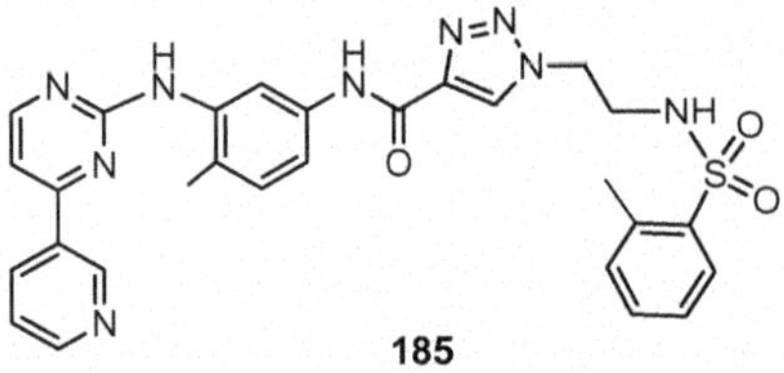

185

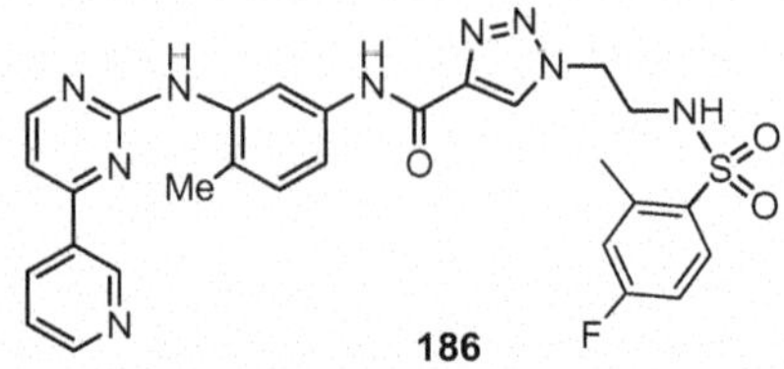

186

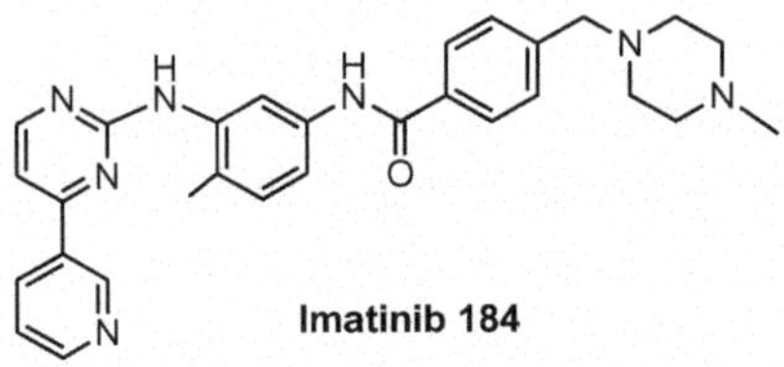

Imatinib 184

	IC_{50} (μM)	
Compound	**Abl**	**Scr**
Imatinib **184**	0.7	26.6
185	1.12	17.9
186	0.8	----

Fig. 15.46 Triazole-bounded imatinib derivatives **185** and **186** as Abl tyrosine kinase inhibitors

15.10 1,2,3-Triazole-Appended Conjugates as Multivalency Effect Through Statistical Rebinding

In addition to the above-mentioned promising pharmacological activities such as anti-diabetic, anti-viral, anti-cancer, anti-tubercular, anti-bacterial, anti-leishmanial, etc. associated with 1,2,3-triazole hybrids and their conjugates, a number of simple to complex dendromeric molecular scaffolds enhancing multivalency effect through statistical rebinding concept and subsequently displayed a notable applications in the study of carbohydrate-protein interaction studies, a very important concept widely applied in drug discovery and development [188]. Interestingly, a click-derived fullerene cored molecular architecture, for example homo-dodecane-valent α-mannosefullerene **187** (Fig. 15.47) displayed a potential multivalency effect with ConA through statistical rebinding [189]. Interestingly, the aggregation deprivation of this mannose-coated fullerene-cored multivalent dendrimer with ConA was found to the highest value of 42 rp/n [189].

Another interesting extension with fullerene cored 1N-ONJ coated with 1,2,3-triazole spacer resulted to the glycodendrimers **188** and **189** (Fig. 15.48) which were displayed their notable inhibitory potency through binding Peanut (*Arachis hypogaea*) agglutinin (PNA) [190]. As a consequence of statistical rebinding, the fullerene dendrimer having 1N-ONJ coat with short spacer (n = 6, dendrimer **188**) exhibited an efficiency with an inhibition factor of 13 rp/n however, dendrimer having 1N-ONJ coat long spacer (n = 9, dendrimer **189**) showed an inhibition factor of 31 rp/n [189]. Through variation of different coating at the end group, a number of diverse molecular systems can be developed simply by applying the similar optimized protocol and the resulting heterocluster effect through consequence of statistical rebinding can further influence the multivalency effect that certainly have an ample opportunities to be explored in this growing field of investigations.

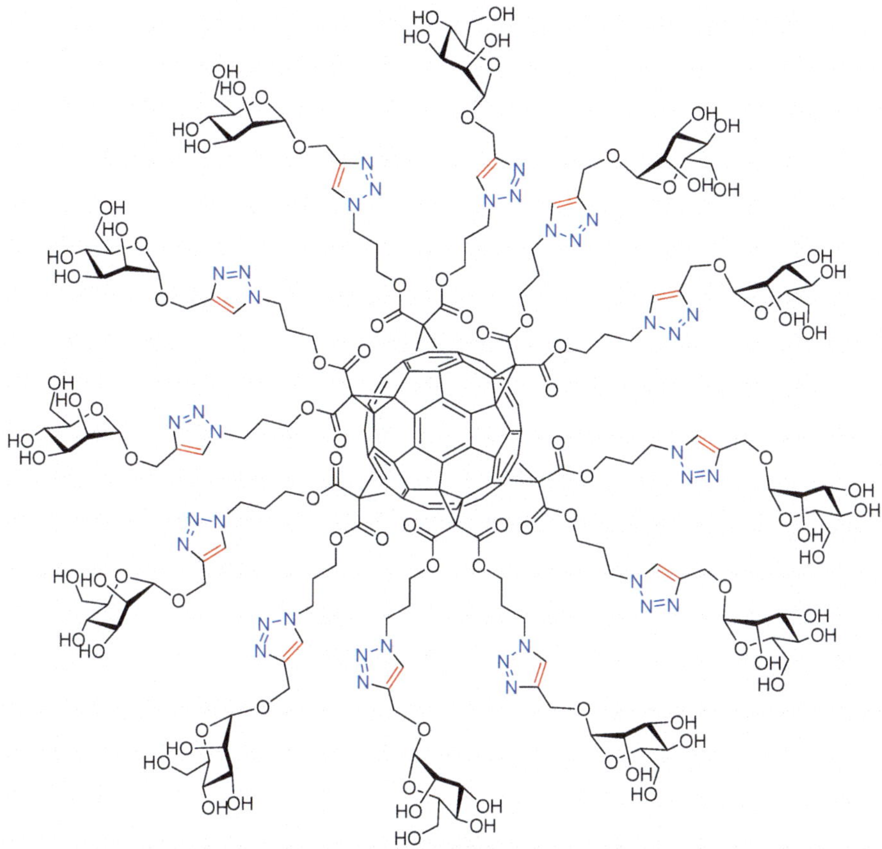

187

Fig. 15.47 Structure of homo-dodecane-valent α-mannosefullerene **187** exhibited promising multivalent effect with ConA (42 rp/n) [189]

Fitting for the binding dendritic cell-specific intercellular adhesion molecule-3-grabbing non-integrin (commonly known as **DC-SIGN**) is used to express in dendritic cells. Such interaction occurs in some important events such pathogen recognition, dissemination, and immune suppression of numerous pathogens. A number of 1,2,3-triazole forming click-mediated rod-like macromolecular architectures of different spacers and ligand unit were developed, where mannoside 1,2,3-triazole-linked glycoconjugate **190** (Fig. 15.49) having two rod-like rigid spacer bivalent analogue displayed notable multivalency effect with DC-SIGN antagonists as a consequence of chelation and statistical rebinding [191]. SPR analysis of rod-based 1,2,3-triazole linked glycodendrimer **191** showed an inhibition factor of 17 rp/n with DC-SIGN (IC_{50} 8 μM) and support the optimized length as suitable spacer [191].

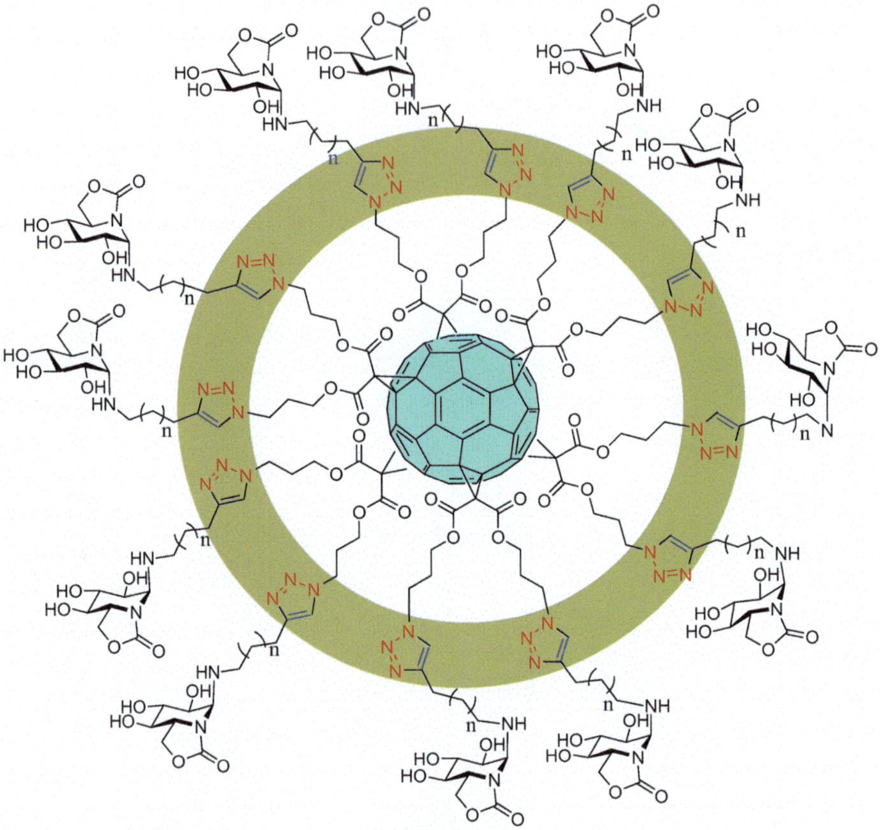

188 (n = 6)
189 (n = 9)

Fig. 15.48 Structure of fullerene core glycodendrimer **188** (13 rp/n) and **189** (31 rp/n) displayed a multivalency effect with PNA [190]

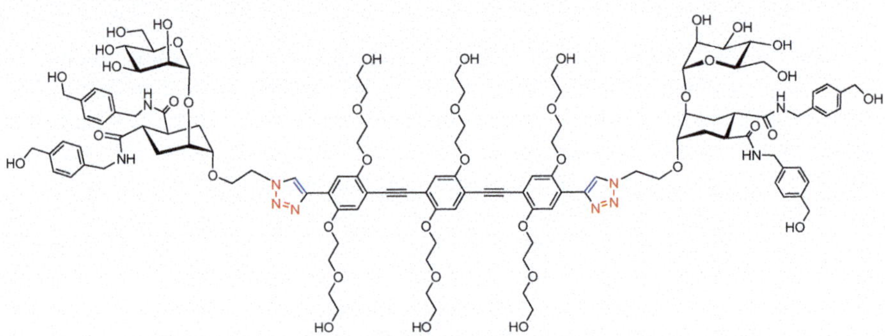

190 (IC_{50} = 8 μM, rp/n = 17)

Fig. 15.49 Rod-based glycodendrimer **190** (multivalency effect with DC-SIGN) [191]

As a consequence of combined statistical rebinding, subsite binding, and aggregation effect, the similar multivalent fullerene core glycodendrimer were evaluated with the binding **Jack bean α-mannosidase (known as Jb-αMan)**. A fullerene glycodendrimer **191** (Fig. 15.50) is able to induce chelation at the binding site of Jb-αMan and exhibited notable multivalency efficiency with inhibitory value of 179 rp/n [192]. The inhibitory activity may be because of the combination of statistical rebinding, aggregation, and subsite binding. The optimized, organized structural motif of Jb-αMan has a vital function that may provide relevant binding sites required for such promising interaction.

In recent years, multivalency effect through interaction of specific ligand pairs with protein is identified as systematized target for the rational drug design and development. When aggregation is prevented, the statistical rebinding is the mode of interaction in ConA. However, aggregation significantly increased the chelation synergism via multivalency effect. Further, PNA also displayed such behavior. In DC-SIGN, the chelation and the statistical rebinding have the combined effect in

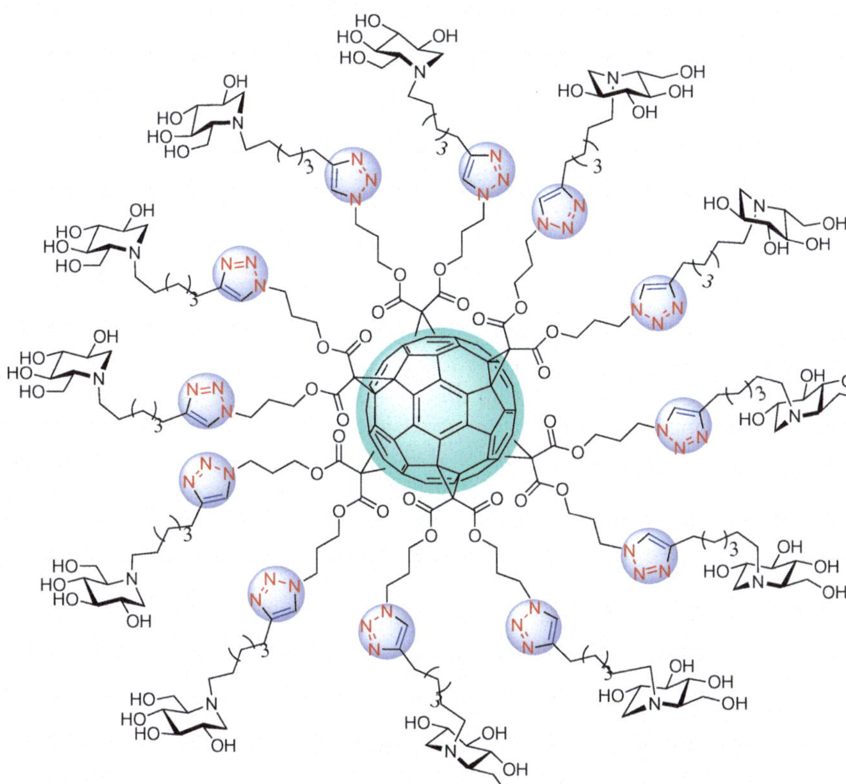

191

Fig. 15.50 Structure of fullerene core glycodendrimer **191** (multivalency effect of 179 rp/n at Jb-αMan) [192]

potency, while the chelation effect has only contributed as multivalency effect in Jb-αMan. Interestingly, the multivalency effect can be proficient through incorporation of the functional residue in fullerene ball or related spherical structure.

15.11 Conclusions and Future Perspectives

1,2,3-Triazole hybrids and their conjugates have been known for a diverse range of pharmaceutical activities such as ant-diabetic, anti-cancer, anti-inflammatory, anti-tubercular, anti-leishmanial, anti-trypanosomal, anti-microbial, anti-bacterial, and anti-viral activities, etc. Notably, molecular hybridization of a variety of pharmacophoric unit with 1,2,3-triazole unit may provide novel, effective and safe drug candidate. This chapter has examined the latest developments in the field of developing carbohydrates-based therapeutics by the application of modular click reaction. The CuAAC of terminal alkynes and azides is a well adaptable coupling method that has assembled a wide range of conjugates with significant biological and pharmacological importance. Compatibility of click reaction with biological systems and carbohydrate chemistry has accelerated lead identification and drug discovery research. It has made it easier to obtain sugar derivatives, ranging from simple to complicated, and has also made it easier to quickly evaluate them in vitro and in vivo. Click reaction mediated 1,4-disubstituted 1,2,3-triazole ring, works well as a linker between sugars and other functional molecules. Many triazolyl conjugates that are synthesized from CuAAC have been found to selectively inhibit enzymes and exhibit a variety of bioactivities, such as immune-stimulating, anti-bacterial, anti-fungal, anti-viral, anti-cancer, and anti-leishmanial properties. Additionally, this magical regioselective reaction has made it easy to construct a variety of multivalent ligands, peptides, macrocycles, clusters, dendrimers, and polymers. These materials are used in sensing required for serum diagnosis, drug delivery system, and many other biomedical applications. Both the metal-induced or metal-free regioselective triazole forming click tool has also been demonstrated to be an excellent method for the successful bioconjugation of biologically relevant systems such as polysaccharides, oligonucleotides, polypeptides, viruses, and also bacteria to a variety of bioactive residue. Despite of several notable feature of CuAAC tool, this virtually flawless coupling method has a significant disadvantage, as copper's toxicity in the cells limits through utilization in the biological systems. This was suggested to perform in vivo imaging experiments in the strain-promoted copper-free azide-alkyne cycloaddition (SPACC) conditions. The bioisosteric relationship of amide functionality with 1,2,3-triazole can be well utilized for the generation of novel leads to several frontline diseases such as cancer, bacterial infections, neurological disorders, and human immunodeficiency.

References

1. Tornøe CW, Christensen C, Meldal M (2002) J Org Chem 67:3057–3062
2. Rostovtsev VV, Green LG, Fokin VV, Sharpless KB (2002) Angew Chem Int Ed 41 2596–2599.
3. Tiwari VK, Mishra BB, Mishra KB, Mishra N, Singh AS, Chen X (2016) Chem Rev 116:3086–3240
4. Agrahari AK, Bose P, Jaiswal MK, Rajkhowa SS, Singh AS, Hotha, Mishra N, Tiwari VK (2021) Chem Rev 121:7638–7956
5. Thirumurugan P, Matosiuk D, Jozwiak K (2013) Chem Rev 113:4905–4979
6. Sharma A, Agrahari AK, Rajkhowa S, Tiwari VK (2022) Eur J Med Chem 238:114454
7. Johansson JR, Beke-Somfai T, Stålsmeden AS, Kann N (2016) Chem Rev 116:14726–14768
8. Kolb HC, Sharpless KB (2003) Drug Discov Today 8:1128–1137
9. Fantoni NZ, Ei-Sagheer AH, Brown T (2021) Chem Rev 121:7122–7154
10. Tripathi RP, Dwivedi P, Sharma A, Kushwaha D, Tiwari VK (2013) Triazolyl glycoconjugates in medicinal chemistry. In: Witczak ZJ, Bielski R (eds) Click chemistry in glycoscience: new development and strategies. Wiley, New York, pp 293–324
11. Gopalan B, Balasubramanian KK (2016) Application of click chemistry in drug discovery and development. In: Chandrasekeran S (ed) Click reactions in organic synthesis. Wiley, New York, pp 25–76
12. Zhang L, Chen XG, Xue P, Sun HHY, Williams ID, Sharpless KB, Fokin VV, Jia GC (2005) J Am Chem Soc 127:15998–15999
13. Jaiswal MK, Gupta A, Ansari FJ, Pandey VK, Tiwari VK (2024) Curr Org Synth 21:513–558
14. John J, Thomas J, Dehaen W (2015) Chem Commun 51:10797–10806
15. Jaiswal MK, Gupta A, Yadav MS, Pandey VK, Tiwari VK (2023) Chem Eur J 29:e202301749
16. Kayet A, Pathak T (2013) J Org Chem 78:9865–9875
17. Jaiswal MK, Tiwari VK Chem Rec e202300167
18. Agard NJ, Prescher JA, Bertozzi CR (2004) J Am Chem Soc 126:15046–15047
19. Codelli JA, Baskin JM, Agard NJ, Bertozzi CR (2008) J Am Chem Soc 130:11486–11493
20. Bird RE, Lemmel SA, Yu X, Zhou QA (2021) Bioconjugate Chem 32:2457–2479
21. Bonandi E, Christodoulou MS, Fumagalli G, Perdicchia D, Rastelli G, Passarella D (2017) Drug Disc Today 22:1572–1581
22. Bozorova K, Zhaoa J, Aisaa HA (2019) Bioorg Med Chem 27:3511–3531
23. Xu Z, Zhao SJ, Liu Y (2019) Eur J Med Chem 1(183):111700
24. Bauman A, Kluba CA, Vomstein S, Walter MA, Mindt TL (2013) Angew Chem In Ed 52:8957–8960
25. Bachl J, Mayr J, Sayago FJ, Cativiela C, Díaz Díaz D (2015) Chem Commun 51:5294–5529
26. Pedersen DS, Abell A (2011) Eur J Org Chem 13:2399–2411
27. Donald BJ, Surani S, Deol HS, Mbadugha UJ, Udeani G (2017) Drug. Des. Devel. Ther. 11:3559–3566
28. Daher SS, Jin X, Patel J, Freundlich JS, Buttaro B, Andrade RB (2019) Bioorg Med Chem Lett 29:1386–1389
29. Platt FM, Neises GR, Reinkensmeier G, Townsend MJ, Perry VH, Proia RL, Winchester BG, Dwek RA, Butters TD (1997) Science 276:428–431
30. Zitzmann N, Mehta AS, Carrouee S, Butters TD, Platt FM, McCauley J, Blumberg BS, Dwek RA, Block TM (1999) Proc Natl Acad Sci USA 96:11878–11882
31. Gross PE, Baker MA, Carver JP, Dennis JW (1995) Clin Cancer Res 1:935–944
32. Rossi LL, Basu A (2005) Bioorg Med Chem Lett 15:3596–3599
33. Anand N, Jaiswal N, Pandey SK, Srivastava AK, Tripathi RP (2011) Carbohydr Res 346:16–25
34. Ferreira SB, Sodero ACR, Cardoso MFC, Lima ES, Kaiser CR, Silva FP, Ferreira VF (2010) J Med Chem 53:2364–2375
35. Diot J, Garcia-Moreno MI, Gouin SG, Mellet CO, Haupt K, Kovensky J (2009) Org Biomol Chem 7:357–363

36. Bonduelle C, Huang J, Barraga TM, Mellet CO, Decroocq C, Etame E, Heise A, Compain P, Lecommandoux S (2014) Chem Commun 50:3350–3352
37. Zhou Y, Zhao Y, O'Boyle KM, Murphy PV (2008) Bioorg Med Chem Lett 18:954–958
38. Howe JD, Smith N, Lee MR, Ardes-Guisot N, Vauzeilles B, Désiré J, Baron A, Blériot Y, Sollogoub M, Alonzi DS, Butters TD (2013) Bioorg Med Chem 21:4831–4838
39. Wang L, Fang Z (2020) J Carbohydr Chem 39:415–436
40. Fletterick RJ, Sprang SR (1982) Acc Chem Res 15:361–369
41. Cheng K, Liu J, Liu X, Li H, Sun H, Xie J (2009) Carbohydr Res 344:841–850
42. Cheng K, Liu J, Sun H, Bokor E, Czifrak K, Konya B, Toth M, Docsa T, Gergely P, Somsak L (2010) New J Chem 34:1450–1464
43. Alexacou KM, Hayes JM, Tiraidis C, Zographos SE, Leonidas DD, Chrysina ED, Archontis G, Oikonomakos NG, Paul JV, Varghese B et al (2008) Proteins: Struct Funct Bioinf 71:1307–1323
44. Cheng K, Liu J, Sun H, Xie J (2010) Synthesis 6:1046–1052
45. Goyard D, Docsa T, Gergely P, Praly JP, Vidal S (2015) Carbohydr Res 402:245–251
46. Bokor E, Docsa T, Gergely P, Somsák L (2010) Bioorg Med Chem 18:1171–1180
47. Goyard D, Baron M, Skourti PV, Chajistamatiou AS, Docsa T, Gergely P, Chrysina ED, Praly JP, Vidal S (2012) Carbohydrate Res 364:28–40
48. Kun S, Nagy GZ, Tóth M, Czecze L, Nguyen van Nhien A, Docsa T, Gergely P, Charavgi M-D, Skourti PV, Chrysina ED et al (2011) Carbohydr Res 346:1427–1438
49. He XP, Deng Q, Gao LX, Li C, Zhang W, Zhou YB, Tang Y, Shi XX, Xie J, Li J et al (2011) Bioorg Med Chem 19:3892–3900
50. Li C, He X-P, Zhang YJ, Li Z, Gao L-X, Shi XX, Xie J, Li J, Chen GR, Tang Y (2011) Eur J Med Chem 46:4212–4218
51. Wang J, He X, Gao L, Sheng L, Shi X, Li J, Chen G (2011) Chin J Chem 29:1227–1232
52. He XP, Li C, Jin XP, Song Z, Zhang HL, Zhu CJ, Shen Q, Zhang W, Sheng L, Xie J et al (2011) New J Chem 35:622–631
53. Song Z, He X-P, Li C, Gao L-X, Wang Z-X, Tang Y, Xie J, Li J, Chen G-R (2011) Carbohydr Res 346:140–145
54. Yang JW, He XP, Li C, Gao LX, Sheng L, Xie J, Shi XX, Tang Y, Li J, Chen GR (2011) Bioorg Med Chem Lett 21:1092–1096
55. Lin L, Shen Q, Chen GR, Xie J (2008) Bioorg Med Chem 16:9757–9763
56. Zhang YJ, He XP, Li C, Li Z, Shi DT, Gao LX, Qiu BY, Shi XX, Tang Y, Guo-Rong C (2010) Chem Lett 39:1261–1263
57. Igual MO, Nunes PSG, da Costa RM, Mantoani SP, Tostes RC, Carvalho I (2018) Carbohyd Res 471:43–55
58. Chrysina ED, Bokor É, Alexacou KM, Charavgi MD, Oikonomakos GN, Zographos SE, Leonidas DD, Oikonomakos NG (2009) Somsák L. Tetrahedron Asymmetry 20:733–740
59. Khan H, Saini KS, Ajay A, Devender N, Bhattacharjee A, Das S, Dwivedi S, Gupt MP, Bora HK, Mitra K, Tripathi RP, Konwar R (2015) Int J Biochem Cell Biol 65:275–287
60. (a) Hou J, Feng C, Li Z, Fang Q, Wang H, Guoxian G, Shi Y, Liu P, Xu F, Zheng Y, Shen J, Wang P (2011) Eur J Med Chem 46:3192–3200; (b) Accenda Tech Co. Ltd. Patent No: WO/2011/003365
61. Suzuki T, Ota Y, Ri M, Bando M, Gotoh A, Itoh Y, Tsumoto H, Tatum PR, Mizukami T, Nakagawa H, Iida S, Ueda R, Shirahige K, Miyata N (2012) J Med Chem 55:9562–9575
62. Tars K, Vullo D, Kazaks A, Leitans J, Lends A, Grandane A, Zalubovskis R, Scozzafava A, Supuran CT (2013) J Med Chem 56:293–300
63. Grandane A, Tanc M, Zalubovskis R, Supuran CT (2014) Biorg. Med. Chem Lett 24:1256–1260
64. Oueis E, Santoni G, Ronco C, Syzgantseva O, Romieu A, Weik M, Jean L, Nachon F, Renard PY (2014) Org Biomol Chem 12:156–161
65. Ronco C, Carletti E, Colletier JP, Weik M, Nachon F, Jean L, Renard PY (2012) Chem. Med. Chem. 7:400–405
66. Gallop MA, Barrett RW, Dower WJ, Fodor SP, Gordon EM (1994) J Med Chem 37:1233–1251

67. Bérubé G (2016) Expert Opin Drug Discov 11:281–305
68. Viegas-Junior C, Danuello A, da Silva Bolzani V, Barreiro EJ, Fraga C (2007) A. Curr Med Chem 14:1829–1852
69. Szumilak M, Wiktorowska-Owczarek A, Stanczak A (2021) Molecules 26:2601
70. Chamberlain SD, Moorman AR, Burnette TC, de Miranda P, Krenitsky TA (1994) Antiviral Chem Chemo 5:64–73
71. Netto GJ, Epstein JI (2009) Med Res Rev 29:419–435
72. Wilkinson BL, Bornaghi LF, Houston TA, Innocenti A, Vullo D, Supuran CT, Poulsen SA (2007) J Med Chem 50:1651–1657
73. Salmon AJ, Williams ML, Maresca A, Supuran CT, Poulsen SA (2011) Bioorg Med Chem Lett 21:6058–6061
74. Moeker J, Peat TS, Bornaghi LF, Vullo D, Supuran CT, Poulsen S-A (2014) J Med Chem 57:3522–3531
75. Murraya AB, Quadrib M, Lib H, McKennaa R, Horenstein NA (2019) Carbohydr Res 476:65–70
76. Parida PK, Sau A, Ghosh T, Jana K, Biswas K, Raha S, Misra A (2014) K. Bioorg Med Chem Lett 24:3865–3868
77. Peterson LB, Blagg BSJ (2010) Bioorg Med Chem Lett 20:3957–3960
78. Zhang J, Garrossian M, Gardner D, Garrossian A, Chang YT, Kim YK, Chang CWT (2008) Bioorg Med Chem Lett 18:1359–1363
79. Freitas LBO, Borgati TF, Freitas RP, Ruiz ALTG, Marchetti GM, Carvalho JE, Cunha EFF, Ramalho TC, Alves RB, Eur J (2014) Med Chem 84:595–604
80. Nandurkar N, Zhang J, Ye Q, Ponomereva L, She Q, Thorson JS (2014) J Med Chem 57:7478–7484
81. Manach CL, Baron A, Guillot R, Vauzeilles B, Beau JM (2011) Tetrahedron Lett 52:1462–1465
82. Wang, G. Q.; Yan, L-l.; Wang, Q. *Heterocycl. Commun.*, **2018**, *24*, 119–124.
83. Halay EE, Ay E, Şalva E, Ay K, Karayıldırım T (2018) Chem Heterocycl Comp 54:158–166
84. Moni L, Marra A, Skotnicki JS, Koehn FE, Gharbia MA, Dondoni A (2013) Tetrahedron Lett 54:6999–7003
85. Nagarsenkar A, Prajapti SK, Guggilapua SD, Birinenib S, Kotapallib S, Ummanni S, Babu BN (2016) Med Chem Commun 7:646–653
86. Krawczyka M, Pastuch-Gawolek G, Mrozek-Wilczkiewiczc A, Kuczak M, Skonieczna M, Musiol R (2019) Bioorg Chem 84:326–338
87. Amdouni H, Robert G, Driowya M, Furstoss N, Metier´C, Dubois A, Dufies M, Zerhouni M, Orange F, Lacas-Gervais S, Bougrin K, Martin AR, Auberger P, Benhida R (2017) J Med Chem 60:1523–1533
88. Panaka S, Trivedi R, Jaipal K, Giribabu L, Sujitha P, Kumar CG, Sridhar B (2016) 813:125–130
89. Lia HN, Wanga H, Wanga ZP, Yana HN, Zhanga M, Liua Y, Cheng MS (2018) Bioorg Med Chem 26:4025–4033
90. Mishra KB, Tiwari N, Bose P, Singh R, Rawat AK, Singh SK, Mishra RC, Singh RK, Tiwari VK (2019) ChemistrySelect 4:2644–2648
91. Zi CT, Yang L, Xu FQ, Dong FW, Yang D, Li Y, Ding ZT, Zhou J, Jiang ZH, Hu JM (2018) Drug Design Develop Therap 12:3393–3406
92. Belmonte-Reche E, Benassi A, Peñalver P, Cucchiarini A, Guédin A, Mergny JL, Rosu F, Gabelica V, Freccero M, Doria F, Morales JC (2022) Eur J Med Chem 232:114183
93. Liu S, Sun Z, Liang M, Song W, Zhang R, Shi Y, Cui Y, Gao Q (2022) Adv Sci 9:2105315
94. Zhang R, Luo Y, Du C, Wu L, Wang Y, Chen Y, Li S, Jiang X, Xie Y (2023) Bioorg Med Chem Lett 81:129128
95. Jordan A, Hadfield JA, Lawrence NJ, McGown AT (1998) Med Res Rev 18:259–296
96. Yang JS, Song D, Ko WJ, Kim B, Kim BK, Park SK, Won M, Lee K, Lee K, Kim HM, Han G (2013) Eur J Med Chem 63:621–628
97. Zi C-T, Li G-T, Li Y, Zhou J, Ding Z-T, Jiang Z-H, Hu, J-M (2015) Nat Prod Bioprospect 5:83–90

98. Wei G, Luan W, Wang S, Cui S, Li F, Liu Y, Yang L, Cheng M (2015) Org Biomol Chem 13:1507–1514
99. Wan Q, Chen J, Chen G, Danishefsky SJ (2006) J Org Chem 71:8244–8249
100. Schauer R (2009) Curr Opin Struct Biol 19:507–514
101. Bose P, Jaiswal MK, Singh SK, Singh RK, Tiwari VK (2023) Carbohydrate Res 527:108804
102. Chen X (2023) Acc Chem Res 57:17618–17619
103. Meindl P, Tuppy H (1969) Hoppe Seylers Z 350:1088–1092
104. Michiels B, Van Puyenbroeck K, Verhoeven V, Vermeire E, Coenen S (2013) PLoS ONE 8:e60348
105. Burch J, Corbett M, Stock C, Nicholson K, Elliot AJ, Duffy S, Westwood M, Palmer S, Stewart L (2009) Lancet. Infect Dis 9:537–545
106. Zou Y, Albohy A, Sandbhor M, Cairo CW (2010) Bioorg Med Chem Lett 20:7529–7533
107. Kai H, Hinou H, Nishimura SI (2012) Bioorg Med Chem 20:2739–2746
108. Li J, Zheng M, Tang W, He PL, Zhu W, Li T, Zuo JP, Liu H, Jiang H (2006) Bioorg Med Chem Lett 16:5009–5013
109. Weiwer M, Chen CC, Kemp MM, Linhardt RJ (2009) Eur J Org Chem 16:2611–2620
110. Yan ZL, Liu AY, Wei XX, Zhang Z, Qin L, Yub Q, Yub P, Lub K, Yang Y (2019) Carbohydr Res 477:32–38
111. Papin C, Doisneau G, Beau JM (2008) Chem Eur J 15:53–57
112. Kushwaha D, Xu P, Kovac P (2017) RSC Adv 7:7591–7603
113. Oliveira IA, Gonçalves AS, Neves LJ, Itzstein MV, Todeschini AR (2014) J Biol Chem 289:423–436
114. Carvalho I, Andrade P, Campo VL, Guedes PMM, Sesti-Costa R, Silva JS, Schenkman S, Dedola S, Hill L, Rejzek M et al (2010) Bioorg Med Chem 18:2412–2427
115. Harrison JA, Kartha KPR, Fournier EJL, Lowary TL, Malet C, Nilsson UJ, Hindsgaul O, Schenkman S, Naismith JH, Field RA (2011) Org Biomol Chem 9:1653–1660
116. Su Y, Meng L, Sun J, Li W, Shao L, Chen K, Zhou D, Yang F, Yu F (2019) Eur J Med Chem 182:111622
117. Li T, Guo L, Zhang Y, Wang J, Li Z, Lin L, Zhang Z, Li L, Lin J, Zhao W et al (2011) Carbohydr Res 346:1083–1092
118. Balzarini J, Pérez-Pérez MJ, San-Félix A, Velazquez S, Camarasa MJ, De Clercq E (1992) Antimicrob Agents Chemother 36:1073–1080
119. Sluis-Cremer N, Hanamouch N, Felix AS, Velazquesz S, Balzarini J, Camarasa MJ (2006) J Med Chem 49:4834–4841
120. Olomola TO, Klein R, Mautsa N, Sayed Y, Kaye PT (2013) Bioorg Med Chem 21:1964–1971
121. Kumar S, Ranjan N, Kellisha P, Gonga C, Watkinsb D, Arya DP (2016) Org Biomol Chem 14:2052–2056
122. Kumar S, Arya DP (2011) Bioorg Med Chem Lett 21:4788–4792
123. Cheng S, Chang X, Wang Y, Gao GF, Shao Y, Ma L, Li X (2015) J Med Chem 58:1372–1379
124. Gorbalenya AE, Baker SC, Baric RS, Groot RJ, Drosten C, Gulyaeva A, Haamans BL, Lauber C, Leontovich AM, Neuman BW et al (2020) Nat Microbiol 5:536–544
125. Kupferschmidt K, Cohen J (2020) Science 367:610–611
126. Łoczechin A, Seron´K, Barras A, Giovanelli E, Belouzard S, Chen Y-T, Metzler-Nolte N, Boukherroub R, Dubuisson J, Szunerits S (2019) ACS Appl Mater Interfaces 11:42964–42974
127. Wang Y, Zhang D, Du G, Du R, Zhao J, Jin Y, Fu S, Ling, Gao L, Cheng Z et al. Lancet 395:1569–1578
128. Merchan A, Ramirez-Lopez P, Martinez C, Suarez JR, Perona A, Hernaiz MJ (2024) Bioconjugate Chem 35:34–42
129. Mishra S, Tripathi N, Mishra S, Kishore N, Singh RK, Tiwari VK (2017) Eur J Med Chem 137:504–554
130. Keri RS, Patil SA, Budagumpi S, Nagaraja BM (2015) Chem Biol Drug Des 86:410–423
131. Zhou B, He Y, Zhang X, Xu J, Luo Y, Wang Y, Franzblau SG, Yang Z, Chan RJ, Liu Y, Zheng J, Zhang ZY (2010) Proc Natl Acad Sci USA 107:4573–4578
132. Isobe H, Fujino T, Yamazaki N, Nieckowski MG, Nakamura E (2008) Org Lett 10:3729–3732

133. Tiwari VK (2020) Carbohydrates in drug discovery and development. Elsevier Inc.
134. Muthukrishnan M, Mujahid M, Yogeeswari P, Sriram D (2011) Tetrahedron Lett 52:2387–2389
135. Menendez C, Christophe A, Rodriguez F, Inard C, Pasca MS, Lherbet C, Baltas M (2012) Eur J Med Chem 52:275–283
136. Kim S, Cho SN, Taegwon O, Kim P (2012) Bioorg Med Chem Lett 22:6844–6847
137. Boechat N, Ferreira VF, Ferreira SB, Ferreira MDLG, Silva FDC, Bastos MDSC, Lourenco MCS, Pinto AC, Krettli AU, Aguiar AC et al (2011) J Med Chem 54:5988–5999
138. Kumar K, Kremer SC, Kremer L, Guerardel Y, Biot C, Kumar V (2013) Organometallics 32:5713–5719
139. Singh BK, Yadav AK, Kumar B, Gaikwad A, Sinha SK, Chaturvedi VR, Tripathi P (2008) Carbohydr Res 343:1153–1162
140. Munagala G, Yempalla KR, Singh S, Sharma S, Kalia NP, Rajput VS, Kumar S, Sawant SD, Khan IA, Vishwakarma RA, Singh PP (2015) Org Biomol Chem 13:3610–3624
141. Ramprasad J, Nayak N, Dalimba U, Yogeeswari P, Sriram D (2015) Bioorg Med Chem Lett 25:4169–4173
142. Reddyrajula R, Dalimba UK (2019) New J Chem 43:16281–16299
143. Suresh A, Naidu K, Srinivasarao MS, Agnieszka N, Ewa AK, Murugesan S, Chander S, Krishnan R, Sekhar KVGC (2017) New J Chem 41:347–359
144. Naik RJ, Kulkarni MV, Pai KSR, Nayak PG (2012) Chem Biol Drug Des 80:516–523
145. Anand A, Naik RJ, Revankar HM, Kulkarni MV, Dixit SR, Joshi SD (2015) Eur J Med Chem 105:194–207
146. Somagond SM, Kamble RR, Bayannavar PK, Shaikh SKJ, Joshi SD, Kumbar VM, Nesaragi AR, Kariduraganavar MY (2019) Arch. Pharm. Chem Life Sci. 352:1900013
147. Huang G-C, Xu Y, Xu Z, Lv Z-S, Zhang J, Guo H-Y, Hu Y-Q, Liu M-L, Guan J-G, Lu Y (2018) J Heterocycl Chem 55:830–835
148. Srivastava S, Bimal D, Bohra K, Singh B, Ponnan P, Jain R, Basil MV, Maity J, Thirumal M, Prasad AK (2018) Eur J Med Chem 150:268–281
149. Ashok D, Gundu S, Aamate VK, Devulapally MG, Bathini R, Manga V (2018) J Mol Struct 1157:312–321
150. Khare SP, Deshmukh TR, Sangshetti JN, Krishna VS, Sriram D, Khedkar VM, Shingate BB (2018) ChemistrySelect 3:13113–13122
151. Ashok D, Chiranjeevi P, Kumar AV, Sarasija M, Krishna VS, Sriram D, Balasubramanian S (2018) RSC Adv 8:16997–17007
152. Jeankumar VU, Reshma RS, Janupally R, Saxena S, Sridevi JP, Medapi B, Kulkarni P, Yogeeswari P, Sriram D (2015) Org Biomol Chem 13:2423–2431
153. Hu YQ, Zhang S, Zhao F, Gao C, Feng LS, Lv ZS, Xu Z, Wu X (2017) Eur J Med Chem 16:255–267
154. Badar AD, Sulakhe SM, Muluk MB, Rehman NNMA, Dixit PP, Choudhary PF, Rekha E, Sriram MD, Havel KPJ (2020) Hetrocycl. Chem. 57:3544–3557
155. Reis WJ, Bozzi IAO, Riberio MF, Halicki PCB, Ferreira LA, Almeida da Silva PE, Ramos DF, de Simone CA, da Silva EN Jr (2019). Bioorg Med Chem 27:4143–4150
156. Kumar D, Beena, Khare G, Kidwai S, Tyagi AK, Singh R, Rawat DS (2014) Eur J Med Chem 81:301–313
157. Boechat N, Ferreira VF, Ferreira SB et al (2011) J Med Chem 54:5988–5999
158. Patil PS, Kasare SL, Haval NB, Khedkar VM, Dixit PP, Rekha EM, Sriram D, Haval KP (2020) Bioorg Med Chem Lett 30:127434
159. Thomas KD, Adhikari AV, Chowdhury IH, Sumesh E, Pal NK (2011) Eur J Med Chem 46:2503–2512
160. Kumar KK, Seenivasan SP, Kumar V, Das TM (2011) Carbohydr Res 346:2084–2090
161. Reddyrajula R, Dalimba U (2019) Chemistry Select 4:2685–2693
162. Marvadi SK, Krishna VS, Sriram D, Kantevari S (2019) Bioorg Med Chem Lett 29:529–533
163. Pogaku V, Krishna VS, Balachandran C, Rangan K, Sriram D, Aoki S, Basavoju S (2019) New J Chem 43:17511–17520

164. Thakare PP, Walunj Y, Chavan A, Bobade VD, Sarkar D, Mhaske PC (2020) J Heterocycl Chem 57:3918–3929
165. Nesaragi AR, Kamble RR, Bayannavar PK, Shaikh SKJ, Hoolageri SR, Kodasi B, Joshi SD, Kumbar VM (2021) Bioorg Med Chem Lett 41:127984
166. Xu Z, Song X-F, Qiang M, Lv Z-S (2017) J Heterocycl Chem 54:3735–3741
167. Xu Z, Song X, Hu Y, Lv MZ (2018) J Heterocycl Chem 55:192–198
168. Xu Z, Lv Z-S, Song X-F, Qiang M (2018) J Heterocycl Chem 55:97–102
169. Chen R, Zhang H, Ma T, Xue H, Miao Z, Chem L, Shi X (2019) Bioorg Med Chem Lett 29:2635–2637
170. Xia X, Zhang Q, Zhao L, Hu Y (2017) Eur J Med Chem 138:66–71
171. Xu Z, Zhang S, Song X, Qiang M, Lv Z (2017) Bioorg Med Chem Lett 27:3643–3646
172. Ding Z, Hou P, Liu B (2019) Arch. Pharm. Chem. Life Sci. 352:1900135
173. Hu YQ, Xu Z, Qiang M, Lv ZS (2018) J Heterocycl Chem 55:187–191
174. Gao F, Chen Z, Ma L, Fan Y, Chen L, Lu G (2019) Eur J Med Chem 180:648–655
175. Kumar K, Singh P, Kremer L, Guerardel Y, Biot C, Kumar V (2012) Dalton Trans 41:5778–5781
176. Singh A, Biot C, Viljoen A, Dupont C, Kremer L, Kumar K, Kumar V (2017) Chem Biol Drug Des 89:856–861
177. Caselli E, Romagnoli C, Vahabi R, Taracila MA, Bonomo RA, Prati F (2005) J Med Chem 58:5445–5458
178. Yadav S, Singh AK, Agrahari AK, Sharma K, Singh AS, Gupta MK, Tiwari VK, Prakah P (2020) Scientific Rep 10:14204
179. Yadav S, Singh A, Agrahari A, Pandey A, Chauhan B, Gupta M, Tiwari VK, Prakash P (2021) Antibiotics 10(4):388
180. Burza S, Croft SL, Boelaert M (2018) Lancet 392:951–970
181. Chappuis F, Sundar S, Hailu A, Ghalib H, Rijal S, Peeling RW, Alvar J, Boelaert M (2007) Nature. Rev Microbiol 5:873–882
182. Dwivedi P, Mishra KB, Mishra BB, Singh N, Singh RK, Tiwari VK (2015) Glycoconj J 32:127–140
183. Wu Y, Jiang Z, Li Z, Gu J, You Q, Zhang X (2018) J Med Chem 61:5332–5349
184. Rajan S, Puri S, Kumar D, Babu MH, Shankar K, Varshney S, Srivastav A, Gupta A, Reddy MS, Gaikwad AN (2018) Eur J Med Chem 143:1345–1360
185. Narsimha S, Battula KS, Nagavelli VR (2018) Synth Commun 48:1220–1226
186. Krishna BR, Thummuri D, Naidu VGM, Ramakrishna S, Mallavadhani UV (2018) Bioorg Chem 78:94–102
187. Kalesh KA, Liu K, Yao SQ (2009) Org Biomol Chem 7:5129–5136
188. Heteren JV, Pieters RJ (2020) Carbohydrate-protein interaction enhancing multivalancy effect through statistical rebinding, In: Twari VK (ed) Carbohydrate in drug discovery and development. Elsevier Inc., pp 383–402
189. Abellán Flos M, García Moreno MI, Ortiz Mellet C, García Fernández JM, Nierengarten JF, Vincent S (2016) P. Chem Eur J 22:11450–11460
190. Rísquez-Cuadro R, García Fernández JM, Nierengarten JF, Ortiz Mellet C (2013) Chem Eur J 19:16791–16803
191. Ordanini S, Varga N, Porkolab V, Thépaut M, Belvisi L, Bertaglia A, Palmioli A, Berzi A, Trabattoni D, Clerici M et al (2015) Chem Commun 51:3816–3819
192. Durka M, Buffet K, Iehl J, Holler M, Nierengarten JF, Vincent SP (2012) Chem Eur J 18:641–651

Chapter 16
Click-Mediated 1,2,3-Triazoles in Catalysis

16.1 Introduction

The process of catalysis involves accelerating a chemical reaction by including a catalyst, in case of transition metal catalysis, transition metals with assistance of ligands and sometimes itself acts as that catalyst to accelerate the rate of various chemical transformations [1–3]. An enormously important and crucial area of chemistry is catalysis. More than 90% of the world's chemical production involves some form of a catalyst [4]. The introduction of asymmetric catalysis in the 1950s brought about a significant change in the approach to organic synthesis, which in turn aided the field of drug discovery and development by offering a practical method for the selective synthesis of a particular isomer of interest in excess. To obtain the desired enantiomer of the result in a metal-catalyzed process, organic molecules were used as chiral ligands [5–8]. However, there were several disadvantages to using metal complexes in asymmetric synthesis, such as the usage of toxic metals, the need for anhydrous/anaerobic reaction conditions, or non-orthogonality with different functional groups [9, 10]. On the other hand, enzymatic catalysis, while providing a non-toxic easy route to produce selected isomers, is very substrate-specific and often only supports one type of reaction per enzyme [11, 12]. Synthetic chemists began developing alternative catalytic systems that could handle diverse reaction conditions, display broad substrate scope like metal catalysts, but were harmless like enzymes to address these problems. The ligand–metal interaction governs the catalytic properties, a ligand is defined as the any species (molecules or ions) having at least one electron pair and that can be donated to a metal atom, and therefore the binding with metal atom takes place through donation of electron pairs from ligand. The interaction between metal atom and ligand may be covalent or ionic; however it stabilizes the oxidation state of the metal during catalytic processes and also modifies its catalytic properties for a reaction [13–15]. The click-derived 1*H*-1,2,3-triazole **1** is a five-membered heterocyclic ring containing three nitrogen atoms at position 1(sp^3), 2(sp^2) and 3(sp^2). The presence of these three nitrogen atoms results in the polarization of π-electrons of

V. K. Tiwari et al., *Click Chemistry*, Materials Horizons: From Nature to Nanomaterials,
https://doi.org/10.1007/978-981-97-4596-8_16

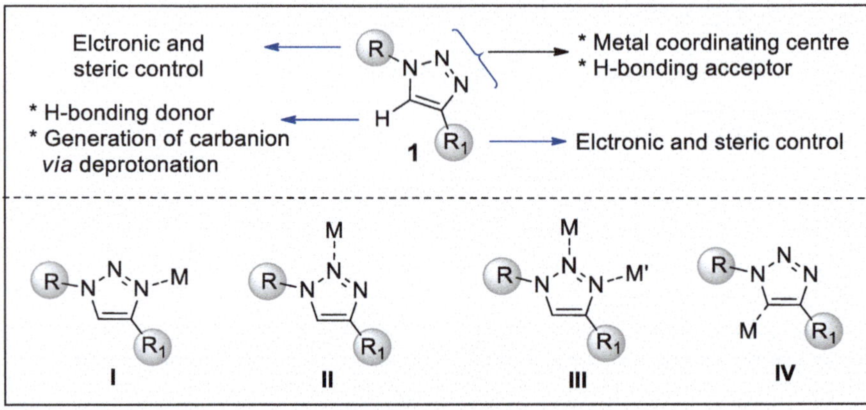

Fig. 16.1 Plausible metal coordination modes (**I–IV**) of 1,4-disubstituted-1,2,3-triazoles

the aromatic system [16]. Presence of loan pairs of electron on ring nitrogen atoms, as well as, polarized π-electrons of the aromatic system makes 1,2,3-triazole ring a suitable moiety to coordinate with metal ions (Fig. 16.1) [17, 18]. Examining the 1,2,3-triazole ring in more detail reveals the appealing properties that this ring, when combined with the appropriate substituent, has to offer to efficient coordination with suitable metal [17, 19–26].

Moreover, the click-derived 1,2,3-triazoles **1** have an array of importance in various fields such as polymer chemistry, drug discovery and development, chelation, dendrimers, surface chemistry, bioconjugation, and many more, which we have been covered in different chapters of the book [27–31]. The importance of ligands and their useful applications in diverse range of organic transformation have led to motivate researchers to look for novel molecules with easily functionalized structures in order to create effective metal–ligand interaction for easy, greener, and selective catalysis [32–34]. It is interesting to note that besides triazole skeleton, carbohydrate-appended triazole also offers an efficient coordinating center for the synthesis of ligands and organocatalysts. Chirality, one of the most interesting features of carbohydrates, entices researchers working in the field of asymmetric synthesis. Furthermore, the large natural abundance of carbohydrate molecules and the existence of many hydroxyl groups as suitable binding sites indicate that they have a promising future in low-cost asymmetric catalysis. Several attempts have been undertaken over the last few decades to employ carbohydrates in asymmetric induction as chiral auxiliaries, organocatalysts, and ligands. Numerous ligands developed from carbohydrate-based triazoles are proved to be efficient and promising catalysts [35–40]. In this chapter, a brief explanation of the structural benefits and application of the click-derived triazoles as ligand with various transition metals has been provided.

16.2 Triazole as Ligand in Metal-Catalyzed Reactions

16.2.1 Click-Derived Triazole/Cu-Assisted Catalysis

The 1,2,3-triazoles have efficient coordinating ability to bind with transition metal and catalyze a particular reaction. Chan et al., reported an example of *N*-ligated 1,2,3-triazole/Cu catalyst for CuAAC reaction in order to explore the click-derived triazole-assisted click reaction (click to click catalysis). The authors employed click-derived tris(benzyltriazolylmethyl)amine (TBTA, **4**) as a ligand to stabilize Cu(I) for the 'click' reaction between phenylacetylene **2** and benzyl azide **3** in their pioneering investigation. The tetradentate ligating nature of triazolyl ligand (TBTA) allows efficient coordination with Cu(I) metal through the basic tertiary amine and triazole units, central *N*-atom expected to be coordinated throughout catalysis, while the 1,2,3-triazole unit temporarily separate from the metal center to allow formation of a Cu(I)-acetylide intermediate, which is essential for the catalytic cycle. The author suggested that the tetradentate nature of TBTA ligand increases catalytic efficacy through preventing disproportionation and oxidation of the catalytic amount of Cu(I) center. This has led to the perception that it is significantly more effective than mono- or bis-triazolyl-based ligands to attain the triazolyl product **1** (Scheme 16.1) [41].

Later, Williams et al. found that Cu in second oxidation state can also be coordinated with TBTA and the resulting Cu(II) complex changed into an active Cu(I) catalyst in the presence of sodium ascorbate, a suitable reducing agent [42]. Moreover, the tris(triazolyl) ligand **5** with apical functional groups at quaternary carbon were consequently regarded as suitable ligands for CuAAC reactions due to their special properties and were further studied for the construction of polymer-immobilized copper catalysts **6**. The Merrifield resin was first used to alkylate the tris(triazolyl) methanol ligand **5** before CuCl was added to form the immobilized catalyst **6**. This catalyst tolerated changes in reaction conditions and shown a very effective 'click' activity toward a variety of substrates in aqueous or methanol/water systems. The great recyclability and simple separation of the product make the protocol highly sustainable and efficient to get triazole derivatives (Scheme 16.2) [43].

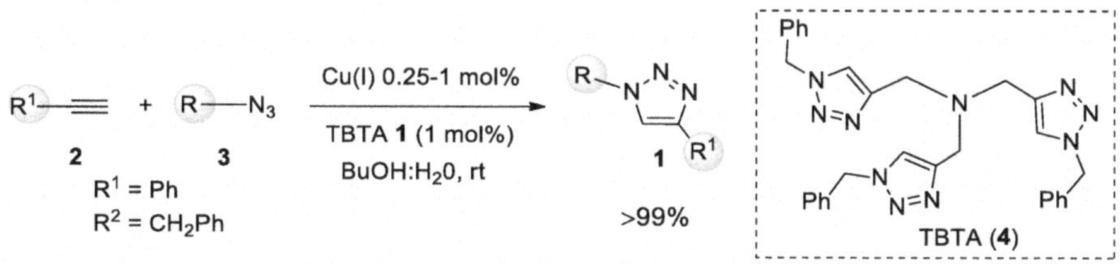

Scheme 16.1 The modular CuAAC reaction in presence of Cu(I)/TBTA catalytic system

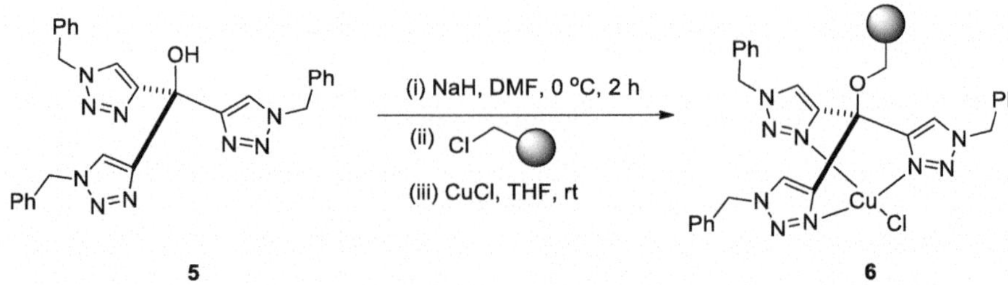

Scheme 16.2 The polymer supported tris(triazolyl)methanol ligand-derived Cu(I) catalyst **6**

Tiwari and coworkers explore the development of triazolyl sugar analogues as the efficient metal ligand to successful synthesis of various biologically important heterocyclic scaffolds through the specific transition metal and developed triazolyl glycohybrids **7** mediated coupling reactions. The Cu(I)-catalyzed intramolecular Ullmann-type coupling reaction of *o*-halo aromatic amides or thioamides or benzamidines **8** that gave high yields of respective benzoxazoles (via C–O cross-coupling) or benzothiazoles (via C–S cross-coupling) or benzimidazoles (via C–N cross-coupling) **9** were successfully investigated by group. In the experimental outcomes, the application of click-derived 1,2,3-triazolyl glycohybrid as the ligand to coordinate Cu(I) center and assist the Cu-catalyzed intramolecular Ullmann coupling reaction were explored. A series of seven novel triazolyl-glycoconjugates **7a–g** were developed from the cost-effective D-glucose, and a benzotriazole-based ligand **7h** is also considered for analyzing the catalytic efficacy with these glycosyl triazole ligands (Fig. 16.2). The CuAAC reaction was employed for the synthesis of these triazole containing carbohydrate-based ligands, where glycosyl azides derived from the D-glucose react with terminal alkynes of choice in presence of $CuSO_4.5H_2O$ and sodium ascorbate in DCM/water (1:1) at room temperature afforded the desired ligands. These carbohydrate triazole hybrid ligands were demonstrated for Cu-catalyzed C-hetero atom (O, S, N atom) cross-coupling to synthesize benz-fused azoles **9**. Among all these triazolyl glycohybrid ligands, ligand **7a** effectively assists the Cu(I)-catalyzed Ullmann-type intramolecular cross-coupling at low catalytic loading to afford benz-fused azoles **9** in excellent yields (up to 99%) (Scheme 16.3). Additionally, the devised protocol was found to be efficient for gram scale practical synthesis of diverse range of heterocyclic scaffolds with recovery of ligand from the reaction media using conventional flash column chromatography. The author suggested the plausible coordination atom in the ligand as one of the *N*-atom from triazole unit and one hydroxyl functionality of sugar analogue [44].

Moreover, the same group extends the application of click-derived sugar containing 1,2,3 triazole ligands in temperature-dependent competitive Sonogashira and Glaser–Hay coupling reaction. In these experiments, previously developed ligand **7a**, **7c** and three new carbohydrate-based 1,2,3-triazole ligands **10a–c** were screened for the Cu-catalyzed C–C coupling reaction Fig. 16.3. The Cu-catalyzed Glaser coupling and Sonogashira coupling afforded the excellent product yields in assistance of ligand **10a**. The carbohydrate containing 1,2,3-triazole ligand **10a** has proven

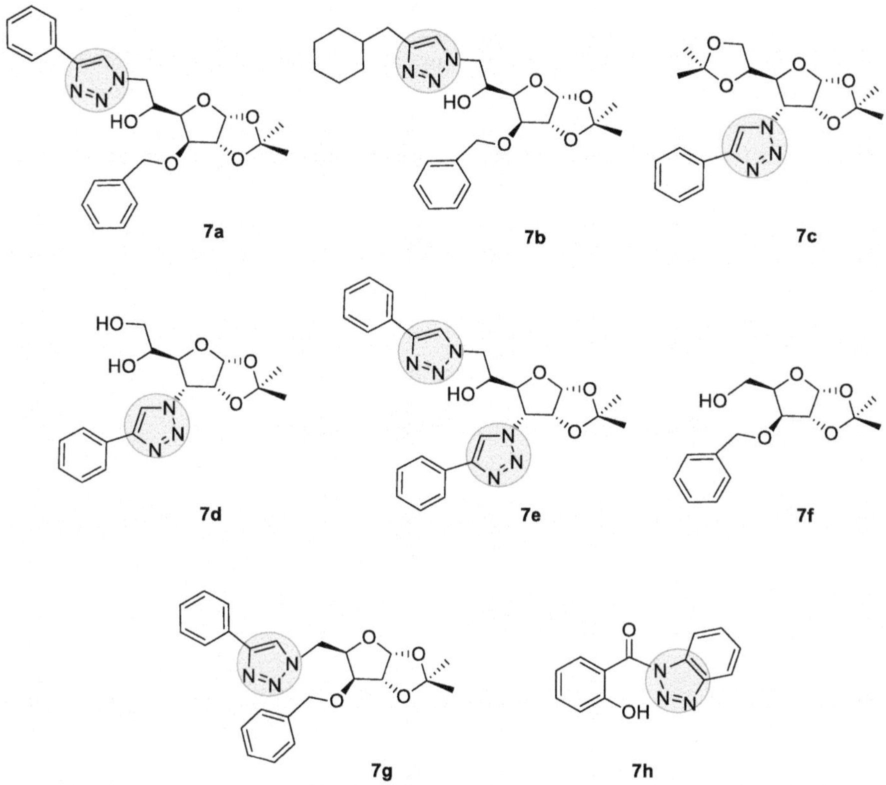

Fig. 16.2 The click-derived 1,2,3-triazolyl-glycoconjugates investigated for the intramolecular Ullmann coupling reaction

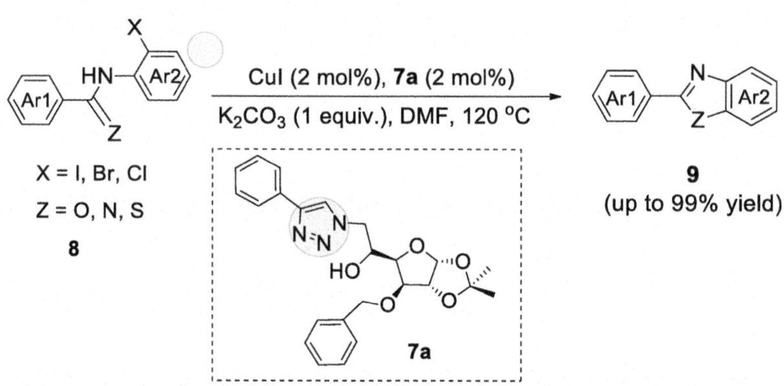

Scheme 16.3 The carbohydrate-based triazole ligand/Cu(I)-assisted intramolecular Ullmann coupling reaction

to be an effective ligand in both the coupling reactions at low catalytic loading. Noteworthy, ligand **10a**/Cu-catalyzed Sonogashira reaction at high temperature (130 °C) affords no side product of homocoupling even without employing inert environment, which is of immense use in this click-derived triazole ligand during catalysis to get beyond the difficulties of Cu-catalyzed Sonogashira reaction (Scheme 16.4) [45].

Kraft et al. explored the click approach to develop various sugar-triazole-pyridine-based bidentate and tridentate glycohybrid ligands and screened them in metal-catalyzed asymmetric synthesis where the click-derived triazole-appended glycohybrid ligands **11** effectively worked for the Cu(II)-catalyzed asymmetric addition of terminal alkyne **2** to *N*-benzylideneaniline **12**. The 1,2,3-triazole-appended glycohybrid ligand **11** with $Cu(OTf)_2$ efficiently catalyzed the enantioselective addition reaction to afford particular enantiomeric product **13** in the good yield (Scheme 16.5) [46].

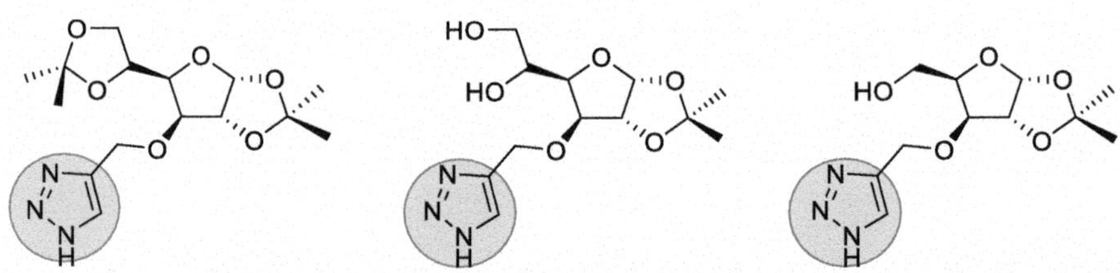

10a **10b** **10c**

Fig. 16.3 Click-derived sugar containing 1,2,3-triazole for the Cu-catalyzed coupling reaction

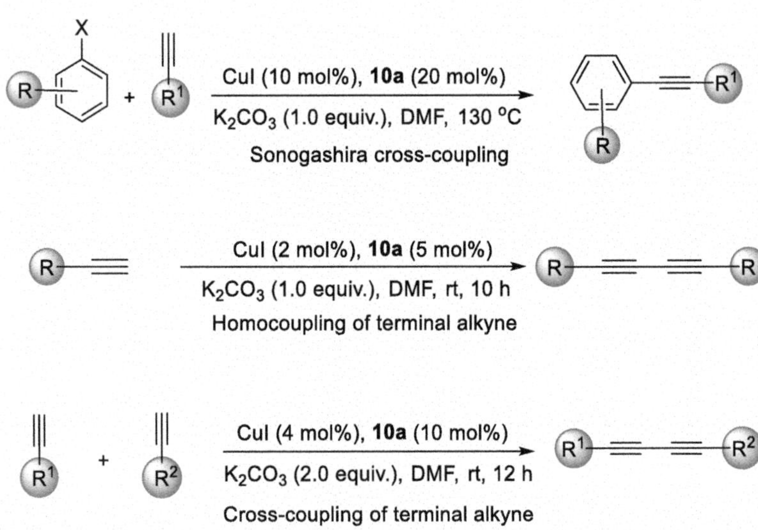

Scheme 16.4 Click-derived sugar-based triazole/Cu-catalyzed competitive C–C coupling reactions

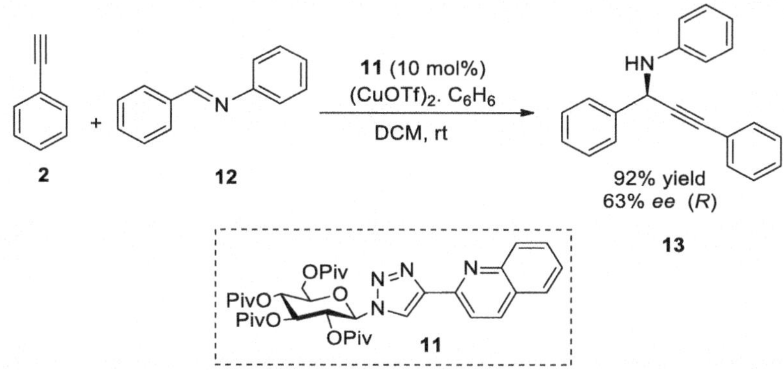

Scheme 16.5 Click-derived sugar based triazole/Cu-catalyzed enantioselective addition reaction

Tiwari et al. recently reported the domino synthesis of substituted quinazolinone analogues through Cu (I)-catalysis in assistance of click-derived pyridyl glycosyl triazole ligands **14a–d** (Fig. 16.4). The pyridyl glycosyl triazole ligands **14a–d** were developed by the Meldal click protocol as we discussed in earlier chapter. The developed ligands were characterized by the presence of pyridine, triazole, and chiral center riched carbohydrate skeleton that act as suitable Cu-stabilizing ligand during catalytic processes. The Cu(I) salt with ligand **14a** effectively assisted the domino synthesis of substituted quinazolinones **15** and **16** at low catalytic loading and mild reaction conditions (Scheme 16.6) [47].

Moreover, the same group explored the application of CuAAC reaction in synthesis of carbohydrate-based mono- and bis-1,2,3-triazolyl ligands as well and used them in catalysis of C–C coupling reaction. The author screened the catalytic activities of previously developed ligands **10a**, **14a–d** and newly developed ligand **17a–c**. These triazole-based ligand in coordination with Cu(I) salt effectively work for the Sonogashira and the Glaser coupling reaction of alkynyl sugars at low catalytic loading. The ligands **17a–c** shown in Fig. 16.5 were developed by the modular click reaction of alkynyl sugar with respective azide in presence of CuI/DIPEA in anhydrous DCM. The previously developed glycosyl triazole **14c**/CuI catalytic system efficiently worked for the Sonogashira coupling reaction to afford various novel glycohybrids including glycosyl rods. However, the macrocyclic bis-triazolyl ligand **17c** with CuI salt assisted the synthesis of sugar containing 1,3-dialkynes [48].

In addition to the successful use of click-derived glycosyl 1,2,3-triazole in Cu catalysis, the same group developed four glycosyl triazole-appended pyridinamide ligands **18a–d** for the Cu-catalyzed C–C bond formation (Fig. 16.6). The authors used modular Cu-catalyzed click reaction of alkynyl pyridinamides and sugar azides in assistance of CuI/DIPEA in anhydrous DCM for 12 h to develop ligands **18a–d** containing carbohydrate framework along with a triazole and pyridinamide unit. The ligand **18b** encapsulates the CuI to provide suitable catalytic environment and stabilize its oxidation state more effectively during coupling reaction of

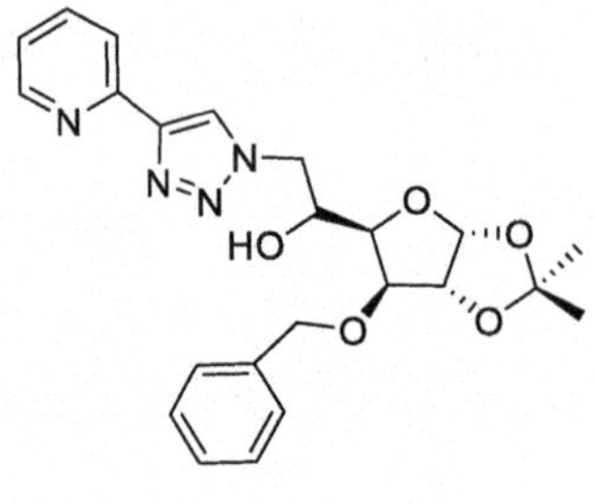

14a

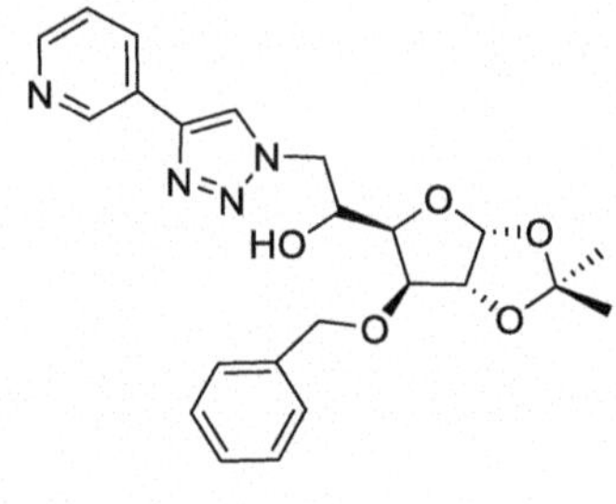

14b

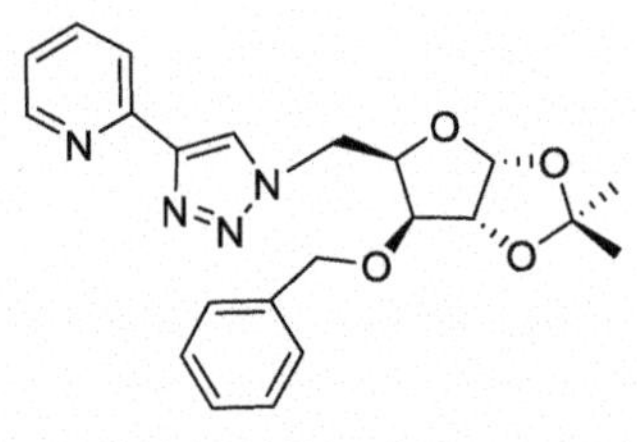

14c

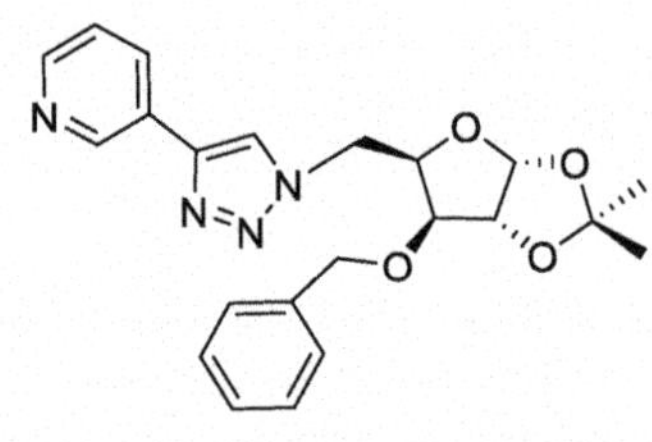

14d

Fig. 16.4 Structure of developed click-derived pyridyl glycosyl triazole ligands **14a–d**

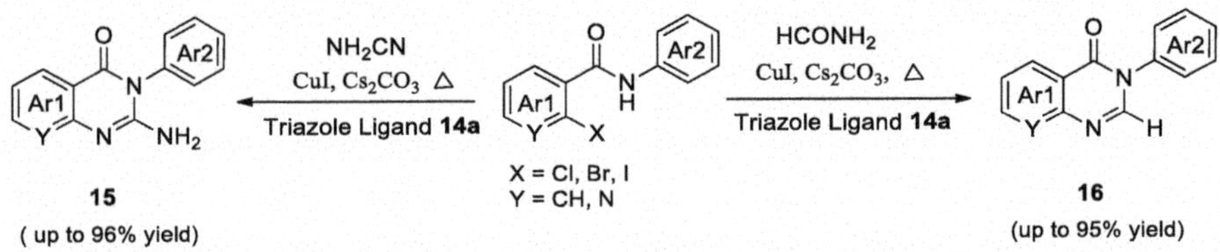

Scheme 16.6 Click-derived pyridyl glycosyl triazole/Cu-catalyzed domino synthesis of quinazolinones

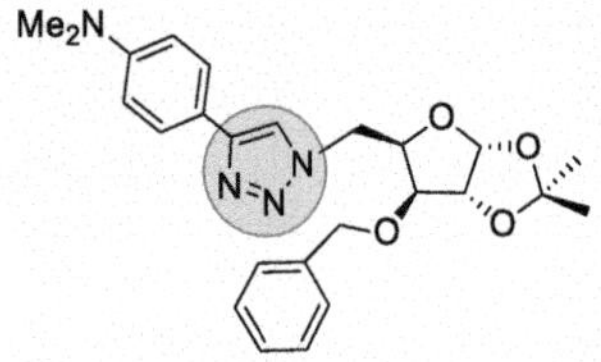

17a

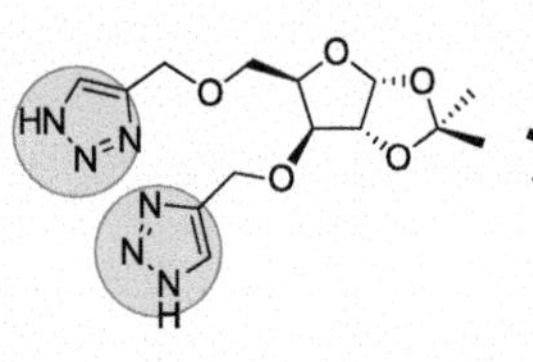

17b

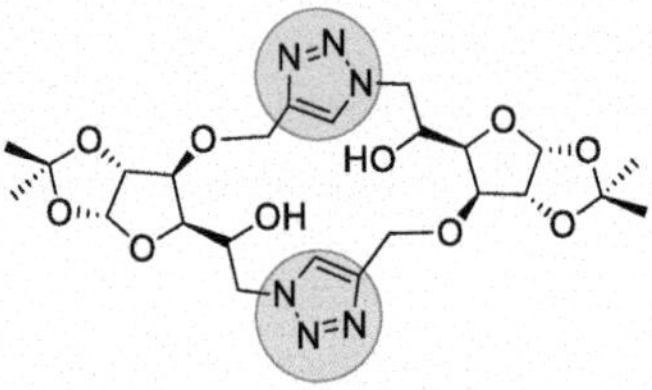

17c

Fig. 16.5 Click-derived sugar containing 1,2,3-triazole **171-c** for the Cu-catalyzed coupling reaction

N-substituted-2-halobenzamides **19** with active methylene compounds to achieve respective isoquinolinones **20** and **21** (Scheme 16.7) [49].

A large number of triazolyl-glycoconjugates synthesized via the CuAAC click reaction of azide with alkyne under Cu(I) catalysis were investigated as the most potent ligands for a rapid CuAAC reaction. The dendrimers can act as a supporting material for Cu(I) nanoparticles. The Astruc's team conducted an intriguing investigation in 2004. They developed a water-soluble catalytic nanoreactor **23** through the reaction of triazole-appended dendrimer **22** with $CuSO_4.5H_2O$

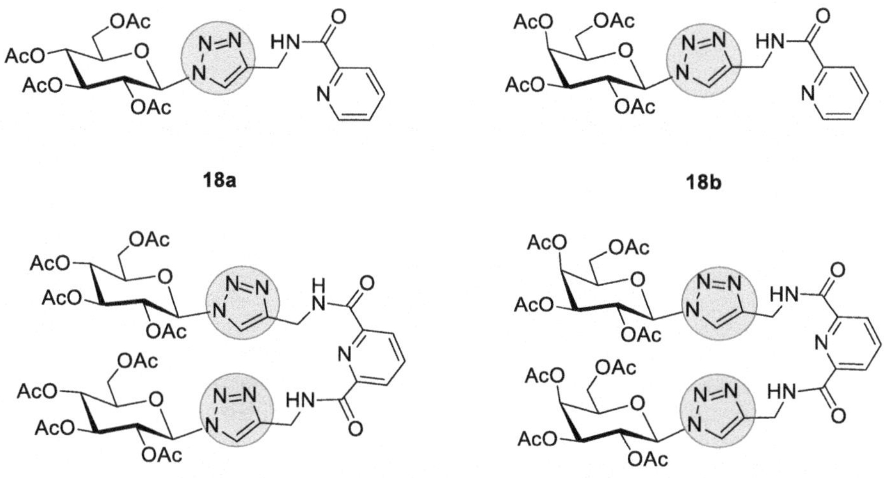

18c **18d**

Fig. 16.6 Click-derived glycosyl triazole-appended pyridinamide ligands **18a–d** designed and screened for Cu-catalyzed C–C coupling reaction

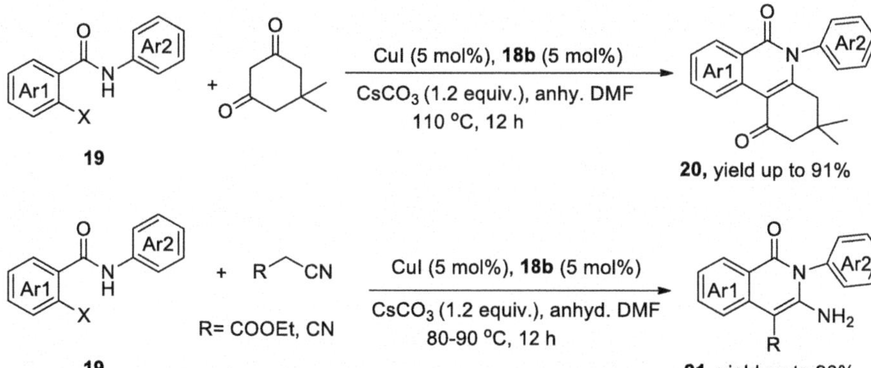

Scheme 16.7 Click-derived triazole-appended pyridinamide/Cu-catalyzed domino synthesis of isoquinolinones

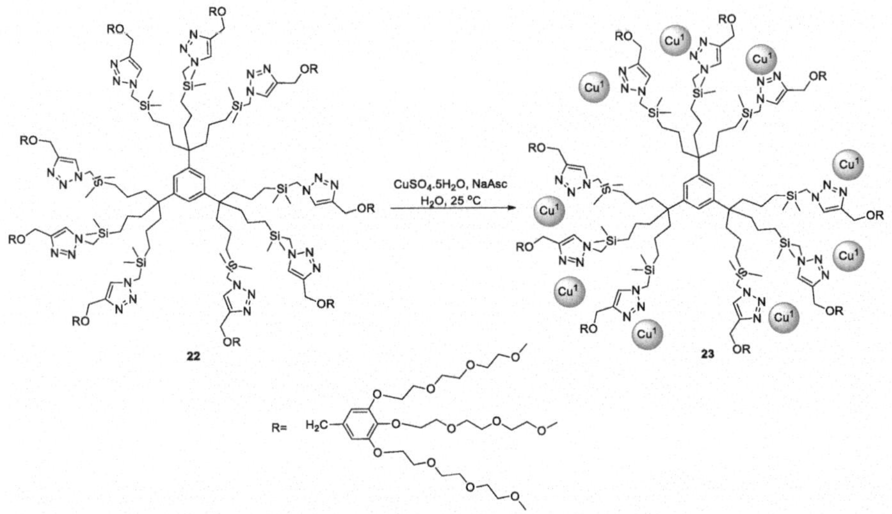

Scheme 16.8 Synthesis of Cu(I) nanoreactor **23** used for the regioselective 1,3-dipolar cycloaddition reaction

and NaAsc (Scheme 16.8). This nanoreactor efficiently promoted the regioselective Cu-catalyzed 1,3-dipolar cycloaddition reaction and was recycled and reused numerous times in the catalytic system without a significant loss in its catalytic effectiveness [50].

Further, the use of click-derived triazoles in click reaction (Click-to-Click catalysis) was successfully achieved by Husain and Bisht in click-inspired triazole-linked resorcin[4]arene glycocodendrimer **26** through the reaction of propargylated resorcin[4]arene **24** with glycosyl azide **25** in the presence of CuI/DIPEA in refluxing chloroform for 12 h (Scheme 16.9). A water-soluble glycodendrimer **27** was developed by deprotection of acetate group on the triazole-appended glycodendrimer **26** in presence of 0.1 M sodium methoxide in methanol. It has an enlarged pseudo-saccharide cavity that serves as an ideal host to be a molecular vehicle for water-insoluble hydrophobic azide and alkyne as clickable partners in aqueous environment because this molecule has eight D-glucopyranoside residues on the upper rim of resorcin[4]-arene. On reaction of azide and alkyne in water as a green medium, the authors were able to effectively investigate the inverse phase transfer molecular host nature of the resulting glycodedrimer **27** to operate as a powerful CuAAC catalyst for the regioselective synthesis of 1,4-disubstituted triazole [51].

In the process of further applicability of 1,2,3-triazole-based metal catalysis, Wang and coworkers designed and developed dendritic glycoconjugates and demonstrated their gelation ability and catalytic activity as transition metal–ligands to assist the rate of CuAAC click reactions. In the catalytic efficacy for modular click reaction (CuAAC), the hexameric glycoclusters of glucose **28**, galactose **29** and glucosamine **30** were found to be the most effective ligands to afford regioselective triazoles with high yields instantly (Fig. 16.7). However, the glycoconjugate containing six-armed

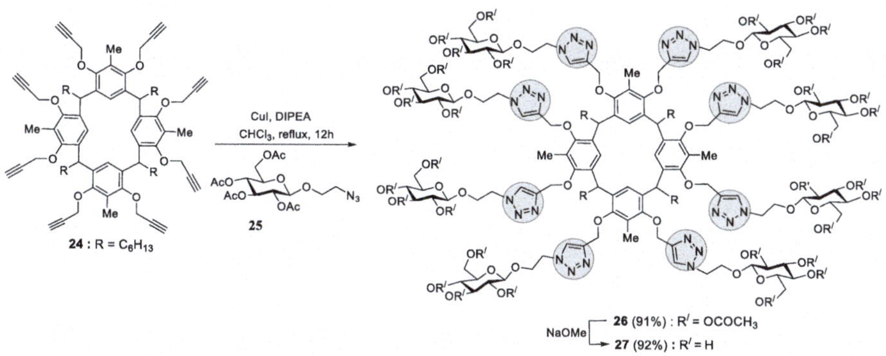

Scheme 16.9 Click-derived resorcin[4]arene–glycoconjugates (**26**,**27**) as potential CuAAC catalyst for assisting the click reaction in aqueous media

triazole-based glucose and galactose analogues in periphery **28** and **29** did not exhibit gelation properties, while glucosamine derivative of triazole-based glycoconjugate **30** was effectively able to form gel in all tested solvents and, thus, 1,2,3-triazole-based glycoconjugate proved to be an effective gelator [52].

In addition to click-derived triazole as ligand or metal catalaysis, the benzotriazole scaffolds were also found to assist various organic transformations such as

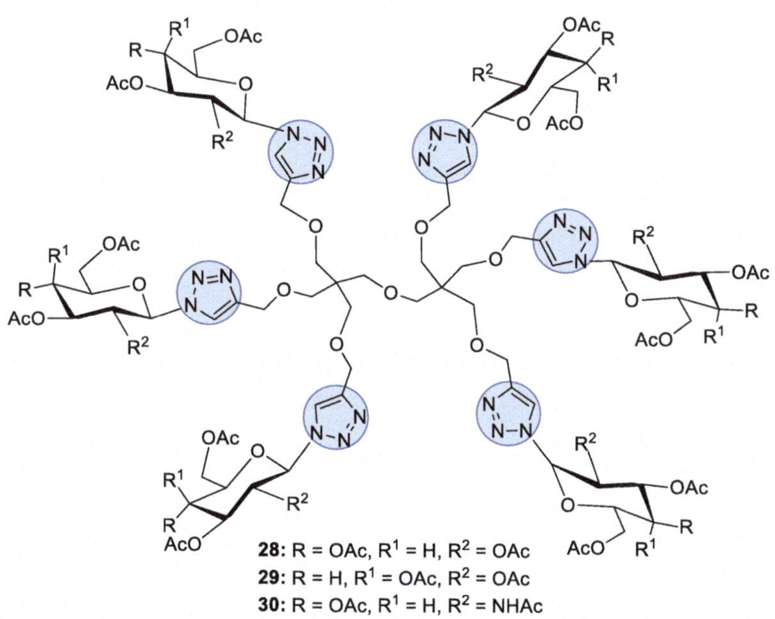

Fig. 16.7 Structure of click-derived 1,2,3-triazole-based dendritic glycoconjugates (**28**–**30**) as ligand to assist CuAAC reaction

Cu-catalyzed *N*-arylation of imidazole [53], electrophilic cyclization of *N*-arylimines [54], Tiwari et al., reported benzotriazole/Cu catalysis for oxidative coupling reaction [55].

16.2.2 Click-Derived Triazole/Pd-Assisted Catalysis

In addition to catalytic coordination of triazole analogue is also reported to be an active catalyst with Cu metal, it is reported with Pd metal as well. A class of click derived triazole-based novel monophosphine ligand **31a–e** (Click Phos) has been developed. These ligands can be obtained easily and be tailor made via 1,3-dipolar cycloaddition of different azide and acetylene. Palladium complexes of these ligands serve as extremely effective catalysts for the Suzuki–Miyaura coupling and amination of aryl chlorides (Scheme 16.10) [56]. Moreover, a class of various phosphine-linked triazole-based ligand reported to assist a range of organic transformations [57].

To extend the efficacy of click derived triazolylmonophosphine as an effective ligand in Pd catalysis, Zhao et al., reported fourteen 5-substituted and twelve 1,5-disubstituted derivatives of 1,2,3-triazoles bearing diaryl or dialkyl phosphine at 5th-position. These developed triazoles were demonstrated as the efficient ligand for the palladium catalyzed Suzuki–Miyaura cross-coupling reactions [58]. The 1,3-dipolar cycloaddition reaction derived triazolyl derivatives of carbohydrates and their metal complexes are particularly prized as extremely effective catalytic systems even at very low loading levels. Zhang et al. presented a very effective example of such a catalytic system containing Pd complex of a 2-pyridyl-1,2,3-triazole ligand generated from β-cyclodextrin. The Cu-catalyzed click of 2-ethynylpyridine with β-cyclodextrin derived azide **32** led to the synthesis of β-cyclodextrin based 2-pyridyl-1,2,3-triazole ligand **33**. A Pd complex **34** was formed by ligation of this ligand with $Pd(OAc)_2$ in toluene. With only 10^{-7} mol% loading, this complex could effectively catalyze the Suzuki–Miyaura coupling with substituted aryl halides and aryl boronic acids, yielding up to 99% of the product in aqueous media (Scheme 16.11) [59].

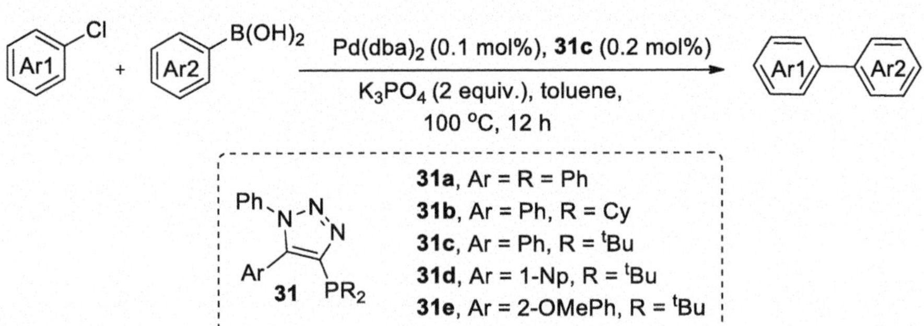

Scheme 16.10 Click-derived monophosphine ligands/Pd complex assisted Suzuki–Miyaura coupling reaction

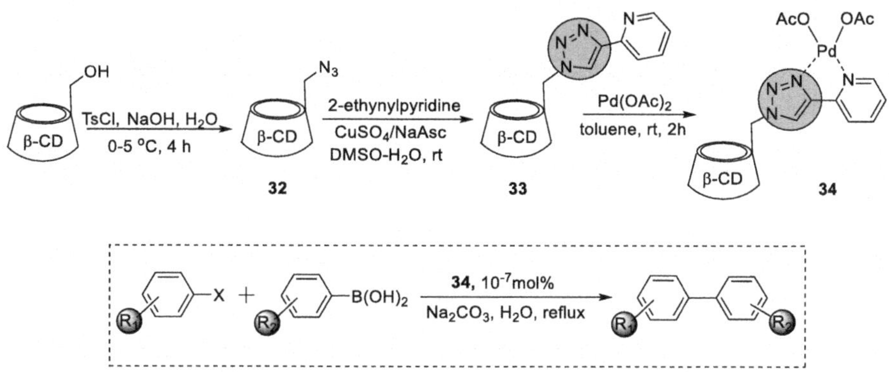

Scheme 16.11 Click-inspired β-cyclodextrin-based 2-pyridyl-1,2,3-triazole ligand/Pd complex in Suzuki–Miyaura coupling

In a very similar fashion to β-cyclodextrin-based pyridyl triazolyl ligand shown in Scheme 16.11, a pyridyl glycosyl traizole Pd complex **35** derived from the D-glucosamine were explored in the solvent-free Heck cross-coupling reaction by Shen et al. Various aryl halides and aryl/hetero aryl halides were underwent in C–C cross-coupling in assistance of this catalyst to afford substituted alkenes in good yields. Additionally, the marketed drug analogue 'Axitinib' **38** was also assembled in presence of devised methodology through the reaction of heteroaryl halide **36** with vinyl pyridine **37** (Scheme 16.12) [60]. In addition, 2-pyridyl-1,2,3-triazole (pyta) ligand were developed using modular click approach and harvested as the ligand to assist various coupling reaction (Suzuki–Miyaura, Sonogashira, and Heck reactions) in the form of palladium based mono- and polymetallic iterative heterogeneous catalysts or metallodendritic catalyst **39** (Fig. 16.8) [61].

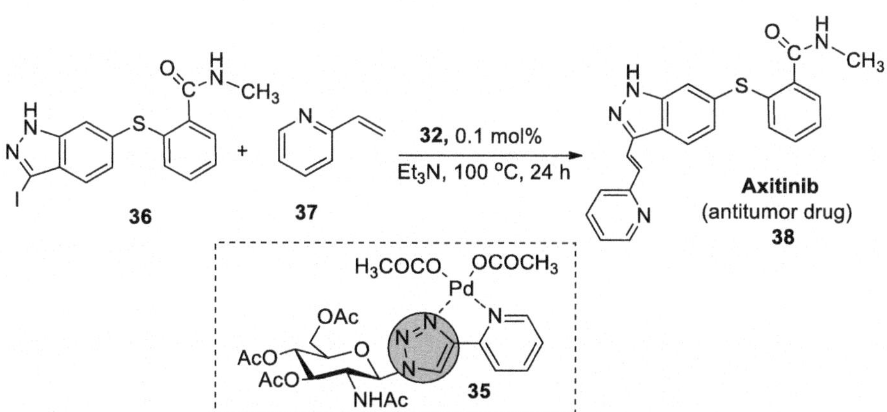

Scheme 16.12 Use of D-glucosamine-based click-derived triazole-based glycohybrid in synthesis of Axitinib via the Heck coupling

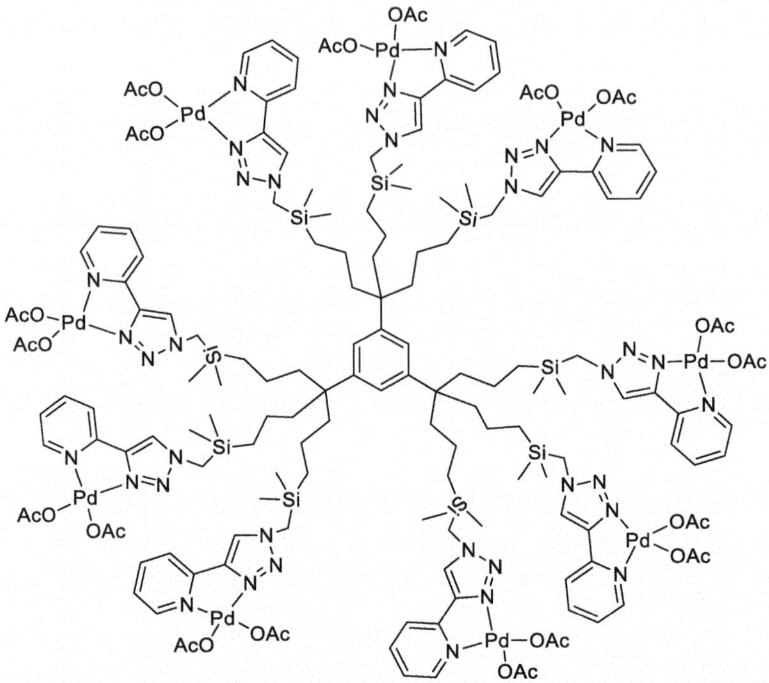

39

Fig. 16.8 Structure of 2-pyridyl-1,2,3-triazole-based metallo dendritic catalyst **39**

The carbo-click approach was utilized to selectively assemble the bidentate and polydentate diphenylphosphino-based triazolyl glycoconjugate ligands **40a**, **40b**, and **40c** by using terminal alkynes of choice with glycosyl azide. When these ligands were screened for the Pd-catalyzed asymmetric reaction of dimethyl malonate to 1,3-diphenylallyl acetate, 100% of the product yield was achieved. However, ligand **40a** exhibits the best asymmetry in the form of an 80/20 S/R ratio at 25 °C in DCM (Scheme 16.13) [62].

Furthermore, click derived pyridyl glycosyl triazole ligand were used to develop nanoparticles of palladium. The pyridyl glycosyl triazole/Pd nanoparticle **42** was achieved by reacting palladium chloride salt with pyridyl glycosyl triazole ligand **41** in toluene for 6 h. The nanoparticle was demonstrated as catalyst in coupling reaction and well characterized by various spectroscopic analyses. The pyridyl glycosyl triazole/Pd nanoparticle catalyst **42** works well for Suzuki–Miyaura cross-coupling reaction (Scheme 16.14) [63].

Moreover to applications of click-derived triazole in Pd-catalyzed coupling reaction, it also acts as directing group for C–H activation of aromatic system. The Ackermann and coworkers reported a simple and incredibly straightforward method

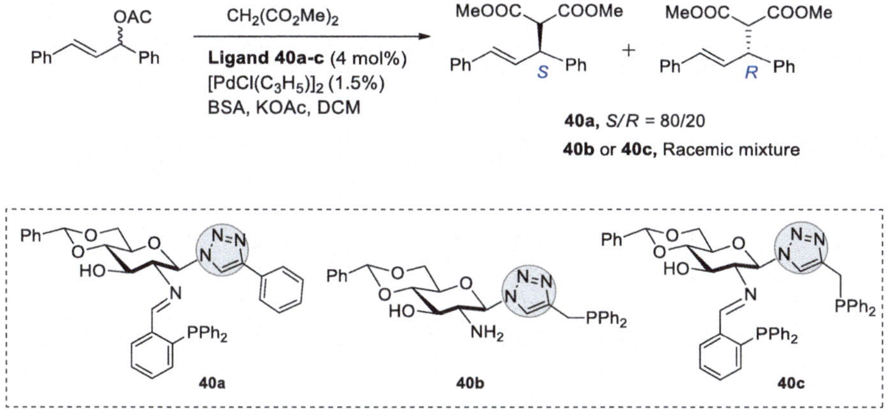

Scheme 16.13 The click-derived diphenylphosphino-based triazolyl ligands in catalysis

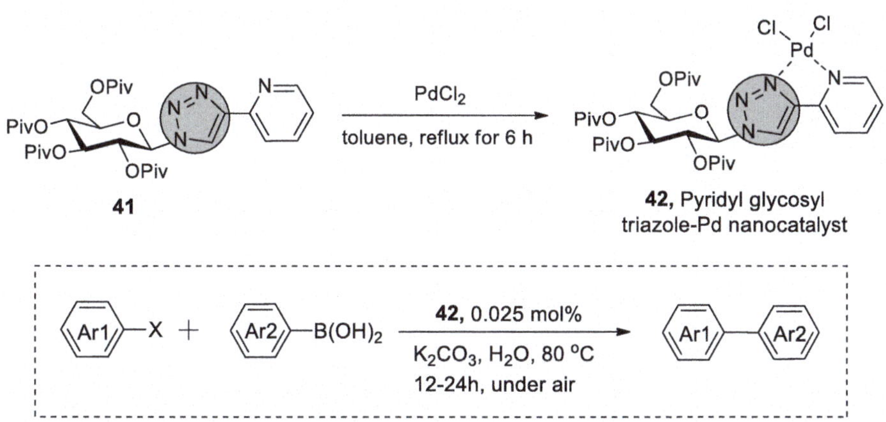

Scheme 16.14 The pyridyl glycosyl triazole-Pd nanocatalyst-assisted Suzuki–Miyaura cross-coupling reaction

for C–H functionalization of triazole substituted amides **43** under palladium catalysis (Scheme 16.15). According to the authors, species **44** is a Pd–ligand complex in which the click-mediated triazole ring functions as the directing group, i.e., necessary to bring about C–H activation under Pd catalysis to form the corresponding product **45**. The authors were able to successfully remove the triazole group from compound **45** in order to obtain a high yield of the appropriate biaryl carboxylic acid while tolerating the ketone functionality (Scheme 16.15) [64].

A versatile and selective C(sp^2)-H halogenation of late-stage diversified click products, such as 4-aryl-1,4-disubstituted triazoles **46** under Pd catalysis, was described by Correa and coworkers (Scheme 16.16). The wide substrate scope and tolerance of different groups present in the molecule are two of the few advantages. To support the mechanism of triazole-assisted C(sp^2)-*H* bromination, the

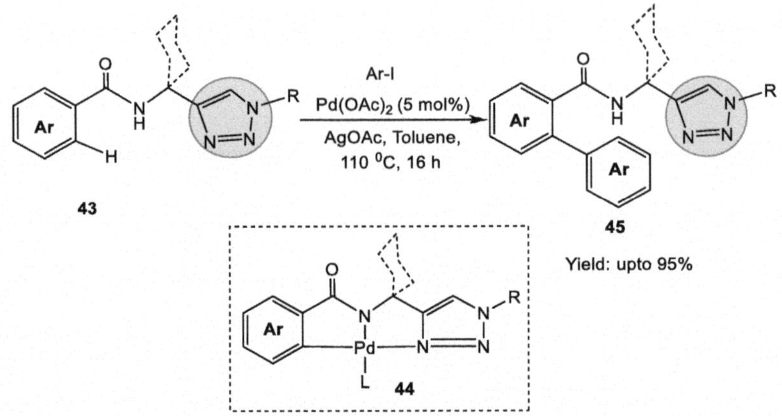

Scheme 16.15 The C–H functionalization of substituted amides **43** under triazole-assisted palladium catalysis

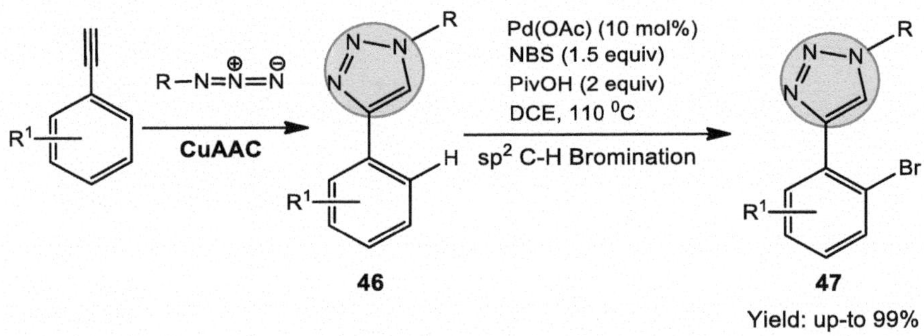

Scheme 16.16 Click-derived triazole-assisted C(sp^2)–H bromination of **46** under palladium catalysis

authors presented DFT analyses and highlighting the path of reaction, triazole **46** is cyclopalladated as the first step in the reaction, followed by oxidative brominationand reductive elimination afford the triazole-aided C(sp^2)-*H* bromination product **47** [65].

16.2.3 Click-Derived Triazole/Ni-Assisted Catalysis

The triazolyl outcome of Click reaction is also exhibited coordination interaction with Ni metal. The authors develop pyridyl glycosyl triazole/Ni nanoparticle **48** using $NiCl_2$ salt with ligand in tetrahydofuran. The pyridyl glycosyl triazole/Ni nanoparticle **48** exists indimeric form and efficiently assists the acylation of amines **49** in

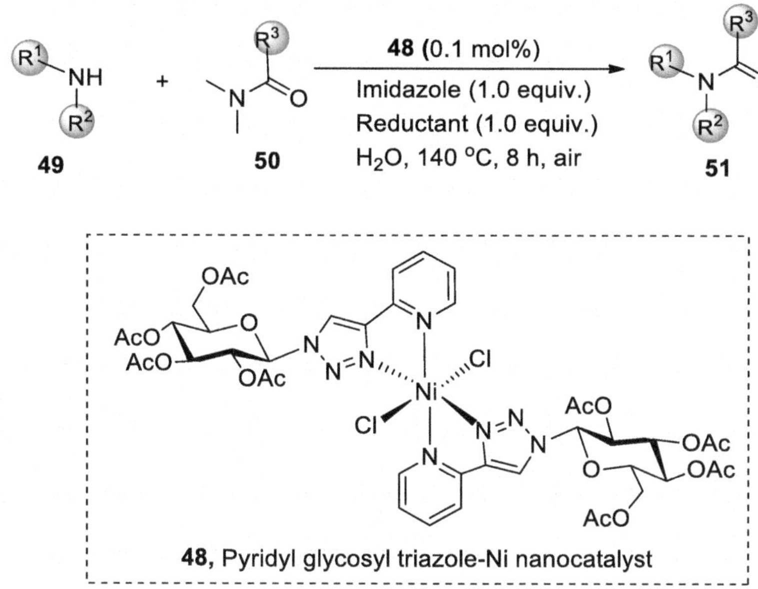

Scheme 16.17 The pyridyl glycosyl triazole–Ni nanocatalyst **48** assisted in acylation of amines

presence of acyl group precursors **50** to afford corresponding amides **51** in aqueous medium at very low catalytic loading (Scheme 16.17) [66].

A few another interesting examples of triazole/Ni catalysis includes a series of new pyridyl–triazole ligand **52a–d** and their corresponding nickel(II) complexes **53a–d** which were made for catalytic uses. The complexes composition was identified as NiL_2Br_2, L = pyridyl–triazole ligand, after investigation using elemental analysis and mass spectrometry. According to structural analysis of one of the complexes, the triazole and pyridine nitrogens from each of the ligands are *trans* to one another, which also revealed that the *bis*-chelating ligands are in the equatorial plane of the octahedron. In terms of activities and turnover rates, the nickel complexes with 1,2,3-triazole ligands, or 'Click' ligands, proved to be fairly effective catalysts for oligomerization of ethylene (Scheme 16.18) [67].

The effective synthesis of an intriguing MNPs-supported nanonickel catalyst was reported which involves the "click" reaction of azide-functionalized magnetic nanoparticles **54** with 2-ethynylpyridine, followed by the immobilization of nickel nanoparticles. The resulting MNPs-supported nanonickel substance **55** was successfully employed as practical and affordable heterogeneous catalyst for fluoride-free Hiyama cross-coupling process (Scheme 16.19) [68].

In addition to MNPs-supported nanonickel catalyst, the uniformly sized spherical NiNPs supported on chitosan surfaces was also reported as the heterogeneous catalyst for Suzuki–Miyaura coupling [69]. The olefin polymerization is the most significant of the nickel-catalyzed reactions due to its wide range of applications in both academic and industrial fields. The use of catalysts, development of this

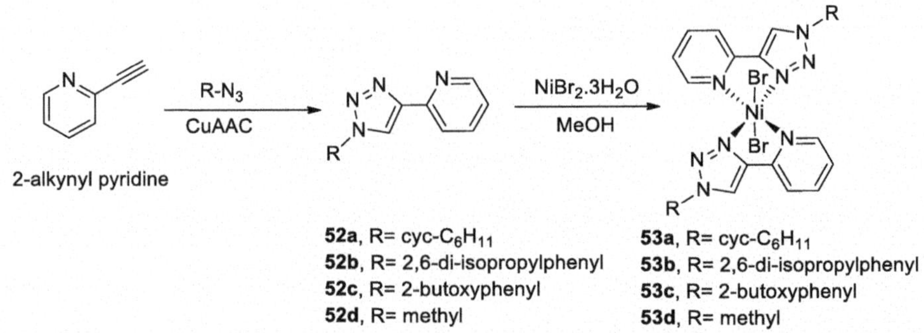

Scheme 16.18 Synthesis of triazole–Ni complex as effective catalysts for the oligomerization of ethylene

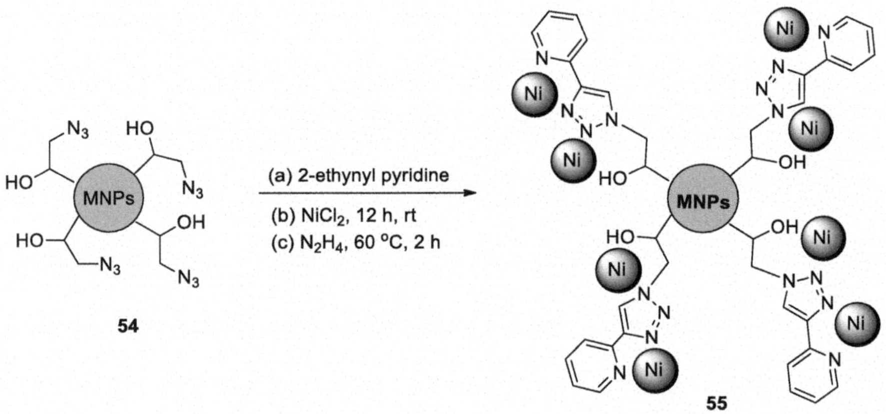

Scheme 16.19 Developed MNPs supported nanonickel catalyst **55** for fluoride-free Hiyama cross-coupling

reaction has advanced significantly on nickel complexes (and other late transition metals) containing Schiff-base ligands as a result of their distinct N-donor characteristics [70]. As a result, over the past few years, researchers have also looked into 1,2,3-triazole-based complexes [67, 71], and it has been found that their catalytic activity is on par with or even superior to that of Schiff-base complexes. The triazolyl-appended *bis*(imino)acenaphthene (triazolyl based-BIAN) **56** and sulfur containing *bis*(imino)acenaphthene Schiff bases **57** were found to be efficient ligand for coordination with Ni and assist the catalysis for polymerization of olefin in presence of methylaluminoxane (MAO) as classical co-catalyst. The triazole-appended BIAN-Ni(II)/MAO catalytic system, triazole moieties act as labile unit for chelation to favor the formation of active system for polymerization unsaturated monomeric unit (Fig. 16.9) [71].

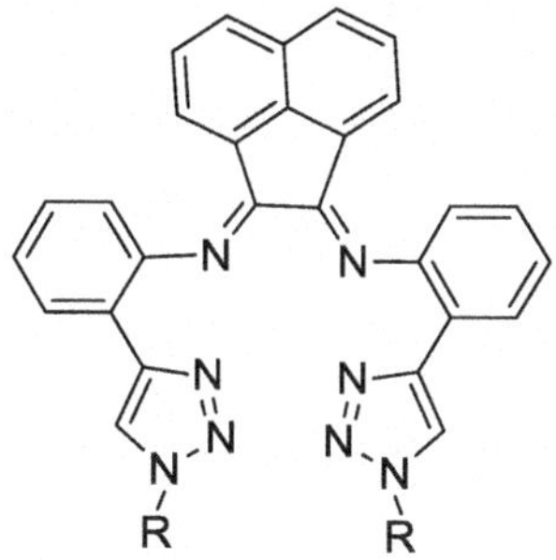

R = Alkyl, aryl

56

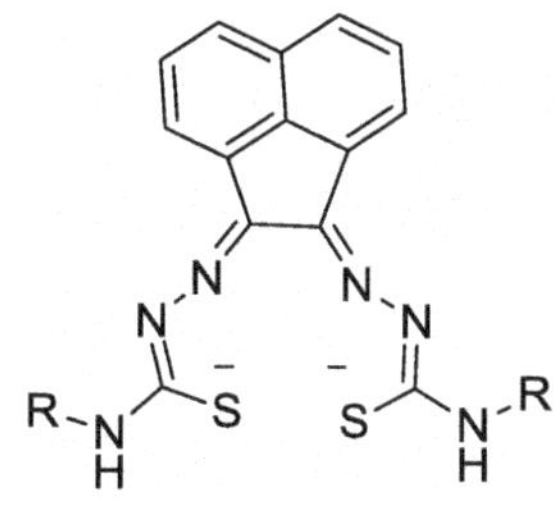

R = Allyl

57

Fig. 16.9 Structure of triazolyl-appended *bis*(imino)acenaphthene and sulfur-containing *bis*(imino)acenaphthene ligand for metal catalysis

16.2.4 Click-Derived Mn/Triazole-Assisted Catalysis

In 3d series transition metal, Mn is the first metal for which transition metal catalysis have been reported, in accordance with the order of elements in the periodic table. For many significant and difficult catalytic processes, triazole-based N_4 tetradentate ligands have recently been widely employed as metal ligands. However, the majority of them are built on a pyridyl framework, which constrained the extent to which they could be altered. A series of N_4 tetradenate ligands based on triazoles have been formed utilizing an effective construction process called ‘Click’ chemistry. Their Mn(II) complex **58** with the click derived tetradentate ligand demonstrated good catalytic activity with low catalyst loading and quick reaction times in the epoxidation of various terminal aliphatic olefins **59** to afford oxirane containing product **60** in assistance of peracetic acid as the oxidant (Scheme 16.20) [72].

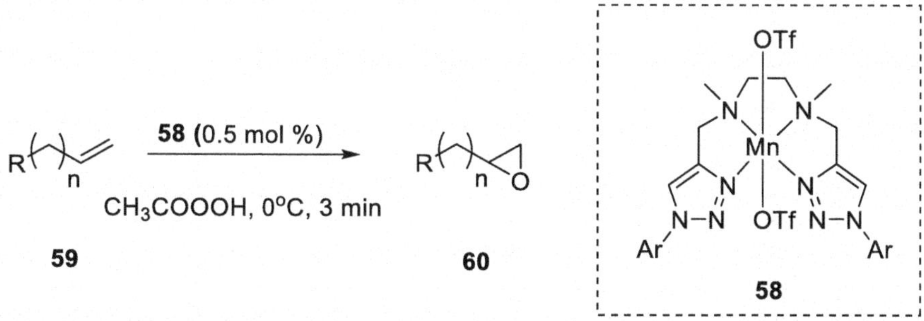

Scheme 16.20 Application of click derived triazole-based tetradentate ligand/Mn complex **58** in epoxidation reaction of olefin

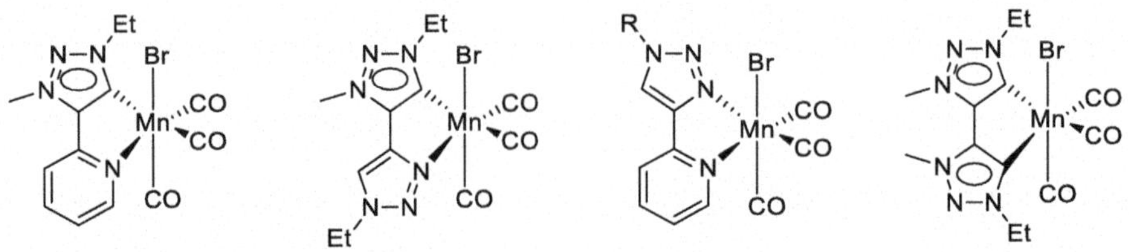

61 62 63 64

Fig. 16.10 Triazolylidene/triazole/pyridine-based Mn complex **61**–**64** for electrocatalytic reduction of CO_2

To explore triazole coordination ability to transition metal, a new family of manganese tricarbonyl complexes with the general formula *fac*-[Mn(L)$(CO)_3$Br] [where L is a bidentate chelating ligand containing mixed mesoionic triazolylidene–pyridine **61**, triazole–pyridine **62**, triazolylidene–triazole **63,** and triazolylidene–triazolylidene **64** have been synthesized for the catalytic use (Fig. 16.10). These complexes contain triazolylidene/triazole/pyridine scaffold. The elemental analysis, IR and NMR spectroscopy were used to characterize each complex. The author looked into the CO_2–electrocatalytic reduction activity of complexes **61**–**64**. The effect of a mesoionic triazolylidene ligand on the electrocatalytic activity of Mn tricarbonyl complexes has been investigated for the first time. Under CO_2 environment, these complexes displayed an improvement in catalytic current and favor reduction of CO_2 to CO [73].

16.2.5 Click-Derived Ru/Triazole-Assisted Catalysis

The triazole was also reported as the directing group for C–H activation in Ru catalysis. Liu et al. have successfully investigated the ortho-alkenylation of arenes **65** with a number of internal alkynes **66** under Ru catalysis by triazole-directed C–H activation to afford the desired product **67** up to 91% yield (Scheme 16.21). It is interesting to note that the technique is effective during alkenylation of triazoles with different terminal alkenes [74].

N-Heterocyclic carbene (NHC) ruthenium complexes have a wide range of catalytic activities, including the oxidation of alcohols or amines, transfer hydrogenation, and olefin-metathesis reactions. The 1,2,3-triazolylidene-based NHC ligands were regarded as being of utmost relevance shortly after the development of 'Click chemistry', particularly due to their capacity to act as stronger sigma donors than Arduengo-type imidazolylidene-based NHC. Taking into account these characteristics, Albrecht and coworkers reported a simple click-inspired synthesis of N-heterocyclic carbene (NHC) glycohybrids, which are ruthenium complexes that contain carbohydrate chiral unit. The NHC ligand leads a stable Ru complex **68** and **69** that can be kept in solution for several hours while being unstable in solid form. The authors looked at the transfer hydrogenation of a few ketones under basic

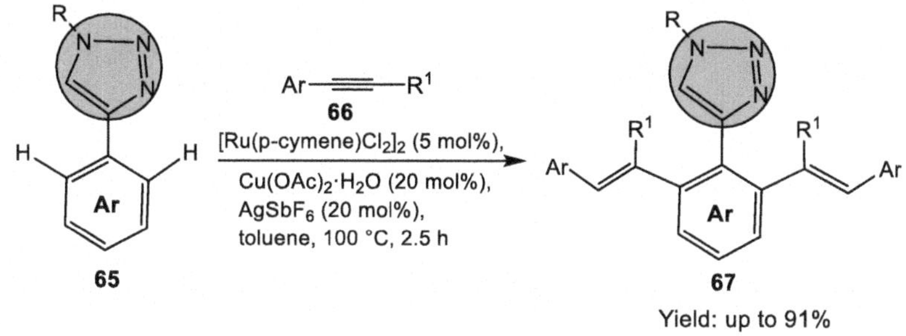

Scheme 16.21 The C–H activation for *ortho*-alkenylation of arenes under click-inspired triazole-directed ruthenium catalysis

circumstances using the deacetylated complexes **68** and **69**. Both ligands showed high catalytic activities and boosted catalytic activity of the ruthenium metal center. However, the ethylene spacer connected triazole NHC reduced the activity of transfer hydrogenation. The directly triazolylidene-bound glucose was discovered to transfer hydrogen at significantly higher rates than other sugar-based Ru-NHC, notably for the oxidation reaction (Scheme 16.22). As a result, the significant positive contribution of the carbohydrate substituent in pyranose or furanose form to provide further broad range of catalyst engineering is well justified and might be utilized in advanced catalysis [75].

The 1,2,3-triazole/Ru complexes also explored as the active catalyst for the oxidation of alcohol. The half sandwich complexes **74–77** of Ru metal coordinated through *N*-atom of click derived triazoles **70–73** in presence of *N*-methylmorpholine *N*-oxide

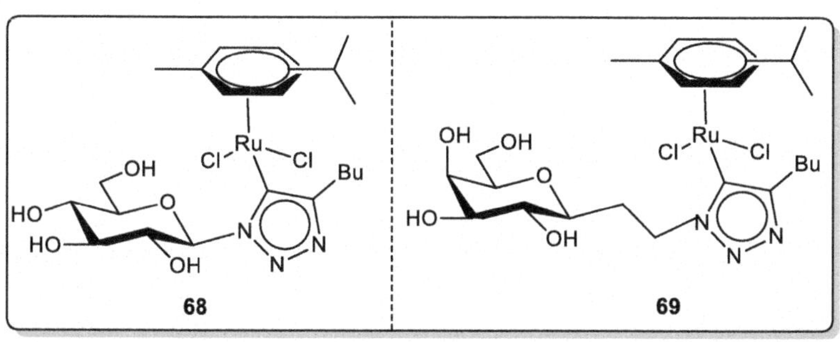

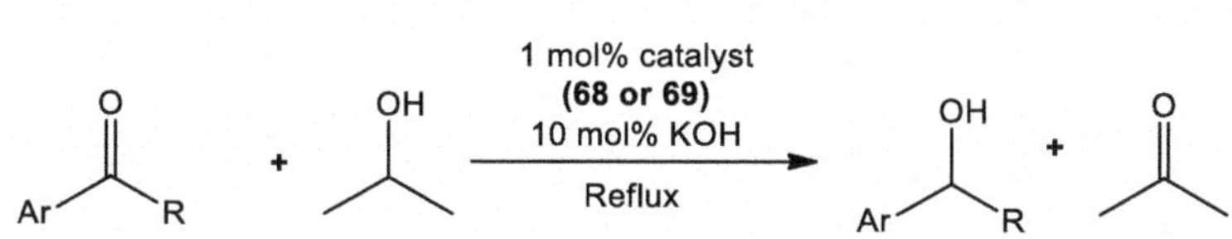

Scheme 16.22 The CuAAC–ruthenium complexes as *N*-heterocyclic carbene (NHC) glycohybrids (**68–69**) and their catalytic activity for the transfer hydrogenation reaction

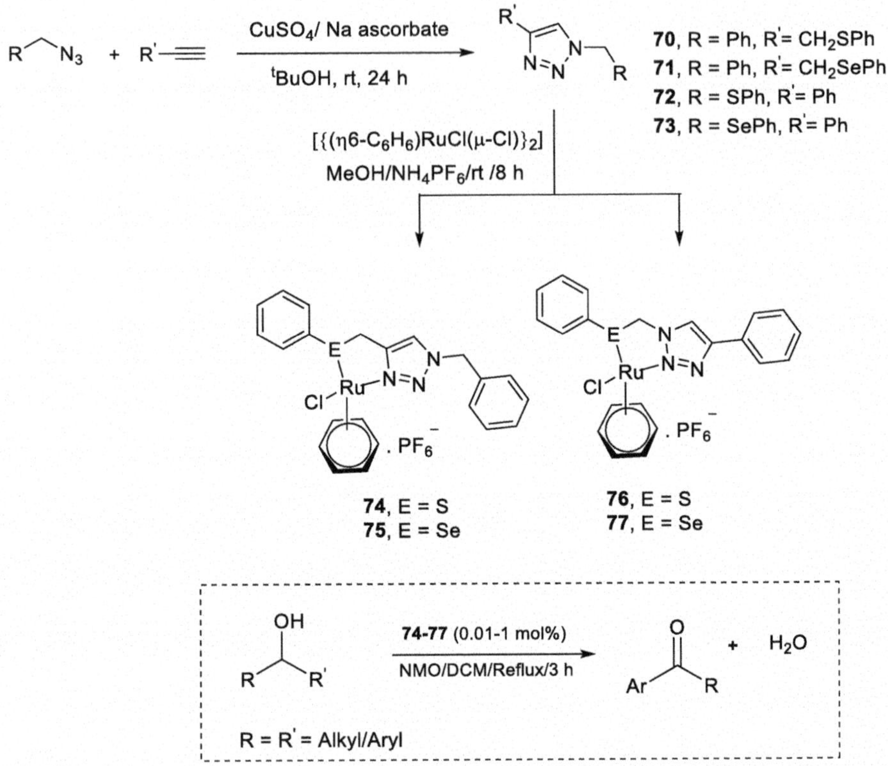

Scheme 16.23 Synthesis and application of 1,2,3-triazole and their *N*-coordinated Ru catalyst

assists the oxidation of an array of primary and secondary alcohols (Scheme 16.23) [76].

16.2.6 Click-Derived Fe/Triazole-Assisted Catalysis

There are very few 1,2,3 triazole-Fe catalysts available in the literature, and nothing is known about how these ligands compare to other *N*-heterocyclic ligands. However, this element has been covered and explored through Shi's research group. The 1,2,3-triazole ligands **78–80** with Fe metal offered acceptable yields of enyne **82** in the propargyl alcohol **81** dehydration reaction among other heteroaromatic ligands such as pyridine, imidazole, tetrazole, and variously substituted triazoles with same binding patterns. This characteristic has been linked to the 1,2,3-triazole's necessary electronic impact in modifying Lewis acidity of Fe^{3+} to accomplish the essential chemoselectivity for the conversion (Scheme 16.24). This result should spur

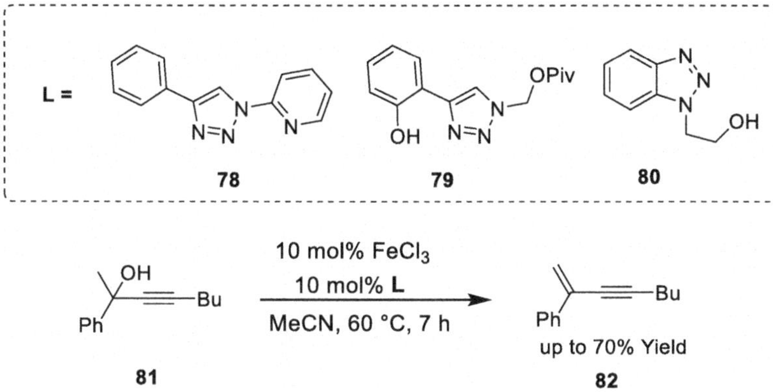

Scheme 16.24 The Fe/triazole coordination in catalysis of dehydration of propargyl alcohol

interest in catalysis using 1,2,3-triazole–Fe complexes in subsequent studies as it demonstrates an advantage of 1,2,3-triazoles over other *N*-heterocycles [77].

16.2.7 Click-Derived Au/Triazole-Assisted Catalysis

Nomiya et al., in 1988 first detailed the synthesis of Au(I)–triazolyl complexes based on anionic triazoles followed by Shi et al. employed the 1,2,3-triazolyl–Au(I) complexes as a new catalyst (Scheme 16.25) [78, 79]. In addition to benzotriazole-assisted Cu catalysis, the cyclization was achieved at room temperature through triazole gold (TA-Au) catalyst **83** as well [80].

With the triazolyl ligand, this Au(I) catalyst **83** demonstrated greatly increased thermal stability and enabled highly efficient intermolecular alkyne hydroamination catalysis at high temperatures (80 °C) without agglomeration. This study expanded the temperature range at which homogeneous Au(I) catalysis is possible [80]. Kannan and Rajakumar explore the modular click reaction to develop gold nanoparticle (Au-NPs) decorated glycodendrimers. The click derived triazolyl glycodendrimers were stabilized or encapsulated the gold nanoparticles (NPs). The first generation

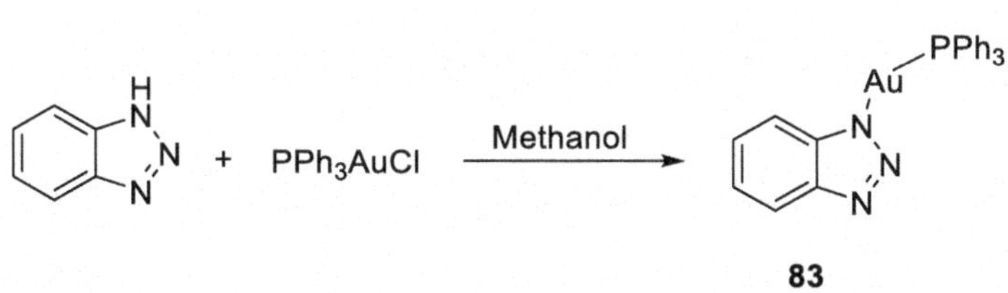

Scheme 16.25 The triazole-coordinated Au-complex **83** as catalyst

dendrimer and the zeroth generation dendrimer enclose and stabilize gold nanoparticles, respectively. The first-generation dendrimer encapsulated gold nanoparticles are termed as Au-dendrimer encapsulated nanoparticles (Au-DENPs), while zero generation stabilize gold nanoparticles are termed as Au-dendrimer stabilized nanoparticles (Au-DSNPs). The high-resolution transmission electron microscopy (TEM) and UV–Vis spectroscopy were used to characterize the Au-DSNPs and Au-DENPs. These gold nanoparticles (Au-NPs) decorated glycodendrimers **84** were found to be efficient in catalyzing the reduction of 4-nitrophenol to 4-amino phenol (Scheme 16.26) [81].

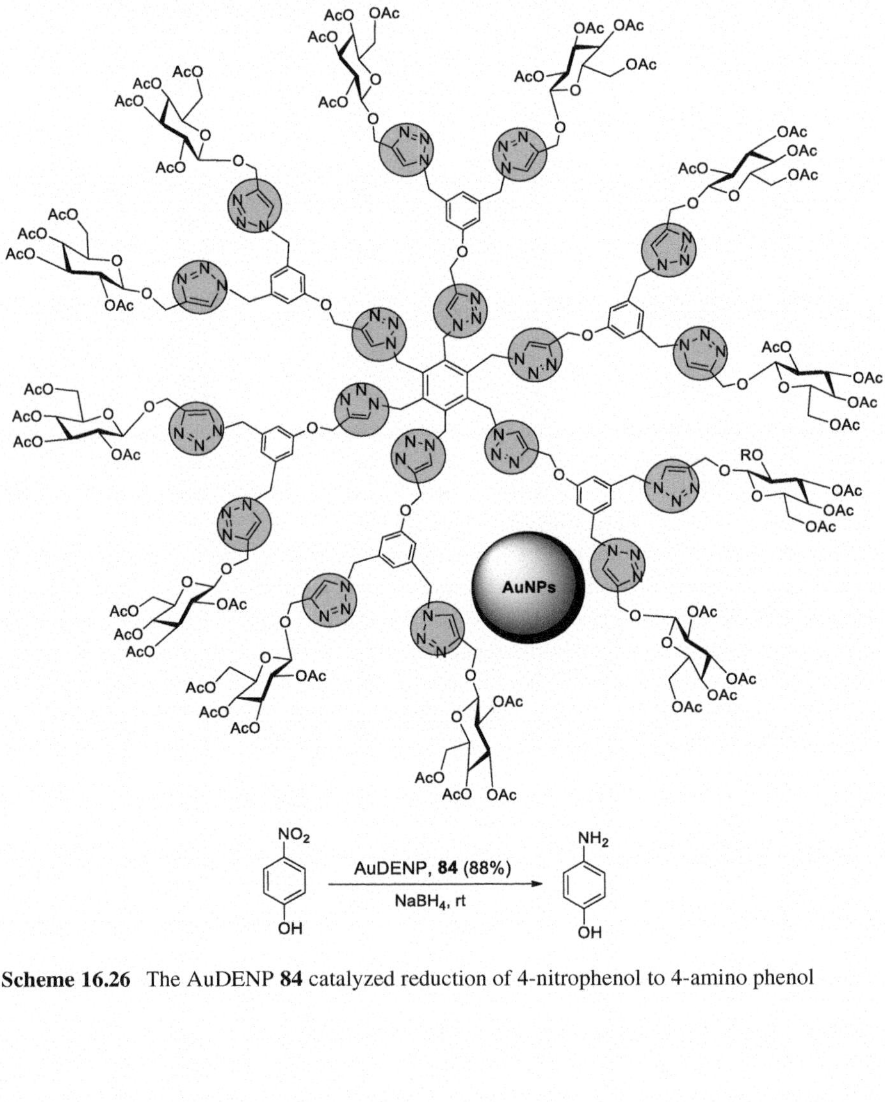

Scheme 16.26 The AuDENP **84** catalyzed reduction of 4-nitrophenol to 4-amino phenol

16.2.8 Click-Derived Rh or Ir/Triazole-Assisted Catalysis

The Rh and Ir transition metal complexes are well-known catalysts to assist alkene hydroamination, hydroformylation, and hydrogenation. The phosphines and NHC-based ligands are frequently used as donor ligands in these reactions. The click-derived 1,2,3-triazoles ligands were also exhibited the efficient coordination with Rh or Ir to catalyze these reactions [82]. The synthesis of the pyrazolyl–triazolyl bidentate ligands PyT and PyS (Ligand **85, 86**) were accomplished by CuI-catalyzed modular click reaction of 1-propargylpyrazole and benzyl azide or phenyl azide, respectively. A series of cationic Rh and Ir complex **87–90** were developed using this ligand. The metal center coordinates through the N_3 atom of the triazolyl moiety and the N_2 atom of the pyrazole moiety, generating six-membered metallacycles, based on single-crystal X-ray diffraction data. The slightly shorter M–N(triazole) bonds in comparison to the M–N(pyrazole) bonds show that the triazolyl donating capacity is stronger than that of the pyrazolyl donor. These novel metal complexes work well as effective catalysts for a variety of alkynamines and alkenamines to undergo intramolecular hydroamination reaction (Scheme 16.27) [83].

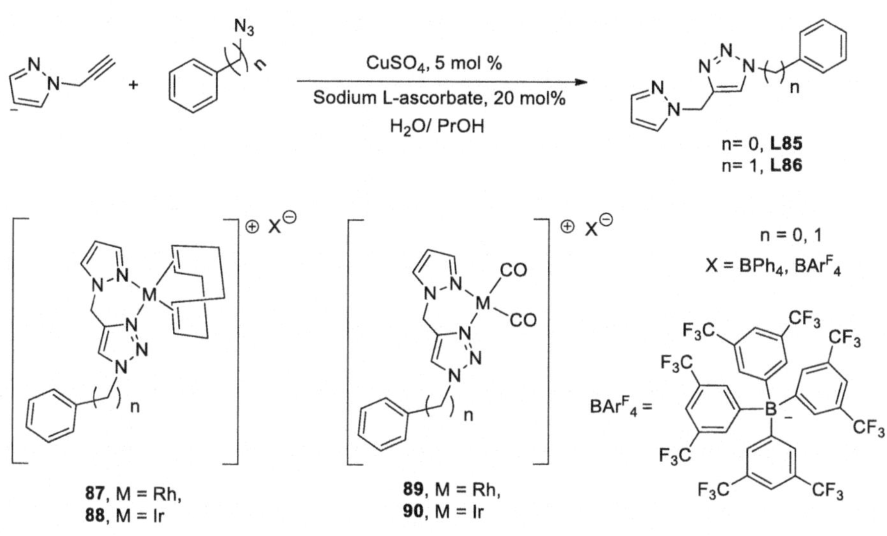

Scheme 16.27 The pyrazolyl–triazolyl bidentate ligand-based Rh(I) and Ir(I) complex

16.3 Triazole-Based Organocatalyzed Reactions

Organocatalysts are preferred over metal ligands to attain enantioselectivity for a chemical transformation because of their ecologically safe, metal-free catalytic activity. Pyrrolidine glycoconjugates have been discovered to be exceptionally effective as organocatalysts for efficient enantioselective Michael addition reaction [84, 85]. The glycosyl triazoles have demonstrated their effectiveness in catalyzing processes as ligands and organocatalysts despite the fact that there are a very few instances of these compounds in the field of catalysis. While the availability of multiple hydroxy groups provides access to modify them to multiple binding sites, which further improves the catalytic efficiency even with very low loading of the catalytic system. The chiral pool of the carbohydrate moiety contributes to the ability of glycosyl triazoles to be a good source to induce asymmetry in a reaction. Additionally, these triazole compounds are non-toxic, making them appropriate for green synthesis and pharmaceutical development. The modular click protocol was utilized to develop a triazolyl-based pyrrolidine glycoconjugates **93** through the reaction of glycosyl alkyne **91** and Fmoc-protected 2-(azidomethyl)pyrrolidineazide **92** in assistance of CuI/DIPEA. The triazole-appended glycohybrids **93**, on Fmoc deprotection of *N*-atom of triazole afforded the final organocatalyst **94**, which was demonstrated for enantioselective Michael addition of nitro olefins **95** (Michael acceptor) with ketones **96** (Michael donor) to afford addition product **97**. 10 mol% of catalytic loading of **94** catalyzed the reaction stereoselectively at room temperature without need of any solvent (Schemes 16.28 and 16.29) [86].

Chandrasekhar et al. reported the asymmetric Michael addition reaction of ketones **95** to nitro olefins **96** at room temperature successfully catalyzed in assistance of click-derived novel hydroxyphthalimide-coupled triazole–pyrrolidine **98** derivative. The authors suggest that when benzoic acid was combined with triazole-based organocatalyst **98** in aqueous medium, excellent yields and high selectivities of product **99** were obtained (Scheme 16.30) [87].

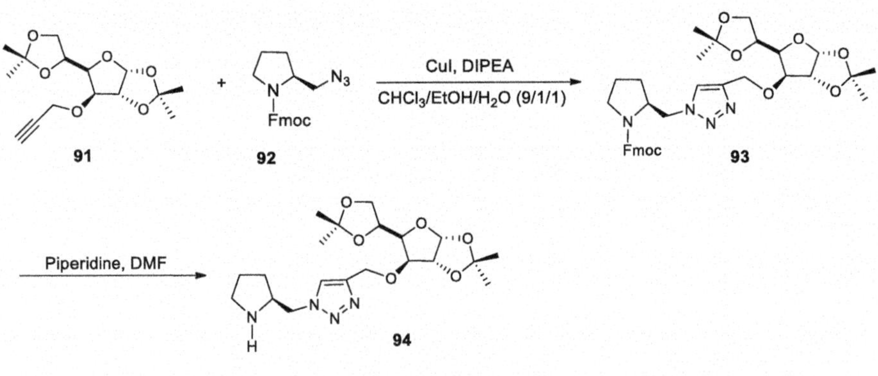

Scheme 16.28 Synthesis of organocatalyst **94** in assistance of CuAAC reaction

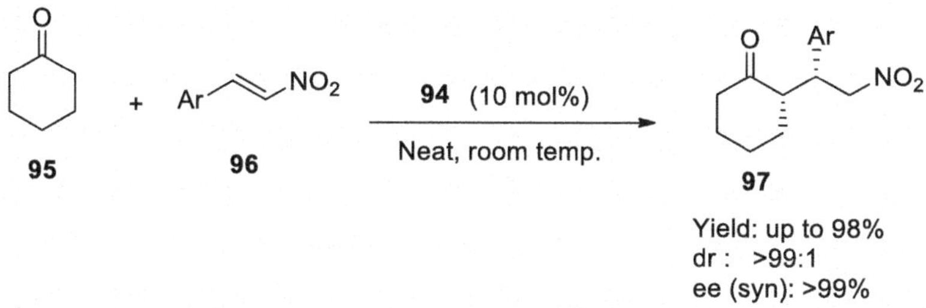

Scheme 16.29 The 1,2,3-triazole-based organocatalyst **94** assisted stereoselective 1,4-conjugate addition reaction

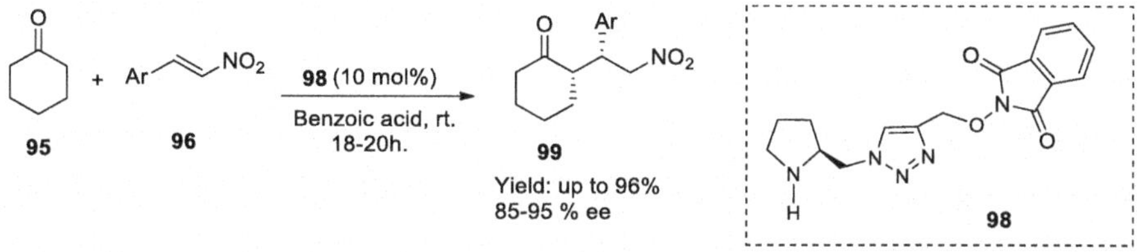

Scheme 16.30 1,2,3-Triazole-based organocatalyst **98** assisted stereoselective 1,4-conjugate addition reaction

Another triazole–pyrrolidine-based organocatalyst was developed and successfully invested in the Michael addition reaction of ketones with nitro olefins. Synthetic approach of organocatalyst **102** involves the click reaction of Boc-protected pyrrolidineazide **100** with terminal alkyne to afford compound **101**, which was on Boc deprotection in presence of trifluoroacetic acid gave the final organocatalyst **102** (Scheme 16.31) [88].

For the enantioselective Michael addition of ketones to nitroalkenes, a silica-supported pyrrolidine–triazole, was also developed. The supported catalyst showed good activity and enantioselectivity, and it can easily be utilized up to four times without reducing its catalytic activity [89].

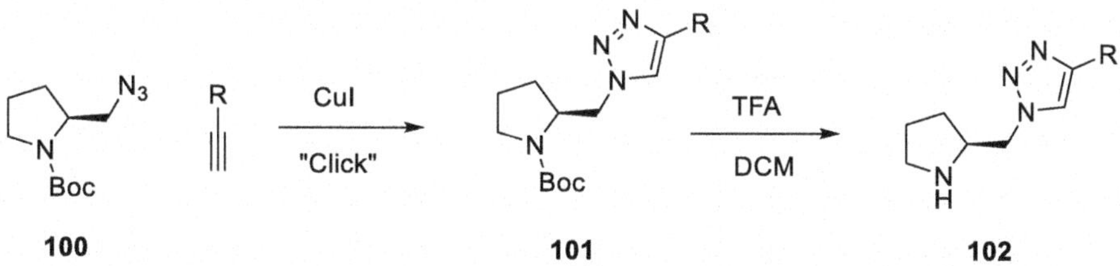

Scheme 16.31 Synthesis of 1,2,3-triazole-based organocatalyst **102** for the Michael addition

16.4 Conclusions and Future Outlook

In this chapter, we have explored the synthesis of triazole analogues in assistance of click reaction and their application as ligand in metal-catalyzed organic transformation with selected examples. In spite of the application of triazolyl moiety in the metal catalysis, we also included a brief description about triazolyl scaffolds as the organocatalyst for the stereoselective organic transformations. Although there are numerous instances of triazolyl metal-ion ligands used in catalysis, more research into metal–triazolyl interactions may lead to the discovery of new ligands that have high efficacy, a wide range of applications, and significant promise for green synthesis.

References

1. Masel RI (2001) Chemical kinetics and catalysis. Wiley Interscience, New York
2. Chiusoli GP, Maitlis PM (eds) (2006) Royal society of chemistry, Cambridge. Chapter 2, 4.a
3. Zhou Q-L (2016) Angew Chem Int Ed 55:5352–5353
4. Sheldon RA, Bekkum HV (eds) (2001) In Fine chemicals through heterogeneous catalysis. Wiley-VCH, Weinheim, Germany, p 538
5. Gladiali S, Alberico E (2006) Chem Soc Rev 35:226–236
6. Ojima I (ed) (2010) Catalytic asymmetric synthesis, 3rd edn. John Wiley and Sons, Inc., Hoboken
7. Carroll MP, Guiry PJ (2014) Chem Soc Rev 43:819
8. Dieguez M, Pamies O, Ruiz A, Diaz Y, Castillon S, Claver Ć (2004) Coord Chem Rev 248:2165–2192
9. Zhou Q-L (2016) Angew Chem 55:5352–5353
10. Krištofíková D, Modrocká V, Mečiarová M, Šebesta R (2020) ChemSusChem 13:2828–2858
11. Reetz MT (2016) Chem Rec 16:2449–2459
12. Benkovic SJ, Hammes-Schiffer S (2003) Science 301:1196–1202
13. Constable EC (1996) Metals and ligand reactivity. Wiley VCH, Weinheim
14. Berrisford DJ, Bolm C, Sharpless KB (1995) Angew Chem Int Ed 34:1059
15. Zhong C, Shi X (2010) Eur J Org Chem 2010:2999–3025
16. Jug K, Chiodo S (2003) J Phys Chem A 107:4172–4183
17. Schulze B, Schubert US (2014) Chem Soc Rev 43:2522–2571
18. McDonald KP, Hua Y, Flood AH (2010) Heterocycl Chem 24:341–366
19. Schweinfurth D, Deibel N, Weisser F, Sarkar B (2011) Nachr Chem 59:937–941
20. Struthers H, Mindt TL, Schibli R (2010) Dalton Trans 39:675–696
21. Liang L, Astruc D (2011) Coord Chem Rev 255:2933–2945
22. Aizpurua JM, Fratila RM, Monasterio Z, Pérez-Esnaola N, Andreieff E, Irastorza A, Sagartzazu-Aizpurua M (2014) New J Chem 38:474–480
23. Crabtree RH (2013) Coord Chem Rev 257:755–766
24. Crowley JD, Lee AL, Kilpin KJ (2011) Aust J Chem 64:1118–1132
25. Donnelly KF, Petronilho A, Albrecht M (2013) Chem Commun 49:1145–1159
26. Aromi G, Barrios LA, Roubeau O, Gamez P (2011) Coord Chem Rev 255:485–546
27. Tiwari VK, Mishra BB, Mishra KB, Mishra N, Singh AS, Chen X (2016) Chem Rev 116:3086–3240
28. Agrahari AK, Bose P, Jaiswal MK, Rajkhowa S, Singh AS, Hotha S, Mishra N, Tiwari VK (2021) Chem Rev 121:7638–7956
29. Kolb HC, Sharpless KB (2003) Drug Discovery Today 8:1128–1137

30. Mishra KB, Tiwari N, Bose P, Singh R, Rawat AK, Mishra RC, Singh RK, Tiwari VK (2019) Chem Sel 4:2644–2648
31. Bose P, Jaiswal MK, Singh SK, Tiwari VK (2023) Carbohydr Res 527:108804
32. Gao J, Bhunia S, Wang K, Gan L, Xia S, Ma D (2017) Org Lett 19:2809–2812
33. Zhao J, Niu S, Jiang X, Jiang Y, Zhang X, Sun T, Ma D (2018) J Org Chem 83:6589–6598
34. Singh SK, Yadav MS, Singh AS, Agrahari AK, Mishra N, Kumar S, Tiwari VK (2021) ACS Omega 6:21125–21138
35. Woodward S, Dieguez M, Pamies O (2010) Coord Chem Rev 254:2007–2030
36. Wojaczynska E, Steppeler F, Iwan D, Scherrmann M-C, Marra A (2021) Molecules 26:7291
37. Singh SK, Mishra N, Kumar S, Jaiswal MK, Tiwari VK (2022) Chem Select 7:e202201314
38. Mishra A, Mishra N, Tiwari VK (2016) Curr Org Synth 1:176–219
39. Mishra N, Agrahari AK, Bose P, Singh SK, Singh AS, Tiwari VK (2020) Sci Rep 10:3586
40. Kushwaha D, Singh SK, Tiwari VK (2020) Carbohydrates in drug discovery and development, pp 403–450
41. Chan TR, Hilgraf R, Sharpless KB, Fokin VV (2004) Org Lett 6:2853–2855
42. Donnelly PS, Zanatta SD, Zammit SC, White JM (2008) Chem Commun 2459–2461
43. Ozkal E, Ozcubukcu S, Jimeno G, Pericas MA (2012) Catal Sci Technol 2:195–200
44. Mishra N, Singh AS, Agrahari AK, Singh SK, Singh M, Tiwari VK (2019) ACS Comb Sci 21:389–399
45. Mishra N, Singh SK, Singh AS, Agrahari AK, Tiwari VK (2021) J Org Chem 86:17884–17895
46. Kraft J, Schmollinger D, Maudrich J, Ziegler T (2015) Synthesis 47:199–208
47. Singh SK, Kumar S, Yadav MS, Tiwari VK (2022) J Org Chem 87(22):15389–15402
48. Singh SK, Kumar S, Yadav MS, Gupta A, Tiwari VK (2023) J Org Chem 88:13440–13453
49. Singh SK, Kumar S, Yadav MS, Bhattacharya S, Tiwari VK (2024) Synthesis 56:975–988
50. Deraedt C, Pinaud N, Astruc D (2014) J Am Chem Soc 136:12092–12098
51. Husain AA, Bisht KS (2019) RSC Adv 9:10109–10116
52. Wang G, Wang D, Bietsch J, Chen A, Sharma P (2020) J Org Chem 85:16136–16156
53. Verma AK, Singh J, Kasi Sankar V, Chaudhary R, Chandra R (2007) Tetrahedron Lett 48:4207–4210
54. Battula S, Vishwakarma RA, Ahmed QN (2014) RSC Adv 4:38375–38378
55. Singh M, Singh AS, Mishra N, Agrahari AK, Tiwari VK (2019) ACS Omega 4:2418–2424
56. Liu D, Gao W, Dai Q, Zhang X (2005) Org Lett 7:4907–4910
57. Dai Q, Gao WZ, Liu D, Kapes LM, Zhang XM (2006) J Org Chem 71:3928–3934
58. Zhao Y, van Nguyen H, Male L, Craven P, Buckley BR, Fossey JS (2018) Organometallics 37:4224–4241
59. Zhang G, Luan Y, Han X, Wang Y, Wen X, Ding C, Gao J (2013) Green Chem 15:2081–2085
60. Shen C, Shen H, Yang M, Xia C, Zhang P (2015) Green Chem 17:225–230
61. Wang D, Denux D, Ruiz J, Astruc D (2013) Adv Synth Catal 355:129–142
62. Khiar N, Navas R, Fernandez I (2012) Tetrahedron Lett 53:395–398
63. Shen H, Shen C, Chen C, Wang A, Zhang P (2015) Catal Sci Technol 5:2065–2071
64. Santrac D, Cella S, Wang W, Ackermann L (2016) Eur J Org Chem 5429−5436
65. Goitia A, Gómez-Bengoa E, Correa A (2017) Org Lett 19:962–965
66. Lin Z, Jin J, Qiao J, Tong J, Shen C (2020) Catalysts 10:230
67. Schweinfurth D, Su CY, Wei SC, Braunstein P, Sarkar B (2012) Dalton Trans 41:12984
68. Hajipour AR, Abolfathi P (2018) Catal Commun 103:92
69. Hajipour AR, Abolfathia P (2017) New J Chem 41:2386–2391
70. Gibson VC, Spitzmesser SK (2003) Chem Rev 103:283–315
71. Li L, Bomes CSB, Bomes PT, Duarte MT, Fan Z (2011) Dalton Trans 40:3365–3380
72. Hao ER, Wang ZY, Jiao LJ, Wang SW (2010) Dalton Trans 39:2660–2666
73. Friaes S, Realista S, Gomes CSB, Martinho PN, Royo B (2021) Molecules 26:6325
74. Li XG, Liu K, Zou G, Liu PN (2014) Eur J Org Chem 7878−7888
75. Badjic JD, Nelson A, Cantrill SJ, Turnbull WB, Stoddart JF (2005) Acc Chem Res 38:723–732
76. Saleem F, Rao GK, Kumar A, Mukherjee G, Singh AK (2013) Organometallics 32:3595–3603
77. Yan W, Ye X, Akhmedov NG, Petersen JL, Shi X (2012) Org Lett 14:2358–2361

78. Nomiya K, Noguchi R, Ohsawa K, Tsuda K (1998) J Chem Soc Dalton Trans 4101–4108
79. Stephen A, Hashmi K, Lothschutz C (2010) ChemCatChem 2:133–134
80. Wang QY, Aparaj S, Akhmedov NG, Peterson JL, Shi XD (2012) Org Lett 14:1334
81. Kannan A, Rajakumar P (2015) RSC Adv 5:46908–46915
82. Astruc D (2008) Nanoparticles and catalysis. Wiley Interscience, New York
83. Wong, Vuong KQ, Gatus MRD, Hua C, Bhadbhade M, Messerle BA (2012) Organometallics 31:7500−7510
84. Agarwal J, Peddinti RK (2011) Tetrahedron Lett 52:117–121
85. Kumar TP, Balaji SV (2014) Tetrahedron: Asymmetry 25:473−477
86. Wang L, Liu J, Miao T, Zhou W, Li P, Ren K, Zhang X (2010) Adv Synth Catal 352:2571–2578
87. Chandrasekhar R, Kumar TP, Haribabu K, Reddy CR (2010) Tetrahedron Asymmetry 21:2372–2375
88. Yan Z-Y, Niu Y-N, Wei H-L, Wu L-Y, Zhao Y-B, Liang Y-M (2006) Tetrahedron: Asymmetry 17:3288–3293
89. Zhao YB, Zhang L-W, Wu L-Y, Zhong X, Li R, Ma J-T (2008) Tetrahedron: Asymmetry 19:1352–1355

Chapter 17
Intramolecular Click Chemistry in Organic Synthesis

17.1 Introduction

The concept of 'Click Chemistry' [1] coined by Prof. K B Sharpless in 2001 and turned out to be one of the most popular protocols soon after the discovery of regioselective 1,2,3-triazole forming tool Cu(I)-catalyzed azide-alkyne cycloaddition (CuAAC) by two independent groups of researchers, Prof. Morten Meldal [2] and Prof. K B Sharpless [3]. This modular and versatile tool has extensive advantages including the excellent regioselectivity, specificity, reliability, biocompatibility, excellent yields, mild reaction conditions, and short reaction time [4]. Thus, the coupling of terminal alkyne **1** with organic azide **2** in the presence of Cu(I) catalyst (CuAAC route) furnished high-to-excellent yields of 1,4-disubstituted triazoles **3** [2, 3], whereas the reaction of terminal alkyne and organic azide under Ru-catalysis (RuAAC route) [5] deliver exclusively 1,5-disubstituted triazoles **4**, the opposite regioisomer (Scheme 17.1a). Both the CuAAC and RuAAC protocols have their advantages because the similar thermally induced cycloaddition reactions produced a mixture of both the regioisomers [6]. In addition to the CuAAC [2, 3] and RuAAC [5], the regioselective triazole forming reactions comprise several other routes of synthesis, for instance, organocatalyzed click cycloaddition [7–10], photocatalyzed click reaction [11], topochemical click cycloaddition [12], and Bertozzi's strain-promoted [3 + 2] azide-alkyne cycloaddition (SPAAC) [13]. The click-mediated 1,2,3-triazole ring acts as amide isosteres; i.e., they are good mimics of the peptide or amide bonds owing to their polar nature and hydrogen bond formation ability [14, 15]. They are a potential pharmacophore, and therefore, triazole-appended carbohydrates, heterocycles, proteins, peptides, lipids, and other scaffolds have found extensive applications in almost all prominent disciplines of chemistry, biology, medicine, technology, agriculture, etc. [16–18]. Also, this illustrated synthetic method found tremendous applications in various emerging fields including glycoscience [19, 20] and drug development [21–23].

V. K. Tiwari et al., *Click Chemistry*, Materials Horizons: From Nature to Nanomaterials,
https://doi.org/10.1007/978-981-97-4596-8_17

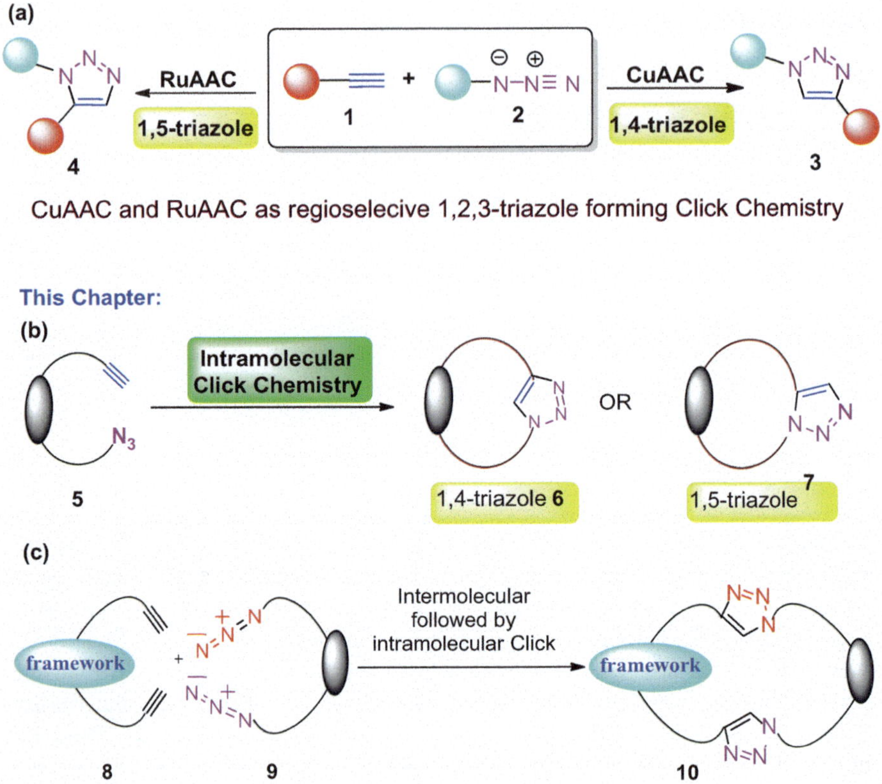

Scheme 17.1 Classical click-mediated azide-alkyne cycloaddition approaches (intermolecular and intramolecular) for the regioselective 1,2,3-triazole synthesis

Despite the enormous scope of organocatalyzed click chemistry and SPAAC bioorthogonal methodology in drug development, the intramolecular variation of either of the established methods has not been well investigated so far in comparison with metal-catalyzed click (CuAAC and RuAAC) and thermally induced click chemistry [24]. There are two common routes for the classical click-mediated macrocyclization reaction from respective azide-alkyne framework, which is either a selective metal-catalyzed or thermally initiated intramolecular azide-alkyne cycloaddition (Scheme 17.1b) or intermolecular click followed by intramolecular click reaction (Scheme 17.1c). In the second route, however both the click steps may proceed simultaneously or can proceed preferentially as the rate of intramolecular cyclization reactions is in general more rapid than the corresponding intermolecular one.

Thus, the triazole forming cycloaddition reactions can be broadly categorized into two major pathways, (i) the intermolecular click reactions in which the cycloaddition reaction is performed between two isolated precursors, i.e., an azide group and an alkyne group and (ii) the intramolecular click reactions in which both the azide and alkyne functionalities are present within same substrate. Over 200 reviews have been published on 'click chemistry' [20]; however, there is still very little literature review and comprehensive reports available on intramolecular azide-alkyne cycloaddition reaction [24–27]. Taking into consideration the above facts, there is a critical need of an inclusive review having emphasis on recent progress on intramolecular regioselective click reactions of diverse azido-alkynes. In summary, this chapter presents the recent approaches on click chemistry-mediated intramolecular azide-alkyne cycloadditions along with their growing applications in organic synthesis.

17.2 CuAAC-Mediated Intramolecular Click Chemistry

The intermolecular and intramolecular CuAAC both can be successfully employed for the macrocyclization reaction which can be further accelerated by including some unconventional methods such as the template-directed synthetic approach or macrocyclization carried out in high-dilution conditions or also by using well-established dynamic combinatorial route. Considering all the aspects, it is evident that the regioselective CuAAC 'Click Chemistry' technique leads to a successful macrocyclization reaction giving rise to a wide range of 1,2,3-triazole-linked macrocyclic compounds including triazole-fused polycycles [28], cyclodextrin mimetics [29], and the biocompatible glycomacrocycles of great utility [30]. Additionally, the 1,2,3-triazoles produced as a result of CuAAC possess the molecular planarity, strong dipole–dipole interactions, chemical stability, and also act as an amide bond isostere due to its hydrogen-bonding capacity [15, 31, 32]. Some non-covalent interactions such as π-stacking interactions and hydrogen-bonding interactions have also been exhibited by the 1,2,3-triazoles.

Meldal's catalytic pair (CuI/DIPEA) [2] and Sharpless catalytic pair ($CuSO_4.5H_2O$/NaAsc) [3] both were firmly established for the easy construction of regioselective 1,4-disubstituted triazole scaffolds, but apart from these two approaches, various other catalytic systems have also been explored for the 1,2,3-triazole formation with great ease [4, 33]. For instance, a high-yielding methodology for the synthesis of 1,4,5-trisubstituted triazole was developed by Sen and co-workers [34]. The reaction path involved a Sonogashira coupling reaction followed by an intramolecular click coupling. In the developed methodology, the Sonogashira coupling was supposed to happen first in which the terminal alkyne of glycosyl azido-alkyne **11** undergoes reaction with aryl iodide **12** in the presence of cuprous oxide-nanoparticle as catalyst Cs_2CO_3 as a base in aqueous media to afford the aryl substituted azido-alkyne sugar intermediate which on subsequent intramolecular click coupling furnished the desired glycohybrid triazole **13** in 85% yield (Scheme 17.2) [35].

Intramolecular click followed by macrocyclization by ring closing metathesis (RCM) [36] or macrolactamization reaction [37] or intramolecular Aldol reaction [38] or intramolecular glycosylation via intramolecular aglycon delivery (IAD) [39] are some of the well-established intramolecular protocols in organic synthesis. Along

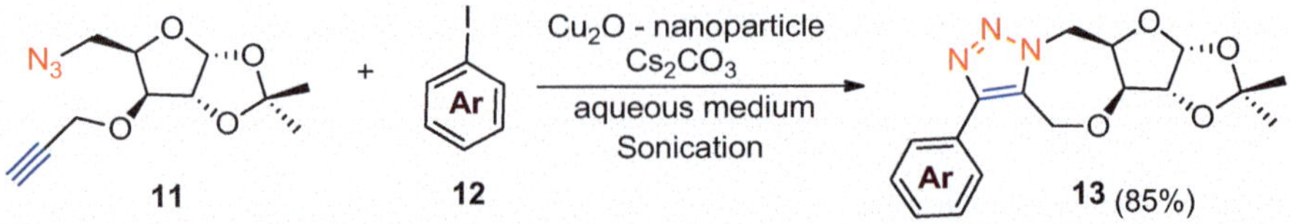

Scheme 17.2 Sonogashira followed by intramolecular Click route for the synthesis of fused glycohybrid triazole from sugar azido-alkyne

with this, intramolecular acylation reactions are also one of the efficient methods in organic synthesis which makes Click-acylation domino reactions of utmost importance in organic synthesis [40]. In addition to the above-mentioned macrocyclization methodologies, numerous cyclic molecular scaffolds including glycomacrocycles and peptidomimetic macrocycles can be readily constructed through the click followed by intramolecular click reaction strategy. Furthermore, another variant of a successful triazole-appended macrocyclization step with high selectivity can also be utilized. Toward this end, Ghorai et al. used CuAAC-mediated macrocyclization route to develop 1,2,3-triazole-based peptidomimetic macrocycle as a class of pseudo cyclo-β-peptides [41]. Linear azido-alkyne **14** derived from D-glucose on treatment with CuI/DIPEA in the presence of *tris*-(benzyltriazolylmethyl)amine (TBTA) underwent intermolecular click followed by intramolecular click macrocyclization and afforded the respective 1,2,3-triazolo-urea containing peptidomimetic macrocycle **15** in 36% yield (Scheme 17.3) [41]. Similar intramolecular click macrocyclization route was successfully used by Nilsson et al. and C_2 symmetry hybrid glycopeptide-containing macrocycles were obtained in good yields [42].

In an another application of intramolecular click cycloaddition, monomeric triazolophane **17** as part of 15-membered strained ring was synthesized through the coupling of the furanoside sugar azide and alkenylated aryl residue linked with unnatural ester tether of compound **16** in the presence of CuI/DIPEA (Scheme 17.4) [43]. The method employed sugar-appended peptide-based azido-alkynes as substrates in intramolecular cycloaddition to access corresponding triazolophane macrocycle as part of a 12- to 17-membered ring system [43].

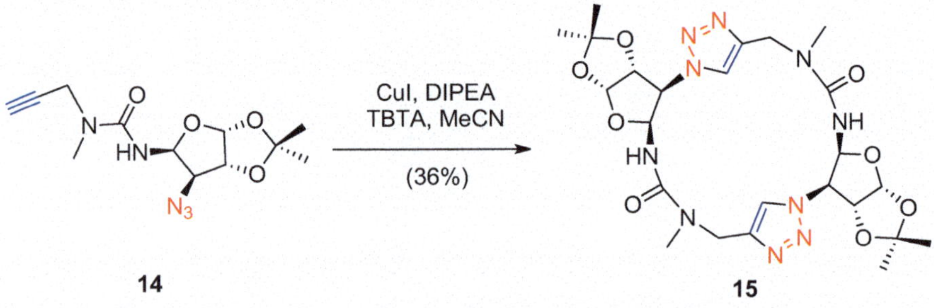

Scheme 17.3 Click-inspired synthesis of triazole/urea-based peptidomimetic macrocycle **15**

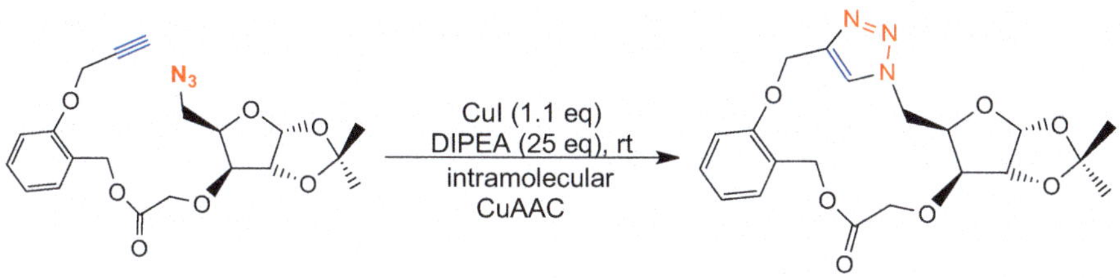

16

17 (32%, part of a 15 memebered ring)

Scheme 17.4 Intramolecular CuAAC for synthesis of monomeric triazolophane from corresponding sugar-tethered azide-alkyne scaffolds

Intramolecular CuAAC was further exploited to synthesize biologically important *C*-spiro-morpholine-fused glycohybrid species **21** by utilizing 2,3,4,6-tetra-*O*-benzyl-D-glucono-1,5-lactone **18** as substrate. First, the compound **18** was subjected to allylation under Barbier conditions followed by ring-closure metathesis (RCM) using Grubbs second-generation catalyst to produce *C*-spiro cyclopentenyl carbohydrate species **19** which further treated with *m*-CPBA followed by *β*-epoxide ring opening using NaN_3/NH_4Cl to furnish azido alcohol **20** (in a 1:1 diastereomeric mixture). Finally, propargylation followed by intramolecular Click cycloaddition of azido alcohol **20** afforded the desired morpholine-fused-spiro-glycohybrids (**21a** and **21b**) (Scheme 17.5) [44].

Sugar amino acids serve as appropriate building blocks for the synthesis of biologically important sugar-appended macrocycles including cyclodepsipeptides and cyclophane-based macrocycles. In this regard, Murphy et al. utilized click conjugation technique for the efficient synthesis of cyclophane-based macrocycle **25** employing glycosyl amino acid **22** as substrate [45]. First, the glycosyl amino acid **22** was transformed into a carbohydrate-derived *bis*-azide **23** which undergoes the consecutive CuAAC click conjugation with *p*-*bis*-propargyloxybenzene **24** to furnish the desired glycotriazolophane macrocycle **25** in good yield (Scheme 17.6)

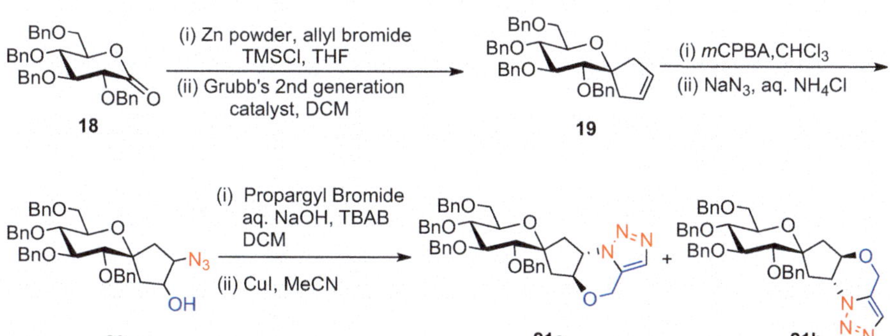

Scheme 17.5 Synthesis of morpholine-fused-spiro-glycohybrid via intramolecular CuAAC

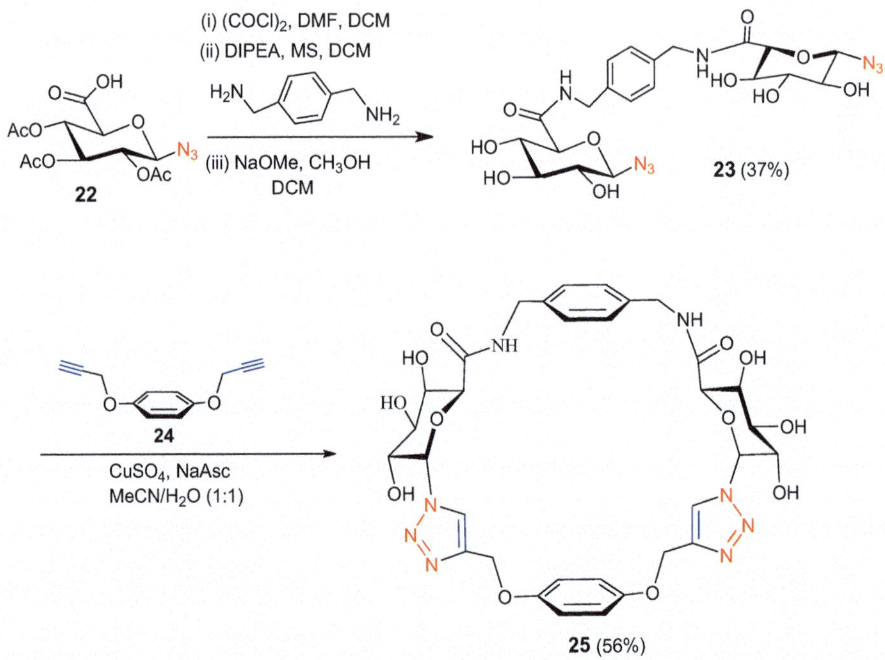

Scheme 17.6 Click conjugation technique for efficient synthesis of glycotriazolophane macrocycle **25** from sugar amino acid

[45]. In a similar fashion, Click chemistry was also utilized for the synthesis of *bis*-triazolophane-appended macrocycle having thioindigo chromophore [46].

Yu et al. developed a series of carbohydrate-appended BODIPY macrocycles (**29**) and (**32**) explored their applicability as metal sensors [47]. BODIPY acid **26** and 6-amino sugar azide **27** were tethered in the presence of EDC/HOBt in DCM at room temperature to obtain the amide analogue **28** which was further subjected to intramolecular Click reaction to afford good yield of the corresponding triazole-appended 23-membered glycosyl macrocycle **29** (Scheme 17.7) [47].

Similarly, Boc-protected glucosyl amino acids (**30a**, and **30b**) were obtained through coupling of 6-amino sugar azide **27** with Boc-l-Asp-OBn, Boc-Gly-OH, or Boc-l-Met-OH, respectively. Boc deprotection of the obtained glycosyl amino acids followed by amide coupling with BODIPY acid **26** provided the azido-alkynes (**31a** and **32b**) which underwent the intramolecular Click reaction to furnish good yields of 26-membered macrocycles **32a** and **32b,** respectively (Scheme 17.8) [47].

Gin et al. further applied the intramolecular CuAAC cycloaddition for the synthesis of *C*-2 or *C*-3 symmetric carbohydrate macrocycles. Orthogonally protected mannose scaffold **33** having azido group at the anomeric position (α-anomer) and counter alkenylated residue (propargyl ether) at the C4-position was employed for the intramolecular cyclo-oligomerization in the presence of CuI/DBU to furnish corresponding carbohydrate-appended macrocycle **34** (Scheme 17.9) [48,

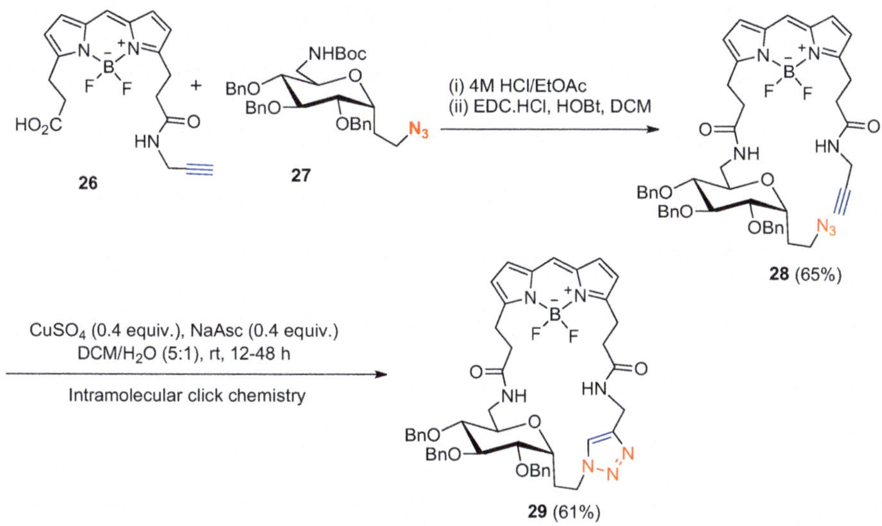

Scheme 17.7 Intramolecular click-mediated synthesis of sugar-based BODIPY-functionalized fluorescent macrocycle **29**

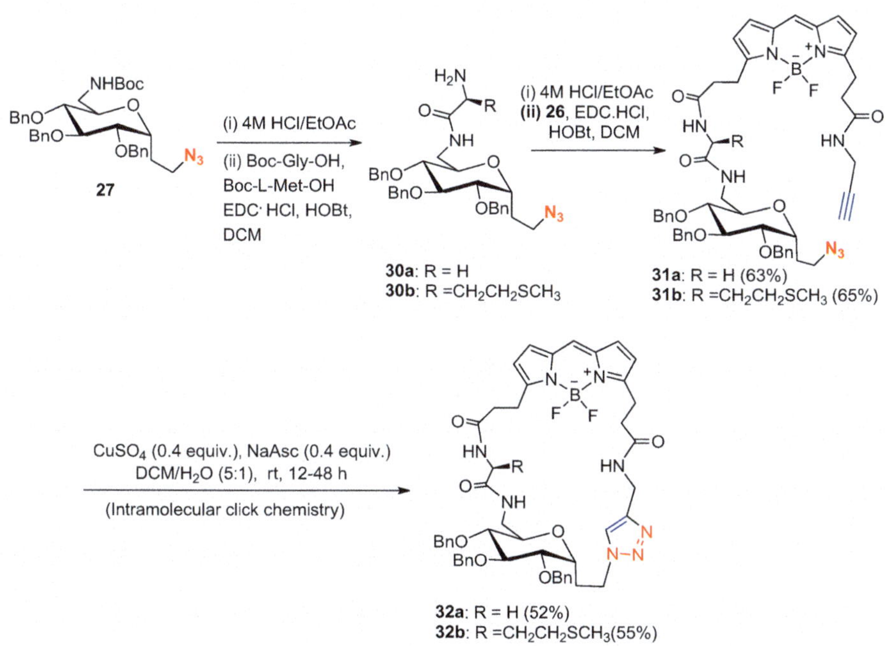

Scheme 17.8 Intramolecular click-mediated synthesis of sugar-based BODIPY-functionalized fluorescent macrocycle **32**

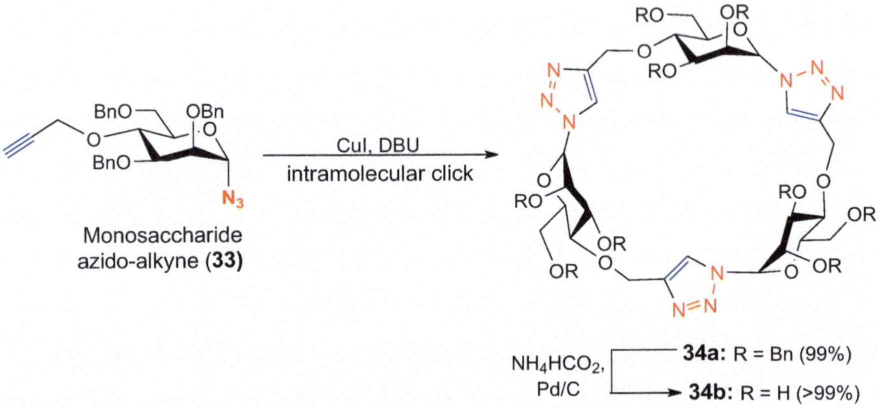

Scheme 17.9 Intramolecular CuAAC in the synthesis of symmetric carbohydrate containing macrocycles **34a** and **34b**

49]. The authors further utilized disaccharide and trisaccharide as substrates under the same reaction conditions to afford corresponding carbohydrate-appended macrocycles. In a similar manner, the intramolecular click macrocyclization method was employed to numerous orthogonally protected glycosyl azido-alkyne derivatives of diverse monosaccharides including D-glucose, D-galactose, and D-mannose for the synthesis of their corresponding macrocycles [49].

An efficient and challenging synthetic methodology for the construction of triazolophane-based fused bicyclic systems has been established by Dondoni et al. by exploiting the intramolecular CuAAC click macrocyclization protocol. The group successfully carried out the synthesis of various *C*n-symmetric cycloglucopyranosides comprising of two, four, or six glucopyranoside frameworks linked together via 1,4-triazole moiety by utilizing the convergent method of synthesis [50]. In a similar investigation, construction of furanoside-containing macrocycle was carried out under the intramolecular CuAAC conditions from a substrate having azido-alkyne functionality comprised of ester (**35a, b**) (Scheme 17.10). The substrates having azido-alkyne functionality comprised of amine and amide were also undergone similar macrocyclization under the similar reaction conditions [51].

Luzzio et al. employed the intramolecular click strategy for the synthesis of endo-fused 1,2,3-triazolo-macrocycles in good-to-excellent yields by using oxazole-containing azidoalkynyl building blocks as substrates and $CuSO_4$/NaAsc as catalyst [26]. An another intramolecular macrocyclization protocol for the high-yielding synthesis of nucleoside macrocycles as part of sixteen-membered ring was demonstrated by Seela et al. [52]. The group carried out the intramolecular coupling of compounds **37** possessing azido group at 5′-position of deoxy ribose and octadiynyl residue at C-5 position of nucleobase under standard Click conjugation conditions to afford the desired nucleoside macrocycle **38** in 71% yield (Scheme 17.11) [52].

Nakane and group successfully synthesized dumbbell oligodeoxynucleotides comprising numerous 1,2,3-triazole units through hybridization of cyclic DNA by

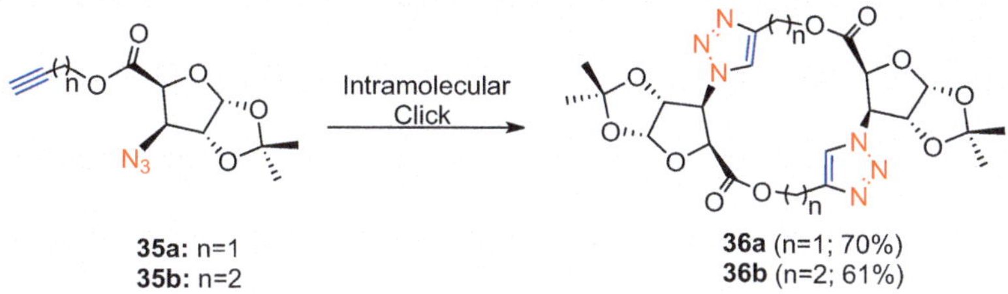

Scheme 17.10 Construction of furanoside-containing macrocycles via intramolecular click reaction

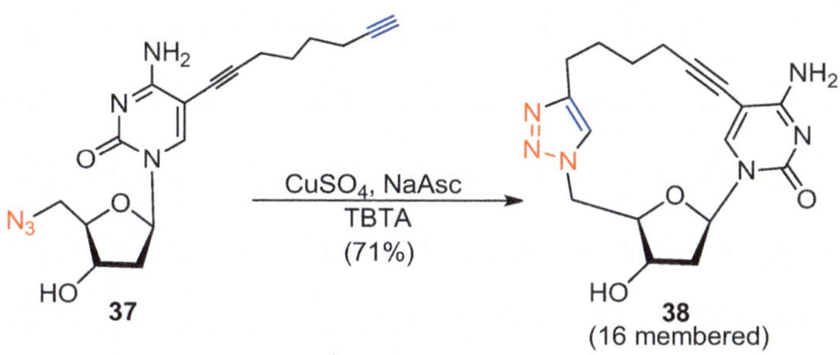

Scheme 17.11 Intramolecular click conjugation-mediated synthesis of nucleoside macrocycle **38**

utilizing intramolecular CuAAC methodology. The developed triazole-containing products exhibited high resistance to exonuclease degradation [53]. In addition, it was observed that the catenated DNA has appreciable binding ability with NF-κB and also inhibits in vitro and *ex-vivo* transcriptions.

Further, the intramolecular CuAAC macrocyclization technique was utilized by Allam et al. for the development of notable macrocycles possessing copper complexation ability using galacturonic acid as starting material [54]. First of all, D-galacturonic acid was converted into the required building block **39** which was then treated with azido amine species **40** in anhydrous THF at 45 °C for 16 h to undergo trans-acylation to furnish galacturonic acid-based bifunctional azido-alkyne intermediate **41** in appreciable yield. Next, intramolecular click cyclization of species **41** under different catalytic conditions gives rise to cyclomonomer product **42** and cyclodimer product **42** (Scheme 17.12) [54]. Different catalytic and solvent systems afford different selectivity and yields of products as depicted in the Scheme 17.12. In addition, interesting Cu(II)complexation ability was displayed by the monomeric macrocyclic product **42**.

Campo et al. further explored the field of intramolecular click and came up with the synthesis of β-1,6-linked-galactose-triazole-linked glycomacrocycles **45a, b** along with linear oligomers from a galactose-derived azido-alkyne framework **44** having an azido functionality with a triethylene spacer at *C*-1 and a 6-*O*-propargyl functionality (Scheme 17.13) [55]. The group reported that the synthesis can be achieved in

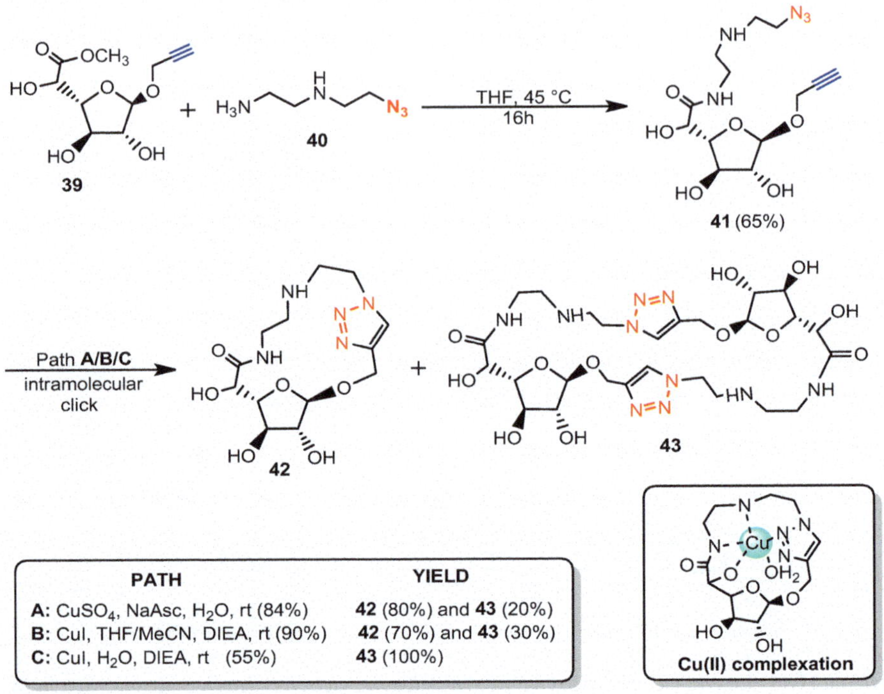

PATH	YIELD
A: $CuSO_4$, NaAsc, H_2O, rt (84%)	**42** (80%) and **43** (20%)
B: CuI, THF/MeCN, DIEA, rt (90%)	**42** (70%) and **43** (30%)
C: CuI, H_2O, DIEA, rt (55%)	**43** (100%)

Scheme 17.12 Synthesis of galacturonic acid-based macrocycles **42** and **43** using intramolecular click reaction

two ways, either through intramolecular click reaction in the presence of $CuSO_4$/Cu turnings in DMF at 110 °C under *MW* condition for 30 min (Method **M1**) or intramolecular click at room temperature for 2 days (Method **M2**). Both the methods furnished glycomacrocyclic products along with linear oligomers. Also, the obtained products **45a, b** were screened for their anti-parasitic activities against *T. Cruzi* in search of a potential lead candidate.

Compain et al. accomplished an efficient synthesis of various cyclodextrin mimetics designed to work as better building blocks for the development of multivalent systems [56]. The developed methodology involved the reaction of orthogonally protected disaccharide **46** with $TMSN_3$ in the presence of catalytic amount of TMSOTf and TEA to produce a sugar intermediate **47** which on treatment with $AgNO_3$ furnished bifunctional carbohydrate framework **48** possessing both clickable azide and alkyne functionalities (Scheme 17.14). Next, the sugar species **48** was subjected to intramolecular click conjugation using $CuSO_4{\cdot}5H_2O$/NaAsc catalytic system in THF/H_2O solvent and produced a major cyclodimer product **49** (42% yield) along with a minor cyclotrimer product **50** (9% yield). Further, the generated carbohydrate-appended macrocycle **49** underwent TIPS deprotection with AgF in MeCN to afford terminal alkyne species further made to react with iminosugar

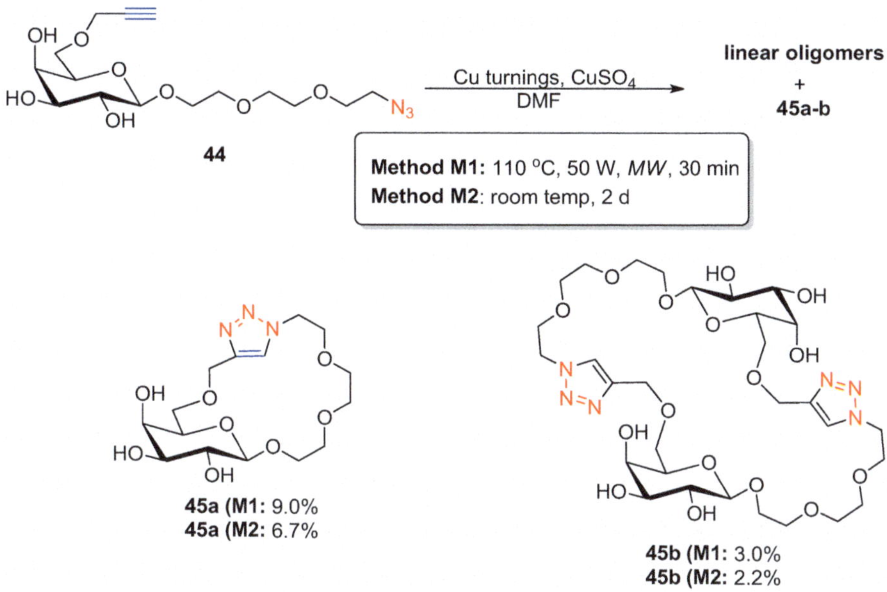

Scheme 17.13 Synthesis of *β*-1,6-linked-galactose-triazole-linked glycomacrocyclic products through intramolecular click protocol

azide under intermolecular click conditions to gain insight into possible multivalent systems.

In similar manner, a facile synthesis of series of *N*-acetyl glucosamine comprising *bis*-triazole macrocycles (**56a–h**) has been established by Wang and group. The reaction proceeds through a successive CuAAC intermolecular click reaction followed by intramolecular click protocol (Scheme 17.15) [57]. *N*-acetyl-D-glucosamine-derived azide **52** underwent intermolecular click coupling with *bis*-terminal alkynes **53** to produce corresponding carbohydrate triazolyl alkynes **54** which was subjected to azidation on reaction with NaN_3 in DMF to afford the azido-alkynes **55a–h.** Intramolecular click reaction in the presence of CuI (20 mol%) and DIPEA at 0 °C in MeCN produced the desired 1,6-linked *N*-acetyl glucosamine-based macrocyclic products **56a–h** in moderate yields.

A modular double-click conjugation technique has been developed and utilized for straightforward construction of a series of *C*2-symmetric macrocycles **59** and **62** (Scheme 17.16). The macrocyclic product **59** was synthesized by employing *bis*-propargyl compound **57** and *bis*-azide **58** as coupling partners and the reaction proceeds through cycloaddition reaction involving double-click macrocyclization in the presence of *S*-phenyl ethyl ammonium acetate in one-pot condition. Using the same approach, a larger macrocycle **62** possessing a phenyl linker between *bis*-triazolyl sugar scaffolds was also synthesized. For this purpose, same substrate **57** was first allowed to react with 4-trimethylsilylethynylphenyl azide **60** underclick

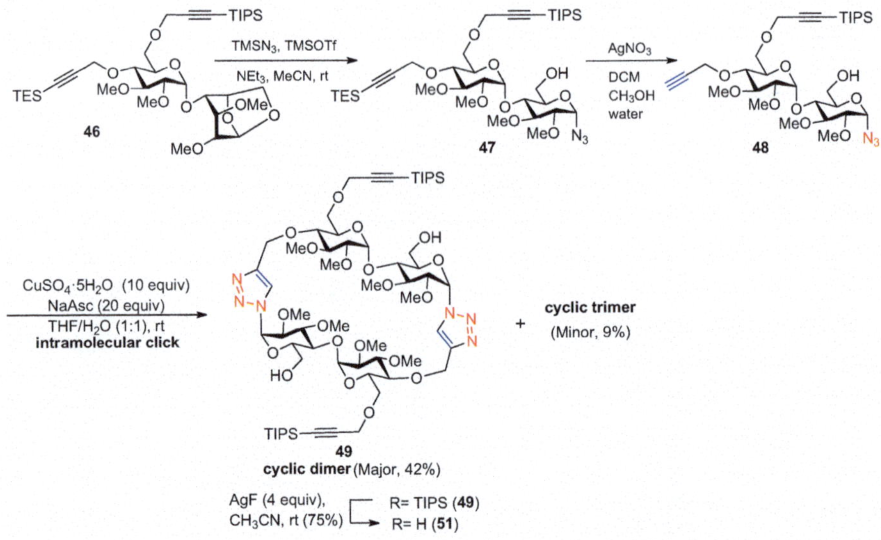

Scheme 17.14 Synthesis of cyclodextrin mimetics under intramolecular click macrocyclization strategy

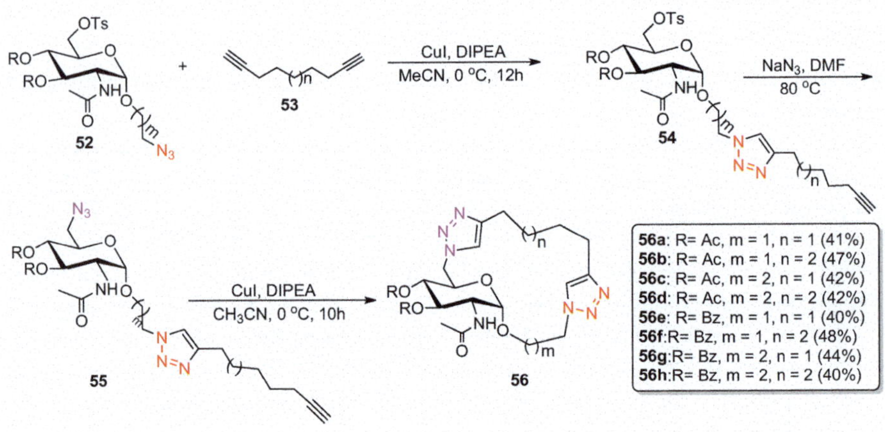

Scheme 17.15 Synthesis of 1,6-linked *N*-acetyl glucosamine-linked macrocycles (**56a–h**) through Click cyclization protocol

conditions followed by trimethylsilyl deprotection to produce a *bis*-triazolyl intermediate **61** which further undergoes same Click macrocyclization coupling with bis-azide **58** to furnish the desired sugar macrocycle **62** (Scheme 17.16) [58]. The developed macrocycles are found to have good application in host–guest interaction and chiral recognition phenomenon.

Cyclodextrin mimic triazole-linked glycomacrocycles are found to have prominent application as guest binding and recognition properties in host–guest

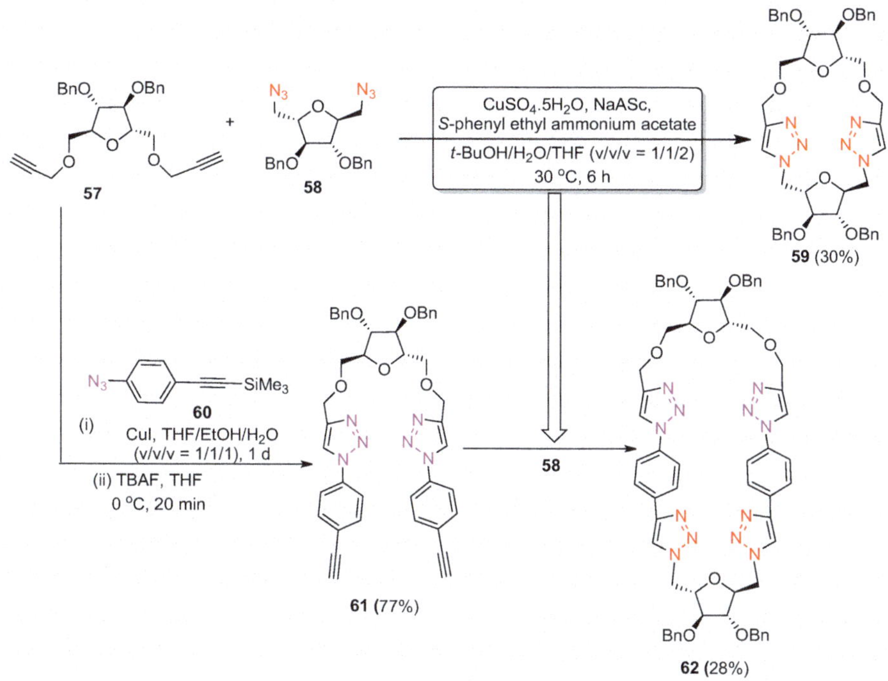

Scheme 17.16 Construction of sugar-tethered macrocycles **59** and **62** under Click macrocyclization condition

supramolecular chemistry. Pathigoolla et al. accomplished the synthesis of reverse cyclodextrin mimetics in appreciable yields using the similar technique of intramolecular CuAAC click reaction (Scheme 17.17) [59]. 1-Azido and 5-acetylene D-galactose **63** was chosen as substrate for this reaction which in turn was synthesized from per-acetylated galactose azide via several steps including acetate removal, 4,6-benzylidene acetal protection, benzylation, reductive ring opening and finally propargylation of the 4-hydroxyl group. Self-macrocyclization of azido-alkyne substrate **63** using CuI/DIPEA catalyst in THF produced corresponding macrocyclic products **64–67** as a mixture of four cyclo-oligomers. The group successfully isolated them and identified as cyclic dimer **64** (15%), trimer **65** (28%), cyclic tetramer **66** (15%), and cyclic pentamer **67** (3%). The group also carried out the debenzylation of obtained compounds and achieved corresponding cyclodextrin mimic triazole-linked glycomacrocyclic products in excellent yield [59].

Several important biomolecules comprising cyclic tetrapyrrole porphyrin rings such as hemoglobin, cytochromes, and Vitamin B-12 are well-known to play crucial roles in various biological events [60]. In addition, a wide variety of porphyrin-tethered triazole molecules developed in recent years have been found to display a range of photophysical, photochemical, electrochemical, and therapeutic properties. Keeping in view the impact of above-described molecules, our group explored

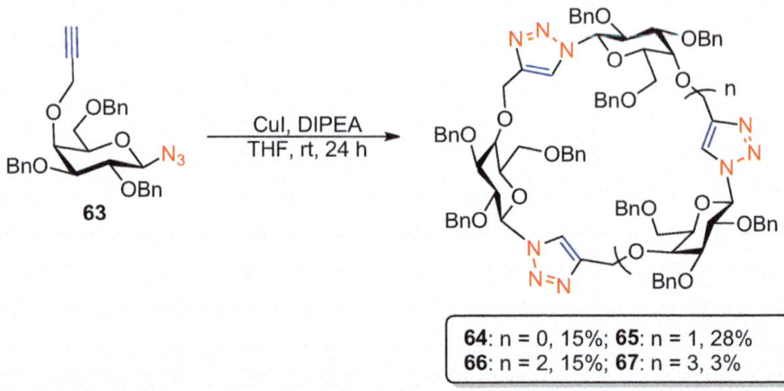

Scheme 17.17 Construction of reverse cyclodextrin mimetics under CuAAC-mediated macrocyclization

the intramolecular click chemistry to synthesize porphyrin-cored glycodendrimer [61] and porphyrin-cored hybrid glycodendrimer [62]. Also, both the developed molecules exhibited potential application for the selective sensing of Cu(II) ions.

Crossley et al. utilized the modular intermolecular click chemistry in an interesting manner and developed a 'capped' porphyrin-triazole-based macrocycle **70** by employing a rigid 1,2,3-triazole as linker to attach the porphyrin to the capping group (Scheme 17.18) [63]. Zinc(II) tetra azido porphyrin **68** and tetra alkyne **69** were utilized as coupling partners under two different and complementary reaction conditions (Method Y1: $CuSO_4·5H_2O$, L-ascorbic acid, TEA, 10% H_2O in THF, reflux, 30 h: Yield 19%) and (Method Y2: $Cu(MeCN)_4PF_6$, TEA, anhydrous DMF, 70 °C, 16 h, Yield 14%) to achieve the desired triazole-linked-capped porphyrin macrocycle **70**. Further, the anion-binding properties of the developed capped porphyrin macrocycle **70** were assessed by Gilday and co-workers and found that it exhibited very strong anion-binding affinities for a number of anions [64].

The CuAAC-inspired intramolecular macrocyclization protocol has also been largely investigated to access other essential macrocyclic systems of extensive applications including temperature-responsive cyclic PNIPAM [65], and cyclic polymer template [66].

Lewandowski has successfully developed a wide range of interesting sucrose-based crown ethers and macrocycles [67–69]. The authors also developed a sucrose-based azide-alkyne derivative **71** by using multiple high-yielding steps [69]. Jarosz and group employed this developed sucrose compound **71** as substrate for the efficient synthesis of sucrose-based macrocyclic architectures under the Click intramolecular reaction conditions [70]. The group also observed that the intramolecular and intermolecular mode selectivity of the reaction depends on the concentration of catalyst as well as starting material. Low concentration of **71** favored intramolecular cyclization and produced triazolyl macrocycle **72** along with its regioisomer1,5-disubstituted triazole **73** whereas high concentration of **71** leads to the intermolecular coupling *bis*-triazolyl macrocycle **74** was obtained as the sole product. It was observed that

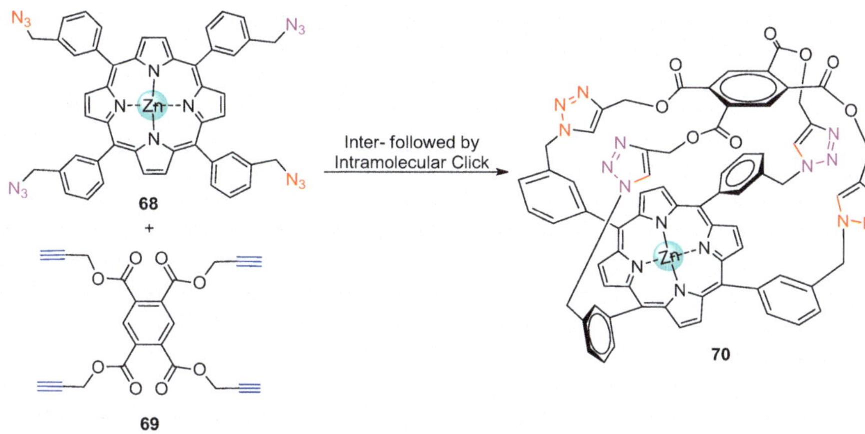

Scheme 17.18 Synthesis of 'capped' porphyrin-triazole-based macrocycle **70**

the steric repulsion in the small cavity of 1,4-regioisomeric macrocycle **74** gives rise to the unexpected formation of isomer **73** (Scheme 17.19) [71].

In a similar fashion, further exploration of CuAAC click conjugation methodology paved the way for an interesting synthesis of sucrose-containing *bis*-triazolyl macrocyclic compounds **77** and **80** [72]. Sucrose-based *bis*-propargyl derivative **75** and bis-azide derivative **78** were synthesized from sucrose in a couple of steps and employed as starting materials for this developed protocol. Next, the sucrose *bis*-alkyne **75** was allowed to react with D-mannitol azide **76** under standard Click

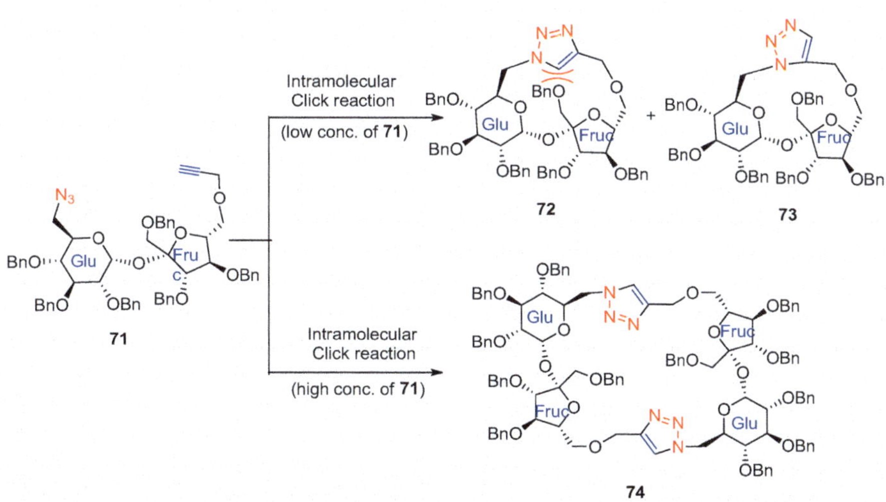

Scheme 17.19 Concentration-dependent click-mediated synthesis of sucrose-based macrocycles **72–74**

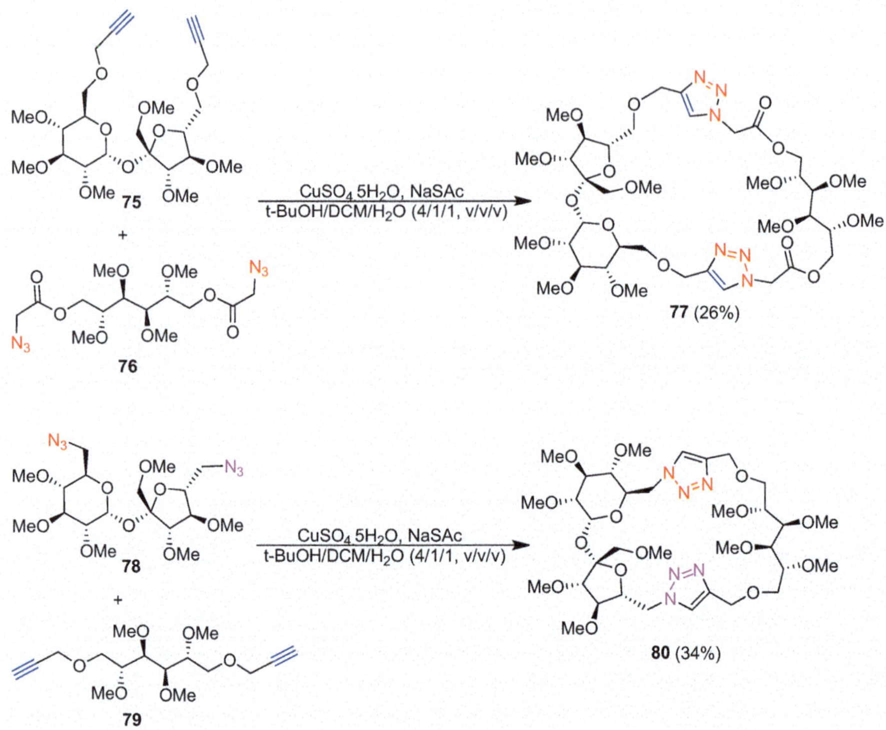

Scheme 17.20 Synthesis of sucrose appended *bis*-triazolyl macrocycles **77** and **80**

treatment and afforded the desired macrocycle **136,** and on the other hand, macrocycle **80** was obtained when sucrose *bis*-azide **78** and *O*-propargylated D-mannitol **79** were coupled using click conjugation protocol (Scheme 17.20) [72].

Sialosides and sialic acid-containing macrocycles are of great importance in the field of drug discovery and development. A review highlighting their significance and their growing impact in biological and medicinal fields has been recently documented by Tiwari et al. recently [73]. Further, keeping in mind the impact of sialosides and sialic acid-linked macrocycles in the domain of glycoscience, the intramolecular CuAAC protocol has been deployed successfully for the development of sialic acid-containing macrocycles of chemotherapeutic potential [74]. A bifunctional galactose azido-alkyne **81** having propargyl group at *C*-1 and azido group at *C*-6 position has been synthesized and used as substrate for the intramolecular macrocyclization in the presence of $CuSO_4$/Cu turnings as catalyst in anhydrous DMF solvent under *MW* irradiation and corresponding sialic acid-linked macrocyclic products **82** and **83** were obtained in excellent yields. Also, the developed products were screened for their therapeutic potential, and it was found that they can serve as a model template for an enzymatic sialylation with *Trypanosoma cruzi trans-sialidase* (TcTS) in the

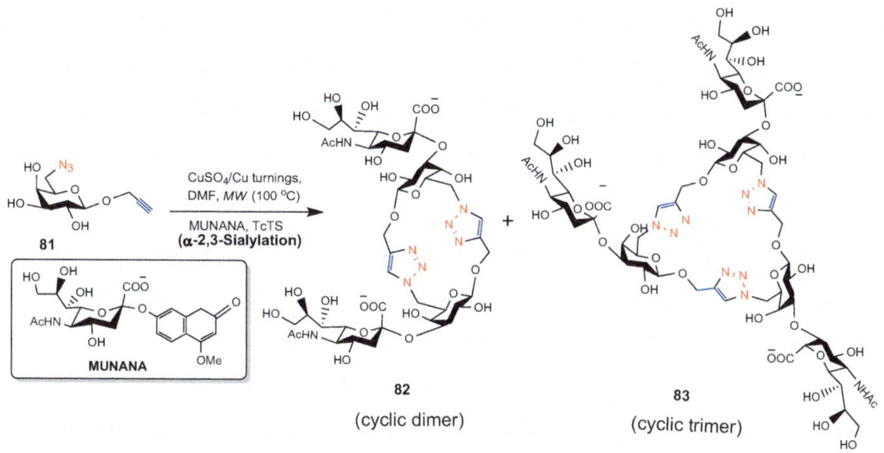

Scheme 17.21 Click-mediated synthesis of sialic acid-containing macrocycles **82** and **83** followed by enzymatic sialylation using TcTS

presence of MUNANA (Scheme 17.21) [74]. It was also observed that the cyclo-oligomerization reaction pathway was concentration dependent. When carried out in dilute solution, the reaction pathway leads to both cyclic dimer and trimer as the major products, whereas the same cyclization reaction when performed in concentrated condition, higher cyclic derivatives, e.g., tetramers or pentamers were produced in trace amounts.

The versatility of the CuAAC click conjugation methodology has been well displayed in its application in the synthesis of an octasaccharide species **84** which was synthesized through one-pot multienzyme synthesis followed by intermolecular CuAAC reaction in good yield. The obtained α-2,6-linked sialoside product **84** having azide at *C*-5 position of sialic acid and alkyne at the terminal end of galactose then serves as the substrate for the synthesis of a 76-membered macrocycle. This step was performed under the intramolecular click macrocyclization conditions using CuI/DIPEA catalytic system in MeCN/H_2O (1:1) mixture to furnish the desired 76-membered macrocyclic product **85** (Scheme 17.22) [35].

The obtained sialic acid-linked macrocycle possesses large number of hydroxyl functional groups, which makes it extremely water soluble, and therefore, it is potentially applicable in the emerging fields such as the host–guest chemistry and the size-dependent interaction study with other hydrophobic molecules. In addition to this, these hydrophilic molecules play crucial role on the cell surfaces and therefore can modulate and facilitate several important cellular interactions in biological systems and they are also useful for the study of the carbohydrate-protein interactions [75]. These characteristics indicate that these molecules will have huge future impact in biological and medicinal fields and their investigation under Click condition including macrocyclization reaction would be very useful to enhance their potential.

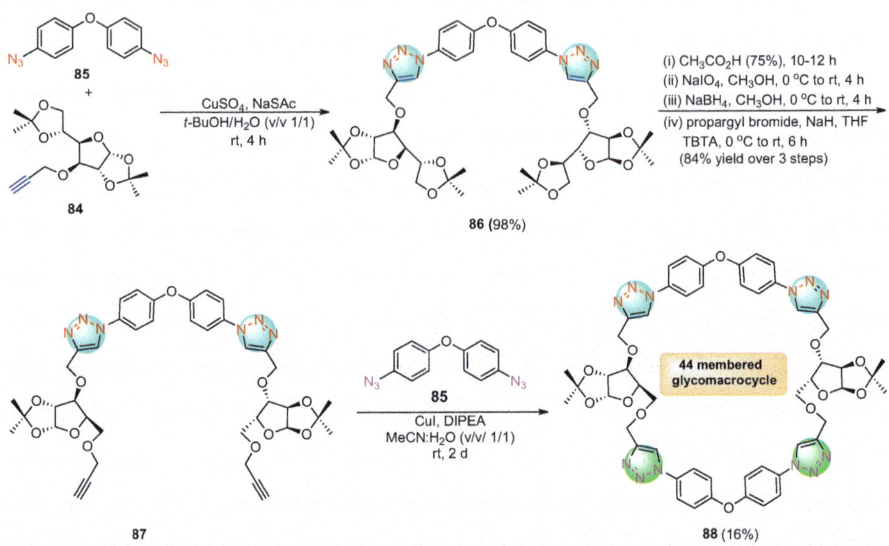

Scheme 17.22 Synthesis of vancomycin-mimic 44-membered glycomacrocycle **88** via CuAAC-mediated macrocyclization approach

A vancomycin-mimic glycomacrocycle possessing potential anti-bacterial activity has been efficiently developed by Tripathi et al. by employing well-established CuAAC macrocyclization approach (Scheme 17.22) [76]. First, *bis*-azide **85** sugar alkyne **84** was allowed to undergo CuAAC coupling in the presence of $CuSO_4 \cdot 5H_2O$ and sodium ascorbate in *t*-BuOH/H_2O (v/v 1:1) to give species **86** which further gets transformed into *bis*-alkyne **87** species via various reactions including isopropylidene deprotection, $NaIO_4$ oxidation, $NaBH_4$ reduction, and propargylation. Further, the obtained compound **87** was utilized as substrate for preparation of the desired final product and therefore underwent intermolecular double-click macrocyclizations with bis-azide **85** under CuI/DIPEA catalytic system in MeCN/H_2O (1:1, v/v) solvent, at room temperature for 2 days to furnish vancomycin-mimic 44-membered glycomacrocycle **88**.

Cryptophycin-52 is a macrocyclic molecule having promising anti-tumor activity because of its endocyclic *trans*-amide linkage. Several significant efforts have been made to mimic its endocyclic *trans*-amide linkage with 1,4-disubstituted 1*H*-1,2,3-triazole ring and ultimately Sewald and group achieved this target through the aid of Click chemistry [77]. The group prepared a linear precursor molecule **89** to accomplish the synthesis of 'clicktophycin-52' **90** which is a cryptophycin-52 triazole analogue, through macrocyclization technique. The azido-alkyne cryptophycin precursor **89** undergoes intramolecular cycloaddition in the presence of CuI/DIPEA as catalyst at room temperature and an inseparable mixture of cyclomonomer and cyclodimer was obtained as a result. Further, the treatment of the obtained mixture with TFA/H_2Oresulted in acidic cleavage of the acetonides group, and free diol of cyclomonomer **90** and cyclodimer **91** were successfully generated (Scheme 17.23).

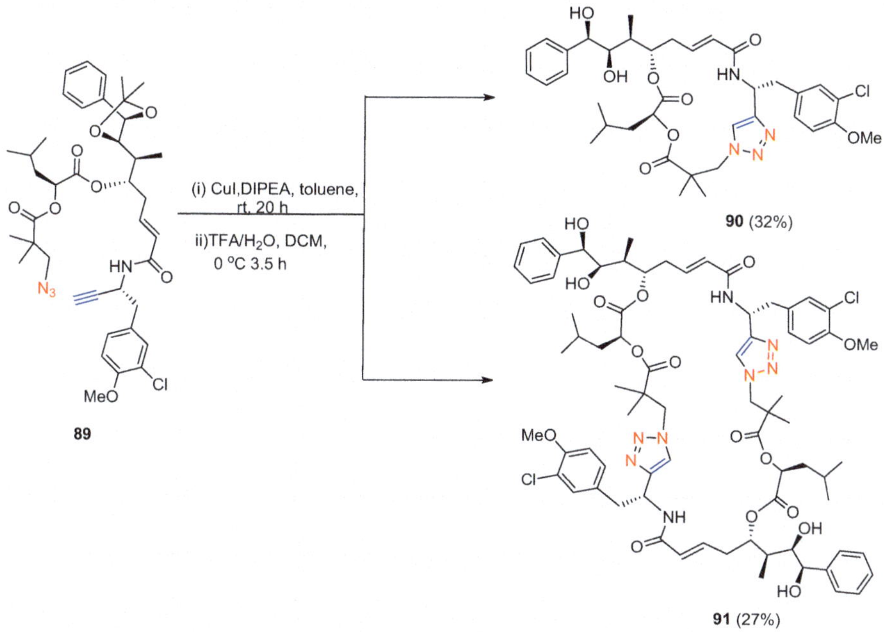

Scheme 17.23 Synthesis of 'clicktophycin-52' **90** (a triazole mimetic analogue of 'cryptophycin-52') with the aid of Click technique

Also, it was observed that the in vitro cytotoxicity of developed 'clicktophycin-52' **91** against the multidrug-resistant human cancer cell line KB-V1 is only slightly reduced as compared to 'cryptophycin-52'.

As observed in the above-illustrated schemes, the traditional intramolecular macrocyclization approach has some limitations mainly because of the simultaneous ongoing competing oligomerization reaction. Extreme high dilution is required for exclusive macrocyclization and to avoid the side reactions [78]. Therefore, various efforts have been made to address the underlying situation, and as a result, click-mediated continuous-flow chemistry came into consideration, which is easy to scale-up, proceed through rate acceleration, and also lies under the umbrella of 'green chemistry' [79].

Under the technique of continuous-flow chemistry, a copper coil continuous-flow technique was efficiently utilized by Bogdan and group for the synthesis of biologically active twelve- to twenty two-membered triazole-tethered macrocycles with the aid of intramolecular CuAAC macrocyclization approach (Scheme 17.24) [80]. Chiral amino alcohol azido-alkynes **92** were employed as starting material in the presence of *tris*[1-*tert*-butyl-1*H*-1,2,3-triazolyl)methyl)amine (TTTA, 0.1 equiv.) and DIPEA (2.0 equiv.), and the reaction was performed in a copper coil flow reactor under continuous-flow condition for 5 min at 150 °C and an excellent yield of the desired macrocyclic product **93** was obtained. Also, it was observed that the reaction

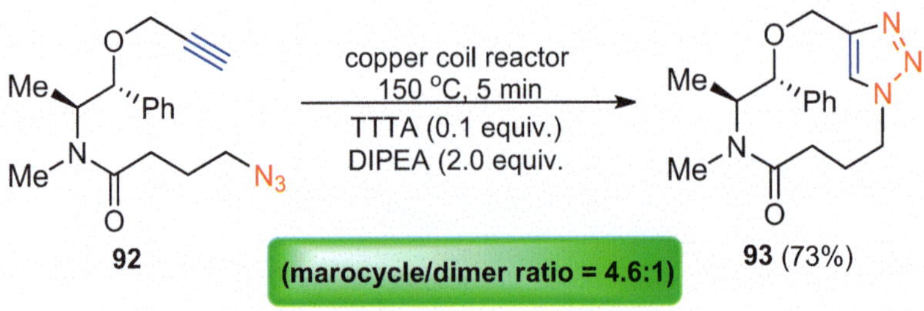

Scheme 17.24 Continuous-flow chemistry supported intramolecular CuAAC macrocyclization for synthesis of biologically active macrocycle **93**

carried out in the presence of suitable ligand and base accelerated the rate of macrocyclization and the competing dimerization products were produced in negligible amounts. The computational study also supported the correlation to the strain energy in the produced macrocyclization product [81].

The technique of continuous-flow chemistry was further explored by Bedard et al. and the group came up with a methodology for the synthesis of a macrocycle **95** based on the click CuAAC macrocyclization. 1-iodo-substituted azido-alkynes **94** underwent conjugation in an intramolecular fashion in the presence of CuI and *N*,*N*,*N*,*N*-tetramethylethane-1,2-diamine (TMEDA) as catalysts at 80 °C with 0.1 mL/min rate to furnish 83% yield of the desired macrocycle **95** (Scheme 17.25) [82]. In addition, the developed macrocycle **95** has suitable applications in the field of metal-catalyzed cross-coupling reactions. For instance, in the Pd-catalyzed Suzuki coupling with aryl boric acids for the synthesis of biologically active macrocycles.

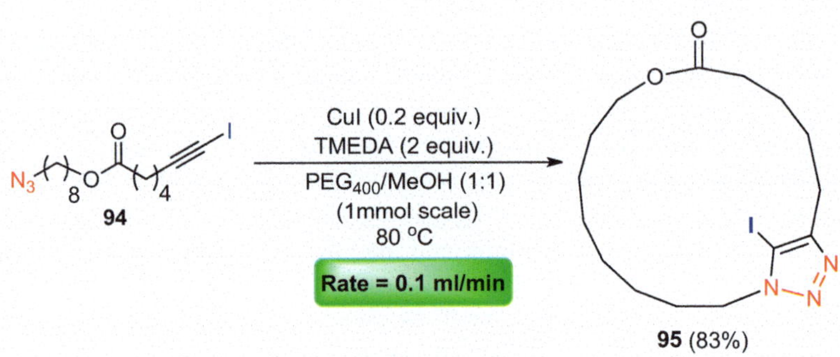

Scheme 17.25 The continuous-flow chemistry for synthesis of a triazole-linked macrocycle **95** based on the intramolecular click macrocyclization approach

17.3 RuAAC-Mediated Intramolecular Click Chemistry

The CuAAC cycloaddition reaction can easily couple two distinct functionalities such as alkynes and azides in a regioselective fashion leading to the preparation of 1,4-triazoles but the drawback of this protocol is that it is limited to terminal alkynes only. Various significant efforts have been made to overcome this limitation, and in this endeavor, Jia and Fokin successfully developed a methodology in which they carried out the regioselective synthesis of 1,5-disubstituted 1,2,3-triazoles from both terminal and internal alkynes using ruthenium complexes as catalysts [5, 83, 84]. Several ruthenium complexes have been studied in search of most efficient catalyst to synthesize various triazolyl molecules in which Cp*RuCl(COD) catalyst was found to be labile and a better choice for cycloaddition at room temperature. The study also revealed that the cyclooctadiene (COD) containing catalyst Cp*RuCl(COD) exhibits excellent catalytic activity at ambient temperature for RuAAC cycloaddition, whereas phosphine complex Cp*RuCl(PPh_3)$_2$ is effective for the coupling of diverse aliphatic azides and internal alkynes at 60 °C [85]. Furthermore, the regioselective 1,5-triazole forming reactions can also be performed using metal catalysts other than Ruthenium. For instance, 1,5-disubstituted triazole can be effectively synthesized through the coupling of azido-alkyne precursor in the presence of Cp_2Ni (10 mol%) catalyst and xantphos (10 mol%) and Cs_2CO_3 in toluene under open-air condition [86].

The one-dimensional ^{13}C NMR spectroscopy can be used to easily distinguish between the structures of regioisomeric 1,2,3-triazoles. The ^{13}C signal at δ = 132.7 ppm corresponding to C_4 carbon reveals the formation of 1,5-isomer, whereas the ^{13}C signal at $\delta = 121.0$ ppm corresponding to C_5 carbon indicates the confirmation of 1,4-regioisomer triazole product [87, 88].

Marcaurelle et al. developed a regioselective approach for the synthesis of macrocyclic triazole units under catalyst control conditions. The group employed a wide range of azido-alkynes of different ring sizes as substrates [89]. The developed protocol proceeds with the construction of a common scaffold that gives rise to a range of macrocyclic triazole products with n or $n + 1$ rings sizes under different reaction conditions. Also, during the catalyst screening, it was found that the ruthenium catalyst [RuClCp*]$_4$ produces a better monomer/dimer ratio than the conventional Cp*RuCl(PPh_3)$_2$ catalyst. It was observed that lower concentrations (2 mM) and higher temperatures (80 °C) favor the reaction and an excellent monomer-to-dimer ratio is obtained under these conditions. The azido-alkyne substrate **96** underwent intramolecular coupling under the optimized RuAAC condition to afford various macrocyclic 1,5-triazole-appended scaffold **97** in appreciable yield (Scheme 17.26).

The same group further extended the previously utilized RuAAC macrocyclization technique and achieved the synthesis of novel macrocyclic histone deacetylase inhibitors (HDACs) based on a three-phase diversity-oriented synthesis (DOS) protocol known as build/couple/pair (B/C/P) [90, 91]. The syn-aldol-derived azido-alkyne substrate (2R,5S,6R) **99** was allowed to undergo intramolecular RuAAC in

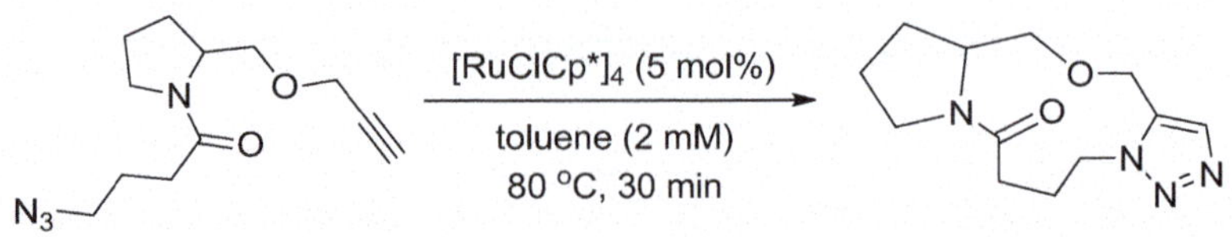

Scheme 17.26 RuAAC-mediated intramolecular macrocyclization reaction

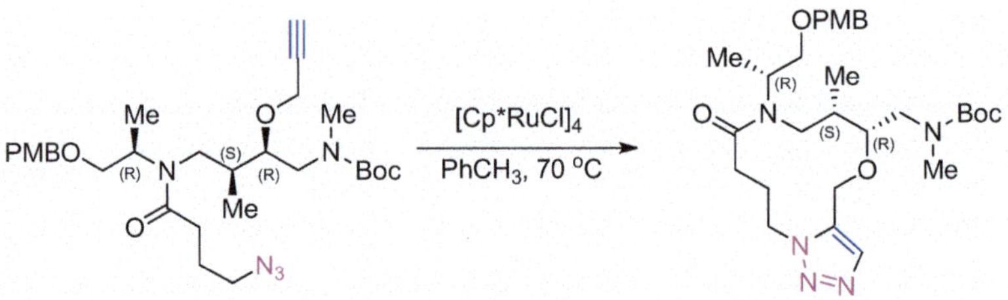

Scheme 17.27 Synthesis of a 12-membered 1,5-triazole-linked chiral macrocycle **99** under standard RuAAC macrocyclization strategy

the presence of $[Cp^*RuCl]_4$ catalyst in toluene at 70 °C to deliver a good yield of the macrocyclic product **99** (Scheme 17.27).

Zhang et al. appreciably exploited both the CuAAC and RuAAC techniques to develop cyclic tripeptides comprised of 1,4- and 1,5-disubstituted triazoles having exceptional resemblance to vancomycin mimetics [92]. The rigidity of the linear azido-alkyne precursor **102** ($n = 1$) leads to a mixture of both dimer and trimer products and thus makes the synthetic path of RuAAC-induced synthesis of corresponding product **104** ($n = 1$) less efficient (Scheme 17.28). The group also performed a BOP-mediated macrocyclization (lactamization) using the substrate having the longest side chain where the monocyclic product was obtained in good yield. It was also observed that the shorter ring size increases the degree of macrocyclization reaction (Scheme 17.28). Further, the same group used $[Cp^*RuCl]_4$ catalyst and exploited the intramolecular RuAAC double macrocyclization strategy and accomplish the synthesis of biologically potent 1,5-triazole-bridged vancomycin bicyclic mimetics [93].

Spring group utilized intramolecular RuAAC macrocyclization protocol to access biologically important novel macrocyclic peptidomimetic scaffolds **111–113** based on three-phase diversity-oriented synthesis (DOS) build/couple/pair method (Scheme 17.29) [94]. The reaction employed azido-alkynes **(108–110)** as substrates and $[Cp^*RuCl]_4$ as catalyst, and it was observed that the reaction progresses in a complete regioselective fashion and produced triazole-appended macrocycles **(111–113)** without epimerization of their respective chiral centers. Also, the proton NMR data revealed a single stereoisomer product under the utilized reaction conditions. In

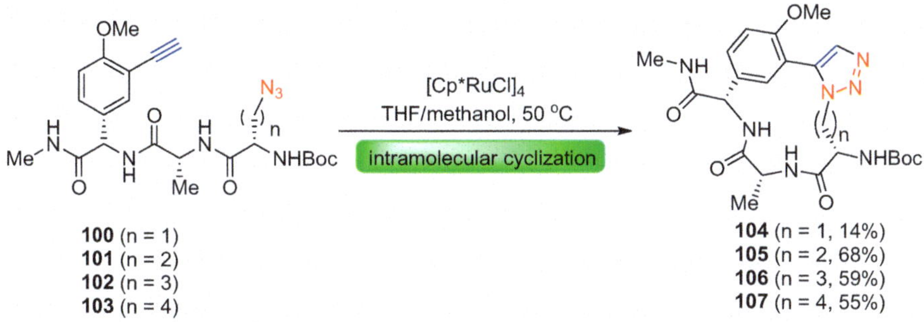

Scheme 17.28 RuAAC-induced macrocyclization and BOP-mediated macrolactamization of peptidic scaffolds **102**

addition, the production of diketopiperazine (DKP) from solid-supported *N*-methyl morpholine, acetic acid, and microwave heating has increased the diversity of the obtained triazolyl macrocyclic products. The biological evaluation of the developed 1,5-triazole-linked macrocyclic product displayed significant anti-microbial activity against *Staphylococcus aureus*.

The synthesis of non-peptidic macrocyclic compounds of chemotherapeutic potential is quite challenging through conventional synthetic methods. The same group in their next work applied the multidimensional coupling approach of two frameworks for the synthesis of Diversity-Oriented Synthesis (DOS)-derived non-peptidic macrocycles and achieved a higher degree of diversification [95]. The group utilized CuAAC and RuAAC reaction protocols for the final macrocyclization step for the construction of the couple phase and prepared a total of 73 macrocyclic products consisting of 59 distinct chemical moieties. Synthesis of compounds **115a–c** through

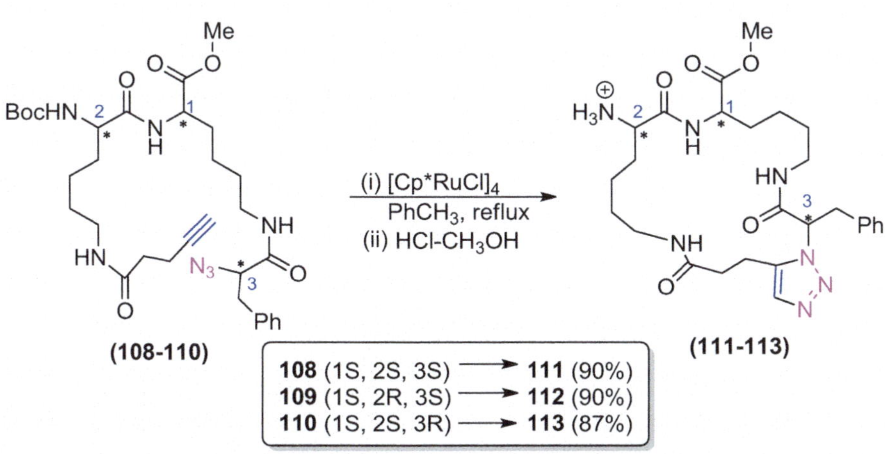

Scheme 17.29 Intramolecular RuAAC-induced synthesis of 1,5-triazole-linked macrocyclic products **111–113** from azido-alkyne precursors **108–110**

RuAAC path at low catalyst loading in the final step depicted in (Scheme 17.30) along with some representative examples.

Fang and group accomplished the direct and easy synthesis of oxazolidinone fused triazolyl macrocycles under standard RuAAC macrocyclization condition in excellent yield by using fused aziridine ring as substrate [96]. The reaction was performed using azido-alkyne oxazolidinone precursor **116** and [Cp*RuCl(PPh_3)$_2$] as ruthenium catalyst in toluene at 80 °C for 20 min to obtain corresponding 11- to 12-membered 1,5-triazolyl fused macrocyclic products **117** and **118** (Scheme 17.31).

Fused bicyclic triazoles, a novel group of pharmacophores possessing potential and specific α-glucosidase inhibition activity have been efficiently synthesized by Shaw et al. through the aid of intramolecular click chemistry [97]. Both thermal and RuAAC routes of azide-alkyne conjugation were utilized by the group for the synthesis of fused bicyclictriazolyl iminosugar and it was observed that the yield of reaction was different in both cases. Under thermal condition, the

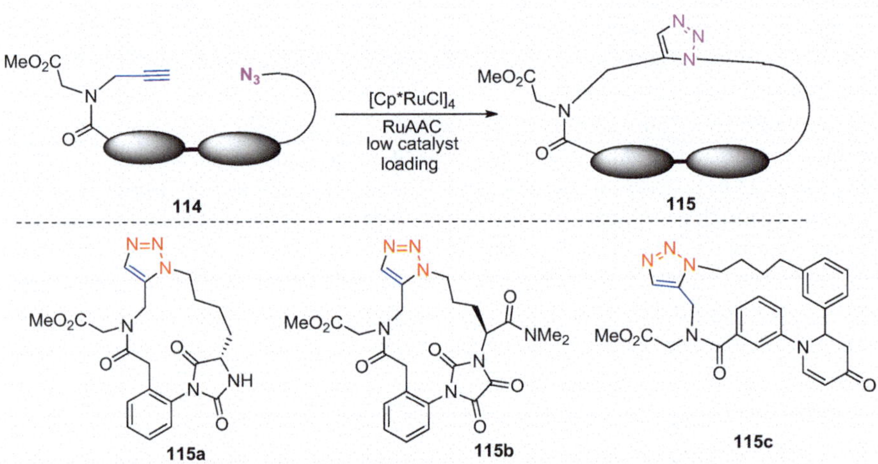

Scheme 17.30 RuAAC-based construction of non-peptidic macrocyclic compounds **115a–c**

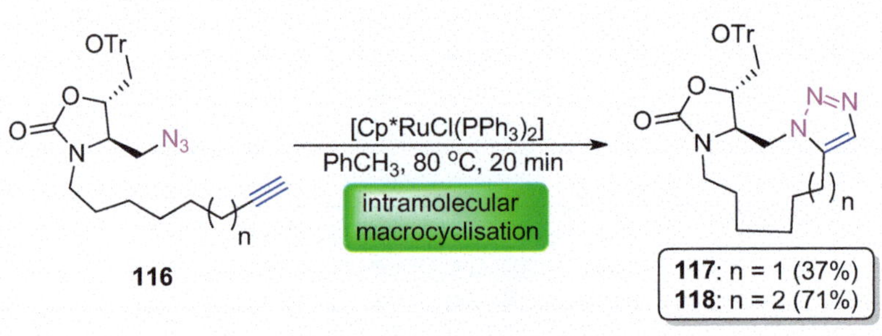

Scheme 17.31 Construction of 11 and 12-membered triazolyl macrocycles via RuAAC intramolecular macrocyclization route

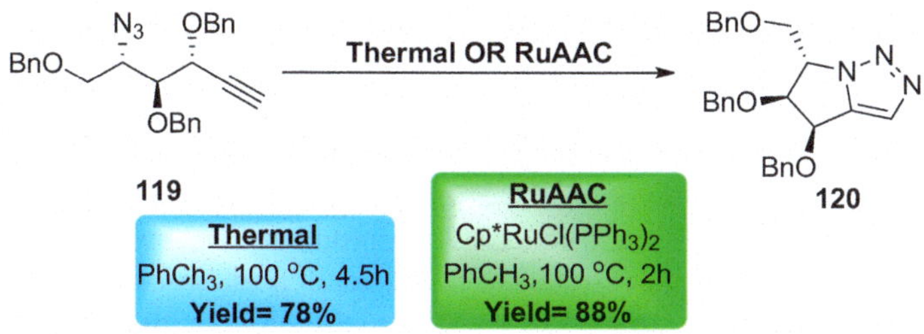

Scheme 17.32 Synthesis of fused bicyclic pyrrolidine iminosugar hybrid **120** through the aid of intramolecular RuAAC approach

cycloaddition reaction of the azido-alkyne substrate **119** furnished 78% yield of fused bicyclic triazolyl product **120**, whereas when reaction was carried out using [Cp*RuCl(PPh_3)$_2$] as catalyst under RuAAC protocol, the yield of reaction increases to 88% (Scheme 17.32).

The above-illustrated schemes and reports identify the intramolecular click chemistry along with the appropriate employment of Ruthenium-based catalysts as a very essential technique for macrocyclization reactions leading to the synthesis of important macrocycles comprising of 1,5-disubstituted triazole linker and the fused ring systems. The described metal-catalyzed approaches, i.e., CuAAC and RuAAC, provide excellent regioselectivity in triazole forming reactions, but the use of copper and ruthenium metals raises the toxicity concerns and limits the applicability of both methods in drug discovery and development. Considering this scenario, the development of a thermal-based metal-free approach becomes a field of critical importance in modern organic synthesis and that will be covered in upcoming sections of this Chapter.

17.3.1 Thermally Induced Intramolecular Click of Diverse Azido-Alkyne Scaffolds

1,3-Dipolar cycloaddition reactions of organic azides as 1,3-dipole and terminal alkynes as dipolarophile or substituted unsaturated molecule under thermally induced conditions have been widely explored to develop diverse range of 1,2,3-triazole conjugates. The similar thermally induced intramolecular cycloaddition of azido-alkyne precursors frequently explored for an access of 1,2,3-triazole-fused benzo-macrocycles [98], functionalized 1,2,3-triazole-fused macrocycles [99–101], as well spiro-fused-1,2,3-triazoles [102]. Thermally induced cycloaddition of azido-alkyne precursors precedes at high temperature and is highly exothermic with high activation barrier. HOMO of the 1,3-dipolar component (e.g., azides) interacts with LUMO of the dipolarophile (e.g., alkynes) to get the regioselective 1,2,3-triazole conjugates. Since the energy differences between the HOMO and LUMO energy levels of azide and terminal alkyne are comparable and thus their interaction often resulted to the formation of nearly equimolar mixture of both the regiomeric 1,2,3-triazoles. Consequently, the uncatalyzed triazole forming cycloaddition reactions proceed with sluggish rate and are typically non-regioselective in triazole formation.

In addition to the expected seven-membered cyclized 1,2,3-triazole-fused molecular system, some of the intramolecular click-mediated cyclization of azido-alkyne resulted to the respective dimeric macrocyclic product [98]. For example, azido-alkyne **121** in refluxing DMF resulted to the formation of seven-membered 1,2,3-triazole-fused monomeric product **122** (in 36% yield) along with a dimeric product **123** (in 24% yield) via intramolecular azide-alkyne reaction (Scheme 17.33).

MW-induced intramolecular click of α-amino acid-derived azido-alkynes gave high yield of functionalized 1,4,5-trisubstiututed triazoles [103]. Likewise, glycosyl azido-alkynes under metal-free and thermally induced condition successfully gave high yield of biologically relevant morpholine-fused [5,1-*c*] 1,2,3-triazole glycoconjugates [104–106].

The sugar azido-alkyne motif **125,** obtained from sugar azido alcohol **124** through its routine propargylation, on regioselective intramolecular click-mediated cyclization under refluxing toluene for 24 h gave biologically relevant spirocyclic triazooxazine nucleoside adduct **126** in 51% yield (Scheme 17.34) [107].

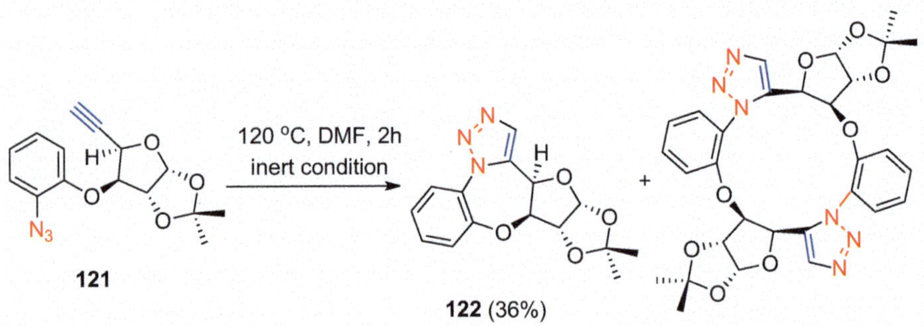

121

122 (36%)

123 (24%, dimer adduct)

Scheme 17.33 Thermally induced Intramolecular click reaction of azido-alkyne **121** for the synthesis of macrocyclic **122** and **123**

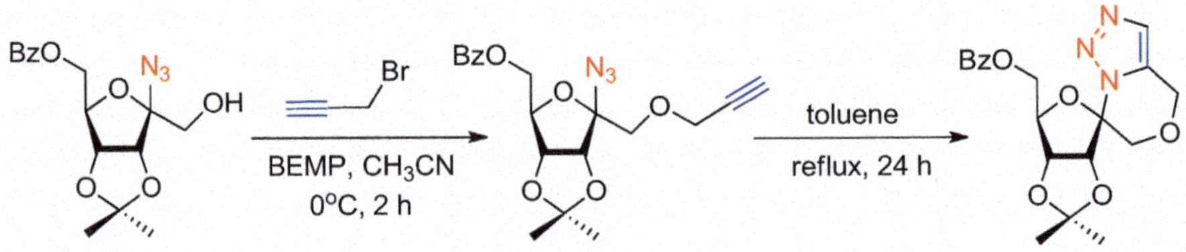

124

125

126 (51%)

Scheme 17.34 Synthesis of spiro-glycohybrid triazole **126** from respective sugar azido alcohol **226**

Likewise, the azido-alkyne derivative **128** obtained from cyclic epoxide **127** through sequential epoxide ring opening followed by propargylation on thermally induced intramolecular triazole forming click cycloaddition in xylene/toluene at 110 °C afforded excellent yield of structurally diverse tricyclic [1,2,3] triazolo[1,5-*d*][1,4] product **128** (Scheme 17.35) [108].

In a similar fashion, intramolecular click cyclization of glycosyl azido-alkyne **130** in toluene at 120 °C afforded the respective tricyclic glycohybrid 1,2,3-triazole **131** in 88% yield [109]. The resulting 1,5-disubstituted triazoles **131** could be used for the development of biologically relevant 1,2,3-triazole-appended spiro-nucleosides. For this purpose, acetylated tricyclic product **132a, b,** obtained in overall 64% yield by treating compound **131** with AcOH/Ac_2O followed by TFA in DCM on glycosylation with thymine, BSA, and catalytic amount of TMSOTf afforded the desired 1,5-disubstituted-1,2,3-triazole-spiro-nucleoside **133** in 85% yield (Scheme 17.36).

Basavaiah group explored intramolecular click chemistry and reported a metal-free expeditious synthesis of triazole-fused benzoxazonines **138** (Scheme 17.37)

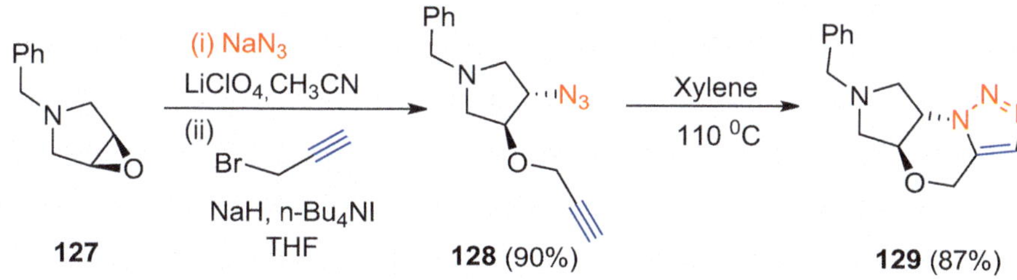

Scheme 17.35 Thermally induced Intramolecular click cycloaddition of azido-alkyne leading to respective tricyclic 1,2,3-triazoles

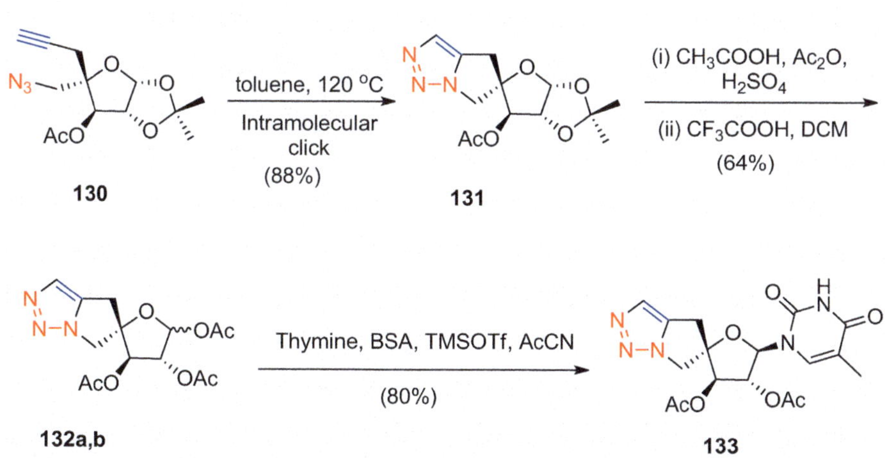

Scheme 17.36 Intramolecular click-mediated synthesis of spiro-nucleoside **133** from azido-alkyne sugar **130**

[110]. Thus, compound **136,** obtained via DABCO-catalyzed Baylis–Hillman–Basavaiah reaction of *O*-propargylated salicylaldehyde **134** with alkyl acrylates **135** followed by treatment with AcCl under basic condition, on azidation reaction afforded the required azido-alkyne **137** that finally underwent thermally induced intramolecular cycloaddition to give the desired biologically relevant triazole-fused benzoxazonines **138**.

Banerji et al. reported silver chloride-catalyzed intramolecular click cyclization of azides **139** and developed the respective triazolo-benzoxazepine scaffolds **140** (Scheme 17.38) [111]. The resulting molecules were evaluated against three cancer cells such as Neura 2a (a neuroblastoma cell line), Hep G2 (a liver cancer cell line), and Hek 293 (a kidney cancer cell line) using MTT assay and displayed in vitro micromolar anti-cancer activity and investigation concluded that they caused apoptosis.

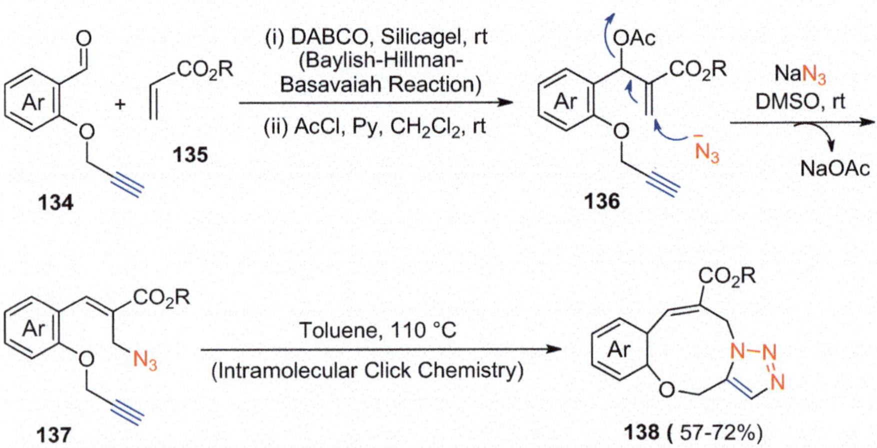

Scheme 17.37 Intramolecular click-mediated synthesis of triazole-fused benzoxazonines **138** from Baylis–Hillman–Basavaiah adduct **136**

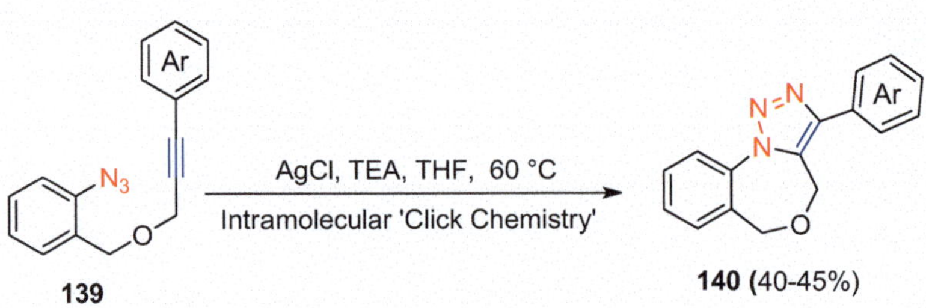

Scheme 17.38 AgCl-mediated intramolecular click cyclization for an easy access of biologically relevant triazolo-benzoxazepines **140**

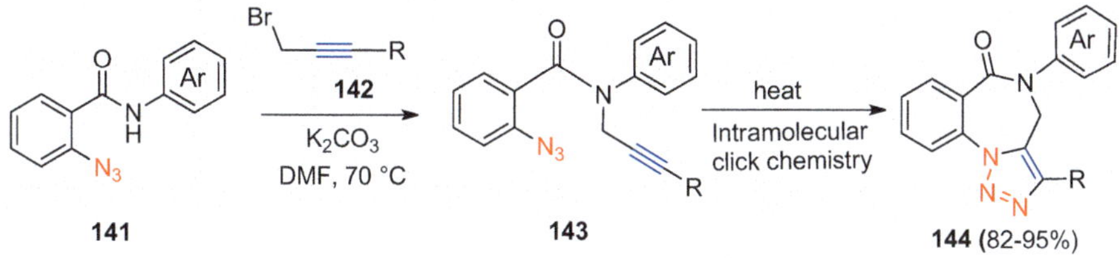

Scheme 17.39 Convenient cyclization of azido-alkyne precursor **143** leading to respective triazole-fused 1,4-benzodiazepinone **144**

The intramolecular click cycloaddition of azido-alkyne precursor **143** obtained from propargylation of *N*-substituted-2-azidobenzamides **141** gave good-to-excellent yield of the desired 1,2,3-triazole-fused 1,4-benzodiazepinones **144** (Scheme 17.39) [112]. The method has some notable features including mild reaction condition, high reaction yield, ease of isolation of the products and moreover avoids the use of catalyst in protocol.

Likewise, *N*-propargylated anilines **146**, obtained via alkenylation of aniline on one-pot Cu-catalyzed Ullmann C-N coupling reaction with *O*-azidobenzylbromides **145** with *O*-azidobenzylbromides **145** using CuI and K_2CO_3 and DMF at 110 °C afforded respective azido-alkyne **147** which on intramolecular click cycloaddition afforded good-to-excellent yield of the desired cyclized triazolo[1,5-a][1,4]benzodiazepinones **148** (Scheme 17.40) [113].

Further intramolecular click chemistry is nicely explored for an easy access of biologically relevant tricyclic glycohybrid 1,2,3-triazoles **151** from corresponding monosaccharide azido-alkyne derivative **149** [114]. The azido-alkyne **149** underwent intramolecular thermally induced cycloaddition and gave high yield of the required 1,2,3-triazole-fused 1,4-oxazepine **150**. The debenzylation was successfully done using Pd/C in MeOH and resulted to the respective diols in almost quantitative yield (Scheme 17.41).

Arbaciauskiene et al. reported a metal-free and thermally induced intramolecular click cycloaddition condition of the azido-alkynes (**152, 154**) to give good yields of respective fused bioactive heterocycles, namely pyrazolo[4,3-*f*][1,2,3]-triazolo-[5,1-*c*][1,4]oxazepines (**153, 155**) (Scheme 17.42) [115].

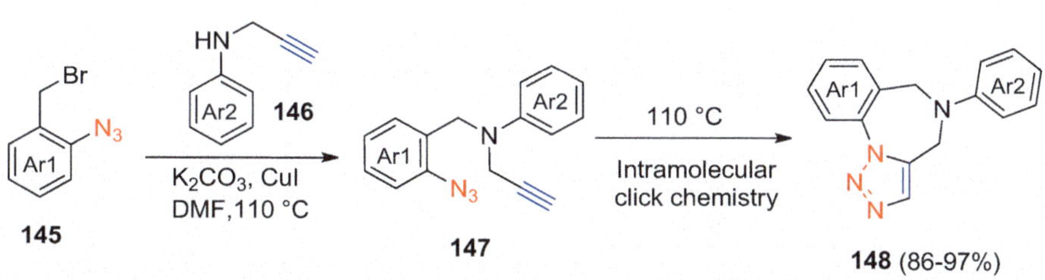

Scheme 17.40 Intramolecular click-inspired synthesis of triazolo[1,5-*a*][1,4]benzodiazepinones **148**

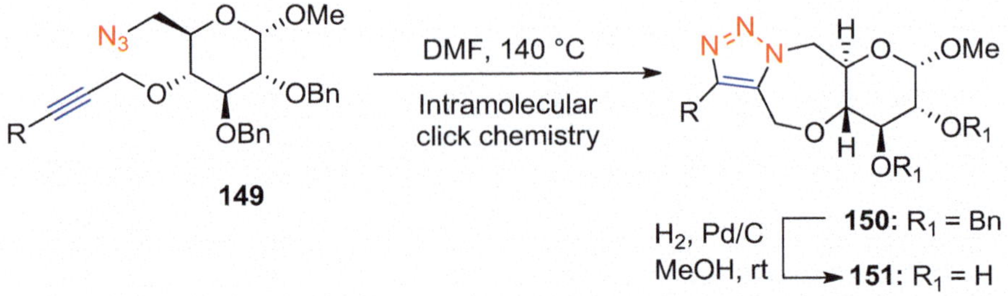

Scheme 17.41 Synthesis of tricyclic 1,2,3-triazole glycohybrids **151** from 6-azido glucopyranoside

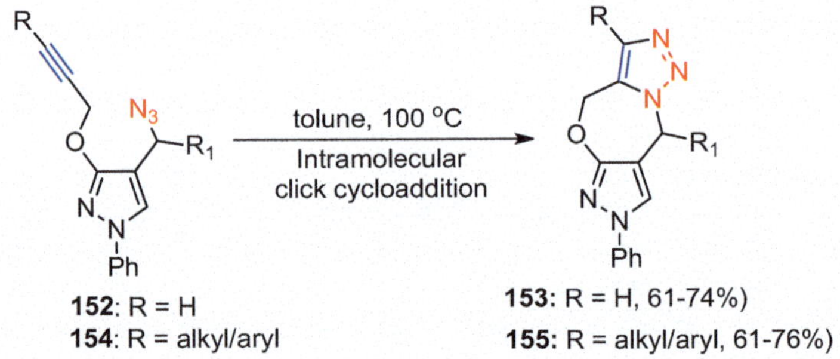

Scheme 17.42 Thermally induced intramolecular click tool for the synthesis of fused heterocycle pyrazolo[4,3-*f*][1,2,3]triazolo[5,1-*c*][1,4]oxazepines **153** and **155**

Sastry and co-workers used cascade Knoevenagel-intramolecular AAC of alkylated aldehyde **156** and phenacyl azides **157** under one-pot condition resulting to the formation of biologically relevant tricyclic heterocycles **159**. Thus, α-keto vinyl azido-alkyne intermediate **158**, obtained by piperidinium acetate-mediated reaction of alkylated aldehyde **156** and phenacyl azides **159** under Knoevenagel condensation condition, on intramolecular click cycloaddition afforded high-to-excellent yield of desired 4*H*-benzo[*g*][1,2,3]triazolo[5,1-*c*][1,4]oxazocines **159** (Scheme 17.43) [116]. The resulting 1,2,3-triazoles fused tricyclic analogues displayed promising in vitro cytotoxic activity ($IC_{50} \leq 13\ \mu M$) against HeLa, MDA-MB-231, DU145, and MCF-7 cancer cell lines. Also, fluorescence titration and viscosity investigation confirmed them to possess high DNA binding affinity.

Hajela et al. reported $Sc(OTf)_3$-catalyzed tandem *C*-2 functionalization intramolecular azide-alkyne cycloaddition route for the diversity-oriented synthesis of annulated tetracyclic diazepines and their spirocyclic benzodiazepinones (Scheme 17.44) [117]. Thus, aryl azides **160** on $Sc(OTf)_3$-mediated *C*-2 functionalization with variety of aryl- or heteroaryl- or aliphatic- or cyclohexanone- or adamantanone- or cyclododecanone-derived propargyl alcohols **161** gave respective azido-alkyne adduct **162**, which finally underwent intramolecular cycloaddition

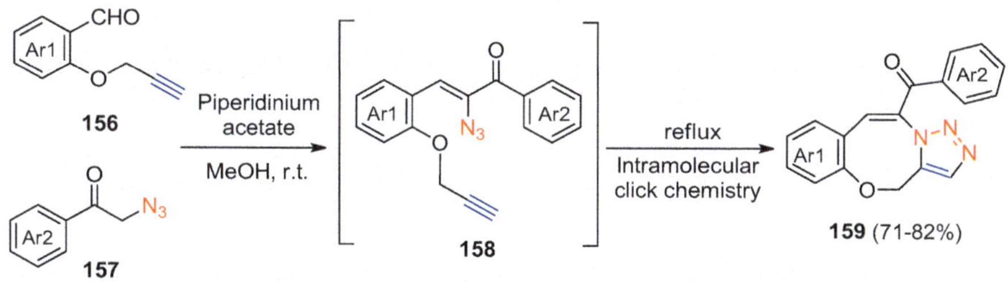

Scheme 17.43 One-pot synthesis of biologically relevant tricyclic heterocycle system **159**

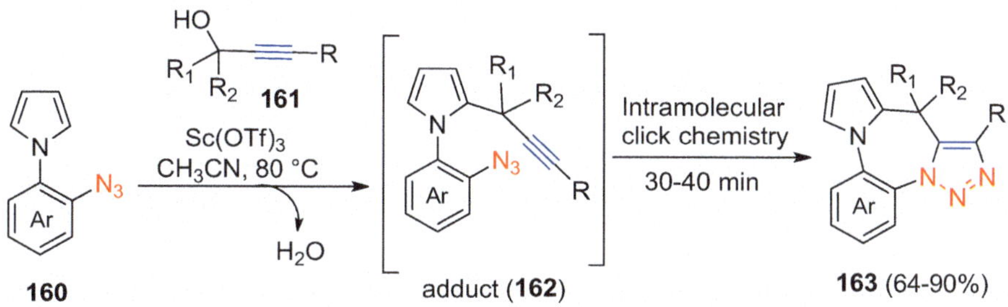

Scheme 17.44 $Sc(OTf)_3$-catalyzed tandem synthesis of benzodiazepinones **163**

under one-pot tandem route gave high-to-excellent yields of biologically relevant tetracyclic annulated benzodiazepinones **163** (Scheme 17.44).

Further in continuation to the synthesis of biologically active benzodiazepinones scaffolds, a straightforward cascade route exemplified with utilizing intramolecular click chemistry is established for an easy access of quinazolino[1,2,3]-triazolo-[1,4]benzodiazepinones **166** [118]. The pentacyclic quinazolino[1,2,3]-triazolo-[1,4]benzodiazepinones **166** was developed by reacting *O*-aminobenzamides **164** with *O*-azidoacetophenone **165** in the presence of 10 mol% of I_2 in MeOH at 50 °C (Scheme 17.45). The molecular iodine-mediated atom economy transformation route incorporates five reactive centers including amide, aniline, carbonyl, azide, and alkyne. The tandem sequence involves the reaction of anilino-keto condensation of *O*-aminobenzamides **164** and azido ketone **165** to form Schiff base **I,** which on subsequent cyclization through amide N-atom to the imine resulted to quinazolinone aminal **(II)** and this finally underwent intramolecular click reaction to give the desired pentacyclic analogues **166**.

Likewise, 2-amino-*N*-benzyl propargylamines **167** on sequential diazotization, azidation, and intramolecular click reactions under one-pot condition afforded high yield of triazole-fused [1,4]benzodiazepinones **168** (Scheme 17.46) [119]. This diversity-oriented reaction proceeds through diazotization of 2-amino-*N*-benzylpropargylamines **167** using $NaNO_2$/HCl at 0–3 °C followed by treatment with NaN_3 produced corresponding azido-alkyne intermediate **II** which on thermally

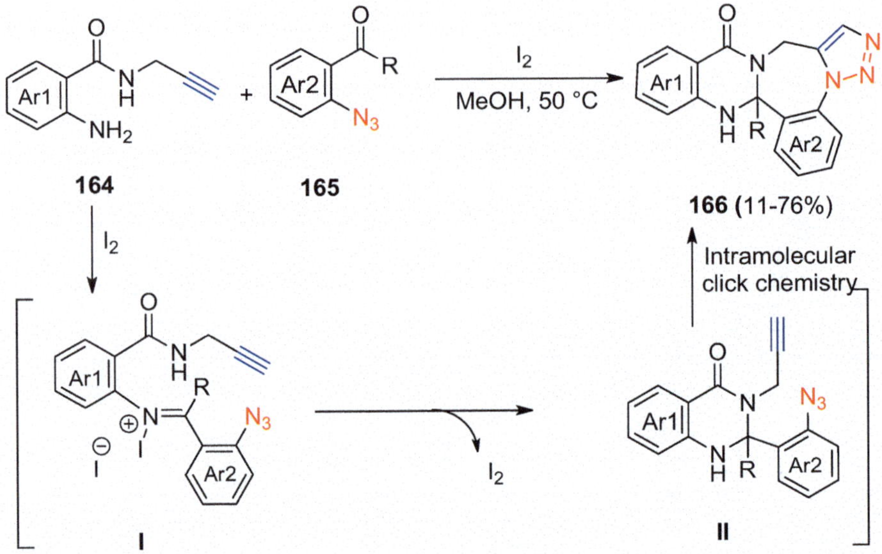

Scheme 17.45 I_2-mediated one-pot cascade route for the synthesis of pentacyclic benzodiazepinones **166**

induced intramolecular click cyclization at 100 C in water resulted to anti-bacterial tricyclic triazole-fused [1,4]benzodiazepinones **168**.

Base-catalyzed one-pot reaction of phenacylazides **169** with *N*-propargyl aldehyde **170** gave good of respective 1,2,3-triazole-fused *N*-heterocycles **173** (Scheme 17.47). This straightforward reaction proceed via piperidine-induced

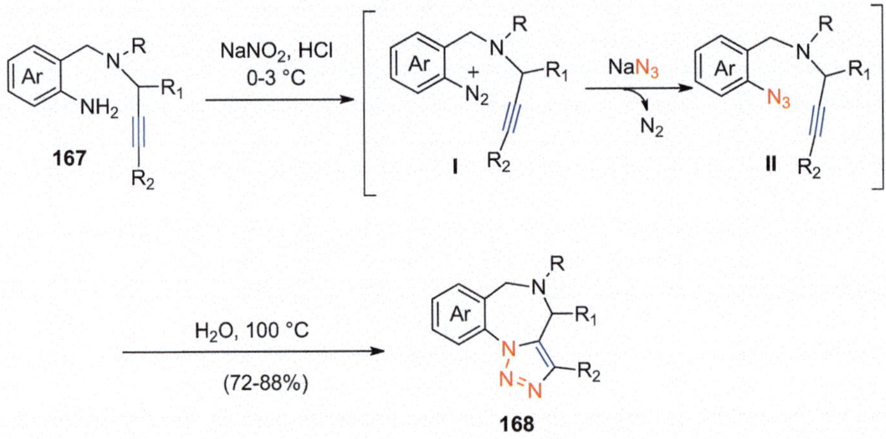

Scheme 17.46 One-pot diversity-oriented synthesis of pharmacologically relevant 1,2,3-triazole-fused [1,4]benzodiazepinones **168**

Knoevenagel condensation then respective product underwent intramolecular triazole forming click reaction at room temperature. Electron donating group present at phenyl ring of phenacylazides **169** resulted to high yields than electron-withdrawing group in compound **177**. Similar reaction of phenacylazides **169** with *N*-propargyl indole-2-carboxaldehyde **174** resulted to 1,2,3-triazole-fused-indolo system **175** (Scheme 17.47) [120]. The resulting fused heterocycles **173** and **175** displayed promising in vitro anti-cancer activities against the NCI 60 cell lines and also possess DNA binding affinity.

3-(Azidoalkoxy)aryne **176**, obtained by Mitsunobu etherification of resorcinols, on intramolecular click cycloaddition afforded high yield of respective 1,2,3-triazole-fused benzotriazoles **177** (Scheme 17.48) [121]. The protocol can also be used for the synthesis of *bis*-1,2,3-triazole-appended benzotriazole skeleton via sequential azide-alkyne and azide-aryne cycloadditions. The cycloaddition required for the formation of 1,2,3-triazole-fused 8-membered heterocyclic skeleton is although less efficient.

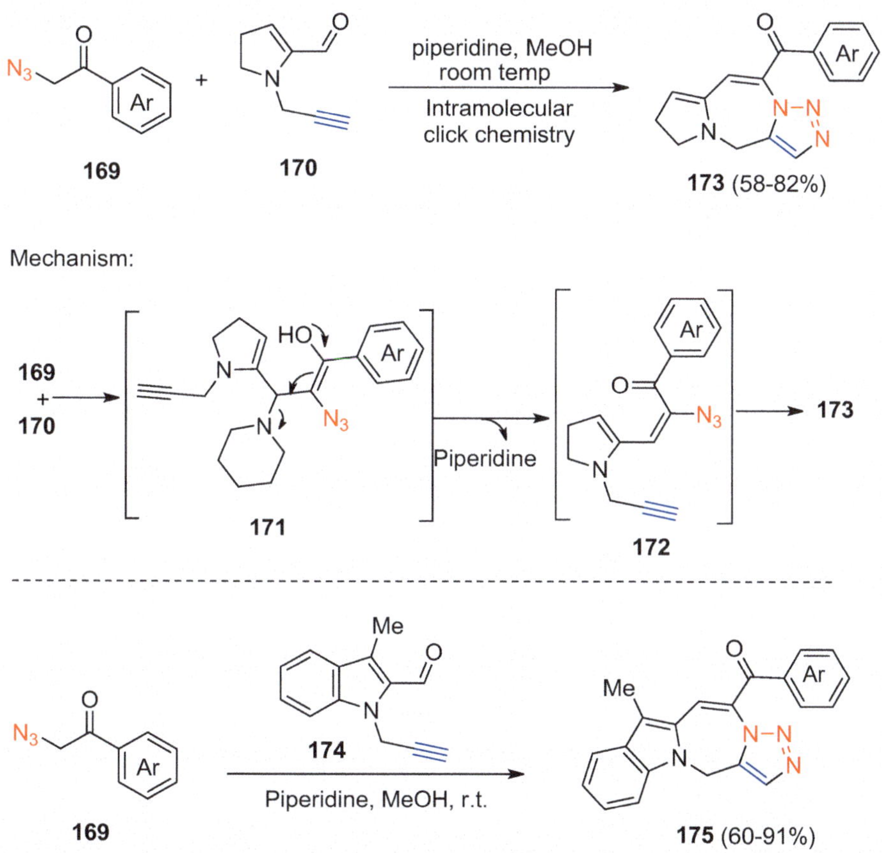

Scheme 17.47 Intramolecular click-mediated synthesis of 1,2,3-triazole-fused pyrrolodiazepine **173** and related indolo-fused heterocycle **175**

Scheme 17.48 Synthesis of fused benzotriazoles **303** via in situ benzyne formation followed by intramolecular click chemistry

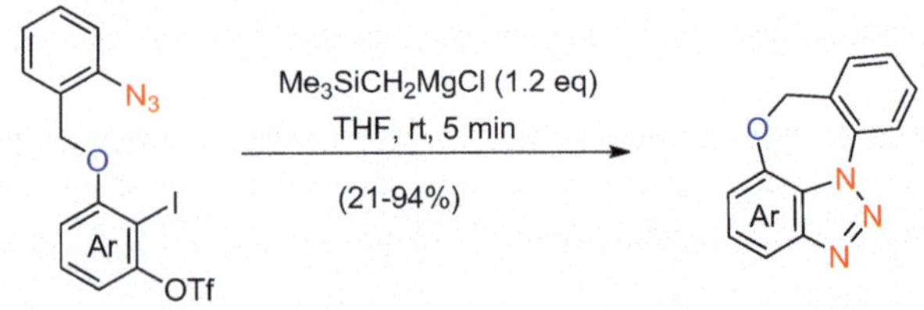

176 **177**

Samala and co-workers reported an expeditious one-pot diversity-oriented synthesis of indole-annulated polyheterocycles such as indolotriazoloquinoxalines **180** and ketoindoloquinoxalines **181** by utilizing 1-(2-nitroaryl)-2-alkynylindoles **178** [122]. The reaction is envisaged to proceed either through thermally induced metal-free intramolecular click cyclization route or through cascade cyclization via aminooxygenation followed by CuI/oxygen-mediated elimination of nitrous oxide. The authors suggested that aromatic nucleophilic substitution of the nitro group with NaN_3 afforded the azido-alkyne intermediate **179** which consequently underwent thermally induced [3 + 2] cycloaddition to give high yield of desired tetracyclic product **180** (Scheme 17.49). On the other hand, compound **178** on treatment with CuI/O_2 and NaN_3 in HMPA gave good-to-high yield of indole-based annulated product **181** via aminooxygenation route.

Soural and co-workers established a catalyst-free solid-supported route for easy access of 4*H*-benzo[*f*][1,2,3]triazolo-[1,5-*a*][1,4]diazepin-6(5*H*)-ones **185** and

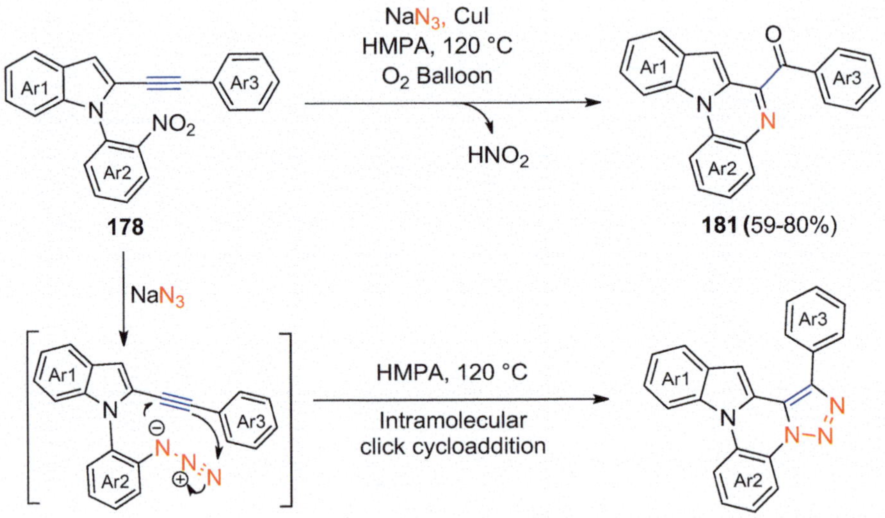

Scheme 17.49 HMPA-mediated thermally induced reaction of 1-(2-nitroaryl)-2-alkynylindoles **304** with NaN_3 for the development of respective annulated tetracyclic analogues **180** and **181**

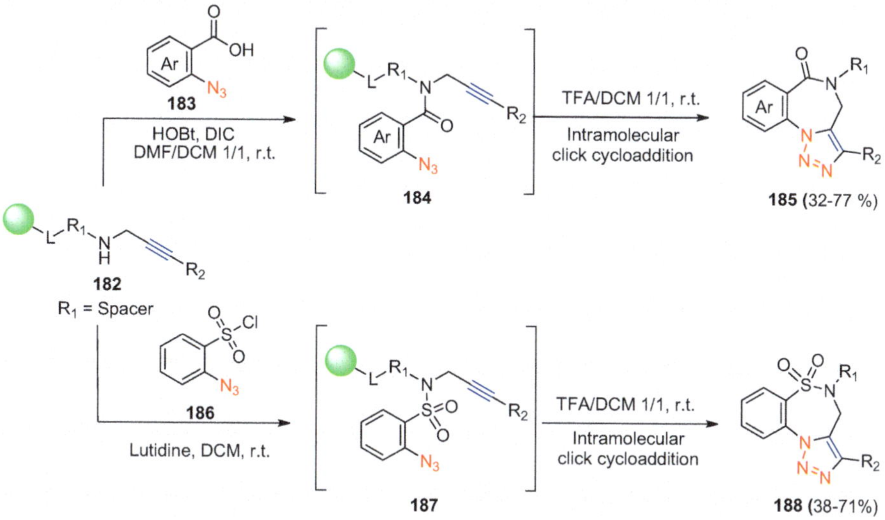

Scheme 17.50 Solid-phase synthesis of 1,4-benzodiazepinones **185** and their sulfonyl analogues **188**

their sulfonyl analogues **188** under mild conditions [123]. A resin bound alkenylated unmasked amine intermediate **182** on standard acid–amine coupling with 2-azidobenzoic acids **183** resulted to azido-alkyne **184** on subsequent removal of polymer resin using TFA/DCM followed by intramolecular click gave the respective 1,2,3-triazolo benzodiazepinones **185** in overall 32% yield (Scheme 17.50). In a similar way, the sulfonyl analogue **188** was obtained through 2,6-lutidine-mediated reaction of amine **182** with 2-azidobenzenesulfonyl chloride **186** followed by intramolecular click of the resulting azido-alkynes **187** and removal of polymer using TFA in DCM (1:1) ratio.

Likewise, reductive amination of *O*-azidobenzaldehydes **189** with salt of 4-aminobut-2-*yn*-1-ol **190** using Hünig's base and $NaBH(OAc)_3$ resulted to the functionalized molecule 321. Respective azido-alkyne **191** on subsequent thermally induced [3 + 2] cycloaddition in refluxing toluene gave good-to-excellent yield of desired tricyclic triazolo-1,4-benzodiazepinone scaffolds **192** (Scheme 17.51) [124].

Piperidinium acetate-catalyzed cascade reaction of 2-alkynyl aryl(heteroaryl) aldehyde **193** and with phenacyl azides **169** in methanol under one-pot condition resulted to the formation of biologically relevant 1,2,3-triazole-fused heterocyclic scaffold **194** [125]. This metal-free cascade is high atom economy and can be scale-up. This is proceed through Knoevenagel condensation of 2-alkynyl aldehydes with phenacyl azides and then consequent intramolecular click reaction (Scheme 17.52) [125].

Likewise, a thermally induced intramolecular click reaction of alkenylated tetrahydro-*β*-carboline intermediate **345** in refluxing toluene gave high yield of 1,2,3-triazolo-fused-tetrahydro-*β*-carbolines **346**. Further, an azido-alkyne **350** on

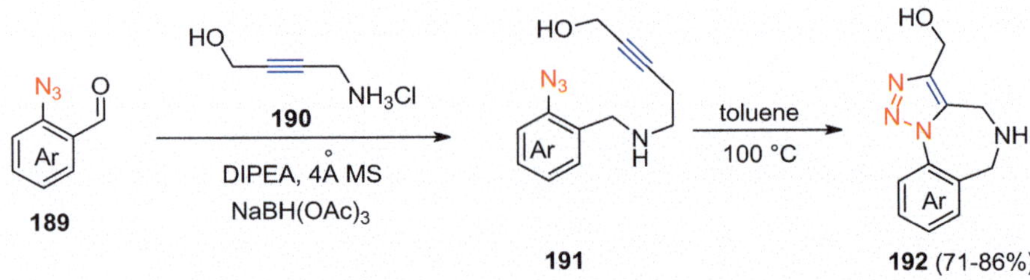

Scheme 17.51 Synthesis of hydroxymethyl substituted triazolo-1,4-benzodiazepinones **192**

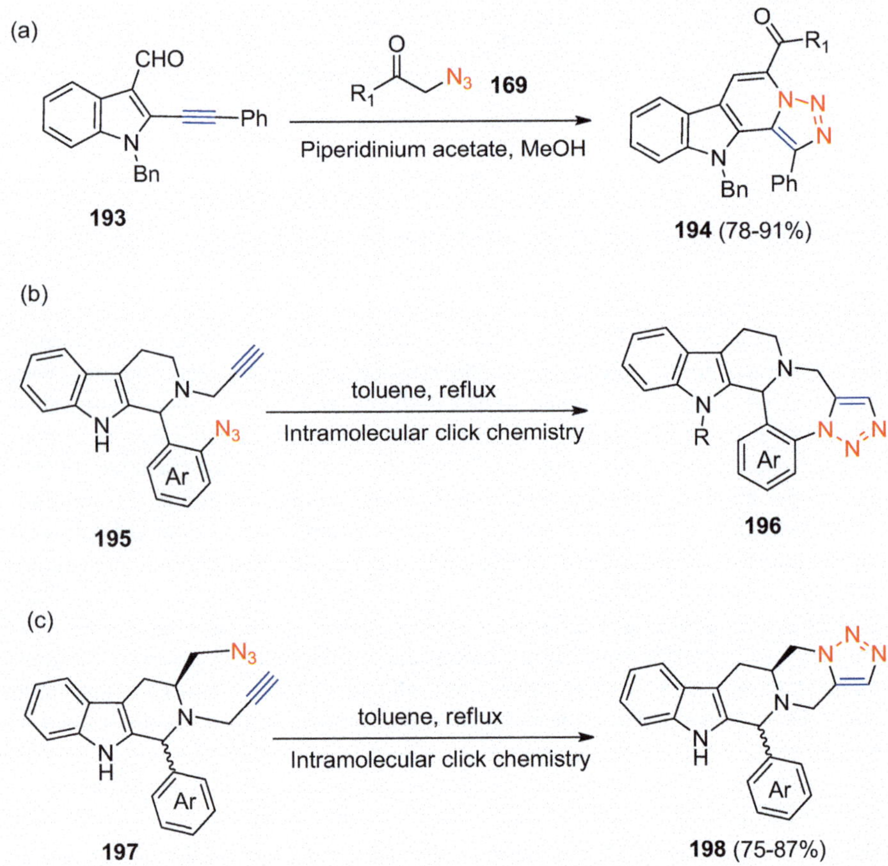

Scheme 17.52 Intramolecular click-mediated one-pot synthesis of 1,2,3-triazolo-fused-tetrahydro-β-carboline

thermally induced metal-free condition afforded high-to-excellent yield of penta-heterocyclic system **351**, a potent cytotoxic agent against B16F10 cell line [126].

Dealing with Cu(I)-catalyzed azide-alkyne cycloaddition (CuAAC) in Chap. 2 and Ru-catalyzed azide-alkyne cycloaddition (RuAAC) in Chap. 3, a number of 1,2,3-triazole-linked molecular architectures has been developed. In earlier section of this chapter, a great number of 1,2,3-triazole-linked macrocycles were developed by applying CuAAC and RuAAC routes. Further, in continuation of click-mediated crown ether-appended macrocycle linked through C-2 and C-3 positions of D-glucose [127], a "click stitching" strategy was successfully explored to develop C1-to-C6 glucose-linked macrocycle architecture [128]. Despite of advantages with metal-catalyzed azide-alkyne cycloaddition as a part of macrocyclization route, a metal-free protocol for similar molecular macrocycle would be positively more profitable to explore them in drug discovery and development. For example, Tripathi et al. devised a metal-free macrocyclization route for the high-yielding synthesis of 14- to 16-memebered macrocycles starting from monosaccharides [129]. Thus, glucose-derived azides **200** was obtained in high yield from glucose butanone **199** via regioselective 6-*O*-tosylation using TsCl/NEt_3 and pyridine followed by acetylation using Ac_2O/pyridine and then azidation by treating the orthogonally protected tosyl analogue with NaN_3 in DMF at 80 °C for 2 h. The resulting azide **355** was subjected to CuAAC click reaction with alkyne tosylates **201** using $CuSO_4$/NaAsc, then azidation to give high yield of the respective triazolyl butenones **202**. Finally, tetrabutylammonium hydrogen sulfate (TBAHS)-mediated macrocyclization of compound **202** gave high yield of respective sugar-containing macrocycle **203**. Removal of acetate groups from macrocycle **203** resulted to the biologically relevant polyhydroxy analogues **204** (Scheme 17.53). The method is equally applicable with other monosaccharides, for example, D-mannose to give high yield of desired D-mannose-based macrocycle.

In addition to the above-described regioselective azide-alkyne cycloaddition route for the 1,2,3-triazoles, the similar scaffolds can also be obtained in good yields starting from nitrostyrenes or related systems like vinyl sulfones [130, 131]. Tiwari et al. explored a metal-free approach for the synthesis of 1,5-disubstituted triazoles **207** by treating nitrostyrene **205** with glycosyl azide **206** in the presence of thiourea-based organocatalyst (Scheme 17.54a) [10]. The similar [3 + 2] cycloaddition of (*E*)-1-(2-nitrovinyl)-2-(alkynyl)benzene **209** with NaN_3 in DMSO at 100 °C for 20–40 min afforded high yields of biologically potent triazolo-fused isoquinolines **210** (Scheme 17.54b) [132]. Despite wide scope of these well-established metal-free synthetic protocols in drug discovery, the intramolecular version is not well explored in literature and certainly further investigation in this promising area is still required.

17.4 Conclusions and Future Outlook

Just after the invention of CuAAC Click Chemistry in 2002, this modular tool was identified as one of the most reliable and widely explored protocol in organic synthesis, particularly for an easy access of diverse molecular architecture of novel

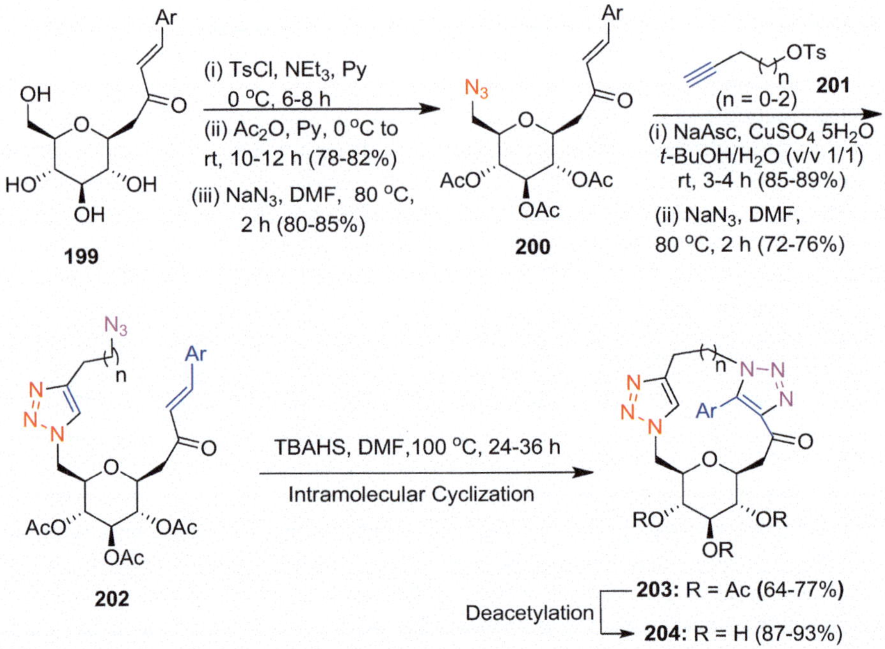

Scheme 17.53 Synthesis of pyran-based 1,4-linked 14- to 16-membered ring macrocycles (**203**, **204**) via intramolecular click chemistry

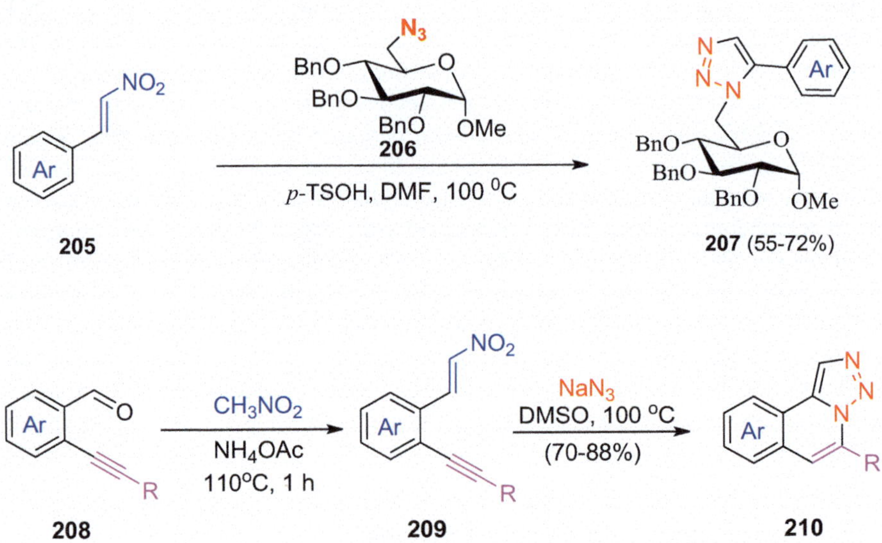

Scheme 17.54 Alternate route for the functionalized 1,2,3-triazoles from nitrostyrenes [131, 132]

applications. Intramolecular variant of triazole forming click is an efficient tool for the successful macrocyclization of azido-alkynes and the resulting cyclic molecular architectures have notable applications in drug development and other emerging fields.

Despite of notable advantages of this modular coupling tool, such macrocyclization procedure is thought-provoking. The challenge is to avoid the undesired cyclo-oligomerization of starting bifunctional precursors in some of the cases. In general, thermally induced intermolecular AAC gave an inseparable mixture of both the regioisomeric 1,2,3-triazoles; however, thermally induced intramolecular AAC of azido-alkyne scaffolds mainly gave the respective 1,5-disubstituted triazolo-fused heterocycles. The intramolecular click-mediated macrocyclization method has notable features in terms of reactivity, selectivity, and yields [24]. The click-mediated regioselective 1,2,3-triazoles has been extensively used in molecular hybridization approach simply through combining of two potent pharmacophores of complementary click partners having either similar or dissimilar units to generate a unique and effective framework with the improved potency. This triazole-linked molecular hybridization tool undoubtedly has enormous scope in the development of effective, safe, and novel mechanism-based drug candidates [22, 133, 134].

The intramolecular click-mediated resulting macrocycles and fused ring systems have widespread applications in medicinal chemistry, sensing, gelation, chelation, catalysis, and supramolecular chemistry and nanotechnology. Also, clicks followed by RCM or IAD or other established cyclization tool were nicely explored for the development of similar molecular scaffolds including macrocycles. The '*Intramolecular Click Chemistry*' can also be explored for the synthesis of similar molecular scaffolds under multicomponent reaction (MCR) condition which is presented in the next chapter. We authors do strongly hope that this chapter may help readers to understand the subject that serves a strong base for future advancements in contemporary procedure for the successful development of diverse 1,2,3-triazole-linked molecular architectures having notable applications in chemistry, biology, medicine, agriculture, and material science. In conclusion, this chapter presents a recent update on azide-alkyne cycloadditions via metal-catalyzed intramolecular cyclization reactions, organocatalyzed intramolecular reactions, and thermally induced intramolecular cycloaddition reactions under metal-free conditions with their emerging applications in organic synthesis with their important features and associated challenges.

References

1. Kolb HC, Finn M, Sharpless KB (2001) Angew Chem Int Ed 40:2004–2021
2. Tornøe CW, Christensen C, Meldal M (2002) J Org Chem 67:3057–3062
3. Rostovtsev VV, Green LG, Fokin VV, Sharpless KB (2002) Angew Chem Int Ed 41:2596–2599
4. Meldal M, Tornoe CW (2008) Chem Rev 108:2952–3015
5. Zhang L, Chen XG, Xue P, Sun HHY, Williams ID, Sharpless KB, Fokin VV, Jia GC (2005) J Am Chem Soc 127:15998–15999

6. Huisgen R (1963) Angew Chem Int Ed 2:565–598
7. Ramachary DB, Shashank AB, Karthik S (2014) Angew Chem Int Ed 53:10420–10424
8. Belkheira M, Abed DE, Pons JM, Bressy C (2011) Chem Eur J 17:12917–12921
9. Wang L, Peng S, Danence LJT, Gao Y, Wang J (2012) Chem Eur J 18:6088–6093
10. Jaiswal MK, Gupta A, Yadav MS, Pandey VK, Tiwari VK (2023) Chem Eur J 29(55):e202301749
11. Kumar GS, Lin Q (2021) Chem Rev 121:6991–7031
12. Hema K, Sureshan KM (2019) Acc Chem Res 52:3149–3163
13. Agard NJ, Prescher JA, Bertozzi CR (2004) J Am Chem Soc 126:15046–15047
14. Ghorai A, Achari B, Chattopadhyay P (2016) Tetrahedron 72:3379–3387
15. Bauman A, Kluba CA, Vomstein S, Walter MA, Mindt TL (2013) Angew Chem In Ed 52:8957–8960
16. Witczak ZJ, Bielski R (2013) Click chemistry in glycoscience: new development and strategies. Wiley-VCH John Wiley & Sons, New York
17. Chandrasekeran S (2016) Click reactions in organic synthesis. Wiley-VCH, John Wiley & Sons, New York
18. Chen MM, Kopittke PM, Jhao FJ, Wang P (2024) Trends Plant Sci 29:167–178
19. Tiwari VK, Mishra BB, Mishra KB, Mishra N, Singh AS, Chen X (2016) Chem Rev 116:3086–3240
20. Agrahari AK, Bose P, Singh AS, Rajkhowa S, Jaiswal MK, Hotha S, Mishra N, Tiwari VK (2021) Chem Rev 121:7638–7956
21. Thirumurugan P, Matosiuk D, Jozwiak K (2013) Chem Rev 113:4905–4979
22. Sharma A, Agrahari AK, Rajkhowa S, Tiwari VK (2022) Eur J Med Chem 238:114454
23. Kolb HC, Sharpless KB (2003) Drug Discov Today 8:1128–1137
24. Jaiswal MK, Tiwari VK (2023) Chem Rec e202300167
25. Tashrifi Z, Khanaposhtani M, Bahadorikhalili S, Larijani B, Mahdavi M (2023) Curr Org Synth 21:166–194
26. Patil PC, Luzzio FA (2017) Tetrahedron 73:4206–4213
27. Pasini D (2016) Synthesis of macrocycles and click chemistry. In: Chandrasekeran S (ed) Click reactions in organic synthesis. Wiley VCH, pp 287–308
28. Dolhem F, Tahli FA, Lievre C, Demailly G (2005) Eur J Org Chem 23:5019–5023
29. Rajesh R, Periyasami G, Raghunathan R (2010) Tetrahedron Lett 51:1896–1898
30. Li H, Aneja R, Chaiken I (2013) Molecules 18:9797–9817
31. Pedersen DS, Abell A (2011) Eur J Org Chem 13:2399–2411
32. Bachl J, Mayr J, Sayago FJ, Cativiela C, Diaz DD (2015) Chem Commun 51:5294–5297
33. Sareen N, Singh AS, Kant R, Tiwari VK, Bhattacharya S (2017) Dalton Trans 46:12705–12710
34. Chatterjee N, Pal R, Sarkar S, Sen AK (2015) Tetrahedron Lett 56:3886–3889
35. Muthana S, Cao H, Yu H, Cheng J, Chen X (2009) J Org Chem 74:2928–2936
36. Dorner S, Westermann BA (2005) Chem Commun 22:2852–2854
37. Ghorai A, Padmanaban E, Mukhopadhyay C, Achari B, Chattopadhyay P (2012) Chem Commun 48:11975–11977
38. Srinivas A, Rao EK (2021) Acta Chim Slov 68:404–413
39. Tiwari VK, Kumar A, Schmidt RR (2012) Eur J Org Chem 2945–2956
40. Larin EM, Lautens M (2019) Angew Chem Int Edn 58:13438–13442
41. Ghorai A, Reddy KS, Achari B, Chattopadhyay P (2014) Org Lett 16:3196–3199
42. Billing JF, Nilsson UJ (2005) J Org Chem 70:4847–4850
43. Ray A, Manoj K, Bhadbhade MM, Mukhopadhyay R, Bhattacharjya A (2006) Tetrahedron Lett 47:2775–2778
44. Lambu MR, Hussain A, Sharma DK, Yousuf SK, Singh B, Tripathi AK, Mukherjee D (2014) RSC Adv 4:11023–11028
45. Leyden R, Murphy PV (2009) Synlett 12:1949–1950
46. Schulz M, Christoffers J (2013) Tetrahedron 69:802–809
47. Bogliotti Y, Yu N, Tang J, Xie J (2013) Eur J Org Chem 7749–7760
48. Bodine KD, Gin DY, Gin MS (2005) Org Lett 7:4479–4482

49. Bodine KD, Gin DY, Gin MS (2004) J Am Chem Soc 126:1638–1639
50. Conte ML, Grotto D, Chambery A, Dondoni A, Marra A (2011) Chem Commun 47:1240–1242
51. Chandrasekhar S, Rao CL, Nagesh C, Reddya CR, Sridhar B (2007) Tetrahedron Lett 48:5869–5872
52. Liu J, Leonard P, Muller SL, Daniliuc C, Seela F (2018) Beil J Org Chem 14:2404–2410
53. Nakane M, Ichikawa S, Matsuda A (2008) J Org Chem 73:1842–1851
54. Allam A, Dupont L, Behr JB, Plantier-Royon R (2012) Eur J Org Chem 4:817–823
55. Campo VL, Ivanova IM, Carvalho I, Lopes CD, Carneiro ZA, Saalbach G, Schenkman S, Santana da Silva J, Nepogodiev SA, Field RA (2015) Tetrahedron 71:7344–7353
56. Lepage ML, Schneider JP, Bodlenner A, Compain P (2015) J Org Chem 80:10719–10733
57. Chen A, Samankumara LP, Dodlapati S, Wang D, Adhikari S, Wang G (2019) Eur J Org Chem 1189–1194
58. Singh A, Khatri V, Malhotra S, Prasad AK (2016) Carbohydr Res 421:25–32
59. Pathigoolla A, Sureshan KM (2014) Chem Commun 50:317–319
60. Li L, Diau EWG (2013) Chem Soc Rev 42:291–304
61. Kushwaha D, Tiwari VK (2013) J Org Chem 78:8184–8190
62. Agrahari AK, Kumar S, Rajkhowa S, Jaiswal MKVK (2022) Tiwari. Chem Sel 7:e202202273
63. Webb JEA, Maharaj F, Blake IM, Crossley MJ (2008) Synlett 2147–2149
64. Gilday LC, White NG, Beer PD (2012) Dalton Trans 41:7092–7097
65. Qiu XP, Tanaka F, Winnik FM (2007) Macromolecules 40:7069–7071
66. Yuan YY, Du JZ, Wang J (2012) Chem Commun 48:570–572
67. Jarosz S, Listkowski A, Lewandowski B, Ciunik Z, Brzuszkiewicz A (2005) Tetrahedron 61:8485–8492
68. Lewandowski B, Jarosz S (2008) Chem Commun 47:6399–6401
69. Jarosz S, Listkowski A (2006) Can J Chem 84:492–496
70. Lewandowski B, Jarosz S (2011) Synth Commun 41:2161–2168
71. Jarosz S, Lewandowski B, Listkowski A (2008) Synthesis 6:913–916
72. Chaciak B, Dabrowa K, Swider P, Jarosz S (2018) New J Chem 42:18578–18584
73. Bose P, Jaiswal MK, Singh SK, Tiwari VK (2023) Carbohydrate Res 527:108804
74. Campo VL, Carvalho I, Silva CHTP, Schenkman S, Hill L, Nepogodiev S, Field RA (2010) Chem Sci 1:507–514
75. Tiwari VK (ed) (2020) Carbohydrates in drug discovery and development. Elsevier Inc., The Netherland
76. Singh K, Sharma G, Shukla M, Kant R, Chopra S, Shukla SK, Tripathi RP (2018) J Org Chem 83:14882–14893
77. Nahrwold M, Bogner T, Eissler S, Verma S, Sewald N (2010) Org Lett 12:1064–1067
78. Driggers EM, Hale SP, Lee J, Terrett NK (2008) Nat Rev Drug Disc 7:608–624
79. Otvos SB, Fulop F (2015) Catal Sci Tech 5:4926–4941
80. Bogdan AR, Jerome SV, Houk KN, James K (2012) J Am Chem Soc 134:2127–2138
81. Bogdan AR, James K (2011) Org Lett 13:4060–4063
82. Bedard A-C, Collins SK (2014) Org Lett 16:5286–5289
83. Boren BC, Narayan S, Rasmussen LK, Zhang L, Zhao HT, Lin ZY, Jia GC, Fokin VV (2008) J Am Chem Soc 130:8923–8930
84. Johansson JR, Beke-Somfai T, Stalsmeden AS, Kann N (2016) Chem Rev 116:14726–14768
85. Himo F, Lovell T, Hilgraf R, Rostovtsev VV, Noodleman L, Sharpless KB, Fokin VV (2005) J Am Chem Soc 127:210–216
86. Kim WG, Kang ME, Lee JB, Jeon MH, Lee S, Lee J, Choi B, Cal PMSD, Kang S, Kee JM, Bernardes GJL, Rohde JU, Choe W, Hong SY (2017) J Am Chem Soc 139:12121–12124
87. Jaiswal MK, Gupta A, Ansari FJ, Pandey VK, Tiwari VK (2024) Curr Org Syn 21:513–558
88. Creary X, Anderson A, Brophy C, Crowell F, Funk Z (2012) J Org Chem 77:8756–8761
89. Kelly AR, Wei JQ, Kesavan S, Marie JC, Windmon N, Young DW, Marcaurelle LA (2009) Org Lett 11:2257–2260
90. Nielsen TE, Schreiber SL (2008) Angew Chem Int Ed 47:48–56

91. Marcaurelle LA, Comer E, Dandapani S, Duvall JR, Gerard B, Kesavan S, Lee MD, Liu H, Lowe JT, Marie JC et al (2010) J Am Chem Soc 132:16962–16976
92. Zhang JQ, Kemmink J, Rijkers DTS, Liskamp RMJ (2011) Org Lett 13:3438–3441
93. Zhang Q, Kemmink J, Rijkers DTS, Liskamp RMJ (2013) Chem Commun 49:4498–4500
94. Isidro-Llobet A, Murillo T, Bello P, Cilibrizzi A, Hodgkinson JT, Galloway W, Bender A, Welch M, Spring DR (2011) Proc Natl Acad Sci USA 108:6793–6798
95. Beckmann HSG, Nie F, Hagerman CE, Johansson H, Tan YS, Wilcke D, Spring DR (2013) Nature Chem 5:861–867
96. Fang F, Vogel M, Hines JV, Bergmeier SC (2012) Org Biomol Chem 10:3080–3091
97. Arora I, Kashyap VK, Singh AK, Dasgupta A, Kumar B, Shaw AK (2014) Org Biomol Chem 12:6855–6868
98. Adhikary ND, Chattopadhyay P (2012) J Org Chem 77:5399–5405
99. Hotha S, Anegundi RI, Natu AA (2006) Tetrahedron Lett 46:4585–4588
100. Arigela RK, Mandadapu AK, Sharma SK, Kumar B, Kundu B (2012) Org Lett 14:1804–1807
101. Ning Y, Wu N, Yu H, Liao P, Li X, Bi X (2015) Org Lett 17:2198–2201
102. Senwar KR, Sharma P, Reddy TS, Jeengar MK, Nayak VL, Naidu V, Kamal A, Shankaraiah N (2015) Eur J Med Chem 102:413–424
103. Balducci E, Bellucci L, Petricci E, Taddei M, Tafi A (2009) J Org Chem 74:1314–1321
104. Mishra KB, Tiwari VK (2014) J Org Chem 79:5752–5762
105. Mishra KB, Agrihari AK, Tiwari VK (2017) Carbohydrate Res 450:1–9
106. Reddy YS, John Pal AP, Gupta P, Ansari AA, Vankar Y (2011) D. J Org Chem 76:5972–5984
107. Dellisola A, McLachlan MMW, Neuman BW, Al-Mullah HMN, Binks AWD, Elvidge W, Shankland K, Cobb AJA (2014) Chem Eur J 20:11685–11689
108. Oliva AI, Christmann U, Font D, Cuevas F, Ballester P, Buschmann EH, Torrens A, Yenes S, Pericas MA (2008) Org Lett 10:1617–1619
109. Rungta P, Mangla P, Maikhuri VK, Singh SK, Prasad AK (2017) Chem Select 2:7808–7812
110. Basavaiah D, Reddy BS, Lingam H (2013) Tetrahedron 69:10060–10067
111. Banerji B, Pramanik SK, Sanphui P, Nikhar S, Biswas SC (2013) Chem Biol Drug Des 82:401–409
112. Majumdar KC, Ganai S (2013) Synthesis 45:2619–2625
113. Majumdar KC, Ganai S (2013) Tetrahedron Lett 54:6192–6195
114. Konda S, Rao P, Oruganti S (2014) RSC Adv 4:63962–63965
115. Arbaciauskiene E, Laukaityte V, Holzer W, Sackus A (2015) Eur J Org Chem 5663–5670
116. Sastry KNV, Routhu S, Datta SG, Nagesh N, Babu BN, Nanubolu JB, Kumar CG, Maurya RA, Kamal A (2016) Org Biomol Chem 14:9294–9305
117. Hussain MK, Ansari MI, Kant R, Hajela K (2014) Org Lett 16:560–563
118. Guggenheim KG, Toru H, Kurth MJ (2012) Org Lett 14:3732–3735
119. Sudhapriya N, Nandakumar A, Arun Y, Perumal P, Balachandran C, Emi N (2015) RSC Adv 5:66260–66270
120. Gour J, Gatadi S, Pooladanda V, Ghouse SM, Malasala S, Madhavi YV, Godugu C, Nanduri S (2019) Bioorg Chem 93:103306
121. Yoshida S, Morita T, Hosoya T (2016) Chem Lett 45:726–728
122. Samala S, Arigela RK, Kant R, Kundu B (2014) J Org Chem 79:2491–2500
123. Fulopova V, Funk P, Popa I, McMaster C, Soural M (2015) Eur J Org Chem 3551–3557
124. Abeykoon GA, Sahn JJ, Martin SF (2021) Tetrahedron Lett 66:152828
125. Maurya RA, Adiyala PR, Chandrasekhar D, Reddy CN, Kapure JS, Kamal A (2014) ACS Comb Sci 16:466–477
126. Nekkanti S, Pooladanda V, Veldandi M, Tokala R, Godugu C, Shankaraiah N (2017) Chem Select 2:7210–7221
127. Salman AA, Tabandeh M, Heidelberg T, Hussen RSD (2015) Carbohydr Res 406:41–45
128. Jogula S, Dasari B, Khatravath M, Chandrasekar G, Kitambi SS, Arya P (2013) Eur J Org Chem 5036–5040
129. Ajay A, Sharma S, Gupt MP, Bajpai V, Hamidullah, Kumar B, Kaushik MP, Konwar R, Ampapathi RS, Tripathi RP (2012) Org Lett 14:4306–4309

130. Kayet A, Dey S, Pathak T (2015) Tetrahedron Lett 56:5521–5524
131. Mishra KB, Tiwari VK (2016) Chem Select 1:3693–3698
132. Arigela RK, Samala S, Mahar R, Shukla SK, Kundu B (2013) J Org Chem 78(20):10476–10484
133. Yadav MS, Rajkhowa S, Singh SK, Jaiswal, MK, Tiwari VK (2024) Chem Sel 9:e202400776
134. Tiwari VK (2021) Chem Rec 21:3029–3048

Chapter 18
Tandem/MCR Click Chemistry in Organic Synthesis

18.1 Introduction

The construction of functionalized 1,2,3-triazoles [1–4] involves stepwise formation that is formation of triazole in the primary step and their functionalization in the successive steps or vice versa. But if similar scaffolds could be formed in a single step without requiring the additional reagents, modification of the reaction conditions, or isolation of the intermediates, that would be certainly far more efficient. Regarding this, one-pot multicomponent reactions are thought to be an indispensable tool for producing various functionalized triazoles with numerous applications in combinatorial as well as in clinical chemistry [5–7]. Along with the CuAAC [8, 9] or RuAAC [10–12], or the thermally triggered intramolecular versions of regioselective azide-alkyne cycloaddition reactions of alkyne **1** and azide **2** to provide the corresponding functionalized 1,2,3-triazoles bound molecular frameworks [13]. It is of utmost importance to investigate the multicomponent reactions (MCRs) of the same transformation in parallel fashions (Fig. 18.1). These reactions are advantageous in terms of reducing time and evading the additional steps involved in the synthesis of targeted molecules. A number of imperative multicomponent reactions, namely Ugi-4CR [14], Passernini 3CR [15], Strecker reaction [16], Hantzsch 3CR dihydropyridine synthesis [17], Biginelli 3CR [18], Kabachnik-Fields 3CR [19], Mannich reaction [20], etc. were established, which offers an efficient route for the construction of a number of biologically significant compounds.

Over 200 reviews have been published on 'click chemistry'; [8] however, there is still very little literature available on Tandem/MCR azide-alkyne cycloaddition reaction 'Click Chemistry' in Organic Synthesis [13]. This chapter precisely the usefulness of multicomponent reaction in click chemistry for the construction of simple to complex triazole-containing compounds with miscellaneous applications in various emerging fields. We focused our aim to discuss the synthesis of various triazole-appended heterocycles, macrocyclization reactions, and construction of glycohybrid triazole architectures.

V. K. Tiwari et al., *Click Chemistry*, Materials Horizons: From Nature to Nanomaterials,
https://doi.org/10.1007/978-981-97-4596-8_18

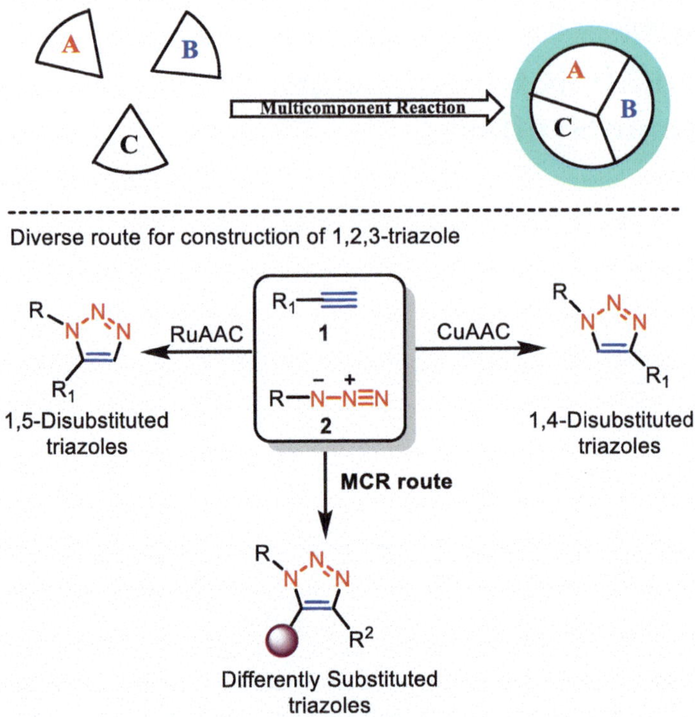

Fig. 18.1 Construction of functionalized 1,2,3-triazoles under different circumstances

18.2 Application of MCR in Click Chemistry

18.2.1 Synthesis of 1,2,3-Triazole-Appended Hybrid Scaffolds

The invention of Click chemistry was proven to be an indispensable tool for the construction of a diverse range of heterocyclic bound as well as fused systems with a number of applications [21–23]. The chemistry of triazole-based heterocycles has attracted a lot of attention. As anti-cancer, anti-bacterial, and anti-tuberculosis medicines, they display a wide variety of biological activities. It has been determined that 1,2,3-triazoles fused with carbo- and heterocycles can treat HIV, cancer, and Alzheimer's disease [24, 25]. Herein we discussed the synthesis of a diverse range of fused heterocycles via MCR Click reaction.

Since Prof. Lieki's laboratory produced isocyanide for the first time in 1859, this functional group has been extensively studied in current organic synthesis, like Ugi multicomponent reaction enables the quick synthesis of densely functionalized frameworks with their prospective uses in drug delivery and development [26, 27]. In order to accomplish the aforementioned reaction, a one-pot, four-component Ugi reaction involving sugar azido-aldehyde, benzylamine, alkyl isocyanide, and phenylpropiolic acid has been documented. This reaction produces a new class of sugar-derived triazolodiazepines **8** with satisfactory yields as well as high selectivity [28]. The approach entails the production of iminium intermediates **I** from sugar azidoaldehyde **3** and benzylamine **4**. The resulting intermediates were then attacked nucleophilically by isocyanides **5** and arylpropiolates **6** to produce intermediate **II,** which was then isomerized to produce Ugi adduct **7** (Scheme 18.1). Subsequently, the sugar-derived triazolodiazepines product **8** is produced by thermally triggered intramolecular click reaction of Ugi adducts **7** in DMF at 120 °C (Scheme 18.1) [28]. The primary advantages of this approach are its easily accessible nature and avoidance of catalysts in order to achieve significant structural diversity in one pot.

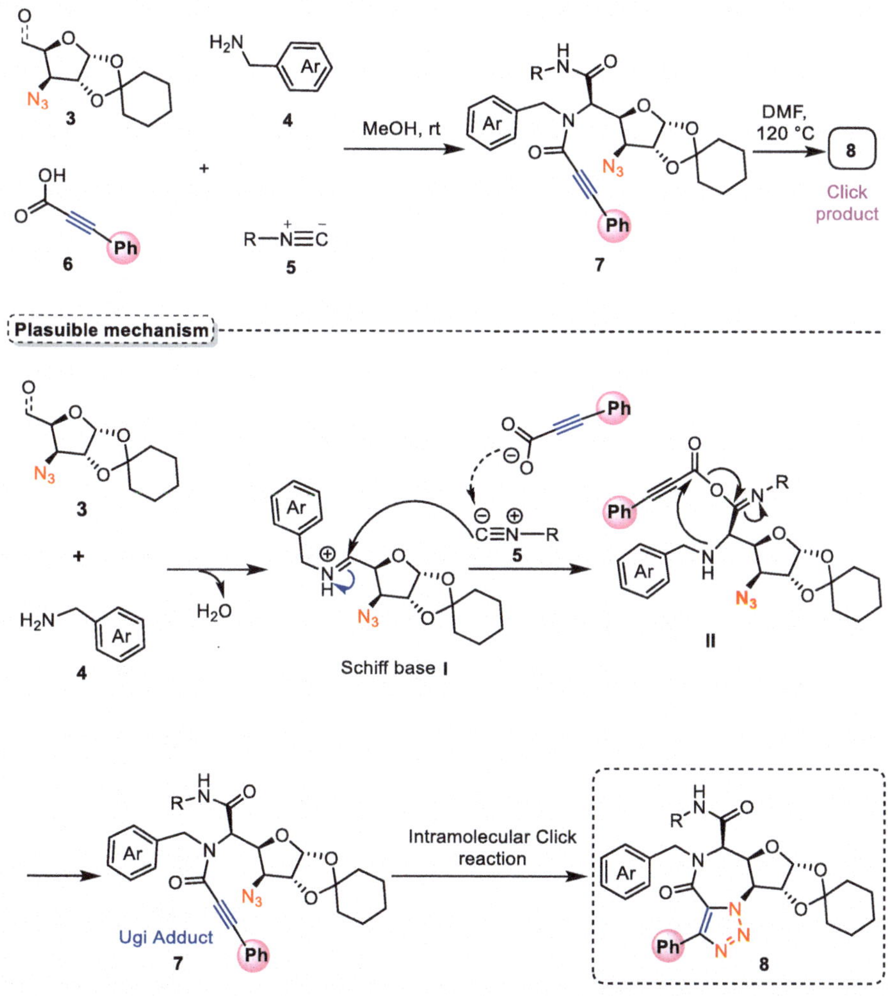

Scheme 18.1 MCR Click reaction for synthesis of sugar-derived triazolodiazepines **8**

In the same way, Barlow and colleagues produced good to outstanding yields of biologically significant amino-triazoloazepinone **13** using the Ugi four-component reaction of *β*-azido-α-amino acid **9**, an isocyanide **5**, propargylamine**10**, and an aldehyde **11**, without the aid of a catalyst [29]. In this one-pot reaction, Ugi adducts **12** are produced, which are then converted to amino-triazoloazepinone **13** at 70 °C by intramolecular [3+2] cycloaddition (Scheme 18.2). The resultant compound **13** was a diastereoisomic combination (1:1) that the authors were unable to isolate in optically pure form. When considering reaction yield and diastereoselectivity in product **13**, the substituent on the phenyl ring of aromatic aldehyde played no discernible influence. Additionally, when using aromatic aldehydes instead of aliphatic aldehydes, MCR was often more effective in the context of a reaction yield.

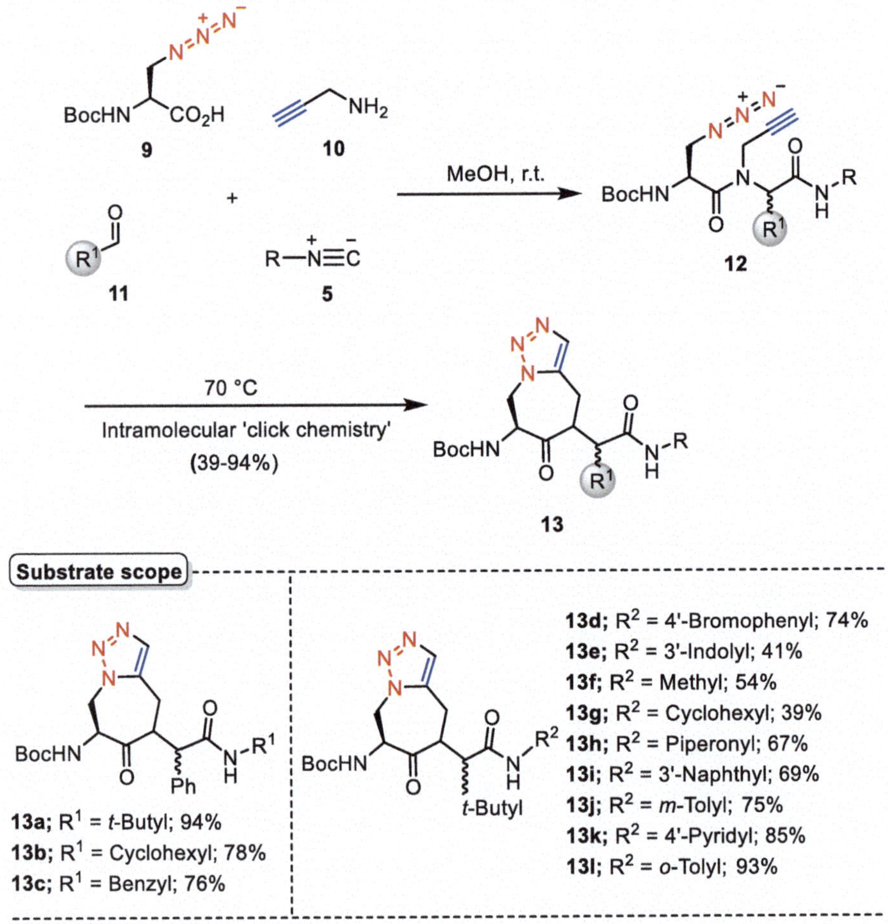

Scheme 18.2 One-pot Ugi four-component reaction for synthesis of bicyclic scaffolds **13**

An analogous four-component high-yielding reaction of isocyanides **5**, 2-azidobenzaldehyde **14**, nitrophenols **15**, and propargyl amines **10**, for the construction of fused-4,5-dihydro-1,2,3-triazolo[1,5-*a*][1,4] benzodiazepinone **16** by utilizing Click-Ugi-Smiles type reaction has been reported by Saeedi et al. in methanol [30]. The procedure involves condensation of propargyl amine **10** with 2-azidobenzaldehyde **14** followed by intramolecular click to produce tricyclic intermediate **II** (Scheme 18.3). Further, the nucleophilic addition of isocyanide and phenoxide ion (from phenol **15**) led to the formation of intermediate **IV**, which on Smiles rearrangement successfully afforded the high yields of functionalized heterocycles **16**. The electron-withdrawing substituent on phenol particularly *O*-nitrophenol significantly improves the reactivity of MCR reaction.

Shafie and the group also established an efficient protocol for the development of a unique class of 1,2,3-triazolobenzodiazepine derivatives **20** and demonstrated their powerful anti-convulsant properties in vivo [31]. When 2-azidobenzoic acids **17** have been put together with benzaldehydes **18** and propargylamine **10** in methanol, an adduct **19** is formed. This adduct then spontaneously undergoes intramolecular click cycloaddition to yield 1,2,3-triazolobenzodiazepine **20** as core component (Scheme 18.4). Moreover, pentylenetetrazole (PTZ) and maximum electroshock (MES)-induced seizures in mice were used to test the in vivo pharmacology of new compounds to determine their anti-convulsant properties. According to the PTZ assay, the majority of the compounds displayed more anti-convulsant effectiveness than the reference medication diazepam (0.5 mg/kg). In PTZ screening, the produced benzodiazepinone **20a** was shown to be more potent than diazepam, while in MES screening, the compounds **20b** and **20c** showed outstanding anti-convulsant efficacy. The compounds **20a** and **20c**, however, appeared to have less neurological impairment than diazepam, according to the rotarod test.

In order to synthesize triazolopyrazinyl acetamides analogue **24** in one-pot fashion, Mahdavi and colleagues in 2020 came up with a new approach and constructed the effective magnetic silica nanoparticle supported copper complex Cu@ABA-Fe_3O_4@SiO_2 **22** [32]. The present investigation stated that Ugi adducts 398 are produced by a four-component reaction of azides **21**, isocyanide **5**, aromatic aldehydes **18**, and propargylamine **10**. Upon intramolecular click cycloaddition, excellent yields of the respective triazolo-fused-pyrazinyl acetamides **24** were isolated (Scheme 18.5). Furthermore, all synthesized compounds were evaluated for anti-bacterial activity against gram-positive and gram-negative bacterial strains such as *E. coli* and *Staphylococcus aureus*. The in vitro assay revealed that the majority of the compounds have significant anti-bacterial activity.

$InCl_3$-catalyzed four-component reactions of α-diketones **26**, propargylic amines **25**, NH_4OAc, and *O*-azidobenzaldehydes **14**, have been utilized by Kurth and colleagues to synthesize imidazotriazolobenzodiazepines **27**in one-pot fashion (Scheme 18.6) [33]. This multicomponent reaction approach is being carried out via an azido-alkyne intermediary **V** that, after thermally induced intramolecular click reaction, yielded the physiologically relevant fused heterocycles scaffold **27** in moderate to good yields. This procedure is renowned for its practicality, one-pot condition, diverse substrate scope, and great atom economy. The mechanistic

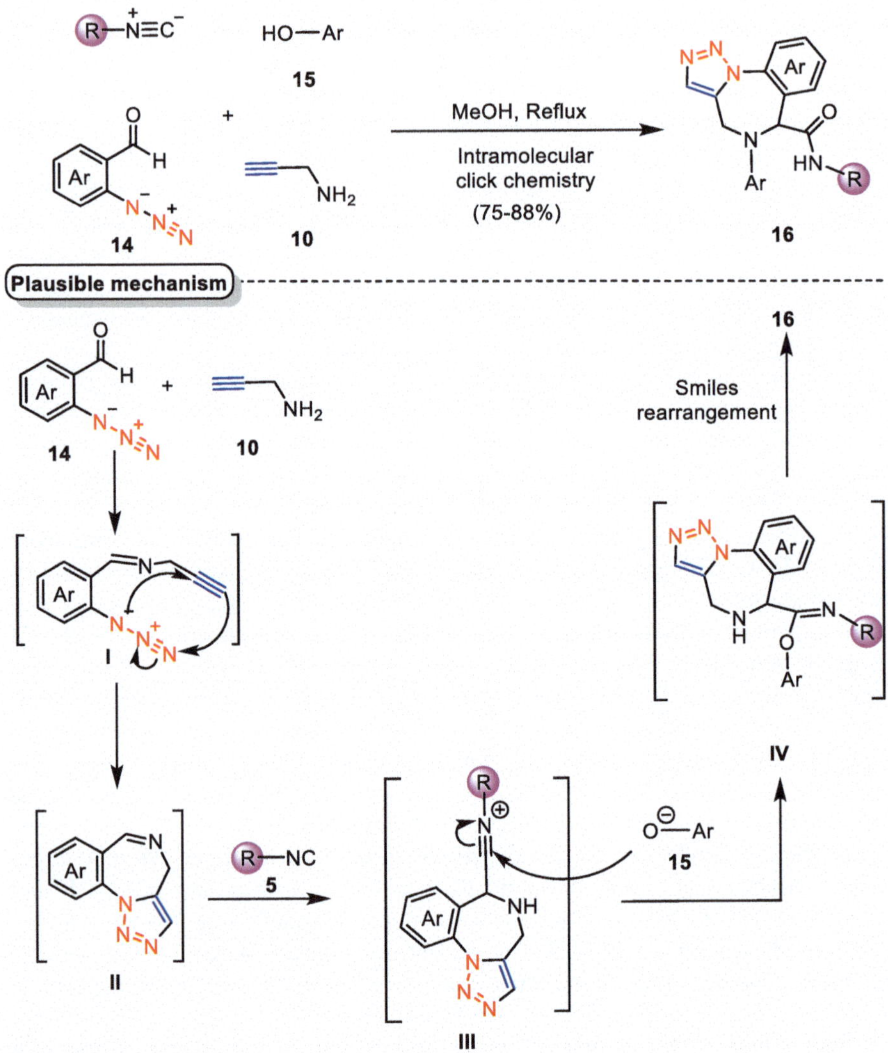

Scheme 18.3 A four-component Ugi-Smiles type reaction for synthesis of 4,5-dihydro-1,2,3-triazolo[1,5-*a*][1,4] benzodiazepinones **16**

pathway for the reaction was also emphasized by Kurth et al. In this case, the tandem reaction begins with the reaction of $InCl_3$with *O*-azidobenzaldehydes **14** to furnish intermediatory **I**, resulting the imine **II**. The nucleophilic addition of propargylamine to imine **II** generated the intermediate **III**. The initial formation of azido-alkyne imidazole intermediary **V**takes place in assistance with **IV,** followed by the intramolecular [3+2] cycloaddition of alkyne and azide under thermal conditions afforded respective imidazotriazolobenzodiazepines **27.** Another route is also

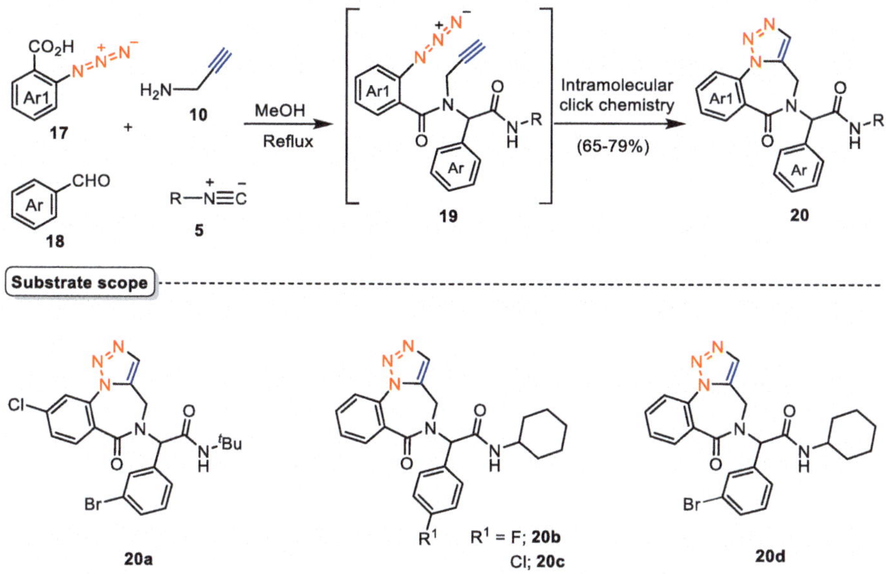

Scheme 18.4 Synthesis of 1,2,3-triazolo-benzodiazepines **20** via Ugi-Click strategy

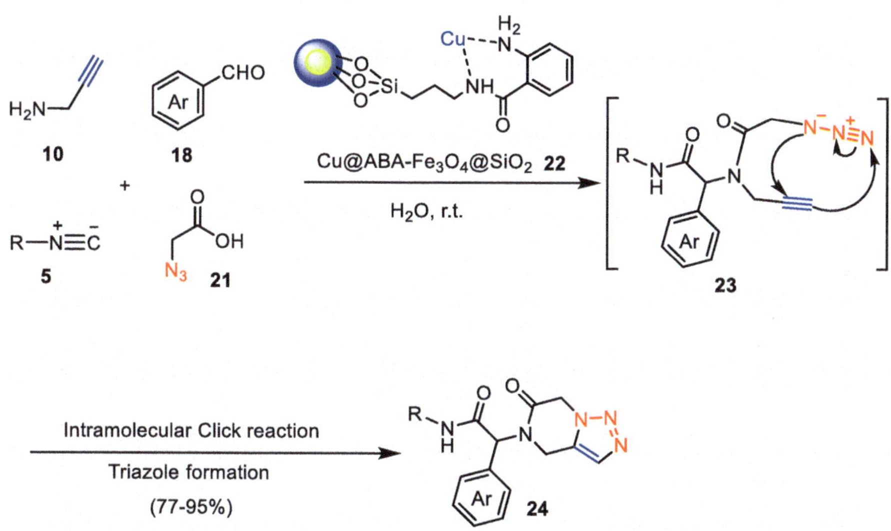

Scheme 18.5 Cu@ABA-Fe_3O_4@SiO_2-induced efficient synthesis of triazolopyrazinylacetamides **24**

possible for the desired compound **V** (not discussed here), in which the formation of triazole takes place before the construction of the imidazole skeleton.

Utilizing a three-component coupling process accompanied by intramolecular cycloaddition, a two-step MCR approach for diastereoselective synthesis of tricyclic-fused pyrazines **368** was disclosed by Lee group [34]. In this procedure, cyclic amines **28,** aldehydes **29,** and terminal alkynes **30** are coupled through the action of $AuBr_3$ to produce propargyl amine analouge**31** in the absence of a solvent. The coupling response of aliphatic aldehydes, such as valeraldehyde, butyraldehyde, cyclohexane carboxaldehyde, and hexanal, was higher than that of aromatic aldehydes. The resulting propargyl amines **31** were then subjected to an intramolecular click cycloaddition process in DMF at 100 °C to form tricyclic-fused pyrazines **32** with outstanding diastereoselectivity (Scheme 18.7).

Similar to this, Basso and colleagues describe a four-step, one-pot method (referred to as the OPHA technique) for the synthesis of tricyclic benzoxazepine

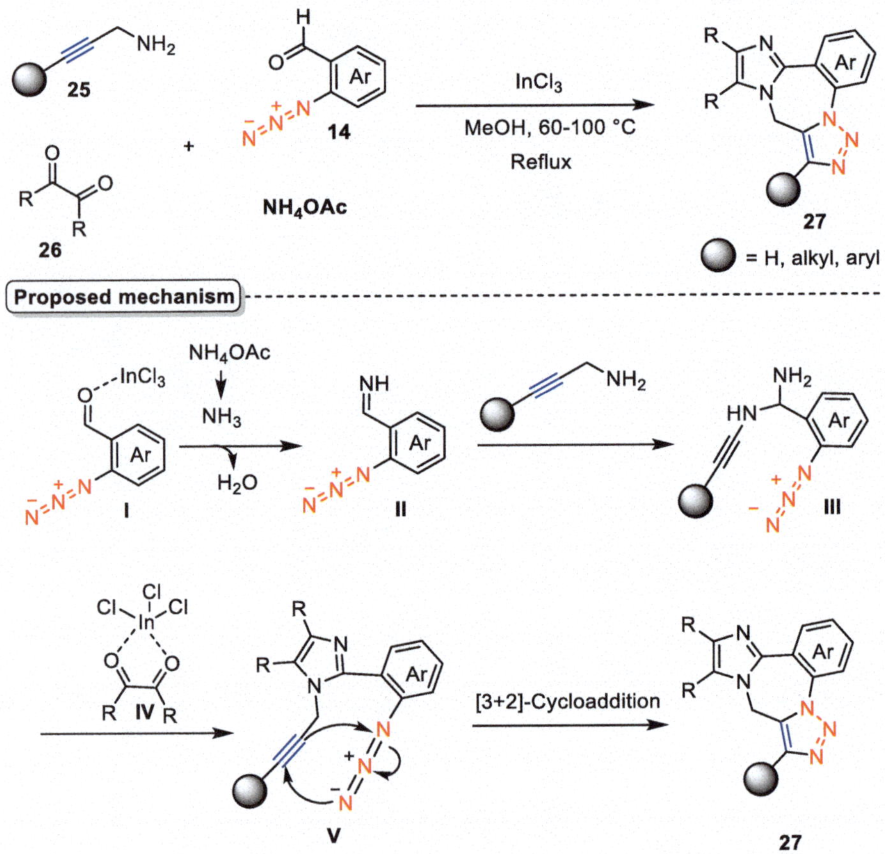

Scheme 18.6 $InCl_3$-induced multicomponent synthesis of imidazo-fuzed-triazolobenzodiazepine scaffold **27**

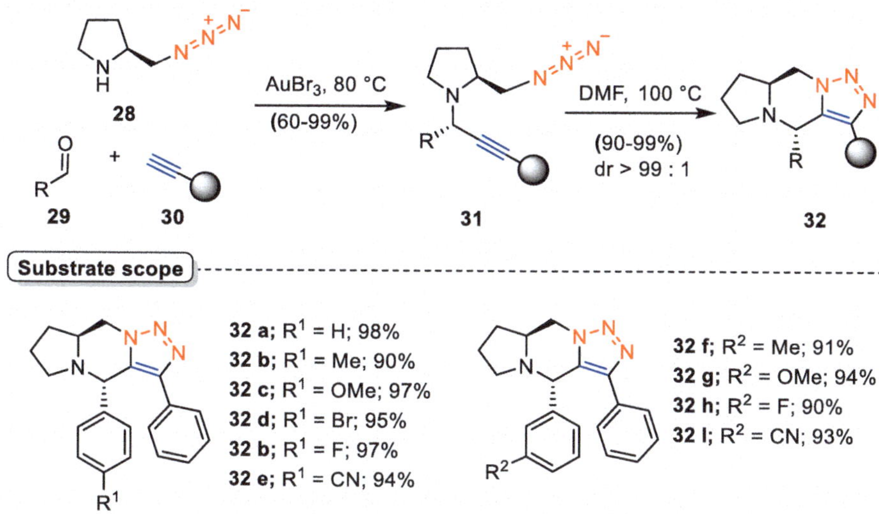

Scheme 18.7 Synthesis of tricyclic-fused pyrazines **32** via $AuBr_3$-promoted MCR approach

36, which was incompatible with an alkylative Passerini reaction. The researchers performed a three-component reaction using propargyl bromides **34**, isocyanide, and 2-azidobenzyl alcohols **33**, with the assistance of TEMPO/PIDA at room temperature, to deliver the adducts **35.** Ultimately, the required tricyclic benzoxazepines **36** were produced in acceptable yields by thermally induced intramolecular click cycloaddition of adduct **35** in toluene at 100 °C upon microwave exposure (Scheme 18.8) [35].

18.2.2 Synthesis of 1,2,3-Triazole-Containing Macrocycles

Macrocyclic scaffolds with a variety of ring systems exhibit an array of clinal properties and are found to be a part of natural products, in perfumery and some representative synthetic macrocyclic compounds, namely clarithromycin and erythromycin show anti-tumor and anti-bacterial properties, respectively [36–38]. Moreover, using the multicomponent reactions (MCR) approach, a varied spectrum of physiologically pertinent carbohydrate-containing macrocycles would be produced [39].

In the year 2006, Tron and colleagues constructed a straightforward and useful method for the synthesis of macrocycles **39**. The macrocyclization achieved in one-pot by associating alkyne bounded acetamide-nitrile **37**, aldehydes and amino azides **38** in tandem fashion to furnish isoxazole-based azido-alkyne followed by intramolecular CuAAC to afford targeted macrocyclic architecture **39** (Scheme 18.9) [40].

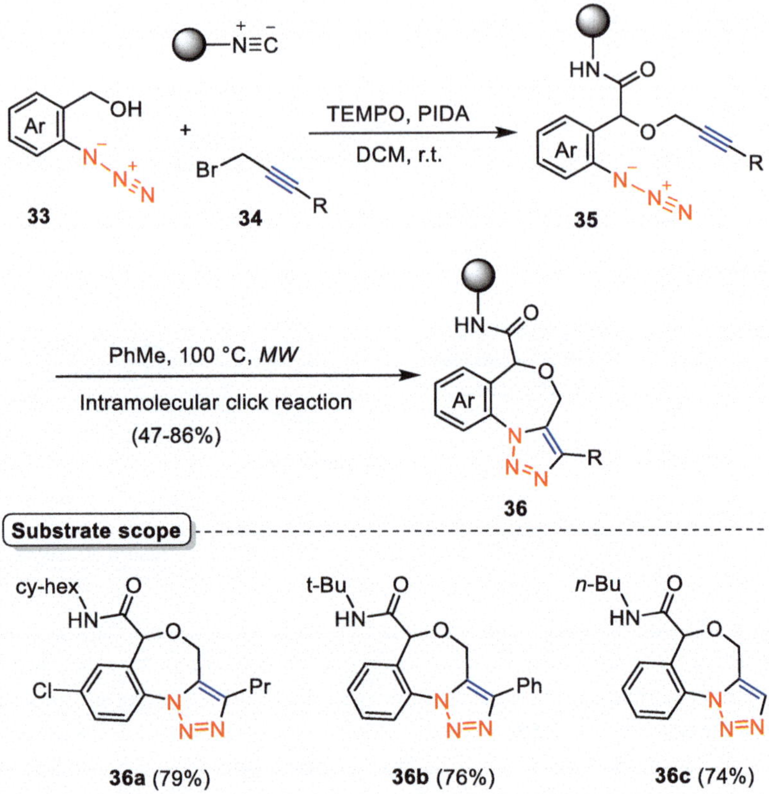

Scheme 18.8 Four-step-one-pot MCR synthesis of benzoxazepines **36**

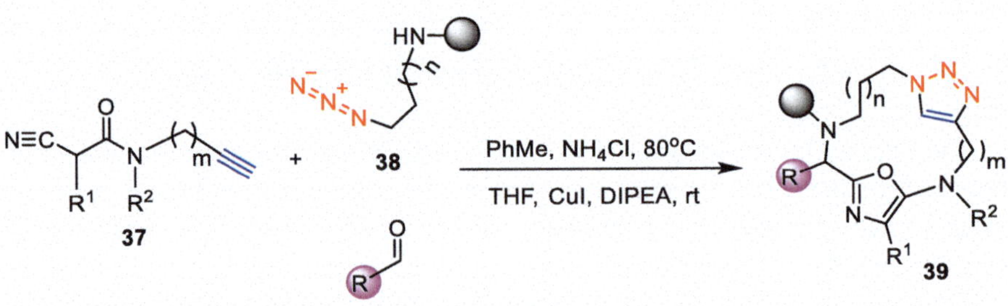

Scheme 18.9 One-pot tandem synthesis of isooxazole-appended triazole-containing macrocycle **39**

In order to achieve the targeted scaffolds, Prasanna and colleagues described an intuitive synthetic pathway for an easy access of sugar-appended macrocycles, where the ultimate molecular scaffolds were achieved in good-to-high yields employing multicomponent reactions (MCR) via the CuAAC-mediated macrocyclization. As a result, the 1,4-triazole-appended macrocycles **43**were produced by the reaction

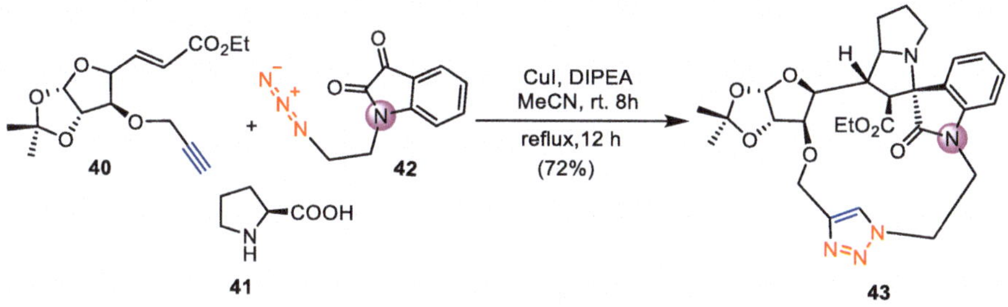

Scheme 18.10 Synthesis of macrocycles **43** via tandem one-pot reaction

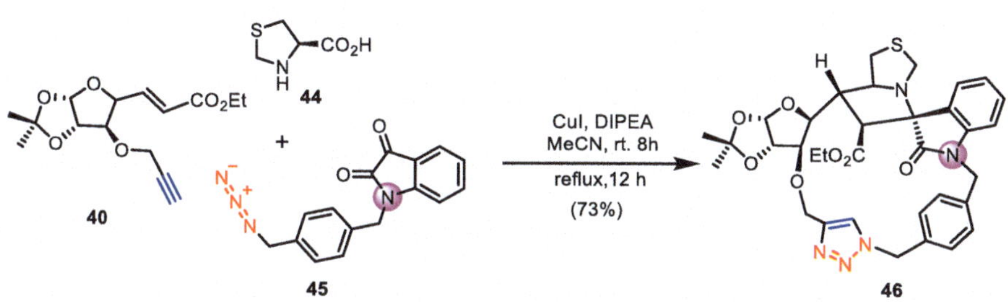

Scheme 18.11 Synthesis of macrocycles **46** via tandem one-pot reaction

of and *N*-alkyl azidoisatin **42** with secondary amino acid **41** and glucose-derived *O*-propargyl enone **40** (Scheme 18.10) [41].

The described approach was also used in the analogous synthesis of a macrocycles **46** employing glucose-derived *O*-propargyl enone **40**, amine **44,** and *N*-benzyl azidoisatin **45** (Scheme 18.11) [41].

A similar work proposed by Rao et al. also showed that azomethine ylides could be macrocyclized in one pot to create a number of biocompatible triazole-linked glycomacrocycles [42]. The methodology involves the one-pot three-component reaction of *O*-propargyl nitro olefin **47**, salicylaldehyde-derived *O*-alkyl azides **48** or **49** and secondary amino acid **50** in the presence of CuI and DIPEA afforded the desired 1,4-triazole-linked glycomacrocycles **51**, and **52** in good yield. The current technique employs a one-pot three-component reaction with salicylaldehyde-derived *O*-alkyl azides **48** or **49**, with *O*-propargyl nitro olefin **47**, and secondary amino acid **50** to furnish the required 1,4-triazole bound glycomacrocycles **51**or **52** respectively in an excellent outcome (Scheme 18.12) [42].

A multicomponent and continuous-flow multistep reaction was developed by Kappe and colleagues for macrocyclization reaction. Initially, linear peptoids **61** were formed through the aforementioned technique, then after intramolecular click chemistry was used as an indispensable tool for their further macrocyclization to furnish macrocyclic compound **62** [43]. For the synthesis of targeted scaffolds **62,** *N*-(triethylammoniumsulfonyl)carbamate) **54**, known as Burgess reagent was used for dehydration of formamides **53** and nucleophilic substitution from bromides **56**

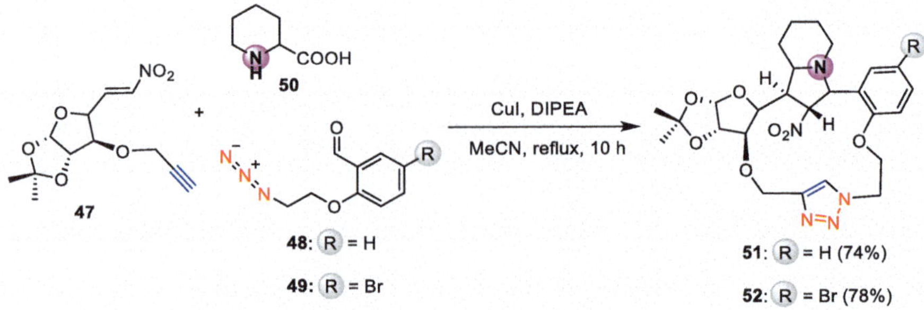

Scheme 18.12 Synthesis of macrocycles **51** and **52** via tandem 1,3-dipolar cycloaddition reaction

with tetrabutylammonium azide (TBAA) **57**, respectively, resulting in the continuous formation of an isocyanide **55** and an azide-functionalized carboxylic acid **58.** Linear peptides **61**, in yields of between 80 and 95%, are produced by combining in situ produced isocyanides and azides with paraformaldehyde **59** and *tert*-butyl amine in a Ugi-4CR at 100 °C. By using the continuous-flow reactor at 140 °C following intramolecular CuAAC macrocyclization, a monomeric or dimeric cyclic peptide macrocycle **62** is obtained in the 80% of yield (Scheme 18.13).

Furthermore, Miranda et al. implemented a multicomponent click cycloaddition reaction under microwave irradiation conditions to prolong the Ugi adducts **66** for the development of new tryptamine-based macrocycles **67.** The indole-derived amine **63** ($n = 1$–5), obtained through a versatile three-step procedure from *N*-Boc tryptamine, amalgamated with isonitriles **5**,O-propargylated benzaldehyde **64**, and acetic or benzoic acid **65**, in a solution of methanol at room temperature to yield the Ugi adducts **66.** An array of complex macrocyclic products **67a–d** were produced in reasonable yields by macrocyclizing Ugi product **66** using 40 mol% CuBr after microwave exposure(Scheme 18.14) [44].

Additionally to the Ugi-click approach, a study on the development of macrocyclic peptidomimetics with 12–28 units, produced from 1,2,3-triazoles was documented extensively [45]. In this Ugi-click approach, azido functionality in isocyanides and acetylenicfunctionality in acid components are fused together to furnish respective bifunctional peptidomimetics, macrocyclization is then made possible by linking these two fragments utilizing modular click reaction. In this study, bifunctional peptidomimetics **72** were produced in MeOH at room temperature by the reaction of *o*/*m*/*p*-substituted ethynylbenzoic acids **68**, isocyanide **69**, DMB-amine**70**, and ketones/ aldehyde **71**. Click macrocyclization produced a combination of 12- and 24-membered cyclomonomers **73**and cyclodimers **74**, respectively, when conducted in the presence of CuI/iPr$_2$NEt/lutidine (Scheme 18.15). The distance between the azide and alkyne groups in the Ugi adduct **72** determined the macrocyclization's selectivity.

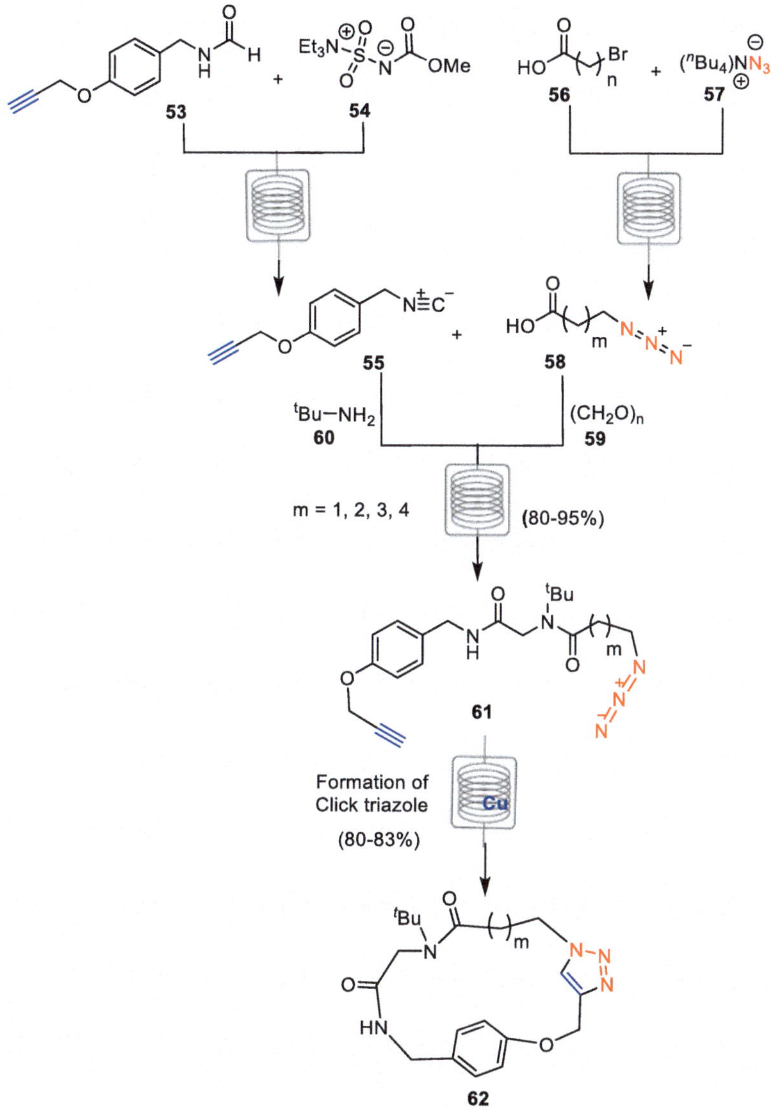

Scheme 18.13 An efficient synthesis of cyclic peptides **62** using Continuous-flow chemistry

18.2.3 Synthesis of Glycohybrid 1,2,3-Triazoles

Triazole-based compounds are significant class of compound toward their biological studies as well as they participation as synthetic precursors in variety of reactions [46–48]. Among them, glycohybrid triazole hybrid analogues are important in many fields, including drug discovery and development, it is necessary to develop

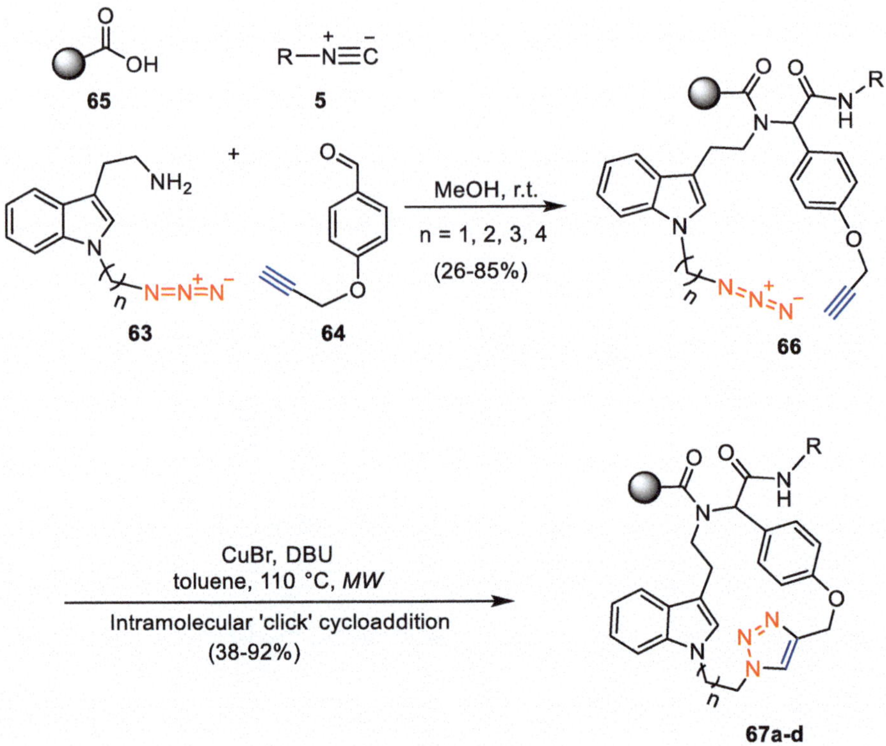

Scheme 18.14 Four-component Ugi reaction toward the synthesis of tryptamine-based macrocycles **67**

quick, simple, high-yielding, and innovative practical synthesis methods. Today, it is acknowledged that the CuAAC-tandem MCR techniques provide a solid framework for quick access to related triazole-fused or isolated systems during multicomponent reactions [49–51].

An alternative strategy for putting the azido functionality at the β-position of a,α,β -unsaturated olefinic ester is to use the Michael addition; [52–54] although in a few circumstances, like the conjugate addition of azido anion to glycosyl, α,β-unsaturated olefinic ester, the approach was unexpectedly complicated. Due to the structural similarity between azide and diazoamines [55], amines might be quickly transformed into the corresponding azido derivatives by using the diazo-transfer reaction. As a result, our group developed a straightforward method for the simple synthesis of glycosyl-β-azido ester that involves Michael addition followed by diazo-transfer in one pot. [56] Because azide and diazoamines have structural similarities. As a result, olefinic ester **75** initially reacted with ethanolic ammonia to produce the corresponding glycosyl β-amino ester as a diastereomeric mixture. The resulting mixture was then used in a subsequent step to initiate a diazotransfer reaction with imidazolyl-1-sulfonyl azide **76** in a combination of K_2CO_3 and $ZnCl_2$, which produced the title glycosyl azide.

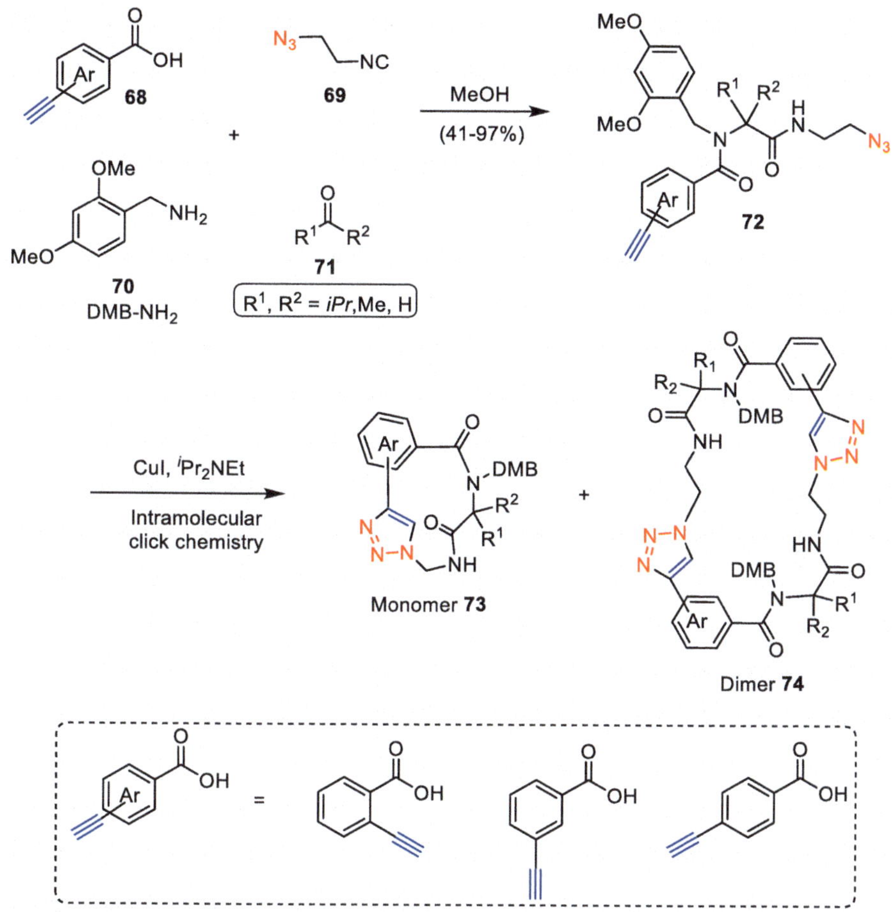

Scheme 18.15 Ugi-Click multicomponent route for the synthesis of macrocyclic peptidomimetics **73** and **74**

Flash column chromatography (SiO_2) made it simple to separate each of the azido esters **77a** and **77b** in their purest form, and NMR (^{1}H and ^{13}C) and single crystal X-ray diffraction investigation helped to clarify their structures. A dianionic tetrazene intermediate **B** was postulated as the pathway for the metal-catalyzed diazotransfer process. After reacting with $ZnCl_2$ and an amino ester in the presence of K_2CO_3, amine complex **A** underwent nucleophilic addition to the electrophilic center of $ImSO_2N_3$ **76**. This was followed by subsequent deprotonation, which produced a Zn-stabilized tetrazene intermediate **B**. This intermediate **B** could then undergo retro 1,3-dipolar cycloaddition to produce the desired glycosyl-β-azidoesters**77a,b** and their diastereomeric ratio was unaffected because the diazotransfer process did not

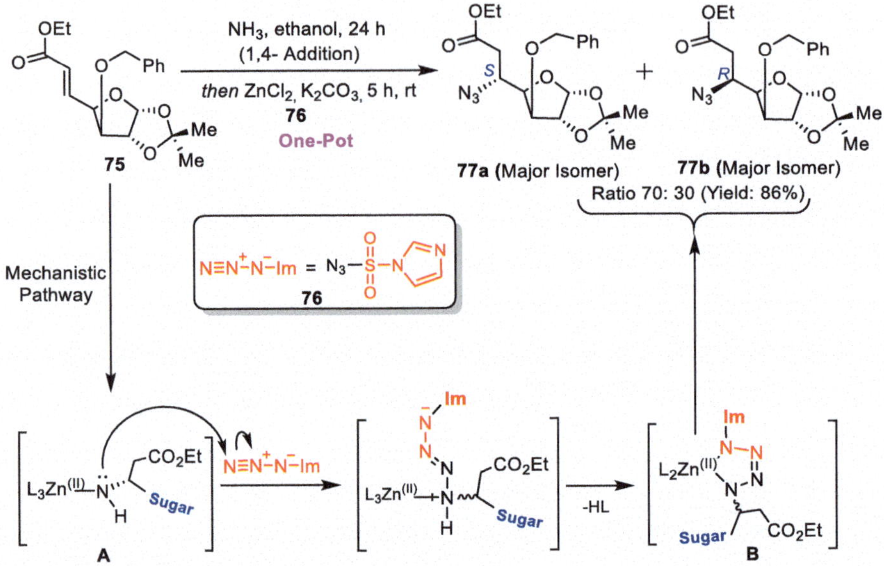

Scheme 18.16 Michael addition was followed by diazotransfer reaction using imidazolyl sulfonyl azide **76** in one pot for the synthesis of glycosylated-β-azido ester **77**

involve the C-5 stereogenic center (Scheme 18.16). These developed azide precursors were employed for the synthesis of a library of triazole-based glycohybrid using modular CuAAC.

It's interesting to note that some azide transfer processes have been successfully investigated in situ allowing quick access to clickable azides for future usage [57, 58]. Recently, Mancheno and colleagues revealed a Cu-Catalyzed one-pot instantaneous method for sp3C-H activation with azidation followed by subsequent CuAAC conjugation for the easy procurement of click-promoted triazole conjugates [59]. In their study, they use unprotected tetrasaccharides **78,** served as the reducing sugar in a similar sp^{3}C-H azidation reaction, which led to the in situ generation of the appropriate azidoglycoside. This azidoglycoside was subsequently utilized for click conjugation with peptide **79**, which contained the clickable propargyl glycine at the side chain, in a one-pot manner, and provided the appropriate complex glycopeptide **80**. (Scheme 18.17) [59].

Ethisterone, which is known to compete with androgen receptor (AR) binding and exhibit lower levels of AR transcriptional activation than dihydrotestosterone [60], was further investigated by the Tiwari group. Taking these considerations into account, the authors commenced toward the production of *bis*-triazolyl ethisterone glycoconjugates **85** in two successive steps. Initially, triazolyl azido alcohols **83** were produced by corresponding glycosyl alkyne **81**, oxirane analogue **82,** and sodium azide. The subsequent reaction of **83** with ethisterone **84** afforded bis-triazolyl ethisterone glycoconjugates **85** employing modular CuAAC tool. (Scheme 18.18) [61].

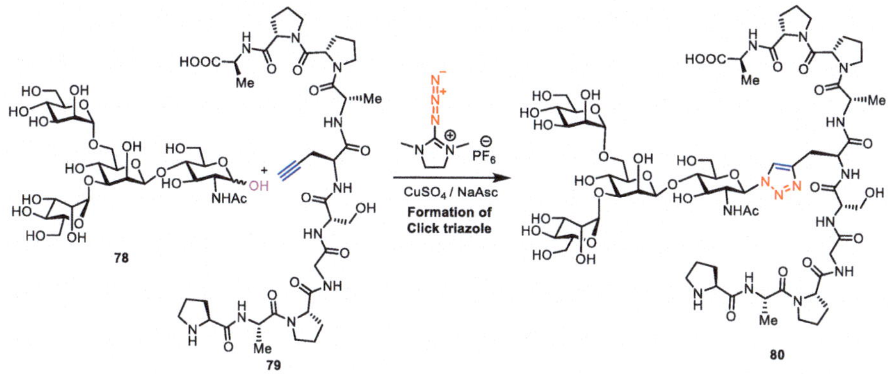

Scheme 18.17 CuAAC conjugation of totally unprotected tetrasaccharides with peptide in one step yields an intriguing glycopeptide **80**

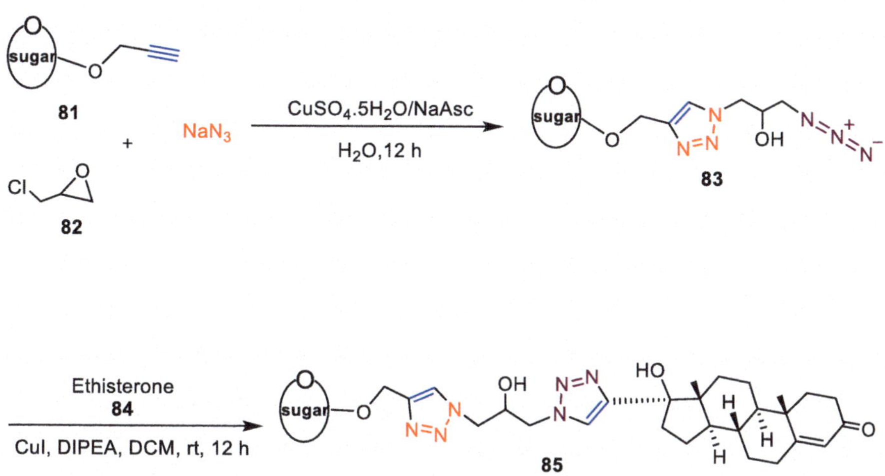

Scheme 18.18 Modular click reaction for the synthesis of bis-triazolyl ethestirone glycoconjugates **85**

As far as, very limited reports have been documented on the diversification of aminocyclitols via CuAAC alteration [62]. In this endeavor, synthesis of triazolyl C7-N aminocyclitol scaffolds was achieved by Qian et al. in two successive steps. In the first step, a diazotransfer reaction between C_7-N aminocyclitolanalogues **86,** namely valienamine, validamine or valiolamine via an imidazole-1-sulfonyl azide hydrochloride (diazotransfer reagent) furnished the corresponding azidocyclitol derivative **87** in assistance with Cu(II)-catalytic system. In step second, under conventional click conditions diverse alkyne functionality undergoes 3 + 2-cycloaddition reaction to furnish the respective C7N aminocyclitol-derivatised 1,2,3-triazoles **88** as a core moiety (Scheme 18.19) [63].

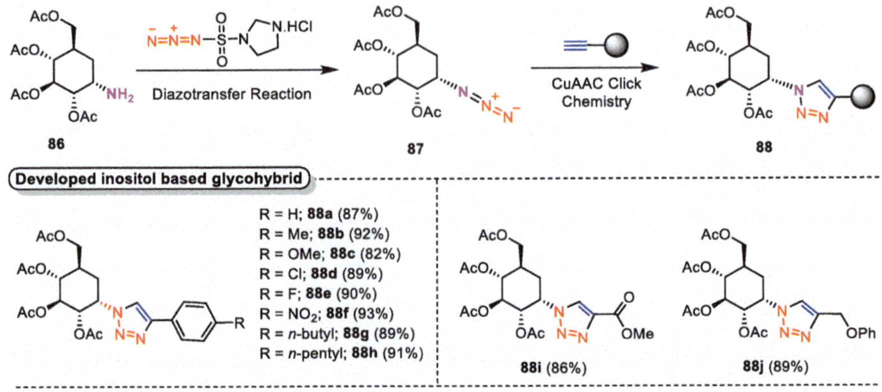

Scheme 18.19 Tandem synthesis of aminocyclitol-derived glycohybrid triazoles **88a-j** via CuAAC click reaction

Furthermore, Choudhary and the group, proposed a tandem synthesis of biocompatible isoindoline-fused triazoles via a Click-Heck coupling reaction using a Pd-Cu catalytic system. Glycosyl derivative **89** with alkyne functionality fused with azido analogue **90** in combination with $Pd(PPh_3)_2Cl_2$ and CuI/Et_3N in DMF solvent at 115 °C led to the formation of corresponding CuAAC product **91**. Afterward, the developed compound **91** undergoes an intramolecular Heck reaction to furnish the desired isoindoline-fused triazoles **92** in excellent yield (Scheme 18.20). An array of terminal alkynes was taken under consideration for synthesis [64].

In similar fashion, a report came with the Sen group for the synthesis of triazole-appended polycyclic glycosides **94** under sonication. Desired glycosides **94** were obtained by the subsequent reaction of glycosyl azido-alkyne analogue **393** with aryl iodide along with Cu_2O-nanoparticle and Cs_2CO_3in aqueous condition through intramolecular CuAAC Sonogashira reaction**.** Additionally, it is known that sonication-induced reactions distribute the catalyst, increasing its surface area.

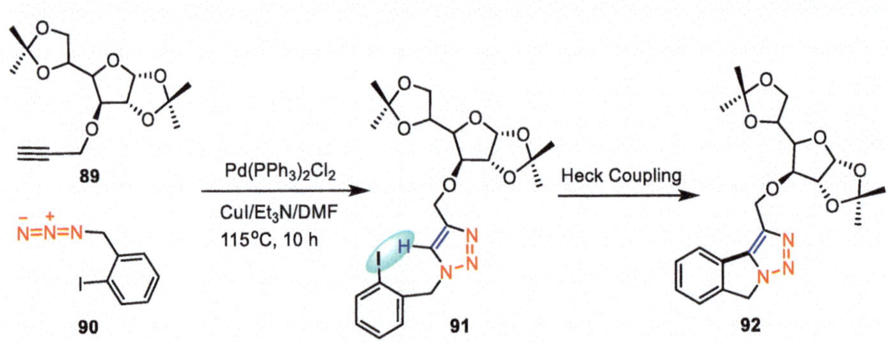

Scheme 18.20 Tandem synthesis of isoindoline-fused triazole **92** under Palladium-Copper Catalysis

Cu_2O, a heterogeneous catalyst with high catalytic activity, was thus recycled for three subsequent processes. The procedure is made even more practical by its wide substrate range and high to excellent reaction yields (72–90%) (Scheme 18.21) [65].

In addition to the simple click technique based on a copper catalytic system, various one-pot multicomponent condensation strategies have been investigated in parallel [66–68]. Mukherjee et al. successfully exploited this challenging tool to synthesize furan-linked triazolyl glycohybrids through a one-pot three-component conversion of D-glucal to respective furan-appended hydroxy triazoles **96** (Scheme 18.22). The click reaction was carried out with the assistance of catalytic amount of $Cu(OTf)_2$ and Cu powder. The reaction was conducted using the precursor **95** in one-pot via successive additions of the reagents under mild reaction conditions, affording an excellent yield of glycoconjugate **96** exhibiting excellent diastereomeric purity [69].

An interesting one-pot synthetic route was developed by Sen group to synthesize an array of triazolyl *N*-carboxamides **100** by the reaction of corresponding amines **97**

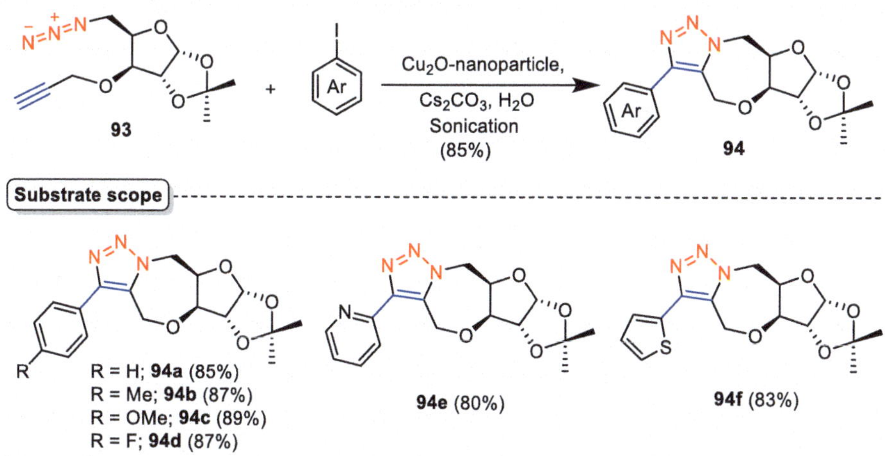

Scheme 18.21 Intramolecular CuAAC-mediated synthesis of triazole-fused polycyclic furanosides **94** in tandem fashion

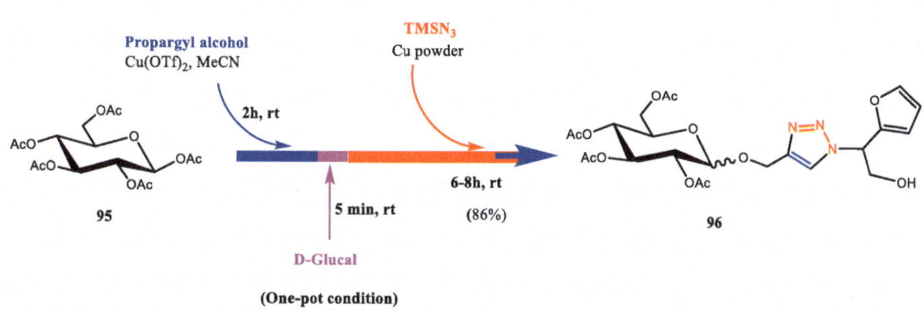

Scheme 18.22 One-pot three-component synthesis of furan-appended triazolyl glycoconjugate **96**

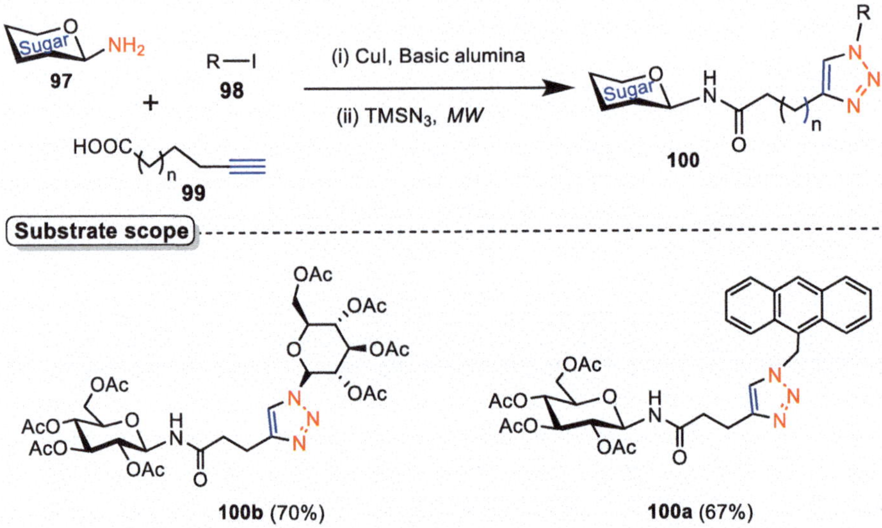

Scheme 18.23 Multicomponent tandem route for the synthesis of triazolyl *N*-Carboxamides **100** using basic alumina solid support

with diverse aryl/alkyl halides **98** and propargyl source **99** in assistance with copper source and basic alumina as solid support. The reaction ignores the use of solvents and bases and no need of ligands to assist the catalytic cycle during course of reaction. Therefore, the current methodology demonstrates the scale up synthesis and practical applicability of reaction (Scheme 18.23) [70]. The extension of same protocol was proposed by Arora group to synthesize biologically relevant unsymmetrical *bis*-1,2,3-triazoles through consecutive domino-Click approach [71].

Jenson in 2019, came with new terminology in multicomponent reaction and named it 'Combining Click Reactions'. They synthesized diverse tri- and tetrasaccharides mimics as only anomer through thiol-ene coupling followed by the regiospecific copper-catalyzed cycloaddition reaction of alkyne and azide. A linear synthesis of trisaccharide mimic was achieved in two steps in one pot. Carbohydrate building block with thiol functionality**101** (X1) and with alkyne functionality **102** (Y1) fused together under given reaction condition to furnish intermediatory **I**, then after subsequent addition of **103** (Z1) under click condition afforded corresponding linear trisaccharide mimic **104** (X1-Y1-Z1) (Scheme 18.24) [72].

18.3 MCR Click Under Multicatalytic Condition

It is very difficult to functionalize the C–H bond of triazole, usually required a high temperature as well as microwave heating [73]. For solving this problem, Xu and co-workers, exemplified the one-pot multicomponent synthesis of trisubstituted

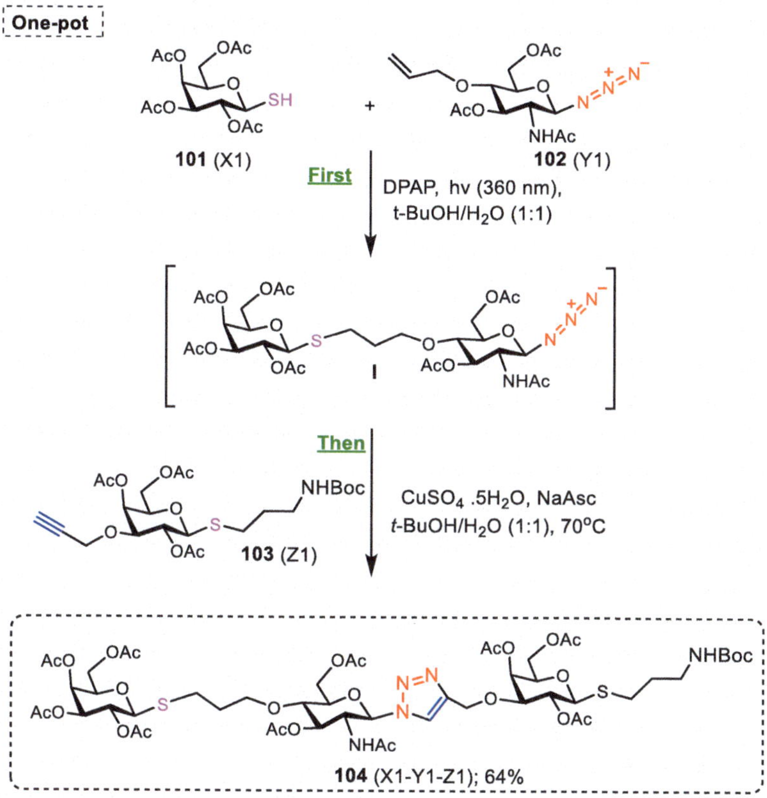

Scheme 18.24 Representative linear synthesis of the Gal-GlcNAc-Gal trisaccharide mimic **104**

triazoles **106**. They developed a new Cu/Pd transmetalation procedure to capture the vinyl copper intermediate **B** to overcome this problem. Thus, it was able to have relatively mild reactions and prevent direct C–H bond breaking. An innovative modular synthesis of trisubstituted triazoles from easily accessible components was achieved by employing an unprecedented Cu/Pd transmetalation relay catalysis. Three distinct substituents can be freely installed into the triazole ring in one step using this process. As shown in Scheme 18.25, the cuprate-triazole intermediate **B** can be formed by the cycloaddition of the copper(I) acetylide **A** with azide **2**. At the very same time, the palladium intermediate **C** gets generated by the oxidative addition of aryl halide **105** to a Pd(0) catalyst. The desirable trisubstituted triazole **106** would be formed through transmetalation involving **B** and **C**, and then followed by reductive elimination and catalyst Pd(0) was regenerated. However, due of the two ineluctable competing reactions, achieving this optimal catalytic cycle is a very difficult goal: (1) The undesirable 1,4-disubstituted 1,2,3-triazole **107** is formed by protonation on the cuprate-triazole intermediate **B** (route I). Additional byproduct **108**, a conventional Sonogashira coupling product is produced by the transmetalation of the palladium intermediary **C** and the copper(I) acetylide **A** (route II). We must have to eliminate all potential proton sources and boost the rate of transmetalation in order to block route I. An increased feasibility of transmetalation offers more space toward route II, which leads to the formation of Sonogashira coupling product **108**, which is undesirable (Scheme 18.25) [74].

The synthesis of chiral triazole-fused pyrazine scaffolds **112** from readily available substrates in a step-economical asymmetric catalytic way is highly appealing. We herein discuss that an efficient Cu/Ag relay-catalyzed protocol employing cascade asymmetric propargylic amination, hydroazidation, and [3 + 2] cycloaddition reaction with high efficiency to access the target enantioenriched 1,2,3-triazolo[1,5-*a*]pyrazine **112** has been accomplished by applying a novel *N,N,P*-ligand through intermediate **A**. The one-pot reaction of three components exhibits high functional group tolerance, excellent enantioselectivities, and a broad substrate scope with readily available starting materials(terminal alkyne **109**, azide source **110** and amine with alkyne functionality **111**) (Scheme 18.26) [75].

A widely useful strain-release glycosylation procedure was developed by Xu and Loh. They used a strained furanosyl donor **113** that produced good to excellent yields of the corresponding glycosylation products **114** with remarkable anomeric selectivities when it was reacted with an array of acceptors· The resulting linker pool molecules **114** with clickable alkyne functionality were modularly coupled to different glycosyl azides (both furanosyl and pyranosyl sugars) in assistance with standard coupling catalyst ($CuSO_4.5H_2O$/NaAsc) to deliver excellent yields of expected 1,4-disubstituted 1,2,3-triazole-appended glycohybrids **115** (Scheme 18.27) [76]. Furthermore, each of the linker diversification and CuAAC click orthogonal domain diversification stages of the multicatalytic diversification technique may be accomplished in a one-pot way to provide very good yields of the respective glycohybrids without isolating the linker pool.

18.4 Miscellaneous Reactions

The copper-catalyzed decarboxylative/click cascade reaction involving propiolic acids **118**, diselenides **119**, and azides**117** for the synthesis of 1,4-disubstituted 5-arylselanyl-1,2,3-triazoles **120** was described by Xu and co-workers. This reaction pathway utilizes $Cu(OAc)_2.H_2O$ (10 mol%), and 1,10-Phen (20 mol%), catalytic system along K_2CO_3 (1.2 equiv.), toluene, at 120 °C for 6 h to enable diverse range of compounds with good to outstanding regioselectivity. The current protocol may be applied to a variety of propiolic acids that contain aromatic, heteroaromatic, and aliphatic groups, as well as alkyl and benzyl azides. Moreover, the synthesized highly functionalized 5-seleno-1,2,3-triazoles **120a** and **120d** displayed remarkable anti-cancer activity in vitro (Scheme 18.28) [77].

Li et al. proposed the regioselective synthesis of 1,4,5-trisubstituted 1,2,3-triazoles **124** in the air is accomplished via a metal-free three-component coupling of α-CF_3 carbonyls **121**, azides **123**, and amines **122**. After using the enolizable α-CF_3 ketones

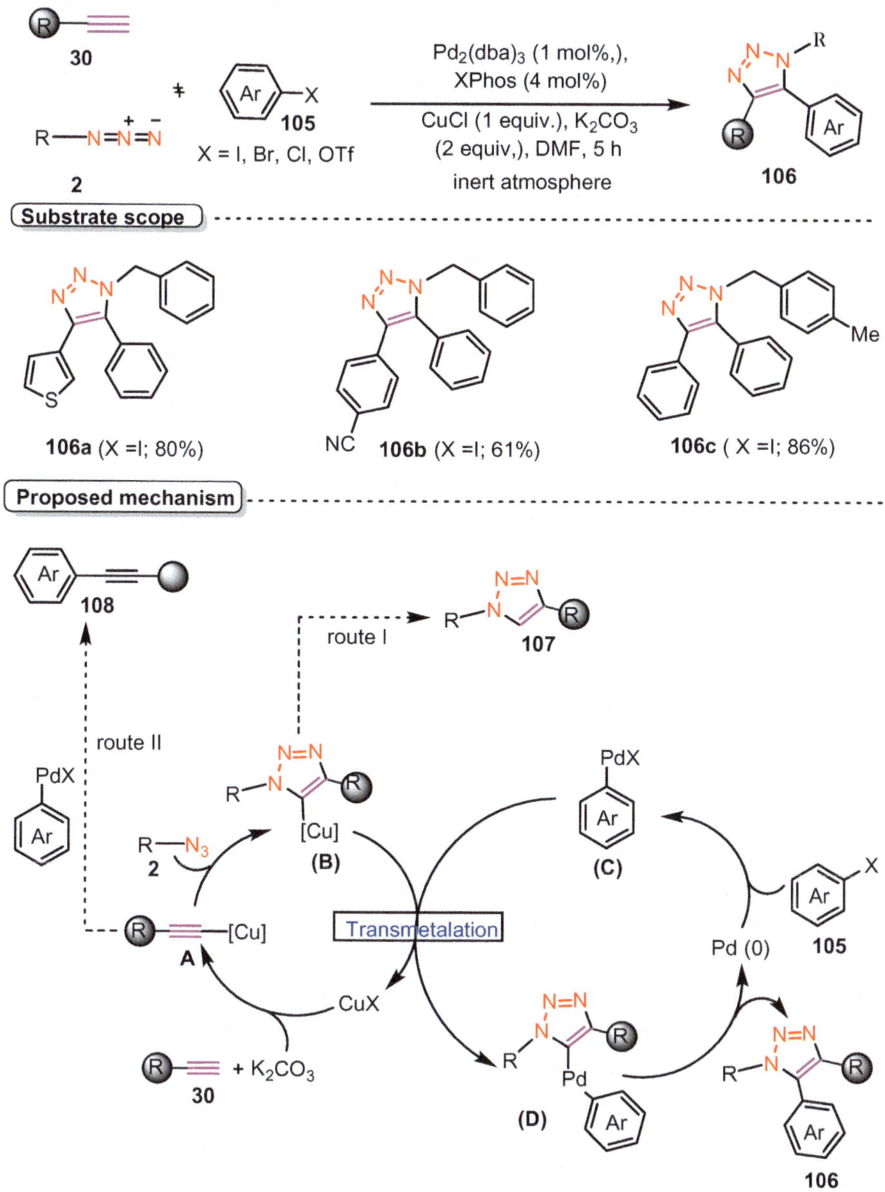

Scheme 18.25 Multicomponent synthesis of trisubstituted triazole under Cu-Pd catalytic system

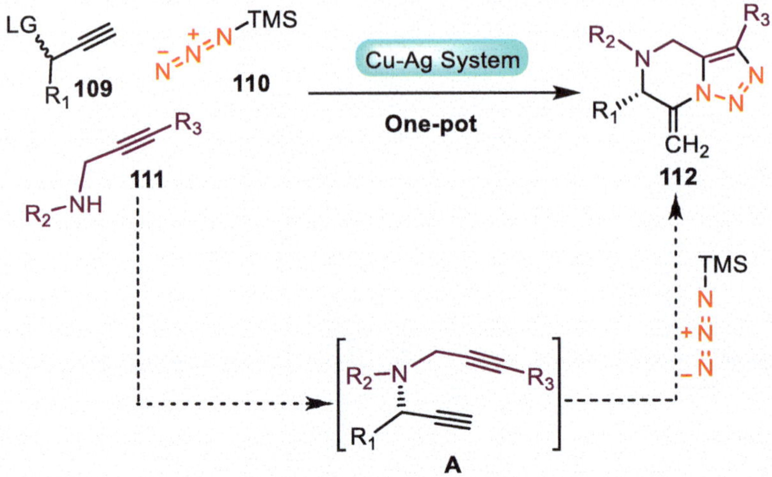

Scheme 18.26 One-pot synthesis of fully substituted 1,2,3-triazolo pyrazine derivative **112**

in an aqueous reaction media, several substituted 4-amide 1,2,3-triazoles were produced. Exclusively, functionalized 5-amino 1,2,3-triazoles were produced in the case of non-enolizable α-CF3 esters, which may allow for the post-functionalization of complex bioactive molecules (Scheme 18.29) [78]. Based on a literature study [79–81], Li proposed the mechanistic route for the formation of **124** and stated that primarily, gem-difluoro vinyl ketone **I** is produced via base-assisted dehydrofluorination of α-CF3 ketone **121**, and it then undergoes nucleophilic addition–elimination with pyrrolidine **122** to produce intermediate **II**. Two conceivable reaction routes are suggested, both of which are based on the active intermediate **II**. When R is an aryl or an alkyl, water reacts with intermediary species **II** to produce the enolizable β-oxo amide **III**, which then undergoes 1,3-dipolar cyclization with aryl azide **123** to furnish the intermediate **V**, and finally removal of pyrrolidine afforded1,4,5-trisubstituted 1,2,3-triazoles **124** (Scheme 18.29).

Furthermore, the intermediate **II** formed in Scheme 18.29 can be instantly reacted with the aryl azide and differently substituted CF_3 carbonyls **125** through 1,3-dipolar cyclization to produce the triazoline intermediate **VI**, which on elimination of HF yielded respective compound **126** (Scheme 18.30) [78].

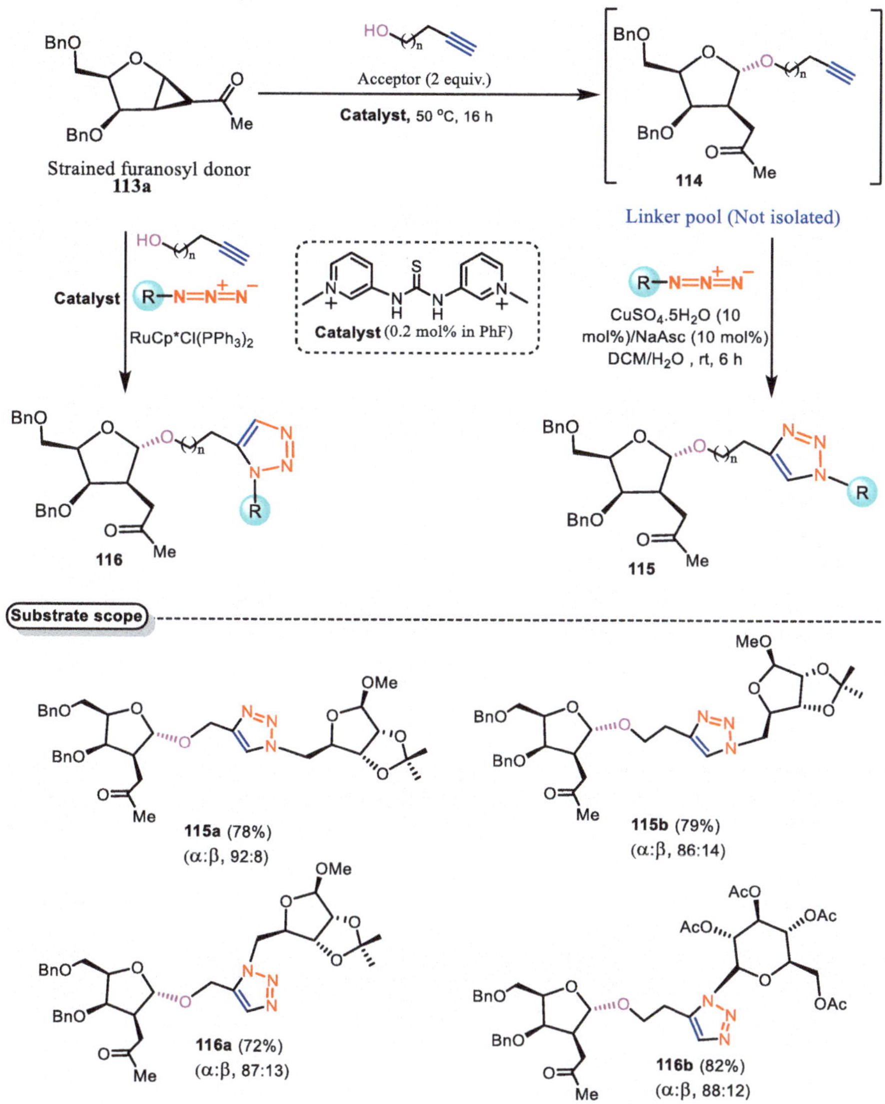

Scheme 18.27 Tandem thiourea-catalyzed one-pot strain-release glycosylation

A three-component regiospecific consecutive click-click approach was developed by Aucagne group for the construction of triazole-peptide ligation in one-pot fashion (Scheme 18.31). Reaction of compound **127** (containing two alkyne functionality that is TMS-protected and another with terminal C-H) with azido derivative **128** led to the regiospecific formation of triazole-linked species **129** on gentle heating. Azido derivative **128** chemoselectively reacts with alkyne functionality with terminal C-H in compound **127**, under given click condition ($CuSO_4$/NaAsc in *t*-BuOH/H_2O). The

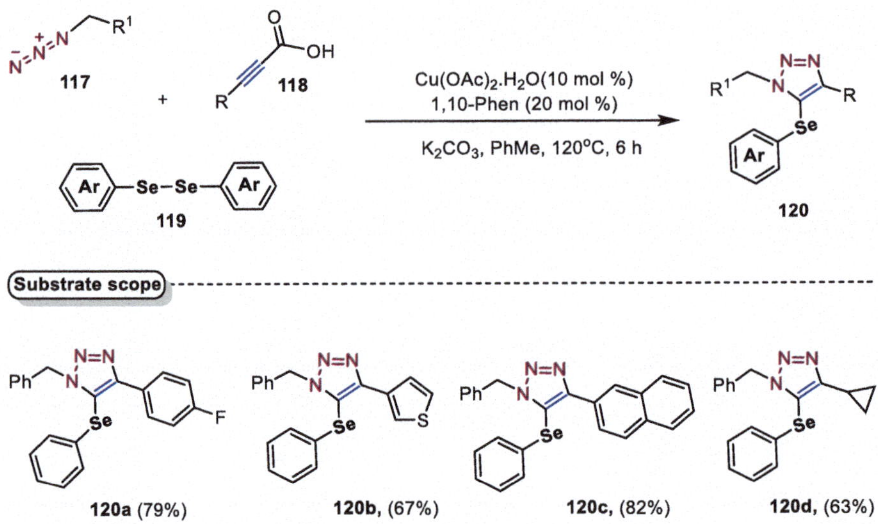

Scheme 18.28 Copper-catalyzed Synthesis of 1,4-disubstituted 5-arylselanyl-1,2,3-triazoles **120** through decarboxylative/click cascade reaction

chemoselectivity was controlled through the catalytic loading, and authors found that the excessive catalytic loading than the standard condition (mentioned in step i) led to the partial TMS deprotection and compound **130** was isolated in 17% yield. The successive addition of second azide **131** to **129** along with catalytic loading of $AgBF_4$ furnished the bis-triazole **132** in excellent yield (Scheme 18.31) [82].

Bis(1,2,3-triazole)s were frequently used as coordinating ligands for diverse range of transition metals [83–86]. Regarding this, in 2011, Simpson, et al. reported an efficient one-pot synthesis of 1,1′-Disubstituted *bis*(1,2,3-triazole)s through successive CuAAC. Benzyl azide reacts with monosilylbutadiynes **133** under click condition furnished TMS-protected ethynyltriazole **134**, along with some undesired symmetrical *bis*(1,2,3-triazole)s and compound **135**. Further, deprotection of ethynyltriazole **134**, using KF in methanol reasonable amount of compound **135** was isolated. The successive addition of diverse azides to **135** afforded unsymmetrical disubstituted bis(1,2,3-triazole)s **136** as a core compound (Scheme 18.32). In place of TMS-protected butadiyne stable monosilylbutadiynes (TIPS-protected) were also taken under consideration for reaction feasibility [87].

Chemical modification of proteins and peptides under multicomponent system offers a new way for drug development [88, 89]. In this regard, the authors demonstrated a primary amine modification process chemoselectively that allows for three-component (mercaptobenzaldehyde **137**, amino peptides **138,** and azide), one-pot bioconjugation toward the synthesis of bifunctional linker **139** (Scheme 18.33). This plan of action is founded on knowledge of the particular reactivity of the thiol group toward imines. The specifically constructed mercaptobenzaldehyde-based bifunctional linker **139** offers very selective and durable amine labeling

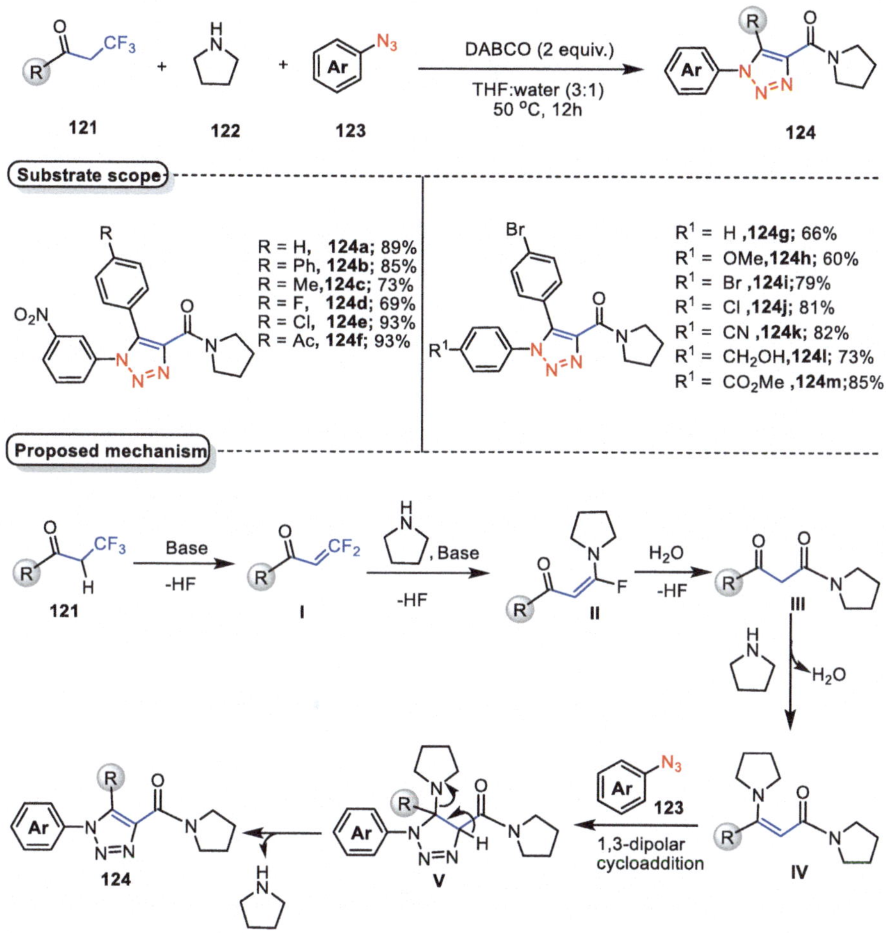

Scheme 18.29 Synthesis of 1,4,5-trisubstituted 1,2,3-triazoles **124** from α-trifluoromethyl carbonyls **121**, and amines **122**, azides **123**

under biocompatible circumstances. This linker has a wide functional group tolerance and is straightforward to construct, allowing for easy payload integration. Finally, research has proven that the addition of a linker does not hinder the function of changed proteins like insulin [90].

An interesting example of one-pot three-component chemoselective reaction was exemplified by Zhu et al. Unsymmetrical bisazides **140** with chelating and non-chelating azido functionality are conjugated to two distinct alkyne molecules **141** and **142** using chemoselective three-component copper(I)-catalyzed azide_alkyne conjugation processes. By combining two alkynes and a bisazide in a single reaction vessel, a bis-triazole molecule **143** can be produced with good regioselectivity, taking advantage of the reactivity gap between aromatic and aliphatic alkynes. This approach

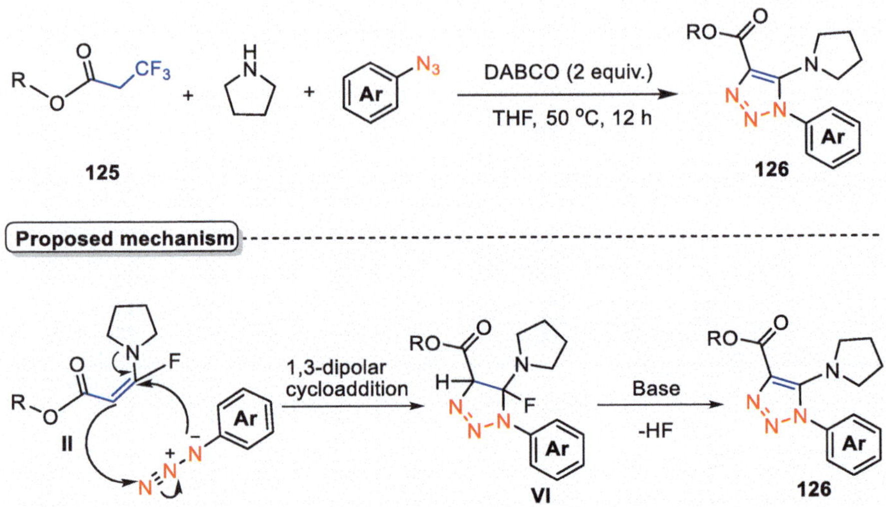

Scheme 18.30 Synthesis of the densely functionalized 5-amino 1,2,3-triazoles **126** from α-CF_3 esters

can be used with aqueous solutions that have a pH of 7, which makes it useful for bioconjugation applications (Scheme 18.34) [91].

18.5 Conclusion and Future Perspectives

Conclusively, we describe the importance of multicomponent reaction in click chemistry and their role in the synthesis of various biologically relevant scaffolds having significant clinical applications. This chapter exclusively highlighted the synthesis of various triazole-appended heterocycles, macrocyclization reactions, and construction of glycohybrid triazole architectures. Moreover, some interesting multicatalytic-promoted MCR and other miscellaneous synthetic pathway including late-stage functionalization of triazole, selenylation, and amidation reaction and protein modification, were also investigated in this chapter. These multicomponent click reactions provide a new vision to chemists toward the development of new methodologies for future perspectives.

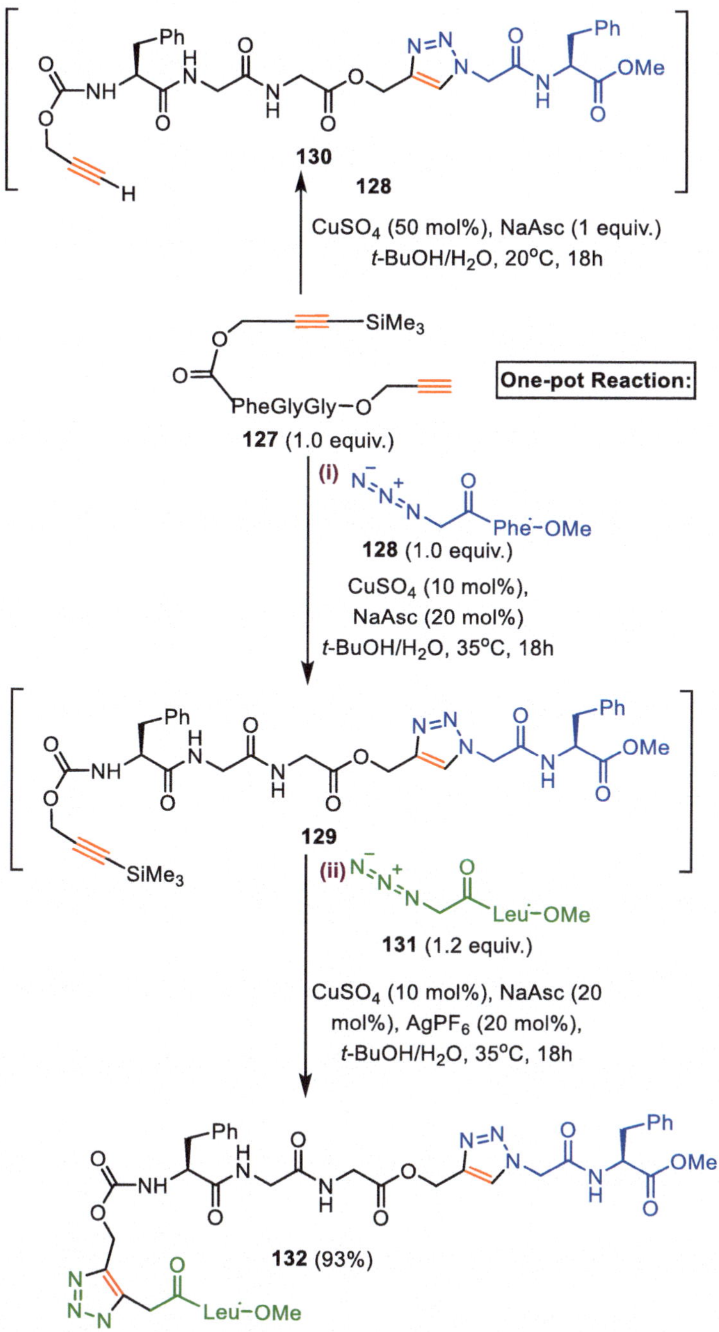

Scheme 18.31 Chemoselective formation of successive *bis*-triazole-linked peptide **131** in one pot

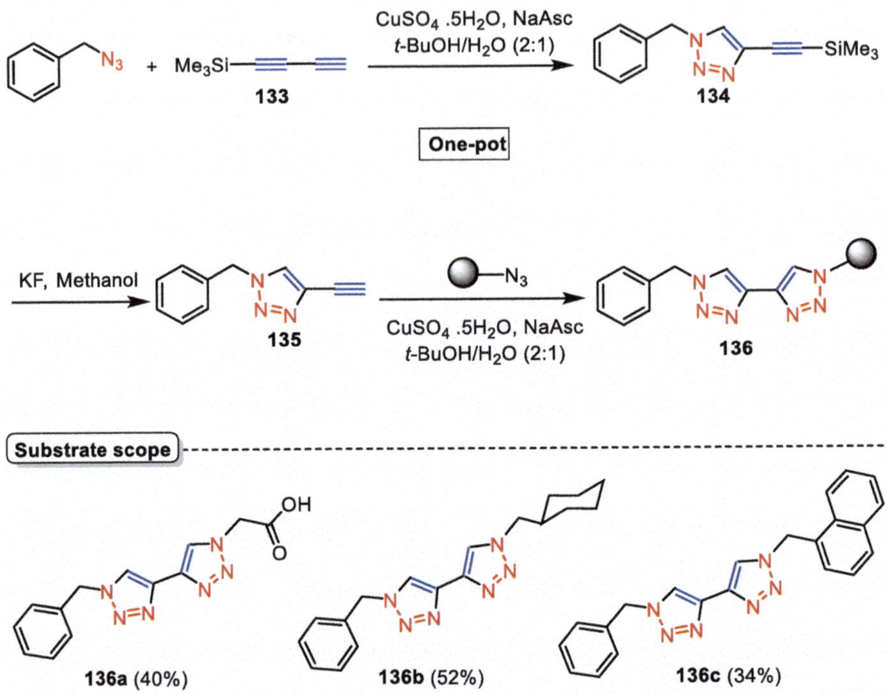

Scheme 18.32 Synthesis of Unsymmetrical 1,1′-Disubstituted *bis*(1,2,3-triazole)s **136**

Scheme 18.33 Three-component one-pot bioconjugation: a conventional click chemistry

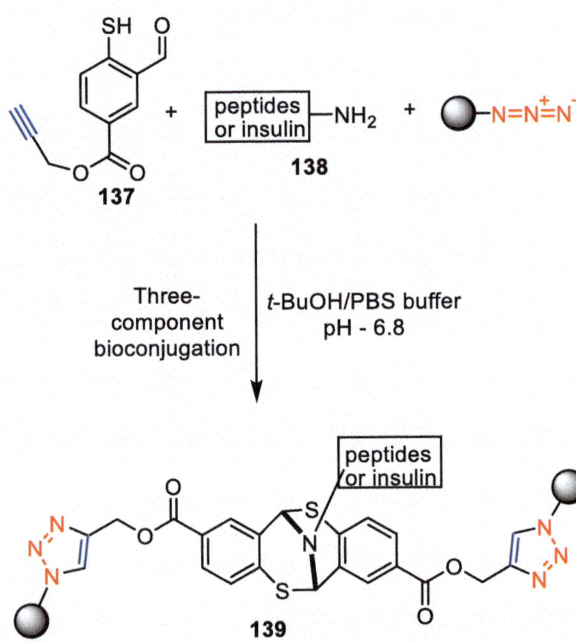

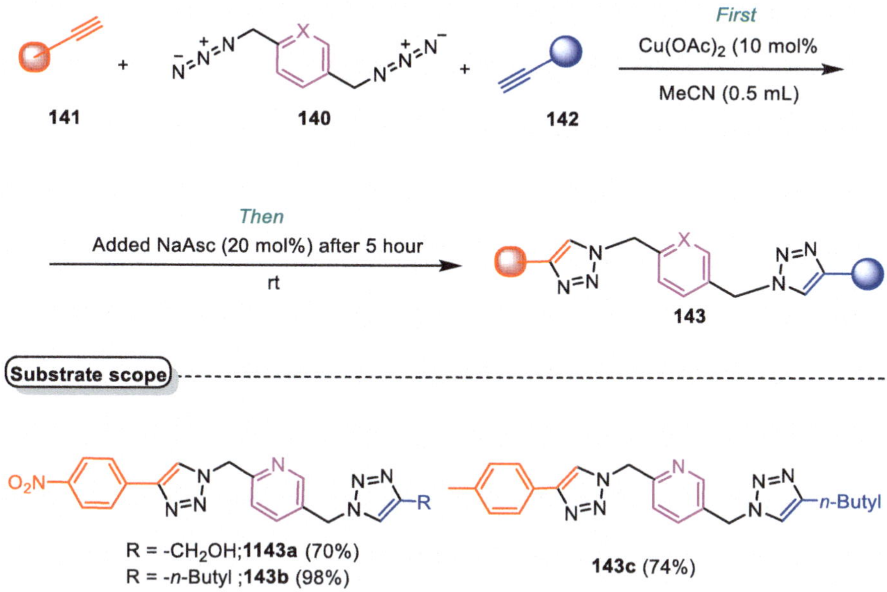

Scheme 18.34 One-pot three-component chemoselective synthesis of bis-triazole **143**

References

1. Meldal M, Tornoe CW (2008) Chem Rev 108:2952–3015
2. Cao Y, Huang H, Wang L, Lin X, Yang JJ (2023) Org Chem 88:4301–4308
3. Opsomer T, Dehaen W (2021) Chem Commun 57:1568–1590
4. Thomas J, Goyvaerts V, Liekens S, Dehaen W (2016) Chem Eur J 22:9966–9970
5. John SE, Gulati S, Shankaraiah N (2021) Org Chem Front 8:4237–4287
6. Domling A, Wang W, Wang K (2012) Chem Rev 112:3083–3115
7. Domling A (2006) Chem Rev 106:17–89
8. Agrahari AK, Bose P, Jaiswal MK, Rajkhowa S, Singh AS, Hotha S, Mishra N, Tiwari VK (2021) Chem Rev 12:7638
9. Tiwari VK, Mishra BB, Mishra KB, Mishra N, Singh AS, Chen X (2016) Chem Rev 116:3086
10. Zhang L, Chen X, Xue P, Sun HY, Williams ID, Sharpless KB, Fokin VV, Jia G (2005) J Am Chem Soc 127:15998–15999
11. Boren BC, Narayan S, Rasmussen LK, Zhang L, Zhao H, Lin Z, Jia G, Fokin VV (2008) J Am Chem Soc 130:8923–8930
12. Johansson JR, Beke-Somfai T, Stalsmeden AS, Kann N (2016) Chem Rev 116:14726–14768
13. Jaiswal MK, Tiwari VK (2023) Chem Rec e202300167
14. Ugi I (1962) Angew Chem Int Ed 1:8–21
15. Passerini M, Simone L (1921) Gazz Chim Ital 51:126–129
16. Jarusiewicz J, Choe Y, Soo K, Yoo C, Park P, Jung KW (2009) J Org Chem 74:2873–2876
17. Ducatti DRB, Massi A, Noseda MD, Duarte MER, Dondoni A (2009) Org Biomol Chem 7:1980–1986
18. Ranu BC, Hajra A, Jana U (2000) J Org Chem 65:6270–6272
19. Bhagat S, Chakraborti AK (2007) J Org Chem 72:1263–1270
20. Mannich C, Krosche W (1912) Arch Pharm 250:647–667
21. Huisgen R, Szeimies G (1967) Möbius. L Chem Ber 100:2494–2507

22. Kolb HC, Finn MG, Sharpless KB (2001) Angew Chem Int Ed 40:2004–2021
23. Tornøe CW, Christensen C, Meldal M (2002) J Org Chem 67:3057–3064
24. Agalave SG, Maujan SR, Pore VS (2011) Chem Asian J 6:2696–2718
25. Donikela S, Mainkar PS, Nayani K, Chandrasekhar S (2019) J Org Chem 84(16):10546–10553
26. Zanze IA (2008) Curr Opin Chem Biol 12:324–331
27. Rocha RO, Rodrigues MO, Neto BAD (2020) ACS Omega 5:972–979
28. Reddy BS, Majumder N, Rao TP (2014) Synthesis 46:3408–3414
29. Barlow T, Jida M, Tourwe D, Ballet S (2014) Org Biomol Chem 12:6986–6989
30. Saeedi M, Mahadavi M, Foroumadi A, Shafiee A (2013) Tetrahedron 69:3506–3510
31. Shafie A, Khanaposhtani MM, Asadi M, Rahimi N, Ranjbar PR, Ghasemi JB, Larijani B, Mahdavi M, Shafaroodi H, Dehpour AR (2020) Mol Div 24:179–189
32. Asgari MS, Sepehri S, Bahadorikhalili S, Ranjbar PR, Rahimi R, Gholami A, Kazemi A, Khoshneviszadeh M, Larijani B, Mahdavi M (2020) Chem Heterocycl Comp 56:488–494
33. Nguyen HH, Palazzo TA, Kurth MJ (2013) Org Lett 15:4492–4495
34. Dhondge AP, Afraj SN, Nuzlia C, Chen C, Lee GH (2013) Eur J Org Chem 4119–4130
35. Moliner FD, Bigatti M, Banfi L, Riva R, Basso A (2014) Org Lett 16:2280–2283
36. Roxburgh CJ (1995) Tetrahedron 51:9767–9822
37. Parenty A, Moreau X, Campagne J-M (2006) Chem Rev 106:911–939
38. Martí-Centelles V, Pandey MD, Burguete MI, Luis SV (2015) Chem Rev 115:8736–8834
39. Witczak ZJ, Bielski R (eds) (2016) Domino and intramolecular rearrangement reactions as advances synthetic methods in glycoscience. Wiley
40. Pirali T, Tron GC, Zhu J (2006) Org Lett 8:4145–4148
41. Prasanna R, Purushothaman S, Raghunathan R (2014) Org Biomol Chem 12:9375–9383
42. Rao JNS, Raghunathan R (2015) Tetrahedron Lett 56:2669–2673
43. Salvador CEM, Pieber BU, Neu PM, Torvisco A, Kleber Z, Andrade C, Kappe CO (2015) J Org Chem 80:4590–4602
44. Chavez-Acevedo L, Miranda LD (2015) Org Biomol Chem 13:4408–4412
45. Zakharova EA, Shmatova OI, Kutovaya IV, Khrustalev VN, Nenajdenko VG (2019) Org Biomol Chem 17:3433–3445
46. Yadav MS, Singh AS, Agrahari AK, Mishra N, Tiwari VK (2019) ACS Omega 4:6681
47. Yadav MS, Singh SK, Agrahari AK, Singh AS, Tiwari VK (2021) Synthesis 53:2494
48. Agrahari AK, Jaiswal MK, Yadav MS, Tiwari VK (2021) Carbohydrate Res 508:108403
49. Tiwari VK (2021) Chem Rec 21:3029
50. Yadav MS, Jaiswal MK, Kumar S, Singh SK, Ansari FJ, Tiwari VK (2022) Carbohyd Res 521:108674
51. Jaiswal, MK, Gupta A, Yadav MS, Pandey VK, Tiwari VK (2023) Chem Eur J e202301749
52. Lakshmipathi P, Rama Rao AV Tetrahedron Lett 38:2551−2552
53. Guerin DJ, Horstmann TE, Miller SJ (1999) Org Lett 1:1107–1109
54. Xu LW, Li L, Xia CG, Zhou SL, Li JW (2004) Tetrahedron Lett 45:1219–1221
55. Witczak Z (2010) J Carbohydr Chem 36:176–193
56. Mishra A, Tiwari VK (2015) J Org Chem 80:4869–4881
57. Meng G, Guo T, Ma T, Zhang J, Shen Y, Sharpless KB, Dong J (2019) Nature 574:86–89
58. Guo T, Meng G, Zhan X, Yang Q, Ma T, Xu L, Sharpless KB, Dong J (2018) Angew Chem Int Ed 57:2605–2610
59. Brandhofer T, Ozdemir A, Gini A, Mancheno OG (2019) Chem Eur J 25:4077–4086
60. Levine PM, Imberg K, Garabedian MJ, Kirshenbaum K (2012) J Am Chem Soc 134:6912–6915
61. Mishra KB, Mishra BB, Tiwari VK (2014) Carbohydr Res 399:2–7
62. Ji L, Zhang DF, Zhao Q, Hu SM, Qian C, Chen XZ (2013) Tetrahedron 69:7031–7037
63. Ji L, Zhou G-Q, Qian C, Chen X-Z (2014) Eur J Org Chem 3622−3636
64. Chowdhury C, Mandal SB, Achari B (2005) Tetrahedron Lett 46:8531–8534
65. Chatterjee N, Pal R, Sarkar S, Sen AK (2015) Tetrahedron Lett 56:3886–3889
66. Qian W, Amegadzie A, Winternheimer D, Allen J (2013) Org Lett 15:286–289
67. Cai Q, Yan J, Ding A (2012) Org Lett 14:3332–3335
68. Alonso F, Moglie Y, Radivoy G, Yus M (2012) Heterocycles 84:1033–1044

69. Yousuf SK, Taneja SC, Mukherjee D (2010) J Org Chem 75:3097–3100
70. Pal R, Sarkar S, Chatterjee N, Sen AK (2013) Tetrahedron Lett 54:5642–5646
71. Arora BS, Shafi S, Singh S, Ismail T, Kumar HMS (2008) Carbohydr Res 343:139–144
72. Brinko A, Risinger C, Lambert A, Blixt O, Grandjean C, Jensen HH (2019) Org Lett 21:7544–7548
73. Ackermann L, Potukuchi HK, Landsberg D, Vicente R (2008) Org Lett 10:3081
74. Wei F, Li H, Song C, Ma Y, Zhou L, Tung C-H, Xu Z (2015) Org Lett 17:2860–2863
75. Geng Z-Q, Zhao C, Qian H-D, Li S-J, Peng H, Xu H (2023) Org Lett 25:4504–4509
76. Xu C, Loh CCJ (2018) Nat Commun 4057:1–12
77. Cui FH, Chen J, Mo Z-Y, Su S-X, Chen Y-Y, Ma X-L, Tang H-T, Wang H-S, Pan Y-M, Xu Y-L (2018) Org Lett 20:925–929
78. Gao G, Kuantao M, Lv L, Lia Z (2022) Adv Synth Catal 364:1402–1408
79. Lv L, Gao G, Luo Y, Mao K, Li Z (2021) J Org Chem 86:17197–17212
80. Yu Y, Zhang F, Peng T, Wang C, Cheng J, Chen C, Houk KN, Wang Y (2021) Science 371:1232–1240
81. Bakulev VA, Beryozkina T, Thomas J, Dehaen W (2018) Eur J Org Chem 262–294
82. Aucagne V, Leigh DA (2006) Org Lett 8:4505–4507
83. Struthers H, Mindt TL, Schibli R (2010) Dalton Trans 39:675
84. Yano M, Tong CC, Light ME, Schmidtchen FP, Gale PA (2010) Org Biomol Chem 8:4356
85. Crowley JD, Bandeen PH, Hanton LR (2010) Polyhedron 29:70
86. Urankar D, Pinter B, Pevec A, De Proft F, Turel I, Kossmrl J (2010) Inorg Chem 49:4820
87. Doak BC, Scanlon MJ, Simpson JS (2011) Org Lett 13:537–539
88. deGruyter JN, Malins LR, Baran PS (2017) Biochemistry 56:3863–3873
89. Tamura T, Hamachi I (2019) J Am Chem Soc 141:2782–2799
90. Dai Y, Weng J, George J, Chen H, Lin Q, Wang J, Royzen M, Zhang Q (2019) Org Lett 21:3828–3833
91. Yuan Z, Kuang G-C, Clark RJ, Zhu L (2012) Org Lett 14:2590–2593

Chapter 19
Post-functionalization of Click-Derived 1,2,3-Triazoles

19.1 Introduction

Click reaction is an important tool for the construction of a diverse range of molecules with interesting biological behavior [1–8]. A notable example of 'click chemistry' is the regioselective 1,2,3-triazole forming Cu(I)-catalyzed azide-alkyne 1,3-dipolar cycloaddition (CuAAC), which has become one of the most popular and legitimate methods for linking two entirely different scaffolds together via biocompatible 1,2,3-triazole linker [9–14]. 1,2,3-Triazole forming click reactions have been frequently explored in organic synthesis for the successful development of entities of small to macromolecular level such as neoglycoconjugates (dendrimers, glycopeptides, and glycoproteins) [15, 16]. A wide range of glycosyl triazolyl ligands of wide applications in chemistry and biology has been developed since last 22 years [17–19]. In recent years, curiosity about the post-modification of click 1,2,3-triazoles, commonly known as interrupted click reaction, has increased beyond the CuAAC, which traps the Cu(I)-triazole entity with appropriate electrophiles [20–22]. Contrarily, the development of intramolecular variations has not drawn much attention. Regarding this, interrupted click cyclization is one of the most widely utilized techniques to create 1,5-fused triazole frameworks [23–26] and it is although uncommon to create 4,5-fused triazole cores using the similar technique (Fig. 19.1) [27, 28]. These fused 1,2,3-triazoles having their distinct role in chemical biology, material chemistry as well as in drug delivery and development got much attention [13, 29–31].

Over 200 reviews have been published on 'triazole forming click chemistry' [1]; however, there is still limited literature available on their post-functionalization. This chapter highlights the victorious development of post-functionalization of click-mediated 1,2,3-triazoles in emerging area of organic synthesis. Using this innovative technique a number of fused heterocycles, substituted sulfino- or selenotriazoles were easily accessed under given reaction condition. Along with synthetic route for some representative 1,2,3-triazole-appended complexes, many other miscellaneous applications of the post-functionalization are also summarizes in this chapter.

V. K. Tiwari et al., *Click Chemistry*, Materials Horizons: From Nature to Nanomaterials,
https://doi.org/10.1007/978-981-97-4596-8_19

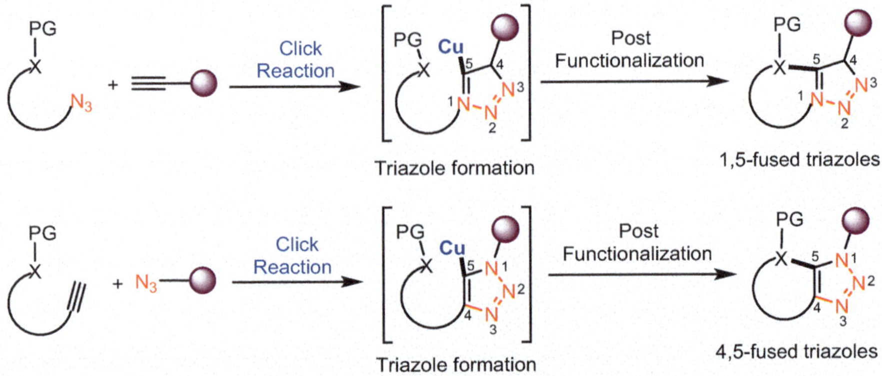

Fig. 19.1 Routes for the construction of fused triazoles via post-functionalization of click 1,2,3-triazoles

19.2 Application of Post-modification Strategy in Modern Organic Synthesis

A number of investigations were devoted for the successful synthesis of biologically relevant 1,2,3-triazole-appended fused heterocyclic scaffolds. Sun and co-workers, proposed an efficient pathway for the tandem synthesis of triazoloquinazolinones **3** under copper catalysis (Scheme 19.1) [32]. A single efficient copper catalyst was used to perform two different and significant reactions in one pot. The synthetic approach involves a consecutive one-pot cycloaddition reaction between azide analogue **1** and alkyne **2** to afford intermediate **I** and then followed by an aerobic intramolecular C–H amidation giving diverse range of triazoloquinazolinones **3** in good-to-excellent yield. The reaction proceeds well in the absence of ligand and is feasible when alkyne and azide functionality was implemented in the reaction.

In year 2016, Lautens came with a new approach for the multicomponent synthesis of fully substituted polycyclic triazoles **4** [33]. This multicomponent reaction included multicatalytic system of copper and palladium. The proposed reaction pathway demonstrated that this programmable multicomponent proceeds with two unique catalytic cycles operating separately from one another. The group projected that the palladium phase would come after the copper-catalyzed step. Foremost, CuAAC reaction between benzyl azide **1** and alkyne **2** led to the formation of intermediate **A**. The intermediate **A** quickly intercepted by the palladium catalyst to furnish intermediate **B**. Carbopalladation reaction on internal alkyne **3**, yielded the ultimate intermediate **C**, and subsequent C–H functionalization of the heterocycle would result in the expected fused polycyclic triazole **4** (Scheme 19.2).

Likewise, Choudhary and co-workers nicely used CuAAC Click-Heck coupling route in tandem fashion and developed a series of biologically relevant isoindoline-fused triazoles in high yields [26]. The authors first consider the click conjugation of glycosyl acetylene **7** with *o*-iodobenzyl azide **8** in presence of $Pd(PPh_3)_2Cl_2$ and

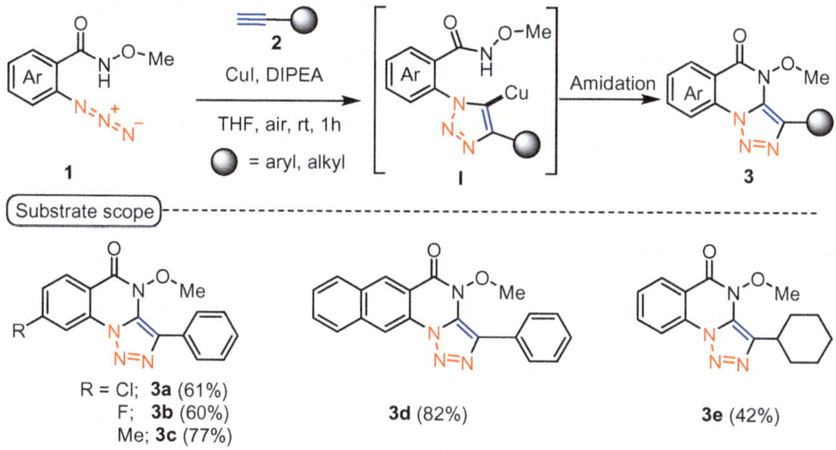

Scheme 19.1 Tandem synthesis of triazoloquinazolinones **3** from azide **1** and alkyne **2**

CuI/NEt$_3$ in refluxing DMF and respective triazle product **9** thus obtained was then underwent palladium-catalyzed Heck coupling as part of post-functionalization of triazole to furnish high yields of desired isoindoline-fused triazole **10** (Scheme 19.3) [26]. The tandem reaction method has some notable features including the compatible with diverse range of terminal alkynes.

In comparison to the synthesis of 1,4-disubstituted-1,2,3-triazoles, the advancement regarding successful placement of substituents at the 5-position has been pointed out in introduction part, is a challenge for modern chemists. Among them, acyl-linked triazoles namely 5-acyl-1,4,5-trisubstituted triazoles are useful frameworks with considerable biological significance [34, 35]. Nevertheless, there has been limited study into the direct synthesis of these functionalities from terminal alkynes using copper catalysis and there is still room for improvement. Lautens et al. in 2019 established an effective approach for the synthesis of 5-acyl triazoles **13** that made use of a basic CuI/2,2'-bipyridine-based catalytic mechanism (Scheme 19.4). The procedure entails an intramolecular acylation onto a carbamoyl chloride accompanied by a copper (I) catalyzed azide-alkyne cycloaddition (CuAAC). Azide **11** and alkyne **12** initially enrolled together to produce the triazole intermediate I, which after intramolecular cyclization through copper catalysis gave rise to the solitary product, 5-acyl triazoles **13**. This click triazole post-modification strategy progresses under benign circumstances, admits a number of functional groups, and is easily scalable. This procedure is an entirely novel approach to synthesis of complicated heterocycles through domino CuAAC/acylation route [36].

In particular, nitrogen- and sulfur-containing heterocycles show broad activities as anti-depressant, anti-cancer, or anti-viral drugs [37–39]. Sulfoximidoyl-containing heterocycles are analogues of sultams, which are investigated as bioactive compounds and functional materials [40]. A novel methodology for the synthesis of fused

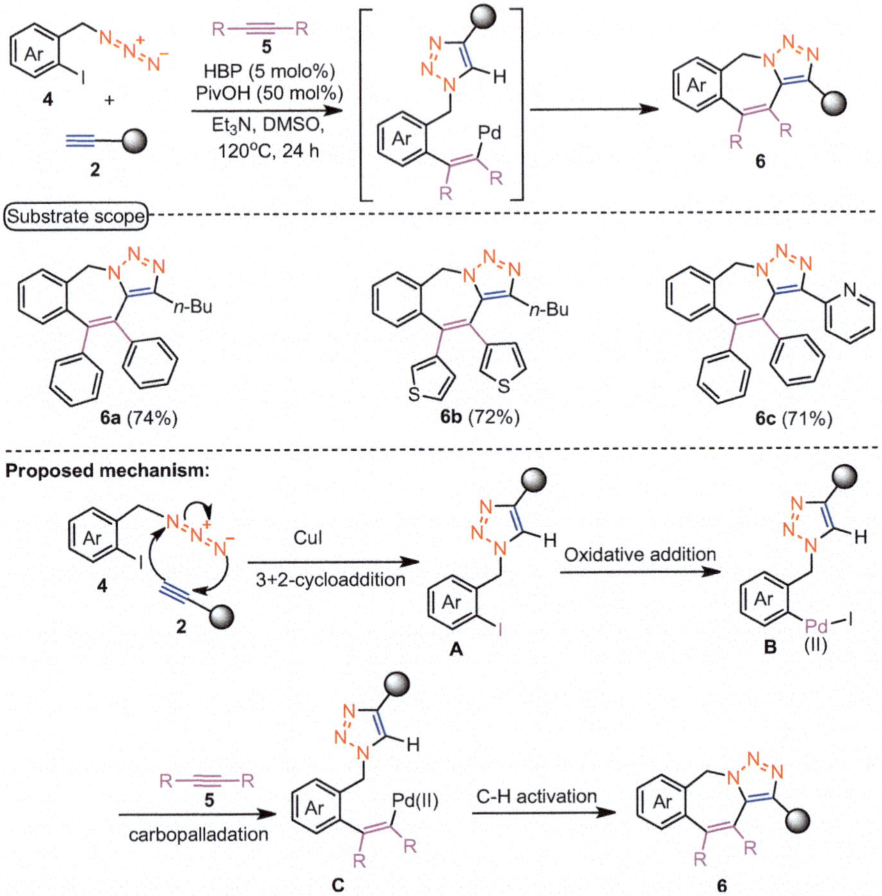

Scheme 19.2 Synthesis of fully substituted polycyclic triazoles **6** using multicatalytic system

triazolyl-containing benzothiadiazine-1-oxides from sulfoximidoyl-containing heterocycles based on the concept of copper-catalyzed interrupted domino reaction has been reported by Bolm and co-workers. 2-azidosulfoximines **15** and 1-iodoalkynes **14** were utilized as substrates in the presence of a copper catalyst and triethylamine base in dichloromethane solvent at 50 °C for 12 h to afford the desired fused triazolylbenzothiadiazine-1-oxide products **16** (Scheme 19.5) [41]. The group also postulated a mechanistic cycle for the conversion. First, the interrupted domino process starts with the reaction of the iodoalkyne **14** with the active copper species to generate a copper π-complex **A**. Subsequently, azide **15** undergoes complexation with species **A** to produce a complex species **B**, which transforms into acyclic vinylidene-type transition state **C**. This complex **C** undergoes a cycloaddition reaction to form 5-iodo-triazole intermediate **D**. An oxidative C–N coupling reaction

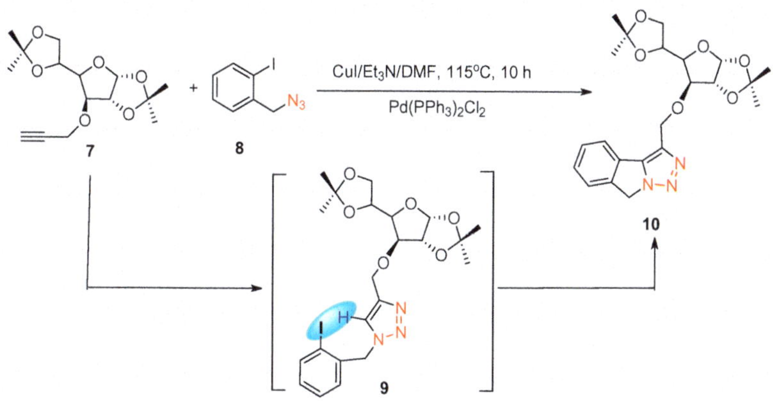

Scheme 19.3 Tandem CuAAC Click-Heck route for the development of biologically relevant isoindoline-fused triazole **10**

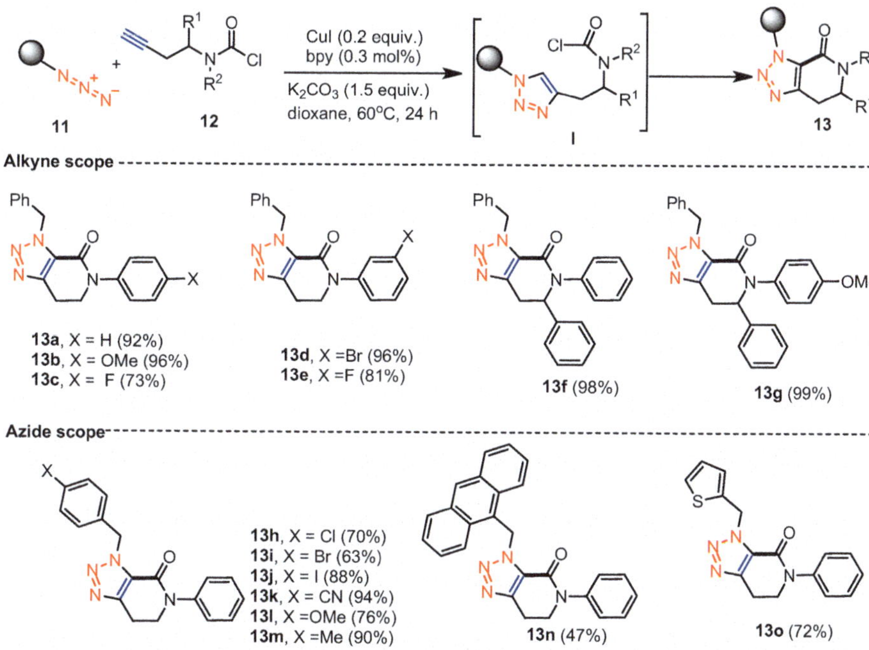

Scheme 19.4 Synthesis of acyl substituted triazoles **13** through interrupted click mechanism

of intermediate **D** to yield the intermediate **E**. Reductive elimination from the intermediate **E** produces the desired product **16** (Scheme 19.5) [41].

Reddy et al. exemplified the post-modification of click-mediated 1,2,3-triazole and established a high-yielding and straightforward click-sulfenylation cascade method

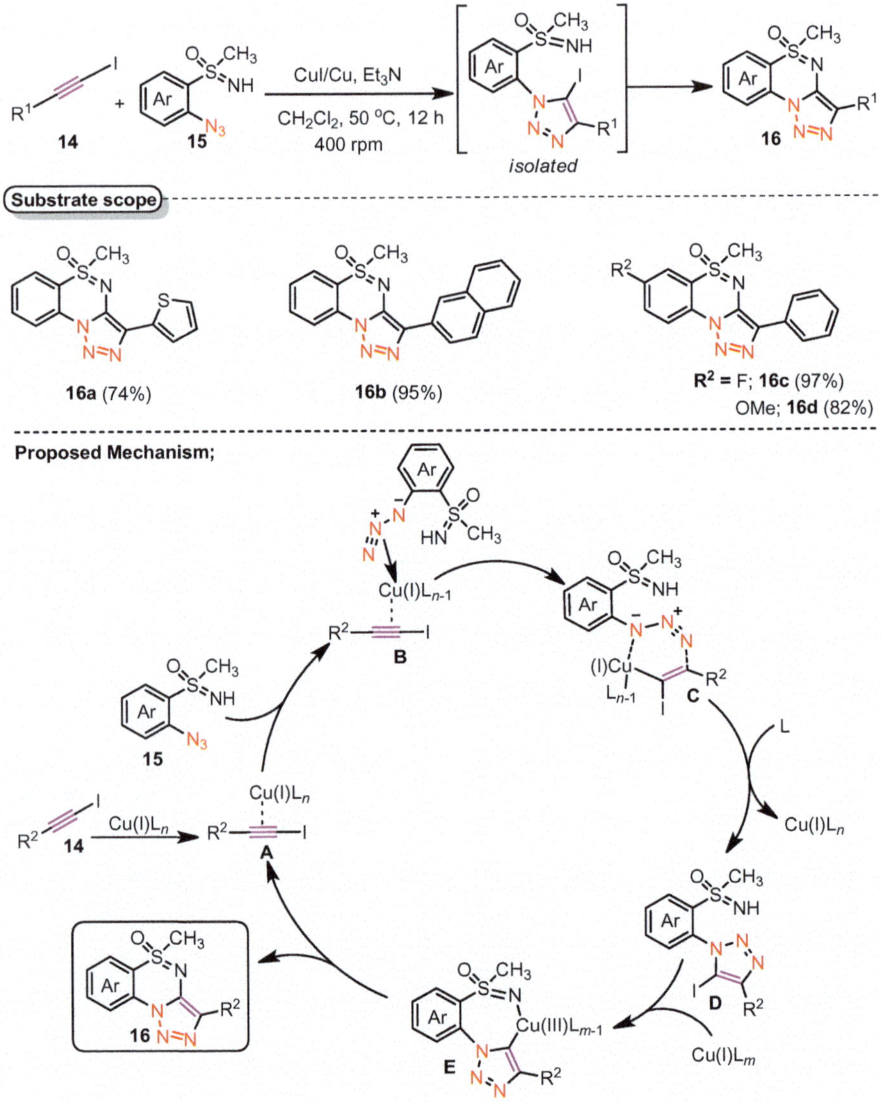

Scheme 19.5 Synthesis of fused triazolyl benzothiadiazine-1-oxides through copper-catalyzed interrupted domino reaction

to deliver high-to-excellent yields of the respective fused heterocycles benzoxathiocino or related benzothiazocino-fused triazole analogues **20** from *O*-propargyl benzyl thiosulfonate **17** (Scheme 19.6) [42].

A reasonable mechanism for this cascade process is provided by Reddy et al., in light of the aforementioned experimental findings and the literature data (Scheme 19.7). At first, CuI-1,10-phen produces the true copper-catalytic species

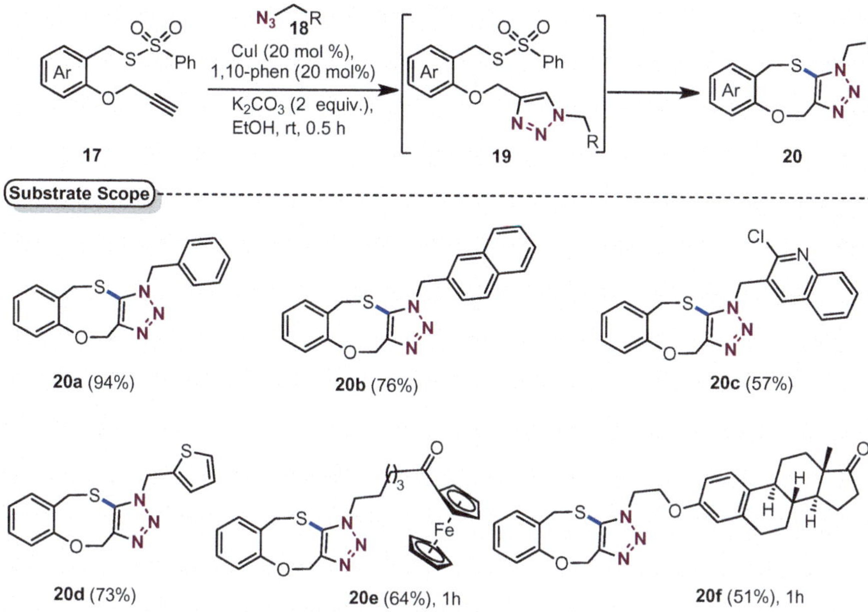

Scheme 19.6 Synthesis of benzothiazocino-fused triazole analogues via post-functionalization of triazole

(CuLn; L:1,10-phen), which may stimulate the π-system of **17** to make the copper(I) acetylide precursor **I**. A significant intermediary **II** would be produced by the [3 + 2]-cycloaddition of **I** with benzyl azide **18** via a nucleophilic β-carbon of Cu(I) acetylide and the electrophilic terminal nitrogen. It is assumed that the cuprate-triazole **II** intermediate is intricate with the sulfur moiety. The intermediate II's oxidative-protonation could result in the uncyclized product **19**. The efficient coordination of Cu–S in **II** may afterward make it easier to give a defined Cu(III) complex **III** through the oxidative-addition reaction. Additionally, the synthesis of the C–S bond would be made possible by the reductive elimination of the organocopper(III) precursor **III**, resulting in the desired scaffold **20** and the regeneration of functioning catalyst (Scheme 19.7) [42].

The method has wide substrate scope and can be utilized for an easy access of diverse range of biologically relevant tri- and tetracyclic frameworks. The mild reaction conditions and excellent functional group tolerance are the additional features. Also, the method works proficiently well via three-component cascade condition between **21**, sodium azide and benzyl bromide **22** to deliver high yield of fused triazoles **23** in gram-scale (Scheme 19.8) [42].

Toward the synthesis of bicyclic 5-thiotriazoles **25** and **28**, Wu and group hypothesized that this moderate domino reaction may successively establish a triazole ring and an intramolecular thioether ring. A fused bicyclic triazole product **25a** was produced in 79% yield when 1- azido-3-chloro-2-(tosyloxy)propane **24a** was coupled

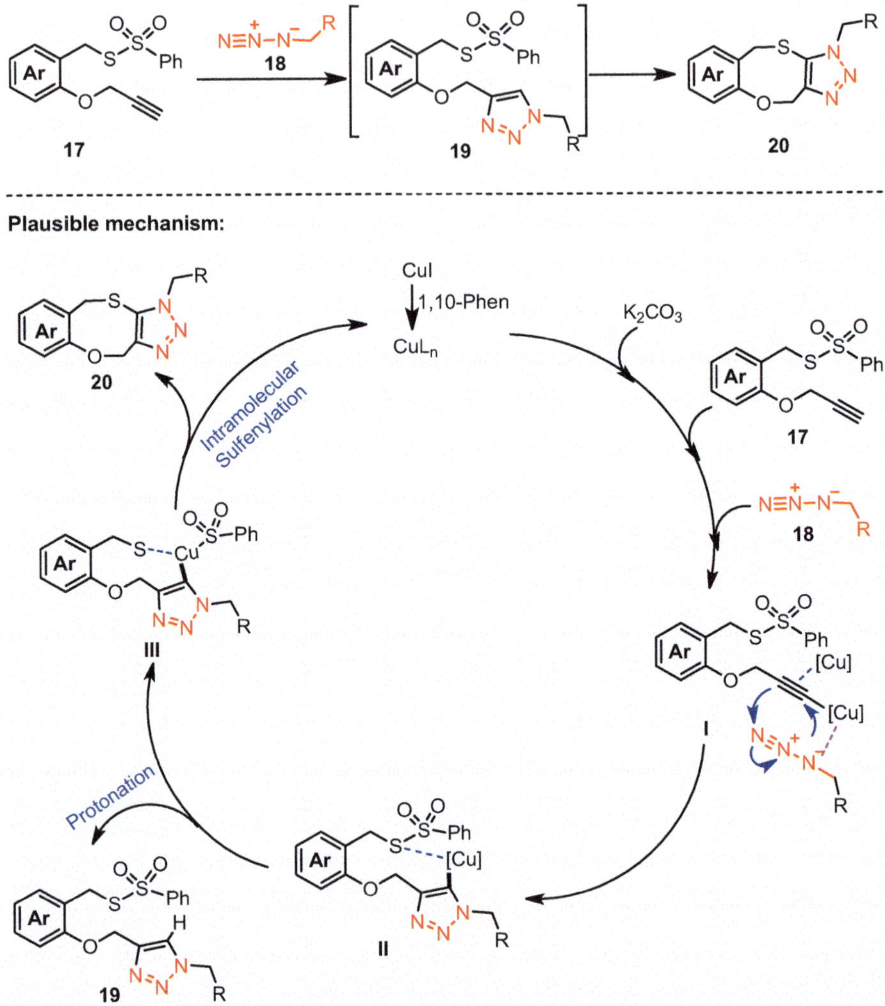

Scheme 19.7 Plausible synthetic route for the formation of benzoxathiocino or related benzothiazocino-fused triazole analogues through post-modification of click-mediated triazole

with phenylacetylene **2a** and S_8 under conventional conditions to verify the validity of this concept. Similar to this, other five-, six-, or eight-membered bicyclic triazoles were synthesized in good to outstanding yields. It was also possible to easily convert alkynes **26** with a tosylate group into bicyclic triazoles **28** likewise **28a** and **28b** in assistance with azides **27** (Scheme 19.9) [43].

A suggested mechanistic route for this protocol is shown in Scheme 19.9 based on the aforementioned experimental findings. Elemental sulfur is changed into copper sulfide **A** in the presence of CuI and base. Intermediate species **B** is produced by oxidative addition to a halide and is then converted into copper(II) thiolate **D** via

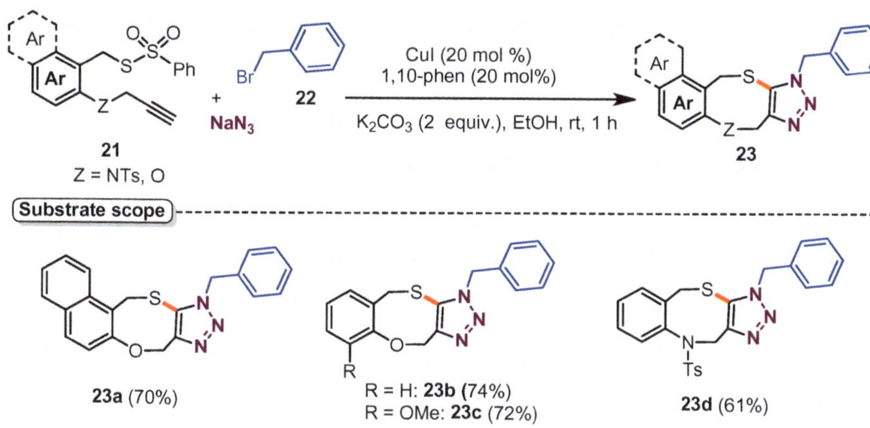

Scheme 19.8 Three-component cascade synthesis of fused triazoles through post-functionalization

reductive organo group transfer and oxidation reaction. Under basic circumstances, a terminal alkyne and CuI have been paired to generate copper(I) acetylide **M⁰**, which on cycloaddition reaction with the azide-generated copper(I) triazol-5-ide intermediate **M¹**. The copper(II) thiolate **D** and **M¹** would come together to generate the significant intermediate **E**. The anticipated 5-thiotriazole results from the reductive elimination of **E** [44, 45].

The formation of medium and large rings is typically difficult and challenging for chemists. In the year 2016, Xu group constructed fused bicyclic triazoles **31** in an intramolecular fashion using copper catalysis (Scheme 19.10). Initially, various functionalized alkyne precursors **29** were synthesized by fusing the thiosulfonyl and alkyne groups together in a single molecule. The reaction was performed with functionalized alkynes **29** and azide **30** to give triazole-appended intermediate species, followed by intramolecular cyclization furnished bicyclic triazoles **31** under standard reaction conditions. It is possible to synthesize several 5- and 6-membered rings, as well as medium- and large-sized rings with 7–14 members, in good yields [46].

19.3 Synthetic Approaches for the Construction of 5-Heterofunctionalized 1,2,3-Triazole

19.3.1 Sulfenylation, Selenylation, and Amination Reactions

Other than above-discussed approaches for the synthesis of fused 1,2,3-triazoles, a particular category of favored triazoles is the 5-heterofunctionalized 1,2,3-triazole, which is found in numerous synthesized compounds with a wide range of biological activities [47–49]. The modular CuAAC reaction, however, is unable to produce

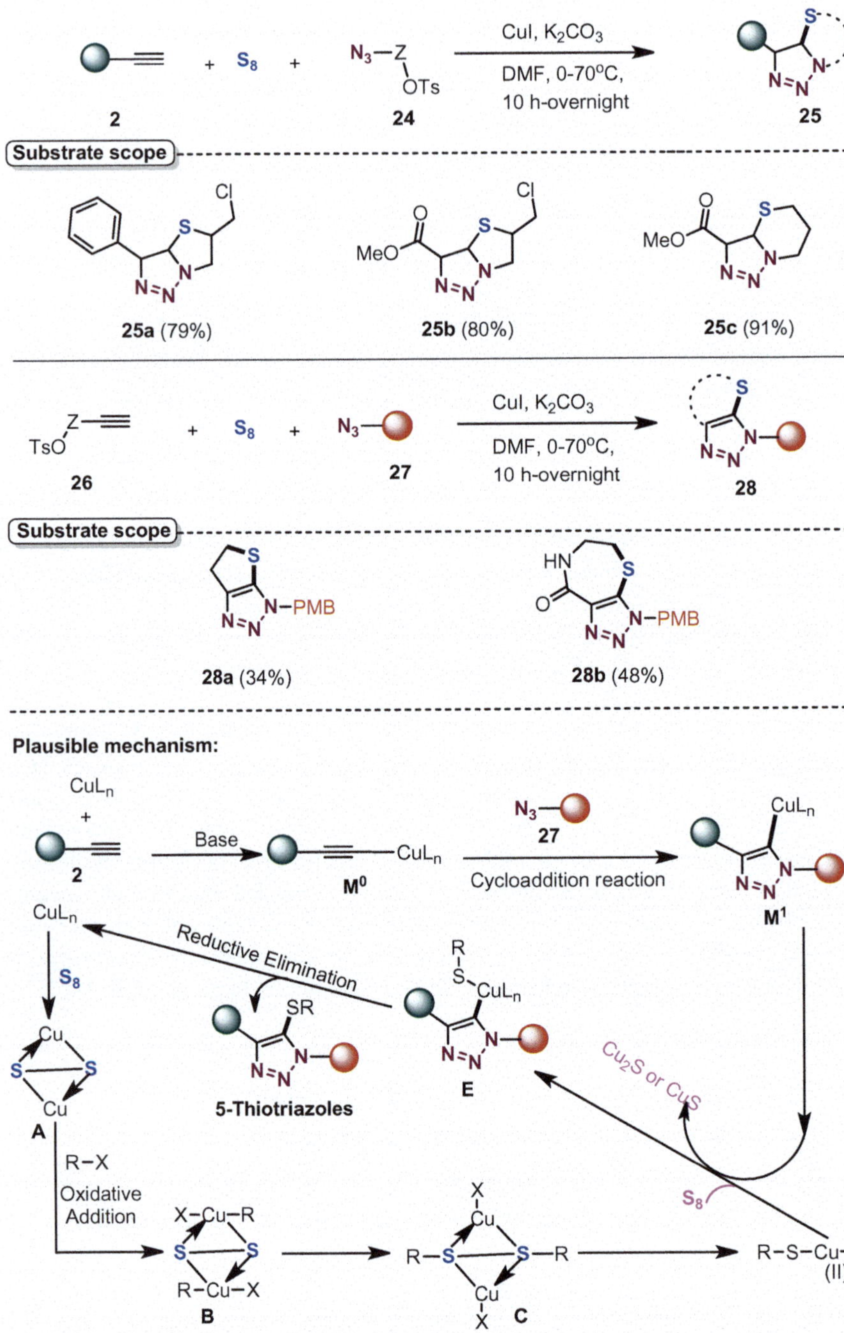

Scheme 19.9 Synthesis of 5-thiotriazole through 1,2,3-triazole post-modification strategies

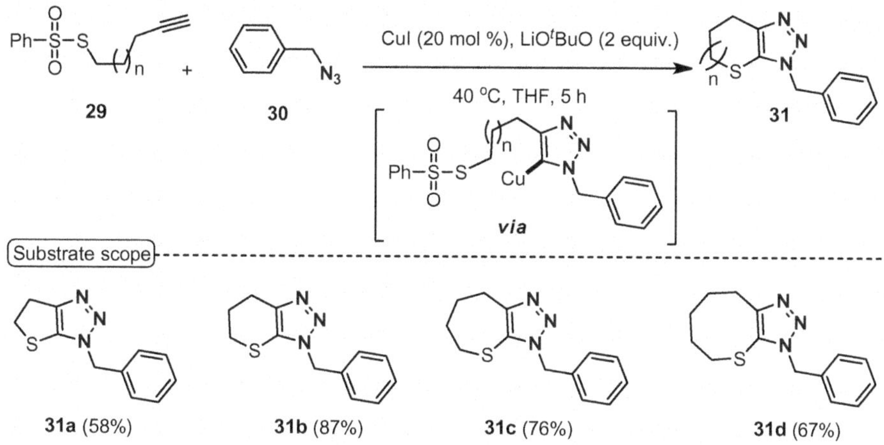

Scheme 19.10 Post-modification of click 1,2,3-triazole for the construction of diverse ring system through intramolecular pathway

these functionalized molecules since it is restricted to terminal alkynes. Therefore, the creation of an effective method to obtain various heterofunctionalized triazoles is crucial. Xu and group exemplified the synthesis of diverse 5-heterofunctionalized triazolesusing an emerging technique known as interrupted click reaction. Diverse alkynes and azides react simultaneously to form the cuprate–triazole intermediate, which was subsequently interrupted by the suitable heteroatom electrophile, afforded respective 5-heterofunctionalized triazoles. In their investigation, Xu group performed sulfenylation, amination, and selenylation reactions using different electrophilic reagents. S-methyl benzenesulfonothioate **34** is a well-known electrophilic sulfenyl agent for the introduction of the sulfenyl group. The sulfenylation reaction was performed by using alkyne **32** (1 equiv.), azides **33** (1.5 equiv.), sulfenylating agent **34** (2 equiv.), CuI (20 mol %), and LiO^{t}Bu as a base. The reaction mixture was allowed to stir in THF for 12 h at room temperature under inert atmosphere to generate the cuprate–triazole intermediate which on interaction with S-methyl benzenethiosulfonate afforded respective 5-thiotriazoles **35** in excellent yield. The amination product **37** was isolated in good yield on reaction with alkyne (1 equiv.), azides (1.5 equiv.), amination reagent **36** (1.5 equiv.), CuI (20 mol %), LiO^{t}Bu in pentane for 12 h at room temperature under inert atmosphere (Scheme 19.11) [46]. The mechanistic pathway describes that this is a competitive reaction. Initially, alkynes **32**on reaction with CuI produces copper acetylide intermediate I, which on subsequent reaction with azide **33** to formcuprate–triazole intermediate **II,** which subsequently reacted with electrophilic reagent **34** or **36**, furnished corresponding 5-substituted triazoles **35** or **37**through intermediate species **III**. A side reaction (protonation at 5th position) was also proceeding simultaneously to form the undesired product 5-*H* triazole **38**. Hence, the reaction condition needs an active electrophilic reagent (Scheme 19.11).

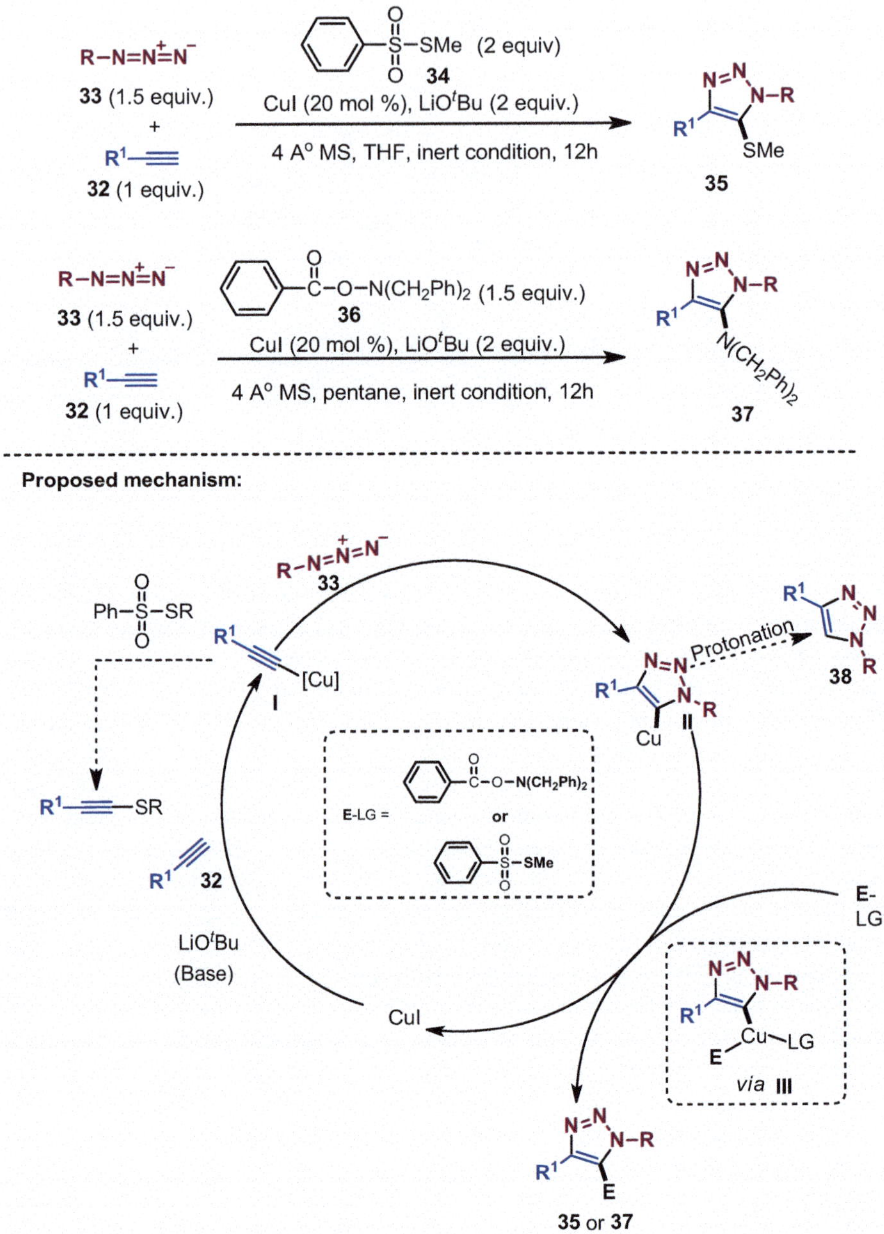

Scheme 19.11 Synthesis of 5-heterofunctionalized triazoles via copper(I)-catalyzed interrupted click reaction (a post-modification strategy)

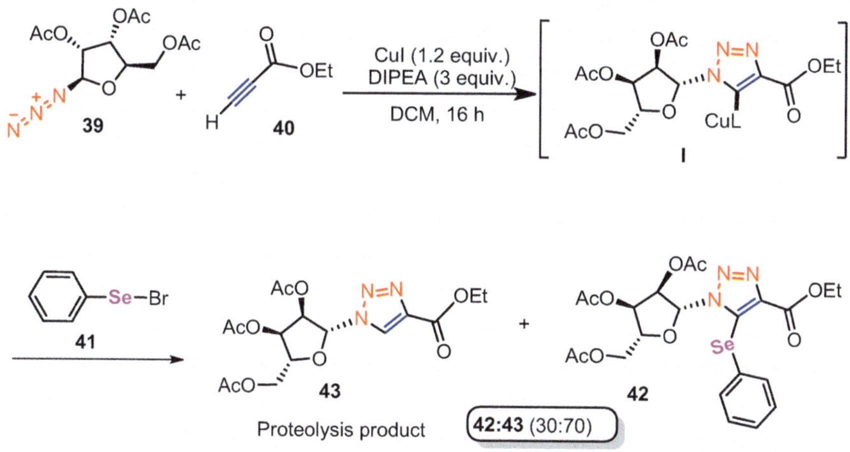

Scheme 19.12 Synthesis of 4,5-functionalized sugar-appended 1,2,3-triazole **42**

Behinda and colleagues devised an approach to synthesize 4,5-functionalized triazolyl-nucleosides **42** by employing an external selenium supply. The key techniques employed in this effective strategy are the tandem reactions of azide-alkyne [1,3]-cycloaddition and electrophilic addition (formation of triazole followed by electrophile addition). Surprisingly, a modest yield of the anticipated 5-phenylselanyl-substituted triazole **42** was obtained when PhSeBr **41** was used as the electrophile source in the reaction among alkyne **40** and azido-nucleoside **39**. The research group attributed the low yield to the weak reactivity of the electrophilic selenium species. The mechanistic route offered that initially, the formation of triazole intermediate **I** takes place by combination of alkyne **40** and azido-nucleoside **39**, which functionalized in the influence of copper followed by the reaction with PhSeBr to furnish 5-phenylselanyl-substituted triazole **42** along with the proteolysis product **43** (Scheme 19.12) [50].

In 2019, Wu and colleagues presented the one-pot multicomponent procedure for synthesizing a variety of 5-selenoor 5-thiotriazoles from elemental selenium or sulfur. Alkynes **2**, benzyl and alkyl azides **44**, variably functionalized arylboronic acids, organohalide **45** or tosylate **46**, and elemental sulfur or elemental selenium engage in this gentle procedure to produce 5-thiotriazoles in a yield of 21–91% or 5-selenotriazoles **47** in a yield of 20–85%, respectively. The processes required for the synthesis of the 5-selenotriazoles **47** were carried out either at 0 °C for 30 min or at 25, 50, or 70 °C for two hours to overnight (Scheme 19.13). It is crucial to emphasize that this practical route is effective for functionalizing the number of pharmaceuticals and bioactive substances. It also has strong functional group compatibility [43].

Yasuike in 2019 generalized an efficient copper-mediated three-component reaction between different organic azides **44**, ethynylstibanes **48** and elemental selenium to furnish diverse range of selenides **49** and diselenides **50**, having 1,2,3-triazole functionality (Scheme 19.14) [51]. For the synthesis of selenides **49**, copper iodide

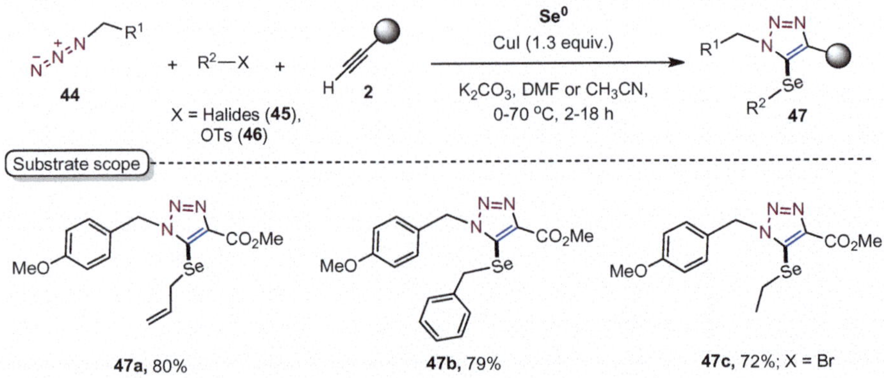

Scheme 19.13 Copper-catalyzed synthesis of 1,4-disubstituted 5-arylselanyl-1,2,3-triazoles **47** through decarboxylative/click cascade reaction

(10 mol %) and1, 10-phenanthroline (10 mol%) accomplished as an effective catalyst and ligand respectively in DMSO at 40 °C for 2–24 h. Different diselenides **40** were synthesized using the same reaction condition except varying the time interval that is from 1–5 h. This one-pot procedure offers regioselective double Se-arylation under straightforward reaction conditions by utilizing an antimony reagent. The established protocol was effectively worked with different alkyl, aryl, and vinyl alkynes as well as differently substituted azides to furnish a series of compound of interest. The proposed mechanism stated that initially organic azides **44**, ethynylstibanes **48** fused together afforded 5-stibanotriazoleintermediate **A**. Copper iodide and 1,10-phenanthroline combine to form an active catalytic species **I.** The intermediate **A** in assistance with active copper-ligand system **I** after transmetallation generated species **B**. The rlectrophilic interaction of selenium with intermediate **A** led to the formation of **C,** then after aromatization and subsequent elimination of Ph_2SbOH from **C** gives the ultimate intermediate **D**. The homocoupling on **D** afforded diselenides **50** while, another route offers the monoselenide **49** through different intermediatary pathway.

The same year, Zheng and Song presented an innovative proposal for the late modification of triazole in order to generate fully substituted triazolyl-organosulfurs. The group performed a reaction with internal thiocyanatoalkynes **51a** or S-(phenylethynyl) 4-methylbenzenesulfonothioate **51b** (1.0 equiv.), with diverse range of azides **23** (1.5 equiv.), in assistance with $Cu(MeCN)_4PF_6$(5 mol%), THF (0.5 M), at 80 °C in a sealed tube for 12–24 h to generate an intermediate species **A**. Nucleophiles such as Grignard reagent or others (2 equiv.) was added to the mixture at 0 °C and allowed the reaction at room temperature for 12 h to furnish respective triazolyl-organosulfurs **52** as sole product. During their investigation, it was noticed that internal thiocyanatoalkynes **51a** may provide greater yields compared to S-(phenylethynyl) 4-methylbenzenesulfonothioate **51b** (Scheme 19.15) [52].

Xu et al. further explored the interrupted click reaction methodology and came up with a novel approach for the synthesis of multisubstituted triazole sulfides by employing elemental sulfur as one of the substrates. The developed protocol was

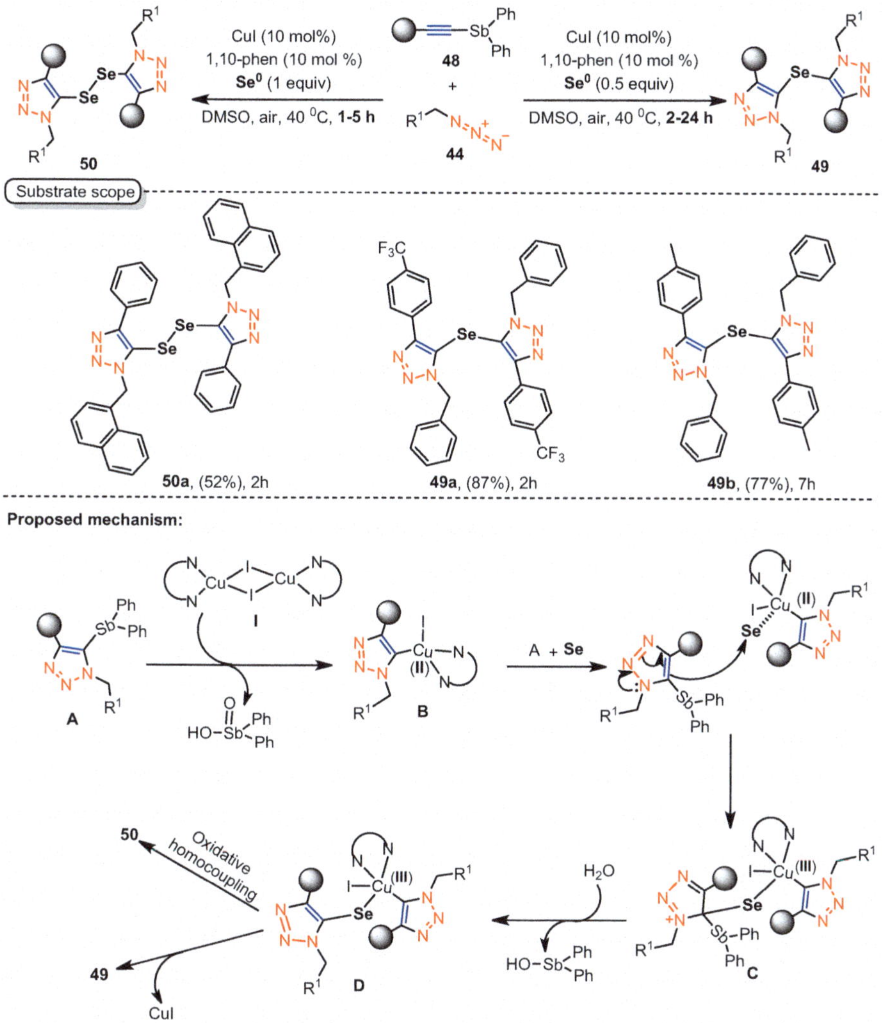

Scheme 19.14 Synthesis of selenides **49** and diselenides **50** via functionalization of click triazole

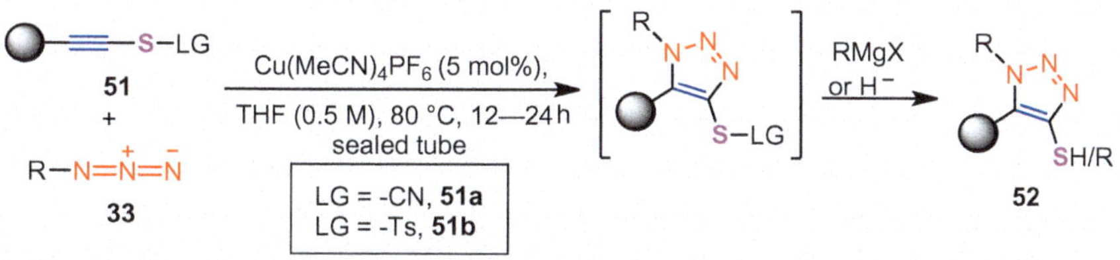

Scheme 19.15 Synthesis of fully substituted triazolyl-organosulfurs **52**

based on the multicomponent copper(I)-catalyzed interrupted click reaction strategy. The one of the most fascinating features of this method was the activation of elemental sulfur to furnish an active electrophilic sulfuration reagent which in situ, intercepts the click intermediate (cuprate-triazole) for the efficient synthesis of 5-thio **54** and 5-selenotriazoles **55** (Scheme 19.16) [53]. The authors also investigated and proposed a plausible mechanism for the developed synthesis. Initially, a copper(I) acetylide intermediate **A** is generated from reaction of alkyne **2**, Cu catalyst, and base. This intermediate **A** undergoes cycloaddition reaction with azide **33** afforded the key cuprate-triazole adduct **B**. On the other hand, aryl iodide **53** undergoes oxidative addition with the Cu(I) catalyst and produces ArCuI species. Further sulfur coordination with ArCuI gives rise to the intermediate **C**, which after reductive elimination produces a reactive electrophilic ArSI species. This species reacts with the cuprate-triazole intermediate **B** to furnish the final triazole product **54**. In an alternate pathway, intermediates **B** and **C** can undergo direct transmetallation reaction to form an intermediate **D**, which transforms into the final product **54** through reductive elimination [53].

The 'Click-to-Chelation' is an exciting, legitimate method for developing unique radiotracers with the anticipated properties for various, discrete clinical purposes. With the help of the modular CuAAC tool, Mindt and colleagues accomplished their goal to synthesize tridentate-inspired triazolyl metal chelators from the corresponding azide and alkyne in a single step, which were then chelated with metal carbonyls furnished desired scaffolds in satisfactory yields [54]. The produced glyco-hybrids were also acknowledged as the potential source of significant biomolecules including bombesin, thymidine, and phospholipids which have many biological uses in emerging fields. Two synthetic approaches 'regular click ligand' and 'inverse click ligand' were mostly used to produce sugar chelate hybrids, depending on the molecular orientation and best conditions for obtaining the ligands. In regular click method, a tridentate ligand system is produced by the combination of sugar azide with a clickable alkyne prochelator, while the inverse method involves combining sugar alkynes with the functionality of an azide prochelator [55].

Mindt et al. highlighted the application of the 'regular click approach' for the one-pot synthesis of ^{99m}Tc-tricarbonyl complex **59.** Primarily, triazole chelator **58** was produced by uniting glycosyl azide **56** and L-propargyl glycine **57**under copper catalysis. Upon addition of [99mTcCO3]$^{+}$ to triazole chelator **58,** ^{99m}Tc-tricarbonyl complex **59** was isolated in a significant yield. (Scheme 19.17) [54].

By applying an approach, similar, to 'Mindt's Click-Chelation' protocol, Fernandez, synthesized ^{99m}Tc(CO)$_3$-glucose-histidine complex **61,** functionalized at the C-2 position from respective azide **60** (Scheme 19.18) [56]. Further investigations show that the compound is approximately 90% radiochemically pure and chemically hydrophilic in nature, yet it has a weak affinity for binding plasma proteins.

An additional successful application of the 'Click-to-Chelation' strategy can be seen in the construction of 1-(2-picolyl)-triazolyl-based homoleptic complexes (**63** and **65**) derived from triazole hybrids (**62** and **64**) with photophysical properties. To generate the desired molecular configuration in high yield, the click-prompted triazolyl ligands are exposed to a metal complexation process (Scheme 19.19) [57].

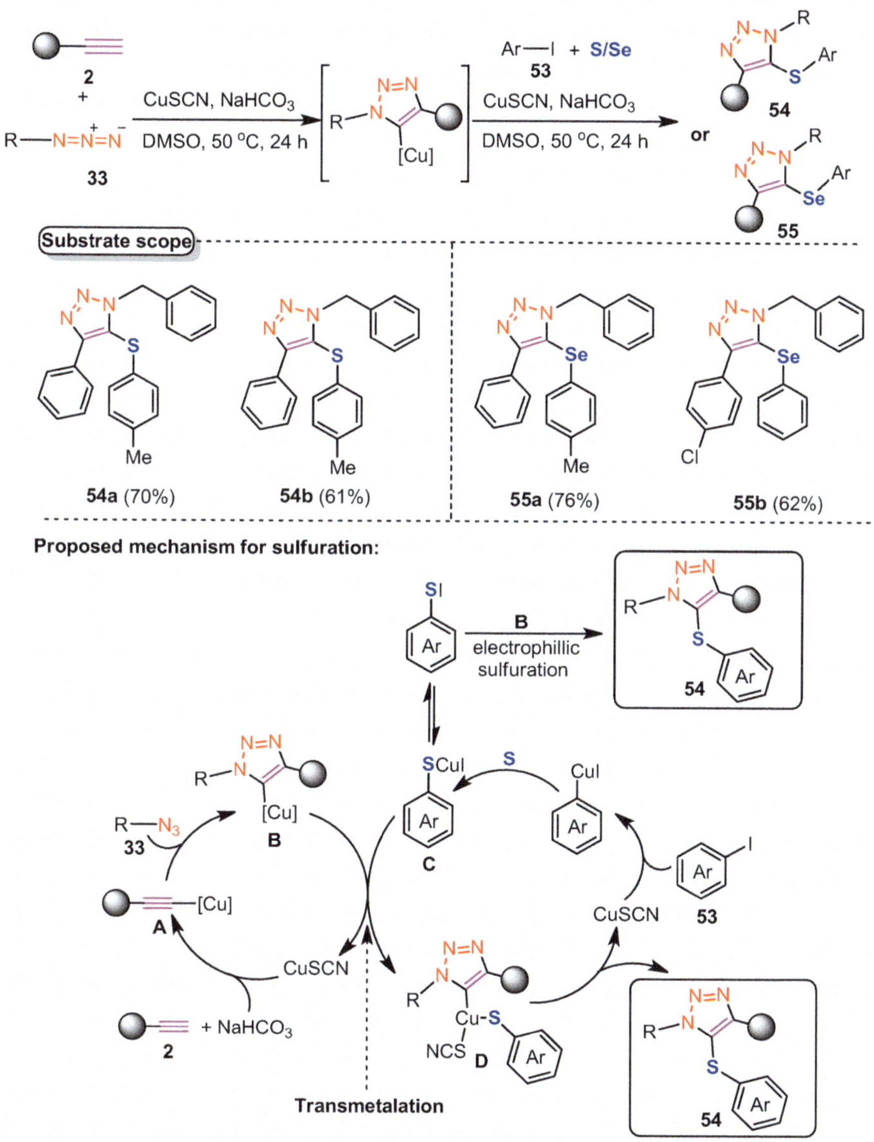

Scheme 19.16 Four-component interrupted click reaction for the synthesis of 5-thio and 5-selenotriazoles

However, using the same synthetic approach, Crowley et al. expanded the production of 2-pyridyl-triazolyl complexes, which had a recurrent band at 280 nm with the same electrical spectrum properties as free ligands [58].

As a result, the 'Click-To-Chelation' tool has significantly altered how glycoconjugate chelates are synthesized and designed over the course of decades of operations

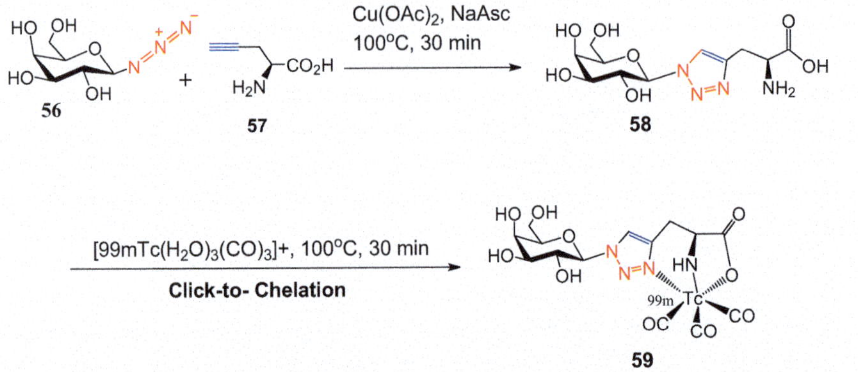

Scheme 19.17 Click-to-chelation approach for the synthesis of ^{99m}Tc-tricarbonyl complex

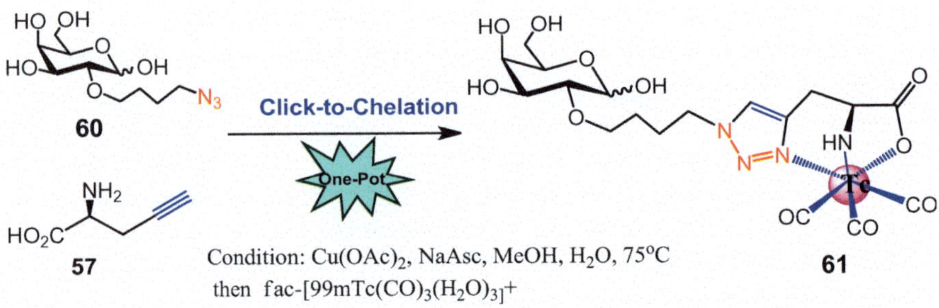

Scheme 19.18 Synthetic strategy for ^{99m}Tc(CO)$_3$-glucose-histidine complex **61** using 'Click-to-Chelate' Protocol

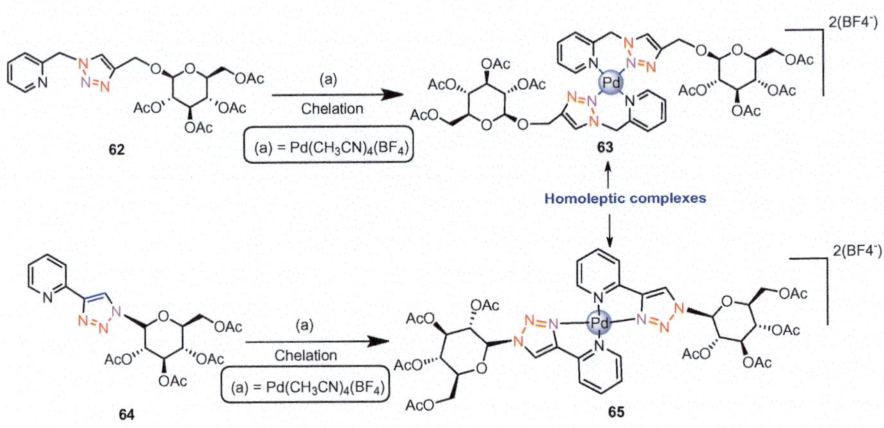

Scheme 19.19 Click triazole-containing 1-(2-picolyl)-triazolyl-based Pd-complexes **63** and **65**

and provided stable chelates with organometallic cores, which are effective labeling agents with the ability to metabolically evaluate tumor activity, thus expanding the skyline of diagnostic tools with enormous biomedical importance. However, there is still a lot of potential for producing more beneficial outcomes by utilizing this simple and effective strategy, which can be set as the everyday diagnostic strategy in routine tasks.

19.3.2 Miscellaneous Reactions

In addition to the above-mentioned post-click approach, the 1,2,3-triazoles could be employed in other important applications including glycosylation methodology. To control the desired regio- and stereoselectivity in glycosylation reactions, the intramolecular glycosylation methodology is found to be quite superior to get the high selectivity in glycosylation product [59, 60]. In the rigid spacer-based intramolecular glycosylations, several rigid spacers including malonyl, succinyl, glutaryl, peptide, *o*- or *m*-xylene, phthaloyl, and isophthaloyl were successfully investigated [59]. Further in 2012, Schmidt and Tiwari used CuAAC-mediated 1,2,3-triazole as the rigid spacer in intramolecular glycosylation protocol, and a number of disaccharide-containing macrocycles were developed in high yield and fair anomeric selectivity [61]. In order to get the better selectivity glycosylation results, Schmidt et al. further extend Click chemistry and used the *o*-azidobenzyl group to generate desired 1,2,3-triazole as rigid spacer required in intramolecular glycosylation methodology [62]. Thus, *O*-linked *ortho*-substituted arylazido-based acceptor **67** was first underwent CuAAC click conjugation with 2-*O*-propargylated glycosyl donor **66** to give excellent yield of desired β-(1–3)-linked triazole **68** as a part of rigid spacer linking together the donor and acceptor (**Step 1**). The intramolecular glycosidation of donor–triazole spacer–acceptor system **68** as a part of fourteen-membered ring formation supports the required intramolecular aglycon delivery and afforded β(1–3)-linked disaccharide **69** as major product (25:1 ratio) in 68% yield (**Step 2**). In this intramolecular aglycon delivery approach, the authors presented a rational approach for the possible anchimeric assistance. There could be a possible stabilization of anomeric carbenium ion as part of six-membered ring through *N*-3 atom of the resulting 1,2,3-triazole. This could approach either from α- and/or the β-side. However, the resulting 2-*O*-linked (1-phenyl-triazol-4-yl)methyl residue nearly does not display anchimeric assistance in the intramolecular glycosidation step, probably due to the lesser electron-donating nature of the triazole *N*-atom [62]. Also, the restricted conformational mobility and interaction between resulting 1,2,3-triazole and phenyl group may overall influence the lower conformational mobility than 2-azidomethylbenzyl residue. At the end, the rigid spacer was removed successfully from the macrocyclic system **69** by reacting it with sodium in presence of liquid ammonia at −75 °C (Birch conditions), and the resulting free disaccharide was then subjected to acylation using Ac_2O in presence of pyridine to give high yield of optically pure disaccharide **70** as glycosylation product (**Step 3**) (Scheme 19.20) [62]. The similar donor–acceptor systems could

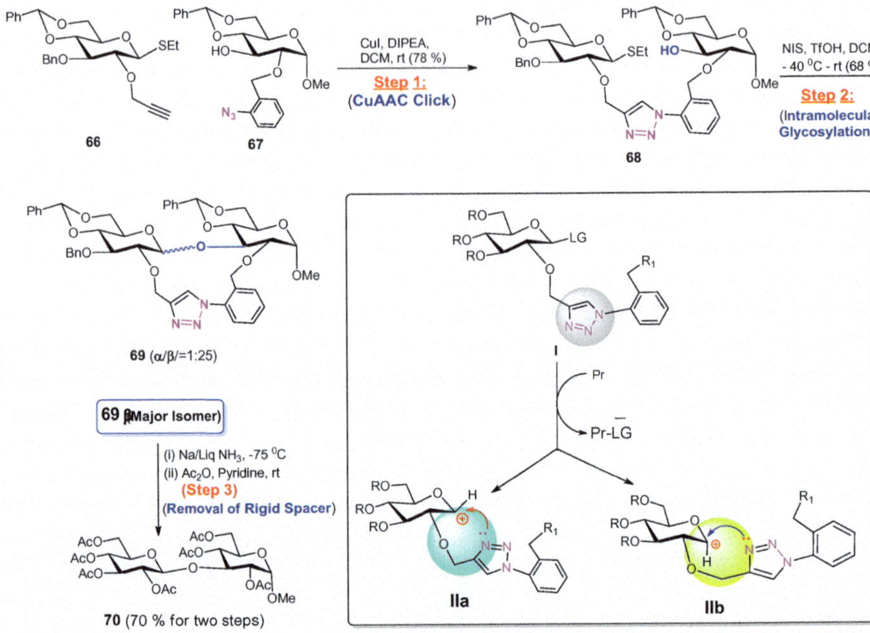

Scheme 19.20 Click-mediated 1,2,3-triazole as rigid spacer and its possible role in anchimeric assistance in intramolecular glycosylation through IAD [62]

be useful for the development of diverse range of carbohydrate-containing macrocyclic systems of notable applications in drug discovery, sensing, complexation, and host–guest chemistry [1, 63].

A novel approach for the regioselective synthesis of 1,4,5-trisubstituted 1,2,3-triazoles has been established by Chit Tsui et al. The group utilized the Cu(I)-catalyzed interrupted click reaction methodology to carry out the synthesis [64]. The reaction was based on the concept of interception of the conventional CuAAC cycle with $TMSCF_3$ as a nucleophilic CF_3 source. Terminal alkynes **71** and azides **33** were employed as the substrates along with (trifluoromethyl)trimethylsilane ($TMSCF_3$) as a nucleophilic CF_3 source for the regioselective synthesis of fully substituted 5-trifluoromethyl 1,2,3-triazoles **72** in single step. The reaction was catalyzed by CuI in the presence of phen as ligand and Ag_2CO_3 was used as an oxidant along with Et_3N for the activation of $TMSCF_3$ (Scheme 19.21). The group also proposed a plausible catalytic cycle for the performed interrupted click reaction. Initially, terminal alkyne **71** reacts with Et_3Nbase to form the copper(I) acetylide intermediate **A**. In the presence of azide **33**, the reaction follows the expected CuAAC pathway leading to the regioselective generation of 5-cuprated 1,2,3-triazole intermediate **B**. An unproductive side reaction would happen via protonolysis of **B** giving rise to triazole **73**. In an alternate pathway, interception of intermediate **B** by nucleophilic CF3-, generated from $TMSCF_3$ and activated by Et_3N, to form the trifluoromethylated copper species **C** which ultimately transforms into the desired 1,4,5-trisubstituted 1,2,3-triazole product **72** [64].

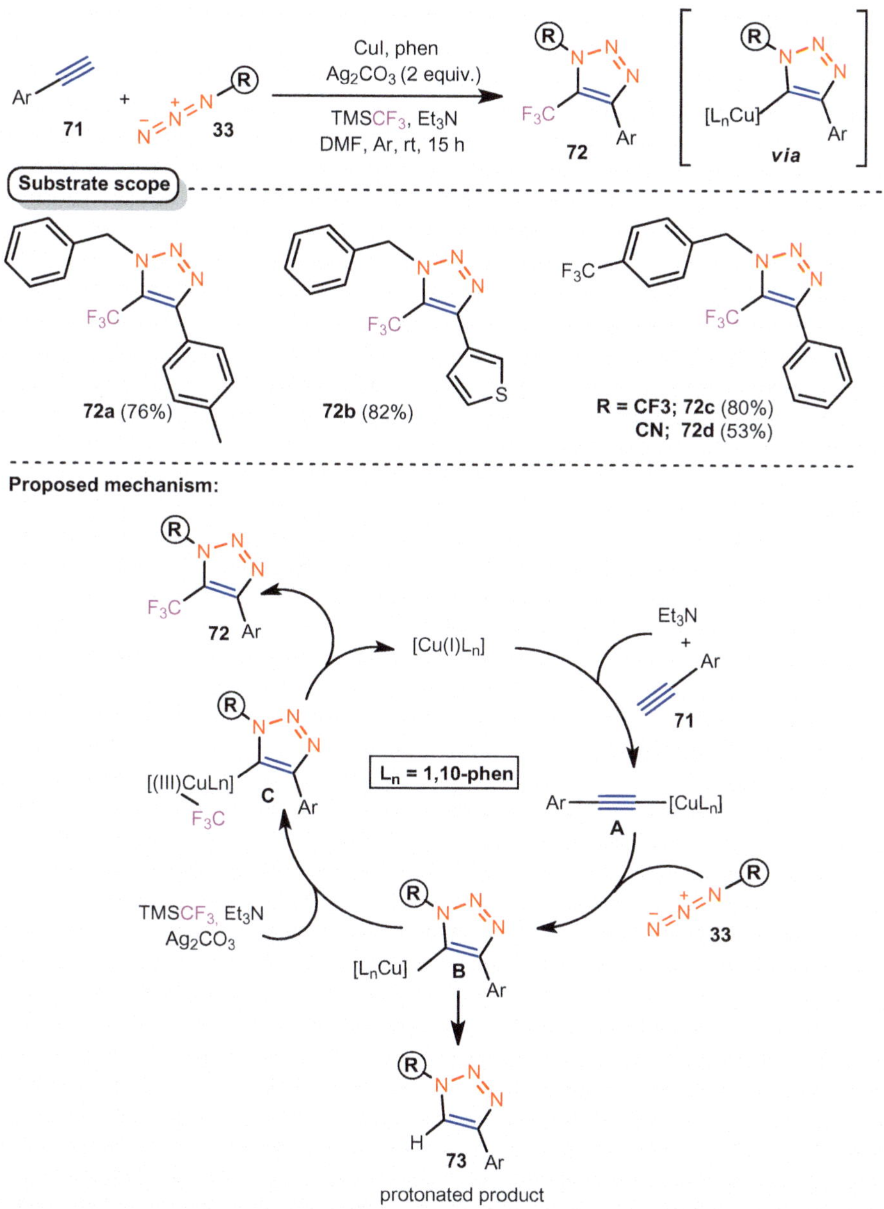

Scheme 19.21 Synthesis of 5-trifluoromethyl 1,2,3-triazoles **72** via copper(I)-catalyzed interrupted click reaction with $TMSCF_3$

Lee et al. documented a multicomponent strategy for the synthesis of fully substituted triazoles **75**. The methodology involved a Cu-catalyzed three-component coupling reaction of alkynes, azides, and diazo compounds for synthesis of fully substituted triazoles. The reaction was performed by employing alkynes **2**, azides **33** and ethyl diazoacetate **74** as substrates in the presence of CuI as catalyst and dtbpy as ligand in acetonitrile solvent at room temperature to afford the desired substituted triazoles (Scheme 19.22) [65]. The mechanistic picture of the developed protocol included the initial formation of Cu-acetylide intermediate **A** on reaction with alkyne **2** and active copper species which, undergoes reaction with azide **33** to give rise to adduct **B**. A [3 + 2] cycloaddition reaction of intermediate **B** produces intermediate **C**. Afterward, ethyl diazoacetate **74** undergoes reaction with intermediate **C** to furnish a Cu carbenoid species **D**, which further transforms into an intermediate **E**. Finally, alkyne **2** coordinates with **E** followed by proton exchange to afford the desired fully substituted triazole **75** and Cu-acetylide **A** is regenerated.

The most straightforward method to create completely substituted triazoles from readily available terminal alkynes **2** and azide **33** in one step and one pot is the recently invented interrupted click reaction. The crucial intermediate in a CuAAC reaction is cuprate-triazole **M1** generated by the reaction of alkynes **2** and azide **33**. The active electrophilic sulfenylating reagent benzenesulfonothioate **76,** in which benzenesulfinate has great leaving properties, has also been used to produce 5-thio triazoles **77** (Scheme 19.23) [66]. A variety of different terminal alkynes both aromatic and aliphatic was assessed, and numerous 5-persulfur-functionalized triazoles were produced in good to outstanding yields and as single regioisomers. This transformation was applicable to aromatic alkynes with either an electron-rich or electron-poor phenyl ring, naphthalene, and heterocycles like indole and thiophene, all of which produced good yields of the desired products. This reaction is equally acceptable for aliphatic alkynes.

19.3.3 Synthesis of Biologically Relevant Sugar Triazoles

In 2020, Sriniwas has been utilized D-glucal in developing pyrene glycoconjugates **83** and **84** with anti-bacterial property. The modular CuAAC technique was implemented to produce the propargyl ether analogue **78**, which was subsequently transformed into the triazole derivative **80** by the association with phenyl azide **79**. The compound **80** was then exposed to oxidation, followed by R-substitution with primary amine in the incidence of $ZnCl_2$, and then after microwave exposure compounds **81a–f** were obtained. Resulting scaffolds **81a–f** on exposure with p-fluorobenzaldehyde in assistance with AcOH and NaOAc furnished triazole-appended thiazolidinone glycosides **82a–f** in excellent yield. Furthermore, resulted compounds **82a–f** were converted into compounds **83a–f** and **84a–f** by cyclocondensation reaction utilizing arylhydrazines and hydroxylamine, respectively (Scheme 19.24).The resultant glycoconjugates **83b**, **84b**, **83c**, **84c**, and **84f**, interestingly have significant anti-bacterial properties toward gram-positive bacteria while the glycohybrid counterparts **83a**, **84a**,

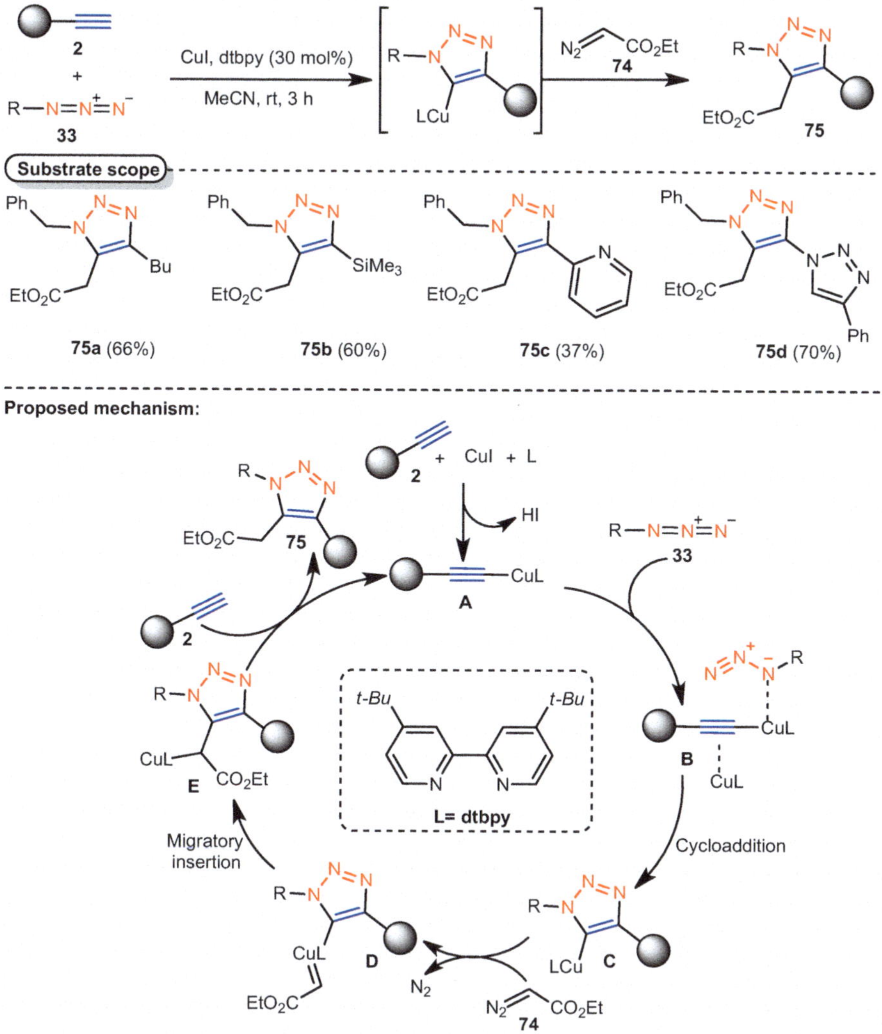

Scheme 19.22 Multicomponent strategy for the synthesis of fully substituted triazoles **75**

83b, **84b**, **83e**, and **84e** exhibit activity against gram-negative pathogens. Beside from that, compounds **83d**, **84d**, **83f**, and **84f** also shown appreciable anti-fungal properties [67].

Another unique series of triazolyl glycohybrids, **87**, was created and developed by Tan et al. by associating different lengths of alkyl moieties with CAR binding components using CuAAC catalysis (Scheme 19.25). Similar to this, the % inhibition of mycelial growth assay was used to evaluate the anti-fungal efficacy of all produced compounds against three different plant-threatening fungus strains. The assay findings showed that the compounds are effective anti-fungal pharmacophores

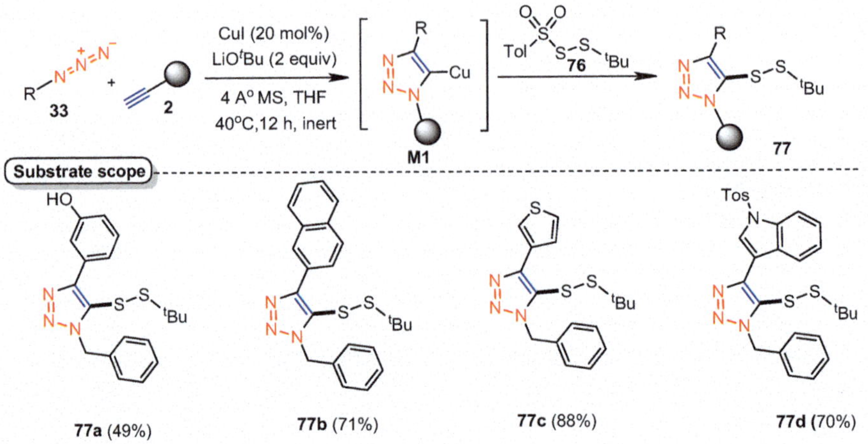

Scheme 19.23 Multicomponent synthesis of 5-substituted per sulfur derivatives **77** through click-mediated 1,2,3-triazole

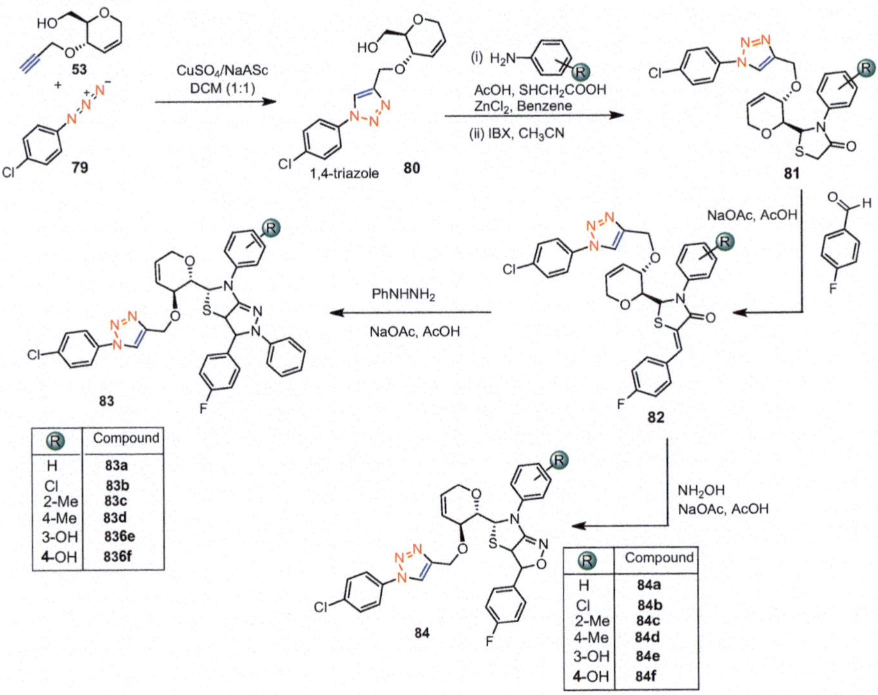

Ⓡ	Compound
H	**83a**
Cl	**83b**
2-Me	**83c**
4-Me	**83d**
3-OH	**836e**
4-OH	**836f**

Ⓡ	Compound
H	**84a**
Cl	**84b**
2-Me	**84c**
4-Me	**84d**
3-OH	**84e**
4-OH	**84f**

Scheme 19.24 Synthesis of CuAAC 'Click Chemistry'-inspired pyrene glycoconjugates (**83a–f** and **84a–f**) as the anti-bacterial motif

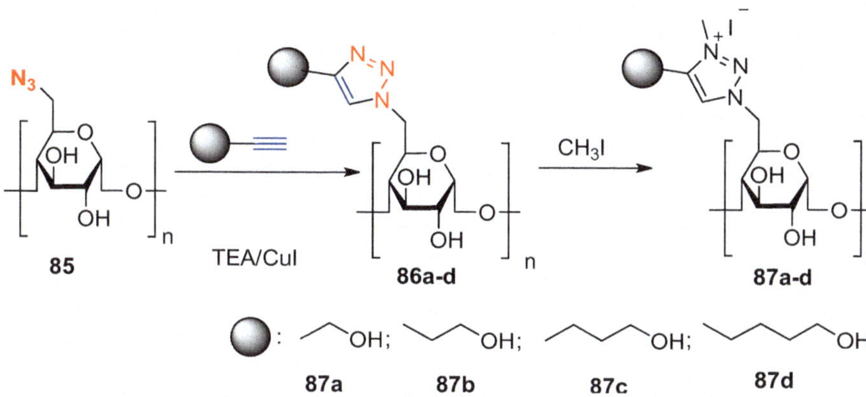

Scheme 19.25 Click-inspired development of triazolyl sugar hybrid (**87a–d**) as anti-fungal agent

in comparison to the standard reference and that specific 1,2,3-triazolium cores are superior anti-fungal functional groups. The length of the alkyl groups is also a key factor in determining the anti-fungal action. It was noticed that as the side-chain length increases, the anti-fungal activity of the resulting series diminishes. Therefore, additional clinical evaluations are necessary before the new leading structures may be validated as anti-fungal [68].

Triazenes have become popular structural scaffolds bringing a wide range of functions in chemistry and medicinal science [69, 70]. They have been utilized for the functionalization and development of numerous carbo and heterocyclic substances and act as concealed diazonium ions [71, 72]. A limited methods for constructing triazenyl-1,2,3-triazoles have been proposed in literatures, which involves the reaction of lithium acetylides and aryl azides or through the azo coupling of amines with 5-diazo-1,2,3-triazole-4-carboxamide [73] and recently triazenyl-functionalized 1,2,3-triazoles was achieved through the Ir(I)-catalyzed cycloaddition of azides and 1-alkynyltriazenes [74]. These procedures still use harsh reaction condition. Recently, Ortiz introduced an interrupted click strategy (post-functionalization of click triazoles) for the synthesis of 1,4-(disubstituted)-5-triazenyl- 1,2,3-triazoles **93**. Under copper (I) catalysis, a broad range of acetylenes reacts with chelating aryl azides having P = O, N–P = O, and SO_3H entities at the ortho position, in a tandem fashion, without ligand. In particular, the reaction between *N*,*N*,*N*,*N*'-tetraethyl-2-azidophenyl-phosphonodiamide **88** and phenylacetylene along with copper source (5 mol%) and sodium ascorbate (1 equiv.) in DMF for 40 min produced intermediate species **A**. This was followed by in situ electrophilic trapping with another equivalent of azide **88** leading to entirely substituted 1,2,3-triazoles **89** regio- and chemoselectively. DFT calculations demonstrate that Cu-chelation is a key factor in the intervention of the CuAAC precursor through azide. From **89**, it is possible to make their individual thermolysis product **93**. The compound **89** tautomerizes into compound **90** when heated, and the breakdown of compound **90** produces diazonium **91** and amine **92**. Diazo group from compound **91** is displaced by amine **92**

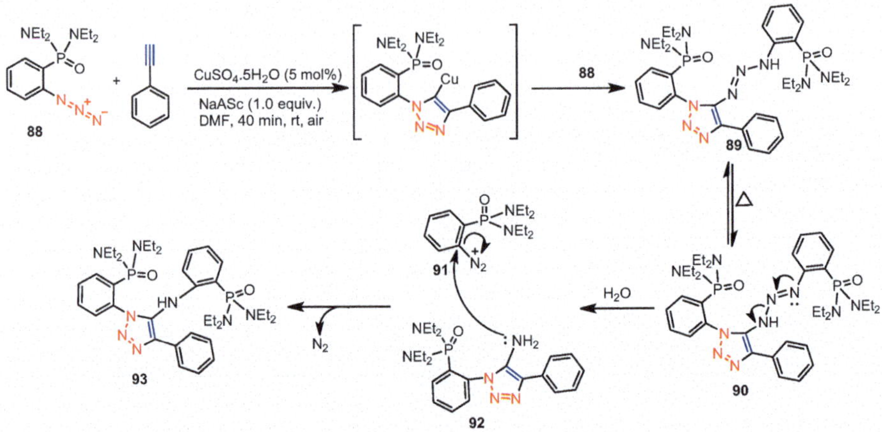

Scheme 19.26 Tandem synthesis of substituted 5-triazenyl-1,2,3-triazoles **93** through interrupted copper(I)-catalyzed azide–alkyne reaction

to produce compound **93** (Scheme 19.26) [75]. In the reaction, trace amount of predictable 1,4-disubstituted 1,2,3-triazole was also isolated.

In recent year, 1,2,3-triazoles have been successfully explored as novel surrogate for the diazo compounds for an easy synthesis of diverse range of pharmaceutically important heterocyclic skeletons [76]. In this regard, an interesting rhodium-supported chemistry was presented by the Anbarsan group, which leads to diverse bioactive scaffolds. An array of polysubstituted pyrroles **96** was successfully achieved through rhodium-catalyzed transannulation of *N*-sulfonyl-1,2,3-triazoles **94** with vinyl ether **95** and significantly acknowledged for the precise synthesis of an anti-cancer drug neolamellarin A (Scheme 19.27a) [77]. From the same group, an amalgamated Rh(II) and Yb(III) catalytic system was inaugurated for the construction of substituted naphthofurans **98** via transannulation of *N*-sulfonyl-1,2,3-triazoles with β-naphthols **97**. The process commences with rhodium-catalyzed insertion of azavinyl carbene into the O–H bond, followed by ytterbium-catalyzed annulative C–C bond formation and simultaneous aerial oxidation (Scheme 19.27b) [78]. Started with the similar substrate that is *N*-sulfonyl-1,2,3-triazoles in assistance with o-quinone methides (o-QMs), generated in situ from 2-hydroxybenzyl alcohols **99**, under rhodium catalysis furnished the synthesis of substituted benzopyrans **100**. In the established process, o-QM attacks α-imino rhodium carbenoid nucleophilically to produce a carbonyl ylide, followed by subsequent 6π-electrocyclization and isomerization reaction (Scheme 19.27c) [79]. More recently, a successful synthesis of dihydro-3,1-benzoxazine derivatives **102** using rhodium catalysis has been achieved using *N*-sulfonyl-1,2,3-triazole and 1,3-amino alcohols **101** generated from aniline. The accomplished reaction opens the door to a variety of substituted dihydro-3,1-benzoxazines and shows off the new reactivity of azavinyl carbenes (Scheme 19.27d) [80]. An interesting chemistry of easily accessible *N*-sulfonyl-1,2,3-triazoles and o-acylaniline **103** derivatives can be used to synthesize 3-aminoquinolines **104** efficiently. In this transition, rhodium azavinyl carbenoid is inserted into a N–H bond, resulting in the formation of C–C and C–N bonds (Scheme 19.27e) [81]. A new substrate system of *N*-sulfonyl-1,2,3-triazoles and 2-alkenylindoles **105** was established to develop an array of 3-aminocarbazoles **106** in two-step process (Scheme 19.27f) [82]. In a similar fashion, reports were documented for the enantioselective synthesis of dihydro-β-carbolines through the Aza Michael reaction [83, 84]. Interestingly, within aromatic C(sp2)–H bonds, rhodium azavinyl carbene is inserted chemo- and regioselectively. During the process, 2,2-diaryl enamides **108** have been achieved through the effective rhodium-catalyzed direct arylation of α-diazoimines, which can be generated from conveniently available 1,2,3-triazole (Scheme 19.27g) [85].

For the purpose of synthesizing different functionalized enamides **110**, an effective Sommelet–Hauser type rearrangement of sulfur ylides produced from α-thioesters **109** and *N*-sulfonyl-1,2,3-triazoles has been effectively carried out using rhodium catalysis. An unusual [2,3]-sigmatropic rearrangement of sulfur ylides with the imine motif has been involved in the devised process (Scheme 19.28a) [86]. Using a similar precursor, a tandem synthesis of substituted tetrahydropyridines **111** was achieved. Primarily, thio-substituted 1-azadiene is formed from thio-bounded *N*-sulfonyl-1,2,3-triazoles and then undergoes a self-Aza-Diels–Alder reaction via tandem rhodium-catalyzed 1,2-sulfur migration to furnish desired tetrahydropyridines **111**. (Scheme 19.28b) [87].

Tiwari and Group involved in similar methodology development via benzotriazole ring cleavage (BtRC) route. The BtRC-mediated functionalization under optimized condition leads to the formation of diverse range of fused heterocycles including benzothiazoles, benzooxazoles, and other biologically relevant molecules (Scheme 19.29) [39, 88–94]. Thus, a synthetic protocol via free-radical intramolecular cyclative cleavage of benzotriazole ring in presence of $(TMS)_3SiH$/AIBN for an easy access of diverse 2-*N*, *S*, and *C*-substituted glycoconjugate benzothiazole was developed through the cleavage of benzotriazole ring (BtRC) [88, 89]. The short reaction time, wide substrate scope, simple workup, good-to-excellent yield under fairly mild conditions, and moreover readily accessibility of the starting materials are the notable features of this BtRC protocol. The methodology demonstrates a significant compatibility under microwave conditions too. Interestingly, the benzotriazole analogues *benzyloxythioacyl benzotriazoles* on reaction with silanes or Bu_3SnH undergo free-radical *β*-scission of *C–O* bond instead of *N–N* bond to afford deoxy product [90]. Also, acyl benzotriazole under similar free-radical condition afforded respective amides, rather than expected benzoxazole [91]. While, a Lewis acid-mediated BtRC of RCOBt resulted to respective benzoxazoles [92]. The BtRC strategy has been successfully explored to get high yield of the targeted bioactive molecules.

In addition to above-mentioned example with triazole post-functionalization methodologies, triazole can also participate in glycosylation reaction [39, 61, 62]. For example, glycosyl triazole or related systems can be obtained in good yields via *N*-glycosylation of thio-glycoside using hypervalent iodine reagent [95]. Thus, glycosyl donor **112** on glycosylation with *N*-triazole or benzotriazole as acceptor in presence

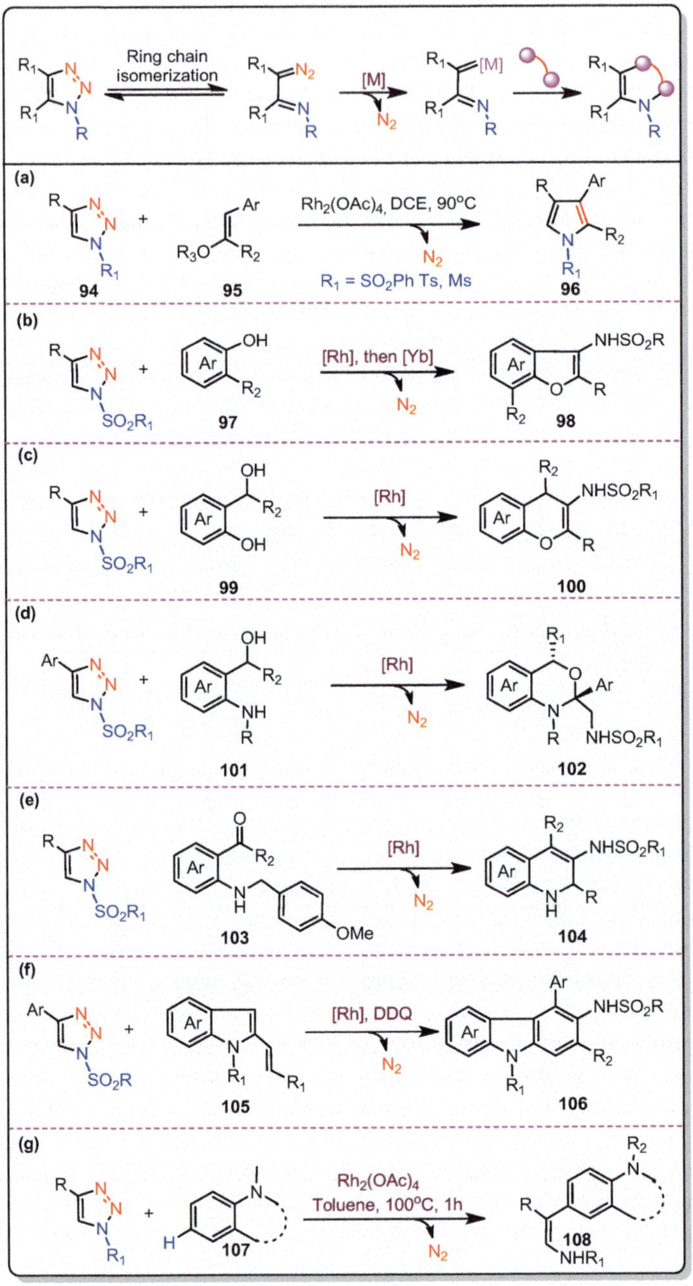

Scheme 19.27 Click-derived 1,2,3-triazoles as novel surrogate for the diazo compounds

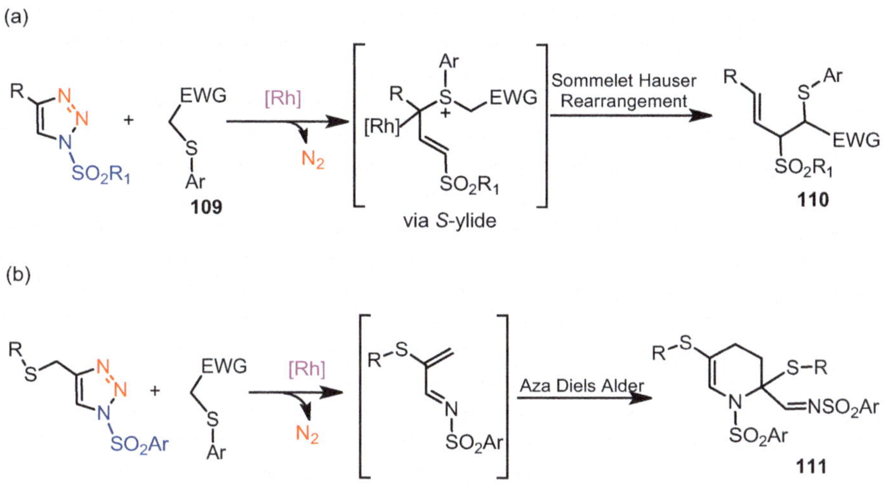

Scheme 19.28 Post-modification of *N*-1 substituted triazole precursor

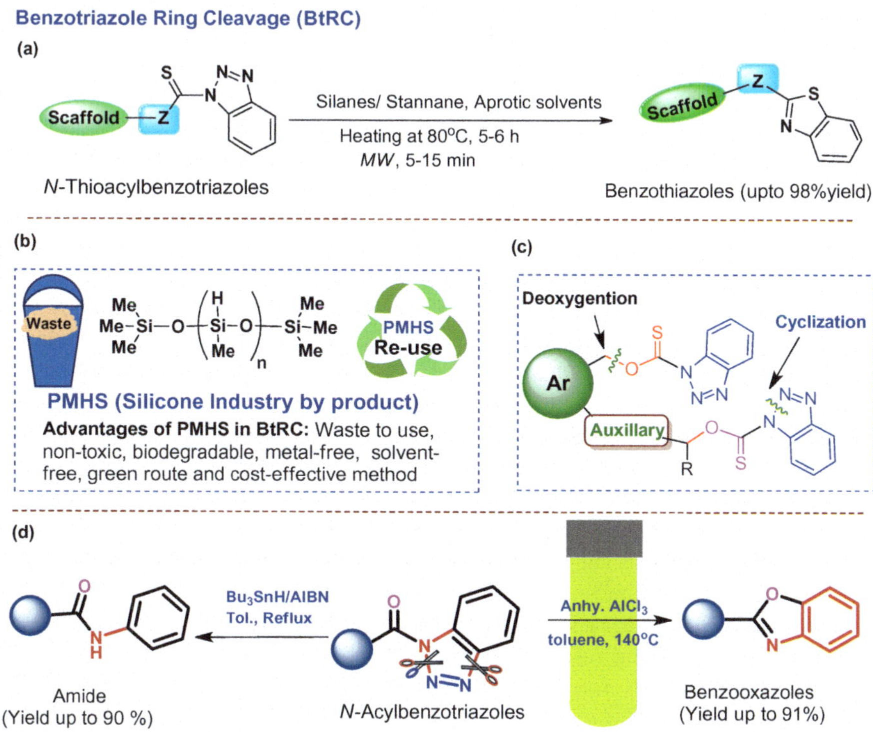

Scheme 19.29 Tiwari's benzotriazole ring cleavage (BtRC) methodology for an easy access of biologically relevant molecules

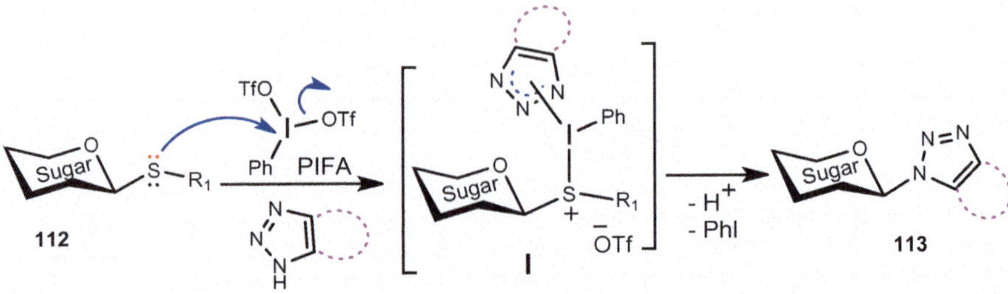

Scheme 19.30 *N*-glycosylation of thio-glycoside via participation with triazole and hypervalent iodine reagent leading to *N*-glycoside with N_1-selectivity

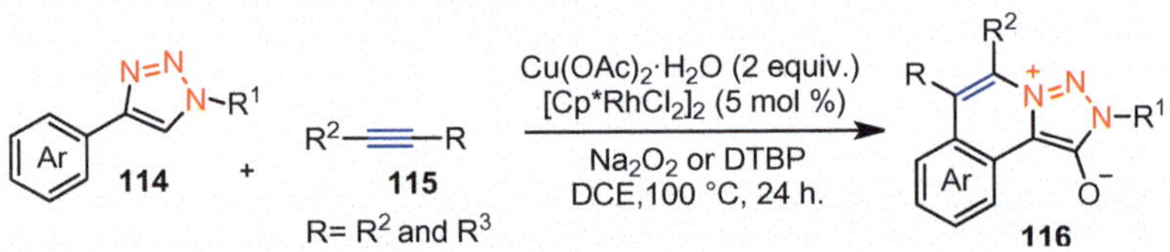

Scheme 19.31 Synthesis of triazolo[5,1-*a*]isoquinolium

of 2.0 equiv. of TfOTMS and 1.0 equiv of $PhI(OTf)_2$ in 1,2-dichloroethane at 80 °C for 24 h afforded respective *N*-glycoside **114** with N_1-selectivity via intermediate **113** (Scheme 19.30) [95].

A straightforward approach for the synthesis of Triazolo[5,1-α] isoquinolium derivatives **116** was proposed by Huayue Wu and co-workers [96]. The ruthenium-copper-catalyzed reaction between triazoles **114** and alkyne **115** to furnished the Triazolo[5,1-*a*] isoquinolium derivatives **3** in the moderate to excellent yield. In this reaction condition, they used oxidizing agent Na_2O_2 or DTBP at 100 °C to sequential triazole-directed C–H activation and C–C, C–N, and C–O bond formation processes in one pot (Scheme 19.31).

Chen and co-workers developed a new protocol for the synthesis of enamine-functionalized 1,2,3-triazoles from terminal 2*H*-azirines **117**, azides, and alkynes. This methodology is grounded on Cu-catalyzed click reaction, in which the triazolyl-Cu was captured by 2*H*-azirine and thus inserted the enamine group on the C-5 position of 1,2,3-triazole **118** via vinyl nitrene transfer reaction. In the reaction mechanism pathway, alkyne first reacts with CuI in the presence of DIPEA and generate copper(I) phenylacetylide **A** which was then react with TsN_3 to give intermediate **B** via CuAAC click. The resulting intermediate **B** was then rapidly reacted with 2*H*-azirine **117** lead to the formation of triazolyl-Cu(I)/2*H*-azirine complex **C** which underwent rearrangement to generate respective Cu (III)-vinyl nitrene **D**. Further via insertion mechanism the resulting Cu–N complex **E** in the last step proceed to the protonation of Cu–N complex and afforded the desired product **118a** in good yield (Scheme 19.32a) [97].

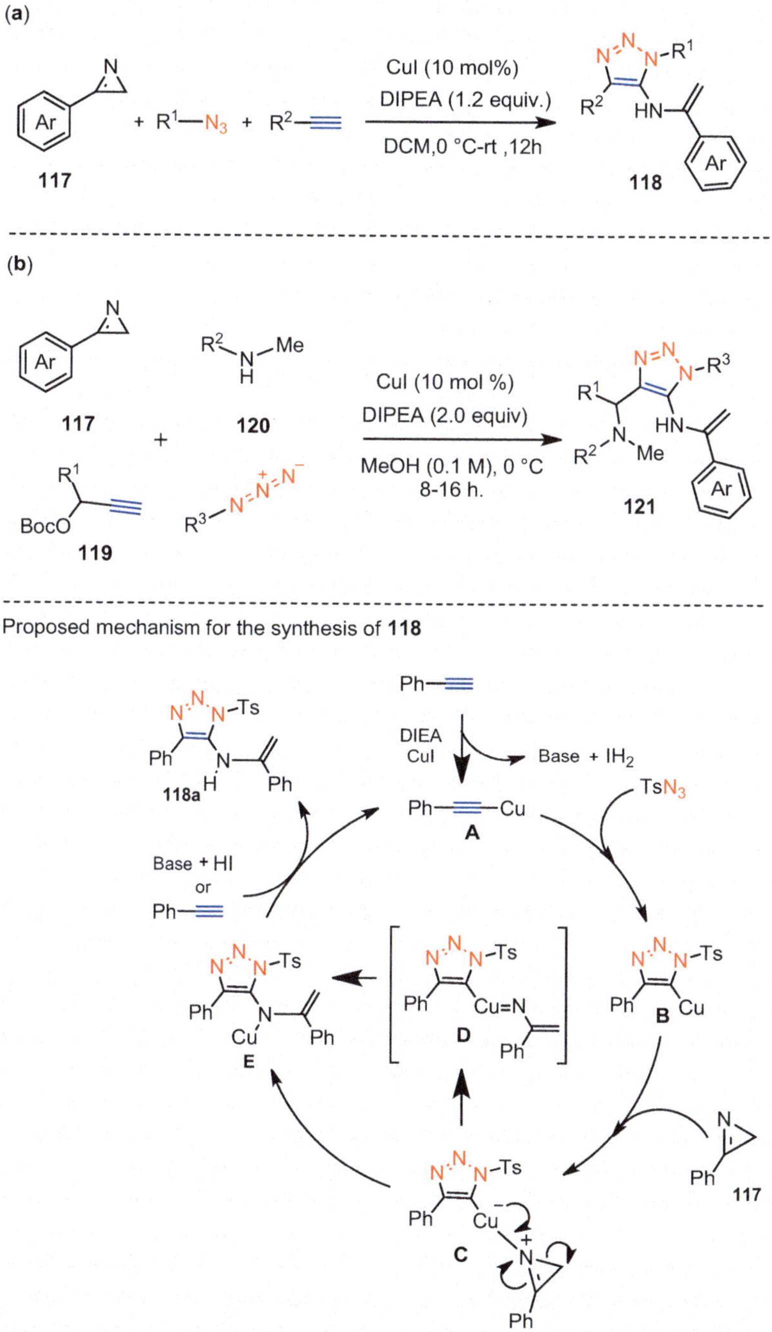

Scheme 19.32 **a** Synthetic protocol of the C–N bond-containing 1,2,3-triazoles derivatives **118** from azirines. **b** Synthesis of polyfunctionalized triazoles **121**

Likewise, Song and co-workers recently introduced a convenient method for the synthesis of polyfunctionalized 1,2,3-triazoles **121** via CuAAC via consecutive formation of four C–N bonds [98]. The reaction proceeds through one-pot strategy under mild reaction condition. Thus, 2*H*-azirines **117,** alkynes **119,** amines **120,** and azides underwent MCR using Cu-catalysis under basic condition and afforded high-to-excellent yield of compound **121** (Scheme 19.32b).

19.4 Conclusions and Future Perspective

The click concept can apply to a wide range of reaction types, usually CuAAC for construction of 1,4-disubstituted 1,2,3-triazole and RuAAC for 1,5-disubstituted 1,2,3-triazole are well explored by different research group. Beyond these conventional approaches, a recent terminology was introduced by different group named interrupted click reaction (postfunctionalization of click triazole) has increased demand. In this chapter, we are emphasized to highlight the synthesis of functionalized 1,2,3-triazole via post-modification of click triazoles in one pot. The application of these post-modification strategies was explored for the construction of diverse range of biologically important scaffolds including triazole-bounded fused heterocycles, sulfenylation, selenylation and amination reactions for development of 5-heterofunctionalized 1,2,3-triazole were briefly discussed. Moreover, the application of post-click approach could be employed in glycosylation methodology for the synthesis of neoglycoconjugates. Hence, these reactions may certainly provide new and ample opportunities for the synthetic and biochemists in the fields of medicinal, agriculture, material, and wide range of synthetic organic chemistry of diverse applications.

References

1. Agrahari AK, Bose P, Jaiswal MK, Rajkhowa S, Singh AS, Hotha S, Mishra N, Tiwari VK (2021) Chem Rev 12:7638–7956
2. Meldal M, Tornoe CW (2008) Chem Rev 108:2952–3015
3. Tiwari VK (2021) Chem Rec 21:3029–3048
4. Sharma A, Agrahari AK, Rajkhowa S, Tiwari VK (2022) Eur J Med Chem 238:11454
5. Kolb HC, Sharpless KB (2003) Drug Discov Today 8:1128–1137
6. Yadav S, Singh AK, Agrahari AK, Sharma K, Singh AS, Gupta MK, Tiwari VK, Prakah P (2020) Sci Rep 10:14204
7. Witczak ZJ, Bielski R (2013) Click chemistry in glycoscience: new development and strategies. Wiley-VCH John Wiley & Sons, New York
8. Chandrasekeran S (2016) Click reactions in organic synthesis. Wiley-VCH, John Wiley & Sons, New York
9. Kolb HC, Finn MG, Sharpless KB (2001) Angew Chem Int Ed 40:2004–2021
10. Tornøe CW, Christensen C, Meldal M (2002) J Org Chem 67:3057-3064
11. Jaiswal MK, Gupta A, Yadav MS, Pandey VK, Tiwari VK (2023) Chem Eur J e202301749

12. Yadav MS, Jaiswal MK, Kumar S, Singh SK, Ansari FJ, Tiwari VK (2022) Carbohydrate Res 521:108674
13. Thirumurugan P, Matosiuk D, Jozwiak K (2013) Chem Rev 113:4905–4979
14. Tiwari VK, Mishra BB, Mishra KB, Mishra N, Singh AS, Chen X (2016) Chem Rev 116:3086–3240
15. Agrahari AK, Jaiswal MK, Yadav MS, Tiwari VK (2021) Carbohydrate Res 508:108403
16. Kumar S, Yadav MS, Singh SK, Rajkhowa S, Tiwari VK (2023) SynOpen 7:145–153
17. Singh SK, Kumar S, Yadav MS, Tiwari VK (2022) J Org Chem 87(22):15389–15402
18. Singh SK, Kumar S, Yadav MS, Tiwari VK (2023) J Org Chem 88:13440–13453
19. Singh SK, Kumar S, Yadav MS, Tiwari VK (2024) Synthesis 56:975–988
20. Wei F, Wang W, Ma Y, Tung CH, Xu Z (2016) Chem Commun 52:14188–14199
21. Wang WF, Wei, Ma Y, Tung C-H, Xu Z (2016) Org Lett 18:4158–4161
22. Fu Z, Yang X, Chit Tsui G (2023) Org Lett 25:4945–4949
23. Wang W, Huang S, Yan S, Sun X, Tung CH, Xu Z (2020) Chin J Chem 38:445–448
24. Brahma K, Achari B, Chowdhury C (2013) Synthesis 45:545–555
25. Hu YY, Hu J, Wang XC, Guo LN, Shu XZ, Niu YN, Liang YM (2010) Tetrahedron 66:80–86
26. Chowdhury C, Mandal SB, Achari B (2005) Tetrahedron Lett 46:8531–8534
27. Larin EM, Lautens M (2019) Angew Chem 131:13572–13576
28. Reddy MN, Swamy KCK (2012) Eur J Org Chem 2013–2022
29. Bozorov K, Zhao J, Aisa HA (2019) Bioorg Med Chem 27:3511–3531
30. Dheer D, Singh V, Shankar R (2017) Bioorg Chem 71:30–54
31. Malik MS, Ahmed SA, Althagafi II, Ansari MA, Kamal A (2020) RSC Med Chem 11:327–348
32. Selvaraju M, Sun CM (2014) Adv Synth Catal 356:1329–1336
33. Qureshi Z, Kim JY, Bruun T, Lam H, Lautens M (2016) ACS Catal 6:4946–4952
34. Cheng H, Wan J, Lin M-I, Liu Y, Lu X, Liu J, Xu Y, Chen J, Tu Z, Cheng Y-SE, Ding K (2012) J Med Chem 55:2144–2153
35. Futatsugi K, Bahnck KB, Brenner MB, Buxton J, Chin JE, Coffey SB, Dubins J, Flynn D, Gautreau D, Guzman-Perez A, Hadcock JR, Hepworth D, Herr M, Hinchey T, Janssen AM, Jennings SM, Jiao W, Lavergne SY, Li B, Li M, Munchhof MJ, Orr STM, Piotrowski DW, Roush NS, Sammons M, Stevens BD, Storer G, Wang J, Warmus JS, Wei L, Wolford AC (2013) Med Chem Commun 4:205–210
36. Larin EM, Lautens M (2019) Angew Chem Int Ed 58:13438–13442
37. Vitaku E, Smith DT, Njardarson JT (2014) J Med Chem 57:10257–10274
38. Yadav MS, Singh AS, Agrahari AK, Mishra N, Tiwari VK (2019) ACS Omega 4:2418–2424
39. Singh AS, Mishra N, Yadav MS, Tiwari VK (2019) Heterocyclic Chem 56:275–280
40. Majumdar KC, Mondal S (2011) Chem Rev 111:7749–7773
41. Hommelsheim R, Bausch S, Selvakumar A, Amer MM, Truong K-N, Rissanen K, Bolm C (2023) Chem Eur J 29:e202203729
42. Reddy RJ, Md. Waheed M, Kumari AH, Krishna GR (2022) Adv Synth Catal 364:319–325
43. Zhang L-L, Li Y-T, Gao T, Guo S-S, Yang B, Meng Z-H, Dai Q-P, Xu Z-B, Wu Q-P (2019) Yynthesis 51:4170–4182
44. Chen C, Xie Y, Chu L, Wang R, Zhang X, Qing F (2012) Angew Chem Int Ed 51:2492–2495
45. Amiri K, Rostami A, Rostami A (2016) New J Chem 40:7522–7528
46. Wang W, Peng X, Wei F, Tung C-H, Xu Z (2016) Angew Chem 128:659–663
47. Biagi G, Calderone V, Giorgi I, Livi O, Scartoni V, Baragatti B, Martinotti E (2000) Eur J Med Chem 35:715
48. Morzherin Y, Prokhorova PE, Musikhin DA, Glukhareva TV, Fan Z (2011) Pure Appl Chem 83:715
49. Oura I, Shimizu K, Ogata K, Fukuzawa S-I (2010) Org Lett 12:1752–1755
50. Malnuit V, Duca M, Manout A, Bougrin K, Behinda R (2009) Synlett 13:2123–2128
51. Yamada M, Matsumura M, Sakaki E, Yen S-Y, Kawahata M, Hyodo T, Yamaguchi K, Murata Y, Yasuike S (2019) Tetrahedron 75:1406–1414
52. Li M, Dong K, Zheng Y, Song W (2019) Org Biomol Chem 17:9933–9941
53. Sun P, Wang W, Lai J, Yan H, Tung C-H, Xu Z (2023) Org Chem Front 10:1890–1896

54. Mindt TL, Struthers H, Brans L, Anguelov T, Schweinsberg C, Maes V, Tourwe D, Schibli R (2006) J Am Chem Soc 9:15096–15097
55. Kluba CA, Mindt TL (2013) Molecules 18:3206–3226
56. Fernandez S, Crócamo N, Incerti M, Giglio J, Scarone L, Rey A (2012) J Label Compd Radiopharm 55:274–280
57. Urankar D, Pinter B, Pevec A, Proft FD, Turel I, Kosmrlj J (2010) Inorg Chem 49:4820–4829
58. Kilpin KJ, Gavey EL, McAdam CJ, Anderson CB, Lind SJ, Keep CC, Gordon KC, Crowley JD (2011) Inorg Chem 50:6334–6346
59. Jung KH, Muller M, Schmidt RR (2000) Chem Rev 100:4423–4442
60. Jaiswal MK, Sharma A, Tiwari VK, Schmidt RR (2023) Intramolecular glycosylation methodology: opportunity and challenges in glycoside formation. In: Synthetic strategies in carbohydrate chemistry, Chapter 2. Elsevier Inc, pp 53–94
61. Tiwari VK, Kumar A, Schmidt RR (2012) Eur J Org Chem 2945–2956
62. Kumar A, Geng Y, Schmidt RR (2012) Eur J Org Chem 35:6846–6851
63. Singh AS, Mishra KB, Mishra A, Tiwari VK (2016) Click chemistry in the synthesis of carbohydrate derivatives. In: Chandrasekaran S (ed) Click chemistry in organic synthesis. Wiley-VCH, pp 161–201
64. Pak K, Cheung S, Chit Tsui G (2017) Org Lett 19:2881–2884
65. Chen F-J, Mamidipalli P, Sabbasani VR, Liu H, Xia Y, Lee D (2021) Org Chem Front 8:6095–6107
66. Wang W, Lin Y, Ma Y, Tung C-H, Xu Z (2018) Org Lett 20:2956–2959
67. Srinivas A, Sunitha M, Shamili S (2020) Acta Chim Sloven 67:1061–1071
68. Tan W, Li Q, Dong F, Qiu S, Zhang J, Guo Z (2017) Carbohydr Polym 160:163–171
69. Dong W, Chen Z, Xu J, Miao M, Ren H (2016) Synlett 27:1318–1334
70. Francisco AP, Mendes E, Santos AR, Perry MJ (2019) Curr Pharm Des 25:1623–1642
71. Zhou J, He J, Wang B, Yang W, Ren H (2011) J Am Chem Soc 133:6868–6870
72. Wang C, Sun H, Fang Y, Huang Y (2013) Angew Chem Int Ed 52:5795−5798
73. Shealy YF, O'Dell CA (1966) J Med Chem 9:733–737
74. Zeng L, Lai Z, Zhang C, Xie H, Cui S (2020) Org Lett 22:2220–2224
75. Navarro Y, Lopez JG, Iglesias MJ, Ortiz FL (2021) Org Lett 23:334–339
76. Akter M, Rupa K, Anbarasan P (2022) Chem Rev 122:13108
77. Rajasekar S, Anbarsan P (2014) J Org Chem 79:8428–8434
78. Kaladevi S, Kamalraj M, Altia M, Rajasekar S, Anbarsan P (2019) Chem Commun 55:4507–4510
79. Yadagiri D, Chaitanya M, Reddy ACS, Anbarsan P (2018) Org Lett 20:3762–3765
80. Rupa K, Yadagiri D, Bagavathi R, Anbarsan P (2023) Org Lett 25:3375–3379
81. Rupa K, Yadagiri D, Anbarsan P (2023) J Org Chem 88:9077–9086
82. Rajasekar S, Ramachandran K, Anbarsan P (2022) J Chem Sci 134:121
83. Rajasekar S, Anbarsan P (2019) J Org Chem 84:7747–7761
84. Rajasekar S, Anbarsan P (2019) Org Lett 21:3067–3071
85. Yadagiri D, Anbarsan P (2014) Org Lett 16:2510–2513
86. Reddy ACS, Reddy PM, Anbarsan P (2020) Chem Commun 56:5649–5652
87. Yadagiri D, Anbarsan P (2015) Chem Sci 6:58470
88. Kumar D, Mishra BB, Bhattacharya S, Tiwari VK (2013) J Org Chem 78:899–909
89. Kumar D, Mishra BB, Tiwari VK (2014) J Org Chem 79:251–266
90. Kumar D, Singh AS, Tiwari VK (2015) RSC Adv 5:31584–31593
91. Singh AS, Kumar D, Tiwari VK (2017) ACS Omega 2:5044–5051
92. Singh M, Singh AS, Bose P, Tiwari VK (2020) Tetrahedron 76:131078
93. Yadav MS, Singh AS, Agrahari AK, Mishra N, Tiwari VK (2019) ACS Omega 4:6681–6689
94. Yu J, Singh AS, Yan G, Yu J, Tiwari VK (2020) Synthesis 52:3781–3800
95. Morimoto K, Yanase K, Ikeda T, Uchikawa C, Kita Y, Kajmoto T (2020) Heterocycles 101:621–630
96. Zhao S, Yu R, Chen W, Liu M, Wu H (2015) Org Lett 17:2828–2831
97. Zhou W, Zhang M, Li H, Chen W (2017) Org Lett 19:10–13
98. Duan X, Zheng N, Liu G, Li M, Wu Q, Sun X, Song W (2022) Org Lett 24:6006–6012

GPSR Compliance

The European Union's (EU) General Product Safety Regulation (GPSR) is a set of rules that requires consumer products to be safe and our obligations to ensure this.

If you have any concerns about our products, you can contact us on ProductSafety@springernature.com

In case Publisher is established outside the EU, the EU authorized representative is:

Springer Nature Customer Service Center GmbH
Europaplatz 3
69115 Heidelberg, Germany

The manufacturer's authorised representative in the EU is Springer Nature Customer Service Centre GmbH, Europaplatz 3, 69115 Heidelberg, Germany. If you have any concerns regarding our products, please contact ProductSafety@springernature.com

Printed and bound by CPI Group (UK) Ltd, Croydon, CR0 4YY
30/09/2026
02225302-0018